Springer

Berlin
Heidelberg
New York
Hong Kong
London
Milan
Paris
Tokyo

Kabat · Claussen · Dirmeyer · Gash · de Guenni · Meybeck
Pielke Sr. · Vörösmarty · Hutjes · Lütkemeier (Eds.)

Vegetation, Water, Humans and the Climate

A New Perspective on an Interactive System

With 246 Figures

 Springer

2004

Editors

Pavel Kabat
Wageningen University and Research Centre
ALTERRA Green World Research
Droevendaalsesteeg 3, 6708 PB Wageningen, The Netherlands
E-mail: *pavel.kabat@wur.nl*

Martin Claussen
Potsdam-Institut für Klimafolgenforschung
Telegrafenberg, 14473 Potsdam, Germany
E-mail: *claussen@pik-potsdam.de*

Paul A. Dirmeyer
Center for Ocean-Land-Atmosphere Studies
4041 Powder Mill Road, Suite 302, Calverton MD 20705-3106, USA
E-mail: *dirmeyer@cola.iges.org*

John H. C. Gash
Centre for Ecology and Hydrology
Maclean Building, Crowmarsh Gifford, Wallingford,
Oxfordshire OX10 8BB, United Kingdom
E-mail: *jhg@ceh.ac.uk*

Lelys Bravo de Guenni
Universidad Simón Bolivar
Departamento de Cómputo Cientifico y Estadistica
P.O. Box 89.000, Caracas 1080-A, Venezuela
E-mail: *lbravo@cesma.usb.ve*

Michel Meybeck
Université de Paris 6/SYSYPHE, Laboratoire de Géologie Appliquée
4 place Jussieu, F-75252 Paris, France
E-mail: *meybeck@ccr.jussieu.fr*

Roger A. Pielke Sr.
Department of Atmospheric Sciences
Colorado State University, Fort Collins CO 80523, USA
E-mail: *pielke@snow.atmos.colostate.edu*

Charles J. Vörösmarty
University of New Hampshire, Institute for the Study of Earth
Oceans & Space (EOS)
Complex Systems Research Center, Water Systems Analysis Group
Morse Hall, 39 College Road, Durham NH 03824-3525, USA
E-mail: *charles.vorosmarty@unh.edu*

Ronald W. A. Hutjes
Wageningen University and Research Centre,
ALTERRA Green World Research
Droevendaalsesteeg 3, 6708 PB Wageningen, The Netherlands
E-mail: *ronald.hutjes@wur.nl*

Sabine Lütkemeier
Potsdam-Institut für Klimafolgenforschung
Telegrafenberg, 14473 Potsdam, Germany
E-mail: *Sabine.Luetkemeier@pik-potsdam.de*

Cover design:
Figures from left to right: *Well in the Sahel, West Africa:* courtesy of John H. C. Gash ©, Centre for Ecology
and Hydrology, Wallingford, Oxfordshire, United Kingdom. *Model cloud:* adapted from Pielke RA, Lee TJ,
Copeland JH, Eastman JL, Ziegler CL, Finley CA (1997) Use of USGS-provided data to improve weather
and climate simulations. Ecol Applications 7:3–21. © printed by permission of Ecological Society of America
and the authors. *Amazon near Manaus:* © Antonio D. Nobre, Instituto Nacional de Pesquisas da Amazônia,
Manaus, Brazil. *Elbe flood:* Jeßnitz (near Bitterfeld) on the river Mulde on 15 August 2002 © M. Zebisch,
Technische Universität Berlin/Potsdam-Institut für Klimafolgenforschung e. V., Germany.

ISSN 1619-2435
ISBN 3-540-42400-8   Springer-Verlag Berlin Heidelberg New York

Library of Congress Control Number: 2004102329

Bibliographic information published by Die Deutsche Bibliothek
Die Deutsche Bibliothek lists this publication in the Deutsche Nationalbibliografie;
detailed bibliographic data is available in the Internet at http://dnb.ddb.de

Springer-Verlag Berlin Heidelberg New York
a member of BertelsmannSpringer Science+Business Media GmbH
http://www.springer.de
© Springer-Verlag Berlin Heidelberg 2004
Printed in Germany

The use of general descriptive names, registered names, trademarks, etc. in this publication does not
imply, even in the absence of a specific statement, that such names are exempt from the relevant protective laws and regulations and therefore free for general use.

Cover Design: Erich Kirchner, Heidelberg
Dataconversion: Büro Stasch (*stasch@stasch.com*), Klaus Häringer · Bayreuth

Printed on acid-free paper – 32/3141/lt – 5 4 3 2 1 0

Preface

This book is the result of an initiative by the *Biospheric Aspects of the Hydrological Cycle* (BAHC), a Core Project of the *International Geosphere-Biosphere Programme* (IGBP). It reports on the more than a decade-long research and findings of a large number of scientists studying the Earth system in terms of the connection between the terrestrial biosphere, the hydrologic cycle and the potential anthropogenic influences. The authors contributing to the five parts of the book have highlighted the research and findings of hundreds of scientists who have worked over the past 15 years on the interface between the hydrological cycle, the terrestrial biosphere and the atmosphere. As you read through the book, it becomes clear that the scientific progress goes well beyond any single international programme: it is interdisciplinary and reflects contributions made towards addressing many of the objectives set forth by a number of projects of IGBP, WCRP (*World Climate Research Programme*), and IHDP (*International Human Dimensions Programme on Global Environmental Change*).

At the programmatic level we often compartmentalise and label research as belonging to a specific named programme, but in reality and at the researchers' level, it is all a seamless process that tackles specific and challenging questions related to the highly interactive processes of vegetation, water and humans within the climate system. In their earliest years, BAHC and GEWEX (the *Global Energy and Water Cycle Experiment* of WCRP) recognised the need for thematic synergies and collaboration between the two research programmes. Both programmes have successfully collaborated in a large number of joint research, observational and modelling activities since their inception (BAHC in 1990 and GEWEX in 1988). The *International Satellite Land Surface Climatology Project* (ISLSCP), a GEWEX project, is perhaps one of the best examples of an excellent collaboration between the two programmes. BAHC and ISLSCP have operated "back to back" since the Tucson aggregation workshop in 1994 (Kabat and Sellers 1997)[1]. ISLSCP is a leading project in producing and consolidating global datasets for global change studies. BAHC and ISLSCP jointly initiated and coordinated an array of land-surface/atmosphere experiments, known as HAPEX and FIFE (for example *Hydrological and Atmospheric Pilot Experiment* in the Sahel and *First ISLSCP Field Experiment*, respectively). Both programmes jointly took the first steps to initiate the largest and most integrative Earth system experiment so far: the *Large Scale Biosphere Atmosphere Experiment in Amazonia* (LBA). It is gratifying to see some of the research and findings resulting from these joint activities presented in this book.

While we are both extremely pleased with the research progress reported in this volume, we are even more excited about the future results of the planned joint activities associated with the recently launched GEWEX Phase II and the new project in IGBP on the land-atmosphere interface, ILEAPS (*Integrated Land Ecosystem – Atmosphere Processes Study*)[2], to which the BAHC community will be a major contributor. For example, both the *Coordinated Enhanced Observing Period* (CEOP) of GEWEX,

---

[1] Kabat P, Sellers PJ (1997) Special issue: Aggregate description of land-atmosphere interactions, foreword. J Hydrol 190/3–4:173–175.
[2] *http://www.atm.helsinki.fi/ILEAPS/*

and the FLUXNET project of world-wide CO_2 flux measurement initiated by BAHC are positioned at the forefront of the Earth system measurement and monitoring approaches. By focusing on a series of reference field sites distributed over all continents (CEOP), on "transect studies" (FLUXNET), and on simultaneous use of satellite and ground observation, these experiments will provide a data set of unprecedented completeness and quality for our scientists to work with.

The *Global Land-Atmosphere System Study* (GLASS) and the *Global Soil Wetness Project* (GSWP) are other examples of successful collaborative activity between the two programmes in modelling land-surface/atmosphere processes and interactions within the climate system (e.g. Feddes et al. 2001)[3]. These projects are promising a new generation of land-surface schemes for Earth system models. The new schemes will evolve into interactive schemes that increasingly incorporate more hydrological, atmospheric, biogeochemical and ecological information.

Finally, while BAHC and GEWEX place much of their emphasis on the physical and biospheric aspects of water, they have also been very much interested in the potential impact of the alteration of the global hydrological cycle on regional water resources and ecosystems. However, despite the reported scenario and case studies (see Part D and E) and the proposed new approach for vulnerability assessments (Part E), at present, specific regional effects continue to be uncertain. This remaining uncertainty is one of the factors that has thus far hindered the effective application of GEWEX and BAHC research results to operational hydrology and water management strategies. Better links to applications in water resources is therefore one of the main priorities of Phase II of GEWEX and of the new joint project GWSP (*Global Water Systems Project*), co-sponsored by IGBP, WCRP, IHDP and DIVERSITAS (*International Programme of Biodiversity Science*). We remain optimistic that within this decade much progress in this area will be made and it will be the subject of a future publication.

Soroosh Sorooshian

Chair,
GEWEX-Scientific Steering Group

Pavel Kabat

Chair and Co-Chair,
BAHC and ILEAPS Science Steering Committees

---

[3] Feddes R A, Hoff H, Bruen M, Dawson TE, de Rosnay P, Dirmeyer P, Jackson RB, Kabat P, Kleidon A, Lilly A, Pitman AJ (2001) Modelling root water uptake in hydrological and climate models. Bull Amer Meteor Soc 82:2797–2809.

Acknowledgements

The editors emphasise that the results reported here are based on the research work of many individual scientists and research teams around the world who have been associated in some way with the objectives of the IGBP-BAHC and WCRP-GEWEX research programmes. Synthesising these results was only achieved through the commitment, and work, often voluntary, of these scientists, and their staff and students. We especially acknowledge the help of all those involved in developing databases, convening/hosting workshops and editing drafts, and all those who participated in the lively discussions which made the work so enjoyable and worthwhile.

We are grateful to the German Federal Ministry of Education and Research (BMBF) for funding the BAHC Project Office at the Potsdam Institute for Climate Impact Research (PIK) to whom thanks are also due. The Dutch National Research Programme on Global Air Pollution and Climate Change (grant 959291) provided support for part of the Science Synthesis process. Additional funding and support have been provided through the following institutions and agencies: the IGBP Central Office in Stockholm, The Netherlands Ministry of Agriculture, Nature Management and Fisheries, The Netherlands Organisation for Scientific Research (NWO), and the WCRP International GEWEX Project Office, Silver Spring, USA.

In particular we would like to thank Will Steffen for his continuous encouragement and support during the entire Synthesis process; Hans-Jürgen Bolle as the first BAHC Chair; Piers Sellers who as the former Chair of ISLSCP helped forge the connections between BAHC and GEWEX; the late Mike Fosberg, former head of the BAHC Core Project Office; and all the former members of the BAHC Scientific Steering Committee. Last but not least we also wish to thank two colleagues at the Potsdam Institute for Climate Impact Research: Dietmar Gibitz-Rheinbay, for helping with the technical side of putting together a book of this size, and Ursula Werner for redrawing many of the figures.

Contents

Introduction 1

Part A Does Land Surface Matter in Climate and Weather? 5

A.1   Introduction 7

A.2   The Climate near the Ground 9
A.2.1   Introduction 9
A.2.2   The Surface Energy Balance 9
A.2.3   The Surface Water Balance 14
A.2.4   Observing the Surface 16

A.3   The Regional Climate 21
A.3.1   Fundamental Mechanism in Land-Atmosphere Interactions 21
A.3.2   Atmospheric Response to Heterogenous Land Forcing 22
        A.3.2.1   Microscale Impact 22
        A.3.2.2   Mesoscale Impact 24
A.3.3   Regional Teleconnections 28
A.3.4   Discussion 31

A.4   The Global Climate 33
A.4.1   Feedbacks, Synergisms, Multiple Equilibria and Teleconnections 33
        A.4.1.1   Feedbacks 33
        A.4.1.2   Synergisms 36
        A.4.1.3   Multiple Equilibria 37
        A.4.1.4   Teleconnections 39
A.4.2   Palaeoclimate 40
        A.4.2.1   Feedbacks in the Arctic Climate System 40
        A.4.2.2   The Sahara 42
        A.4.2.3   Historical Land-cover Change 44
A.4.3   Sensitivity to Decadal Biogeochemical Feedbacks 45
        A.4.3.1   The Land Surface and Climate Change 45
        A.4.3.2   Biogeochemical Feedbacks 46
        A.4.3.3   Transient Experiments 48
A.4.4   Seasonal Variability 48
A.4.5   Impact of Land Surface on Weather 52
        A.4.5.1   Brief Literature Survey 52
        A.4.5.2   European Centre for Medium-Range Weather Forecast
                  (ECMWF) Examples 53

A.5   The Sahelian Climate 59
A.5.1   Introduction 59
        A.5.1.1   Background 59
        A.5.1.2   Climate Anomalies and Climate Change in the Sahel 60
        A.5.1.3   The Complex Processes of Land-use Change in the Sahel 62

A.5.2 Observational Studies of Sahelian Land-surface/Atmosphere
      Interactions 63
      A.5.2.1 The Sahelian Energy Balance Experiment (SEBEX) 63
      A.5.2.2 The Hydrological and Atmospheric Pilot Experiment
              in the Sahel (HAPEX-Sahel) 64
      A.5.2.3 Coupling Tropical Atmosphere and Hydrological Cycle
              (CATCH) 65
      A.5.2.4 Savannas in the Long Term (SALT) 65
      A.5.2.5 Satellite Data 66
A.5.3 Coupled Modelling of Sahelian Land-Atmosphere Interactions ... 66
      A.5.3.1 Brief Overview 66
      A.5.3.2 Large Scale Force-Response Studies of the Sahelian
              Climate Anomaly: The Relative Importance of Land-surface
              Processes and Sea-surface Temperatures 68
      A.5.3.3 Mesoscale Interactions between Sahelian Precipitation
              and Land-surface Patterns 71
      A.5.3.4 Climate System Interactions in the Sahel 71
A.5.4 Understanding Mechanisms 73
A.5.5 Conclusion 75

A.6   **The Amazonian Climate** 79
A.6.1 Introduction 79
A.6.2 Future Climates in Amazonia 81
A.6.3 Observations of Land-surface/Atmosphere Interactions 82
A.6.4 The General Characteristics and Variability
      of Water and Energy Balances in the Amazon Basin 85
      A.6.4.1 Introduction 85
      A.6.4.2 Water Balance 86
      A.6.4.3 Energy Balance 87
A.6.5 Deforestation and Climate 88

A.7   **The Boreal Climate** 93
A.7.1 The Boreal Ecosystem, Boreal Climate and High-latitude Climate
      Change and Variability 93
      A.7.1.1 Climate and Boreal Vegetation 94
      A.7.1.2 Effects of Fire and Insects on Vegetation and Land Cover ... 95
      A.7.1.3 High-latitude Climate Change 96
      A.7.1.4 Changes in Snow Extent, Depth and Duration 96
A.7.2 Energy Dissipation and Transport by the Boreal Landscape 101
      A.7.2.1 Effect of Soil Type on Surface Energy Partitioning ... 102
      A.7.2.2 Effect of Land-cover Type on Seasonal Variation
              in Relative Humidity 103
      A.7.2.3 Role of Stomatal Control 103
      A.7.2.4 Role of Latitudinal Gradient 104
      A.7.2.5 Role of Moss 106
      A.7.2.6 Role of Albedo of Forests, Wetlands and Lakes 106
      A.7.2.7 Role of Fire-induced Atmospheric Aerosols 107
      A.7.2.8 Role of Surface Hydrology 107
      A.7.2.9 Scaling Energy and Water Flux from the Plot to the Region ... 108
A.7.3 Biospheric Carbon Exchange: Carbon Dioxide and Methane 109
      A.7.3.1 Measurement Methods 109
      A.7.3.2 Controlling Factors (above and below Ground) 110
      A.7.3.3 Geographic Variations in Carbon Flux 110
      A.7.3.4 Seasonal and Interannual Variations in Carbon Flux ... 111
      A.7.3.5 Methane 112

A.7.3.6 The Effect of Landscape Patterns, Disturbance
         and Succession on Carbon Cycling 112
A.7.4   Sensitivity Experiments 113
A.7.4.1 Snow Albedo and Climate Feedback 113
A.7.4.2 Carbon Sequestration and Radiative Feedbacks 113
A.7.4.3 Effects of Climate Change on Land Cover 114
A.7.4.4 Hydrological Feedbacks 114

A.8   The Asian Monsoon Climate 115
A.8.1   Introduction 115
A.8.2   Role of Human-induced Large-scale Land-use/cover Change
         on the Water Cycle and Climate in Monsoon Asia 115
A.8.2.1 Atmospheric Water Cycle over Monsoon Asia 115
A.8.2.2 Is Monsoon Rainfall Decreasing?
         The Impact of Deforestation in Thailand on the Water Cycle 116
A.8.2.3 Do Water-fed Rice Paddy Fields Increase Rainfall
         in Monsoon Asia? 118
A.8.2.4 Conclusions 120
A.8.3   Can Human-induced Large-scale Land-cover Changes
         Modify the East Asian Monsoon? 120
A.8.3.1 History of Land-cover/Land-use Changes over East Asia 121
A.8.3.2 Design of the Numerical Experiments 121
A.8.3.3 Changes of Surface Dynamic Parameters
         under Two Vegetation Coverages 123
A.8.3.4 Changes of the East Asia Monsoon
         by Human-induced Land-cover Changes 123
A.8.4   Summary 127

A.9   Summary, Conclusion and Perspective 129

        References 137

Part B How Measurable is the Earth System? 155

B.1   Introduction 157
B.1.1   The Need for Integrated Experiments 157
B.1.2   The Experimental Design 157
B.1.3   Guide to Part B 158

B.2   The Energy Balance Closure Problem 159
B.2.1   Examples of Energy Balance Closure in Field Experiments 159
B.2.2   Reasons for Poor Closure 161
B.2.3   Experiment Design 163
B.2.4   Calculation and Analysis Errors 163
B.2.5   Overall Accuracy – What Can Be Expected 164
B.2.6   Pushing Eddy Covariance past Its Limits 165
B.2.7   Coping with Poor Energy Balance Closure When Modelling 166
B.2.8   Summary and Conclusions 166

B.3   Radiation Measurements in Integrated Terrestrial Experiments 167
B.3.1   Introduction 167
B.3.2   Radiometry in Integrated Terrestrial Experiments 167
B.3.3   Available Radiometer Designs 169
B.3.4   Overview of Radiometry in the Last Two Decades 170
B.3.5   Summary 171

B.4   **Surface Turbulent Fluxes** 173
B.4.1   Where Are We Now? 173
B.4.2   How Did We Get Here? 173
B.4.3   The Main Micrometeorological Methods 174
    B.4.3.1   The Aerodynamic Gradient Method 174
    B.4.3.2   The Bowen Ratio (or Energy Balance) Method 176
    B.4.3.3   The Eddy Covariance (or Correlation) Method 176
    B.4.3.4   Flux Footprint 180
    B.4.3.5   The Nocturnal "CO_2 Loss" Problem 181
B.4.4   The Future for Surface Fluxes 182

B.5   **Accuracy and Utility of Aircraft Flux Measurements** 183
B.5.1   Introduction 183
B.5.2   Technology of Airborne Flux Measurement 184
B.5.3   Accuracy of Airborne Measurements 185
B.5.4   Utility of Airborne Flux Measurement 186
B.5.5   Conclusions 187

B.6   **Boundary Layer Budgeting** 189
B.6.1   Introduction 189
B.6.2   Characteristic Structure of the Boundary Layer 189
B.6.3   Budget Relations 190
B.6.4   More Recent Budget Studies Conducted over Land 191
B.6.5   Conclusions and Suggestions for Ongoing Work 196

B.7   **Vegetation Structure, Dynamics and Physiology** 199
B.7.1   Introduction 199
B.7.2   Measurement of Vegetation Structure, Dynamics and Physiology 199
B.7.3   Vegetation Measurements in the Integrated Terrestrial Experiment 202
B.7.4   Synergy of Vegetation Measurements with Other Measurement
    Programmes in the Integrated Terrestrial Experiment 204
B.7.5   Summary 204

B.8   **Remote Sensing and Land-surface Experiments** 207
B.8.1   Introduction 207
B.8.2   Remote Sensing Input to the Experiments 208
B.8.3   Impact of Field Experiments on Remote Sensing of the Land Surface 208
B.8.4   The Future 212

B.9   **The Water Balance Concept – How Useful Is It as a Guiding**
    Principle for the Design of Land-Atmosphere Field Experiments? 213
B.9.1   Background 213
B.9.2   The Water Balance Concept 213
    B.9.2.1   Snow-free Land Surface 213
    B.9.2.2   Snowpack Water Balance 214
    B.9.2.3   Atmospheric Water Balance 214
B.9.3   Application of the Water Balance Concept 214
    B.9.3.1   Intensive Field Campaigns 214
    B.9.3.2   GEWEX Continental Scale Experiments 218
B.9.4   Assessment and Recommendations 219

B.10   **Use of Field Experiments in Improving the Land-surface Description**
    in Atmospheric Models: Calibration, Aggregation and Scaling 221
B.10.1   Introduction 221

B.10.2 Benefit of Datasets from Integrated Terrestrial Experiments
        for Surface Modelling in Atmospheric and Hydrological Models 221
B.10.3 A Modelling Strategy for Upscaling from the Plot Scale
        to the Size of a GCM Grid Box 222
B.10.4 Use of a Macroscale Hydrological Model to Investigate Aggregation
        of Hydrological Processes 225
B.10.5 Concluding Remarks 227

**B.11   Further Insight from Large-scale Observational Studies
          of Land/Atmosphere Interactions** 229
B.11.1 Introduction 229
B.11.2 International Co-operation 229
B.11.3 The Use of Land-surface Data to Validate Global Models 229
B.11.4 Surface Flux Measurements 230
B.11.5 Hydrological Catchment Measurements 232
B.11.6 Aggregation and Models 233
B.11.7 Future for Large-scale Integrated Experiments 233
B.11.8 Concluding Remarks 233

        References 235

Part C The Value of Land-surface Data Consolidation 245

C.1   Motivation for Data Consolidation 247
C.1.1   The Volume of Data 248
C.1.2   The Breadth of Data 250
C.1.3   The Trend toward Interdisciplinary Science 251
C.1.4   Who Needs Consolidated Data? 251
C.1.5   What Is Consolidation? 252

C.2   Existing Degrees of Consolidation 255
C.2.1   Project-specific Data Collections 255
C.2.2   Subject-specific Archives 258
C.2.3   Broad Data Archives 259
C.2.4   Co-registration 261
C.2.5   Tools and Data 264

C.3   Achieving Full Data Consolidation 267
C.3.1   Necessary Elements 267
C.3.2   Tools 268
C.3.3   Data Maintenance 270
C.3.4   Motivating Data Providers 271

C.4   Terrestrial Data Assimilation 273
C.4.1   Topographic Coherence of Weather –
        The Role of Statistical Assimilation 275
        C.4.1.1 Stochastic Weather Models for Scenario Generation 275
        C.4.1.2 Scheme for Generating Stochastic Weather Scenarios ... 276
        C.4.1.3 Stochastic Models 277
        C.4.1.4 Topographic Dependent Interpolation
                 of Weather Model Parameters 278
        C.4.1.5 Spatial Structure and Topographic Dependencies
                 of Weather Anomalies 278
C.4.2   Terrestrial Model Prediction 279

C.4.3   Terrestrial Observations 280
C.4.4   Data Assimilation Concepts and Methods 281
C.4.5   Current Projects 283
C.4.6   Future Opportunities 286
        C.4.6.1   Terrestrial Observation 286
        C.4.6.2   Terrestrial Simulation 286
        C.4.6.3   Terrestrial Data Assimilation 287

C.5   Conclusions 289

        References 291

Part D The Integrity of River and Drainage Basin Systems:
        Challenges from Environmental Change 297

D.1   Introduction 299

D.2   Responses of Hydrological Processes to Environmental Change
        at Small Catchment Scales 301
D.2.1   Introduction 301
D.2.2   Terrestrial Hydrological Processes –
        Overview, Definitions, Classification 301
        D.2.2.1   Fundamental Hydrological Processes 301
        D.2.2.2   Spatial Differentiation of Vertical Hydrological Processes 302
        D.2.2.3   Runoff Generation and Runoff Components 305
        D.2.2.4   Time Behaviour of Runoff Components 307
        D.2.2.5   Unresolved Understanding of Processes of Subsurface Flow
                  and Limitations in Assessing the Controlling Subsurface
                  Characteristics and Parameters 308
D.2.3   The Unsaturated Zone and Its Interaction with the Atmosphere
        through the Biosphere 310
        D.2.3.1   The Role of Soil Moisture
                  in Coupled Land-surface/Atmosphere Modelling 310
        D.2.3.2   Soil Water Flow and Root Water Uptake at the Field Scale 311
        D.2.3.3   Effects of Frozen Soil Moisture 312
        D.2.3.4   Representation of Available Soil and Root Information
                  in Land-surface Models 313
        D.2.3.5   Conclusions 316
D.2.4   Overland Flow, Erosion and Associated Sediment
        and Biogeochemical Transports 317
        D.2.4.1   Impact of Climate Change 317
        D.2.4.2   Impact of Land-use Change 319
D.2.5   Subsurface Stormflow and Lateral Flow Processes 322
        D.2.5.1   Rapid, Shallow Subsurface Stormflow Processes 322
        D.2.5.2   Separating Event Water and Subsurface Stormflow
                  in the Storm Hydrograph 324
        D.2.5.3   Modelling Lateral Flow at the Catchment Scale 326
        D.2.5.4   Subsurface Flow and Catchment-scale Nutrient Dynamics 326
        D.2.5.5   Conclusions 328
D.2.6   Integrated Ecohydrological Modelling Considering Nutrient Dynamics
        in River Catchments 328
        D.2.6.1   General issues 328
        D.2.6.2   Structure of Integrated Ecohydrological Models 329
        D.2.6.3   Assessment of Land-surface Heterogeneity in Modelling 330
        D.2.6.4   GIS-based Estimation of Land-surface Characteristics
                  and Related Model Parameters 331

D.2.6.5 Calibration and Verification of Component
Models (Modules) within Integrated Models 332
D.2.6.6 Examples of Integrated Ecohydrological Models:
SWIM and ACRU 332
D.2.7 Conclusions 337

**D.3   River Basin Responses to Global Change
and Anthropogenic Impacts** 339
D.3.1 Introducing the River Basin Scale and Its Response
to Anthropogenic Change 339
D.3.2 Natural Landscape Processes at the River Basin Scale 341
D.3.2.1 Introduction 341
D.3.2.2 Natural Watercourses and Aquatic Ecosystems 341
D.3.2.3 Evaporation and Transmission Losses
from Riverine Systems 343
D.3.2.4 Wetlands 345
D.3.3 Anthropogenic Modifications of the River Basin Landscape 347
D.3.3.1 Land-use Change and Its Impacts on Hydrological Responses 347
D.3.3.2 Plantation Afforestation Effects 349
D.3.3.3 Urban Influences on Hydrological Responses 352
D.3.3.4 Water Quality Degradation Resulting
from Agricultural Pollution by Nitrogen and Phosphorus 354
D.3.3.5 Salinisation 357
D.3.4 Water and River Engineered Landscape 362
D.3.4.1 River Channel Modification 362
D.3.4.2 Dams and Their Impacts 363
D.3.5 The Road Ahead 1:
Integrated Water Resources Management (IWRM) 365
D.3.5.1 Integrated Water Resources Management as a Response
to an Inheritance of Damaged River Basins 365
D.3.5.2 What Is Integrated Water Resources Management (IWRM)? 367
D.3.5.3 The River Basin as the Fundamental Unit for IWRM 369
D.3.5.4 At What Space and Time Scales Should IWRM
Be Carried Out? 370
D.3.5.5 Differences in IWRM between Developing Countries
and Developed Countries 371
D.3.5.6 Conditions for the Success of IWRM 372
D.3.5.7 Conclusions 372
D.3.6 The Road Ahead 2: Restoration of Riverine Ecosystems 373
D.3.7 Conclusions 374

**D.4   Responses of Continental Aquatic Systems at the Global Scale:
New Paradigms, New Methods** 375
D.4.1 Introduction 375
D.4.2 Terms of Reference 376
D.4.2.1 Relevant Time and Space Scales Associated with
Global Change and Continental Aquatic Systems 376
D.4.2.2 Emerging Techniques for Analysing
Continental Aquatic Systems and Global Change 377
D.4.3 Changes in River Connectivity and Basin Characteristics:
Palaeo to Present 377
D.4.3.1 A Global Classification System for Flow Connectivity
in River Systems 377
D.4.3.2 Major Earth System Processes Controlling Land-to-Ocean
Coupling: Glaciation/De-glaciation, Climate Variability
and Recent Tectonics 381

D.4.4   Human Conditioning of Continental Runoff 385
        D.4.4.1   A Focus on Reservoirs 386
        D.4.4.2   Impacts of Land-cover Change on Water Budgets 390
D.4.5   Global Sediment Flux 390
        D.4.5.1   The Continuum of Fluxes from Field Erosion
                  to River Mouth Export 391
        D.4.5.2   Approaches toward Estimating Basin Fluxes 391
        D.4.5.3   Additional Temporal Complexities 396
        D.4.5.4   Globally, What Is the Net Change
                  in Riverborne Sediment Flux due to Humans? 397
D.4.6   Global River Transfer of Carbon and Its Alteration and Storage 397
        D.4.6.1   Sources, Sinks and Re-cycling 397
        D.4.6.2   Estimates of Riverborne Carbon Flux to the Oceans 399
D.4.7   Global Riverine Nutrient Flux to the Oceans 402
        D.4.7.1   Inventory Methods 402
        D.4.7.2   New Regression and Multiple Regression Models 405
        D.4.7.3   "Hot Spots" at the Global Scale 407
        D.4.7.4   Stoichiometric Changes of $N:P:Si$ 408
D.4.8   Future Trends 409
        D.4.8.1   Pressure on Inland Water Systems 409
        D.4.8.2   The Future State of Riverine Carbon Loads 411
        D.4.8.3   The Future State of Inorganic Nitrogen Loads in Rivers 411
D.4.9   Future Research 412

D.5   Case Study 1: Integrated Analysis of a Humid Tropical Region –
        The Amazon Basin 415
D.5.1   Towards an Integrated Analysis of the Amazon Basin 415
D.5.2   Coupling Hydrology, Organic Matter and Nutrient Dynamics
        in Large River Basins 415
D.5.3   The Amazon Basin: Vargem Grande to Óbidos 417
D.5.4   Hydrology of the Amazon River System: A Mainstem Perspective 419
        D 5.4.1   Patterns of Rainfall 419
        D.5.4.2   Mainstem and Tributary Hydrographs 419
        D.5.4.3   Models of Amazon Water Movement 421
D.5.5   River Chemistry 422
        D.5.5.1   A Synoptic View of Chemical Profiles 422
        D.5.5.2   In-river Dynamics 424
        D.5.5.3   Organic Geochemical Signatures 425
        D.5.5.4   Dynamics of Floodplains 425
D.5.6   Potential Impact of Anthropogenic Change on the River System 427
D.5.7   Towards a Synthetic Model of Drainage Basins 427

D.6   Case Study 2: Integrated Ecohydrological Analysis of a Temperate
        Developed Region: The Elbe River Basin in Central Europe 429
D.6.1   General Outline of the Elbe River Basin 429
D.6.2   Integrated Analysis of Hydrological Processes and Nitrogen
        Dynamics 431
        D.6.2.1   Comparison of Nitrogen Dynamics in the Lowland
                  and Mountainous Sub-regions of the Elbe 431
        D.6.2.2   Regional Nitrogen Dynamics across the German Part
                  of the Elbe Basin 433
D.6.3   Agricultural Land-use Change and Its Impact
        on Water Resources 435
D.6.4   Climate Change Impacts on Hydrology and Crop Yields 437
D.6.5   Conclusions 439

D.7   Case Study 3: Modelling the Impacts of Land Use
       and Climate Change on Hydrological Responses in the Mixed
       Underdeveloped/Developed Mgeni Catchment, South Africa 441
D.7.1 Setting the Scene 441
D.7.2 Attributes of the Mgeni Catchment and Human Pressures 443
D.7.3 Configuration of the Mgeni System for Simulation Modelling 444
D.7.4 Verification of Simulated Hydrological Outputs 446
D.7.5 Modelling Impacts of Contrasting Land Uses
       on Streamflow Generation 446
D.7.6 Modelling Impacts of Land Uses on Water Quality Indicators 447
D.7.7 Scenario Studies on Impacts of Land Use
       on Water Quantity and Quality 450
       D.7.7.1 Effects of Individual Land Uses on Runoff
                at the Management Catchment Level 450
       D.7.7.2 Impacts of Subsistence Farming and Informal Settlements
                on Water Quality and Quantity 452
       D.7.7.3 Impacts of Potential Climate Change on Streamflows 452
D.7.8 Conclusions 453

D.8   Conclusions: Scaling Relative Responses
       of Terrestrial Aquatic Systems to Global Changes 455
D.8.1 Terrestrial Aquatic Systems and the Earth System under Pressure 455
D.8.2 Spatial Organisation of Terrestrial Aquatic Systems
       and Their Responses to Anthropogenic Change 457
D.8.3 Spatial Scale of Drivers Operating on Terrestrial Aquatic Systems 459
       D.8.3.1 Natural Drivers 459
       D.8.3.2 Anthropogenic Drivers 460
       D.8.3.3 Integrated Water Management and Governance 461
D.8.4 Time Scales of Responses of Continental Aquatic Systems (CAS)
       to Imposed Changes 461
D.8.5 Continental Aquatic Systems and Emergence of the Anthropocene 461
D.8.6 Continental Aquatic Systems Shared by Social Systems and the
       Biogeophysical Earth System: An Extension of the DPSIR Approach 463

       References 465

Part E How to Evaluate Vulnerability
        in Changing Environmental Conditions? 481

E.1   Introduction 483

E.2   Predictability and Uncertainty 485

E.3   Contrast between Predictive and Vulnerability Approaches 491
E.3.1 Societal Needs 493
E.3.2 Quantifying Uncertainty Using a Bayesian Approach 494

E.4   The Scenario Approach 497

E.5   The Vulnerability Approach 499
E.5.1 Risk, Hazard and Vulnerability: Concepts 499
E.5.2 Anthropogenic Land-use and Land-cover Changes 502
E.5.3 Procedures to Assess the Effect of Environmental Conditions
       on Water Resources: Natural Landscape Changes 506
E.5.4 An Example of the Vulnerability Approach: Ecosystem Vulnerability 509

E.6   Case Studies 515
E.6.1   Population and Climate 515
E.6.2   Water Resources in the Lake Erhai Basin, China 523
E.6.3   Yellow River: Recent Trends 526
E.6.4   Examples of Hazard Determination and Risk Mitigation
        from South Africa 528

E.7   Conclusions 537

        References 539

        Index 545

Contributors

Simon J. Allen

CECS, University of Edinburgh
John Muir Building, The King's Buildings
Mayfield Road, Edinburgh, EH9 3JK, United Kingdom
E-mail: s.allen@ed.ac.uk

Roni Avissar

Department of Civil and Environmental Engineering
Duke University
123 Hudson Hall, Durham NC 27708-0287, USA
E-mail: avissar@duke.edu

Dennis Baldocchi

Ecosystem Science Division and Berkeley Atmospheric
Science Center, Department of Environmental Science,
Policy and Management
University of California, Berkeley
151 Hilgard Hall, Berkeley CA 94720, USA
E-mail: baldocchi@nature.berkeley.edu

Brad Bass

Adaptations and Impacts Research Group
Environment Canada at the University of Toronto
33 Willcocks Street, Toronto, Ontario, Canada M5S 3E8
E-mail: brad.bass@ec.gc.ca

Alfred Becker

Potsdam-Institut für Klimafolgenforschung
Telegrafenberg, 14473 Potsdam, Germany
E-mail: Alfred.Becker@pik-potsdam.de

Anton C. M. Beljaars

European Centre for Medium-Range Weather Forecasts (ECMWF)
Shinfield Park, Reading, Berkshire RG2 9AX, United Kingdom
E-mail: beljaars@ecmwf.int

Alan K. Betts

Atmospheric Research
58 Hendee Lane, Pittsford VT 05763-9405, USA
E-mail: akbetts@aol.com

Richard A. Betts

Met Office
Hadley Centre for Climate Prediction and Research
Fitzroy Road, Exeter, EX1 3PB, United Kingdom
E-mail: richard.betts@metoffice.com

Hans-Jürgen Bolle

Stücklenstraße 18c, 81247 München, Germany
E-mail: hansj.bolle@lrz.badw-muenchen. de

Mike Bonell

Section: Hydrological Processes and Climate
UNESCO Division of Water Sciences
1 rue Miollis, 75732 Paris Cedex 15, France
E-mail: m.bonell@unesco.org

Lelys Bravo de Guenni

Universidad Simón Bolivar
Departamento de Cómputo Cientifico y Estadistica
P.O. Box 89.000, Caracas 1080-A, Venezuela
E-mail: lbravo@cesma.usb.ve

Ross Brown

Climate Research Branch, Meteorological Service of Canada
2121 Trans Canada Highway, Dorval, QC, Canada H9P 1J3
E-mail: Ross.Brown@ec.gc.ca

Jake Brunner

Conservation International
1919 M Street, Washington DC 20036, USA
E-mail: J.brunner@conservation.org

Thomas Chase

Cooperative Institute for Research in Environmental Sciences
(CIRES) and Department of Geography
Campus Box 216, University of Colorado, Boulder, CO 80309, USA
E-mail: tchase@cires.colorado.edu

Jing M. Chen

Department of Geography and Program in Planning
University of Toronto
100 St. George St., Toronto, Ontario, Canada M5S 3G3
E-mail: chenj@geog.utoronto.ca

Wenjun Chen

Environmental Monitoring Section
Canada Centre for Remote Sensing
588 Booth St., Ottawa, Ontario, Canada K1A 0Y7
E-mail: wenjun.chen@ccrs.nrcan.gc.ca

Martin Claussen

Potsdam-Institut für Klimafolgenforschung
Telegrafenberg, 14473 Potsdam, Germany
E-mail: claussen@pik-potsdam.de

Peter M. Cox

Met Office
Hadley Centre for Climate Prediction and Research
Fitzroy Road, Exeter, EX1 3PB, United Kingdom
E-mail: peter.cox@metoffice.com

Timothy L. Crawford †

Field Research Division NOAA/ARL
1750 Foote Drive, Idaho Falls, ID 83402, USA
E-mail: use: *dobosy@atdd.noaa.gov*

Alistair D. Culf

Centre for Ecology and Hydrology, Maclean Building, Crowmarsh
Gifford, Wallingford
Oxfordshire OX10 8BB, United Kingdom
E-mail: use: *jhg@ceh.ac.uk*

Paul A. Dirmeyer

Center for Ocean-Land-Atmosphere Studies
4041 Powder Mill Road, Suite 302, Calverton MD 20705-3106, USA
E-mail: *dirmeyer@cola.iges.org*

Ronald J. Dobosy

Atmospheric Turbulence and Diffusion Division NOAA/ARL
Post Office Box 2456, Oak Ridge TN, 37831-2456, USA
E-mail: *dobosy@atdd.noaa.gov*

A. J. (Han) Dolman

Dept. Geo-Environmental Sciences
Free University
de Boelelaan 1085, 1081 HV Amsterdam, The Netherlands
E-mail: *han.dolman@geo.falw.vu.nl*

Hervé Douville

Météo-France/CNRM, GMGEC/UDC
42 avenue Coriolis, 31057 Toulouse, France
E-mail: *herve.douville@meteo.fr*

Reinder A. Feddes

Wageningen University, Environmental Sciences,
Sub-Department Water Resources,
Chair Soil Physics, Agrohydrology and Groundwater Management
Nieuwe Kanaal 11, 6709 PA Wageningen, The Netherlands
E-mail: *reinder.feddes@wur.nl*

Balazs Fekete

University of New Hampshire
Institute for the Study of Earth, Oceans & Space (EOS)
Complex Systems Research Center, Water Systems Analysis Group
Morse Hall, 39 College Road, Durham NH 03824-3525, USA
E-mail: *balazs.fekete@unh.edu*

David R. Fitzjarrald

Atmospheric Sciences Research Center
University at Albany, SUNY
251 Fuller Road, Albany NY 12203, USA
E-mail: *fitz@asrc.cestm.albany.edu*

Thomas Foken

Abteilung Mikrometeorologie
Universität Bayreuth
Universitätsstraße 30, D-95440 Bayreuth, Germany
E-mail: *thomas.foken@uni-bayreuth.de*

Burkhard Frenzel

Institut für Botanik
Universität Hohenheim
Garbenstr. 30, D-70599 Stuttgart, Germany
E-mail: *bfrenzel@uni-hohenheim.de*

Steve Frolking

Institute for the Study of Earth, Oceans, and Space
University of New Hampshire
39 College Rd., Durham NH 03824-2622, USA
E-mail: *steve.frolking@unh.edu*

Congbin Fu

START Regional Center for Temperate East Asia, c/o Institute of
Atmospheric Physics, Chinese Academy of Sciences,
Qi Jia Huo Zi, De Sheng Men Wai Street, Beijing 100029, China
E-mail: *fcb@ast590.tea.ac.cn*

John H. C. Gash

Centre for Ecology and Hydrology, Maclean Building, Crowmarsh
Gifford, Wallingford
Oxfordshire OX10 8BB, United Kingdom
E-mail: *jhg@ceh.ac.uk*

Pamela Green

University of New Hampshire
Institute for the Study of Earth, Oceans & Space (EOS)
Complex Systems Research Center, Water Systems Analysis Group
Morse Hall, 39 College Road, Durham NH 03824-3525
USA, E-mail: *pam.green@unh.edu*

Vijay Gupta

Dept. of Civil and Environmental Engineering and Coopreative
Institute for Research in Environmental Sciences, Campus Box 216,
University of Colorado, Boulder CO 80309-0216, USA
E-mail: *guptav@cires.colorado.edu*

Florence Habets

Meteo-France - CNRM
42, avenue Gustave Coriolis, F-31057 Toulouse, France
E-mail: *florence.habets@meteo.fr*

Forrest G. Hall

Goddard Space Flight Center
National Aeronautic and Space Administration (NASA)
Code 923, Greenbelt MD 20771, USA
E-mail: *fghall@ltpmail.gsfc.nasa.gov*

Sven Halldin

Department of Earth Sciences, Uppsala University
Villavägen 16, SE-75236 Uppsala, Sweden
E-mail: *sven.halldin@hyd.uu.se*

Niall Hanan

Natural Resource Ecology Laboratory
Colorado State University
Fort Collins CO 80521, USA
E-mail: *niall@nrel.colostate.edu*

Richard J. Harding

Centre for Ecology and Hydrology, Maclean Building
Crowmarsh Gifford, Wallingford
Oxfordshire OX10 8BB, United Kingdom
E-mail: *rjh@ceh.ac.uk*

Holger Hoff

Potsdam-Institut für Klimafolgenforschung
Telegrafenberg, 14473 Potsdam, Germany
E-mail: *hhoff@pik-potsdam.de*

Paul Houser

Hydrological Sciences Branch
Goddard Space Flight Center (GSFC)
National Aeronautics and Space Administration (NASA)
Code 974, Bldg. 33, Greenbelt MD 20771, USA
E-mail: *houser@hsb.gsfc.nasa.gov*

Gordon H. Huang

Environmental Systems Engineering
University of Regina
Regina, Sask., Canada S4S 0A2
E-mail: *gordon.huang@uregina.ca*

Michael F. Hutchinson

Centre for Resource and Environmental Studies
Australian National University
GPO Box 4, Canberra ACT 0200, Australia
E-mail: *hutch@cres.anu.edu.au*

Ronald W. A. Hutjes

Wageningen University and Research Centre
ALTERRA Green World Research
Droevendaalsesteeg 3, 6708 PB Wageningen, The Netherlands
E-mail: *ronald.hutjes@wur.nl*

Roy Jenne

National Center for Atmospheric Research
1850 Table Mesa Drive, Boulder, CO 80305-3000, USA
E-mail: *jenne@ucar.edu*

Pavel Kabat

Wageningen University and Research Centre
ALTERRA Green World Research
Droevendaalsesteeg 3, 6708 PB Wageningen, The Netherlands
E-mail: *pavel.kabat@wur.nl*

Yann H. Kerr

Centre National d'Etudes Spatiales (CNES), CESBIO
18 Avenue Edouard Belin, 31401 Toulouse, France
E-mail: *yann.kerr@cesbio.cnes.fr*

Stefan W. Kienzle

Department of Geography, The University of Lethbridge
4401 University Drive, Lethbridge, Alberta, Canada T1K 3M4
E-mail: *stefan.kienzle@uleth.ca*

Timothy Kittel

National Center for Atmospheric Research
1850 Table Mesa Drive, Boulder, CO 80305-3000, USA
E-mail: *kittel@ucar.edu*

Randal Koster

Global Modeling and Assimilation Office
Goddard Space Flight Center (GSFC)
National Aeronautics and Space Administration (NASA)
Code 900.3, NASA/GSFC, Greenbelt, MD 20771, USA
E-mail: *randal.d.koster@nasa.gov*

Bart Kruijt

Wageningen University and Research Centre
ALTERRA Green World Research
Droevendaalsesteeg 3, 6708 PB Wageningen, The Netherlands
E-mail: *bart.kruijt@wur.nl*

Valentina Krysanova

Potsdam-Institut für Klimafolgenforschung
Telegrafenberg, 14473 Potsdam, Germany
E-mail: *Valentina.Krysanova@pik-potsdam.de*

Yumiko Kura

World Resources Institute (WRI)
10 G Street, NE, Washington DC 20002, USA
E-mail: *yumiko@wri.org*

Pierre Lacarrère

Meteo-France - CNRM
42 avenue Gustave Coriolis, F-31057 Toulouse, France
E-mail: *pierre.lacarrere@meteo.fr*

Eric F. Lambin

Department of Geography
Université Catholique du Louvain
3 place Louis Pasteur, B-1348 Louvain La Neuve, Belgium
E-mail: *lambin@geog.ucl.ac.be*

Thierry Lebel

Laboratoire d'étude des Transferts en Hydrologie et
Environnement (LTHE), UMR 5564
BP 53, F-38041 Grenoble cedex 9, France
E-mail: *Thierry.Lebel@inpg.fr*

John Leese

GEWEX Continental Scale International Project (GCIP) Office
Office of Global Programs
National Oceanic and Atmospheric Administration (NOAA)
1100 Wayne Avenue, Silver Spring MD 20901, USA
E-mail: *leese@ogp.noaa.gov*

Rik Leemans

Environmental Systems Analysis Group
Department of Environmental Sciences, Wageningen University
De Drijenborch, Ritzema Bosweg 32a
6703 AZ Wageningen, The Netherlands
E-mail: *rik.leemans@wur.nl*

Dennis P. Lettenmaier

Department of Civil and Environmental Engineering
University of Washington
Seattle, WA, USA
E-mail: *lettenma@ce.washington.edu*

Changming Liu

Institute of Geographic Science and Natural Resources Research
Chinese Academy of Sciences, Anwai
Datun Road, Building 917, Beijing 100101, P. R. China
E-mail: *liucm@igsnrr.ac.cn*

Lei Liu

Department of Civil Engineering, Dalhousie University
1360 Barrington St., Halifax, NS, Canada B3J 1Z1
E-mail: *lei.liu@dal.ca*

Simon Lorentz

School of Bioresources Engineering and Environmental Hydrology
University of KwaZulu-Natal, Pietermaritzburg
Carbis Road, Scottsville, 3201 Pietermaritzburg, RSA
E-mail: *lorentz@ukzn.ac.za*

Sabine Lütkemeier

Potsdam-Institut für Klimafolgenforschung
Telegrafenberg, 14473 Potsdam, Germany
E-mail: *Sabine.Luetkemeier@pik-potsdam.de*

José A. Marengo

Centro de Previsão de Tempo e Estudos Climáticos (CPTEC)
Instituto Nacional de Pesquisas Espaciais (INPE)
Rodonia Presidente Dutra, km 40,
Cachoeira Paulista, SP 12630-000, Brazil
E-mail: *marengo@cptec.inpe.br*

Luis Antonio Martinelli

Centro de Energia Nuclear na Agricultura
Universidade de Sao Paulo/Piracicaba
Avenida Centenario 303, Piracicaba SP, Brazil
E-mail: *lamartin@pintado.ciagri.usp.br*

Emilio Mayorga

School of Oceanography, University of Washington
P. O. Box 357940, Seattle WA 98195-7940, USA
E-mail: *emiliom@u.washington.edu*

Jeffrey J. McDonnell

Department of Forest Engineering, Oregon State University
Corvallis OR 97331-5706, USA
E-mail: *jeff.mcdonnell@orst.edu*

Robert H. Meade

U.S. Geological Survey, Denver Federal Center
Lakewood CO 80225-0046, USA
E-mail: *rhmeade@usgs.gov*

Blanche W. Meeson

Goddard Space Flight Center
National Aeronautics and Space Administration (NASA)
Greenbelt MD 20771, USA
E-mail: *meeson@see.gsfc.nasa.gov*

Michel Meybeck

Université de Paris 6/SYSYPHE
Laboratoire de Géologie Appliquée
4 place Jussieu, F-75252 Paris, France
E-mail: *meybeck@ccr.jussieu.fr*

John Moncrieff

Institute of Ecology and Resource Management
University of Edinburgh, Mayfield Rd., Darwin Building
Edinburgh EH9 3JU, United Kingdom
E-mail: *j.moncrieff@ed.ac.uk*

Bart Nijssen

Departments of Hydrology and Water Resources / Civil Engineering and Engineering Mechanics, The University of Arizona
Tucson, AZ 85721, USA
E-mail: *nijssen@u.arizona.edu*

Carlos A. Nobre

Centro de Previsão de Tempo e Estudos Climáticos (CPTEC)
Instituto Nacional de Pesquisas Espaciais (INPE)
Rodovia Presidente Dutra, km 40
Cachoeira Paulista, SP 12630-000, Brazil
E-mail: *nobre@cptec.inpe.br*

Joel Noilhan

Meteo-France - CNRM
42, avenue Gustave Coriolis, F-31057 Toulouse, France
E-mail: *joel.noilhan@meteo.fr*

Dennis Shoji Ojima

Natural Resource Ecology Laboratory, Colorado State University
Fort Collins CO 80523, USA
E-mail: *dennis@nrel.colostate.edu*

Taikan Oki

Institute of Industrial Science, University of Tokyo
4-6-1 Komaba, Meguro-ku, Tokyo 153-8505, Japan
E-mail: *taikan@iis.u-tokyo.ac.jp*

Richard J. Olson

Oak Ridge National Laboratory
2105 Driftwood Drive, Stevens Point, Wi 54481, USA
E-mail: *olson627@juno.com*

Lucille Perks

School of Bioresources Engineering and Environmental Hydrology
University of KwaZulu-Natal, Pietermaritzburg
Carbis Road, Scottsville, 3201 Pietermaritzburg, RSA
E-mail: *perksl@ukzn.ac.za*

Gerhard Petschel-Held

Potsdam-Institut für Klimafolgenforschung
Telegrafenberg, 14473 Potsdam, Germany
E-mail: *Gerhard.Petschel@pik-potsdam.de*

Thomas J. Phillips

Program for Climate Model Diagnosis and Intercomparison
(PCMDI), Lawrence Livermore National Laboratory
P.O. Box 808, L-103, Livermore, CA 94551-0808, USA
E-mail: *phillips14@llnl.gov*

Roger A. Pielke Sr.

Department of Atmospheric Sciences, Colorado State University
Fort Collins CO 80523, USA
E-mail: *pielke@snow.atmos.colostate.edu*

Roger A. Pielke Jr.

Center for Science and Technology Policy Research
University of Colorado/CIRES, UCB 488
1333 Grandview Ave., Boulder CO 80309-0488, USA
E-mail: *pielke@cires.colorado.edu*

Andrew J. Pitman

Macquarie University, Division of Environmental and Life Sciences
North Ryde, NSW 2109, Australia
E-mail: *apitman@penman.es.mq.edu.au*

Jan Polcher

Laboratoire de Météorologie Dynamique du CNRS
Université Pierre et Marie Curie, 4 pl Jussieu, F-75252 Paris, France
E-mail: *jan.polcher@lmd.jussieu.fr*

Steven D. Prince

Department of Geography, University of Maryland
1113 Lefrak Hall, College Park MD 20742-8225, USA
E-mail: *sp43@umail.umd.edu*

Robert Rabin

National Severe Storms Laboratory
National Oceanic and Atmospheric Administration
Norman, OK 73072, USA
E-mail: rabin@ssec.wisc.edu

Carmen Revenga

Senior Associate, Information Program
World Resources Institute (WRI)
10 G Street, NE, Washington DC 20002, USA
E-mail: carmenr@wri.org

Jeffrey E. Richey

School of Oceanography, University of Washington
Seattle, WA 98195-7940, USA
E-mail: jrichey@u.washington.edu

Steven W. Running

Montana Forest and Conservation Experiment Station
School of Forestry, University of Montana
Missoula MT 59812-1063, USA
E-mail: swr@ntsg.umt.edu

Joel Schafer

Biospheric Sciences Branch, Science Systems & Applications
NASA/Goddard Space Flight Center
Greenbelt, MD 20771, USA
E-mail: Joel.S.Schafer.1@gsfc.nasa.gov

Roland E. Schulze

School of Bioresources Engineering and Environmental Hydrology
University of KwaZulu-Natal
Pietermaritzburg
Carbis Road, Scottsville, 3201 Pietermaritzburg, RSA
E-mail: schulzeR@ukzn.ac.za

Maria Assuncão Silva Dias

IAG, Universidade de São Paulo
Rua do Matao 1226, São Paulo 05508-900 SP, Brazil
E-mail: mafdsdia@model.iag.usp.br

Thomas J. Stohlgren

U.S. Geological Survey
Fort Collins Science Center
Natural Resource Ecology Laboratory
Colorado State University, Fort Collins CO 80523, USA
E-mail: Tom_Stohlgren@USGS.gov

Kiyotoshi Takahashi

Meteorological Research Institute (MRI)
Japan Meteorological Agency (JMA)
Nagamine 1-1, Tsukuba, Ibaraki, 305-0035, Japan
E-mail: ktakahas@mri-jma.go.jp

Kirsten Thompson

World Resources Institute (WRI)
10 G Street, NE, Washington DC 20002, USA
E-mail: kirsten@wri.org

Christian Valentin

IRD, Institut de Recherche pour le Développement
32 rue Henri Varagnat, 93143 Bondy, France
E-mail: Christian.Valentin@bondy.ird.fr

Riccardo Valentini

Department of Forest Science and Environment
Università degli Studi della Tuscia
Via S. Camillo de Lellis, I-01100 Viterbo, Italy
E-mail: rik@unitus.it

Kristine Verdin

EROS Data Center
47914 252nd Street, Sioux Falls, SD 57198-0001, USA
E-mail: kverdin@edcsgw6.cr.usgs.gov

Reynaldo Luiz Victoria

Centro de Energia Nuclear na Agricultura
Universidade de Sao Paulo/Piracicaba
Avenida Centenario, 303, Piracicaba, SP 13416-000, Brazil
E-mail: reyna@mail.cena.usp.br

Pedro Viterbo

European Centre for Medium-Range Weather Forecasts (ECMWF)
Shinfield Park, Reading, Berkshire RG2 9AX, United Kingdom
E-mail: p.viterbo@ecmwf.int

Charles J. Vörösmarty

University of New Hampshire
Institute for the Study of Earth, Oceans & Space (EOS)
Complex Systems Research Center, Water Systems Analysis Group
Morse Hall, 39 College Road, Durham NH 03824-3525, USA
E-mail: charles.vorosmarty@unh.edu

Christopher P. Weaver

Center for Environmental Prediction and Department of
Environmental Sciences, Rutgers University
New Brunswick, NJ 08901, USA
E-mail: weaver@cep.rutgers.edu

Frank Wechsung

Potsdam-Institut für Klimafolgenforschung
Telegrafenberg, 14473 Potsdam, Germany
E-mail: Frank.Wechsung@pik-potsdam.de

David Werth

Department of Environmental Sciences, Cook College
Rutgers University
New Brunswick, NJ 08903, USA
E-mail: werth@cep.rutgers.edu

Yongkang Xue

Department of Geography
University of California, Los Angeles
1255 Bunche Hall, Los Angeles CA 90095-1524, USA
E-mail: yxue@geog.ucla.edu

Tetsuzo Yasunari

Frontier Research System for Global Change (FRSGC), and
Hydrospheric Atmospheric Research Center (HyARC)
Nagoya University, Nagoya, Aichi 464-8601, Japan
E-mail: yasunari@ihas.nagoya-u.ac.jp

Xubin Zeng

Department of Atmospheric Sciences
University of Arizona
PAS Building 81, Tucson AZ 85721, USA
E-mail: zeng@atmo.arizona.edu

Introduction

This volume is a synthesis of the research undertaken by the Biospheric Aspects of the Hydrological Cycle (BAHC) Core Project of the International Geosphere-Biosphere Programme (IGBP) since its inception in 1990. Before reading about the wealth of new insights that are presented in this volume, it is important to return to the origins of BAHC and our level of understanding of global change and the nature of the Earth system at that time.

The original aim of BAHC was to improve our knowledge of how terrestrial ecosystems and their components affect the water cycle, freshwater resources and the partitioning of energy on Earth. To address this overall goal, the project had four quite specific objectives:

- Development, testing and validation of models representing the transfer of water through the soil, vegetation and the atmosphere;
- Progress on how to aggregate the land-surface properties and fluxes from varied landscapes from local to regional scales;
- A study of the temporal and spatial diversity of biosphere-hydrosphere interactions;
- Development of a weather generator to both characterise an important part of the hydrological cycle and to provide time-varying boundary conditions to models.

These objectives reflect the state of understanding in the late 1980s. The dominant paradigm then was that the Earth's environment was largely controlled by the coupled dynamics of the planet's two great fluids – the atmosphere and the oceans. There was much debate in the global change community on whether biology had any significant role at all to play in Earth system dynamics. The terrestrial vegetation was considered to be a passive recipient of the impacts of changes in ocean-atmosphere dynamics; it was a spectator rather than a player in the functioning of the Earth system. Water and energy exchange between the land and the atmosphere was regarded in the same way. Indeed, the specific BAHC objectives listed above are highly suggestive of the terrestrial vegetation being thought of as a lower boundary condition for the atmosphere, where all of the action occurred.

Now, a decade later, one of the most important overall findings of IGBP research, highlighted at the Global Change Open Science Conference in Amsterdam in July 2001 and in the Amsterdam Declaration on Global Change, is that biology is a much more important player in Earth system dynamics than was earlier thought. More than any other Core Project in IGBP, BAHC research led to that conclusion. The achievements of BAHC over the past decade, as presented in this volume, thus go well beyond its original, more narrow remit; rather, the authors provide a new perspective on the interplay between two important components of the Earth system – the hydrological cycle and the terrestrial biosphere, i.e. vegetation.

Since 1990, the BAHC project has also gradually evolved to encompass additional questions, such as:

- To what extent are lateral fluxes of water, nutrients and sediments by rivers dependent on climate variability and/or direct human activities such as land use, pollution or river engineering?
- How can the dynamic context of Earth system changes and rapid human development – and their associated uncertainties – be expressed as vulnerabilities and risks that ultimately need to be dealt with?

Through more than a decade of BAHC activities, from local-scale experiments in areas comparable in size to a General Circulation Model grid square, large, regional field campaigns such as the LBA (Large Scale Biosphere-Atmosphere Experiment in Amazonia) or Asian GAME (GEWEX Asian Monsoon Experiment), to developing regional and global models, another important BAHC task gradually developed:

- How to consolidate the land-use and water datasets at different spatial and temporal resolutions.

All these stages of the BAHC project have been synthesised in the five parts of this book.

In *Part A* evidence is presented which convincingly demonstrates that the land surface does matter in weather and climate. This evidence comes from a whole range of spatial scales, from point and local measurement all the way up to global scale, multi-century modelling. The vegetation-atmosphere interactions that regulate local weather and hydrological balances and the regional climate have been fully demonstrated through a set of major international field campaigns, e.g. in the Sahel, Canada, China, Thailand and the Amazon, which were co-organised by BAHC scientists. These experiments brought about a step-by-step improvement in our knowledge on basic energy- and water-exchange processes such as precipitation, evaporation, transpiration, infiltration, surface runoff and on the related fluxes of trace gases and aerosols. These studies were carried out in major world biomes from the tropical wet forest to the boreal forest and thus have yielded important insights into the richness of response evoked by different land surfaces across the globe. These experiments also pointed out the importance, in ways not appreciated a decade ago, of feedbacks within the climate system caused by land-use and land-cover change such as deforestation and agriculture.

Model-based interpretation of palaeobotanic evidence of large-scale vegetation change – for example the greening of the Sahara and the shift of the boreal tree line several thousand years ago – leads to the conclusion that vegetation not only follows changes in the atmosphere, but it amplifies climate changes via its interaction with the other components of the natural Earth system. It is concluded that without consideration of biogeochemical, biogeophysical and biogeographical feedbacks, attribution of recent climate change as well as assessment of climate change in the upcoming decades are most likely to be incomplete.

As discussed in more detail in *Part B*, BAHC has played a major role in the planning and coordination of a series of international field experiments. The aim of these experiments has been to quantify how the land-surface functions at the regional scale and to provide the data to allow the fluxes of heat, water vapour and carbon to be represented in regional and global scale models. The experimental philosophy has been to measure all the components of the land-surface/atmosphere interaction over areas of around 100 km by 100 km. Experiments have been completed, or are under way, in key biomes of the world – concentrating on those which are likely to be vulnerable to climate change, or where land-use change may influence climate (please see Table C.1 for a list of field experiments).

Part B critically examines the progress made by this series of integrated land-surface experiments and assesses the development of our ability to measure each of the components of the land-surface/atmosphere interaction at different time and space scales. As a recommendation for future integrated experiments, the early involvement of social science is stressed and a need for a regional extension of research into the tropics is emphasized.

Arguably, the scientific efficacy of Earth system science today greatly depends on data co-registration, standardisation, assimilation, consolidation, maintenance and distribution. *Part C* of the book presents the scientific and technical tools needed to achieve the land-surface and water-related data consolidation at the interdisciplinary level. Data consolidation is the bringing together of related datasets from disparate sources, in differing spatial and temporal resolutions and in various formats, into an organised, standardised, co-registered and fully documented database for redistribution. Consolidation has two main facets. The first is the synthesis of disparate datasets across space and time to produce continuous, high-quality data. The other aspect is the standardisation of formats, documentation, attribution, and tools to generate access and distribute the consolidated datasets. Whereas synthesis makes the data useful, standardisation makes it useable.

The best path to data consolidation is a joint effort from both the largest and smallest data providers and users, including institutions that have historically restricted the free dissemination of data. Consolidation can only work if the providers of data are rewarded for their extra efforts, in the form of recognition and citation whenever their data are used. Success of data consolidation also requires a change in the thinking of scientific funding organisations. Non-operational data streams from individual scientific projects with little infrastructure must have motivation and support to participate in consolidation. The interdisciplinary and inter-project exchange that today's programme managers advocate would be much easier to attain if they specifically supported data management in each grant. Regardless of the approach, it is also necessary that all programmes cooperate and support a common set of standards across both national and disciplinary boundaries.

The global change research community has mostly focused its attention on the question of climate change and variability. A primary goal of *Part D* of this synthesis has been to assess the importance of additional factors shaping the character of rivers and associated drainage basins. This work has emphasized the nature of interactions among the physical, biological, and social dimensions of the land-based water cycle, expressed through the conceptual framework of the *drainage basin* as a functioning hydrological unit. A variety of spatial and temporal scales was considered, ultimately the full spectrum from patch-to-globe, from literally minutes-to-millennia.

These complex interactions between humans and the Earth system are illustrated in a set of three case studies: the River Amazon (Brazil), the River Elbe (Germany) and the Mgeni River (South Africa), representing multiple climatic, population density and economic development gradients. Humans exert an influence on the water cycle that goes beyond greenhouse warming and includes land-cover change (deforestation for agriculture and timbering, urbanisation), industrialisation, pollution, and water resources development.

Major human-derived impacts also include large increases in the residency time of river waters on land, large increases in erosion followed by interception of a substantial fraction of sediment destined for the world's oceans, order-of-magnitude increases in riverine nutrient flux in industrialised regions, severe loss of discharge to the coastal zones of river with large-scale irrigation works and flow diversion, virtually instantaneous changes in discharge regime due to reservoir operations, water balance feedbacks through widespread land-cover change. Semi-arid river basins such as the Colorado, Nile, Murray, Amu Darya, and Huang He are the most sensitive ones, combining high water use and regulation with extreme climate variability and they bear important consequences for freshwater resources, for ecosystems and for humans living in the coastal zone.

Part D promotes the use of integrated water resource management (IWRM) which should now integrate basin-wide, water-demand management within a broader Earth system perspective. Management of water systems must increasingly take into account not only their natural hydrology and biogeochemical cycles, but also their hydraulics and water management systems. IWRM is essential to the wise management of what are becoming, in many parts of the world, scarce water resources and can be used in fruitful ways to help manage regional water systems, including the politically sensitive issue of transboundary water systems.

Part E concludes that the involvement of land-surface processes as a major influence on regional and global climate variability and change significantly complicates the ability to project future climate in response to human disturbance. As a result of this lack of skillful predictive ability, we discuss the use of vulnerability assessments as the primary methodology to evaluate risks associated with environmental variability and change, including those risks associated with climate. Even when skillful projections are possible, starting with vulnerability assessments provides the quantification of the greatest threats to a resource.

The vulnerability paradigm is illustrated with several specific examples. A mathematical framework is introduced, and definitions of terms used are provided. This more holistic framework with which to evaluate environmental risk associated with human-caused and natural Earth system variability and change offers a more complete framework for policymakers than the reliance on projections which provide only a subset of future conditions.

This synthesis makes important contributions to the development of a holistic Earth system science approach. One of the strengths of BAHC has been its ability to place its work into the broader context of the Earth system and to engage and collaborate with others to generate new insights into the workings of our life support system. The research presented here on the coupled ocean-atmosphere-land dynamics shows how the nonlinear dynamics of the Earth system, revealed through the increasingly rich palaeo-record, can only be understood by considering the interactive coupling of components of the Earth system, some of which are now directly managed by humans, such as, for example, the biogeo-chemical cycle of nutrients within impacted and/or regulated river basins at the regional to global scale.

The past decade has been one of achievement, excitement and surprise, and it has raised new questions and challenges that have caused us to reflect on the nature of global change science and its ability to rise to these new challenges. The outstanding success of BAHC encourages us to meet these challenges. The science presented in this volume provides a critical underpinning for, and acts as a bridge to, the new approaches and structures needed to build a more integrative Earth system science. The BAHC community will now split; some parts are spearheading the development of a new IGBP project focused on the land-atmosphere interface, others are developing a project on global change and the water system, with a much stronger emphasis on terrestrial water resources and links to the socioeconomic and biodiversity sciences, while others will continue to contribute to the GAIM (Global Analysis, Integration and Modelling) goals and its reformulation, and to the emergence of an integrated Earth system perspective in IGBP.

These developments would not have been possible thirteen years ago. This volume, built on the efforts of a large international community of scientists, attested by the great number of contributors to this volume and of related papers, describes the exciting journey of understanding that has changed our view of the way in which the hydrological cycle interacts with terrestrial ecosystems and humans, from a lower boundary condition on a physical system to a central, dynamic feature of a single, integrated Earth system including the present-day Human dimension.

The Editors

Part A

Does Land Surface Matter in Climate and Weather?

Edited by Martin Claussen

Chapter A.1

Introduction

Martin Claussen

Traditionally, climate has been considered as the mean state of the atmosphere (e.g. Hann 1908), or averaged weather, including its statistical ensemble properties (WMO 1984). On the other hand, classifications of climate have often been developed on the basis of conditions essential to life, primarily for plants. Hence in many classical considerations of climate, its interaction with the biosphere played a dominant role (Bolle 1985). For example, Köppen (1936) described vegetation as "crystallised, visible climate" and referred to it as being an indicator of climate that is much more accurate than our instruments.

We interpret Köppen's statement in the sense that he considered vegetation as being so completely determined by climate as to be a perfect climate indicator. If Köppen (1936) had taken into account the possibility that vegetation could affect atmospheric and oceanic circulation, then he certainly would have sought a more "objective" parameter. In fact until recently, some climate researchers doubted that vegetation had a strong and significant impact on the large climate changes of the past. For example, when examining different theories of ice ages, deMarchi (1885) concluded that the occurrence of glacial epochs does not depend on changes in the "covering of the Earth's surface (vegetation)". Likewise, Alexander von Humboldt (1849) imagined the desertification of North Africa to be caused by an oceanic impact. He argued that somewhere in the "dark past", the subtropical Atlantic gyre was much stronger and flooded the Sahara, thereby washing away vegetation and fertile soil. Today, we are convinced that vegetation itself affects the development of deserts strongly – as we shall demonstrate in Chapt. A.5.

When considering energy storage capacities, it seems reasonable to identify the ocean as the main driver of climate changes on time scales longer than a year (Peixoto and Oort 1992). Accordingly, coupled atmosphere-ocean models were regarded as state-of-the-art climate models (Cubasch et al. 1995) and many modellers would still hold this view (Grassl 2000). The nature and the distribution of global vegetation patterns in these models are kept constant in time. Only short-term plant physiology and, to some extent, fractional vegetation and leaf area is allowed to change with meteorological conditions.

More recently, climate system models have been built in which not only atmospheric and oceanic but also vegetation dynamics are simulated explicitly (e.g. Petoukhov et al. 2000; Cox et al. 2000; Friedlingstein et al. 2001). This development is in line with a more physical, not merely a statistical, view of climate in terms of a dynamic system. In the new concept (e.g. Peixoto and Oort 1992), the climate system is defined to encompass not only the abiotic world or, as it is sometimes called, the physical climate system – the atmosphere, hydrosphere (mainly the oceans but also rivers), the cryosphere (inland ice, sea ice, permafrost and snow cover), the pedosphere (the soils) and the lithosphere (the Earth's crust and the more flexible upper Earth's mantle) – but also the living world, the biosphere (see Fig. A.1). Hence in the new concept, vegetation appears as a dynamic state variable of the (global) climate system.

In this part of the book, we present evidence that forces us to believe that land surface does matter in weather and climate. This evidence comes from a whole range of spatial scales, from point and local measurement all the way up to global scale multi-century modelling. Accordingly, this part is organised in the following way (see Fig. A.2).

In the next three chapters, we summarise our knowledge on the interaction between vegetation and atmosphere at the local, the regional and the global scale. In Chapt. A.2, we discuss the (physical) concept of land-surface/atmosphere interaction. We will not derive a complete mathematical theory. Instead, we will present an outline of the basic ideas. Moreover, we will highlight contributions that various international projects have made to this problem. Because Chapt. A.2 deals mainly with the direct land-surface/atmosphere interaction which takes place within a few decametres of the surface, we have chosen "the climate near the ground" as a title.

Land-surface/atmosphere interaction is not only a local phenomenon, nor is the land surface homogeneous. Hence, one might ask: does the heterogeneity of land surfaces matter in weather and climate? And if it does,

Fig. A.1.
Sketch of the climate system as a dynamical system encompassing the abiotic world, sometimes referred to as the physical climate system, and the living world, the biosphere. The subcomponents of the climate system interact via fluxes of energy, momentum, water, and biogeochemical substances such as carbon, nutrients etc.

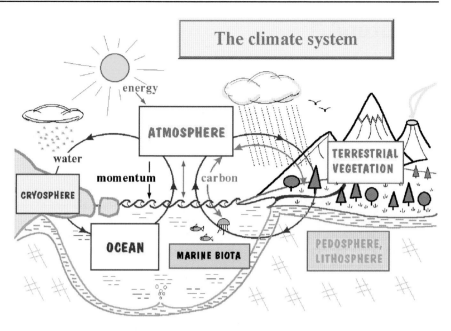

to what extent? These questions will be tackled in Chapt. A.3, "the regional climate", where we demonstrate that changes in the regional structure of landscapes can indeed affect global atmospheric dynamics.

In Chapt. A.4 we pursue the global perspective. We will present evidence that the (dynamic) climate system cannot be described properly if vegetation dynamics are ignored. The discussion focuses on theoretical considerations of global atmosphere/land-surface interaction, on interpretation of palaeoclimatic records, on sensitivity studies of biogeochemical processes (i.e. mainly the carbon cycle), on global weather forecasts and, finally, on the seasonal dynamics of soil moisture.

In many regions on Earth, changes in land cover affect local climate. Nonetheless, changes in the large-scale circulation seem to be dominant, as seen in the Mediterranean region for example (Bolle 2003). Some regions on Earth, however, appear to be "hot spots", i.e. regions of strong atmosphere/land-surface interaction, and these regions are discussed in detail in Part A. The Sahelian (Chapt. A.5) and the boreal (Chapt. A.7) regions are transition zones from a hot desert to tropical forests and from polar deserts to boreal forests, respectively. In these regions, changes in the land surface induce changes in the atmosphere, and both changes amplify each other. The Amazon forest (Chapt. A.6) is important not only for the global climate system as a very active region of the water cycle but also as a major source of oxygen and a major pool of biomass and biodiversity. The Asian monsoon is perhaps the largest circulation driven by land-atmosphere-ocean interaction. Any changes in this system, for example large-scale changes of the land surface owing to land use, can potentially affect the life of more than 1.5 billion humans (Chapt. A.8).

The reader should not be surprised if some (actually only a few) redundancies occur. This is intended. All chapters can be read separately; however, only the whole of Part A yields the complete picture.

Fig. A.2.
Flow chart of Part A

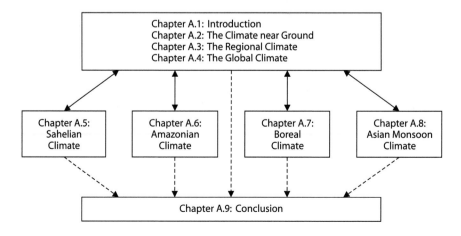

Chapter A.2

The Climate near the Ground

Andrew J. Pitman · Han Dolman · Bart Kruijt · Riccardo Valentini · Dennis Baldocchi

A.2.1 Introduction

This chapter provides a brief theoretical framework to explain how and why the land surface can affect weather and climate. The key equations that describe how energy and water are exchanged between the atmosphere and the land surface are explained. The rest of Part A builds on this framework to illustrate examples of where evidence exists to support the role of the land surface in weather and climate.

A.2.2 The Surface Energy Balance

The amount of energy available to the climate is controlled by the global energy cycle which is largely dominated by atmospheric processes (see Rosen 1999). Of 100 units of energy entering the global climate system, 46 are absorbed by the surface and 31 are exchanged

as sensible and latent heat (Fig. A.3, from Rosen 1999). The land surface influences significantly the way that these 31 units of energy are partitioned between sensible and latent heat, and acts as a significant medium to store energy on both the diurnal and seasonal time scale. The land surface is also a major store of carbon in both the vegetation and the soil. These roles combine to make the land surface a key component of the climate system.

Net Radiation

The driving force for the climate system is the Sun. The Sun emits shortwave radiation which is reflected, absorbed or transmitted by the atmosphere. An amount of energy (S) reaches the Earth's surface and some is reflected (depending on the albedo, α) (see Fig. A.4). Infrared radiation is also received ($L\downarrow$) and emitted ($L\uparrow$) by the Earth's surface (depending on the temperature

Fig. A.3.
Schematic diagram of the annual mean global energy balance. Units are percent of incoming solar radiation. The solar (shortwave) fluxes are shown on the *left-hand side*, and the terrestrial (longwave) fluxes are on the *right-hand side* (after Rosen 1999)

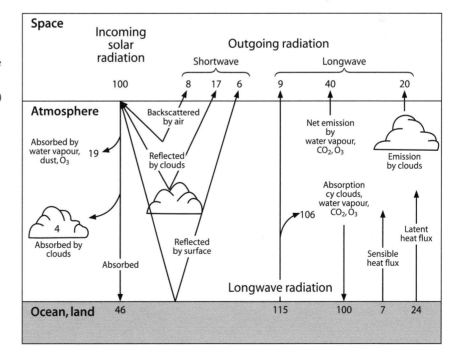

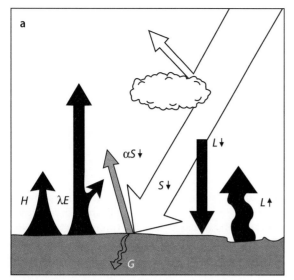

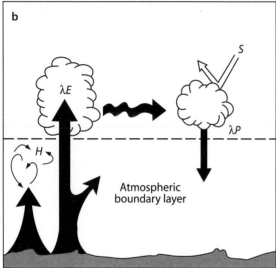

Fig. A.4. Interactions between the land surface and the atmosphere that have direct impacts on the climate system. **a** Surface radiation budget; **b** effect of heat fluxes on the atmosphere (after Sellers 1992)

The calculation of G has generally been approached as a simple heat diffusion problem (Sellers et al. 1997) using a range of approaches from force-restore (e.g. Deardorff 1977) to multilevel soil models that account for phase changes within a heterogeneous soil profile. A difficulty in updating R_n is that as the fluxes on the right hand side of Eq. A.2 change, the temperature of the surface changes. This affects R_n through $L\uparrow$ as well as through the subsequent evolution of the fluxes.

In order to obtain a reasonable estimate for R_n, the emissivity of the surface must be known (it is usually assumed to be 1.0). The albedo also needs to be calculated and this can be difficult since it can vary diurnally with solar insolation angle, seasonally with vegetation changes and stochastically with rain or snowfall.

In terms of the climate system, it is important to model the partitioning of available energy between sensible and latent heat as well as possible since more latent heat contributes water vapour to the atmosphere and tends towards increasing cloudiness and precipitation, while increases in sensible heat tends to warm the planetary boundary layer (Fig. A.4). Given the key role that latent and sensible heat play in the climate system, it is necessary to simulate the diurnal, seasonal and longer term variations in these fluxes as well as possible and this has become a key focus in how land-surface processes are represented in numerical models of weather and climate. Basically, the sensible heat flux can be represented as a quasi-diffusive process which can be written in the potential difference resistance form as:

$$H = \frac{T_s - T_r}{r_a} \rho c_p \tag{A.3}$$

In order to solve this equation, the surface temperature is required which can be complicated to obtain in the presence of sparse vegetation or other surface heterogeneity. In addition, the aerodynamic resistance (r_a) must be represented.

The Aerodynamic Resistance

The aerodynamic resistance is a turbulent diffusion term which impedes the transfer of momentum and scalar properties, such as heat, water vapour and CO_2, away from the vegetation surface to the free atmosphere. It is inversely dependent on the wind speed and the logarithm of the surface roughness length. This is, in turn, a function of the drag properties of the land surface. Stability corrections need to be applied to account for the effects of convection on r_a. The aerodynamic resistance can be written as

$$r_{am} = \int_{d+z_0}^{h} \frac{dz}{K_m} \tag{A.4}$$

and emissivity of the land and atmosphere). The net balance of the incoming and reflected shortwave radiation, and the net balance of the downwelling and emitted long-wave radiation at the Earth's surface is called net radiation.

$$R_n = S\downarrow (1 - \alpha) + L\downarrow - L\uparrow \tag{A.1}$$

The primary role of the land surface is to partition this net radiation into the turbulent energy fluxes (sensible heat, H and latent heat, λE) and the soil heat flux (G) shown schematically in Fig. A.4:

$$R_n = H + \lambda E + G \tag{A.2}$$

Fig. A.5.
Schematic of a leaf cross section, showing links between stomatal gas exchange and photosynthesis (after Sellers 1992)

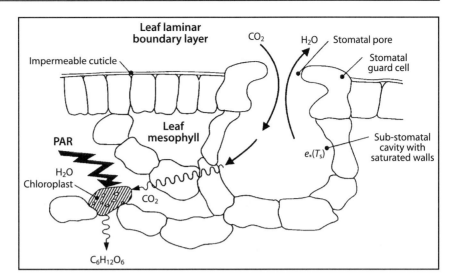

where d is the zero plane displacement height, z_0 the aerodynamic roughness length, and K_m the eddy diffusivity for momentum (Thom 1972). There are two noteworthy features concerning this equation. Firstly, the equation appears to describe momentum transport only and secondly, it uses a mean eddy diffusivity.

Momentum transport in real canopies takes place via eddy diffusion and pressure forces. It is the latter process that makes momentum transfer different from transport of scalar quantities. In this case, for instance for heat transport, an extra resistance (quasi laminar resistance, r_b, Verma et al. 1989) has to be introduced that effectively describes diffusion across the quasi laminar boundary layer around the canopy:

$$r_{ah} = \int_{d+z_0}^{h} \frac{dz}{K_m} + r_b \qquad (A.5)$$

The parameterisation of r_b is difficult since it depends on the ratio of the roughness lengthes of heat and momentum.

The problem in modelling the aerodynamic transfer from plant canopies arises when the sources of latent and sensible heat are not the same (e.g. as in the case for sparse canopies). Here, dual source models are required to describe the energy transfer from vegetation and soil to the atmosphere (Shuttleworth and Wallace 1985). The use of K-theory to describe the aerodynamic transfer from and in canopies breaks down when the length scale of the eddies is no longer comparable to the size of the mean gradient and becomes comparable to the height of the canopy. This gives rise in forests, for example, to the phenomenon of counter-gradient fluxes (Denmead and Bradley 1985). This is caused by eddies, responsible for the transport, operating at larger length scales than the scales at which the mean gradients change. In practice, these constraints on the use of K-theory play a less

important role, as for forests, the mean resistance under dry canopy conditions is within the stomata and for less dense canopies the foliage is less dense and distributed more uniformly (Dolman and Wallace 1991).

Where K-theory fails, a different framework to model canopy turbulence can be used (e.g. Raupach 1989) based on Lagrangian (fluid following) theory or using a plane mixing layer rather than as a boundary layer. Kruijt et al. (2000) found that for a tall Amazonian rainforest the length scales were smaller than suggested by the plane mixing layer analogy, suggesting that the aerodynamic resistance is also smaller than predicted.

Representing Evaporation

The representation of evaporation within the land surface has undergone profound changes over the last few decades as our understanding of the physical and biological controls on evaporation have increased. The calculation of evapotranspiration was represented traditionally by an equation of the form:

$$\lambda E = \beta \frac{e^\star(T_s) - e_r}{r_s} \frac{\rho c_p}{\gamma} \qquad (A.6)$$

where β is a moisture availability term which ranges non-linearly between zero and unity, and r_s is a resistance to the transfer of mass from the surface to the air. In order to represent β, information regarding the soil moisture is required. A second major difficulty is the representation of r_s which is relatively complex and couples the flow of water between the leaf and the atmosphere with the flow of CO_2 (Fig. A.5). In early models r_s was ignored, assumed to be a constant, or made dependent on just vegetation type. Sellers et al. (1997) refer to these as "first generation" models (Fig. A.6). This sim-

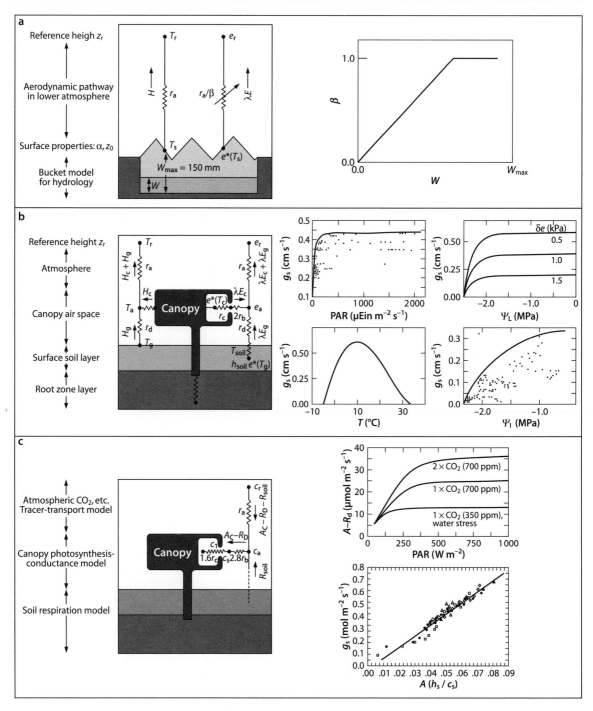

Fig. A.6. Development of land-surface parameterisations in atmospheric general circulation models. **a** First-generation "bucket" model. *Left*: The sensible and latent fluxes flow from the surface to the atmosphere through an aerodynamic resistance r_a. *Right*: The latent heat flux is regulated by the level of moisture W in the bucket through the moisture availability function β. **b** Second-generation model with separated vegetation canopy and soil. Moisture fluxes from the (dry) vegetation canopy escape to the canopy air space and free atmosphere through a bulk canopy surface resistance ($r_c = 1/g_c$). Sensible and latent heat fluxes from the ground (subscript "g") and canopy (subscript "c") combine in the canopy air space to give the total fluxes passed to the atmosphere. *Right*: Terms that relate to leaf conductance (g_s), light intensity (PAR), vapour pressure deficit (δe), temperature (T), and leaf water potential (ψ_l). The *points* denote observations; the *lines* denote a hypothetical limit to the $f(\psi_l)$ relation. The canopy resistance r_c can be calculated by integrating g_s over the depth of the canopy and inverting it. **c** Third-generation model. A carbon flux pathway is added to the moisture and heat flux pathways shown in (**b**). *Left*: The canopy A-g_s model controls canopy carbon and water fluxes simultaneously and consistently. *Upper right*: Dependence of leaf assimilation A on light (PAR) intensity, CO_2 concentration, and water stress. *Lower right*: Relation between g_s and A for a number of species, with a line of best fit (from Sellers et al. 1997, © American Association for the Advancement of Science, reprinted with permission; see there for additional references)

ple approach was adequate in these early models, but the evolution in our understanding of the role of the vegetation in climate has led to the recognition that for the land surface to play a realistic role in the evolution of the Earth's climate, the biophysical feedback of vegetation on climate must be represented. This biophysical feedback links changes in the climate with changes in evapotranspiration which result from changes in the CO_2 concentration in the atmosphere. In order to represent this in models of weather and climate, the interactions between CO_2 and the vegetation must be represented. Simple models of r_s fail to do this and thus more biophysically realistic approaches have been developed.

Monteith (1965) was the first to suggest that the control vegetation exerts over transpiration can be expressed as a canopy resistance in the flow path of water evaporation from the saturated inner surface of the stomata to the outside atmosphere. The crucial development was to conceive the vegetation as a "single big leaf" that exerts control over evaporation in much the same way as single leaves control their evaporation. This approach has become the cornerstone in the development of the "second generation" of biophysical models (Sellers et al. 1997, Fig. A.6), such as BATS (Biosphere-Atmosphere Transfer Scheme, Dickinson et al. 1986), SiB (Sellers et al. 1986) and a series of similar models. Jarvis (1976) developed a stomatal conductance model that predicted the response of stomata to environmental variables, such as atmospheric humidity deficit, temperature, solar radiation (or PAR, Photosynthetically Active Radiation), leaf water potential and ambient CO_2 concentration. Although in the case of atmospheric humidity deficit, the cause of the response is still disputed, by and large, individual leaves exhibit similar behaviour when subjected to these environmental stresses. Stewart (1988) and Dolman et al. (1988) showed that at canopy level, forests exhibit similar responses and hence the behaviour of canopy conductance could be predicted well, when a model was calibrated previously on micrometeorological measurements. Unfortunately, using a temperature function from one forest with another for humidity deficit from another forest, is not advisable hence a new canopy conductance model needs to be calibrated each time.

A different approach is to take advantage of the close relationship of photosynthesis and stomatal conductance as is observed at leaf level (Ball et al. 1987). These models are based on the premise that stomata have evolved to serve the conflicting roles (e.g. Cowan and Farquhar 1977) of supporting photosynthesis (the uptake of CO_2) and restricting evaporation (release of H_2O). The resulting model, for which experimental evidence is increasing, divides the response of stomata into two broad categories: those that depend on photosynthesis directly, and those that are independent of photosynthesis. Factors of the first category generally maintain a constant proportionality of conductance to pho-

tosynthesis (g_s/A), while factors of the other category, those that affect the diffusion of CO_2 and H_2O, generally cause g_s/A to change. These considerations lead to the following equation:

$$g_s = \frac{mAh_s}{c_s} + b \qquad (A.7)$$

where m and b are the slope and intercept respectively of the linear curve relating g_s to A, and h_s and c_s respectively the relatively humidity and CO_2 concentration at the surface. This model is generally known as the Ball-Berry model and Sellers (1992) used it to generate a canopy scale model that effectively linked canopy conductance and photosynthesis. The advantage of this "third generation" model (Sellers et al. 1997, Fig. A.6) is that the biochemistry of photosynthesis is part of such a model, and the model provides a direct output of CO_2 uptake that can be used to derive net primary production and other derived quantities. The photosynthesis models are generally based on Farquhar and Caemmerer (1982) that describe the rate of CO_2 assimilation as the minimum of a Rubisco regeneration limited rate and an electron transport limited rate.

To scale the Ball-Berry model to canopy level, assumptions need to be made about the distribution of photosynthetic capacity and radiation. Sellers et al. (1989) and most people after them, assume that the distribution of nitrogen and maximum carboxylation capacity (V_{max}) is in proportion to the distribution of absorbed radiation. Radiation is assumed to follow Beer's law, but more complicated schemes may be used that distinguish between dark and sun lit and clumped leaves. This to a large extent is a matter of personal preference and has to be weighed against other uncertainties in the model parameters.

When such a conductance model, based on generally valid biochemical relations and properties is applied, the number of parameters used decreases considerably, compared to the stress function approach. This may be in appearance only: in practice there is uncertainty also about the specific values used in the photosynthesis models, such as the temperature response. However, the clear bonus of using this approach with photosynthesis is that the land-surface model can also be used to predict CO_2 uptake. This is particularly relevant with the increasing importance of the biospheric uptake of CO_2 (e.g. Cox et al. 2000) to land-surface modelling.

Summary

One of the main roles of the land-surface parameters in weather and climate modelling is therefore to simulate the partitioning of net radiation between latent and sensible heat fluxes. It has now been realised that in order

Fig. A.7.
Schematic illustration of the
Earth's water cycle (after Oki
1999)

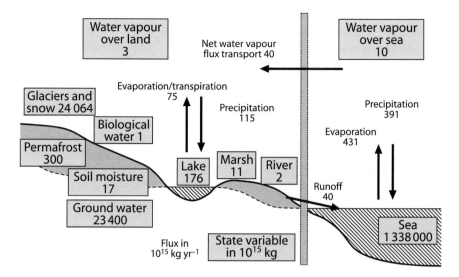

to do this in a way that is consistent with the role vegetation plays in the carbon cycle, a biophysical approach to the representation of the evapotranspiration process is required, linked tightly to the exchange of carbon dioxide between the surface and the atmosphere. As this understanding has evolved, a second major area of research, the role that the land surface plays in the partitioning of available water between evaporation and runoff, has also developed.

A.2.3 The Surface Water Balance

While most water is stored in the oceans rather than on the land surface (see Oki 1999 for a detailed discussion of the global water cycle) the proportion of the total water stored in the soil (0.0012%, Oki 1999) plays a vital role in the provision of food and freshwater. The hydrological cycle (Fig. A.7) is linked to the surface energy balance through the latent heat flux, the energy flux associated with the evaporation of water. The focus on soil moisture within climate modelling has increased in the 1990s following major internationally coordinated efforts within the International Geosphere-Biosphere Programme (in particular, the core project BAHC) and the World Climate Research Program (in particular GEWEX, the Global Energy and Water Cycle Experiment). These efforts, most identifiable by the Project for the Intercomparison of Land-Surface Parameterization Schemes (PILPS; Henderson-Sellers et al. 1995) and the Global Soil Wetness Project (GSWP; Dirmeyer et al. 1999) have led to significant improvements in our understanding of the role of the surface water balance in weather and climate.

Water is lost from the surface through two major mechanisms. Water falling as precipitation may be intercepted by a canopy and re-evaporated or may reach the soil surface, infiltrate and then be absorbed by roots,

and subsequently transpired. The Earth's surface may also lose water via runoff. Runoff used to be considered relatively unimportant to weather and climate but, in an elegant illustration of the complexities and nonlinearities within the system, recent work has highlighted the role of freshwater input into the oceans in influencing the thermohaline circulation which in turn may be a significant factor in explaining long-term climate changes. Runoff also has a significant role to play in large scale hydrology validation (see Chahine 1999).

The basic role of the land-surface model is to partition available water (assumed here to be precipitation, P, but it could also include snow melt), between evaporation (E) and runoff which is usually split between a fast component (R_{surf}) and a slow component (R_{drain}). Soil moisture is modelled as a change in the storage component (ΔS). Overall, this balancing of incoming and outgoing fluxes of water is called the surface water balance:

$$P = E - R_{drain} - R_{surf} - \Delta S \qquad (A.8)$$

Climate models solve this equation using a representation of the land surface which might range from a simple representation of water storage in a globally constant moisture holding capacity (e.g. Manabe 1969), to extremely complex models with many levels for soil moisture. Irrespective of the complexity of the land-surface scheme, or the philosophy underpinning the model, all land-surface schemes must simulate a quantity representing runoff and a quantity representing soil moisture.

Soil Moisture

Most land-surface schemes used in climate models solve the one-dimensional mass conservation equation based on Darcy's Law or Richards' (1931) equation. For example, the change in water within the soil with time can be expressed by:

$$\frac{\partial w}{\partial t} = -\frac{1}{\rho_w}\frac{\partial Q}{\partial z} + s_w \qquad (A.9)$$

where w is the volumetric soil moisture content, ρ_w is the density of liquid water, z is depth, Q is the vertical soil water flux and s_w is a moisture sink term which includes evaporation and runoff. From this equation, differences between models evolve from choices resulting from decisions on how to model individual aspects of s_w, how to split z into layers and how to assign parameters (e.g. hydraulic conductivity). At the most basic level, Manabe (1969) parameterised available soil moisture by assuming a 15 cm soil moisture holding capacity globally. Robock et al. (1995) have shown that the model works quite well in comparison to observations and in comparison to a more complex model. PILPS has shown that the soil moisture simulated by the Manabe (1969) scheme is within the range simulated by other models (e.g. Shao and Henderson-Sellers 1996), and provided E is calculated properly, the model cannot be distinguished from more complex schemes at longer timescales (Desborough 1999).

However, there is a majority feeling in land-surface modelling that a more complex parameterisation of soil moisture is required. There is a long history of more complex schemes, but one of the first was developed by Deardorff (1977) and is used by, for example, Noilhan and Planton (1989). This model is known as the "force-restore" model and consists of two soil layers: a thick upper layer of the order of 10 cm and a deeper second layer of some 1–2 m which account for two time scales of moisture variability (see Noilhan and Mahfouf 1996). Many of the more recent land-surface models include more complex multi-layer diffusion models. In general, most land-surface schemes choose to use three layers, but some have now employed five or more (e.g. Boone and Wetzel 1996).

There are major difficulties in simulating soil moisture in climate models and various land-surface models simulate very different amounts for a given state of the climate. These are not believed to affect the reliability of the climate models for simulating the climate, or changes in the climate. This is because soil moisture is simulated to provide information for calculating evaporation so even if a large difference in soil moisture exists between two models, this need not be important provided the evaporation parameterisation is adjusted. This does cause major difficulties, however, in comparing modelled soil moisture with observed soil moisture, or in using model-derived soil moisture in impact studies.

Runoff

Runoff is usually modelled very simply in climate models even though it is known to affect climate model simulations (Viterbo and Illari 1994). At the simplest level, Manabe (1969) assumed that runoff was zero unless water (precipitation or snow melt) was added to a reservoir where the soil wetness (W) was above a critical value (W_c):

$$R = \begin{cases} 0 & \text{if } W < W_c \\ P - E & \text{if } W \geq W_c \end{cases} \qquad (A.10)$$

There is an increasing consensus that in order to simulate runoff, schemes need to differentiate between the two basic types or time scales of runoff – a quick (infiltration excess) component and a slow (drainage) component (Dooge 1982, Viterbo 1996).

Surface runoff can be parameterised very simply (i.e. all incident water infiltrates until saturation and all remaining water becomes runoff) or in complex ways based on a parameterisation of the rate at which water infiltrates into the soil, or in a wide variety of ways in between including calculating infiltration as the residual of the surface water balance. Unless routing is employed, the surface runoff is usually assumed to be lost into the ocean *immediately*. This appears to be a gross simplification, but at the spatial scales of climate models, and for the purposes of simulating the climate, it may be reasonable. The problem develops, however, when attempts are made to interpret this quantity for regional scale impacts assessment. Further, the work of Chen et al. (1997), Liang et al. (1998) and many others have demonstrated the strong interaction between the surface water balance and the surface energy balance. This leads to the obvious conclusion that systematic errors in the modelling of surface runoff lead inevitably to systematic errors in the modelling of the partitioning of available energy between latent and sensible heat, and presumably impacts on the simulation of the climate.

Drainage is usually assumed to occur at a rate defined by the hydraulic conductivity (k) using Darcy's Law. This is achieved following a method such as Clapp and Hornberger (1978) such that:

$$k = k_{sat}\left[\frac{\theta_1(z)}{\theta_p}\right]^{(2b+3)} \qquad (A.11)$$

where k_{sat} is the saturated hydraulic conductivity and θ_p is the pore volume fraction. Cosby et al. (1984) derived statistical relationships between b, θ_p and k_{sat} which are commonly used within climate models (Viterbo and Beljaars 1995, discuss the implementation of these equations in more detail). Global datasets for k_{sat}, θ_p, and b

must be available for several soil texture classes to use these parameterisations or single global constants must be used. It is usual to ignore variation in these parameters with depth (z) given the scarcity of data.

The lack of any reliable data on the spatial heterogeneity in these parameters at the scale of a climate model grid square, combined with the lack of any quality data for a reasonable fraction of the globe, introduces uncertainty in our ability to model the surface water balance. Pitman et al. (1999), Chen et al. (1997), Wetzel et al. (1996a) and Liang et al. (1998) have all shown that a suite of land-surface schemes partition rainfall between evaporation and runoff very differently. Some of this variability is driven by the parameterisation of processes which generate runoff and the differences in the storage characteristics of the models (Gedney et al. 2000).

While the simulation of runoff is difficult in weather and climate models, effort in this direction is very worthwhile and progress is being made. Runoff is probably the only quantity simulated by land-surface models that can be validated at the large scale. One of the problems is that runoff, simulated by land-surface schemes, represents the water that is lost *locally* by the climate model grid square. This cannot be compared in its original form to observed runoff at a measurement location remote from the grid box, because there is a time factor for the transfer of local runoff across a landscape. Some land-surface schemes now therefore include river routing (e.g. Sausen et al. 1994; Hagemann and Dümenil 1998) in order to simulate the annual cycle of river discharge into the ocean. This appears to improve the modelling of runoff from some large drainage basins (Dümenil et al. 1997) although water storage and runoff in regions of frozen soil moisture remain outstanding problems (Arpe et al. 1997; Pitman et al. 1999).

It is possible to take the runoff component of some hydrological models (e.g. ARNO, TOPMODEL) and incorporate them into the physics component of the climate model (see for example Dümenil and Todini 1992; Stamm et al. 1994). Global fields of the parameters required by the hydrological component then need to be provided at the resolution of the climate model. A good example of this being achieved is the VIC model (Wood et al. 1992). Tightly linked with the attempts to couple hydrological models of soil moisture and runoff into climate models is the need to include river routing. Routing models in climate models cannot rely on parameters tuned on the basis of inflow and outflow data which are generally unavailable on the global scale (Todini and Dümenil 1999) and may change fundamentally as ecosystems respond to increasing atmospheric carbon dioxide. Liston et al. (1994) and Vörösmarty et al. (1989) have developed independently river routing schemes for use within climate models based on a linear reservoir. Miller et al. (1994), and Hagemann and Dümenil (1998)

have developed models to be applied globally. All these models offer potential and are useful developments but weaknesses in their application are linked to major difficulties in parameter estimation and even availability of appropriate databases for simple things such as vegetation, soil types, land use, topography (Todini and Dümenil 1999).

Summary

The partitioning of available water between runoff and evaporation is a key role that the land surface plays in weather and climate. A major effort in how to model soil moisture, how to model runoff and how to represent other key aspects of the land surface with regards to the surface water balance has been made in the last decade. Some of the progress achieved in the last decade is reported elsewhere in this book (see Chapt. B.9 and D.2).

The development of land-surface models over the last couple of decades has been a very major international collaborative effort. In the last decade, two major intercomparison exercises have developed that use a range of methods to explore the relationship between model parameterisation and performance. PILPS (Henderson-Sellers et al. 1995, 2002; see also special issues of Global and Planetary Change, volumes 13 (1996) and 19 (1998)) and the Global Soil Wetness Project (GSWP; Dirmeyer et al. 1999, and the Special Issue of the J. Meteorological Society of Japan, volume 77 (1999)) have led point-based and catchment-based intercomparisons (in the case of PILPS) and a global intercomparison in the case of GSWP. These substantial efforts have led to increased recognition of the nature of land-surface models. The recent integration of PILPS and GSWP into the new Global Land-Atmosphere System Study (GLASS) offers considerable potential to further improve land-surface modelling in the future.

A.2.4 Observing the Surface

Study of the Earth's biogeochemistry and hydrology involves quantifying the flows of matter in and out of the atmosphere with an array of methods (Canadell et al. 2000). At the global scale, scientists assess carbon dioxide sources and sinks using inversion modelling of CO_2, ^{13}C and O_2 concentration and wind fields (Ciais et al. 1995; Denning et al. 1996). At regional and continental scales, this approach is limited by the sparseness of the measurement network and their biased placement in the marine boundary layer (Denning et al. 1996). Instruments mounted on satellite platforms view the Earth in fine detail (1 km to 30 m resolution) and offer the potential to evaluate surface carbon fluxes on the basis of algorithms that are driven by reflected and emitted radiation measurements (Running et al. 1999; Cramer et al. 1999), but these need ground truth data to validate the algorithms. One measure of carbon flux "ground truth" can be provided by biomass surveys (Kauppi et al. 1992; Gower et al. 1999). However, these surveys provide information on decadal time scales and do not provide information on shorter-term physiological forcings and mechanisms that are needed by satellite-driven algorithms. Forest inventory studies are labour intensive and inferential estimates of net carbon exchange and rarely measure growth of small trees and below-ground carbon.

The eddy covariance method (see Chapt. B.4), a nonintrusive automated micrometeorological technique, provides a direct measure of net carbon and water fluxes between vegetated canopies and the atmosphere (Baldocchi et al. 1988; Aubinet et al. 2000). The eddy covariance method is able to measure fluxes over hours, days, seasons and years with minimal disturbance to the underlying vegetation with a relatively large area of land (footprints have longitudinal length scales of 100 to 2 000 m).

With the eddy covariance method we can measure, at the stand scale, how ecosystems respond to environmental forcings. Response functions, generated by a network of carbon flux measurement sites can be used to validate and improve upon algorithms being used by remote sensors and global and regional scale modellers to scale carbon and water fluxes from landscape to regional and continental scales. Direct flux measurements also identify new properties that emerge as we transcend scales from the leaf to canopy and canopy to landscape scales.

Technological advances during the 1980s enabled the eddy covariance method to be applied over crops (Anderson et al. 1984; Desjardins et al. 1984; Ohtaki 1984), forests (Verma et al. 1986) and native grass (Verma et al. 1989). As technology became more reliable, micrometeorologists started to conduct integrated studies over selected portions of the growing season. The most well-known of such investigations that involved CO_2 flux measurements are the FIFE (First ISLSCP Field Experiment; Sellers et al. 1988), HAPEX-Sahel (Hydrological Atmospheric Pilot Experiment; Goutourbe et al. 1997a), and ABRACOS (Gash et al. 1996) experiments. By 1990, technology allowed eddy fluxes to be measured for extended periods. Wofsy et al. (1993) and Vermetten et al. (1994) in the Netherlands were among the first investigators to attempt to measure CO_2 and water vapour fluxes continuously over a forest over the course of a year with the eddy covariance method. Spurred by these two pioneering studies, other towers were soon established and operating by 1993 in North America (Oak Ridge, TN, Greco and Baldocchi 1996; Prince Albert, Saskatchewan, Canada, Black et al. 1996) and Japan (Yamamoto et al. 1999) and in Europe by 1994 (Valentini et al. 1996). This era also heralded the start of longer, quasi-continuous integrated experiments, as noted by the BOREAS (Boreal Ecosystem-Atmosphere Study; Sellers et al. 1997) and NOPEX (Northern Hemisphere Climate-Processes Land-surface Experiment; Halldin et al. 1999) experiments.

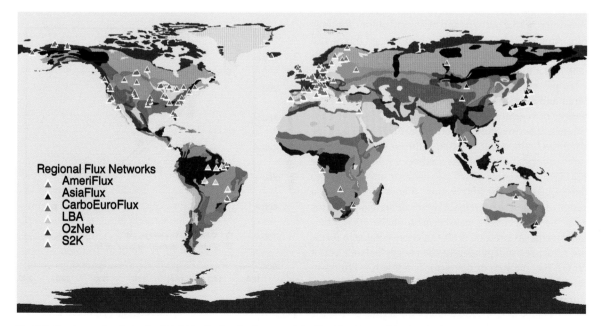

Fig. A.8. Flux tower sites from the Oak Ridge National Laboratory Distributed Active Archive Center (ORNL DAAC). FLUXNET web page available on-line [*http://public.ornl.gov/fluxnet/ecoregions.cfm*] from ORNL DAAC, Oak Ridge, Tennessee, USA, accessed January 10, 2002

Fig. A.9.
Annual course of net ecosystem CO_2 exchange for three temperate deciduous forests, one in Oak Ridge, TN, Petersham, MA, and Denmark

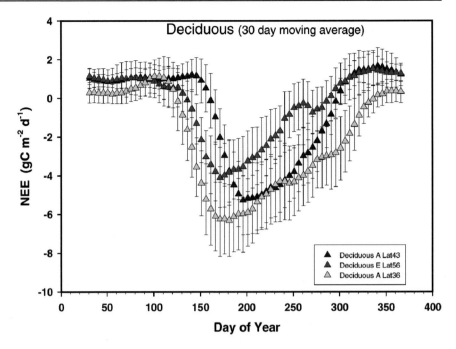

Recently a great advance in continuous long-term measurements of biospheric exchanges has been achieved. FLUXNET is a global network of long-term micrometeorological flux measurement sites at which the exchanges of carbon dioxide, water vapour and energy between the biosphere and atmosphere are measured. Through FLUXNET, the data streams from various regional flux measurement networks (AmeriFlux, CarboEurope, AsiaFlux, and OzFlux) are combined and a mechanism for synthesising data across sites and biomes is realised. The goals of FLUXNET include quantifying temporal dynamics, spatial patterns and biotic and abiotic forcings of carbon dioxide, water vapour and energy fluxes and providing data for the validation of remote sensing products that are inferring fluxes of carbon dioxide.

At present over 140 flux measurement sites are registered and operating on a continuous basis (Fig. A.8). A variety of vegetation is under study at sites on five continents. Information derived from FLUXNET can be combined with satellite-based information to improve our understanding of the carbon cycle. While the FLUXNET initiative offers tremendous opportunity to develop data on the functioning of ecosystems aross a range of scales, there remain methodological problems with the independent measurement of the energy balance terms. At present, if all the energy balance terms are measured independently, the energy balance is not closed (i.e. the sum of the terms is not zero). Efforts to minimise this error and thereby improve the quality of flux measurements is ongoing. Multi-year carbon flux records allow us to examine interannual time scales; the timing of leaf expansion may be advanced or delayed by a month due to large-scale climatic features that can be associated with El Niño-La Niña cycles (Myneni et al.

1997; Keeling et al. 1995). The FLUXNET project can produce direct information on the impact of changing growing season length on net ecosystem carbon dioxide exchange. Initial data are indicating that each additional day of growing season length affects net ecosystem CO_2 exchange of temperate deciduous forests by 6 g m^{-2} (Fig. A.9). Combining the data in Fig. A.9 with remote sensing data (e.g. Myneni et al. 1997) will allow us to understand how CO_2 exchange is perturbed across larger space scales.

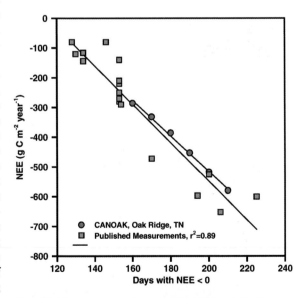

Fig. A.10. Measurements and computations of the relation between length of growing season and net ecosystem CO_2 exchange of broadleaved deciduous forests. The relation is linear, it accounts for over 80% of the variance and a coupled biophysical model is able to capture this behaviour (after Baldocchi et al. 2001)

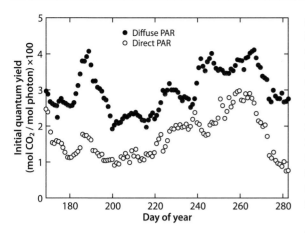

Fig. A.11. Seasonal trend of initial quantum yield of canopy scale CO_2 exchange as a function of clear and cloudy conditions. The data are from an aspen stand and were acquired by T A Black and colleagues (after Gu et al. 2002)

The second example involves the assessment of light use efficiency, as used by many satellite algorithms to determine gross primary productivity (Cramer et al. 1999). The accumulating body of FLUXNET data is showing that light use efficiency varies by a factor of two whether the sky is clear or cloudy (Gu et al. 1999a; Fig. A.10). This type of information is not widely utilised by the remote sensing community, hence their current estimates of regional carbon exchange may be in error.

The third example relates to the seasonal forcing of biophysical models for computing carbon exchange. At present many models may vary weather inputs and leaf area index seasonally on the basis of remote sensing measurements but they assume static values for features such as photosynthetic capacity. FLUXNET data are being used to examine how well model calculations, based on certain parameterisation schemes, reproduce the power spectrum of measured carbon fluxes (Fig. A.11). Identifying periods of low and high coherence between measured and modelled fluxes across an array of plant functional types will enable us to refine how we model carbon fluxes across a spectrum of time scales. Global assessments of carbon fluxes using satellite-based sensors cannot proceed without this necessary intermediate step.

Overall Summary

The role of the land surface in partitioning available water between evaporation and runoff, and in partitioning available energy between sensible and latent heat fluxes is significant and affects the state of the atmosphere directly. In the remainder of this book evidence will be presented to support the role of the land surface and highlight key contributions made by BAHC to this understanding.

Acknowledgement

We wish to thank Lydia Dümenil-Gates for her useful suggestions on a draft of this chapter.

Chapter A.3

The Regional Climate

Roni Avissar · Christopher P. Weaver · David Werth · Roger A. Pielke Sr. · Robert Rabin
Andrew J. Pitman · Maria Assuncão Silva Dias

Regional climate is typically affected by atmospheric, oceanic and land processes at all scales (Pielke 2001a). However, this chapter focuses on the issue "Does the horizontal heterogeneity of the soil-vegetation-atmosphere system affect the regional climate?"

A.3.1 Fundamental Mechanism in Land-Atmosphere Interactions

During cloud-free day-time conditions, turbulent heat fluxes near the ground surface are affected strongly by the ability of the surface to redistribute the radiative energy absorbed from the sun and the atmosphere into sensible and latent heat. As discussed in Pielke (2002) and reproduced in part here, the surface energy and moisture budgets for bare and vegetated soils (i.e. snow and ice effects are not considered in this discussion) are illustrated schematically in Fig. A.12 and Fig. A.13. On bare dry land, the absorption of this energy results in a relatively strong heating of the surface, which usually generates a strong turbulent sensible heat flux in the atmospheric surface layer and a large soil heat flux. In that case, there is no evaporation (i.e. no latent heat flux) and the Bowen ratio (i.e. the ratio of sensible to latent heat flux) is infinite. By contrast, on bare wet land, as is common in irrigated agricultural areas and/or after rain events, the incoming radiation is mostly used for evaporation. In that case, the turbulent sensible heat flux and the soil heat flux are usually much smaller than the latent heat flux. As a result, the Bowen ratio is close to zero. When the ground is covered by dense vegetation, water is extracted mostly from the plant root zone by transpiration. Thus, latent heat flux is dominant, even if

Fig. A.12.
Schematic illustration of the surface heat budget over (*top*) bare soil, and (*bottom*) vegetated land. The roughness of the surfaces (and for the vegetation, its displacement height), will influence the magnitude of the heat flux. Dew and frost formation and removal will also influence the heat budget (adapted from Pielke and Avissar 1990)

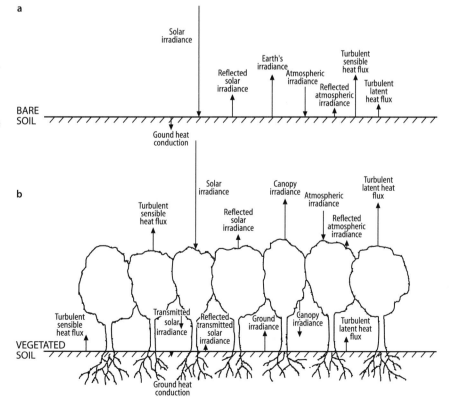

Fig. A.13.
Schematic illustration of the surface moisture budget over (*top*) bare soil, and (*bottom*) vegetated land. The roughness of the surface (and for the vegetation, its displacement height), will influence the magnitude of the moisture flux. Dew and frost formation and removal will also influence the moisture budget (adapted from Pielke and Avissar 1990)

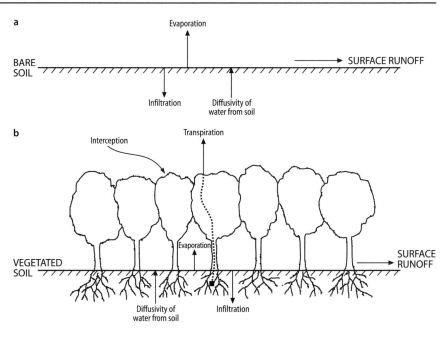

the soil surface is dry, but only as long as there is enough water available in the root zone and plants are not stressed.

Due to these differences in surface fluxes, the characteristics of the atmosphere above dry and wet (or vegetated) land are significantly different, as illustrated in Fig. A.14 (e.g. Avissar 1992). The faster heating rate produced on dry land generates a vigorous turbulent mixing and an unstable stratified atmospheric planetary boundary layer (PBL), which often extends up to a height of 2 000–3 000 m or more during the afternoon. The slower heating rate of wet (vegetated) land limits the development of the PBL, typically, to a height of less than 1 000 m. However, evapotranspiration provides a supply of moisture which significantly increases the amount of water in this shallow PBL. During the afternoon, the temperature of the PBL above dry land is considerably warmer than above wet land. During night-time, however, the strong cooling of bare and vegetated surfaces creates almost the same atmospheric inversion.

Since clouds and precipitation generated in these different PBLs can be quite different, the system of interactions and feedbacks that develop between the land and the atmosphere is very complex. A model of this system needs to account for the movement and storage of water in the ground, the ability of the vegetation to extract the water from the ground, and the dynamic response of the PBL to varying boundary conditions. For instance, Betts et al. (1993, 1996) have shown a close coupling between errors in a model's surface parameterisation and its PBL, clouds, and moist convection. More evapotranspiration can provide a cooler, shallower BL, and positive equivalent potential temperature anomalies (because of less entrainment), which can act to destabilise the PBL to moist convection and, in favourable condi-

tions, might promote heavier precipitation. Precipitation at one location results in stabilising subsidence patterns elsewhere. Because of surface heterogeneities, substantial variations will occur in the instability of PBL air, which may act to reduce the area covered by moist convection.

A.3.2 Atmospheric Response to Heterogenous Land Forcing

It is well known that the land characteristics vary extensively even at a scale of a few metres. The impact of their spatial variability on the atmosphere at the microscale and mesoscale is discussed in the following subsections. The purpose of this discussion is to evaluate the importance of the dynamic response of the atmosphere to land-surface heterogeneities on land-atmosphere interactions at the regional scale and to evaluate whether or not current parameterisations used in GCMs and other large-scale models are acceptable.

A.3.2.1 Microscale Impact

Li and Avissar (1994), and Rodriguez-Camino and Avissar (1999) investigated the impact of microscale (or "patch" scale) variability of the most important land-surface characteristics (as identified by Henderson-Sellers 1993; Collins and Avissar 1994; and Rodriguez-Camino and Avissar 1998): leaf area index, stomatal conductance, roughness, soil-surface wetness, and albedo on the atmospheric turbulent heat fluxes near the ground surface. They found that, in general, the variability of a particular characteristic affects the long-wave radiation emitted by the sur-

Fig. A.14.
Diurnal variation of profiles of (**a** and **b**) potential temperature and (**c** and **d**) specific humidity in the atmospheric planetary boundary layer above (**a** and **c**) a bare, dry land and above (**b** and **d**) a densely vegetated, moist land, simulated for a cloudless midsummer day (August 15) at latitude 37° N. Graphs on the right-hand side of the figure represent two selected profiles, at 0300 LST (*dashed line*) and at 1500 LST (*solid line*) (from Avissar 1992, © American Geophysical Union)

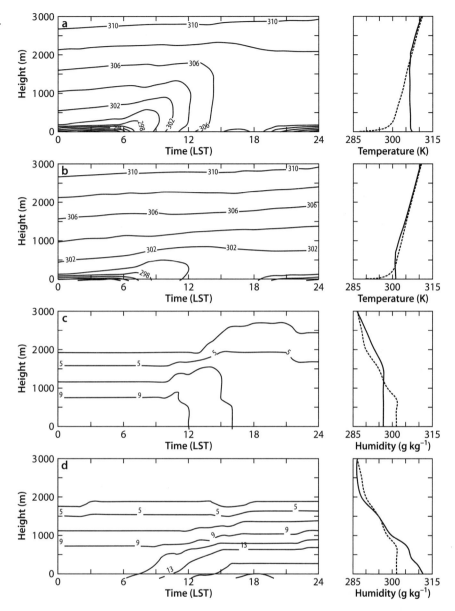

face and the heat fluxes simultaneously, yet not with the same intensity. On average, the latent heat flux is the most sensitive and the radiative flux the least sensitive to spatial variability. For instance, a difference as large as 150 W m^{-2} was found between the latent heat flux calculated with a log-normal distribution of leaf area index and the mean value corresponding to this distribution. By comparison, the corresponding difference of longwave radiation emitted by the surface was 90 W m^{-2}. They also noted that unlike for the other land-surface characteristics, the spatial variability of albedo created differences of not more than 20 W m^{-2}, indicating a relatively linear relationship between this characteristic and land-surface energy fluxes.

A detailed analysis of the land-surface energy fluxes integrated at the patch scale using explicitly the spatial variability of land-surface characteristics or their corresponding mean indicated that, under unstable atmospheric conditions, stomatal conductance and leaf area index variability have the most significant effect on spatially integrated energy fluxes from vegetated land. On bare land, soil-surface wetness strongly affected the surface fluxes. Under neutral and stable atmospheric conditions, surface roughness appeared to have the predominant effect on the surface fluxes. Holtslag and Ek (1996), in a study of the interaction of the atmospheric boundary layer with the heterogeneous pine forest in HAPEX-MOBILHY (Hydrological Atmospheric Pilot Experiment – Modélisation du Bilan Hydrique) on a scale of about 10 km also found that the coupled atmosphere-vegetation system is sensitive to the roughness length. They concluded that this parameter affects in particular the sensible heat flux and, as a result, the boundary layer height and profiles of mean variables.

In these studies, however, it was assumed that the impact of the variability of land-surface characteristics did not extend beyond the atmospheric surface layer. This, of course, simplified the land-atmosphere models being used but limited the studies to the patch scale. The impact of the spatial variability of land-surface characteristics at a scale larger than the patch scale, though still considered microscale, was studied with large-eddy simulations (LES). This modelling technique, which was pioneered by Deardorff (1972), is accurate and, therefore, very attractive to study the turbulence activity in the PBL. For instance, Deardorff (1974) showed that turbulence statistics obtained with an LES model compared nicely with those observed in the 1967 Wangara Experiment (Clarke et al. 1971). Note that with this modelling technique, only small (less energetic) eddies need to be parameterised and largest-size eddies are resolved. Numerical experiments with LES by Moeng (1984, 1986, and 1987), Hechtel et al. (1990), Hadfield et al. (1991, 1992), Walko et al. (1992), Avissar et al. (1998), Avissar and Schmidt (1998), and Gopalakrishnan et al. (2000) confirmed the high potential of this technique, especially with the powerful computers readily available today.

Until recently, only a few investigations had been conducted to understand better the impacts of spatial variability of land-surface characteristics on the PBL (Hechtel et al. 1990; Hadfield et al. 1992; and Walko et al. 1992). But more complete LES studies have contributed additional insights on this issue. For instance, Avissar and Schmidt (1998), Gopalakrishnan et al. (2000), and Baidya Roy and Avissar (2000) investigated the scale of forcing at which vertical profiles of horizontally-averaged atmospheric variables in the convective boundary layer (CBL) are affected significantly by the spatial variability of surface heat fluxes or the presence of hills. Their results indicate that as long as the length scale of the surface sensible heat flux perturbation is smaller than about 5 km and the topographical features are not higher than about 400 m, it seems that the dynamics of the CBL are not affected by landscape heterogeneity. Furthermore, these studies emphasized that the response of the atmosphere to the scale of forcing is quite non-linear. They attributed this non-linearity to the ratio of the horizontal pressure gradient created by the heat flux heterogeneity to the buoyancy intensity, which is related to the mean surface sensible heat flux.

It is interesting to note that, based on theoretical considerations and a simple slab model of the CBL, Raupach (1993) claimed that the CBL should not be affected by landscape heterogeneity in flat terrain at a horizontal length scale of 1–5 km. Claiming that turbulence is very efficient at mixing the boundary layer, Shuttleworth (1988a) suggested that this scale could be as large as 10 km. Linear analysis studies (Dalu and Pielke 1993; and Dalu et al. 1996) have reached a similar conclusion.

A.3.2.2 Mesoscale Impact

The contrast between land and water generates breezes (i.e. sea, lake, and land breezes), which are mesoscale circulations. Several papers and textbooks describe at length the mechanism involved in the generation of these circulations, and their impact on weather and climate (e.g. Yan and Anthes 1987; Pielke 2001c, 2002 among many others). More recent investigations (Lawton et al. 2001) have indicated that other landscape discontinuities (e.g. irrigated land in arid areas, deforestation, and afforestation) also provide an environment appropriate for the development of mesoscale circulations. These studies include those of Ookouchi et al. (1984), Mahfouf et al. (1987), Segal et al. (1988), Pielke et al. (1991), Segal and Arritt (1992), Avissar and Chen (1993), Chen and Avissar (1994a), Goutorbe et al. (1994), Mahrt et al. (1994), Lynn et al. (1995a), Avissar and Liu (1996), Stohlgren et al. (1998), Taylor et al. (1998), Wang et al. (1996, 1998) and Chase et al. (1999).

Studies by Hammer (1970), Barnston and Schikedanz (1984) and Otterman et al. (1990) provide evidence that increased convective precipitation occurs over extended crop areas during favourable atmospheric conditions. Land-use features are believed to play a role in the intensity and position of drylines[1] in the southern plains of the US. In a case study of localised thunderstorm development along a dryline in Oklahoma, Hane et al. (1997) observed a significant "bulge" or distortion in the dryline formed downwind from a swath of sparse vegetation cover. It was suspected that enhanced sensible heat flux and vertical mixing in this swath was a factor in the local displacement of the dryland position.

Pielke et al. (1997) present a sensitivity experiment to evaluate the importance of land-surface conditions on thunderstorm development. Using identical lateral boundary and initial values, two model simulations for 15 May 1991 were performed for the Oklahoma-Texas Panhandle region. One experiment used the current landscape (which includes irrigated crops and shrubs as well as the natural short-grass prairie), while the second experiment used the natural landscape in this region (the short-grass prairie). Figure A.15 provides the results at 15:00 local standard time for both experiments. The simulation with the current landscape (Fig. A.15, *top*) produced a thunderstorm system along the dryline, while only a shallow line of cumulus clouds were produced using the natural landscape (Fig. A.15, *bottom*). A thun-

---

[1] Dryline: The borderline between hot, dry air moving down the east slopes of the Rocky Mountains and humid air coming from the Gulf of Mexico.

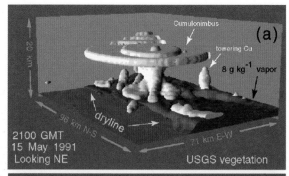

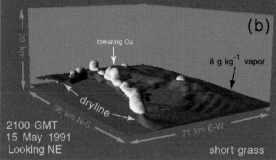

Fig. A.15. Simulation of cloud and water vapour mixing ratio fields at 21 GMT on 15 May 1991 obtained with (**a**) current vegetation, and (**b**) natural vegetation. The clouds are depicted by white surfaces with $q_c = 0.01$ g kg^{-1} with the sun illuminating the clouds from the west. The vapour mixing ratio in the planetary boundary layer is depicted by the grey surface with $q_v = 8$ g kg^{-1}. The flat surface is the ground. Areas formed by the intersection of clouds or the vapour field with lateral boundaries are flat surfaces, and visible ground implies $q_v < 8$ g kg^{-1}. The vertical axis is height, and the backplanes are the north and east sides of the simulated domain (from Pielke et al. 1997)

derstorm was observed in this region on 15 May 1991 with the other meteorological quantities also simulated realistically (Shaw et al. 1997; Ziegler et al. 1997; Grasso 2000). The thunderstorm developed when the current landscape was used since the enhanced vegetation coverage (higher leaf area) permitted more transpiration of water vapour into the air than would have occurred with the natural landscape. The result was higher convective available potential energy with the current landscape. Ziegler et al. (1995), Shaw et al. (1997) and Grasso (2000) also concluded that the spatial distribution of vegetation was very important in determining the location of the dryline and the intensity of cumulus convection which developed in it. Lyons et al. (1993, 1996) and Huang et al. (1995), in a contrasting result, found that the replacement of native vegetation with agriculture reduced sensible heat flux, with a resultant decrease in rainfall. Wetzel et al. (1996b), in a study in the Oklahoma area, found that cumulus clouds form first over hotter, more sparsely vegetated areas. Over areas covered with deciduous forest, clouds were observed to form one to two hours later due to the suppression of vertical mixing.

Rabin et al. (1990) also found from satellite images that cumulus clouds form earliest over regions of large sensible heat fluxes and are suppressed over regions with large latent heat flux during relatively dry atmospheric conditions. The sensitivity of small cumulus clouds to the land-surface has become evident since the first routine satellite imagery in the 1960s. In addition to the suppressive effect of the Great Lakes, early observations of the Mississippi basin suggested that relatively gentle topography can affect the distribution of these clouds (Oliver and Oliver 1963; Anderson et al. 1966). Through composite images of Colorado, later studies by Klitch et al. (1985) identified a match between terrain slope and cumulus cloud formation. Sequences of hourly visible and infrared satellite image pairs of the south-east United States revealed the sensitivity of day-time cloud frequency to small terrain features (Gibson and Vonder Haar 1990). Over monthly periods, the land surface within the Mississippi Valley was found to impose a distinctive signature on day-time cumulus clouds (Rabin and Martin 1996). Cumulus appeared to be modulated by even modest elevation features. Locally, slope and aspect may also modulate cumulus. At times, soils and vegetation may be more important than elevation in controlling the distribution of cumulus. This was the case over southern Illinois during the drought month of July 1988. There, cumulus frequency tended to be associated inversely with plant cover and available soil moisture (Fig. A.16 and Fig. A.17). Similarly, high-resolution geostationary satellite observations of Oklahoma indicated that shallow cumulus clouds formed earlier and become more numerous over landscapes of wheat stubble (where surface air temperature and Bowen ratio was highest) than over adjacent landscapes of pasture and row crops (Rabin et al. 1990). A similar effect was observed from satellite-derived monthly cumulus frequency during the dry season in Brazil (Cutrim et al. 1995). For the first time, an effect of deforestation on cloud cover was noted. Enhancement of cumulus cloud cover was evident over deforested areas in Rondônia which was similar to that observed over natural grasslands in nearby regions. They also noted enhancement of shallow cumulus over small mountain ranges in the Amazon region. Souza et al. (2000) explained the thermodynamics of local circulation between forest and pasture, including the effect of sloping terrain, as a heat engine. They suggested that topographical features enhance convective systems because of the anomalous high sensible heat flux injected into the updrafts moving upslope.

Pielke et al. (1991) used a set of numerical simulations to demonstrate that the surface heterogeneity created by alternating strips of land and water generates mesoscale heat fluxes often more significant than turbulent fluxes. Further analysis by Avissar and Chen (1993), Chen and Avissar (1994a,b), Lynn et al. (1995a,b), Weaver and

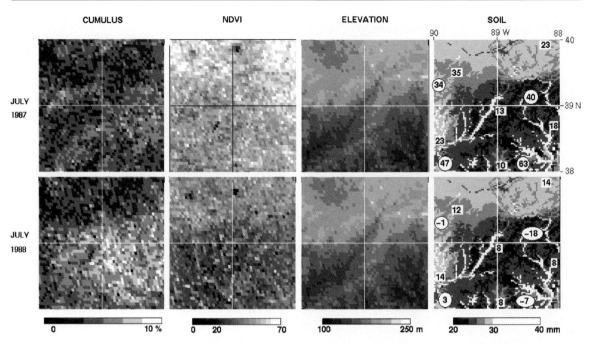

Fig. A.16. Maps of shallow cumulus frequency, NDVI, elevation and potential plant-available water over southern Illinois. Values plotted on the plant-available water maps are measured soil moisture content in mm (from Rabin and Martin 1996, © American Geophysical Union)

Avissar (2001) and Baidya Roy and Avissar (2002) indicated that mesoscale heat fluxes are created by various types of landscape discontinuity, and are affected by different background conditions e.g. wind velocity, thermal stratification and humidity profile of the atmosphere, latitude, and day of the year.

These mesoscale circulations and associated fluxes can generate clouds which affect the radiation and the precipitation regimes and, as a result, the hydrological cycle (Chen and Avissar 1994a; Avissar and Liu 1996;

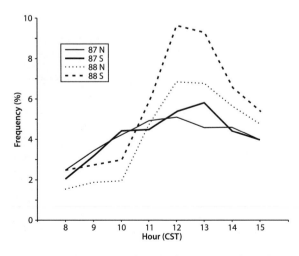

Fig. A.17. Hourly variations of cumulus frequency over the northern half (**N**, *thin curves*) and southern half (**S**, *thick curves*) of the region in Fig. A.16. *Solid curves* are for July 1987. *Dashed curves* are for July 1988 (from Rabin and Martin 1996, © American Geophysical Union)

Weaver and Avissar 2001; Baidya Roy and Avissar 2002). For instance, Bougeault et al. (1991) simulated with the French Weather Service limited-area numerical model the formation of clouds next to the boundary between a forest and a crop area. Their results were supported by satellite images and observations of various meteorological parameters in the CBL. Similarly, Lyons et al. (1993) presented a satellite photograph of convective clouds which had clearly developed at the boundary between irrigated land and native vegetation in Southwestern Australia. Weaver and Avissar (2001) used the Regional Atmospheric Modelling System (RAMS) to simulate several case-study days observed at the Atmospheric Radiation Measurement (ARM) Cloud and Radiation Testbed (CART) (Stokes and Schwartz 1994). This Department of Energy (DOE) testbed is located in Oklahoma and Kansas and is part of the Global Energy and Water Cycle Experiment (GEWEX) Continental Scale International Project (GCIP) Enhanced Seasonal Observing Period for 1995 (ESOP-95). Because of a lack of suitable observations of the complete atmospheric effects of mesoscale landscape heterogeneity, the only reasonable tool for a comprehensive investigation of this phenomenon is a numerical model. Satellite imagery and composited rainfall measurements collected as part of ESOP-95 make possible the proper interpretation of the model results. The results shown in Weaver and Avissar (2001) illustrate how model simulations, coupled with complementary observations, provide a powerful method for answering complex questions of this type.

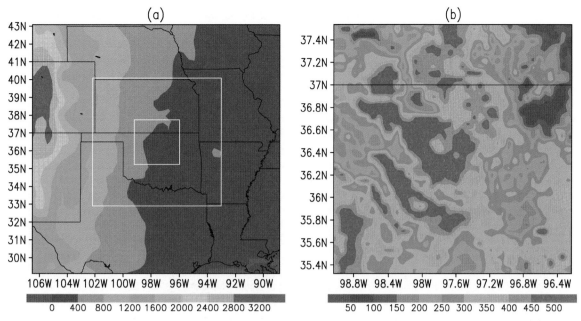

Fig. A.18. a The full RAMS simulation domain (Grid 1) showing surface topographic height (m) and the locations of the nested Grids 2 and 3; **b** Grid 3 surface sensible heat flux (W m^{-2}) at 13:00 local time (1900 UTC) on 13 July, 1995 (from Weaver and Avissar 2001)

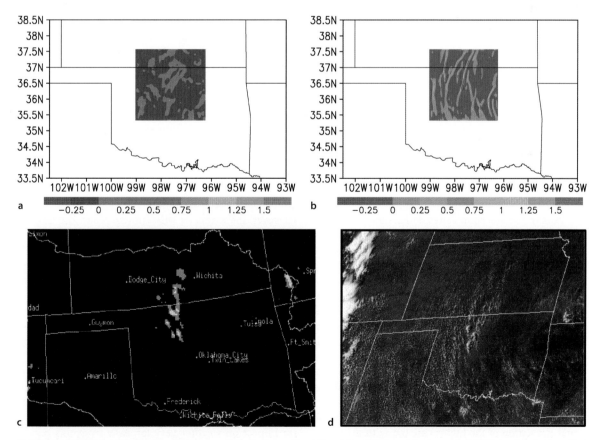

Fig. A.19. a RAMS simulated w (m s^{-1}) at 1100 m altitude at noon local time (1800 UTC) on July 13, 1995. **b** RAMS simulated w (m s^{-1}) at 1100 m altitude at 15:30 local time (2130 UTC) on 13 July, 1995. **c** ABRFC accumulated precipitation composite for 16:00 to 17:00 local time (2200–2300 UTC) on 13 July, 1995. **d** GOES-8 satellite visible image at 15:15 local time (2115 UTC) on 13 July, 1995 (from Weaver and Avissar 2001)

The ARM/CART site measures approximately 300 km by 350 km in Oklahoma and Kansas, and this heavily-instrumented area is a source of frequent measurements of surface sensible and latent heat fluxes, as well as other meteorological quantities, throughout the year (Stokes and Schwartz 1994). The human impact on this region is readily apparent. During summer, large differences in land use across this site, particularly the contrast between already-harvested winter wheat in the central portion of the domain and actively growing vegetation elsewhere, result in significant gradients in surface fluxes, as explained by Zhong and Doran (1998) and illustrated in Fig. A.18b. Sensible heat flux contrasts of approximately 200 W m^{-2} exist between areas (patches) with characteristic scales of 50–100 km.

Figure A.19 shows the mid-PBL vertical velocity (w) field at noon and 15:30 local time for 13 July, 1995. At noon (Fig. A.19a), relatively weak upward motion of a few cm s^{-1} is located over patches of large sensible heat flux (recall Fig. A18b) with scales in the order of 20–50 km. Over the course of the afternoon, a circulation with strong upward motion on the order of 1 m s^{-1} gradually evolves around larger surface scales, with convergence of air flowing from high latent heat (cooler) to high sensible heat (warmer) areas (Fig. A.19b). Over some locations, this circulation penetrates from the surface to a height of more than 3 km.

That this circulation pattern is undoubtedly a real-world feature is demonstrated by the excellent correspondence between the simulated w field and the composited rainfall measurements (Fig. A.19c) and satellite cloud observations (Fig. A.19d). The clouds and precipitation are oriented along the linear, north-south convergence lines of strong upward motion. The model is able to capture not only the snapshot agreement presented here, but reproduces correctly the evolution in time of the convergence field as diagnosed by successive hourly rainfall composites. Furthermore, Weaver and Avissar (2001) emphasize that such a good agreement between the location and orientation of the clouds, the patterns of upward motion, and the satellite imagery, is also found on other days in July 1995.

To demonstrate the important role these circulations play in transporting heat and moisture vertically, Weaver and Avissar (2001) also calculated the sensible and latent heat fluxes by both the mesoscale and microscale (turbulent) motions. The mesoscale fluxes are one measure of the intensity of the landscape-induced circulations. To highlight the impact of the surface variability on the fluxes, rather than simply the mean surface conditions, Weaver and Avissar (2001) performed an additional simulation of the July 13 case but with the realistic surface sensible and latent heat fluxes replaced with their mean values, thus eliminating any mesoscale heterogeneity. This mean flux case, rather than developing

any organised circulation, has a flow field (not shown) characterised by essentially random updrafts and downdrafts at a small horizontal and vertical scale which, at the 2 km grid increment of the model, are very weak (of the order of a few cm s^{-1} compared to the 1–2 m s^{-1} upward motion of the realistic July 13 simulation as shown in Fig. A.19b).

Some model and analytical studies have indicated that, except under very weak large-scale background wind conditions (e.g. of the order of 2–3 m s^{-1} or less), landscape-induced mesoscale flows are inhibited and the corresponding mesoscale fluxes are therefore also very weak (Wang et al. 1996; Avissar and Schmidt 1998; Liu et al. 1999a). Hence the most common argument against their climatological importance, i.e. the percentage of time these landscape heterogeneity effects are significant, is expected to be small (Zhong and Doran 1997, 1998). This view has been challenged recently (Vidale et al. 1997; Wang et al. 1998), and the results of Weaver and Avissar (2001) demonstrate for the first time for a range of cases that increasing large-scale wind does not necessarily inhibit these mesoscale circulations (Fig. A.20). Indeed, the day with the largest mesoscale flux (July 7) is also the day with the highest winds (over 12 m s^{-1} in the lower PBL), and all but one of the remaining days have winds of 5–10 m s^{-1} in the lowest 2 km. This means that the conditions under which such effects are significant probably occur more often than assumed previously.

The results shown here demonstrate that, by changing the scale and properties of the naturally occurring land-surface elements, human influences can affect local weather and climate significantly, at least in the short term. Since both natural and human-modified heterogeneous landscapes are ubiquitous around the globe (e.g. snow/soil, land/water, pasture/forest, city/country), such atmospheric effects must influence global climate today. However, the extent of this influence, and the longer-term impacts of accelerating land-use changes on future regional and global climate, are currently unknown. Evaluating these impacts is complicated by complex vegetation dynamics (e.g. Lu et al. 2001; Eastman et al. 2001a; and Eastman et al. 2001b).

A.3.3 Regional Teleconnections

Additional discussion on regional teleconnections can be found in Sect. A.4.1.4. The effect of above-average temperatures in the eastern and central Pacific Ocean, referred to as "El Niño", has been shown to have a major impact on weather very far away from this region (Shabbar et al. 1997). The warm ocean surface provides appropriate conditions needed for the development of thunderstorms, which export large amounts of heat,

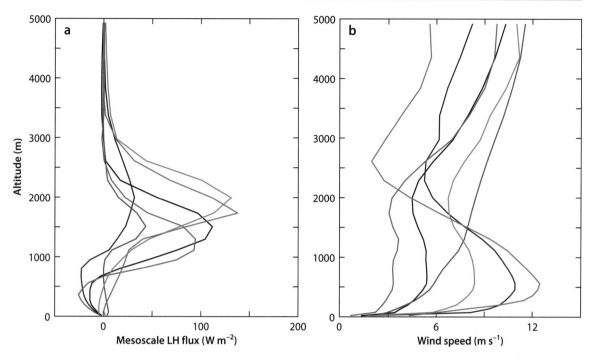

Fig. A.20. a Grid 3 averaged mesoscale latent heat flux (W m^{-2}) at 16:00 local time (22:00 UTC) for each of the sixe case study days (6, 7, 10, 12, 13, and 21 July 1995). **b** Grid 3 averaged wind speed (m s^{-1}) at 8:00 local time (14:00 UTC) for each of the six case study days (6 July: *black line*; 7 July: *red line*; 10 July: *green line*; 12 July: *dark blue line*; 13 July: *light blue line*; 21 July: *purple line*). We show early morning wind here because it is illustrative of the large-scale wind conditions existing prior to the development of any landscape-generated mesoscale circulations. The winds later in the day include the effects of these circulations, and thus they cannot properly be considered to reflect the ambient synoptic-scale environment in which such circulations must form

moisture and kinetic energy to the middle and higher latitudes. This transfer, which alters the ridge and trough pattern associated with the polar jet stream (Hou 1998), is referred to as a "teleconnection" (Glantz et al. 1991; Namias 1978; Wallace and Gutzler 1981). Almost two-thirds of the global precipitation is associated with mesoscale cumulonimbus and stratiform cloud systems located Equatorward of 30° (Keenan et al. 1994). In addition, much of the world's lightning occurs over tropical continents, with maxima also over mid-latitude continents in the warm seasons (Lyons 1999; Rosenfeld 2000a). These tropical regions are also undergoing rapid landscape change, which is believed to alter the frequency and intensity of thunderstorms (Baidya Roy and Avissar 2002; Werth and Avissar 2002).

As shown in the pioneering study by Riehl and Malkus (1958), and Riehl and Simpson (1979), 1 500–5 000 thunderstorms (which they refer to as "hot towers") are the conduit to transport this heat, moisture, and wind energy to higher latitudes. Since thunderstorms only occur in a relatively small percentage of the area of the tropics, a change in their spatial patterns would be expected to have global consequences.

Wu and Newel (1998) concluded that sea-surface temperature variations in the tropical eastern Pacific Ocean have three unique properties that allow this region to influence the atmosphere effectively: large magnitude, long persistence and spatial coherence. Since land-use change has the same three attributes, a similar teleconnection is expected to occur as a result of man-made landscape changes in the tropics (Pielke 2001b).

Figure A.21, from Werth and Avissar (2002), illustrates how the mean precipitation of a few regions is affected by heavy deforestation in Amazonia. One notices a significant reduction in precipitation over central North America during the summer season. Other regions affected by this deforestation include Central America, the Gulf of Mexico, the Tropical West Pacific and the Indian Ocean. Using different GCM configurations, Chase et al. (1996, 2000) and Zhao et al. (2001) also found precipitation teleconnections as a result of landscape change. Other studies support the result that there is a significant effect on the large-scale climate due to land processes (e.g. Betts et al. 1997; Broström et al. 1998; Brubaker and Entekhabi 1995; Claussen 1997, 1998; Costa and Foley 2000; Dirmeyer and Zeng 1999; Eltahir 1996; Pitman and Zhao 2000). An important conclusion from these studies is that landscape change directly alters local and regional heat and moisture fluxes in two ways. First, the local and regional convective available potential energy is modified since the Bowen ratio and the albedo are changed as the surface heat and moisture

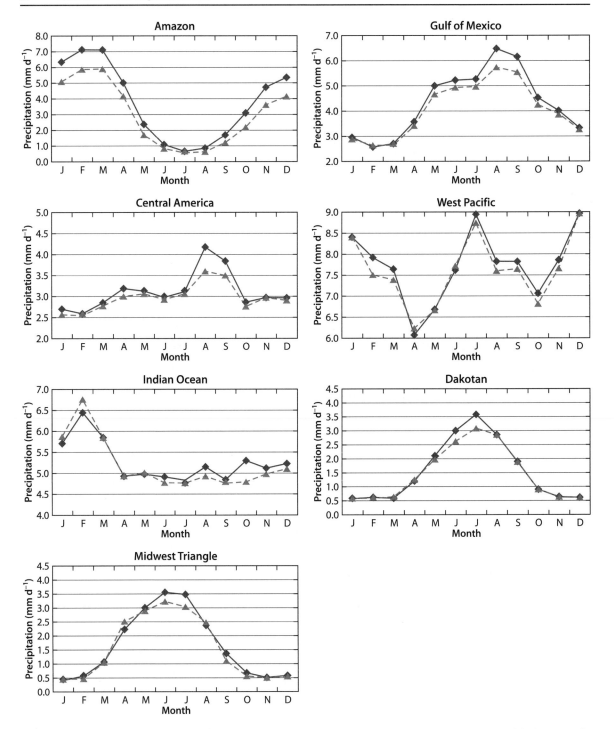

Fig. A.21. Annual variation in precipitation in Amazonia and other regions of the world produced by the GISS (Goddard Institute for Space Studies) GCM with current vegetation (*blue line*) and a deforested Amazonia (*red line*). Each point represents the average of 6 realisations of 8-years each (from Werth and Avissar 2002, © American Geophysical Union)

budgets are altered. Second, larger-scale heat and moisture convergence, and associated large-scale circulations, can be modified as a result of changes in the large-scale atmospheric pressure field due to the landscape alterations.

Based on an extensive review of the various studies performed on this issue, Pielke (2001b) concluded that "regional and global model studies indicate that the spatial patterning of deep cumulus convection, particularly in the tropics and mid-latitude summers, are altered significantly as a result of landscape changes. These alterations in cumulus convection teleconnect to middle and higher latitudes, which alters the weather in those regions. This effect appears to be most clearly defined in the Winter Hemisphere." Werth and Avissar (2002) also emphasized that thunderstorms are not simulated explicitly with GCMs and current parameterisations in these models do not account for thunderstorms generated by mesoscale landscape heterogeneity. Therefore, the major impact of man-made landscape modification is not represented properly in these models, and the modelling results obtained so far probably underestimate the real effects of landscape alterations locally, regionally, and by teleconnection to other regions. Better parameterisations and/or model resolution will provide additional insights on this issue in the foreseeable future.

A.3.4 Discussion

The various studies reviewed here lead to the following conclusions:

1. The land-surface fluxes are non-linearly and strongly related to key land-surface characteristics (i.e. stomatal conductance, roughness, leaf area index, and surface wetness). Therefore, to calculate these fluxes properly, it is important to account for the heterogeneity of these characteristics at the canopy scale;
2. When the heterogeneity of the surface fluxes generated by landscape patchiness is larger than about 5 km, or topographical features are larger than about 400 m, mesoscale circulations (or "organised turbulence") develop in the planetary boundary layer, affecting significantly the transport of heat, moisture, and momentum in the atmosphere. In particular, this has a major impact on the generation of clouds and precipitation, including thunderstorms. Therefore, this heterogeneity must be either resolved or properly parameterised in GCMs and other large-scale models;
3. Man-made landscape alterations (e.g. deforestation, agricultural development, etc.) affect the atmosphere locally and, when such alterations are done (at least) at the mesoscale, they also affect the regional hydrometeorology. Furthermore, the impact of such extensive alterations can propagate by teleconnection to other regions.

These conclusions have significant implications for the representation of land processes in GCMs and other weather/climate models. First, parameterisations based on the "big leaf" approach, in which each plant-covered grid element of the model is assumed to be represented

by a single leaf, without consideration for heterogeneity, is not recommended. This is because important nonlinear relations are omitted with this approach.

To circumvent this problem, a few alternatives have been proposed. For instance, Avissar (1992), Famiglietti and Wood (1994), and Giorgi (1997a,b) proposed to use the probability density functions of the most important land characteristics in so-called "statistical-dynamical" parameterisations. Using either analytical (Giorgi 1997a,b) or numerical (Avissar 1992) solutions, these parameterisations account explicitly for the spatial heterogeneity of the land characteristics.

Effective (or aggregated) parameters also aim at accounting for the non-linear relations between land characteristics and fluxes but, unlike in the "statistical-dynamical" approach, these relations are calculated implicitly through "effective" functions and/or parameters (e.g. Noilhan and Lacarrère 1995; Kabat et al. 1997a). For instance, an effective surface roughness was proposed by André and Blondin (1986), Mason (1988) and Claussen (1990), and an effective stomatal resistance was proposed by Claussen (1990) and by Blyth et al. (1993). Rodriguez-Camino and Avissar (1999) considered different types of aggregating functions (linear, trigonometric, parabolic, square root and logarithmic) on the accuracy of various effective parameters. They found that non-linear functions are generally more useful than linear functions to calculate the effective value of those parameters which affect the surface heat fluxes independently of the atmospheric stability (e.g. leaf area index and soil surface wetness).

With the "mosaic of tiles" approach (e.g. Avissar and Pielke 1989; Claussen 1991a; Koster and Suarez 1992; Ducoudre et al. 1993; Walko et al. 2000), each land-use patch or "tile" of a grid element is coupled independently to the atmosphere of the model, and patches affect each other only through the atmosphere. The grid-average surface fluxes are obtained in this case by flux aggregation, i.e. by averaging the surface fluxes over each tile weighted by their fractional area. For tiles of a typical length of up to several hundred metres, the so-called blending-height method has proven to be useful (Mason 1988; Claussen 1991a, 1994). This method is very similar to the "mosaic of tiles" approach, but it includes the effect of local advection, i.e. the influence of tiles on the near-surface air flow upstream of the tile under consideration. Even extreme cases, e.g. very stably stratified air flow of sea-ice with some patches of open water which yield an areally-averaged heat flux counter to the averaged temperature gradient, can be treated with this method (Claussen 1991b; Stössel and Claussen 1993).

While these various techniques (i.e. statistical-dynamical, effective/aggregation, mosaic-of-tile) address more or less correctly the issue summarised in Conclusion (1), none of them is equipped to address the impacts of landscape heterogeneity at scales larger than about 5 km, as needed for addressing Conclusion (2). In other words, these parameterisations are not designed to be used in models at a resolution larger than about 5 km.

To address these larger scales of heterogeneity, a few preliminary parameterisations have been proposed (Lynn et al. 1995b; Zeng and Pielke 1995a,b; Arola 1999; Liu et al. 1999a). For instance, Lynn et al. (1995b) and Liu et al. (1999a) were able to develop such a parameterisation based on synoptic conditions (resolved by the GCMs), a characteristic length scale of the land-surface heterogeneity, and the variance of surface heat fluxes. Based on their findings, one can infer that an important challenge for hydrometeorologists is to develop a parameterisation that could provide the characteristic length scale of the land-surface patchiness and the distribution of surface heat fluxes. Avissar (1998) has proposed such a land parameterisation, but its application in GCMs and other models remains to be demonstrated.

From this review of state-of-the-art studies, we conclude that there is little doubt that the horizontal heterogeneity of the soil-vegetation-atmosphere system *does* affect the regional climate. While a significant effort has been made to address the microscale heterogeneity at the canopy scale, only very few studies have been conducted to address the great challenge of representing correctly the coupled land-planetary boundary layer (including clouds and precipitation) system in heterogeneous landscape. Since precipitation is a key parameter of the hydrological cycle (including its biological aspect), it is essential that progress be achieved on this issue.

Chapter A.4

The Global Climate

Martin Claussen · Peter M. Cox · Xubin Zeng · Pedro Viterbo · Anton C. M. Beljaars
Richard A. Betts · Hans-Jürgen Bolle · Thomas Chase · Randall Koster

In this chapter, we present evidence that changes in land-surface, in particular those relating to vegetation-atmosphere interaction, affect weather and climate at the global scale. We also show that without understanding vegetation dynamics we are unable to describe long-term climate variability, nor can we fully interpret palaeoclimatic reconstructions or predict global weather properly. This chapter is organised as follows (see also Table A.1):

In Sect. A.4.1, we present theoretical considerations. We show how vegetation interferes with the atmosphere and the other components of the climate system at the global scale, and we explore the consequences of the non-linear character of atmosphere-vegetation dynamics.

In Sect. A.4.2, we interpret palaeoclimatic and palaeobotanic reconstructions in the light of the theory being developed in Sect. A.4.1.

In Sect. A.4.3, emphasis is given to the interaction of atmosphere-vegetation dynamics and the global carbon cycle. We mainly focus on numerical sensitivity experiments which highlight potential changes in the climate system under the condition of increased emissions of CO_2.

In Sect. A.4.4, we discuss the memory effects which soil moisture imposes on the global climate system. This memory effect becomes important where seasonal climate variability is concerned.

In Sect. A.4.5, we present evidence that improved representation of atmosphere-vegetation interaction in a model of global weather forecast has substantially improved prediction skill.

Martin Claussen

A.4.1 Feedbacks, Synergisms, Multiple Equilibria and Teleconnections

A.4.1.1 Feedbacks

According to the modern definition of climate (e.g. Peixoto and Oort 1992), the land surface is considered to be a state variable and not just a boundary condition of the climate system. In this respect it is important to realise that the climate system encompasses several components: the abiotic world (the atmosphere, hydrosphere, cryosphere, pedosphere) and the living world (the biosphere). The components of the climate system are connected via fluxes of energy, substances and momentum which carry feedbacks among them. These feedbacks are non-linear. Hence, we can expect feedbacks in the climate system to lead to abrupt changes in the system through subtle changes in the external forcing, and to yield multiple states under the same, constant external forcing. Vegetation-climate interaction can be interpreted in terms of biogeochemical and biogeophysical feedbacks.

Biogeochemical feedbacks are associated with changes in terrestrial biomass and, therefore, with changes in the chemical composition of the atmosphere. When considering atmospheric CO_2, biogeochemical feedbacks are generally negative, i.e. they dampen any initial perturbation in the system. For example, an increase in biomass yields a stronger CO_2 uptake, which could lead to a global near-surface cooling and which in turn would reduce biomass. Vice versa, a decrease in vegetation tends to lower biomass and to enhance atmospheric CO_2, thus yielding conditions favourable for vegetation growth via global-scale surface warming. This sketch of a negative biogeophysical feedback loop is of course grossly simplified. Any change in atmospheric CO_2 could trigger other feedbacks by, for example, affecting the efficiency of CO_2 uptake at enhanced CO_2 concentrations (e.g. Sellers et al. 1997) and plant growth (e.g. Eastman et al. 2001b). Which feedback dominates depends on the time scale under consideration and on its synergism with other, biogeophysical feedbacks (Betts et al. 1997; Woodward et al. 1998). A more thorough discussion of biogeochemical feedbacks will be given in Sect. A.4.3.

Table A.1. Structure of Chapter A.4

Section	Focus	Time scales
A.4.1	Theory	Years to millennia
A.4.2	Palaeoclimate	Millennia/centuries
A.4.3	Carbon cycle	Decades
A.4.4	Soil moisture	Seasons
A.4.5	Weather forecast	Days

Biogeophysical feedbacks – or *biogeographical feedbacks*, if vegetation displacement is emphasized – directly affect near-surface (i.e. the lowest few metres of the atmosphere above ground) energy, moisture, and momentum fluxes via changes in surface structure, such as albedo, roughness, leaf area, etc. Biogeophysical feedbacks can conveniently be categorised by analysing the surface-energy budget.

$$S(1-\alpha) + \varepsilon R - \varepsilon\sigma T_G^4 \pm H \pm \lambda E = \pm G \qquad (A.12)$$

where S refers to insolation, α, ε, and T_G to the albedo, the emissivity and the temperature, respectively, of the atmosphere-surface interface. R is the atmospheric radiation, σ the Stefan-Bolzmann constant, and H and λE represent the turbulent fluxes of sensible and latent heat respectively (the positive sign applies to downward turbulent heat fluxes, which often occur during night time, and the negative sign to upward, mostly day-time, fluxes). On the right hand side of the equation, G symbolises the heat flux into the ground. (Here, the cartesian co-ordinate system is used: the positive sign refers to an upward soil heat flux, and the negative sign, to a downward flux – in contrast to Chapt. A.2 in which a positive sign is allocated to flux towards the atmosphere-soil interface.)

A decrease in vegetation which causes a decrease in leaf mass and leaf area often results in an increase in surface albedo. Exceptions are found in regions with dark, wet soil. An increase in surface albedo favours near-surface cooling, thereby reducing vegetation mass even further. Vice versa, a vegetation increase leads to a reduction in surface albedo and thus to a near-surface warming and more favourable conditions of vegetation growth. Hence in many cases the *surface-albedo feedback*, i.e. the feedback related to changes in the first term of the energy budget equation, tends to amplify an initial perturbation. The situation could be reversed, however, if vegetation is replaced by dark, wet soil.

The surface-albedo feedback is particularly strong in areas with snow cover, as first discussed by Otterman et al. (1984) and Harvey (1988, 1989a,b). The albedo of snow-covered vegetation is much lower for forests, such as taiga, than for low vegetation such as tundra. For snow-covered grass, albedo values of some 0.75 were measured (Betts and Ball 1997) and for snow-covered forests, 0.2 to 0.4 (see also Sect. A.4.5). Hence the darker, snow-covered taiga receives more solar energy than snow-covered tundra which, in turn, favours the growth of taiga. This feedback, sometimes called *taiga-tundra feedback*, is sketched in Fig. A.22a, (full arrows, left-hand branch) where ΔA refers to a change in absorption of solar radiation, ΔT to a change in near-surface temperatures, and ΔV to a change in vegetation structure, i.e. leaf area or living biomass. The arrows in this figure should be interpreted as denoting relationships. For example, the arrow pointing from ΔT to ΔV in Fig. A.22a

should be interpreted as $\Delta V \sim + \Delta T$ (leaf area or living biomass increases with increasing temperature). The taiga-tundra feedbacks plays a key role in explaining palaeoclimatic changes in the mid-Holocene and glacial inception (see Sect. A.4.2).

In the subtropics, we find a positive feedback – the *desert-albedo feedback* proposed by Charney (1975). Charney argued that the high albedo of sand deserts

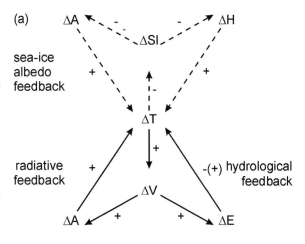

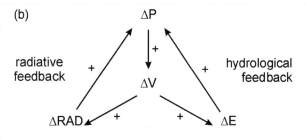

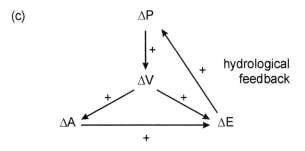

Fig. A.22. Sketch of atmosphere-vegetation interactions at (**a**) northern high latitudes, (**b**) subtropics, and (**c**) tropics. In the sketch, not all possible feedbacks are drawn, but only those considered to be the most important on a large scale. The *arrows*, for example from ΔT to ΔV, or ΔT to ΔSI in (**a**), have to be interpreted as relations, e.g. $\Delta V \sim + \Delta T$ or $\Delta SI \sim -\Delta T$. The *dashed arrows* in (**a**) indicate the sea-ice albedo feedback which enhances the biogeophysical feedback (see Sect. A.4.2.1). T refers to temperature or temperature sum; V, to vegetation density or area covered by vegetation; P, to annually or seasonally averaged precipitation; A, to absorption of solar energy at the surface; E, to evaporation and transpiration; SI, to sea-ice area; H, to heat fluxes from the ocean into the atmosphere; RAD, to the divergence of radiative fluxes in the atmosphere

causes a negative deviation of the atmospheric radiation budget. Above sand deserts, more solar radiation is reflected and more thermal radiation leaves the surface into space because of reduced cloudiness over the dry deserts than over darker, vegetated surfaces. This is indicated in Fig. A.22b, left-hand branch, by $\Delta RAD \sim \Delta V$. Hence above sand deserts, the atmosphere cools more strongly than over vegetation, and this radiative cooling is compensated by adiabatic sinking and heating of air just as in Föhn situations in the leeward side of mountains. Sinking motion further reduces cloudiness and the possibility of convective rain (ΔP in Fig. A.22b). As demonstrated by Dickinson (1992), Charney's theory of albedo-induced desertification is not wholly valid. It is incomplete as it ignores any interaction with changes in soil moisture, as shown in Fig. A.22b, right-hand branch. (In a later paper, Charney et al. (1977) prescribed large soil moisture for vegetation and little soil moisture for desert in order to simulate hydrological feedbacks to a first approximation.) However, Charney's theory can easily be extended to include the interaction with soil moisture and evaporation (Claussen 1997). It appears that the combined subtropical albedo-vegetation-precipitation feedback loop operates in an amplifying, i.e. positive way (see right-hand branch in Fig. A.22b with ΔE as change in evaporation).

When considering the desert-albedo effect in West Africa in detail the feedback processes are, of course, more complex. Eltahir (1996), and Eltahir and Gong (1996) suggested that the dynamics of the monsoon circulation over West Africa are regulated by the meridional distribution of boundary-layer entropy (i.e. the logarithm of the equivalent potential temperature). In their theory the dynamics of the moist atmosphere across the land-ocean boundary are the dominant factor which determines the strength of the African summer monsoon. This theory contrasts with Charney's (1975) theory and simulations by Xue and Shukla (1993) which emphasize the dynamics of a dry atmosphere across the savanna-desert border. One important implication of the land-atmosphere-ocean interaction mechanism proposed by Eltahir and Gong (1996) is that the variability in the sea-surface temperatures off the coast of West Africa and/or changes in land cover in the coastal region of West Africa may play a more important role in the regional monsoon climate than do changes in land cover near the desert border. By using a zonally symmetric model of West Africa, Zheng and Eltahir (1998) and Zheng et al. (1999) found that, consistently, land-cover changes along the border between Sahara and Sahel have a minor impact on the simulated monsoon circulation, while deforestation along the West African coast – which more strongly affect boundary-layer entropy – triggered a collapse of the monsoon circulation. On the other hand, Claussen (1997) demonstrated that land-cover change and subsequent vegetation-atmosphere interaction in the

Sahara can indeed strongly modify the African summer monsoon. He showed that changes in the atmospheric energy budgets and large-scale zonal atmospheric circulation arising from the land-cover change are consistent with Charney's theory. However, this result does not directly conflict with Eltahir's theory, because a change in land cover always yields a change in boundary-layer entropy. In his three-dimensional, global atmosphere-biome model, Claussen (1997) found a strong increase in West African summer monsoon and an increased gradient in moist static energy between land and ocean, but a weaker temperature gradient between land surface and ocean surface, when the western Sahara was covered by savanna. Claussen's (1997) results and the earlier numerical experiments by Xue and Shukla (1993) indicate that besides the regional atmosphere-ocean contrast which Eltahir's theory focusses on, large-scale processes can affect the African summer monsoon as suggested by Charney (1975). The African summer monsoon appeared to be closely linked to the Asian summer monsoon and the strength of the tropical easterly jet and the African easterly jet. In simulations with reduced Sahara in present-day climate (Claussen 1997), as well as in palaeo-simulations of a mid-Holocene greening of the Sahara (Claussen and Gayler 1997), the tropical easterly jet at 200 hPa was found to be stronger and the African easterly jet at 700 hPa near 15° N weaker than in present-day climate and present-day African deserts – in agreement with Xue and Shukla's (1993) desertification experiments (see also Chapt. A.5).

Returning to the classification of biogeophysical feedbacks, Levis et al. (1999a) proposed the existence of an *emissivity-vegetation feedback*, i.e. feedbacks associated with changes in the emissitiy ε and, thereby, in the absorption of atmospheric radiation and emission of infra-red radiation (second and third term of the energy budget equation). Presumably, the emissivity-vegetation feedback is important for the near-surface climate only in the absence of insolation, for example during the long polar night at high northern latitudes. Levis et al. (1999a) assumed that the emissivity of snow is smaller than that of vegetated surfaces which is not undisputed, because the emissivity of fresh snow approaches unity. Hence in their model, an increase in leaf area owing to an increase in taiga, for example, strengthens the infrared heat loss during winter leading to a near-surface cooling. This could put a stress on taiga which retreats, thereby leaving the ground for the more cold-resistant tundra.

Hydrological feedbacks are effective mainly in the growing season. They work both in negative and positive directions (therefore, both signs, + and –, are shown in Fig. A.22a, right-hand branch). A reduction in vegetation which is caused by a reduction in leaf area and/or in roots yields a reduction in transpiration E and thus a reduction in evaporative cooling. Keeping the available energy, i.e. the sum of energy fluxes through solar radiation, $S(1 - \alpha)$, thermal radiation, $\varepsilon R - \varepsilon \sigma T_G^4$, and the soil heat flux G, constant, then the sensible heat H increases at the cost of the latent heat flux λE, thereby warming the near-surface atmosphere. Warming at least in high latitudes tends to be favourable for plant regrowth. On the other hand, a reduction in transpiration could reduce local rainfall, which might hamper the growth of vegetation limited by precipitation, e.g. tropical rainforest. Thus, the sign of this feedback strongly depends on the region under consideration.

Hydrological feedbacks also affect near-surface fluxes indirectly. For example, near-surface warming caused by a reduction in evaporation can be amplified further by a reduction in cloudiness and thus an increase in insolation. Hence hydrological feedbacks can oppose surface-albedo feedbacks by changing the near-surface heat fluxes and the *planetary albedo*, i.e. the albedo of the surface and the atmosphere, including clouds (e.g. Betts 1999b). The problem becomes even more complicated, because a reduction in low clouds causes a reduction in long-wave atmospheric radiation, R, which induces a near-surface cooling, mainly at night. Moreover, a reduction in evaporation could, all other conditions being equal, reduce condensation and the release of condensation heat into the lower atmosphere with a consequent cooling near the ground. In summary, hydrological feedbacks act in various ways. Both directions, positive and negative, may occur depending on the time of day and on the season. On an annual average basis and at high northern latitudes, climate models indicate that the negative feedback dominates and therefore, we have put the positive sign on the lower right-hand branch in Fig. A.22a in parentheses.

Meehl (1994) outlined the competition between two branches of hydrological feedbacks of the Indian summer monsoon, both being positive and negative. He demonstrated that an increase in surface moisture leads to stronger precipitation and thereby better vegetation growth and increased surface moisture. On the other hand, stronger evaporation from moister surfaces induce weaker temperature contrasts between the continent and the Indian ocean. Hence moister land surfaces tend to weaken the monsoon, thereby attenuating large-scale moisture advection.

Hydrological feedbacks can be modified by albedo feedbacks as shown by Lofgren (1995a,b). He showed by use of numerical simulations that in the tropics, a change in albedo yields a change in the net radiation budget and thus, in the energy available for evaporation and transpiration (see Fig. A.22c). This clearly demonstrates that surface-albedo feedbacks and hydrological feedbacks cannot be treated separately, particularly if both feedbacks are triggered by changes in vegetation.

It seems likely that the sign and strength of biogeophysical feedbacks change with the geographical location on a large scale, as indicated in Fig. A.22. In mid-latitudes, the situation is presumably quite complex as surface-albedo feedbacks and hydrological feedbacks compete with each other. However, at high northern latitudes the strong and positive taiga-tundra feedback presumably dominates, as does the positive vegetation-albedo-precipitation feedback (the "extended" Charney's feedback) in the subtropics. In the tropics, hydrological feedback seems to dominate, as indicated by deforestation experiments (e.g. Polcher and Laval 1994a; Henderson-Sellers et al. 1993).

Biogeophysical and biogeochemical feedbacks do not operate independently in nature. Since models which are capable of describing both processes are under continuous development (see also Sect. A.4.3), we can discuss only preliminary results. By using a coupled atmosphere-ocean-vegetation model with a simple carbon cycle Claussen et al. (2001) quantified the relative magnitude of biogeophysical and biogeochemical processes as well as their synergisms (see below). Their sensitivity studies show that biogeochemical and biogeophysical processes being triggered by large-scale land-cover changes oppose each other on the global scale. Tropical deforestation tends to warm the planet because the increase in atmospheric CO_2 and hence atmospheric radiation, outweighs the biogeophysical effects. In mid and high northern latitudes, however, biogeophysical processes, mainly the taiga-tundra feedback through its synergism with the sea-ice-albedo feedback, win over biogeochemical processes, thereby eventually leading to a global cooling in the case of deforestation and to a global warming in the case of afforestation (see Fig. A.23). Synergisms between biogeophysical and biogeochemical processes appear to be weak on the global scale.

A.4.1.2 Synergisms

Although the analysis of feedbacks is very useful to identify and isolate biospheric processes, it is often the interaction between feedbacks, the synergism, which becomes important for climate system dynamics. Indeed, climate modellers sometimes describe synergisms, rather than feedbacks, albeit the opposite has been claimed (e.g. Braconnot et al. 1999). For example, to isolate biospheric feedbacks, differences between simulations with a coupled atmosphere-ocean-vegetation and a coupled atmosphere-ocean model are interpreted as the results of biospheric feedbacks. However, implicitly hidden in the differences between model results are not only feedbacks, but also synergisms of biospheric feedbacks with other, atmosphere-ocean feedbacks. A proper feedback analysis requires a number of simulations in which the feedback processes are switched on and off individually in all possible combinations. Examples of "proper" feedback analyses are given by Berger (2001) who uses the factor-separation technique by Stein and

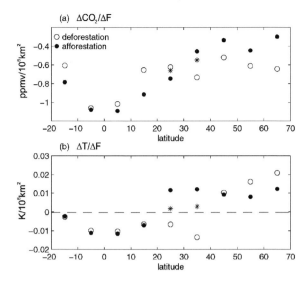

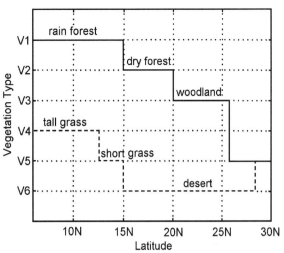

Fig. A.23. Ratio of changes of global atmospheric CO_2 concentration (**a**) and global mean near-surface temperature T (**b**) to changes in forest area F owing to deforestation and afforestation in latitudinal belts of 10 degrees width. Units given in ppmv CO_2 per 10^6 km^2 and K per 10^6 km^2 of tree area directly affected by deforestation (*open circles*) or afforestation (*full circles*), respectively. An *asterix* indicates a simulation in which deserts, i.e. grid cells with marginal or no vegetation cover, are left unchanged in the case of afforestation. Sensitivities are computed after the coupled atmosphere-ocean-vegetation model has approached a new equilibrium state, in which the land surface in a zonal belt is completely deforested or afforested. Regrowth of grassland is allowed in the deforested area (taken with modifications from Claussen et al. 2001)

Alpert (1993) and Claussen (2001a) who extends the classical feedback analysis used in electrical engineering (see Peixoto and Oort 1992).

An example of a strong synergism between a biogeophysical feedback and an atmosphere-ocean feedback is the so-called "biome paradox" which will be discussed in Sect. A.4.2. Another type of synergism was detected by Ganopolski et al. (2001). They analysed the effect of boreal and tropical deforestation on global climate, and they find that their results strongly depend on the inclusion of ocean-atmosphere interaction. If an atmosphere-only model is used, then tropical deforestation tends to warm the near-surface atmosphere, even on a global average. Hence the hydrological feedback dominates the surface-albedo feedback. But letting the ocean interact as part of the climate system turns global warming into a global cooling with some marginal warming in the tropics only. The reason for this difference is differences in evaporation: in the experiments with fixed sea surface temperatures, evaporation from the ocean increases owing to a drier atmosphere and approximately compensates for the reduction of evaporation over land surfaces. The interactive ocean, however, cools in response to a decreased atmospheric water content, thereby diminishing evaporation from the ocean surfaces and decreasing the water-vapour greenhouse effect further.

Fig. A.24. Vegetation types at the equilibrium state into which a zonally-symmetric coupled biosphere-atmosphere model of West Africa develops into when it starts with two extreme initial conditions. *Upper curve*: the model starts from a rainforest-covered West Africa; *lower curve*: the model starts from a desert-covered West Africa. The vegetation types include rainforest (*V1*), dry forest (*V2*), woodland (*V3*), short grass (*V5*), and desert (*V6*). (This figure is taken with modifications from Wang and Eltahir 2000b)

A.4.1.3 Multiple Equilibria

As the interaction between components of the climate system is non-linear one might expect multiple equilibrium solutions. Gutman et al. (1984) and Gutman (1984, 1985) found only unique, steady-state solutions in their zonally averaged model. (Actually, they regarded their results as "tentative and merely as an illustration of the suggested approach", because of the simplicity of their model.) The possibility of multiple equilibria in the three-dimensional atmosphere-vegetation system was discovered later by Claussen (1994) and subsequently analysed in detail by Claussen (1997, 1998) for present-day climate, i.e. present-day insolation and sea surface temperature. Two solutions of the atmosphere-vegetation system appear: the arid, present-day climate and a humid solution resembling more the mid-Holocene climate, i.e. with a Sahara greener than today, albeit less green than in the mid-Holocene. The two solutions differ mainly in the subtropical areas of northern Africa and, but only slightly, in Central East Asia. The possibility of multiple equilibria in the atmosphere-vegetation system of North-West Africa has recently been corroborated by Wang and Eltahir (2000a–c; compare Fig. A.24) and Zeng and Neelin (2000) by using completely different models of the tropical atmosphere and dynamic vegetation. The latter studies indicate that the existence of multiple states of the atmosphere-vegetation system in West Africa depends also on the inclusion of other perturbation, such as changes in sea-surface temperatures. Moreover, more than two stable states can be attained depending on the initial conditions chosen (see also Sect. A.5.3.4).

Interestingly, the stability of the atmosphere-vegetation system seems to change with time: experiments with mid-Holocene vegetation yield only one solution, the green Sahara (Claussen and Gayler 1997), while for the Last Glacial Maximum (LGM), two solutions exist (Kubatzki and Claussen 1998) (Fig. A.25).

So far, no other regions on Earth have been identified in which multiple equilibria could evolve on a large scale. Levis et al. (1999b) have sought multiple solutions of the atmosphere-vegetation-sea-ice system at high northern latitudes. Their model converges to one solution in this region, corroborating the earlier assertion (Claussen 1998) that multiple solutions manifest themselves in the subtropics, mainly in northern Africa.

Why do we find multiple solutions in the subtropics, but none at high latitudes and why for present-day and LGM climates, but not for the mid-Holocene climate? Claussen et al. (1998) analysed large-scale atmospheric patterns in present-day, mid-Holocene, and LGM climates. They found that velocity potential patterns, which indicate divergence and convergence of large-scale atmospheric flow, differ between arid and humid solutions mainly in the tropical and subtropical regions. It appears that the Hadley-Walker circulation shifts slightly to the west. This is consistent with Charney's (1975) theory of albedo-induced desertification in the subtropics. Moreover, changes in surface conditions directly influence vertical motion, and thereby large-scale horizontal flow, in the tropics (Eltahir 1996), but hardly at all at middle and high latitudes (e.g. Lofgren 1995a,b). For the mid-Holocene climate, the large-scale atmospheric flow is already close to the humid mode, even if one prescribes present-day land-surface conditions. This is caused by differ-

Fig. A.25.
Multiple equilibria computed for present-day climate (**a**), and for the climate of the last glacial maximum (**c**). For mid-Holocene conditions, only one solution is obtained (**b**). This figure summarises the results of Claussen (1997) (**a**), Claussen and Gayler (1997) (**b**) and Kubatzki and Claussen (1998) (**c**)

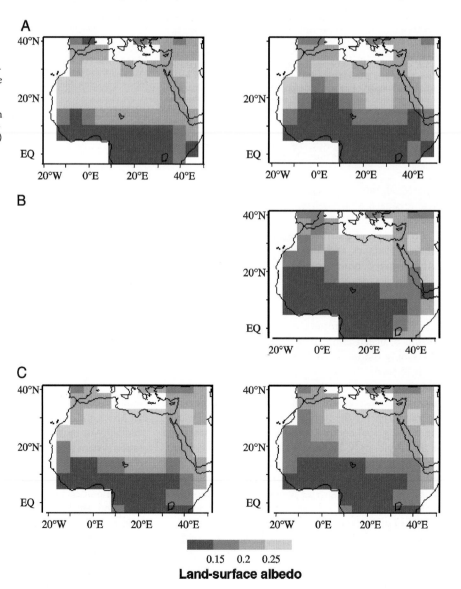

ences in insolation: in the mid-Holocene boreal summer, the Northern Hemisphere received up to 40 W m^{-2} more energy than today, thereby strengthening African and Asian summer monsoon (Kutzbach and Guetter 1986). During the LGM, insolation was quite close to present-day conditions.

A more ecological interpretation of multiple equilibria was given by Brovkin et al. (1998). They developed a conceptual model of vegetation-precipitation interaction in the western Sahara which is applied to interpret the results of comprehensive models. The conceptual model finds three solutions for present-day and LGM climate; one of these, however, is unstable to infinitesimally small perturbations. The humid solution is shown to be less probable than the arid solution, and this explains the existence of the Sahara desert as it is today. For mid-Holocene climate, only one solution is obtained. Application of the conceptual model to biospheric feedbacks at high latitudes (Levis et al. 1999b) yields only one solution for present-day conditions.

Are multiple equilibria only relevant to the atmosphere-vegetation system, or do they occur also in the atmosphere-ocean-vegetation system? So far, multiple solutions have not yet been found in the model of Ganopolski et al. (1998, 2001) and Petoukhov et al. (2000). (The model attains multiple solutions associated with multiple states of the thermohaline convection.) The authors blame this deficit on the coarse resolution of this model because northern Africa is represented by just three grid boxes, Sahara, Sudan and tropical northern Africa. Subsequently, Saharan precipitation in the coarse model of Ganopolski et al. (1998) is less sensitive to changes in land-surface conditions than the West-Saharan precipitation in the model used by Claussen (1997, 1998). On the other hand, the study of Ganopolski et al. (1998) shows that the biogeophysical feedback in northern Africa is mainly a vegetation-atmosphere feedback. Therefore, the conclusion from coupled vegetation-atmosphere models should generally be valid, i.e. vegetation-atmosphere-ocean models (with finer horizontal resolution, however) should also exhibit multiple equilibria in the north-western African region.

The discussion of multiple equilibria seems to be somewhat academic. However the existence of these can explain abrupt transitions in vegetation structure (Claussen et al. 1998; Brovkin et al. 1998). If global stability changes in the sense that one equilibrium solution becomes less stable to finite amplitude perturbations than the others, then an abrupt change in the system from the less stable to a more stable equilibrium is to be expected. This will be discussed in more detail in Sect. A.4.2. Likewise, the existence of multiple equilibria can be used to interpret rainfall variability in the Sahel region, suppression of interannual variability and enhancement of decadal variability, see Chapt. A.5.

Thomas Chase

A.4.1.4 Teleconnections

From theoretical considerations, it should be expected that large-scale changes in land cover will have remote climatic effects, just like the El Niño phenomenon in the easterly tropical Pacific does. Here we consider teleconnections arising from large-scale land-cover change, particularly in the tropics. Teleconnections owing to changes in the carbon cycle are discussed in Sect. A.4.3, and teleconnections associated with changes in riverine transport owing to changes in land cover are highlighted in Part D.

The three major tropical convective heating centres are associated with the land surfaces of Africa, Amazonia, and the maritime continent of Indonesia, Malaysia, New Guinea and surrounding regions (e.g. Krueger and Winston 1973; Chen et al. 1988; Berbery and Nogues-Paegle 1993). Significant precipitation recycling occurs due to the presence of dense, transpiring vegetation in tropical regions (e.g. Lettau et al. 1979; Burde and Zangvil 1996). Changes in this vegetation structure have major impacts on the momentum and radiant energy absorbed at the surface and its partitioning into latent and sensible forms which affect surface temperatures, boundary layer energetics (see Chapt. A.2) and structure and the strength and positioning of convective storms (e.g. Dickinson and Kennedy 1992; Pielke et al. 1997; Nobre et al. 1991; Eltahir 1996; Polcher and Laval 1994b, see also Chapt. A.3). Small changes in the magnitude and spatial pattern of tropical convection may then alter the magnitude and pattern of upper-level tropical outflow which feeds the higher latitude zonal jet (e.g. Bjerknes 1969; Krishnamurti 1961; Chen et al. 1988; Oort and Yienger 1996).

Aside from affecting the mean zonal flow, changes in tropical convection may also force anomalous Rossby waves which can propagate to higher latitudes in a westerly background flow (e.g. Wallace and Gutzler 1981; Tiedtke 1984; James 1994; Tribbia 1991; Berbery and Nogues-Paegle 1993). Therefore, land-cover changes which result in changes in tropical convection may affect weather and climate regimes remotely both in the tropics and at high latitudes. This process is analogous to the well documented remote effects attributed to the opposing phases of the El Niño-Southern Oscillation (ENSO).

Finally, changes in tropical and mid-latitude vegetation cover appear to play a significant role in the strength and positioning of tropical monsoon circulations (e.g. Eltahir 1996; Meehl 1994). Because tropical monsoon circulations have global effects and interactions which extend far beyond the tropics, any change in these circulations would be expected to alter weather and climate remotely both by affecting the mean flow and the waves superimposed on that flow.

Relatively few studies have commented directly on teleconnections resulting from land-cover changes. Franchito and Rao (1992) found a reduction in large-scale tropical circulations in a land-cover change experiment using a zonally averaged climate model. Such changes in global-scale circulations would be expected to generate remote climatic effects though that study did not specifically examine such a possibility. McGuffie et al. (1995) noted weak, remote, high latitude effects in a complete tropical deforestation simulation (i.e. all tropical rainforests converted to grasslands), while Chase et al. (1996) hypothesised that large, winter hemisphere climate anomalies in an atmospheric general circulation model (GCM) simulation of the effects of realistic global green leaf area change were due to tropical deforestation affecting low latitude convection and therefore global-scale circulations. Zhang et al. (1996) further explored simulated mid-latitude teleconnections due to complete tropical deforestation with a linear wave model and found that appropriate conditions existed for the propagation of tropical waves into the extratropics in that case.

More recently, specific efforts towards identifying teleconnection patterns resulting from vegetation change have been mounted. Chase et al. (2000) examined GCM model simulations of the effect of observed levels of land-cover change globally and found strong evidence of changes in global-scale circulations and for the propagation of Rossby waves into the mid-latitudes. Pitman and Zhao (2000) and Zhao et al. (2001) again demonstrated that the remote effects of observed levels of land-cover change were prevalent in a variety of models under a variety of configurations and model assumptions and that remote temperature anomalies resulting from land-cover change could be similar in magnitude to the effects of present levels of CO_2. Gedney and Valdes (2000), also using a GCM, specifically examined the effects of a wholesale removal of the Amazonian rainforest on remote climates and found significant evidence for a reduction in large scale circulations generated by tropical convection and for propagating Rossby waves which affected rainfall in Northern Hemisphere winter. They further strengthened their results with the aid of a simple wave model which produced a similar pattern when driven by the GCM-simulated diabatic heating pattern.

Martin Claussen

A.4.2 Palaeoclimate

A.4.2.1 Feedbacks in the Arctic Climate System

Palaeobotanic evidence indicates that during the early to middle Holocene, some 9 000 to 6 000 years ago, boreal forests extended north of the modern tree line (Frenzel et al. 1992; Kutzbach et al. 1996b; Cheddadi et al. 1997). It is assumed that this migration was triggered by changes in the Earth's orbit which led to stronger insolation in Northern Hemisphere summer, i.e. the growing season. Insolation during northern hermisphere winter, however, was weaker than today. Hence, when assuming that changes in insolation are the dominant cause for climate change, one would expect warmer summers but colder winters than today during the early and middle Holocene. However, not only summers but also winters were presumably warmer than today in many regions of the Northern Hemisphere, as reported, for example, for Europe by Cheddadi et al. (1997). This winter-time warming is assumed to be caused by the taiga-tundra feedback (see Sect. A.4.1), which is supposed to have amplified initial summer warming to be felt all year round. It is, therefore, sometimes called the biome paradox (e.g. Berger 2001). For example, by imposing an increase in forest area by some 20% as a surface condition, Foley et al. (1994) found that changes in land-surface conditions give rise to an additional warming of some 4 °C in spring and about 1 °C in the other seasons. The additional warming is mainly caused by a reduction of snow and sea-ice volume by nearly 40% and subsequent reduction in surface albedo. Further simulations using similar experimental set-ups but different models (Kutzbach et al. 1996b) corroborated the earlier results. These studies clearly point to the importance of vegetation-atmosphere interaction at high northern latitudes in amplifying climate change triggered by some external forcing. However, these sensitivity studies did not isolate the effect of a decrease in vegetation-snow albedo and the sea-ice-albedo feedback. It was thus not clear how much the biospheric process actually contributes to the mid-Holocene warming at high northern latitudes.

As mentioned in Sect. A.4.1, the taiga-tundra feedback is a positive one: a reduction in surface albedo increases near-surface temperatures which, in turn, favour growth of taller vegetation, reducing surface albedo further (see Otterman et al. 1984). The feedback is limited by topographical constraints, e.g. coast lines, or by the insolation. The studies of Claussen and Gayler (1997) and Texier et al. (1997), when using different atmospheric models but the same biome model of Prentice et al. (1992), confirm this assertion. However, both models show a rather small northward expansion of boreal forests. This is not surprising, as the annual cycle of sea-surface temperatures (SSTs) and Arctic sea-ice volume is kept constant. Obviously the synergism between terrestrial and marine feedbacks (see Fig. A.22a) is missing. This has clearly been demonstrated by Ganopolski et al. (1998) who used a coupled atmosphere-ocean-vegetation model. They found a summer warming over Northern Hemisphere continents of some 1.7 °C (in comparison with present-day climate) owing to orbital forcing on the atmosphere alone. Inclusion of ocean-atmosphere feedbacks (but keeping the vegetation structure constant in time) reduces this signal to some 1.2 °C, whereas the vegetation-snow-albedo feedback (but now without any oceanic feedback) enhances the summer warming to 2.2 °C. In the full system (including all feedbacks) this additional warming is not reduced, as one would expect from linear reasoning, but is increased to 2.5 °C as a result of a synergism between the vegetation-snow-albedo feedback and the Arctic sea-ice-albedo feedback. Likewise, orbital forcing alone induces a wintertime cooling of some –0.8 °C. The biogeophysical feedbacks alone reduce this cooling to –0.7 °C, and the atmosphere-ocean interaction, to –0.5 °C. The synergism between the two feedbacks, however, causes a winter warming of some 0.4 °C (see Fig. A.26). Therefore, the warming of Northern Hemisphere winters does not seem to be a pure biospheric feedback, but it is caused mainly by the synergism between this feedback and the oceanic feedback. Hence the so-called biome paradox does not seem to be a pure biome paradox, but rather a biome-sea-ice paradox.

During the mid-Holocene, orbital forcing triggered a warming of the Northern Hemisphere in summer, whereas the opposite is valid for the end of the Eemian warm period some 115 ka BP (ka BP = 1 000 years before

present) as pointed out by Harvey (1989b) and subsequently by Gallée et al. (1992), Berger et al. (1992, 1993) and Gallimore and Kutzbach (1996). These studies show that the vegetation-snow-albedo feedback contributes significantly to the temperature response to orbital forcing. Gallimore and Kutzbach (1996) state that even a prescribed increase in surface albedo which is deduced from a biome model estimate of tundra expansion at 115 ka BP is sufficient to induce glaciation over northeastern Canada. (Actually, Gallimore and Kutzbach (1996) did not simulate glacial inception, just the occurrence of permanent snow cover). DeNoblet et al. (1996) supported this hypothesis by using a coupled atmosphere-biome model, although they obtain just a substantial increase in snow depth but no large-scale perennial snow cover over north Canada. Moreover, they restrict themselves to the biospheric feedbacks, ignoring any synergism between land surface and sea ice (which presumably could help to obtain perennial snow cover).

Presumably, the taiga-tundra feedback and its synergism with other feedbacks play an important role also in the warm world of the late Cretaceous period, some 100 to 64 million years ago. Otto-Bliesner and Upchurch (1997) showed that their model in contrast to palaeocli-

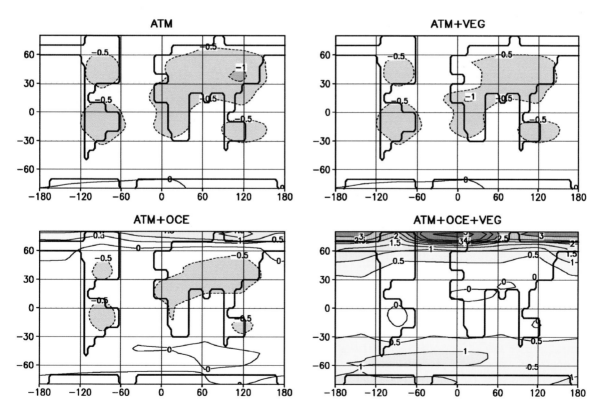

Fig. A.26. Differences in near-surface temperatures during Northern Hemisphere winter (December, January, February) between mid-Holocene climate and present-day climate simulated by Ganopolski et al. (1998). The authors used different model configurations: the atmosphere-only model (labelled *ATM*), the atmosphere-ocean model (*ATM + OCE*), the atmosphere-vegetation model (*ATM + VEG*) and the fully coupled model (*ATM + OCE + VEG*). In *ATM*, *ATM + OCE*, and *ATM + VEG*, present-day land-surface and ocean-surface conditions depending on the model configuration are used (taken from Wasson and Claussen 2002, © reprinted with permission from Elsevier Science)

mate reconstructions does not inhibit growth of inland-ice masses in the Northern Hemisphere. They demonstrated that only by prescribing a forest-type vegetation in high northern latitudes does their model yield a summer climate warm enough to melt any ice sheets.

A.4.2.2 The Sahara

Palaeoclimatic reconstructions indicate that during the so-called Holocene climatic optimum some 9 000–6 000 years ago, the summer in the Northern Hemisphere was warmer than today. The summer monsoon in northern Africa was stronger than today according to lake level reconstructions (Yu and Harrison 1996), estimates of aeolian dust fluxes (deMenocal et al. 2000), and distribution of sand dunes (Sarnthein 1978). Moreover, palaeobotanic data (Jolly et al. 1998) reveal that the Sahel reached at least as far north as 23° N. (The present boundary extends up to 18° N.) Hence, there is an overall consensus that during the Holocene optimum, the Sahara was greener than today – albeit reconstructions of mid-Holocene vegetation distribution in northern Africa vary (Frenzel et al. 1992; Hoelzmann et al. 1998; Anhuf et al. 1999, Prentice et al. 2000) with respect to vegetation coverage and geographical detail.

The greening was tentatively attributed to changes in orbital forcing (e.g. Kutzbach and Guetter 1986) that amplified the northern African summer monsoon. However, this amplification in summer rain did not seem to be large enough to explain a large-scale greening (Harrison et al. 1998; Joussaume et al. 1999). In their model, Texier et al. (1997) yielded a positive feedback between vegetation and precipitation in this region which is, however, much too weak to get any substantial greening (see Fig. A.27, *upper figure*). They suggested an additional (synergistic) feedback between SST and land-surface changes. Braconnot et al. (1999) found indeed such a synergism which however is also not strong enough to produce a strong northward shift of vegetation. By modifying surface conditions in northern Africa (increased vegetation cover, increased area of wet-lands and lakes) Kutzbach et al. (1996a) obtained some change in their model which leads to an increase in precipitation in the south-eastern part of the Sahara but almost none in the western part. Broström et al. (1998) obtained a stronger, but still unrealistically small northward migration of savanna. Claussen and Gayler (1997) found a strong feedback between vegetation and precipitation and an almost complete greening in the western Sahara and some in the eastern part (see Fig. A.27, *middle figure*). Claussen and Gayler (1997) and Claussen et al. (1998) explained the positive feedback by an interaction between high albedo of Saharan sand deserts and atmospheric circulation, as hypothesised by Charney (1975). They extended

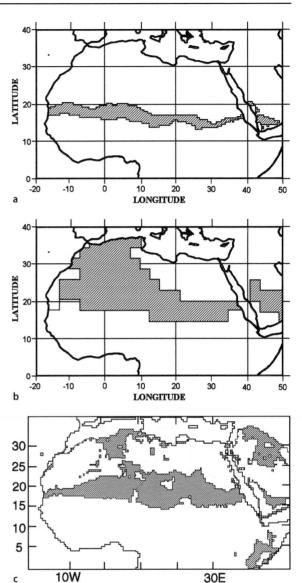

Fig. A.27. Reduction of desert from present-day climate to mid-Holocene climate simulated by (**a**) the model of Texier et al. (1997), (**b**) Claussen and Gayler (1997), and (**c**) Doherty et al. (2000). *Shading with lines* from the lower left to the upper right indicate a reduction of desert, and the *opposite shading*, an increase

Charney's theory by accounting for atmospheric hydrology, i.e. moisture convergence and associated convective precipitation, and interactive soil moisture. Recently, Doherty et al. (2000) used a coupled atmosphere dynamic vegetation model to assess the problem of the green Sahara. They report a reduction of Saharan desert by some 50% (see Fig. A.27, *lower figure*) which seems to agree best with Anhuf's reconstruction, but is too low in comparison with the estimate by Hoelzmann et al. (1998). However, the result by Doherty et al. (2000) is not taken directly from the coupled atmosphere-vegetation model. It is an interpretation of model results by using an equilibrium vegetation model BIOME-3 (Haxeltine and Prentice 1996). The original coupled model has a rather strong bias towards a green Sahara with present-day climate.

Which model is correct? This question cannot yet satisfactorily be answered given the uncertainty of palaeobotanic reconstructions. To tackle the problem from a more theoretical point of view, deNoblet-Ducoudre et al. (2000) compared the extreme with regard to the magnitude of Saharan greening models of Claussen and Gayler (1997) and Texier et al. (1997). Both groups used the same biome model, but different atmospheric models. Moreover, the atmospheric model and biome model were asynchronously coupled in different manners: Claussen and Gayler (1997) used the output of the climate model directly to drive the biome model, while Texier et al. (1997) took the difference between model results and a reference climate as input to the biome model. The latter, so-called anomaly approach, prevents the coupled model from drifting to an unrealistic climate which could be induced by some positive feedbacks between biases in either model. Hence this method is similar to the flux correction in coupled atmosphere-ocean models. It turns out that the difference between coupling procedures affects the results of the coupled atmosphere-biome model only marginally. DeNoblet-Ducoudre et al. (2000) concluded therefore that the differences in northern Africa greening cannot be attributed to the coupling procedure rather it can be traced back to different representations of the atmospheric circulation in the tropics. The atmospheric model of Claussen and Gayler (1997) somewhat overestimates the duration of the African monsoon, while the other model of Texier et al. (1997) yields an unrealistic near-surface pressure distribution and, therefore, a too zonal circulation. The authors demonstrated how the one model yielded an unrealistically arid climate and why they believe more in the other model as far as the existence of a strong biogeophysical feedback in northern Africa is concerned. But they could not prove that the latter model is completely trustworthy. Hence this issue certainly needs further consideration.

A second argument concerns the missing interaction with the ocean. Kutzbach and Liu (1997) provided simulations using an asynchronously and partially coupled atmosphere-ocean model (no freshwater fluxes, no dynamic sea-ice model). They found an increase in African monsoon precipitation as a result of increased SST in late summer, bringing the model in closer agreement with palaeo data. Similarly Hewitt and Mitchell (1998), using a fully coupled atmosphere-ocean model, observed an increase in precipitation over northern Africa, but still not as intense as the data suggest. They assumed that missing biospheric feedbacks were the cause of their model failure. Ganopolski et al. (1998) have re-addressed this issue using a coupled atmosphere-vegetation-ocean model in different combinations (as an atmosphere-only model, an atmosphere-vegetation model, an atmosphere-ocean model, and a fully coupled model). They concluded that the biospheric feedback dominates in the subtropics, while the synergism between this feedback and an increase in monsoon precipitation owing to increased SST adds only little. This result is in qualitative agreement with most model experiments on palaeomonsoon (Kutzbach et al. 2001). Braconnot et al. (1999) found a stronger impact of SST than did Ganopolski et al. (1998), albeit that the biospheric feedback seems to dominate. Unfortunately, Braconnot et al. (1999) did not perform a systematic feedback and synergism analysis or factor-separation technique so the relative magnitude of feedbacks could not be quantified.

How does the biogeophysical feedback in northern Africa change with time? As mentioned in Sect. A.4.1, this feedback could lead to multiple equilibrium solutions which could explain abrupt transitions in vegetation structure (Claussen et al. 1998). Brovkin et al. (1998) found in their conceptual model that the green solution becomes less stable around 3.6 ka BP. Keeping in mind that the variability of precipitation is larger in humid regions than in arid regions of northern Africa (e.g. Eischeid et al. 1991), one would expect a transition roughly in between 6 ka BP and 4 ka BP.

In fact, there is evidence that the mid-Holocene wet phase in northern Africa ended around 5.0–4.5 ka BP, even in the high continental position of the east Sahara (Pachur and Wünnemann 1996; Pachur and Altmann 1997). Petit-Maire and Guo (1996) presented data which suggest that the transition to present-day arid climate did not occur gradually but in two steps with two arid periods, at 6.7 to 5.5 ka BP and 4 to 3.6 ka BP. Other reconstructions indicate that freshwater lakes in the eastern Sahara began to disappear from 5.7 to 4 ka BP, when recharge of aquifers ceased at the end of the wet phase (Pachur and Hoelzmann 1991). Pachur and Hoelzmann (1991) suggest that climate change at the end of the mid-Holocene was faster in the western than in the eastern Sahara. Recently, deMenocal et al. (2000) found an abrupt change in terrigenous material dated at 5 500 years ago. The record was obtained from marine cores drilled near the Atlantic coast of northern Africa.

Claussen et al. (1999) have analysed the transient structures in global vegetation patterns and atmospheric characteristics using the coupled atmosphere-ocean-vegetation model of Ganopolski et al. (1998), but with a dynamic vegetation module. Their simulations clearly show (see Fig. A.28) that subtle changes in orbital forcing triggered changes in northern African climate which were then strongly amplified by biogeophysical feedbacks in this region. The timing of the transition, which started at around 5.5 ka BP in the model, was governed by a global interplay between atmosphere, ocean, sea ice,

Fig. A.28.
Simulation of transient development of precipitation (**b**)
and vegetation fraction in
the Sahara (**c**) as response to
changes in insolation (**a** depicts insolation changes on
average over the Northern
Hemisphere during boreal summer). Results from Claussen
et al. (1999) are compared with
data of terrigenous material
in North Atlantic cores off the
northern African coast (**d**)
(this figure is taken with modifications from deMenocal et al.
2000)

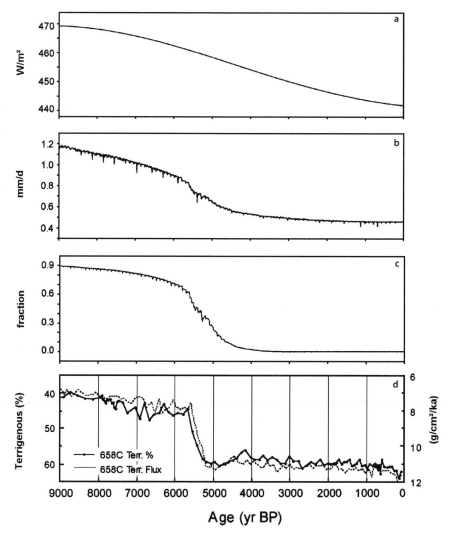

and vegetation. The latter is affected by a change in tropical SST and by the synergisms between biospheric and
oceanic feedbacks, mentioned in the previous chapter,
which influence the large-scale meridional temperature
gradient. Hence the abrupt desertification, abrupt in
comparison with the subtle change in orbital forcing, is
a regional effect, the timing of which depends, however,
on global processes.

Martin Claussen · Hans-Jürgen Bolle

A.4.2.3 Historical Land-cover Change

Historical land-cover change has been a long-term process. It started with the first human settlements which grew
in conjunction with population density, and has accelerated over the last century (see Sect. E.5.2). For example,
northern Africa in Roman times was considered one of
the most prosperous and rich areas of the western world.
The production of wheat, olive oil, and wine was greater

than in Mediterranean Europe. Reorganisation of land
use by the Romans started some 2 100 years ago when
marginal land was cultivated also and the environment
modified by drainage systems, river damming and aq-
ueducts. Associated with agricultural expansion and further deforestation around 2 ka B.P. the vegetation
changed as documented in and reconstructed from historical sources (Reale and Dirmeyer 2000).

These large-scale changes of land use may have contributed to long-term climate change in this area. To test
this hypothesis, a coupled land-atmosphere climate
model was re-run with the present distribution of vegetation exchanged against the land cover as suggested
by Reale and Dirmeyer with the vegetation for the end
of the Roman classical period (Reale and Shukla 2000).
Many of the properties of summertime precipitation
suggested by both the historical and archeological evidence were recreated in the simulations. For example,
northern Africa from the Nile valley to the Atlas mountains was simulated as significantly wetter with the veg-

etation at the time of the Roman classical period. Much of southern Europe (e.g. Greece and the Iberian Peninsula) were also wetter, with reduced rainfall over the Mediterranean Sea itself. The model results also underline the interaction between the Mediterranean area and the Hadley circulation. According to the model simulations there was a stronger northward propagation of the Inter-Tropical Convergence Zone and its associated rains into the Sahel during boreal summer in the Roman classical period; rainfall over sub-Saharan Africa was also affected. In these computations all other parameters except the vegetation cover remained unchanged. In view of the complex interactions between the surface and the atmosphere this may not be a fully adequate procedure to really *prove* that Mediterranean climate is changed by changes of land use, because changes in large-scale atmospheric dynamics, for example changes in the North Atlantic Oscillation, together with changes in sea-surface temperatures also affect Mediterranean climate (Bolle 2003). Nonetheless the study by Reale, Dirmeyer and Shukla provides a valuable first step into more detailed investigations.

Brovkin et al. (1999) have focused on the biogeophysical effect of the land-cover change on climate during the last millennium. A dynamic scenario of deforestation, derived from rates of land conversion to agriculture, leads to a global cooling of 0.35 °C with a more notable cooling of the Northern Hemisphere (0.5 °C), according to their atmosphere-ocean-vegetation model. The cooling is most pronounced in the northern middle and high latitudes, especially during the spring season because of an increase in surface albedo and the tundra-taiga feedback. This is in qualitative agreement with studies by Betts (1999a) of the effect of present-day land use on climate. Betts found a local cooling of some 2 K in winter in the Eurasian agricultural region which amounts to a cooling of 1 K in the annual mean. Global changes were small, however, presumably owing to prescribed, i.e. fixed, sea-surface temperatures.

Chase et al. (1996, 2000) found from their model simulations that historical global land-cover changes could have led to an increase in large regional temperature changes with particularly strong warming and cooling regions at higher latitudes in the Northern Hemisphere winter. Because anthropogenic land-cover changes are largest in the tropics, but relatively small for high latitudes, the authors argue that these changes are mainly caused by tropical to high-latitude teleconnections (see Sect. A.4.1.4). These results are surprising in that previous experiments involving land-cover changes (e.g. Amazon deforestation experiments, see Chapt. A.6) did not indicate a strong impact on global climate. The results of Chase et al. (2000) are supported by Zhao et al. (2001) who used a different land-cover change pattern, but similar climate model configuration. Moreover, the experiments performed by Zhao et al. (2001) suggest that land-cover change over South-east Asia appears to cause larger global impacts than those over South America. Hence the discrepancy between the studies by Chase et al. (2000) and Zhao et al. (2001) on the one hand and by Brovkin et al. and Betts on the other awaits further consideration.

From their study, Brovkin et al. (1999) concluded that anthropogenic land-cover change, at least partially, contributed to the so-called Little Ice Age in the 16–19th centuries. However, the authors admitted that their conclusion is preliminary as they have not taken into account changes in the terrestrial carbon budget. The release of carbon into the atmosphere owing to historical global land-cover change is quite significant. For example, carbon emission due to land-cover change for the years 1860–1980 was estimated by Houghton et al. (1982) to have been 180 Gt (Gt = 10^{15} g). A large portion of the carbon released by deforestation has presumably been taken up by the ocean. Investigation of this problem using a coupled atmosphere-ocean-carbon cycle model is underway.

Peter M. Cox · Richard A. Betts

A.4.3 Sensitivity to Decadal Biogeochemical Feedbacks

A.4.3.1 The Land Surface and Climate Change

The continued increase in the atmospheric concentration of greenhouse gases, primarily due to anthropogenic emissions, is expected to result in significant climate change over the next 100 years (e.g. Houghton et al. 2001). The land surface will also play a role in determining the climate change. The climate response to anthropogenic disturbances, including emissions of greenhouse gases, can affect many things such as changes in crop productivity, or the frequency and intensity of floods and droughts. Over the last 15 years, a broader recognition of this role has driven significant work designed to improve the representation of the land surface in climate models. As a result, we now have a greater number and range of candidate land-surface schemes (LSSs) than ever before. Intercomparison projects such as PILPS (see also Sect. A.2.3 and Henderson-Sellers et al. 1996) have demonstrated that differences amongst these LSSs produce a spread of responses to a given atmospheric forcing. However, we are only just beginning to assess how these differences contribute to uncertainties in simulations of climate change. In fact, the current generation of LSSs have explored only a subset of biogeophysical and hydrological feedbacks, and have not yet considered biogeochemical feedbacks.

During a recent collaborative project, Polcher et al. (1998) carried out a series of climate change experiments using four separate Atmospheric General Circulation

Models (AGCMs). Each AGCM simulated the current climate ($1 \times CO_2$) and a climate with doubled CO_2 ($2 \times CO_2$) using two different land-surface representations. Although the nature of the imposed LSS changes was allowed to vary amongst the AGCMs, Crossley et al. (2000) were able to estimate the importance of the land surface by comparing the between-AGCM variation with the between-LSS variation. The land-surface representation was found to play a role in determining differences between the modelled $1 \times CO_2$ climates, but made a much larger contribution to differences in climate sensitivity ($2 \times CO_2 - 1 \times CO_2$).

Gedney et al. (2000) used results from the same numerical experiments to understand the hydrological response of each LSS to climate change. They adapted a methodology developed by Koster and Milly (1997), in which each LSS is mapped on to a generic bucket model. Gedney et al. (2000) defined this generic model by the intercept and slope of just two straight lines, describing how evaporative fraction and runoff vary with scaled soil moisture. The gradients and intercepts of these lines were estimated using monthly mean variables taken from each of the $1 \times CO_2$ control runs. Although the models displayed similar dependencies of evaporative fraction on soil moisture, their runoff responses differed markedly, resulting in varying degrees of soil moisture

limitation under the control climates. As a consequence, the models also produced different hydrological responses to a given climate change. For example, in Amazonia all AGCMs predicted a decrease in annual mean precipitation and an increase in annual mean available energy (Fig. A.29a). However the LSSs did not produce similar hydrological responses to this climate change, with evaporation increasing in some cases and decreasing in others (Fig. A.29b). The simulation of runoff over the active soil moisture range appears to be a critical factor influencing the hydrological sensitivity to climate and climate change.

A.4.3.2 Biogeochemical Feedbacks

Changes in plant structure affect atmospheric CO_2 if these changes are associated with changes in biomass. In addition, plants respond physiologically to carbon dioxide through increased water use efficiency and increased rates of photosynthesis under elevated CO_2 concentrations. There is therefore even greater potential for vegetation change and feedbacks under CO_2-induced climatic change. Despite this, most GCM climate change simulations have used prescribed fixed vegetation distributions which are insensitive to both climate and CO_2.

Fig. A.29.
Differing land-surface responses to GCM-simulated climate change in Amazonia from Gedney et al. (2000).
a Changes in precipitation, dP, and available energy, dA;
b hydrological responses in terms of changes in evaporation, dE, and runoff, dY. The *capital letters* each represent a given AGCM, and the *a* and *b* denote the two distinct LSSs coupled to each.
Figures **c** and **d** demonstrate that the simulated changes in evaporation, dE, and soil moisture, dM, can be accurately estimated using the generic bucket model (aside from models *Hb* and *Cb* which include direct effects of CO_2 on transpiration)

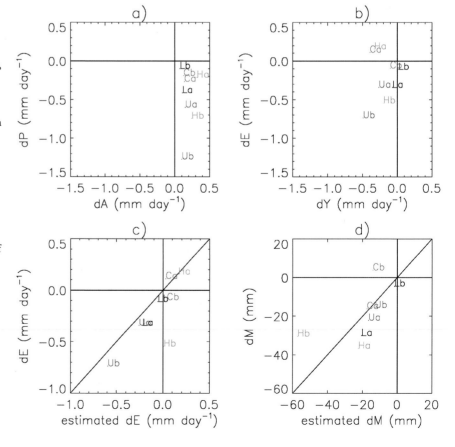

The first attempts to remedy this deficiency in GCMs have concentrated on the introduction of more realistic representations of plant physiology (Sellers et al. 1996a, Cox et al. 1998). These approaches are based on links between transpiration and photosynthesis. Stomatal pores on plant leaves are central to both processes, providing the means by which CO_2 can diffuse into the leaf to be fixed during photosynthesis, and the means by which water escapes during transpiration. Empirical

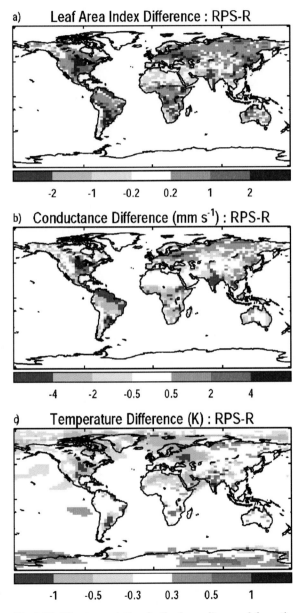

a) **Leaf Area Index Difference : RPS-R**

-2   -1   -0.2   0.2   1   2

b) **Conductance Difference (mm s^{-1}) : RPS-R**

-4   -2   -0.5   0.5   2   4

c) **Temperature Difference (K) : RPS-R**

-1   -0.5   -0.3   0.3   0.5   1

Fig. A.30. Climate-vegetation feedbacks as diagnosed from the modelling study of Betts et al. (1997). **a** Change in leaf area index; **b** change in canopy conductance for water vapour; **c** change in near-surface air temperature, resulting from the combined effect of physiological ("P") and structural ("S") responses of the vegetation to a doubling of CO_2 and the associated climate change

evidence also suggests stomatal conductance varies approximately linearly with net photosynthesis (Ball et al. 1987), which facilitates the construction of coupled stomatal conductance-leaf photosynthesis models. Simple upscaling algorithms have been developed which enable these models to be used directly within GCM LSSs, where they can provide a more physiologically-based treatment of stomatal conductance, as a function of environmental variables (e.g. solar radiation, temperature, humidity deficit, CO_2).

An interesting outcome of this approach is that stomatal conductance becomes a decreasing function of atmospheric CO_2 concentration, which is consistent with most (but not all) field and laboratory experiments (Field et al. 1995). As a consequence Sellers et al. (1996a) suggest that CO_2 induced-stomatal closure alone could have appreciable local impacts on CO_2 induced climate warming. In a similar set of numerical experiments, Cox et al. (1999) found that stomatal closure results in an increase in annual mean runoff under $2 \times CO_2$, rather than the reduction which occurs in the absence of this effect. They also report enhanced warming over tropical forests, which are coupled to the atmosphere through a small aerodynamic resistance and thus are especially sensitive to the increase in stomatal resistance.

Betts et al. (1997) coupled the "DOLY" vegetation model (Woodward et al. 1995) to the Hadley Centre AGCM (Johns et al. 1997) in order to estimate the role of vegetation feedbacks in climate change. This study included both the direct effects of CO_2 on photosynthesis and transpiration, as well as the indirect effects arising from the associated climate change. A number of $2 \times CO_2$ climate-vegetation simulations were carried out to isolate the impact of CO_2-induced stomatal closure (i.e. the physiological response), from the impact of changes in vegetation structure and distribution (i.e. the structural vegetation feedback). In broad agreement with Sellers et al. (1996a), Betts et al. (1997) found that stomatal closure alone warmed the surface under a $2 \times CO_2$ atmosphere, especially in moist forested regions. However, on including changes in vegetation structure, they found that changes in leaf area index (LAI) counteracted much of this response. LAI increased at the global scale as a result of enhanced photosynthesis at high CO_2 (Fig. A.30a) and this acted to maintain similar or higher canopy conductance values, even though the stomatal conductance of each leaf was reduced (Fig. A.30b) However, in some regions (e.g. north-east Brazil), LAI decreased owing to reductions in precipitation and the stomatal effect was amplified by the structural change. Other biophysical feedbacks were also found to be important at the regional scale. For example, increased LAI in the Siberian boreal forests produced an additional surface warming due to the masking of the bright snowcover by darker vegetation (Fig. A.30c).

A.4.3.3 Transient Experiments

The studies described above have been critical in highlighting the role of vegetation in climate and climate change, but they each have limitations which need to be addressed. The models using interactive plant physiology have the advantage of consistency between the water and CO_2 fluxes, but lack the longer-term feedbacks associated with vegetation structural change. The asynchronous coupling technique enables these structural feedbacks to be estimated, but hides a potential inconsistency since the GCM and the vegetation model each carry out different internal calculations of the same fluxes (e.g. evaporation) and stores (e.g. soil moisture). Also, these experiments are limited to the study of climate-vegetation equilibria.

In reality, vegetation responds on a range of timescales, so that the slower components (e.g. forests) are unlikely to be in equilibrium with the changing environment of the next century. Thus there is a requirement for Dynamic Global Vegetation Models (DGVMs) which treat fast and slow responses within a consistent framework. A number of DGVMs are under development (Cramer et al. 2000), and some have already been coupled consistently to GCMs (Foley et al. 1998; Cox et al. 2000) and to so-called EMICs (Earth System Models of Intermediate Complexity) (Claussen 2001b; Claussen et al. 2002). On the regional scale, Eastman et al. (2001b) and Lu et al. (2001) have completed season-long coupled atmosphere-biogeochemical model simulations on the relative importance of land-use change and direct, local CO_2 forcing. Examples of transient atmosphere-vegetation simulations were already given in the previous chapter, and these are complemented by examples from Sahelian climate variability in Chapt. A.5.

Most examples of important atmosphere-vegetation feedbacks have dealt with biogeophysical processes. The more critical land-atmosphere interactions over the next century, however, will probably involve biogeochemical feedbacks, in particular, carbon and water cycle feed-backs. Currently only about half of the anthropogenic emissions from fossil fuel burning and deforestation remain airborne, with the remainder being absorbed by the oceans and the terrestrial biosphere. These components of the Earth system are thus buffering us from the full climatic effect of our emissions. However, the atmosphere-ocean and atmosphere-land exchanges of carbon are both known to be sensitive to climate, so how will the airborne fraction of emissions vary in the future?

The greatest uncertainties are associated with the terrestrial biosphere. The cause of the present-day carbon sink is still in doubt, with CO_2-fertilisation, nitrogen deposition and forest regrowth all implicated in certain regions. The location of the sink is even more debatable, perhaps because this is subject to great interannual variability. However, despite these uncertainties, there are good reasons to believe that the terrestrial carbon cycle will play a key role in climate over the next century. Whilst increases in atmospheric CO_2 are expected to enhance photosynthesis (and reduce transpiration), the associated climate warming is likely to increase plant and soil respiration. There is therefore a competition between the direct effect of CO_2, which tends to increase terrestrial carbon storage, and the indirect effect, which may reduce carbon storage.

The outcome of this competition has been seen in a range of DGVMs (Cramer et al. 2000), each of which simulate reduced land carbon under atmospheric changes (based on GCM sensitivity experiments) alone and increased carbon storage with CO_2 increases only. In most DGVMs, the combined effect of the CO_2 and associated changes in the atmosphere results in a reducing sink towards the end of the 21st century, as CO_2-induced fertilisation begins to saturate but soil respiration continues to increase with temperature. The manner in which soil respiration responds in the long-term to temperature is thus a key uncertainty in the projections of CO_2 in the 21st century.

One of the first fully coupled climate-carbon cycle simulations using comprehensive models also highlights the importance of the land (Cox et al. 2000; Friedlingstein et al. 2001). Cox et al. included the HadOCC ocean carbon cycle model and the TRIFFID DGVM within the Hadley Centre coupled ocean-atmosphere GCM (Gordon et al. 2000). The coupled system produced a stable pre-industrial state displaying significant interannual variability in the carbon cycle, which was realistically driven by the model's own ENSO cycle. A first climate change experiment using the IS92a emissions was found to produce reasonable natural carbon sinks for the current day, but a terrestrial carbon source after about 2050 in the coupled simulation. This sink-to-source transition was the combined result of a regional climate change in the Amazon, which resulted in forest dieback (Fig. A.31, *top panel*), and a more widespread release of soil carbon via the mechanism identified above (Fig. A.31, *lower panel*). Much work is needed to identify the uncertainties in this experiment, but it indicates that feedbacks between climate and the land are likely to be critical determinants of CO_2 increases and climate change.

Xubin Zeng · Randal Koster

A.4.4 Seasonal Variability

After highlighting the role of biogeophysical and biogeochemical feedbacks in climate, we focus here on seasonal weather. Most of the early studies on the role of land-surface in weather and climate focused on soil moisture on seasonal time scales, because of the relatively long

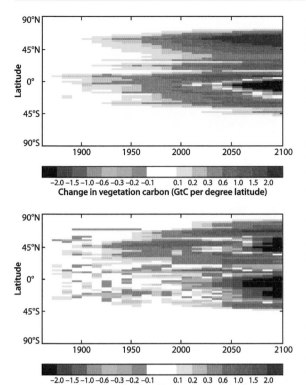

Fig. A.31. Time-latitude plots showing the changes in vegetation carbon (*top panel*) and soil carbon (*bottom panel*), using the coupled climate-carbon cycle simulations given in Cox et al. (2000)

persistence of soil moisture compared with the atmospheric persistence of about two weeks. In fact, some four decades ago, Namias (1959) had already proposed a possible mechanism for the impact of soil moisture on seasonal prediction: a dry spring would result in a relatively large sensible heat flux into the atmosphere to maintain anticyclonic circulations and this, in turn, would lead to reduced summer rainfall. Soil moisture was also the only prognostic variable in the first generation land-surface model (i.e. bucket model; Manabe 1969). Significant progress has been made since then, and only the most recent results will be summarised here for brevity.

Through a series of global simulations spanning a total of several thousand years, Koster et al. (2000) systematically assessed the impact of land-surface *versus* ocean boundary conditions on the seasonal-to-interannual variability and predictability. The influence of sea surface temperature was removed through the application of climatological sea surface temperatures, while the influence of an interactive land surface was removed through the prescription of climatological evaporation efficiency, β, defined as the ratio of actual evaporation to the potential evaporation. The prescription of β does not imply a prescription of the evaporation rate; instead, it implies a prescription of the land surface's ability to deliver moisture to the atmosphere in response to the stated demand. In simple (though not precise) terms, prescription of β can be thought of as prescribing soil moisture contents and canopy interception amounts. Koster et al. demonstrated that the contribution of ocean, atmosphere, and land processes to the precipitation variance can be characterised, to first order, with a linear model, as shown in Fig. A.32. The variance in the absence of land-atmosphere feedback (*top left panel*) has the same overall structure as the total variance (not shown) but is generally smaller. This implies that land-surface processes primarily affect the amplitude of precipitation variance rather than determine its spatial pattern. The *bottom right panel* shows that land-atmosphere feedback increases this variance everywhere except in deserts, high latitudes, and the very wet area of the tropics. The oceanic contribution (*top right panel*) is high in the tropics and in parts of the Sahara but remains below 30% throughout midlatitudes, where atmospheric contribution is high (*bottom left panel*). The *top and bottom right panels* also show that the areas where the oceanic contribution is large generally coincide with areas for which the land amplification is small, implying that the land and ocean generally affect precipitation variance in different regions. In other words, the synergy discussed in Sect. A.4.1 between oceanic and land-surface feedbacks for each region is weak.

The potential for seasonal-to-interannual predictability of precipitation was also discussed in Koster et al. (2000), under the assumption that all relevant ocean and land boundary conditions are themselves perfectly predictable; i.e. the absolute upper limit of precipitation predictability was addressed. Although the chaotic nature (e.g. Lorenz 1963; Zeng et al. 1993) of the atmosphere imposes fundamental limits on precipitation predictability in many regions, foreknowledge of sea surface temperatures was found to contribute significantly to predictability throughout much of the tropics, but not in mid-latitudes and high-latitudes, in agreement with other studies (e.g. Shukla 1998). In contrast, foreknowledge of the land-surface moisture state was found to contribute significantly to predictability in transition zones between dry and humid climates, particularly over the tropics and summer extratropics.

Dirmeyer (1999) examined how the insertion of soil moisture data generated under the Global Soil Wetness Project (GSWP) affects the climate simulated by an atmospheric model. These data were generated by the off-line integration of a land-surface model forced by observed and assimilated meteorological data. He found that the GSWP soil wetness was significantly different from that of the land-atmosphere coupled model's own climatology, and produced a better simulation of precipitation anomaly patterns over monsoon regions and the summer hemisphere extratropics. When the soil wetness data from the wrong year were used, the simulation of precipitation anomaly patterns was significantly degraded, indicating the importance of inter-an-

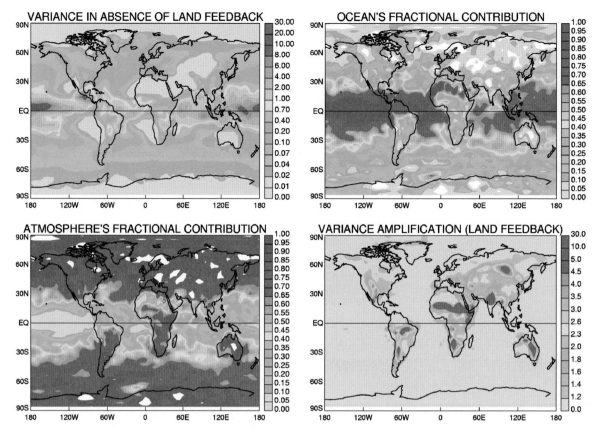

Fig. A.32. Breakdown of the contributions of oceanic, atmospheric and land-surface processes to precipitation variance, assuming a linear framework. *Top left*: precipitation variance in the absence of land-atmosphere feedback; *top right*: fraction of precipitation variance induced by variable sea surface temperatures; *bottom left*: fraction of precipitation variance induced by chaotic atmospheric dynamics; *bottom right*: amplification of variance due to land-atmosphere feedback (adapted from Koster et al. 2000)

nual variability in soil wetness to the interannual variability of summer precipitation. However, differences in precipitation due to interannual sea surface temperature variability generally dominated those apparently caused by soil wetness variances, which is not consistent with the above conclusion of Koster et al. (2000) that the the land and ocean generally affect precipitation variance in different regions. This inconsistency could be caused by the small ensemble size of two years of simulations in Dirmeyer (1999) versus the much larger ensemble size of decades in Koster et al. (2000), because of the different approaches to specifying soil moisture and/or of the use of different global atmospheric and land models.

Instead of prescribing soil moisture throughout the model integration, Fennessy and Shukla (1999) addressed the impact of initial soil wetness on seasonal atmospheric prediction in summer. Compared with several previous studies that prescribed initial soil wetness arbitrarily, they initialised soil wetness with climatological values and with those derived from the output of an operational analysis-forecast model. The impact of initial soil wetness was found to be largely local and was largest on near-surface fields, in agreement with previ-

ous studies. Significant impacts were found in several tropical and extratropical regions where a wet initial state typically increases the seasonal mean evaporation and decreases surface air temperature. While half of the regions have significant increases in seasonal mean precipitation in response to increased initial soil wetness, the precipitation response is overall more variable and highly dependent on the response of the moisture flux convergence to the initial soil wetness anomaly. Consistent with previous studies (e.g. Yeh et al. 1984), they also found that the above results depend on several factors, such as the areal extent and magnitude of the initial soil wetness anomaly, the persistence of the anomaly, the strength of the regional dynamical circulations, etc. Addressing the same issue but with a focus on the Mississippi River basin, Bonan and Stillwell-Soller (1998) also demonstrated that soil moisture feedbacks amplify the severity and persistence of floods and droughts on monthly to seasonal time scales and that the manner in which soil water limits evaporation and transpiration is a key control of this response. On an even smaller spatial scale, using observational data, Findell and Eltahir (1997) confirmed the soil moisture-rainfall feedback over Illinois, USA.

Persistence of the soil moisture anomaly has also been addressed in Liu and Avissar (1999a,b). Their analysis of model output indicated a significant persistence, on time scales of months to seasons, in soil moisture and temperature with the former being stronger than the latter. Both model output and observations showed the dependence of soil moisture persistence on latitude and regional climatology. Their sensitivity analysis also indicated that the soil moisture persistence on a seasonal time scale is most affected by surface hydrological processes (e.g. evaporation, runoff, and soil moisture diffusion), while the thermal characteristics of the land-atmosphere system mostly affect the monthly time scale.

Snow cover and depth directly affect springtime soil moisture and hence affect summer temperature and precipitation prediction. Walker (1910) used Himalayan snow as one of the predictors for the summer monsoon rainfall in India. Using satellite snow data, Bamzai and Shukla (1999) found that composites for high and low snow cover over Eurasia show spatially homogeneous large-scale patterns of snow cover. However, when individual regions are analysed, western Eurasia becomes the only geographical region for which a significant inverse correlation exists between winter snow cover and subsequent summer monsoon rainfall in India. This reverse relationship holds especially in those years when snow was anomalously high or low for both the winter and the subsequent spring season. Also, contrary to previous findings, no significant relationship is found between the Himalayan seasonal snow cover and subsequent monsoon rainfall.

Soil moisture and evaporation and transpiration are directly related to the vegetation root distribution. Using the comprehensive global root database (Canadell et al. 1996; Jackson et al. 1996, 1997), Zeng et al. (1998) and Zeng (2000) have derived vertical root distribution data (including rooting depth) for various vegetation types that are significantly different from those specified in most land models. Using a variety of observational datasets, Zeng et al. (1998) found that observed root distribution significantly improves the off-line simulation of surface water and energy balance. Multi-year simulations further demonstrated the impact of observed root distribution (versus that specified in current land models) on global modelling over tropics and mid-latitudes. In particular, over the Northern Hemisphere mid-latitudes, inclusion of observed root distribution changes simulated soil moisture in the root zone, precipitation, and surface air temperature in summer. These results are also consistent with the above studies involving soil moisture anomalies.

In addition to soil moisture, land-use and land-cover change also play a role in atmospheric seasonal to interannual variability. Land-cover change directly changes surface albedo and roughness lengths and affects soil moisture as well as the land-atmosphere exchanges of energy, water, and carbon. Here we focus on the continental United States on seasonal time scales because the land-use and land-cover change over other regions and on different time scales have been covered in the following chapters.

Through numerical modelling, Bonan (1997) showed that the climate of the United States with present-day vegetation is significantly different from that with natural vegetation. The primary difference between natural and modern vegetations is the replacement by crops of much of the natural needleleaf evergreen, broadleaf deciduous, and mixed forests of the eastern United States, and the natural grasslands in the central US. The climate signals caused by modern vegetation in the Bonan results include significant near-surface temperature and humidity changes in spring and summer (e.g. summer cooling of up to 2 °C over a wide region of the central United States and moistening of the near-surface atmosphere by 0.5 to 1.5 g kg^{-1} over much of the United States). These changes in near-surface temperature and humidity extend well into the atmosphere, up to 500 hPa, and affect the atmospheric boundary layer and atmospheric circulation. Compared with other anthropogenic climate forcings, the summer cooling from deforestation could offset more than 100 years of warming over the United States at the current, observed rate.

The importance of land cover to the summer climate over the US. has also been demonstrated by Xue et al. (1996c) from a different perspective. In their effort to investigate their model's significant positive surface air temperature biases of up to 4 °C in summer over the central United States, they found that the temperature biases were coincident with the agricultural region of the central US. where negative precipitation biases also occurred. A series of numerical experiments further showed that these biases were largely caused by the erroneous prescription of crop-vegetation phenology in their land-surface model. They also demonstrated that, for US summer monthly and seasonal predictions, the vegetation seasonality plays a crucial role. The effect of land cover on near-surface atmospheric variables during the North American summer is very pronounced and persistent in their results but is largely limited to the area of the anomalous land-surface forcing.

On a smaller spatial scale, Pielke et al. (1999) addressed the influence of land-cover change on summer surface air temperature in south Florida, USA. There has been a widespread conversion of natural vegetation classes to urban and agricultural land, and into grassy shrubland over this region during the past 100 years. Using land-cover data in 1993 and 1973, their modelling studies showed a 9% decrease in rainfall averaged in July and August with the 1973 landscape and an 11% decrease with the 1993 landscape, as compared with the model results using 1900 land-cover data (when the region was close to its natural state). Along with the decrease of precipitation (primarily deep cumulus rainfall), surface temperature increases in response to the land-cover change (compare Fig. A.15 as well).

In addition to the above work, several studies have also addressed the general role of land surface in seasonal to interannual variability. Pielke et al. (1999) demonstrated that soil moisture distribution and landscape could cause a significant non-linear interaction between vegetation growth and precipitation on seasonal time-scales, and hence suggested that seasonal weather prediction is an initial value problem. Koster and Suarez (1999) derived an equation that relates the interannual variability of evaporation to gross characteristics of the atmospheric forcing. This is complementary to the original Budyko equation that describes the partitioning of annual mean precipitation into annual mean evaporation and runoff. These equations suggested, and the modelling results in Koster and Suarez (1999) confirmed, that the ratio of an evaporation anomaly (e.g. that caused by land-surface processes) to the corresponding precipitation anomaly tends to be significantly less than the ratio of mean evaporation to mean precipitation. Using a quasi-isentropic back-trajectory algorithm, Dirmeyer and Brubaker (1999) showed that about 41% of precipitation over the Mississippi River basin originated as evaporation from the same basin during April–July 1988 and 33% during April–July 1993.

Pedro Viterbo · Anton C. M. Beljaars

A.4.5 Impact of Land Surface on Weather

The importance of land surface in numerical weather prediction stems from a blend of practical considerations and basic physical principles. First and foremost, accurate forecasts of near-surface weather parameters are requested by the numerical weather prediction users' community. The quality of such products as the diurnal cycle of near-surface air temperature and humidity, winds, low-level cloudiness and precipitation is to a large extent determined by the physical realism of the model representation of the surface/atmosphere interactions. Secondly, the model needs to represent correctly the processes that control the essential surface-atmosphere feedback mechanisms, in as much as they affect the forecasts for the lower troposphere. Thirdly, the partitioning of the available energy at the surface between sensible and latent heat fluxes determines the soil wetness, which acts as one of the forcings – or, at the very least, a modulator – of low frequency atmospheric variability, e.g. extended drought periods (see Sect. A.4.4). Finally, remote sensing observations of the atmosphere are increasingly important in characterising the state of the atmosphere over land. The sensors that are sensi-

tive to the lower troposphere can only be used in the presence of a good quality background field of skin temperature.

It is important to emphasize that, despite the fact that deterministic forecasts of numerical weather prediction do not extend beyond two weeks, the model land-surface controls much longer time-scales. The soil variables in the model are initialised every six hours based on indirect, imperfect and very sparse observations (see Part C). Data assimilation combines that information with a short-term forecast to get an optimal estimate of the land-surface initial conditions. Cycling the creation of initial conditions is tantamount to an extended model integration nudged by observations and extends the memory of the surface variables beyond the forecast duration. The realism of the representation of land-surface processes is crucial for handling correctly the *memory* properties, represented by soil moisture in sub-tropics and mid-latitudes spring and summer and by the snow mass in higher latitudes and mountainous areas.

Despite all the above considerations, the impact of land surface on weather was only recently recognised, as can be seen by the sparse literature available. Therefore, surface-atmosphere interaction mechanisms that are relevant for the atmospheric simulation at the regional to continental scales are briefly reviewed, based on numerical or observational studies. Moreover, examples based on recent experience at the European Centre for Medium-Range Weather Forecasts will illustrate when and where the land surface can affect the weather.

A.4.5.1 Brief Literature Survey

Betts et al. (1996) review the impact of land surface in the context of global numerical weather prediction: Diurnal and seasonal feedback loops are discussed as well as feedback loops controlling the boundary layer evolution. We will highlight here typical mechanisms controlling the interaction between land surface and the atmosphere over the US, Europe, and the tropics.

Early modelling efforts (Benjamin and Carlson 1986; Lanicci et al. 1987) have shown the sensitivity of precipitation in the US Great Plains to evaporation upstream, in the Mexican Plateau. The characteristic storm environment, leading to heavy precipitation over the Midwest, involves the breakdown of a capping inversion formed by an overlying pre-existing boundary layer from the Mexican plateau, which overlies the cool, moist, boundary layer originating in the Gulf of Mexico. This complex pattern of differential advection impacts on the strength of the capping inversion, and the strength of the inversion is controlled by evaporation *upstream* of the precipitation area. Lower values of evaporation lead to a stronger capping inversion, and the low level flow from the Gulf will not break through the inversion until much further north. The location and extent of the heavy precipitation associated with the July 1993 US floods was found to be highly sensitive to the correct representation of these mechanisms in the European Centre for Medium-Range Weather Forecasts model (Beljaars et al. 1996; Viterbo and Betts 1999a).

Spatial gradients of soil moisture can also enhance the differential heating maintaining and reinforcing the surface front in a pre-storm environment and intensifying the thermally direct (ageostrophic) circulation (Chang and Wetzel 1991; Fast and McCorcle 1991). A similar mechanism is active in a cold front associated with a severe squall line developing explosively (Koch et al. 1997).

The seasonality of leaf area index impacts on the systematic errors of US Midwest lower tropospheric temperature in summer (Xue et al. 1996c; Yang et al. 1994). At a smaller scale, there are many observational and modelling studies (see Chapt. A.3) demonstrating the importance of mesoscale fluxes and smaller-scale heterogeneity, specially at low wind speeds.

Over Europe, Rowntree and Bolton (1983) demonstrated the role of local and non-local response of medium-range rainfall forecasts to anomalies in the initial soil. The mechanisms relevant to this soil moisture-precipitation feedback were scrutinised in Schär et al. (1999) in a study demonstrating the impact of idealised (and large) anomalies of soil water on the European summer circulation. Unlike the desert-albedo feedback hypothesis (Charney 1975; see also Sect. A.4.1 and A.4.4), the European soil-precipitation feedback is not of a large-scale dynamical nature, i.e. it is not associated with changes in large-scale flow. Precipitation recycling has also a small role in Europe. Three main feedback loops have been identified. Wet soils, associated with low Bowen ratios, lead to the build-up of a shallow boundary layer, concentrating moist entropy at those low levels and giving higher values of convective available potential energy. Additionally, lower Bowen ratios lead to higher relative humidity, lowering the level of free convection. Finally, a positive feedback of radiative origin, with increased cloud cover but larger net radiative flux, leads to larger moist entropy and convective instability.

All the examples above are from the extratropics in spring and summer snow-free situations. Spring examples in the presence of snow are presented below. There is little or no impact of land surface on the atmospheric circulation in winter (see Giorgi 1990).

Despite an extensive list of publications on the role of land surface in the tropical climate (see next section and Chapt. A.5, A.6, and A.8), there is scant evidence of its impact for short- and medium-range forecasts. A nota-

ble exception is the work of Walker and Rowntree (1977) who demonstrated the role of enhanced soil moisture gradients on the short-range (1–2 days ahead) forecast of the generation of easterly waves, using a simplified model version over West Africa.

A.4.5.2 European Centre for Medium-Range Weather Forecast (ECMWF) Examples

A.4.5.2.1 *Impact of Soil Moisture*

July 1993 showed anomalously high precipitation over the central USA, with exceptional flooding of the Mississippi (Changnon 1996). During this month, the new model version of the European Centre for Medium-Range Weather Forecasts (ECMWF) model (CY48) and the then operational version (CY47), were running in parallel at full resolution (spectral truncation T213, grid-point spacing ~ 60 km), including data assimilation. Beljaars et al. (1996) compared the performance of the two schemes, looking at the average of all one-, two- and three-day forecasts verifying between 9 and 25 July. While the day one precipitation of the two systems is very similar, and similar to the observed precipitation, the forecasts at day 3 are markedly different. In the new system, the location and intensity of the maximum precipitation is similar to the observations (40° N, 95° W), while the old system has less than half the precipitation amount in the area of the observed maximum and has a spurious maximum of precipitation displaced 800 km north east. In the old system, there is a gradual reduction in precipitation from day 1 to day 3, while the new system is able to better maintain the intensity. However, evaporation at the area of maximum precipitation is similar for the old and new system, and in both systems there is no evidence of forecast spin down, strongly suggesting that the local evaporation is not responsible for the differences in precipitation. It turns out that the maximum of the evaporation difference is located over the Mexican Plateau, 1 000 km south west of the precipitation maximum, two to three days upstream as suggested by backwards trajectories ending up at 750 hPa, 40° N, 95° W. The mean thermodynamic profiles, similar for day 1 forecasts, are very different for day 3 forecasts. The old model shows a too strong capping inversion above the boundary layer, with air too warm and too dry and much lower values of convective available potential energy. It is clear that the differential advection mechanism characteristic of the US monsoon is responsible for the differences in precipitation. When compared to the new model, the soil on the Mexican Plateau has much lower values of soil moisture in CY47, giving a much reduced evaporation, which in turn produces a warm and dry air mass that caps the boundary layer downstream, inhibiting convection. In CY47, the soil model values are strongly forced to an erroneous, too dry, climatology: in such a data-dense area, atmospheric profiles are initiated to correct values but during the forecast they slowly feel the influence of the erroneous soil moisture values. In CY48 the soil moisture values were initialised to field capacity at the beginning of July, consistent with values of June precipitation in the area much above normal. There is no forcing to climatology in CY48 (Viterbo and Beljaars 1995) and the model is capable of maintaining high values of moisture throughout July. Monthly integrations performed with CY47 and CY48 suggest the importance of the memory associated with idealised soil moisture anomalies in the initial conditions (Beljaars et al. 1996). The monthly precipitation fields with CY48 compare much better with operations than those of CY47.

CY48 surface model ran with predicted soil moisture throughout the first half of 1994. It was clear that a very large near-surface warm and dry bias developed over the Northern Hemisphere continental areas at the end of spring and beginning of summer. A scheme to initialise soil water based on the short-term forecast errors of near-surface atmospheric humidity was developed to overcome that problem (Viterbo 1996). In order to test the new scheme, three complete data assimilation-forecast experiments were ran at T213 for the month of June 1994: (*a*) Control (CY48, no assimilation of soil moisture); (*b*) as Control, but using the initialised soil moisture values, and; (*c*) as in (*b*), but using a prognostic cloud scheme with much more realistic cloud cover over land (Tiedtke 1993). In response to a wetter soil, the near-surface warm and dry bias reduce from Control to the initialised soil water experiment. A lower tropospheric warm bias develops in the Control model and is greatly reduced when initialisation of soil water is used. Both experiments have too little cloud cover over land with too large surface shortwave radiative fluxes, but the wetter soil conditions of experiment (*b*) manage to maintain evaporation in the face of excessive net radiation at the surface. The third simulation displays even smaller biases, associated with a larger, more realistic cloud cover and smaller radiative biases.

The algorithm to initialise soil water is successful in controlling model drifts but dampens the seasonal cycle and interannual variations of evaporation and soil moisture. Viterbo and Betts (1999a) revisited the July 1993 simulation, using the initial soil water algorithm and the prognostic cloud scheme. The new system gives poorer results for precipitation than Beljaars et al. (1996). Although much better than CY47, there is a suggestion of northward displacement and reduction in the precipitation maximum in day 2 forecasts. It appears that the initialisation of soil moisture at field capacity at the beginning of July in Beljaars et al. was crucial to obtain a good simulation of the excessive rainfall events.

A.4.5.2.2 *Boreal Forests*

Surface albedo is the prime regulator of the net energy available at the surface. The albedo of snow-free land-surfaces ranges from values of 0.1 in forests to values of 0.35 over deserts. For areas seasonally covered with snow, that range can extend up to 0.85. Betts and Ball (1997; see also Sect. A.7.2) analysed the annual cycle of albedo in the BOREAS (Boreal Ecosystem-Atmosphere Study) experiment, Canada, focusing on the snow season, comparing several measurement sites located over grass, aspen and coniferous forest. Representative values for daily averaged albedo of snow-covered grass sites are 0.75, while corresponding values for the aspen and conifer sites are 0.21 and 0.13, respectively, with values as high as 0.4, one to two days after snowfall. The lower albedo of the boreal forests in the presence of snow corroborates data from other observational studies and the few attempts at making a hemispheric-satellite based estimate of albedo.

The European Centre for Medium-Range Weather Forecasts model version of 1994, at the time of the BOREAS experiment, treated the albedo of snow covered areas with no regard to the land cover: beyond a critical value for snow depth, the albedo of snow covered areas was rarely outside the 0.7–0.8 range. As a result, when compared to experimental results, net radiation was too low and near-surface air temperatures were too cold. A modification to the scheme was designed such that the albedo of snow-covered surfaces tended to the asymptotic value of 0.2 in the presence of forests and 0.7 otherwise (Viterbo and Betts 1999b). The modified scheme has much reduced biases in temperature and radiation in the high latitudes. The cold bias in temperature in the control scheme extended in the vertical to the whole troposphere, increasing with forecast range and affecting most continental areas. The *bottom panel* of Fig. A.33 shows the day 5 forecast error of 850 hPa temperature, averaged for March and April in 1996. The very high albedo induces cooling errors exceeding –3 K in North America and –7 K in Asia. The *top panel* shows the corresponding figure for 1997. In spring 1997, with the new snow albedo scheme, the cold bias over the boreal forest has been almost eliminated. The new scheme also improved the quality of the medium-range forecasts, as evidenced by better scores of 500 hPa geopotential fields.

The forecast results above corroborate the study of Thomas and Rowntree (1992) on the role of the boreal forests in conditioning the climate at high latitudes. The spring months of two five-year experiments, the first with a (realistic) snow albedo and the second with the high latitude forests removed, are compared. The latter experiment is colder than the former in the continental areas north of 50° N. Pielke and Vidale (1995) suggested that the boundary between tundra and boreal forests is a re-

Fig. A.33.
Average 5-day forecast temperature errors at 850 hPa in the European Centre for Medium-Range Weather Forecasts model for March–April 1996 (*bottom panel*) and 1997 (*top panel*) (from Viterbo and Betts 1999b, © American Geophysical Union)

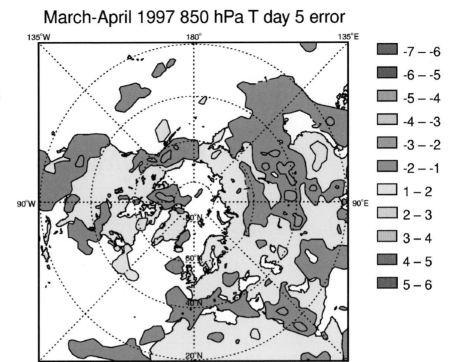

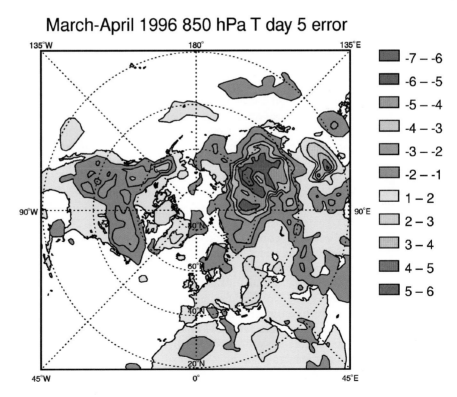

gion of enhanced horizontal temperature gradients, acting as a pre-conditioner for baroclinic instability and "locking" the climatological position of the polar front. In analysing further refinements to the European Centre for Medium-Range Weather Forecasts snow model,

van den Hurk et al. (2000) show that (*a*) simulating the boreal forest control on evaporation in spring (reduced transpiration from frozen soils) and (*b*) increasing the runoff over frozen soils, further improves the agreement of model results with observations.

Fig. A.34.
History of monthly biases (*thick solid lines*) and standard deviations (*thin solid lines*) with respect to observations of the day-time (72-hour: *red*) and night-time (60-hour: *blue*) operational two-metre temperature forecasts, averaged for all available surface stations in the European area of 30 to 72° N and 22° W to 42° E

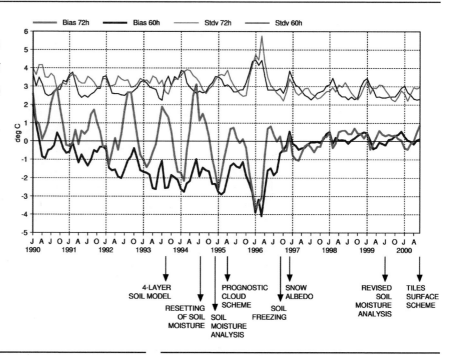

A.4.5.2.3 Cold Climates and Soil Water Freezing

The operational ECMWF forecasts for the winters of 1993–94 through to 1995–96 showed a tendency to produce a cold bias over continental areas in winter. This error was particularly severe during the winter of 1995–96 for Scandinavia, a year characterised by reduced snow amounts and, consequently, larger thermal coupling between the soil and the atmosphere above. Viterbo et al. (1999) diagnosed two main problems contributing to that error. Firstly, the energy involved in phase changes in the soil was not taken into account. When positive temperatures approach 0 °C, a substantial amount of the external cooling demand (i.e. infrared cooling) is used to freeze the soil water, thereby decreasing the rate of soil cooling; a similar effect occurs in melting. Soil water phase transitions act as a thermal barrier, increasing the soil inertia at temperatures close to 0 °C. Secondly, the downward sensible heat flux, prevailing in winter conditions, was too small, leading the model into a positive feedback loop where cooling reduced the heat flux and made the soil even colder. Model changes were designed to incorporate the missing physical mechanisms. Separate seasonal integrations with both model schemes revealed a greatly alleviated soil and near-surface atmospheric cooling drift: screen-level temperatures reduced from –10 °C to close to zero. Despite the considerable warming in the model soil and near-surface winter climate in continental areas, there was negligible impact on the free atmosphere temperature and the atmospheric flow. In winter, stable situations, the atmosphere is decoupled from the surface and changes at the surface do not propagate upwards, unless they affect the momentum budget.

A.4.5.2.4 History of Biases

Probably the best summary of the impact of land surface on weather can be shown on Fig. A.34, displaying the history of ECMW Forecasts operational short-range forecast errors of 2 m temperature over Europe as a time-series of monthly averages. These errors show a large annual cycle, are different for night and day (72 and 60 hour forecasts verifying at 12 and 00 UT, respectively), and have a rich history of the many model changes that were made over the years. We will discuss only the model changes made from 1993 onwards.

In August 1993, a surface scheme with a climatological deep-soil boundary condition for temperature and moisture was replaced by the free-running four-layer scheme (Viterbo and Beljaars 1995), but the impact is not very obvious. The summer day-time bias of August 1993 was smaller than that of the previous year but, at that time, the soil scheme has been running freely for only two months (including the July parallel test described earlier). The next summer showed a pronounced warm bias related to a gradual drying out of the soil which was reduced in July 1994 by resetting the soil moisture to field capacity over vegetated areas. A simple soil moisture analysis scheme was introduced in December 1994 (Viterbo 1996) with a clear beneficial impact on the day-time bias for summer 1995. The night-time temperatures have been biased cold for many years, related to an overly large amplitude of the diurnal cycle. The winter of 1995/1996 was particularly bad, mainly because the European area was blocked for most of the winter with easterly winds and very cold temperatures, although changes to the cloud scheme or the orographic drag might have had a negative impact on night-time temperatures. It is interesting that the reduction of the day-time bias actually increased the night-time bias by displacing the entire diurnal cycle to colder temperatures. Soil freezing and increased boundary layer diffusion in stable layers, introduced in September 1996, improved the monthly error statistics considerably. The winter-time bias was largely eliminated and the amplitude of the diurnal cycle was down to a reasonable level. The snow albedo reduction described in Sect. A.4.5.2.2 was introduced in December 1996, but its impact is not clear over Europe due to the relatively small area covered by snow and the overall magnitude of the errors linked to the excessive soil cooling, corrected three months earlier. Finally, a much more selective way of initialising soil moisture (Douville et al. 2000), introduced in April 1999, might be responsible by a slight reduction in the standard deviation of temperature in that year. A new surface scheme was introduced in June 2000 (van den Hurk et al. 2000), but it is too early to assess its operational performance. The statistics presented in Fig. A.34 are averaged over a month and over a large area. The errors on a day-to-day basis can still be large but are less systematic, and are often related to errors in the forecast clouds or the presence of snow.

Chapter A.5

The Sahelian Climate

Yongkang Xue · Ronald W. A. Hutjes · Richard J. Harding · Martin Claussen · Steven D. Prince
Thierry Lebel · Eric F. Lambin · Simon J. Allen · Paul A. Dirmeyer · Taikan Oki

A.5.1 Introduction

The Sahel has experienced several drought periods in the past 500 years, however no available records show a drought as persistent and severe as the one that started in the 1960s (Nicholson 1978). Many thousands of people died and many more suffered severe disruption of their lives in the severe phases of this drought (e.g. in 1984). The human dramas and socio-economic consequences resulting from drought-induced famines in the Sahel region have presented a strong motivation for research into the causes of the drought. Ever since, the causes for this prolonged drought have been sought somewhere between two extreme views. One view considers land-surface degradation resulting from population pressure in excess of the region's carrying capacity as the main driver. This implies the existence of positive land-surface/atmosphere feedbacks mostly internal to the region, and in principle could incite the development of mitigation strategies to reverse the trend. The other view attributes the drought to unfavourable anomalous patterns in sea-surface temperature (SST) in the oceans. This implies the existence of a driver external to the region and by nature beyond human control, and requires the development of adaptation strategies to make the region's societies less vulnerable to droughts.

To position ourselves in this scientific debate, several important questions must be answered. We focus on the questions related to the first view, as it is still the more controversial of the two.

1. Can land-surface dynamics in principle lead to a drought of the magnitude witnessed over the last decades?
2. What is the role of soil moisture dynamics? Does soil degradation alter these dynamics?
3. What is the role of internal vegetation dynamics? Is vegetation a passive victim of the drought or an active modulator?

If the answer to the first question is positive, and with knowledge of issues raised in the other two, we can go a step further:

4. Are observed land-surface anomalies, if any, sufficiently strong to induce climate anomalies of the observed magnitude?
5. If so, do anthropogenic activities indeed form a significant driver in the observed land-surface dynamics?

Only with sufficient knowledge of all these issues may we be able to assess the relative importance of land surface *versus* SST forcing as the cause of droughts in the region, and may we hope to answer questions such as:

6. Is the current drought a temporary variation or a persistent and perhaps irreversible trend?

We will address aspects of all these questions in the next sections. First we will characterise the Sahelian region in ecological and climatological terms on one hand, and socio-political terms on the other. Then we will review the scientific efforts that have built the current understanding of land-surface/hydrology/climate interactions in the region, which includes discussion of the role of past and present field experiments in the region, and the results and evidence from modelling studies. Finally, after analysing the mechanisms, we will come back to the questions formulated above and assess where we stand.

A.5.1.1 Background

The Sahel is a tropical, semi-arid region (approximately 1.5×10^6 km^2) along the southern margin of the Sahara desert formed from large parts of six African countries and smaller parts of three more. The word "Sahel" was derived from the Arabic for a "shore", and this was presumably how the vegetation cover seemed to early traders who entered the region from the Sahara desert to the south. Today it is a bioclimatic zone of predominantly annual grasses with shrubs and trees, receiving a mean annual rainfall of between 150 and 600 mm y^{-1}. There is a steep gradient in climate, soils, vegetation, fauna, land use and human utilisation, from the almost lifeless Sahara desert in the north to savannas to the south. The similarity of climate and land cover in the

E-W direction contrasts dramatically with the strong N-S gradient. The uniformity of this geographical pattern is partly a result of an absence of rapid changes in topography but also because of the zonal nature of the climate with desert to the north and ocean or humid forests to the south. The geographical patterns of rainfall, vegetation cover, soils, human settlement and land use all share this zonal arrangement and are strongly correlated, so that cause and effect are hard to disentangle. The continental scale land mass and relative flat orography (excluding eastern Sahel) warrant that land-surface/atmosphere interactions play a major role in the regional climate and also allow such interactions to be detected relatively easily in model simulations. In fact, tropical northern Africa is the first region where continental scale land-surface/atmospheric interaction studies were conducted (Charney 1975).

The monsoon climate system increases further the potential for interaction between the lower atmosphere and the land surface. The Sahel summer climate is dominated by the West African monsoon system. The basic driver of the monsoon circulation is provided by thermal contrast between the continent and adjacent oceanic regions (Holton 1992). Any changes in the contrast may affect substantially the monsoon flow. Furthermore, the region is located in the tropics. In the tropics, the dynamical flow instabilities are relatively weak (compared to mid-latitudes). The boundary conditions modify the diabatic heating (latent and radiative heating) which in turn changes circulation and rainfall at seasonal and interannual time scales.

Soils in this region are dominated by a sand sheet of varying depth, usually resulting in unstructured, free-draining soils with low nutrient content and with a strong tendency to form an impervious "cap". This low soil nutrient content is sometimes considered a more serious constraint on rangeland quality and production than low rainfall (Breman and de Wit 1983). As a result of the crust formation, much rainfall runs off rather than infiltrating and is subsequently concentrated in local depressions.

These pools provide surface water for livestock allowing traditionally nomadic herdsmen to take their cattle, sheep, goats, camels and donkeys away from permanent water points such as boreholes, wells and rivers during the rainy season. As the people of the Sahel are dependent almost entirely on subsistence livestock production and agriculture they are therefore particularly vulnerable to the frequent droughts that occur in the region.

A.5.1.2 Climate Anomalies and Climate Change in the Sahel

Rainfall is the controlling factor in the life of the Sahel. A brief rainy season occurs, caused by the northward movement of the Intertropical Convergence Zone (ITCZ) in the northern summer, which causes humid air from the Gulf of Guinea to undercut the dry north-easterly air. West-moving squall lines and local convective activity cause mixing of the two air masses and results in short, torrential thunderstorms that increase in frequency and rainfall amount towards the south where the humid air mass is deeper. The rainy season varies in length from two to four months, but "rain days" can be separated by weeks of no rain.

Rainfall in the Sahel is characterised by high spatial and temporal variability, both within and between seasons, and by a high north-south gradient in the region. The climatological annual rainfall gradient over the Sahel is about 1 mm km^{-1} (Lebel et al. 1992). Not only is year to year variability high but also longer dryer or wetter periods may continue over a number of years.

Although the Sahel is usually defined with reference to a range of annual average rainfall, it is not always the amount of rainfall that is the most important characteristic, rather the spatial and temporal variation in rainfall, especially interannual variations, which determines much of the type and pattern of vegetation, fauna and human utilisation. Using satellite remote sensing the Sahel can be delimited by the zone of high interannual variations in vegetation activity. These characteristics have been discussed by a number of studies (e.g. Lamb 1978a,b; Nicholson and Palao 1993; Lebel et al. 1997).

Since the late 1960s, a persistent summer drought has lasted for more than 30 years. Although in the 1990s the rainfall was not as scarce as the 1980s, it was still below the climatological average (Fig. A.35). One of the important features of sub-Sahara drought is the displacement

Fig. A.35.
July-August-September rainfall anomalies for the Sahel region (10–20° N, 15° W to 40° E)

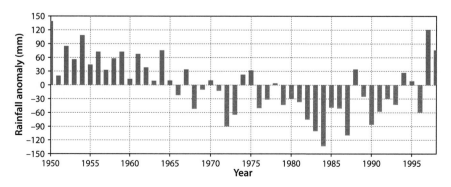

of isohyets in the 1980s, retreating southward in the northern Sahel during dry years and also shifting slightly to the south in the southern Sahel. This redistribution reduces the rainfall in the Sahel and increases rainfall to the south of 10° N. Figure A.36b shows the zonal average rainfall over the African continent from Nicholson's observational data. It indicates that the maximum rainfall was not reduced very much and it was the shifting alone that gave rise to the rainfall deficiency in the Sahel.

At small scales the data, even though relatively scarce by standards of temperate countries, do allow for detailed statistical analysis of spatial and temporal dynamics in certain areas. Le Barbé and Lebel (1997) have shown that the drought in Niger is associated mostly with a decrease of the number of rain events at the height of the rainy season (July/August). The mean event rainfall and the length of the rainy season hardly contributed to the decrease in annual rainfall. The rainy season duration and timing between dry (1970–1990) and wet (1950–1970) periods are similar.

A number of studies has also shown that many rivers in the region exhibit substantial reduction in discharge between the 1950s through the 1980s (for example, Savenije 1995; Oki and Xue 1998) as a consequence of the reduced rainfall. Annual discharge declined to less than one-third at some stations during the 1980s. The discharge of the River Niger at Niamey was reduced from an annual average of $1\,060\ m^3\,s^{-1}$ over the period 1929–1968 to $700\ m^3\,s^{-1}$ over the period 1969–1994, e.g. a decrease of 34%. At the same time the average discharge during the month of low water decreased from $64\ m^3\,s^{-1}$ to $11\ m^3\,s^{-1}$. Figure A.37 shows the river runoffs during 1950–1970 and 1971–1990

from the Koulikoro station based on the data from the Global Runoff Data Centre. This station, with a basin size of $12\,104\ km^2$, is located in the upstream region of the River Niger, and represents the mean runoff from the south-western Sahel region. The figure shows that the river runoff is substantially reduced during 1971–1989, and this reduction occurs mainly during the summer monsoon season.

In addition, several investigations have found that the temperature has increased in the Sahel during dry years (Tanaka et al. 1975; Schupelius 1976; Kidson 1977). Tanaka et al. (1975) reported a surface air temperature increase of about 2 K. The observed temperature data, provided by the Oak Ridge National Laboratory (Vose et al. 1992), has also shown that the summer surface temperature over the Sahel region is greater by 1–2 °C during the summers of the 1980s compared with the 1950s (Xue 1997).

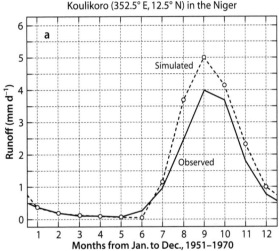

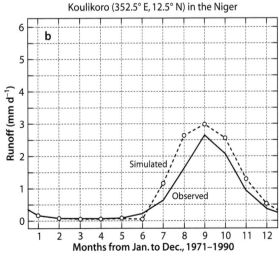

Fig. A.37. Comparison of simulated and observed runoff. **a** Control run and observed mean from 1951–1970; **b** desertification run and observed mean from 1971–1990 (after Oki and Xue 1998)

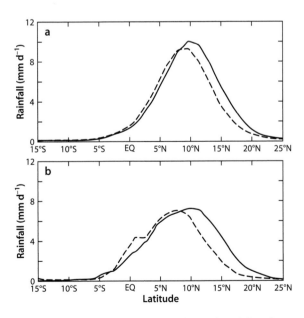

Fig. A.36. Zonally averaged rainfall distribution (mm d^{-1}). **a** Three-month mean for control run (*solid line*) and desertification run (*dashed line*); **b** *solid line* is the mean for 1950 and 1958, *dashed line* is for 1983 and 1984 (from Xue and Shukla 1993)

In prehistoric times, the Sahelian region had a much more dramatic desertification episode than that which persists today. The climate was humid in the early and mid-Holocene from 10 000 to *c.* 4 000 years B.P. (Alexandre et al. 1997; Claussen and Gayler 1997). During that period, swamp forest vegetation was established in the interdune depression in the Nigerian Sahel. The vegetation limit in the region reached at least 23° N. A dry phase during the late Holocene (between *c.* 4 000 and 1 200 years B.P.) followed and came to be replaced by wetter climate conditions from *c.* 1 000 years B.P. The modern shrubs and savanna were developed *c.* 700 years B.P. Subtle changes in the Earth's orbit, strongly amplified by regional and global climate-vegetation interactions, are currently believed to be the main causes for this change in the region's climate 4 000 years B.P. (Kutzbach et al. 1996a).

Historical data reveal several other drought periods from the 16th century in the Sahelian area (data from before the 16th century are very scare), but no records show a drought as persistent and severe as the one that started in the 1960s (Nicholson 1978). Only the drought during the 1840s might be as severe as the one in the 1980s.

A.5.1.3 The Complex Processes of Land-use Change in the Sahel

The Sahel region has a variety of land uses that generate goods and services for the population: fuelwood in natural vegetation areas, food for subsistence and market demands in croplands, livestock in the pastoral area. Pastoralism is particularly extensive. Biomass production relies on the natural productivity of grasslands. Burning is used in most rangelands to preserve savannas against woody invasion and to reduce weeds and pests. Production from shrubs and trees, and crop residues is also important for consumption by livestock. In extensive farming systems in African savannas, four years of cultivation is the usual time limit before the land is put into fallow.

The Sahelian region has undergone significant changes in land cover over the past decades. These changes result from interactions between short-term climate fluctuations and longer-term anthropogenic impacts, in particular, agricultural expansion, agricultural intensification, and rangeland modifications (Stephenne and Lambin 2001). Expansion of cultivation into previously uncultivated areas or migration into unsettled areas takes place with the help of newly developed technology. Agricultural expansion thus leads to deforestation and to a reduction in pastoral land. Pastoral land also expands into natural vegetation areas. Note that grazing livestock move into fallows and cropland during the dry season. Expansion of cropland and pastoral land is thus driven by changes in human and animal populations that increase the demand for food crops and forage, through variability in rainfall that modifies land productivity, and policy changes that modify the rules of access to resources and market changes. The rate and reversibility of change are, however, much higher for rainfall compared to demographic, institutional or economic changes.

Once specific land thresholds are reached, additional demand for food crops results in agricultural intensification. In Sahelian agriculture, intensification mostly takes place as a shortening of the fallow cycle, compensated by the use of labour, agricultural inputs (mainly natural fertilisers) and selection of seeds with, so far, only a minor use of mechanisation and irrigation (Diop 1992; Gray 1999). A shortening of the fallow cycle without input rapidly depletes soil fertility.

The increase in livestock population combined with shrinking pastoral land results in an increasingly sedentary society, where livestock relies more on crop residues for consumption. This puts increasing grazing pressure on pastoral land. Conflicts between pastoralists and sedentary farmers sometimes take place when land scarcity becomes acute.

State policies throughout sub-Saharan Africa are framed in the belief that rangelands are over-stocked by pastoralists, leading to rangeland degradation (Oba et al. 2000). The resulting management strategies aim to control, modify, and even obliterate traditional patterns of pastoralism (Ellis and Swift 1988). Rangelands are maintained by the interaction of human and biophysical drivers, and reducing or excluding the human use may trigger significant changes in these ecosystems. Indeed, reduced grazing is associated with a loss of species diversity, a decline in vegetation cover and reduced plant production. A weakened indigenous pastoral system may lead to economic decline or to urban migration with rural remittances.

There has been rapid population growth in the region over the past 50 years. For example, the population of five Sahelian countries (Senegal, Niger, Mali, Sudan and Chad) rose from 20 million in 1950 to 55 million in 1990 and is projected to rise to over 135 million by 2025 (World Resources Institute 1994). This rise in population, as well as political constraints on nomadism, will continue to cause extensive land-cover conversion. This population growth has occurred at a time of generally reduced rainfall but the proposition that this has led to widespread desertification is not fully established.

Desertification is a loosely defined concept but it is taken here to mean the degradation associated with increased soil erosion caused by wind and water, soil compaction and reduced water holding capacity, reduced infiltration of rainfall and vegetation production, and loss of palatable species, all leading to a decline in crop yields, livestock production, and fuel-wood supply (Le Houérou and Hoste 1977; Dregne 1983; Rapp 1987;

Le Houérou 1989; Goudie 1990; Barrow 1991; Verstraete and Schwartz 1991). Semi-arid rangelands are highly dynamic and resilient, moving through multiple vegetation states by shifting chaotically in response to human and biophysical drivers. This intrinsic variability of rangeland ecology makes it difficult to distinguish directional change (e.g. loss of biodiversity, soil degradation) from readily reversible fluctuations, such that interpretations of "degradation" and "desertification" must be viewed cautiously.

The distinction between desertification and degradation on the one hand, and drought on the other, is not always clearly drawn. Although there is consensus on local degradation in the Sahel (e.g. Lindqvist and Tengberg 1993; Mabbutt and Floret 1980), opinions on widespread degradation in the Sahel are quite controversial. A number of studies showed that overgrazing of the natural rangelands by livestock, northwards extension of cultivation and increased fuel-wood extraction, coupled with severe drought, are leading to widespread land degradation (Gornitz 1985; Dregne and Tucker 1988; Skoupy 1987; Lanly 1982; Dregne and Chou 1992; Akhtar-Schuster 1995; Benjaminsen 1993; Gray 1999). However, some studies, based on the satellite data from the late 1970s, have challenged views on large scale degradation (e.g. Watts 1987; Forse 1989; Tucker et al. 1991; UNCED 1992; Nicholson et al. 1998; Prince et al. 1998). Measuring rates of dryland degradation is thus a complex challenge and requires long time-series of rainfall data, remote sensing-based indicators of surface conditions and field observations of soil attributes, floristic composition, etc.

The issue is in urgent need of reconsideration with more appropriate data (Helldén 1991). The actual extent and severity of regional and subcontinental-scale degradation is an important issue that affects policy with respect to economic development strategies and aid programmes, and also the degree to which changes in the land cover need to be considered in the context of global climate change modelling (Rasmusson 1987; Schlesinger et al. 1990). Prince et al. (1998) have used the ratio of net primary production (NPP) to precipitation – the rain-use efficiency (RUE) – to map degradation. In the Sahel

the results suggested that NPP was remarkably resilient, a fact that was reflected in only little variation in the RUE during the period of study. Thus, in much of the region, NPP seems to be in step with rainfall, recovering rapidly following drought and not supporting the fears of widespread, subcontinental scale desertification taking place in the nine-year period that is studied (Nicholson et al. 1998). In fact, the results show a small but systematic increase in RUE for the Sahel as a whole from 1982 to 1990, although some of the areas contained within the region did have persistently low values.

A.5.2 Observational Studies of Sahelian Land-surface/Atmosphere Interactions

To investigate interactions of land surfaces with the atmosphere and impact of land-cover change in the Sahel region, requires regional measurements of land-surface energy and water balances, and of boundary layer and free-atmosphere dynamics. They provide useful information about Sahelian surface hydro-meteorology and are essential to validate models.

Two large field campaigns were conducted during the late 1980s (Sahelian Energy Balance Experiment, SEBEX) and 1991–1992 (HAPEX-Sahel). Building on these, two additional studies are currently operational (SALT and CATCH). Most of these projects incorporate substantial remote sensing programmes. Figure A.38 shows the locations of these four field experiments. Table A.2 lists the web sites for these experiments.

A.5.2.1 The Sahelian Energy Balance Experiment (SEBEX)

This experiment was conducted by the U.K. Institute of Hydrology and the International Crops Research Institute for the Semi-Arid Tropics (ICRISAT) Sahelian Center at two sites near Niamey, Niger, from 1988 through to 1991 (Wallace et al. 1992) and covering three contrasting Sahelian land types. One site was a fenced area of fallow sa-

Table A.2. Relevant websites relating to Sahelian studies

HAPEX-Sahel online database (Hydrological Atmospheric Pilot Experiment)	*http://www.orstom.fr/hapex/index.htm#1*
CATCH project (Couplage de l´atmosphère tropicale et du cycle hydrologique)	*http://www.lthe.hmg.inpg.fr/catch/welcomefr.html*
SALT project (Savannas in the Long Term)	*http://medias.obs-mip.fr/www/francais/lettre/10/ Dossiers/pf2225/pf2225.htm*
WAMP project (West African Monsoon Project)	*http://www.met.rdg.ac.uk/~swsthcri/tropical.html*
CLD project (Climate and Land Degradation)	*http://www.nwl.ac.uk/ih/cld/*
Sahel standardised rainfall index (Mitchell)	*http://tao.atmos.washington.edu/data_sets/sahel/#values*
Sahel standardised rainfall index (Hulme)	*http://www.cru.uea.ac.uk/tiempo/floor2/data/sahel.htm*
El Niño impacts in Africa	*http://www.essc.psu.edu/~osei/west/westaf1.html*

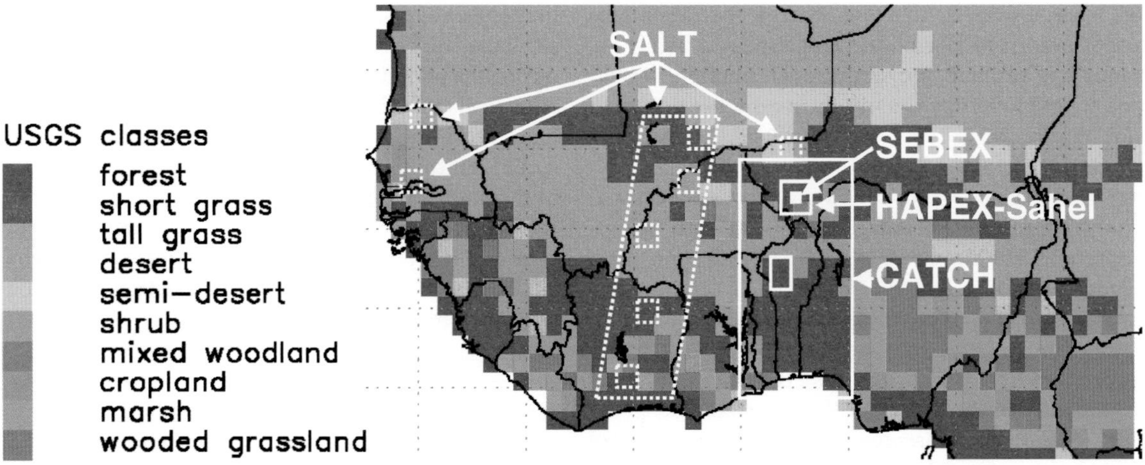

Fig. A.38. Location of past and present land-surface experiments in the Sahel region, projected on USGS land-cover classification. SEBEX comprised two sites near Niamey (Niger), HAPEX Sahel a 1° gridbox in the same area. CATCH includes the latter, while adding a catchment of similar area in Benin, both nested inside two larger regions. SALT includes eight 40 km² along two gradients

vanna, which supported a comparatively high density of natural vegetation, and a second was an area of degraded natural forest, which consisted of a mosaic of dense vegetation and large areas of bare soil. There was also agricultural land that was used for growing traditional crops such as millet. One major objective of the project was to obtain direct measurements of available energy, evaporation and sensible heat flux from different land surfaces, and to improve understanding of the impact of changes in vegetation on the energy and water balance. The measurements covered both the Sahelian dry and rainy seasons. Although there were minor differences among measured variables, all datasets include hourly evaporation rates, sensible heat flux, soil heat flux, friction velocity, short-wave and long-wave radiation flux, net radiation at surface, and meteorological data.

A.5.2.2 The Hydrological and Atmospheric Pilot Experiment in the Sahel (HAPEX-Sahel)

The HAPEX-Sahel experiment was executed in Niger, West Africa during 1991–1992 (Goutorbe et al. 1994, 1997a). The objective of this experiment was to improve the parameterisation of surface hydrological processes in semi-arid areas at scales consistent with general circulation models, and to develop methods for monitoring the surface hydrology at such large scales. With objectives similar to the SEBEX experiment, the experiment was carried out at a much larger scale, i.e. in a 1° square (2–3° E, 13–14° N) in Niger. The vegetation within the experimental area was typical of the southern Sahelian zone: arable crops, fallow savanna and sparse dryland forest. The observations included a period of intensive measurement during the transition

period of the rainy to the dry season, complemented by a series of long-term measurements. Analyses of data from satellite remote sensing combined with ground measurements were carried out to estimate energy fluxes for the GCM grid scale.

Three super-sites were instrumented with a variety of hydrological and meteorological equipment to provide detailed information of surface energy exchange at the local scale. Two less heavily instrumented sites were also established. These were to extend the spatial coverage over the square. In addition to these sites, a network of automated weather stations was established over the full 1° square. They provided information on the regional-scale variability of the main climatic variables. Meanwhile, balloon and aircraft measurements were also carried out to provide information about the development of the atmospheric boundary layer.

The experiments in the Sahel were extensive and complex and the main experimental results are presented in a Special Issue of the Journal of Hydrology (Goutorbe et al. 1997a). The main contributions from these experiments of interest to the climate modelling community are:

- Measurements of the energy balance of the main cover types were made, including mean surface albedo, skin temperature and evaporation, for millet cropland and natural vegetation. Millet is the dominant crop grown by subsistence farmers, in areas cleared of natural savanna vegetation. The millet crop had a significantly higher surface albedo and skin temperature and lower evaporation rate (Table A.3; Gash et al. 1997), indicating some of the changes in land-surface properties were likely to have been a result of the extension of agriculture.

Table A.3.
Mean surface albedo, surface skin temperature and evaporation at the HAPEX-Sahel Southern Super-Site, through August and September, 1992 and simulated differences between degraded and control experiments over the test area

Vegetation type	Surface albedo	Surface skin temperature (°C)	Evaporation (mm m⁻¹)
Millet	0.27	29.5	77
Fallow savanna	0.19	28.4	115
Observed difference	0.08	1.1	−38
Simulated differences	0.08	0.8	−24

- Detailed measurements of the bare soil patches within the vegetation covers were made. The energy and water balance of the bare soil areas turned out to be crucial to the response of the overall landscape.
- The extensive raingauge network has yielded considerable insight into spatial rainfall and soil moisture patterns in the Sahel (see e.g. Lebel et al. 1997; Taylor and Lebel 1998). These data are beginning to be used to understand the aggregate description of the Sahelian landscape (e.g. Taylor and Blyth 2000).
- Extensive site and spatial datasets have been produced which are providing an invaluable resource for current and future Sahelian research.

A.5.2.3 Coupling Tropical Atmosphere and Hydrological Cycle (CATCH)

While HAPEX-Sahel produced key results, some important shortcomings made it difficult to place those results in a broad climatic context. First, the lack of appropriate atmospheric measurements at the regional scale prevented a link between the mesoscale and the large circulation. Second, intensive experiments for a limited time are not on their own sufficient to describe the interannual and decadal variabilities of the water cycle. Finally, the mechanisms that control the variability of West African climate and hydrology must be studied.

Therefore, CATCH addresses scales larger than those in HAPEX-Sahel. A nested approach encompasses the following:

- West Africa as a whole, to study the structure and the variability of large atmospheric entities (e.g. easterly waves);
- the so-called CATCH regional window (0–5° E; 6–15° N) will be used as a reference area to compare the outputs of various atmospheric models (global to mesoscale) with observations;
- two focus areas were selected for fine resolution measurements and process studies: the HAPEX-Sahel square (2–3° E; 13–14° N) and the upper Ouémé catchment (1.5–2.5° E; 9–10° N);
- super-sites, covering in the order of a hundred km², allow for small-scale studies, especially fluxes at the soil-atmosphere interface.

CATCH is organised around four themes:

1. an atmospheric theme addresses climatic variability related to east African waves and global teleconnections aiming at improving seasonal precipitation forecasts, and spatio-temporal dynamics of convective processes in relation to planetary boundary layer (PBL) dynamics;
2. a biospheric theme focusing on the role of small-scale spatial and structural (trees *v.* herbaceous vegetation) heterogeneity in evapotranspiration and carbon cycling;
3. a hydrospheric theme focusing on precipitation/runoff dynamics, as well as on non-saturated zone moisture dynamics at larger scales;
4. a land-atmosphere interaction theme focusing on an integrated (modelling) analysis of coupling of the other themes.

Recently, CATCH has evolved into a larger international project centered on the study of the whole West African monsoon system and of its links with the water cycle. This project, named AMMA, is endorsed by CLIVAR, GEMEX and GCOS.

A.5.2.4 Savannas in the Long Term (SALT)

While AMMA is a project rather strongly oriented towards atmospheric and hydrological processes, SALT addresses more ecological issues. Set up as a transect (and as such endorsed as an IGBP-GCTE Transect), its observational programme includes eight sites covering the two major eco-climatic gradients in the region: the south to north aridity gradient, and a west to east continentality gradient. SALT's primary objective is to understand and predict current and future dynamics and changes in West African savanna ecosystems under climatic and anthropogenic pressures. It addresses savanna dynamics in terms of species composition and spatial structure in relation to the cycles of water, energy, carbon and nutrients. A special focus is on how these are affected by land degradation and erosion.

Apart from ground measurements, remote sensing plays an important role in SALT to scale up vegetation dynamics, to monitor fire dynamics and to study aerosol dynamics.

A.5.2.5 Satellite Data

Validation and calibration studies using the SEBEX and HAPEX-Sahel data have helped to improve the models' simulations in semi-arid areas and to understand the important features and mechanisms of land-surface/atmosphere interactions in the Sahel area (Goutorbe et al. 1997b; Kabat et al. 1997b; Xue et al. 1996a, 1998). However, considering the high spatial variability in land-surface conditions in the Sahel, and given that these two experiments cover relatively small areas and short periods of time, both the observed and derived land data from HAPEX-Sahel are not wholly representative of vegetation conditions over the entire Sahel area.

During the past two decades, new satellite data have brought the Earth systems science community to the brink of a new era for land-surface data information (Shuttleworth 1998). In the earlier stage, vegetation maps were the main products from satellite data (DeFries and Townshend 1994). Recently, vegetation parameters with more physical meaning, such as leaf area index (LAI) and biomass, and surface meteorological, ecological, and hydrological variables, such as surface temperature, albedo, soil moisture, can be derived with increasing accuracy from space (for example, Los et al. 2000; Goetz 1997; Goward et al. 1999), due to more optimised sensor characteristics, better atmospheric correction procedures and better algorithms. Such datasets will increasingly feed directly to models for land-atmosphere interaction studies.

During HAPEX-Sahel, a very wide range of aircraft and satellite measurements was made (Prince et al. 1995). Recently, a subset of a global biophysical land-surface dataset at 8 km spatial and 15-day temporal resolutions for 1982–1998 derived from AVHRR by NASA Goddard Space Flight Center (GSFC) has been introduced to Sahel studies by Kahan (2002). This dataset (hereafter referred to as Los data) is an extension of their previous 1° dataset (Los et al. 2000), and includes LAI and fraction of photosynthetically active radiation (FPAR) absorbed by vegetation. Using this dataset and vegetation cover information, other vegetation properties, such as green leaf fraction, surface roughness length, vegetation cover percentage, are obtained.

This dataset was used to specify the land-surface conditions over a 1° square area in the Sahel. With a forcing dataset from ISLSCP Initiative I CD-ROM (International Satellite Land Surface Climatology Project; Meeson et al. 1995), uniform over the one degree box, the off-line version of the Simplified Simple Biosphere Model (SSiB; Xue et al. 1991, 1996b) was integrated from 1 January 1987 through 31 December 1988 at every 8 km pixel. Figure A.39 shows the July, August, September (JAS) means for latent heat and surface air temperature for the both years 1987 and 1988 from this 1° area, respectively. The figures show that even with a uniform meteorological

forcing in a 1° box, vegetation distribution (with 8 km resolution) alone can produce a north-south temperature gradient due to significant spatial variation in sensible and latent heat fluxes. Furthermore, using the same Los dataset, the model also simulated the interannual variations well. The year 1987 (a dry year) has higher surface temperature and lower latent heat flux than 1988 (a normal year).

A.5.3 Coupled Modelling of Sahelian Land-Atmosphere Interactions

A.5.3.1 Brief Overview

A number of studies with different models have been conducted over the past 20 years to examine the role of biospheric feedbacks in the Sahel drought of the same period. In the pioneering work by Charney et al. (1977) on the effects of albedo on the climate, he found that increases in albedo caused a reduction in precipitation. This discovery has been confirmed by a number of studies with different models (for example, Chervin and Schneider 1976; Sud and Fennessy 1982; Laval and Picon 1986). The effects of soil moisture and evaporation have also been investigated (e.g. Walker and Rowntree 1977; Sud and Fennessy 1984). Most studies showed that less initial soil moisture leads to less precipitation. Furthermore, the combined effects of surface albedo and soil moisture have been studied (e.g. Sud and Molod 1988; Kitoh et al. 1988). Kitoh et al. (his Table 2) found that the combined effects almost equalled the sum of the albedo or soil moisture effects alone. Besides these two parameters, other factors have also been investigated (Cunnington and Rowntree 1986). These modelling studies consistently demonstrated that the land surface may have a significant impact on the Sahel climate.

In the studies that were carried out during the 1970s and 1980s, simple surface layer models of the bucket type were used for sensitivity studies. In most cases, only a single land-surface parameter was tested each time. The primary factors of the desertification under investigation were surface albedo and soil moisture. In most of these sensitivity studies, the area associated with land-surface changes and the extents of the changes in the surface characteristics, such as albedo and soil moisture, were somewhat arbitrary. Atmosphere-biosphere interactions are, however, much more complex in the real world and involve many more parameters and processes.

Therefore, sophisticated land-surface models are needed to assess realistically the impact of desertification on the Sahel drought. A proper evaluation of the surface feedback to climate can be obtained only when all comparable components of the energy and water balances are considered. A number of surface schemes that include a realistic representation of the vegetation

Fig. A.39.
$1° \times 1°$ fields of latent heat ($\mathrm{W\,m^{-2}}$) and surface air temperature (°C) in the Sahel in two successive years (1987: dry; 1988: wet) as simulated by SSiB. The images show the impact of the land-surface variability obtained from satellite observations (at 8 km resolution). Climate forcings were uniform for the entire grid cell

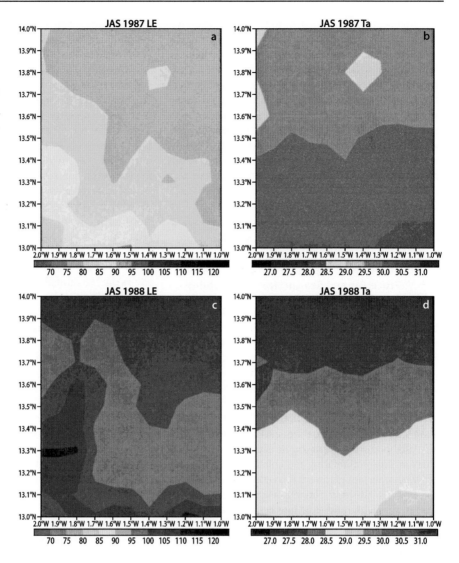

responses have been applied to the Sahel (e.g. the Biosphere-Atmosphere Transfer Scheme, BATS, Dickinson et al. 1986; the Simplified Simple Biosphere Model, SSiB, Xue et al. 1991; the Meteorological Office Surface Energy Scheme, MOSES, Cox et al. 1998). They have been coupled with atmospheric models to conduct a number of Sahel drought studies (Xue et al. 1990; Xue and Shukla 1993; Xue 1997; Clark et al. 2001). Different from the studies in the early 1980s, some of these studies not only tested the sensitivity of the regional climate to land-surface processes but were also intended to link the land-surface processes to observed decadal climate anomalies and, therefore, explored the possible mechanisms involved more realistically.

Most recently, an additional level of complexity is being explored by the coupling of dynamic vegetation models to atmospheric models. Though still in its early development stage, this field of activity has already produced a number of interesting results regarding land-surface climate interactions in the Sahel. Some studies

investigated the climate equilibrium states (e.g. Claussen and Gayler 1997; Wang and Eltahir 2000b). Using a GCM with an interactive vegetation model, Claussen and Gayler (1997) found multiple equilibrium states in climate/vegetation dynamics in northern Africa which could not be found in other parts of the world. This discovery was confirmed by Wang and Eltahir (2000a) with a coupled two-dimensional climate/dynamic vegetation model. In another study using a coupled atmosphere and dynamic vegetation model, Zeng et al. (1999) reproduced the decadal precipitation variability in Sahel, which could not be simulated if interactive vegetation processes were not included in their model.

Mesoscale models are also introduced to Sahel land-atmosphere interaction studies. Thermal anomalies in a landscape may trigger convection and lead to rainfall. The causes of these contrasts in surface energy partitioning can be attributed to contrasts in soil moisture (Taylor et al. 1997b) and/or to contrasts in vegetation density (Hutjes and Dolman 1999). Whilst models illustrate this possibility, it remains to be proven from observations that mesoscale heat flux gradients are strong enough to influence rainfall patterns by this mechanism.

In addition to land-surface effects, observational and model studies have revealed that SST anomalies are highly relevant to the seasonal and interannual rainfall variability in northern Africa (Lamb 1978a,b; Palmer 1986; Semazzi et al. 1988; Folland et al. 1991; Lamb and Peppler 1991; Shinoda and Kawamura 1994; Rowell et al. 1995; Xue and Shukla 1998). Some of the modelling studies on this subject will be briefly introduced in Sect. A.5.3.2 and A.5.3.4 for comparison.

A.5.3.2 Large Scale Force-Response Studies of the Sahelian Climate Anomaly: The Relative Importance of Land-surface Processes and Sea-surface Temperatures

Starting in the late 1980s, numerical experiments with coupled biosphere-atmosphere models have been conducted to derive a better understanding of the mechanisms of land-atmosphere interactions in northern Africa. More realistic simulations of climate anomalies have been achieved by using land-surface descriptions which can be changed to represent changes closer to the real world than was the case in the earlier experiments. In the following we will use the results obtained by Xue and collaborators to describe in detail the up-to-date achievement in simulating the climate anomalies and the most important pathways through which land-cover change affects Sahelian climate. Meanwhile, other recent Sahel studies with models of interactive vegetation and mesoscale models will also be presented.

To explore the impact of land degradation in the Sahel on seasonal climate variability and the water balance, the coupled Center for Ocean-Land-Atmosphere Study's GCM (COLA GCM, Kinter et al. 1988) and SSiB land-surface model (Xue et al. 1991, 1996b) was integrated using a "normal" vegetation map (control simulation) and several vegetation maps where savanna or shrubs with ground cover were changed to shrubs with bare soil in a specified area (degraded simulation). Climatological SSTs were used as the lower atmospheric boundary conditions over the oceans. Because of the internal variability in GCMs, the model in Xue's study (1997) was integrated from 1 June, 2 June, and 3 June for three years, respectively, and ensemble means were used to identify climate impacts.

The world vegetation map was read into the coupled surface-atmosphere model to provide the land-surface conditions required by SSiB. Twelve vegetation types are recognised by SSiB, including trees, short vegetation, arable crops and desert. Different vegetation and soil properties, including surface albedo, LAI, soil hydraulic conductivity and surface roughness length, are defined for each vegetation type. For land-surface degradation simulations, the normal vegetation types in the Sahel area were changed to the types that would result from land degradation, altering the prescribed vegetation and soil properties. Since there are no quantitative data available for the 1950s' land-cover information in the Sahel, the selections of the degradation areas were based on some available information (e.g. Dregne and Chou 1992) but do not necessarily represent the reality that occurred during the past 50 years. As a result, these simulations must be regarded as sensitivity studies.

The differences between the ensemble mean July-August-September (JAS) rainfall of the degraded and control simulations are shown in Fig. A.40b. The rainfall is reduced in the degraded area but increases slightly to the south. This dipole pattern is consistent with the observed pattern for dry climate anomalies in Fig. A.40a, which shows the JAS rainfall differences between the 1980s and the 1950s. The simulated rainfall in a test area (from 9° N to 17° N, and 15° W to 43° E), including most of the specified degraded area, is reduced by 39 mm month^{-1}, close to the 45 mm month^{-1} observed reduction.

At the beginning of the Sahelian dry season (October-November-December, OND), when the Intertropical Convergence Zone (ITCZ) moves to the south, little rainfall is observed or simulated in the Sahel. However, the areas of reduced rainfall related to land degradation shift to the south of the Sahel, with a positive anomaly over eastern Africa (Fig. A.40d), consistent with the observed OND rainfall anomaly (Fig. A.40c). Thus, the effect of land degradation is not limited to the summer rainy season within the Sahel but extends to the autumn and into East Africa.

The JAS surface air temperature is higher in the degraded simulation than the control, consistent with the observed JAS temperature difference between the 1980s and the 1950s. In the test area, the simulated surface air temperature increases by 0.8 K, close to the observed increase over the same area, 1.1 K, and consistent with the difference measured in the field (Table A.3).

The impact of land-surface degradation on river discharge variability in tropical northern Africa has also been investigated. The simulated soil moisture, surface runoff and subsurface drainage in degradation experiments decrease, consistent with the reduction in rainfall. To compare the runoff at grid points in GCM simulations with observed river discharge, a linear river routing scheme is applied. The lateral flow direction is taken from Total Runoff Integrating Pathways (TRIP; Oki and Sud 1998). The simulated river discharges from control and degradation experiments at the Koulikoro station, located at the top end of the River Niger, represent the mean runoff from south western Sahel region and are used to compare with the observed 1951–1970 mean and the 1971–1990 mean, respectively (Oki and Xue 1998). Some stations in the Sahel cannot be used because their

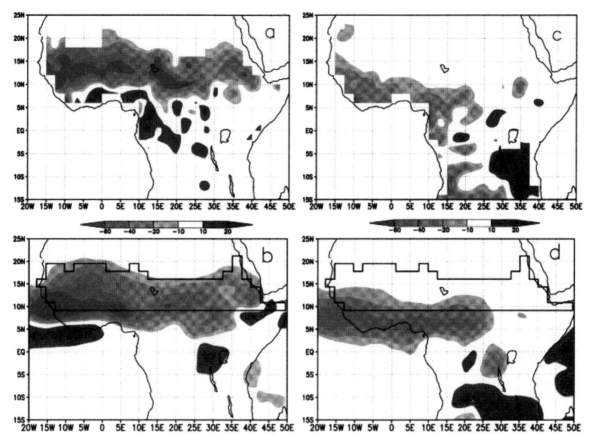

Fig. A.40. a Observed JAS rainfall difference (mm month^{-1}) between 1980s and 1950s; **b** JAS rainfall difference in ensemble mean, degraded minus control; **c** same as (**a**) but for OND; **d** same as (**b**) but for OND. In the degraded simulations the vegetation types in the *heavy lines* were changed to shrubs with bare soil (adapted from Xue 1997)

data are severely affected by irrigation. The mean monthly discharges are simulated fairly well and the contrast between control and degraded simulations corresponds to the observed difference between 1951–1970 and 1971–1990 (Fig. A.37).

Since we may never know the exact extent and degree of real land degradation in the last 50 years, another set of tests was designed to investigate how the extent of specified land-surface changes may affect the results. Five subregions were degraded in turn: northern Sahel, southern Sahel, West Africa, East Africa, and the coast area along the Gulf of Guinea (Clark et al. 2001). In general, degradation results in reduced rainfall over the changed surface. But considerable differences between the areas indicate that the location of the degraded area is important. Degradation in northern Sahel and West African Sahel results in the largest and most significant reductions of rainfall. In particular, degradation of northern Sahel causes widespread reduction of rainfall across tropical northern Africa, both within and outside the degraded area. Meanwhile, the deforestation in the coast area produces little rainfall variation.

The simulated climate anomalies are caused by the specified land degradation, which affects the atmosphere mainly through modulating the hydrological processes and energy balance at the surface. There are about 20 parameters in a biosphere model. To investigate the effects of each parameter in this degradation study, numerical degradation experiments have also been conducted, in which not the vegetation type but rather the individual parameter was changed in the degraded area (Xue et. al. 1997). The results show that surface albedo, surface aerodynamic resistance, stomatal resistance, LAI, and hydraulic conductivity of the soil have the largest impact on the simulation results. The surface albedo change caused by land degradation has a strong effect on the surface energy and water balances, so it must be specified realistically in Sahel studies. The albedo changes used in the Sahel study were reasonable (0.09–0.1) and close to the observed albedo difference between the cropland and natural vegetation in the Sahel (Table A.3). Despite the importance of albedo, changes in other vegetation and soil properties (such as LAI) also contribute significantly to rainfall and surface temperature anomalies.

Land degradation also manifests itself in the changes in large-scale circulation and easterly wave propagation. One of the important features of sub-Saharan drought is the displacement of isohyets in the dry years. The isohyets retreat southward in northern Sahel during dry years, which reflects the changes in ITCZ position. In southern Sahel, the isohyets also shift slightly to the south. This movement reduces the rainfall in the Sahel and increases rainfall to the south of 10° N. In the model simulations, both shifting and reduction of the maximum rainfall play roles in reducing the rainfall in the Sahel (Fig. A.36). Both the strength and the depth of the monsoon flow may change.

Reed (1986), based upon the observed data from 1967 to 1982, concluded that the sub-Saharan baroclinic zone and near equatorial rain belt jointly spawned about the same number of disturbances over Africa each year, but that fewer strong and/or highly convective disturbances occurred near the coast during periods of extended droughts. In the degraded experiments (Xue and Shukla 1993), the intensity of disturbances was greatly reduced but not the number of the disturbances. However, recently available observational data from Niger (Le Barbé and Lebel 1997) show that the drought in Niger is mostly associated with a decrease in the number of rain events in July and August. The mean event rainfall and the length in the rainy season hardly contributed at all to the decrease in annual rainfall.

At the continental scale, the oceans play an important role, and the relative strength of forcing of land surface *versus* SSTs has been a subject of some studies. In particular, the relationship between SST and seasonal to interannual rainfall variations in the Sahel region has long been an important scientific subject. The first comprehensive observational evidence for a possible relationship between the Atlantic SST anomalies and the Sahelian rainfall anomalies was presented by Lamb (1978a,b). Lamb showed that a relationship exists between displacements of Atlantic SST patterns and rainfall patterns. Tropical Atlantic SST patterns are known to accompany extreme sub-Saharan rainfall conditions, especially towards the west. This discovery was further confirmed, and extended by several subsequent studies (Hastenrath 1984, 1990; Druyan 1991; Lamb and Peppler 1991), to explain long-term changes. In addition, weaker but significant correlation was found between El Niño/ Southern Oscillation indices and sub-Saharan rainfall especially towards eastern (horn of Africa) and southeastern Africa. But there is considerable disagreement in the literature on the ENSO signal in the Sahel (e.g. Wolter 1989; Nicholson and Kim 1997).

Several other observational and modelling studies suggested that the global SST anomalies also played an important role in producing rainfall anomalies over the Sahel and the adjoining regions (Folland et al. 1986; Pal-

mer 1986; Folland et al. 1991; Palmer et al. 1992; Rowell et al. 1995). Folland et al. (1991) found that the relatively modest variations observed in the large-scale patterns of SST had a substantial impact on the variations of Sahel rainfall. Tropical oceans, on the whole, had considerably more influence than extra-tropical oceans. They also found that warmer SST in the Southern Hemisphere relative to that in the Northern Hemisphere had been associated with Sahel drought. Rowell et al. (1995), using realistic SST as the boundary condition for the UK Meteorological Office GCM, found that the model was able to simulate the interannual and decadal variations of Sahelian rainfall very well.

These results were not confirmed by the simulations by other modelling groups (Sud and Lau 1996). In an experiment with the COLA GCM and the same SSTs used by Folland et al. (1991), Xue and Shukla (1998) found that, although the results generally supported the conclusion of Folland et al., they were quite model-dependent. For example, in the 1958 case, the COLA GCM could not produce the observed rainfall anomaly pattern as did the U.K. Meteorological Office model. The simulated anomalies had relatively larger sensitivity to the initial conditions in this study than those in their desertification study (Xue and Shukla 1993). Furthermore, the simulated rainfall anomalies were smaller than observed ones. The results from the desertification experiments (Xue and Shukla 1993; Xue 1997) showed that despite the dramatic changes in prescribed land-surface conditions in the desertification experiment, the simulated rainfall anomaly was still smaller than observations. In the COLA GCM, neither SST nor land anomaly forcing alone could reproduce the observed rainfall anomalies. It is likely, therefore, that both SST and land-surface anomalies play a role in Sahelian rainfall.

The teleconnection between Sahel climate and other parts of the world is an important scientific subject. Observational evidence has shown a correlation between the rainfall anomalies in the Sahel and the rainfall anomalies in other tropical regions (e.g. India) and subtropical regions (e.g. Europe, Ward 1995). But thus far, there are few studies on this subject.

Xue's study (1997, Fig. A.40d) shows there may be a potential link between Sahelian and East African climate, and some impact of Sahel desertification on circulation in some tropical regions. Further study is necessary to investigate this issue. Another example of possible teleconnections influencing Sahelian rainfall comes from a palaeoclimate study in which vegetation changes associated with agricultural expansion and deforestation in Mediterranean Europe and North Africa toward the end of the Roman Classical Period (some 2 000 years ago) appear to contribute to long-term climate change in the Sahel. Reale and Dirmeyer (2000) and Reale and Shukla (2000) reconstructed the distribution of vegetation in lands surrounding the Mediterranean Sea from a variety of historical and palynological sources. The Mediterranean forests of southern Europe, the Middle East, and the mountains of North Africa and the Roman agricultural belt in North Africa were restored. That distribution was substituted for present-day vegetation in the COLA GCM and the model has been integrated for both cases. The most striking difference in the simulations is that rainfall over sub-Saharan Africa is greatly affected. There is a stronger northward propagation of the ITCZ and its associated rain into the Sahel during boreal summer in the Roman Classical Period case.

A.5.3.3 Mesoscale Interactions between Sahelian Precipitation and Land-surface Patterns

The heterogeneity of vegetation distribution within a GCM grid box has not been taken into account in previous GCM simulations. Observational evidence showed that spatial heterogeneity of surface types in the HAPEX-Sahel area was high (Prince et al. 1995), with a distinct north-south gradient. Taylor et al. (1997a) found soil moisture variations to be the important cause of variations in surface energy partitioning. Hutjes and Dolman (1999) also found variations in vegetation density to be more dominant in surface energy partitioning than soil moisture in certain cases. The later work also focused on issues of aggregation of surface characteristics. Based on a vegetation classification map with 30 m resolution (Prince et al. 1997), Arain et al. (1997) aggregated vegetation parameters to the model resolution of 2 km, using either those of the dominant class or averaging each parameter appropriately weighted by subgrid relative areas of each vegetation class. Either method produced similar domain-averaged heat fluxes but the variation in fluxes differed significantly. Moreover, only aggregated vegetation parameters could, at least qualitatively, reproduce observed rainfall, whereas using dominant class parameters at the land surface did not produce any rainfall at all.

Observational evidence suggests positive surface feedback leads to persistent rainfall gradients, which are much stronger and more localised than the large-scale average. Taylor and Lebel (1998) found strong positive correlation between daily and antecedent rainfall differences at a range of time scales. In semi-arid regions this mechanism may lead to preferred and seasonally persistent rainfall patterns. The influence of antecedent rainfall on storm development is particularly noticeable when intense large-scale storms passed over areas of marked gradients in evaporation. Preliminary 2D-simulations of a single storm with resolved convective processes by Taylor and co-workers (pers. comn.) could reproduce rainfall enhancement over wet soils surrounded

by dry ones. In a similar 3D-study of two consecutive storms, Hutjes and co-workers found that (pers. comn.) rain did not fall over the area wetted by the first storm, but fell instead on the still dry land.

Both preliminary studies need to have a larger number of ensembles for statistical confidence favouring either mechanism. These are two forms of a land-surface/atmosphere feedback that occurs at a much smaller scale than in previous studies, and therefore needs to be included in future regional studies of climate and mesoscale change.

A.5.3.4 Climate System Interactions in the Sahel

In all the studies described above, vegetation types were fixed during the period of model integration. This may be realistic for time periods of days up to a few years but obviously becomes less so for longer, decadal to centennial or even millennial scale studies, although we do not know yet how much the impact could be. The logical conclusion from the studies discussed above is that there is a need for a realistic representation of land-surface conditions in land-atmosphere models. At the same time it stresses the need to test the reverse link: responses of land-surface conditions to environmental changes.

In the past decade, a number of groups attempted to quantify the missing climate-vegetation feedbacks by coupling equilibrium biogeography models "asynchronously" to GCMs and regional models (for example, Prentice et al. 1992; Neilson and Marks 1994; Woodward et al. 1995). These vegetation models attempt to simulate global vegetation patterns and predict the potential change in vegetation distribution associated with changes in other components of the climate system. This kind of study involves an iterative procedure in which the GCM calculates a climate implied by a given land cover; and the vegetation model calculates the land cover implied by a given atmospheric-ocean condition. This process is repeated until a mutual climate-vegetation equilibrium is reached (for example, Claussen 1994).

In another type of model, the land cover is treated as an interactive element, by incorporating dynamic global vegetation models (DVGMs) within climate models in a physically consistent way (for example, Friend et al. 1993). During recent years, several studies were conducted to develop physically consistent biophysical/biogeochemical models (for example, Sellers et al. 1996c; Dickinson et al. 1998). Attempts have also been made to develop dynamic biosphere models (Foley et al. 1998) and to use interactive vegetation models for regional climate prediction (for example, Pielke et al. 1999; Lu et al. 2001; Eastman et al. 2001a,b).

Claussen and Gayler (1997) used the first type of vegetation model developed by Prentice et al. (1992) cou-

Fig. A.41.
The role of the land surface in controlling precipitation variability in the Sahel, as inferred from the study of Zeng et al. (1999). *Top panel*: the observed anomalies in rainfall (*black bars*) and NDVI (*green dots*); *lower three panels*: modelled anomalies in rainfall (*black bars*) and vegetation (*green dots*) arising from ocean-atmosphere interactions ("*AO*"), ocean-atmosphere-land interactions ("*AOL*") and ocean-atmosphere-land-vegetation interactions ("*AOLV*")

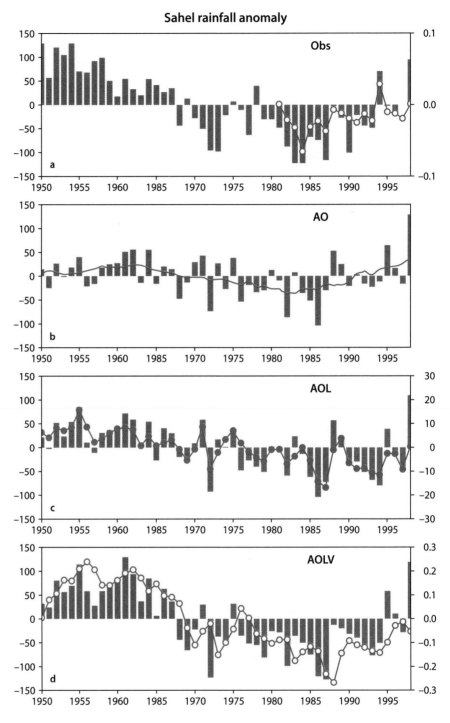

pled to ECHAM (climate model based on the ECMWF model, version 3.2) to simulate climate evolution over the long time scales. As discussed in Sect. A.5.1.2, the climate in the Sahara during the early and middle Holocene was much wetter than today, and vegetation was found far north of the present desert border. Using an interactively coupled model, this shift in vegetation has been recaptured. On the other hand, when only an atmos-

pheric model and present-day land-surface conditions are employed, the enhancement of simulated precipitation was insufficient to diagnose an encroachment of xerophytic woods/shrub into the Sahara north of approximately 20° N.

Using the coupled model, Claussen (1994, 1997, 1998) also found that, starting from different initial conditions, the system came to multiple equilibrium states in climate/vegetation dynamics in northern Africa. Under present-day conditions of the Earth's orbital parameters and SST, two stable equilibria of vegetation patterns were possible. One solution corresponds to present-day sparse vegetation in the Sahel and desert in the Sahara (desert equilibrium) while the second solution yielded savanna, which extended far into the western part of the Sahara (green equilibrium).

Wang and Eltahir (2000a,b) used a different modelling concept to show how different initial vegetation conditions may lead to different equilibrium climate patterns. They used a zonally symmetric two-dimensional regional model, synchronously coupled to the Integrated Biosphere Simulator (IBIS, Foley et al. 1996) biosphere model. Their surface conditions outside the tropics were fixed to NCEP (National Centers for Environmental Prediction) re-analysis climatology. IBIS is a DVM of the second type which models vegetation growth and competition explicitly. Using this model they found that the equilibrium climate pattern, as in Claussen (1994), was sensitive to the initial vegetation distribution. Starting with desert covering all of west Africa, the vegetation at equilibrium varied from tall grass near the coast to short grass and desert northward. In contrast, starting with forest all over west Africa, the equilibrium vegetation consisted mostly of forests covering most of west Africa, with a narrow grassland band in the north and a much higher productivity and rainfall than in the first case.

In addition, they showed that the new equilibrium states depended on the initial disturbances in model simulation. The system was able to recover completely in terms of vegetation distribution and rainfall when the grass biomass removal was less than 60% between 12.5 and 17.5° N. With 60–75% biomass removed, the system converged to a new equilibrium, about 40% drier and 65% less productive. When even more biomass was removed the system collapsed to a 60% drier and nearly 100% less productive equilibrium. These new drier equilibria were associated with both weaker large-scale circulation and suppressed local convection. The reason for this weakening/suppression was found to differ between wet and dry seasons. In the dry season, a degradation process that leads to higher albedo, reduced net radiation, and cooler land surface was the dominant process, while in the wet season reduced evaporation and warm land surface was the dominant process. Because the lack of the zonal wave disturbance in a 2-D model may have contributed to the existence of multiple equilibrium solutions, further studies with a 3-D model are necessary.

Recent work by Zeng and Neelin (2000) shows how interannual variability in forcing (SST) affects the multiplicity of stable climate patterns. Using a simple DVM, conceptually of the second type, they showed how in absence of SST variability (using the same SST clima-tology each year again) the equilibrium vegetation was sensitive to initial vegetation distribution, consistent with the findings of both Claussen and Wang. However, when real SSTs, exhibiting variability on various time scales, were used to force the model, the equilibrium climate pattern becomes nearly independent of initial conditions.

In another study, Zeng et al. (1999) also showed that the interaction between all three vegetation, soil and oceans components, best reproduces observed climate in terms of interannual precipitation variability over the Sahel. In a model run with soil moisture and vegetation fixed, precipitation showed little variability compared to the climatology of the last 50 years. Decadal variability is nearly absent and the amplitude of interannual variability is small. A run with soil moisture feedback, but with each year having the same imposed vegetation dynamics, improves precipitation variability somewhat in both respects (timescale and amplitude). Allowing vegetation to respond to precipitation/soil moisture reproduced the best-observed precipitation decadal variability and with an amplitude of the correct magnitude (Fig. A.41).

A.5.4 Understanding Mechanisms

All the experiments described above demonstrate the importance of land-surface processes in Sahel climate. Here, we analyse the mechanisms involved in the atmosphere/land-surface interactions and their relative importance for the Sahel region. The following discussion of the dominant physical mechanisms is outlined in Fig. A.42, with heavy lines representing the main processes. The major differences in the energy balance between degradation and control experiments are also listed in Table A.4 for comparison.

Vegetation degradation and soil drying increase the albedo. In the southern Sahel, when the albedo increases, more solar radiation would be reflected back into the atmosphere. However, there are fewer clouds as a result of less evaporation associated with the higher albedo. Therefore, a negative feedback occurs, and the solar radiation that reaches the ground is not significantly changed. In the northern Sahel, however, although a higher albedo increases the short-wave radiation that reflects into the atmosphere, it cannot reduce clouds any further because there are not many clouds in the first place. The values given in Table A.4 show that in the northern Sahel, where clouds are scarce, a cloud/radiation feedback is, therefore, not important and the increased albedo leads to reduced surface heating. In the southern Sahel, in contrast, the cloud/radiation interaction is particularly strong and the net short-wave radiation absorbed by the surface does not change. The net

Fig. A.42.
Schematic diagram of land-surface degradation interactions and feedback processes. The *dark lines* represent the main process. The *dashed line* between the surface and MFC indicates the uncertainty of their interaction

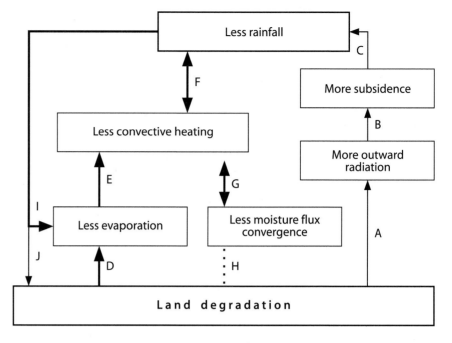

long-wave radiation at the surface is reduced because the higher surface temperature increases outgoing long-wave radiation. The reduced cloud and water vapour in the degradation simulations also reduce the incoming long-wave radiation at the surface. However, this reduction is smaller compared with other energy components.

The changes in sensible heat flux are associated with the changes in short-wave radiation. The sensible heat flux is reduced when the short-wave radiation decreases substantially, such as in the northern Sahel. Otherwise, it increases to balance the reduction of latent heat flux in the surface budget such as in the southern Sahel. Evaporation in Table A.4 is substantially lower in both regions, correlating with the simulated rainfall reductions. Evaporation decreases partially because of the reduced net ra-

diation, but more importantly because of lower LAI, surface roughness length, higher stomatal resistance, and changes in other vegetation and soil properties in the degradation simulations. The reduced evaporation is similar to that observed (Table A.3) when replacing the denser natural vegetation by thinner crops. Again, there is a difference between northern and southern Sahel. When there is more evaporation initially (southern Sahel), the effect is stronger.

Because of the large reduction in evaporation (see the letter E in Fig. A.42), less moisture is transferred to the atmosphere through the boundary layer. Table A.4 shows that this results in less convection and lower atmospheric latent heating rates at 500 mb, where the largest reduction occurs. (Fig. A.42, F). The lower convective latent heating rate is responsible for more than 70% of the re-

Table A.4.
Simulated ensemble mean differences between control and degradation experiments over different regions

Parameter	South Sahel[a] (JAS)[b]	North Sahel[c] (JAS)
Precipitation (mm month^{-1})	−56	−29
Latent heat flux / evaporation (W m^{-2} / mm month^{-1})	−23 / −26	−18 / −21
Sensible heat flux (W m^{-2})	10	−8
Net short-wave flux (W m^{-2})	−1	−16
Net long-wave flux (W m^{-2})	−12	−10
Total diabatic heating at 500 mb (K d^{-1})	−0.69	−0.59
Convective latent heating at 500 mb (K d^{-1})	−0.47	−0.41
Moisture flux convergence (mm month^{-1})	−29	−12
Surface temperature (K)	1.1	0.4
Albedo (W m^{-2})	0.10	0.09

[a] From 9 to 13° N and 15 to 43° E.
[b] *JAS:* July, August, September.
[c] From 13 to 17° N and 15 to 43° E.

duction in the total atmospheric diabatic heating rate in these experiments. The reduced total diabatic heating rate in the atmosphere is associated with relative subsidence, which in turn weakens the African monsoon flow, reduces moisture flux convergence (Fig. A.42, G), and lowers rainfall (Fig. A.42, F). How precisely monsoon flow and moisture convergence are affected depends on the geographical position of the land-surface anomaly as shown by Clark et al. (2001).

Early Sahel studies considered albedo as the major factor causing the long-term drought. But the dramatic albedo changes assigned in these studies and the cooling caused by high surface albedo are not supported by observations, and as a result the credibility of the land-surface effect has long been challenged (for example, Ripley 1976). In the experiments discussed above, the albedo changes were reasonable (close to the Nicholson et al. (1998) estimation). The dominant factor in diabatic cooling of the upper troposphere and enhanced subsidence appears to be the reduction of convective heating rates as a result of reduced latent heat flux and moisture flux convergence. The change in radiative heating rate was rather a secondary factor, although it may trigger other anomalous processes.

Once rainfall is affected additional feedbacks start playing a role. Reduced rainfall reduces evaporation producing a positive feedback (Fig. A.42). These feedbacks exhibit different temporal scales associated with different "memory time scales" of the processes involved. Very fast response time scales, such as hours or less, are normally associated with evaporation of moisture in the top soil layer and water intercepted by the vegetation. This process could be active during a storm (Taylor and Lebel 1998). Longer time scales, such as weeks to a season, are associated with plant transpiration, which reduce the deeper soil moisture. Yet longer time scales, such as seasons to years, are related to plant development. The study by Zeng et al. (1999) illustrates how the combination of multiple feedbacks, each with its own time constant, recreates a realistic interannual rainfall variability. When a drought is persistently long enough, the ecosystems may change or "move". This brings up issues such as ecosystem resilience as explored by Wang and Eltahir (2000b). At the longest time scales, such as centuries and longer, the additional interaction may result in "step changes" in response to gradually imposed forcings (Claussen et al. 1999).

In the above discussion, the surface evaporation is a dominant factor. However, in some cases, even surface temperature change alone could also produce a robust feedback process, such as in Charney's study (1975) and Wang and Eltahir's study (2000b). We shall not discuss this process further here since the hydrological processes are dominant in the desertification experiments presented in this chapter.

By and large, the above analysis shows that land degradation, and in particular vegetation degradation, modifies the water and energy balance as well as the partitioning of available energy over sensible or latent heat fluxes. These are first-order effects and their relative importance may vary spatially and temporally. In the atmosphere, differences in radiation, latent and sensible heat inputs lead to altered heat and moisture contents within the atmospheric boundary layer. This effect produces a feedback to the surface through atmospheric stability conditions and other environmental factors, which control stomatal behaviour of plants, thus creating a first potential loop. Meanwhile, the turbulence fluxes and thermal structure within the boundary layer also affect convective heating, total diabatic heating, subsidence, and moisture convergence. All these processes affect cloud formation, precipitation and the strength of the monsoon flow. Clouds strongly affect radiation flux transfer and create the second potential feedback loop. When precipitation is affected, additional feedback loops are activated through soil moisture storages, vegetation growth and phenology, and eventually ecosystem changes.

A.5.5 Conclusion

Research on Sahelian land-surface/atmosphere interactions has been continuing for several decades. It started with a very simple model with even no water in it (Charney 1975). During the early 1980s, a number of studies were conducted by simple bucket type of land models coupled with GCMs. The vegetation models were introduced from the late 1980s. All this research showed a consistent result: the Sahelian climate is sensitive to the local land-surface processes.

In the introduction, we have raised several issues related to land-surface/atmosphere interaction in the Sahel region. This Chapt. A.5, by reviewing the relevant research during the past two decades, has aimed at understanding and clarifying these issues. The answer to question one in the introduction is apparently positive. In principle, land-surface changes can result in rainfall reductions of the observed magnitude. The studies discussed in Sect. A.5.3.2 illustrate this. If changes in vegetation properties between the 1950s and the 1980s were comparable to the specified control/desertification differences in that study, the results indicate that the degradation could lead to regional climate changes of the order of the differences found between the 1950s and the 1980s, with increases in the surface air temperature, and reductions in the summer rainfall, runoff, and soil moisture over the Sahel region. The impact is not only limited to the specified desertification areas and the JAS period but it also affects the region south of this area and continues into the OND period in East Africa. All these results are in line with the observed climate anomalies, which suggest that the climate anomaly during the 1980s could be due to land-surface processes. While early studies attributed the drought primarily to a albedo/radiation driven process chain, we now realise that an evaporation/convection driven process chain is probably more important.

An apparent shortcoming in sensitivity studies was that the specified land-cover changes and simulated land-surface energy and water balances at the surface were not verified by observations. The SEBEX and HAPEX-Sahel field experiments provided very important information of land-surface processes in the Sahel region. These field measurements have led to a considerable increase in our understanding of the surface energy balance at the Sahelian land surface and in particular the interplay between the hydrological responses of soil and vegetation.

In general, these experiments have resulted in improved calibration and validation of the surface biosphere models. Our knowledge of physical and biological characteristics of soil and vegetation and energy and hydrological processes has improved significantly (e.g. Taylor and Blyth 2000; Xue et al. 1996a). These spatial analyses are leading to a strategy for the aggregate description of the landscape in these regions where both the rainfall and the vegetation have high spatial variability. However, since all these measurements were conducted in one region, it is difficult to apply the results to the continental scale. Therefore, new field measurements should cover much larger areas and should include more typical vegetation. In addition, long-term measurements, which include at least a few growing seasons, are necessary to understand the seasonal and interannual variability in the region. Both the SALT and CATCH projects seek to fill these needs.

Satellites have provided land-surface conditions since the late 1970s. The various results discussed before (Sect. A.5.2.5) demonstrate the great potential in applying the satellite dataset for Sahel studies. These data provide temporal and spatial variations of land-surface conditions over the entire globe and cannot be obtained from any other method. They can also be and have been used to specify land-surface conditions more realistically in model simulations, although more research is needed to develop and validate the methodology used to convert satellite signals to vegetation characteristics and other land-surface information. In addition, they provide useful information for potential validations of dynamic vegetation models, which we believe has not yet been attempted. The full potential of satellite measurements should be explored further.

Our knowledge of the role of vegetation dynamics in the Sahelian drought has also greatly improved recently by the use of dynamic vegetation models. Although such models are still at the preliminary stage, several studies with African climate as a prime subject have demonstrated their promising potential. Thus far, these studies focused on time scales of decades and centuries. Further development in coupled dynamic vegetation-atmosphere models and more realistic simulations at seasonal-interannual-decadal scales are necessary.

The field experiments, satellite data, off-line land-model validations, dynamic vegetation models and coupled models greatly help us to understand land-surface processes and the roles of soil and vegetation in these processes, and the influence and mechanisms of land-surface/atmosphere interactions (Questions 2 and 3 in the introduction).

As to the real land-cover change in the Sahel, we may never know the real scope of the changes that occurred during the past half century. However, it is undeniable that the human population in Sahel has increased substantially over the last 50 years. The rise in population and projected further increases have caused, and will continue to cause, extensive conversion of land from natural vegetation to agriculture, and increasing demand for fuelwood and grazing (Stephenne and Lambin 2001). Currently, work is under way to use this type of information to produce more realistic reconstructions of land-surface dynamics over the past 40 years and to make projections in to the future (Taylor pers. comn.). This line of research will eventually provide a better answer to the question of whether it is really population pressure leading to land degradation that leads to desertification. We now know it could; we may soon know whether it did, and what might happen in the future. This should partially answer the Questions 4 and 5.

At this stage, we still cannot decide whether the current drought is a temporary phenomenon, what is the exact role of humans in its cause, or what kind of possible remedy should be taken (Question 6). One area that remains unclear is how well the normal and degraded vegetation scenarios specified in the Sahel represent the real land-surface conditions in the 1950s and the 1980s, as reliable information is unavailable. With the dynamic vegetation models, we may be able to assess more realistically the role of SST, land-surface processes, and other processes in the long-term African drought.

However, we might come closer to answering this question by looking at other regions undergoing similar desertification trends, such as Inner Mongolia (see Chapt. A.8 and Xue 1996). To decide whether such droughts are "permanent", at least on human time scales, we may look back into the distant past that some groups have begun to explore (Claussen et al. 1999).

The present chapter demonstrates that changes in surface properties caused by intensification of land use may have had serious consequences for the regional climate. To avoid such consequences, the implementation of sustainable resource management policies must be a priority. Xue and Shukla (1996) demonstrated that widespread afforestation of the Sahel might reverse the regional climate changes caused by land degradation. In this chapter we mainly presented the biophysical process studies for the Sahel area. We believe other dimensions of the Sahelian droughts, e.g. water resources, food security issues, and human activity (socio-economic and political drivers of land-use change), should be much more tightly integrated with physical process studies. Thus new links in the process chains may be discovered, leading to better reconstruction of past and better predictions of future climate changes and anomalies in this sensitive region.

Chapter A.6

The Amazonian Climate

Carlos A. Nobre · Maria Assuncão Silva Dias · Alistair D. Culf · Jan Polcher · John H. C. Gash
José A. Marengo · Roni Avissar

Carlos A. Nobre

A.6.1 Introduction

The Amazon Basin contains the largest extent of tropical forest on Earth, over 5×10^6 km^2, and accounts for a large proportion of the planet's animal and plant species. It also contains over 2×10^6 km^2 of savanna vegetation to the north and to the south of the domain of the tropical forest. The annual discharge of the River Amazon into the Atlantic Ocean of more than 200 000 m^3 s^{-1} contributes about 18% of the global flow of fresh water into the oceans. It is thought that humans have inhabited tropical South America for at least 12 000 years, but their impact on the vegetation cover was quite small from immemorial times to about 1950. By the time the Europeans arrived in this continent 500 years ago, there were several million indigenous people living in Amazonia. Their impact on the vegetation cover was negligible because they practised small-scale shifting and burning agriculture. Because many tribes selected useful plant species to grow near their dwellings (for instance, the Kayapo "gardening"), it is thought that the species composition and distribution have gradually changed from pre-historical times but with little overall effect on the vegetation cover. For most of the last 500 years, the forests of Amazonia were left untouched, including the period known as the "rubber boom" (second half of the 19th century to about 1910). During that period, the river system was used to bring hundreds of thousand of workers to the far-distant portions of the forest to tap the rubber tree, but land clearing was not significant. In fact, after the demise of the rubber boom by 1910 many workers settled along the river corridors, and borrowed many of the habits from the indigenous tribes that they had displaced such as adopting small-scale shifting and burning agriculture, and having fish as their main protein source. In summary, they caused only negligible disturbance to the forest and by the 1950s only about 1% of the forest of Amazonia had been cleared.

The occupation and development of Amazonia since the 1950s and its impact on the vegetation cover and on the composition of the atmosphere have changed dramatically, with most of the changes taking place over the last 30 years. During the 1950s and 1960s many countries of South America started to implement government plans to develop and integrate the economy of Amazonia. The plans called for a network of roads criss-crossing the region and put emphasis on agricultural development. For Brazil, that was accomplished through government incentives and tax benefits to attract large companies for cattle ranching, and a programme to settle in Amazonia millions of landless peasants from other parts of Brazil. As a result the population of Brazilian Amazonia has grown rapidly from 3.5 million in 1970 to almost 20 million in 2000. This rapid development has led to the deforestation of over 500 000 km^2 in Brazil alone. Current rates of annual deforestation range from 15 000 to 20 000 km^2 in Brazilian Amazonia (Instituto Nacional de Pesquisas Espaciais – INPE, 2001). About 75% of land clearing occurs within 50 km of the roads (Alves 2002). There are basically two spatial patterns of deforestation. On the one hand, there are the rectangular-shaped large cattle ranches as contiguous or semi-contiguous clearings ranging from a few kilometres to 10 or 20 km along each side. On the other hand, there are the spatially complex patterns associated with the settlement projects known as the "fish backbone" pattern, that is, a patchy deforestation pattern of clear plots on secondary roads 4 or 5 km apart which exit from a main road. Except over south-western Amazonia (Rondônia), where there is an almost continuous area cleared (with dimensions of approximately 40 km across by 200 km along the paved highway BR 364), most of the remaining deforestation is widely scattered over the eastern, southern and south-western parts of the Basin in a region known as the "arc of deforestation". As a consequence of these deforestation patterns, the land cover is gradually becoming less homogeneous over small up to the mesoscale compared with the more homogeneous forest cover.

A number of field studies carried out over the last 15 years (the results of some of those field experiments are reviewed in Sect. A.6.3 below) showed *local* changes in the water, energy, carbon and nutrient cycling, and atmospheric composition caused by deforestation and biomass burning. Large-scale and regional changes in the water cycle have not been detected so far. Regional air surface temperatures have increased over Amazonia by about 0.4 °C during the 20th century, but this seems to be associated with global climate change and not to land-cover changes (Victoria et al. 1998). In addition to land-cover changes, it is becoming ever more important to consider also the changes in atmospheric composition brought about by biomass burning. Large concentrations of aerosols in the boundary layer during the biomass burning season (July through October in most of Amazonia south of the Equator) may change radiative properties of the lower troposphere due to absorption of solar radiation by soot aerosols. This, in turn, may even change the dynamics of the boundary layer and affect cloud microphysical processes as has been suggested recently by the Large Scale Biosphere-Atmosphere Experiment in Amazonia (LBA; Tarasova et al. 1999, 2000; Andreae et al. 2002; Silva Dias et al. 2002).

It is quite difficult to predict future land-use and land-cover change for Amazonia in the 21st century. On one hand, current development plans for the region call for opening up new and paving existing roads which may lead to continued and even increasing land clearing. On the other hand, pilot initiatives towards sustainable development, minimising deforestation and biomass burning, have been implemented successfully in the last few years over a number of areas in Amazonia. Some early positive results of such initiatives are already apparent. For instance, the number of AVHRR-derived "hot spots" for Amazonia have decreased by 35% from 1999 to 2000 in Mato Grosso (CPTEC 2002) – the part of Amazonia with the highest rates of biomass burning – most likely due to measures to prevent biomass burning that have been implemented in that area.

The average annual precipitation over the Amazon Basin ($\sim 7 \times 10^6$ km^2) is about 2.3 m, which makes that region one of the important heat sources for the tropical atmosphere. Basin rainfall is affected by seasonal to interannual climate variability associated with planetary-scale ocean-atmosphere interactions, e.g. ENSO. Basin-scale annual rainfall anomalies linked with SST anomalies in the tropical Pacific Ocean are of the order of 10 to 15% of the total rainfall, but can be as high as 40% for specific subregions within the Basin. That is the case for northern and eastern Amazonia, which experiences negative (positive) rainfall anomaly during El Niño (La Niña) events. Numerical simulation of total substitution of vegetation in Amazonia (pasture replacing forest), as will be discussed in detail in Sect. A.6.5 below, indicated a general trend for warmer temperatures (0.5 to 2.5 °C warmer) and drier conditions (5 to 15% reduction of Basin-wide precipitation). These reductions are of the same order of magnitude as those associated with remote lower boundary forcing (e.g. SST anomalies). However, the former will lead to unidirectional changes, that is, towards dryness, which could possibly reduce permanently the strength of the Amazonia tropospheric heat source. The potential effect of such permanent change in affecting

climate patterns has not, as yet, been investigated in detail, but could in principle affect the climate of distant regions due to the propagation of planetary-scale atmospheric waves (Rossby waves) emanating from the heated region.

The rainfall over Amazonia shows distinct geographical patterns. Precipitation maxima in excess of 3 m annually are found over western Amazonia, with a secondary elongated maximum towards the south-east over the South Atlantic Convergence Zone (SACZ). Another maximum in excess of 3 m is found along the Atlantic coast which owes its existence to rainfall systems linked to sea breeze circulation and westward propagating instability lines (Cohen et al. 1995). The reason for the maximum over western Amazonia is not clear, but may be related to the proximity to the Andes Cordillera and the concave shape of that mountain range (Salati and Vose 1984) possibly creating the conditions for low tropospheric moisture convergence (Figueroa and Nobre 1990). Modelling studies of the role of the Andes Cordillera on the climate of South America have shown that the main aspects of geographical distribution would be maintained in Amazonia in the absence of the mountain, but with some minor differences. The summer maximum of precipitation over southern Amazonia and its extension to the south-east associated with the SACZ would be displaced to the north but would still be present (Figueroa et al. 1995; Rocha 2001). In the absence of the Andes Cordillera, the largest effect would take place over subtropical South America, including parts of the Cerrado south of Amazonia. Those areas would be semi-arid because transient, synoptic-scale frontal systems would not extend as far north as they do with the Andes (Satyamurty at al. 1998). Straddled by these two rainfall maxima, there is a region of minimum precipitation as low as 1.6 m annually linking the regions of lower precipitation of northern South America to the Cerrado regions of Brazil, south-east of the Amazon forest. The annual cycle of rainfall shows a distinctive feature. Unlike Equatorial Africa, where the ITCZ-like precipitation band crosses the Equator twice a year in its seasonal march, in the Amazon the annual rainfall has only one peak a year. That has to do with the fact that, during Northern Hemisphere summer, the maximum precipitation is found in northern South America and Central America. However, during the Southern Hemisphere spring, September and October, the rainfall does not migrate slowly towards the Southern Hemisphere. Instead, the rainfall reappears rather abruptly in Central Brazil and southern Amazonia, associated with the seasonal establishment of the SACZ (Horel et al. 1989; Rao and Hada 1990; Liebmann et al. 1999). The transition from dry conditions over southern Amazonia during winter is interrupted when a frontal system in September or beginning of October breaks the winter conditions and organises convection along the SACZ. Although large-scale controls play a strong role in spatial and temporal variability of rainfall in Amazonia, the land-surface forcing is important. The strong diurnal cycle of precipitation over tropical South America, with maximum convection and precipitation in the afternoon and night hours (Garreaud and Wallace 1997) stresses the importance of this land-surface forcing.

A number of water balance studies for the Amazon (reviewed in Sect. A.6.3 below) indicate that for the Basin as a whole and for long-term averages, total evaporation is about 50% of total precipitation. There has been a suggestion that the forest is efficient in recycling water vapour. Isotopic-based studies carried out during the 1970s and 1980s (Salati and Vose 1984) across the Basin from the Atlantic coast to western Amazonia, indicated that water vapour is recycled several times before exiting the Basin to the south or to the north-west. However, other studies challenge that view and maintain that most of the water vapour used for precipitation in the region comes from the Atlantic Ocean, and that evapotranspiration plays a secondary role (Eltahir and Bras 1994). That would happen because precipitating systems during the rainy season are organised by synoptic- and sub-synoptic scale perturbations and the main moisture source is external to Amazonia. On the other hand, during the dry season, forest evapotranspiration continues to be high throughout the Basin, between 3 and 4 mm d^{-1}, but that water vapour mostly exits the Basin without taking part in the precipitation processes. There is little doubt that recycling of water vapour is important since total precipitable water increases westward from the Atlantic coast and reaches a maximum over western Amazonia, 2 000 to 3 000 km inland, as shown during the GTE-ABLE 2B Experiment (Nobre et al. 1991). It is likely that the forest evapotranspiration may play an important role for dry season rainfall. ABRACOS observations have shown that forest evapotranspiration is about twice as large as that of grass vegetation in pastures during the dry season (Gash and Nobre 1997). This is even more important when one considers that dry season rainfall originates from isolated deep convection or mesoscale convective complexes distant from oceanic moisture sources. A more complete understanding of the water cycle of Amazonia, including the role of the land surface in recycling water vapour, is still to be accomplished. The LBA Experiment is contributing towards that goal.

As mentioned above, the observed patterns of land-cover change over the last two decades in Amazonia have led to increase in the heterogeneity of the land cover, which might in fact lead to more vigorous mesoscale circulations if the scale of the heterogeneity is 10 km or larger. Observations of boundary layer growth over pasture and forest (Fisch et al. 1996) have shown higher mixed layer heights for pastures consistent with higher surface sensible heating fluxes (Culf et al. 1996), which results in more boundary layer clouds over pasture (Cutrim et al. 1995).

Carlos A. Nobre

A.6.2 Future Climates in Amazonia

In terms of potential future climate change due to human-caused global climate change, the IPCC (Intergovernmental Panel on Climate Change) sensitivity simulations for the family of emissions scenarios (IPCC 2000), ranging from a low emissions scenario (CO_2 concentration of 550 ppmv by 2100) to a high emissions ("business as usual") scenario (830 ppmv by 2100) indicate significant climate change for Amazonia. Given the emissions scenarios, Carter and Hulme (2000) used the results of ten global climate models to calculate the range of projected temperature and rainfall anomalies through to 2100. For Amazonia, projected temperature increases range from about 1 °C (low emission scenario) to about 6 °C (high emission scenario) by 2080. Uncertainties in the projections of precipitation changes are still very high. Some GCM simulations show decreases in precipitation whereas other GCM simulations show increases over most of Amazonia as also seen in the latest IPCC report (IPCC 2001). In terms of magnitudes of precipitation changes, these are projected as 0 to ±3% for the low emissions scenario and 0 to ±10% for the high emissions scenario. Such changes are of the same order as the Basin-wide rainfall changes associated with SST influences and of large-scale deforestation. However, the critical aspect for maintenance of the tropical forests is the duration of the rainy season. The uncertainty of projections on changes in the seasonal cycle of precipitation is even larger than that in the annual rainfall changes.

The uncertainty with respect to projected rainfall changes in the future makes it extremely difficult to "predict" the impact of climate change on Amazonian ecosystems. If there is a reduction in rainfall, this will be in addition to that expected to take place as a response to large-scale deforestation. The result will be a significant increase in the susceptibility of the ecosystems to fire and a reduction of species not tolerant to drought or fire. On the other hand, if precipitation increases as a result of global climate change, this might counteract the deforestation-induced rainfall reductions. The final result would be a smaller change in rainfall. With respect to temperature changes, global climate change would add to the temperature increase due to large-scale deforestation. As seen in Sect. A.6.3 below, observations show temperature increases between 1 and 2 °C due to replacement of forest by grass vegetation in pastures. This increase is of the same order as that projected to take place for the low emissions scenario but it is substantially smaller than the projected increase for the high emissions scenario of up to 6 °C warming over Amazonia. The temperature effect due to changes in the energy balance at the surface due to deforestation would be added to that of the GHG radiative forcing and the result would be a large increase in the risk of forest fires since the drying-out of vegetation during the dry season and its flammability are higher with increasing temperatures (Nepstad et al. 1994).

Based on a global climate change simulation of the Hadley Centre GCM, which explicitly represents the terrestrial biota carbon cycle and particularly showed a scenario of significant precipitation reductions in Amazonia, Cox et al. (2000) speculated about the possibility of an Amazonian forest dieback due to severe drought in parts of the Basin. In that calculation, the forest dieback would release a large amount of carbon stored in the biomass into the atmosphere over a period of one to two decades. The CO_2 thus released would act as a positive feedback for the GHG radiative forcing. Although the positive feedback between forest dieback and CO_2 radiative forcing may be of scientific interest, the uncertainty in estimates of regional rainfall changes for Amazonia due to global climate change makes it difficult to establish the likelihood of such a natural "disaster" occuring in the near or distant future.

In summary, these two effects on temperature would be a positive feedback on the susceptibility of Amazonia ecosystems due to global climate change and to regional deforestation. However, the uncertainty on how precipitation will change in Amazonia due to global climate change makes it impossible to determine whether the final climatic feedback will be positive or negative. It is hoped that this uncertainty in estimating likely regional climate changes for Amazonia will be improved substantially during this decade through the use of more complex climate models which take into account explicitly the regional climate and the two-way interaction between the biosphere and the atmosphere.

Alistair D. Culf · John H. C. Gash
Maria Assuncão Silva Dias

A.6.3   Observations of Land-surface/Atmosphere Interactions

The forest of the Amazon Basin has long been recognised as an important component of the global climate system and is often referred to in the popular press as the "lungs of the world". Until recently, however, the actual interactions between this huge tropical forest and the atmosphere were poorly understood, largely because of the inaccessibility of the region and the consequential difficulty of making measurements there. With the increasing power of computers throughout the 1980s, it became possible to run sophisticated climate models and make predictions of the effect of Amazonian deforestation on the climate. The need to improve the physics of such models and to check their performance against reality has been the driving force behind a succession of land-atmosphere interaction studies in the region during the last 20 years.

ARME

The first micrometeorological measurements in Amazonia were made during the 1980s in the Amazon Regional Micrometeorology Experiment (ARME) (Shuttleworth 1988a). ARME, which was a collaboration between Brazilian and British scientists, made routine measurements of the near surface climate, the proportion of rainfall intercepted and subsequently re-evaporated from the forest canopy, and the soil moisture status over a 25-month period at Reserva Ducke, close to Manaus in central Amazonia. Additional measurements of the surface energy balance, including radiation, and sensible and latent heat flux (Shuttleworth et al. 1984b; Shuttleworth 1988a; Moore and Fisch 1986), vertical gradients of temperature, humidity and windspeed (Shuttleworth et al. 1984b; Viswanadham et al. 1990), and plant physiological measurements (Roberts et al. 1990, 1993) were made during four intensive field campaigns.

The albedo of the Reserva Ducke forest was found to be around 12% (Shuttleworth et al. 1984b), similar to the limited results obtained from other tropical forests, while the loss of incoming rainfall due to interception by the forest canopy was found to be less than anticipated, at 10–15% (Lloyd and Marques 1988; Lloyd et al. 1988). It was shown that there was extreme spatial variability in throughfall and frequent random relocation of collection gauges beneath the forest canopy is a necessity if accurate results are to be obtained. The ARME data were used to construct and calibrate a one-dimensional micrometeorological model of the forest-atmosphere interaction for the Reserva Ducke site. The model was then applied to the climate and rainfall record to predict the evaporation over the entire study period of 25 months (Shuttleworth 1988a). This analysis predicted a fairly uniform evaporation rate throughout the year of about 110 mm month^{-1}, accounting for half of the incoming precipitation.

As the first micrometeorological dataset from Amazonian rainforest, the ARME data were used extensively to calibrate various models of surface conductance (Dolman et al. 1991) and comprehensive GCM land-surface sub-models of the rainforest-atmosphere interaction (e.g. Sellers et al. 1989) The data were also useful in checking the ability of GCMs to predict area average surface exchange. These validation exercises (e.g. Shuttleworth and Dickinson 1989) revealed many areas in the models that required improvement, in particular the description of convective cloud and rainfall interception.

ABLE-2

Whilst ARME concentrated on the surface, the Amazon Boundary Layer Experiment (ABLE-2) focused on atmospheric transport. The experiment was designed to study the rate of exchange of material between the Earth's surface and the atmospheric boundary layer, and the processes by which gases and aerosols are moved between the boundary layer and the "free" troposphere. ABLE-2 was a collaboration between US and Brazilian scientists consisting of two phases: the first in the Amazonian dry season (ABLE-2A, July to August 1985); and the second in the wet season (ABLE-2B, April to May 1987). The core research data were gathered from a base in Manaus by NASA Electra aircraft flights that extended from Belém, at the mouth of the River Amazon, west to Tabatinga, on the Brazil-Colombia border. These observations were supplemented by ground-based chemical and meteorological measurements in the dry forest, the Amazon floodplain, and the tributary rivers through the use of enclosures, the ARME tower, a large tethered balloon, and weather and ozone sondes. The air above the Amazon forest was found to be extremely clean during the wet season but deteriorated dramatically during the dry season as the result of biomass burning. Amazonian ozone deposition rates were found to be 5 to 50 times higher than those previously measured over pine forests and water surfaces. The measurements from the ARME tower in Reserva Ducke included the first eddy correlation measurements of CO_2 flux from tropical forest (Fan et al. 1990). A summary of the results from ABLE-2 is given by Garstang et al. (1990), whilst detailed results can be found in a special issue of the Journal of Geophysical Research (February 1988).

ABRACOS

At the end of the 1980s there was growing international concern about the possible effects of Amazonian deforestation on climate, but although the models' representation of tropical forest was now relatively well parameterised, there was little knowledge about the properties of the landscape which followed forest clearance. The Anglo-Brazilian Amazonian Climate Observation Study (ABRACOS) was set up to address this issue and also to extend the spatial and temporal scale of forest measurements which had previously been concentrated in central Amazonia for periods of, at most, a few months.

In ABRACOS, measurements were concentrated in three locations, Manaus, Rondônia, in south-west Amazonia, and Marabá in the east (see Gash et al. 1996). At each of these locations, paired forest-pasture sites were set up and climate and soil moisture monitored continuously for several years. A number of short intensive measurement campaigns involving micrometeorology (e.g. Wright et al. 1992, 1995), plant physiology (e.g. McWilliam et al. 1993) and detailed studies of the soil moisture changes (e.g. Tomasella and Hodnett 1997) were also conducted at these sites between 1990 and 1994. Initially, the

intensive measurements were concentrated at Fazenda Dimona, close to Manaus, but in 1992 the focus was changed to Rondônia on the south-west periphery of the Amazon Basin, an area with a more seasonal rainfall climatology and substantial areas of deforestation.

The value for tropical forest albedo determined from the ARME data was confirmed by the ABRACOS data, with a range of 0.11 to 0.13 being observed. However, the long time series of data now available showed that at all the forest sites there was a seasonal trend in albedo which was correlated with soil moisture. The albedo values of the other forest sites were found to be slightly higher than the Manaus site, with an average for the three forest sites of 0.13. On average, the albedo of the pasture sites was found to be about 0.18 (Culf et al. 1995), ranging from 0.16 at low leaf area index to 0.2 at high leaf area index (Wright et al. 1996a). The radiation balance is also affected by the surface temperature, which is much higher at the pasture sites than the forest, leading to higher values for long-wave emission. The combined effect of increased albedo and increased long-wave emission at the pasture sites is to reduce the net radiation received by the pasture sites by an average of 11% compared to the forests.

As with the ARME data, micrometeorological studies showed a strong diurnal decline in surface conductance at the forest sites (Shuttleworth 1988a; Wright et al. 1996b), with a strong correlation between canopy surface conductance and air humidity deficit. This pattern was also found in the physiological behaviour of foliage in the upper half of the canopy but less so in the lower half of the canopy (Roberts et al. 1990; McWilliam et al. 1996; Sá et al. 1996). The evaporation measurements showed that, in the dry season, the shallower rooting depth of the pasture leads to a reduction in the evaporation in response to the development of a soil moisture deficit. In contrast, the deeper rooting depth of the forest allows the evaporation to continue unabated throughout the dry season, with no reduction in the evaporation being observed. These results are confirmed by the observations of Nepstad et al., (1994) who showed that evergreen forests in north-eastern Amazonia maintained evaporation rates during five-month dry periods by absorbing water from the soil at depths of more than 8 m. Hodnett et al. (1996) used a simple daily soil water balance model, calibrated with ABRACOS measurements and long-term rainfall records to show that a deep rooting strategy is essential if trees on the low water availability soils in the Manaus area are to survive the dry years which occur relatively regularly. The model indicated that uptake from below 2 m occurred in all years of the 27-year rainfall record, with an average dry season uptake from below 2 m of 72 mm. There is little evidence from either micrometeorological studies of surface conductance (Shuttleworth 1988a) or physiological studies (Roberts et al. 1990) that reductions in soil moisture cause a lowering of conductances. The main results of ABRACOS are summarised by Nobre et al. (1996b), and Gash and Nobre (1997) and presented in detail in a book, Amazonian Deforestation and Climate (Gash et al. 1996).

Rondônian Boundary Layer Experiment (RBLE)

The site in Rondônia was deliberately chosen because there is a large area of forest juxtaposed with an area of partially deforested land large enough to develop its own characteristic atmospheric boundary layer. To test this hypothesis, the ongoing surface measurements being made under ABRACOS were enhanced between 1992 and 1994 by a series of three atmospheric boundary layer sounding campaigns known as the Rondônian Boundary Layer Experiment (RBLE) carried out during the dry season (Nobre et al. 1996a). The first RBLE concentrated on the forest site, while the second made measurements at both forest and pasture sites, although not simultaneously. In the third and final phase, simultaneous radiosonde and tethersonde measurements were made at both sites several times a day for a two-week period. The results from RBLE showed that the convective boundary layer grew to different depths over the two surface types during the day in response to the different partitioning of the net radiation at the surface. A further feedback was noted in that the enhanced boundary layer growth over the pasture in the dry season seemed to result in the formation of cumulus clouds earlier in the day over the pasture than over the forest, systematically reducing the incoming solar radiation (Culf et al. 1996). This result was confirmed by Cutrim et al. (1995) who developed an algorithm for determining the existence of shallow cumulus clouds from satellite imagery. Satellite images analysed using this technique showed the corridor of deforestation, which follows the main road through Rondônia, clearly marked out by shallow cumulus clouds.

Fisch et al. (1996) attempted to model the Rondônian convective boundary layer by using a slab model approach. They were able to simulate the observations over the forest but could not reproduce the observed rapid growth over the deforested area. They suggested that the juxtaposition of small patches of forest and pasture might lead to enhanced turbulence in the early morning, causing a faster breakdown of the nocturnal inversion than would occur over a uniform surface. The RBLE data were also used by Dolman et al. (1999), in a mesoscale modelling study designed to highlight deficiencies in the understanding of land-surface/atmosphere interactions in the run up to the Large Scale Biosphere-Atmosphere Experiment in Amazonia (LBA). Mesoscale models were run in both one and three-dimensional modes for the Rondônia area. The study showed that the current models were unable to simulate accurately the atmospheric boundary layer over the partially deforested areas. The models' underestimate of temperature was put down to the neglect in the models of the radiative effects of aerosols, and a poor understanding of the entrainment of air into the boundary layer from above. The work illustrated the need for further measurements in these areas. Silva Dias and Regnier (1996) showed, through mesoscale model simulations, that local circulations developed between pasture and surrounding forest in Rondônia during the dry season with enhanced low level convergence during the day-time which would help shallow cumulus to develop in spite of a deep mixed layer. Souza et al. (2000) show that not only the difference in vegetation which induces different mixed layer temperatures and depths, is important in establishing the local circulations but also that even a small elevation difference, such as that between the RBLE pasture and forest sites, enhances their intensity.

Carbon Dioxide Flux Measurements

During the 1990s, the climate change debate made a subtle shift away from the direct influence of Amazonian deforestation on global climate through changes in albedo, aerodynamic roughness and other surface properties and began to concentrate on the effects of greenhouse gases and carbon dioxide in particular. In this debate, the role of the Amazonian forest is still critical and yet few measurements of carbon fluxes over this important biome existed at that time (the short run of measurements made by Fan et al. (1990) during the ABLE-2 campaign being the notable exception). In 1992, the scope of ABRACOS was extended further by the addition of carbon dioxide flux measurements at the Rondônia sites. The first short campaign (Grace et al. 1995) showed the forest to be a strong sink for CO_2, a surprising result at that time when many thought the forest to be in equilibrium. Further CO_2 flux measurements were made at the site in succeeding years, confirming the initial result. In 1995 a flux measurement site was established close to Manaus in central Amazonia and run for a year (Malhi et al. 1998). This site was found to be an even stronger sink of CO_2. The measurements were in agreement with biomass measurements made at nearby forest plots (Phillips et al. 1998), but a short campaign to measure profiles of CO_2 in the nocturnal boundary layer above the forest (Culf et al. 1999), suggested that the nocturnal, respirative flux may have been underestimated by the eddy correlation system, leading to an overestimate of the sink strength.

LBA – The Large Scale Atmosphere-Biosphere Experiment in Amazonia

The LBA (LBA science planning group, 1996) is a Brazilian-led, international research initiative designed to create the new knowledge needed to understand the climatological, ecological, biogeochemical, and hydrological functioning of Amazonia, the impact of land-use change on these functions, and the interactions between Amazonia and the Earth system.

In the last decade, flux measurement technology has developed to a point where it is now feasible to monitor fluxes continuously for periods of several years. This allows studies of interannual variability and the detailed response of the landscape to extreme climatic events such as El Niño and La Niña. The LBA measurement strategy is to use this new-found ability to make long-term measurements in Rondônia, Manaus, Santarém and Caxiuanã. The initial results from these sites again point to large uptakes of CO_2 by the forest ecosystem, although there is still much debate about the size of the errors inherent in the measurement technique and the problems of making accurate measurements by eddy correlation at night (see Chapt. B.4).

Wet AMC

In early 1999, the pioneering work of ABRACOS and RBLE in linking observed changes in the surface energy balance to changes in the atmospheric boundary layer was significantly extended by the Wet Atmospheric Mesoscale Campaign (WETAMC/LBA) (Silva Dias et al. 2002), probably the most ambitious campaign of land-surface/atmosphere interaction studies carried out in the tropics to date. The WETAMC/LBA focused on both the local effects of deforestation and on the regional response to the larger scale forcing. The campaign was a joint venture between Brazilian and European scientists who joined forces with scientists from NASA's Tropical Rainfall Measurement Mission (TRMM) (Simpson et al. 1996). A major ground validation programme within TRMM was co-located with the WETAMC/LBA, bringing the benefits of additional observational capacity and increased spatial coverage to both studies. The field phase took place in Rondônia during January and February 1999. Some 250 scientists participated in the WETAMC/LBA and operational instrument systems included: four high frequency radiosonde launch sites, four tethered balloons, two forest and four pasture flux measurement towers, a network of automatic weather stations, a dense raingauge network, two doppler radars and two instrumented aircraft.

Early analysis of the wealth of data generated during the experiment has produced some interesting examples of the effect of surface features on convective processes. Although on average, during this wet season campaign, there was little difference in mixed layer heights over forest and pasture, for individual cases there are important differences. The radar facility enabled the location of cloud formation to be mapped onto the un-

derlying vegetation and it was found that the formation of cumulus clouds occurred earlier over forest areas than over pasture (the reverse of the case in the dry season). Rain showers then occurred over the forest, inhibiting further growth of the mixed layer. It therefore appears that the timing of convection initiation during the Amazonian wet season may be sensitive to changes in the land surface. The possible effect of this timing on the further development of deep convection is currently under investigation. The reproduction of the proper timing of convection by mesoscale numerical models is a major challenge for the future.

Data Availability

The data obtained during ARME, ABLE-2, ABRACOS and RBLE have been brought together and published on CD-ROM as part of the Pre-LBA Initiative. In addition to the data from studies described here, the CD-ROM contains satellite imagery, soil survey data and hydrological measurements. Information about the Pre-LBA CD-ROM can be obtained from LBA Central Office at CPTEC/INPE (Centro de Previsão de Tempo e Estudos Climáticos/ Instituto Nacional de Pesquisas Espaciais), Cachoeira Paulista, São Paulo, Brazil (E-mail: *lba@cptec.inpe.br*).

José A. Marengo

A.6.4 The General Characteristics and Variability of Water and Energy Balances in the Amazon Basin

One of the goals of the LBA experiment is to evaluate the land-surface and atmospheric parameterisations for regional numerical weather prediction models and their associated four-dimensional data assimilation systems. Surface water and energy budgets are parameterised in models, and a host of new parameters that results from physical parameterisations has become available as well. Soil moisture, evaporation, net radiation are some examples. A first look at the main components of these budgets from point measurements has indicated differences with model results.

A.6.4.1 Introduction

The present document is an up-to-date review of issues regarding the water and energy balances of the Amazon Basin, using results from the early isolated field observations from the 1970s through to the recent LBA measurements, including the recently available global re-analysis (Roads et al. 2002; Marengo 2004; Costa and Foley 1998; Curtis and Hastenrath 1999). For a careful review, the reader is referred to Marengo and Nobre (2001), Fisch et al. (1998), and Marengo (2004) for studies on the water balance of Amazonia. Results from recent regional energy budget studies that used instrumental, satellite and model data from the recent LBA-WET AMC (Wet Atmospheric Mesoscale Campaign) and LBA/TRMM field campaigns from January to April 1999, can be found in Silva Dias et al. (1999, 2002) and Gu et al. (2004).

A.6.4.2 Water Balance

The water balance of the Amazon Basin is of great importance because it is the world's largest hydrographic system. Hence, there is a concern that large-scale land-use changes may change significantly the flow regimes of rivers within the region and the land-atmosphere exchange of moisture.

The lack of long-term and continuous sounding observations of the atmosphere, precipitation and evaporation data and of measurements of river discharge along the Amazon and its main tributaries has forced many scientists to use indirect methods for determining the water balance for the region. Since the early 1980s several studies have been devoted to the water balance of the Amazon Basin, using few observations from the sounding network in Brazilian Amazonia, rainfall and river data from some individual stations, or from regional experiments carried out in the Basin over the last 20 years (see reviews in Matsuyama 1992; Oki et al. 1995; Chu 1982; Villa Nova et al. 1976; Marques et al. 1980a; Milliman and Meade 1983; Salati and Marques 1984; Vörösmarty et al. 1989, 1996; Marengo 2004).

Early studies by Salati and Marques (1984) have attempted to quantify the components of the water balance: Average precipitation was estimated as 11.9×10^{12} m^3 y^{-1} (Villa Nova et al. 1976); discharge of the Amazon at Obidos was quantified as 5.5×10^{12} m^3 y^{-1}, from Oltman (1967); and evapotranspiration was estimated by the Penman method as 6.4×10^{12} m^3 y^{-1}. The latter is in good agreement with the difference between precipitation and river discharge. The Amazon rivers drains an area of approximately 5.8×10^6 km^2, with an average discharge of 5.5×10^{12} m^3 y^{-1}, as indicated above (compare Chapt. D.5). Most of these estimates are based on the records of the Amazon in Obidos (1° 55' S, 55° 28' W).

For the Amazon Basin, meteorological studies of the water balance have ranged from large-scale atmospheric water budgets based on the European Centre for Medium-Range Weather Forecasts re-analysis or the FGGE gridded data (Matsuyama 1992; Oki et al. 1995; Rao et al. 1996) and previously, based on climatic observations of precipitation, a few radiosonde stations, and river discharges (Marques et al. 1980a,b; Chu 1982; Marengo et al. 1994). Rao et al. (1996) used the European Centre for Medium-Range Weather Forecasts data from 1985–89 and their results suggested that: (a) the Amazon Basin

is the principal source of moisture for Central Brazil for the period September–February, and (b) the water vapour flux from the equatorial Atlantic associated with trade winds is the main moisture source for the Amazon Basin. However, the interannual variations of these fluxes and their associations with El Niño-La Niña have not been studied yet.

Table A.5 shows the estimates of the annual water balance across scales in the Basin, from point measurements to Basin-wide hydrological budgets – in other words, from a few point radiosonde, rainfall and streamflow observations in Amazonia to the Basin-wide rainfall data available from CMAP (Xie and Arkin 1997, 1998; Legates and Willmott 1990), and the NCEP/NCAR (National Centers for Environmental Prediction/National Center for Atmospheric Research), ECMWF, and NASA/DAO reanalyses. Most of them show that the rate between evapotranspiration and precipitation varies from 40 to 75%.

The availability of world-wide gridded re-analysis (NCEP, ECMWF, NASA-DAO) have allowed assessments of the water vapour transport into Amazonia, and in fact some recent works have shown contradictory results in terms of trends in input moisture into the Amazon Basin.

Preliminary studies using climate modelling at CPTEC/INPE show some evaluations of continental-hydrology data using observed and GCM generated stream-flow at the main channel of the Amazon. The model runoff is produced by the Simplified SiB scheme incorporated in the formulation of the CPTEC/COLA

Table A.5. Annual water balance of the Amazon Basin as deduced from previous studies (P = precipitation, ET = evapotranspiration, R = runoff, in mm y^{-1}) (Marengo and Nobre 2001)

Study	P	ET	R
Baumgartner and Reichel (1975)	2 170	1 185	985
Villa Nova et al. (1976)	2 005	1 080	925
Marques et al. (1979)	2 083	1 000	1 083
Marques et al. (1977)	2 328	1 261	1 067
Jordan and Heuveldop (1981)	3 684	1 985	1 759
Leopoldo et al. (1982)	2 076	1 676	400
Franken and Leopoldo (1984)	2 510	1 641	869
Shuttleworth (1988a,b)	2 636	1 329	1 317
Vörösmarty et al. (1989)	2 260	1 250	1 010
Russell and Miller (1990)	2 010	1 620	380
Nizhizawa and Koike (1992)	2 300	1 451	849
Matsuyama (1992)	2 153	1 139	849
Marengo et al. (1994)	2 888	1 616	1 272
Vörösmarty et al. (1996)	2 301	1 221	1 080
Costa and Foley (1998)	2 160	1 360	1 106
Oki (1999)	2 076	1 023	1 053
Zeng (1999)	2 044	1 879	365
Marengo et al. (2000)	2 146	1 581	93

(Center for Ocean-Land-Atmosphere) Atmospheric General Circulation Model. The observed and modelled streamflow are integrated for the whole area of the Amazon Basin, and then assessed at its mouth near the Obidos gauge station. According to ANEEL scientists, the discharge at Obidos is 6.7% greater than the Amazon River at its mouth. Climatological monthly river runoff at the mouth of the Amazon is obtained at each grid point of the model river basin and then integrated; the model estimates have then been compared with the observations. The discussion here includes the rainfall evaluation from previous chapters.

For the Amazon Basin, the ensemble mean appears to mimic the phase and the interannual variability of the observed streamflow quite well. The model shows a build-up in the discharge during late summer to autumn. However, the timing of the modelled peak discharge tends to occur earlier than that which is observed at Obidos. There is a lag of 1–2 months between the seasonal peaks of the observed and modelled flows and, in general, the model underestimates the streamflow, with an observed mean average of 3.2 mm d^{-1} against a mean of 1.2 mm d^{-1}. This is in accordance with the underestimation of rainfall by the CPTEC/COLA model, especially in northern and central Amazonia. However, the model does reproduce the observed low discharges during 1983 and the largest values during 1982, such as the abundant late 1982 pre-El Niño 1983 rainfall (Marengo et al. 1998), and the rainfall in the 1988/89 La Niña event.

Comparison of the several models of the Atmospheric Model Intercomparison Project (AMIP) experiment by Lau et al. (1996) indicates that for the Amazon Basin several models show zero or very little discharge during autumn, while the observations show a background of 2.1 mm d^{-1}, due in part to the inadequate parameterisation of underground water storage in the land-surface schemes. Our estimates here point at model discharges that are approximately 61% below the measurements made at Obidos for the period 1982–91, and the autumn low season discharges reach between 0.5 and 0.8 mm d^{-1}, as compared to the observed 2.5 mm d^{-1}.

A.6.4.3 Energy Balance

As part of the LBA effort to improve the parameterisation of the land-surface processes in short and long-range forecasts, long-term continuous direct measurements of the components of the surface energy balances were initiated in 1990 at three locations in Amazonia, during the ABRACOS Experiment. Since 1998, the flux measurement systems part of the LBA research projects from Brazil, the US and Europe, began to be deployed within the Brazilian Amazon Basin, and later on in the Venezuelan Llanos. The flux towers were deployed in forested and agricultural areas, as well as in some swamps. These sites exhibit typical land-surface characteristics for different regions of Amazonia and will provide detailed information on the seasonal and annual cycle of the local surface energy budget. It is especially important for LBA to obtain the warm season energy budget, which is affected by both land-surface and hydrological processes.

Despite the great importance of the Amazon Basin to the regional and global climate, measurements of the energy balance terms in the Basin are very sparse, and the few observations available are mostly at local scale, made during field experiments such ARME, ABLE-2B, and FLUAMAZON (Machado 2000), ABRACOS (Culf et al. 1996), and the recent LBA-WET AMC campaign (Silva Dias et al. 1999, 2000, 2002). These few measurements are basically photosynthetically active radiation (PAR), net radiation measured above and inside the forest canopy, and terrestrial radiation above the canopy. Continued observations of the energy budget terms have been collected since the 1980's at the forest Duke Reserve site near Manaus (Shuttle-worth 1988a).

Previous studies using the ABRACOS data (Culf et al. 1996) have shown long-term measurements of radiation components at the surface, showing that on average the effect of replacing forest by pasture in Amazonia is to reduce the net radiation at the surface by about 11% as a result of both changes in albedo and in the long-wave radiation budget. Wright et al. (1992) found that for the Manaus site the ratio of evaporation to net radiation was similar at both the forest and pasture sites during the wet season. Evaporation differences between sites (Marabá and Ji-Paraná, in eastern and western Amazonia respectively) were attributable to differences in net radiation.

Results from ABLE-2B and FLUAMAZON for a few locations in Amazonia (Machado 2000) documented that for the Amazon Basin during those two experiments (April–May 1987 and November–December 1989, for both ABLE-2B and FLUAMAZON, respectively), the surface energy seems to be in equilibrium. However, the quantity of energy stored at the surface seems to be limited, defining a time scale at which the surface and the atmosphere need to export or receive energy in order to control their deficit or gain of energy.

During the 1999 LBA-WET AMC campaign (Silva Dias et al. 2002), measurements of radiation balance and sensible and latent heat fluxes were taken at a micrometeorological tower at the forest site near Ji-Paraná. From these measurements, the penetration of radiation inside the canopy indicate that less than 40% of the PAR and incoming solar radiation above the forest reaches the 30 m level and only 8% of the PAR and 14% of the solar radiation reaches the 15-metre level. Studies by Gu et al. (1999b) found that solar radiation at the ground surface on a grassland site in Rondônia during January–February 1999 could exceed the solar constant under cloudy conditions. The biophysical consequences of the surface radiation enhancement by clouds depend on the pace of biological responses, which are largely controlled by the responses of the stomata to changes in light. Further studies are needed in order to understand the impacts of the bimodal roles of clouds on mass and energy exchanges between terrestrial ecosystems and the atmosphere.

The short and medium range forecast models, which simply parameterise the surface energy balance components as seasonally dependent fractions of the available energy, would be challenged to reproduce the surface energy fluxes thus observed.

On the Basin-wide scale, studies using the global re-analysis for estimating regional aspects of the energy balance are lacking. Few studies using the ISCCP (International Satellite Cloud Climatology Project) cloud products have been devoted to studying the diurnal cycle and interannual variations of deep convection in South America (Marengo 1995; Garreaud and Wallace 1997).

Jan Polcher · Roni Avissar · José A. Marengo

A.6.5 Deforestation and Climate

Tropical Deforestation Experiments

Tropical forests in South-America, Africa and Southeast Asia are being deforested to make land available for agriculture. The destruction of one of the richest ecosystems on Earth not only reduces the biodiversity but through the changes of the physical characteristics of the surface, could also impact on climate. Since complex land-surface schemes are coupled to GCMs, one can attempt to model in a realistic manner the changes at the surface and study the impact on climate. The first deforestation experiment was conducted by Dickinson and Henderson-Sellers in 1988. Since then the experiments have been repeated with nearly all GCMs coupled to complex land-surface schemes. Some of the recent experiments are presented in Table A.6.

Sensitivity of Convection to Land-surface Changes

Two major field campaigns have been carried out recently in the Amazon Basin in order to understand the surface processes in the untouched forest and how they differ from those over a re-grown pasture. Their results are presented in Sect. A.6.3 and it suffices here to recall that the main impacts of deforestation on surface conditions have been quantified as follows:

- *Albedo* will increase upon deforestation from about 0.12 to about 0.18.
- *Evaporation characteristics* will be altered and result in a reduced net evaporation from the surface.
- *Surface roughness* will be reduced from about 2 m to 0.02 m.

Most deforestation experiments used the data from the ARME campaign to choose parameters for the tropical forest and validate the model over this region. Parameter values for pasture were deduced from other observations. A number of experiments used parameter modifications from previous studies to enable comparisons. Currently, the ABRACOS dataset has only been used to set up and validate one deforestation experiment (Lean and Rowntree 1997). In all the cases presented in Table A.6 the entire Amazon Basin was deforested. In three of these experiments the deforestation was also performed over Africa and South-east Asia.

The complexity of the surface change renders the analysis of the modifications of the surface fluxes difficult. Indeed, the shift in the balance of fluxes cannot be attributed to a single cause. Furthermore, the combination of the changes in the radiation, turbulent fluxes and evaporative fraction can compensate or amplify each other. Nevertheless, a few observed modifications should be reproduced in GCM experiments. The albedo increase should reduce the net radiative flux at the surface, as observed by Culf et al. (1996). This impact can however be

Table A.6. Some of the recent tropical deforestation experiments

Reference	Model	Resolution	Surface scheme	Length (y)
Nobre et al. (1991)	NMC	R40	SiB	1
Dickinson and Kennedy (1992)	COMI	R15	BATS	3
Lean and Rowntree (1993)	UKMO	2.5° × 3.75°	Warrilow et al. (1986)	3
Polcher and Laval (1994a)	LMD-3	2.0° × 5.6°	SECHIBA	1
Polcher and Laval (1994b)	LMD-6	2.0° × 5.6°	SECHIBA	11
Manzi and Planton (1996)	Emeraude	T42	ISBA	3
Zang et al. (1996)	CCM1	R15	BATS	11
Lean and Rowntree (1997)	UKMO	2.5° × 3.75°	Revised Warrilow et al. (1986)	10
Hahmann and Dickinson (1997)	CCM2	T42	BATS	8

amplified by an increase in cloud cover as indicated by two studies (Chu et al. 1994; Cutrim et al. 1995). The reduction in net radiation can also be enhanced by a larger long-wave radiation caused by a warmer surface. Evaporation will be reduced by the combination of two effects: first, a reduction in interception loss resulting from the smaller canopy capacity and aerodynamic roughness of pasture and second, the annual cycle of transpiration for both biomes will be different as pasture has a shallower root distribution than forest. The first effect will be dominant during the rainy season while the second one will prevail as the dry season sets in and the soil dries out. It is believed that the impact on transpiration will be larger than that on interception (Wright et al. 1992).

Annually averaged results from a few recent deforestation experiments are given in Table A.7. All models display a reduction in net surface radiation (ΔR_n), except for one where no significant change was found, caused by the increase in albedo imposed in the deforestation scenario and partly compensated by increase in temperature and thus long-wave emission. An interesting fact is that in all GCMs, except in the one which had fixed clouds (Nobre et al. 1991), cloud cover is reduced and counteracts the higher albedo in the deforestation scenario. In Manzi and Planton (1996), the albedo increase was cancelled by the combination of long-wave radiation and cloud feedbacks. A reduction in evaporation (ΔE) is a common feature in all experiments, although a wide range of values is covered. The annual evolution of this change varies from a maximum during the dry season (e.g. Lean and Rowntree 1997), to nearly constant reduction throughout the year (e.g. Zang et al. 1996). This result depends on the ability of the land-surface scheme to represent properly the contrast in soil moisture control on transpiration for pasture and forest and the difference in interception loss for both biomes. Surface temperature (ΔT) is warmer in the deforestation experiment except in two cases where no significant changes could be detected in the annual mean. As surface temperature is the result of the balance of fluxes and its variation is a complex combination of the modification experienced by the fluxes, it is not the variable best suited to evaluate

the impact of deforestation. When comparing the first and penultimate experiments presented in Table A.7, it appears that their temperature changes are comparable but they differ in all other aspects and especially in their impact on the hydrological cycle.

The most striking contradictions between these deforestation experiments are certainly the changes in moisture convergence ($\Delta(P - E)$). If for a moment, one is to follow the reasoning exposed by Mintz in his review paper (Mintz 1984) on the impact of evaporation reductions over extratropical continents, one would expect a lower moisture convergence. It was an unexpected result when for the first time a deforestation experiment (Polcher and Laval 1994a) displayed an increase in moisture convergence. However, if one considers for instance the results presented by Shukla and Mintz (1982) or Milly and Dunne (1994) and the fact that the fraction of recycled water in the Amazon Basin is estimated to be between 50% (Salati 1987) and 25% (Eltahir and Bras 1994), it appears that this outcome is possible. This result demonstrates that the sensitivity to surface processes is different in the tropics and extra-tropics.

The divergences of the model's results at the surface may have different causes. Three main categories can be distinguished.

- The imposed changes in surface parameters are different in all models. This is due to the fact that values are obtained from different sources but also because, by design, the land-surface schemes need different parameters. For instance, a scheme with an explicit representation of the sub-grid variability at the surface will only change albedo on the tropical forest tile which is being deforested, not over the entire gridbox. Most schemes also have parameters which cannot be measured directly or for which no observations are available and thus their modification has to be estimated.
- The GCMs used are different and it is not known if they possess the same sensitivities to surface processes. For instance, cloud schemes used in these models vary from fixed clouds to highly complex prog-

Table A.7.
The impact of deforestation on the surface fluxes and the hydrological cycle for some of the recent experiments. Results are annual means averaged over an area within the Amazon Basin. Changes in the following variables are presented: R_n = net surface radiation, ET = evapotranspiration, T = surface temperature, and $P - E$ = moisture convergence

Reference	ΔR_n [W m^{-2}]	ΔET [mm y^{-1}]	ΔT [K]	$\Delta(P - E)$	
Nobre et al. (1991)	−26.0	−500.0	+2.0	Decrease	↘
Dickinson and Kennedy (1992)	−11.0	−25.5	+0.6	Decrease	↘
Lean and Rowntree (1993)	−18.5	−198.0	+2.1	Decrease	↘
Polcher and Laval (1994a)	−12.0	−985.0	+3.8	Increase	↗
Polcher and Laval (1994b)	−14.2	−127.8	+0.1	Decrease	↘
Manzi and Planton (1996)	+1.0	−113.0	+1.3	Increase	↗
Zang et al. (1996)	−16.3	−402.0	+0.3	Decrease	↘
Lean and Rowntree (1997)	−12.7	−157.0	+2.3	Increase	↗
Hahmann and Dickinson (1997)	−10.0	−149.0	+1.0	Decrease	↘

nostic ones. Thus the compensation in the net short-wave flux of the albedo increase can result from different processes. Manzi and Planton (1996) have shown that when the convection scheme is changed in the Météo-France GCM the sensitivity is altered. The same may also be true for the land-surface schemes.

- The climates simulated by the GCMs are different and therefore have different sensitivities. For instance the length of the rainy season will vary from one GCM to another, and thus the reduction in interception loss will be more or less important.

From the experiments carried out so far it is not clear what the impact will be of tropical deforestation on the atmospheric branch of the hydrological cycle. From the academic land-surface change experiments presented in Sect. A.4.1 no dominant process can be identified. The lower net surface radiation should by itself tend to reduce moisture convergence; on the other hand the lower roughness of pasture should favour the converging low level fluxes in the region. Finally, the atmosphere has the potential to compensate for the reduced evaporation with an increase in moisture convergence. At this point it is not clear how these processes combine when the three surface properties are changed simultaneously. Lean and Rowntree (1997) put forward the hypothesis that the linear combination of the roughness length and albedo changes explains the largest fraction of the atmospheric impact. However, no experiment is presented for the sensitivity of the atmosphere to evaporation changes only to prove that a simple linear approximation is valid.

Surface processes and tropical climate are characterised by short time scales and the forcing to the atmosphere is local, thus it is difficult to analyse their interactions with space-time averaged data. One may view the evolution of the tropical convergence zone as a rapid alternation between convective events, in which energy from the surface is carried upward and transformed, and situations of subsidence where the surface input is dominated by divergence in the lower layers. As the surface energy is used differently in each of these situations, the atmosphere will also display contrasting sensitivities. Thus it appears difficult to judge the responses of the different GCMs using large regional or long temporal averages which neglect these differences. Indeed, the responses of the GCMs might appear different only because the locations or the temporal characteristics of the simulated convergence zones are not the same. Thus devising a method for analysing the sensitivity of the atmosphere to land-surface processes which is independent from the way the tropical climate is simulated will help to understand the mechanisms at work and compare the results between various models.

Sensitivity of Convection to Land-surface Changes

One possible refinement in the analysis of the interactions between land surfaces and the tropical atmosphere is to separate the different synoptic situations and to study them separately. To discriminate between these situations one may use the intensity of the energy cycle of convection. The method introduced by Polcher (1995) will be described in this section.

The atmospheric column is characterised by its vertically integrated energy balance. In the tropics, this simple model is very useful as it helps to identify convective systems and measure their intensity. Convection may be viewed (Fig. A.43) as a thermodynamical engine that takes energy received from the surface, from large-scale convergence of enthalpy and latent energy and from absorbed radiation, to transform it into potential energy. On the other hand, subsidence is the transformation of potential energy into enthalpy which, combined with the surface fluxes, produces a divergence of enthalpy and latent energy. The equations describing this system are discussed in detail by Polcher (1995). This simple model of convection can be applied to single convective systems as well as to the Hadley-Circulation. To study the surface processes, the measure best suited to characterise the intensity of convection is potential energy divergence (PED). Three main reasons can be given. (*i*) PED is independent of the surface fluxes. (*ii*) It is the energy which is exported from the convective regions and transported to the areas of subsidence. (*iii*) PED is a very

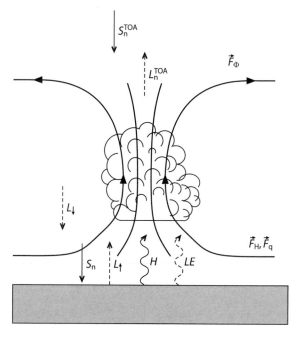

Fig. A.43. A simple model for the energetics of tropical convection

sharp measure of convection as it is the largest single flux in the balance of energy.

This measure of convection is computed with daily averaged data for each grid-box in the region of interest. The data points are then binned according to their PED and the relations to the other energy fluxes are studied. This diagnostic can also be applied at higher frequencies as long as the variations in the storage of energy within the column is small compared to the fluxes. Individual peaks in the PED can be tracked if the time evolution of convection needs to be studied.

The water cycle can be analysed over the spectrum of values of potential energy divergence in the same way as it was done for the energy cycle of convection. The water balance equation for the atmospheric column may be written in terms of energy as:

$$L\frac{\partial Q}{\partial t} = LE - LP - \nabla \vec{F}_q \qquad (A.13)$$

where L is the latent heat of evaporation and Q precipitable water. Figure A.44 shows the four terms of the water balance as a function of the intensity of convection. One may note that all three regions shown here have a similar behaviour and that characteristic properties can be defined for each class of convective events.

For subsidence, local evaporation feeds precipitation and the divergence of latent heat. Depending on the region, the fraction of water recycled or exported will vary. As convection starts it can either feed on local evaporation or moisture convergence. When it intensifies it is dominated by the convergence of moisture and local

evaporation plays a limited role. Precipitation is proportional to $-\nabla \vec{F}_q$. Using Fig. A.44, a first estimation can be made of the sensitivity of precipitation to local evaporation depending on the convective regime. Precipitation of subsequent events will be sensitive to local evaporation while when the PED exceeds 2 kW m^{-2} a reduction in evaporation can be compensated by an increase in moisture convergence.

Using the tool described above, the deforestation experiment presented in Polcher and Laval (1994b) was analysed. In the present summary we will only describe the conclusions reached on the precipitation change which was observed in this experiment. The first result, and the most important one, is that the precipitation decrease is associated with a reduction in the number of intense convective events. Furthermore this decrease is statistically significant during the months of May and June, which could not be shown on the regionally averaged precipitation. The second result revealed by this study is that in a situation of subsidence and weak convection, the recycling rate of water was reduced as a direct consequence of the reduction of evaporation caused by deforestation.

An interesting aspect of this result is that Le Barbé and Lebel (1997) found a similar behaviour in the precipitation record in the sub-Saharian region. A statistical method was applied to the daily rainfall observations available over the region to distinguish between the number of precipitating systems and the amount of rainfall they produce. The authors found that the interannual variation of precipitation is dominated by the variation of the number of systems and that the change in char-

Fig. A.44.
The water cycle of convection as a function of potential energy divergence (PED)

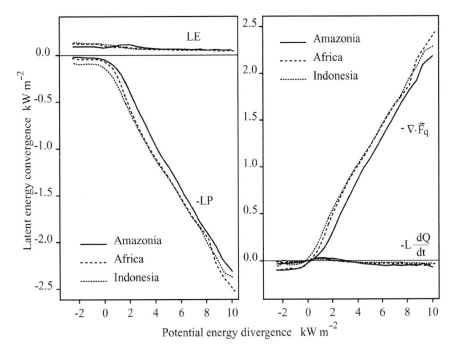

acteristic rainfall is only minor. Thus it appears that the response of the GCM described above is associated to processes that can also be found in the real world.

To study the sensitivity of the number of convective events, ten variations of the deforestation experiment reported in Polcher and Laval (1994b) were conducted. The parameters of pasture in these experiments were varied according to the known uncertainties (Polcher 1995). These experiments, performed with the same GCM, cover a range of results nearly as large as the one presented in Table A.7. Some simulations showed an increase in moisture convergence over Amazonia and others a reduction. It could be shown that the change in sensible heat flux at the surface determined the change in the number of intense convective events and thus the large-scale moisture convergence over the region. The evaporation change on the other hand affected the re-

cycling rate and thus the precipitation brought by the weaker convective events but without modifying the moisture convergence.

An interesting consequence of this mechanism is that the total result on regional precipitation will depend on the distribution of convective and subsequent events. The climate of the control experiment will be a determining factor for the outcome of the deforestation experiment. This might explain the diversity of results for the deforestation experiments described in Table A.7.

An important caveat to this result is that the correlation was performed with the changes in surface fluxes of the grid-box where convection occurs. In the case of a homogeneous deforestation this is not a major restriction as it can be assumed that the local flux variations are similar to those of the neighbouring grid-boxes.

Chapter A.7

The Boreal Climate

Forrest G. Hall · Alan K. Betts · Steve Frolking · Ross Brown · Jing M. Chen · Wenjun Chen
Sven Halldin · Dennis P. Lettenmaier · Joel Schafer

The boreal ecosystem encircles the Earth above about 48° N, covering Alaska, Canada, and Eurasia. It is second in areal extent only to the world's tropical forests and occupies about 21% of the Earth's forested land surface (Whittaker and Likins 1975). Nutrient cycling rates are relatively low in the cold wet boreal soils. Whittaker and Likins (1975) estimate the annual net primary productivity of the boreal forest at 800 g C m^{-2}y^{-1} and its tundra at 140 g C m^{-2}y^{-1}, in contrast to tropical forests averaging 2 200 g C m^{-2}y^{-1} and temperate forests at 1 250 g C m^{-2}y^{-1}. However, the relatively low nutrient cycling rates at high latitudes result in relatively high long-term boreal carbon storage rates averaging roughly 30 to 50 g C m^{-2}y^{-1} (Harden et al. 1992), a result of relatively high root turnover from trees, shrubs and mosses with relatively low decomposition rates. Over the past few thousand years, these below-ground storage processes have created a large and potentially mobile reservoir of carbon in the peats and permafrost of the boreal ecosystem. Currently, the boreal ecosystem is estimated to contain approximately 13% of the Earth's carbon, stored in the form of above-ground biomass and 43% of the Earth's carbon stored below-ground in its soils (Schlesinger 1991). Meridional gradients in atmospheric CO_2 concentrations suggest that forests above 40° N sequester as much as 1 to 2 gigatons of carbon annually (Denning et al. 1995; Randerson et al. 1997), or nearly 15 to 30% of that injected into the atmosphere each year through fossil fuel combustion and deforestation. Given the enormous areal extent of the ecosystem, roughly 20 Mkm^2 (Sellers et al. 1996b; Fig. A.45), shifts in carbon flux of as little as 50 g C m^{-2}y^{-1} can contribute or remove one gigaton of carbon annually from the atmosphere. The size of the boreal forest, its sensitivity to relatively small climatic variations, its influence on global climate and the global carbon cycle, therefore, make it critically important to better understand and represent boreal ecosystem processes correctly in global models.

In Sect. A.7.1 we will first describe the global boreal ecosystem, its landscape structure and composition and the extant factors that have shaped this landscape; its palaeo- and modern climate, and its fire disturbance history. We will pay particular attention to the effects of high-latitude climate change on snow extent and depth.

In Sect. A.7.2 we will review research relevant to energy dissipation and transport between the boreal land-surface and atmospheric boundary layer. We will discuss how the boreal landscape, its albedo and biophysical control on the surface energy and water budget affects atmospheric circulation in the short term and climate at longer time scales. In Sect. A.7.3 we will consider biogeochemical cycling by the boreal landscape, focusing on carbon uptake and release for both carbon dioxide and methane; how the cycling rates are affected by land-cover type, climate, soils and surface hydrology. Finally, in Sect. A.7.4 we will examine the projected impacts of future climate change on the land surface and potential feedbacks, including the effects of variation in snow cover extent and duration, carbon sequestration and release by the surface to the atmosphere and the effects of land-cover change.

A.7.1 The Boreal Ecosystem, Boreal Climate and High-latitude Climate Change and Variability

The boreal ecosystem, or taiga, has no precise definition beyond the meaning of its name, cold northern ecosystem. By about 7 000 years ago, the great ice fields and glacial lakes that had covered Canada, part of the US and Eurasia during the last glacial maximum, had withdrawn from what now constitutes the boreal ecosystem, leaving behind in their retreat bare rock, sand, gravel and clay. Following that, precursors to modern boreal species recolonised the boreal landscape from their ice-age refuges in mountains and warmer coastal regions, and from the warmer southern regions as far south as 30° N (Larsen 1980). Over time, these soil horizons were overlain by mineral soils from wind erosion, and organic soils and peats as the boreal soils accumulated detritus from forests and mosses. Harden et al. (1992) place historical carbon accumulation rates in these peat soils in the range of 10 to 50 g C m^{-2}y^{-1}.

Topography, frigid long winters, followed by hot dry summers, in combination with frequent large-scale lightning-induced fires are the dominant forces that structure the patchiness and composition of the vast majority of the boreal landscape. For a majority of the boreal

Fig. A.45.
Polar boreal vegetation map at 1° spatial resolution derived from ISLSCP I (Sellers et al. 1996b) land-cover dataset based on AVHRR data (DeFries and Townshend 1994). *White* latitude line is Arctic Circle; *red* latitude lines drawn at 60 and 40° N

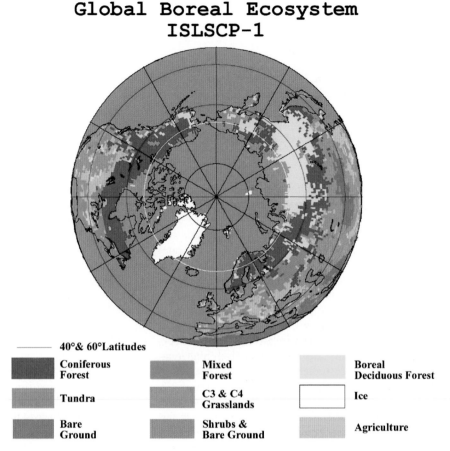

Global Boreal Ecosystem ISLSCP-1

——— 40°& 60°Latitudes

■ Coniferous Forest	■ Mixed Forest	■ Boreal Deciduous Forest
■ Tundra	■ C3 & C4 Grasslands	□ Ice
■ Bare Ground	■ Shrubs & Bare Ground	■ Agriculture

ecosystem, its species have limited commercial value and deforestation from logging is small. Where logging is of economic importance and fire is controlled, such as in the north-European boreal forests of Sweden and Finland, patchiness is largely determined by property boundaries.

These conditions have selected for a few dominant species, including stands of spruce, tamarack, fir and pine as well as aspen, birch and poplar. The ecosystem is also populated by vast wetland complexes (Tarnocai et al. 2000) consisting of lakes, bogs, fens, marshes, swamps and muskeg. For the purposes of this chapter, we will consider the boreal ecosystem to consist of the sub-arctic forests (predominantly coniferous) and wetlands (muskeg) lying primarily above 48° N as well as the tundra, primarily above about 60° N and below the Arctic Circle. Above roughly 60° N, the forest degrades gradually to forest shrubland, then tundra. The southern limit is less clearly defined, transitioning gradually into prairies and croplands, and more temperate deciduous and coniferous forests and grasslands.

To understand the first order interactions between the atmosphere and land, boreal vegetation types can be lumped into the broad categories shown in Fig. A.45, which indicates that coniferous, deciduous and mixed coniferous/deciduous boreal forests occupy roughly 20 Mkm² between 60 and 40° N. It is within these latitudes that tracer model studies indicate forests may be a strong sink for carbon dioxide of approximately 1 to 2 gigatons annually (Denning et al. 1995; Randerson et al. 1997).

These wetlands are generally unresolvable from either AHVRR or Landsat Thematic Mapper data (Hall et al. 1997; Beaubien et al. 1999), although ground survey maps do exist (Tarnocai et al. 2000). As will be seen in Sect. A.7.3.5, wetlands play an important role in methane flux to the atmosphere and thus the future development of a remote sensing capability to estimate their areal extent and type (fen, marsh, bog or swamp) is important.

A.7.1.1 Climate and Boreal Vegetation

Larsen (1980) observes of the boreal forest, "It is bounded on the north over most of Canada and Eurasia approximately by the position of the July 13 °C isotherm with marked departures in regions possessing montane or oceanic climatic influences. The southern limit of the boreal region in central and eastern Canada is bounded roughly by the position of the July 18 °C isotherm. But where drier conditions prevail, the southern edge of the forest border lies to the north of this isotherm, trending into regions where annual precipitation is greater. The same general relationships between the boreal forest and summer isotherms hold also in Eurasia." The harsher climate at the northern transition zone, accompanied by lower soil temperatures, results in an open wooded lichen landscape, which slowly merges into the tundra.

The boreal ecosystem so defined, consists of three major climatic zones: largely flat continental interiors, montane and coastal regions, including the Pacific coast, Hudson Bay, and the Arctic coast in Eurasia and the East Siberian seas. The climate of Scandinavia is maritime to the west and gradually shifts over to continental in the eastern parts of Finland. The climate in the North American boreal zone is typified by short, warm, moist summers and long, cold, dry winters. Mean annual temperatures range from –10 to –4.5 °C in the interior Yukon and Alaska, 1 to 5.5 °C in northern coastal British Columbia and –4 to 5.5 °C in the extensive boreal shield ecozone (Ecological Stratification Working Group (ESWG) 1996). The northern continental plain of the boreal forest retains ground snow cover for up to eight months in a year resulting in a short growing season, in comparison with the Avalon peninsula which experiences milder, wetter weather due to the buffering effect of the Atlantic ocean (ESWG 1996). The North American boreal forest region is characterised by large, often monotypic stands of white and black spruce, jackpine, balsam fir, and tamarack. Historically, this predominantly coniferous forest had a minor component of trembling aspen and balsam poplar (Rowe 1972). An increase in the black spruce and tamarack species component occurs as the tree line is approached (ESWG 1996). From north to south, the southern transition zone in the west incorporates an increasing amount of trembling aspen, balsam poplar, and willow. The transition to steppe-like vegetation is made across a positive 2 °C mean temperature gradient (Singh and Wheaton 1991). In the east, the southern transition zone is characterised by intermingled temperate species of eastern white and red pine, yellow birch, sugar maple, and black ash (Rowe 1972). The appearance, structure and function of the boreal forest is remarkably similar at all longitudes, consisting primarily of needleleaf evergreen and needleleaf deciduous conifer, underlain by low shrubs and herbacious plants, atop a thick layer of mosses and lichens. From the functional viewpoint of energy balance and biogeochemical cycling, the most important aspect of species variation within the boreal ecosystem is the variation in structural characteristics from one species to the next, i.e. deciduous versus evergreen and broad versus needleleaf.

In addition to climate change, fire and insect, defoliation plays a dominant role in shaping boreal forest landscapes (Suffling 1995; Hogg 1999), see Sect. A.7.1.2.

A.7.1.2 Effects of Fire and Insects on Vegetation and Land Cover

Fire is a central part of the life cycle of boreal ecosystems except where logging is of economic importance and fire is controlled in areas such as Northern Europe, for example. Although fires burn the boreal forest regularly, new trees quickly emerge in most burnt areas. The history of Canada's boreal forest has been mostly cycles of destruction and renewal by wild fires. Lightning caused 35% of Canada's forest fires since 1930 but was disproportionately responsible for 85% of the burned area. Most fires were small, with 2–3% of the fires growing to more than 200 ha and eventually contributing to ~ 98% of the total burned area (Weber and Flannigan 1997). The fire season in Canada ranges from April through October. Typically, there is virtually no winter activity but a flurry of spring fires after snow melt, followed by a decline as green-up progresses northward. Lightning-induced fires are most active in mid-summer and cease in the autumn.

As shown in Fig. A.46, the period of 1930–1979 saw relatively low forest fire activity in Canada, with the total burned area of 1.1 Mha y⁻¹ on average and a range from 0.2 to 3.8 Mha y⁻¹ (Weber and Flannigan 1997). During the last two decades since 1980, the average total burned area increased to 2.6 Mha y⁻¹, with a maximum 7.6 Mha y⁻¹ in 1989. Another high fire activity period occurred during the period 1850–1920, with mean burned area of ~ 2.6 Mha y⁻¹ as inferred by Chen et al. (2000) based on their analyses of forest age structure information from 1920 (Kurz et al. 1995). While the high fire activity period in the recent decades might be related to climate change, the large fire disturbance rates in late 19th century and early 20th century were probably induced by human settlement.

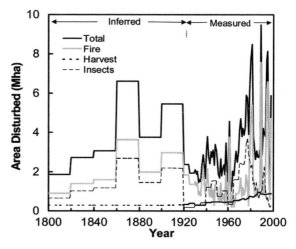

Fig. A.46. Measured and inferred areal extent of disturbance in Canada's forests since 1800 (after Chen et al. 2000). Forest age structure data for 1920 (Kurz et al. 1995) were used to inferred disturbed areas before 1920

In addition to forest fire, insect-induced mortality and harvest affect the carbon cycle of forest ecosystems. The total disturbed area was large during 1860–1920 and 1980–1998, and small during the other periods in the 19th and 20th century (Weber and Flannigan 1997; Kurz et al. 1995; Canadian National Forestry Database Program, *http://www.nrcan.gc.ca/cfs*; Chen et al. 2000).

This temporal distribution of disturbance rates resulted in significant variations in the percentages of young and less productive stands (i.e. < 20 years old), productive stands (i.e. 20–100 years old), and old and less productive stands (i.e. > 100 years old). From 1800–1820, fire and insect-induced mortality rates were lower than their pre-industrial averages. Consequently, the percentage of young stands decreased while that of productive stands increased. The increasing disturbance rates from 1820 to 1880 reduced the percentage of productive stands to a minimum of 41% around 1880. The decreasing trends in the percentage of productive stands has reversed since then and reached a maximum of 71% in 1940, as fire and insect-induced mortality rates decreased in this period. From 1940–1970, the fire and insect-induced mortality rates were still low, but the stands disturbed during the 1860–1920 period entered into old and less productive ages, reducing the percentage of productive stands. The increases in fire and insect-induced mortality rates during the last two decades increased the percentage of young stands substantially and reduced the percentage of productive stands further.

A.7.1.3 High-latitude Climate Change

At the present rate of increase, atmospheric CO_2 concentration will double before the end of the next century (Houghton et al. 1995). Sensitivity experiments made with AGCMs point to large temperature increases in the northern high latitude continental interiors (Schlesinger and Mitchell 1987; Houghton et al. 1995; Sellers et al. 1996a), partly due to projected changes in the polar sea ice climatology and snow-albedo feedbacks. A significant surface warming trend was observed in the boreal zone during the 1980s and early 1990s, reaching 1.25 °C per decade within the Canadian interior (Chapman and Walsh 1993). Serreze et al. (2000) estimated that approximately half the pronounced recent rise in surface Northern Hemisphere temperatures reflected changes in atmospheric circulation. Thompson and Wallace (1998) demonstrated that the predominantly positive phase of the Arctic Oscillation over the last three decades of the 20th century had a major role in the recent warming observed over Eurasia. Correlation of the Thompson and Wallace Arctic Oscillation index with Northern Hemisphere monthly snow cover extent over the 1972 to 2000 period revealed statistically significant correlations over

the January–March period, with a maximum correlation in March (–0.583). Further analysis by sub-regions revealed that this link was concentrated over western Eurasia with no significant associations over North America. Dai et al. (1997) have determined that since 1900 zonal precipitation is increasing at all latitudes, but especially above 40° N. Within the troposphere, the arctic atmosphere has retained nearly the same temperatures since at least the late 1970s (Chase et al. 2002), which is consistent with regional surface changes being due to circulation pattern changes, rather than a uniform tropospheric warming. Arctic sea ice trends (Pielke et al. 2000; *http://faldo.atmos.uiuc.edu/CT*) also document a more complex pattern of temporal variability, with total arctic 2002 sea ice areal coverage returning to near the extent observed at the beginning of the satellite record in the early 1980s.

A.7.1.4 Changes in Snow Extent, Depth and Duration

Although the available data have a number of uncertainties and shortcomings for documenting snow cover variability and trends over the Northern Hemisphere boreal forest zone, they provide evidence that this region has experienced a significant reduction in spring snowpack over the second half of the 20th century. The reduction is in response to enhanced spring warming over Northern Hemisphere high latitudes (Fig. A.47). There are some indications that winter snow depths have increased over northern Eurasia (Ye et al. 1998) which is consistent with observed trends of increased precipitation. However, further analysis of *in situ* snow depth and snow cover extent data, particularly the updated Soviet snow-depth dataset, is required to confirm the continuation of this trend. Winter snow depths decreased over much

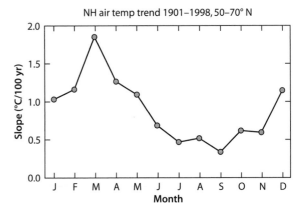

Fig. A.47. Linear trend in Northern Hemisphere land temperatures averaged over 50–70° N for the 1901–1998 period. Computed using an updated version of the Jones (1994) gridded land-surface air temperature anomaly dataset. Trends are statistically significant for the February–September period

of Canada in the mid-1970s in response to a change in atmospheric circulation (Brown and Braaten 1998).

Trends of snow cover extent (1972–1998) from satellite data show significant reductions in spring snow cover over high-latitude and mountainous regions of the Northern Hemisphere, but less change over the boreal forest zones of North America and central Eurasia. A recent climate modelling study by Fyfe and Flato (1999) reported a clear elevational dependency in simulated temperature warming linked to the rise in the snow line and amplified warming from the snow-albedo feedback. It is speculated that the lagged response over the boreal zone is related to the deeper snowpack in these areas, and to the reduced snow cover-albedo feedback over forest.

While there is a clear negative correlation between Northern Hemisphere temperatures and snow cover extent (Robinson and Dewey 1990), surface snow cover conditions represent an integrated response to temperature and precipitation. Thus, atmospheric circulation plays an important role in snow cover variability and change. Dominant modes of atmospheric circulation such as the Pacific-North America (PNA) and North Atlantic Oscillation (NAO) have been shown to exert strong influences on North America and Eurasian winter snow cover respectively.

A.7.1.4.1 Changes in Snow Extent

The National Oceanic and Atmospheric Administration's (NOAA) weekly snow cover dataset provides the longest record of *hemispheric-wide* snow cover extent and has been used extensively to monitor trends in snow cover extent (e.g. Robinson et al. 1993) and to investigate snow cover-climate relationships (e.g. Gutzler and Rosen 1992; Karl et al. 1993; Groisman et al. 1994). The data consist of digitised weekly charts of snow cover derived from visual interpretation of NOAA visible satellite imagery by trained meteorologists. The charts were digitised on an 89 × 89 polar stereographic grid for the Northern Hemisphere with a 190.5-km resolution at 60° N. Regular monitoring of Northern Hemisphere snow cover with visible satellite imagery began in November 1966 (Dewey and Heim 1982), but the sub-point resolution of the pre-1972 satellites was ~ 4.0 km compared to 1.0 km for the VHRR launched in 1972 (Robinson et al. 1993). Pre-1972 data have recently been re-charted at Rutgers University to extend the NOAA weekly dataset back to 1966. These data were unavailable at the time this chapter was being prepared.

NOAA weekly snow data from 1972–1998 were used to investigate snow cover trends over the boreal forest region of the Northern Hemisphere. This was a period of widespread warming with significant warming trends in most months of the year over the Northern Hemi-

sphere boreal zone (Fig. A.47). This warming was accompanied by significant reductions in Northern Hemisphere snow cover extent over the May to August period (Fig. A.48).

To characterize the spatial and temporal characteristics of snow cover changes over this period, regression analysis was carried out at each NOAA gridpoint of the number of days with snow cover in the first and second halves of the snow year.

The results (Fig. A.49 and Fig. A.50) show evidence of a marked seasonal difference in snow cover trends, with snow cover increases in the autumn half of the snow season, and widespread snow cover reductions in the spring over many areas of the Northern Hemisphere.

The trend toward reduced spring snow cover is consistent with the spring snow cover feedback mechanism observed by Groisman et al. (1994), i.e. decreased snow cover reduces surface albedo, increases surface absorption of radiation, raises surface temperatures, and hence increases snow melt. The areas experiencing the largest significant reductions in spring snow cover were western mountain regions of North America (particularly the Pacific NW), the Canadian Arctic Islands and Ungava peninsula, the Central Siberian Plateau and Yablonovyy Range, and a broad swath running from the Caucasus Mountains to Tibet. Over these areas spring snow cover reductions were of the order of 1.0 d y^{-1}. The boreal forest region of NA, heavily forested by evergreen conifers, showed relatively little change in spring snow cover perhaps because the snow cover feedback is weak, due to strong shadowing by the evergreen conifers which reduces markedly the difference between snow-on and snow-off albedo. In Eurasia, where deciduous conifer is abundant, snow-on versus snow-off albedos differ significantly and most of this region experienced spring snow cover reductions (with the exception of the area between 40 and 80° E where spring snow cover increased slightly).

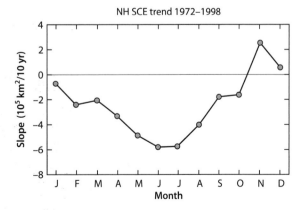

Fig. A.48. Trends in Northern Hemisphere snow cover extent from 1972–1998 from NOAA weekly data. Reductions in the May–August period are statistically significant

Fig. A.49.
Contoured values of the slope
t-statistic for change in snow
cover duration in the first half
of the snow year (August–
January) over the 1972–1998
period. Values greater than ±2
indicate locally significant
trends

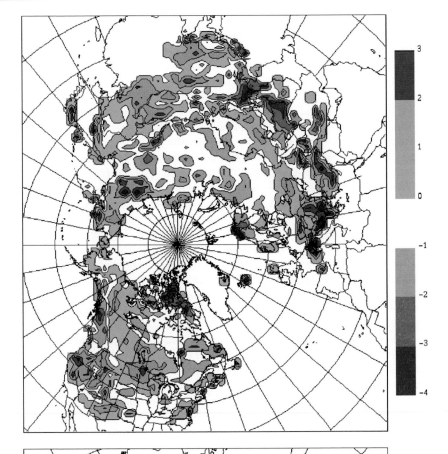

Fig. A.50.
Contoured values of the slope
t-statistic for snow cover duration in the second half of
the snow year (February to
July) over the 1972–1998 period. Values greater than ±2
indicate locally significant
trends

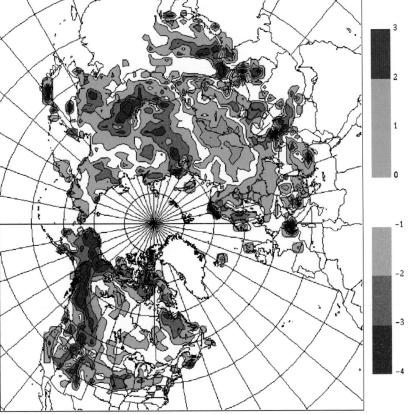

A.7.1.4.2 Trends in Snow Depth and Duration

In situ observations of daily snow depth, which provide the longest record of snow cover depth over individual plots, can be used to investigate satellite snow cover trends, and to extend the satellite snow cover record back in time. These data also provide information on the timing and duration of the snow cover season. The two countries with the largest fraction of the Northern Hemisphere boreal forest zone, Canada and Russia, have historical snow depth data going back to the early 1900s. However, the observation programmes and the spatial and temporal distributions of the data are quite different with respect to monitoring snow cover in the boreal forest. Both countries also have *in situ* snow water equivalent observations from snow courses, but these do not cover extended periods and are less suitable for investigating long-term snow cover variability.

Canadian snow depth observations are made by ruler at open locations (often airports) and may not be fully representative of snow conditions in forested areas. This contrasts with snow depth measurements in the Former Soviet Union which were made from snow stakes located to represent the area surrounding a climate station. A description of Canadian snow depth data is provided by Brown and Braaten (1998) and the data are available on CD-ROM (MSC 2000). A CD-ROM compilation of Former Soviet Union historical snow depth data for the period 1881–1995 is provided by the US National Snow and Ice Data Service (NSIDC 1999).

There are few long-term daily snow depth observations in Canada north of ~ 55° N before World War II. Analysis of trends in Canadian snow depth data over the 1946–1995 period (Brown and Braaten 1998) revealed widespread decreases in late winter and early spring snow depths over much of Canada, with the greatest depth decreases occurring in February and March (Fig. A.51). This decrease was associated with significant decreases in spring and summer snow cover duration over most of western Canada and the Arctic (Fig. A.52). Increases are shown as cross-hatched contour lines and decreases as regular lines. The filled squares show the locations of the stations included in the analysis (which were required to have at least 40 years of data in the period). Dark shading is used to highlight areas where changes in snow depth were locally significant ($p < 0.05$). Light shading is used to highlight areas where stations exhibited marginally significant ($0.05 < p < 0.10$) snow depth changes (Brown and Braaten 1998). The snow depth changes were characterised by a rather abrupt transition to lower snow depths in the mid-1970s that coincided with a well-documented shift in atmospheric circulation in the Pacific-North America sector of the Northern Hemisphere that had widespread ecological impacts (Ebbesmeyer et al. 1991).

An indication of longer-term variability in spring snow cover over the southern boreal forest zone of North America can be inferred from reconstructed information on April snow cover extent (Brown 2000). Analysis of NOAA weekly snow cover information showed that during April, snow cover variability is concentrated over

Fig. A.51.
Average change in March mean monthly snow depth over the 1946–95 period (cm y^{-1}) (Brown and Braaten 1998)

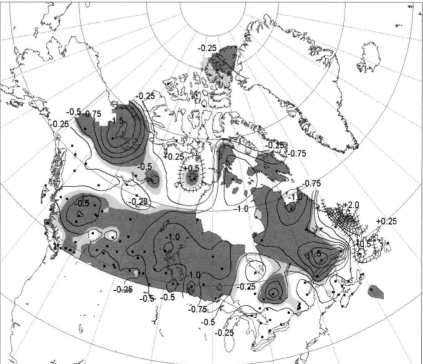

Fig. A.52.
Average change in spring (MAM) snow cover duration over the 1946–95 period (d y⁻¹). *Shading* same as previous figure (Brown and Braaten 1998)

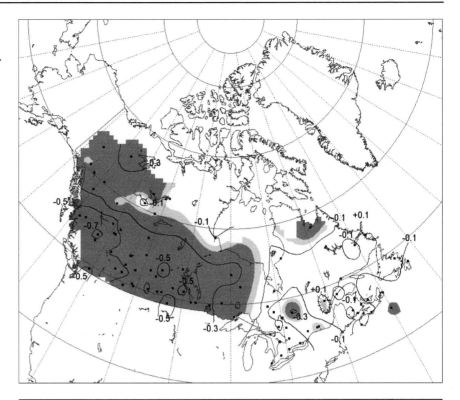

Fig. A.53.
Interannual variability in north American monthly snow cover extent for April. *Open diamonds* are anomalies for the station-derived snow cover index; *solid diamonds* are snow cover extent anomalies from the NOAA satellite data; and the *heavy solid lines* are the result of passing a 9-year binomial filter to the combined station (1915–1971) and satellite (1972–2000) anomaly series. Anomalies were standardised with respect to a common 1972–1992 reference period (updated from Brown 2000)

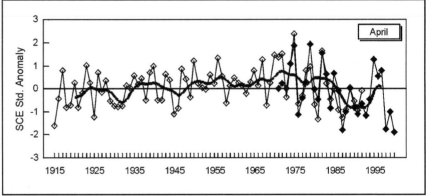

a relatively narrow band from 50–60° N that covers the southern limits of the boreal forest zone. Figure A.53 suggests that reduction in North America spring snow cover is a relatively recent phenomenon, and that snow cover exhibited a gradual increase from the early 1900s to the early 1970s. However, analysis of area-averaged snow depth information for April showed evidence of a significant decreasing trend that began ~ 1945. Mean snow depths were estimated to have approximately halved over the 1945–1997 period. These results suggest that major changes have likely occurred in spring snowpack conditions in the North American boreal zone over the last five decades. The difference between the snow cover extent and depth results can be explained by the relatively deep spring snowpack in the boreal forest regions (i.e. depth changes have little impact on snow cov-

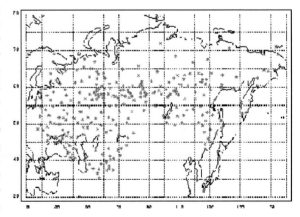

Fig. A.54. Spatial distribution of observing sites in the Historical Soviet Daily Snow Depth dataset (see NSIDC 1999)

Fig. A.55.
Reconstructed variation in
Eurasian snow cover extent
for April (from Brown 2000).
Lines and symbols follow the
explanation given in Fig. A.53

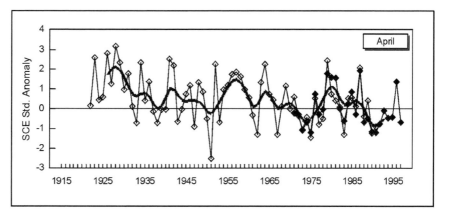

ered area), and a reduced albedo feedback (Betts and Ball 1997). According to Harding and Pomeroy (1996), the reduced radiation received at the forest floor plus the shelter from vigorous turbulent exchanges act to suppress snowmelt and sublimation. One mechanism explaining the observed decrease in spring snow depths is a reduction in the solid fraction of precipitation accompanying spring warming. Zhang et al. (2000) show that western and central regions of Canada have experienced significant spring (MAM) decreases in the solid fraction of precipitation over the 1950–1994 period (the only season to exhibit this trend).

Figure A.54 shows the spatial distribution of snow depth observed in the Former Soviet Union dataset. Analysis of the first release of the dataset (up to 1984) by Ye et al. (1998) concluded that winter snow depth had increased over most of northern Russia (60–70° N – the boreal forest zone), and had decreased over southern

Russia. The rate of increase over northern Russia was estimated to be 4.7% per decade, which is consistent with estimates of precipitation increases over northern mid-latitudes (Bradley et al. 1987; Vinnikov et al. 1990). The Former Soviet Union snow depth dataset was recently extended to 1995. Reconstruction of snow cover extent over Eurasia by Brown (2000) showed evidence of a significant reduction in April snow cover over the 1922–1997 period (Fig. A.55). During April, the snow line is usually located just to the south of the Eurasian boreal forest zone. Brown (2000) determined that increasing winter season snow depth and reduced spring snow cover extent were consistent with concurrent trends of warming and increased precipitation.

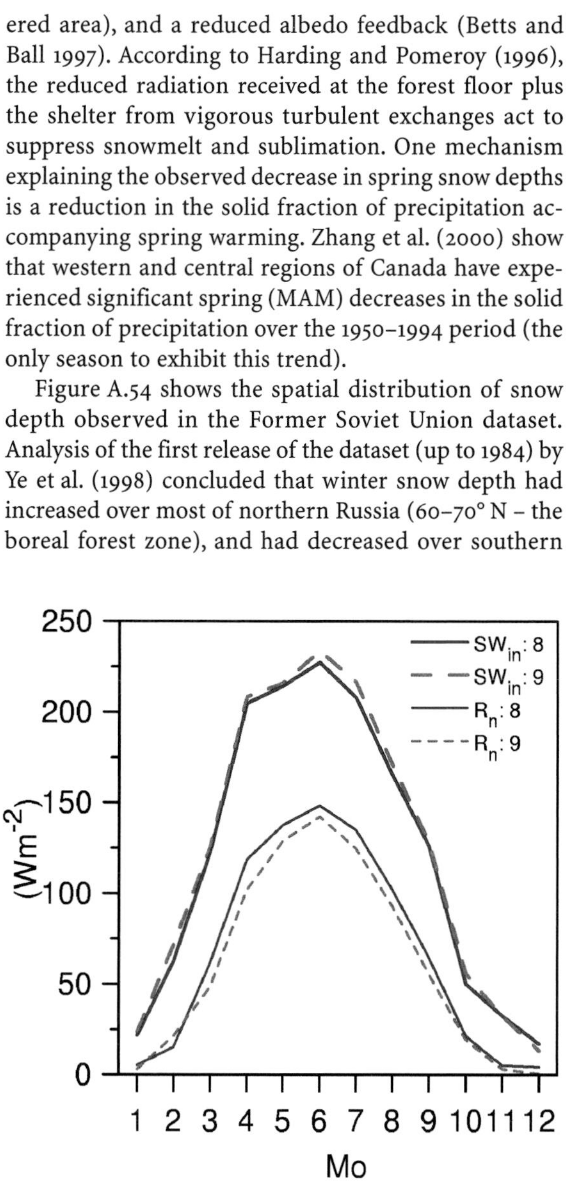

A.7.2 Energy Dissipation and Transport by the Boreal Landscape

The amount of solar energy reaching the land surface during the day varies seasonally and is determined by daylength, solar elevation, cloud cover and atmospheric opacity. The fraction of the sun's incident energy absorbed by the land surface is in turn determined by surface reflectivity (albedo). The manner in which the surface partitions the absorbed energy into sensible and latent heat is a major determinant of boundary layer dynamics, weather and climate patterns (see Fig. A.3). The partitioning of the net radiation into the sensible and latent heat fluxes and ground heat flux is controlled by the availability of water for evaporation (which is dramatically affected by the seasonal cycle of soil freezing), and by the biophysical controls on transpiration in the warmer months.

Figure A.56 shows the annual cycle of incoming short-wave radiation (SW_{in}) and net radiation (R_n) at the surface for two BOREAS (Boreal Ecosystem-Atmosphere Study) mesonet sites (Thompson, site #8 and the old jackpine, site #9) in the BOREAS northern study area (from Betts et al. 2001). The variation in the net radiation seasonal cycle is very large at this latitude and it

Fig. A.56. Annual cycle of incoming short-wave and net radiation (*1*: January)

drives the very large seasonal cycle of temperature. There is also a visible and important seasonal asymmetry with respect to the summer solstice in Fig. A.56. In April, both SW_{in} and R_n are considerably larger than in August.

This must be a consequence of lower cloud cover in April, associated with the relative humidity minimum (shown later in Fig. A.58), since the mean solar zenith angle is greater in April than in August.

A.7.2.1 Effect of Soil Type on Surface Energy Partitioning

Figure A.57 shows the three-year (1994–1996) average annual cycle of temperature. The solid lines are mean air temperature at about 6 m above the canopy for Thompson, site #8 and NOJP, site #9. The differences in temperature at this level are very small on the seasonal scale. The dotted and dashed lines, however, labelled with the site numbers, are soil temperatures (at 10 cm and 50 cm depth respectively). Here below ground, the intersite differences are very large. The annual cycle of soil temperature at the jackpine site (with a sandy soil and only a thin lichen ground cover) is much larger than at the mixed spruce site at Thompson, where there is an organic soil with a thick overlying moss layer. At this site, the annual cycle of temperature at 50 cm is small, and the soil at this depth is still frozen in June. This subsurface heterogeneity is an important feature of the landscape and, along with the vegetation type itself, affects surface fluxes since, for example, water is unavailable for transpiration until the ground melts in spring (as we shall see later). However, the sub-surface heterogeneity of temperature is not reflected in the temperature cycle above the canopy.

Figure A.58 shows the annual cycle (1994–1996) of relative humidity above the canopy, showing the remarkable seasonal asymmetry over the boreal forest. In spring (April is month 4) when temperatures are still below freezing but the solar elevation and the net radiation are increasing rapidly, the relative humidity is at a minimum. The minimum day-time relative humidity is very low, often below 25%, as water is generally not available for evaporation at the surface. Evaporation is so low in spring that the forest could be thought of as a "green desert". Consequently the dry boundary layer in spring can be as much as 2 000 m deep, driven by very large sensible heating with very little evaporation (Betts et al. 1996, 1999, 2001). Throughout the spring, summer and early autumn, mean relative humidity rises to an October peak as the surface becomes warm relative to the atmosphere. In October, the ground has not frozen at either site and water is readily available for evaporation. This seasonal asymmetry in relative humidity is in sharp contrast to the seasonal thermal cycle with a mid-summer maximum, and is an important characteristic of the northern latitudes. The seasonal phase lag of ground temperature with respect to air temperature, which is sharply increased by the phase change of soil water, severely restricts evaporation in spring, when the ground is frozen, while the reverse is true in the autumn, until the ground freezes.

The freezing and thawing of the soil moderates winter temperatures because during the freeze process the effective heat capacity of the soil is increased by a factor of 20 (Viterbo et al. 1999). It introduces a significant lag into the system. In spring a significant part of the net radiation goes into melting the ground, and lakes (and in warming them). This energy becomes available in the

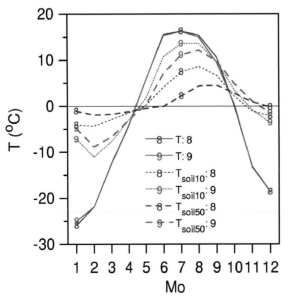

Fig. A.57. Annual cycle of air and soil temperature (*1*: January)

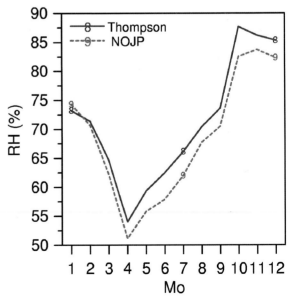

Fig. A.58. Annual cycle of relative humidity (*1*: January)

early winter, when the surface refreezes. Water is not available for transpiration in spring until the ground melts or for evaporation until the ground, wetlands and lakes warm with respect to the atmosphere. As discussed above, this leads to very low evaporative fraction in spring, with very large sensible heat fluxes from the forest canopy, which in turn produce deep dry boundary layers in spring. In autumn, when the lakes and ground are warm relative to the cooling atmosphere, the situation reverses. Evaporative fraction is high from the conifers and lakes, but low from deciduous species following senescence. Net radiation is much lower by the time the surface freezes, so sensible heat fluxes are very low and autumn boundary layers become very shallow, often capped by stratocumulus.

A.7.2.2 Effect of Land-cover Type on Seasonal Variation in Relative Humidity

At the BOREAS sites the dominant conifer species, black spruce, has a characteristic increase in evaporative fraction (EF), the ratio of latent heat to the sum of sensible and latent heat, from a minimum in spring to a maximum in late summer and autumn. Figure A.59 shows two sites for 1996 from the southern study area (Jarvis et al. 1997) and the northern study area (Goulden et al. 1997; Betts et al. 1999). Indeed this rise of evaporative fraction is responsible for the steady rise of mean relative humidity during the year as shown in Fig. A.58.

It can be thought of as a positive feedback between the surface ecological control and the relative humidity of the BL, since for black spruce, inferred vegetative resistance decreases almost linearly with increasing relative humidity (Betts et al. 1999). At deciduous and wetland sites there are both similarities and differences.

Figure A.60 shows the old aspen site in the southern study area (SOA: Blanken et al. 1997) and the fen site in the northern study area (NFen: Lafleur et al. 1997; Joiner et al. 1999b) for the same year. Once temperature climbs near the end of May (Day 150) (when the 10-day average temperature reaches 5 °C), latent heat flux rises rapidly at SOA, as leaf-out occurs. In mid-summer, evaporation is higher at the deciduous aspen site than at the spruce sites, but it is lower in the autumn than at the spruce sites, after the leaves die and fall. At the wetland site, NFen, the spring rise of evaporative fraction is also steep (there are some missing data), and there is a fall as the fen dries out in the summer. The leaf area index of the wetlands is lower and so transpiration (which can draw on a deeper root zone than surface evaporation) plays a less important role than at the aspen site.

A.7.2.3 Role of Stomatal Control

The well-known importance of the stomatal control of transpiration by a coniferous forest (e.g. Lindroth 1985; Jarvis and McNaughton 1986) was confirmed by BOREAS and NOPEX (Jarvis et al. 1997; McCaughey et al. 1997; Joiner et al. 1999a; Goulden et al. 1997; Betts et al. 1999; Cienciala et al. 1997, 1998, 1999; Grelle et al. 1999; Morén 1999). The diurnal cycle of evaporative fraction at the five flux sites in Fig. A.61 are quite different, even though there is little difference in the atmospheric forcing. At the Nfen, evaporative fraction climbs from mid-morning to mid-afternoon as evaporation increases with falling relative humidity. At the black spruce site, evaporative fraction does not rise as relative humidity falls to its afternoon minimum, indicating that stomatal control is playing a role. At this site evaporative fraction has a broad day-time minimum. At the jackpine site, where water is

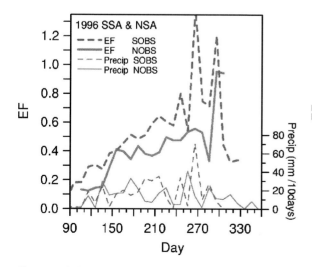

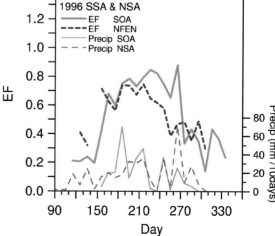

Fig. A.59. Black spruce evaporative fraction (*EF*) with corresponding 10-day precipitation

Fig. A.60. As Fig. A.59 for aspen and fen sites

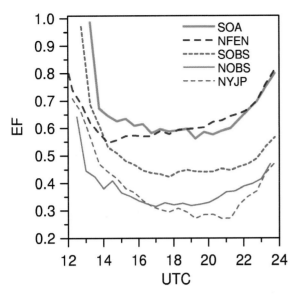

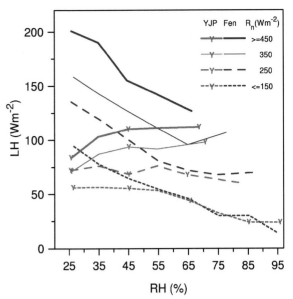

Fig. A.61. Summer day-time evaporative fraction (*EF*) for five flux sites (JJA 1996)

Fig. A.62. Distinct dependence of surface evaporation on relative humidity and net radiation at fen and jackpine sites

most limited because of the sandy soil, the solid curve shows that evaporative fraction drops to a minimum in mid-afternoon about the time of the relative humidity minimum. Here stomatal control must be the strongest. Figure A.62 (from Betts et al. 2001) illustrates this important climatic role even more clearly. We compare the link between measured evaporation from the NYJP and NFen sites, relative humidity and net radiation, by averaging the day-time data (1200–2400 UTC) from the months June to October of 1994 and 1996 into 10% bins in relative humidity, and 100 W m^{-2} bins of net radiation. During the day-time, the atmospheric forcing variables of temperature and humidity differ little between the sites, as discussed previously, but evaporation has a quite different behaviour. Evaporation increases with R_n as expected for both sites. At the fen, evaporation increases with decreasing relative humidity, consistent with the high availability of water. However, at the jackpine site, at low radiation levels, evaporation is flat with decreasing relative humidity, and at high radiation levels, evaporation actually falls as relative humidity decreases, showing that the sensitivity of stomatal control to relative humidity (or vapour pressure deficit) is the dominant process. So we see that at low relative humidity, the evaporation curves diverge widely, even though the atmospheric drivers are almost identical. Because the landscape is dominated by a forest with strong stomatal control, the climatic equilibrium of the diurnal cycle is controlled by the forest, not the fens. When evaporation is low, despite high Rn, relative humidity is low and the diurnal cycle of lifting condensation level (LCL) is large. The fens on the other hand are not in equilibrium with the overlying dry BL and have a much higher surface evapora-

tion. We show the dependence of LH on relative humidity rather than vapour pressure deficit (VPD), which has been widely used in the study of canopy physiology, because VPD has a strong temperature dependence and the important atmospheric variables, relative humidity and LCL are tightly linked (Betts et al. 2001). In addition, Betts et al. (1999) found that vegetative resistance, inferred from data at the northern study area old black spruce site, was more tightly correlated with T and relative humidity than with T and VPD. The theoretical link between equilibrium evaporation, equilibrium BL depth and vegetative resistance is discussed more extensively in Betts (2000a). Other studies using branch bags (Rayment and Jarvis 1999) have shown that assimilation of CO_2 increases with light levels, decreases with increasing VPD, and has an optimum temperature range.

A.7.2.4 Role of Latitudinal Gradient

BOREAS had two study areas, at the north-east and south-west ends of a transect across the boreal forest. Figure A.63 from Barr et al. (2001) contrasts the BOREAS flux sites in the southern and northern study area. It compares the diurnal cycles (an average from 24 May to 19 Sept, 1994, corresponding essentially to the growing season) of (*a*) air temperature and relative humidity, (*b*) evaporative fraction, (*c*) solar and net radiation and (*d*) surface conductance. The meteorological variables are averaged over all sites. The *upper panels* show an afternoon maximum of temperature and a minimum of relative humidity. The *evaporative fraction panels* show a day-time minimum of evaporative fraction at all sites,

Fig. A.63.
Diurnal variation in air temperature, T_a (*red lines*), relative humidity: RH (*blue lines*), evaporative fraction, EF, solar radiation, R_s (*black lines*), net radiation, R_n (*blue lines*) and surface conductance to water vapour, g_s from BOREAS tower flux sites in the southern study area (*SSA*) and northern study area (*NSA*) averaged between 24 May and 19 Sept 1994. We first averaged the fluxes by time of day before calculating EF. The mean surface conductance was calculated from measured data only (e.g. with no gap filling) after excluding values below the 10th and above the 90th percentiles. The line style is the same for each land-cover type. Local time in the southern and northern study area is 6 hours less than UTC (*OA:* old aspen; *OBS:* old black spruce; *OJP:* old jackpine; *YJP:* young jackpine)

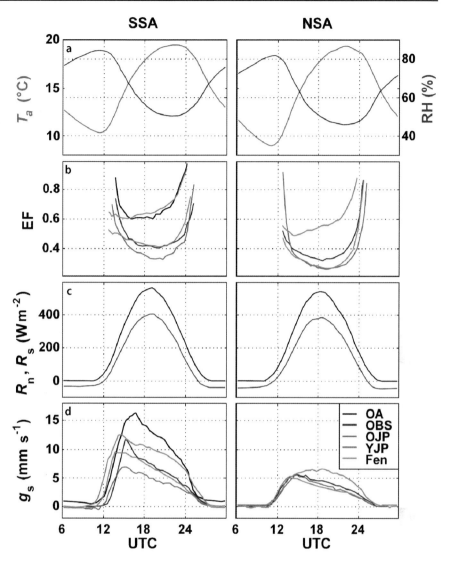

with evaporative fraction at the southern study area exceeding evaporative fraction at the northern study area for each paired land cover. (Gaps in the data were filled using a model: see Barr et al. 2001.) In addition, the figure shows the characteristically different diurnal patterns of evaporative fraction for each land-cover type, independent of geographic location. At the fen sites, evaporative fraction increases as relative humidity falls from an early-morning maximum to a mid-afternoon minimum. At the aspen and old black spruce sites, the day-time pattern is relatively flat. At the jackpine sites, evaporative fraction falls the most and reaches the lowest afternoon minimum. These differences reflect the decreasing availability of water for evaporation and transpiration and the strongest stomatal control on transpiration at the jackpine sites. Mean relative humidity is a little lower for the northern than the southern study area, consistent with the uniformly lower evaporative fraction.

The southern-northern study area difference in evaporative fraction is consistent with the earlier results of

Barr and Betts (1997), who analysed the boundary-layer budgets of the BOREAS radiosondes. They reported mean mid-day Bowen ratios for the BOREAS northern and southern study area during the 1994 intensive field campaigns that correspond to evaporative fractions of 0.53 and 0.45, respectively. These values are intermediate between the lower conifer values and the higher aspen and fen values in Fig. A.63, and are about 30% higher than the mean mid-day values for mature black spruce (0.43 in the southern study area and 0.34 in the northern study area). These differences illustrate that, although the boreal forest landscape is dominated by conifers, particularly black spruce (Betts et al. 2001), its energy balance at the landscape scale is also influenced significantly by other land covers with higher evaporative fractions.

The *lower panels* in Fig. A.63 show the mean diurnal cycles of solar and net radiation (*middle*) and the derived surface conductance to water vapour (*bottom*). The southern study area aspen site has the highest surface conductance and the jackpine sites have the lowest. Unlike the fen sites, where conductance is nearly symmetric with radiation (more so in the north than the south), the diurnal pattern of conductance is markedly asymmetric at the forest sites. The high forest conductances in the early-to-mid morning reflect the maximum daytime stomatal opening as a result of low-in-magnitude leaf water potentials, high relative humidity and low vapour pressure deficit (Margolis and Ryan 1997). At some sites and times, they may also reflect the presence of early-morning dew on the canopy. The fall of conductance between mid-morning and late afternoon at the forest sites reflects stomatal control as vapour pressure deficit increases (Margolis and Ryan 1997).

As was observed with EF, the paired sites (fen, old black spruce, old jackpine and young jackpine) each have higher mean conductance in the southern than in the northern study area. There is rather little difference in incoming short-wave radiation or in temperature between the northern and southern study area, and the 5% lower mid-day relative humidity in the northern study area is not sufficient to explain the significantly lower g_s. The higher rainfall in 1994 in the southern study area may explain part of the higher conductance, particularly at the jackpine sites. However, Betts et al. (2001) showed a similar difference between the black spruce sites in 1996, when rainfall was similar in both the southern and the northern study area. Thus, it is not likely that the seasonal atmospheric and soil water constraints between south and north are entirely responsible for the lower surface conductance in the north. Since g_s is a computed bulk stomatal conductance of the vegetated surface (rather than needle-level) the lower g_s in the north may be a result of lower canopy leaf area index.

A.7.2.5 Role of Moss

The wet conifer sites in BOREAS were characterised by extensive moss layers (Goulden et al. 1997; Price et al. 1997), which play an important role in the surface energetics and hydrology. The moss layer, typically tens of centimetres thick, is a thermal insulator, which limits the exchange of heat with the underlying mineral soil, and greatly reduces the annual range of soil temperature (see Fig. A.57 for site #8). In addition, the moss stores surface water after rainfall events, some of which drains and some of which is evaporated to the atmosphere. Price et al. (1997) estimated that about 23% of the precipitation throughfall was evaporated from the moss layer within a few days. Betts et al. (1999) found that evaporation for a black spruce site depended strongly on the wetness of the moss and canopy reservoirs. In contrast, variations in soil water in the underlying soil layer of the spruce root zone had no measurable effect on transpiration, probably because volumetric soil water never fell below 35% in this organic soil layer.

A.7.2.6 Role of Albedo of Forests, Wetlands and Lakes

The high latitude coniferous forests have a very low surface albedo in summer, around 0.08–0.09 for spruce and 0.09–0.13 for jackpine, and rather low values in winter, typically < 0.2 (Betts and Ball 1997), since the canopy shades the snow at low solar elevation angles. In contrast, more open wetland sites have much higher albedos in winter (Joiner et al. 1999b), as do the frozen lakes. The northern forests and wetlands are regions of sharply contrasting albedo, so this plays an important role on both short timescales and the annual timescale.

Figure A.64 shows daily average albedo from one year of the BOREAS mesonet data. In summer, the conifer sites have a very low albedo < 0.1, comparable to the albedo of the lakes, while deciduous and grass sites have a significantly larger albedo. In winter the contrast is very large between forest (where the snow is beneath the canopy and shaded from the low sun angles, and albedo only rarely exceeds 0.2) and open sites and frozen lakes which can have albedos in the range 0.6–0.8, depending on the age of the snow. Errors in the mean albedo of the forest, wetland and lake system within Numerical Weather Prediction (NWP) models can have a large impact on surface temperature forecasts before snowmelt in spring (Viterbo and Betts 1999a). The length of the snow-free or unfrozen period has a marked impact on the annual net radiation balance of open areas and lakes at high latitudes, but less so for the forested areas, where the annual range of albedo is smaller. The presence of the low albedo forest may maintain a higher mean annual temperature than in a region at the same latitude without forest, even though snow melts earlier at open sites.

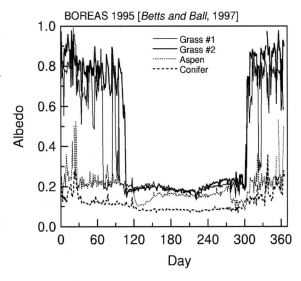

Fig. A.64. Daily average albedo, showing two grassland sites, an aspen site and an average of seven conifer sites (from Betts and Ball 1997, © American Geophysical Union)

Other studies have also indicated the importance of properly representing surface albedo in the snow-covered winter. For example, NWP simulations by Melas et al. (2001), incorporating a land-surface scheme based on Deardorff (1978), were used to study the diurnal variation of the boundary layer structure and surface fluxes during four consecutive days with air temperatures well below zero, snow-covered ground and changing synoptic forcing. Model results were evaluated against *in situ* measurements performed during the NOPEX winter field campaign in the northern NOPEX study area in Sodankylä, Northern Finland, in March 1997 (Gryning et al. 1999). The results show that the land-surface parameterisation employed in the NWP did not estimate day-time sensible heat fluxes correctly, in particular underestimating a pronounced afternoon maximum. This drawback was to a large extent removed by the implementation of a shading factor, as discussed in Gryning et al. (1999), to more accurately account for canopy heating resulting from increased interception of solar radiation at low solar elevations and the implementation of an objective heat storage scheme.

A.7.2.7 Role of Fire-induced Atmospheric Aerosols

Smoke aerosols produced by forest fires modify regional climate (Anderson et al. 1996; Charlson 1997) by (*i*) increasing cloud albedo, (*ii*) reducing surface insolation and (*iii*) influencing cloud formation and precipitation tendencies. The introduction of small particles such as those produced by the biomass burning in the boreal forest, common during the Canadian summer, have been shown to increase cloud albedo by acting as auxiliary cloud condensation nuclei, (Facchini et al. 1999). For the same water content, the average droplet size will diminish as the available number of activated cloud condensation nuclei increases, and the reflectivity of the cloud is consequently enhanced. This same effect also decreases the likelihood that the size threshold for precipitation will be encountered, potentially diminishing or suppressing precipitation in the affected regions (Rosenfeld 1998, 2000b). In addition, the heating of the troposphere as incident solar energy is absorbed by suspended smoke particles typically decreases the environmental lapse rate, and inhibits convective cloud development. This suppression of warm rain processes by heavy smoke conditions has been demonstrated conclusively by the Tropical Rainfall Measuring Mission (TRMM) observations (Rosenfeld 1999).

Fire-induced atmospheric aerosols also can reduce irradiance received at the surface. The single scatter albedo is a measure of the bulk absorption of an aerosol layer that ranges from 0.0–1.0 with lower values representing greater absorption by the aerosol. Observed single scatter albedo values derived from Cimel sunpho-

tometers at five sites in the Canadian boreal forest during 1994 and 1995 averaged 0.95 (standard deviation of 0.03). While these single scatter albedo values are somewhat higher than those typical for biomass burning, they still indicate a substantial reduction in received insolation due to aerosol absorption.

As the background (non-burning) aerosol loadings were typically of the order of 0.1 (aerosol optical thickness at 500 nm), the burning season produces a major modification of the radiation balance with bi-weekly average aerosol optical thicknesses as high as 1.0 for some intervals. This represents a reduction in received photosynthetically active radiation (PAR) irradiance at the surface of approximately 20% for mid-day conditions for a prolonged duration.

Gu et al. (2003) found that volcanic aerosols from the 1991 Mount Pinatubo eruption greatly increased diffuse radiation worldwide for the following 2 years and estimated that this increase in diffuse radiation alone enhanced noontime photosynthesis of a deciduous forest by 23% in 1992 and 8% in 1993 under cloudless conditions. They concluded that "the aerosol-induced increase in diffuse radiation by the volcano enhanced the terrestrial carbon sink and contributed to the temporary decline in the growth rate of atmospheric carbon dioxide after the eruption."

A.7.2.8 Role of Surface Hydrology

Boreal hydrology plays a significant role in both surface energy balance and carbon cycling (see Sect. A.7.3). Evapotranspiration is limited by energy and water availability over much of the year (see Fig. A.56). This constraint tends to increase annual runoff ratios (ratio of annual runoff to annual precipitation). Even though the boreal forest would be a desert on the basis of annual precipitation alone, evaporative demand is low enough that runoff ratios are quite high. The excess of precipitation saturates large regions of clay soils with relatively little topography, creating the great boreal wetlands: lakes, bogs, fens, marshes and forested peatlands. These forested peatlands grow atop relatively shallow water tables (mostly < 40 cm) capped from beneath by an impermeable clay horizon, with black spruce and tamarack rooted in peat soils, kept moist throughout the growing season by a thick moss layer and low evapotranspiration rates (~ 2 mm d^{-1}). Nutrient cycling in this cool, moist root layer is slow, supporting only a sparse above-ground vegetation canopy (low leaf area index) with concomittantly low photosynthetic and evapotranspiration rates. Combined with limited energy availability, this results in a relatively low summertime evaporative fraction and relative humidity (see Fig. A.58–A.62), even during the growing season. Because of the low evapotranspiration rates over land, and the extensive area of lakes, open water plays an important role in the energy balance of the region. Venäläinen et al. (1999), found that while latent heat fluxes from the forest exceeded that from lakes in late spring and early summer, in the boreal region of Sweden and Finland in late summer, lake values exceeded those of the forest. Furthermore, whereas night-time latent heat flux was essentially zero from the forest, it was significant from lakes throughout the summer. These features, along with relatively large areal coverage by lakes, wetlands and fens (especially in the North American boreal forest) also result in a highly damped stream flow from snowmelt runoff and occasional intense summer precipitation. Representation of these effects of surface storage on water and energy balances is a key feature of boreal hydrology that is not presently well represented in land-surface models.

Boreal hydrology also supports an important carbon storage generation mechanism in the spruce/tamarack wetlands, bogs and fens. Runoff storage in the wetland peats leads to low decomposition rates below ground with high root turnover, resulting in carbon storage rates ranging from 30 to 160 g C $m^{-2} y^{-1}$ (e.g. Gorham 1991; Trumbore et al. 1999; Alm et al. 1997). Depending on the degree of soil saturation, anaerobic decomposition of organic materials in the boreal wetlands can generate methane (see Sect. A.7.3.5). Very wet sites, such as peatlands and beaver ponds, are generally sources of methane, ranging from from 0–20 mg $CH_4 m^{-2} d^{-1}$ for frozen palsas[2] and peat plateaus to ~ 380 mg $CH_4 m^{-2} d^{-1}$ for open, graminoid fen sites (Bubier et al. 1995). Forested peatlands are generally associated with drier conditions and lower methane emission rates; permafrost collapse scars are associated with wet conditions and higher emission rates (Bubier et al. 1995). Beaver ponds emit both methane and carbon of about 50–150 mg $CH_4 m^{-2} d^{-1}$, and carbon dioxide fluxes of about 2–10 g $CH_4 m^{-2} d^{-1}$ (Roulet et al. 1997).

In wintertime, snow plays a central role in boreal hydrology. Even although the total fraction of annual precipitation that falls as snow in boreal forests is often well less than half, spring snow accumulations commonly account for much more than half of the annual runoff (see Bowling et al. 2000, for a map of this fraction over the Arctic drainage basin). Winter snowpacks, although present for much of the year, are usually quite thin, because temperatures are so low that relatively little precipitation occurs during the coldest mid-winter months. Because the snowpacks are quite thin, snow redistribution by wind, and accompanying sublimation, are important processes that affect peak snow accumulations at the onset of spring snowmelt. By some estimates (Pomeroy

et al. 1997) sublimation can result in a loss of one-third or more of the winter snow accumulation.

BOREAS results suggest that the fraction of runoff originating as snow is highest near the northern boundary of the boreal forest. At the gauged streams in the BOREAS northern study area, for instance, there was always a well-defined spring snowmelt peak in the seasonal hydrographs, whereas in the southern study area near the southern boreal forest boundary, the spring peak was less distinct, and in some years accounted for only a modest fraction of the annual runoff.

A.7.2.9 Scaling Energy and Water Flux from the Plot to the Region

The last few years of field measurements of energy, water and heat fluxes at scales of a kilometre or less, over relatively homogeneous surfaces, have resulted in the evaluation and refinement of soil-vegetation-atmosphere transfer schemes (SVATs) that permit reasonably accurate computation of energy, water and heat fluxes from the surface to the atmosphere. These models use as input, incident radiation, land cover structural characteristics and vegetation biophysics. SVAT models that represent these surface processes correctly are critical to the performance of boundary layer dynamics models that compute boundary layer properties as a function of surface conditions. These boundary layer properties are in turn crucial to parameterising general circulation models of the atmosphere, GCMs. Accurate GCMs are necessary to understand the relationships between climate and vegetation at global scales.

Even though the resolution of GCMs is steadily falling (the ECMWF 10-day forecast model has a current resolution of 40 km), GCMs must somehow either incorporate or ignore the effects of a heterogeneous landscape on surface fluxes. However, questions arise as to whether the average properties of a heterogeneous surface, used as input to a one-dimensional SVAT model, will compute energy, water and heat fluxes properly over the heterogeneous surface. A number of scaling studies have investigated this question. One strategy that has been employed is tiling (Koster and Suarez 1992; Van den Hurk et al. 2000).

Batchvarova et al. (2001) have shown that even a sparse canopy coverage by trees may have an influence on the area-averaged fluxes of momentum out of proportion to the area covered by the sparse trees, indicating that area averaging of surface roughness is a process that requires attention. The heat fluxes may be influenced to such a degree that the area-average heat flux may run counter to the area-averaged potential temperature gradient.

The regional sensible heat flux over the southern NOPEX experimental area, an inhomogeneous region with patches of forest, agricultural fields, mires and lakes

---

[2] Palsa is a type of wetland created when permafrost melts beneath the soil and creates a collapse of the land above the permafrost. These collapsed "scars" generally are circular, sunken beneath their surroundings, are poorly drained, thus support a unique type of vegetation such as brown moss.

at the southern boundary of the boreal zone, was estimated for three days of the campaign in 1994 (Gryning and Batchvarova 1999). It was found to be lower than the heat flux over forest and higher than the heat flux over agricultural fields. The regional sensible heat flux estimated by the mixed-layer evolution method was compared to a land use-weighted average sensible heat flux. The two independent estimates of the regional heat flux were found to be in general agreement. This result is supported by the findings of Gottschalk et al. (1999).

Based on measurements from the NOPEX winter experiment Batchvarova et al. (2001) estimated the regional (aggregated) momentum and sensible heat fluxes for two days over a site in Finnish Lapland during the late part of the winter. The study shows that the forest dominates and controls the regional fluxes of momentum and sensible heat in different ways. The regional momentum flux was found to be only slightly smaller than the measured momentum flux over the forest, although the forest (deciduous and coniferous) covered only 49% of the area. The regional sensible heat flux was estimated to be 30 to 50% of the values measured over a coniferous forest. This percentage corresponds roughly to the areal coverage of coniferous forest in the area (see Chapt. B.10 for further discussion).

A.7.3 Biospheric Carbon Exchange: Carbon Dioxide and Methane

The major terms in the net carbon balance of a boreal ecosystem are carbon uptake via photosynthesis (gross primary productivity, GPP), carbon loss by the autotrophic respiration of vegetation (Ra), and carbon loss by heterotrophic respiration (Rh) of decomposers. The net carbon assimilation by the vegetation is called net primary production (NPP = GPP – Ra). In boreal forest ecosystems Ra is typically more than 50% of GPP (Ryan et al. 1997). Net ecosystem exchange (NEE) is equal to net uptake by vegetation minus release by decomposition:

$$NEE = NPP - Rh = GPP - Ra - Rh \qquad (A.14)$$

NEE is generally a small difference between much larger gross photosynthesis and respiration, and can be near zero for some mature forest stands (e.g. Goulden et al. 1997). Additional pathways of carbon loss from boreal ecosystems are generally either small or very large but episodic. The smaller pathways include volatile organic carbon (VOC) emissions from foliage (Lerdau et al. 1997) and dissolved organic and inorganic carbon (DOC and DIC) losses with groundwater flow (e.g. Aitkenhead and McDowell 2000). The larger but episodic pathways include harvest and fire.

A.7.3.1 Measurement Methods

Ecosystem carbon balances can be measured in three fundamentally independent ways (Table A.8): (*i*) eddy correlation methods, (*ii*) carbon stock methods or (*iii*) the inverse method.

The eddy correlation method places high frequency sonic anemometers and gas sampling inlets above the canopy, and correlates variance in CO_2 concentrations with variance in vertical air motion to assess net CO_2 flux between the canopy and the atmosphere (e.g. Wofsy et al. 1993). In theory, near-continuous measurements of CO_2 exchange can be integrated to determine NEE over a measurement period. In practice, there are a number of challenges and uncertainties associated with the eddy correlation method (e.g. Goulden et al. 1996; Grelle and Lindroth 1994, 1996). The eddy correlation method provides a very good picture of how the ecosystem responds to weather at hourly to seasonal time scales, and is an important tool for developing and evaluating process-based ecosystem models. See Chapt. B.4 for further description of this methodology.

The carbon stock method requires actual measurement of the carbon in soils and vegetation at a number of stands. In principle, net carbon sequestration at a site can be measured directly if a site is visited repeatedly over a number of years. However, net sequestration is often a very small fraction of the standing carbon stocks, and thus sequestration can be difficult to detect. Often,

Table A.8. Methods for assessing ecosystem carbon balance

	Eddy correlation method	Carbon stock method	Inverse method
Observation	Net carbon fluxes	Carbon in biomass and soils	Atmospheric [CO_2], transport, sources
Inference	Net C sequestration	Net C flux	Net C flux and sequestration
Measurement effort	Intensive	Extensive	Very extensive
Characterise ecophysiology	Yes	No	No
Disaggregate components	Possible with enough measurements	Yes	No
Other uses	Process model validation and parameterisation	Chronosequences: ecosystem evolution	Regional view, large-scale evaluation

stand-age chronosequences are measured and carbon sequestration is inferred from these data under the assumption that all sites have followed or will follow a similar successional/developmental trajectory (e.g. Rapalee et al. 1998). This carbon stock method can provide information on the contributions of the different components of the ecosystem (trees, mosses and ground vegetation, soils) to total carbon pools and carbon sequestration.

The inverse method combines observed gradients in atmospheric CO_2 concentration and isotopic ratios, modelled or measured atmospheric transport, and known anthropogenic CO_2 sources to infer net CO_2 exchange at the land surface (e.g. Heimann and Kaminsky 1999). As the current network of atmospheric data is very large-scale (e.g. continental or larger), only inferences can be made (e.g. Fan et al. 1998). However, both the other methods are quite localised, so the inverse method provides a meaningful constraint on regional carbon fluxes. Ultimately, the three methods should be consistent with, and constrain, each other.

A.7.3.2 Controlling Factors (above and below Ground)

Rates of photosynthesis, respiration, and decomposition have a number of biophysical and ecological controls. Controls on seasonal and interannual variability are discussed in Sect. A.7.3.4. Base photosynthetic rates for foliar tissue are generally correlated with foliar nitrogen content (e.g. Field and Mooney 1986; Reich et al. 1997; Dang et al. 1997). Needle-leaf evergreen trees, which dominate the boreal forest, generally have low tissue nitrogen content and low photosynthetic rates (e.g. Reich et al. 1997). *GPP* and *NEE* rates for boreal peatlands are generally lower than for temperate ecosystems (Frolking et al. 1998). Incident PAR has a very strong control on photosynthetic rates, both over a diurnal cycle, and on day-to-day variability (e.g. Goulden et al. 1997; Suyker et al. 1997; Jarvis et al. 1997; Joiner et al. 1999a). Vapour pressure deficits greater than about 1 kPa have been shown to cause stomatal closure and limit photosynthetic and transpiration rates for boreal trees (Dang et al. 1997; Hogg and Hurdle 1997; Hogg et al. 1997; see also Sect. A.7.2.3). However, a strong vapour pressure deficit control on NEE and GPP of black spruce (*Picea mariana* (Mill.) BSP) stands was not observed (e.g. Jarvis et al. 1997; Goulden et al. 1997).

Lavigne et al. (1997) estimated that, for six coniferous boreal stands, soil respiration (roots plus heterotrophs) was ~ 50–70% of total ecosystem respiration; foliar respiration was ~ 25–40% of total ecosystem respiration, and above-ground woody tissue respiration was ~ 5–15% of total ecosystem respiration. Ryan et al. (1997) estimated

that root respiration was about half of total autotrophic respiration. All respiration rates showed temperature dependence with Q10 values (i.e. rate increases for a 10 °C temperature increase) of ~ 1.5 for wood, ~ 1.9 for roots, and ~ 2.0 for foliage (Ryan et al. 1997). Goulden et al. (1997) fit an exponential function to total ecosystem respiration as measured by eddy correlation in a boreal black spruce stand that is equivalent to a Q10 value of about 2.0. Savage et al. (1997) reported a Q10 value of 2.6 for temperature dependence of soil respiration measured with manual chambers in the same black spruce stand. Goulden and Crill (1997) reported a Q10 value of about 2.0, based on automated chamber measurements in this stand. Bubier et al. (1998) reported Q10 values of 3.0 to 4.1 for a range of peatland sites using chamber measurements. Water table is also an important control on respiration in boreal wetlands because of slower rates of CO_2 emission in anaerobic conditions (Moore and Dalva 1993; Silvola et al. 1996; Bubier et al. 1998). Peatlands can switch from being a net sink of carbon to being a net source in exceptionally dry periods due to increased aerobic respiration rates (Alm et al. 1999; Bellisario et al. 1998).

A.7.3.3 Geographic Variations in Carbon Flux

Above-ground NPP (ANPP) rates of 120–350 g C $m^{-2} y^{-1}$ were measured for forest stands in the BOREAS study area in central Canada, within the range of data from a number of boreal sites around the world (Gower et al. 1997). There are fewer estimates of below-ground NPP for boreal forests, but Steele et al. (1997) and Ruess et al. (1996) estimated fine root NPP to be roughly 50–100% of above-ground litterfall for deciduous stands and 100–200% of above-ground litterfall for evergreen stands in North America. Analysing eddy correlation carbon flux data from a network of towers ranging from northern to southern Europe, Valentini et al. (2000) concluded that much of the observed variation in NEE between these sites was due to differences in net ecosystem respiration, while GPP was fairly uniform across the sites. Ecosystem respiration generally increased, south to north, from roughly 50% to nearly 100% of GPP. Valentini et al. (2000) postulated that this may be due to enhanced Rh as a result of a disequilibrium in soil organic matter pools at the more northern sites, due both to climatic warming and site disturbances. At an evergreen boreal stand in Canada, Goulden et al. (1998) observed high ecosystem respiration during a warm summer, which they attributed primarily to anomalous warming/thawing of soil organic matter about 0.5–1.0 m below the surface. However, at a deciduous boreal stand in Canada, Black et al. (2000) observed very little variation in the annual total ecosystem respiration over four years, despite ~ 3.5 °C variation in mean annual temperature.

Lindroth et al. (1998) and Morén (1999) measured CO_2 fluxes using eddy correlation above a mixed pine and spruce forest in central Sweden during a two-year period, from June 1994 to May 1996. During that period, the forest was observed to be a net source of carbon. Analysis of night-time fluxes showed that 79% of the variation in night-time CO_2 respiration could be explained by an exponential temperature function. Using this function to estimate respiration during day-time allowed gross uptake to be estimated as well. In 1995, the gross uptake was ~ 1250 g C m^{-2} y^{-1} and the total ecosystem respiration was ~ 1500 g C m^{-2} y^{-1}. Typical eddy correlation measurements of growing season, mid-day NEE for boreal ecosystems are an uptake of 3–25 g CO_2 m^{-2} y^{-1} (e.g Sellers et al. 1997; Lindroth et al. 1998).

Millennial-scale carbon sequestration rates in peatlands have been estimated by dividing total carbon content by basal age, and average about 30 g C m^{-2} y^{-1} (e.g. Gorham 1991). Using three different methods Trumbore et al. (1999) estimated C sequestration rates of 0–160 g C m^{-2} y^{-1} for different components of a boreal peatland in central Manitoba, Canada. Alm et al. (1997) used the static chamber method to measure a mean C sequestration rate of about 70 g C m^{-2} y^{-1} for a boreal peatland in Finland. Estimates of boreal peatland ANPP range from about 100 to 500 g C m^{-2} y^{-1} (e.g. Trumbore et al. 1999; Thormann and Bayley 1997). There are few measurements of below-ground NPP for peatlands (e.g. Backéus 1990; Saarinen 1996), and estimates must be considered extremely uncertain.

Regional-scale rates of NEE, NPP, and methane flux are very difficult to measure, due to the large spatial and temporal domains needed. Typical eddy correlation towers generate a long time series but sample only a relative small area (< 1 km^2) (see Sect. B.4.3.4). Eddy correlation instrumentation on aircraft can cover larger spatial domains but generally have very poor temporal coverage (see Chapt. B.5). Atmospheric sampling of CO_2 concentrations, combined with inverse modelling, can represent large area and time domains but with poor temporal and spatial resolution. Combining land-cover maps derived from remote sensing (e.g. Steyaert et al. 1997) with ecosystem models has been used to generate regional flux estimates of CO_2 (e.g. Liu et al. 1999b). Similarly, combining remote sensing land-cover maps with site level flux data has been used to generate regional methane flux estimates (Roulet et al. 1994; Reeburgh et al. 1998). In a model scaling exercise, Kimball et al. (1999) showed that results were sensitive to the scale of resolution of the landscape complexity. Small landscape features that can have large fluxes (e.g. beaver ponds for CO_2 and CH_4, peatlands for CH_4) can be difficult to resolve and quantify at the landscape scale, and can thus lead to significant uncertainty in regional analyses. Rates and intensity of disturbance of boreal ecosystems by fire,

insect defoliation, harvest, or permafrost degradation will also have a significant impact on the regional carbon balance in any year, and on the long-term carbon balance of any site (e.g. Harden et al. (1997); also see papers in Apps and Price (1996), and Apps et al. (1995)). These effects are not well quantified at this time.

A.7.3.4 Seasonal and Interannual Variations in Carbon Flux

Ecosystem carbon fluxes show a strong seasonal signal, with maximum uptake rates in early and mid-summer, and maximum loss rates in early autumn and late spring. Boreal field sites with year-round eddy correlation observations show low but persistent and significant net loss of carbon throughout the long, cold winter months (Goulden et al. 1997; Lindroth et al. 1998; Black et al. 2000). Winter soil flux measurements show that much of this respiration comes from below-ground and is highly variable, both spatially and temporally (Winston et al. 1997). Dormant season Ra accounts for about one-quarter of yearly total plant respiration (Ryan et al. 1997).

Eddy correlation measurements of CO_2 exchange have been made at a number of boreal stands, but only a few sites have reported near-continuous, year-round data for more than one year. After two years of eddy correlation measurements at a mature pine and spruce forest in central Sweden, Lindroth et al. (1998) estimated a net loss of about 200 g C m^{-2} y^{-1}. They postulated that the forest cannot sustain a persistent loss of C at these rates, and that the two years of observation may have been anomalous due to a dry summer in one year reducing photosynthesis, and to some anomalously warm months enhancing respiration. After nearly two and a half years of eddy correlation measurements at a mature spruce stand in central Manitoba, Canada, Goulden et al. (1997) estimated that the gross photosynthesis and total respiration were nearly in balance at about 800 g C m^{-2} y^{-1} each. Despite being an evergreen forest, gross photosynthesis was quite low and insensitive to air temperatures ranging from –10 to +10 °C during April, before the ground had thawed. Gross photosynthesis increased with increasing air temperature during May–October. After four complete years of eddy correlation measurements at a boreal deciduous forest in central Saskatchewan, Canada, Black et al. (2000) reported net carbon sequestration of 80 to 290 g C m^{-2} y^{-1}. A warm, early spring in 1998 caused an earlier leaf-out, and these additional two to four weeks of active photosynthesis led to the large NEE, more than double that of the other three years. Total ecosystem respiration in 1998 was similar to the other three years (Black et al. 2000). Measurements at the black spruce forest in Manitoba also showed an early start to the active growing season in 1998, and annual NEE at the site switched from a weak carbon source to a weak carbon sink for that year (Steve Wofsy, pers. comn.). Because the annual air temperature wave lags the annual insolation wave by about a month at these latitudes, by the time boreal ecosystems thaw in spring, light levels are quite high. In addition, the ecosystems warm rapidly and water is abundant, so net productivity can be high early in the active growing season. By contrast, when ecosystems freeze-up in the autumn (October–November) light levels and photosynthesis rates are low. Thus gross productivity is much more sensitive to variation in the timing of spring thaw than autumn freeze-up (Frolking 1997).

A.7.3.5 Methane

Four primary controls on methane fluxes have been identified. Net methane emission generally increases with increasing soil temperature, with increasing soil moisture and/or shallower water tables, with increasing conductance through plant aerenchymous tissue, and with increasing substrate availability (e.g. Crill et al. 1991; Chanton and Dacey 1991). For upland soils that are net consumers of atmospheric methane, the primary control of flux rates is gas diffusion rates, which are closely related to soil moisture content. For peatlands that are net emitters of methane, Bubier et al. (1995) found that mean seasonal peat temperature at the average position of the water table explained most of the spatial variability in mean seasonal CH_4 fluxes for a diverse peatland

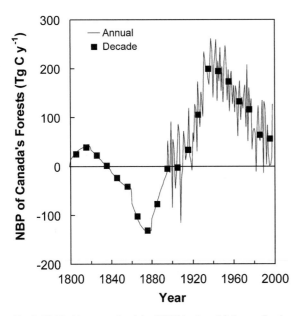

Fig. A.65. Net biome productivity (NBP) in Canada's forests for the last two centuries (expanded after Chen et al. 2000). In the estimate, disturbance (fire, insect-induced mortality, harvest) and non-disturbance (temperature, growing season length, nitrogen deposition, CO_2 fertilisation) factors are considered

complex in Manitoba, Canada. Vegetation assemblages are generally correlated with the biophysical controlling factors, and can be useful for scaling flux estimates to larger regions (Bubier et al. 1995). Net ecosystem production is a good predictor of methane emission in flooded wetland soils (Whiting and Chanton 1993), but this relationship weakens when the water table is below the surface of the peat (Bellisario et al. 1998).

Methane is both produced and consumed by microbial processes in boreal soils. Boreal ecosystem methane uptake rates measured with static chambers in drier upland soils (~ 0.1–1.0 mg CH_4 m^{-2} d^{-1}; e.g. Moosavi and Crill 1997; Savage et al. 1997) are generally comparable to rates in other ecosystems. Continuous methane flux measurements over a five-month growing season at an aspen stand in Saskatchewan, Canada, using a micrometeorological tower and the flux-gradient method, averaged an uptake of about 1.4 mg CH_4 m^{-2} d^{-1} (Simpson et al. 1997). Wetter sites can exhibit episodes of methane emission and methane uptake, generally depending on degree of soil saturation (e.g. Moosavi and Crill 1997; Simpson et al. 1997). Very wet sites, such as peatlands and beaver ponds, generally are net sources of methane (e.g. Bubier et al. 1995; Roulet et al. 1997; Moore and Roulet 1993). Seasonal average emission rates from a peatland complex in Manitoba, Canada ranged from 0–20 mg CH_4 m^{-2} d^{-1} for frozen palsas and peat plateaux to ~ 380 mg CH_4 m^{-2} d^{-1} for open, graminoid fen sites (Bubier et al. 1995). Forested peatlands are generally associated with drier conditions and lower methane emission rates; permafrost collapse scars are associated with wet conditions and higher emission rates (Bubier et al,. 1995). Beaver ponds emit both methane and carbon dioxide; Roulet et al. (1997) using a micrometeorological tower and the flux-gradient method, measured methane emission rates of about 50–150 mg CH_4 m^{-2} d^{-1}, and carbon dioxide fluxes of about 2–10 g CO_2 m^{-2} d^{-1}.

A.7.3.6 The Effect of Landscape Patterns, Disturbance and Succession on Carbon Cycling

As shown in Fig. A.65, the high disturbance rates in late 19th century caused Canada's forests to release C at a rate of ~ 30–120 Tg C y^{-1} (Chen et al. 2000). During 1930–1970 the low disturbance rates, in combination with forest regrowth in areas disturbed in the late 19th century, allowed Canada's forests to uptake C at a rate of ~ 100–200 Tg C y^{-1}. The increased disturbances during 1980–1996 released 60 Tg C y^{-1} from Canada's forests. Similar results were reported by Kurz et al. (1995). However, the positive effects on forest growth of nitrogen deposition, enhanced nitrogen mineralisation and increased growing season length with increasing temperature and increased atmospheric CO_2 concentration might have outweighed the negative effects of the increase in disturbance rates in recent decades, making Canada's forests currently a small sink of ~ 50 Tg C y[-1] on average (Chen et al. 2000). Among the four disturbance factors, fire had the most important contribution to the inter-annual and decadal variations of net biome productivity (NBP), followed by insect-induced mortality.

A.7.4 Sensitivity Experiments

Climate model sensitivity simulations of the response to increasing atmospheric CO_2 concentrations indicate that the rapid warming could occur at these high latitudes (43–65° N) with the most marked effects within the continental interiors, see Houghton et al. (1995), Mitchell (1983), Schlesinger and Mitchell (1987), Sellers et al. (1996a). Analyses of long-term, ground-measured surface temperature records are consistent with these model simula-tions, showing increasing temperature trends of as much as 1.25 °C per decade) (Chapman and Walsh 1993) with much of this warming occurring in the spring and fall.

Such perturbations in northern continental climates could well lead to significant changes in the carbon cycle and the ecological functioning of the boreal forest, which could feedback onto the global climate. The boreal ecosystem may be responding to such warming by increasing vegetation growth. Studies by Keeling et al. (1996) have found recent increases in the amplitude of the atmospheric CO_2 concentration and a broadening of the seasonal cycle by as much as six days, suggesting increased vegetation growth. Results from BOREAS indicate that an earlier onset of spring photosynthetic activity leads to large increases in annual vegetative carbon uptake (Sellers et al. 1997). Satellite studies of high latitude vegetation show concomitant increases in both leaf area duration and vegetation density (Myneni et al. 1997) over the past decade.

However, an increasingly warm, dry boreal climate could potentially shift the ecosystem from being a net sink of carbon to a net source. Results from BOREAS (Hall 1999) and NOPEX (Halldin et al. 1999) also show that if climate warming leads to drying and warming of the boreal soils, accelerated decomposition of the organic matter stored in the peat soils and organic soil horizons could potentially release huge quantities of stored soil carbon to the atmosphere (Sellers et al. 1997; Lindroth et al. 1998).

A.7.4.1 Snow Albedo and Climate Feedback

A number of studies also indicate the influence of boreal land-surface/atmosphere interactions on global weather patterns and interannual variations in climate. For example, Viterbo and Betts (1999a) using the European Centre for Medium-Range Weather Forecasts (ECMWF) model, showed that varying assumptions about the albedo of the snow-covered boreal ecosystem can lead to wide differences in spring temperature forecasts across North America and, in fact, throughout the Northern Hemisphere, Asia and the western Pacific.

The spring snow cover feedback mechanism observed by Groisman et al. (1994) in which decreased snow cover reduces surface albedo, increases surface absorption of radiation, raises surface temperatures, and hence increases snow melt, affects seasonal atmospheric circulation as well as long-term climate. As air temperatures warm, the snow cover feedback acts to either amplify the rate of retreat of the snow extent, and its seasonal duration, or mitigate it.

A.7.4.2 Carbon Sequestration and Radiative Feedbacks

Bounoua et al. (1999) used a coupled biosphere-atmosphere model to examine the radiative and physiological effects on climate from a doubling of the atmospheric carbon dioxide concentration. Their study demonstrated the importance of accounting for vegetation response to climate change. When only the radiative effects from doubled CO_2 were taken into account, a mean temperature increase of 2.6 K over land was predicted with a 7% increase in precipitation. However, the changes varied with latitude, with an increase of 4 K at boreal latitudes (> 50° N), but only 1.7 K over the tropics. As the physiological effects of vegetation were added, the predicted temperature increase over land increased only slightly, but a smaller difference of 0.7 K was observed between the boreal and tropical latitudes. This smaller difference decreased evaporation and transpiration rates of land vegetation, a result of increased atmospheric CO_2 concentrations.

Other climate change sensitivity experiments also indicate that boreal latitudes could undergo warming, resulting in the extension of the frost-free season by one to nine weeks in North America (Brklacich et al. 1997). For the Prairies, Ontario and Québec, the model results suggest an extension of three to five weeks, while for Atlantic Canada near the current agricultural margin the season is three to four weeks longer. Air temperature during the frost-free season is also simulated to increase. Estimates related to agricultural moisture regimes show a broad range compared to the thermal regimes. For instance, in the Prairies and Peace River regions, simulated precipitation changes range from decreases of about 30% to increases of about 80%. As the frost-free season becomes longer and warmer, most model simulations suggest higher potential evaporation and transpiration rates and thus the larger seasonal soil moisture deficits (Brklacich et al. 1997).

A.7.4.3 Effects of Climate Change on Land Cover

Based on pollen records, species would shift northward with increasing temperatures (Oechel and Vourlitis 1994). In the northern regions, species richness would be reduced through the loss of less abundant species and a move to species characteristic of other biomes (Chapin et al. 1995). The area of boreal forest in Québec could decrease by as much as 20% (Singh 1988). Solomon and West (1987) simulated tree dynamics in north-western Ontario and found that the dominance of spruce slowly gave way to an increasing proportion of sugar maple and white pine that entered the canopy in a $2 \times CO_2$ sensitivity experiment. Plöchl and Cramer (1995) used an ecosystem model to locate tundra and taiga distributions under a $2 \times CO_2$ sensitivity experiment and suggested that both biomes will diminish in extent when forced to shift northward. Many of these simulations are based on equilibrium states, which may never be realised. More realistic estimates for land-cover changes (e.g. since industrialisation) in response to the past climate are not yet found in the literature.

In assessing the potential for agricultural expansion to areas in north-west Canada (i.e. north of 55° N and west of 110° W) and Alaska, Mills (1994) identified 57 Mha of potentially arable land. This estimate drops to 39 Mha when climatic limitations are imposed, but under a sensitivity experiment of $2 \times CO_2$ it increases to 55 Mha. Similar conclusions have been reached for the northern agricultural regions in Québec and Ontario (Brklacich et al. 1997). Agricultural land potential north of 60° N is generally considered not to be sensitive to these simulated climate changes.

The studies suggest that warmer frost-free seasons under the full range of global climatic change sensitivity experiments would increase the rate of development of grain crops, reducing the time between seeding and harvesting (e.g. by about one-and-one-half to three weeks in most regions of Canada for spring-seeded cereals and coarse grains) (Brklacich et al. 1997). In northern regions, the simulated decrease in maturation time would reduce the risk of frost-induced crop injury, a decided benefit for these regions. In the Peace River region, the positive and negative influences tend to counteract each other, reducing the magnitude of the positive impacts of climatic change on cereal yields. In the Prairies, the effects of climatic change on grain yields are more pronounced and variable, with spring-seeded cereal yields in the western Prairies reducing by as much as 35% and those in the eastern Prairies increasing by as much as 66%. Similar results are suggested for Ontario and Québec, except for the more generally positive impacts on grain production, especially corn, in northern Ontario and Québec.

A.7.4.4 Hydrological Feedbacks

The hydrology of boreal regions is particularly susceptible to climate (Bowling et al. 2000; Nijssen et al. 2001). Near the southern border of the boreal forest, winter snow accumulations, and hence spring runoff, might well decrease. However, parts of the boreal forest that now have extremely cold winters could see increased winter precipitation according to some climate models. This, in conjunction with earlier onset of spring, would change the seasonal flow of the northern rivers, most likely shifting peak runoff associated with spring snow-melt to earlier in the year, and possibly reducing summer flows. Furthermore, both short-term feedbacks in the climate – due to changes in albedo associated with variations in the length of snow cover, especially near the northern limit of the boreal forest, and tundra areas where vegetation effects on albedo are least – may well occur due to changes in the surface radiative balance, especially in spring. Likewise, changes in the amount of discharge, and the seasonality of runoff of the major north-flowing rivers, could affect the global thermohaline circulation, due to both direct effects on the freshwater balance and to indirect effects on albedo of ice-free portions of the Arctic Ocean. More local effects would almost certainly occur due to changes in depth of the active (seasonally thawed) layer in permafrost areas, which are particularly important near the northern boundary of the boreal forest. These fundamental changes in growing season length, and soil properties, would almost certainly affect the northern and southern boundaries of the boreal forest, which would have accompanying changes on surface hydrology.

During the summertime, increases in precipitation at northern latitudes (Dai et al. 1997) could also have significant impacts on boreal hydrology and, through the mechanisms discussed in Sect. A.7.2.8, surface energy balance and carbon cycling and storage rates. For example, if increased precipitation trends lead in the long-term to a shallower water table overall, nutrient cycling rates could potentially decrease further, leading to decreased summertime evapotranspiration. On the other hand, drying and warming of the boreal climate could lead to increases in the water table depth, increased above-ground productivity, but also increased decomposition rates of the peat soils. The balance between these two processes could potentially change the boreal ecosystem from a net sink to a net source of carbon, or vice versa. This potential coupling between subsurface hydrology, carbon, energy and water cycling needs to be explored further.

Chapter A.8

The Asian Monsoon Climate

Congbin Fu · Tetsuzo Yasunari · Sabine Lütkemeier

A.8.1 Introduction

Temperate East Asia comprises a major portion of the Earth's largest continent; and it is bordered by the planet's highest mountains and largest ocean. Mainly due to strong land-ocean thermal contrast and the dynamic and thermal effects of the Tibetan Plateau, East Asia has a well-developed monsoon climate system. The life of a human population of more than 1.5 billion in East Asia depends on monsoon rainfall and also suffers from drought and flood disasters related to the high variability of monsoon climate. Moreover, East Asia is the homeland of some of the world's oldest, most advanced and most rapidly evolving human civilisations and human activities have powerfully influenced the changes in every aspect of the environment, including the monsoon climate, as will be shown in this chapter.

The Asian Monsoon is well known as a land-atmosphere-ocean system. The land-ocean heating contrast between the Tibetan Plateau and the Indian Ocean produces strong south-westerly monsoon flow across the Equator, which transports huge amounts of water vapour over south, South-east and East Asia. Another water vapour channel is located from the tropical Pacific Ocean toward east and South-east Asia. Water vapour transport from the Indian and Pacific Ocean through these two main channels plays an essential role in maintaining large-scale cumulus convection and precipitation over humid Asia, or "monsoon Asia", which reinforce the monsoon circulation system through latent heat release. The Asian monsoon is thus itself a huge water recycling system resulting in a humid climate and considerable vegetation dominating the eastern half of the Eurasian continent. This humid monsoon climate maintains a meridionally-oriented dense vegetation zone called the "green belt" from tropical South-east Asia to sub-polar Siberia, which is one of the centres of biodiversity on Earth, and a centre of world population. In addition to the moisture transport from the oceans, evapotranspiration from the land surface of the continent also plays an essential role in the water cycle, which is controlled by surface conditions, e.g. vegetation types,

soil moisture, albedo, surface roughness, etc. These surface conditions are easily modified by human activities, and may change the water cycling process between land and atmosphere which, in turn, may affect the local to regional climate of monsoon Asia.

The first section of this chapter focuses on the basic time-space characteristics of the water cycling processes over the monsoon Asian and Eurasian continent, based on observed data. Two recent modelling studies will then be introduced that investigate the possible weakening and intensifying of monsoon rainfall due to human-induced land-cover/land-surface change.

The second section emphasizes the actual land-use and land-cover changes and their impact on the East Asian summer monsoon. Historical changes introduced by hu-man activities over the last 3 000 years will be taken into account as well.

Tetsuzo Yasunari

A.8.2 Role of Human-induced Large-scale Land-use/cover Change on the Water Cycle and Climate in Monsoon Asia

Land-surface conditions are easily modified by human activities and may change the water cycling process between land and atmosphere, which, in turn, may affect the local to regional climate of monsoon Asia. These human-induced effects on the monsoon climate may, however, differ in strength and spatial extent, depending upon the basic atmospheric conditions at the surface. In other words, the human-induced change in monsoon climate may occur (strengthened or weakened) selectively from one specific season to another, and from one geographical region to another.

A.8.2.1 Atmospheric Water Cycle over Monsoon Asia

Precipitation is the most important factor for life on land, i.e. for vegetation as well as water resources for human beings. What, then, contributes to the origin of precipitation (*P*) in monsoon Asia and the interior of the Eura-

Fig. A.66.
Seasonal atmospheric water balance over various regions in monsoon Asia (India, Indo-China, Middle China Plain) by using the objectively-analysed global meteorological data for 15 years (1979–1993) from ECMWF (Yasunari et al. 2000a). *Blue*: Precipitation (*P*); *black*: water vapour flux convergence (*C*); *green*: evapotranspiration (*E*)

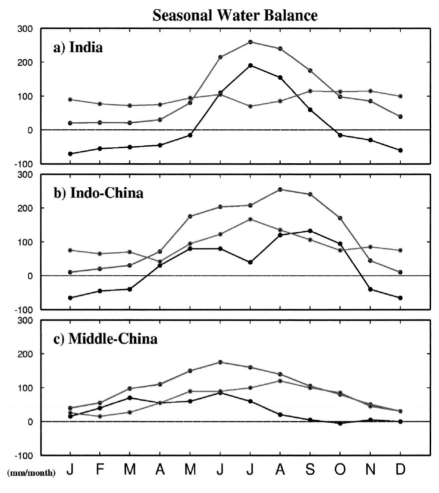

sian continent? In tropical monsoon Asia, water vapour transport (with its convergence (*C*)) should be a major moisture source for precipitation. Evapotranspiration (*E*) is another source for precipitation. The analysis of the atmospheric water balance over various regions using the objectively-analysed global meteorological data for 15 years (1979–1993) from ECMWF forecasts reveals time-space characteristics of $P - C - E$ (Yasunari et al. 2000a) as shown in Fig. A.66. In the humid tropics, both *C* and *E* contribute nearly equally to *P*. In fact, over the Indian subcontinent and the Indo-China peninsula, the contribution of *C* is larger than that of *E*. However, this feature changes from month to month. In the mature phase of the monsoon season, *C* plays a major role in *P*, but at the beginning and ending stages, the contribution of *E* is relatively large. In the East Asian monsoon region (central China plain), the contribution of *E* tends to become larger, presumably due to a sub-seasonal change in the atmosphere (from the Meiyu (or Baiu in Japanese) frontal rain period to hot summer under the subtropical high) and surface wetness condition. The interannual variability of *P* during the mature phase of the rainy season is highly correlated with *C*, both in the

south-east and east monsoon region, associated with the variability of the large-scale monsoon and atmospheric circulation. However, the large contribution of *E* to *P* at the beginning and end of the rainy season demonstrates the essential roles of evaporation from wet surfaces and transpiration from vegetation.

A.8.2.2 Is Monsoon Rainfall Decreasing? The Impact of Deforestation in Thailand on the Water Cycle

The long-term record of rainfall in Thailand since 1950 shows not only the interannual variability but also the long-term decreasing trend, particularly in September, when the monthly maxima appear in most parts of the country. Figure A.67 includes two time series of rainfall for September in Sakhon Nakhon and Phitsanuloc, showing considerable decrease since the 1970s. One possible cause of this decreasing trend may be related to the decadal-scale change in the frequency of rain-bearing weather disturbances (e.g. typhoon and tropical depressions) coming from the tropical western Pacific, as-

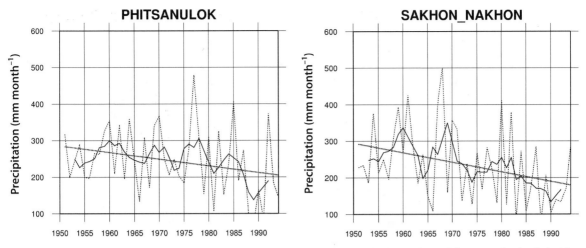

Fig. A.67. Interannual variation of rainfall amount in September in Sakhon Nakhon and Phitsanuloc, with linear trends. The *dashed line* denotes annual values, the *solid line* denotes the trend by linear regression (from Kanae et al. 2001)

sociated with the change in the atmosphere-ocean system (Yasunari et al. 2000b). Presumably, this decadal or longer-term change may be attributed to the anthropogenically-induced "global warming". Another possible cause may be due to the more local anthropogenic effect of land-cover change, i.e. deforestation in Thailand: indeed, deforestation accelerated in the 1970s as shown in Fig. A.68.

By using RAMS (Regional Atmospheric Modelling System; Pielke 2002), Kanae et al. (2001) successfully simulated this considerable rainfall decrease in September, by changing the land surface from fully forested to completely deforested conditions under the atmospheric circulation of September (when the near-surface westerly wind is weaker than that in other monsoon months) as shown in Fig. A.69. They concluded that the decreas-

ing trend in rainfall in this month may be at least partly attributed to the recent deforestation and associated surface vegetation conditions in Thailand through the changes in surface energy and water balance. It is noteworthy that this effect of deforestation to decrease rainfall appeared remarkably prominent in September, when the south-west monsoon current has become weak. In July and August, when the monsoon current is still strong, the change in land-surface condition by deforestation tends to change the distribution of rainfall, rather than to decrease the overall rainfall amount over the deforested area.

The recent intensive observational studies under the GAME (GEWEX Asian Monsoon Experiment)-Tropics have indeed revealed that the seasonal course of surface energy and water balance is completely different be-

Fig. A.68.
Changes of forested area in the whole of Thailand and north-east Thailand (from Kanae et al. 2001)

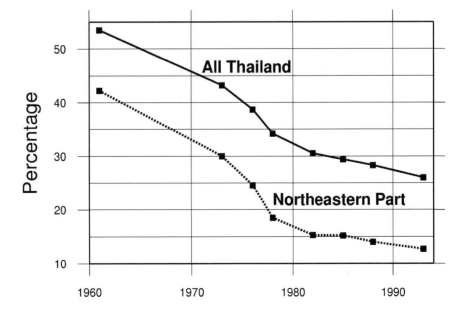

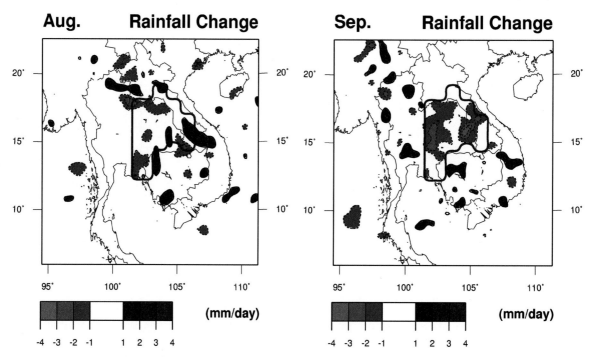

Fig. A.69. Simulated rainfall change over Thailand with surface boundary conditions of the land surface for the full forested and the full deforested condition under the mean atmospheric circulation of (*left*) August and (*right*) September. *Blue colour* shows positive, *red colour* shows negative rainfall difference (of fully deforested minus fully forested case). The *thick line* denotes the area of land-surface change from past to current, e.g. after deforestation (from Kanae et al. 2001)

tween the fully forested and the deforested or mixed shrub/paddy field areas. In the dense monsoon forest within the Kog-Ma basin in northern Thailand, annual maximum evapotranspiration appeared in March, i.e. in the middle of the dry season (Suzuki et al. 2001), while in the mixed shrub/paddy field area in the central plain it appeared in the middle of the rainy season (Toda et al. 2002). The recent simulation by using a multi-layer canopy model has shown that the evapotranspiration maximum in March in the forested area is caused mainly by the water storage and physiology characteristics of forest (Tanaka et al. 2003).

Water availability for agriculture is basically related to the inventory of water resources, i.e. precipitation (P) minus evapotranspiration (E). In this sense, the water availability ($P - E$) tends to be smaller in the deforested paddy or shrubby area than the forested area, particularly in the rice-growing (rainy) season, where E also increases as P increases. Thus, the vulnerability of agriculture to rainfall variability should be intensified whenever and wherever forests are changed to agricultural fields, particularly in the tropical or subtropical monsoon regions. If the deforested area becomes large enough, such as in the plains of Thailand, P itself becomes affected by the deforested surface condition through changes in surface roughness, albedo and atmospheric moisture convergence, as suggested by Kanae et al. (2001).

A.8.2.3 Do Water-fed Rice Paddy Fields Increase Rainfall in Monsoon Asia?

Monsoon Asia is characterised as a traditional rice-cultivation society, which actually has enabled this region to support more than half of the population in the world. Rice paddy fields occupy a huge area in South-east and East Asia. In Thailand, we have suggested that deforestation and wide-spread mixed shrub/paddy fields may reduce rainfall through the effect of a smoother surface and a decrease in evaporation efficiency compared to the forest zone. In this case, the Bowen ratio (ratio of sensible heat to latent heat) at the surface has been increased by deforestation, and the moist, static stability of the atmosphere has increased due to reduced rainfall. On the other hand, the water-fed paddy fields may function to increase evaporation compared with drier farm lands and fields. This may be the case in the southern part of the China plain. In June and July, the China plain is strongly affected by the Meiyu frontal activity, which is a major part of the Asian monsoon system in east Asia. The south-west monsoon outflow from south and South-east Asia is a main moisture source for rainfall in the Meiyu frontal zone.

In addition, most (more than 80%) of the area is occupied by water-fed paddy fields in this rainy season. An interesting result from a recent model study, based

Fig. A.70.
Change of cumulus convection by using the surface condition of water-fed paddy field (with large latent heat flux) and dry farmland (with large sensible heat flux) (Shinoda and Uyeda 2001)

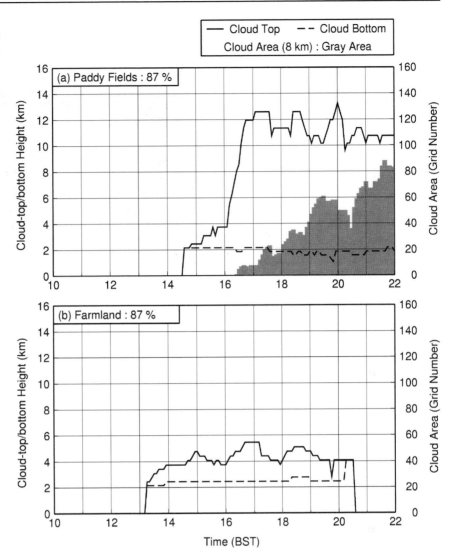

on observations of surface energy and water fluxes, is that the large area of water-fed rice paddy fields in the southern and central China plain may play a key role in the enhancement of deep cumulus convection and rainfall in the region.

Based upon the observational data of the GAME-HUBEX (Huai-he River Basin Experiment) mesoscale and surface hydro-meteorological experiment, Shinoda and Uyeda (2001) conducted numerical simulations of cloud and precipitation systems using a cloud-resolving mesoscale model with two different boundary conditions, i.e. with water-fed paddy fields, and with dry farmland. In addition, they performed a set of sensitivity tests with different atmospheric conditions, deducing two factors for the development of deep convective clouds in this region. One is a large amount of latent heat flux from the surface, and the other is a moist environment in the middle troposphere. Paddy fields are widely distributed over this region, especially in the

Yangtze and Huai-he River Valleys. Since water-fed paddy fields can supply large amounts of latent heat flux into the lower atmosphere, this surface boundary condition has been proved to be responsible for developing deep convective clouds as shown in Fig. A.70. On the other hand, when the environment in the middle troposphere is moist, a shallow convective cloud can develop into a deep one without losing its positive buoyancy because the negative buoyancy by evaporative cooling of an entrained air mass is less. Interestingly, this moist environment in the middle troposphere is maintained not only by the monsoon southerly (horizontal advection), but also is reinforced by vertical transport of water vapour from the lower atmosphere through shallow convective clouds (vertical advection). These features of the development of deep convective clouds over this region are summarised in Fig. A.71. The development process of convective clouds, which is induced by large amounts of water vapour supply from the water-fed

Fig. A.71.
Schematic view of the moisture transport of monsoon flow, shallow convection, Meiyu front with deep convective activity over south/central China plain during early summer (Shinoda and Uyeda 2001)

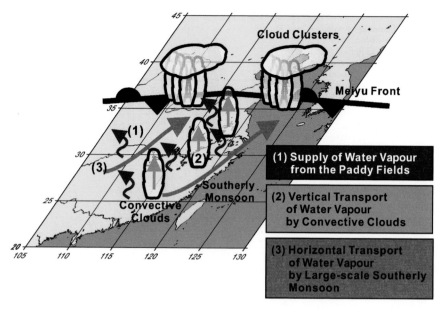

paddy fields to the south of the front, reinforces the transport of water vapour into the Meiyu front. Through this process, the moist subtropical air mass to the south of the front is formed and maintained over the China plain, which eventually strengthens the Meiyu frontal activity.

This result suggests that the rain-fed rice paddy fields in east and South-east Asia form a landscape well harmonised with the monsoon climate through the positive hydrological feedback between the surface and atmosphere. These atmospheric moisture environments are different from those over other continental plains, e.g. the Great Plains of North America and the Amazon Basin in South America.

A.8.2.4 Conclusions

This study first overviewed the basic seasonal and regional characteristics of the atmospheric water balance of monsoon Asia, and suggested that rainfall change may be sensitive to surface conditions when and where there is a large contribution of evapotranspiration to the water balance.

The two model studies on the impact of land-cover/land-use changes on convective activity and rainfall in the tropics and subtropics of monsoon Asia, i.e. Thailand and the south China plain have strongly suggested that the sensitivity of rainfall and the water cycle may change drastically, depending upon seasonal and regional atmospheric wind and moisture conditions.

In Thailand, the trend of decreasing rainfall in September was attributed to the recent deforestation and associated surface vegetation conditions through the change in surface energy and water balance. This effect

of deforestation to decrease rainfall appeared most remarkably in September, when the south-west monsoon current has become weak. In July and August, when the monsoon current is still strong, the change of land-surface condition by deforestation tends to change the distribution of rainfall, rather than to decrease the overall rainfall amount over the deforested area.

In the subtropical China plain, where the Meiyu frontal activity is dominant in early summer, it is suggested that the widely distributed water-fed rice paddy fields play an important role in the maintenance and reinforcement of moistening air masses to the south of the front through moistening the lower and middle troposphere by cumulus convection. These results may imply that the rain-fed rice paddy fields in east and South-east Asia constitute a landscape well synchronised with the monsoon climate through the positive hydrological feedback between the surface and atmosphere.

Thus, these model studies, together with observational atmospheric water balance studies, suggest that the human-induced changes in surface conditions in monsoon Asia would affect the regional monsoon climate and water balance, in either a positive or negative manner, depending upon the basic climate conditions.

Congbin Fu

A.8.3 Can Human-induced Large-scale Land-cover Changes Modify the East Asian Monsoon?

Both observational as well as theoretical studies have proved that human-induced large-scale land-cover changes, such as destructive forest harvesting and over-cultivation or over-grazing of grassland, have been one of the major causes for the deterioration of regional cli-

Fig. A.72.
Expansion of agriculture land over China since 11th century B.C. (Deng et al. 1983)

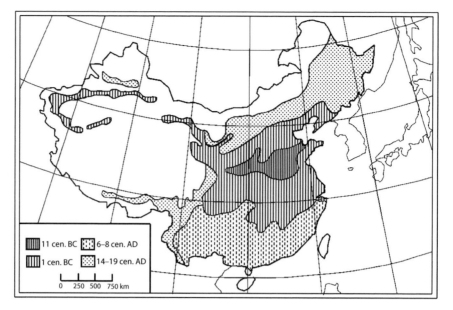

11 cen. BC 6–8 cen. AD
1 cen. BC 14–19 cen. AD

0 250 500 750 km

mate and environment (e.g. Charney 1975; Lean and Warrilow 1989; Nobre et al. 1991; Wei and Fu 1998; Pielke 2001c; Pielke et al. 1991; Xue 1996). The interaction between climate and land cover over the Asia monsoon region is particularly strong in two respects:

1. In terms of natural processes, the highly variable monsoon climate with regard to precipitation and temperature, forces changes in function and structure of terrestrial ecosystems on various time scales through changing their physiological processes (e.g. Fu and Wen 1999). Long-term monsoon climate changes can even alter the biogeographic distributions of the ecosystems (An et al. 1990) and such changes bring about feedback effects on the monsoon climate itself (Xue 1996).

2. On the other hand, the long history of civilisation has caused significant changes in land cover over Asia. The human-induced land-use and land-cover changes in East Asia in particular, one of the largest regions of anthropogenic activities in the world is striking: in the past 3 000 years, more than 60% of the region has been affected by the conversion of various categories of natural vegetation into farmland, grassland into semi-desert and widespread land degradation. Such human-induced land-cover changes result in significant changes to surface dynamic parameters, such as albedo, surface roughness, leaf area index and fractional vegetation coverage. The variation in the East Asian monsoon presented here is the result of land-cover changes only. It is likely that anthropogenic modification of the monsoon system would occur through changing the surface fluxes of energy, water and greenhouse gases under different land-use patterns.

A.8.3.1 History of Land-cover/Land-use Changes over East Asia

Owing to increasing population, industrialisation and urbanisation, the natural ecosystems within Asia such as forest, grassland and wetlands, have been encroached upon by farmland and other man-made ecosystems on a large scale. For example, there has been a significant expansion of farmlands throughout China since the 11th century B.C. (Fig. A.72, Deng et al. 1983). In the early days, all the country was covered by various kinds of natural ecosystems except for a very narrow band of farmland along the lower reaches of the Yellow River basin. The agricultural area was gradually expanded, both southward and northward, and into the upper reaches of the major river basins. Around the late 19th century, the land-use pattern was set up much as it is today, although the land cover was continuously changing until recently. Now, man-made ecosystems cover nearly 80% of the total terrestrial area. Such human-induced land-cover changes have been as great in Asia as in any other part of the world, if not greater. To what extent has the monsoon system itself been modified by such changes? In this chapter numerical experiments using a high-resolution regional climate model are presented which examine the most likely response of the Asian monsoon system to human-induced land-cover change.

A.8.3.2 Design of the Numerical Experiments

The natural vegetation has been altered in East Asia over the last millennia such that its reconstruction other than by modelling is rather difficult. However, it is feasible to give an estimate of the equilibrium, or potential vegetation (Vp) that could exist today by using the so-called biome approach (e.g. Prentice et al. 1992; Ojima 2000). Potential vegetation is computed from atmospheric and soil conditions only, and neglects any anthropogenic influence on the biosphere.

Remote sensing information is now widely used to provide maps of vegetation cover at quite high temporal resolution. The current vegetation used here is based on the global land-cover classification data as part of the global datasets for land-atmosphere models, ISLSCP Initiative I, developed in the International Satellite Land Surface Climatology Project (ISLSCP, see also Sect. C.2.4) (Meeson et al. 1995). Although one has to realise that the current vegetation cover will be somewhat different to the data for 1987–1988, these differences are relatively smaller than the differences with the potential vegetation as described in the previous paragraph. Therefore, we assume that the vegetation cover data (Vc) can be looked upon as approximating the actual current vegetation cover. The human-induced land-cover change is defined as the difference between the potential and current vegetation, as presented by Fig. A.73. More than 80%

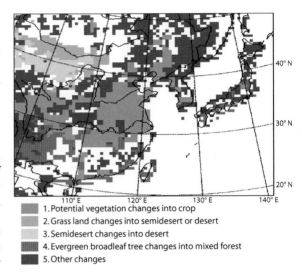

1. Potential vegetation changes into crop
2. Grass land changes into semidesert or desert
3. Semidesert changes into desert
4. Evergreen broadleaf tree changes into mixed forest
5. Other changes

Fig. A.73. Changes of vegetation cover from potential (Vp) to current (Vc) vegetation

of the region has been affected by conversion of various categories of natural vegetation into farmland, grassland into semi-desert and widespread land degradation. The most pronounced changes occur in north-west China

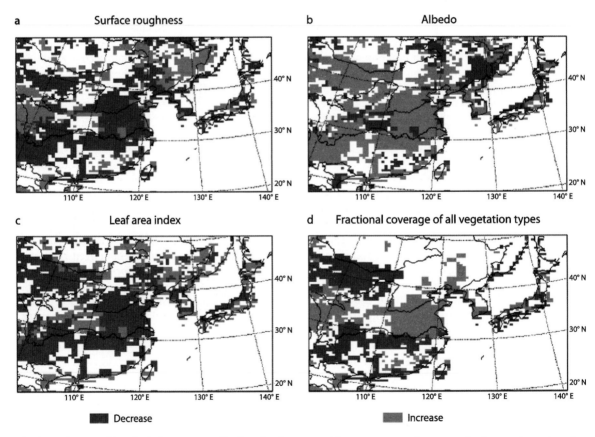

Fig. A.74. Changes of four physical parameters of land-surface from potential to current vegetation cover. **a** Surface roughness; **b** surface albedo; **c** leaf area index; **d** fractional vegetation coverage

where the grassland has been changed into semi-desert or desert and in East China where the forest has been replaced by crop lands. There are also significant changes over Japan.

A pair of numerical experiments is performed for the above two land-cover conditions by using the Regional Integrated Environmental Model System (RIEMS) version 1 (Fu et al. 2000). This model consists of mainly three components of the climate system: a mesoscale atmospheric dynamic model, a radiation scheme and a land-surface scheme; besides two components of major regional anthropogenic forcing factors: land-cover changes and changes in greenhouse gases and aerosol emissions. The validation study on the model performance has shown reasonably good results in its capacity to simulate the regional climate in the Asia monsoon region (Fu et al. 1998).

The differential fields of integration by the two vegetation cover datasets (current minus potential vegetation) are used to exemplify the impacts of a change in natural vegetation, since the two simulations are identical to each other for all conditions, including the large-scale driving fields used as the initial and lateral boundary conditions, as are the parameters of all physical processes except for the vegetation cover. In order to maintain the linkages between the large-scale environment and the simulated region, a relaxation scheme with ten buffer zones is applied for nesting at the lateral boundary (Wei et al. 1998).

A.8.3.3 Changes of Surface Dynamic Parameters under Two Vegetation Coverages

Changes in four main surface parameters, resulting from comparisons from potential to current vegetation distribution, are shown in Fig. A.74: surface roughness, albedo, leaf area index and total vegetation fractional coverage. In those areas where the natural vegetation (mainly forests) have been turned into farmland or where grasslands have been turned into semi-desert or desert (the *green area* shown in Fig. A.73), the significant decrease in surface roughness and leaf area index observed is shown in *blue* (Fig. A.74a,c); increase in albedo is shown in *red* (Fig. A.74b). The total fractional vegetation coverage is higher in the farmland area than in natural forests (the *red area* in Fig. A.74d), but is lower in the semi-desert and desert areas in comparison with grassland and in areas of mixed forests in comparison with evergreen broadleaf forests (the *blue area* in Fig. A.74d). Changes in these surface parameters would surely modify the exchanges of energy and water between the land surface and the atmosphere and result in the changes in atmospheric circulation as shown in the next section.

A.8.3.4 Changes of the East Asia Monsoon by Human-induced Land-cover Changes

To examine the potential modification of the East Asian monsoon system, the analyses focus here on the changes in the monsoon circulation and related surface climate. Fig. A.75a presents the mean changes of vector wind and

(a) Vector wind and geopotential height

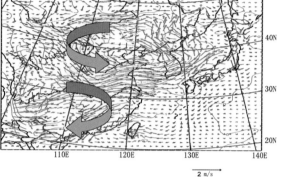

b Meridional circulation

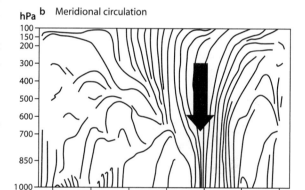

c Zonal vertical circulation

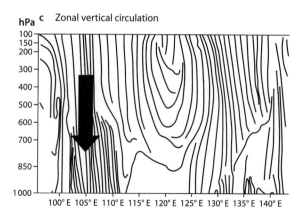

Fig. A.75. Changes of summer monsoon circulation over East Asia under two vegetation covers (current minus potential) during summer (June, July, August). **a** Vector wind and geopotential height at 850 hPa in m (*blue*: north wind, *red*: south wind, *pink*: positive, *green*: negative); **b** mean meridional circulation along 100–120° E; **c** mean zonal circulation along 25–40° N

Fig. A.76.
Changes of specific humidity at 850 hPa (g kg^{-1}) over East Asia under two vegetation covers (current minus potential) in summer (June, July, August)

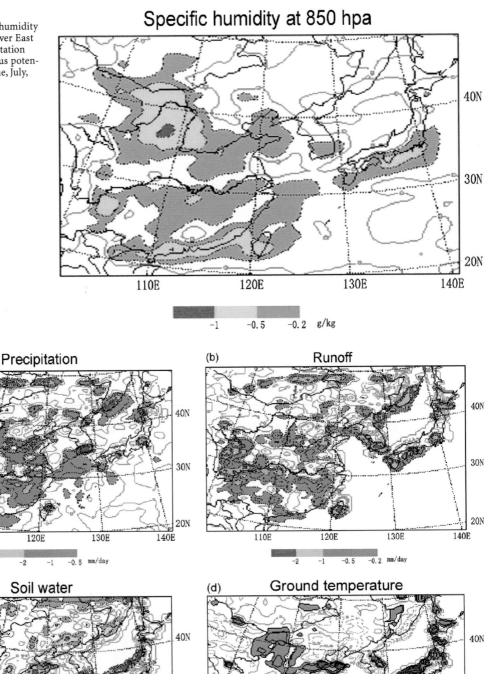

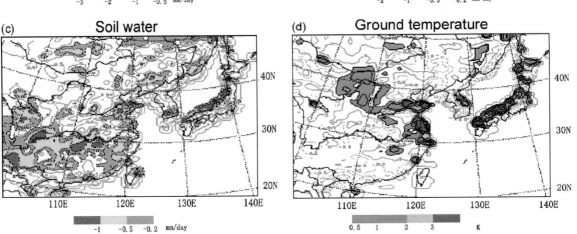

Fig. A.77. Changes of surface climate over East Asia under two vegetation covers (current minus potential). **a** Precipitation (mm d^{-1}); **b** runoff (mm d^{-1}); **c** soil moisture (mm d^{-1}); **d** ground temperature (K)

geopotential height in the lower atmosphere (850 hPa) in summer over East Asia. The weakening of the monsoon depression is shown by the positive anomalies in the region to the south of 30° N and the weakening of summer monsoon is shown by the northerly anomalous flow. There is a negative departure in height over the northern part of the domain, representing the development of a low-pressure system over there, which brings about the anomalous north-west flow.

The changes in mean meridional circulation and zonal circulation are shown in Fig. A.75b,c with major characteristics being enhancement of descending motion flows over 35–40° N and 100–115° E respectively, which would prevent the development of the summer monsoon circulation.

Both these two northerly anomalous flows and the enhancement of descending motion over East Asia would prevent moisture transport northward and the development of convective activities, resulting in more dry conditions in the atmosphere over most of the domain, as indicated by the differential field of the specific humidity at 850 hPa (Fig. A.76).

Figure A.77 presents the changes in surface climate related to summer monsoon changes. All components of the surface water cycle, such as precipitation (*a*), runoff (*b*) and soil moisture (*c*) are reduced over most of the region. It indicates the weakening of the water cycle through the deterioration of the natural vegetation. There are no significant changes in surface temperature except for a relative warming of an area in the northern China plain (Fig. A.77d), mainly related to the significant reduction in surface evaporation.

In contrast, the winter monsoon over East Asia becomes stronger with the deterioration in the natural vegetation cover, as shown by the strong anomalous northerly flow in the differential fields of vector wind and geopotential height at 850 hPa shown in Fig. A.78a. This circulation pattern would bring dry and cold air masses from inland down to all regions of East Asia, resulting in changes in the surface climate, such as the reduction of atmospheric humidity (Fig. A.78b) and precipitation (Fig. A.78c), mostly in the southern part of the region, and cold temperatures over almost the whole region (Fig. A.78d).

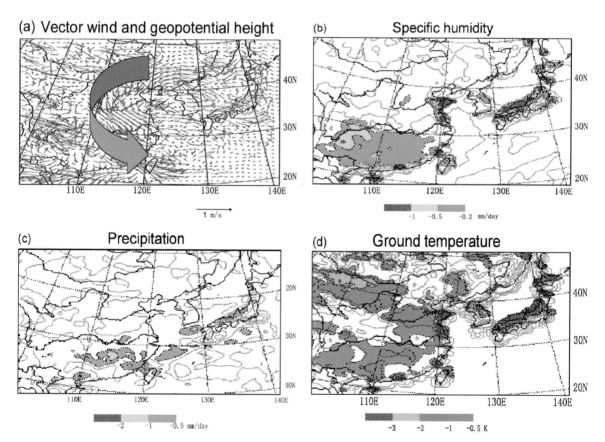

Fig. A.78. Changes of winter (December, January, February) monsoon circulation and related surface climate. **a** Vector wind and geopotential height at 850 hPa in m (*blue:* north wind, *red:* south wind, *pink:* positive, *green:* negative); **b** surface humidity (g kg⁻¹); **c** precipitation (mm d⁻¹); **d** ground temperature (K)

Conclusions

According to the above analysis, human-induced land-cover changes have modified the monsoon circulation by the weakening of the summer monsoon and the enhancement of the winter monsoon over East Asia, which result in related changes to the surface climate over the region. The conclusions derived from the numerical experiments require observational evidence to support them.

Figure A.79 is the time evolution of the aridity index over East China since 1880, showing a significant trend in aridification during the last 120 years, with a 36-year period of oscillation (the Brucker period) superimposed on it (Fu 1994). Since the moisture condition over East China is mainly related to the intensity of the summer monsoon, it is a reflection of the weakening of the summer monsoon during that period.

On a longer time scale, a 25 000-year lake level dataset for the Daihai Lake in Inner-Mongolia shows a signifi-

cant reduction in its level, beginning about 3 000 years ago (Fig. A.80) (S. M. Wang, pers. comn.). This is also an indication of an aridification trend over Northern China and therefore the weakening of the summer monsoon since then. Over a much longer period, the proxy datasets of winter and summer monsoon indices from loess deposition show an overall dry trend since 2.5 Myears ago.

It seems that the deterioration in natural vegetation due to development of human society is perhaps one of the anthropogenic factors superimposed on the natural variability of the monsoon system.

Acknowledgment

This research took place within the project "Predictive study of aridification in Northern China in association with life-supporting environment changes", G1999043400, National Key Basic Research Development Programme under the support of the Chinese Ministry of Science and Technology.

Fig. A.79.
Variations in aridity index of East China since 1880 (from Fu 1994, © reprinted by permission of Wiley & Sons, Inc.)

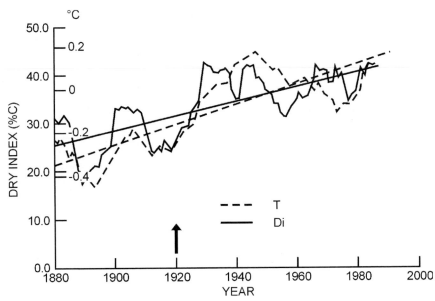

Fig. A.80.
Changes of lake level of Daihai Lake in the Inner-Mongolia since 25 000 BP (m) (S. M. Wang, pers. comn.)

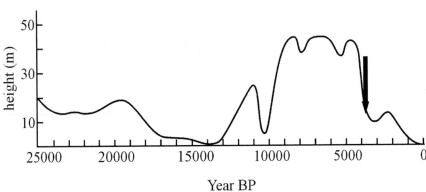

A.8.4 Summary

Those parts of the Asian continent which are influenced by the monsoon are a prime example of how the land-surface is affecting weather and in the long run climate. Model calculations have shown that the human-induced changes of land surfaces – like deforestation or establishing rice paddies – have a decipherable impact on albedo, energy fluxes and precipitation. Although palaeorecords show that aridification is a long-term trend, for example, in Inner Mongolia since about 2000 B.C., the anthropogenical influences reinforce this trend, as can be seen in the decreasing rainfall in Thailand since 1950. The precipitation distribution is most probably linked to evaporation as can be seen in Thailand for rice paddies in the summer months while in September the weaker monsoon results in reduced overall precipitation. Model calculations of human-induced changes in China compare potential with current vegetation showing that the present land surface results in decreasing precipitation in summer and winter.

Chapter A.9

Summary, Conclusion and Perspective

Martin Claussen · Andrew J. Pitman

Before we provide an overall conclusion of Part A, we would like to summarise the most important results of Part A chapter by chapter:

Chapter A.2: The Climate near Ground

There has been a remarkable improvement in our understanding in the last two decades of how the land surface of the Earth interacts with the atmosphere (see Sect. A.2.2 and A.2.3). This can be traced back to both developments in modelling as well as the increased availability of sophisticated equipment to study land-surface processes *in situ*. In terms of the climate system, it is important to model the partitioning of available energy between sensible and latent heat as well as possible (Sect. A.2.2) since more latent heat contributes water vapour to the atmosphere and tends towards increasing cloudiness and precipitation, while increases in sensible heat tends to warm the planetary boundary layer. The modelling of this partitioning is closely linked with the modelling of carbon within the biosphere. Given the fundamental role that latent and sensible heat play in the climate system, it is necessary to simulate the diurnal, seasonal and longer term variations in these fluxes as well as possible (Sect. A.2.2) and this has become a key focus in how land-surface processes are represented in numerical models of weather and climate. The modelling of carbon exchange and storage at the surface is an integral component in the simulation of climate variability and change.

The latent heat flux is the energy flux associated with the evaporation of water. While most water is stored in the oceans rather than on land, the land-surface moisture stored in the soil plays a vital role in the provision of food, fibre and fresh water (Sect. A.2.3). There are major difficulties in simulating soil moisture in climate models and various land-surface models simulate very different amounts for a given state of the climate. Runoff is usually modelled very simply in climate models even though it is known to affect climate model simulations. There is an increasing consensus that in order to simulate runoff, schemes need to differentiate between the two basic types or time scales of runoff (dealt with in more detail in Part D) – a quick (infiltration excess) component and a slow (drainage) component (Sect. A.2.3). While the simulation of runoff is difficult in weather and climate models, effort in this direction is very worthwhile and progress is being made, driven in part by the realisation that runoff is probably the only quantity simulated by land-surface models that can be validated at the large scale.

Major initiatives (e.g. FLUXNET) are providing spatially distributed networks of surface observations which offer considerable potential to validate and thereby improve the modelling of surface processes (Sect. A.2.4).

Chapter A.3: The Regional Climate

Man-made landscape alterations significantly affect the atmosphere at the microscale and, when such alterations expand to the mesoscale, they also affect the regional hydrometeorology. Furthermore, the impact of such extensive alterations can propagate by teleconnection to other regions.

The above conclusion has significant implications for the representation of land processes in atmospheric simulation models. Indeed, parameterisations based on the "big leaf" approach, in which each plant-covered grid element of the model is assumed to be represented by a single leaf, without consideration for heterogeneity, is not recommended. This is because important non-linear relations are omitted with this approach.

While a significant effort has been made to address the microscale land heterogeneity, very few studies have been conducted to address the great challenge of representing correctly the coupled land-PBL system (including clouds and precipitation) in a heterogeneous landscape. Since precipitation is a key parameter of the hydrological cycle (including its biological aspect), it is essential that progress be achieved on this issue.

Chapter A.4: The Global Climate

Humankind has altered the Earth's surface; it is estimated that over 45% of the land surface is directly affected by human-induced land-cover change. Over the last few centuries, the intensity and scale of these modifications has increased significantly, and model experiments suggest that this land-cover change has an impact not only on regional climate, but also on the global scale (Sect. A.4.1.4 and A.4.2.3).

Changes in vegetation pattern affect the chemical composition of the atmosphere, and thereby, the greenhouse effect, via changes in photosynthesis and respiration. In turn, changes in the chemical composition of the atmosphere influence plant physiology and thus vegetation pattern. These biogeochemical feedbacks act globally because of rapid (usually within one or two years) global mixing of gaseous substances within the atmosphere. Sensitivity studies indicate that positive feedbacks can arise if CO_2 emission continues to increase. These feedbacks, associated with a strong increase in heterotrophic respiration and dieback of tropical rainforest, could turn the land surface from a carbon sink into a carbon source (Sect. A.4.3.3).

Theoretical work demonstrates that the atmosphere-vegetation system is an intransitive system, i.e. the system yields multiple equilibria depending on the initial conditions from which the system is started. For example, several models show that under present-day insolation and ocean temperatures, two solutions are found for North Africa and central Asia: a dry climate and a more humid climate (Sect. A.4.1.3). Moreover, the atmosphere-vegetation system can react to external forcing in a disproportional way which leads to abrupt climate and vegetation change in some regions.

Interpretation of palaeoclimatic and palaeobotanic records by using climate system models suggest that atmosphere-vegetation interaction in the past has amplified climate changes which are triggered by orbital forcing, i.e. changes in the Earth's orbit around the sun. For example, the synergism between the sea-ice albedo feedback in the Arctic and the so-called taiga-tundra feedback led to warmer winters than today in the mid-Holocene (some 6 000 years ago) despite less insolation at this time (Sect. A.4.1.2 and A.4.2.1). This synergism also contributed to the last glacial inception some 115 000 years ago in a significant way. (Most models cannot explain the last glacial inception without taking into consideration biogeophysical effects.) Likewise, atmosphere-vegetation interaction is likely to be the major amplification factor for the greening of northern Africa during the early and mid-Holocene. The non-linearity of this feedback helps to explain the rather abrupt aridification of North Africa some 5 500 years ago (Sect. A.4.2.2).

Global weather forecasts have improved substantially in part because of more realistic representations of near-surface processes (Sect. A.4.5 and 4.5.2.4). Changes in albedo in snow-covered areas during spring and changes in soil moisture have been shown to be particularly important. Most numerical weather forecast models have their poorest skill in predicting precipitation in the tropics year round and the spring/summer extratropics. Improvements in the understanding of the interaction between soil moisture and convection will most likely alleviate deficiencies.

Because of its memory effect, soil moisture is one of the main agents in seasonal climate variability (Sect. A.4.4). Soil moisture primarily affects the amplitude of precipitation variance rather than determines its spatial structure over global land except in deserts, high latitudes, and the very wet area of the tropics. Foreknowledge of soil moisture also contributes significantly to long-term predictability of precipitation in transition zones between dry and humid climates, particularly over the tropics and summer extratropics.

Chapter A.5: The Sahelian Climate

The Sahel is a tropical, semi-arid region along the southern margin of the Sahara desert (Sect. A.5.1.1). Although five hundred years of available records have revealed several drought periods in this region, none have been as persistent and severe as the current one that began in the 1960s (Sect. A.5.1.2). So far, there have been two views regarding the surface boundary forcing on the cause of the drought: the first group believes that the main cause of the drought is land-surface degradation resulting from population pressure in excess of the region's carrying capacity (Sect. A.5.1.3). This view implies the existence of a positive land-surface/atmosphere feedback, primarily internal to the region. In principle, it favours the development of mitigation strategies to reverse this dry trend. The second group attributes the drought to unfavourable anomalous patterns in sea surface temperature in the oceans (Sect. A.5.3.2). This view implies that the cause of the current drought is external to the region and is thus not a consequence of local human activities. It encourages the development of adaptation strategies to make the region less vulnerable to the effects of droughts.

The first land-surface/atmosphere interaction study investigating the albedo feedback in the northern African region was conducted in 1975 to determine whether land-surface dynamics could lead to severe droughts. Since then, numerous sensitivity studies using different General Circulation Models (GCM) and various types of land-surface models have demonstrated consistently that the land surface can have a significant impact on the Sahel climate (Sect. A.5.3.1). In particular, studies that use a coupled atmosphere-biosphere model have shown that drastic land-cover degradation in Sahel could cause summer rainfall reductions of the observed magnitude in northern Africa. In these studies, the maximum summer rainfall region shifted to the south and synoptic atmospheric easterly waves became weaker. The land-cover change not only influenced the summer rainfall in the

Sahel, but also the autumn rainfall in East Africa. In addition, these studies also simulated increases in the surface air temperature, and reductions in the runoff and soil moisture over the Sahel region (Sect. A.5.3.2). These changes are consistent with the observed decadal climate anomalies in the northern African region. Therefore, if changes in vegetation properties between the 1950s and the 1980s were comparable to the specified land degradation in that study, the results indicate that the degradation can lead to regional climate changes of the order of the differences between the 1950s and the 1980s.

Early land-surface/atmosphere interaction studies attributed the drought to a primarily albedo/radiation driven process chain. However, observations did not support the dramatic albedo changes used in these studies and the simulated surface cooling caused by high surface albedo. These apparent discrepancies challenged the credibility of the land-surface effect. With the coupled atmosphere-biosphere model, we now realise that an evaporation/convection driven process chain is probably more important (Sect. A.5.4). The large reduction in evaporation due to land degradation results in less convection and lowered atmospheric heating rates. The reduced diabatic heating rate in the atmosphere is associated with less surface pressure decreases in this region, which in turn weakens the African monsoon flow, reduces moisture flux convergence, and lowers the rainfall and evaporation, which produces a positive feedback.

The lack of observed field data for the validation of model-simulated surface energy and water balance has hampered earlier Sahel sensitivity studies. However, two field experiments, SEBEX (1988–1991) and HAPEX-Sahel (1991–1992) have provided very important information on land-surface processes in the Sahel region (Sect. A.5.2.1 and A.5.2.2). These field measurements have led to a considerable increase in our understanding of the surface energy balance at the Sahelian surface, in particular the interplay between the hydrological responses of soil and vegetation. In addition, these two field experiments have resulted in improved calibration and validation of the surface biosphere models. However, it is difficult to place these results in a broad climatic context since the experiments were conducted in only one region. To apply results from field measurements at a continental scale, more comprehensive, long-term, and large-scale measurements are needed (Sect. A.5.2.3 and A.5.2.4).

Since the late 1970s, satellites have provided data on land-surface conditions, including temporal and spatial variations of land-surface conditions at a global and continental scale, which cannot be obtained by any other method. Studies have demonstrated the great potential in applying satellite data for Sahel studies (Sect. A.5.2.5). Additional research is needed to develop and validate the methodology used to convert satellite signals to vegetation characteristics and other land-surface information.

Presently, we cannot determine with certainty whether the current drought is a temporal phenomenon or whether it is primarily caused by human activities. This problem arises mainly because we do not yet know how well the normal and degraded vegetation scenarios specified in the Sahel studies represent the actual land-surface conditions between the 1950s and the 1980s, as the reliability of the information currently available is uncertain. With the development of dynamic vegetation models (Sect. A.5.3.4), we may be able to assess more realistically the role of sea surface temperatures, land-surface processes, and other processes in the long-term Sahel drought.

Chapter A.6: The Amazonian Climate

The forest of the Amazon Basin has long been recognised as an important component of the global climate system and is often referred to in the popular press as the "lungs of the world". The undisturbed forests may play an important role as a sink of excess atmospheric CO_2.

The Amazon Basin contains the largest extent of tropical forest on Earth, over 5×10^6 km^2, and accounts for a large proportion of the planet's animal and plant species (perhaps as large as 1/3 of all species). It contains also over 2×10^6 km^2 of savanna vegetation to the north and to the south of the domain of the tropical forest. Average annual rainfall in the Basin is 2.3 m and the annual discharge of the Amazon River into the Atlantic Ocean of more than 200 000 m^3 s^{-1} contributes about 18% of the global flow of fresh water into the oceans.

The occupation and development of Amazonia since the 1950s and its impact on the vegetation cover and on the composition of the atmosphere have changed dramatically, where most of the changes took place over the last 25 years. The rapid development has led to the deforestation of over 500 000 km^2 in Brazil alone. Current rates of annual deforestation range from 15 000 to 20 000 km^2 in Brazilian Amazonia. The spatial patterns of land-cover change are highly heterogeneous.

A number of field studies carried out over the last 20 years showed local changes in the water, energy, carbon and nutrient cycling, and atmospheric composition caused by deforestation and biomass burning. Diurnal amplitude of surface air temperature is 3 to 5 °C higher and dry season forest transpiration is reduced by 30% to 50% over deforested areas in comparison to forested ones.

However, large-scale and regional changes in the water cycle have not been detected so far. Regional air surface temperatures have increased over Amazonia by about 0.4 °C during the 20th century, but this seems to be associated with global climate change and not to land-cover changes.

In addition to land-cover changes, it is becoming ever more important also to consider the changes in atmospheric composition brought about by biomass burning.

Large concentrations of aerosols in the boundary layer during the biomass burning season (July through October in most of Amazonia south of the Equator) may change radiative properties of the lower troposphere due to absorption of solar radiation by soot aerosol. This, in turn, may even change boundary layer dynamics and affect cloud microphysical processes.

It is likely that transpiration of forests and evaporation from forests may play an important role for dry season rainfall. ABRACOS observations have shown that forest transpiration and evaporation can be up to twice as large as that of grass vegetation in pastures during the dry season. This is even more important when one considers that dry season rainfall originates from isolated deep convection or mesoscale convective complexes distant from oceanic moisture sources. A more complete understanding of the water cycle of Amazonia, including the role of the land surface in recycling water vapour, is still to be accomplished. The LBA Experiment is contributing towards that goal.

The observed patterns of land-cover change over the last two decades in Amazonia have led to increase of the heterogeneity in the land cover, which might in fact lead to more vigorous mesoscale circulations if the scale of the heterogeneity is 10 km or larger. Dry season observations of boundary layer growth over pasture and forest have shown higher mixed layer heights for pastures consistent with higher surface sensible heating fluxes, which results in the observations of more boundary layer clouds over pasture.

A number of numerical simulations of complete deforestation of Amazonia and replacement by pastures indicate a different post-deforestation climate: 0.5 to 2 °C increase in surface temperature, 20 to 30% reduction in evaporation and transpiration and 5 to 25% decrease in precipitation. The contrasts in pre- and post-deforestation climates are more pronounced during the dry season. That has led to the hypothesis of a future tendency towards "savannisation" of the southern and northern boundaries of the tropical forest due to lengthening of the dry season.

Sensitivity experiments with increased greenhouse gas (GHG) emissions given in the IPCC Third Assessment Report indicate changes in temperature by 2100 ranging from 1–1.5 °C (low GHG emissions scenario) to 4–6 °C (high GHG emission scenario) for the regional climate of Amazonia. Estimates of rainfall changes are still highly uncertain, but generally indicate drier climate over eastern Amazonia. The synergism between climate changes due to atmospheric composition changes and land-use changes could combine to produce warmer temperatures and dryness in the coming decades. That would pose an increasing threat to the maintenance of the tropical forest due to increased risk of forest fires and disappearance of species less tolerant to climate extremes.

Chapter A.7: The Boreal Climate

The boreal ecosystem encircles the Earth above about 48° N, covering Alaska, Canada, and Eurasia. It is second in areal extent only to the world's tropical forests and occupies about 21% of the Earth's forested land surface. Model studies indicate that the boreal ecosystems are likely to have amplified palaeoclimatic changes triggered by external forcing (Sect. A.4.1.2 and A.4.2.1).

Meridional gradients in atmospheric CO_2 concentrations suggest that boreal forests above 40° N sequester as much as 1 to 2 gigatons of carbon annually or nearly 15 to 30% of that injected into the atmosphere each year through fossil fuel combustion and deforestation. Given the enormous areal extent of the ecosystem, roughly 20 Mkm^2, shifts in carbon flux of as little as 50 g C m^{-2} y^{-1} can remove (or add) one gigaton of carbon annually from (to) the atmosphere

The boreal ecosystem can be generally divided roughly into uplands and peatlands, the uplands supporting tree growth on mineral soils with net ecosystem exchange in the range of 100 to 300 g C m^{-2} y^{-1}. The uplands burn once every hundred years on the average, thus can only store carbon over multi-decadal time scales. Peatlands however, store carbon below ground with long-term (centuries to millennia) carbon accumulation rates in the range of 10 to 50 g C m^{-2} y^{-1}. Thus, the rate at which the global boreal ecosystem stores carbon has a secular component of about 30 g C m^{-2} y^{-1} or about 0.6 gigatons annually, with seasonal and decadal fluctuations driven by atmospheric variation and land-surface disturbance. For example, in Canada the total area disturbed was large during 1860–1920 and 1980–1998 leading to increased uplands carbon uptake in the decades between.

The high-latitude ecosystem differs dramatically from temperate and tropical ones. The surface energy balance can change in just a few days as the boreal snow cover melts, the frozen peats beneath begin to thaw allowing photosynthesis, transpiration and evaporation to occur. The boreal ecosystem is for the most part, a wetland ecosystem composed of nutrient-limited conifers growing on cold, moisture-saturated peats. Tower and radiosonde observations of the surface and boundary layer show a strong diurnal coupling between the surface energy budget, the atmospheric boundary layer and the diurnal exchange of carbon dioxide with the atmosphere. The observations (Sect. A.7.2.4) show an average diurnal cycle in which surface relative humidity increases to mid-morning then drops rapidly within the expanding and warming boundary layer acting to limit canopy stomatal conductance and transpiration and evaporation. Reduced surface moistening increases surface temperature and sensible heating accelerating warming and drying of the boundary layer. This diurnal feedback mechanism acts also to restrict canopy photosynthetic uptake of carbon after mid morning. Thus in spite of this ecosystem's water-saturated surface, the atmospheric boundary layer is often dry and deep, more characteristic of an arid ecosystem. The deep, dry boundary layers overlying a water-saturated surface led to the apt description "the green desert".

Since 1970, the boreal ecosystem has experienced a significant reduction in spring snowpack and an overall decrease in the seasonal duration of snow cover. Autumn and winter snow cover have, however, recovered to their early extent. Moreover, there were regional variations. The boreal forest region of North America, heavily forested by evergreen conifers, showed relatively little change in spring snow cover perhaps because the snow cover feedback is weak, due to strong shadowing by the evergreen conifers which reduces markedly the difference between snow-on and snow-off albedo. In Eurasia, where deciduous conifer is abundant, snow-on versus snow-off albedos differ significantly and most of this region experienced spring snow cover reductions.

Significant warming of the atmosphere near the ground has occurred in many regions at high northern latitudes, with much of this warming occuring in winter and spring resulting in an increasing growing season length. These regions include the boreal forests of western North America, Northern Europe and large areas in Northern Asia. The near-surface warming in some locations exceeds 1 K per decade. By contrast, the lower troposphere in the Arctic atmosphere has retained nearly the same temperatures since at least the late 1970s. With respect to near-surface warming, there are strong regional variations, and some regions have experienced a cooling. Tower flux and chamber measurements of above and below ground photosynthesis and respiration have helped to elucidate the dynamics and ecophysiology of boreal carbon exchange and how climate changes might alter the source/sink relationships within this ecosystem. Tower flux measurements show that the wetlands fluctuate between being a weak source to a weak sink of carbon with source/sink strengths of about ±50 g C cm^{-2}. Tower measurements also showed that this small net ecosystem exchange was the net difference between two much larger carbon flux rates of about 1 kg of carbon uptake from net primary production and about 1 kg of carbon loss from heterotrophic respiration, primarily as a result of soil decomposition. Over those years annual net primary production was rather more stable than heterotropic respiration.

Other detailed studies showed that net ecosystem exchange in the boreal ecosystem is enhanced by early snow melt and subsequent soil thaw which initiates early photosynthetic uptake of carbon while the soil is still relatively cool and heterotrophic respiration is low. For the same reason, cool summers and late falls also enhance net carbon uptake. Heterotrophic respiration is fundamentally a function of soil temperature. Thus years with longer growing seasons and cool summer soil temperatures should in general be associated with increased carbon uptake. Shorter growing seasons on the other hand, with hot summers and warmer soil temperatures should, in general, be associated with increased carbon release. If the strong high-latitude warming trend continues, leading to warmer soils and a reduction in the extent of the boreal permafrost zone, the resultant increases in soil organic matter decomposition could switch the boreal ecosystem from a long-term carbon sink to a significant carbon source.

Chapter A.8: The Asian Monsoon Climate

Temperate East and South-east Asia is part of the largest continent, situated between high elevation (e.g. the Tibetan Plateau) and the largest ocean (the Pacific). As this area has for a long time been home to an ever increasing human population human-induced changes to the land cover have had a huge influence on the resulting parameters like surface roughness, albedo and leaf-area index (Sect. A.8.3.4).

These anthropogenically induced changes are superimposed on a natural aridification which has been shown by observing and interpreting lake-level reductions since 2 000 BP in Inner Mongolia.

Precipitation depends on evaporation and transpiration and convergence of water vapour transport, for which respective contributions are highly variable from month to month. An analysis of monthly data for 15 years (1979–1993) has shown that, for example, the large contribution from evaporation, especially at the beginning and end of a monsoon, makes it highly probable that the evaporation from wet surfaces and transpiration from vegetation is a major factor in the resulting precipitation during that period. By contrast, there is a high correlation between precipitation in the mature phase of a monsoon and vapour convergence (Sect. A.8.2.1).

Deforestation in Thailand since the 1970s seems to be correlated with reduced annual precipitation in September when the monsoon is already weakening. For the time of the maximum monsoon (i.e. July and August) model simulations show a change in the distribution of rainfall but no decrease of rainfall (Sect. A.8.2.2).

Observations (through GAME Tropics) show (Sect. A.8.2.2) that the change from forest to mixed shrub/paddy field in Thailand shifts the time of maximum evaporation and transpiration from the middle of the dry season to the middle of the rainy season. This is increasing vulnerability to variability in rainfall, which means the possibility of reduced monsoonal precipitation in the growing season.

Model results seem to indicate that wet rice paddies with increased evaporation in the southern China plain may enhance deep cumulus convection and thus rainfall in the region (Sect. A.8.2.3). This would suggest a positive hydrological feedback between the land surface and the atmosphere.

A comparison of regional cimate simulations using current vegetation cover with those using potential vegetation (here regarded as natural vegetation cover) as surface-boundary condition reveals a reduction of summer monsoon with current vegetation, which results in reduced precipitation. Current vegetation, on the other hand, enforces winter monsoon but as the winds during the winter monsoon originate from the land and not the ocean, the result is the same, namely a reduction in precipitation (Sect. A.8.3.4).

Overall Conclusion

- Humans have altered a significant fraction of the Earth's surface. It is estimated that over 45% of the Earth's land surface is currently affected by human-induced land-cover change.
- All the land surface is affected indirectly by anthropogenic changes in the chemical composition of the atmosphere.
- Over the last few centuries, the intensity and scale of these modifications has increased significantly.
- The land surface affects the partitioning of water and energy and changes in the state of the surface cause changes in fluxes of momentum, energy, moisture and substances between surface and atmosphere. There has been a remarkable improvement – through empirical and theoretical studies – in our understanding of how the land surface interacts with the atmosphere. New models of land-surface/atmosphere interaction include not only biophysical processes but also their feedback with carbon uptake and respiration.
- Global weather forecasting has substantially improved in part because of more realistic representations of land-surface processes. Soil moisture was shown to play a key role in seasonal forecasting because of the memory effect, i.e. the long (in comparison with weather dynamics) time scales involved in soil hydrology.
- Modelling experiments have identified regional scale impacts due to land-cover change (deforestation, desertification and change in land use, in general). Observational evidence from several regions – including Amazonia, Sahel, boreal ecosystems, and the Asian monsoon region discussed in detail in Part A – all provide evidence that land-cover change can influence the atmospheric boundary layer and thereby convection and cloudiness over large regions.

- Modelling studies also indicate that land-cover change (mainly deforestation), is likely to have influenced and to continue to influence the evolution of global climate.
- Theoretical studies show that the interaction between atmosphere and vegetation interaction is non-linear, presumably at all spatial scales, i.e. the atmosphere-vegetation system can react in a disproportional way to subtle external forcing. Moreover, the atmosphere-vegetation system can reveal multiple equilibria with the possibility of abrupt changes between different equilibria under varying external forcing. For example, it has been hypothesised that the abrupt vegetation change that happened some 5 500 years ago in northern Africa was a non-linear response of the subtropical atmosphere-vegetation system to steady, subtle changes in astronomically-induced insolation changes.
- Further theoretical studies on the sensitivity of the climate system to large-scale perturbation show the potential of strong changes in terrestrial carbon fluxes. The uncertainty in biogeochemical feedbacks are substantial and deserve considerable attention since such feedback mechanisms could contribute very large sources of greenhouse gases in the future.
- Palaeoclimate simulations have been improved considerably by explicit inclusion of terrestrial biosphere dynamics. In general, model-based interpretation of palaeobotanic evidence of large-scale vegetation change during the last several thousand years, including historical human land use force us to conclude that land-surface/atmosphere interaction does matter in global climate change.
- As a corollary, it is concluded that without consideration of biogeochemical, biogeophysical and biogeographical feedbacks, attribution of recent climate change as well as assessments of climate change in the upcoming decades will be incomplete.

Martin Claussen · Andrew J. Pitman

Perspective

Remarkable progress in exploring the role of the land surface in the climate system has been made over the last decades – a progress that led to the overall conclusion summarised above. That, however, does not imply that all questions with respect to land-surface/atmosphere interaction have been solved. Some open points were listed in the summary of the chapters, and more can be found in the text. Here we would like to address a more general problem.

In Fig. A.81 we have depicted selected observational and theoretical studies – just a subset of papers cited in the chapters of Part A – which provide strong support for the role of the land surface at various time and space

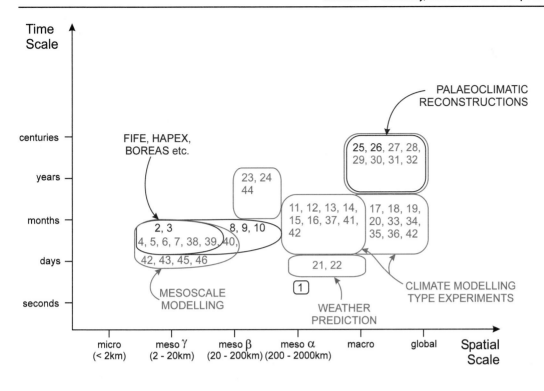

Fig. A.81. Selected observational and modelling evidence providing support for the role of the land-surface at various time and space scales. The *blue* refers to observational studies (including field experiments such as FIFE, HAPEX and BOREAS) while the *red* refers to evidence acquired via modelling. The list of papers cited here is just a subset of a more comprehensive list discussed in detail in the chapters of Part A

scales. Observational and theoretical studies are indicated by different colours, and these markers point at a conceptual problem.

Observational and theoretical studies do not overlap at all scales. Measurements are generally on far too short a time scale to determine the role of the surface in climate and perhaps also in weather, and modelling tends to focus on sensitivity studies (because modelling tends to focus on exploration of the stability of the climate system instead of a realistic description of the past). Climate modelling is on too coarse a scale to answer questions about weather, while mesoscale modelling is on too short a time scale to answer questions about climate.

Therefore, it would be worthwhile to close the gap between observational and theoretical evidence. On the one hand, we need long-term measurements over large spatial scales. This need not be expensive, and in fact it is already being done through FLUXNET (see Chapt. A.2, as well as Part B). These data may provide the long time scale data we need, and when aggregated, are approaching being global in extent. From the modelling perspective, realistic perturbation experiments, with a suite of climate models, would help answer the climate question. Perhaps the final answer will come from simulations of the climate system of the last decades and centuries (i.e. coupled atmosphere-ocean-biosphere simulations including marine and terrestrial carbon cycle). If we can produce high quality simulations of the climate system at these time scales, then we will be able to quantify the effect of land-surface dynamics in the climate system. There already are first model experiments which tackle this problem, and we are confident that further studies will strengthen our knowledge on land-surface dynamics and quantify some of our above conclusions.

Last but not least, we recognise that not only the land surface itself, but also the feedbacks between the land surface and the other components of the climate system are modified by humans. This calls for a more integrated view in which the dynamics of the land surface in the climate system are described not only as a biogeochemical and biogeophysical system, but in which the interference with humankind is taken account interactively. First steps in this direction which lead to the so-called vulnerability approach are discussed in Part E.

References Part A

Aitkenhead JA, McDowell WH (2000) Soil C:N ratio as a predictor of annual riverine DOC flux at local and global scales. Global Biogeochem Cy 14:127–138

Akhtar-Schuster M (1995) Degradadationsprozesse und Desertifikation im semiariden randtropischen Gebiet der Butana/Rep. Sudan. Göttinger Beiträge zur Land- und Forstwirtschaft in den Tropen und Subtropen, Heft 105, 166 S

Alexandre A, Meunier J-D, Lezine A-M, Vincerns A, Schwartz D (1997) Phytoliths: indicators of grassland dynamics during the late holocene in intertropical Africa. Palaeogeogr Palaeocl 136: 213–229

Alm J, Talanov A, Saarnio S, Silvola J, Ikkonen E, Aaltonen H, Nykänen H, Martikainen PJ (1997) Reconstruction of the carbon balance for microsites in a boreal oligotrophic pine fen, Finland. Oecologia 110:423–431

Alm J, Schulman L, Walden J, Nykanen H, Martikainen PJ, Silvola J (1999) Carbon balance of a boreal bog during a year with an exceptionally dry summer. Ecology 80(1):161–174

Alves DS (2002) Space-time dynamics of deforestation in Brazilian Amazônia. Int J Remote Sens 23:2903–2908

An ZS, Xihao W, Yanchou L, De'er Z, Xiangjun S, Guangrong D (1990) A preliminary study on Paleoenvironment change of China during the last 20 000 years. In: Liu (ed) Loess, Quaternary geology, global change. Science Press, Beijing, pp 1–26

Anderson RK, Ferguson EW, Oliver VJ (1966) The use of satellite pictures in weather ananlysis and forecasting. Tech Note 75, World Meteor. Org., Geneva, 184 pp

Anderson DB, Verma SB, Rosenberg NJ (1984) Eddy correlation measurements of CO_2, latent heat and sensible heat fluxes over a crop surface. Bound-Lay Meteorol 29:167–183

Anderson BE, Grant WB, Gregory GL, Browell EV, Collins JE, Sachse GW, Bagwell DR, Hudgins CH, Blake BR, Blake NJ (1996) Aerosols from biomass burning over the tropical South Atlantic region: distributions and impacts. J Geophys Res 101(IssD19): 24117–24137

André J-C, Blondin C (1986) On the effective roughness length for use in numerical three-dimensional models. Bound-Lay Meteorol 35:231–245

André J-C, Bougeault P, Mahfouf J-F, Mascart P, Noilhan J, Pinty J-P (1989) Impact of forests on mesoscale meteorology. Philos T Roy Soc B 324:407–422

Andreae MO, de Almeida SS, Artaxo PE, Brandão C, Carswell FE, Ciccioli P, Culf AD, Esteves JL, Gash JHC, Grace J, Kabat P, Lelieveld J, Malhi Y, Manzi AO, Meixner FX, Nobre A, Nobre C, de Lourdes Ruivo MA, Silva-Dias MA, Stefani P, Valentini R, von Jouanne J, Waterloo M (2002) Biogeochemical cycling of carbon, water, energy, trace gases and aerosols in Amazonia: the LBA-EUSTACH experiment. J Geophys Res 107(D20), 8066, doi:10.1029/2001JD000524, 2002

Anhuf D, Frankenber P, Lauer W (1999) Die postglaziale Warmphase vor 8 000 Jahren. Geol Rundsch 51:454–461

Apps MJ, Price DT (eds) (1996) Forest ecosystems, forest management and the global carbon cycle. NATO ASI Series Vol. I-40. Springer-Verlag, Berlin, 452 pp

Apps MJ, Price DT, Wisniewski J (eds) (1995) Boreal forests and global change. Kluwer Academic, Dordrecht, 540 pp

Arain MA, Michaud JD, Shuttleworth WJ, Yang Z-L, Dolman A-J (1997) Mapping surface cover parameters using aggregation rules and remotely sensed cover classes. Q J Roy Meteor Soc 123(B):2325–2348

Arola A (1999) Parameterisation of turbulent and mesoscale fluxes for heterogeneous surfaces. J Atmos Sci 56:584–598

Arpe K, Behr H, Dümenil L (1997) Validation of the ECHAM4 climate model and re-analyses data in the Arctic region. In: Proc. Workshop on the Implementation of the Arctic Precipitation Data Archive (APDA) at the Global Precipitation Climatology Centre (GPCC), Offenbach, Germany, World Climate Research Programme WCRP-98, WMO/TD No. 804, pp 31–40

Aubinet M, Grelle A, Ibrom A, Rannik Ü, Moncrieff J, Foken T, Kowalski AS, Martin PH, Berbigier P, Bernhofer C, Clement R, Elbers J, Granier A, Grünwald T, Morgenstern K, Pilegaard K, Rebmann C, Snijders W, Valentini R, Vesala T (2000) Estimates of the annual net carbon and water exchange of European forests: the EUROFLUX methodology. Adv Ecol Res 30:113–175

Avissar R (1992) Conceptual aspects of a statistical-dynamical approach to represent landscape subgrid-scale heterogeneities in atmospheric models. J Geophys Res 97:2729–2742

Avissar R (1998) Which type of soil-vegetation-atmosphere transfer scheme is needed for general circulation models: a proposal for a higher-order scheme. J Hydrol 212:136–154

Avissar R, Chen F (1993) Development and analysis of prognostic equations for mesoscale kinetic energy and mesoscale (subgrid-scale) fluxes for large scale atmospheric models. J Atmos Sci 50:3751–3774

Avissar R, Liu Y (1996) A three-dimensional numerical study of shallow convective clouds and precipitation induced by land-surface forcing. J Geophys Res 101:7499–7518

Avissar R, Pielke RA (1989) A parameterisation of heterogeneous land-surface for atmospheric numerical models and its impact on regional meteorology. Mon Weather Rev 117:2113–2136

Avissar R, Schmidt T (1998) An evaluation of the scale at which ground-surface heat flux patchiness affects the convective boundary layer using a large-eddy simulation model. J Atmos Sci 55(16):2666–2689

Avissar R, Eloranta EW, Gurer K, Tripoli GJ (1998) An evaluation of the large-eddy simulation option of the regional atmospheric modeling system in simulating a convective boundary layer: a FIFE case study. J Atmos Sci 55(7):1109–1130

Backéus I (1990) Production and depth distribution of fine roots in a boreal open bog. Ann Bot Fenn 27:261–265

Baidya Roy S, Avissar R (2000) Scales of response of the convective boundary layer to land-surface heterogeneity. Geophys Res Lett 27:533–536

Baidya Roy S, Avissar R (2002) Impact of land-use/land-cover change on regional hydrometeorology in Amazonia. J Geophys Res 107(D20), 8037, doi: 10.1029/2000JD000266, 2002

Baldocchi DD, Hicks BB, Meyers TP (1988) Measuring biosphere-atmosphere exchanges of biologically related gases with micrometeorological methods. Ecology 69:1331–1340

Baldocchi DD, Falge B, Wilson K (2001) A spectral analysis of biosphere-atmosphere trace gas flux densities and meteorological variables across hour to multi-year time scales. Agr Forest Meteorol 107:1–27

Ball JT, Woodrow IE, Berry JA (1987) A model predicting stomatal conductance and its contribution to the control of photosynthesis under different environmental conditions. In: Biggins J (ed) Progress in photosynthesis research. Nihjoff, Dordrecht, pp 221–224

Bamzai AS, Shukla J (1999) Relation between Eurasia snow cover, snow depth, and the India summer monsoon: an observational study. J Climate 12:3117–3132

Barbé L le, Lebel T (1997) Rainfall climatology of the HAPEX-Sahel region during the years 1960–1990. J Hydrol 188–189:43–73

Barnston AG, Schikedanz PT (1984) The effect of irrigation on warm season precipitation in the southern Great Plains. J Appl Meteorol 23:865–888

Barr AG, Betts AK (1997) Radiosonde boundary-layer budgets above a boreal forest. J Geophys Res 102:29205–29212

Barr AG, Betts AK, Black TA, McCaughey JH, Smith CS (2001) Intercomparison of BOREAS northern and southern study area surface fluxes in 1994. J Geophys Res 106(D24):33543–33550

Barrow CJ (1991) Land degradation: development and breakdown of terrestrial environments. Cambridge University Press, 295 pp

Batchvarova E, Gryning SE, Hasager CB (2001) Regional fluxes of momentum and sensible heat over a sub-arctic landscape during late winter. Bound-Lay Meteorol 99(3):489–507

Baumgartner A, Reichel R (1975) The world water balance: mean annual global, continental and maritime precipitation, evaporation and run-off. Elsevier, Amsterdam, 167 pp

Beaubien J, Cihlar J, Simard G, Latifovic R (1999) Land cover from multiple Thematic Mapper scenes using a new enhancement-classification methodology. J Geophys Res 104(D22):27909–27920

Beljaars ACM, Viterbo P, Miller MJ, Betts AK (1996) The anomalous rainfall over the United States during July 1993: sensitivity to land-surface parametrization and soil moisture anomalies. Mon Weather Rev 124:362–383

Bellisario L, Moore TR, Bubier J (1998) Net ecosystem exchange of CO_2 in peatlands, Thompson, Manitoba. Ecoscience 5:534–41

Benjamin SG, Carlson TN (1986) Some effects of surface heating and topography on the regional severe storm environment. Part I: Three-dimensional simulations. Mon Weather Rev 114:307–329

Benjaminsen TA (1993) Fuelwood and desertification: Sahel orthodoxies discussed on the basis of field data from the Gourma region in Mali. Geoforum 24/4:397–409

Berbery EH, Nogues-Paegle J (1993) Intraseasonal interactions between the tropics and extratropics in the Southern Hemisphere. J Atmos Sci 50:1950–1965

Berger A (2001) The role of CO_2, sea-level and vegetation during the Milankovitch-forced glacial-interglacial cycles. In: Bengtsson LO, Hammer CU (eds) Geosphere-biosphere interactions and climate. Cambridge University Press, New York, pp 119–146

Berger A, Fichefet T, Gallée H, Tricot C, van Ypersele JP (1992) Entering the glaciation with a 2-D coupled climate model. Quaternary Sci Rev 11/4:481–493

Berger A, Gallée H, Tricot C (1993) Glaciation and deglaciation mechanisms in a coupled two-dimensional climate-ice-sheet model. J Glaciol 39/131:45–49

Betts RA (1999a) The impact of land use on the climate of the present day. In: Ritchie H (ed) Research activities in atmospheric and oceanic modelling. CAS/JSC WGNE Report No. 28. WMO, Geneva, pp 7.11–7.12

Betts RA (1999b) Self-beneficial effects of vegetation on climate in an ocean-atmosphere general circulation model. Geophys Res Lett 26(19):1457–1460

Betts AK (2000a) Idealized model for equilibrium boundary layer over land. J Hydrometeorol 1:507–523

Betts RA (2000b) Offset of the potential carbon sink from boreal forestation by decreases in surface albedo. Nature 408:187–190Betts AK, Ball JH (1997) Albedo over the boreal forest. J Geophys Res 102D:28901–28909

Betts AK, Ball JH, Beljaars ACM (1993) Comparison between the land-surface response of the European Centre model and the FIFE-1987 data. Q J Roy Meteor Soc 119:975–1001

Betts AK, Ball JH, Beljaars ACM, Miller MJ, Viterbo P (1996) The land surface-atmosphere interaction: a review based on observational and global modeling perspectives. J Geophys Res 101D:7209–7225

Betts RA, Cox PM, Lee SE, Woodward FI (1997) Contrasting physiological structural vegetation feedbacks in climate change simulations. Nature 387:796–799

Betts AK, Goulden ML, Wofsy SC (1999) Controls on evaporation in a boreal spruce forest. J Climate 12:1601–1618

Betts AK, Ball JH, McCaughey JH (2001) Near-surface climate in the boreal forest. J Geophys Res 106(D24):33529–33541

Bjerknes J (1969) Atmospheric teleconnections from the equatorial Pacific. Mon Weather Rev 97:163–172

Black TA, DenHartog G, Neumann HH, Blanken PD, Yang PC, Russell C, Nesic Z, Lee X, Chen SG, Staebler R, Novak MD (1996) Annual cycles of water vapour and carbon dioxide fluxes in and above a boreal aspen forest. Global Change Biol 2:219–229

Black TA, Chen WJ, Barr AG, Arain MA, Chen Z, Nesic Z, Hogg EH, Neumann HH, Yang PC (2000) Increased carbon sequestration by a boreal deciduous forest in years with a warm spring. Geophys Res Lett 27:1271–1274

Blanken PD, Black TA, Yang PC, Neumann HH, Nesic Z, Staebler R, den Hartog G, Novak MD, Lee X (1997) Energy balance and canopy conductance of a boreal aspen forest: partitioning overstory and understory components. J Geophys Res 102(D24): 28915–28927

Blyth EM, Dolman AJ, Noilhan J (1993) Sensitivity of mesoscale circulations to surface wetness: an example from HAPEX-MOBILHY. J Appl Meteorol 33:445–454

Bolle H-J (1985) What is climate? Adv Space Res 5(6):5–14

Bolle H-J (ed) (2003) Mediterranean climate – variability and trends. Springer-Verlag, Berlin, 320 pp

Bonan GB (1997) Effects of land use on the climate of the United States. Climatic Change 37:449–486

Bonan GB, Stillwell-Soller LM (1998) Soil water and the persistence of floods and droughts in the Mississippi River basin. Water Resour Res 34:2693–2701

Bonan GB, Pollard D, Thompson SL (1992) Effects of boreal forest vegetation on global climate. Nature 359:716–718

Boone A, Wetzel PJ (1996) Issues related to low resolution modeling of soil moisture: experience with the PLACE model. Global Planet Change 13:161–181

Bougeault P, Bret B, Lacarrere P, Noilhan J (1991) An experiment with an advanced surface parameterisation in a mesobeta-scale model. Part II: the 16 June 1986 simulation. Mon Weather Rev 119:2374–2392

Bounoua L, Collatz GJ, Sellers PJ, Randall DA, Dazlich DA, Los SO, Berry JA, Fung I, Tucker CJ, Field CB, Jensen TG (1999) Interactions between vegetation and climate: radiative and physiological effects of doubled atmospheric CO_2, J Climate 12: 309–324

Bowling LC, Lettenmaier DP, Matheussen BV (2000) Hydroclimatology of the Arctic drainage basin. In: Lewis L (ed) Freshwater budget of the Arctic Ocean. Kluwer, Dordrecht, pp 57–90

Braconnot P, Joussaume S, Marti O, deNoblet-Ducoudre N (1999) Synergistic feedbacks from ocean and vegetation on the African monsoon response to mid-Holocene insolation. Geophys Res Lett 26(16):2481–2484

Bradley RS, Diaz HF, Eischeid JK, Jones PD, Kelly PM, Goodess CM (1987) Precipitation fluctuations over Northern Hemisphere land areas since the mid-19th century. Science 237:171–275

Breman H, de Wit CT (1983) Rangeland productivity and exploitation in the Sahel. Science 221:1341–1347

Brklacich M, Bryant C, Veenhof B, Beauchesne A (1997) Implications of global climatic change for Canadian agriculture: a review and appraisal of research from 1984 to 1997, chapter 4 Canada country study: climate impacts and adaptations. Environment Canada

Broström A, Coe M, Harrison SP, Gallimore R, Kutzbach JE, Foley J, Prentice IC, Behling P (1998) Land-surface feedbacks and paleomonsoons in northern Africa. Geophys Res Lett 25:3615–3618

Brovkin V, Claussen M, Petoukhov V, Ganopolski A (1998) On the stability of the atmosphere-vegetation system in the Sahara/Sahel region. J Geophys Res 103(D24):31613–31624

Brovkin V, Ganopolski A, Claussen M, Kubatzki C, Petoukhov V (1999) Modelling climate response to historical land-cover change. Global Ecol Biogeogr 8(6):509–517

Brown RD (2000) Northern Hemisphere snow cover variability and change, 1915-1997. J Climate 13:2339-2355

Brown RD, Braaten RO (1998) Spatial and temporal variability of Canadian monthly snow depths, 1946-1995. Atmos Ocean 36:37-54

Brubaker KL, Entekhabi D (1995) An analytic approach to modeling land-atmosphere interaction. 1: Construct and equilibrium behavior. Water Resour Res 31:619-632

Bubier JL, Moore TR, Bellisario L, Comer NT, Crill PM (1995) Ecological controls on methane emissions from a northern peatland complex in the zone of discontinuous permafrost, Manitoba, Canada. Global Biogeochem Cy 9:455-470

Bubier JL, Crill PM, Moore TR, Savage K, Varner RK (1998) Seasonal patterns and controls on net ecosystem CO_2 exchange in a boreal peatland complex. Global Biogeochem Cy 12:703-714

Burde GI, Zangvil A (1996) Estimating the role of local evaporation in precipitation in two dimensions. J Climate 9:1328-1338

Canadell J, Jackson RB, Ehleringer JR, Mooney HA, Sala DE, Schulze ED (1996) Maximum rooting depth of vegetation type at the global scale. Oecologia 108:583-595

Canadell J, Mooney H, Baldocchi D, Berry J, Ehleringer J, Field CB, Gower T, Hollinger D, Hunt J, Jackson R, Running S, Shaver G, Trumbore S, Valentini R, Yoder B (2000) Carbon metabolism of the terrestrial biosphere. Ecosystems 3:115-130

Carter T, Hulme M (2000) Interim characterizations of regional climate and related changes up to 2100 associated with the provisional SRES marker emissions scenarios. IPCC Secretariat, c/o WMO, Geneva, Switzerland

Chahine (1999) Integrated land-atmosphere-ocean hydrology. In: Browning KA, Gurney RJ (eds) Global energy and water cycles. Cambridge University Press, pp 282-285

Chang J-Y, Wetzel PJ (1991) Effect of spatial variations of soil moisture and vegetation on the evolution of a prestorm environment: a numerical case study. Mon Weather Rev 119:1368-1390

Changnon SA (ed) (1996) The great flood of 1993. Westview Press, pp xii, 324

Chanton JP, Dacey JWH (1991) Effects of vegetation on methane flux, reservoirs and carbon isotopic composition. In: Sharkey TD, Holland EA, Mooney HA (eds) Trace gas emissions from plants. Academic Press, San Diego CA, pp 65-92

Chapin III FS, Shaver GR, Giblin AE, Nadelhoffer KJ, Laundre JA (1995) Responses of arctic tundra to experimental and observed changes in climate. Ecology 76:649-711

Chapman WL, Walsh JE (1993) Recent variations of sea ice and air temperature in high latitudes. B Am Meteorol Soc 74:33-47

Charlson R (1997) Direct climate forcing by anthropogenic sulfate aerosols: the Arrhenius paradigm a century later. Ambio 26(1):25-31

Charney JG (1975) Dynamics of deserts and drought in the Sahel. Q J Roy Meteor Soc 101:193-202

Charney JG, Quirk WK, Chow S-H, Kornfield J (1977) A comparative study of the effects of albedo change on drought in semiarid regions. J Atmos Sci 34:1366-1387

Chase TN, Pielke RA, Kittel TGF, Nemani RR, Running SW (1996) Sensitivity of a general circulation model to global changes in leaf area index. J Geophys Res 101:7393-7408

Chase TN, Pielke RA Sr, Kittel TGF, Baron JS, Stohlgren TJ (1999) Potential impacts on Colorado Rocky Mountain weather due to land-use changes on the adjacent Great Plains. J Geophys Res 104:16673-16690

Chase TN, Pielke RA, Kittel TGF, Nemani R, Running SW (2000) Simulated impacts of historical land-cover changes on global climate in northern winter. Clim Dynam 16(2-3):93-105

Chase TN, Herman B, Pielke RA Sr, Zeng X, Leuthold M (2002) A proposed mechanism for the regulation of mid-tropospheric temperatures in the Arctic. J Geophys Res 107(D14):4193, doi: 10.1029/2001JD001425

Cheddadi R, Yu G, Guiot J, Harrison SP, Prentice IC (1997) The climate of Europe 6000 years ago. Clim Dynam 13:1-9

Chen F, Avissar R (1994a) The impact of land-surface wetness heterogeneity on mesoscale heat fluxes. J Appl Meteorol 33:1323-1340

Chen F, Avissar R (1994b) Impact of land-surface moisture variability on local shallow convective cumulus and precipitation in large-scale models. J Appl Meteorol 33:1382-1401

Chen T-C, Tzeng R-Y, Van Loon H (1988) Study on the maintenance of the winter subtropical jet streams in the Northern Hemisphere. Tellus 40A:392-397

Chen TH, Henderson-Sellers A, Milly PCD, Pitman AJ, Beljaars ACM, Polcher J, Abramopoulos F, Boone A, Chang S, Chen F, Dai Y, Desborough CE, Dickinson RE, Dumenil L, Ek M, Garratt JR, Gedney N, Gusev YM, Kim J, Koster R, Kowalczyk EA, Laval K, Lean J, Lettenmaier D, Liang X, Mahfouf JF, Mengelkamp HT, Mitchell K, Nasonova ON, Noilhan J, Robock A, Rosenzweig C, Schaake J, Schlosser CA, Schulz JP, Shao Y, Shmakin AB, Verseghy DL, Wetzel P, Wood EF, Xue Y, Yang ZL, Zeng Q (1997) Cabauw experimental results from the project for intercomparison of land-surface parameterisation schemes. J Climate 10:1194-1215

Chen JM, Chen WJ, Liu J, Cihlar J, Gray S (2000) Annual carbon balance of Canada's forest during 1895-1996. Global Biogeochem Cy 14:839-850

Chervin RM, Schneider SH (1976) On determining the statistical significance of climate experiments with general circulation models. J Atmos Sci 33:405-412

Chu P-S (1982) Diagnostics of climate anomalies in tropical Brazil. Ph. D. Dissertation. Department of Meteorology, University of Wisconsin, Madison, USA

Chu P-S, Yu Z-P, Hastenrath S (1994) Detecting climate change concurrent with deforestation in the Amazonian basin: which way has it gone. B Am Meteorol Soc 75(4):579-583

Ciais P, Tans PP, Trolier M, White JWC, Francy RJ (1995) A large North Hemisphere terrestrial CO_2 sink indicated by the $^{13}C/^{12}C$ ratio of atmospheric CO_2. Science 269:1098-1102

Cienciala E, Kucera J, Lindroth A, Cermak J, Grelle A, Halldin S (1997) Canopy transpiration from a boreal forest in Sweden during a dry year. Agr Forest Meteorol 86:157-167

Cienciala E, Kucera J, Ryan MG, Lindroth A (1998) Water flux in a forest during two hydrologically contrasting years; species specific regulation of canopy conductance and transpiration. Ann Sci Forest 55:47-61

Cienciala E, Kucera J, Lindroth A (1999) Long-term measurements of stand water uptake in Swedish boreal forest. Agr Forest Meteorol 98:1-4, 99:547-554

Clapp RB, Hornberger GM (1978) Empirical equations for some soil hydraulic properties Water Resour Res 14:601-604

Clark D, Xue Y, Harding RJ, Valdes PJ (2001) Modeling the impact of land-surface degradation on the climate of tropical North Africa. J Climate 14(8):1809-1822

Clarke RH, Dyer AJ, Brook RR, Reid DG, Troup AJ (1971) The Wangara experiment: boundary layer data. Div. Meteor. Phys. Tech. Paper No. 19, CSIRO, Melbourne, Australia, 341 pp

Claussen M (1990) Area-averaging of surface fluxes in a neutrally stratified, horizontally inhomogeneous atmospheric boundary layer. Atmos Environ 24a:1349-1360

Claussen M (1991a) Estimation of areally-averaged surface fluxes. Bound-Lay Meteorol 54:387-410

Claussen M (1991b) Local advection processes in the surface layer of the marginal ice zone. Bound-Lay Meteorol 54:1-27

Claussen M (1994) On coupling global biome models with climate models. Climate Res 4:203-221

Claussen M (1997) Modeling biogeophysical feedback in the Africa and Indian monsoon region. Clim Dynam 13:247-257

Claussen M (1998) On multiple solutions of the atmosphere-vegetation system in present-day climate. Global Change Biol 4:549-560

Claussen M (2001a) Biogeophysical feedbacks and the dynamics of climate. In: Schulze ED, Harrison SP, Heimann M, Holland EA, Lloyd J, Prentice IC, Schimel D (eds) Global biogeochemical cycles in the climate system. Academic Press, San Diego, pp 61-71

Claussen M (2001b) Earth system models. In: Ehlers E, Krafft T (eds) Understanding the Earth system: compartments, processes and interactions. Springer-Verlag, Heidelberg, pp 145-162

Claussen M, Gayler V (1997) The greening of Sahara during the mid-Holocene: results of an interactive atmosphere-biome model. Global Ecol Biogeogr 6:369-377

Claussen M, Brovkin V, Ganopolski A, Kubatzki C, Petoukhov V (1998) Modeling global terrestrial vegetation-climate interaction. Philos T Roy Soc B 353:53-63

Claussen M, Kubatzki C, Brovkin V, Ganopolski A, Hoelzmann P, Pachur HJ (1999) Simulation of an abrupt change in Saharan vegetation at the end of the mid-Holocene. Geophys Res Letters 26(14):2037–2040

Claussen M, Brovkin V, Petoukhov V, Ganopolski A (2001) Biogeophysical versus biogeochemical feedbacks of large-scale landcover change. Geophys Res Lett 26(6):1011–1014

Claussen M, Mysak LA, Weaver AJ, Crucifix M, Fichefet T, Loutre M-F, Weber SL, Alcamo J, Alexeev VA, Berger A, Calov R, Ganopolski A, Goosse H, Lohman G, Lunkeit F, Mokhov II, Petoukhov V, Stone P, Wang Zh (2002) Earth system models of intermediate complexity: closing the gap in the spectrum of climate system models. Clim Dynam 18(7):579–586

Cohen JCP, Silva Dias MAF, Nobre CA (1995) Environmental conditions associated with Amazonian squall lines: a case study. Mon Weather Rev 123:3163–3174

Collins D, Avissar R (1994) An evaluation with the Fourier amplitude sensitivity Test (FAST) of which land-surface parameters are of greatest importance for atmospheric modelling. J Climate 7:681–703

Cosby BJ, Hornberger GM, Clapp RB, Ginn TR (1984) A statistical exploration of the relationships of soil moisture characteristics to the physical properties of soils. Water Resour Res 20:682–690

Costa MH, Foley J (1998) Trends in the hydrological cycle of the Amazon basin. J Geophys Res 104:14189–14198

Costa MH, Foley JA (2000) Combined effects of deforestation and doubled atmospheric CO_2 concentrations on the climate of Amazonia. J Climate 13:35–58

Cowan IR, Farquhar GD (1977) Stomatal function in relation to leaf metabolism and environment. Sym Soc Exp Biol 31:471–505

Cox PM, Huntingford C, Harding RJ (1998) A canopy conductance and photosynthesis model for use in a GCM land-surface scheme. J Hydrol 212–213:79–94

Cox PM, Betts RA, Bunton CB, Essery RLH, Rowntree PR, Smith J (1999) The impact of new land-surface physics on the GCM simulation of climate and climate sensitivity. Clim Dynam 15:183–203

Cox PM, Betts RA, Jones CD, Spall SA, Totterdell IJ (2000) Acceleration of global warming due to carbon-cycle feedbacks in a coupled climate model. Nature 408:184–187

CPTEC (Centro de Previsão de Tempo e Estudos Climáticos) (2002) Climanálise-Boletim de Monitoramento Climático. Outubro de 2000. Cachoeira Paulista, Brazil (available at *www.cptec.inpe.br*)

Cramer W, Kicklighter DW, Bondeau A, Moore III B, Churkina G, Nemry B, Ruimy A, Schloss AL, Kaduk J (1999) Comparing global models of terrestrial primary productivity (NPP): overview and key results. Global Change Biol 5:1–15

Cramer W, Bondeau A, Woodward FI, Prentice IC, Betts RA, Brovkin V, Cox PM, Fisher V, Foley JA, Friend AD, Kucharik C, Lomas MR, Ramankutty N, Stitch S, Smith B, White A, Young-Molling C (2000) Global response of terrestrial ecosystem structure and function to CO_2 and climate change: results from six dynamic global vegetation models. Global Change Biol 7:357–373

Crill PM, Harriss RC, Bartlett KB (1991) Methane fluxes from terrestrial wetland environments. In: Rogers JE, Whitman WB (ed) Microbial production and consumption of greenhouse gases: methane, nitrogen oxides, and halomethanes. Am Soc Micro, pp 91–110

Crossley JF, Polcher J, Cox PM, Gedney N, Planton S (2000) Uncertainties linked to land-surface processes in climate change simulations. Clim Dynam 16:949–961

Cubasch U, Santer BD, Hegerl GC (1995) Klimamodelle – wo stehen wir? Phys Bl 51:269–276

Culf AD, Fisch G, Hodnett MG (1995) The albedo of Amazonian forest and ranchland. J Climate 8:1544–1554

Culf AD, Esteves JL, Marques Filho A de O, da Rocha HR (1996) Radiation, temperature and humidity over forest and pasture in Amazonia. In: Gash JHC, Nobre CA, Roberts JM, Victoria RL (eds) Amazonian deforestation and climate. John Wiley & Sons, Chichester, pp 175–191

Culf AD, Fisch G, Malhi Y, Costa RC, Nobre A, Marques Filho A de O, Gash JHC, Grace J (1999) Carbon dioxide measurements in the nocturnal boundary layer over Amazonian forest. Hydrol Earth Syst Sc 3:39–53

Cunnington WM, Rowntree PR (1986) Simulations of the Saharan atmosphere-dependence on moisture and albedo. Q J Roy Meteor Soc 112:971–999

Curtis S, Hastenrath S (1999) Trends of upper-air circulation and water vapour over equatorial South America and adjacent oceans. Int J Climatol 19:863–876

Cutrim E, Martin DW, Rabin RM (1995) Enhancement of cumulus clouds over deforested lands in Amazonia. B Am Meteorol Soc 76:1801–1805

Dai A, Fung IY, DelGenio AD (1997) Surface observed global land precipitation variations during 1900–1988. J Climate 10(11):2943–2962

Dalu GA, Pielke RA (1993) Vertical heat fluxes generated by mesoscale atmospheric flow induced by thermal inhomogeneities in the PBL. J Atmos Sci 50:919–926

Dalu GA, Pielke RA, Baldi M, Zeng X (1996) Heat and momentum fluxes induced by thermal inhomogeneities. J Atmos Sci 53:3286–3302

Dang Q-L, Margolis HA, Sy M, Coyea MR, Collatz GJ, Walthall CL (1997) Profiles of PAR, nitrogen and photosynthetic capacity in the boreal forest: implications for scaling from leaf to canopy. J Geophys Res 102:28845–28860

Deardorff JW (1972) Numerical investigation of neutral and unstable planetary boundary layers. J Atmos Sci 29:91–115

Deardorff JW (1974) Three-dimensional numerical study of the height and mean structure of a heated planetary boundary layer. Bound-Lay Meteorol 7:81–106

Deardorff JW (1977) A parameterisation of ground-surface moisture content for use in atmosphere prediction models. J Appl Meteorol 16:1182–1185

Deardorff JW (1978) Efficient prediction of ground surface temperature and moisture, with inclusion of a layer of vegetation. J Geophys Res 83(C4):1889–1903

DeFries RS, Townshend JRG (1994) NDVI-derived land-cover classification at global scales. Int J Remote Sens 15(17):3567–3586

deMarchi L (1885) In: Arrhenius S (1896) On the influence of carbonic acid in the air upon the temperature of the ground. The London Edinburgh and Dublin Philosophical Magazine and Journal of Science 41:237–276

deMenocal PB, Ortiz J, Guilderson T, Adkins J, Sarnthein M, Baker L, Yarusinski M (2000) Abrupt onset and termination of the African humid period: rapid climate response to gradual insolation forcing. Quaternary Sci Rev 19:347–361

Deng JZ, Chuangjun W, Zhikang X (1983) General view of agriculture geography of China. Science Press, 334 pp

Denmead OT, Bradley EF (1985) Flux-gradient relationships in a forest canopy. In: Hutchison BA, Hicks BB (ed) The forest-atmosphere interaction. D. Reidel Publishing Company, Dordrecht, pp 421–442

Denning AS, Fung I, Randall DA (1995) Strong simulated meridional gradient of atmospheric CO_2 due to seasonal exchange with the terrestrial biota. Nature 376:240–243

Denning AS, Collatz JG, Zhang C, Randall DA, Berry JA, Sellers PJ, Colello GD, Dazlich DA (1996) Simulations of terrestrial carbon metabolism and atmospheric CO_2 in a general circulation model. Part 1: Surface carbon fluxes. Tellus 48B:521–542

deNoblet N, Prentice IC, Jousaume S, Texier D, Botta A, Haxeltine A (1996) Possible role of atmosphere-biosphere interactions in triggering the last glaciation. Geophys Res Lett 23(22):3191–3194

deNoblet-Ducoudre N, Claussen M, Prentice IC (2000) Mid-Holocene greening of the Sahara: first results of the GAIM 6000 year BP experiment with two asynchronously coupled atmosphere/biome models. Clim Dynam 16(9):643–659

Desborough CE (1999) Surface energy balance complexity in GCM land-surface models. Clim Dynam 15:389–403

Desjardins RL, Buckley DJ, St Amour G (1984) Eddy flux measurements of CO_2 above corn using a microcomputer system. Agr Forest Meteorol 32:257–265

Dewey KF, Heim G Jr (1982) A digital archive of Northern Hemisphere snow cover, November 1966 through December 1980. B Am Meteorol Soc 63:1132–1141

Dickinson RE (1992) Changes in land use. In: Trenberth KE (ed) Climate system modeling. Cambridge University Press, Cambrige, pp 698–700

Dickinson RE, Henderson-Sellers A (1988) Modelling tropical deforestation: a study of GCM land-surface parametrizations. Q J Roy Meteor Soc 114:439–462

Dickinson RE, Kennedy P (1992) Impact on regional climate of Amazon deforestation. Geophys Res Lett 19(19):1947–1950

Dickinson RE, Henderson-Sellers A, Kennedy PJ, Wilson MF (1986) Biosphere-atmosphere transfer scheme (BATS) for the NCAR community climate model. NCAR Technical Note TN-275 + STR, 69 pp

Dickinson RE, Shaikh M, Bryant R, Graumlich L (1998) Interactive canopies for a climate model. J Climate 11:2823–2836

Diop AB (1992) Les paysans du bassin arachidier. Conditions de vie et comportements de survie. Politique Africaine 45:39–61

Dirmeyer PA (1999) Assessing GCM sensitivity to soil wetness using GSWP data. J Meteorol Soc Jpn 77:367–385

Dirmeyer PA, Brubaker KL (1999) Contrasting evaporative moisture sources during the drought of 1988 and the flood of 1993. J Geophys Res 104:19,383–19,397

Dirmeyer PA, Zeng FJ (1999) Precipitation infiltration in the simplified SiB land-surface scheme. J Meteorol Soc Jpn 77:1–13

Dirmeyer PA, Dolman AJ, Sato N (1999) The pilot phase of the global soil wetness project. B Am Meteorol Soc 80:851–878

Doherty R, Kutzbach J, Foley J, Pollard D (2000) Fully coupled climate/dynamical vegetation model simulations over northern Africa during the mid-Holocene. Clim Dynam 16:561–573

Dolman AJ, Wallace JS (1991) Lagrangian and k-theory approaches in modeling evaporation from sparse canopies. Q J Roy Meteor Soc 117:1325

Dolman AJ, Stewart JB, Cooper JD (1988) Predicting forest transpiration from climatological data. Agr Forest Meteorol 42:339–353

Dolman AJ, Gash JHC, Roberts J, Shuttleworth WJ (1991) Stomatal and surface conductance of tropical rainforest. Agr Forest Meteorol 54:303–318

Dolman AJ, Dias MAS, Calvet JC, Ashby M, Tahara AS, Delire C, Kabat P, Fisch GA, Nobre CA (1999) Meso-scale effects of tropical deforestation in Amazonia: preparatory LBA modelling studies. Ann Geophys-Atm Hydr 17. 1095–1110

Dooge JCI (1982) Parameterisation of hydrologic processes. In: Eagleson PS (ed) JSC study conference on land-surface processes in atmospheric general circulation models. World Climate Research Program, Cambridge University Press, Cambridge, pp 243–288

Douville H, Viterbo P, Mahouf J-F, Beljaars ACM (2000) Evaluation of the optimum interpolation and the nudging techniques for soil moisture analysis using FIFE data. Mon Weather Rev 128: 1733–1756

Dregne HE (1983) Desertification of arid lands. Hardwood Academic Publications, New York, 242 pp

Dregne HE, Chou NT (1992) Global desertification dimensions and costs. Degradation and restoration of arid lands. Texas Tech. University, pp 249–282

Dregne HE, Tucker CJ (1988) Desert encroachment. Desertification Control Bulletin 16:16–19

Druyan LM (1991) The sensitivity of sub-Saharan precipitation to Atlantic SST. Climatic Change 18:17–36

Ducoudre NI, Laval K, Perrier A (1993) SECHIBA, a new set of parameterisations of the hydrologic exchanges at the land-atmosphere interface within the LMD atmospheric general circulation model. J Climate 6:248–273

Dümenil L, Todini E (1992) A rainfall-runoff scheme for use in the Hamburg model. In: O'Kane JP (ed) Advances in theoretical hydrology: a tribute to James Dooge. European Geophysical Society Series of Hydrological Sciences, 1., Elsevier

Dümenil L, Hagemann S, Arpe K (1997) Validation of the hydrological cycle in the Arctic using river discharge data. Proc. Workshop on Polar Processes in Global Climate (13–15 November 1996, Cancun, Mexico), AMS, Boston, USA

Eastman JL, Pielke RA, McDonald DJ (1998) Calibration of soil moisture for large eddy simulations over the FIFE area. J Atmos Sci 55:1131–1140

Eastman JL, Coughenour MB, Pielke RA (2001a) Does grazing affect regional climate? J Hydrometeorol 2:243–253

Eastman JL, Coughenour MB, Pielke RA Sr (2001b) The regional effects of CO_2 and landscape change using a coupled plant and meteorological model. Global Change Biol 7:797–815

Ebbesmeyer CC, Cayan DR, McLain DR, Nichols FH, Peterson DH, Redmond KT (1991) 1976 step in the Pacific climate: forty environmental changes between 1968–1975 and 1977–1984. Proc. Seventh Annual Pacific Climate (PACLIM) Workshop, Pacific Grove, CA, pp 115–126

Eischeid JD, Diaz HF, Bradley RS, Jones PD (1991) A comprehensive precipitation dataset for global land areas. DOE/ER-6901T-H1, TR051. United States Dep. of Energy, Carbon Dioxide Research Program, Washington, D.C

Ellis JE, Swift DM (1988) Stability of African pastoral ecosystems: alternate paradigms and implications for development. J Range Manage 41(6):450–459

Eltahir EA, Bras RL (1994) Precipitation recycling in the Amazonian basin. Q J Roy Meteor Soc 120:861–880

Eltahir EAB (1996) Role of vegetation in sustaining large-scale atmospheric circulation in the tropics. J Geophys Res 101(D2):4255–4268

Eltahir EAB, Gong C (1996) Dynamics of wet and dry years in West Africa. J Climate 9(5):1030–1042

ESWG (1996) A national ecological framework for Canada. Agriculture Canada and Agri-Food Canada, Ottawa, 125 pp

Facchini MC, Mircea M, Fuzzi S, Charlson RJ (1999) Cloud albedo enhancement by surface-active organic solutes in growing droplets. Nature 401(6750)257–259

Famiglietti J, Wood EF (1994) Multiscale modeling of spatially variable water and energy balance processes. Water Resour Res 30: 3061–3078

Fan SM, Wofsy SC, Bakwin PS, Jacobs DJ (1990) Atmosphere-biosphere exchange of CO_2 and O_3 in the central Amazon forest. J Geophys Res 95:168851–16854

Fan S, Gloor M, Mahlman J, Pacala S, Sarmiento J, Takahashi T, Tans P (1998) A large terrestrial carbon sink in North America implied by atmospheric and oceanic carbon dioxide data and models. Nature 282:442–446

Farquhar GD, Caemmerer von S (1982) Modeling of photosynthetic response to environmental conditions. In: Lange OL, Nobel PS, Osmond CB, Ziegler H (eds) Encyclopedia of plant physiology, vol. 12B, Physiological plant ecology. II. Water relations and carbon assimilation. Springer-Verlag, New York, pp 549–587

Fast JD, McCorcle MD (1991) The effect of heterogeneous soil moisture on a summer baroclinic circulation in the central United States. Mon Weather Rev 119:2140–2167

Fennessy MJ, Shukla J (1999) Impact of initial soil wetness on seasonal atmospheric prediction. J Climate 12:3167–3180

Field CB, Mooney HA (1986) The photosynthesis-nitrogen relationship in wild plants. In: Givinish TJ (ed) On the economy of plant form and function. Cambridge University Press, Cambridge, pp 25–56

Field C, Jackson R, Mooney H (1995) Stomatal responses to increased CO_2: implications from the plant to the global scale. Plant Cell Environ 18:1214–1225

Figueroa SN, Nobre CA (1990) Precipitation distribution over central and western tropical South América. Climanálise 5:35–46

Figueroa SN, Satyamurty P, Silva Dias PL (1995) Simulations of the summer circulation over the South American region with an Eta coordinate model. J Atmos Sci 52:1573–1584

Findell KL, Eltahir EAB (1997) An analysis of the soil moisture-rainfall feedback based on direct observations from Illinois. Water Resour Res 33:725–735

Fisch G, Culf AD, Nobre CA (1996) Modelling convective boundary layer growth in Rondonia. In: Gash JHC, Nobre CA, Roberts JM, Victoria RL (eds) Amazonian deforestation and climate. John Wiley & Sons, Chichester, pp 425–435

Fisch G, Marengo J, Nobre C (1998) Uma revisao geral sobre o Clima da Amazonia. Acta Amazonica 28:101–126

Fitzjarrald DR, Acevedo OA, Moore KE (2001) Climatic consequences of leaf presence in the eastern United States. J Climate 14(4):598–614

Foley J, Kutzbach JE, Coe MT, Levis S (1994) Feedbacks between climate and boreal forests during the Holocene epoch. Nature 371: 52–54

Foley JA, Prentice IC, Ramankutty N, Levis S, Pollard D, Sitch S, Haxeltine A (1996) An integrated biosphere model of land-surface processes, terrestrial carbon balance and vegetation dynamics. Global Biogeochem Cy 10:603–628

Foley JA, Levis S, Prentice IC, Pollard D, Thompson SL (1998) Coupling dynamic models of climate and vegetation. Global Change Biol 4:561–579

Folland CK, Palmer TN, Parker DE (1986) Sahel rainfall and worldwide sea temperatures, 1901–1985. Nature 320:602–607

Folland CK, Owen J, Ward MN, Colman A (1991) Prediction of seasonal rainfall in the Sahel region using empirical and dynamical methods. J Forecasting 1:21–56

Forse B (1989) The myth of the marching desert. New Sci 4:31

Franchito SH, Rao VB (1992) Climatic change due to land-surface alterations. Climatic Change 22(1):1–34

Franken W, Leopoldo P (1984) Hydrology of catchment areas in Central-Amazonian forest streams. In: Soili H (ed) The Amazon, limnology and landscape ecology of a mighty tropical river and its basin. pp 501–519

Frenzel B, Pesci M, Velichko AA (1992) Atlas of paleoclimates and paleoenvironments of the Northern Hemisphere: Late Pleistocene–Holocene. Geographical Research Institute, Budapest

Friedlingstein P, Bopp L, Ciais P, Dufresne J-L, Fairhead L, LeTreut H, Monfray P, Orr J (2001) Positive feedback between future climate change and the carbon cycle. Geophys Res Lett 28:1543–1546

Friedrich K, Mölders N (2000) On the influence of surface heterogeneity on latent heat fluxes and stratus properties. Atmos Res 54:59–85

Friend AD, Shugart HH, Running SW (1993) A physiology-based model of forest dynamics. Ecology 74:792–797

Frolking S (1997) Sensitivity of spruce/moss boreal forest net ecosystem productivity to seasonal anomalies in weather. J Geophys Res 102:29053–29064

Frolking S, Bubier JL, Moore TR, Ball T, Bellisario LM, Bhardwaj A, Carroll P, Crill PM, Lafleur PM, McCaughey JH, Roulet NT, Suyker AE, Verma SB, Waddington MJ, Whiting GJ (1998) Relationship between ecosystem productivity and photosynthetically-active radiation for northern peatlands. Global Biogeochem Cy 12:115–126

Fu CB (1994) An aridity trend in China in association with global warming. In: Zepp RG (ed) Climate-biosphere interaction: biogenic emission and environmental effects of climate change. John Wiley & Sons, Chichester, pp 1–17

Fu CB, Wen G (1999) Variation of ecosystems over East Asia in association with seasonal, interannual and decadal monsoon climate variability. Climatic Change 43:477–494

Fu CB, Wei HL, Chen M (1998) Evolution of summer monsoon rainbelts over East China in a regional climate model. Sci Atmos Sin 22:522–534

Fu CB, Wei HL, Qian Y, Chen M (2000) Documentation on a Regional Integrated Environmental Model System (RIEMS, version 1). TEACOM Science Report No. 1:1–26. START Regional Committee for Temperate East Asia, Beijing, China

Fyfe JC, Flato GM (1999) Enhanced climate change and its detection over the Rocky Mountains. J Climate 12:230–243

Gallée H, van Ypersele JP, Fchefet T, Marsiat I, Tricot C, Berger A (1992) Simulation of the last glacial cycle by a coupled, sectorial averaged climate-ice sheet model. 2. response of insolation and CO$_2$ variations. J Geophys Res 97(D14):15713–15740

Gallimore RG, Kutzbach JE (1996) Role of orbitally-induced vegetative changes on incipient glaciation. Nature 381:503–505

Ganopolski A, Kubatzki C, Claussen M, Brovkin V, Petoukhov V (1998) The influence of vegetation-atmosphere-ocean interaction on climate during the mid-Holocene. Science 280:1916–1919

Ganopolski A, Petoukhov V, Rahmstorf S, Brovkin V, Claussen M, Eliseev A, Kubatzki C (2001) CLIMBER-2: a climate system model of intermediate complexity. Part II: Validation and sensitivity tests. Clim Dynam 17:735–751

Garreaud R, Wallace JM (1997) The diurnal march of convective cloudiness over the Americas. Mon Weather Rev 125:3157–3171

Garstang M, Ulanski S, Greco S, Scala J, Swap R, Fitzjarrald D, Browell E, Shipman M, Connors V, Harriss R, Talbot R (1990) The Amazon Boundary-Layer Experiment (ABLE 2B): a meteorological perspective. B Am Meteorol Soc 71:19–31

Gash JHC, Nobre CA (1997) Climatic effects of Amazonian deforestation: some results from ABRACOS. B Am Meteorol Soc 78(5):823–830

Gash JHC, Nobre CA, Roberts JM, Victoria RL (eds) (1996) Amazonian deforestation and climate. John Wiley & Sons, Chichester, 611 p

Gash JHC, Kabat P, Monteny BA, Amadou M, Bessemoulin P, Billing H, Blyth EM, deBruin HAR, Elbers JA, Friborg T, Harrison G, Holwill CJ, Lloyd CR, Lhomme JP, Moncrieff JB, Puech D, Soegaard H, Taupin JD, Tuzet A, Verhoef A (1997) The variability of evaporation during the HAPEX-Sahel intensive observation period. J Hydrol 188/189:385–399

Gedney N, Valdes PJ (2000) The effect of Amazonian deforestation on the Northern Hemisphere circulation and climate. Geophys Res Lett 27:3053–3056

Gedney N, Cox PM, Douville H, Polcher J, Valdes PJ (2000) Characterising GCM land-surface schemes to understand their responses to climate change. J Climate 13(17):3066–3079

Gibson HM, Vonder Haar TH (1990) Cloud and convection frequencies over the southeast United States as related to small-scale geographic features. Mon Weather Rev 118:2215–2227

Giorgi F (1990) Sensitivity of wintertime precipitation and soil hydrology simulation over the western United States to lower boundary specification. Atmos Ocean 28:1–23

Giorgi F (1997a) An approach for the representation of surface heterogeneity in land-surface models: Part I: Theoretical framework. Mon Weather Rev 125:1885–1899

Giorgi F (1997b) An approach for the representation of surface heterogeneity in land-surface models: Part II: Validation and sensitivity experiments. Mon Weather Rev 125:1900–1919

Glantz MH, Katz RW, Nicholls N (eds) (1991) Teleconnections linking worldwide climate anomalies. Cambridge University Press, Cambridge, England

Goetz SJ (1997) Multi-sensor analysis of NDVI, surface temperature and biophysical variables at a mixed grassland site. Int J Remote Sens 18(1):71–94

Gopalakrishnan S, Baidya Roy S, Avissar R (2000) An evaluation of the scale at which topographical features affect the convective boundary layer using large-eddy simulations. J Atmos Sci 57:334–351

Gordon C, Cooper C, Senior CA, Banks H, Gregory JM, Johns TC, Mitchell JFB, Wood RA (2000) The simulation of SST, sea ice extents and ocean heat transport in a version of the Hadley Centre coupled model without flux adjustments. Clim Dynam 16:147–168

Gorham E (1991) Northern peatlands: role in the carbon cycle and probable responses to climatic warming. Ecol Appl 1:182–195

Gornitz V (1985) A survey of anthropogenic vegetation changes in west Africa during the last century – climate implication. Climatic Change 7:285–235

Gottschalk L, Batchvarova E, Gryning SE, Lindroth A, Melas D, Motovilov Y, Fresh M, Heikinheimo M, Samuelsson P, Grelle A, Persson T (1999) Scale aggregation – comparison of flux estimates from NOPEX. Agr Forest Meteorol 98–99

Goudie AS (1990) Desert degradation. In: Goudie AS (ed) Techniques for desert reclamation. John Wiley & Sons, Chichester, pp 1–33

Goulden ML, Crill PM (1997) Automated measurements of CO$_2$ exchange at the moss surface of a black spruce forest. Tree Physiol 17:537–542

Goulden ML, Munger JW, Fan S-M, Daube BC, Wofsy SC (1996) Measurements of carbon sequestration by long-term eddy covariance: methods and a critical evaluation of accuracy. Global Change Biol 2:169–182

Goulden ML, Daube BC, Fan S-M, Sutton DJ, Bazzaz A, Munger JW, Wofsy SC (1997) Physiological responses of a black spruce forest to weather. J Geophys Res 102:28987–28996

Goulden ML, Wofsy SC, Harden JW, Trumbore SE, Crill PM, Gower ST, Fries T, Daube BC, Fan S-M, Sutton DJ, Bazzaz A, Munger JW (1998) Sensitivity of boreal forest carbon balance to soil thaw. Science 279:214–217

Goutorbe J-P, Lebel T, Tinga A, Bessemoulin P, Brouwer J, Dolman AJ, Engman E, Gash JHC, Hoepffner M, Kabat P, Kerr YH, Monteny B, Prince S, Saïd F, Sellers P, Wallace JS (1994) HAPEX-Sahel: a large-scale study of land-atmosphere interactions in the semi-arid tropics. Ann Geophys 12:53–64

Goutorbe J-P, Dolman AJ, Gash JHC, Kerr YH, Lebel T, Prince SD, Stricker JNM (eds) (1997a) HAPEX-Sahel. Elsevier, Amsterdam, 1079 pp

Goutorbe J-P, Noilhan J, Lacarrere P, Braud I (1997b) Modelling of the atmospheric column over the central sites during the HAPEX-Sahel. J Hydrol 188/189:1017–1039

Goward SN, Xue Y, Czaykowski K (1999) Evaluating land-surface moisture conditions from remotely sensed temperature/vegetation index measurement: an exploration employing the simplified simple biosphere model. Remote Sens Environ 79: 225–242

Gower ST, Vogel JG, Norman JM, Kucharik CJ, Steele SJ, Stow TK (1997) Carbon distribution and aboveground net primary production in aspen, jack pine, and black spruce stands in Saskatchewan and Manitoba, Canada. J Geophys Res 102:29029–29041

Gower ST, Kucharik CJ, Norman JM (1999) Direct and indirect estimation of leaf area index, fpar and net primary production of terrestrial ecosystems. Remote Sens Environ 70:29–51

Grace J, Lloyd J, McIntyre J, Miranda A, Meir P, Miranda H, Nobre C, Moncrieff J, Massheder J, Malhi Y, Wright I, Gash J (1995) Carbon dioxide uptake by an undisturbed tropical rain forest in South-West Amazonia, 1992–1993. Science 270:778–780

Grassl H (2000) Status and improvements of coupled general circulation models. Science 288:1991–1997

Grasso LD (2000) A numerical simulation of dryline sensitivity to soil moisture. Mon Weather Rev 128:2816–2834

Gray LC (1999) Is land being degraded? A multi-scale investigation of landscape change in south-western Burkina Faso. Land Degrad Dev 10:329–343

Greco S, Baldocchi DD (1996) Seasonal variations of CO_2 and water vapour exchange rates over a temperate deciduous forest. Global Change Biol 2:183–198

Grelle A, Lindroth A (1994) Flow distortion by a solent sonic anemometer: wind tunnel calibration and its assessment for flux measurements over forest and field. J Atmos Ocean Tech 11: 1529–1542

Grelle A, Lindroth A (1996) Eddy-correlation system for long-term monitoring of fluxes of heat, water vapour and CO_2. Global Change Biol 2:297–307

Grelle A, Lindroth A, Mölder M (1999) Seasonal variation of boreal forest surface conductance and evaporation. Agr Forest Meteorol 98/99:563–578

Groisman PY, Karl TR, Knight RW (1994) Observed impact of snow cover on the heat balance and the rise of continental spring temperatures. Science 263:198–200

Gryning S-E, Batchvarova E (1999) Regional heat flux over the NOPEX area estimated from the evolution of the mixed layer. Agr Forest Meteorol 98/99:159–168

Gryning S-E, Batchvarova E, De Bruin HAR (1999) Energy balance of sparse coniferous high-latitude forest under winter conditions. Bound-Lay Meteorol 99(3):465–488

Gu L, Fuentes JD, Shugart HH, Staebler RM, Black TA (1999a) Responses of net ecosystem exchanges of carbon dioxide to changes in cloudiness: results from two North American deciduous forests. J Geophys Res 104:31421–31434

Gu L, Fuentes J, Garstang M, Heitz R, Sigler J (1999b) The clouds induced bimodal characteristics and enhancement of surface irradiance as observed over a Brazilian grassland. In: Preprints of the 15th Conference of Hydrology. Long Beach, California, January 9–14 1999, pp 394–397

Gu L, Baldocchi D, Verma SB, Black TA, Vesala T, Falge EM, Dowty RP (2002) Advantages of diffuse radiation for terrestrial ecosystem productivity. J Geophys Res 107(D6), 10.1029/2001JD001242

Gu L, Baldocchi DD, Wofsy SC, Munger JW, Michalsky JJ, Urbanski SP, Boden TA (2003) Response of a deciduous forest to the Mount Pinatubo eruption: enhanced photosynthesis. Science 299:2035–2038

Gu J, Cooper H, Grose A, Liu G, Norman J, Waterloo M, Nobre A, Araujo A, Manzi A, Oliveira P, Von Randow C, Silva Dias P, Marengo J, Merritt J, Smith E (2004) Modeling carbon sequestration over large scale Amazon basin aided by satellite observations. Part 1: wet and dry season SRB flux and precipitation based on GOES retrieval. J Appl Meteorol (in press)

Gutman G (1984) Numerical experiments on land-surface alterations with a zonal model allowing for interaction between the geobotanic state and climate. J Atmos Sci 41:2679–2685

Gutman G (1985) On modeling dynamics of geobotanic state-climate interaction. J Atmos Sci 43:305–306

Gutman G, Ohring G, Joseph JH (1984) Interaction between the geobotanic state and climate: a suggested approach and a test with a zonal model. J Atmos Sci 41:2663–2678

Gutzler DS, Rosen RD (1992) Interannual variability of wintertime snow cover across the Northern Hemisphere. J Climate 5: 1441–1447

Hadfield MG, Cotton WR, Pielke RA (1991) Large-eddy simulations of thermally-forced circulations in the convective boundary layer. Part I: A small-scale circulation with zero wind. Bound-Lay Meteorol 57:79–114

Hadfield MG, Cotton WR, Pielke RA (1992) Large-eddy simulations of thermally forced circulations in the convective boundary layer. Part II: The effect of changes in wavelength and wind speed. Bound-Lay Meteorol 58:307–328

Hagemann S, Dümenil L (1998) A parameterisation of the lateral water flow for the global scale. Clim Dynam 14:17–31

Hahmann AN, Dickinson RE (1997) RCCM2-BATS model over tropical South America: application to tropical deforestation. J Climate 10(8):1944–1964

Hall FG (1999) BOREAS in 1999: experiment and science overview. J Geophys Res-Atmos 104(D2):27627–27639

Hall FG, Knapp D, Huemmerich KF (1997) Physically based classification and satellite mapping of biophysical characteristics in the southern boreal forest. J Geophys Res 102(D24):29569–29580

Halldin S, Gryning S-E, Gottschalk L, Jochum A, Lundin L-C, Van de Griend AA (1999) Energy, water and carbon exchange in a boreal forest landscape – NOPEX experiences. Agr Forest Meteorol 98/99:1–4, 5–29

Hammer RM (1970) Cloud development and distribution around Khartoum. Weather 25:411–414

Hane CE, Bluestein HB, Crawford TM, Baldwin ME, Rabin RM (1997) Severe thunderstorm development in relation to along-dryline variability: a case study. Mon Weather Rev 125:246–266

Hann J (1908) Handbuch der Klimatologie, Band I, 3. Auflage. Engelhorn, Stuttgart, 394 S

Harden JW, Sundquist E, Stallard R, Mark R (1992) Dynamics of soil carbon during deglaciation of the Laurentide ice sheet. Science 258:1921–1924

Harden JW, O'Neill KP, Trumbore SE, Veldhuis H, Stocks BJ (1997) Moss and soil contributions to the annual net carbon flux of a maturing boreal forest. J Geophys Res 102:28,805–28,816

Harding RJ, Pomeroy JW (1996) The energy balance of the winter boreal landscape. J Climate 9:2778–2787

Harrison SP, Jolly D, Laarif F, Abe-Ouchi A, Dong B, Herterich K, Hewitt C, Joussaume S, Kutzbach JE, Mitchell J, de Noblet N, Valdes P (1998) Intercomparison of simulated global vegetation distributions in response to 6 kyr BP orbital forcing. J Climate 11:2721–2742

Harvey LDD (1988) A semianalytic energy balance climate model with explicit sea ice and snow physics. J Climate 1:1065–1085

Harvey LDD (1989a) An energy balance climate model study of radiative forcing and temperature response at 18 ka. J Geophys Res 94(D10):12873–12884

Harvey LDD (1989b) Milankovitch forcing, vegetation feedback, and North Atlantic deep-water formation. J Climate 2:800–815

Hastenrath S (1984) International variability and annual cycle: mechanisms of circulation and climate in the tropical Atlantic sector. Mon Weather Rev 112:1097–1107

Hastenrath S (1990) Decadal-scale changes of the circulation in the tropical Atlantic sector associated with Sahel drought. Int J Climatol 10:459–472

Haxeltine A, Prentice IC (1996) BIOME3: an equilibrium biosphere model based on ecophysilogical constraints, resource availability and competition among plant funtional types. Global Biogeochem Cy 10:693–709

Hechtel LM, Moeng C-H, Stull RB (1990) The effects of nonhomogeneous surface fluxes on the convective boundary layer: a case study using large-eddy simulation. J Atmos Sci 47:1721–1741

Heimann M, Kaminski T (1999) Inverse modelling approaches to infer surface trace gas fluxes from observed atmospheric mixing ratios. In: Bouwman AF (ed) Approaches to scaling of trace gas fluxes in ecosystems. Elsevier, Amsterdam, pp 277–295

Helldén U (1991) Desertification – time for an assessment? Ambio 20:372–383

Henderson-Sellers A (1993) A factorial assessment of the sensitivity of the BATS land-surface parameterisation scheme. J Climate 6:227–247

Henderson-Sellers A, Dickinson RE, Durbridge TB, Kennedy PJ, McGuffie K, Pitman AJ (1993) Tropical deforestation: modeling local- to regional-scale climate. J Geophys Res 98(D4):7289–7315

Henderson-Sellers A, Pitman AJ, Love PK, Irannejad P, Chen T (1995) The project for intercomparison of land-surface parameterisation schemes (PILPS) Phases 2 and 3. B Am Meteorol Soc 76:489–503

Henderson-Sellers A, McGuffie K, Pitman A (1996) The project for the intercomparison of land-surface parametrization schemes (PILPS): 1992–1995. Clim Dynam 12:849–859

Henderson-Sellers A, Pitman AJ, Irannejad P, McGuffie K (2002) Land-surface simulations improve atmospheric modeling. EOS Trans Am Geophys Union 83:145, 152

Hewitt CD, Mitchell JFB (1998) A fully coupled GCM simulation of the climate of the mid-Holocene. Geophys Res Lett 25(3): 361–364

Hodnett MG, Tomasella J, Marques A de O, Oyama MD (1996) Deep soil water uptake by forest and pasture in central Amazonia: predictions from long-term daily rainfall data using a simple water balance model. In: Gash JHC, Nobre CA, Roberts JM, Victoria RL (eds) Amazonian deforestation and climate. John Wiley & Sons, Chichester, pp 79–99

Hoelzmann P, Jolly D, Harrison SP, Laarif F, Bonnefille R, Pachur H-J (1998) Mid-Holocene land-surface conditions in northern Africa and the Arabian peninsula: a dataset for the analysis of biogeophysical feedbacks in the climate system. Global Biogeochem Cy 12:35–51

Hogg EH (1999) Simulation of interannual response of trembling aspen stands to climate variation and insects defoliation in western Canada. Ecol Model 114:175–193

Hogg EH, Hurdle PA (1997) Sap flow in aspen: implications for stomatal responses to vapour pressure deficit. Tree Physiol 17: 501–509

Hogg EH, Black TA, den Hartog G, Neumann HH, Zimmermann R, Hurdle PA, Blanken PD, Nesic Z, Yang PC, Staebler RM, McDonald KC, Oren R (1997) A comparison of sap flow and eddy fluxes of water vapour from a boreal deciduous forest. J Geophys Res 102(D24):28929–28939

Holton JR (1992) An introduction to dynamic meteorology. Academic Press, 507 p

Holtslag AAM, Ek M (1996) Simulation of surface fluxes and boundary layer development over the pine forest in HAPEX-MOBILHY. J Appl Meteorol 35:202–213

Horel J, Hahmann A, Geisler J (1989) An investiation of the annual cycle of convective activity over the tropical Americas. J Climate 2:1388–1403

Hou AY (1998) Hadley circulation as a modulator of the extra-tropical climate. J Atmos Sci 55:2437–2457

Houghton RA, Hobbie JE, Melillo JM, Moore B, Peterson BJ, Shaver GR, Woodwell GM (1982) Changes in the carbon content of terrestrial biota and soils between 1860 and 1980: a net release of CO_2 to the atmosphere. Ecol Monogr 53:235–262

Houghton JT, Meiro Filho LG, Callander BA, Harris N, Kattenburg A, Maskell K (eds) (1995) Climate change 1995: the science of climate change, technical summary. Cambridge University Press, New York, pp 9–97

Houghton JT, Ding Y, Griggs DJ, Noguer M, van der Linden P, Dai X, Maskell K, Johnson CI (eds) (2001) Climate change 2001: the scientific basis. Contribution of Working Group I to the Third Assessment Report of the Intergovernmental Panel on Climate Change. Cambridge University Press, 881 p

Huang X, Lyons TJ, Smith RCG (1995) Meteorological impact of replacing native perennial vegetation with annual agricultural species. In: Kalma JD, Sivapalan M (eds) Scale issues in hydrological modelling. Advanstar Communications, Chichester, pp 401–410

Humboldt A von (1849) Ansichten der Natur, 3. Aufl. J.G. Cotta, Stuttgart, Tübingen, Reprint 1969. P. Reclam, Stuttgart

Hutjes RWA, Dolman AJ (1999) The effects of subgrid variability in vegetation cover and soil moisture on regional scale weather: a case study for the Sahel. In: Harding RJ (ed) Modelling the effect of land degradation on climate. Climate and Land Degradation first annual report, Centre for Ecology and Hydrology, Wallingford, UK

Instituto Nacional de Pesquisas Espaciais – INPE (2001) Monitoring of the Brazilian Amazon forest by satellite 1999–2000. Separata. Instituto Nacional de Pesquisas Espaciais, São José dos Campos, SP, Brazil. (available at *http://www.inpe.br/ Informacoes_Eventos/amazonia.htm*)

IPCC (Intergovernmental Panel on Climate Change) (2000) Emissions scenarios – IPCC special report. IPCC Secretariat, c/o WMO, Geneva, Switzerland

IPCC (Intergovernmental Panel on Climate Change) (2001) Climate change 2001: the scientific basis. Contribution of Working Group I to the IPCC Third Assessment Report. Cambridge University Press, 980 p

Jackson RB, Canadell J, Ehleringer JR, Mooney HA, Sala OE, Schulze ED (1996) A global analysis of root distributions for terrestrial biomes. Oecologia 108:389–411

Jackson RB, Mooney HA, Schulze ED (1997) A global budget for fine root biomass, surface area, and nutrient contents. P Natl Acad Sci USA 94:7362–7366

James IN (1994) Introduction to circulating atmospheres. Cambridge University Press, Cambridge

Jarvis PG (1976) The interpretation of the variations in leaf water potential and stomatal conductance found in canopies in the field. Philos T Roy Soc B 273:593–610

Jarvis PG, McNaughton KG (1986) Stomatal control of transpiration. Adv Ecol Res 15:1–49

Jarvis PG, Massheder JM, Hale SE, Moncrieff JB, Rayment M, Scott SL (1997) Seasonal variation of carbon dioxide, water vapour, and energy exchanges of a boreal black spruce forest. J Geophys Res 102:28953–28966

Johns TC, Carnell RE, Crossley JF, Gregory JM, Mitchell JFB, Senior CA, Tett SFB, Wood RA (1997) The second Hadley Centre coupled ocean-atmosphere GCM: model description, spinup and validation. Clim Dynam 13:103–134

Joiner DW, McCaughey JH, Lafleur PM, Bartlett PA (1999a) Water and carbon dioxide exchange at a boreal young jack pine site in the BOREAS northern study area. J Geophys Res 104:27641–27652

Joiner DW, Lafleur PM, McCaughey J, Bartlett PA (1999b) Interannual variability in carbon dioxide exchanges at a boreal wetland in the BOREAS northern study area. J Geophys Res 104:27663–27672

Jolly D, Harrison SP, Damnati B, Bonnefille R (1998) Simulated climate and biomes of Africa during the late Quarternary: comparison with pollen and lake status data. Quaternary Sci Rev 17(6–7):629–657

Jones PD (1994) Hemispheric surface air temperature variations: a reanalysis and an update to 1993. J Climate 7:1794–1802

Jordan C, Heuveldop J (1981) The water balance of an Amazonian rain forest. Acta Amazon 11:87–92

Joussaume S, Taylor KE, Braconnot P, Mitchell JFB, Kutzbach JE, Harrison SP, Prentice IC, Broccoli AJ, Abe-Ouchi A, Bartlein PJ, Bonfiels C, Dong B, Guiot J, Herterich K, Hewit CD, Jolly D, Kim JW, Kislov A, Kitoh A, Loutre MF, Masson V, McAvaney B, McFarlane N, deNoblet N, Peltier WR, Peterschmitt JY, Pollard D, Rind D, Royer JF, Schlesinger ME, Syktus J, Thompson S, Valdes P, Vettoretti G, Webb RS, Wyputta U (1999) Monsoon changes for 6000 years ago: results of 18 simulations from the Paleoclimate Modeling Intercomparison Project (PMIP). Geophys Res Lett 26(7):859–862

Kabat P Hutjes, RWA Feddes, RA (1997a) The scaling characteristics of soil parameters: from plot scale heterogeneity to subgrid parameterisation. J Hydrol 190:363–396

Kabat P, Prince SD, Prihodko L (eds) (1997b) Hydrologic Atmospheric Pilot Experiment in the Sahel (HAPEX-Sahel). Report 130, Department of Agrohydrology, DLO Winand Staring Center for Integrated Land, Soil and Water Research, Wageningen, The Netherlands, 380 p

Kahan DS (2002) The impact of land-cover change and land condition on climate in West Africa and the Arctic. Master Thesis, Department of Geography, University of California, Los Angeles

Kanae S, Oki T, Musiake K (2001) Impact of deforestation on regional precipitation over the Indochina Peninsula. J Hydromet 2:51–70

Karl TR, Groisman PY, Knight RW, Heim RR Jr (1993) Recent variations of snow cover and snowfall in North America and their relation to precipitation and temperature variations. J Climate 6:1327–1344

Kauppi PE, Mielikainen K, Kuuseia K (1992) Biomass and carbon budget of European forests, 1971 to 1990. Science 256:70–74

Keeling CD, Whorf TP, Wahlen M, vd Plicht J (1995) Interannual extremes in the rate of rise of atmospheric carbon dioxide since 1980. Nature 375:666–670

Keeling CD, Chin JFS, Whorf TP (1996) Increased activity of northern vegetation inferred from atmospheric CO_2 measurements, Nature 382:146–149

Keenan TD, Ferrier B, Simpson J (1994) Development and structure of a maritime continent thunderstorm. Meteorol Atmos Phys 53:185–222

Kidson J (1977) African rainfall and its relation to the upper air circulation. Q J Roy Meteor Soc 103:441–456

Kimball JS Running SW, Saatchi SS (1999) Sensitivity of boreal forest regional water flux and net primary production simulations to subgrid-scale land-cover complexity. J Geophys Res 104:27789–27802

Kinter III JL, Shukla J, Marx L, Schneider EK (1988) A simulation of the winter and summer circulation with the NMC global spectral model. J Atmos Sci 45:2486–2522

Kitoh A, Yamazaki K, Tokioka T (1988) Influence of soil moisture and surface albedo changes over the African tropical rain forest on summer climate investigated with the MRI-GCM-I. J Meteorol Soc Jpn 66:65–85

Klitch MA, Weaver JF, Kelly FP, Vonder Haar TH (1985) Convective cloud climatologies constructed from satellite imagery. Mon Weather Rev 113:326–337

Koch SE, Aksakal A, McQueen JT (1997) The influence of mesoscale humidity and evapotranspiration fields on a model forecast of a cold-frontal squall line. Mon Weather Rev 125:384–409

Köppen W (1936) Das geographische System der Klimate. In: Köppen W, Geiger R (eds) Handbuch der Klimatologie, Band 5, Teil C. Gebrüder Bornträger, Berlin

Koster RD, Milly PCD (1997) The interplay between transpiration and runoff formulations in land-surface scheme used with atmospheric models. J Climate 10:1578–1591

Koster RD, Suarez MJ (1992) Modeling the land-surface boundary in climate models as a composite of independent vegetation stands. J Geophys Res 97:2697–2715

Koster RD, Suarez MJ (1999) A simple framework for examining the interannual variability of land-surface moisture fluxes. J Climate 12:1911–1917

Koster RD, Suarez MJ, Heiser M (2000) Variance and predictability of precipitation at seasonal-to-interannual time scales. J Hydrometeorol 1:26–46

Krishnamurti TN (1961) The subtropical jet stream of winter. J Meteorol 18:358–369

Krueger, AF, Winston JS (1973) A comparison of the flow over the tropics during two contrasting flow regimes. J Atmos Sci 31(2):358–370

Kruijt B, Malhi Y, Lloyd J, Nobre AD, Culf A, Miranda AC, Pereira MGP, Grace J (2000) Turbulence above and within two Amazon rain forest canopies. Bound-Lay Meteorol 94:297–311

Kubatzki C, Claussen M (1998) Simulation of the global biogeophysical interactions during the last glacial maximum. Clim Dynam 14:461–471

Kurz WA, Apps MJ, Bekema SJ, Lekstrum T (1995) 20th century carbon budget of Canadian forests. Tellus 47:170–177

Kutzbach JE, Guetter PJ (1986) The influence of changing orbital parameters and surface boundary conditions on climate simulations for the past 18 000 years. J Atmos Sci 43:1726–1759

Kutzbach JE, Liu Z (1997) Response of the African monsoon to orbital forcing and ocean feedbacks in the middle Holocene. Science 278:440–443

Kutzbach JE, Bonan G, Foley J, Harrison SP (1996a) Vegetation and soil feedbacks on the response of the African monsoon to orbital forcing in the early to middle Holocene. Nature 384:623–626

Kutzbach JE, Bartlein PJ, Foley JA, Harrison SP, Hostetler SW, Liu Z, Prentice IC, Webb T (TEMPO) (1996b) Potential role of vegetation feedback in the climate sensitiviy of high-latitude regions: a case study at 6 000 years BP. Global Biogeochem Cy 10(4):727–736

Kutzbach JE, Harrison SP, Coe MT (2001) Land-ocean-atmosphere interactions and monsoon climate change: a paleoperspective. In: Schulze ED, Harrison SP, Heimann M, Holland EA, Lloyd J, Prentice IC, Schimel D (eds) Global biogeochemical cycles in the climate system. Academic Press, San Diego, pp 73–86

Lafleur PM, McCaughey JH, Joiner DW, Bartlett PA, Jelinski DE (1997) Seasonal trends in energy, water and carbon dioxide fluxes at a northern boreal wetland. J Geophys Res 102:29009–29020

Lamb PJ (1978a) Large-scale tropical Atlantic surface circulation patterns associated with sub-Saharan weather anomalies. Tellus 30:240–251

Lamb PJ (1978b) Case studies of tropical Atlantic surface circulation patterns during recent Subsaharan weather anomalies: 1967 and 1968. Mon Weather Rev 106:482–491

Lamb PJ, Peppler RA (1991) West Africa. In: Glantz M, Katz RW, Nicholls N (eds) Teleconnections linking worldwide climate anomalies. Cambridge University Press, pp 121–189

Lanicci JM, Carlson TN, Warner TT (1987) Sensitivity of the Great Plains severe-storm environment to soil-moisture distribution. Mon Weather Rev 115:2660–2673

Lanly JP (1982) Tropical forest resources. FAO Forestry Paper, 30, Rome, 106 p

Larsen JA (1980) The boreal ecosystem. Academic Press, New York

Lau K-M, Kim J, Sud Y (1996) Intercomparison of hydrologic processes in AMIP GCMs. B Am Meteorol Soc 10:2209–2227

Laval K, Picon L (1986) Effect of a change of the surface albedo of the Sahel on climate. J Atmos Sci 43:2418–2429

Lavigne MB, Ryan MG, Anderson DE, Baldocchi DD, Crill PM, Fitzjarrald DR, Goulden ML, Gower ST, Massheder JM, McCaughey JH, Rayment M, Striegl RG (1997) Comparing nocturnal eddy covariance measurements to estimates of ecosystem respiration made by scaling chamber measurements at six coniferous boreal sites. J Geophys Res 102:28977–28986

Lawton RO, Nair US, Pielke RASr, Welch RM (2001) Climatic impact of tropical lowland deforestation on nearby cloud forests. Science 294:584–587

LBA Science planning group (1996) The large scale biosphere-atmosphere experiment in Amazonia (LBA). Winand Staring Centre, PO Box 125, NL-6700, AC Wageningen, The Netherlands. 44 P

Lean J, Rowntree P (1993) A GCM simulation of the impact of Amazonian deforestation on climate using an improved canopy representation. Q J Roy Meteor Soc 119:509–530

Lean J, Rowntree P (1997) Understanding the sensitivity of a GCM simulation of Amazonian deforestation to the specification of vegetation and soil characteristics. J Climate 10:1216–1235

Lean J, Warrilow DA (1989) Simulation of the regional climate impact of Amazon deforestation. Nature 342:411–413

Le Barbé L, Lebel T (1997) Rainfall climatology of the HAPEX-Sahel region during the years 1950–1990. J Hydrol 188/189:43–73

Lebel T, Sauvageot H, Hoepffner M, Desbois M, Guillot B, Hubert P (1992) Rainfall estimation in the Sahel: the EPSAT-Niger experiment. Hydrolog Sci J 37:201–215

Lebel T, Taupin JD, D'Amato N (1997) Rainfall monitoring during HAPEX-Sahel: 1. general conditions and climatology. J Hydrol 189:74–96

Legates D, Wilmott C (1990) Mean seasonal and spatial variability in gauge corrected, global precipitation. Int J Climatol 10:111–127

Le Houérou HN (1989) The grazing land ecosystems of the African Sahel. Springer-Verlag, Berlin, 282 p

Le Houérou HN, Hoste CH (1977) Rangeland production and annual rainfall relations in the Mediterranean Basin and in the African Sahelo-Sudanian zone. J Range Manage 30(3): 181–189

Leopoldo P, Franken W, Matsui E, Roibeiro M (1982) Estimativa da evapotranspiracao da floresta Amazonica de terra firne. Acta Amazon 12:23–28

Lerdau M, Litvak M, Palmer P, Monson R (1997) Controls over monoterpene emissions from boreal forest conifers. Tree Physiol 17:563–569

Lettau H, Lettau K, Molion LCB (1979) Amazonia's hydrologic cycle and the role of atmospheric recycling in assessing deforestation effect. Mon Weather Rev 107:227–238

Levis S, Foley JA, Pollard D (1999a) Potential high-latitude vegetation feedbacks on CO_2-induced climate change. Geophys Res Lett 26(6):747–750

Levis S, Foley JA, Brovkin V, Pollard D (1999b) On the stability of the high-latitude climate-vegetation system in a coupled atmosphere-biosphere model. Global Ecol Biogeogr 8:489–500

Levis S, Foley JA, Pollard D (2000) Large scale vegetation feedbacks on a doubled CO_2 climate. J Climate 13:1313–1325

Li B, Avissar R (1994) The impact of spatial variability of land-surface characteristics on land-surface heat fluxes. J Climate 7:527–537

Liang X, Wood EF, Lettenmaier DP, Lohmann D, Boone A, Chang S, Chen F, Dai Y, Desborough CE, Dickinson RE, Duan Q, Ek M, Gusev YM, Habets F, Irranejad P, Koster R, Mitchell KE, Nasonova ON, Noilhan J, Schaake J, Schlosser CA, Shao Y, Schmakin AB, Verseghy D, Warrach K, Wetzel P, Xue Y, Yang Z-L, Zeng Q-C (1998) The project for intercomparison of land-surface parameterisation schemes (PILPS) Phase 2(c) Red-Arkansas River basin experiment: 2. Spatial and temporal analysis of energy fluxes. Global Planet Change 19(1–4):137–159

Liebmann B, Kiladis G, Marengo J, Ambuzzi T, Glick J (1999) Submonthly convective variability over South America and the South Atlantic Convergence Zone. J Climate 10:1877–1891

Lindqvist S, Tengberg A (1993) New evidence of desertification from case studies in northern Burkina Faso. Geogr Ann A 75A:127–135

Lindroth A (1985) Canopy conductance of coniferous forests related to climate. Water Resour Res 21(3):297–304

Lindroth A, Grelle A, Morén A-S (1998) Long-term measurements of boreal forest carbon balance reveal large temperature sensitivity. Global Change Biol 4:443–450

Liston GE, Sud YC, Wood EF (1994) Evaluating GCM land-surface hydrology parameterisations by computing river discharges using a runoff routing model: applicatioon to the Mississippi basin. J Appl Meteorol 33:394–405

Liu Y, Avissar R (1999a) A study of persistence in the land-atmosphere system using a general circulation model and observations. J Climate 12:2139–2153

Liu Y, Avissar R (1999b) A study of persistence in the land-atmosphere system with a fourth-order analytical model. J Climate 12:2154–2168

Liu Y, Weaver CP, Avissar R (1999a) Toward a parameterisation of mesoscale fluxes and moist convection induced by landscape heterogeneity. J Geophys Res 104:19515–19553

Liu J, Chen JM, Cihlar J, Chen W (1999b) Net primary production distribution in the BOREAS region from a process model using satellite and surface data. J Geophys Res 104:27735–27754

Lloyd CR, Marques AO (1988) Spatial variability of throughfall measurements in Amazonian rainforest. Agr Forest Meteorol 42:63–73

Lloyd CR, Gash JHC, Shuttleworth WJ, Marques AO (1988) The measurements and modelling of rainfall interception by Amazonian rainforest. Agr Forest Meteorol 43:277–294

Lofgren BM (1995a) Sensitivity of land-ocean circulations, precipitation, and soil moisture to perturbed land-surface albedo. J Climate 8(10):2521–2542

Lofgren BM (1995b) Surface albedo-climate feedback simulated using two-way coupling. J Climate 8(10):2543–2562

Lorenz EN (1963) Deterministic nonperiodic flow. J Atmos Sci 20:130–141

Los SO, Collatz GJ, Sellers PJ, Malmstrom CM, Pollack NH, DeFries RS, Bounoua L, Parris MT, Tucker CJ, Dazlich DA (2000) A global 9-year biophysical land-surface dataset from NOAA AVHRR data. J Hydrometeorol 1:183–199

Lu L (1999) Implementation of a two-way interactive atmospheric and ecological model and its application to the central United States. PhD thesis, Colorado State University, Fort Collins, Colo.

Lu L, Pielke RA Sr, Liston GE, Parton WJ, Ojima D, Hartman M (2001) Implementation of a two-way interactive atmospheric and ecological model and its application to the central United States. J Climate 14:900–919

Lynn BH, Rind D, Avissar R (1995a) The importance of mesoscale circulations generated by subgrid-scale landscape-heterogeneities in general circulation models. J Climate 8:191–205

Lynn BH, Abramopolous F, Avissar R (1995b) Using similarity theory to parameterize mesoscale heat fluxes generated by subgrid-scale landscape discontinuities in GCMs. J Climate 8:932–951

Lyons WA (1999) Lightning. In: Pielke RA Sr, Pielke RA Jr (eds) Storms. Hazard and Disaster Series. Routledge Press, pp 60–79

Lyons TJ (2002) Clouds prefer native vegetation. Meteorol Atmos Phys 80(1–4):131–140

Lyons TJ, Schwerdtfeger P, Hacker JM, Foster IJ, Smith RCG, Xinmei H (1993) Land-atmosphere interaction in a semiarid region: the bunny fence experiment. B Am Meteorol Soc 74:1327–1334

Lyons TJ, Smith RCG, Xinmei H (1996) The impact of clearing for agriculture on the surface energy budget. Int J Climatol 16:551–558

Mabbutt JA, Floret C (1980) Case studies on desertification. In: Natural resources research. UNESCO, Paris, 279 p

Machado LAT (2000) The Amazon energy budget using field experiment data. In: Preprints of the 6th International Conference on Southern Hemisphere Meteorology and Oceanography, Santiago, Chile, April 3–7, 2000, pp 328–329

Mahfouf J-F, Richard E, Mascart P (1987) The influence of soil and vegetation on the development of mesoscale circulations. J Clim Appl Meteorol 26:1483–1495

Mahrt L, Sun JS, Vickers D, MacPherson JI, Pederson JR, Desjardins RL (1994) Observations of fluxes and inland breezes over a heterogeneous surface. J Atmos Sci 51:2484–2499

Malhi Y, Nobre AD, Grace J, Kruijt B, Pereira MGP, Culf A, Scott S (1998) Carbon dioxide transfer over a central Amazonian rain forest. J Geophys Res 103:31593–31612

Manabe S (1969) Climate and ocean circulation. Part I: The atmospheric circulation and the hydrology of the Earth's surface. Mon Weather Rev 97:739–774

Manzi AO, Planton S (1996) Calibration of a GCM using ABRACOS and ARME data and simulation of Amazonian deforestation. In: Gash JHC, Nobre CA, Roberts JM, Victoria RL (eds) Amazonian Deforestation and Climate. John Wiley & Sons, Chichester, pp 505–529

Marengo J (1995) Interannual variability of deep convection in the tropical South American sector as deduced from ISCCP C2 data. Int J Climatol 15:995–1010

Marengo J (2004) Characteristics and space-time variability of the atmospheric water balance in the Amazon basin. Clim Dynam (accepted)

Marengo J, Nobre C (2001) The hydroclimatological framework in Amazonia. In: Richey J, McClaine M, Victoria R (eds) Biogeochemistry of Amazonia. Oxford University Press, pp 17–42

Marengo J, Miller J, Russell G, Rosenzweig C, Abramopoulos F (1994) Calculations of river-runoff in the GISS GCM: impact of a new land-surface and runoff routing model in the hydrology of the Amazon River. Clim Dynam 10:349–361

Marengo J, Nobre CA, Sampaio G (1998) On the associations between hydrometeorological conditions in Amazonia and the extremes of the Southern Oscillation. In: Extended abstracts of Memorias Tecnicas, Seminario internacional consecuencias climaticas e hidrologicas del evento El Niño a escala regional y local. 26–29 noviembre 1997, Quito, Ecuador. pp 257–266

Marengo J, Tomasella J, Uvo CRB (1998) Long-term streamflow and rainfall fluctuations in tropical South America: Amazonia, Eastern Brazil and Northwest Peru. J Geophys Res 103:1775–1783

Margolis HA, Ryan MG (1997) A physiological basis for biosphere-atmosphere interactions in the boreal forest: an overview. Tree Physiol 17:491–499

Marques J, dos Santos M, Villa Nova N, Salati E (1977) Precipitable water and water wapor flux between Belem and Manaus. Acta Amazon 7:355–362

Marques J, Salati E, dos Santos M (1980a) Cálculo de evapotranspiração na bacia Amazônica a traves do método aerológico. Acta Amazon 10:357–361

Marques J, Salati E, dos Santos M (1980b) A divergencia do campo do fluxo de vapour d'água e as chuvas na região Amazônica. Acta Amazon 10:133–140

Mason PJ (1988) The formation of areally-averaged roughness lengths. Q J Roy Meteor Soc 114:399–420

Matsuyama H (1992) The water budget in the Amazon River Basin during the FGGE Period. J Meteorol Soc Jpn 70:1071–1083

McCaughey JH, Lafleur PM, Joiner DW, Bartlett PA, Costello AM, Jelinsky DE, Ryan MG (1997) Magnitudes and seasonal patterns of energy, water and carbon exchanges at a boreal young jack pine site in the BOREAS northern study area. J Geophys Res 102(D24):28997–29007

McGuffie K, Henderson-Sellers A, Zhang H, Durbidge TB, Pitman AJ (1995) Global climate sensitivity to tropical deforestation. Global Planet Change 10:97–128

McWilliam A-LC, Roberts JM, Cabral OMR, Leitao MVBR, de Costa ACL, Maitelli GT, Zamparoni CAGP (1993) Leaf area index and above-ground biomass of terra firme rain forest and adjacent clearings in Amazonia. Funct Ecol 7:310–317

McWilliam A-LC, Cabral OMR, Gomes BM, Esteves JL, Roberts J (1996) Forest and pasture leaf-gas exchange in south-west Amazonia. In: Gash JHC, Nobre CA, Roberts JM, Victoria RL (eds) Amazonian deforestation and climate. John Wiley & Sons, Chichester, pp 265–285

Meehl GA (1994) Influence of the land surface in the Asian summer monsoon: external conditions versus internal feedbacks. J Climate 7:1033–1049

Meeson BW, Corprew FE, McManus JMP, Myers DM, Closs JW, Sun KJ, Sunday DJ, Sellers PJ (1995) ISLSCP Initiative I – Global datasets for land-atmosphere models, 1987–1988. Volumes 1–5, published on CD-ROM by NASA(USA_NASA_GDDAC_ISLSCP_001-USA_NASA_GDAAC_ISLSCP_005)

Melas D, Persson T, DeBruin H, Gryning S-E, Batchvarova E, Zerefos C (2001) Numerical model simulations of boundary-layer dynamics during winter conditions. Theor Appl Climatol 70(1–4):105–116

Miller JR, Russell GL, Caliri G (1994) Continental-scale river flow in climate models. J Climate 7:914–928

Milliman J, Meade R (1983) World-wide delivery of river sediment to the oceans. J Geol 91:1–21

Mills PF (1994) The agricultural potential of northwestern Canada and Alaska and the impact of climatic change. Arctic 47:115–123

Milly PCD, Dunne KA (1994) Sensitivity of the global water cycle to the water-holding capacity of land. J Climate 7:506–526

Mintz Y (1984) The sensitivity of numerically simulated climates to land-surface boundary conditions. In: Houghton JT (ed) The global climate. Cambridge University Press, pp 79–105

Mitchell JFB (1983) The seasonal response of a general circulation model to changes in CO_2 and sea temperature. Q J Roy Meteor Soc 109:113–152

Moeng C-H (1984) A large-eddy simulation model for the study of planetary boundary-layer turbulence. J Atmos Sci 41:2052–2062

Moeng C-H (1986) Large-eddy simulation of a stratus-topped boundary layer, Part I: structure and budgets. J Atmos Sci 43:2886–2900

Moeng C-H (1987) Large-eddy simulation of a stratus-topped boundary layer, Part II: implications for mixed-layer modeling. J Atmos Sci 44:1605–1614

Mölders N (2000) Similarity of microclimate as simulated in response to landscapes of the 1930s and the 1980s. J Hydrometeorol 1:330–352

Monteith JL (1965) Evaporation and environment. In: Fogg GE (ed) The state and movement of water in living organisms. Soc. Exptl. Biol. Symp. No. 19, Cambridge University Press, Cambridge, pp 205–234

Moore T, Dalva M (1993) Influence of temperature and water table position on carbon dioxide and methane emissions from columns of peatland soils. J Soil Sci 44:651–664

Moore CJ, Fisch G (1986) Estimating heat storage in Amazonian tropical forest. Agr Forest Meteorol 38:147–169

Moore TR, Roulet NT (1993) Methane flux: water table relations in northern wetlands. Geophys Res Lett 20:587–590

Moosavi SC, Crill PM (1997) Controls on CH_4 and CO_2 emissions along two moisture gradients in the Canadian boreal zone. J Geophys Res 102:29261–29278

Morén A-S (1999) Modelling branch conductance of Norway spruce and Scots pine in relation to climate. Agr Forest Meteorol 98/99:1–4, 579–593

MSC (2000) Canadian snow data CD-ROM. CRYSYS Project, Climate Processes and Earth Observation Division, Meteorological Service of Canada, Downsview, Ontario, January 2000

Myneni RB, Keeling CD, Tucker CJ, Asrar G, Nemani RR (1997) Increased plant growth in the northern high latitudes from 1981–1991. Nature 386:698–702

Namias J (1959) Persistence of mid-tropospheric circulations between adjacent months and seasons. In: Bolin B (ed) The atmosphere and the sea in motion: Rossby memorial volume. Rockefeller Institute Press and Oxford University Press, pp 240–248

Namias J (1978) Multiple causes of the North American abnormal winter 1976–1977. Mon Weather Rev 106:279–295

Neilson RP, Marks D (1994) A global perspective of regional vegetation and hydrologic sensitivities from climate change. J Veg Sci 5:715–730

Nepstad DC, Carvalho CR, Davidson EA, Jipp PH, Lefebvre PA, Negrelros GH, da Silva ED, Stone TA, Trumbore SE, Vieira S (1994) The role of deep roots in the hydrological and carbon cycles of Amazon forests and pastures. Nature 372:666–669

Nicholson SE (1978) Climate variations in the Sahel and other African regions during the past five centuries. J Arid Environ 1:3–24

Nicholson SE, Kim J (1997) The relationship of the El Niño-Southern Oscillation to African rainfall. J Climatol 17:117–135

Nicholson SE, Palao IM (1993) A re-evaluation of rainfall variability in the Sahel. Part I. Characteristics of rainfall fluctuations. Int J Climatol 13:371–389

Nicholson SE, Tucker CJ, Ba MB (1998) Desertification, drought, and surface vegetation: an example from the west African Sahel. B Am Meteorol Soc 79:815–829

Nijssen B, O'Donnell GM, Hamlet AF, Lettenmaier DP (2001) Hydrologic sensitivity of global rivers to climate change. Climatic Change 50(1–2):143–175

Nizhizawa T, Koike Y (1992) Amazon-ecology and development. Iwanami, Tokyo, 221 p

Nobre C, Sellers P, Shukla J (1991) Amazonian deforestation and the regional climate change. J Climate 4:957–988

Nobre CA, Fisch G, da Rocha HR, Lyra RF, da Rocha EP, Ubarana VN (1996a) Observations of the atmospheric boundary layer in Rondônia. In: Gash JHC, Nobre CA, Roberts JM, Victoria RL (eds) Amazonian deforestation and climate. John Wiley & Sons, Chichester, pp 413–424

Nobre CA, Gash JHC, Roberts JM, Victoria RL (1996b) The conclusions from ABRACOS. In: Gash JHC, Nobre CA, Roberts JM, Victoria RL (eds) Amazonian deforestation and climate. John Wiley & Sons, Chichester, pp 577–595

Noilhan J, Lacarrère P (1995) GCM grid-scale evaporation from mesoscale modelling. J Climate 8:206–223

Noilhan J, Mahfouf J-F (1996) The ISBA land-surface parameterisation scheme. Global Planet Change 13:145–159

Noilhan J, Planton S (1989) A simple parameterisation of land-surface processes for meteorological models. Mon Weather Rev 117:536–549

NSIDC (1999) Historical Soviet daily snow depth data – Version 2.0. National Snow and Ice Data Center, Cooperative Institute for Research in Environmental Sciences, University of Colorado, Boulder, CO, CD-ROM data product

Oba G, Stenseth NC, Lusigu WJ (2000) New perspectives on sustainable grazing management in arid zones of sub-Saharan Africa. Bioscience 50(1):35–51

Oechel WC, Vourlitis GL (1994) The effects of climate change on land-atmosphere feedbacks in arctic tundra regions. Trends Ecol Evol 9:324–329

Ohtaki B (1984) Application of an infrared carbon dioxide and humidity instrument to studies of turbulent transport. Bound-Lay Meteorol 29:85–107

Ojima D (2000) Integrated analysis of climate change impacts on land use in Temperate East Asia (LUTEA): integration of ecosystem and economic factors determing land use under climate change. In: Phelan A, Virji H, Sobti M (eds) Land-use/land-cover change in temperate East Asia: current status and future trend. International START secretariat in Washington, D.C. USA, pp 1–18

Oki T (1999) The global water cycle. In: Browning KA, Gurney RJ (eds) Global energy and water cycles. Cambridge University Press, pp 10–29

Oki T, Sud YC (1998) Design of Total Runoff Integrating Pathways (TRIP) – a global river channel network. Earth Interactions 2(1): 1–37

Oki T, Xue Y (1998) Investigation of river discharge variability in a Sahel desertification experiment. Preprint of Ninth Symposium on Global Change Studies, pp 259–260

Oki T, Musiake K, Matsuyama H, Masuda K (1995) Global atmospheric water balance and runoff from large river basins. Hydrol Process 9:655–678

Oliver VJ, Oliver MB (1963) Cloud patterns, 1: some aspects of the organization of cloud patterns. Mon Weather Rev 91:621–629

Oltman R (1967) Reconnaissance investigations of the discharge and water quality of the Amazon, in Atas do Simp. sobre Biota Amazônica, Rio de Janeiro, CNPq, Vol. 3, pp 163–185

Ookouchi Y, Segal M, Kessler RC, Pielke RA (1984) Evaluation of soil moisture effects on the generation and modification of mesoscale circulations. Mon Weather Rev 112:2281–2292

Oort AH, Yienger JJ (1996) Observed interannual variability of the Hadley circulation and its connection to ENSO. J Climate 9: 2751–2767

Otterman J, Chou M-D, Arking A (1984) Effects of nontropical forest cover on climate. J Clim Appl Meteorol 23:762–767

Otterman J, Manes A, Rabin S, Alpert P, Starr DO'C (1990) An increase of early rains in southern Israel following land-use change. Bound-Lay Meteorol 53:333–351

Otto-Bliesner BL, Upchurch GR (1997) Vegetation-induced warming of high-latitude regions during the Late Cretaceous period. Nature 385:804–807

Pachur H-J, Altmann N (1997) The Quaternary (Holocene, ca. 8 000 a BP). In: Schandelmeier H, Reynolds P-O (eds) Palaeogeographic-palaeotectonic atlas of North-Eastern Africa, Arabia, and adjacent areas Late Neoproterozoic to Holocene. pp 111–125

Pachur H-J, Hoelzmann P (1991) Paleoclimatic implications of Late Quaternary lacustrine sediments in Western Nubia, Sudan. Quaternary Res 36:257–276

Pachur H-J, Wünnemann B (1996) Reconstruction of the palaeoclimate along 30° E in the eastern Sahara during the Pleistocene/Holocene transition. In: Heine K (ed) Palaeoecology of Africa and the surrounding islands. pp 1–32

Palmer TN (1986) Influence of the Atlantic, Pacific and Indian Oceans on Sahel rainfall. Nature 322:251–253

Palmer TN, Brankovic C, Viterbo P, Miller MJ (1992) Modeling interannual variations of summer monsoons. J Climate 5:399–417

Peixoto JP, Oort AH (1992) Physics of climate. American Institute of Physics, New York

Petit-Maire N, Guo Z (1996) Mise en evidence de variations climatiques holocenes rapides, en phase dans les deserts actuels de Chine et du Nord de l'Afrique. Cr Acad Sci II A 322:847–851

Petoukhov V, Ganopolski A, Brovkin V, Claussen M, Eliseev A, Kubatzki C, Rahmstorf S (2000) CLIMBER-2: a climate system model of intermediate complexity. Part I: Model description and performance for present climate. Clim Dynam 16(1):1–17

Phillips OL, Malhi Y, Higuchi N, Laurance WF, Núñez PV, Vásquez RM, Laurance SG, Ferreira LV, Stern M, Brown S, Grace J (1998) Changes in the carbon balance of tropical forests: Evidence from long-term plots. Science 282:439–442

Pielke RA (2001a) Earth system modeling – an integrated assessment tool for environmental studies. In: Matsuno T, Kida H (eds) Present and future of modeling global environmental change: toward integrated modeling. Terrapub, pp 311–337

Pielke RA (2001b) Influence of the spatial distribution of vegetation and soils on the prediction of cumulus convective rainfall. Rev Geophys 39:151–177

Pielke RA Sr (2002) Mesoscale meteorological modeling. 2nd Edition, Academic Press, San Diego, CA, 676 p

Pielke RA, Avissar R (1990) Influence of landscape structure on local and regional climate. Landscape Ecol 4:133–155

Pielke RA, Vidale PL (1995) The boreal forest and the polar front. J Geophys Res 100(D):25755–25758

Pielke RA, Dalu GA, Snook JS, Lee TJ, Kittel TGF (1991) Nonlinear influence of mesoscale land use on weather and climate. J Climate 4:1053–1069

Pielke RA, Lee TJ, Copeland JH, Eastman JL, Ziegler CL, Finley CA (1997) Use of USGS-provided data to improve weather and climate simulations. Ecol Appl 7:3–21

Pielke RA, Liston GE, Eastman JL, Lu LX, Coughenour M (1999) Seasonal weather prediction as an initial value problem. J Geophys Res-Atmos 104:19463–19479

Pielke RA Sr, Liston GE, Robock AG (2000) Insolation-weighted assessment of Northern Hemisphere snow-cover and sea-ice variability. Geophysical Research Letters 27(19):3061–3064

Pitman AJ, Zhao M (2000) The relative impact of observed change in land cover and carbon dioxide as simulated by a climate model. Geophys Res Lett 27:1267–1270

Pitman AJ, Henderson-Sellers A, Yang Z-L, Abramopoulos F, Boone A, Desborough CE, Dickinson RE, Gedney N, Koster R, Kowalczyk E, Lettenmaier D, Liang X, Mahfouf J-F, Noilhan J, Polcher J, Qu W, Robock A, Rosenzweig C, Schlosser C, Shmakin AB, Smith J, Suarez M, Verseghy D, Wetzel P, Wood E, Xue Y (1999) Key results and implications from phase 1(c) of the Project for Intercomparison of Land-surface Parameterisation Schemes. Clim Dynam 15:673–684

Plöchl M, Cramer WP (1995) Coupling global models of vegetation structure and ecosystem processes – an example from Arctic and boreal ecosystems. Tellus 47B:240–250

Polcher J (1995) Sensitivity of tropical convection to land-surface processes. J Atmos Sci 52 (17) :3143–3161

Polcher J, Laval K (1994a) The impact of African and Amazonian deforestation on tropical climate. J Hydrol 155:389–405

Polcher J, Laval K (1994b) A statistical study of the regional impact of deforestation on climate in the LMD GCM. Clim Dynam 10: 205–219

Polcher J, Crossley J, Bunton C, Douville H, Gedney N, Laval K, Planton S, Rowntree PR, Valdes P (1998) Importance of land-surface processes for the uncertainties of climate change; a European project. GEWEX News 8(2):11–13

Pomeroy JW, Marsh P, Gray DM (1997) Application of a distributed blowing snow model to the Arctic. Hydrol Process 11:1451–1464

Prentice IC, Cramer W, Harrison SP, Leemans R, Monserud RA, Solomon AM (1992) A global biome model based on plant physiology and dominance, soil properties and climate. J Biogeogr 19: 117–134

Prentice IC, Jolly D, and BIOME 6000 members (2000) Mid-Holocene and glacial-maximum vegetation geography of the northern continents and Africa. J Biogeogr 27(3):507–519

Price AG, Dunham K, Carleton T, Band L (1997) Variability of water fluxes through the black spruce (*Picea mariana*) canopy and feather moss (*Pleurozium schreberi*) carpet in the boreal forest of Northern Manitoba. J Hydrol 196:310–323

Prince SD, Kerr YH, Goutorbe J-P, Lebel T, Tinga A, Bessemoulin P, Brouwer J, Dolman AJ, Engman ET, Gash JHC, Hoepffner M, Kabat P, Monteny B, Saïd F, Sellers P, Wallace JS (1995) Geographical, biological and remote sensing aspects of the Hydrologic Atmospheric Pilot Experiment in the Sahel (HAPEX-Sahel). Remote Sens Environ 51:215–234

Prince SD, Brown de Colstoun EC, Strand H (1997) Land-cover characterisation of the HAPEX Sahel west central supersite with remote sensing data. In: Kabat P, Prince SD, Prihodko L (eds) Hydrologic atmospheric pilot experiment in the Sahel (HAPEX-Sahel) methods, measurements and selected results from the west central supersite. Report 130, DLO Winand Staring Centre, Wageningen, Netherlands

Prince SD, Brown de Colstoun, EC, Kravitz LL (1998) Evidence from rain-use efficiencies does not indicate extensive Sahelian desertification. Global Change Biol 4: 359–374

Rabin RM, Martin DW (1996) Satellite observations of shallow cumulus coverage over the central United States: An exploration of land-use impact on cloud cover. J Geophys Res 101(D3):7149–7155

Rabin RM, Stadler S, Wetzel PJ, Stensrud DJ (1990) Observed effects of landscape variability on convective clouds. B Am Meteorol Soc 71:272–280

Randerson JT, Thompson MV, Conway TJ, Fung I, Field C (1997) The contribution of terrestrial sources and sinks to trends in the seasonal cycle of atmospheric carbon dioxide. Global Biogeochem Cy 11(4):535–560

Rao VB, Hada K (1990) Characteristics of rainfall over Brazil: annual variations and connections with the Southern Oscillation. Theor Appl Climatol 42:81–91

Rao VB, Cavalcanti I, Hada K (1996) Annual variation of rainfall over Brazil and water vapour characteristics over South America. J Geophys Res 101:26539–26551

Rapalee G, Trumbore S, Davidson E, Harden J, Veldhuis H (1998) Soil carbon stocks and their rates of accumulation and loss in a boreal forest landscape. Global Biogeochem Cy 12:687–702

Rapp A (1987) Desertification. In: Gregory KJ, Walling DE (eds) Human activity and environmental processes. John Wiley & Sons, Chichester, pp 425–443

Rasmusson EH (1987) Global climate change and variability: effects of drought and desertification on Africa. In: Glantz MH (ed) Drought and hunger in Africa. Cambridge University Press, Cambridge, pp 3–22

Raupach MR (1989) Applying Lagrangian fluid mechanics to infer scalar source distributions from concentration profiles in plant canopies. Agr Forest Meteorol 47:85–108

Raupach MR (1993) The averaging of surface flux densities in heterogeneous landscapes. In: Bolle H-J, Feddes RA, Kalma JD (eds) Exchange processes at the land surface for a range of space and time scales. International Association of Hydrological Sciences (IAHS) Publication No. 212, pp 343–355

Rayment MB, Jarvis PG (1999) An improved open chamber system for measuring soil CO2 effluxes in the field. J Geophys Res 102:28779–28784

Reale O, Dirmeyer PA (2000) Modeling the effects of vegetation on Mediterranean climate during the Roman classical period. Part I: Climate history and model sensitivity. Global Planet Change 25:163–184

Reale O, Shukla J (2000) Modeling the effects of vegetation on Mediterranean climate during the Roman classical period. Part II: Model simulation. Global Planet Change 25:185–214

Reeburgh WS, King JY, Regli SK, Kling GW, Auerbach NA, Walker DA (1998) A CH4 emission estimate for the Kuparuk River basin, Alaska. J Geophys Res 103:29005–29013

Reed RJ (1986) On understanding the meteorological causes of Sahelian drought. In: Chagas C, Puppi G (eds) Persistent meteoceanographic anomalies and teleconnections. Pontificia Academia Scientiarvm, Ex Aedibcs Academicis, pp 179–213

Reich PB, Walters MB, Ellsworth DS (1997) From tropics to tundra: global convergence in plant functioning. P Natl Acad Sci USA 94:13730–13734

Richards LA (1931) Capillary conduction of liquids through porous mediums. Physics 1:318–333

Riehl H, Malkus JS (1958) On the heat balance in the equatorial trough zone. Geophysica 6:504–537

Riehl H, Simpson JM (1979) The heat balance of the equatorial trough zone, revisited. Contrib Atmos Phys 52:287–297

Ripley EA (1976) Drought in the Sahara: insufficient biogeophysical feedback? Science 191:100–101

Roads J, Kanamitsu M, Stewart R (2002) CSE water and energy budgets in the NCEP-DOE reanalyses. J Hydrometeorol 3:227–248

Roberts J, Cabral OMR, de Aguiar LF (1990) Stomatal and boundary-layer conductances measured in a terra firme rainforest, Manaus, Amazonas. Brazil. J Appl Ecol 27:336–353

Roberts JM, Cabral OMR, Fisch G, Molion LCB, Moore CJ, Shuttleworth WJ (1993) Transpiration from Amazonian rainforest calculated from stomatal conductance measurements. Agr Forest Meteorol 65:175–196

Robinson D, Dewey KF (1990) Recent secular variations in the extent of Northern Hemisphere snow cover. Geophys Res Lett 17:1557–1560

Robinson DA, Dewey KF, Heim RR (1993) Global snow cover monitoring: an update. B Am Meteorol Soc 74:1689–1696

Robock A, Vinnikov KY, Schlosser CA, Speranskaya NA, Xue Y (1995) Use of midlatitude soil moisture and meteorological observations to validate soil moisture simulations with biosphere and bucket models. J Climate 8:15–35

Rocha EP (2001) Moisture balance and influence of surface boundary conditions over Amazônia precipitation. Ph.D. Thesis in Meteorology, Instituto Nacional de Pesquisas Espaciais, São José dos Campos, Brazil, 210 p (in Portuguese)

Rodriguez-Camino E, Avissar R (1998) Comparison of three land-surface schemes with the Fourier amplitude sensitivity test (FAST). Tellus 50A:313–332

Rodriguez-Camino E, Avissar R (1999) Effective parameters for surface heat fluxes in heterogeneous terrain. Tellus 51A:387–399

Rosen RD (1999) The global energy cycle. In: Browning KA, Gurney RJ (eds) Global energy and water cycles. Cambridge University Press, pp 1–9

Rosenfeld D (1998) Satellite-based insights into precipitation formation processes in continental and maritime convective clouds. B Am Meteorol Soc 79(11):2457–2476

Rosenfeld D (1999) TRMM observed first direct evidence of smoke from forest fires inhibiting rainfall. Geophys Res Lett 26(20):3105–3108

Rosenfeld J (2000a) Sentinels in the sky. Weatherwise 53:24–29

Rosenfeld D (2000b) Suppression of rain and snow by urban and industrial air pollution. Science 287(5459):1793–1796

Roulet NT, Jano A, Kelly CA, Klinger LF, Moore TR, Protz R, Ritter JA, Rouse WR (1994) Role of the Hudson Bay lowland as a source of atmospheric methane. J Geophys Res 99:1439–1454

Roulet NT, Crill PM, Comer NT, Dove AE, Bourbonniere RA (1997) CO2 and CH4 between a boreal beaver pond and the atmosphere. J Geophys Res 102:29313–29320

Rowe JS (1972) Forest regions of Canada. Canadian Forestry Service Publication No. 1300. Information Canada, Ottawa. 172 p

Rowell DP, Folland CK, Maskell K, Ward MN (1995) Variability of summer rainfall over tropical north Africa (1906–1992): observation and modeling. Q J Roy Meteor Soc 121:669–704

Rowntree PR, Bolton JA (1983) Simulation of the atmospheric response to soil moisture anomalies over Europe. Q J Roy Meteor Soc 109:501–526

Ruess RW, Van Cleve K, Yarie J, Viereck LA (1996) Comparitve estimates of fine root production in successional taiga forests of interior Alaska. Can J Forest Res 26:1326–1336

Running SW, Baldocchi DD, Turner D, Gower ST, Bakwin P, Hibbard K (1999) A global terrestrial monitoring network, scaling tower fluxes with ecosystem modeling and EOS satellite data. Remote Sens Environ 70:108–127

Russell G, Miller J (1990) Global river runoff calculated from a global atmospheric general circulation model. J Hydrol 117:241–254

Ryan MG, Lavigne MB, Gower ST (1997) Annual carbon costs of autotrophic respiration in boreal forest ecosystems in relation to species and climate. J Geophys Res 102:28871–28884

Sá TD de A, da Costa J de PR, Roberts JM (1996) Forest and pasture conductances in southern Pará, Amazonia. In: Gash JHC, Nobre CA, Roberts JM, Victoria RL (eds) Amazonian deforestation and climate. John Wiley & Sons, Chichester, pp 242–263

Saarinen T (1996) Biomass and production of two vascular plants in a boreal mesotrophic fen. Can J Bot 74:934–938

Salati E (1987) The forest and the hydrological cycle. In: Dickinson RE (ed) Geophysiology of Amazonia. John Wiley & Sons, New York, pp 273–296

Salati E, Marques J (1984) Climatology of the Amazon region. In: Sioli H (ed) The Amazon: limnology and landscape ecology of a mighty tropical river and its basin. W. Junk, Dordrecht

Salati E, Vose P (1984) Amazon basin: a system in equilibrium. Science 225:129–138

Sarnthein M (1978) Sand deserts during glacial maximum and climatic optimum. Nature 272:43–46

Satyamurty P, Nobre CA, Silva Dias PL (1998) Tropics-South América. In: Karoly DJ, Vincent DG (ed) Meteorology of the Southern Hemisphere. AMS 3:119–140

Sausen R, Schubert S, Dümenil L (1994) A model of the river-runoff for use in coupled atmosphere-ocean models. J Hydrol 155:337–352

Savage K, Moore TR, Crill PM (1997) Methane and carbon dioxide exchange between the atmosphere and northern boreal forest soils. J Geophys Res 102:29279–29288

Savenije HHG (1995) New definitions for moisture recycling and the relationship with land-use changes in the Sahel. J Hydrol 167:57–78

Schär C, Lüthi D, Beyerle U, Heise E (1999) The soil-precipitation feedback: a process study with a regional climate model. J Climate 12:722–741

Schlesinger WH (1991) Biogeochemistry: an analysis of global change. Academic Press, San Diego Ca

Schlesinger ME, Mitchell JFB (1987) Climate model calculations of the equilibrium climatic response to increased carbon dioxide. Rev Geophys 25(4):760–798

Schlesinger WH, Reynolds JF, Cunningham GL, Huenneke LF, Jarrell WM, Virginia RA, Whitford WG (1990) Biological feedbacks in global desertification. Science 247:1043–1048

Schupelius GD (1976) Monsoon rains over west Africa. Tellus 28:533

Segal M, Arritt RW (1992) Non-classical mesoscale circulations caused by surface sensible heat-flux gradients. B Am Meteorol Soc 73:1593–1604

Segal M, Avissar R, McCumber MC, Pielke RA (1988) Evaluation of vegetation effects on the generation and modification of meso-scale circulations. J Atmos Sci 45:2268–2292

Sellers PJ (1992) Biophysical models of land-surface processes. In: Trenberth KE (ed) Climate system modelling. Cambridge University Press

Sellers PJ, Mintz Y, Sud YC, Dalcher A (1986) A simple biosphere model (SiB) for use with general circulation models. J Atmos Sci 43:505–530

Sellers PJ, Hall FG, Asrar G, Strebel DE, Murphy RE (1988) The First ISLSCP Field Experiment (FIFE). B Am Meteorol Soc 69:22–27

Sellers PJ, Shuttleworth WJ, Dorman JL, Dalcher A, Roberts JM (1989) Calibrating the simple biosphere model for Amazonian tropical forest using field and remote sensing data: Part 1, average calibration with field and remote sensing data. J Appl Meteorol 28: 727–759

Sellers PJ, Bounoua L, Collatz GJ, Randall DA, Dazlich DA, Los SO, Berry JA, Fung I, Tucker CJ, Field CB, Jensen TG (1996a) Comparison of radiative and physiological effects of doubled atmospheric CO_2 climate. Science 271:14

Sellers PJ, Meeson BW, Closs J, Collatz J, Corprew F, Dazlich D, Hall FG, Kerr Y, Koster R, S Los, Mitchell K, McManus J, Myers D, Sun KJ, Try P (1996b) The ISLSCP Initiative I global datasets: surface boundary conditions and atmospheric forcings for land-atmosphere studies. B Am Meteorol Soc 77(9):1987–2005

Sellers PJ, Randall DA, Collatz GJ, Berry JA, Field CB, Dazlich DA, Zhang C, Collelo GD, Bounoua L (1996c) A revised land-surface parameterisation (SiB2) for atmospheric GCMs. Part I, model formulation. J Climate 9:676–705

Sellers PJ, Dickinson RE, Randall DA, Betts AK, Hall FG, Berry JA, Collatz GJ, Denning AS, Mooney HA, Nobre CA, Sato N, Field CB, Henderson-Sellers A (1997) Modeling the exchanges of energy, water and carbon between continents and the atmosphere. Science 275:502–509

Semazzi FHM, Mehta V, Sud YC (1988) An investigation of the relationship between sub-Sahara rainfall and global sea surface temperatures. Atmos Ocean 26:1471–1485

Serreze MC, Walsh JE, Chapin FS, Osterkamp T, Dyurgerov M, Romanovsky V, Oechel WC, Morison J, Zhang T, Barry RG (2000) Observational evidence of recent change in the northern high-latitude environment. Climatic Change 46:159–162

Shabbar A, Bonsal B, Khandekar M (1997) Canadian precipitation patterns associated with the Southern Oscillation. J Climate 10: 3016–3027

Shao Y, Henderson-Sellers A (1996) Validation of soil moisture in landsurface parameterisation schemes with HAPEX data. Global Planet Change 13:3–11

Shaw BL, Pielke RA, Ziegler CL (1997) A three-dimensional numerical simulation of a Great Plains dryline. Mon Weather Rev 125: 1489–1506

Shinoda M, Kawamura R (1994) Tropical rainbelt, circulation, and sea surface temperatures associated with the Sahelian rainfall trend. J Meteorol Soc Jpn 72:341–357

Shinoda T, Uyeda H (2001) Factors in the development of deep convection in the southern region far from the Mei-yu front over Eastern China during the GAME/HUBEX IOP'98. Proc. the Fifth International Study Conference on GEWEX and GAME (Volume 1) Nagoya, Japan, Oct. 3–5, 2001, pp 185–190

Shukla J (1998) Predictability in the midst of chaos: a scientific basis for climate forecasting. Science 282:728–731

Shukla J, Mintz Y (1982) Influence of land-surface evaporation on the Earth's climate. Science 215:1498–1501

Shuttleworth WJ (1988a) Evaporation from Amazonian rainforests. Philos T Roy Soc B 233:321–346

Shuttleworth WJ (1988b) Macrohydrology – the new challenge for process hydrology. J Hydrol 100:31–56

Shuttleworth WJ (1998) Combining remotely sensed data using aggregation algorithms. Hydrol Earth Syst Sc 2(2–3):149–158

Shuttleworth WJ, Dickinson RE (1989) Comments on 'Modelling tropical deforestation: a study of land-surface parameterisations' by R. E. Dickinson and A. Henderson-Sellers (January 1988, 114, 439–462). Q J Roy Meteor Soc 115:1177–1179

Shuttleworth WJ, Wallace JS (1985) Evaporation from sparse crops – an energy combination theory. Q J Roy Meteor Soc 111:839–855

Shuttleworth WJ, Gash JHC, Lloyd CR, Roberts JM, Marques A de O, Fisch G, de Silva P, Ribeiro MNG, Molion LCB, de Abreu Sa LD, Nobre CA, Cabral OMR, Patel SR, de Moraes JC (1984a) Eddy correlation measurements of energy partition for Amazonian forest. Q J Roy Meteor Soc 110:1143–1162

Shuttleworth WJ, Gash JHC, Lloyd CR, Moore CJ, Roberts J, Marques A de O, Fisch G, Silva V de P, Ribeiro MNG, Molion LCB, de Abreu Sa LD, Nobre JCA, Cabral OMR, Patel SR, de Moraes JC (1984b) Observations of radiation exchange above and below Amazonian forest. Q J Roy Meteor Soc 110:1163–1169

Shuttleworth WJ, Yang Z-L, Arain MA (1997) Aggregation rules for surface parameters in global models. Hydrol Earth Syst Sc 2: 217–226

Silva Dias MAF, Regnier P (1996) Simulation of mesoscale circulations in a deforested area of *Rondonia* in the dry season. In: Gash JHC, Nobre CA, Roberts JM, Victoria RL (eds) Amazonian deforestation and climate. John Wiley & Sons, Chichester, pp 531–547

Silva Dias MAF, Dolman A, Silva Dias P, Rutledge S, Zipser E, Fisch G, Artaxo P, Manzi A, Marengo J, Nobre C, Kabat P (1999) An overview of the WET AMC/LBA January and February 1999. In: Preprints of the 15th Conference on Hydrology, Long Beach, California, January 9–14 2000. pp 305–306

Silva Dias MAF, Dolman A, Silva Dias P, Rutledge S, Zipser E, Fisch G, Artaxo P, Manzi A, Marengo J, Nobre C, Kabat P (2000) Rainfall and surface processes in Amazonia during the WET AMC/LBA – An overview. In: Preprints of the 6th International Conference on Southern Hemisphere Meteorology and Oceanography, Santiago, Chile, April 3–7 2000, pp 249–250

Silva Dias MAF, Rutledge S, Kabat P, Silva Dias PL, Nobre C, Fisch G, Dolman AJ, Zipser E, Garstang M, Manzi A, Fuentes JD, Rocha H, Marengo J, Plana-Fattori A, Sá L, Alvalá R, Andreae MO, Artaxo P, Gielow R, Gatti L (2002) Cloud and rain processes in a biosphere-atmosphere interaction context in the Amazon Region. J Geophys Res LBA Special Issue 107(D20), 8072, doi:10.1029/2001JD000335,2002

Silvola J, Alm J, Ahlholm U, Nykänen H, Martikainen PJ (1996) CO_2 fluxes from peat in boreal mires under varying temperature and moisture conditions. J Ecol 84:219–228

Simpson J, Kummerow C, Tao WK, Adler RF (1996) On the tropical rainfall measuring mission (TRMM). Meteorol Atmos Phys 60:19–36

Simpson IJ, Edwards GC, Thurtell GW, den Hartog G, Neumann HH, Staebler RM (1997) Micrometeorological measurements of methane and nitrous oxide exchange above a boreal aspen forest. J Geophys Res 102:29331–29342

Singh B (1988) The implications of climate change for natural resources in Quebec. Climate Change Digest Series CCD88-08. Atmospheric Environment Service, Downsview, Ontario, 11 p

Singh T, Wheaton EE (1991) Boreal forest sensitivity to global warming: implications for forest management in western interior Canada. Forest Chron 67:342–348

Skoupy J (1987) Desertification in Africa. Agric. Met. Prog., Proc. Reg. Train. Sem. on Drought and Desertification in Africa. Addis Ababa, 33–45, Geneva, WMO

Smith EA, Hsu AY, Crosson WL, Field RT, Fritschen LJ, Gurney RJ, Kanemasu ET, Kustas WP, Nie D, Shuttleworth WJ, Stewart JB, Verma SB, Weaver HL, Wesley ML (1992) Area-averaged surface fluxes and their time-space variability over the FIFE experimental domain. J Geophys Res 97:18599–18622

Solomon AM, West DC (1987) Simulating forest ecosystem responses to expected climate change in eastern North America: applications to decision making in the forest industry. In: Shands W, Hoffmann JS (eds) The greenhouse effect, climate change, and the U.S. forests. The Conservation Foundation, Washington D.C., pp 189–217

Souza EP, Rennó NO, Silva Dias MAF (2000) Convective circulations induced by surface heterogeneities. J Atmos Sci 57:2915–2922

Stamm JF, Wood EF, Lettenmaier DP (1994) Sensitivity of a GCM simulation of global climate to the representation of land-surface hydrology. J Climate 7:1218–1239

Steele S, Gower ST, Vogel JG, Norman JM (1997) Root mass, net primary production and turnover in aspen, jack pine and black spruce forests in Saskatchewan and Manitoba, Canada. Tree Physiol 17:577–588

Stein U, Alpert P (1993) Factor separation in numerical simulations. J Atmos Sci 50:2107–2115

Stephenne N, Lambin EF (2001) A dynamic simulation model of land-use changes in the African Sahel (SALU). Agr Ecosyst Environ 85:145–162

Stewart JB (1988) Modelling surface conductance of pine forests. Agr Forest Meteorol 43:19–35

Steyaert LT, Hall FG, Loveland TR (1997) Land-cover mapping, fire-regeneration, and scaling studies in the Canadian boreal forest with 1-KM AVHRR and Landsat TM Data. J Geophys Res 102: 29581–29598

Stohlgren TJ, Chase TN, Pielke RA, Kittel TGF, Baron J (1998) Evidence that local land-use practices influence regional climate and vegetation patterns in adjacent natural areas. Global Change Biol 4:495–504

Stokes GM, Schwartz SE (1994) The Atmospheric Radiation Measurement (ARM) program: programmatic background and design of the cloud and radiation test bed. B Am Meteorol Soc 75:1201–1220

Stössel A, Claussen M (1993) On the momentum forcing of a large-scale sea-ice model. Clim Dynam 9:71–80

Sud YC, Fennessy M (1982) A study of the influence of surface albedo on July circulation in semiarid regions using the GLAS GCM. J Climatol 2:105–125

Sud YC, Fennessy M (1984) A numerical study of the influence of evaporation in semiarid regions on the July circulation. J Climatol 4:383–398

Sud YC, Lau WK-M (1996) Comments on "Variability of summer rainfall over tropical north Africa (1906–1992): observation and modeling" by Rowell, Folland, Maskell, Ward (1995). Q J Roy Meteor Soc 122:1001–1006

Sud YC, Molod A (1988) A GCM simulation study of the influence of Saharan evapotranspiration and surface-albedo anomalies on July circulation and rainfall. Mon Weather Rev 116:2388–2400

Suffling R (1995) Can disturbance determining vegetation distribution during climate warming? A boreal test. J Biogeogr 22:501–508

Suyker AE, Verma SB, Arkebauer TJ (1997) Season-long measurement of carbon dioxide exchange in a boreal fen. J Geophys Res 102:29021–29028

Suzuki M, Takizawa H, Tanaka N, Yoshifuji N, Tangtham N (2001) Energy and water budget in hill evergreen forest, northern Thailand. Proceedings of the International Workshop on GAME-AAN/Radiation, Phuket, Thailand, 7–9 March (2001) Bulletin of the Terrestrial Environment Research Center, University of Tsukuba, No. 1 Supplement March 2001, p 48

Tanaka K (2002) Multi-layer model of CO_2 exchanges in a plant community coupled with the water budget of leaf surfaces. Ecol Model 147(1):85–104

Tanaka M, Weare BC, Navato AR, Newell RE (1975) Recent African rainfall patterns. Nature 255:201

Tanaka K, Takizawa H, Tanaka N, Kosaka I, Yoshifuji N, Tantasirin C, Piman S, Suzuki M, Tangtham N (2003) Transpiration peak over a hill evergreen forest in northern Thailand in the late dry season: Assessing the seasonal changes in evapotranspiration using a multilayer model. J Geophys Res 108(D17): Art. No. 4533

Tarasova T, Nobre CA, Holben BN, Eck TF, Setzer AW (1999) Assessment of smoke aerosol impact on surface solar irradiance measured in the Rondônia region of Brazil during SCAR-B. J Geophys Res 104:161–170

Tarasova T, Nobre CA, Eck TF, Holben BN (2000) Modeling of gaseuous, aerosol, and cloudiness effects on surface solar irradiance measured in Brazil's Amazônia. J Geophys Res 105:961–971

Tarnocai C, Kettles IM, Lacelle B (2000) Peatlands of Canada (1 : 6 500 000 map and digital data base of Canadian Peatlands). Geological Survey of Canada, Ottawa, Open File 3834

Taylor CM, Blyth EM (2000) Rainfall controls on evaporation at the regional scale: an example from the Sahel. J Geophys Res 105: 15469–15479

Taylor CM, Clark DB (2001) The diurnal cycle and African easterly waves: a land-surface perspective. Q J Roy Meteor Soc 127 (573): 845–867

Taylor CM, Lebel T (1998) Observational evidence of persistent convective-scale rainfall patterns. Mon Weather Rev 126:1597–1607

Taylor CM, Harding RJ, Thorpe AJ, Bessemoulin P (1997a) A mesoscale simulation of land-surface heterogeneity from HAPEX-Sahel. J Hydrol 188/189:1040–1066

Taylor CM, Saïd F, Lebel T (1997b) Interactions between the land-surface and mesoscale rainfall variability during HAPEX-Sahel. Mon Weather Rev 125:2211–2227

Taylor CM, Harding RJ, Pielke RA Sr, Vidale PL, Walko RL, Pomeroy JW (1998) Snow breezes in the boreal forest. J Geophys Res 103: 23087–23101

Texier D, de Noblet N, Harrison SP, Haxeltine A, Jolly D, Joussaume S, Laarif F, Prentice IC, Tarasov P (1997) Quantifying the role of biosphere-atmosphere feedbacks in climate change: coupled model simulations for 6000 years BP and comparison with palaeodata for northern Eurasia and northern Africa. Clim Dynam 13:865–882

Thom AS (1972) Momentum, mass and heat exchange of vegetation. Q J Roy Meteor Soc 98:124–134

Thom AS (1976) Momentum, masses and heat exchange of plant communities. In: Monteith JL (ed) Vegetation and the atmosphere, Volume 1. Principles. Academic Press, London, pp 57–109

Thomas G, Rowntree PR (1992) The boreal forests and climate. Q J Roy Meteor Soc 118:469–497

Thompson DWJ, Wallace JM (1998) The Arctic Oscillation signature in the wintertime geopotential height and temperature fields. Geophys Res Lett 25:1297–1300

Thormann MN, Bayley SE (1997) Above-ground net primary production along a bog-fen-marsh gradient in southern boreal Alberta, Canada. Ecoscience 4:374–384

Tiedtke M (1984) The effect of penetrative cumulus convection on the large-scale flow in a general circulation model. Beitr Phys Atmos 57:216–224

Tiedtke M (1993) Representation of clouds in large-scale models. Mon Weather Rev 121:3040–3061

Toda M, Nishida K, Ohte N, Tani M, Musiake K (2002) Observation of energy fluxes and evapotranspiration over terrestrial complex land covers in the tropical monsoon environment. J Meteorol Soc Jpn 80(3):465–484

Todini E, Dümenil L (1999) Estimating large-scale runoff. In: Browning KA, Gurney RJ (rds) Global energy and water cycles. Cambridge University Press, pp 265–281

Tomasella J, Hodnett MG (1997) Estimating soil water retention characteristics from limited data in Brazilian Amazonia. Soil Sci 163:190–202

Tribbia JJ (1991) The rudimentary theory of atmospheric teleconnections associated with ENSO. In: Glantz MH, Katz RW, Nicholls N (eds) Teleconnections linking worldwide climate anomalies. Cambridge University Press, pp 285–307

Trumbore SE, Bubier JL, Harden JW, Crill PM (1999) Carbon cycling in boreal wetlands: a comparison of three approaches. J Geophys Res 104:27673–27682

Tucker CJ, Dregne HE, Newcomb WW (1991) Expansion and contraction of the Sahara desert from 1980 to 1990. Science 253:299–301

UNCED (1992) Managing fragile ecosystems: combating desertification and drought. United Nations Conference on Environment and Development

Valentini R, Matteucci H, Dolman AJ, Schulze E-D, Rebmann C, Moors EJ, Granier A, Gross P, Jensen NO, Pilegaard K, Lindroth A, Grelle A, Bernhofer C, Grünwald T, Aubinet M, Ceulemans R, Kowalski AS, Vesala T, Rannik U, Berbigier P, Loustau D, Gudmundsson HT, Ibrom A, Morgenstern K, Clement R, Moncrieff J, Montagnani L, Minerbi S, Jarvis PG (2000) Respiration as the main determinant of carbon balance in European forests. Nature 404:861–865

Valentini De Angelis R, Matteucci P, Monaco G, Dore R, Scarascia S, Mugnozza GE (1996) Seasonal net carbon dioxide exchange of a beech forest with the atmosphere. Global Change Biol 2:199–208

Van den Hurk BJJM, Viterbo P, Beljaars ACM, Betts AK (2000) Offline validation of the ERA40 surface scheme. ECMWF Tech Memo, No. 295, available from ECMWF, Shinfield Park, Reading RG2 9AX, UK, 43 p

Venäläinen A, Frech M, Heikinheimo M, Grelle A (1999) Comparison of latent and sensible heat fluxes over boreal lakes with concurrent fluxes over a forest: implications for regional averaging. Agr Forest Meteorol (98/99)1–4:535–554

Verma SB, Baldocchi DD, Anderson DF, Matt DR, Clement RF (1986) Eddy fluxes of CO, water vapour, and sensible heat over a deciduous forest. Bound-Lay Meteorol 36:71–91

Verma SB, Kim J, Clement RJ (1989) Carbon dioxide, water vapour and sensible heat fluxes over a tall grass prairie. Bound-Lay Meteorol 46:53–67

Vermetten AWM, Ganzeveld L, Jeuken A, Hofschreuder P, Mobren GMJ (1994) CO_2 uptake by a stand of Douglas fir: flux measurements compared with model calculations. Agr Forest Meteorol 72:57–80

Verstraete MM, Schwartz SA (1991) Desertification and global change. Vegetatio 91:3–13

Victoria RL, Matinelli LA, Moraes JM, Ballester MV, Krusche AV, Pellegrino G, Almeida RMB, Richey JE (1998) Surface air temperature variations in the Amazon region and its borders during this century. J Climate 11(5):1105–1110

Vidale PL, Pielke RA, Barr A, Steyaert LT (1997) Case study modeling of turbulent and mesoscale fluxes over the BOREAS region. J Geophys Res 102:29167–29188

Villa Nova N, Salati E, Matsui E (1976) Estimativa da evapotranspiração na bacia Amazônica. Acta Amazon 6:215–228

Vinnikov KY, Groisman PY, Lugina KM (1990) Empirical data on contemporary global climate changes (temperature and precipitation). J Climate 3:662–677

Viswanadham Y, Molion LCB, Manzi AO, Sa LDA, Silva VP, Andre RGB, Nogueira JLM, Dos Santos RC (1990) Micrometeorological measurements in Amazon forest during GTE ABLE-2A mission. J Geophys Res 95:13669–13682

Viterbo P (1996) The representation of surface processes in general circulation models. Ph.D. Thesis, University of Lisbon, 201 p, available from the author, ECMWF, Shinfield Park, Reading RG2 9AX, England

Viterbo P, Beljaars CM (1995) An improved land-surface parameterisation scheme in the ECMWF model and its validation. J Climate 8:2716–2748

Viterbo P, Betts AK (1999a) The impact of the ECMWF reanalysis soil water on forecasts of the July 1993 Mississippi flood. J Geophys Res 104D:19361–19366

Viterbo P, Betts AK (1999b) The forecast impact of the albedo of the boreal forests in the presence of snow. J Geophys Res 104D: 27803–27810

Viterbo P, Illari L (1994) The impact of changes in the runoff formulation of a general circulation model on the surface and near surface parameters. J Hydrol 155:325–336

Viterbo P, Beljaars ACM, Mahfouf J-F, Teixeira J (1999) The representation of soil moisture freezing and its impact on the stable boundary layer. Q J Roy Meteor Soc 125:2401–2426

Vörösmarty CJ, Moore B III, Grace AL, Gileda MP, Melillo JM, Peterson B, Jrastetter EB, Steudler PA (1989) Continental scale models of water balance and fluvial transport: an application to South America, Global Biogeochem Cy 3:241–256

Vörösmarty CJ, Willmott C, Choudhury B, Schloss A, Stearns T, Roberson S, Dorman T (1996) Analyzing the discharge regime of a large tropical river through remote sensing, ground-based climatic data, and modeling. Water Resour Res 32:3137–3150

Vose RS, Keim R, Schmoyer RL, Karl TR, Steuer PM, Eischeid JK, Peterson TC (1992) The global historical climatology network: long-term monthly temperature, precipitation, sea level pressure, and station pressure data. ORNL/CDIAC-53, NDP-041, 189 p

Walker GT (1910) Correlation in seasonal variations of weather. Memo India Meteor Dept 21:22–45

Walker J, Rowntree PR (1977) The effect of soil moisture on circulation and rainfall in a tropical model. Q J Roy Meteor Soc 103:29–46

Walko RL, Cotton WR, Pielke RA (1992) Large-eddy simulations of the effects of hilly terrain on the convective boundary layer. Bound-Lay Meteorol 58:133–150

Walko RL, Band LE, Baron J, Kittel TGF, Lammers R, Lee TJ, Ojima DS, Pielke RA, Taylor C, Tague C, Tremback CJ, Vidale PL (2000) Coupled atmosphere-biophysics-hydrology models for environmental modeling. J Appl Meteorol 39:931–944

Wallace JM, Gutzler DS (1981) Teleconnections in the geopotential height field during the Northern Hemisphere winter. Mon Weather Rev 109:784–812

Wallace JS, Allen SJ, Culf AD, Dolman AJ, Gash JHC, Holwill CJ, Lloyd CR, Stewart JB, Wright IR, Sivakumar MVK, Renard C (1992) SEBEX: the Sahelian energy balance experiment. Final report on ODA project T06050C1, Institute of Hydrology, U.K.

Wang G, Eltahir EAB (2000a) Biosphere-atmosphere interactions over West Africa. Part I. Development and validation of a coupled dynamic model. Q J Roy Meteor Soc 126:1239–1260

Wang G, Eltahir EAB (2000b) Biosphere-atmosphere interactions over West Africa. 2. Multiple equilibira. Q J Roy Meteor Soc 126: 1261–1280

Wang G, Eltahir EAB (2000c) Role of vegetation dynamics in enhancing the low-frequency variability of Sahel rainfall. Water Resour Res 36(4):1013–1021

Wang J, Bras RL, Eltahir EAB (1996) A stochastic linear theory of mesoscale circulation induced by the thermal heterogeneity of the land surface. J Atmos Sci 53:3349–3366

Wang J, Eltahir EAB, Bras RL (1998) Numerical simulation of nonlinear mesoscale circulations induced by the thermal heterogeneities of land surface. J Atmos Sci 55:447–464

Ward MK (1995) Analyzing the boreal summer relationship between worldwide sea-surface temperature and atmospheric variability. In: Stroch H von, Navarra A (eds) Analysis of climate variability. Springer-Verlag, Heidelberg, pp 95–117

Warrilow DA, Sangster AB, Slingo A (1986) Modelling of land-surface processes and their influence on European climate. Technical Report No. 38, 92 p, U.K. Meteorological Office, London Road, Bracknell, Berkshire RG12 2SZ U.K.

Wasson RJ, Claussen M (2002) Earth system models: a test using the mid-Holocene in the Southern Hemisphere. Quaternary Sci Rev 21(7):819–824

Watts M (1987) Drought, environment and food security: some reflections on peasants, pastoralists and commoditization in dryland West Africa. In: Glantz HH (ed) Drought and hunger in Africa. Cambridge University Press, Cambridge, pp 171–211

Weaver CP, Avissar R (2001) Atmospheric disturbances caused by human modification of the landscape. B Am Meteorol Soc 82: 269–281

Weber MG, Flannigan MD (1997) Canadian boreal forest ecosystem structure and function in a changing climate: impact on fire regimes. Environ Rev 5:145–166

Wei HL, Fu CB (1999) Study of the sensitivity of a regional model in response to land-cover change over Northern China. Hydrol Process 12:2249–2265

Wei HL, Fu CB, Wang WC (1998) Impacts of lateral boundary condition treatment of a regional climate model on the East Asia summer monsoon precipitation. Sci Atmos Sin 22:779–790

Werth D, Avissar R (2002) The local and global effects of Amazon deforestation. J Geophys Res 107(D20): Art. No. 8087

Wetzel PJ, Liang X, Irannejad P, Boone A, Noilhan J, Shao Y, Skelly C, Xue Y, Yang Z-L (1996a) Modeling the vadose zone liquid water fluxes: infiltration, runoff, drainage, interflow. Global Planet Change 13:57–71

Wetzel PJ, Argentini S, Boone A (1996b) Role of land surface in controlling daytime cloud amount: two case studies in the GCIP-SW area. J Geophys Res 101:7359–7370

Whiting GJ, Chanton JP (1993) Primary production control of methane emission from wetlands. Nature 364:794–795

Whittaker RH, Likins GE (1975) Primary production of the biosphere. Springer-Verlag, New York

Winston GC, Sunquist ET, Stephens BB, Trumbore SE (1997) Winter CO_2 fluxes in a boreal forest. J Geophys Res 102:28795–28804

WMO (1984) Scientific plan for the World Climate Research Programme. WCRP Publ. Series No. 2, Geneve

Wofsy SC, Goulden ML, Munger JW, Fan SM, Bakwin PS, Daube BC, Bassow SL, Bazzaz FA (1993) Net exchange of CO_2 in a midlatitude forest. Science 260:1314–1317

Wolter K (1989) Modes of tropical circulation, southern oscillation, and Sahel rainfall. J Climate 2:149–172

Wood EF, Lettenmaier DP, Zartarian VG (1992) A land-surface hydrology parameterisation with subgrid variability for general circulation models. J Geophys Res 97(D3):2717–2728

Woodward FI, Smith TM, Emanuel WR (1995) A global land primary productivity and phytogeography model. Global Biogeochem Cy 9:471–490

Woodward FI, Lomas MR, Betts RA (1998) Vegetation-climate feedback in a greenhouse world. Philos T Roy Soc B 353:29–39

World Resources Institute (1994) World resources 1994–1995. Oxford University Press, New York

Wright IR, Gash JHC, Da Rochas HR, Shuttleworth WJ, Nobre CA, Maitelli GT, Zamparoni CAGP, Carvalho PRA (1992) Dry season micrometeorology of central Amazonian ranchland. Q J Roy Meteor Soc 118:1083–1099

Wright IR, Manzi AO, da Rocha HR (1995) Canopy surface conductance of Amazonian pasture:model application and calibration for canopy climate. Agr Forest Meteorol 75:51–70

Wright IR, Nobre CA, Tomasella J, da Rocha HR, Roberts JM, Vertamatti E, Alvala RC, Hodnett MG, Culf AD (1996a) Towards a GCM surface parameterization for Amazonia. In: Gash JHC, Nobre CA, Roberts JM, Victoria RL (eds) Amazonian deforestation and climate. John Wiley & Sons, Chichester, pp 473–504

Wright IR, Gash JHC, da Rocha HR, Roberts JM (1996b) Modelling surface conductance for Amazonian pasture and forest. In: Gash JHC, Nobre CA, Roberts JM, Victoria RL (eds) Amazonian deforestation and climate. John Wiley & Sons, Chichester, pp 437–457

Wu Z-X, Newell RE (1998) Influence of sea surface temperature on air temperature in the tropic. Clim Dynam 14:275–290

Xie P, Arkin P (1997) Global precipitation: a 17-yr monthly analysis based on gauged observations, satellite estimates and numerical model outputs. B Am Meteorol Soc 78:2539–2558

Xie P, Arkin P (1998) Global monthly precipitation estimates from satellite-observed outgoing longwave radiation. J Climate 11:137–164

Xue Y (1996) The impact of desertification in the Mongolian and Inner Mongolian grassland on the regional climate. J Climate 9:2173–2189

Xue Y (1997) Biosphere feedback on regional climate in tropical north Africa. Q J Roy Meteor Soc 123B:1483–1515

Xue Y, Shukla J (1993) The influence of land-surface properties on Sahel climate. Part I: desertification. J Climate 6:2232–2245

Xue Y, Shukla J (1996) The influence of land-surface properties on Sahel climate. Part II: afforestation. J Climate 9:3260–3275

Xue Y, Shukla J (1998) Model simulation of the influence of global SST anomalies on the Sahel rainfall. Mon Weather Rev 126:2782–2792

Xue Y, Liou K-N, Kasahara A (1990) Investigation of the biophysical feedback on the African climate using a two-dimensional model. J Climate 3:337–352

Xue Y, Sellers PJ, Kinter JL III, Shukla J (1991) A simplified biosphere model for global climate studies. J Climate 4:345–364

Xue Y, Allen SJ, Li Q (1996a) Sahel drought and land-surface processes – a study using SEBEX and HAPEX-Sahel data. Preprint of Second International Scientific Conference on the Global Energy and Water Cycle, pp 11–12

Xue Y, Zeng FJ, Schlosser CA (1996b) SSiB and its sensitivity to soil properties – a case study using HAPEX-Mobilhy data. Global Planet Change 13:183–194

Xue Y, Fennessy MJ, Sellers PJ (1996c) Impact of vegetation properties on U.S. summer weather prediction. J Geophys Res 101:7419–7430

Xue Y, Elbers J, Zeng FJ, Dolman AJ (1997) GCM parameterization for Sahelian land-surface processes. In: Kabat P, Prince S, Prihodko L (eds) HAPEX-Sahel west central supersite: methods, measurements and selected results. Rep 130, HM/07.97. The Winand Staring Center for Integrated Land, Soil and Water Research, The Netherlands, pp 289–297

Xue Y, Zeng FJ, Schlosser CA, Allen S (1998) A simplified Simple Biosphere Model (SSiB) and its application to land-atmosphere interactions. (Chinese Journal of Atmospheric Sciences) Sci Atmos Sin 22:575–586

Xue Y, Juang HH, Kanamitsu M, Hansen M (1999) Asian monsoon and vegetation interactions. GEWEX News 9:8–9

Yamamoto S, Murayama S, Saigusa N, Kondo H (1999) Seasonal and interannual variation of CO_2 flux between a temperate forest and the atmosphere in Japan. Tellus 51B:402–413

Yan H, Anthes RA (1987) The effect of variations in surface moisture on mesoscale circulation. Mon Weather Rev 116(1):192–208

Yang R, Fennnessy MJ, Shukla J (1994) The influence of initial soil wetness on medium-range surface weather forecasts. Mon Weather Rev 122:471–485

Yasunari T, Yatagai A, Masuda K (2000a) Time-space characteristics of atmospheric water balance in monsoon areas based on ECMWF reanalysis data. Proceedings of the second WCRP international conference on reanalysis. WMO/TD-No. 985, pp 261–265

Yasunari T, Yatagai A, Arakawa O (2000b) Seasonal and interannual variabilities of rainfall and atmospheric water balance in Thailand and associated large-scale atmospheric circulation. Extended abstract of the International GAME-Tropics workshop, March 6–7, 2000, Cha-am Bay, Thailand

Ye H, Cho H-R, Gustafson PE (1998) The changes in Russian winter snow accumulation during 1936–1983 and its spatial patterns. J Climate 11:856–863

Yeh TC, Wetherald RT, Manabe S (1984) The effect of soil moisture on the short-term climate and hydrology change – a numerical experiment. Mon Weather Rev 112:474–490

Yu G, Harrison SP (1996) An evaluation of the simulated water balance of Eurasia and northern Africa at 6000 yr BP using lake status data. Clim Dynam 12:723–735

Zang H, Henderson-Sellers A, McGuffie K (1996) Impacts of tropical deforestation. Part I: Process analysis of local climate change. J Climate 9:1497–1517

Zeng N (1999) Seasonal cycle and interanual variability in the Amazon hydrological cycle. J Geophys Res 104:9097–9106

Zeng X (2000) Global vegetation root distribution for land modeling. J Hydrometeorol 2:525–530

Zeng N, Neelin JD (2000) The role of vegetation-climate interaction and interannual variability in shaping the African savanna. J Climate 13:2665–2670

Zeng X, Pielke RA (1995a) Landscape-induced atmospheric flow and its parameterisation in large-scale numerical models. J Climate 8:1156–1177

Zeng X, Pielke RA (1995b) Further study on the predictability of landscape-induced atmospheric flow. J Atmos Sci 52:1680–1698

Zeng X, Pielke RA, Eykholt R (1993) Chaos theory and its application in the atmosphere. B Am Meteorol Soc 74:631–644

Zeng X, Dai Y-J, Dickinson RE, Shaikh M (1998) The role of root distribution for land climate simulation. Geophys Res Lett 25:4533–4536

Zeng N, Neelin JD, Lau K-M, Tucker CJ (1999) Enhancement of interdecadal climate variability in the Sahel by vegetation interaction. Science 286:1537–1540

Zhang H, Henderson-Sellers A, McGuffie K (1996) Impacts of tropical deforestation. Part II: the role of large scale dynamics. J Climate 10:2498–2521

Zhang X, Vincent LA, Hogg WD, Nitsoo A (2000) Temperature and precipitation trends in Canada during the 20th century. Atmos Ocean 38(3):395–429

Zhao M, Pitman AJ (2002) The impact of land-cover change and increasing carbon dioxide on the extreme and frequency of maximum temperature and convective precipitation. Geophys Res Lett 29:21–24

Zhao M, Pitman AJ, Chase TN (2001) The impact of land-cover change on the atmospheric circulation. Clim Dynam 17:467–477

Zheng X, Eltahir EAB (1998) The role of vegetation in the dynamics of west African monsoons. J Climate 11:2078–2096

Zheng X, Eltahir EAB, Emanuel KA (1999) A mechanism relating tropical Atlantic spring sea surface temperature and west African rainfall. Q J Roy Meteor Soc 125, 1129–1163

Zhong S, Doran JC (1997) A study of the effects of spatially varying fluxes on cloud formation and boundary layer properties using data from the southern Great Plains cloud and radiation testbed. J Climate 10:327–341

Zhong S, Doran JC (1998) An evaluation of the importance of surface flux variability on GCM-scale boundary-layer characteristics using realistic meteorological and surface forcing. J Climate 11:2774–2788

Ziegler CL, Martin WJ, Pielke RA, Walko RL (1995) A modeling study of the dryline. J Atmos Sci 52:263–285

Ziegler CL, Lee TJ, Pielke RA Sr (1997) Convective initiation at the dryline: a modeling study. Mon Weather Rev 125:1001–1026

Part B

How Measurable Is the Earth System?

Edited by John H. C. Gash and Pavel Kabat

This part of the book is dedicated to the memory of
Tim Crawford,
who tragically died of a stroke while flying an ocean
boundary-layer experiment off Massachusetts.

Chapter B.1

Introduction

John H. C. Gash · Pavel Kabat

B.1.1 The Need for Integrated Experiments

Twenty years ago, the numerical weather prediction models on which today's Global Circulation Models (GCMs) are based could afford to have relatively inaccurate representations of the surface fluxes of energy and water: the average residence time for water vapour in the atmosphere is about a week, so the accuracy of weather forecasts a day ahead is not highly dependent on a good representation of the evaporation. However, as is evident from Part A, when these models are used for climate prediction, if the transfer of energy and water through the land surface is not correct, then the accuracy of the climate prediction will suffer. Yet the practicalities of global scale modelling mean that within these models the land surface of the Earth is divided into grid squares several hundred kilometres across. At each grid point the land surface must be represented by a simple, and computationally efficient, set of equations, with only a few parameters to represent the soil and vegetation. This need to represent all the biomes of the Earth, but at a coarse scale, created a demand for both new land-surface models and new data to inform those models.

The demand for data at the scale of the GCM grid square and the demand for data from all the major biomes on Earth created a new challenge to the land-surface experimentalists (see Shuttleworth 1988b). Previous data had almost all been collected in small plots, mostly in response to the demands of agriculture and hydrology in the developed world, while whole regions of the world were almost scientifically unexplored, particularly in the tropics and at high latitudes. The challenge was largely met by a series of coordinated international experiments (see Table C.1 and Shuttleworth 1991; André et al. 1999) in which all the components of the energy, water and, latterly, the carbon balance were measured simultaneously over an area comparable to that of a GCM grid square.

B.1.2 The Experimental Design

During the 1960s and '70s, there were a number of major international meteorological experiments held over the ocean, with a concentration of effort on the tropical Atlantic (see Garstang and Fitzjarrald 1999). These experiments were based on intensive measurement campaigns in which multiple teams made coordinated measurements from buoys, ships, balloons and aircraft. The design of the first Integrated Terrestrial Experiments (ITEs) held over the land surface in the 1980s owed much to the design philosophy of these atmosphere-ocean experiments, but naturally with an added emphasis on the vegetation and soil. The ABLE campaigns in Amazonia (see Sect. A.6.3 and Garstang and Fitzjarrald 1999) and HAPEX-MOBILHY (Hydrological Atmospheric Pilot Experiment – Modélisation du Bilan Hydrique; André et al. 1988) were pioneers.

HAPEX-MOBILHY was designed to address the scaling issue presented above: how to represent an area of mixed vegetation with a set of parameters simple enough to be used in a GCM? The experimental design was to measure all the components of the energy and water budget, and the meteorological and hydrological controls, feedbacks and outcomes. Other major ITEs which followed expanded on this issue, while at the same time adding data from different biomes. FIFE (First ISLSCP Field Experiment), in an area of temperate grassland, was designed to develop the methodologies of remote sensing (Sellers et al. 1992). EFEDA (European Field Experiment in a Desertification-Threatened Area; Bolle et al. 1993) and then HAPEX-Sahel (Goutorbe et al. 1997a, see also Chapt. A.5) addressed the issue of desertification, in Spain and the Sahel respectively. These regions were thought to be vulnerable to the effects of climate change induced by human action, either directly through land-use change or indirectly through global warming. The emphasis on global warming and the feedback between vegetation and climate through the uptake and release of CO_2 resulted in the carbon balance being incorporated into GCMs, also creating the need to measure CO_2 fluxes. Knowledge of the boreal forest of Canada (Sellers et al. 1995a; Hall 1999) and Europe (Halldin et al. 1998, 1999, 2001) was added by the BOREAS (Boreal Ecosystem-Atmosphere Study) and NOPEX (Northern Hemisphere Climate-Processes Land-surface Experiment) field studies respectively (see also Chapt. A.7). In these experiments the advances in technology, which now allow continuous operation of flux measuring systems, were exploited to move from campaign-based experiments to routine monitoring. In the most ambitious experiment to date, the Large Scale Biosphere-Atmosphere Experiment in Amazonia (LBA) has introduced a holistic approach to the study of Amazonia, measuring the functioning of the Amazon tropical forest as an integrated whole (see Chapt. A.6).

B.1.3 Guide to Part B

In this part of the book we examine the progress made by this series of integrated land-surface experiments, assess our current ability to measure the components of the land surface at different space and time scales, and identify the unsolved problems. No attempt has been made to produce a consensus; the individual chapters should be regarded as critical reviews from the personal viewpoints of the authors. Some opinions may be regarded as controversial, and some issues may be raised which we often prefer to ignore – but we believe that these issues are best presented and discussed openly.

There are still unsolved measurement problems and limitations in our ability to measure surface fluxes. Energy closure, the difference between the net radiation entering the surface and the fluxes of energy leaving, is a critical test of surface flux measurements. How this test should be applied is discussed in Chapt. B.2. Our ability to measure net radiation itself is discussed in Chapt. B.3, which looks at how net radiation is measured and the shortfalls in the accuracy of sensors and their calibration. Chapter B.4 charts the progress we have made in applying micrometeorological techniques to the routine measurement of surface fluxes. At the larger scale, atmospheric boundary layer techniques are used to integrate the components of a varied landscape: the problems and pitfalls of using aircraft are discussed in Chapt. B.5, and of using boundary layer budgeting in Chapt. B.6. The way in which vegetation controls the fluxes of energy, water and carbon makes understanding the structure and behaviour of plants the key to producing accurate models of these fluxes. Chapter B.7 discusses the role of vegetation measurements and the contribution they have made. At the regional scale only remote sensing can give the large two-dimensional pictures we need both in designing the experiments and in scaling up the results: application of remote sensing is discussed in Chapt. B.8. Chapter B.9 assesses the role of the catchment hydrology in these experiments. Chapter B.10 reviews how far and how successfully the results have been applied in large scale modelling. Finally we assess the lessons that have been learned (Chapt. B.11).

Chapter B.2

The Energy Balance Closure Problem

Alistair D. Culf · Thomas Foken · John H. C. Gash

The equation representing the energy balance of the Earth's surface is a fundamental component of all models of land-surface/atmosphere interaction. In its most common, simplest form, as applied over bare soil or short vegetation, this equation states that the available energy at the surface (the net radiation (R_n) minus the ground heat flux (G)) is equal to the sum of the sensible heat flux (H) and the latent heat flux (λE, where λ is the latent heat of vaporisation)

$$R_n - G = H + \lambda E \qquad (B.1)$$

(In this chapter the commonly used convention of treating radiative fluxes as positive downwards, and other fluxes as positive when directed away from the surface has been adopted). In some situations, when significant amounts of energy are stored in the biomass, or within the canopy space of tall vegetation, or when snow melts, extra terms may be added on the left hand side of Eq. B.1 (e.g. McGregor and Gellatly 1996), but the principal that energy is neither created nor destroyed at the Earth's surface remains the same. The so-called energy balance closure problem is that when the components are measured separately in the field, Eq. B.1 rarely, if ever, balances, indicating either an inability to measure or calculate the individual terms accurately, or an incomplete understanding of the physics of the system. Experimen-

talists often use the surface energy balance as a measure of the quality of their measurements, sometimes by examining the size of the ratio $(H + \lambda E)/(R_n - G)$, often referred to as the "recovery ratio", or by calculating the sum $(R_n - G - H - \lambda E)$. A dataset in which the energy balance does not close to within ±10% is often considered to be unreliable, although, of course, an apparently good recovery ratio can mask large, compensating errors in the individual components of the balance. Atmospheric modellers, trying to use field data for calibration or validation have to deal with the inconsistency between the perfect energy closure demanded by the models and the poor closure often delivered by the data.

B.2.1 Examples of Energy Balance Closure in Field Experiments

There are many publications giving field results of energy balance closure. The differences in instrumentation, software, quality control, etc. used in the different experiments make a systematic review extremely difficult but a selection from various studies are presented here to illustrate the wide range of results that are obtained.

The results from the first energy balance measurement campaigns were reported by Russian expeditions (Elagina et al. 1978; Elagina et al. 1973; Orlenko and Legotina 1973; Tsvang et al. 1987). They closed the energy balance to within 80% and suggested that problems with the eddy covariance measurements and horizontal advection were the probable reasons for the shortfall. Later, Bolle et al. (1993), Braud et al. (1993) and Foken (1990) showed that the storage of energy in the canopy and the upper soil layer were also contributing factors. Panin et al. (1996) combined the results of the first international land-surface experiments FIFE (Kanemasu et al. 1992), KUREX-88 (KURsk EXperiment; Tsvang et al. 1991) and TARTEX-90 (TARTu EXperiment; Foken et al. 1993) and proposed that the difference in energy closure ranging from 90 to 60% might be caused by heterogeneities in the surrounding area.

Recently, it has become possible to make energy balance measurements for much longer periods of several months to years. Greco and Baldocchi (1996) present a

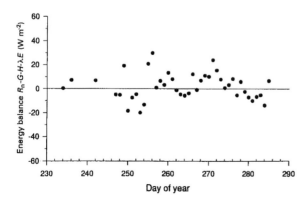

Fig. B.1. The residual of the energy balance on a daily basis for measurements made over savanna vegetation in the Sahel (after Kabat et al. 1997a)

year of energy balance measurements over temperate deciduous forest. From a statistical test on the summation of the net radiation and energy balance components on days with valid data, they concluded that there was no significant difference between net radiation and the sum of the components. However, a later study at the same site (Wilson and Baldocchi 2000) obtained worse results, with the sum of λE and H being only about 80% of the net radiation on an annual basis. Kabat et al. (1997a) present energy closure results over an area of patterned woodland ("tiger bush") in the Sahel. The results (see Fig. B.1) are presented as the daily residual ($R_n - G - H - \lambda E$) of the energy balance and, although there is scatter of up to 30 W m^{-2}, the average residual over a 48-day period is just 11 W m^{-2} in a system where the daily average net radiation may be as large as 150 W m^{-2}. Measurements of the surface energy balance of a boreal black spruce forest were made by Jarvis et al. (1997). Over the entire growing season of 120 days, they closed the energy budget to within 3%, with the energy stored by photosynthesis accounting for 1% of the net radiation and the soil heat flux for 3%. Recently, Aubinet et al. (2000) presented some energy closure data from six EUROFLUX forest sites as plots of the sum of the turbulent fluxes ($H + \lambda E$) against the available energy ($R_n - G$) (see Fig. B.2). The slope of the regression lines on these plots varied from 0.99 in the best case to 0.70 in the worst. The authors were unable to find any systematic difference in the sites or instrumentation to account for the variation. The root mean square error of these hourly data was of the order of 60 W m^{-2}. These results are fairly typical of those currently being obtained from many long-term monitoring sites.

Energy balance results over tropical forest have been presented by a number of authors, beginning with Shuttleworth et al. (1984) who obtained satisfactory closure of the energy balance (93% on average for the eight days of data presented). Wright et al. (1996) used the Institute of Hydrology Mk 2 Hydra (Shuttleworth et al. 1988) over tropical forest and generally obtained recovery ratios higher than 100%, while Grace et al. (1995b) using an Edisol system (Moncrieff et al. 1997) at the same site obtained values of less than 100%. When averaging with a short time constant digital filter Malhi et al. (2002) were initially unable to close the energy balance to a satisfactory degree, but when they abandoned detrending they obtained good closure. Results from the Hydra have been presented by a number of authors over vari-

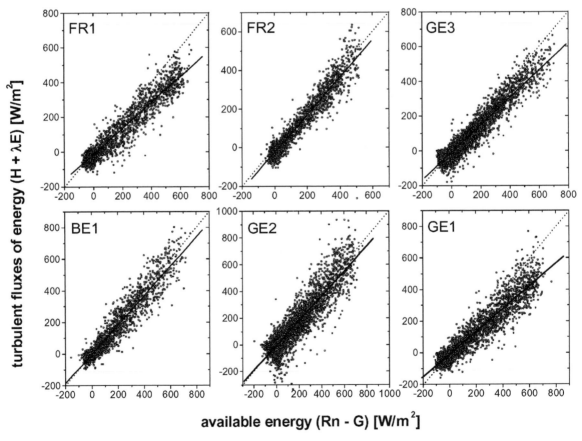

Fig. B.2. Half hourly measurements of the energy balance made at six forest sites during the EUROFLUX long-term measurement programme illustrating the wide range of results obtained, even with relatively standardised instrumentation, but also the general underestimation of the turbulent fluxes (as presented by Aubinet et al. 2000, © reproduced with permission from Academic Press)

ous vegetation types. Finch and Harding (1998) measured the energy balance components over pasture for a year and obtained good energy closure, with the sum of the net radiation being within 5% of the sum of the turbulent fluxes over the entire year and with good agreement between the acccumulated evaporation and independent measurements of soil moisture depletion.

Oncley et al. (2000) carried out an experiment over irrigated cotton in California specifically to address the energy closure problem. The experiment was designed to tackle several aspects of the problem, from turbulent flux measurement errors to the influence of heterogeneities and the storage term. Initially only 80% closure was obtained, but when the storage term was taken into account the energy closure was much improved.

B.2.2 Reasons for Poor Closure

There are many possible reasons for obtaining poor energy closure when making measurements of the surface energy balance, ranging from measurement errors associated with the individual instruments to the inability of the methods used to measure certain physical phenomena. Some of these effects are discussed in the following sections.

Net Radiation Sensors

Although net radiation is a fundamental parameter in the land-surface/atmosphere interaction, its accurate measurement is problematic and is discussed in some detail by Halldin (Chapt. B.3, this book). In fact, the problem of poor energy closure may seem to be greater now than 15 years ago partly because some of the net radiometers used in those early experiments underestimated the radiative flux substantially – an error which compensated for underestimation of the turbulent fluxes by eddy covariance. In FIFE a side-by-side comparison of net radiometers used at 22 surface sites (Field et al. 1992) revealed differences of 5 to 7% for instruments of the same type and of 10 to 15% for different types. Hodges and Smith (1997) compared the net radiometers used in BOREAS and found that the most commonly used radiometer in that study underestimated net radiation by about 5% in the day-time and 45% at night. Halldin and Lindroth (1992) compared the performance of six types of net radiometer and found considerable variation in performance and in the accuracy of manufacturers' calibrations. The response of the instruments in the longwave range was particularly variable and they concluded that a standard calibration procedure in this range was required before progress could be made. At the same time the World Meteorological Organisation (WMO) made an effort to reduce the errors of all radiation components

and to establish the Basic Surface Radiation Network, (Gilgen et al. 1994). In the case of long-wave radiation this was realised by a pyrgeometer (Eppley, USA) with additional measurements of the dome temperature made by the World Radiation Centre in Davos, Switzerland (Philipona et al. 1995). The error of this device should be less than 3%. Presently available net radiometers, such as the CRN 1 from Kipp & Zonen (Netherlands), underestimate the net radiation by about 3–5%, a much smaller error than older types such as the Radiation Energy Balance Systems Q-series or the Schenk sensor which commonly underestimated by 10–20%. However, practicalities of project finance mean that many older instruments are still in service.

The best designed instrument can still perform badly in the field if it is not well maintained. Dirty, scratched or otherwise damaged domes on net radiometers will adversely affect measurements. As with other types of instruments, regular maintenance is essential if high quality measurements are required.

Windspeed and Temperature Measurements

In the past 10 to 15 years the commercial availability of reasonably priced sonic anemometers has largely eliminated the need for individual research groups to develop their own. Whilst this ready availability has undoubtedly led to increased research activity, it has also led to an increased reliance on quoted specifications. The results of calibrations of anemometer windspeed or sonic temperature measurements are rarely reported in current energy balance studies, yet Mortensen (1994) for example, showed in an intercomparison experiment that a commonly used anemometer (Solent 1012R2, Gill Instruments, Lymington, UK) underestimated the vertical windspeed by 12%. This is a typical error for sonic anemometers which have a mounting below the measuring path of the vertical wind velocity (Foken et al. 1997). Distortion of the flow by the sonic anemometer itself can cause large errors in flux measurements, mostly due to cross-talk between the horizontal and vertical velocity components (Wyngaard 1988). Flow distortion has been shown to be a problem for all sonic anemometers, especially omni-directional probes like the Solent which are used in long-term measurement programmes. For fundamental research, probes with a selected sector of small flow distortion, like the Kaijo Denki Probe A or the CSAT 3, should be used (Foken and Oncley 1995). Obviously, the addition of extra bulky instrumentation to measure water vapour, carbon dioxide or other scalars, close to the anemometer head, is likely to exacerbate the flow distortion problem.

Use of the so-called "sonic temperature", routinely provided by the commercial instruments, is now relatively commonplace, and has the advantage over thermocouples or other sensors in that it requires no additional instrumentation which would obstruct the flow and also makes the system more robust for long-term unattended measurement. However, the accuracy of the measurement often remains unchallenged. In theory, the sonic temperature is nearly equal to the virtual temperature (Kaimal and Gaynor 1991). Therefore, the calculated flux is the buoyancy flux and not the smaller sensible heat flux, which has to be calculated (Schotanus et al. 1983).

Water Vapour Fluctuation Measurements

Water vapour measurements are usually made with an open path fast-response hygrometer, or by using a closed path system in which air is pumped from the anemometer head to a laboratory infra-red gas analyser. The relative merits of both types of system are discussed in detail by Leuning and Judd (1996). Drift in the measurement by several open path sensors has been shown to be a problem (Foken and Oncley 1995). In some instruments (Shuttleworth et al. 1988) this problem is avoided by making a separate measurement of background humidity with a slower response, but more stable instrument.

The need to measure carbon dioxide fluxes in addition to water vapour and sensible heat has led to increased use of closed path systems to measure rapid fluctuations in water vapour and carbon dioxide (e.g. Moncrieff et al. 1997 and Chapt. B.4). Whilst these systems have the advantage that they are created from readily available components and the Webb correction (which accounts for the effects of density fluctuations) is minimised by damping out temperature fluctuations before the measurement is made, they do introduce other uncertainties into the calculated flux. For example, the delay time of the air in the tube must be calculated accurately. This time may depend on the cleanliness of the sampling tube, and there is a general consensus that the tube should be changed or cleaned regularly. The frequency response of the system to water vapour fluctuations has also been shown to deteriorate rapidly with dirty tubing (Leuning and Judd 1996). However, tube cleaning or renewal is rarely reported and, in many experiments, it is not carried out regularly if indeed at all. Calibration practices also vary widely between groups, with some using sophisticated automatic calibration procedures several times a day (e.g. Goulden et al. 1996) and others preferring to calibrate the analyser on a weekly basis only (e.g. Malhi et al. 1998) or longer (Valentini et al. 1996).

Both open and closed path systems perform badly during and immediately after rain. With open path systems rain droplets obscure the optical path during rain and remain on the lens of the hygrometer after rainfall.

In closed path systems water may be sucked into the sampling tube during rainfall, resulting in severe damping of the humidity fluctuations (Moors 1999, pers. comn.) until the tubing dries out. Energy balance closure for periods including rainfall is therefore likely to be poor and these periods should be excluded from any assessment of the dataset based on energy balance criteria.

Soil Heat Flux and Heat Storage Measurements

When compared to the effort devoted to the measurement of the turbulent fluxes, the effort put into the accurate measurement of the soil heat flux or heat storage is often minimal. In percentage terms, the error in the soil heat flux is probably the largest of any of the components of the energy budget under most conditions, and can be as large as 50%. However, because the soil heat flux is typically only about 10% of the net radiation, the impact of this error on the budget is relatively small and the returns on investing heavily in improving the measurement are likely to be small.

If the soil heat flux is not measured with soil heat flux plates very close to the surface, soil heat storage should be included in the energy balance equation, since the energy storage in even the upper 20 mm of the soil can be as large as 40 $W m^{-2}$ (~ 20% of the net radiation). Kukharets and Tsvang (1999) used soil heat storage to partially explain the non-closure of the surface energy balance by taking into account the fact that soil heat capacity is dependant on soil moisture (Peters-Lidard et al. 1998). The change in energy stored in the soil was investigated during an experiment in 1997 at the Lindenberg Meteorological Observatory in Germany (Kukharets et al. 1998). Soil heat storage was calculated for five minute intervals at a depth of about 20 mm and was found to be nearly equal to the residual of the surface energy balance. On a day of strongly changing cloudiness, about two-thirds of the residual was explained by soil heat storage. The heat storage and energy balance residual terms showed non-steady state changes, which are not seen using 30-minute time series, and a phase shift between the two parameters. This phase shift was investigated explicitly during the solar eclipse on 11 August 1999 in Germany (Foken et al. 2001). In that experiment there was a phase shift between the changing radiation conditions and the increase of turbulent exchange processes after the totality of about 20 minutes. When the timing of the measured turbulent fluxes was adjusted to remove this phase difference, the energy balance closure was improved. Such large phase differences do not occur normally, but phase shifts of a few minutes would explain some uncertainties in the energy balance closure.

The problems of measuring soil heat flux and soil heat storage are sometimes circumvented, for the purpose of evaluating the quality of turbulent flux measurements, by assuming that the net amount of energy going into storage over a period of 24 hours or several days will be negligible and can be ignored. This assumption is another possible source of error. There may be large changes from day to day unless weather conditions are particularly stable. Even in areas of the tropics in which conditions are very similar from day to day, the soil may follow a warming cycle after rainfall which takes several days. To evaluate the accuracy of fluxes at shorter timescales, the storage terms certainly need to be evaluated carefully.

In forests with dense canopies, little solar radiation reaches the forest floor and so the soil heat flux and the soil heat storage are much less important than under short or sparse vegetation. However, considerable storage of energy can occur between the forest floor and the height of the turbulent flux measurements, both in the air and in the biomass. If eddy covariance measurements are made at a height of 20 m above a 30 m forest canopy, the energy stored in the air below the level of the eddy flux measurements between early morning and late afternoon could be as much as 3% of the net radiation. Again, as with soil heat storage, this storage term should be considered if accurate measurement of the energy budget is required.

B.2.3 Experiment Design

In addition to the measurement errors associated with each of the different components of the energy balance equation, the experiment design can have a significant impact on the energy closure. The most significant problem in most cases is reconciling the different horizontal scales for which the different component measurements of the budget are representative. An overview of these scales is given in Table B.1.

Sensible and latent heat fluxes measured by eddy covariance are not point measurements but averages over an area known as the fetch or flux footprint (see Chapt. B.4). The size of the fetch, usually defined as the distance from within which x% of the measured flux emanates, can be calculated using the scheme of Schmid (1997) or Schuepp et al. (1990), a modification of the method derived by Gash (1986a). For typical measurement heights over short vegetation of 2 to 5 m, the footprint is tens of metres long in the unstable case and up to 1 km in the stable case. Over forests, flux measurements are often made as high as 15 to 20 m above the canopy (Malhi et al. 1998, Aubinet et al. 2000) in which case the 90% effective fetch will extend to several kilometres even in unstable conditions, and to tens of kilometres at night. In some situations, when the vegetation cover is mixed or patchy, for example, the measurement height is deliberately increased so that the flux obtained will be an average over a larger area. However, as discussed below, this practice can lead to increases in the size of the corrections which must be applied to the data if sensor response and data processing algorithms are not also adjusted appropriately. In contrast to the turbulent flux measurements, the net radiation, soil heat flux and storage measurements (if appropriate) are often made close to the mast on which the eddy covariance instrumentation is mounted, and refer to a very small area.

Efforts are sometimes made to increase the spatial representivity of the net radiation, soil heat flux and storage measurements. For net radiation over short crops, several instruments distributed across the flux footprint can be used to assess the spatial variability. This approach was followed by Cain et al. (2001) who installed five carefully inter-calibrated net radiometers in an attempt to measure spatially averaged fluxes over a barley field. Differences in net radiation of up to 15% were recorded at different points in the field. In systems with separate component vegetation types, making separate measurements over each of the components and then creating an average, weighted according to the proportion of each type in the fetch, may be sufficient (Culf et al. 1993). These approaches are often impractical over forests however, because of the need for additional towers to get the instrumentation above the canopy. A moving trolley system has been employed in some studies to determine spatially averaged net radiation.

B.2.4 Calculation and Analysis Errors

Averaging Times and Running Means

In the eddy covariance method, fluctuations in the windspeed and various scalars are calculated by subtracting

Table B.1.
Typical horizontal scales of the terms of the surface energy balance

Term of the energy balance equation	Horizontal scale (m)	Measurement height (m)
Latent heat flux	10^2	$2\ldots\ 10 \times 10^0$
Sensible heat flux	10^2	$2\ldots\ 10 \times 10^0$
Net radiation	$10^0 \ldots 10^1$	$1\ldots\ 2 \times 10^0$
Soil heat flux	10^{-1}	$-2\ldots -10 \times 10^{-2}$
Storage term	$10^{-1} - 10^1$	

the mean value of the measurement in question from its instantaneous value. Multiplication and subsequent averaging of these fluctuations gives the various covariances required. The value to use for the mean in this process is a subject of some debate. For example, some researchers use an autoregressive running mean with a time constant chosen to eliminate slow variation due to diurnal changes or sensor drift, while other researchers favour the use of a linear detrend over blocks of data typically of 20 minutes to one hour in length. Rannik and Vesala (1999) compared several of these different methods and concluded that the use of the linear detrend is preferable. This is because using a fast response running mean – which is perfectly acceptable over short vegetation – can lead to underestimation of the flux if used over tall forest vegetation where the low frequency eddies play a larger role in the turbulent transport. Coordinate rotation is commonly applied to the wind velocity measured by three dimensional anemometers. This rotation is designed to compensate for non-level terrain and forces the mean vertical velocity to be zero, but this procedure can also act as a crude filter on long period eddies – too short an averaging time results in the calculated fluxes being too low (Malhi et al. 2002). The choice of running mean or linear detrend average length also impacts on the size of the frequency response corrections discussed below.

Although eddy covariance is the most direct way of measuring the turbulent fluxes, there is still a raft of corrections which must be applied to the raw fluxes (see Chapt. B.4). These corrections themselves introduce scope for further errors if applied blindly or incorrectly. The most complex and generally poorly understood of all the corrections which should be applied are those for frequency response. These corrections are applied since eddy covariance systems only respond to a limited band of the wide spectrum of atmospheric turbulence. At the high frequency end the response of the systems is limited by the frequency response of the various component sensors, while low frequencies are excluded by the choice of moving average or other detrending method. It is common practice to calculate an approximation to the flux which is expected to occur at the high and low frequencies missed by the system and to add this to the measured flux. The method generally used was first described by Moore (1986). For the non-specialist, the account given by Moncrieff et al. (1997) is more easily penetrable. Recently Horst (2000) pointed out some errors in previously published works on frequency response corrections which, although discussed in the literature, are still commonly made. In essence, the method involves calculating the theoretical spectrum for the particular instrument height, atmospheric stability and windspeed for the period under consideration, and determining the proportion of the spectrum which is outside the system range of response. The meas-

ured flux is then increased according to that proportion. The spectra commonly used for this procedure were in general derived from the Kansas experiment over grassland (Kaimal et al. 1972) and yet, because the correction routine is often embedded deeply in many eddy covariance systems, they are applied in every situation where the systems are used. Over tall, forest vegetation, the actual spectra may be shifted towards much lower frequencies.

Measuring at large heights in low windspeed conditions whilst using a relatively quickly responding running mean can increase substantially the size of the corrections which are calculated by the method described above. The theoretical spectrum in this situation shifts to very low frequencies and the correction factors generated by the frequency response routine become very large. For a month of recent measurements in Amazonia, under these types of conditions, 40% of the 30-minute average values of sensible heat flux included a frequency response correction factor of greater than 1.3, with low frequency corrections making a 24% contribution to the energy budget for that month. Malhi et al. (2002) have shown the sensitivity of calculated tropical forest fluxes to the analysis procedure. Energy balance closure changed from only 70% to over 100% by including transport at long timescales. Since the correction process is only an approximation, such large contributions can only increase the uncertainty in the final values obtained for the turbulent fluxes. Measurement height, expected windspeed ranges, and the frequency responses of the various sensors must all be considered together when planning a campaign to ensure that the size of the frequency response corrections is kept to a minimum.

B.2.5 Overall Accuracy – What Can Be Expected

As discussed above, and as with any measurement, each component of the energy budget equation has an uncertainty associated with it. Conservatively estimating these uncertainties to be 5% on the turbulent fluxes, 3% on the net radiation and 20% on the soil heat flux, we find – for typical values of the fluxes for temperate summer conditions – that the uncertainty in the value of the recovery ratio is around 9%. With a more pessimistic estimation of the errors as 10% on the turbulent fluxes, 5% on the net radiation and 30% on the soil heat flux, the expected error on the recovery ratio is of the order of 16%. A large scatter on the energy balance is therefore certainly not unexpected given the uncertainties in the individual measurements. Here, the errors on the individual terms have been assumed to be symmetrical. In reality, the measurements, especially of the turbulent fluxes and the net radiation, are more likely to be low than high – a tendency which may partly account for any bias towards low recovery ratios in field data.

B.2.6 Pushing Eddy Covariance past Its Limits

Up until now we have considered the aspects of the energy closure problem concerned with our inability to measure the terms of the equation with sufficient accuracy. In this section we will consider the possibility that the commonly used methods are inappropriate in some conditions.

Non-homogeneous Terrain

It is clear that non-homogeneous terrain should have an influence on energy balance closure. Firstly, a non-homogeneous surface should affect the eddy covariance flux measurements, a problem which can be resolved partly by a quality check of the data (Foken and Wichura 1996). Secondly, in a non-homogeneous landscape, horizontal heat exchange by advection will occur. It is not clear why this advection should always be an energy gain but numerical studies (Friedrichs et al. 2000) have shown that the turbulent fluxes increase over surfaces with small scale heterogeneties. If this flux is generated in the longwave part of the turbulent spectra, the eddy covariance technique is unable to measure it.

It is possible to re-write Eq. B.1 as

$$R_n - G = k(H + \lambda E) \qquad (B.2)$$

where k is a factor characterising the non-closure of the energy balance and the heterogeneity of the surface

(Panin et al. 1996). From analysis of the results of studies over various types of surface cover there is some evidence to suggest that k is a function of the shelter coefficient and has its largest values, in the region of 1.6, when measurements are made over bare soil or short grass. Over tall vegetation (e.g. Aubinet et al. 2000; Lee and Black 1993) better closure is normally achieved.

Non-turbulent Conditions

It is questionable whether the eddy covariance method should be used at all during the night when windspeeds are low and there are stable conditions which suppress the eddies (see Fig. B.3). However, many researchers do apply the method routinely in such conditions with varying degrees of success. One problem with application of the method at night, is that intermittent events, associated with the periodic break-up of the nocturnal boundary layer, can transport large amounts of energy even when the atmosphere is strongly stably stratified (Handorf and Foken 1997). The energy transport in such individual events may be measured more accurately by using wavelet analysis (Collineau and Brunet 1993a,b), rather than by traditional eddy covariance techniques. The problem of making accurate night-time flux measurements has, traditionally, been of little importance given the relative sizes of the day-time and night-time sensible and latent heat fluxes. Recently, however, it has become more critical with the need to measure carbon dioxide fluxes accurately in both day-time and night-time to determine net ecosystem carbon exchange. The

Fig. B.3.
High frequency vertical velocity data recorded by an eddy covariance system over tropical forest during (a) nocturnal, stable conditions and (b) day-time, unstable conditions to illustrate how turbulence at night is typically severely damped out by the atmospheric stability (data from Culf 2000)

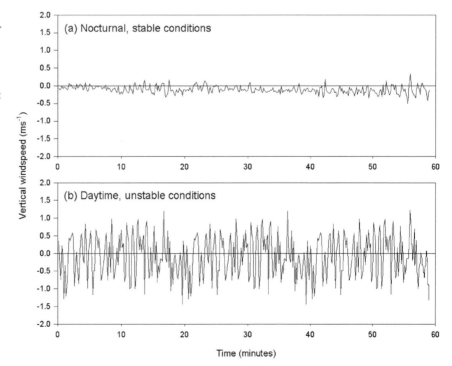

common underestimation of nocturnal carbon dioxide fluxes in calm conditions has prompted much research activity in this area (see also Chapt. B.4) and solutions may soon be forthcoming.

Large-scale Turbulence Structures

The high aerodynamic roughness of forests can lead to the generation of larger scale turbulent structures than typically occur over short vegetation. These coherent structures may be of a scale which is too large to be detected by the usual eddy covariance algorithms, but they may transport significant amounts of heat and momentum. The structures or "rolls" are often orientated along the mean wind direction and, therefore, a measurement system on a fixed tower may sample only part of the structure for long periods of time, possibly introducing additional systematic errors into the flux measurements.

B.2.7 Coping with Poor Energy Balance Closure When Modelling

The energy balance will not close perfectly at every timescale all the time even in the very best dataset from an ideal experiment site. The uncertainty inherent in every measurement makes this the case and, as we have seen above, even with conservative estimates of the errors in the components in the energy budget, there may still be a residual in the energy balance of ±10% of the net radiation which has to be dealt with in a sensible way by the modelling community. Sometimes, in a final dataset derived from an experimental campaign, the re-sidual of the energy balance is distributed, according to the Bowen ratio, to the sensible and latent heat fluxes. Another approach would be to accept that the measurements have an associated error and to run model calibrations or validations on a range of values encompassing these uncertainties to give an average model output which itself has a range of uncertainty. In an attempt to calibrate the Simple Biosphere (SiB) model for tropical rainforest, Sellers et al. (1989) tackled the problem in their optimisation routine by comparing the calculated and observed values of the evaporative fraction ($\lambda E / (\lambda E + H)$) rather than simply the sensible and latent heat flux alone.

B.2.8 Summary and Conclusions

The law of energy conservation is one of the most fundamental and absolutely certain physical laws and, therefore, its application in micrometeorology should not be abandoned lightly. It is important, in the face of apparent increasing ease of making complex surface energy balance measurements, not to neglect the rigour required to achieve good results. Equally, it should be realised that the techniques developed for uniform fetches and stationary conditions cannot be expected to perform as well over the complex terrain and in the rapidly changing conditions in which they are now commonly used. The size of uncertainties and possible errors should be quoted more frequently in the literature, whilst the modelling community must accept the existence of experimental error in the measurements that it feeds on and adopt sensible procedures when propagating these uncertainties through the models and into the output and predictions they produce.

Chapter B.3

Radiation Measurements in Integrated Terrestrial Experiments

Sven Halldin

B.3.1 Introduction

The radiation budget at the Earth's surface is central to all discussions about a possible global climate change. Such a change is believed to be triggered by the well-established increase in atmospheric content of radiatively active gases. The physical consequence of this increase is an increased longwave radiation of a few W m^{-2} from the sky to the surface. The direct detection of this effect would require longwave radiometers – pyrgeometers – with an accuracy better than 1 W m^{-2} maintained over a period of several decades. Such instruments and observational programmes do not exist and are unlikely to do so in the foreseeable future.

Radiation measurements in experimental studies of climate change have instead relied on an increased understanding of the processes governing the climate, e.g. the energy and water balances at the surface, and the functioning of the vegetation cover. All of the major integrated terrestrial experiments sponsored by IGBP and WCRP (World Climate Research Programme) have utilised radiation measurement for three main purposes. The most important has been to quantify the radiation part of the energy balance. Another purpose has been the measurement of radiation as a driving variable for photosynthesis, evaporation, and transpiration. Radiometry has also played an important role in tracing the atmospheric content of aerosols and trace gases for corrections of images from satellite sensors.

Development of radiometric techniques within the integrated terrestrial experiments has been a rather subordinate task in spite of the major role played by radiation, both as a driver of climate change and many exchange processes, and as the main component of the land-surface energy balance. Measurement problems have been larger for other factors influencing land-surface exchange. There have, however, been considerable uncertainties in integrated terrestrial experiment radiometry, with a growing awareness of these problems among the researchers involved. This chapter describes radiometry in the major integrated terrestrial experiments and focuses on the successes and failures in broadband radiometry aimed at understanding the energy budget at the land surface.

B.3.2 Radiometry in Integrated Terrestrial Experiments

In all the major integrated terrestrial experiments, the primary radiation measurements have been performed with commercially available equipment. No experiment has dealt with radiometer development. Some of the experiments have had a focus on remote sensing and these have stressed the importance of photometric monitoring to provide atmospheric corrections of satellite scenes. All the integrated terrestrial experiments have aimed at an understanding of fluxes over an area approximately the size of a GCM grid cell and have deployed radiometers over land surfaces ranging from 100 km^2 to 10 000 km^2. The spatial variability of radiation over such a region is of little importance if the problem is, for example, to feed assimilation models with radiation levels controlling the processes at individual study sites. When, on the other hand, the purpose is to understand the spatial variability of radiation fluxes, such a deployment is only meaningful if the radiometers have an accuracy better than that of the variability. This has been inherently assumed in some integrated terrestrial experiments. In others it has led to intercomparison projects where radiometers have been calibrated against a local or global standard. Such intercomparison projects have sometimes been seen as technical-support projects and not always worthy of publishing, so information about them is sometimes hidden in internal reports.

The amount of evidence from these different investigations (Table B.2) points in the same direction as presented by Culf et al. in Chapt. B.2, who elaborate the different reasons behind the non-closure of the surface energy balance in different integrated terrestrial experiments. Even if the relative errors are often smaller in net radiometry than in other energy-balance components, they result in an unacceptably large error in the net radiation, which is a dominant component of the energy balance.

Table B.2. Main results from calibration, comparison, and radiometer correction experiments in integrated terrestrial experiments

Integrated terrestrial experiment	Main results
FIFE (Field et al. 1992)	Significant site-to-site variation in observed radiation, attributable to systematic differences between instruments of different design.
NOPEX (Halldin and Lindroth 1992)	(*i*) No major progress will be possible before there is agreement on a longwave radiometric reference; (*ii*) there are large differences in the long- and shortwave responsivities between net radiometers of different designs; and (*iii*) manufacturer calibration factors are generally unreliable.
EFEDA (Van den Hurk and Malhi 1993)	Four net-radiometer instrument types compared relatively well between each other but with two unexplained exceptions. Since the net radiometers also differed considerably from the independently measured component-sum radiation, it was concluded that "a careful evaluation of the thermopile-instrument is certainly needed."
HAPEX-Sahel (Lloyd et al. 1997)	Two net-radiometer designs had an unexplained difference of 12%.
BOREAS (Hodges and Smith 1997)	"A standard make of net pyrradiometer, used at 15 of the 21 measuring sites, underestimates net radiation by about 5% in the day-time and 45% at night" and "The distribution of measuring sites and intercalibration errors are shown to produce distinctive patterns of bias errors in the space-time fields."
BOREAS (Smith et al. 1997)	A 25% range of variation in measured net radiation "presumably related to different methods of calibration used by different manufacturers"; "Fundamental deficiencies in the two-sided thermopile design used in most net pyrradiometers under non-equilibrium conditions … [that] cannot be corrected for in the standard calibration procedures…". The conclusion is that "there are such large intrinsic uncertainties with net pyrradiometer-type measurements because of the design of the instrument and the inapplicability of the calibration procedure to the actual measuring procedure, that such devices should not be used when conducting measurement campaigns requiring accuracy and precision levels consistent with climate monitoring." NASA is recommended to support development of an international standard and an internationally-sanctioned calibration facility for devices used to measure net radiation "to seek to improve the measuring technology."

Fig. B.4.
Dome transmissivities for net radiometers (*top*), pyranometers and pyrgeometers (*bottom*) along with short- and longwave, MODTRAN-derived (Moderate Resolution Transmittance Code), spectra from a clear midsummer sky at noon in the NOPEX southern research area (*middle*; courtesy of Bruno Dürr, Physikalisch-Meteorologisches Observatorium Davos/World Radiation Center). The spectral irradiances (normally given as $W\,m^{-2}$ per mm) are multiplied by the wavelength (mm; shown on a logarithmic scale) in order to keep the area under the curves proportional to the total irradiation

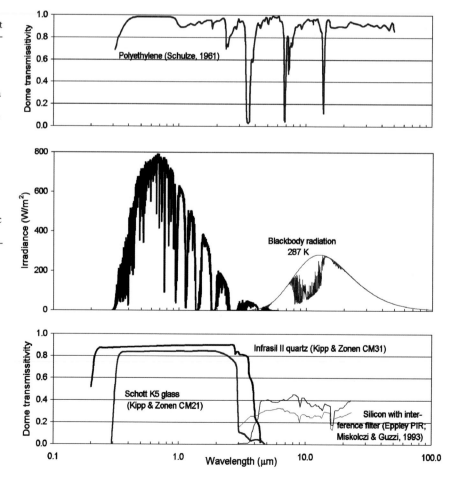

B.3.3   Available Radiometer Designs

Sun photometers are instruments used for spectral investigations of direct radiation from the sun. Depending on their purpose, they are manufactured with different selections of wave lengths and widths of wavelength bands. Photometers used for atmospheric corrections are manufactured in small numbers and few intercomparisons or evaluations have been published where many photometer designs have been involved. Intercomparison of photometers is a regular duty of the World Radiation Centre (Fröhlich et al. 1995) in Davos, but these investigations are concentrated on photometers that follow WMO wave-band guidelines.

There are basically two types of broadband radiometers aimed at measuring the whole shortwave and/or longwave radiation at the surface (Fig. B.4). The majority of commercially marketed instruments are bolometric, i.e. they measure the total amount of radiative energy falling on them. The majority of the bolometric radiometers use thermopile sensors of different design even if other sensors, e.g. Peltier elements could be conceived. The detailed sensor design is often a manufacturer's secret, so sensor principles are not known for all radiometers. A minority of broad-band radiometers is based on semi-conductor sensors. These are extremely rapid and have a wave-length-dependent response. They must, therefore, be provided with wave-length-specific filters for the type of application they are aimed at. If a radiometer of this type is used in an environment with a different spectrum than it was designed for, it will produce unpredictable results. The LICOR LI-200 pyranometer and LI-190 PhAR sensor are well-known examples of semi-conducting sensors.

Broadband, bolometric radiometers are available as pyrheliometers (narrow-angle instruments aimed directly at the sun), pyranometers (shortwave sensors), pyrgeometers (longwave sensors), and net radiometers (also called [net] pyrradiometers and radiation-balance meters). Pyrheliometers are used for two main purposes: to record the direct solar radiation and to make pyranometer calibrations traceable to the World Radiation Reference (WRR). Many pyrheliometers are, thus, designed as reference instruments and are not intended for long-term monitoring. The variability in output between high-quality pyrheliometers participating in the operational definition of WRR is normally better than four or five digits after the decimal point. The Eppley NIP is one instrument commonly deployed for monitoring purposes in a variety of weather conditions.

Pyranometers are available from a variety of manufacturers, e.g. Kipp & Zonen (a division of SCI-TEC since 1996), Eppley, Ph. Schenk, McVan (Middleton), Swissteco, and Astrodata. The Kipp & Zonen CM11/CM21 and Eppley PSP fulfill the WMO requirements of "Secondary Standards". Both have been thoroughly investigated by a number of independent researchers with respect to their accuracy in terms of e.g. time constant, drift, linearity of output, directionality, spectral selectivity, temperature response, and zero offset. They have, therefore, achieved a widespread use as high-quality reference instruments in both the integrated terrestrial experiment community and among other groups dealing with radiometry. The agreement on the World Radiometric Reference in 1981 was a prerequisite for the establishment of these two pyranometers as de facto standards. Before WRR there was a competition between the Smithsonian and the Ångström scales (differing ~ 5%) followed by the International Pyrheliometric Scale 1956 (IPS56) which was poorly related to the "true" SI value (Fröhlich et al. 1995).

Pyrgeometers are available from a few manufacturers, but only the Eppley PIR has received widespread acceptance. The Eppley PIR is a development from, and shares many features with, the Eppley PSP pyranometer. The major difference between the two lies with the domes. The PSP is covered by double domes of Schott optical glass (similar to the Kipp & Zonen CM11/CM21), which transmits shortwave radiation between 0.3 µm and 3 µm whereas the PIR is covered by a single silicon dome, covered on the inside by a vacuum-deposited interference filter that transmits longwave radiation between 4 µm and 50 µm. Because of the different nature of short- and longwave radiation, there is a general problem with the physically inconsistent way in which long- and shortwave calibration factors are commonly defined. Whereas the shortwave responsivity is a true quotient between the incident radiation and the voltage output of the instrument, the responsivity of the pyrgeometer is an approximation of the energy balance of the sensor surface. The lack of a "WRR for longwave radiation" and a generally accepted method for calibrating pyrgeometers have been serious drawbacks for the measurement of longwave radiation. There have, furthermore, been problems with the Eppley PIR. The most important one deals with the shortwave interference. Although the silicon dome only transmits insignificant amounts of shortwave radiation it is heated by it. This heating creates an internal convection which influences the energy balance of the sensor surface, creating a shortwave interference error that may reach 50–100 W m^{-2}, according to a variety of investigators. There is also a small but measurable amount of solar radiation above the traditional 3-µm threshold distinguishing short- and longwave radiation (Fig. B.5). This "longwave solar radiation" is normally small, but may in special circumstances amount to 20 W m^{-2} (Marty 2000). Depending on the quality of the pyrgeometer dome (see Fig. B.4) this may be a significant error that is best avoided by shading the dome from direct solar radiation. Another problem relates to the battery-driven, electrical circuit of the Eppley PIR, which is designed to mimic the longwave radiation

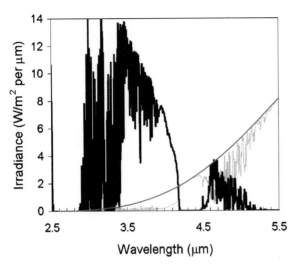

Fig. B.5. Spectral overlap of solar and atmospheric radiation for a clear midsummer sky at noon at the NOPEX southern research area (courtesy of Bruno Dürr, Physikalisch-Meteorologisches Observatorium Davos/World Radiation Center)

emitted from the sensor such that the instrument can give a voltage directly proportional to the incident longwave radiation. Several investigators have shown that this battery circuit is inapplicable for high-precision measurements. There is, furthermore, a bewildering variety of options for operating the Eppley PIR and many of its users are unaware of the right way to use the instrument to provide high quality data (Fairall et al. 1998).

Net radiometers are aimed at the simultaneous recording of incoming minus outgoing radiation in the combined long- and shortwave bands. There are two main types of net radiometers. The first type, seldom used in integrated terrestrial experiments, is constructed with separate upward- and downward-facing sensors, combined through a body with high thermal capacity and conductivity, the temperature of which is measured separately. This type of instrument is manufactured by Mario Käseberg and Ph. Schenk. The Mario-Käseberg (previously Bruno-Lange) instrument has been used as a high-precision pyrgeometer in some international intercomparisons and was suggested by Halldin and Lindroth (1992) as a *de facto* standard in net radiometry. The second type consists of one sensor exposed at both ends to radiation. Instruments of this type are available from a number of (mostly small-scale) manufacturers, e.g. REBS (a continuation of Micromet), Swissteco, Kipp & Zonen, Ph. Schenk, McVan (Middleton), and Astrodata. Most of these instruments are protected by a dome of polyethylene, which transmits both long- and shortwave radiation. Depending on its thickness, it may or may not be self-supporting. Thin domes need pressurisation and dry air/nitrogen to function properly. Polyethylene domes are vulnerable to bird attacks and cracking at low temperatures. Manufacturers recommend that they be changed

regularly but no serious investigation has been performed to study their directional properties, degradation because of weathering and long-term exposure to radiation, and other factors that are vital to their performance. The Kipp & Zonen instrument uses a conical-shaped teflon-coated sensor surface instead of a dome. The Astrodata sensors are not covered by anything but their highly absorbing paint. Calibration of net radiometers is a problem, not only because manufacturer calibrations are commonly unreliable but also because of the inherent sensor problems with differing responsivity in the long- and shortwave for covered instruments. This problem may possibly be solved with an internationally sanctioned calibration facility.

B.3.4 Overview of Radiometry in the Last Two Decades

Whereas the integrated terrestrial experiment community has primarily just been using radiometers, sometimes rather uncritically, there has been considerable work on development and evaluation of radiometers and radiometric practices within the International Energy Agency (IEA) and the Baseline Surface Radiation Network (BSRN) of WCRP. The key scientists in this community are the ones that developed and still maintain the WRR through regular intercomparisons (IPCs) at the World Radiation Centre in Davos. As strongly suggested by both Halldin and Lindroth (1992), and Smith et al. (1997), there is a need for an international standard for longwave radiation. Work on such a standard has been ongoing within the IEA/BSRN community during the 1990s. Early investigations showed that calibration factors for Eppley PIRs achieved with different methods at different laboratories could differ by 30–40%, but later developments of calibration procedures have demonstrated a much smaller scatter in responsivity (< 2%) of five roving reference instruments when calibrated at six of eleven laboratories worldwide during a two-year period (Philipona et al. 1998). This agreement relates to the responsivity of the sensor and not to the shortwave interference correction. There is now a concensus that high-quality longwave radiometry requires instruments that are ventilated and shaded from direct sun irradiation. Even with these precautions, it is difficult to reach an accuracy of ±6–7 W m^{-2} as requested by the TOGA/COARE (Tropical Ocean Global Atmosphere/Coupled Ocean-Atmosphere Experiment) community (Fairall et al. 1996) as a necessary requirement to correctly determine the surface energy balance over the ocean. Since the main problem seems to depend on the "dome correction", Philipona et al. (1995) have suggested a modification of the Eppley PIR, with additional measurements of dome temperature at three azimuthal angles and at 45° elevation, together with a new calibration procedure. These steps are assumed to give an accuracy of atmospheric and terrestrial longwave measurements down to ±2 W m^{-2}. Support for this statement has come through evaluation of longwave sky radiation measured with standard and modified instruments against calculations with LOWTRAN7 (Low Resolution Transmittance Code) and MODTRAN (Moderate Resolution Transmittance Code) (Ohmura et al. 1998).

B.3.5 Summary

The integrated terrestrial experiment community has identified measurement of net radiation as a weak point in the determination of the energy balance over different land surfaces. Even if many investigations have been carried out rather uncritically, the community has realised that there is a major need (*i*) for an internationally accepted longwave-radiometry standard, and (*ii*) that many of the existing net radiometer designs are not providing the accuracy and precision needed for the measurement of a satisfactory surface energy balance. There has also been a trend to favour one specific brand of net radiometers. This has been unlucky since this design has been shown to have inherent design problems that cannot be overcome by simple recalibration. This one-mark predominance has aggravated the problems of lacking energy closure, for example, since this brand has been used as a practical radiometrical reference in some intercalibrations.

The TOGA/COARE community, working with similar problems to those over land surfaces, have identified the need to measure the surface net radiation with an accuracy of ±6–7 W m^{-2} to determine correctly the total surface heat input to the ocean mixed layer. A similar accuracy would also be needed over land. To achieve this accuracy there is a need to establish a longwave standard. It is difficult to argue for a different standard than that proposed by the BSRN community, i.e. the modified Eppley PIR calibrated at the World Radiation Centre. If this standard is accepted (and this should not prevent it from being challenged and developed further), the development of net-radiometer technology would be based on more solid ground. Until a breakthrough has occurred in this field, it is suggested that the proper way of measuring net radiation is through the measurement of the individual components. Evaluation and intercalibration of instruments for doing this should receive high priority in the future.

Chapter B.4

Surface Turbulent Fluxes

John Moncrieff

B.4.1 Where Are We Now?

Micrometeorological methods to measure surface fluxes have several key properties which make them useful in global change research. The methods do not disturb the vegetation, they can operate for extended periods, and they provide a spatial average of fluxes over typically a few hundred to a few thousand m^2. Developments in both hardware and theory have moved the subject forward considerably over the past 40 years. We have moved from a position where micrometeorology was used in small research-led experiments, usually of quite short duration and in fair weather, to a point where fluxes are measured continuously and routinely at a number of sites around the globe as part of a flux network. In addition, surface flux measurements are no longer seen just as an interest in themselves but rather as a component of much wider research. Surface fluxes are required to validate process-based or empirical models operating at smaller and greater scales and they provide essential ground-truth observations for remote sensing. The changes in instrumentation and methodology are easily demonstrated by considering the series of large-scale experiments supported by BAHC and others. In the early HAPEX-MOBILHY experiment (1986) for instance, the dominant surface flux methodology was the aerodynamic gradient technique (with the exception of a Hydra eddy covariance station over forest, see Gash et al. 1989); by the time of HAPEX-Sahel (1992), eddy covariance systems were as important as gradient and Bowen ratio techniques and by the time of the BOREAS (Boreal Ecosystem-Atmosphere Study) experiments in the mid-1990s, surface fluxes were almost entirely measured by eddy covariance.

Although much progress has been made in the past 30 years in making surface flux measurements as accurate and precise as possible, it is accepted that there are limits as to what can be achieved by taking measurements under the relatively restricted set of conditions which are considered optimum for surface flux measurement – flat terrain, homogeneous surface vegetation and sub-surface conditions, with measurements being averaged over statistically stable time periods. Such ideals are rare if not impossible to meet and there is an increasing realisation that some fundamental re-thinking of the way in which we analyse our data is essential. This will probably mean that future measurements of surface flux will require another level of quite sophisticated analysis to operate in conjunction with the field measurements. This review summarises the major milestones in the path to our present understanding of surface fluxes, emphasising the main practical discoveries which have got us to where we are now. The view we shall reflect is that of practising experimentalists faced with making flux measurements for one of the flux networks.

B.4.2 How Did We Get Here?

Figure B.6 is a schematic to represent the progress made thus far and over the past decade. The first part of the diagram shows a typical daily course of the flux of carbon dioxide measured over a forest site. Such a diagram has commonly been published for many years and the observations could have been made by any one of a number of micrometeorological methods. Such information would have been interesting to plant physiologists researching the functioning of vegetation and its response to environmental factors. The second diagram represents an annual course of CO_2 fluxes and was obtained by an eddy covariance system. Such information would be useful to test larger scale models of C sequestration and plant growth. The third part of the diagram shows that we are now capable of making multi-year measurements to capture interannual variability in C source and sink strengths. This sort of information, in turn, is of immense interest to global change modellers who can use it to test their long-term simulations of the global climate.

To arrive at the point where we are now, there had to be a continual process of advance and refinement, a mixture of theory and technology-led advances in a number of disciplines. Table B.3 presents a subjective and highly selective list of some of "the usual suspects", the authors and their papers which have had the greatest influence on most flux groups who have started up in the past decade or so – these are the papers most often men-

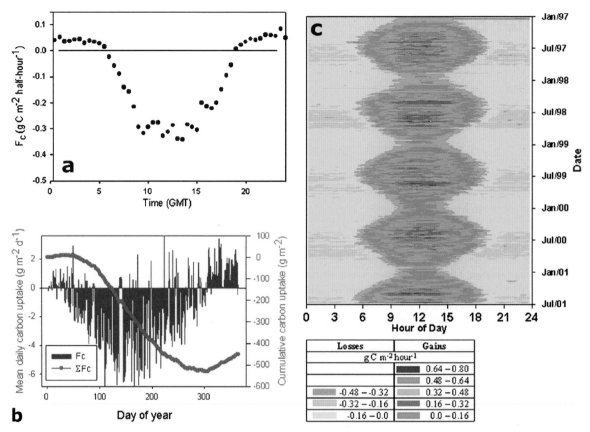

Fig. B.6. Carbon uptake at Griffin Forest, Aberfeldy, Scotland. An illustration of how progress has been made over the years in measuring CO_2 fluxes in the surface layer. (**a**) shows a typical daily curve of CO_2 flux as might have been measured 20 years ago by Bowen-ratio or gradient methods. The data come from an eddy co-variance system in May 1997 at a field site in Scotland. (**b**) shows a season's measurements at the same site and is the type of data-set being generated in the early to mid 1990s by the first automatic flux stations. The *narrow bars* represent the daily carbon uptake (g m^{-2} d^{-1}) from the start of 1997. The *heavy solid line* shows the cumulative carbon uptake (g m^{-2}) over the course of the year. The growing season was 255 days long at this site in 1997 and the forest sequestered 466 g C m^{-2} over the whole year. (**c**) shows that we now have the capability to make long-term flux measurements, capable of distinguishing interannual variability in C flux. In (**c**), each daily curve (as in (**a**)) has been colour-coded to reveal the seasonal change in net C flux at this site in Scotland

tioned in the flux literature. There are of course many others not mentioned here who have contributed greatly to the development of theory and instrumentation. The author suggests this as his personal "shorthand" when describing the major signposts and it is definitely not exhaustive, simply reflecting the author's own bias.

In addition, there are a number of excellent reviews of the methodology published previously and which cover the subject in more detail than space permits here (e.g. Raupach and Legg 1984; Baldocchi et al. 1988; Verma 1990; Lenschow 1995).

B.4.3 The Main Micrometeorological Methods

Three micrometeorological methods have dominated surface flux measurement over the past four decades. Two are indirect methods which relate the vertical flux to the gradient of the scalar of interest e.g. CO_2, H_2O or O_3. The aerodynamic gradient method combines scalar gradients

with the gradient of wind speed whereas the energy balance or Bowen ratio method uses the available energy, i.e. net radiation minus soil heat flux ($R_n - G$), in conjunction with scalar gradients to obtain the flux. Both these methods dominated the first three decades since the 1960s. In the past decade, the eddy covariance (or correlation) technique has grown in importance to the point where it is now far and away the predominant technique, particularly in long-term continuous flux studies. It is a direct method which measures turbulent fluctuations in wind velocity and scalars.

B.4.3.1 The Aerodynamic Gradient Method

The flux, F_χ, of scalar χ by the aerodynamic gradient method can be written in the generalised form

$$F_\chi = -K \frac{\partial \chi}{\partial z} \tag{B.3}$$

Table B.3. Some of the major papers on the practical aspects of surface flux measurements

Year	Lead author	Why it was influential
1951	Swinbank	First description of device to measure the fine scale of turbulence – leading to the development of CSIRO's 'evapotron' (Dyer and Maher 1965) and 'fluxatron' (Dyer et al. 1967).
1963	Kaimal and Businger	Described a practical sonic anemometer.
1970	Webb; Paulson	Described the main working equations in the aerodynamic gradient method and the corrections required for diabatic influence on wind profiles.
1972	Kaimal et al.	Quantification of some 'universal' spectral and co-spectral shapes.
1972	McBean	Produced working guidelines for eddy covariance measurements.
1975	Hyson and Hicks	Described an open-path H_2O instrument for eddy covariance.
1975	Thom	Gave a comprehensive treatment of the aerodynamic gradient and Bowen ratio methods and placed them in the context of flow above and within a vegetation canopy.
1977	Desjardins	First description of the eddy accumulation technique.
1980	Webb, Pearman and Leuning	Gave equations which would correct gas concentration measurements for the effects of changes in air density. Relevant in all surface flux measurements where density rather than mixing ratio is measured.
1982	Ohtaki and Matsui	Described an open-path fast-response CO_2 and H_2O analyser for eddy covariance.
1983	Coppin and Taylor	Described a 3-dimensional sonic anemometer which, although itself never went into commercial production, provided the model for most subsequent commercial developments.
1986	Moore	Described a system of transfer functions which could be used to correct for the inevitable inadequacies of all parts of a complete measuring system.
1986a	Gash	Described how fetch was influenced by transition over surfaces evaporating at different rates.
1986	McMillen	Probably first readily available description of a working eddy covariance system with emphasis on real-time processing including coordinate rotation and the influence of the length of averaging period on flux loss.
1988	Shuttleworth et al.	Described their highly successful 'Hydra' eddy covariance system which went on to be used in a large number of land-surface experiments over a range of ecosystems. One of the first continuous flux systems for (relatively) unattended operation.
1990	Schmid and Oke; Schuepp et al.	Following on from Gash (1986a), published equations which were easily adopted by the community to identify the source regions for scalar and flux footprints.
1990	Businger and Oncley	'Relaxed' Desjardin's eddy accumulation technique and introduced the simplification that air was sampled at constant flow rate into reservoirs simply according to the direction of the vertical wind, rather than at a rate proportional to it. Opened the way for the rapid adoption of the relaxed eddy accumulation method.
1990	Leuning and Moncrieff	The first of a series of papers which quantified the influence of ducted tubing in a closed-path eddy covariance system (Leuning and King 1992, Leuning and Judd 1996).
1993	Wofsy et al.	The culmination of much of what had gone before in that this paper was the first to report long-term (2 y), near-continuous measurement of the surface flux of CO_2 made by the eddy covariance method.
1994	Lenschow et al.; Horst and Weil	Established some ground rules for flux footprint ('How far is far enough?') and sampling strategies ('How long is long enough?') in surface flux measurements.
1996	Baldocchi et al.	Not really a methodology paper but one which pointed out that the future importance of surface fluxes lay within the context of observations made at a range of scales from cell to remote sensing and co-ordinated centrally.
1996	Foken and Wichura	Suggested various measures which could be used to assess the quality of data collected.
1998	Lee	The first of what is likely to be a series of papers from the community (e.g. Baldocchi et al. 2000) as surface fluxers are forced to re-assess their measurements in the light of the realisation that we are analysing our data taken at one point in space (the tower) yet the transport processes are three-dimensional.
2000	Aubinet et al.	The first of the FLUXNET networks to be established, this paper described the methodology adopted for eddy covariance measurements within the EuroFlux network.
2000	McNaughton and Laubach	Probably initiated a re-evaluation of Monin-Obukhov similarity theory, which, along with the work of Lee above, is likely to lead to a period of re-assessment of how surface flux data should be treated and analysed and what additional, if any, measurements need to be made in a complete eddy covariance instrumentation system.
2001	Falge et al.	An important paper in that it addressed the crucial question of how to 'fill in' for the inevitable gaps in data caused by system failure or inappropriate field conditions. Recognised that substantial differences in annual totals of carbon and water fluxes even within FLUXNET might be due to differences in how individual operators fill in their data gaps. Made some suggestions as to how to proceed.

where K is an eddy diffusivity for scalar χ. A considerable body of literature has been devoted to establishing how the eddy diffusivities for scalars and momentum are related and how they are affected by atmospheric stability (Dyer 1974; Kaimal and Finnigan 1994). In general, the method has been most successfully applied in campaign-mode (as opposed to long-term studies) and over short vegetation such as agricultural fields. Over tall, rough surfaces such as forests, gradients of scalars can be difficult to resolve, placing extreme demands on the precision of scalar analysers; in addition, the roughness elements themselves influence eddy diffusivites and the underlying theory may no longer hold. The layer of the atmosphere most affected by the elements of the canopy is known as the roughness sub-layer and extends to at least two to three times canopy height (Garratt 1978). For a tall forest, this may well be several tens of metres into the lower atmosphere. The practical difficulty of establishing towers or masts at such heights becomes an issue, as does the importance of staying within the inertial sub-layer of the atmosphere and not running out of upwind fetch. It is possible, however, to develop empirical corrections which permit the aerodynamic method to be used in the roughness sub-layer (Cellier and Brunet 1992; Mölder et al. 1999). There are also a number of corrections which have to be applied to wind and scalar profiles to allow for the influence of atmospheric stability on profile shape, but these seem to have been well established over a number of years (Paulson 1970; Garratt 1978). Nonetheless, the aerodynamic gradient method continues to be developed to measure fluxes of gases for which no fast-response sensor exists. Hall et al. (1999) use the gradient technique to measure fluxes of hydrogen peroxide over forest. Griffith and Galle (2000) describe a dual-beam Fourier Transform Infrared spectrometer (FTIR) which, when used in a profile system, could resolve differences in concentration to 0.2% of ambient and over a 20-minute period, and could resolve fluxes of 500 and 20 ng N m^{-2} s^{-1} for NH_3 and N_2O respectively. Whilst much of their instrumentation could be left unattended for up to a day, some technical and logistical issues remain but solutions to these promise to increase these resolutions even further.

B.4.3.2 The Bowen Ratio (or Energy Balance) Method

This is also an indirect method for obtaining surface fluxes and can be written as:

$$F_\chi = (R_n - G)\frac{d\chi}{dT_e} / \rho c_p \tag{B.4}$$

where R_n is the net all-wave radiation, G is soil heat flux and T_e is the equivalent temperature $= (T + e/\gamma)$ with

e being the vapour pressure and γ the psychrometric constant, 66 Pa K^{-1} (Monteith and Unsworth 1990).

There are many different formulations in the literature for this method. The method requires accurate measurements of both R_n and G, but this can still be a problem (see Chapt. B.3). The method fails when evaporation and heat flux are equal and opposite (often at dusk) and is difficult to apply when the available energy ($R_n - G$) is small (e.g. at night), but because it equates the diffusivities for scalars and does not use wind profile information, it can be used in the roughness sub-layer. A further advantage of the technique is that it does not require the semi-empirical stability corrections of the aerodynamic gradient method. Because sources and sinks of scalars may not be co-incident within a canopy, neither the Bowen ratio nor aerodynamic gradient method can be applied within canopies.

B.4.3.3 The Eddy Covariance (or Correlation) Method

When the second edition of Monteith and Unsworth's 'Principles of Environmental Physics' was published in 1990, the authors noted that the eddy covariance technique was still in its relative infancy, awaiting the development of suitable durable and weatherproof sensors, particularly for CO_2 and H_2O. In the decade since, this method has truly come of age and is now the predominant flux methodology, particularly for long-term flux measurements of carbon and water. In general terms,

$$F_\chi = \overline{w'\chi'} \tag{B.5}$$

where w' is the instantaneous departure of vertical wind speed from the mean value and χ' is the instantaneous departure of scalar concentration from its mean value. The overbar indicates that the product of these two terms on an instantaneous basis has been averaged over a suitable averaging period, typically 20–30 minutes. The equation is thus very simple but there is complexity involved in ensuring that the measuring system has fully recorded all the eddies – large and small – that have passed by the sensors during the averaging period. A number of correction terms have been identified and published and the community is reasonably satisfied with this procedure. Major steps to solve the problems were made by workers such as Kaimal et al. (1972), Webb et al. (1980), Moore (1986) and Leuning and co-workers, as described in Table B.3.

The reasons for the relatively late adoption of complete eddy covariance systems are largely technical-suitable sensors for measuring the rapid fluctuations in vertical wind speed and scalar concentration only became commercially available to the whole community at quite a late date. The period up to the late 1980s/early 1990s might be regarded as the period of "eddy-correlation specials" (e.g. Auble and Meyers 1992) in that available instruments were essentially "one-offs" and would often require quite a lot of user intervention. Nonetheless they mark a significant advance in the subject and instruments such as the "Lawrence Livermore National Laboratory Sensor" which was based on an original design by Bingham et al. (1980) were widely used (e.g. Anderson et al. 1984; Verma et al. 1986). After this period, commercial companies began to produce suitable instruments for the whole community. Instruments of particular note are the sonic anemometers from companies such as Gill Instruments (R2 and R3 models, Lymington, UK), ATI (e.g. "K"-series, Longmont, CO) and Campbell Scientific (CA27T, CSAT3, Logan, Utah); infra-red gas analysers also became available either in an open-(Advanced Systems, Japan, model E009, based on the design of Ohtaki and Matsui (1982) or closed-configurations (e.g. LI-COR 6252/6262 series). The open-path analysers are simpler to operate but present a bulkier profile to the turbulent wind near the sonic path; they also need to be corrected for density fluctuations caused by simultaneous heat and moisture transfer in the measured eddies (Webb et al. 1980).

Ducting the wind via a narrow tube placed near the sonic anemometer to a gas analyser at some metres distant has been the preferred method in most of the flux networks largely because no really suitable open-path analyser was available. For species such as carbon dioxide and water vapour, a number of systems have been described in the literature and their deployment is now almost routine (e.g. Grelle and Lindroth 1996; Moncrieff et al. 1997; Aubinet et al. 2000). Figure B.7 is a schematic of a typical eddy covariance system. The choice of LI-COR6262 has been almost universal in the flux stations up to about this year given that the instrument has a fast-enough response and relative stability in gain and offset. The disadvantage of the ducting method is that pumps are required and this will add to the power consumption of the system, an issue where tower locations are not near a main supply. LI-COR have recently introduced two new gas analysers specifically for the eddy covariance market, an open-path (Model 7500) and a closed-path (Model 7000) analyser.

Only a relatively restricted number of chemical species have suitably fast-response gas analyers available for them to be used in eddy covariance. There are many

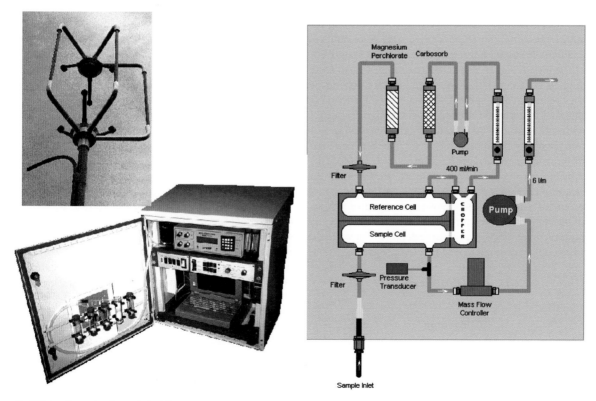

Fig. B.7. A schematic of a typical eddy covariance system. A sonic anemometer above the canopy measures the turbulent fluxes of horizontal and vertical wind speeds. Air is sucked down an inlet tube near the sonic head to a fast-responding infra-red gas analyser at the base of the tower. The expanded schematic shows the gas path within the gas analyser. A mass flow controller and pressure transducer can be used to maintain a constant rate of flow down the sample tube (and hence constant lag of gas sample between the sonic head and optical bench of the infra-red gas analyser). Gas concentrations in the sample cell are measured relative to a reference cell I in which air is dried and scrubbed of carbon dioxide

other species for which "eddy correlation specials" remains the only solution e.g. CH_4 (Verma et al. 1992), O_3 (Massman and Grantz 1995) NO_2 (Wesely et al. 1982; Coe and Gallagher 1992) and isoprene (Guenther and Hills 1998). It is possible to use devices such as tunable diode lasers to measure a large number of trace gases (e.g. the TGA100 from Campbell Scientific Inc. can measure concentrations of N_2O, CH_4 and NH_3 and others, at 10 Hz), but these instruments are still research-grade and expensive (Kim et al. 1999).

Just as important as the technology, however, is the progress made in analysing turbulence data and, in particular, correcting flux estimates for the inevitable shortcomings in the measurement system.

B.4.3.3.1 *Total System Losses*

It is not possible to create an eddy covariance system which has no effect whatsoever on the fluxes it is designed to measure. For example, instruments do not have an infinitesimally small response time, nor can nearby co-located instruments sample exactly the same parcel of air simultaneously (with the exception of sonic temperature and wind speed in the calculation of sensible heat flux). As for actually digitising and recording the signal, the sampling frequency and time constraints of analogue or digital filters also influence the magnitude of the loss of signal. The corrections which need to be applied are frequency-dependent, i.e. they depend on the spectral and co-spectral shapes. A small fraction of the total flux may be carried by small, high-frequency eddies that cannot be detected by the measuring system. Typically, the approach is to ascribe this "flux loss" to its component origins by a series of transfer functions. A transfer function is simply a multiplier in the range 0–1 which varies with frequency. Moore (1986) proposed a scheme whereby a series of transfer functions could be defined for each of the correction terms required in an eddy covariance system. We can write the fractional error in measured flux ($\Delta F_\chi / F_\chi$) as

$$\frac{\Delta F_\chi}{F_\chi} = 1 - \frac{\int_0^\infty T_w \rho\chi(n) C_w \rho\chi(n)\,dn}{\int_0^\infty C_w \rho\chi(n)\,dn} \qquad (B.6)$$

where $T_{w\rho\chi}(n)$ is the convolution of all the transfer functions applicable to the measurement and $C_{w\rho\chi}(n)$ is the cospectrum of the flux F_χ. n is the natural frequency. Having obtained the correction terms we now have to obtain representative co-spectral models which can be applied to estimate the whole system losses. The most widely used cospectral models are those from Kaimal et al. (1972) and we use normalised versions in which the numerical integrals of the normalised co-spectra equal

unity over an unspecified frequency range as described by Moore (1986). There are situations when applying a universal "Kaimal" spectrum is inappropriate and then it is necessary to define unique spectral and co-spectral shapes, sometimes even making on-line corrections using these tailor-made models (Eugster and Senn 1995).

The eddy covariance method is now used almost universally to produce routine and continuous fluxes of CO_2, H_2O, heat and momentum. Under the umbrella of FLUXNET, about 85 stations world-wide are using very similar technology and software to log and analyse data. Within each of the flux networks, there are programmes of intercalibration and standardisation that quantify the degree of uniformity across the networks. It can be argued that the technology is now mature enough such that measurements are routine and dependent on funding rather than babysitting instrumentation. This is a solution we have reached literally only within the past few years and as such is a bit of a departure for micrometeorologists. We have moved from a "fair-weather science" (in the main) to one where long-term fluxes can be obtained within the protocols of a network.

B.4.3.3.2 *Relaxed Eddy Accumulation*

For the scalars for which no fast-response gas analyser exists, the relaxed eddy accumulation method can be used. This is a technique related to eddy covariance in which air is sampled into two sampling reservoirs, one for up-draughts, the other for down-draughts. After a suitable interval of time, say 30 minutes, one sampling bag will contain air which is proportional to $w^+\chi^+$, the other containing air proportional to $w^-\chi^-$. The difference between the two (large) quantities gives the net flux. The vertical wind speed measured by a sonic anemometer is linked to two valves which open and close in response to the direction of the vertical wind. The requirement for a fast sensor to measure vertical wind speed remains, but there is a great relaxation in the speed requirement for the chemical analyser. A further advantage is that by accumulating gas into reservoirs the difference between the bags is enhanced and it is then possible to use high-precision gas analyses in the laboratory to determine the differences.

The method was first suggested by Desjardins (1977) and given the name eddy accumulation. Although attractive in principle, these early studies used the method whereby air was sampled at a rate proportional to the magnitude of the vertical wind speed and technically this was difficult to achieve with the required accuracy. The idea was taken up again by Businger and Oncley (1990) who wrote the flux for a gas with concentration χ, as

$$F_\chi = \overline{w'\chi'} = \beta(\zeta)\sigma_w(\chi^+ - \chi^-) \qquad (B.7)$$

where $\beta(\zeta)$ is a coefficient usually determined by experiment and a weak function of atmospheric stability (ζ), σ_w is the standard deviation of the vertical wind speed, and ($\chi^+ - \chi^-$) is the gas concentration difference between the two sampling bags at the end of the sampling period. Businger and Delaney (1990) discussed the resolution required in a chemical sensor to make it feasible to be deployed as part of a relaxed eddy accumulation system.

As Businger and Oncley point out, the measurement of ($\chi^+ - \chi^-$) is a direct difference of concentration not weighted by w as in the original method. Thus in this method they *relax* the conditions for the original eddy accumulation and the method is based simply on sampling air at constant volumetric flow rate into different reservoirs on the condition of either upward or downward moving air. The coefficient $\beta(\zeta)$ was found to be approximately 0.6 over a wide range of stability. The use of the semi-empirical beta factor worried some but it has since been put on a much sounder footing by its interpretation with the statistics of turbulence (Milne et al. 1999; Baker 2000). It is interesting to note that the number of papers being published in the year 2000 is almost an order of magnitude more than in the first couple of years of the 1990s. The relaxed eddy accumulation method has been used to measure a range of trace gas and other fluxes including, for example, herbicides (Pattey et al. 1995), methane and nitrous oxide (Fowler et al. 1995), isoprene (Bowling et al. 1999a) and terpenes (Christensen et al. 2000).

An interesting development of the relaxed eddy accumulation technique has recently been made by Bowling et al. (1999b) in which the usual vertical wind speed threshold used to open and close the sampling valves is increased, effectively permitting the system to collect only the strongest updrafts and downdrafts which contribute to the flux. The threshold is defined as a "hyperbolic hole" (H) and can be written:

$$H = \left| \frac{w'}{\sigma_w} \frac{c'}{\sigma_c} \right| \tag{B.8}$$

where w' and c' are the turbulent fluctuations of vertical wind speed and CO_2 concentration respectively; σ_w and σ_c are the standard deviation of vertical wind speed and CO_2 concentration respectively. A hole size is decided upon before sampling begins and the sampling valves are only energised when

$$\left| \frac{w'}{\sigma_w} \frac{c'}{\sigma_c} \right| > H$$

The advantage of this method is that differences in scalar concentration are maximised, thus increasing the likelihood of achieving a resolvable difference between the up- and down-sampling reservoirs. It is necessary

to obtain information on the scalar independently of the relaxed eddy accumulation method e.g. c' (or a similar scalar with similar probability distribution) could be obtained by simultaneous measurements by eddy covariance. Wichura et al. (2000) have shown how to use hyperbolic relaxed eddy accumulation to measure fluxes of the stable isotope ^{13}C over spruce forest.

B.4.3.3.3 Chamber Methods

Eddy covariance measurements of surface fluxes are usually made at some height above the canopy. What they sense is the sum of the individual contributions to the total flux arising from a variety of sources and sinks on and within the canopy and including the ground surface. In the sense of Fig. B.8, we are measuring above the upper system boundary. A flux measurement therefore cannot be used to partition the measured flux back into these sources and sinks – there is no way of knowing after the event what proportion of the flux comes from which part of the system. Information on the functioning of the vegetation must come from methods other than by eddy covariance. Chamber methods have a long history of being able to provide measurements of trace gas exchange, for example, on the scale of individual leaves, branches, trunks and even whole trees (Livingston and Hutchinson 1995). Figure B.8 shows schematically the fluxes which can be measured on the scale of the leaf by various leaf chambers such as porometers or at the scale of the branch in branch bags. Information on the respiration from the bole can be measured by stem chambers and the combined autotrophic and heterotrophic respiration from the soil (commonly just soil CO_2 efflux) by soil chambers of varying design (Norman et al. 1997). Such chamber methods have immense value in that they can be used to scale up to the whole canopy and can be used to verify fluxes of water vapour made by eddy covariance e.g. Saugier et al. (1997). Whilst it is difficult to conceive of catchment-scale closure for carbon, it is possible to integrate observations of evapotranspiration at a number of scales to attempt to close the water balance at this scale (Wilson et al. 2001).

Instrumentation for chamber methods is in general much less expensive to build than for micrometeorological techniques such as eddy covariance and it is easier to use (Rayment and Jarvis 1999). Because chambers are relatively small in physical size (typically 100–500 mm in diameter), the question of spatial variability and adequate sampling is problematic with all chamber methods and to achieve an adequate representative sample, these methods can be labour intensive. In addition, because the chambers are placed directly on or around the plant material, they may directly influence the energy and radiation balance of the system under study and hence affect the result. Of course such influences have been stud-

Fig. B.8.
Fluxes above and within the
canopy. The Upper System
Boundary (USB) marks the
level through which the vegetation exchanges carbon and
water with the atmosphere.
Direct micrometeorological
methods such as eddy covariance operate at this level. Below the USB A_1 and R_1 are the
net assimilatory and respiratory exchanges by the leaves.
The subscripts w, s and r refer
to respiratory fluxes from
wood, soil and roots respectively

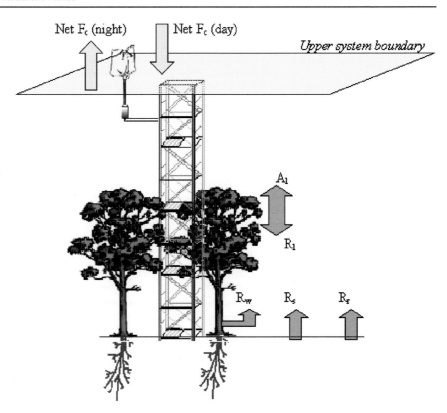

ied in great detail and practical methodologies have been
published which minimise these effects (Livingston and
Hutchinson 1995; Fang and Moncrieff 1998).

B.4.3.4 Flux Footprint

Fluxes measured by micrometeorological sensors are effectively the integration of fluxes from a variety of sources
and sinks in the landscape for a distance of several hundred metres upwind from the measuring point. The height
at which the measurements are chosen to be made must
be determined both by consideration of the frequency
response of the instrumentation and also the "fetch" or
extent of the upwind area from which the signal comes.
Eddies become progressively larger with height up to the
depth of the planetary boundary layer, typically 1–2 km
by day, and this means that instrumentation with a slower
response can be used successfully at heights well above
the vegetation. As the surface is approached, however, the
spectrum of turbulence includes a larger proportion of
smaller eddies that actively exchange mass and momentum between the surface and the atmosphere. The instrumentation must therefore be capable of sampling
high frequency eddies, typically up to 10 Hz. In principle
one could use an eddy covariance system well above the
canopy to avoid the problem of frequency response of
analysers. However, as we move up in height the area of
flux integration becomes larger and the requirements of

surface homogeneity become more and more stringent.
In fact if the instruments are placed too high above the
surface it is possible that they could extend out of the
boundary layer representative of the nearby vegetation
and be measuring some component of fluxes from a different type of vegetation further upwind. A convenient
rule-of-thumb suggests a fetch:height ratio of about
100 : 1; thus a fetch of 500 m would allow instruments to
be placed up to a height of about 5 m above the surface.
The fetch : height ratio depends on atmospheric stability
and surface roughness in so far as they influence the degree of mixing of internal boundary layers as they are
advected over different types of surface (Mulhearn 1977;
Gash 1986b; Grelle and Lindroth 1996).

One promising approach is to define a "footprint" or
source region that is a measure of the relative importance of sources upwind which contribute to the measured flux. This area can be regarded as contributing most
of the flux measured, and its areal extent and position
can be calculated from a knowledge of surface roughness, atmospheric stability, wind speed and direction. A
number of models are available concerning footprint
calculation based either on Lagrangian or particle trajectory models (Schuepp et al. 1990; Finn et al. 1996;
Flesch 1995) or analytical dispersion models (Horst and
Weil 1994; Schmid 1997). Figure B.9 shows a comparison between the analytic model of Horst and Weil (1992)
and a typical Lagrangian stochastic model (*bFlat*, used
at Edinburgh). Both models show the characteristic peak

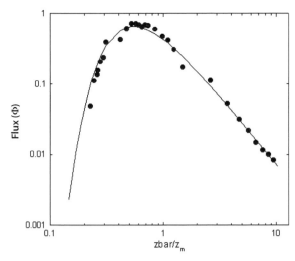

Fig. B.9. A typical "flux footprint" diagram. The *solid line* represents the analytical solution of Horst and Weil (1992) in which the relative contribution to the flux (Φ) varies with distance upwind from the source, here represented in the dimensionless form by the ratio of the mean plume height (zbar) to measurement height (z_m). The *points* are results from a Lagrangian stochastic simulation under the same conditions of neutral stability and parameterised to correspond to the simulation of Horst and Weil (1992)

relatively close to the origin (under the atmospheric stability and surface conditions chosen) with a tail extending out to greater distances upwind. At present, however, most flux footprint models are of limited application in the roughness sub-layer. Over heterogeneous surfaces, models based on Lagrangian rather than Gaussian dispersion are preferred. The question of flux footprint within the canopy remains unclear (Baldocchi et al. 1996) as does the size of the concentration footprint which applies to the gradient and Bowen-ratio methods of flux measurement (Horst 1998).

Ideally, the measurement site should be extensive, flat and horizontally uniform to ensure that fluxes are measured in the surface layer. A longer fetch is needed for stable conditions but the requirement may be relaxed under unstable stratification (Kaimal and Finnigan 1994). With an increase in measurement height, the peak of the flux footprint becomes more and more distant. Similarly, the peak contribution moves closer to the point of measurement as the atmosphere becomes more unstable. With increasing stability, the peak contribution moves further from the point of measurement. Formulations such as those of Schuepp et al. (1990) were easy to code and it became trivial to apply them to field configurations. One problem, however, with the flux footprint models based on analytical solutions such as the FSAM (Flux Source Area Model; Schmid 1997) is that they are generally restricted to use within the surface layer and also in a relatively restricted range of atmospheric stability. Quite often, they are used well outside these limits of applicability, simply on the grounds that

little else is readily available. The need to extend these limits e.g. for flux footprints for tall towers or for observations from aircraft, has spurred on the development of Lagrangian dispersion models which can cope with being "stretched" in this way e.g. the Lagrangian Stochastic Particle Dispersion Model of Rotach et al. (1996), and de Haan and Rotach (1998). This would appear to be the way forward for flux footprint modelling.

The benefits of being able to quantify source and sink regions for scalar concentrations and fluxes have been several-fold. One of the driving forces for such a theory was when the first International Satellite Land Surface Climatology Project (ISLSCP) experiments started and remote sensing equipment was placed on board airborne platforms such as helicopters; the observations they made of the surface had to be related to what the tower flux teams were measuring at the same time. Also, by defining an area on the ground from which the fluxes were emanating, meant that others could enter the flux footprint to make observations of plant physiology and soil efflux, thus bridging the scale gap.

B.4.3.5 The Nocturnal "CO$_2$ Loss" Problem

A cloud does exist on the horizon for surface fluxers, however. For some time it has been clear that some of the long-term flux sites have difficulty closing the energy balance during the day and also reporting "flux loss" at night. The latter point is best illustrated on graphs of measured eddy flux and soil efflux against friction velocity (Grace et al. 1995a). At low friction velocities, u^* (as a measure for turbulent mixing), the eddy flux underestimates the soil efflux yet the efflux is almost always accounted for at higher wind speeds. Apparently, the eddy flux tower is "missing" this flux. It is accepted that the eddy covariance instrumentation is not malfunctioning in any of the studies which report this phenomenon. There seems to be a wind speed dependency, usually expressed in a graph of CO$_2$ flux by chamber and eddy covariance on the y-axis and friction velocity u^* on the x-axis. Divergence between the two methods is greatest at the low u^* end and decreases as u^* increases, typically reaching parity at u^* values of 20–30 cm s^{-1}. Not all sites show this phenomenon and just what to do about it is guaranteed to cause an argument when micrometeorologists gather to discuss it. Some make no correction to their night-time data whilst others substitute the values of CO$_2$ flux in calm conditions with values estimated by a simple temperature function in well-mixed conditions (Goulden et al. 1996; Aubinet et al. 2000). Since sinks and sources for carbon are usually finely balanced (net (eddy) flux being the small resultant of the much larger fluxes of net photosynthesis and soil respiration), any errors in the method would affect the certainty which could be attributed to the annual carbon balance (Moncrieff et al. 1996). The search is on for a framework which could be used to correct the observations. The issue boils down to the fact that we are trying to interpret scalar budgets in a 2- or 3-D flow field yet our observations come from observations at a single point (Finnigan 1999).

Two principal arguments have been put forward to explain the phenomenon – that EC does not sample the intermittent turbulent events that can occur at any time of the day (and which can carry the majority of the flux) or that local drainage flows carry away the CO_2 respired from the soil before it can be transported up to the eddy flux sensor. This is unquestionably one of the "hottest topics" in the field at the moment. There has been an empirical approach to the drainage idea (e.g. Baldocchi et al. 2000) used the solution of Lee (1998) to improve their annual flux estimates) but it does not appear to be a generic solution. Staebler et al. (2001) have recently attempted to measure the drainage at night of airflow around their tower site but their early results demonstrate just how difficult the experiments and interpretation are going to be.

Solving the issue of nocturnal flux loss is essential as any error such as this has the potential to alter dramatically the annual net sums for the exchange of trace gases such as CO_2. Falge et al. (2001) used the first results from 28 datasets in the FLUXNET project and examined the implication of different methods of gap filling, including the nocturnal flux loss issue. Data coverage over a yearly period averaged out at 65%, thus indicating the scale of the problem. They also showed that different methods of gap filling, including "correcting" for nocturnal flux loss, could make substantial differences to annual totals for carbon and water exchange. For carbon, replacing nocturnal flux estimates with those obtained during windier periods, resulted in an average increase of carbon sequestered by about $75 \text{ g C m}^{-2} \text{y}^{-1}$ but it could be as large as $+200 \text{ g C m}^{-2} \text{y}^{-1}$ which, put in context, is about the same as an average net ecosystem exchange rate for a temperate forest. The implications of this study are that the community needs to find a solution to this problem quickly. The path to our salvation is likely to lie in developing our understanding of how re-processing of raw data by the various de-trending and co-ordinate rotation algorithms influences the trace gas and energy budgets. This is most likely to result in real low frequency contributions to the signal being rediscovered (see also Culf et al., Chapt. B.2).

B.4.4 The Future for Surface Fluxes

The above analysis is meant to suggest that, by and large, surface fluxes have "come of age" and can be measured with quantifiable levels of accuracy. It is probably fair to say that the uncertainties associated with surface flux measurements now depend more on our data analysis and interpretation rather than on the instrumentation used to obtain that data. Sonic anemometers and fast infra-red gas analysers for CO_2 and H_2O are now stable and accurate enough and their response characteristics are well enough quantified. The next few years will be an exciting time for micrometeorologists making flux measurements. It is likely that we will have to reconsider our analysis methods to establish "best practice"; the capability to store all raw data means that we can perform sensitivity analysis on all our stored data and re-process as and when required. It is possible that our measurements will have to be constrained by models of fluid flow and biology. It is also an exciting time for surface fluxers as they become involved in multi-disciplinary field experiments and their data are used in studies at a range of scales. It is becoming clear that such an integrated approach is required to resolve issues associated with global change, environmental pollution, etc. Making that sort of effort more routine is what is being planned now. One example of a recent call was Canadell et al. (2000) who argued for a top-down (air sampling network, inverse numerical methods and satellite data) and a bottom-up approach (surface flux at tower and aircraft, physiological process studies etc) to be integrated. Micrometeorology has gone through a number of step-like changes in our ability to measure and understand the atmosphere-biosphere interactions in the past four decades and it is likely we are heading for another decade of exciting advances and developments.

Chapter B.5

Accuracy and Utility of Aircraft Flux Measurements

Timothy L. Crawford · Ronald J. Dobosy

B.5.1 Introduction

Strategies to assess long-term atmosphere-ecosystem exchange of CO_2 and H_2O must deal not only with time trends but also with spatial variability. Flux-towers, always limited in number, efficiently measure time trends but the representativeness of a tower site – or the significance of spatial variation between sites – is best addressed through flux measurements from small aircraft (see Fig. B.10). Recent technological advances in aircraft and instruments allow airborne flux measurements to be made with enhanced precision, greater ease and lower cost. Challenges remain, however, in all aspects of the activity: instrumentation, data processing, and data interpretation.

Airborne eddy-flux observations obtained with modern instruments properly installed and operated on appropriate aircraft, will give results no less accurate than from a careful flux-tower operation. The primary difference is in how the data must be interpreted (Mahrt 1998). Tower data form a time series relying on mean wind to advect the turbulence past the sensors. An airplane, because of its speed, experiences turbulence more as a space series. The computed fluxes match best (as we will show) when conditions are homogeneous and

Fig. B.10. Atmospheric exchange with Arctic tundra north of the Brooks Range, Alaska. The nose boom supports a temperature and wind-gust probe, while CO_2 and H_2O are sampled by infra-red absorption in the open chamber (appears Π-shaped in the picture) just aft and offset from the boom's root (photograph by Edward Dumas)

stationary. However, spatial and temporal variations are the rule and thus drive spatial and temporal averages apart. It follows, therefore, that the best use of airborne data is to examine the rich spatial structure between and beyond towers, and to estimate non-local terms, such as advection, in the energy and mass budgets.

With the recent availability of low-cost systems on light aircraft, airborne measurements promise to become more prevalent. Nevertheless, they are practical primarily during daylight hours in brief, intensive field campaigns; extrapolation of day-time spatial structure to night-time conditions or as long-term information remains for now the province of models. The development and validation of such models will be enhanced greatly by the increased availability of airborne observations of spatial structure.

B.5.2 Technology of Airborne Flux Measurement

As with towers, sampling of air-surface exchange from aircraft requires an accurate, undistorted, high-frequency record of horizontal (u, v) and vertical (w) components of the wind velocity and scalars (temperature, pressure, mass concentrations) just above the canopy. However, airborne sensors are constantly in rapid complex motion, both linear and rotational, in an environment of flow distortion. Wind velocity and scalar parameters that can be measured directly on a tower have to be derived from multiple measurements on an airplane. Errors in magnitude or timing of these feeder data propagate through derived winds and scalars into computed fluxes. Though sensors of sufficient accuracy for this work are readily available commercially, their proper installation and use are major issues. In finding the optimum configuration, vehicle costs, flexibility, imposed flow distortion, frequency response, flight speed and operational altitudes all have to be considered. Considerable complexity in data processing can be avoided through careful selection of airframe and sensor configurations. An accurate common time reference for all data is critical.

Wind measurement from an airplane is simple in concept, difficult in detail, and vital to the flux computation. Mathematically, the central problem is to convert airflow, measured from a moving airplane, to winds in fixed Earth coordinates. Most of the sensed airflow arises from the airplane's own motion, which must be measured and removed to determine the wind. Strictly, we need to know the motion of the sensors, not of the airplane. This realisation can lead to important optimisation. In the 1980s, inertial navigation systems (INS) were the most accurate way to measure the sensors' motion. The INS achieved good wind accuracy, but availability was limited because of cost and size. Further, the INS measured the motion of that part of the aircraft where

it was mounted. The airflow sensors' velocity was found by extrapolation, a complex, error-inflating process. With the introduction of small, low-cost differential GPS (Global Positioning System) and micro-accelerometer technology, the sensors' motion can now be measured directly. The mathematics has become simpler and more robust.

Flow distortion is a universal consideration in accurate flux measurement. On a tower, flow is distorted by the blockage and drag of the sensors and their support structure. On an airplane such incidental distortion is augmented by powerful flow distortion generated intentionally to provide propulsion and lift. Good airborne installations, like good tower installations, minimise distortion by minimising the disturbance and then placing the sensors as far away from it as is practical. A small, low-drag airframe, rear-mounted engine, and long instrument boom have clear advantages (Crawford and Dobosy 1992). A rear-mounted engine not only removes the propeller's disturbance from the nose, but also shifts the centre of mass aft, moving the wings aftward as well.

The farther aft the wings, the weaker the upwash at the nose. Upwash is the forward part of the circulation generated by the wings in producing lift. Its magnitude is positively correlated with the vertical wind velocity being measured (Crawford et al. 1996). Thus, upwash contamination, if unaccounted, causes fluxes to be overestimated. Characteristic upwash ranges from 0.5 to 2.5 m s^{-1}, depending on the wing loading, flight speed, and forward distance from the wing to the measurement location. Pressure-radome installations, being generally close to the wing, experience strong upwash relative to sensors mounted on long probes. Smaller airplanes with light wing loading generate less upwash.

Flight speed, being at least a factor of ten greater than the wind being measured, imposes strict accuracy limits. A 1% error in the sensors' or the relative airflow's velocity produces at least a 10% error in wind. Fortunately, modern technology has facilitated these measurements and associated computations, increasing their accuracy while greatly reducing their cost.

Flight speed also affects turbulence measurements in other ways. The faster the flight, the more the turbulent information is compressed in time, improving the sample. This is useful close to the surface, where the turbulence is rich with character driven by surface forcing. It must, however, be accompanied by proportionally faster and more accurate sensors. Further, fast aircraft close to the surface are less manoeuvrable and more intrusive to humans. For low-altitude work (10–15 m above ground), 50 m s^{-1} is a practical airspeed, collecting 50 samples per second. With increased altitudes, the surface signal becomes obliterated by turbulent blending. Also, measured flux departs from its surface value as storage and advection beneath the aircraft become more significant (Betts et al. 1990). Further, the horizontal scale of the turbulence increases with height above ground, requiring longer flight tracks to obtain a statistically stable covariance (Lenschow et al. 1994). Fast, higher-flying aircraft are more suited to larger regions where surface detail is less resolved (e.g. Oncley et al. 1997).

The airplane's aerodynamic characteristics correlate its flight speed with vertical wind velocity. For example, when an airplane enters an updraught, constant altitude is maintained by lowering the nose. As the airplane pitches downward, it accelerates. The opposite occurs when descending air is encountered. The airplane thus travels more rapidly through updraughts and more slowly through downdraughts. A constant-rate time series provides a biased sample, with more observations during downdraughts. Constant altitude sampling of other organised flow structures (roll vertices, microfronts, slope flows, etc.) may also modulate the airplane's speed, introducing bias into time averages relative to the space average. Such bias can be as much as 15% on small aircraft, though much less on large aircraft. Estimating the ensemble-average eddy flux from observed airborne time series thus requires conversion to a space series, as discussed by Crawford et al. (1993).

B.5.3 Accuracy of Airborne Measurements

Typical GPS technology can now define sensors' attitude, velocity, and position in Earth coordinates to an accuracy of 0.05°, 0.02 m s^{-1} and 0.01 m respectively. The better GPS receivers report ten times per second but achieve the stated accuracy up to about 1 Hz. Extension to higher frequencies is readily accomplished by measuring accelerations, which increase in amplitude with increasing frequency. Pitch, for example, is found as the second integral of its angular acceleration. This is measured as the difference between vertical accelerations at a known separation along the longitudinal axis of the airplane. Error accumulates rapidly in these integrals, but not in the first two seconds with accelerometers of ordinary good quality. This is adequate because of the high accuracy of GPS at frequencies up to 1 Hz. Figure B.11 shows the quality of the match between pitch angles determined by GPS and by accelerometers. The raw GPS measurement (dashed line through trough) is filtered to remove frequencies above 0.5 Hz (circles). Integrated accelerations are filtered to remove frequencies below 0.5 Hz (solid line about zero). The sum of these filtered signals (solid line through the trough) is more accurate over the whole frequency range than the GPS alone. The two curves, which would obliterate each other, are mutually offset for visibility. The raw GPS trace is noisy in comparison, but the noise is generally within about 0.1° (dashed line about zero). At an airspeed of 50 m s^{-1} an error of only 0.1 m s^{-1} would result from straight use of GPS for pitch.

GPS accuracies are still improving with the adoption of dual frequency receivers, more powerful embedded microprocessors and advanced firmware. Airborne wind measurements have the potential accuracy of 0.02 m s^{-1} horizontally and 0.03 m s^{-1} vertically. Unfortunately, adoption of this new technology has been slow. For various reasons, none of the current airborne wind systems achieves this accuracy in mean wind observations. However, mean wind accuracy is rapidly improving. We believe the residual contamination due to unresolved platform motion to be less than 0.2 m s^{-1} horizontally and much less for vertical winds. The accuracy of turbulent wind should be greater than that for the mean wind because its energy occurs in spectral regions higher than most aircraft motion.

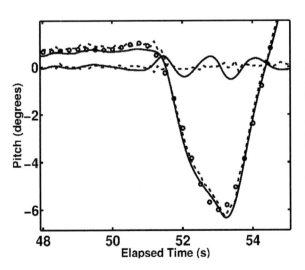

Fig. B.11. Pitch angle by GPS, extended by accelerometers

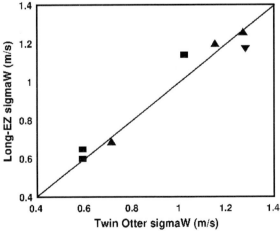

Fig. B.12. Variance of vertical wind component, measured from two airplanes

Assessment of the overall accuracy of airborne flux measurements can be made by intercomparisons in the field. Comparison among airborne systems, as during the BOREAS experiment (Dobosy et al. 1997) show the overall precision of measurement. A more recent comparison, after upgrades to the Long-EZ's GPS receivers, yielded a further improved match. In particular, the variance of vertical velocity measured from the Long-EZ agreed closely with that measured from the Twin Otter of the Canadian National Research Council, as can be seen in Fig. B.12.

Compared to surface fluxes reported from towers, airborne measurements initially produced flux estimates systematically low, by 15% or more (Shuttleworth 1991). Through multiple passes over the same track at lower altitudes (30 m or less) the match has been markedly improved. The surface must, of course, be sufficiently homogeneous to ensure similar footprint character. Figure B.13 shows the quality of match that was achieved by the Long-EZ/Twin Otter combination about a tower operated by S. Verma as part of the AmeriFlux programme funded by the Southern Great Plains Regional Office of the US Department of Energy's NIGEC programme.

B.5.4 Utility of Airborne Flux Measurement

Some important lessons have been learnt as the technique of airborne flux measurement has evolved dur-

ing its application in a series of integrated terrestrial experiments. The utility of the technique derives from both transit speed and freedom of track. Being versatile in space but limited in temporal coverage, airborne flux measurements complement naturally the measurements from fixed towers. The best experiment designs deploy airborne flux measurements between tower sites or along paths passing over at least one fixed tower comparable in height to the flight altitude. Shorter paths traversed frequently are better than long paths traversed rarely. Several important experiments illustrate airborne deployment.

The HAPEX-MOBILHY (André et al. 1988) observed the hydrological budget on a 100 km square in southwest France, with an intensive observation period (IOP) in 1986. The heterogeneity of the landscape was covered by measuring micrometeorological and hydrological parameters at locations representative of the major vegetation communities. The King Air of the US National Center for Atmospheric Research (NCAR) flew a 150 km flight track during the IOP at several depths in the boundary layer, estimating surface fluxes from passes at 100 m altitude. These fluxes, computed over 10-km segments of the path were somewhat low compared to tower measurements. But they showed internal consistency and documented clearly the change in energy partition between agriculture and forest.

The European Field Experiment in a Desertification-Threatened Area (EFEDA) occurred in June 1991 in eastern Spain. The Falcon airplane of the Deutsches Zentrum

Fig. B.13.
Flux airplanes and tower under homogeneous conditions in Oklahoma

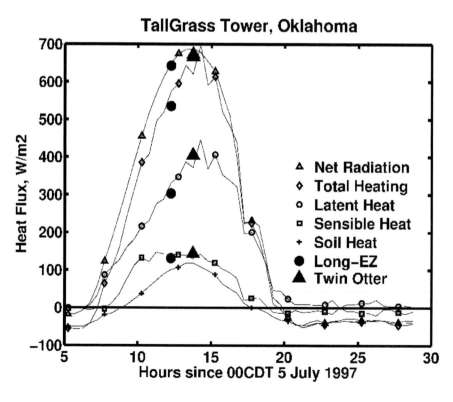

für Luft- und Raumfahrt (DLR) flew L-shaped patterns at three altitudes from 400 m to 2 500 m above ground. Michels and Jochum (1995) found links between the pattern of surface fluxes and the boundary layer's character over its whole depth in this arid and partially irrigated agricultural region. Fluxes sampled at 400 m were, however, sometimes significantly different from those at the surface. The latent heat flux on 23 June 1991 tended to increase at 400 m between Barrax on the east and Tomelloso on the west, opposite to the surface pattern. Modelling reproduced this result and related it to mesoscale moisture advection from the Mediterranean Sea, 200 km to the east (Noilhan et al. 1997).

HAPEX-Sahel extended the observations to the Sahel region of Africa in 1992. The Météo-France Merlin IV aircraft flew 50-km overlapping rectangles over the surface array (Saïd et al. 1997). Of interest to airborne flux measurements, they determined the behaviour of the latent heat flux to be quite variable, compared with that of the sensible heat flux. For the drydown season, an averaging length of 30 km was required to achieve stable statistics, but only 7 km for sensible heat.

In the Boreal Ecosystem-Atmosphere Study (BOREAS) of 1994, surface towers were deployed similarly to HAPEX-MOBILHY. Two study regions of 100-km scale were defined in the Canadian boreal forest, including disturbed areas, lakes, and mixed forest stands. Four flux airplanes flew predetermined transects connecting the fixed towers and extending over a broader range of the heterogeneity of the region (Dobosy et al. 1997; Desjardins et al. 1997). Figure B.14 shows how the uptake of CO_2 varies

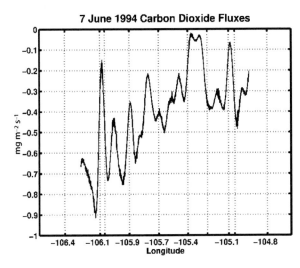

Fig. B.14. CO_2 uptake in Saskatchewan during BOREAS, from Long-EZ. Negative values indicate uptake, so smallest uptake is at the top of the plot. Indicated longitude delineates surface cover. Halkett Lake at 106.1° W is flanked by aspen on both sides. Candle Lake at 105.4° W has mixed deciduous and coniferous trees on the west but coniferous trees on the east. White Gull Lake at 105.1° W is flanked on both sides by coniferous trees

strongly between active aspens to the west and less productive pine and fir to the east. The three lakes (106.1° W, 105.4° W, and 105.1° W) are marked by their minimal uptake of CO_2. It was found important to the design that these transects be flown repeatedly, since heterogeneity implies multiple populations from which samples are drawn. A single 100 km pass over such a surface may represent an averaging time of 30 minutes, but the sample time over each constituent of the path is considerably less.

Measurements were made during the Southern Great Plains Experiment of 1997 (SGP97) using two primary flux airplanes. The goal of this experiment was to examine the influence of heterogeneities in soil moisture, as observed by an innovative passive microwave radiometer intended ultimately for satellite use. One important aspect of the study was the effect on the development of the convective boundary layer. The two flux airplanes flew missions at low levels to sample the influence of the surface moisture heterogeneity on the low-level fluxes. This allowed interpolation between measurements at surface towers, located as appropriate in homogeneous subregions. The airplanes, however, were also able to sample near the base of the entrainment zone at the top of the mixed layer, as had Michels and Jochum (1995). Again, repeated passes were important because of the large spatial scale of the motions near the top of the mixed layer (Dobosy and MacPherson 1999; MacPherson et al. 1999).

Airborne measurements are well suited to remote land areas. Arctic work has been greatly expanded by their flexibility and range (Brooks et al. 1996; Oechel et al. 1998).

Secondary circulations, forming on scales from 10 km to 100 km, can have profound influence on dispersion of admixtures to the air and on the transport of moisture, heat, and momentum. Airborne turbulence measurements are uniquely capable of sampling the flow structures in such circulations as sea breezes and atmospheric frontal structures. Again, it is necessary to make repeated passes over defined lines to ensure proper sampling. (Sun et al. 1997; Eckman et al. 1999).

B.5.5 Conclusions

The advent of small specialised flux airplanes has increased the capabilities and greatly reduced the cost of airborne flux measurement. We have found it best to fly as low as possible, repeating the same track as often as possible, when relating observed fluxes to the underlying surface. Fluxes are thus better determined along paths than over areas. Proper choice of the path(s) maximises effectiveness. Today's smaller, cheaper airplanes also make multiple deployments more tractable. Such deployments have been found to be highly effective in the BOREAS and SGP97 experiments.

Table B.4, not exhaustive, shows a growing number of organisations making airborne flux measurements. The number of new entries with small airplanes shows a potential for a greater availability of low-cost airborne flux measurements. Larger airplanes will remain important for the more unusual or complex measurements, but we expect a growing set of capabilities to be derived from small airplanes as technologies mature.

Table B.4. Organisations making airborne flux measurements

Organisation	Airplane	Size	Example reference
Airborne Research Australia	Grob 109	S	Lyons et al. (1993)
MetAir (Switzerland)	Stemme S10	S	Neininger et al. (1999)
National Research Council (Italy)	Sky Arrow	S	Delivery September 2000
Air Resources Lab (NOAA, USA)	LongEZ	S	Crawford et al. (1996)
San Diego State University	Sky Arrow	S	Brooks et al. (2001)
University of Lund (Sweden)	Sky Arrow	S	Delivery July 2000
University of Manchester (UK)	Cessna 182	S	Wood et al. (1999)
Airborne Research Australia	Cessna 404	M	Matthews et al. (2000)
Institut National des Sciences de l'Univers (France)	Fokker 27	M	Durand et al. (1998)
Météo-France	Merlin IV	M	Durand et al. (1998)
National Research Council (Canada)	Twin Otter	M	Mailhot et al. (1998)
NOAA/ARL	Twin Otter	M	Luke et al. (1998)
University of Wyoming	King Air	M	Dobosy et al. (1997)
National Center for Atmospheric Research (USA)	Electra C130	L L	Oncley et al. (1997) Wang et al. (1999)

Chapter B.6

Boundary Layer Budgeting

David R. Fitzjarrald

B.6.1 Introduction

Vigorous mixing in the lower atmosphere is often confined within a relatively shallow planetary boundary layer (PBL), into which the surface fluxes of atmospheric constituents converge. The horizontal scale of mixing – and hence the degree of spatial averaging performed by the boundary layer turbulence – is proportional to the thickness of the layer h. This thickness ranges from 100s of metres for stable boundary layers (SBLs) to 1–3 kilometres for convective boundary layers (CBLs). The intensity of the mixing depends on the surface buoyancy flux, and this in turn determines the thickness of the boundary layer. Turbulent mixing at these scales blurs local gradients set up by contrasting land-surface types. By using boundary layer (BL) budgets to estimate surface fluxes one exploits the horizontal averaging property of this turbulent mixing. The surface flux is found as a residual; all transport is estimated through the sides and top of the box. The BL budget method is an approach with more advocates than true practitioners. Its potential inspires many to instruct the community on the virtues of the continuity equation, as it is applied to field observations.

The budget method represents an alternative to the recently popular eddy flux tower approach (e.g. Ameriflux, Euroflux) or chamber methods. Fluxes measured from towers are constrained to represent a time-varying area whose diameter in most cases is often less than a kilometre or two, and chambers, which are used to sample tiny surface areas (approximately 1 m^2) (e.g. Mosier 1989). Artificial stirring is needed to homogenise even such small chambers. Useful results continue to be found using these approaches, but they are always plagued by doubts over the representativeness of the measurement.

Ideally, accumulation within the reference box is due to vertical flux convergence alone, so that the BL budget approach is then just another chamber method; but although the stirring is natural, the chamber leaks at the top and on three sides. As with any residual method, errors congregate with the signal sought. But it is not the presence of such error that is the major weakness of the method; rather it is that few studies to date have properly accounted for all of the terms in the budget. Authors take pains to explain why certain terms are unimportant, often dismissing cases if the method fails spectacularly. In many cases, studies billed as boundary layer budgets are really hybrids, with some observations feeding into a model that supplies estimates of the missing terms, such as horizontal advection, that are hard to characterise with observations alone. The trick seems to be to select environmental conditions for which several terms in the continuity equation are not significant.

B.6.2 Characteristic Structure of the Boundary Layer

Garstang and Fitzjarrald (1999) give details of CBL history and structure. At least since Rotch's 1893 kite soundings in New England (and probably before) the structure of the CBL has been known. A layer nearly well mixed in potential temperature (Θ), specific humidity (q) and other scalars (C) lies beneath a capping inversion. Montgomery (1947) presented the modern version of these profiles. During the day, surface buoyancy fluxes provide turbulent energy to promote turbulent mixing. The characteristic velocity scale, w^*, depends on the surface buoyancy flux (in conventional notation):

$$w^\star \equiv \left[\frac{g}{\Theta_0} \overline{w'\Theta'_v}\, h \right]^{1/3}$$

Eddies are effectively bounded by the CBL thickness, h, so that characteristic mixing distances in the CBL are Uh/w^* (see Garratt (1992) for examples). A capping inversion structure indicates synoptic scale subsidence of the air above the inversion, evident on cloudless days or between clouds on other days.

Over land, the surface layer becomes stable in late afternoon, and the former mixing layer is left aloft as a rapidly decaying fossil mixed layer, atop a shallow (100–300 m) and evolving SBL. The thickness of the SBL is harder to quantify, but alternative definitions lead to estimates that are within approximately 30% of each other (e.g. André and Mahrt 1982). The stable boundary layer may never reach approximate equilibrium over the night. Frequent appearance of drainage circulations, nocturnal jets, and intermittent mixing, for example, mean that the periods over which budgets are taken must be selected carefully.

Inferring surface processes from changes in the lower atmosphere is hardly a new approach. G. I. Taylor's kite soundings off Nova Scotia (Taylor 1914) were part of an early quantification of air-mass modification. Burke (1946) cited Sverdrup's University of California Los Angeles lectures during World War II as inspiration when presenting a graphical method to forecast continental air mass modification. Interest in the downstream evolution of the trade-wind boundary layer ("the trades") over the tropical ocean prompted some of the first boundary layer heat and moisture budgets. For budgeting purposes, this situation represents a "best-case" scenario (a degree of surface homogeneity unrealisable over land), and it offers important lessons for the terrestrial case. Analyses of data obtained over the tropical oceans focused on inferring the cloud-base water vapour flux F_{qh}. LeMone (1995) presented four estimates of $\partial F_q / \partial z$ in the CBL over the tropical ocean made at different spots over a 30-year period. Consensus points to small (approximately $1 \, \text{g kg}^{-1} \, \text{d}^{-1}$), but significant, moistening of the layer. Nichols and LeMone (1980) compared aircraft-measured heat and moisture fluxes in GATE with Brümmer's (1978) C-scale triangle (50 km on a side) budget estimates. They attributed important differences in flux estimates to indirect influences of nearby large clouds. Fitzjarrald (1982; see Garstang and Fitzjarrald 1999, p. 278), used composite CBL soundings from GATE and an equilibrium slab model (see below) to infer enviromental subsidence and cloud-base mass flux. Results were comparable to those found using a large-scale budget method based on horizontal convergence seen from a radiosonde array. When a similarly-sized mesoscale triangle was set up over land in the Amazon rainforest during ABLE-2B (see Chapt. A.6.3), researchers found surface divergent conditions on average, despite the location in the convergent tropics (Oliveira 1990; Greco et al. 1990). In conclusion:

a mixed layer structure is evidence of subsidence; and
b the local subsidence reflects both synoptic-scale sinking and compensating subsidence between clouds. Even fair weather clouds alter mass budgets.

Over land, large-scale water budgets based solely on precipitation, runoff, and observed advective components have led to disappointing results. Rasmusson (1967) found large residuals in the observed water balance for North America, and a similar recent study (Gutowski et al. 1997) that used the NCEP (National Centers for Environmental Prediction) reanalysis data products, with their greatly expanded data sources, ended up missing water balance closure by 40%. One should not place too much faith in data just because it is gridded! To date, BL budgets performed over mesoscale, or smaller areas, on land have largely skirted the issue of the importance of clouds.

B.6.3 Budget Relations

How one might use the budget method to estimate surface fluxes using the boundary layer budget method has been explained repeatedly in recent years (e.g. Raupach et al. 1992; Denmead et al. 1998, 1999). A brief explanation will suffice here. After Reynolds averaging to find turbulent fluxes (F_x), a continuity equation for the mean value of constituent C is:

$$\frac{\partial C}{\partial t} + \nabla_{xy}(C\vec{V}) = \frac{\partial F_c}{\partial z} - S_c - \overline{w}\frac{\partial C}{\partial z} \tag{B.9}$$

where the second term on the left denotes horizontal advection and S_c denotes a source or sink term. The thickness h of the budget box may be fixed or moving. In the meteorological literature, it has been common to identify h with an active turbulent region (the boundary layer) whose thickness varies diurnally. The continuity equation for a box with a moving lid is:

$$h\frac{\partial C_m}{\partial t} = F_0 - F_i \left\langle \nabla_{xy}(\vec{V}C) \right\rangle + \left[\frac{\partial h}{\partial t} - \overline{w}\right][C_+ - C_m] \tag{B.10}$$

where $\left\langle \nabla_{xy}(\vec{V}C) \right\rangle$ represents the layer-averaged horizontal divergence of C, and subscript m refers to the layer mean. The first term in square brackets on the right is referred to as the *entrainment velocity*, $w_{en} = \partial h / \partial t - \overline{w}$, where $\overline{w}$ is the mean subsidence rate in the environment; C_+ and C_m represent the ambient concentration just above the budget box and the average of C over the layer, respectively. For many authors, it describes the growth rate of the turbulent layer relative to the less turbulent layer above. When the inversion slopes, then $h(t, x)$ and an additional term $U_m \nabla_{xy} h$ must be included, (U_m is the mean horizontal wind speed and xy is the horizontal plane) and one must assume that the mean airflow intersects the inversion at PBL top. For details consult Garratt (1992, p. 153) or Betts (1992).

The elementary boundary layer budget approach is quite difficult to realise using field observations alone. Accurate estimates of the turbulent and mean transports through the top of the box are rarely made. Directly measured turbulent flux at h is found principally in aircraft missions of limited duration. To derive the mean transport requires $h(t)$ and C_+, found from frequent profiles of the atmospheric constituent considered, but also accurate estimates of $\overline{w}$, typically orders of magnitude smaller than the horizontal wind speed. One can designate the box top above the maximum height of the active turbulent layer. This eliminates the need to find $\partial h/\partial t$ and the turbulent flux F_h (e.g. Wofsy et al. 1988). This advantage comes at the expense of increased reliance on obtaining very accurate estimates of $\overline{w}$, the "jump" $C_+ - C_m$, and $\partial C_+/\partial z$, the environmental scalar gradient above the budget layer. Subsidence $\overline{w}$, can be found as $-D h$, where D is the observed horizontal divergence in the boundary layer, though this is not often done. Researchers often turn to model output statistics to find subsidence, a practice that holds more promise for composites than it does for individual cases. Subsidence can also be estimated by following persistent features in series of sequential soundings (Carlson and Stull 1986; Freedman and Fitzjarrald 2001). Finding transport through the sides of the box requires accurate estimate of often-small horizontal gradients. The best way to do this is to have frequent direct soundings, an ideal application for continuous remote sensing of winds and concentration profiles. The many limitations assure that few studies to date have succeeded in finding all of the important terms in the budgets for the "leaky box" of the boundary layer. Fewer still include data for any substantial part of a seasonal cycle.

Hybrid Method

In many cases, data are interpreted using a slab model (Businger 1954; Betts 1973; Tennekes 1973; Garratt 1992, p. 145ff). In addition to Eq. B.9, the latter authors introduced a *closure assumption* to relate the buoyancy flux at h to the surface value: $F_{bh} = -\beta F_{bo}$. Estimates of the entrainment coefficient, β, vary between 0.2 and 0.5, with the former, until recently, being a consensus value. The considerable variability results from the difficulty in obtaining direct measurements of F_{bh}, a measurement usually made from aircraft platforms (see Grossman 1992 and Betts 1992 for discussion). Raupach (1991, 2000) argued that the CBL cannot reach equilibrium in a single day, but Fitzjarrald (1997; Garstang and Fitzjarrald 1999, p. 173) showed that the importance of β diminishes when sequences longer than a single day are considered. Freedman and Fitzjarrald (2001) verified the diminished role of entrainment after the first day of frontal recovery sequence. In a common version of the hybrid method, one uses β to help close the heat budget, and then applies the resulting w_{en} to estimate F_h. As more model elements enter the budget estimation, the distinction between observation and assertion is blurred. In many situations, the slab model serves to interpolate CBL thickness h for times between soundings.

BL Gradient Methods

The "mixed" layer of the CBL exhibits gradients (cf. Garstang and Fitzjarrald 1999, p. 174ff) and these can be exploited to infer fluxes from profiles. Brutsaert (1999) argued for use of PBL flux-gradient relations, similar to those of the Monin-Obukhov hypothesis for the surface layer. Data presented to support Brutsaert's empirical functions exhibit large scatter. Lack of accurate profiles limits the utility of the CBL gradient approach. This lack of data probably results from difficulties in obtaining quality ensemble averages of CBL mean and flux quantities. For many scalars, fluxes peak during convective conditions (day-time), just when buoyant mixing acts to minimise CBL gradients. Wyngaard and Brost (1984) introduced a similarity hypothesis that accounts for the effects of entrainment as well as surface fluxes on CBL gradients, an idea validated using large eddy simulations (LES; e.g. Moeng and Wyngaard 1984). Mesoscale effects, which are common in observations but not realisable in the LES, limit the applicability of the model-derived similarity expressions for field use (Kiemle et al. 1998). A related complication is that real-world profiles are frequently altered in baroclinic or cloudy conditions.

Aircraft Studies

In many aircraft studies (see also Chapt. B.5) surface fluxes were found by assuming a perfectly mixed CBL (linear turbulent F_c with height) and extrapolating downward from measurements made at several flight levels. This is not strictly a boundary layer budget approach. Lenschow and Johnson (1968) showed how to do this, but Moore et al. (1992) showed that large (approximately 30%) errors can result from the extrapolation to the surface. With data from repeated flight legs, the local change term is usually found by comparing concentration at common points on the leg; the horizontal advective terms are found using the spatial gradient along the flight line in the direction of the mean wind. Sometimes the tendency is obtained from surface concentration measurements (e.g. Lenschow et al. 1980a; Raupach et al. 1992).

B.6.4 More Recent Budget Studies Conducted over Land

Budget terms actually observed for a number of recent boundary-layer budget studies are shown in Table B.5 and Table B.6. If the budget method is to be valuable for inferring surface fluxes, one must demonstrate that the terms in Eq. B.9 or Eq. B.10 can be found accurately or

Table B.5. Selected boundary layer heat and water vapour budgets over land (Y = term observed; M = model-supplied term; – = not observed)

Reference	Variables	Stability	Surface	Platform	$\delta C/\delta t$	F_h	$\nabla_{xy}(VC)$	$h(t)$	$\delta QR/\delta z$	Obs period	Test: Sfc flux	Error (%)	Anecdotal
Betts et al. (1990)	q,Θ	C	Prairie/FIFE	A/C	Y	Y	Y	Y	–	6 2-h periods	Eddy	30	Y
Betts et al. (1992)	q,Θ	C	Prairie/FIFE	A/C	Y	Y	Y	Y	–	8 2-h periods	Eddy	20	Y
Grossman (1992)	q,Θ	C	Prairie/FIFE	A/C	Y	Y	Y	Y	–	2 2-h periods	Eddy	20	Y
Kustas and Brutsaert (1987)	q	C	Hills near Alps	Sounding, 7/day	Y	–	–	·	–	11 days	Lysimeter	300	N
Dolman et al. (1997a)	q,Θ	C	Arid/HAPEX-S	Soundings, model	Y	M	–	Y	–	3 days	Eddy	150	Y
Frech et al. (1998)	q,Θ	C	B-forest-NOPEX	A/C	Y	Y	–	Y	Y	2 2-h periods	Eddy	30	N
Barr and Betts (1998)	q,Θ	C	B-forest/BOREAS	Soundings, ?/day	Y	–	M	Y	–	39 days	A/C	25	N
André et al. (1990)	q,Θ	C	M-forest-HAPEX-M	Soundings, model	Y	Y	M	Y	–	?	Eddy?	26	Y
LeMone and Grossman (1999)		C	Prairie/CASES-97	A/C	Y	Y	Y	Y?	–	2 3-h periods	Eddy	30	N
Moore et al. (1992)	q,Θ	C	Tundra/ABLE-3A	A/C	–	Y	–	Y	–	6 d, 3-h periods	Eddy	30–50	N
Freedman and Fitzjarrald (2001)		C	NE USA	Soundings, 2/day	Y	M	M	Y	–	17 6-day seq	Eddy	20	N

Table B.6. Selected trace gas budgets over land (Y = term observed; M = model-supplied term; – = not observed; NA = insufficient information in the reference to determine this term)

Reference	Variables	Stability	Surface	Platform	$\delta C/\delta t$	F_h	$\nabla_{xy}(VC)$	$h(t)$	$\delta QR/\delta z$	Obs period	Test: Sfc flux	Error (%)	Anecdotal
Denmead et al. (2000)	CH_4	S	Cattle lot	Profile	Y	Y	Y	–	–	1 night	Literature	30	Y
Denmead et al. (2000)	CH_4	C	Agric. sfc	22 m concen.	Y	Y	–	–	–	16 days	A/C flask sequence	10–100	N
Choularton et al. (1995)	CH_4	S	Peatlands	Teth. balloon	Y	–	–	Y	–	1 night, 2 profiles	Eddy	20	Y
Choularton et al. (1995)	CH_4	C	Peatlands	A/C	–	–	–	–	–	4 days	Sfc gradient	50	Y
Levy et al. (1999)	CO_2	C	Boreal forest	Tower, sounding 5/day	Y	Y	Y	Y	–	4 days	Eddy sfc	10–300	N
Culf et al. (1999)	CO_2	S	Trop. forest	Teth. balloon	Y	–	–	–	–	10 days	Eddy	250	Y
Wofsy et al. (1988)	CO_2	C	Trop. forest	A/C	Y	–	–	–	–	9 days	Literature	NA	N
Potasnak et al. (1999)	CO_2	C	ML forest	Sfc obs, stat. Analysis	Y	–	–	–	–	Seasonal	Eddy	NA	N
Helmig et al. (1998)	NMHC	C	Trop. forest	Teth. balloon, 5 prof.	Y	–	–	–	–	1 day	ML gradient	NA	Y!
Guenther et al. (1996)	Isoprene	C	ML forest	Teth. balloon, 5 prof.	Y	–	–	Y	–	11 days	REA	50	Y
Raupach et al. (1992)	CO_2	C	Wheat	Sfc data	Y (sfc)	Y	–	–	–	11 days	Eddy	30	N
Davis et al. (1994)	NMHC	C	Trop. temp. forests	Teth. balloon	Y	–	–	Y	–	11 days	Literature	NA	N
Greenberg et al. (1999)	Isoprene	C	Many forested sites	Teth. balloon	Y	–	–	–	–	Many prof., 11 y	Model	NA	Y
Ritter et al. (1990)	Θ,q,O_3,CO	C	ABLE-2B trop forest	A/C	Y	Y	Y	–	–	4 h on 1 day	Eddy q, analysis	>20	Y
Ritter et al. (1992)	O_3,CH_4,CO	C	ABLE-3A	A/C	–	Y	Y	–	–	10 2-h flts, 4 days	Eddy	40–60	N
Lenschow et al. (1980b)	Θ,q,O_3	C	Front range Colorado	A/C	Y (sfc)	–	Y	–	–	2 2-h flts, 2 days	Literature	100	NA
Graber et al. (1998)	CO_2,q	C	ECOMONT	A/C	Y	–	Y	–	–	11-h period	None	NA	Y

shown to be negligible. In extremely few cases have most of the budget terms in Eq. B.9 been found, and the bulk of these studies encompass very short times.

A conceptual difficulty in assessing the quality of the budget method is that one often has for comparison only the very point flux measurements that the budget method, with its larger areal representativeness, aims to supplant. When two methods do not yield comparable results, it is hard to tell which is valid. It is common over land to seek nearly horizontally homogeneous environments (e.g. FIFE, CASES, OASIS); over such surfaces better agreement among methods occurs. This simplifies analysis of errors, but it is precisely in areas of heterogeneity where the budget method is most promising. Percentage errors in the Table are rough estimates, found in most cases by comparison with tower-based eddy covariance fluxes. For very short-term observations, errors cited in Table B.5 and Table B.6 represent a subset of best-quality results, and are not directly comparable to those with longer datasets or large composites. When comparisons with independent flux estimates were made, very large errors persisted. Smaller error estimates resulted when authors made estimates based on instrument and data processing characteristics alone.

Heat and Water Vapour Budgets Found Using Aircraft with Flux-measuring Capability

An *ideal* box budget consists of the simple situation in which F_{Co} converges totally into the budget box, and one can infer it from $\overline{\partial C_m}/\partial t$. Such is the unspoken expectation of researchers that events with significant advective influence are obscured, though everyone knows that advection dominates heat and moisture budgets at certain stages of the synoptic cycle. Lenschow et al. (1980a) found horizontal advection to play a key role in limiting the accuracy of assessing the energy budget for one case; whereas Betts et al., (1992) argued in another case that it can be ignored. In FIFE (Kansas, USA), Betts et al. (1992, 1996c) found large residuals in the heat and moisture budgets, which Grossman (1992) attributed in part to inadequate sampling of long-wavelength flux contributions. However, similar large residuals were later reported by LeMone and Grossman (1999) in CASES (in Kansas again). They stressed the importance of estimating the radiative flux divergence in the mixed layer.

André et al. (1990, HAPEX-MOBILHY) claimed 30% agreement between aircraft and surface energy balance estimates; but larger scatter was found for similar comparisons of evaporation. Subsequent budget analyses in that project were done using model outputs. This experiment design is notable because of the use of flights every three days, in contrast to the commonly found case study approach for aircraft observation (e.g. FIFE, BOREAS). Moore et al. (1992) compared regional fluxes found by aircraft in ABLE-3a over a simple tundra-lake surface with surface data scaled up using a simple two-component "mosaic". They concluded that 30% errors are unavoidable in making regional flux estimates.

A hybrid budget method for four days of data obtained in HAPEX-Sahel reported by Dolman et al. (1997a) also exhibited large residuals. They too emphasized the importance of capturing advective and subsidence contributions to the budgets. In NOPEX, Frech et al. (1998) reported that advection was important to heat and moisture budgets on two flight days. The study was one of the few to measure radiative flux divergence in the CBL. However, they excluded half of their data (one afternoon), noting: "Since the residual is quite large on June 13, we refrain from inverting the budget to estimate a surface flux." The difficulty was attributed to cloudiness over part of the flight path. In the same experiment, Levy et al. (1999) reported CO_2 fluxes on two pairs of days. They estimated CO_2 above the BL by looking at upstream concentrations using trajectories. While CBL budget estimates of surface flux agreed with tower-based estimates during the first period, there was very poor agreement on the second day (Fig. B.15). These authors have been notably open about the difficulties in applying the budget method. Less public self-selection of data by others probably makes the error estimates in the Table highly optimistic.

Lenschow et al. (1980b) showed that the ozone deposition velocity (F_c/C_0) could be measured from aircraft observations. With time, improved instrumentation was developed and deployed in the NASA ABLE experiments in the Amazon and at high latitudes (Ritter et al. 1990, 1992, 1994). They reported local change, flux divergence and advective change terms for methane, O_3, and CO. Estimated errors for surface fluxes lay in the 30–80% range.

These aircraft budget studies have been very valuable in quantifying many of the terms in the boundary layer budget under convective conditions, but they have emphasized isolated cases on ideal ("golden"), often cloudless days – a consequence of the tremendous effort required for aircraft observations – and they yield what must be considered "anecdotal results". Studies encompassing a few hours on an occasional day do not provide a basis for making strong inferences about regional fluxes. After all these years of effort, CBL budgets made using aircraft observations are still being presented as "proof of concept" case studies.

Sounding Sequences

Especially for the more readily observed Θ and q, sequences of soundings in the convective and stable boundary layers have been used to find $\partial C/\partial t$. Kustas and Brutsaert (1987) argued that good estimates of sur-

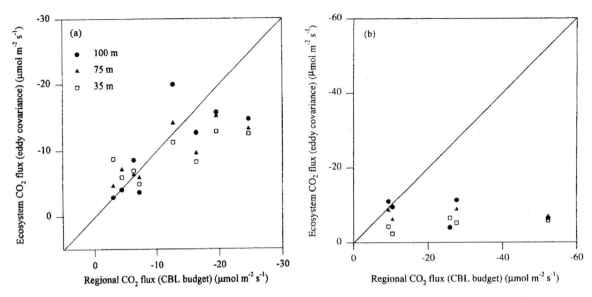

Fig. B.15. CBL budget CO_2 flux compared with tower eddy covariance fluxes (**a**) on "good" days 27–28 June and (**b**) on "bad" days 1–2 July in Sweden during NOPEX (from Levy et al. 1999, © reprinted with permission from Elsevier Science)

face fluxes in rolling terrain could be found, though the evidence they presented indicated that there were very large uncertainties. CBL heat and water vapour budgets were found from soundings in FIFE and in BOREAS (Barr et al. 1997; Barr and Betts 1998); ECMWF model output was used to estimate the horizontal advection terms (Barr et al. 1997). In BOREAS, they found vapour flux divergence in the CBL, a prediction of persistent drying over several days, but persistent drying over many days was not observed at the surface. Freedman and Fitzjarrald (2001) formed composites from 17 six-day sounding sequences, each in the aftermath of a cold front passage in the north-eastern US. The advective terms in the heat and vapour budgets were found from model estimates. They argued that the composite estimates of the advective terms are better than those found for individual cases. Half of F_{q0} converged into the CBL during the sequences, but advection associated with the approach of the next synoptic event means that little net moistening occurred over the sequence in midsummer. Entrainment was only important to the afternoon heat and water vapour budgets on the first day in the sequence (Fig. B.16). We should not take days in isolation and pronounce the results to be of general validity.

Sounding by Aircraft

Wofsy et al., (1988) estimated the CO_2 flux in the Amazon Basin (ABLE-2a) using vertical soundings from the NASA Electra. They avoided problems finding the entrainment velocity $\partial h/\partial t$ by designating the box top h to be well above the maximum altitude of the CBL. This proved to be highly effective; CO_2 above the active mixed layer remained nearly constant over a day, providing a reference against which to watch the uptake and respiration cycles of the rainforest below. No effort was made to estimate horizontal or vertical advection, or to estimate the importance of deep convection in the region in vertical mass redistribution (e.g. Scala et al. 1990). More recently, this approach was used to design a regional CO_2 surface flux estimation project in the midwestern US (COBRA, Stephens et al. 2000). They outline an innovative effort to track air mass origin by defining fingerprints from ratios of other trace gases (Potasnak et al. 1999). The weakness of this approach is always how to find subsidence, and this is particularly true near mesoscale convective activity, a condition under which the horizontal gradient of w can be large. A second limitation is the assertion that aircraft trajectories can characterise mean concentrations adequately over large volumes in the troposphere; this is rather like trying to fill a room with thread. Advective terms are meant to be found using mesoscale model output (e.g. MM5, Eta, RAMS see Pielke 2002), but these also are questionable in the presence of clouds. These limitations diminish as the number of cases increases, especially if there is sufficient data for careful composites to be constructed.

Fluxes of important non-methane hydrocarbons have been inferred from soundings using tethered balloons (e.g. Davis et al. 1994; Greenberg et al. 1999), but to date only a very few days have been analysed anywhere. With reactive gases, such as isoprene, there is the additional complication that the chemical sink term cannot be ignored. Guenther et al. (1996) found the diurnal accumulation of isoprene in the CBL. To find the surface emission, they assumed that all surface flux converged into

Fig. B.16.
Thermodynamic diagram for boundary layer cumulus frontal composite. *Thick short dashed* (automatic weather station, Orange MA airport) and *long dashed lines* represent afternoon CBL conditions (Harvard Forest flux tower) for maximum temperature T (K) and mean q (g kg^{-1}). *Thin curved solid lines* sloping upwards are surface-based lifting condensation levels (LCL, km); *curved dotted lines* are relative humidity RH (%). *Arrows* emanating from HF tendency represent contributions from surface (*solid*), transport (*dotted*), and entrainment (*dashed*) terms. *Arrows* in lower right hand corner represent tendency Bowen ratios $B' \equiv [c_p \partial T/\partial t] / [(L_v \partial q)/(\vartheta \partial t)]$ (from Freedman and Fitzjarrald 2001)

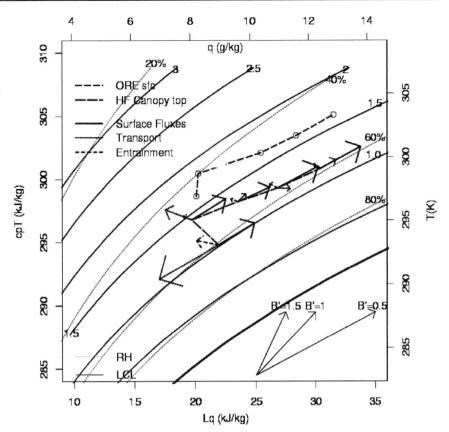

the mixed layer and reacted there. Flux estimates thus obtained vary tremendously, and convergence to within a factor of three is still considered good. Helmig et al. (1998) presented flux estimates for a number of biogenic nonmethane hydrocarbons emitted from the forest in the western Amazon, using a simple accumulation method. These estimates were based on only six soundings from a single day. No one who knows the difficulty of operating in the field in remote areas would discount the importance of these field observations of important atmospheric constituents. However, neither should the very real limitations of applying the BL budget approach on an isolated dataset be obscured. 20–40% errors in surface fluxes are the best we can expect for many atmospheric constituents, and this applies only in ideal situations.

SBL Budgets

Choularton et al. (1995) estimated methane flux from wetlands using tethered balloon measurements. SBL thickness was estimated from sodar measurements; surface flux was assumed to converge totally into the layer. Fluxes found using the SBL budget approach and the conventional surface layer gradient method differed by up to a factor of three. The authors noted that this outcome was "probably as a result of the heterogeneity in

methane source strengths and effective measurement fetch for the different techniques". Culf et al. (1999) also used tethered balloon soundings to estimate nocturnal CO_2 fluxes over the Amazon forest for ten consecutive nights. They found that budget estimates of surface CO_2 flux were much larger than surface eddy flux measurements on six of the nights; they agreed on four nights. They argued that horizontal inhomogeneity in CO_2 led to the nights of disparity and attempted to estimate advective effects using a limited study of BL trajectories. Unfortunately, this analysis did not take into account important river-land breeze circulations in the region (Oliveira and Fitzjarrald 1993). How can we know if the budget method yields better fluxes than the eddy flux approach at night? These two studies point to the way out of the conundrum – i.e. regular sequential observations, but in future for much longer periods. In this way, the regular behaviour of local circulations can be quantified and used to good effect in making composites.

Very Small Scale Observations

Denmead et al. (1999, 2000) described a profile technique to find methane emission from a 24 × 24 m cattle lot. Net transport into a volume 3.5 m high was found by differencing profiles on the up- and downwind sides of the volume. The authors estimated that the error in the surface flux estimation was 30–40% and opined that it would be extremely difficult with existing instrumentation to measure weak fluxes (e.g. CO_2 from soil respiration) over small areas using this approach. Wojcik and Fitzjarrald (2001) were the first (and probably the last) to use the boundary budget method to estimate turbulent and radiation fluxes above and below a newly poured concrete bridge deck (10 × 30 × 0.5 m). Comparisons with laboratory studies verified that they were able to infer the thermal source term from the exothermic concrete reaction from environmental measurements alone.

CBL Gradient Approaches

Many authors advocate inferring surface fluxes from gradients in the CBL, using the Wyngaard-Brost CBL similarity hypothesis cited above. Davis et al. (1994) estimated surface isoprene fluxes over the tropical and midlatitude forests. Direct surface fluxes to compare with were not available, but their resulting estimates were in line with results in the literature. A troubling finding was that differences of up to 40% rested on the choice of displacement height. Kiemle et al. (op.cit.) estimated Fqh in BOREAS by interpreting mean q profiles obtained with airborne DIAL (Differential Absorption Lidar). They found that q entrainment rate estimates were strongly dependent on the h estimate. The CBL gradient approach is also natural for analysing data from "tall towers" (e.g. Bakwin et al. 1998), but this application too will require careful sorting of data to avoid periods with dominating baroclinic influences.

Using Surface Data for Approximate CBL Budgets

Regular sounding for most trace gases is not yet feasible. Raupach (1991) presented an innovative approach to combine sounding and surface mean data to estimate the heat and water vapour fluxes. Taking advantage of occasional upstream CO_2 observations and an onshore sea breeze to estimate free atmosphere concentrations, Raupach et al. (1992) compared CO_2 fluxes from a hybrid CBL budget fed by surface data with direct flux observations (Fig. B.17). They argued that finding the agreement to within 30% represents "a very satisfactory result, given the crudeness of the approximations we have used to estimate h". Potasnak et al. (1999) took advantage of the CO/CO_2 ratio for tower observations to identify air masses originating in industrial areas and infer mesoscale carbon uptake at a site in the NE US.

Fitzjarrald et al. (2001) used daily changes in surface temperature and humidity to infer seasonal changes owing to phenology in the same region. They found that

average surface flux convergence into the CBL peaks in springtime, reaching no more than 5% when 30-yr daily averages are formed. Freedman and Fitzjarrald (2001) showed that this jumps to 50% when only "accumulation" periods following cold fronts are considered. Denning et al. (1999) recently introduced the "rectifier" concept, arguing that "photosynthesis and thermally driven buoyant convection in the atmosphere are both driven by solar radiation, and therefore "beat" on the same diurnal, synoptic, and seasonal frequencies". At the time of leaf emergence, sensible heat flux H decreases dramatically and latent heat flux LE jumps upward. In regions with widespread deciduous forest cover (e.g. north-eastern US), mixed layer thickness h is largest just prior to leaf emergence (Freedman and Fitzjarrald 2001) which leads to the summer maximum in net carbon uptake. Mixed layer thickness and CO_2 are closer to being in quadrature, rather than in phase (Fitzjarrald et al. 2003).

B.6.5   Conclusions and Suggestions for Ongoing Work

Despite the many advantages of using a boundary layer budget method, few of the many published reports have been thorough enough to tell if the approach works. Being based on a small number of cases, most published work is in the "anecdotal" category. What is to be done? We cannot discount the budget method; we cannot expect everything from flux tower networks or a small number of tall towers. It seems that the best approach is to perform regular observations over long periods using multiple measurement platforms. Suggestions for future work include:

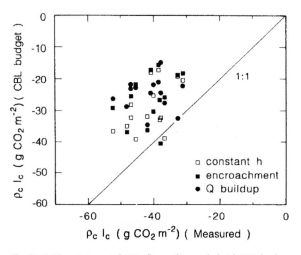

Fig. B.17. Time integrated CO_2 fluxes from a hybrid CBL budget approach compared with direct measurements in Australia. Results from using different mixed layer model growth assumptions are indicated (from Raupach et al. 1992, reprinted from the Australian Journal of Botany with permission from CSIRO Publishing)

- Use data compositing creatively. Seek $\overline{w}$ not only from model outputs but also from measured horizontal divergence on boundaries. Most results get better when averaged. This suggests a slightly different aircraft deployments strategy. Look for "synoptic events" and make sequential flux and sounding runs. Sampling by aircraft along likely upwind trajectories should be routine.

- Deploy remote sensing instruments to get the BL winds (e.g. profilers and acoustic sounders) and concentrations (e.g. DIAL) in horizontal arrays. Operate these networks for several seasons continuously. Month-long intensive observations will not be sufficient. Coulter and Kallistratova (1999) and Coulter et al. (1999) show that integrated sodar/RASS/radar arrays are now capable of obtaining horizontal divergences over mesoscale areas continuously. Ground-based DIAL sensors can return time series of O_3, CO_2 and water vapour. Furger et al. (1995) showed that horizontal heat flux divergence could be obtained in such networks in ideal conditions, though further study is needed to understand errors on many days.

- A less expensive approach is to develop hybrid schemes using continuous time series from surface data (Raupach et al. 1992; Potasnak et al. 1999). Se-lect data for transient periods of weak turbulence and estimate surface flux from $\partial C/\partial t$; use bounding value (envelope) analyses for guidance; take advantage of the early evening, morning transitions (Acevedo and Fitzjarrald 2001).

- Using light aircraft for repeated soundings may often be more efficient than deploying fleets of larger, more highly instrumented planes.

- Do not ignore the role of clouds. Freedman and Fitzjarrald (2001) demonstrate that fair-weather cumulus perforate the top of the budget box, releasing q from the mixed layer in reducing H and LE by intercepting sunlight. Their presence drives the BL toward constant relative humidity equilibrium.

- A continuing challenge is to move beyond CBL budgeting and to exploit the nocturnal case. One can use the regularity of local winds to find the advection terms. Such "natural" boxes as alpine valleys provide predictable circulations ideal for making composites (e.g. Graber et al. 1998). Drainage flows from coastal areas carry respiratory CO_2 from airsheds over water. Measuring the volume of the outflow and its excess CO_2 allows an independent estimate of the forest respiration rate.

Chapter B.7

Vegetation Structure, Dynamics and Physiology

Niall Hanan

B.7.1 Introduction

Vegetation measurements have evolved significantly during the series of integrated terrestrial experiments, with a broadening of interest in plant and soil processes that affect not only the canopy water and energy balance but, increasingly, ecosystem carbon balance and biogeochemistry. This evolution was related to increasing recognition of the importance of the terrestrial biosphere in climate, growing interest in the components of the global carbon cycle, and recognition of the importance of interactions between biosphere, atmosphere, management and biogeochemical cycles. An explicit focus within the integrated terrestrial experiment on the contribution of stand-scale processes to regional-scale processes means that consideration of spatial scale and heterogeneity, in both measurement and modelling activities, has been a necessary common theme.

This section discusses the role of measurements of vegetation structure, dynamics and physiology in integrated terrestrial experiments. In this context vegetation measurements are taken to include measurements of physical, structural and physiological state variables as well as measurements of vegetation processes, including plant and leaf-level fluxes of heat, water and carbon dioxide. The section will focus on the nature and scope of "vegetation" studies during past and current integrated terrestrial experiments, and how such measurements contribute to the larger goal of understanding the complex interactions between biological, biophysical and physical processes, both within ecosystems and at the interface between ecosystem and atmosphere.

B.7.2 Measurement of Vegetation Structure, Dynamics and Physiology

Vegetation Structure

Vegetation structural parameters include height, density, size and spatial distribution of trees, shrubs and herbs as well as the standing biomass of various above ground fractions (e.g. wood, leaves and reproductive structures) and below ground fractions (coarse and fine roots). Leaf area index (LAI) is commonly measured, as are various aspects of canopy architecture, including leaf angle distribution and the spatial distribution of leaves and branches within the canopy volume. In most cases, measurements are separated by broad physiognomic or physiological type (e.g. trees versus herbs, grasses versus forbs, C_3 versus C_4), although there has been less emphasis within the integrated terrestrial experiment on measurement of structural parameters to species level. Gap fraction and the interception of solar and photosynthetically active radiation are frequently measured since these relate to the energy absorption by the canopy and thus to energy balance, transpiration and photosynthesis.

Point-in-time measurements of less dynamic structural parameters, including height and density of trees and shrubs, canopy architecture and, in perennial vegetation without marked seasonality, leaf area indices are useful to define the basic characteristics of the vegetation. However, structural quantities such as LAI, biomass and gap fraction are often markedly seasonal, requiring repeat measurements to fully characterise intra- and interannual variations.

A variety of methods can be used to survey vegetation structure, including quadrat count and line transect methods to determine cover, size and density of woody and herbaceous species and harvest techniques for biomass estimation. With knowledge of the specific leaf area (leaf area / leaf mass, $m^2 kg^{-1}$), biomass estimates can be used to estimate LAI. Field methodologies and theoretical considerations for these common ecological field techniques are reviewed in numerous publications (e.g. Coombs et al. 1987; Krebs 1999). Harvest measurements of above ground biomass fractions, using quadrats or selecting individual plants, can be accurate and convenient in short-growing vegetation where the samples are incorporated in an appropriate experimental design. Measurements at randomised or pre-selected harvest locations seek either to obtain a good estimate of the average conditions across a site or, more frequently, seek to estimate the average in several classes based on biomass, geomorphology or species. In the latter "stratified" sampling design the relative importance of each class is estimated by line transect or other methodology and then combined with the dry mass measurements to determine spatially averaged site biomass.

The problems with harvest measurements arise when it is necessary to consider root biomass, and in woodland and forested sites where the size of the trees makes it difficult to obtain a sufficient sample. Root biomass estimation is difficult for several reasons: (*i*) horizontal and vertical variability in root mass associated with trees, shrubs and herbaceous species is often significant, meaning that sampling methods (e.g. soil cores) may require a large number of samples to obtain a meaningful average; (*ii*) complete measurement of the roots of individual trees, shrubs (or even perennial grasses) requires the slow and arduous excavation of a soil pit surrounding the individual whose roots may extend both horizontally and vertically; and (*iii*) field and laboratory identification of live roots, and separation of fine roots, can be very difficult and the various methods that have been developed are often soil species specific or difficult to apply (Boehm 1979). In HAPEX-Sahel, while several groups devoted considerable effort to obtaining good measurements of above-ground biomass, relatively few measurements of below-ground root dynamics were made (Monteny 1993; Hanan et al. 1997c). In woodland and forest sites, the sheer mass of individual trees means that harvest measurements can be few in number at best, and repeat measurements to estimate increments become impractical and of doubtful accuracy at anything less than one-to-five year intervals, depending on site heterogeneity and age class distributions.

For experiments in forest sites where harvest estimates are difficult, allometric relationships between more easily measured parameters (density, diameter at breast height, etc.) are frequently employed to estimate quantities such as stem biomass, total wood mass, root mass and leaf mass/area. Such calibration relationships can be established on a restricted number of trees harvested expressly for the purposes of the experiment, while in some cases more general forest mensuration measurements for the region are employed. However, while allometric relationships are useful to estimate less dynamic silvicultural parameters related to diameter at breast height, such as timber volume, they may not be so well established for the parameters of interest in the integrated terrestrial experiment (e.g. leaf mass and leaf area). At best, where allometric relationships are available for leaf mass/area, they may assume no interannual variation in leaf dynamics and offer little or no information on the seasonal leaf dynamics of deciduous species.

Indirect measurements of vegetation structure based on radiative transfer through the canopy are frequently employed because of their ease of use and non-destructive nature. The methods include hemispherical photography (Bonhomme and Chartier 1972; Chen et al. 1991; Rich et al. 1993; Fournier et al. 1996), measurements of angular gap fraction using directional radiation sensors (e.g. Gower and Norman 1991; Cohen et al. 1995; Hanan and Bégué 1995; Chen 1996; Stenberg 1996), and measurements of solar radiation penetration using light bars or small sensors set out in measurement arrays (e.g. Evans et al. 1960; Gower and Norman 1991; Bégué et al. 1996b; Chen 1996). The photographic methods require postprocessing of the gap fraction in the image while the radiometric methods require measurements of incident radiation to compare with the radiative flux at the measurement height. Critical analyses and comparisons of these methods are presented in several publications (e.g. Chason et al. 1991; Fassnacht et al. 1994).

Light penetration methods can be used to estimate a number of canopy characteristics, including gap fraction, leaf area, leaf angle distribution, and canopy cover. Each of the methods has advantages and drawbacks and provides direct measurements of a limited set of parameters only (e.g. diffuse gap fraction, proportional penetration of solar radiation). Measurements of fractional light interception are directly useful in energy balance and photosynthesis research (e.g. Hanan and Bégué 1995; Bégué et al. 1996a,b). However, a common application of light interception techniques is in the estimation of derived quantities, particularly LAI. LAI from light penetration measurements relies on assumptions about the nature of radiative transfer in vegetation canopies related to the spatial distribution of leaves and the degree of departure from random. The latter "clumping index" is complicated by the size of individual leaves and needles (which being non-zero immediately impart a degree of clumping), the spatial distribution of leaves and leaf angles, the angular distribution of incident radiation and spectral reflectance of canopy materials in the wavelengths to which the instrument is sensitive. Thus the clumping index cannot be measured directly and correction of LAI estimates for leaf clumping requires inverse calibration using independent (usually harvest) estimates of leaf area indices. While much progress has been made in this field, the accuracy of LAI measurements using light penetration techniques depends on extensive ancillary information.

Vegetation Productivity

The net productivity of plant canopies is related to the temporal dynamics of vegetation structure. Production measurements thus frequently employ similar techniques to those discussed in the previous section. In annual and deciduous vegetation, where seasonal net production is of similar magnitude to biomass, repetitive measurements of standing biomass through the season is sometimes used to approximate net production of leaves, stems and roots. Losses through herbivory and senescence should be measured (or small) for this technique to capture net production dynamics. However, in perennial and non-seasonal vegetation, net production of leaves, wood and roots may be only a small fraction of the standing biomass. In these situations, the census of biomass change, with the inherent errors involved, may not be of sufficient accuracy to determine net production within reasonable error margins. Leaf production on an annual to multi-annual basis can be estimated using litter fall techniques and woody stem growth can be estimated by measurement of radial increment using a model of stem and branch architecture. In many cases, particularly in tall vegetation where harvest measurements are difficult to apply, net production of leaves, stem and roots relies on allometric relationships derived from a limited number of destructive samples on-site or in supposedly similar situations of vegetation, climate and soils at nearby or distant sites.

In the case of roots the development of minirhizotron and root in-growth methods means that field estimation of root production is now more tractable than measurement of total root mass, using relatively non-intrusive techniques (Bragg and Cannell 1983; Aber et al. 1985; Caldwell and Virginia 1991). However, rapid production and turnover of fine roots is still not fully captured by these techniques, resulting in a tendency to underestimate total production.

Vegetation Physiology and Function

Measurement of vegetation physiological and functional properties at leaf and plant levels allows a better understanding of how the component parts, the different functional and physiological groups (e.g. forbs, grasses, shrubs, seedlings and mature trees, C_3 and C_4 species), contribute to the behaviour of the whole system. Plant and leaf level physiology varies spatially: with differences in soil and canopy microclimate, seasonally with weather and leaf age, and interannually as perennial plants grow and mature. Within the integrated terrestrial experiment, measurements of stomatal conductance, sap flow and leaf water status have contributed to an understanding of the controls on transpiration and surface energy balance, while measurements of leaf and branch photosynthesis and respiration contribute to understanding canopy photosynthesis and net ecosystem exchange.

Measurements using leaf chambers and branch bags provide information from individual leaves and leaf-assemblages on stomatal conductance and photosynthetic rate that can be compared to canopy-level equivalents. Measurements across a range of light, humidity and CO_2 concentrations in environmentally controlled chambers allow stomatal and photosynthesis parameters to be estimated. These include the empirical relationships describing stomatal behaviour and photosynthesis parameters such as maximum photosynthetic rate, Rubisco capacity, dark respiration rates, apparent quantum yield and light and CO_2 compensation points (Farquhar and Sharkey 1982; Farquhar and von Caemmerer 1982; Brooks and Farquhar 1985; Ball 1987; Ball et al. 1987).

A variety of techniques to measure the flux of water in the stems of plants is now available, including collar and probe systems using heat pulse, constant power or constant heat flux methods (e.g. Cermak et al. 1973; Granier 1985; Pearcy et al. 1989). Sap flow measurements allow the assessment of contributions by different species and functional groups to overall canopy transpiration. Most applications of the method are on shrubs and trees, although some systems are now small enough to measure larger herbaceous species. Using relationships between sap flow in an individual, and stem size, leaf area or other suitable allometric descriptor, sap flow measurements can be scaled up to estimate the contribution of populations to ecosystem-level transpiration.

Leaf water content and leaf water potential are often measured in the integrated terrestrial experiment since they relate to plant water status and overall physiological condition. Leaf water content can be measured with minimal extra effort during collection of biomass samples for drying. These measurements are also useful in assaying the water status of the vegetation for remote sensing applications such as the interpretation of radar back-scatter measurements (Magagi and Kerr 1997). Pre-dawn leaf water potential measurements can be used as a surrogate for root-zone integrated soil water availability and thus as an index of potential transpiration rates during the day following measurements. However, although the common assumption in pre-dawn water potential measurements is that the vegetation comes to equilibrium with the soil water potential during the course of the night, plant water potentials are often much more negative than soil water potentials across the rooting depth (e.g. Hanan et al. 1997b). This indicates that plants are unable to take up sufficient water during the night to reach equilibrium or, perhaps more likely, that rhizosphere water potentials are not well represented by bulk soil water measurements.

Leaf nitrogen concentration is measured as an index of the physiological capacity of the plants and related to phenological changes in photosynthetic activity as well as spatial variations within the canopy in response to variations in microclimate (particularly light climate). The influence of light climate at different locations in a canopy on leaf nitrogen concentration, photosynthetic capacity and specific leaf area is of great interest to studies of the physiological adaptations of plants to their environment and the ways in which individual leaves contribute to canopy-level fluxes. In particular, models describing canopy processes from leaf-level require information on how light and physiological capacity covary at different positions within the canopy in order to develop leaf-to-canopy scaling methodologies.

B.7.3 Vegetation Measurements in the Integrated Terrestrial Experiment

Measurements of vegetation structure, dynamics and physiology are key to understanding landscape-level processes and biosphere-atmosphere interactions. In the integrated terrestrial experiments vegetation measurements contribute to a wide range of studies aimed at (*i*) understanding the contribution of individual leaves and plants to canopy-scale function, (*ii*) remote sensing of vegetation structure and function, and (*iii*) parameterisation and validation of simulation models of ecosystem processes. Some examples of these studies are given below.

Numerous studies in the integrated terrestrial experiment have examined the photosynthetic capacity and stomatal behaviour of dominant plant species, and related chamber measurements to canopy-scale evapotranspiration and CO_2 flux (Polley et al. 1992; Dang et al. 1997; Flanagan et al. 1997; Hanan and Prince 1997; Levy et al. 1997; Middleton et al. 1997). A number of studies measured sap flow in the main tree or shrub species (Allen and Grime 1995; Soegaard and Boegh 1995; Hogg et al. 1997; Tuzet et al. 1997; Boegh et al. 1999) and discussed methods for comparison of sap flow measurements to net ecosystem fluxes measured by eddy covariance. While scaling stomatal conductance to estimate canopy con-

ductance is complicated by variations in physiology and environment within the canopy, scaling sap flow measurements from individual stems to estimate stand-level transpiration is a convenient way to determine the contribution of trees or shrubs to overall ecosystem evaporation (e.g. Tuzet et al. 1997; Fig. B.18).

Vegetation measurements in the integrated terrestrial experiment have been used to interpret and apply remotely sensed optical, thermal and microwave measurements. Optical remote sensing has focused on remote estimation of photosynthetically active radiation interception (e.g. Demetriades-Shah et al. 1992; Hall et al. 1992; Walter-Shea et al. 1992; Brown de Coulston et al. 1995; Hanan et al. 1997a), LAI and biomass (Turner et al. 1992; Fassnacht et al. 1994; Chen et al. 1997), because of the importance of these structural properties in surface albedo, canopy photosynthesis and transpiration. Measurements of thermal emissions have been used to estimate surface resistance and sensible heat flux (e.g. Friedl 1995; Boegh et al. 1999) (see also Chapt. B.8) which, with estimates of net radiation, allow inference of energy partition and latent heat fluxes. Microwave measurements in certain frequencies show promise for estimation of canopy water content and, by implication, canopy leaf area and green biomass (Magagi and Kerr 1997).

The wealth of complementary measurements in the integrated terrestrial experiment at leaf and canopy scales have provided invaluable data for the development, parameterisation and validation of simulation models describing physical, biophysical and geochemical processes in terrestrial ecosystems. Examples include biophysical models describing the interactions controlling transfers of heat, water and carbon dioxide within the canopy-soil complex and between the vegetation and atmosphere (Hope 1992; Blyth and Harding 1995; Sellers et al. 1995b; Braud et al. 1997; Colello et al. 1998; Hanan et al. 1998; Grant et al. 1999; Pauwels and Wood

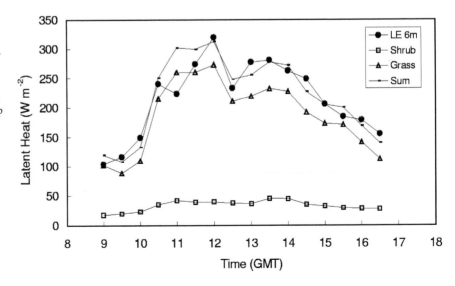

Fig. B.18.
Latent heat (LE) partitioning in Sahelian fallow-savanna between grass and shrub components. Net ecosystem and grass-layer LE were measured by eddy covariance systems at 6.0 m and 0.75 m, respectively. Shrub LE was measured by sap flow on 5 stems, scaled to estimate overall shrub transpiration by stem basal area. Data are from HAPEX-Sahel East Central site on 31 August 1992 reported in Tuzet et al. 1997, © reprinted with permission from Elsevier Science

1999); models of soil and vegetation biogeochemistry and productivity (e.g. Kimball et al. 1999; Peng et al. 1998; Parton et al. 1996); and models of radiative transfer and surface reflectance (Bégué et al. 1994; Bégué et al. 1996a; Van Leeuwen and Huete 1996; Ni et al. 1997; Nijssen and Lettenmaier 1999; Roujean 1999). The radiative transfer studies show that canopy architectures in savannas

(Bégué et al. 1996b) and boreal forests (Ni et al. 1997) confer sufficient heterogeneity in the light environment that simple diffusive media radiative transfer is not able to capture the patterns of light interception. Hybrid models, that combine canopy geometric-optics with diffusive attenuation, represent average canopy light climate better in a one-dimensional framework (Fig. B.19).

Fig. B.19.
Comparison of modelled and measured vertical transmission of photosynthetically active radiation (PAR) in boreal forests in (**a**) a jackpine stand (BOREAS southern old jackpine) and (**b**) black spruce (BOREAS southern old black spruce). Data are from Ni et al. 1997. Transmittances are relative to above canopy incoming PAR

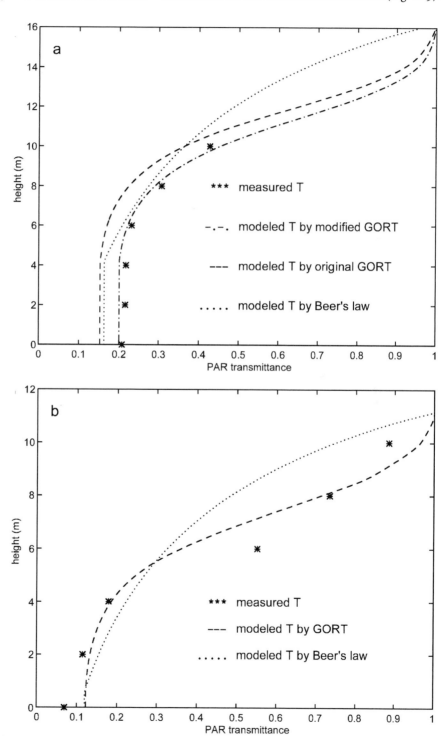

B.7.4 Synergy of Vegetation Measurements with Other Measurement Programmes in the Integrated Terrestrial Experiment

Vegetation measurements in the integrated terrestrial experiment are designed to be part of the larger integrated experiment. As such they are of great value in interpreting and understanding the fine-scale processes at plant and leaf scales that contribute to canopy and landscape behaviour observed by, for example, eddy covariance towers, aircraft or satellite remote sensing. However, the combination of measurements can result in synergies that lead to new understanding or allow the assessment of ecosystem properties not easily measured directly. In a modelling study Hanan et al. (1998) analysed net ecosystem CO_2 fluxes in a fallow-savanna site during HAPEX-Sahel, and partitioned the total measured flux into photosynthetic uptake, above ground, root and soil respiration. An example of synergy is shown in Fig. B.20a, where net photosynthetic CO_2 uptake (expressed as CH_2O production) and harvest measurements of above ground biomass production are used to estimate, by difference, the production and turnover of roots during the growing season. While dependent on the accuracy with which net uptake rates and above ground production are estimated (Hanan et al. 1997c; Hanan et al. 1998), this combination of analyses can provide first-order estimates of root production, which is notoriously difficult to estimate by other means. The analysis shows the full phenological cycle, from leaf initiation to senescence, of this highly seasonal environment. Root production during most of the season is more than double the combined above ground leaf and wood production.

The combination of net uptake and measured/inferred allocation to above ground and below ground biomass production also allows allocation coefficients to be estimated (Fig. B.20b). Allocation coefficients are commonly used in models of vegetation growth and dynamics, and can be difficult to determine empirically. These data show the variation through the season of allocation to above ground and below ground production. Root allocation in the early season is negative because leaf flush of the shrub component (*Guiera senegalensis*) depends on stored carbohydrate reserves. Thus early growth is not observed as a net ecosystem CO_2 uptake until the leaves of the canopy switch from being net consumers to net producers. However, after only 10 days of growth, photosynthetic production is sufficient for allocation to the roots to be positive and, during most of the growing season, allocation to root growth, turnover and replenishment of carbohydrate reserves accounts for more than 50% of net photosynthesis.

B.7.5 Summary

Vegetation measurements, defined here as measurements of vegetation structure, dynamics and physiology, are an essential component of the integrated terrestrial experiment. They provide a general understanding of the nature of the ecosystem and the contribution of species and functional groups, leaves and plants, to canopy-scale function. They are also central to parameterisation and validation of ecosystem process models, including canopy radiative transfer schemes, biophysical models of surface water and energy balances, and models of carbon exchange and production, and vegetation dynamics.

While vegetation measurements may involve difficult and arduous sampling schemes, particularly in forest ecosystems where biomass and production are hard to measure accurately, the benefit of these measurements to the overall success of the integrated experiments is such that the effort is well repaid. In tall and high bio-

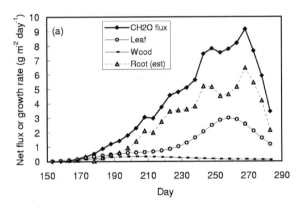

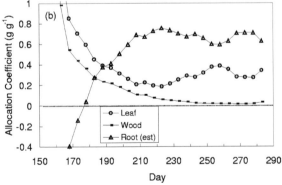

Fig. B.20. Estimation of root production and allocation dynamics of a Sahelian fallow-savanna using measured/modelled net ecosystem CO_2 exchange and harvest measurements of above ground net production. **a** Net CH_2O flux, measured leaf and above ground wood production and inferred root production. **b** Allocation coefficients for leaf, wood and roots. Daily net photosynthetic uptake was estimated from eddy covariance measurements minus estimated heterotrophic respiration. This analysis derived from data and analyses presented in Hanan et al. 1998

mass systems, development of strategies for assessment of vegetation biomass, leaf area and production, and the spatial variation in these characteristics is crucial. In so far as is practicable, allometric relationships describing intra- and interannual variations in these ecosystem properties should be developed directly and locally within the context of the integrated experiment to avoid reliance on relationships developed elsewhere or with less precision.

While outside the scope of this chapter, the increasing emphasis on soil and canopy biogeochemistry during the more recent integrated terrestrial experiments is a positive development. The biogeochemical cycles of nutrients, including nitrogen and phosphorous and other macro- and micro-nutrients, play an important role in ecosystem processes, including controls on productivity, structure and physiology of vegetation, decomposition and turnover of organic material in soils, and the production and consumption of radiatively and chemically active trace gases. Consideration of biogeochemical processes is essential as we aim for fuller understanding of the Earth system and the interactions between the biosphere, geosphere, hydrosphere and atmosphere that determine the role of the terrestrial system within the climate system and the response to anthropogenic land-use change.

Chapter B.8

Remote Sensing and Land-surface Experiments

Forrest G. Hall · Yann H. Kerr

B.8.1 Introduction

Global change studies require coupled global carbon, energy and water models that simulate the dynamics of forests, grasslands and agricultural regions and their response to climate change. Global datasets of land-use and land-cover change are needed, so that these models can be used to investigate the response of various ecosystems to climate variation and to analyse the associated feedbacks (Sellers et al. 1996a). Satellite remote sensing is the only feasible method of obtaining many of the parameters required on a global scale over long time periods. As shown in Fig. B.21 coupled carbon, energy and water models require as input land cover and biophysical parameters such as: vegetation type, total and green leaf area index (LAI), surface aerodynamic roughness, snow-free albedo, background reflectance, soil properties, topography and river routing, and variables such as: canopy temperature, FPAR, precipitation, runoff, soil moisture, snow and ice cover, atmosphere-surface radiation balance, cloud data, and the near surface meteorology.

Satellite data series for land monitoring began with Landsat in 1972. Landsat was followed by others such as the NOAA-AVHRR satellite/sensor series, which provided continuous, global coverage from about 1982. Unfortunately these early satellites suffered from unstable calibrations and the data could not be corrected for the radiative effects of the atmosphere. Therefore, it was difficult to use the data from these satellites to provide a meaningful record of land-cover change. Further, although algorithms existed for identifying land-cover type at local scales, none were available to reliably map land-cover type and the biophysical parameters needed to understand how the Earth's land surface and atmosphere interacted at seasonal, annual and interannual timescales. Only a series of coordinated field experiments, conducted over the major global biomes involved in global change, could provide a quantitative global observational capability. Such experiments were necessary to identify the precise measurement requirements

a) **Surface Energy Budget**

Net Radiation Absorbed = R_n =

Net Short Wave $\{Sw_\downarrow(1-\alpha)\}$ + Net Long Wave $\{Lw_\downarrow - Lw_\uparrow\}$

Remote Sensing Inputs

Atmospheric Aerosols and Clouds, Surface Albedo $\Longrightarrow$ Net Short Wave $\{Sw\ (1-\rho)\}$

Atmospheric
Water Vapor, Temperature Profiles, Cover Type (ε) $\Longrightarrow$ Net Long Wave $\{Lw - Lw\}$

b) **Surface Heat and Mass Budget**

Net Radiation Absorbed = R_n =

Latent Heat (LE) + Sensible Heat (H) + Ground Heat Flux (G)

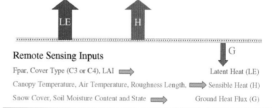

Remote Sensing Inputs

Fpar, Cover Type (C3 or C4), LAI $\Longrightarrow$ Latent Heat (LE)

Canopy Temperature, Air Temperature, Roughness Length, $\Longrightarrow$ Sensible Heat (H)

Snow Cover, Soil Moisture Content and State $\Longrightarrow$ Ground Heat Flux (G)

c) **Surface Carbon Budget**

Net Ecosystem Exchange = NEE = Gross Primary Production (GPP) -

Autotrophic Respiration (R_a) - Heterotrophic Respiration (R_h)

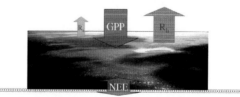

Remote Sensing Inputs

LAI, Fpar, Cover Type, Soil Moisture, Temperature $\Longrightarrow$ Gross Primary Production (GPP)

Biomass, Cover Type, Temperature, $\Longrightarrow$ Autotrophic Respiration (R_a)

Soil Moisture, Soil Temperature, Snow Cover $\Longrightarrow$ Heterotrophic Respiration (R_h)

Fig. B.21. Physical and biophysical inputs obtainable from satellite remote sensing to initialise, force and validate hydrometeorological and biogeochemical cycling models. **a** Surface-atmosphere radiation exchange; **b** surface-atmosphere energy exchange; **c** surface-atmosphere carbon exchange

and acquire the datasets needed to develop and validate the remote sensing algorithms. To be generally applicable the algorithms needed to be based on canopy radiative transfer theory, requiring that the theory itself had to be developed and validated.

Under the auspices of the International Satellite Land Surface Climatology Project (ISLSCP), a series of field experiments focused on establishing:

1. the process model input requirements (parameters and accuracies);
2. the development and validation of the remote sensing algorithms that relate top-of-the-atmosphere observed radiance fields to the required surface parameters for a homogenous surface;
3. the accuracy with which the surface-reflected and emitted radiance fields could be observed over long time periods (calibration and atmospheric correction);
4. the development and validation of canopy radiative transfer theory (models) underlying the algorithm physics and,
5. the scale invariance of the algorithms as surface and atmospheric heterogeneity come into play at coarser geographic scales.

A series of experiments (see Chapters A.5–A.7) HAPEX-MOBILHY, FIFE, BOREAS, EFEDA, HAPEX-Sahel, NOPEX, SALSA (Semi-Arid Land-Surface-Atmosphere) and LBA were designed and implemented, sponsored by a number of US, Canadian, European and South American agencies.

Field experiments have thus been pivotal in the development of global change models, the development and validation of remote sensing algorithms, and in improving sensor calibration and atmospheric correction techniques.

B.8.2 Remote Sensing Input to the Experiments

An important criterion in selecting the locations of the field experiments was that they should be representative samples of the Earth's major biomes – particularly those that are experiencing large-scale change in land use (the Sahel, Amazonia) or likely to be sensitive to the effects of climate change (the Sahel and boreal regions). The timing of the experiments reflected a graduated series of increasingly difficult challenges. The experiments were designed to coordinate process studies with remote sensing investigations, using satellite, airborne, and surface-based instruments. In the initial stages of experiment design, remote sensing images provided local and regional land-cover maps to select study sites within biomes and to pinpoint measurement locations representing the important biome vegetation communities.

As components of the experiments, remote sensing studies were essential to scaling up process models from leaf and plot levels, to regional and global scales. Large-scale validation techniques were incorporated in the field experiments to test scale-integration methods directly; these techniques included airborne flux and profile measurements (see Chapt. B.5), meteorological observations, and modelling. For example, in the BOREAS experiment (see Table B.7), efforts were implemented to characterise surface component optical and microwave properties from the leaf level (using laboratory spectrometers), to the canopy and stand level (using tower and helicopter-mounted spectrometers), and to the study area and regional level (using aircraft and satellite platforms).

These measurements permitted linkages between optical and microwave remote sensing, and vegetation biophysical parameters at scales that include leaf, canopy and regional levels using field, aircraft and satellite-borne sensors and radiative transfer models. A wide range of biometric and radiometric data were collected and archived by these experiments.

Remote sensing science investigators developing the algorithms produced multi-year site-level, study area level and regional scale maps of radiation and biophysical parameters to provide inputs for ecosystem or climate models and to evaluate the expected performance of future space-borne sensors. These datasets, when combined with the wealth of multi-disciplinary field measurements, provide a diverse and unique database for advancing the study of the various subject ecosystems and their interactions with the atmosphere, as well as for developing and testing new algorithms for mapping terrestrial biophysical parameters. Specifically, they prepare the way for the next generation of Earth-observing satellites. The data have been archived in a number of Earth science data archives and are generally available electronically or on CD-ROM (see also Part C).

B.8.3 Impact of Field Experiments on Remote Sensing of the Land Surface

A major focus of the field campaigns was the development of remote sensing algorithms for deriving seasonal, annual and decadal maps of vegetation type and biophysical properties at regional and global scales. These parameters are important in modelling the photosynthetic uptake of carbon, and the physiologically coupled release of water and its effects on the surface energy budget. Algorithms were also developed and tested for measuring incident short- and long-wave radiation, and Photosynthetically Active Radiation (PAR), the fraction of these absorbed by the vegetation, and the subsequent release of energy back to the atmosphere in the form of reflected short-wave and emitted long-wave radiation, and latent and sensible heat.

Table B.7. BOREAS ground, aircraft and satellite instruments with measurements and related products (FPAR: Fraction of photosynthetically active radiation)

	Measurements	Products
Ground-based sensors		
PARABOLA	3-band bi-directional radiance	Bidirectional reflectance distribution function, spectral albedo
Laboratory and field	Spectral radiance spectroradiometers	Understorey and tree component reflectances
Remote sensing aircraft/sensor		
NASA DC-8/AIRSAR	Radar backscatter	Forest type, biomass density, canopy moisture
NASA ER-2/AVIRIS, MAS	Spectral radiance	Forest type, canopy moisture, atmospheric properties
Piper Chieftain/CASI	Spectral radiance	Bidirectional reflectance distribution function
NASA C-130	TMS, ASAS, MAS, POLDER	Spectral radiance, bidirectional radiance, bidirectional reflectance distribution function, albedo, forest type, FPAR
NASA helicopter	MMR, SE-590, POLDER	Spectral radiance
Radar backscatter	C band scatterometer	Bidirectional reflectance distribution function, FPAR, canopy scattering profiles
Remote sensing satellites		
AVHRR	Spectral reflectance, emittance	Land cover, FPAR
Landsat TM	Spectral reflectance, emittance	Land cover, FPAR, LAI, biomass density
SPOT	Spectral reflectance	Land cover
ERS-1, 2 SAR	Radar backscatter	Freeze/thaw
JERS-1 SAR	Radar backscatter	Land cover
Radarsat	Radar backscatter	Land cover
SIR-C/XSAR	Radar backscatter	Land cover, biomass density
GOES	Spectral radiance	PAR, albedo, downwelling irradiance

Normalised Difference Vegetation Index

At the outset of these experiments, the Normalised Difference Vegetation Index (NDVI) and its derivatives were widely used for continental to global monitoring, but with limited understanding and validation. Global images of composited NDVI from the AVHRR sensor (AVHRR NDVI) corresponded well with known surface patterns of vegetation type and their variation with climate, but the quantitative use of vegetation indices to monitor surface energy, water and carbon exchange had not been developed. Exactly what were the vegetation indices measuring and what was their dependence on extraneous effects such as atmospheric and sun angle variations? By combining ground and aircraft-based measures of surface reflectance, vegetation biophysical properties, and atmospheric effects for a number of different vegetation types, under a wide range of seasonal and meteorological conditions, field experiments quantified the behaviour and utility of a variety of vegetation indices and stimulated their wide use.

For example, it was found that NDVI responds nearly linearly to the fraction of incident photosynthetically active radiation absorbed by the photosynthetically active tissue in the canopy (fAPAR) and not that absorbed by non-photosynthetic elements (Hall et al. 1992). Thus

NDVI provides a direct measure of vegetation photosynthetic capacity. These findings stimulated wider use of NDVI and other similar vegetation indices as a driving variable for global monitoring of photosynthesis and evapotranspiration and thus the development of global-scale, surface-vegetation atmosphere transfer models (Sellers et al. 1996a).

The coordinated model and algorithm development activities also led to a quantitative understanding of the input accuracy requirements of carbon, water and energy process models, with consequent implications for remote sensing algorithms, sensor calibration and atmospheric correction requirements. For example, by establishing that an accuracy of 0.05 for NDVI is required to track climate-induced variations in vegetation, it was identified that atmospheric and calibration errors need to be kept to less than 1% in absolute reflectance for the red band and 4% for the near-infrared. Tests conducted during the field experiments, utilising a dense network of ground-based observations of atmospheric optical depth, combined with atmospheric correction algorithms, demonstrated that satellite observations could be corrected to this accuracy (Hall et al. 1992).

Detailed measurements in a number of field experiments also found the relationship between NDVI and photosynthetic capacity to be dependent on background, view and sun angle (Deering et al. 1992). These findings led to the development of global algorithms that correct composited AVHRR NDVI time series for sun angle and other variations to track more accurately climate and disturbance-induced variations in vegetation photosynthetic capacity. These algorithms made possible the use of multi-year NDVI time series to estimate vegetation net primary production from seasonally accumulated models (Sellers et al. 1996b).

Surface Temperature and the Energy Balance

Another important development resulting from the field experiments was the realisation that the partly physical, partly physiological chain linking NDVI, FPAR, photosynthesis, and latent heat flux provided a powerful way for inferring the partitioning of the surface energy budget – more effective even than the computation of surface sensible heat flux from thermal remote sensing. The thermal remote sensing measurements in FIFE were the first to show that surface temperature could not be used to compute surface sensible heat flux to a useful accuracy when the aerodynamic surface temperature is not equal to the radiative surface temperature (Hall et al. 1992). This condition occurs in sparse vegetation where the observed radiative temperature of the surface may be dominated by the temperature of the bare soil, which does not contribute a proportional amount to the sensible heat flux. This result stimulated a number of research efforts aimed at using the thermal band more reliably for monitoring surface energy exchange, including the definition of an "excess" resistance in the aerodynamic resistance term of diffusion theory. More fruitful avenues have included combining time series of thermal measurements with atmospheric boundary layer theory to estimate surface-atmosphere convective energy exchange.

Surface Roughness, Soil Moisture and Moisture State

As shown in Fig. B.21 soil moisture, moisture state (frozen or liquid) and surface roughness are important variables in both surface radiation and surface energy and surface carbon budgets. Variations in soil moisture and surface roughness over space and time create variations in surface structure and electromagnetic properties, causing in turn variations in the amount of microwave radiation emitted or scattered from the Earth's surface. These variations provide an entrée for using microwave remote sensing to infer surface moisture and roughness using algorithms rooted in canopy radiative transfer theory. While truck-mounted, aircraft and space-borne instruments have been developed over the past 40 years to measure variations in emitted or scattered microwave radiation, operational inference of soil moisture, moisture state, vegetation opacity, and surface roughness remains problematical at high frequency or for active systems. Both theoretical and practical challenges have slowed the development of a space capability to measure surface moisture and roughness. The microwave signal emitted or scattered from the surface responds in a complicated way to surface roughness, vegetation structure, vegetation moisture content and soil moisture. For example in FIFE, Wang (1995; Wang et al. 1992) showed that L band microwave brightness temperatures varied as much with variations in topography and vegetation as with soil moisture. To estimate the magnitude and interactions of these separate effects, a robust scattering theory is needed, which in turn specifies the need for algorithms using muliple polarisations and bands or view angles to sort out the various effects. During HAPEX-Sahel and FIFE several studies focused on the use of both multisensor and multifrequency (L and C bands) measurements to infer surface parameters and variables showing that synergistic approaches, multiangular or multifrequency observations, must be used to retrieve both soil moisture and vegetation amount reliably. Experiments in HAPEX-Sahel and FIFE showed that the moisture contribution of the overlying vegetation can in some cases be accounted for using both L and C band sensors and the relationship between the optical thickness at L and C bands. However, the L band single polarisation and the C band dual polarisation passive microwave data are complementary only if additional information on soil moisture measurements are available over the L and C band signal penetration depths. In the future, it will be necessary to test, compare and analyse the different concepts, including L band multi-angular polarised data, in order to assess which concept is more suitable for inferring surface characteristics over natural areas where the surface parameters are closely dependent.

Sensor resolution from space (> 25 km) at passive microwave frequencies also presents a challenge when translating coarse, average soil moisture measurements into meaningful variation at metre scales. Plants respond to local, not average moisture and in some cases in a very non-linear fashion. Moisture, topography and roughness can vary considerably within a 25 km grid. Such variation could present difficult up-scaling problems, particularly in an extreme case where many landscape elements were near moisture saturation and the remainder at or below wilting point. Little data exist on the spatial variation of soil moisture or the scale invariance of soil moisture algorithms. During FIFE, Charpentier and Groffman (1992) studied surface soil moisture variability within an 80 m footprint and found that soil moisture variability was not affected by topography and that variability was lower when soils were wet than when dry, suggesting greater scaling issues during drought conditions. However, Sellers et al. (1995b), showed for a much larger (2×15 km^2) area within the FIFE study site that while the relationships linking surface and root-zone soil wetness to the surface transpiration rates were non-linear, soil moisture variability decreased significantly as the study site is dried out, which partially cancels out the effects of these non-linear functions. Clearly, more studies are needed to examine the scale invariance as a function of moisture level, scale and region. To reduce the effects of spatial variations in surface roughness on soil moisture measurements, change detection techniques have been used to detect changes in soil moisture.

But perhaps the major factor inhibiting widespread use of remotely sensed soil moisture data is the lack of long-term, regional and global datasets and appropriate satellite systems to acquire them. For the most part passive microwave data have been collected only from short duration aircraft campaigns, or from the Scanning Multichannel Microwave Radiometer (SMMR) and Special Sensor Microwave/Imager (SSM/I) passive microwave satellites, inadequate for observing soil moisture through most vegetation. However, even with this restriction, global soil moisture estimates have been made using these satellites. Theory shows that data from the SSMR passive microwave system is better for soil moisture estimates than the SSM/I data, because SMMR is more sensitive to soil moisture (due to the lower frequencies available); however, its period of record is limited to 1982 to 1987. In both cases the footprint is rather large, varying from about 25 km for the SSM/I to about 150 km for the C band of the SMMR. The 150 km footprint, however, limits the utility of the soil moisture data for carbon cycling modelling. Investigations of more advanced satellite systems are under way, such as passive microwave systems using aperture synthesis to obtain higher spatial resolution.

Beginning with the free flyer SEASAT in 1978, and the Shuttle Imaging Radar (SIR) missions SIR-A, SIR-B and and SIR-C in 1981, 1984 and 1994 respectively, NASA has launched a number of active microwave (RADAR) space missions for biomass estimation, land-cover classification, change detection (burned area), flooding and inundation, and soil moisture. NASA has also developed and flown a number of aircraft active microwave instruments. Microwave sensing has at least two advantages over optical frequencies: (1) cloud penetration and (2) as the wavelength increases there is also increasing penetration of the vegetation to the soil. Thus, microwave sensing provides an all-weather imaging capability over a wider range of vegetation types, particularly important in regions with high amounts of cloud cover such as the tropics and the high-latitude boreal ecosystem. One main drawback to the existing Synthetic Aperture Radar (SAR) systems is that there are no existing algorithms for the routine determination of soil moisture from single frequency, single polarisation radars. A second limitation comes from long period between repeat passes; for the most part 35 to 46 days although the RADARSAT has a

three-day capability for much of the globe in a SCANSAR (wide swath, 500 km) mode.

Currently, no NASA radar missions are in orbit. As opposed to the dominance of the US in passive optical space missions, active microwave space missions have been dominated by others: Europe (ERS-1, ERS-2), Japan (JERS-1, ALOS), Canada (RADARSAT) and also Russia (Mir-Priroda and Almaz-1). Only limited regional vegetation data products have resulted from NASA space and aircraft microwave missions. SEASAT lasted only three months and the Shuttle missions a few days, limiting coverage to non-contiguous swaths; however, a global rainforest mapping project and global boreal forest mapping project, involving aircraft synthetic aperture radar SAR scenes from 1995 and 1996, are under way. NASA also plans to buy radar and elevation data from an airborne Interferometric Synthetic Aperture Radar for Elevation (INSARE) system. The data will be useful for a wide range of applications involving land-use, land-cover, and terrain modelling.

SAR systems offer perhaps the best opportunity to measure soil moisture routinely over the next few years. Currently, the European Remote Sensing (ERS-1) C band and Japanese Earth Resources Satellite (JERS-1) L band SARs are operating, as is the Canadian RADARSAT (also C band). Although it is believed that an L band system would be optimum for soil moisture, the preliminary results from the ERS-1 demonstrate its capability as a soil moisture instrument by using the wind scatterometer over land (spatial resolution 50 km, temporal resolution about four days).

Radiative Exchange at the Surface

Field experiments also led to the development and validation of remote sensing techniques to measure the surface-atmosphere radiative exchange. For example, by comparing observations of short-wave, long-wave and net radiation from the Geostationary Operational Environmental Satellite (GOES) with measurements from an extensive network of ground-based radiometers. Efforts in FIFE (see the FIFE I (Sellers and Hall 1992) and II (Hall and Sellers 1995) Special Issues of the Journal of Geophysical Research) quantified the error associated with the surface-atmosphere radiation budget, showing that:

- PAR can be inferred to an accuracy of 8.2 W m^{-2} with satellite algorithms;
- Solar insolation can be inferred to an accuracy of 21.6 W m^{-2};
- Surface albedo to about 3% absolute, about 15% relative;
- Downwelling longwave radiation to about 20 W m^{-2} and
- Net radiation can be inferred to roughly 50 W m^{-2}.

B.8.4 The Future

BOREAS catalysed several advances in remote sensing algorithm development that permitted boreal vegetation to be monitored by type and state, and to track changes that may be due to fire, direct human activity, or climate change. Algorithm developments during the various field experiments have led to the production of AVHRR-derived global vegetation maps spanning 1981 to 2000, time-series fields of land cover, biophysical parameters, phenology and snow cover. All these can be compared with the physical climate record and to seasonal and interannual variations in atmospheric CO_2 concentration. AVHRR data will also be used to monitor changes in the fire disturbance regime over the same period of record. The radiometric quality of the AVHRR data series will have to be enhanced to meet these tasks; this requires the development of techniques for improving long-term calibration and atmospheric correction of the data. The MODIS/MISR and other sensors that are better calibrated, atmospherically corrected and with more optimum bands, should soon provide significant additional capability for monitoring land. Finally, radar satellites such as ERS-1 and JERS-1 have been used to monitor the interannual variability in the freeze-thaw boundary in the boreal ecosystem, shown to be a key factor in the interannual variability of the carbon flux. To take advantage of the different attributes of optical and radar sensors, further remote sensing research and development is required; in particular, data fusion algorithms that combine optical and microwave sensors, as well as other data such as topographic data, could be developed to provide richer information about the biome.

Enhancement of land-atmosphere process models and large-scale parameter quantification using satellite data would be considerable achievements. Success in these two areas is almost certain, based on the early results and work in progress. However, the ultimate goal is to incorporate the improved process models and remote sensing datasets within large-scale energy-water-carbon models to calculate surface-atmosphere fluxes of these quantities for the biome over the period of record of the Earth-observing satellites, say from 1980 to the present. For this calculation, the surface state should be constrained by satellite data while the atmospheric conditions are specified from meteorological analyses or via direct coupling with a Global Circulation Model.

The interannual variations in carbon flux inferred from tracer and isotopic analyses could be compared with estimates generated by the integrated models to shed light on which processes are responsible for perturbations in the terrestrial carbon budget and where, geographically and biologically, they operate. In this way, field experiments should help bridge the huge gap that currently exists between global-scale inferences about the changing terrestrial carbon budget and local-scale understanding of the controlling ecophysiological processes. When this is done, it should be possible to construct useful predictive models that can anticipate future changes in the global physical climate system and the carbon cycle. In this context, the main challenges for land remote sensing are:

1. remote sensing of the spatial and temporal distributions of atmospheric CO_2 concentration to constrain model-based estimates of regional, continental and global variations of land/ocean/atmospheric carbon exchange.
2. remote sensing of land biomass and biomass change resulting from land-use change, logging, fire and changes in rates of growth resulting from climate change, CO_2 fertilisation and nitrogen deposition.
3. more direct measurements of vegetation productivity using remote sensing to observe rates of vegetation photosynthetic activity. Additional coordinated field experiments of the type mounted in FIFE, BOREAS, HAPEX-MOBILHY, EFEDA, NOPEX and HAPEX-Sahel will be needed to develop the remote sensing capabilities in concert with the modelling capability.

Chapter B.9

The Water Balance Concept – How Useful Is It as a Guiding Principle for the Design of Land-Atmosphere Field Experiments?

Dennis P. Lettenmaier · Bart Nijssen

B.9.1 Background

The series of land-atmosphere field experiments described in this chapter have been conducted, over the last decade and a half, with the overall goal of improving the understanding of the controls exerted by the land surface on fluxes of moisture and energy to and from the atmosphere. Among these experiments are: (1) HAPEX-MOBILHY conducted in southern France in the mid 1980s (André et al. 1986); (2) the First ISLSCP Field Experiment (FIFE) conducted in the grasslands of central Kansas at about the same time (Sellers and Hall 1992; Sellers et al. 1992); (3) HAPEX-Sahel in the early 1990s (Goutorbe et al. 1994, 1997a); (4) NOPEX (Halldin et al. 1998, 1999) and later WINTEX (WINTer EXperiment) in Scandinavia beginning in the early 1990s (Halldin et al. 2001); (5) BOREAS in the boreal forest of north central Canada from 1994 to 1996 (Sellers et al. 1995a, 1997; Hall 1999), as well as the various WCRP-GEWEX continental scale experiments. Among the latter, all of which are currently ongoing, LBA, MAGS, GAME (GEWEX Asian Monsoon Experiment) have had the most extensive field activities. All of these experiments were designed to measure the exchanges of energy and moisture between the land surface and the atmosphere over scales ranging from a few kilometres up to the size of large continental river basins. Some of the experiments also measured the exchange of CO_2 and other radiatively active trace gases between the land and atmosphere.

B.9.2 The Water Balance Concept

An underlying concept in most of these experiments has been to "close the water balance", i.e. to provide an estimate, based as directly as possible on observations, of each of the terms in the surface (or atmospheric) water balance. These water balance terms could then be used to diagnose model predictions and ultimately to improve model parameterisations. This chapter reviews the application of the water balance concept in land-atmosphere field campaigns, assesses how it has been implemented, and how useful it has been as a guiding principle for experimental design.

B.9.2.1 Snow-free Land Surface

At the land surface (and assuming the absence of snow cover) the water balance at a point, and at an instant in time, can be written:

$$P - E = Q + \nabla \int \rho \, \Theta(z) \, dz \tag{B.11}$$

where P and E are precipitation and evaporative fluxes at the surface, Q is surface runoff, $\int \rho \, \Theta(z) \, dz$ is depth-integrated soil moisture storage, and ∇ is the divergence. Eq. B.11 is not particularly useful unless it is applied over an area, as the divergence of soil moisture is difficult or impossible to measure at a point. Instead, the logical unit for observation is the river or stream catchment, over finite time, resulting in:

$$P_a - E_a = Q_s + \frac{dS_a}{dt} + Q_g \tag{B.12}$$

where P_a and E_a are spatially averaged (over the catchment) precipitation and evapotranspiration, respectively, Q_s is streamflow at the catchment outlet, dS_a/dt is the change in surface and subsurface (lakes and wetlands, soil moisture, and groundwater) storage, and Q_g is the net groundwater flux across the catchment boundary. Usually, an attempt is made to select catchments such that the net groundwater flux across the basin boundary is minimal, which allows the last term on the right-hand side of Eq. B.12 to be neglected. Spatially averaged precipitation is typically observed via a network of rain gauges. Although these are point measurements, there are well-established methods for estimating the errors in mean areal precipitation (see e.g. Rodriguez-Iturbe and Mejia 1974). Evapotranspiration is usually measured from flux towers, which although not technically point measurements because they "sample" an upwind footprint, usually represent areas considerably smaller than the size of catchments (usually several tens of kilometres, see

Chapt. B.4) for which runoff observation errors can be considered comparatively small. Therefore, diagnosis of water balance errors requires some method of transferring tower evapotranspiration (or more correctly latent heat flux) observations to the larger catchment.

Estimation of soil (and groundwater) storage change is likewise problematic because soil moisture changes occur over much smaller spatial scales (e.g. order of metres) than typical catchment sizes. In principal, storage change could be estimated via a sufficiently dense network of soil moisture profiling instruments (e.g. neutron probe or *in situ* time-domain reflectometry), but in practice this is usually cost-prohibitive. For a purely observationally based diagnosis, the problem can be avoided if time averages are taken over one or more annual cycles. Alternatively, one could use a modelling approach to estimate the storage change, which makes it easier to compute water balances over short time periods. Any such estimates have the disadvantage, of course, of being model-dependent.

B.9.2.2 Snowpack Water Balance

For experiments like NOPEX, which focus – at least in part – on the winter season, diagnosis of the energy and moisture budget of the snowpack is one of the primary objectives. In this case, the primary components of the water balance are precipitation, sublimation and evaporation, melt, and storage change. Unlike the snow-free surface condition, it is in principal possible to measure all terms at a point, as well as over an area. Notwithstanding wind-related effects on the catch efficiency of precipitation gauges, point observations are possible. Likewise, melt can be estimated at a point (or over small areas) using a variety of snow lysimeters, and storage change in the snowpack can be estimated either by manual measurements of depth and density profiles (e.g. snow pits), or with automated devices like snow pillows. Latent heat flux, the sum of sublimation and evaporation of liquid water in the snowpack, can be estimated via flux towers, despite complications with low temperature measurements. In some cases, expanding from the point to a defined area may reduce errors. For instance, combination of spatial networks of depth measurements with density information at a limited number of locations can help to reduce errors in estimation of storage and hence storage change. On the other hand, melt is difficult – if not impossible – to estimate over an area, so local observations must be extrapolated.

B.9.2.3 Atmospheric Water Balance

For larger, continental scale domains, like the GEWEX continental scale experiments, water balances can be computed for the atmosphere using the water balance equation. Specifically,

$$\vec{\nabla} \cdot \int \vec{v} q \frac{dp}{g} + \frac{d}{dt} \int q \frac{dp}{g} = E - P \tag{B.13}$$

where q is specific humidity, v is the horizontal velocity vector, p is atmospheric pressure, and g is the gravitational acceleration. The first term is the moisture flux convergence, the second is the change in moisture storage in the atmosphere, and P and E are precipitation and evapotranspiration at the land surface. Although Eq. B.13 is valid at a point, in practice it must be applied over an area although, unlike the surface moisture balance, the area can be of arbitrary geometry. The main practical constraint in application of Eq. B.13 is that accurate estimates of moisture flux convergence can only be made over very large areas, typically of order 10^5–10^6 km^2. Various methods have been used to estimate moisture flux convergence; purely data-based methods typically use the so-called "picket fence" approach, which makes use of the equivalence between a line integral around the boundaries of the spatial domain and the spatial average of the divergence. Change in atmospheric moisture storage is usually small over timescales larger than about a month but in any event, the second term in Eq. B.13 is computed using essentially the same data that are needed for the first. Precipitation estimates can be obtained using the same methods (and inheriting the same complications) as are present for surface budgets. As for the surface budgets, evapotranspiration is difficult to estimate (and in fact Eq. B.13 is sometimes used to compute evapotranspiration). In principal, local (e.g. tower) observations can be transferred to the larger area (see e.g. Nijssen and Lettenmaier 2002) although such an approach has yet to be tested over large areas, and may be inappropriate if flux observations are sparse, as is usually the case when synoptic observations form the primary data source. Alternatively, model estimates could be applied.

B.9.3 Application of the Water Balance Concept

B.9.3.1 Intensive Field Campaigns

HAPEX-MOBILHY

The field phase of the Hydrological Atmospheric Pilot Experiment – Modélisation du Bilan Hydrique (HAPEX-MOBILHY, André et. al. 1986), took place during 1985–1987, with a special observing period from 1 May to 15 July, 1986. The study area was a 100×100 km^2 area in south-west France, which included several gauged catchments. Most of the flux measurements took place during the special observing period, while precipitation, discharge and soil moisture measurements were gathered during the entire period.

Habets et al. (1999c) (see also Sect. B.10.4) used a macroscale hydrological model to simulate streamflow and the annual water budget for selected gauged catchments within the HAPEX-MOBILHY area. An estimate of the annual water balance terms was made by transferring model results, verified using observed latent heat flux during the special observing period, to the longer two-year period 1986–1987. Anomalies between observed and simulated soil moisture for six sites were likewise transferred to the larger gauged catchments over the simulation period. The simulated water budget showed that soil moisture storage increased by about 65 mm in 1986 and decreased by about 27 mm in 1987. The stream-flow simulations were poorest in that part of the study area where groundwater contributed significantly to river discharge. The initial water table level was obtained as a result of a multi-year spin-up to dynamic equilibrium, which could be another source of error in the water balance estimates, as data were not available to estimate directly the initial soil moisture and groundwater storage states.

FIFE

The International Satellite Land Surface Climatology Project (ISLSCP) sponsored an experiment similar in nature to HAPEX-MOBILHY at a 15×15 km² prairie grassland site near Manhattan, Kansas, with field phases in 1987 and 1989 (Sellers et al. 1992). The scientists involved with the First ISLSCP Field Experiment (FIFE), recognised that "... an annual cycle is the minimum interval required to observe the range of surface conditions that would allow an adequate evaluation of interpretive and predictive algorithms" (Sellers et al. 1992). Nonetheless, the primary acquisition of field data was during four Intensive Field Campaigns (IFCs) of several weeks each in the warm seasons of 1987 and 1989. Although there was a long-term gauged catchment (King's Creek) within the study area, flow during the summer season when most flux measurements were taken was very low, and the gauged area represented less than 5% of the study area. Continuous observations of surface meteorological data at a few sites were made over the entire 1987–1989 study period, which provided a record of the slowly varying processes associated with vegetation phenology and hydrology. In addition to automated weather stations, surface energy fluxes were observed at fourteen Bowen ratio and six eddy correlation stations, and short and longwave radiation observations were made at most of the flux tower locations. Although some of the flux stations were operated during the entire summer of 1987, most only operated during the IFCs. No specific mention of the water balance was made in the overview paper by Sellers and Hall

(1992) who summarised the major results and explored future research directions.

Famiglietti et al. (1992) used a spatially distributed hydrological model to simulate the water balance for the King's Creek catchment for the period 1 June to 9 October, 1987. Although modelled evapotranspiration compared well with observations, no comparison was made between modelled and observed soil moisture, river discharge or water table level. Duan et al. (1996a, 1996b) calculated a water balance for the FIFE area based on observations for the period from late May to mid-October 1987. They concluded that the accumulated difference between observed and estimated soil moisture changes (estimated as the residual of measured precipitation, evapotranspiration and runoff) was less than 20 mm, or about 5% of the precipitation during the period (Fig. B.22). The water budgets at individual sites did not balance as well as did the mean areal water budget. Duan et al. (1996a) therefore concluded that "... it is recommended that the spatially averaged time series of rainfall, fluxes, and soil moisture be used for the development and testing of land-surface models. This means that future field experiments should include more than a single flux site and that several supporting precipitation and soil moisture measurements should be located near each flux site within the area that influences evaporation measured at the flux site." In addition they noted that the FIFE experiment was conducted during the dry period when only about 10% of the annual runoff typically occurs. Consequently, the experiment missed valuable information that would have been captured by wet period observations.

It is noteworthy that despite the explicit recognition of the importance of long-term observations by Sellers et al. (1992), neither of the two hydrological studies reviewed covered a period longer than about five months. The focus of the studies on the warm season can almost certainly be tracked to differences in data density and completeness over the full annual cycle.

HAPEX-Sahel

During 1991 and 1992 a second HAPEX experiment was conducted in Sahelian West Africa (Goutorbe et al. 1994, 1997a). An Intensive Observation Period was carried out from 15 August to 9 October, 1992, during the transition from the wet to the dry season. The hydrology in the Sahel region is dominated by endorheic (inward draining) areas, which typically have quite small surface areas (rarely exceeding a few km²). Ponds and pools form in the beginning of the dry season and dry out – two to three months after the last rainfall. Outward draining streams tend to be ephemeral and "it is therefore generally impossible to rely on measured streamflow at the catchment outlet to provide spatially integrated values of the residual (evaporation) term in the water balance." (Goutorbe et al.

Fig. B.22.
Water balance for the FIFE
area during 1987 (from Duan
et al. 1996a, © American Geophysical Union). The *IFC-1* to
IFC-4 indicate the periods
when the intensive field campaigns took place. *P_cum* is
cumulative precipitation,
E_cum is cumulative evapotranspiration, *Q_cum* is spatially averaged cumulative
discharge, *Est SM Chg* is the
estimated soil moisture change
and *Obs SM Chg* is the observed soil moisture change

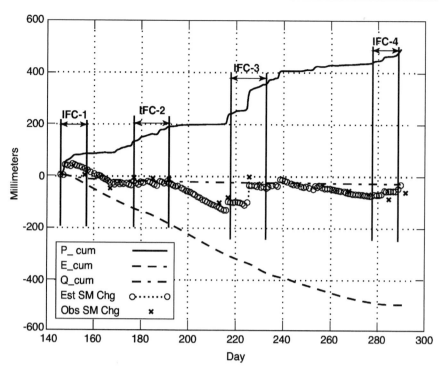

1994). For this reason, the experimental design focused on estimation of spatially integrated evaporation.

The observational network consisted of an outer 1° × 1° grid box (longitude 2–3° E and latitude 13–14° N), within which were three intensively monitored 10–20 km² "Supersites", two less-instrumented sites, and a network of weather stations. Spatial variability of rainfall was measured via a dense network of gauges (107 gauges on a 12 km² square mesh with additional gauges at the supersites) combined with a weather radar (C band, 500 m radial resolution, 1.5 degree angular resolution). Long-term input of water to the aquifer was estimated using existing wells and boreholes at several hundred sites. Soil moisture profiles were measured at the supersites using neutron probes.

Dolman et al. (1997b) identified a number of key results of the experiment that bear on the water balance of the region. Among these are the interannual variability of precipitation processes, the role of the spatial variability of precipitation as it interacts with soil characteristics, and the importance of discontinuous surface flow and groundwater characterisation. Dolman et al. also note that endorheic pools are the only readily observable hydrological integrators controlling infiltration and recharge. Reliable estimates of the water balance over the study area would not be possible without exhaustive consideration of local characteristics.

Because of these complications, no explicit attempt was made to close the water balance of the HAPEX-Sahel region. Leduc et al. (1997) studied water table fluctuations and aquifer recharge using observations from several hundred wells in the study area. Like surface runoff, groundwater recharge is dominated by infiltration from temporary drainage networks (pools and streams). Although estimates of regional recharge were produced, which was estimated to be about 10% of annual rainfall, no comparison was made with the residual of areal rainfall and evapotranspiration over the same period. In addition to the role of local processes on regional fluxes, estimation of the long-term water balance would be complicated by the fact that micrometeorological observations at the supersites were made primarily during the Intensive Observing Period, although this complication is mitigated somewhat by the strong seasonality of precipitation in the Sahel, virtually all of which occurs during the Intensive Observing Periods. Nonetheless, direct estimation of key elements of the long-term water balance such as evapotranspiration over multiple seasonal cycles is not possible from HAPEX-Sahel observations.

BOREAS

In the early and mid 1990s, two separate experiments were initiated to study the interaction between the boreal forest and the climate system. The Boreal Ecosystem Atmosphere Study (BOREAS, Sellers et al. 1995a, 1997; Hall 1999) was conducted in the boreal forest of central Canada, while the Northern Hemisphere Climate-Processes Land-surface Experiment (NOPEX) was initiated in the boreal forest of northern Europe (Halldin et al. 1998, 1999). Most of the BOREAS field work took place from 1994 to 1996.

The primary scientific issues to be addressed by BOREAS were the likely sensitivity of the boreal forest biome to climate change, including the role of feedbacks, and the magnitude of, and controls on, fluxes of carbon between the boreal forest and the atmosphere. To study these issues a multi-scale experiment strategy was used, in which most of the field studies were conducted within two study areas near the southern and northern edge of the boreal forest. These areas, defined by rectangles 100–200 km a side, lay within a much larger 1 000 × 1 000 km^2 region. The 1995 overview article by Sellers et al. does not mention the water balance explicitly. However, the hydrological observation system did support representation of the surface water balance. It consisted of dense networks of precipitation gauges at the northern and southern study areas, a rain radar which was operated during the 1994 growing season, and a network of soil moisture, snow surveys, moss moisture surveys, water level measurements and stream discharge measurements, as well as turbulent heat, radiation and micrometeorological observations at tower sites throughout the region.

Nonetheless, there were no papers in the 1997 BOREAS special issue of the Journal of Geophysical Research that dealt explicitly with the water balance over any of the observation scales within the BOREAS region. Metcalfe and Buttle (1999) studied the water balance of the fen site in the Northern Study Area, which has an area of 2.4 km^2. They commented on the difficulty in scaling-up from such small-scale observations, which is essential to determine the sensitivity of the boreal forest water balance to climate change. In this study, water storage change was estimated explicitly from observations of soil moisture, the depth of the water table, and the depth of the active layer, to identify a residual term in the water balance. During the snow melt period this error term was large and visual observation suggested that it was dominated by surface depression storage in areas not classified as wetlands. This water was subsequently lost, either through evaporation or surface drainage, or contributed to interannual changes in the storage balance.

Over the study period, snowmelt was found to account for about one-third of the water input to the system but for a much higher fraction (almost two-thirds) of the observed runoff. Over the three years of the study, about half of the precipitation fell during the summer but only three summer precipitation events produced significant hydrograph responses. Evapotranspiration was the dominant component in the surface water balance, and was near its maximum around the summer solstice when the net solar radiation was also at a maximum, the vegetation was fully grown and moisture was readily available. Partitioning of snowmelt (which occurs when evapotranspiration is small) into storage and runoff was strongly dependent on antecedent moisture

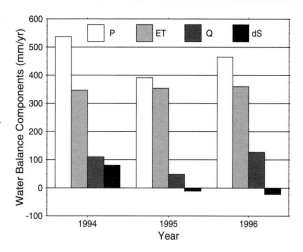

Fig. B.23. Water balance of the White Gull Creek basin in the BOREAS Southern Study Area (from Nijssen and Lettenmaier 2002). P – Precipitation, ET – Evapotranspiration, Q – discharge, and dS – estimated storage change. The storage change was estimated as the residual of the other three terms

conditions. During summer, evapotranspiration balanced precipitation, and the source of runoff was subsurface storage depletion. During late summer and early autumn, evapotranspiration decreased as a result of reduced net radiation, subsurface storage replenishment and vegetation moisture stress.

Metcalfe and Buttle estimated that the error in the water balance was about 15–20%, which was dominated by the accuracy of the latent heat flux observations at the flux towers. A number of secondary sources of water balance errors was identified. They concluded that "… small surface storages, and their contribution to evapotranspiration are a key component in boreal forest water balance dynamics."

Nijssen and Lettenmaier (2002) studied the water balance of the White Gull Creek basin (about 603 km^2) in the BOREAS southern study area for the period 1994–1996. Multiple linear regression equations of canopy resistance as a function of environmental conditions were developed for each vegetation type. The canopy resistances were scaled by the LAI and combined with the Penman-Monteith equation to derive a spatially distributed estimate of evapotranspiration. Their study showed that, based on such an estimate, combined with measurements of precipitation and discharge, the estimated change in storage was –11 mm in 1995 and –22 mm in 1996, which are only a few percent of annual precipitation. However, water balance calculations suggested that the storage change in 1994 was 80 mm (or almost 15% of annual precipitation), and that the residual could be traced largely to heavy precipitation in July of 1994 (Fig. B.23). On the other hand, soil moisture and fen water level observations did not support the computed 1994 storage change value. It was therefore concluded that most of this "missing" water was likely to have been either evaporated or contributed to recharge of a deeper lying aquifer. In any event, the observation network, especially for soil moisture and other subsurface storage, was inadequate to determine with certainty the fate of the excess water.

One interesting water balance study (Barr et al. 2000) performed as part of the Canadian BERMS programme (Boreal Ecosystem Research and Monitoring Sites, a follow-on to BOREAS) used a piezometer installed at a depth of 34.6 m near the BOREAS mature aspen site to monitor pressure changes in the aquifer, integrating the mass loading effectively over an area of 10 hectares. The piezometric and gauged estimates of precipitation agreed to within 5%. Maximum accumulated snow in winter 1997–1998 was 60 mm as estimated from the piezometer, and 55 mm from surface based surveys, evapotranspiration agreed to within about 10% with the tower flux measurements.

NOPEX

The Northern Hemisphere Climate-Processes Land-surface Experiment (NOPEX) studied land-surface/atmosphere interaction in a northern European forest-dominated landscape (Halldin et al. 1998, 1999). The main NOPEX region represented the southern edge of the boreal zone, while the second site in northern Finland represented conditions at the northern edge of the boreal forest. Unlike BOREAS, the experimental design focused on long-term monitoring, which started in 1994. Focused field data collection was conducted in May–June 1994, April–May 1995 and winter 1998–1999. According to Halldin et al. (1998), the long-term monitoring was conducted explicitly to assure the ability to evaluate water balances predicted by models. They note the dependence of the accuracy of the computed water balance on the length of the observing period. Long-term studies within NOPEX were conducted in four small, intensely studied catchments. For the larger region a weather radar complemented data from a synoptic meteorological network and a supplemental precipitation gauge network. Groundwater levels were observed by an extended network of stations operated by the Swedish Geological Survey and discharge records were available at 11 long-term stream gauges with upstream areas of 7–950 km^2.

To date, no studies appear to have been conducted on the long-term water balance at the NOPEX catchments. Grelle et al. (1997), in a study of evaporation at the central NOPEX Norunda site where a large flux tower is operated above a coniferous forest canopy, noted the difficulty of extrapolating flux tower observations of latent heat to larger areas. With respect to the water balance at the Norunda tower site, they argued that flux aggregation was the main problem, due to variations in atmospheric stability over varying source areas.

B.9.3.2 GEWEX Continental Scale Experiments

The GEWEX continental scale experiments represent regions, typically with areas of the order of 10^6 km^2 and mostly defined by river basin boundaries, that are the basis for research activities designed to better represent water and energy fluxes in weather prediction and climate models. Of the five continental scale experiments, three (GCIP, MAGS, LBA) are large continental river basins (Mississippi, Mackenzie, and Amazon). The fourth, GAME, includes multiple continental river basins (e.g. Lena) as well as areas (Tibetan Plateau) not defined by river basin boundaries. The fifth (BALTEX) includes a number of smaller rivers that constitute the drainage area to the Baltic Sea.

The GEWEX observational strategy for the continental scale experiments is somewhat different to that for the intensive field experiments summarised in Sect. B.2.1. The continental scale experiments have attempted to make use of existing long-term datasets wherever possible, supplemented where necessary by additional observations that allow leveraging from historic observations. In many cases they have used data from sources other than *in situ* observations, for example, satellite data, analysis or re-analysis fields from Numerical Weather Prediction (NWP) models. Furthermore, they have tended to focus more on estimation of the atmospheric branch of the water (Eq. B.13) and energy cycles, rather than the land branch (Eq. B.11). For this reason, the continental scale experiments have attempted to produce the most accurate possible estimates of atmospheric convergence of moisture, either directly from radiosonde profiles, or via analysis or re-analysis fields. In contrast to estimation of the surface water balance, which is often complicated by spatial variability at other than very small spatial scales, the relative accuracy of atmospheric moisture convergence estimates increases with area. For this reason, attempts to close the atmospheric water balance have been most successful at the scale of very large river basins, e.g. the entire Mississippi. A few studies have made use of the commonality of evapotranspiration, E, in Eq. B.11 and B.13 to test consistency of the estimated surface and atmospheric budgets over large river basins. Abdulla et al. (1996), for instance, compared estimates of monthly evapotranspiration over the Arkansas-Red River basin (one of the GCIP Large Scale Study Areas). As shown in Fig. B.24, the estimates matched quite well. It should be noted that the two estimates of evapotranspiration are not independent, as precipitation, P, in both Eq. B.11 and B.13 was identical. On the other hand, atmospheric storage changes over monthly time steps are quite small, so that (given the same precipitation P) temporal variability in the atmospheric estimate is dominated by the convergence, while in the surface estimate it is dominated by storage changes. Other studies (e.g. Yarosh et al.

Fig. B.24.
Water balance of the Arkansas-Red River basin as estimated from the atmospheric water balance and estimated using a macroscale hydrological model (VIC – Variable Infiltration Capacity Model). **a** Monthly time series 1972–1986; **b** mean monthly series (from Abdulla et al. 1996, © American Geophysical Union)

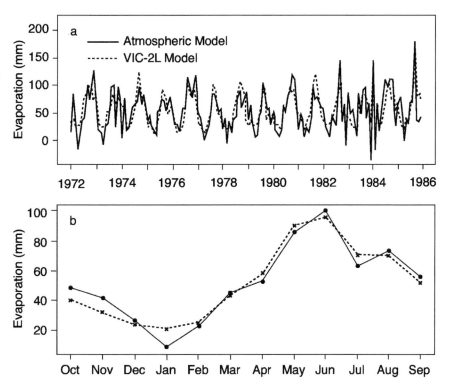

1999) have shown that atmospheric convergence over the Mississippi is strongly dependent on the diurnal variability of atmospheric moisture flux, which is poorly represented by the twice-daily observations such as those used by Abdulla et al. (1996). Considering these observational difficulties, the close agreement of the two moisture budgets shown by Abdulla et al. (1996) may be fortuitous. In any event, it is surprising that more large-scale analyses that exploit the commonality of evapotranspiration in the surface and atmospheric water balances have not been attempted.

An exception to the general nature of the continental scale experiments noted above is the LBA, the study region for which is the Amazon Basin. LBA includes a focused observational programme, which consists primarily of two tower transects, and intensive field observations in the Rondônia region. Observations at the towers and within the Rondônia region began in early 1999 and in most cases are ongoing at the time of writing, although some field activities started earlier. Dunne et al. (1997) identified the guiding questions for hydrologists in LBA as "What are the spatial and temporal patterns of water storage, water flux, and the transport of bioactive materials in the soils and river corridors of the Amazon Basin, and how are they influenced by variations in climate and land use?" As originally conceptualised, a nested design within the Rondônia region with experimental units ranging from individual hillslopes and ponds at the microscale (< 10 km^2) small catchments, up to the mesoscale (10^4 km^2) was intended to provide the observational basis for addressing these questions.

The observational design was intended explicitly to provide water budget closure across the range of spatial scales represented by the nested catchments. Up to the mesoscale (Rondônia) catchments, the design should allow direct computation of the change in storage, as all other terms in Eq. B.11 would be observed directly. At the continental (Amazon) scale, the storage change would be calculated by coupling the atmospheric and land-surface water balance, as in Abdulla et al. (1996). However, funding for key elements of this design is not yet in place and it is not clear at the time of writing how successful the strategy will be.

B.9.4 Assessment and Recommendations

With a decade and a half of experience with land-atmosphere field experiments, the time is opportune to take stock of the success – or lack thereof – in the various experimental designs. As discussed in Chapt. B.10, these experiments have certainly provided the observational basis for improvements in the representation of the land-surface in NWP and climate models, which in one form or another is a stated objective of most of the experiments. Most of the overview articles cited in Sect. B.9.1 and B.9.2 make note of some of the key successes, which do not bear repetition here. On the other hand, the results in general indicate less success in closure of the water balance over the various range of spatial scales represented by (and in some cases, within) the various studies. We summarise here some of the problems that have been encountered and then conclude with recommendations that should help improve subsequent experiments.

Problems with Water Budget in Previous Studies:

- The short-term focus of most field experiments has precluded estimation of the role of storage change (or methods of analysis that can avoid having the budget dominated by this term). To date, there has been minimal success in direct observation of storage change, although Barr et al. (2000) have shown intriguing results in the BOREAS follow-on, albeit for relatively small areas.
- Some studies have failed to focus on catchments, and even when gauged catchments have been included, the experimental design has often tended not to deploy resources in a manner that would allow use of the catchment to close the water balance.
- There is an inconsistency between current methods of observing surface energy fluxes, which are based primarily on towers and hence represent relatively small areas with varying footprints, as opposed to observation of runoff, which requires gauged catchments, and for which the accuracy of measurements generally increases with the drainage area. At present, a theoretical basis for transferring tower estimates of surface fluxes (especially evapotranspiration) to larger areas (e.g. whole catchments) is lacking.
- Scaling of results over a range of spatial scales is often argued as the conceptual basis for experimental design, typically taking tower-scale fluxes, in combination with aircraft transects and satellite observations over the region. However, mechanisms for merging such disparate data sources are lacking, and to date there has been no convincing demonstration of implementation of this strategy to produce water budget closure.

Recommendations:

- If closure of the water budget is desired, this needs to be considered explicitly in the experimental design stage – including type of observations, space-time resolution, duration of the observation programme and the analysis strategy. In general, "closure" is usually a misnomer, as it is rarely possible to observe directly all terms in the water balance (especially storage change). Closure of the water budget may not be appropriate or possible in some studies. However, if water balance closure is an objective, all terms must be considered – including storage change.
- Direct estimation of storage change is a problem in most cases, but averaging over long periods eliminates information about the dynamics of the land system that is provided by high frequency observations. More attention needs to be given to methods of estimating directly storage change over areas large enough that streamflow can be measured accurately (typically tens to hundreds of km^2 and upwards). Monitoring of groundwater levels, and soil moisture observations designed with the objective of integrating over the study area, are two areas that need more attention.
- Given the difficulty of estimating all terms in the surface water budget, there is a need to consider alternative observation and analysis strategies, such as land-surface data assimilation, combination of surface and atmospheric budgets, and new observation strategies.
- There is a need for longer-term observation strategies that can leverage from shorter term intensive observations. Year-round measurement of surface fluxes is now feasible (e.g. NOPEX/WINTEX, BERMS, AmeriFlux). More attention should be paid to how spatially intensive observations over short periods can be integrated with long-term limited area observations, and the implications of the integration strategy for design of observations.

Chapter B.10

Use of Field Experiments in Improving the Land-surface Description in Atmospheric Models: Calibration, Aggregation and Scaling

Joel Noilhan · Pierre Lacarrère · Florence Habets · Richard J. Harding

B.10.1 Introduction

When the first integrated terrestrial experiments were designed in the early 1980s, land-surface processes in atmospheric models were poorly described, as compared to present-day parameterisations. There was also very little data available that was suitable for the local calibration of surface schemes or few datasets for the large-scale implementation of these schemes within mesoscale Numerical Weather Prediction (NWP) or climate models. At that time, two main tasks faced the community. First, to collect enough information from representative land uses to develop and calibrate advanced land-surface parameterisations of the water and energy budgets (in place of classical simple schemes inspired from the bucket model of Manabe, 1969). Second, to develop an upscaling strategy that would allow those surface schemes to be used at a grid scale of several tens of kilometres, while keeping complexity and computing cost compatible with the other physical parameterisations of the atmosphere. The integrated terrestrial experiments HAPEX-MOBILHY, FIFE, EFEDA, NOPEX, HAPEX-Sahel and BOREAS opened the way for a more realistic parameterisation of land-surface changes in NWP and GCM models and thus solved some of the problems outlined above. The successes and failures of this strategy are summarised briefly.

B.10.2 Benefit of Datasets from Integrated Terrestrial Experiments for Surface Modelling in Atmospheric and Hydrological Models

The observations from integrated terrestrial experiments have been invaluable to the modelling community. They have been used in a wide range of ways, notably:

1. the use of point observations (e.g. sensible, latent heat, carbon and momentum fluxes) to develop, test and calibrate land-surface models.
2. The use of a suite of measurements to improve our understanding of the dominant processes, thus leading to improved formulations.
3. The combined use of surface and boundary layer measurements to improve models, particularly NWP and mesoscale models.
4. The use of area average fluxes to test model output directly and indirectly through aggregation routines, often linked to single column versions of the atmospheric models.

A distinction can be made between the methodologies adopted for climate and NWP models. For the NWP models it is possible to simulate one or more "golden" days within the field experiment. This allows a comparison of not just the surface fluxes but also many aspects of the model, such as the development of the boundary layer from radiosonde and aircraft data, and their spatial variability (see e.g. Taylor et al. 1997a; Bringfelt et al. 1999; Betts et al. 1998b; Bougeault et al. 1991). For GCM land-surface schemes, it is less easy to make use of the whole suite of observations and recourse is usually made to tests of the "stand alone" land-surface scheme driven by near-surface observations. While valuable, such comparisons miss out many of the interactions between the surface and the lower atmosphere and provide no opportunity to test the boundary layer scheme. The increasing tendency to use the same land-surface scheme in the NWP and climate models as, for example, in MO-SES (the Meteorological Office Surface Energy Scheme) in the UK Meteorological Office Unified Model or the ISBA surface scheme in the ARPEGE NWP model of Météo-France (Giard and Bazile 2000) provides an excellent opportunity to combine these two approaches.

The use of point data to develop land-surface models is probably the most direct use of the datasets as, for example, by most of the major modelling groups for most of the major experiments (e.g. Cox et al. 1998; Chen et al. 1996; Noilhan et al. 1991, and many more). While an integrated terrestrial experiment is not essential to produce such data, the infrastructure of these experiments has ensured that such point datasets are made available to the community in a timely and well documented manner. The presence of supporting information on vegetation characteristics, soil moisture, forcing data, etc. associated with the experimental structure has increased the value of the data. Such tests and calibrations are, of course, only part of the answer: we need to know the mean values and variability across a landscape (or grid cell) and aggregation algorithms are also necessary. This is where the more extensive measurements of the integrated terrestrial experiment are important.

Synthetic or representative datasets have been assembled and widely distributed to the scientific community for SVAT scheme calibration and intercomparisons. This was done for one year of observations for a soybean crop in HAPEX-MOBILHY (PILPS phase 2b; Shao and Henderson-Sellers 1996) and for four land covers representative of the EFEDA area (Linder et al. 1996). These intercomparisons proved very fruitful since many improvements were brought to the Soil-Vegetation-Atmosphere Transfer (SVAT) schemes involved as, for instance, the inclusion of gravitational drainage in ISBA by Mahfouf and Noilhan (1996). Important limitations in the current parameterisations of water and energy partitioning were clearly identified, for example the link between transpiration and the soil moisture (Mahfouf et al. 1996) or the partitioning between surface evaporation and plant transpiration (Linder et al. 1996). For this aspect, the data collected over various types of sparse vegetation canopies during EFEDA (vineyard, natural vegetation) and HAPEX-Sahel (savanna, millet crop) proved to be very valuable (Braud et al. 1997).

Important developments in land-surface schemes have been achieved using data from the integrated terrestrial experiments in this stand-alone mode. Notable examples are the inclusion of carbon exchanges in the models (see e.g. Cox et al. 1998; Calvet et al. 1998) and the improvement of the representation of bare soil evaporation and soil water description (e.g. Braud et al. 1993).

In the FIFE experiment Betts and Ball (1994 and 1998) produced a dataset, which is an average over all available meteorological and surface flux stations, contained within the 15×15 km^2. This is obviously a considerable improvement on single station datasets, such datasets have not yet been produced for the other experiments. These experiments cover a wider region and larger diversity of land-cover type and thus any averaged dataset would be dependent on the land-cover database and aggregation scheme used; they may therefore be less useful. There is, however, still a need for improved, standardised and easily accessible datasets from many experiments for use by the modelling community.

The impact of the HAPEX experiments on the operational atmospheric forecasts has been significant. An important step was the development of an assimilation method for soil moisture based on optimum interpolation developed and tested with the HAPEX-MOBILHY database (Mahfouf 1991; Bouttier et al. 1993). Given the importance of soil moisture initialisation for short to medium range forecasts, the assimilation of soil mois-

ture is still a pre-requisite for an operational implementation of a land-surface scheme. Such a scheme whereby soil moisture is nudged with forecast errors of air temperature and humidity at screen level has been in operation at Météo-France since March 1998 and at ECMWF since 1999 (Douville et al. 2000; Viterbo and Beljaars 1995). Real improvement of the 6 hour forecast of low level fields in a global NWP model was obtained (Giard and Bazile 2000) as a result of the surface scheme itself including the aggregation of surface properties as well as the assimilation scheme for the soil and surface variables. Another approach to analyse soil moisture is being developed now in the Land-Data Assimilation System (LDAS) (Chen and Mitchell 1999). It takes advantage of continuous measurements comparable to those made in the HAPEX and GCIP campaigns. Atmospheric observations, satellite data and surface flux measurements available in near real time are used to produce both high-resolution atmospheric forcing and validation data to perform continuous simulation of SVAT schemes. The soil moisture is then simulated using all available information and is supposed to be the best guess. It can then be used to initialise soil moisture in NWP models.

Direct comparison of time series data from the field experiments and model simulations from nearby gridpoints from data assimilation and forecast models has proved very useful in identifying systematic error in model parameterisations (Betts et al. 1993, 1996a,b, 1997, 1998a,b,c) and in developing improved schemes (Viterbo and Beljaars 1995; Hong and Pan 1996; Chen et al. 1996). Particularly well documented is the work by Betts and his co-workers on the FIFE data and more recently on BOREAS data to improve ECMWF and NCEP/NCAR models. Typical improvements include the soil hydrology, evaporation, soil heat flux and boundary layer parameterisations. Recently the impacts of frozen soil and snow have been assessed. The sum of these improvements has led to significant improvements in the skill of these forecast models (see Sect. A.4.5.2).

B.10.3 A Modelling Strategy for Upscaling from the Plot Scale to the Size of a GCM Grid Box

The experimental framework set up during HAPEX-MOBILHY was to be used in combination with mesoscale modelling of the atmosphere and the hydrology to bridge the gap between surface exchanges at the plot scale up to the scale of a GCM grid box (André and Bougeault 1988). The broad lines of this multi-scale strategy were also followed during EFEDA 91 (Bolle et al. 1993) and HAPEX-Sahel (Goutorbe et al. 1997a) in semi-arid areas. The experiments took place over relatively flat areas, large enough to be significant with respect to the scales used in GCMs and with enough heterogeneities in land-surface cover to make upscaling studies possible (forest and agricultural patches in MOBILHY, irrigated/non-irrigated crops and vineyards in EFEDA, bare ground, savanna and tiger bush in HAPEX-Sahel).

The experimental strategy was well adapted to give adequate information on physical processes involved:

1. at the point scale: micrometeorological observations, soil moisture down to 2 m depth, surface fluxes and biometric parameters;
2. within the boundary layer: frequent vertical soundings made during special observing periods; and
3. for the upscaling issue: instrumented aircraft flown close to the surface for providing estimates of areal average fluxes and links between surface observations or satellite remote sensing.

It can be estimated that this network performed particularly well on certain golden days, as described for instance by Noilhan et al. (1991) for the 16 June of HAPEX-MOBILHY, by Noilhan et al. (1997) for the case of 26 June of EFEDA (1991) and by Taylor et al. (1997a) for the case of 8 October of HAPEX-Sahel (1992). However, some criticisms can be made: for instance, the somewhat limited period of the year (one month in the case of EFEDA) in which the field campaigns were fully operational prevented the testing of aggregation methods for the whole annual hydrological cycle. In fact, this had implications on the number of situations actually considered in the modelling analysis if one also takes into account problems associated with instrumental deficiencies (for instance aircraft operations) or too complex atmospheric situations encountered in that period (particularly for frontal intrusion during HAPEX-MOBILHY or heavy convection in the HAPEX-Sahel experimental domain).

The combination of field observations at various scales with mesoscale atmospheric modelling for selected days can be considered as a major success of the HAPEX experiments. This is particularly true for upscaling processes from scales of a few to hundreds of kilometres. The success of the strategy was largely because it was possible to validate the simulated turbulent fluxes at the surface and at different levels within the atmospheric boundary layer. At this point, the careful selection of surface measurement sites and the definition of aircraft operations had a crucial importance. Summarised briefly, the key aspects of the interplay between the special observing period and mesoscale modelling were:

1. the calibration of a soil vegetation atmosphere transfer scheme against observations collected at each representative land-cover site within the experimental domain: a number of agricultural sites and the pine forest site in HAPEX-MOBILHY (Noilhan and Planton

1989), vineyard, natural shrub lands and irrigated crops in EFEDA (Linder et al. 1996), bare ground, savanna and tiger bush in the HAPEX-Sahel square (Goutorbe et al. 1997a; Braud et al. 1993). Here, it is essential to perform the calibration using a single column model with the same SVAT and physical package that ultimately will be used in the mesoscale model for the upscaling investigation. Indeed, the simulation of the full interaction between the surface and the atmosphere will ensure the relevance of calibration parameters. Here, it can be said that the complementary nature of the dataset collected at the surface (soil moisture, surface fluxes and biological parameters) and in the atmosphere (radiosounding at least every six hours, including night-time periods) were particularly well suited to this purpose.

2. the mapping of the surface parameters in a large domain for the hydrostatic mesoscale modelling at a grid-size of a few kilometres. The size of the domain has to be centred on the $1° \times 1°$ experimental region: it has to be considerably larger (more than ten times larger) to avoid possible perturbations by lateral boundary conditions. It must also allow for the explicit simulation of possible mesoscale circulations generated by land variability at a length scale of hundreds of kilometres (e.g. vegetation breeze between a dry forest and adjacent wet crops in HAPEX-MOBILHY, or a sea-breeze in EFEDA), which could affect the atmospheric boundary layer within the HAPEX square. Surface parameters were inferred from the local calibrations and from classifications of vegetation and soil types (Mascart et al. 1988). At this stage, satellite remote sensing of the NDVI both for instantaneous values and seasonal-to-annual evolution allowed the classification of vegetative cover into main types (Champeaux et al. 2000).

3. the careful validation of a mesoscale simulation of selected contrasting days using detailed comparisons with all the data available (meteorological variables and fluxes at the surface, boundary layer evolution and aircraft fluxes). The initialisation of soil prognostic variables, particularly the soil moisture in the root zone, has to be done carefully because of their strong influence on the simulation of the structure of the PBL for short time simulations. Full advantage has been taken of long-term observations of soil moisture (Goutorbe et al. 1997b; Cuenca et al. 1997) by using a simple model of the water budget to provide soil moisture fields to initialise the mesoscale model (Taylor et al. 1997a). This also underlines another success of the HAPEX field data which was able to monitor the soil moisture on a few annual cycles for representative vegetation and soil types. To ensure a correct driving of the lateral boundaries of the mesoscale model, re-analysis proved necessary for the

three experiments. This was attempted by adding extra radiosoundings taken during the experiment to the conventional network and was successful for EFEDA and HAPEX-MOBILHY while still underway for HAPEX-Sahel. For this last experiment, the operational large-scale analysis available at the time of the experiment revealed important deficiencies, particularly for the PBL humidity analysis in the Sahelian region. Detailed evaluation of the performance of mesoscale modelling in reproducing accurate simulations both of surface exchanges and PBL evolution have been shown by Noilhan et al. (1991, 1997) and Taylor et al. (1997a). Particularly important for the upscaling issue was the good simulation of the turbulent fluxes at the surface and within the boundary layer. For the first time, the impact of large horizontal differences in surface fluxes on the mesoscale circulation triggering cumulus clouds at the top of the boundary layer was observed and simulated (Bougeault et al. 1991). One difficulty identified in this calibration phase was related to the link between aircraft and surface fluxes. Besides the difficulty of flying close enough to the surface (in particular during EFEDA, see Noilhan and Lacarrère 1995), the underestimation of aircraft turbulent fluxes (eddy correlation measurements) as compared with surface fluxes was a drawback for the simulation validation. However, the analysis revealed that the underestimation affected mainly the evaporation flux, and the spatial pattern of the sensible heat flux was used successfully to diagnose the organisation of a vegetation breeze (HAPEX-MOBILHY) or regional gradient in response to surface heating (HAPEX-Sahel).

4. once validated, the reference mesoscale simulations were used to investigate the upscaling issue. Three possible ways of describing the subgrid scale variability of land surface for large areas were investigated. The first method provided by the 3-D mesoscale model corresponded to the so-called tile approach, whereby the surface exchanges for each land-use fraction were computed explicitly and the reference arealy-averaged surface fluxes were estimated by averaging the whole spatial distribution. This tile approach was used to check the aggregated methods where the subgrid scale variability is characterised through effective parameters. In that case, the main idea was to define the best way to combine the surface parameters in such a way that only one effective flux calculation for the whole large area provides an excellent approximation of the reference "tile approach" averaged flux. The aggregation methods were tested by comparing effective surface fluxes computed with a single column model with the 3-D arealy-averaged fluxes. Noilhan et al. (1997) described the results obtained for selected clear days of HAPEX-MOBILHY and EFEDA, as well as a rainy day where only one part on the domain was covered

by the rainfall. Noilhan and Lacarrère (1995) proposed simple rules to estimate effective parameters by averaging each of them differently, in such a way that conserved quantities (surface fluxes) were indeed conserved. Accordingly, one should average linearly albedos, leaf area indices and vegetation cover values, as this is the way these parameters influence the various physical processes described in many SVAT schemes. On the other hand, one would average the roughness lengths logarithmically, as the friction velocity, while averaging stomatal resistances inversely. Despite the non-linear relationships between surface fluxes and surface parameters, Noilhan et al. (1997) showed that the effective surface fluxes matched the area-averaged 3-D fluxes with a relative error of less than 10%. However, these aggregation studies were based on real conditions with moderated land-surface variability in soil water. For the EFEDA cases in dry-vegetation conditions, the variability of stomatal resistance was relatively small as compared with the extreme variability presented by a partially wet mesoscale canopy. On the other hand, if the surface scheme was fed by surface parameters corresponding to the dominant vegetation type, which corresponded to what was usually done in GCMs, the "dominant" surface fluxes departed significantly from the reference surface fluxes.

A similar analysis for HAPEX-Sahel (Harding et al. 1996) and Hall et al. (1992) for FIFE confirm these results. The success of the use of suitably averaged parameters and driving variables arises from the fact that the turbulent fluxes are linear functions of the driving variables within the ranges found in these landscapes, and the driving variables themselves do not vary in a systematic way across the domain. While transpiration can be estimated with acceptable accuracy using mean parameters, some other processes show strongly non-linear behaviour and must be treated differently. Rainfall interception is an example that was identified very early. Dolman and Gregory (1992) and Taylor (1995), using data from the ARME and ABRACOS experiments, highlighted both these issues and the potential solutions (see Sect. A.6.3).

More recently the problems associated with the representation of snow have been discussed by Harding and Pomeroy (1996), Viterbo and Betts (1999) and Essery (1998) with reference to the BOREAS data. Solutions have been suggested using tile representations for forest and patchy snow. The representation of runoff remains an unresolved problem. It is frequently dealt with, to some extent, by assuming a distribution of rainfall across a grid cell. Wood et al. (1992) describe models (such as VIC), which incorporate some statistical description of the subgrid variability of soil moisture. Although these types of model have been used successfully within climate models (e.g. Liang et al. 1994) the full integration between hydrological and climate models is beset with problems and has yet to be accomplished. The comparatively short experimental periods of integrated terrestrial experiments and tendency to use flat regions (important for good flux measurements) means that the structure of the integrated terrestrial experiments does not lend itself to hydrological investigations (see also Chapt. B.9). Despite this, however, some useful hydrological simulations have been made of the integrated terrestrial experiment study areas (e.g. Famiglietti et al. 1992 for the FIFE region). Notably the HAPEX-MOBIHLY data was used in Phase 2b of the PILPS intercomparison (Desborough et al. 1996). Significantly this latter study illustrated graphically the failings of the current hydrological parameterisations contained within the land-surface models, with no evidence that the more complex parameterisations produced improved results (Pitman et al. 1999).

A further issue is the role of meso-scale circulations, driven by sub-grid variability, on the energy and water exchanges at the surface and within the boundary layer. These have been studied extensively in FIFE, BOREAS and HAPEX-Sahel. Eloranta and Forrest (1992) suggest that the heteorogeneity at the surface enhances the mixing at the top of the boundary layer by a factor of two. Detailed rainfall measurements taken as part of HAPEX-Sahel show a persistence of rainfall patterns throughout the wet season (Taylor et al. 1997b; Fig. B.25), with analysis of the surface and boundary layer measurements leading to the suggestion that squall lines are enhanced when passing over previously wetted surfaces (Taylor 2000). Evidence for a land/atmosphere interaction at the scale of 10×10 km^2 was totally unexpected and an outcome of the measurement within HAPEX-Sahel of a range of

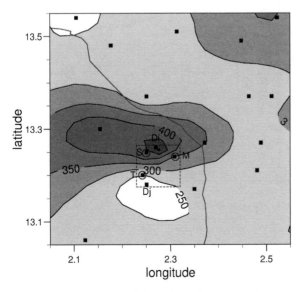

Fig. B.25. Accumulated rainfall (mm) 31 July–18 September 1992 at the Southern Super Site during HAPEX-Sahel. Rain gauge sites, *Di*: Diokoti, *S*: savanna, *M*: millet, *T*: tiger bush and *Dj*: Djakindji (from Taylor et al. 1997b)

spatially distributed surface and atmospheric parameters. For further discussion of land/atmosphere interactions at this scale see Sect. A.3.2.2).

Recently, the impact of aggregate parameters in a GCM was analysed by Burke et al. (2000) who showed that sub-grid vegetation information may affect global-scale circulation patterns. Using the same methodology, the effect of sub-grid scale precipitation on arealy-averaged evaporation was analysed for a case of frontal intrusion in the HAPEX-MOBILHY area (Blyth et al. 1993). Indeed, sub-grid precipitation and its relation to interception loss is an important issue in GCM modelling (Shuttleworth 1988a) and it was shown that the partitioning of evaporation between wet and dry canopy conditions was incorrect if the precipitation was uniformly spread in the grid box. The mesoscale model provided the reference areal evaporation as well as the fractional area covered by rain within the large domain. This information was used to calibrate the results of a single column model where the heterogeneity of rainfall is allowed for within the surface scheme. Noilhan et al. (1997) showed that effective evaporation was simulated satisfactorily if the fraction of vegetation covered by intercepted rainfall was multiplied by the fractional area covered by precipitation. If the sub-grid distribution of rain was not taken into account in the single column model, the evaporation flux was clearly overestimated because the interception reservoir was maintained at its saturated value.

In semi-arid regions, such as that studied in HAPEX-Sahel, the sub-grid variability of rainfall can be important. Taylor and Blyth (2000) showed how bare soil evaporation in the HAPEX-Sahel square displayed a strong spatial variation, due to rainfall variability. Again, transpiration was well represented by mean parameters but differential growth patterns could influence the spatial heterogeneity of transpiration. The key problem in calculating area-averaged variables were caused by interception and evaporation from the surface soil moisture. Failure to assume a sub-grid distribution results in errors in evaporation of up to 30%.

B.10.4 Use of a Macroscale Hydrological Model to Investigate Aggregation of Hydrological Processes

Aggregation of soil processes received less attention than aggregation of vegetation within the HAPEX experiments because of the difficulty in monitoring regional hydrology over a long period of time during the experiments. This is probably one of the weaknesses of these experiments although a hydrological component was included in each of them (see also Chapt. B.9). A first guess for aggregating soil processes was proposed by Noilhan and Lacarrère (1995) who examined the impact of ef-

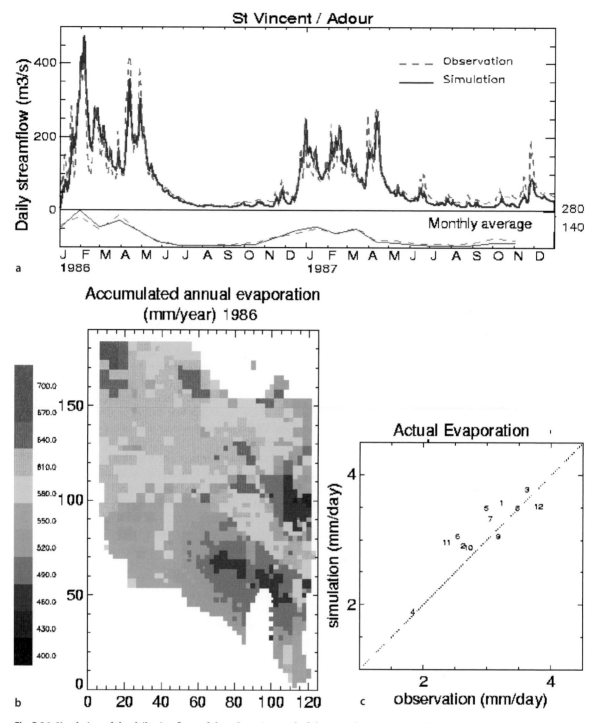

Fig. B.26. Simulation of the daily riverflows of the Adour river and of the annual evaporation in the HAPEX-MOBILHY area with the coupled ISBA-MODCOU model (Habets et al. 1999b, © American Geophysical Union). **a** Comparison of observed and simulated riverflows at the outlet of the Adour basin during 1986 and 1997); **b** simulated annual evaporation in the HAPEX-MOBILHY area; **c** comparison of observed and simulated evaporation for the special observing period at the 12 flux stations

fective soil parameters computed from a linear average of the soil texture. For the HAPEX-MOBILHY square composed of loamy and sandy soils, they showed that the aggregation of soil leads to a better simulation of effective evaporation and soil moisture than those com-

puted from the dominant soil texture. However, they noted some limitations in the linear approach to overcome the non-linear transfers in the soil. Similar conclusions were reached by Kabat et al. (1997b) and Braud (1998) who suggested that aggregated soil parameters failed to predict grid-average drainage below the root zone and the total runoff. Braud et al. (1995) examined the impact of the spatial variability of the soil hydraulic conductivity on the simulation of surface fluxes from a dataset of 73 soil samples taken in the EFEDA central site, showing that the median value of the distribution of hydraulic conductivity led to an acceptable simulation of effective surface fluxes.

Recently, Habets et al. (1999a,b) tested an original method for investigating soil upscaling using a distributed hydrological model coupled with a land-surface scheme within the HAPEX-MOBILHY area. The main idea was to use the four catchments in the area as natural integrators of surface runoff and deep drainage. The method was similar to that for the atmosphere but over a longer time (two years) including the special observing period of the HAPEX experiment. Firstly, the coupled model was implemented in the area with a spatial resolution reaching 1 km. Full advantage was taken of the dense surface atmospheric network (for instance more than 200 raingauges within the 100 × 100 km² experimental area) to interpolate in time (three hours) and space (five kilometres) all the atmospheric quantities required by the surface scheme. Using realistic fields of vegetation and soil types in the area, the coupled model was integrated continuously with the observed forcing to simulate the daily streamflows of the four catchments. Habets et al. (1999b) showed that the model simulated correctly the regional hydrology (streamflow and evaporation, Fig. B.26) provided that a sub-grid scale parameterisation of surface runoff was included in the surface scheme. In this parameterisation, it is assumed that a fraction of the grid-box is saturated when the soil water content exceeds a threshold value (taken as the wilting point). This contributing area generates surface runoff before the entire grid-box is saturated. If this sub-grid scale runoff scheme was not activated, the total runoff, and consequently the streamflows, were strongly underestimated. Similar conclusions were found during the PILPS Phase 2e Red-Arkansas River experiment. In HAPEX-MOBILHY, the observations of soil moisture at 12 sites during two years and the evaporation during two months were particularly important for checking the model outputs at the different stages of the hydrological cycle. Once the distributed hydrological model was calibrated, an aggregation test was made to study how a GCM can simulate the hydrological budget of a large area considered as a single grid-cell (compare Fig. B.27). Habets et al. (1999b) showed that the aggregated values of the water budget (evaporation, surface runoff and drainage) were comparable to the reference water budget, provided that sub-grid scale surface runoff and interception losses were parameterised. If these sub-grid processes were not described, the arealy-averaged evaporation was over-estimated and the total runoff and the river flows were consequently under-estimated.

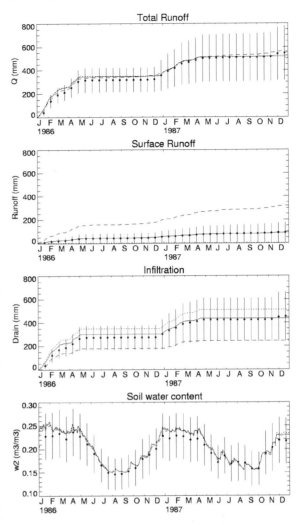

Fig. B.27. Aggregation of evaporation, surface runoff and drainage for the HAPEX-MOBILHY area: The area-averaged evaporation, runoff and drainage fluxes (solid points and vertical bars) are computed by the reference distributed hydrological model. The aggregated fluxes are simulated with stand alone 1-D tests including a sub-grid scale parameterisation for interception losses and for surface runoff: no sub-grid scale runoff scheme (.....), standard sub-grid surface runoff (——), and high value of the parameter B controlling the sub-grid runoff scheme (- - -)

B.10.5 Concluding Remarks

The experimental framework deployed during the HAPEX experiments provided a valuable database to test advanced land-surface parameterisations and upscaling methods now in operation in many atmospheric models. In particular, the combination of measurements of surface fluxes, the atmospheric boundary layer and land-surface properties have provided an important opportunity to implement and test models of the surface and boundary layer. In addition, the monitoring of the long-term evolution of the water budget at many sites, as well as the riverflows during HAPEX-MOBILHY, proved to be essential for the implementation of a distributed hydrological model. Significant numerical results were obtained at the mesoscale, showing that for moderate stress conditions, effective parameters for the vegetation led to a fair estimate of aggregated surface fluxes for a large area with a sub-grid variability of land cover. When the annual hydrological cycle is considered in temperate climatic conditions, hydrological modelling indicates that aggregated fluxes of evaporation, drainage and surface runoff can be simulated satisfactorily provided that sub-grid scale surface runoff and interception losses are taken into account in the aggregation procedure. For the cases examined, the aggregation of soil processes received less attention than vegetation aggregation although the impact of non-linear processes are probably stronger.

Comparitively little has been done on scaling at small scales from the patch to a few kilometres. Blyth (1993) and Blyth and Harding (1995) used the measurements at the tiger bush site in HAPEX-Sahel to investigate this scale using a two-dimensional boundary layer model. This study showed that the horizontal transfer of heat between hot bare soil and cooler bushes was important and was not described by a simple patch, or tile, model. Such analysis requires the use of non-hydrostatic mesoscale models and a high resolution database for the soil and the vegetation. Nowadays, these tools are becoming available and will open the way for a detailed analysis of the complex interaction between the circulation within the atmospheric boundary layer and heterogeneity at the surface (Bélair et al. 1997, discussed such an example).

The increasing demand for improving the representation of surface processes in both meteorological and hydrological models has led to a strong tendency to resolve explicitly the sub-grid variability at the surface (in place of aggregated parameters), following the tile approach first introduced by Koster and Suarez (1992). Such an approach is now used in the revised land-surface scheme of the ECMWF model (Viterbo and Beljaars 1995). Recently, van den Hurk et al. (2000) show the benefit of the tile approach for the description of snow processes and surface fluxes in the BOREAS area but suggested that the tiling approach does not give major improvements in the case of grass and cropland. The authors concluded that the tile-model performance becomes increasingly dependent on a proper choice of vegetation parameters, giving rise to an increased number of degrees of freedom. On the other hand, a correct treatment of very important physical processes at the surface such as the interception of snow by forest canopies, the CO_2 assimilation, the urban and open water, lake and irrigated areas will probably necessitate a tile approach in the next generation of land-surface schemes, although very few experimental data are now available for model verification (particularly for the soil heterogeneity). Here, the key issues are related to the ability to monitor the spatial and temporal evolutions of an increasing number of surface parameters that largely determine the response of the land surface. Now, only indirect verification of the simulation of energy and water partitioning is performed in global atmospheric models and the development of coupling with distributed hydrological models should offer a valuable and additional source of validation of the water cycle through the comparison of observed river flows (Habets et al. 1999a).

All the previous pioneering work has to be checked and improved for larger basins in the world. That is one of the objectives of GEWEX, through an ensemble of continental scale experiments which will collect long series of comprehensive observations with an integrated analysis of the dataset, including coupled modelling and data assimilation. This is particularly important if GEWEX is to meet one of its more critical objectives, namely: "to develop our ability to predict the variation of global and regional hydrological processes and water resources and their response to environmental change".

Chapter B.11

Further Insight from Large-scale Observational Studies of Land/Atmosphere Interactions

John H. C. Gash · Pavel Kabat

B.11.1 Introduction

In 1991 Jim Shuttleworth published a paper (Shuttleworth 1991) overviewing the major international observational studies of the 1980s and discussing the form and function of those to come. In this chapter we deliberately plagiarise the title and format of that paper as we look at the progress that has been made in the past 10 years. The research within the series of integrated land-surface experiments described in this part of the book represents an enormous effort by the scientific community. Has this effort been worth it? In particular, has there been sufficient extra benefit from having large experiments, with a range of simultaneous measurements, to justify the cost, in time and money, of coordinating these operations? We would argue that it has been worth this effort. We also argue that, by nature of their international status, IGBP and WCRP have been able to make a contribution to the initiation, prioritisation and implementation of these experiments, which national organisations would have found difficult to achieve. The same applies to the dissemination of the results.

The reviews in this part of the book make it clear that our ability to measure the Earth system has now advanced to the point where we have the ability to study not just the individual components but also large areas of the complete system itself. It is now possible to monitor the interactions between components of the system, as well as the behaviour of the surface itself at the landscape, or ecosystem, scale. Techniques have evolved to meet the science questions, and the datasets collected have driven the development of better and more realistic models. Equally important is the intellectual coming together of those experimentalists who previously worked at the plot scale, but who now think and measure at the landscape scale; and in the other direction those global modellers who no longer regard the land surface as "wet green blotting paper" but now appreciate the small-scale problems of soil and plant behaviour.

Despite this progress, we cannot claim complete success – some components of the system and some techniques have been neglected. In this chapter we attempt to summarise the views in the preceding chapters, in places adding our own opinions, but while emphasing the successes also pointing out the outstanding problems.

B.11.2 International Co-operation

Writing today it might seem strange that Shuttleworth found it necessary to include a section on international collaboration in his 1991 paper. Yet, at the time that paper was written it was far from clear what the responsibilities and areas of interest of the different international programmes and projects were. There was definitely tension amongst those involved in building the programmes, and this was diverting energy and resources away from the task of driving the science. There was apparent duplication and competition between WCRP projects such as HAPEX-MOBILHY, ISLSCP (which included FIFE) and the emerging GEWEX, and the new IGBP with its core project BAHC. The scientific community found this annoying and confusing: Shuttleworth in fact called for some "discipline" to be imposed. Perhaps as a result of Shuttleworth's paper, or perhaps as a result of a few key, personal interventions by leading scientists, the necessary "discipline" was imposed. The inevitable areas of overlap between international programmes are now easily accommodated, and major experimental projects such as the LBA comfortably form part of IGBP, IHDP and WCRP – indeed, the whole emphasis of the new agendas being developed by these programmes is now on interdisciplinary, collaborative research.

B.11.3 The Use of Land-surface Data to Validate Global Models

Shuttleworth (1991) pointed out the value of field data in checking the accuracy of Global Circulation Model (GCM) predictions of the current climate. The classic example is that of the large error in the predicted radiation balance of Amazonia identified by Shuttleworth and Dickinson (1989). It is satisfying that subsequent work by Culf et al. (1998) comparing Amazonian field data with the predictions from several GCMs found generally quite good agreement. Predicting the correct mean rainfall is one of the most rigorous, and commonly made, tests of model performance. However, impact studies now require not just accurate predictions of average conditions, but also of extremes. It was thus an important finding by Huntingford et al. (2002) that the Hadley Centre Regional Climate model, nested in their GCM, made credible predictions of extreme rainfall return periods for three sites in Britain. However studies such as this are rarely reported and there is the potential for much more use to be made of the growing amounts of surface data (see Part C) for validating GCMs.

B.11.4 Surface Flux Measurements

Net Radiation

Shuttleworth (1991) noted the poor performance of net radiometers and observed that there were differences between designs of instrument. Some instruments suffered from a non-linearity in response and errors of 10% were possible. The situation has improved only slightly. The design of one particular instrument, which was found to behave badly in FIFE, was changed, but more importantly, radiometers which measure all four of the radiation components separately are now commercially available. These latter instruments are inherently more accurate than all-wave instruments. It has become increasingly clear (see Chapt. B.3) that most commercially available net radiometers fall short of the accuracy expected of them and the community is, in general, still far too uncritical of net radiometers and the data they produce. The net radiation measurement thus remains a weak point in the derivation of the land-surface energy balance.

As is explained in Chapt. B.3, there is an urgent need for an internationally accepted long-wave radiometry standard to enable the development of new net radiometers to be based on more solid ground. Until then it is best to avoid sensors which measure the net (short- and long-wave balance) radiation with a single sensor, but to use those instruments which measure the components separately.

Surface Energy Partition and CO$_2$ Flux

We have progressed from a position where micrometeorological measurements of surface fluxes were made by specialists in small, research-led campaigns, to the point where the techniques are now applied routinely by ecologists and hydrologists in measurement networks and integrated terrestrial experiments (see Chapt. B.4).

Technological developments have allowed methods to progress from gradient techniques to closed path eddy covariance systems, which are run to give long-term measurements of carbon and water fluxes. Development continues with the deployment of commercially-available, open path sensors designed specifically for field use. These instruments have lower power consumption and require less maintenance – they make an important next step in the process of moving eddy flux measurement from being a research technique to one of operational monitoring. Other developments are expanding the variety of surface-based flux measurements: for example with the scintillometry technique it is now possible to measure aggregated sensible heat fluxes over long path lengths of up to several kilometres (de Bruin et al. 1995).

Continuous measurements give much greater insight into the water and carbon cycles. Shuttleworth (1991) pointed out the rather small variation in evaporation among the measurement sites in FIFE. A somewhat similar result was found in HAPEX-Sahel: comparing the evaporation between sites resulted in plots from which it was difficult to see any signal amongst the noise of day to day variation (Gash et al. 1997). Only when cumulative plots of energy use and evaporation were plotted for the whole season did the systematic differences between the three major vegetation types become obvious. The implication of these systematic differences – particularly the comparison between bush savanna and agricultural fields – has been explored in GCM modelling studies (Taylor et al. 2002), demonstrating a link between actual land-use change and climate.

Continuous measurements also capture the extremes which are usually missed by campaign mode observations. These extremes fill the modellers' data space and thus increase the likelihood of accurate model predictions. Measuring over the whole range of the soil freezing and thawing cycle (as in NOPEX and BOREAS) or capturing extreme events (such as El Niño-driven droughts) are examples of how continuous data capture can contribute to understanding.

Progress has also been made on the interpretation of the measurements: flux footprint theory now allows micrometeorological flux measurements to be related to the source area of vegetation in the up-wind fetch (see Chapt. B.4). It is also now realised that this footprint is dynamic and varies according to the atmospheric conditions. Flux footprint analysis is particularly useful in the interpretation of data from sites with natural vegetation which is inherently heterogeneous (e.g. Schmid and Lloyd 1999), and where soil or topographic effects are thought to produce spatially variable fluxes (e.g. Soegaard et al. 2000; Araújo et al. 2002).

Assessing the sensitivity of flux calculations to data-processing procedures, particularly the question of averaging times and the removal of background trends in the basic variables, is an area of active research (see Chapt. B.2). Long held beliefs are being questioned and an open-minded approach is being adopted. Raw data are now easily stored, so re-processing of existing data-sets would be straightforward if new standard procedures were to be derived and published.

The frequent failure of the sum of the outgoing energy fluxes to meet the net incoming radiational energy, i.e. the energy balance closure problem, remains unsolved. We can list a number of reasons why energy closure may not be occurring (see Chapt. B.2): e.g. the peculiarities of the site, the instruments' response, or the data-processing procedures, but we do not understand why the same equipment and procedures sometimes produce good closure, but sometimes not. Solving this enigma is a high priority on the micrometeorological research agenda.

In the efforts to achieve continuous data, eddy correlation is often used in calm, stable night-time conditions, where the method is much more prone to under-measurement of fluxes. While this is relatively unimportant for water vapour fluxes, which are low at night, the CO_2 fluxes from respiration, which continue through the night, are inaccurately measured. As a result the "measured" night-time fluxes often include huge and poorly known corrections and the daily totals are biassed in favour of the day-time photosynthesis. Almost every reported measurement of CO_2 defines a net sink of CO_2 because of this, but can they all be true? Nor is measuring the carbon balance just an academic excercise – these data are likely to influence policy at international level, and perhaps to direct huge resources between countries in carbon exchange agreements. There are large investments being made in "FluxNet" arrays on the basis that eddy flux measurements of net CO_2 are made with a proven technology; but that is not the case. At good sites with uniform terrain, and in dry, daylight conditions the method can produce good results; but currently we cannot say the same about night-time measurements. This is a very serious issue which is undermining the credibility of the flux measuring community, and it must be resolved.

Aircraft Flux Measurements

Aircraft flux measurements were a central part of all the integrated land-surface experiments of the 1980s and 1990s. Their objective was to give large-scale, average fluxes. This was done using the eddy correlation method, but by flying through the boundary layer sampling the turbulence along the flight path. Initially they were presented as an alternative to the tower-based approach which relies on the turbulence blowing past the tower to sample vegetation less than a kilometre up-wind. It is now accepted that the two approaches are complementary rather than competing alternatives: aircraft give the variation in space, towers give the variation in time.

Measuring fluxes from aircraft is difficult. The platform is moving and it is not stable, it in fact responds to the turbulence which it is trying to measure. Crawford and Dobosy (Chapt. B.5) point out the little appreciated fact that aircraft respond to an updraft by lowering the nose, which accelerates the aircraft. If not corrected for, this will give the worst nightmare of practitioners of eddy correlation – correlated errors.

Almost all of the attempts to use aircraft to measure fluxes have failed to some degree by under-measuring fluxes. But slow, low-flying, light aircraft that make minimal distortion to the air flow and use modern GPS technology to monitor every movement of the aircraft have shown that the method can be used successfully. If the measurement approach advocated here is followed, we now know enough to measure fluxes accurately.

Boundary Layer Budgets

The boundary layer budgeting method uses the mixing power of boundary layer turbulence to derive surface fluxes at the scale of boundary layer development – several tens of kilometres. The method is an attractive component of an integrated land-surface experiment because many of the necessary measurements will already be being made for other purposes. However, as Fitzjarrald points out (Chapt. B.6), although the method at first sight looks both straightforward and rigorous, it is fraught with difficulties. The boundary layer doesn't always behave as the text books would have us believe. It is not a sealed box; it is a box which leaks at the top and at the sides, its size increases during the day and, if clouds form, it has a component which is changing state. Thus in deriving the surface flux from the change in concentration in the boundary layer and its rate of growth we have to make a series of difficult measurements or uncertain assumptions. Specifically we have to quantify the advective and entrainment terms in the continuity equation. This is not easy and it is perhaps not surprising that Fitzjarrald reports that only in extremely few cases has the method been used successfully – and then only over very short time periods. It is indicative of the state of development of this method that the results of boundary layer budgeting studies are almost always presented as tests of the method, rather than as giving insight into the large-scale surface fluxes. Nevertheless efforts continue to use the averaging power of the boundary layer to give large area fluxes – if application of the method is to be successful it is vital that we build on the past experience outlined in this chapter.

Soil Moisture

Although there is no chapter in this book uniquely dedicated to soil moisture studies, soil measurements have been ubiquitous throughout all the studies. The soil moisture deficit is the most important, long-term control on transpiration and the evaporation of water directly from the soil surface has been shown to be critical in defining the short-term variation of evaporation. However, although soil moisture content may be relatively straightforward to measure, soil hydraulic properties are not. Soil is inherently variable and the presence of macropores and fissures makes parameterising the flow of water through soil problematic. Measurements made on small soil cores simply do not translate to larger scales and there is still no easy method of measuring large-scale soil hydraulic properties. On the other hand measuring soil moisture content has become easier and more automated, and a variety of new instruments are now available and in routine use. These instruments derive the variation in soil moisture content from the measured electrical properties of the soil.

Vegetation Measurements

The main contribution of vegetation measurements in the land-surface experiments has been to measure the leaf and plant level fluxes of heat, water and carbon (see Chapt. B.7). These measurements provide the basic information for parameterising and calibrating the land-surface models. Stem, root and leaf measurements have been used to separate the total canopy scale flux into the components from the different vegetation types and to separate transpiration from evaporation from the soil. The balance between these two fluxes has been shown to be critical in defining the evaporation from sparse vegetation, and under those conditions it is vital to model them separately. The development of the sap flow technique has played an important part in collecting the data which provided this insight.

Micrometeorological measurements of CO_2 flux give the net exchange, but this is unconstrained. Without additional photosynthesis and respiration measurements on the vegetation and soil, there is no check on the accuracy of eddy flux carbon measurements and no data for the modellers to test the two components of their carbon balance models – photosynthesis and respiration. Of the two processes respiration has without doubt been the poor relation: while the process of photosynthesis is relatively well understood and modelled, most respiration models are crude and there are little data against which to test them.

The capacity of remote sensing (see Chapt. B.8) to produce distributed fields of surface variables is being increasingly, if rather slowly, exploited by the global modelling community and much of the basic work behind this has been done within the integrated terrestrial experiments. Because it is primarily the physical properties of plants which determine the reflectance and emission of radiation from the Earth's surface, vegetation measurements are needed to quantify and model plant structure. This work has gone on in parallel with the development and validation of remote sensing algorithms.

Propagation of Errors

It is important to recognise that field data will never be perfect, and the modelling community which uses the data must come to terms with the inevitable errors. Errors in the data feed through to become errors in the model parameters derived from those data. These errors are a source of uncertainty which should be allowed to propagate through the model and be quantified in terms of their effects on model predictions. One interesting recent development is the "climate*prediction*.net" project (Allen 1999). This project aims to utilise the idle time of thousands of Personal Computers to run massive ensembles of climate model simulations. Plans for this project include investigating the effects of uncertainty in model parameters on the climate predictions. It is planned that the land-surface parameterisation will be one of the first to be investigated – this should quantify the effect of measurement errors on model predictions.

B.11.5 Hydrological Catchment Measurements

Most of the land-surface experiments have concentrated on campaign mode operation with little attention being paid to experimental networks designed to test closure of catchment water balances. As Lettenmaier and Nijssen point out (in Chapt. B.9), this is unfortunate because closing the water balance at the catchment scale is a constraint on the measurements which could, and should, have been exploited. Unfortunately, this mistake continues to be made and there are few examples of sites where an integrated experimental design has incorporated continuous eddy correlation flux measurements into a catchment-based measurement framework. A good example of how this should be done is the tropical rainforest study in a catchment near Manaus (Araújo et al. 2002). In this study above-canopy measurements of carbon and water fluxes are being made as part of a comprehensive catchment carbon and water balance, which includes the export of carbon by streamflow.

B.11.6 Aggregation and Models

The experimental data from the series of integrated terrestrial experiments have been widely used for the point calibration of SVATS used in weather prediction and climate models. Positioning the experiments over different and important biomes of the world has greatly added to their value in this respect. The policy of making the data easily available (see Part C) has also been a key factor. However, much of the initial justification of the experiments was to learn how to model large areas of mixed vegetation types, i.e. to improve our understanding of the dominant processes, and to test and develop aggregation schemes. These have been developed (see Chapt. B.10) and are in use, but a tile approach still seems more sensible to some. The aggregate parameter approach is computationally more efficient, but is more problematic in areas with very high contrasts in fluxes. In our view there is room for both approaches, but as computing power increases and the size of model grid squares decreases the difference between the two methods diminishes.

B.11.7 Future for Large-scale Integrated Experiments

The second phase of IGBP is now addressing the challenge of how to make the results of global change research more relevant and useful. There is much emphasis on integrated research: both in terms of understanding the interactions between components of the Earth system and in working with scientists from other global change programmes, such as the social scientists in the International Human Dimensions Programme on Global Environmental Change (IHDP). The land-surface experiments described in this part of the book have of course been at the interface of the land and atmospheric systems and in that respect they act as a model as to how different communities of scientists can work together. Current experiments, such as the Large Scale Biosphere-Atmosphere Experiment in Amazonia (LBA, see also Sect. A.6.3), now include social science as well: studies, such as those on the causes of deforestation and the sustainability of different land management systems, are being added to those of the physical and biological functioning of the Amazon Basin. Thus LBA also points to the way in which large experiments can contribute to the new agenda. However this integration of physical and social scientists in LBA was attempted only late in the planning process and has not been easy. One lesson learnt is that the integration must start at the beginning, with shared proposal development and project design. This human-centred research requires a bottom-up approach with those who will be using the results involved in formulating the issues to be researched.

LBA is also a component of the World Climate Research Programme's Coordinated Enhanced Observing Period (CEOP). This initiative combines intensive satellite measurements and a global set of surface observations, with a simultaneous global atmospheric modelling effort. The objective is to increase understanding of climate variability and our ability to predict variations in the water cycle at the seasonal time scale. A project like CEOP requires very different design and management to that being advocated for the human-centred research discussed above. A top-down approach is needed, in which the issues are defined and the project designed by a group of scientists operating at a high level.

The challenge for the international programmes is thus to design and initiate experiments and observational programmes which meet the needs of integrated research across this wide spectrum of scientific methodologies. This will not be easy to achieve – it requires open minds and new thinking. A new approach is particularly needed in the tropics, which are still neglected in terms of global change research. Research agendas drawn up in the developed world rarely take account of the different priorities of developing countries. In much of the tropics there is a dynamic situation with both landscapes and social structures undergoing rapid change. In developing countries the problems of global change are inseparable from the problems of development. Experimental projects and observational programmes must recognise that.

B.11.8 Concluding Remarks

At the time of the first integrated land-surface experiment, only 15 years ago, the interactions and feedbacks between the land surface and the climate were of interest to a handful of visionary scientists. Today climate change is at the top of the scientific agenda and we have moved from arguing about whether or not climate change will occur, to when it will occur and to what degree. GCM predictions are increasingly regarded as credible forecasts. Within GCMs, the land-surface fluxes are now modelled, for the whole of the Earth's land surface, with a degree of realism and accuracy which far exceeds that produced by the simplistic descriptions generally used 15 years ago. The series of the international experiments discussed in this part of the book have contributed to this progress by anchoring the land-surface descriptions of the GCMs in the reality of actual data. There is no room for complacency – there is still plenty to do – but the fieldworkers who have collected these datasets are entitled to feel pride in a job well done!

References Part B

Abdulla FA, Lettenmaier DP, Wood EF, Smith JA (1996) Application of a macroscale hydrologic model to estimate the water balance of the Arkansas-Red River Basin. J Geophys Res 101:7449–7459

Aber JD, Melillo JM, Nadelhoffer KJ, Mcclaugherty CA, Pastor J (1985) Fine root turnover in forest ecosystems in relation to quantity and form of nitrogen availability: a comparison of two methods. Oecologia 66:317–321

Acevedo OC, Fitzjarrald DR (2001) The early evening transition: temporal and spatial variability. J Atmos Sci 58(17):2650–2667

Allen M (1999) Do-it-yourself climate prediction. Nature 401: 642

Allen SJ, Grime VL (1995) Measurements of transpiration from savannah shrubs using sap flow gauges. Agr Forest Meteorol 75:23–41

Anderson DE, Verma SB, Rosenberg NJ (1984) Eddy correlation measurements of CO_2, latent heat and sensible heat over a crop surface. Bound-Lay Meteorol 29:263–272

André J-C, Bougeault P (1988) On the use of HAPEX data for the validation and development of parameterization schemes of land-surface processes. WCRP publication series No. 126, Appendix B

André J-C, Mahrt L (1982) The nocturnal surface inversion and influence of clear-air radiative cooling. J Atmos Sci 39:864–878

André J-C, Goutorbe J-P, Perrier A (1986) HAPEX-MOBILHY: a hydrologic atmospheric experiment for the study of water budget and evaporation flux at the climatic scale. B Am Meteorol Soc 67:138–144

André J-C, Goutorbe J-P, Perrier A, Becker F, Bessemoulin P, Bougeault P, Brunet Y, Brutsaert W, Carlson T, Cuenca R, Gash J, Gelpe J, Hildebrand P, Lagouarde J-P, Lloyd C, Mahrt L, Mascart P, Mazaudier C, Noilhan J, Ottlé C, Payen M, Phulpin T, Stull R, Shuttleworth J, Schmugge T, Taconet O, Tarrieu C, Thepenier R-M, Valencogne C, Vidal-Madjar C, Weill A (1988) Evaporation over land-surfaces: first results from HAPEX-MOBILHY special observing period. Ann Geophys 6:477–492

André J-C, Bougeault P, Goutorbe J-P (1990) Regional estimates of heat and evaporation over non-homogeneous terrain. Examples from the HAPEX-MOBILHY Programme. Bound-Lay Meteorol 50:77–108

André J-C, Noilhan J, Goutorbe J-P (1999) Use of small-scale models and observational data to investigate coupled processes. Mesoscale field experiments and models. In: Browning KA, Gurney RJ (eds) Global energy and water cycles. Cambridge University Press, pp 183–200

Araújo AC, Nobre AD, Kruijt B, Culf AD, Stefani P, Elbers J, Dallarosa R, Randow C, Manzi AO, Valentini R, Gash JHC, Kabat P (2002) Comparative measurements of carbon dioxide fluxes from two nearby towers in a central Amazonian rain forest: the Manaus LBA site. J Geophys Res 107(D20): Art. No. 8090 SEP–OCT 2002

Aubinet M, Grelle A, Ibrom A, Rannik Ü, Moncrieff J, Foken T, Kowalski AS, Martin PH, Berbigier P, Bernhofer Ch, Clement R, Elbers J, Granier A, Grünwald T, Morgenstern K, Pilegaard K, Rebmann C, Snijders, W, Valentini R, Vesala T (2000) Estimates of the annual net carbon and water exchange of forests: the EUROFLUX methodology. Adv Ecol Res 30 113–175

Auble DL, Meyers TP (1992) An open path, fast response infrared-absorption gas analyzer for H_2O and CO_2. Bound-Lay Meteorol 59:243–256

Baker JM (2000) Conditional sampling revisited. Agr Forest Meteorol 104:59–65

Bakwin PS, Tans PP, Hurst DF, Zhao C (1998) Measurements of carbon dioxide on very tall towers: results of the NOAA/CMDL program. Tellus 50B:401–415

Baldocchi DD, Hicks BB, Meyers TP (1988) Measuring biosphere-atmosphere exchanges of biologically related gases with micrometeorological methods. Ecology 69:1331–1340

Baldocchi DD, Valentini R, Running S, Oechel W, Dahlman R (1996) Strategies for measuring and modelling carbon dioxide and water vapour fluxes over terrestrial ecosystems. Global Change Biol 2:159–168

Baldocchi D, Finnigan J, Wilson K, Paw UKT, Falge E (2000) On measuring net ecosystem carbon exchange over tall vegetation on complex terrain. Bound-Lay Meteorol 96:257–291

Ball JT (1987) Calculations related to gas exchange. In: Zeiger E, Farquhar GD, Cowan IR (eds) Stomatal function. Stanford University Press, Stanford, pp 446–476

Ball JT, Woodrow IE, Berry JA (1987) A model predicting stomatal conductance and its contribution to the control of photosynthesis under different environmental conditions. In: Biggens J (ed) Progress in photosynthesis research. Martinus Nijhoff, Dordrecht, IV, pp 221–224

Barr AG, Betts AK (1998) Radiosonde boundary layer budgets above a boreal forest. J Geophys Res 102:29205–29212

Barr AG, Betts AK, Desjardins RL, MacPherson JI (1997) Comparison of regional surface fluxes from boundary-layer budgets and aircraft measurements above boreal forest. J Geophys Res 102:29213–29218

Barr AG, van der Kamp G, Schmidt R, Black TA (2000) Monitoring the moisture balance of a boreal aspen forest using a deep groundwater piezometer. Agr Forest Meteorol 102:13–24

Bégué A, Hanan NP, Prince SD (1994) Radiative transfer in shrub savanna sites in Niger: preliminary results from HAPEX-Sahel. 2. Photosynthetically active radiation interception of the woody layer. Agr Forest Meteorol 69:247–266

Bégué A, Prince SD, Hanan NP, Roujean JL (1996a) Shortwave radiation budget of Sahelian vegetation. 2. Radiative transfer models. Agr Forest Meteorol 79:97–112

Bégué A, Roujean JL, Hanan NP, Prince SD, Thawley M, Huete A, Tanré D (1996b) Shortwave radiation budget of Sahelian vegetation. 1. Techniques of measurement and results during HAPEX-Sahel. Agr Forest Meteorol 79:79–96

Bélair S, Lacarrère P, Noilhan J, Masson V, Stein J (1997) High resolution simulation of surface and turbulent fluxes during HAPEX-MOBILHY. Mon Weather Rev 126:2234–2253

Betts AK (1973) Non-precipitating cumulus convection and its parameterization. Q J Roy Meteor Soc 99:178–196

Betts AK (1992) FIFE atmospheric boundary layer budget methods. J Geophys Res 97:18523–18531

Betts AK, Ball JH (1994) Budget analysis of FIFE-1987 sonde data. J Geophys Res 99 3655–3666

Betts AK, Ball JH (1998) FIFE surface climate and site-average dataset 1987–1989. J Atmos Sci 55, 1091.1108

Betts AK, Desjardins RL, MacPherson JI, Kelly RD (1990) Boundary-layer heat and moisture budgets from FIFE. Bound-Lay Meteorol 50:109–138

Betts AK, Desjardins RL, MacPherson JI (1992) Budget analysis of the boundary layer grid flights during FIFE 1987. J Geophys Res 97:18533–18546

Betts AK, Ball JH, Beljaars ACM (1993) Comparison between the land-surface response of the ECMWF model and the FIFE-1987 data. Q J Roy Meteor Soc 119:975–1001

Betts AK, Hong SY, Pan HL (1996a) Comparison of NMC/NCAR reanalysis with 1987 FIFE data. Mon Weather Rev 124:1480–1498

Betts AK, Ball FH, Beljaars ACM, Miller MJ, Viterbo P (1996b) The land-surface-atmosphere interaction: a review based on observational and global modeling perspectives. J Geophys Res 101:7209–7225

Betts AK, Desjardins RL, MacPherson JI, Kelly RD (1996c) Boundary-layer heat and moisture budgets from FIFE. Bound-Lay Meteorol 50:109–137

Betts AK, Chen F, Mitchell KE, Janijic ZI (1997) Assessment of the land-surface and boundary layer models in two operationnal versions of the NCEP Eta model using FIFE data. Mon Weather Rev 125:2896–2916

Betts AK, Viterbo P, Wood E (1998a) Surface energy and water balance for the Arkansas-Red River Basin from the ECMWF Reanalysis. J Climate 11:2881–2897

Betts AK, Viterbo P, Beljaars ACM (1998b) Comparison of the land-surface interaction in the ECMWF reanalysis model with the 1987 FIFE data. Mon Weather Rev 126:186–198

Betts AK, Viterbo P, Beljaars ACM, Pan HL, Hong SY, Goulden ML, Wofsy SC (1998c) Evaluation of the land-surface interaction in the ECMWF and NCEP/NCAR reanlyses over grassland (FIFE) and boreal forest (BOREAS). J Geophys Res 103:23079–23085

Bingham GE, Gillespie CH, McQuaid JH, Dooley DF (1980) A miniature, battery-powered, pyroelectric detector-based differential infrared absorption sensor for ambient concentrations of carbon dioxide. Ferroelectrics 34:15–19

Blyth EM (1993) Using a simple SVAT scheme to describe the effect of scale on aggregation. Bound-Lay Meteorol 72:267–285

Blyth EM, Harding RJ (1995) Application of aggregation models to surface heat flux from the Sahelian tiger bush. Agr Forest Meteorol 72:213–235

Blyth EM, Dolman AJ, Noilhan J (1993) Sensitivity of mesoscale circulations to surface wetness: an example from HAPEX-MOBILHY. J Appl Meteorol 33:445–454

Boegh E, Soegaard H, Hanan NP, Kabat P, Lesch L (1999) A remote sensing study of the NDVI-Ts relationship and the transpiration from sparse vegetation in the Sahel based on high-resolution satellite imagery. Remote Sens Environ 69:224–240

Boehm W (1979) Methods of studying root systems. Springer-Verlag, New York

Bolle HJ, André J-C, Arrue JL, Barth HK, Bessemoulin P, Brasa A, de Bruin HAR, Cruces J, Dugdale G, Engman ET, Evans DL, Fantechi R, Fiedler F, van de Griend A, Imeson AC, Jochum A, Kabat P, Kratzsch T, Lagouarde JP, Langer I, Llamas R, Lopez-Baeza E, Melia Miralles J, Muniosguren LS, Nerry F, Noilhan J, Oliver HR, Roth R, Saatchi SS, Sanchez Dias J, de Santa Olalla M, Shuttleworth WJ, Soegaard H, Stricker J, Thornes J, Vauclin M, Wickland D (1993) EFEDA: european field experiment in a desertification-threatened area. Ann Geophys 11:173–189

Bonhomme R, Chartier P (1972) The interpretation and automatic measurement of hemispherical photographs to obtain sunlit foliage area and gap frequency. Israel J Agr Res 22:53–61

Bougeault P, Noilhan J, Lacarrère P, Mascart P (1991) An experiment with an advanced surface parameterization in a meso-beta-model. Part I: Implementation. Mon Weather Rev 119:2358–2373

Bouttier F, Mahfouf JF, Noilhan J (1993) Sequential assimilation of soil moisture from atmospheric low level parameters. J Appl Meteorol 32:1335–1352

Bowling DR, Baldocchi DD, Monson RK (1999a) Dynamics of isotopic exchange of carbon dioxide in a Tennessee deciduous forest. Global Biogeochem Cy 13:903–922

Bowling DR, Delany AC, Turnipseed AA, Baldocchi DD, Monson RK (1999b) Modification of the relaxed eddy accumulation technique to maximize measured scalar mixing ratio differences in updrafts and downdrafts. J Geophys Res 104:9121–9133

Bragg PL, Cannell RQ (1983) A comparison of methods, including angled minirhizotrons, for studying root growth and distribution in a spring oat crop. Plant Soil 73:435–440

Braud I (1998) Spatial variability of surface properties and estimation of surface fluxes of a savannah. Agr Forest Meteorol 89(1):15–44

Braud I, Noilhan J, Bessemoulin P, Mascart P, Haverkamp R, Vauclin M (1993) Bareground surface heat and water exchanges under dry conditions – observations and parameterization. Bound-Lay Meteorol 66:173–200

Braud I, Dantasantonino AC, Vauclin M (1995) A stochastic approach to studying the influence of the spatial variability of soil hydraulic properties on surface fluxes, temperature and humidity. J Hydrol 165(1–4):283–310

Braud I, Bessemoulin P, Monteny B, Sicot M, Vandervaere JP, Vauclin M (1997) Unidimensional modelling of a fallow savannah during the HAPEX-Sahel experiment using the SiSPAT model. J Hydrol 188/189:912–945

Bringfelt B, Heikinheimo M, Gustafsson N, Perov V, Lindroth A (1999) A new land-surface treatment for HIRLAM – comparisons with NOPEX measurements. Agr Forest Meteorol 98/99(1–4):239–256

Brooks A, Farquhar GD (1985) Effect of temperature on the CO_2/O_2 specificity of ribulose-1,5-biphosphate carboxylase/oxygenase and the rate of respiration in the light. Planta 165:397–406

Brooks SB, Crawford TL, McMillen RT, Dumas EJ (1996) Airborne measurements of mass, momentum, and energy fluxes. Arctic Landscape Flux Survey (ALFS) 1994, 1995. NOAA Technical Memorandum ARL/ATDD-216

Brooks SB, Dumas EJ, Verfaillie J (2001) Development of the sky arrow surface/atmosphere flux aircraft for global ecosystem research. International Aerospace Abstracts, American Institute of Aeronautics and Astronautics, August 2001, paper AIAA 2001-0544

Brown de Coulston EC, Walthall CL, Russell CA, Irons JR (1995) Estimating the fraction of absorbed photosynthetically active radiation (fAPAR) at FIFE with airborne bidirectional spectral reflectance data. J Geophys Res 100:25523–25535

Brümmer B (1978) Mass and energy budgets of a 1-km high atmospheric box over the GATE C-scale triangle during undisturbed and disturbed weather conditions. J Atmos Sci 35:997–1011

Brutsaert W (1999) Aspects of bulk atmospheric boundary layer similarity under free-convective conditions. Rev Geophys 37:439–451

Burke CJ (1946) Transformation of polar continental air to polar maritime air. J Meteor 2:94–112

Burke EJ, Shuttleworth WJ, Yang Z-L, Mullen SL, Arain AM (2000) Impact of heterogeneous vegetation on the modeled large scale circulation in CCM3-BATS. Geophys Res Lett 27(3):397–400

Businger JA (1954) Some aspects of the influence of the earth's surface on the atmosphere. Proefschrift, Rijkuniversiteit te Utrecht, The Netherlands, Chapt. 3

Businger JA, Delaney AC (1990) Chemical sensor resolution required for measuring surface fluxes by three common micrometeorological techniques. J Atmos Chem 10:399–410

Businger JA, Oncley SP (1990) Flux measurement with conditional sampling. J Atmos Ocean Tech 7:349–352

Cain JD, Rosier PTW, Meijninger W, De Bruin HAR (2001) Spatially averaged sensible heat fluxes measured over barley. Agr Forest Meteorol 107:307–322

Caldwell MM, Virginia RA (1991) Root systems. In: Pearcy RW, Ehleringer J, Mooney HA, Rundel PW (eds) Plant physiological ecology. Chapman & Hall, London, pp 367–398

Calvet J-C, Noilhan J, Roujean JL, Bessemoulin P, Cabelguenne M, Olioso A, Wigneron JP (1998) An interactive vegetation SVAT model tested against data from six contrasting sites. Agr Forest Meteorol 92. 73–95

Canadell JG, Mooney HA, Baldocchi DD, Berry JA, Ehleringer JR, Field CB, Gower ST, Hollinger DY, Hunt JE, Jackson RB, Running SW, Shaver GR, Steffen W, Trumbore SE, Valentini R, Bond BY (2000) Carbon metabolism of the terrestrial biosphere: a multitechnique approach for improved understanding. Ecosystems 3:115–130

Carlson MA, Stull RB (1986) Subsidence in the nocturnal boundary layer. J Appl Meteorol 25:1088–1099

Cellier P, Brunet Y (1992) Flux-gradient relationships above tall plant canopies. Agr Forest Meteorol 58:93–117

Cermak J, Deml M, Penka M (1973) A new method of sap flow rate determination in trees. Biol Plantarum 15(3):171-178

Champeaux J-L, Arcos D, Bazile E, Giard D, Goutorbe J-P, Habets F, Noilhan J, Roujean J-L (2000) AVHRR-derived vegetation mapping over Western Europe for use in numerical weather prediction models. Int J Remote Sens 21:1183–1200

Charpentier MA, Groffman PM (1992) Soil moisture variability within remote sensing pixels. J Geophys Res 100(D12): 18955–18959

Chason JW, Baldocchi DD, Huston MA (1991) A comparison of direct and indirect ways for estimating forest canopy leaf area. Agr Forest Meteorol 57:107–128

Chen JM (1996) Optically-based methods for measuring seasonal variation of leaf area index in boreal conifer stands. Agr Forest Meteorol 80:135–163

Chen F, Mitchell K (1999) Using the GEWEX/ISLSCP forcing data to simulate global soil moisture fields and hydrological cycle for 1987–1988. J Meteorol Soc Jpn 77:167–182

Chen JM, Black TA, Adams RS (1991) Evaluation of hemispherical photography for determining plant area index and geometry of a forest stand. Agr Forest Meteorol 56:129–143

Chen FK, Mitchell J, Schaake J, Xue Y, Pan H-L, Koren V, Duan Q, Betts BK (1996) Modeling land-surface evaporation by four schemes and comparison with FIFE observations. J Geophys Res 101:7251–7268

Chen JM, Rich PM, Gower ST, Norman JM, Plummer S (1997) Leaf area index of boreal forests: theory, techniques, and measurements. J Geophys Res 102D:29429–29443

Choularton TW, Gallagher MW, Bower KN, Fowler D, Zahniser M, Kaye A (1995) Trace gas flux measurements at the landscape scale using boundary-layer budgets. Philos T Roy Soc A 351:357–369

Christensen CS, Hummelshøj P, Jensen NO, Larsen B, Lohse C, Pilegaard K, Skov H (2000) Determination of the terpene flux from orange species and Norway spruce by relaxed eddy accumulation. Atmos Environ 34:3057–3067

Coe H, Gallagher MW (1992) Measurements of dry deposition of NO2 to a Dutch heathland using the eddy correlation technique. Q J Roy Meteor Soc 118:767–786

Cohen S, Mosoni P, Meron M (1995) Canopy clumpiness and radiation penetration in a young hedgerow apple orchard. Agr Forest Meteorol 76:185–200

Colello GD, Grivet C, Sellers PJ, Berry JA (1998) Modeling of energy, water and CO_2 flux in a temperate grassland ecosystem with SiB2: May-October 1987. J Atmos Sci 55:1141–1169

Collineau S, Brunet Y (1993a) Detection of turbulent coherent motions in a forest canopy. Part I: Wavelet analysis. Bound-Lay Meteorol 65:357–379

Collineau S, Brunet Y (1993b) Detection of turbulent coherent motions in a forest canopy. Part II: Time-scales and conditional averages. Bound-Lay Meteorol 66:49–73

Coombs J, Hall DO, Long SP, Scurlock JMO (1987) Techniques in bioproductivity and photosynthesis. Pergamon Press, Oxford

Coppin P, Taylor R (1983) A three-component sonic anemometer for general micrometeorological research. Bound-Lay Meteorol 27:27–42

Coulter RL, Kallistratova MA (1999) The role of acoustic sounding in a high-technology era. Meteorol Atmos Phys 71:3–13

Coulter RL, Klazure G, Lesht BM, Martin TJ, Shannon JD, Sisterson DL, Wesely ML (1999) The Argonne boundary layer experiments facility: using minisodars to complement a wind profiler network. Meteorol Atmos Phys 71:53–59

Cox PM, Huntingford C, Harding RJ (1998) A canopy conductance and photosynthesis model for use in a GCM land-surface scheme. J Hydrol 212/213:79–94

Crawford TL, Dobosy RJ (1992) A sensitive fast-response probe to measure turbulence and heat flux from any airplane. Bound-Lay Meteorol 59:257–278

Crawford TL, McMillen RT, Dobosy RJ, MacPherson I (1993) Correcting airborne flux measurements for aircraft speed variation. Bound-Lay Meteorol 66:237–245

Crawford TL, Dobosy RJ, McMillen RT, Vogel CA, Hicks BB (1996) Air-surface exchange measurement in heterogeneous regions: extending tower observations with spatial structure observed from small aircraft. Global Change Biol 2:275–285

Cuenca RH, Brouwer J, Chanzy A, Droogers P, Galle S, Gaze SR, Sicot M, Stricker H, Angulo-Jaramillo R, Boyle SA, Bromley J, Chebhoumi AG, Cooper JD, Dixon AJ, Flies JC, Gandah M, Guadu J-C, Laguerre L, Lecocq J, Soet M, Stewart HJ, Vandervaere J-P, Vauclio M (1997) Soil measurements during HAPEX-Sahel intensive period. J Hydrol 188–189:224–266

Culf AD (2000) Examples of the effects of different averaging methods on carbon dioxide fluxes calculated using the eddy correlation method. Hydrol Earth Syst Sc 4:193–198

Culf AD, Allen SJ, Gash JHC, Lloyd CR, Wallace JS (1993) The energy and water budgets of an area of patterned woodland in the Sahel. Agr Forest Meteorol 66:65–80

Culf AD, Fisch G, Lean J, Polcher J (1998) A comparison of Amazonian climate data with general circulation model simulations. J Climate 11:2764–2773

Culf AD, Fisch G, Malhi Y, Costa RC, Nobre AD, Marques Filho A, Gash JHC, Grace J (1999) Carbon dioxide measurements in the nocturnal boundary layer over Amazon forest. Hydrol Earth Syst Sc 3:39–53

Dang QL, Margolis HA, Sy M, Coyea MR, Collatz GJ, Walthall CL (1997) Profiles of photosynthetically active radiation, nitrogen and photosynthetic capacity in the boreal forest: implications for scaling from leaf to canopy. J Geophys Res 102D:28845–28859

Davis KJ, Lenschow DH, Zimmerman PR (1994) Biogenic non-methane hydrocarbon emissions estimated from tethered balloon observations. J Geophys Res 99:125587–25598

de Bruin HAR, van den Hurk BJJM, Kohsiek W (1995) The scintillation method tested over a dry vineyard area. Bound-Lay Meteorol 76:25–40

De Haan P, Rotach MW (1998) A novel approach to atmospheric dispersion modeling: the Puff-Particle Model (PPM). Q J Roy Meteor Soc 124:2771–2792

Deering DW, Middleton EM, Irons JR, Blad BL, Walter Shea EA, Hays CJ, Walthall CW, Eck TF, Ahmad SP, Banerjee BP (1992) Prairie grassland bidirectional reflectances measured by different instruments at the FIFE site. J Geophys Res 97(D17):18887–18904

Demetriades-Shah TH, Kanemasu ET, Flitcroft ID (1992) Comparison of ground- and satellite-based measurements of the fraction of photosynthetically active radiation intercepted by tallgrass prairie. J Geophys Res 97D:18947–18950

Denmead OT, Harper LA, Freney JR, Griffith DWT, Leuning R, Sharpe RR (1998) A mass balance method for non-intrusive measurements of surface-air trace gas exchange. Atmos Environ 32:3679–3688

Denmead OT, Leuning R, Griffith DWT, Meyer CP (1999) Some recent developments in trace gas flux measurements techniques. In: Bouwman AF (ed) Approaches to scaling trace gas fluxes in ecosystem. Elsevier Science

Denmead OT, Leuning R, Griffith DWT, Jamie IM, Esler MB, Harper LA, Freney JR (2000) Verifying inventory predictions of animal methane emissions with meteorological measurements. Bound-Lay Meteorol 96:187–209

Denning AS, Takahashi T, Friedlingstein P (1999) Can a strong atmospheric CO_2 rectifier effect be reconciled with a "reasonable" carbon budget? Tellus 51B:249–253

Desborough CE, Pitman AJ, Irannejad P (1996) Analysis of the relationship between bare soil evaporation and soil moisture simulated by 13 land-surface schemes for a simple non-vegetated site. Global Planet Change 13:47–56

Desjardins RL (1977) Description and evaluation of a sensible heat flux detector. Bound-Lay Meteorol 11:147–154

Desjardins RL, MacPherson JI, Mahrt L, Schuepp P, Pattey E, Neumann H, Baldocchi D, Wofsy S, Fitzjarrald D, McCaughey H, Joiner DW (1997) Scaling up flux measurements for the boreal forest using aircraft-tower combinations. J Geophys Res 102D:29125–29133

Dobosy RJ, MacPherson JI (1999) Intercomparison between two flux airplanes at SGP97 Dallas TX. Abstracts, 14th Conf on Hydrology, Am Meteorol Soc, Dallas, Texas, pp 137–140

Dobosy RJ, Crawford TL, MacPherson JI, Desjardins RL, Kelly RD, Oncley SP, Lenschow DH (1997) Intercomparison among the four flux aircraft at BOREAS in 1994. J Geophys Res 102D:29101–29111

Dolman AJ, Gregory D (1992) A parameterization of rainfall interception in GCMs. Q J Roy Meteor Soc 118:455–467

Dolman AJ, Culf AD, Bessmoulin P (1997a) Observations of boundary layer development during the HAPEX-Sahel intensive observation period. J Hydrol 188/189:998–1016

Dolman AJ, Gash JHC, Goutorbe JP, Kerr Y, Lebel T, Prince SD, Stricker JNM (1997b) The role of the land surface in the Sahelian climate: HAPEX-Sahel results and future research needs. J Hydrol 188/189:1067–1079

Douville H, Viterbo P, Mahfouf JF, Anton C, Beljaars M (2000) Evaluation of the optimum interpolation and nudging techniques for soil moisture analysis using FIFE data. Mon Weather Rev 128:1733–1756

Duan QY, Schaake JC, Koren VI (1996a) FIFE 1987 water budget analysis. J Geophys Res 101:7197–7207

Duan QY, Schaake JC, Koren VI (1996b) Correction to FIFE 1987 water budget analysis by Q. Y. Duan, J. C. Schaake, and V. I. Koren. J Geophys Res 101:29603

Dunne T, Richey JE, Vörösmarty CJ, Becker A, Bonell M, Hodnett MG, Kabat P, Marengo JA, Martinelli LA, Melack J, O'Connell PE, Tomasella J, Victoria RL (1997) Land-surface hydrology and water chemistry. In: Nobre CA, Artaxo P, Becker A, Brown IF, Dolman H, Dunne T, Gash JHC, Grace J, Janetos AC, Kabat P, Keller M, Krug T, Marengo JA, McNeal RJ, Prince SD, Silva Dias PL, Tomasella J, Victoria CJ, Wickland DE. LBA extended science plan, LBA Project Office, Brazil, October 1997

Durand, P, Dupuis H, Lambert D, Benech B, Druilhet A, Katsaros K, Taylor PK, Weill A (1998) Comparison of sea surface flux measured by instrumented aircraft and ship during SOFIA and SEMAPHORE experiments. J Geophys Res 103C:25125–25136

Dyer AJ (1974) A review of flux-profile relationships. Bound-Lay Meteorol 7:363–372

Dyer AJ, Maher FJ (1965) Automatic eddy-flux measurement with the 'evapotron'. J Appl Meteorol 4:622–625

Dyer AJ, Hicks BB, King KM (1967) The fluxatron – a revised approach to the measurement of eddy fluxes in the lower atmosphere. J Appl Meteorol 6:408–413

Eckman RM, Crawford TL, Dumas EJ, Birdwell KR (1999) Airborne meteorological measurements collected during the Model Validation Program (MVP) field experiments at Cape Canaveral, Florida. NOAA Technical Memorandum ARL/ATDD-233, 54 p

Elagina LG, Zubkovskii SL, Kaprov BM, Sokolov DY (1973) Experimental investigations of the energy balance near the surface. Trudy Glavny Geofiziceskij Observatorii 296:38–45 (in Russian)

Elagina LG, Kaprov BM, Timanovskii DF (1978) A characteristic of the surface air layer above snow. Izvestia AN SSSR Fizika Atmosfery I Oceana 14:926–931 (in Russian)

Eloranta EW, Forrest EW (1992) Volume-imaging lidar observations of the convective structure surrounding the flight path of a flux-measuring aircraft. J Geophys Res 97:18383–18394

Essery R (1998) Boreal forests and snow in climate models. Hydrol Process 12:1561–1567

Eugster W, Senn W (1995) A cospectral correction model for measurement of turbulent NO$_2$ flux. Bound-Lay Meteorol 74:321–340

Evans GC, Whitmore TC, Wong YK (1960) The distribution of light reaching the ground vegetation in a tropical rain forest. J Ecol 48:193–204

Fairall CW, Bradley EF, Rogers DP, Edson JB, Young GS (1996) Bulk parameterization of air-sea fluxes for tropical ocean-global atmosphere coupled ocean-atmosphere response experiment. J Geophys Res-Oceans 101:3747–3764

Fairall CW, Persson POG, Bradley EF, Payne RE, Anderson SP (1998) A new look at calibration and use of Eppley precision infrared radiometers. Part I: Theory and applications. J Atmos Ocean Tech 15:1229–1242

Falge E, Baldocchi DD, Olson R, Anthoni P, Aubinet M, Bernhofer C, Burba G, Ceulemans R, Clement R, Dolman H, Granier A, Gross P, Grunwald T, Hollinger D, Jensen N-O, Katul G, Keronen P, Kowalski A, Lai CT, Law BE, Meyers T, Moncrieff J, Moors E, Munger JW, Pilegaard K, Rannik U, Rebmann C, Suyker A, Tenhunen J, Tu K, Verma S, Vesala T, Wilson K, Wofsy S (2001) Gap filling strategies for defensible annual sums of net ecosystem exchange. Agr Forest Meteorol 107:71–77

Famiglietti JS, Wood EF, Sivapalan M, Thongs DJ (1992) A catchment scale water balance model for FIFE. J Geophys Res 97:18997–19007

Fang C, Moncrieff JB (1998) An open-top chamber for measuring soil respiration and the influence of pressure difference on CO$_2$ efflux measurement. Funct Ecol 12:319–326

Farquhar GD, Caemmerer S von (1982) Modelling of photosynthetic response to environmental conditions. In: Physiological plant ecology II. Lange OL, Nobel PS, Osmond CB, Ziegler H (eds). Springer-Verlag, Berlin, 12B, pp 549–587

Farquhar GD, Sharkey TD (1982) Stomatal conductance and photosynthesis. Annu Rev Plant Phys 33:317–345

Fassnacht KS, Gower ST, Norman JM, McMurtrie RE (1994) A comparison of optical and direct methods for estimating foliage surface area index in forests. Agr Forest Meteorol 71:183–207

Field RT, Fritschen LJ, Kanemasu ET, Smith EA, Stewart JB, Verma SB, Kustas WP (1992) Calibration, comparison, and correction of net radiation instruments used during FIFE. J Geophys Res-Atmos 97:18681–18695

Finch JW, Harding RJ (1998) A comparison between reference transpiration and measurements of evaporation for a riparian grassland site. Hydrol Earth Syst Sc 2:129–136

Finn D, Lamb B, Leclerc MY, Horst TW (1996) Experimental evaluation of analytical and lagrangian surface-layer flux footprint models. Bound-Lay Meteorol 80:283–308

Finnigan J (1999) A comment on the paper by Lee (1998): "On micrometeorological observations of surface-air exchange over tall vegetation". Agr Forest Meteorol 97:55–64

Fitzjarrald DR (1982) New applications of a simple mixed layer model. Bound-Lay Meteorol 22:431–453

Fitzjarrald DR (1997) Surface exchanges and air-mass modification. Proc. 12th AMS Symp. Boundary Layers and Turbulence, Vancouver, BC, Canada, 587 p

Fitzjarrald DR, Acevedo OC, Moore KE (2001) Climatic consequences of leaf presence in the eastern United States. J Climate 14(4):598–614

Fitzjarrald DR, Freedman JM, Sakai RK, Moore KE (2003) The seasonal rectifier isn't right. (in preparation)

Flanagan LB, Brooks JR, Ehleringer JR (1997) Photosynthesis and carbon isotope discrimination in boreal forest ecosystems: a comparison of functional characteristics in plants from three mature forest types. J Geophys Res 102D:28861–28869

Flesch TK (1995) Backward-time lagrangian stochastic dispersion models and their application to estimate gaseous emissions. J Appl Meteorol 34:1320–1332

Foken T (1990) Probleme bei der Bestimmung vertikaler turbulenter Feuchtetransporte im Rahmen von ISLSCP-Experimenten (Ergebnisse von KUREX-88 und TARTEX-90). Erste Deutsch-Deutsche Klimatagung, Berlin, 8 p

Foken T, Oncley SP (1995) Results of the workshop 'Instrumental and methodical problems of land-surface flux measurements'. B Am Meteorol Soc 76:1191–1193

Foken T, Wichura B (1996) Tools for quality assessment of surface-based flux measurements. Agr Forest Meteorol 78:83–105

Foken T, Gerstmann W, Richter SH, Wichura B, Baum W, Ross J, Sulev M, Mölder M, Tsvang LR, Zubkovskii SL, Kukharets VP, Aliguseinov AK, Perepelkin VG, Zelený J (1993) Study of the energy exchange processes over different types of surfaces during TARTEX-90. Deutscher Wetterdienst, Forschung und Entwicklung, Arbeitsergebnisse, 4, 34 p

Foken T, Weisensee U, Kirzel H-J, Thiermann V (1997) Comparison of new-type sonic anemometers, 12th Symposium on Boundary Layer and Turbulence. American Meteorological Society, Boston, Vancouver, BC, pp 356–357

Foken T, Wichura B, Klemm O, Gerchau J, Winterhalter M, Weidinger T (2001) Micrometeorological conditions during the total solar eclipse of August 11, 1999. Meteorol Z 10(3):171–178

Fournier RA, Landry R, August NM, Fedosejevs G, Gauthier RP (1996) Modelling light obstruction in three conifer forests using hemispherical photography and fine tree architecture. Agr Forest Meteorol 82:47–72

Fowler D, Hargreaves KJ, Skiba U, Milne R, Zahniser MS, Moncrieff JB, Beverland IJ, Gallagher MW (1995) Measurement of CH$_4$ and N$_2$O fluxes at the landscape scale using micrometeorological methods. Philos T Roy Soc A 351:339–356

Frech M, Samuelsson P, Tjernström M, Jochum AM (1998) Regional surface fluxes over the NOPEX area. J Hydrol 212/213:155–171

Freedman JM, Fitzjarrald DR (2001) Postfrontal air mass modification. J Hydrometeorol 2(4):419–437

Friedl MA (1995) Modeling land-surface fluxes using a sparse canopy model and radiometric surface temperature measurements. J Geophys Res 100:25435–25446

Friedrichs K, Mölders N, Tetzlaff G (2000) On the influence of surface heterogeneity on the Bowen-ratio: a theoretical case study. Theor Appl Climatol 65:181–196

Fröhlich C, Philipona R, Romero J, Wehrli C (1995) Radiometry at the Physikalsch-Meteorologisches Observatorium Davos and World Radiation Centre. Opt Eng 34:2757–2766

Furger M, Whiteman CD, Wilczak JM (1995) Uncertainty of boundary layer heat budgets computed form wind profiler-RASS networks. Mon Weather Rev 123:790–799

Garratt JR (1978) Flux profile relationships above tall vegetation. Q J Roy Meteor Soc 104:199–211

Garratt JR (1992) The atmospheric boundary layer. Cambridge University Press, New York, 316 p

Garstang M, Fitzjarrald DR (1999) Observations of surface to atmosphere interactions in the tropics. Oxford University Press, New York, 405 p

Gash JHC (1986a) A note on estimating the effect of a limited fetch on micrometeorological evaporation measurements. Bound-Lay Meteorol 35:409–413

Gash JHC (1986b) Observations of turbulence downwind of a change in surface roughness. Bound-Lay Meteorol 36:227–237

Gash JHC, Shuttleworth WJ, Lloyd CR, André J-C, Goutorbe J-P, Gelpe J (1989) Micrometeorological measurements in Les Landes Forest during HAPEX-MOBILHY. Agr Forest Meteorol 46:131–147

Gash JHC, Kabat P, Monteny BA, Amadou M, Bessemoulin P, Billing H, Blyth EM, deBruin HAR, Elbers JA, Friborg T, Harrison G, Holwill CJ, Lloyd CR, Lhomme J-P, Moncrieff JB, Puech D, Sögaard H, Taupin JD, Tuzet A, Verhoef A (1997) The variability of evaporation during the HAPEX-Sahel intensive observation period. J Hydrol 188/189:385–399

Giard D, Bazile E (2000) Implementation of a new assimilation scheme for soil and surface variables in a Global NWP model. Mon Weather Rev 128:997–1015

Gilgen H, Whitlock CH, Koch F, Müller G, Ohmura A, Steiger D, Wheeler R (1994) Technical plan for BSRN data management. WRMC, Techn. Rep. 1, 56 p

Goulden ML, Munger JW, Fan S-M, Daube BC, Wofsy SC (1996) Measurements of carbon sequestration by long-term eddy covariance: methods and a critical evaluation of accuracy. Global Change Biol 2:169–182

Goutorbe J-P, Lebel T, Tinga A, Bessemoulin P, Brouwer J, Dolman AJ, Engman ET, Gash JHC, Hoeppfner M, Kabat P, Kerr YH, Monteny B, Prince S, Saïd F, Sellers P, Wallace JS (1994) HAPEX-Sahel: a large scale study of land-atmosphere interactions in the semi-arid tropics. Ann Geophys 12:53–64

Goutorbe J-P, Lebel T, Dolman AJ, Gash JHC, Kabat P, Kerr YH, Monteny B, Prince SD, Stricker JNM, Tinga A, Wallace JS (1997a) An overview of HAPEX-Sahel: a study in climate and desertification. J Hydrol 188/189:4–17

Goutorbe J-P, Noilhan J, Lacarrère P, Braud I (1997b) Modelling of the atmospheric column over the central sites during HAPEX-Sahel. J Hydrol 188/189:1017–1039

Gower ST, Norman JM (1991) Rapid estimation of leaf area index in conifer and broad-leaf plantations. Ecology 72:1896–1900

Graber WK, Siegwolf R, Furger M (1998) Exchange of CO_2 and water vapour between a composite landscape and the atmosphere. In: Hydrology, Water Resources and Ecology in Headwaters, IAHS Publ 248:107–114

Grace J, Lloyd J, McIntyre J, Miranda A, Meir P, Miranda H, Nobre C, Moncrieff JB, Massheder J, Malhi Y, Wright I, Gash J (1995a) Net carbon dioxide uptake by undisturbed tropical rain forest in 1992/1993. Science 270:778–780

Grace J, Lloyd J, McIntyre J, Miranda A, Meir P, Miranda H, Moncrieff J, Massheder J, Wright I, Gash J (1995b) Fluxes of carbon dioxide and water vapour over an undisturbed tropical forest in southwest Amazonia. Global Change Biol 1:1–12

Granier A (1985) Une nouvelle methode pour la mesure du flux de seve brute dans le tronc des arbres. Ann Sci Forest 42:193–200

Grant RF, Black TA, den Hartog G, Berry JA, Neumann HH, Blanken PD, Yang PC, Russell C, Nalder IA (1999) Diurnal and annual exchanges of mass and energy between an aspen-hazelnut forest and the atmosphere: testing the mathematical model Ecosys with data from the BOREAS experiment. J Geophys Res 104D:27699–27717

Greco S, Baldocchi DD (1996) Seasonal variations of CO_2 and water vapour exchange rates over a temperate deciduous forest. Global Change Biol 2:183–197

Greco S, Swap R, Garstang M, Ulanski S, Shipham M, Harriss RC, Talbot R, Andreae MO, Artaxo P (1990) Rainfall and surface kinematic conditions over central Amazonia during ABLE 2B. J Geophys Res 95:17001–17014

Greenberg JP, Guenther A, Zimmerman P, Baugh W, Geron C, Davis K, Helmig D, Klinger LF (1999) Tethered balloon measurements of biogenic VOCs in the atmospheric boundary layer. Atmos Environ 33:855–867

Grelle A, Lindroth A (1996) Eddy-correlation system for long-term monitoring of fluxes of heat, water vapour and CO_2. Global Change Biol 2:297–307

Grelle A, Lundberg A, Lindroth A, Moren A-S, Cienciala E (1997) Evaporation components of a boreal forest: variations during the growing season. J Hydrol 197:70–87

Griffith DWT, Galle B (2000) Flux measurements of NH_3, N_2O and CO_2 using dual beam FTIR spectroscopy and the flux-gradient technique. Atmos Environ 34:1087–1098

Grossman RL (1992) Convective boundary layer budgets of moisture and sensible heat over an unstressed prairie. J Geophys Res 97:18425–18438

Guenther A, Baugh W, Davis K, Hampton G, Harley P, Klinger L, Vierling L, Zimmerman P (1996) Isoprene fluxes measured by enclosure, relaxed eddy accumulation, surface layer gradient, mixed layer gradient, and mixed layer mass balance techniques. J Geophys Res 1017:18555–18567

Guenther AB, Hills AJ (1998) Eddy covariance measurement of isoprene fluxes. J Geophys Res 103:13145–13152

Gutowski WJ, Chen Y, Ötles Z (1997) Atmospheric water vapor transport in NCEP-NCAR reanalyses: comparison with river discharge in the central United States. B Am Meteorol Soc 78: 1967–1970

Habets F, Etchevers P, Golaz C, Leblois E, Ledoux E, Martin E, Noilhan J, Ottlé C (1999a) Implementation of the ISBA surface scheme in a distributed hydrological model, applied to the HAPEX-MOBILHY area. J Hydrol 217:45–118

Habets F, Etchevers P, Golaz C, Leblois E, Ledoux E, Martin E, Noilhan J, Ottlé C (1999b) Simulation of the water budget and the river flows of the Rhone basin. J Geophys Res 104:31145–31172

Habets F, Noilhan J, Golaz C, Goutorbe JP, Lacarrère P, Leblois E, Ledoux E, Martin E, Ottlé C, Vidal-Madjar D (1999c) The ISBA surface scheme in a macroscale hydrological model applied to the HAPEX-MOBILHY area. Part II: Simulation of streamflows and annual water budget. J Hydrol 217:97–118

Hall FG (1999) Introduction to special section: BOREAS in 1999: experiment and science overview. J Geophys Res 104:27627–27639

Hall FG, Sellers PJ (1995) First International Satellite Land Surface Climatology Project (ISLSCP) Field Experiment (FIFE) in 1995. J Geophys Res 100(D12):25383–25395

Hall FG, Huemmrich KF, Goetz SJ, Sellers PJ, Nickeson JE (1992) Satellite remote sensing of surface energy balance: success, failures and unresolved issues in FIFE. J Geophys Res 97D:19061–19089

Hall B, Claiborn C, Baldocchi DD (1999) Measurement and modeling of the dry deposition of peroxides. Atmos Environ 33:577–589

Halldin S, Lindroth A (1992) Errors in net radiometry: comparison and evaluation of six radiometer designs. J Atmos Ocean Tech 9:762–783

Halldin S, Gottschalk L, van de Griend AA, Gryning SE, Heikinheimo M, Hogstrom U, Jochum A, Lundin LC (1998) NOPEX – a Northern Hemisphere climate processes land-surface experiment. J Hydrol 212/213:172–187

Halldin S, Gryning S-E, Gottschalk L, Van de Griend AA, Jochum A (eds) (1999) NOPEX special issue boreal forests and climate. Agr Forest Meteorol 98/99:696pp

Halldin S, Gryning S-E, Lloyd CR (2001) Land/atmosphere exchange in high latitude landscapes. Theor Appl Climatol 70:1–3

Hanan NP, Bégué A (1995) A method to estimate instantaneous and daily intercepted photosynthetically active radiation using a hemispherical sensor. Agr Forest Meteorol 74:155–168

Hanan NP, Prince SD (1997) Stomatal conductance of west central supersite vegetation in HAPEX-Sahel: measurements and empirical models. J Hydrol 188/189:536–562

Hanan NP, Bégué A, Prince SD (1997a) Errors in remote sensing of intercepted photosynthetically active radiation: an example from HAPEX-Sahel. J Hydrol 188/189:676–696

Hanan NP, Kabat P, Prince SD, Prihodko L (1997b) Vegetation production and water relations. In: Kabat P, Prince SD, Prihodko L (eds) Hydrologic Atmospheric Pilot Experiment in the Sahel (HAPEX-Sahel): methods, measurements and selected results from the west central supersite. Wageningen, The Netherlands, DLO Winand Staring Centre, pp 88–105

Hanan NP, Prince SD, Bégué A (1997c) Modelling vegetation primary production during HAPEX-Sahel using production efficiency and canopy conductance model formulations. J Hydrol 188/189:651–675

Hanan NP, Kabat P, Dolman AJ, Elbers JA (1998) Photosynthesis and carbon balance of a Sahelian fallow savanna. Global Change Biol 4:523–538

Handorf D, Foken T (1997) Analysis of turbulent structure over an Antarctic ice shelf by means of wavelet transformation. 12th Symposium on Boundary Layer and Turbulence, Vancouver BC, Canada, 28 July–1 August 1997, American Meteorological Society, pp 245–246

Harding RJ, Pomeroy JW (1996) The energy balance of a winter boreal landscape. J Climate 9:2778–2787

Harding RJ, Taylor CM, Finch JW (1996) Areal average fluxes from mesoscale meteorological models: the application of remote sensing. In: Stewart JB, Engman E, Kerr Y (eds) Scaling up in hydrology using remote sensing. John Wiley & Sons, Chichester, pp 59–76

Helmig D, Balsley B, Davis K, Juck LR, Jensen M, Bognar J, Smith T, Arrieta RV, Rodriguez R, Birks JW (1998) Vertical profiling and determination of landscape fluxes of biogenic nonmethane hydrocarbons within the planetary boundary layer in the Peruvian Amazon. J Geophys Res 103:25519–25532

Hodges GB, Smith EA (1997) Intercalibration, objective analysis, intercomparison and synthesis of BOREAS surface net radiation measurements. J Geophys Res-Atmos 102:28885–28900

Hogg EH, Black TA, den Hartog G, Neumann HH, Zimmermann R, Hurdle PA, Blanken PD, Nesic Z, Yang PC, Staebler RM, McDonald KC, Oren R (1997) A comparison of sap flow and eddy fluxes of water vapor from a boreal deciduous forest. J Geophys Res 102D:28929–28937

Hong S-Y, Pan H-L (1996) Implementing a nonlocal boundary layer vertical diffusion scheme for the NCEP medium range forecast model. Mon Weather Rev 124:2322–2339

Hope AS (1992) Estimating the daily course of Konza prairie latent heat fluxes using a modified Tergra model. J Geophys Res 97D:19023–19031

Horst TW (1998) The footprint for estimation of atmosphere-surface exchange fluxes by profile techniques. Bound-Lay Meteorol 90:171–188

Horst TW (2000) On frequency response corrections for eddy covariance flux measurements. Bound-Lay Meteorol 94:517–520

Horst TW, Weil JC (1992) Footprint estimation for scalar flux measurements in the atmospheric surface layer. Bound-Lay Meteorol 59:279–296

Horst TW, Weil JC (1994) How far is far enough? The fetch requirements for micrometeorological measurements of surface fluxes. J Atmos Ocean Tech 11:1018–1025

Huntingford C, Jones RG, Prudhomme C, Lamb R, Gash JHC, Jones DA (2003) Regional climate model predictions of extreme rainfall for a changing climate. Q J Roy Meteor Soc 129(590):1607–1621

Hyson P, Hicks BB (1975) A single-beam infrared hygrometer for evaporation measurement. J Appl Meteorol 14:301–307

Jarvis PG, Massheder JM, Hale SE, Moncrieff JB, Rayment M, Scott SL (1997) Seasonal variation of carbon dioxide, water vapour and energy exchanges of a boreal black spruce forest. J Geophys Res 102:28953–28966

Kabat P, Dolman AJ, Elbers JA (1997a) Evaporation, sensible heat and canopy conductance of fallow savannah and patterned woodland in the Sahel. J Hydrol 188–189. 494–515

Kabat P, Hutjes RWA, Feddes RA (1997b) The scaling characteristics of soil parameters: from plot scale heterogeneity to subgrid parameterization. J Hydrol 190:337–396

Kaimal JC, Businger JA (1963) A continuous wave sonic anemometer-thermometer. J Appl Meteorol 2:156–164

Kaimal JC, Finnigan JJ (1994) Atmospheric boundary layer flows: their structure and measurement. Oxford University Press, New York, vii+242 p

Kaimal JC, Gaynor JE (1991) Another look at sonic thermometry. Bound-Lay Meteorol 56:401–410

Kaimal JC, Wyngaard JC, Izumi Y, Coté OR (1972) Spectral characteristics of surface layer turbulence. Q J Roy Meteor Soc 98:563–589

Kanemasu ET, Verma SB, Smith EA, Fritschen LJ, Wesely M, Field RT, Kustas WP, Weaver H, Stewart JB, Gurney R, Panin G, Moncrieff JB (1992) Surface flux measurements in FIFE – an overview. J Geophys Res 97:18547–18555

Kiemle C, Ehret G, Giez A, Davis KJ, Lenschow DH, Oncley SP (1998) Estimation of boundary layer humidity fluxes and statistics from airborne differential absorption lidar (DIAL). J Geophys Res 102:29189–29204

Kim J, Verma SB, Billesbach DP (1999) Seasonal variation in methane emission from a temperate *Phragmites*-dominated marsh: effect of growth stage and plant-mediated transport. Global Change Biol 5:433–440

Kimball JS, Running SW, Saatchi SS (1999) Sensitivity of boreal forest regional water flux and net primary production simulations to sub-grid-scale land-cover complexity. J Geophys Res 104(D22):27789–27801

Koster R, Suarez MJ (1992) Modeling the land-surface boundary in climate models as a composite of independent vegetation stands. J Geophys Res 97:2697–2715

Krebs CJ (1999) Ecological methodology. Addison Wesley Longman Inc., Menlo Park, CA

Kukharets VP, Tsvang LR (1999) Variations of the surface temperature and the problem of heat balance on surface. Izv RAN Fiz Atmos I Okeana 35:207–214 (in Russian)

Kukharets VP, Perepelkin VG, Tsvang LR, Richter SH, Weisensee U, Foken T (1998) Energiebilanz an der Erdoberfläche und Wärmespeicherung im Boden. In: Foken T (ed) Ergebnisse des LINEX-97/1-Experimentes. Deutscher Wetterdienst, Forschung und Entwicklung, Arbeitsergebnisse 53, pp 19–26

Kustas WP, Brutsaert W (1987) Budgets of water vapor in the unstable boundary layer over rugged terrain. J Clim Appl Meteorol 26:607–620

Leduc C, Bromley J, Schroeter P (1997) Water table fluctuation and recharge in semi-arid climate: some results of the HAPEX-Sahel hydrodynamic survey (Niger). J Hydrol 188/189:123–138

Lee X (1998) On micrometeorological observations of surface-air exchange over tall vegetation. Agr Forest Meteorol 91:39–49

Lee X, Black TA (1993) Atmospheric turbulence within and above a Douglas-fir stand; Part II: Eddy fluxes of sensible heat and water vapour. Bound-Lay Meteorol 64:369–389

LeMone MA (1995) The cumulus-topped atmospheric boundary layer. In: Moeng C-H (ed) The planetary boundary layer and its parameterization. 1995 Summer Colloquium, available from NCAR, Boulder CO, USA

LeMone MA, Grossman RL (1999) Evolution of potential temperature and moisture during the morning: CASES-97. Abstract J7.1, 13th AMS Symposium on Boundary Layers and Turbulence, Dallas, TX

Lenschow DH (1995) Micrometeorological techniques for measuring biosphere-atmosphere trace gas exchange. In: Matson PA, Harriss RC (eds) Biogenic trace gases: measuring emissions from soil and water. Methods in ecology series. Blackwell Science, pp 126–163

Lenschow DH, Johnson WB Jr (1968) Concurrent airplane and balloon measurements of atmospheric boundary-layer structure over a forest. J Appl Meteorol 7:79–89

Lenschow DH, Wyngaard JC, Pennell WT (1980a) Mean-field and second-moment budgets in a baroclinic, convective boundary layer. J Atmos Sci 37:1313–1326

Lenschow DH, Delany AC, Stankov BB, Stedman DH (1980b) Airborne measurements of the vertical flux of ozone in the boundary layer. Bound-Lay Meteorol 119:249–265

Lenschow DH, Mann J, Christens L (1994) How long is long enough when measuring fluxes and other turbulence statistics? J Atmos Ocean Tech 11:661–673

Leuning RL, Judd M (1996) The relative merits of open- and closed-path analysers for measurement of eddy fluxes. Global Change Biol 2:241–254

Leuning RL, King KM (1992) Comparison of eddy covariance measurements of CO_2 fluxes by open- and closed-path CO_2 analysers. Bound-Lay Meteorol 59:297–311

Leuning RL, Moncrieff JB (1990) Eddy covariance CO_2 flux measurements using open- and closed-path CO_2 analysers: corrections for analyser water vapour sensitivity and damping of fluctuations in air sampling tubes. Bound-Lay Meteorol 53:63–76

Levy PE, Moncrieff JB, Massheder JM, Jarvis PG, Scott SL, Brouwer J (1997) CO_2 fluxes at leaf and canopy scale in millet, fallow and tiger bush vegetation at the HAPEX-Sahel southern supersite. J Hydrol 188/189:612–632

Levy PE, Grelle A, Lindroth A, Mölder M, Jarvis PG, Jriujt B, Moncrieff JB (1999) Regional-scale CO_2 fluxes over central Sweden by a boundary layer budget method. Agr Forest Meteorol 98/99: 169–180

Liang X, Lettenmaier DP, Wood EF, Burges SJ (1994) A simple hydrologically-based model of land-surface water and energy fluxes for general circulation models. J Geophys Res 99:14415–14428

Linder W, Noilhan J, Berger M, Bluemel K, Blyth E, Boulet G, Braud I, Dolman A, Fiedler F, Grunwald J, Harding R, vd Hurk B, Jaubert G, Mueller A, Ogink M (1996) Intercomparison of surface schemes using EFEDA flux data. Technical Report No. 39, Météo-France, CNRM/GMME

Livingston GP, Hutchinson GL (1995) Enclosure-based measurement of trace gas exchange: applications and sources of error. In: Matson PA, Harriss RC (eds) Biogenic trace gases: measuring emissions from soil and water. Methods in Ecology Series. Blackwell, pp 14–51

Lloyd CR, Bessemoulin P, Cropley FD, Culf AD, Dolman AJ, Elbers J, Heusinkveld B, Moncrieff JB, Monteny B, Verhoef A (1997) A comparison of surface fluxes at the HAPEX-Sahel fallow bush sites. J Hydrol 188/189:400–425

Luke WT, Watson TB, Olszyna KJ, Gunter RL, McMillen RT, Wellman EL, Wilkison SW (1998) A comparison of airborne and surface trace gas measurements during the Southern Oxidants Study (SOS). J Geophys Res 103D, 22317–22337

Lyons TJ, Hacker JM, Foster IJ, Schwerdtfeger P, Smith RCG, Xinmei H, Bennett JM (1993) Land-atmosphere interaction in a semi-arid region: the Bunny Fence Experiment. B Am Meteorol Soc 74:1327–1334

MacPherson JI, Dobosy RJ, Verma S, Justas WP, Prueger JH, Williams A (1999) Intercomparisons between flux aircraft and towers in SGP97. Abstracts, 14th Conf on Hydrology, Am Meteorol Soc, Dallas, TX, pp 125–128

Magagi RD, Kerr YH (1997) Retrieval of soil moisture and vegetation characteristics by use of ERS-1 wind scatterometer over arid and semi-arid areas. J Hydrol 188/189:361–384

Mahfouf JF (1991) Analysis of soil moisture from near-surface parameters: a feasibility study. J Appl Meteorol 30:1534–1547

Mahfouf J-F, Noilhan J (1996) Inclusion of gravitational drainage in a land-surface scheme based on the force restore method. J Appl Meteorol 35:987–992

Mahfouf JF, Ciret C, Ducharne A, Irannejad P, Noilhan J, Shao Y, Thornton P, Xue Y, Yang ZL (1996) Analysis of transpiration results from the RICE and PILPS workshop. Global Planet Change 13:73–88

Mahrt L (1998) Flux sampling errors for aircraft and towers. J Atmos Ocean Tech 15:416–429

Mailhot J, Strapp JW, MacPherson JI, Benoit R, Belair S, Donaldson NR, Froude F, Benjamin M, Zawadzki I, Rogers RR (1998) The Montreal-96 Experiment on Regional Mixing and Ozone (MERMOZ): an overview and some preliminary results. B Am Meteorol Soc 79:433–442

Malhi Y, Nobre AD, Grace J, Kruijt B, Pereira MGP, Culf A, Scott S (1998) Carbon dioxide transfer over a central Amazonian rain forest. J Geophys Res 103:31593–31612

Malhi Y, Pegoraro E, Nobre AD, Pereira MGP, Grace J, Culf AD, Clement R (2002) The energy and water dynamics of a central Amazonian rain forest. J Geophys Res 10.1029/2001JD000623

Manabe S (1969) The atmospheric circulation and the hydrology of the Earth's surface. Mon Weather Rev 97:739–774

Marty CA (2000) Surface radiation, cloud forcing and greenhouse effect in the Alps. PhD thesis, ETHZ, Zürich. 139 p

Mascart P, Gelpe J, Pinty JP (1988) Etude des caractéristiques texturales des sols dans la zone HAPEX-MOBILHY (study of the soil texture in the HAPEX MOBILHY area). Technical report, OPGC, 95

Massman WJ, Grantz DA (1995) Estimating canopy conductance to ozone uptake from observations of evapotranspiration at the canopy scale and at the leaf scale. Global Change Biol 1: 183–198

Matthews S, Hacker JM, Williams A, Hutley L (2000) Heat fluxes over Northern Australia: results from aircraft measurements during the NATT experiment. Proceedings of the AMOS 2000 Conference, February 2000, Melbourne, Australia

McBean GA (1972) Instrument requirements for eddy correlation measurements. J Appl Meteorol 11:1078–1084

McGregor GR, Gellatly AF (1996) The energy balance of a melting snowpack in the French Pyrenees during warm anticyclonic conditions. Int J Climatol 16:479–486

McMillen (1986) A BASIC program for eddy correlation in non-simple terrain. NOAA Technical Memorandum ERL-147, Air Resources Lab., Maryland

McNaughton KG, Laubach J (2000) Power spectra for wind and scalars in a disturbed surface layer at the base of an advective inversion. Bound-Lay Meteorol 96:143–185

Metcalfe RA, Buttle JM (1999) Semi-distributed water balance dynamics in a small boreal forest basin. J Hydrol 226:66–87

Michels BI, Jochum AM (1995) Heat and moisture flux profiles in a region with inhomogeneous surface evaporation. J Hydrol 166:383–407

Middleton EM, Sullivan JH, Bovard BD, Deluca AJ, Chan SS, Cannon TA (1997) Seasonal variation in foliar characteristics and physiology for boreal forest species at the five Saskatchewan tower sites during the 1994 Boreal Ecosystem-Atmosphere Study. J Geophys Res 102D:28831–28844

Milne R, Beverland IJ, Hargreaves K, Moncrieff JB (1999) Variation of the beta coefficient in the relaxed accumulation method. Bound-Lay Meteorol 93:211–225

Miskolczi F, Guzzi R (1993) Effects of nonuniform spectral dome transmittance on the accuracy of infrared radiation measurements using shielded pyrradiometers and pyrgeometers. Appl Optics 32:3257–3265

Moeng C-H, Wyngaard JC (1984) Statistics of conservative scalars in the convective boundary layer. J Atmos Sci 41:3161–3169

Mölder M, Grelle A, Lindroth A, Halldin S (1999) Flux-profile relationships over a boreal forest - roughness sublayer corrections. Agr Forest Meteorol. 98/99:645–658

Moncrieff JB, Malhi Y, Leuning R (1996) The propagation of errors in long-term measurements of carbon and water. Global Change Biol 2:231–240

Moncrieff JB, Massheder JM, de Bruin H, Elbers J, Friborg T, Heusinkveld B, Kabat P, Scott S, Soegaard H, Verhoef A (1997) A system to measure surface fluxes of momentum, sensible heat, water vapour and carbon dioxide. J Hydrol 189(1–4):589–611

Monteith JL, Unsworth MH (1990) Principles of environmental physics, 2nd Edition. Arnold

Monteny BA (1993) HAPEX-Sahel 1992: Campagne de mesures Supersite Central Est. ORSTOM, Montpellier

Montgomery RB (1947) Introduction: problems covering convective layers. Ann NY Acad Sci 48:707–713

Moore CJ (1986) Frequency response corrections for eddy correlation systems. Bound-Lay Meteorol 37:17–35

Moore KE, Fitzjarrald DR, Ritter JA (1992) How well can regional fluxes be derived from smaller-scale estimates? J Geophys Res 98:7187–7198

Mortensen NG (1994) Wind measurements for wind energy applications - a review. Proceedings of the 16th British wind energy association conference, Stirling, Scotland, 15–17 June, 1994

Mosier AR (1989) Chamber and isotope techniques. In: Andreae MO, Schimel DS (ed) Exchange of trace gases between terrestrial ecosystems and the atmosphere. John Wiley & Sons, Chichester, 346 p

Mulhearn PJ (1977) Relations between surface fluxes and mean profiles of velocity, temperature and concentration, downwind of a change in surface roughness. Q J Roy Meteor Soc 103:785–802

Neininger B, Bäumle MB, Liechti O, Lehning M (1999) Airborne measurements of air pollution in the regions of Geneva and Berne, 1996–1997. Final report for projects AirOBsGeneva and BOPS, Part I. Swiss Agency for the Environment, Forests and Landscape, Berne, 32 p

Ni W, Li X, Woodcock CE, Roujean J-L, Davis RE (1997) Transmission of solar radiation in boreal conifer forests: measurements and models. J Geophys Res 102D:29555–29566

Nicholls S, LeMone MA (1980) The fair weather boundary layer in GATE: the relationship of subcloud fluxes and structure to the distribution and enhancement of cumulus clouds. J Atmos Sci 37:2051–2067

Nijssen B, Lettenmaier DP (1999) A simplified approach for predicting shortwave radiation transfer through boreal forest canopies. J Geophys Res 104D:27859–27868

Nijssen B, Lettenmaier DP (2002) Water balance dynamics of a boreal forest watershed: White Gull Creek basin, 1994–1996. Water Resour Res 38(11): Art. No. 1255

Noilhan J, Lacarrère P (1995) GCM grid-scale evaporation from mesoscale modelling. J Climate 8:206–223

Noilhan J, Planton S (1989) A simple parameterization of land-surface processes for meteorological models. Mon Weather Rev 117:536–549

Noilhan J, Lacarrère P, Bougeault P (1991) An experiment with an advanced surface parameterization in a mesoscale model. Part 3: Comparison with the HAPEX-MOBILHY dataset. Mon Weather Rev 119:2393–2413

Noilhan J, Lacarrère P, Dolman AJ, Blyth EM (1997) Defining area-average parameters in meteorological models for land-surfaces with mesoscale heterogeneity. J Hydrol 190:302–316

Norman JM, Kucharik CJ, Gower ST, Baldocchi DD, Crill PM, Rayment M, Savage K, Striegl RG (1997) A comparison of six methods for measuring soil-surface carbon dioxide fluxes. J Geophys Res-Atmos 102(D24):28771–28777

Oechel WC, Vourlitis GL, Brooks SB, Crawford TL, Dumas EJ (1998) Intercomparison between chamber, tower, and aircraft net CO_2 exchange and energy fluxes measured during the Arctic system sciences land-atmosphere-ice interaction (ARCSS-LAII) flux study. J Geophys Res 103:28993–29003

Ohmura A, Dutton EG, Forgan B, Fröhlich C, Gilgen H, Hegner H, Heimo A, König-Langlo G, McArthur B, Müller G, Philipona R, Pinker R, Whitlock CH, Dehne K, Wild M (1998) Baseline Surface Radiation Network (BSRN/WCRP): new precision radiometry for climate research. B Am Meteorol Soc 79:2115–2136

Ohtaki E, Matsui T (1982) Infrared device for simultaneous measurement of fluctuations of atmospheric carbon dioxide and water vapor. Bound-Lay Meteorol 24:109–119

Oliveira AP (1990) Planetary boundary dynamics over the Amazon rain forest. Ph. D. dissertation, Dept. of Atmospheric Sciences, University at Albany, SUNY, 296 p

Oliveira AP, Fitzjarrald DR (1993) The Amazon River breeze and the local boundary layer: I. Observations. Bound-Lay Meteorol 63:141–162

Oncley SP, Lenschow DH, Campos TL, Davis KJ, Mann J (1997) Regional-scale surface flux observations across the boreal forest during BOREAS. J Geophys Res 102D:29147–29154

Oncley SP, Foken T, Vogt R, Bernhofer C, Liu H, Sorbjan Z, Pitacco A, Grantz D, Riberio L (2000) The EBEX 2000 field experiment. 14th Symposium on Boundary Layer and Turbulence. American Meteorological Society, Boston Aspen, CO, pp 322–324

Orlenko LR, Legotina SI (1973) The energy balance over the underlying surface during KENEX-71. Trudy Glavny Geofiziceskij Observatorii 296:46–56 (in Russian)

Panin GN, Tetzlaff G, Raabe A, Schönfeld H-J, Nasonov AE (1996) Inhomogeneity of the land surface and the parametrization of surface fluxes – a discussion. Wiss. Mitt. aus dem Institut für Meteorologie der Universität Leipzig und dem Institut für Troposphärenforschung e. V. Leipzig 4:204–215

Parton WJ, Coughenour MB, Scurlock JMO, Ojima DS, Gilmanov TG, Scholes RJ, Schimel DS, Kirchner TB, Menaut J-C, Seastedt TR, Garcia Moya E, Kamnalrut A, Kinyamario JI, Hall DO (1996) Global grassland ecosystem modelling: development and test of ecosystem models for grassland systems. In: Breymeyer AI, Hall DO, Melillo JM, Agren GI (eds) Global change: effects on coniferous forests and grasslands. John Wiley & Sons, Chichester, SCOPE 56:229–266

Pattey E, Cessan AJ, Desjardins RL, Kerr LA, Rochette P, St-Amour G, Zhu T, Headrick K (1995) Herbicides volatilization measured by the relaxed eddy accumulation technique using two trapping media. Agr Forest Meteorol. 76:201–220

Paulson CA (1970) The mathematical representation of wind speed and temperature profiles in the unstable atmospheric surface layer. J Appl Meteorol 9:857–861

Pauwels VRN, Wood EF (1999) A soil-vegetation-atmosphere transfer scheme for the modeling of water and energy balance processes in high latitudes; 2. application and validation. J Geophys Res 104D:27823–27839

Pearcy RW, Schulze E-D, Zimmermann R (1989) Measurement of transpiration and leaf conductance. In: Pearcy RW, Ehleringer JR, Mooney HA, Rundel PW (eds) Plant physiological ecology. Chapman & Hall, London, pp 137–160

Peng C, Apps MJ, Price DT, Nalder IA, Halliwell DH (1998) Simulating carbon dynamics along the Boreal Forest Transect Case Study (BFTCS) in Central Canada. Global Biogeochem Cy 12: 381–392

Peters-Lidard CD, Blackburn E, Liang X, Wood EF (1998) The effect of soil thermal conductivity parameterization on surface energy fluxes and temperatures. J Atmos Sci 55:1209–1224

Philipona R, Fröhlich C, Betz Ch (1995) Characterization of pyrgeometers and the accuracy of atmospheric long-wave radiation measurements. Appl Optics 34:1598–1605

Philipona R, Fröhlich C, Dehne K, DeLuisi J, Augustine J, Dutton E, Nelson D, Forgan B, Novotny P, Hickey J, Love SP, Bender S, McArthur B, Ohmura A, Symour JH, Foot JS, Shiobara M, Valero FPJ, Strawa AW (1998) The baseline surface radiation network pyrgeometer round-robin calibration experiment. J Atmos Ocean Tech 15:687–696

Pielke RA Sr (2002) Mesoscale meteorological modeling. 2nd Edition, Academic Press, San Diego, CA, 676 p

Pitman AJ, Desborough CE, Henderson-Sellers A (1999) Uncertainty in the parameterisation of land-surface schemes: experience from PILPS. In: ECMWF and WRCP/GEWEX workshop on modeling and data assimilation for land-surface-processes, 29 June–2 July 1998, ECMWF, Reading, UK

Polley HW, Norman JM, Arkebauer TJ, Walter-Shea EA, Greegor DH, Bramer B (1992) Leaf gas exchange of Andropogon geradii Vitman, Panicum virgatum L., and Sorghastrum nutans (L.) nash in a tallgrass prairie. J Geophys Res 97:18837–18844

Potasnak MJ, Wofsy SC, Denning AS, Conway TJ, Muner JW, Barnes DH (1999) Influence of biotic exchange and combustion sources on atmospheric CO_2 concentrations in New England from observations at a forest flux tower. J Geophys Res 194:9561–9569

Rannik U, Vesala T (1999) Autoregressive filtering versus linear detrending in estimation of fluxes by the eddy covariance method. Bound-Lay Meteorol 91. 259–280

Rasmusson EM (1967) Atmospheric water vapor transport and the water balance of North America. Part I. Characteristics of the water vapor flux field. Mon Weather Rev 95:403–426

Raupach MR (1991) Vegetation-atmosphere interaction in homogeneous and heterogeneous terrain: some implications of mixed-layer dynamics. Vegetatio 91:105–120

Raupach MR (2000) Equilibrium evaporation and the convective boundary layer. Bound-Lay Meteorol 96:107–141

Raupach MR, Legg BJ (1984) The uses and limitations of flux-gradient relationships in micrometeorology. Agr Water Manage 8:119–131

Raupach MR, Denmead OT, Dunin FX (1992) Challenges in linking atmospheric CO_2 concentrations to fluxes at local and regional scales. Aust J Bot 40:697–716

Rayment MB, Jarvis PG (1999) Seasonal gas exchange of black spruce using an automated branch bag system. Can J Forest Res 29:1528–1538

Rich PM, Clark DB, Clark DA, Oberbauer SF (1993) Long-term study of solar radiation regimes in a tropical wet forest using quantum sensors and hemispherical photography. Agr Forest Meteorol 65:107–127

Ritter JA, Lenschow DH, Barrick JDW, Gregory GL, Sachse GW, Hill GF, Woerner MA (1990) Airborne flux measurements and budget estimates of trace species over the Amazon basin in the GTE/ABLE2B expedition. J Geophys Res 96:16875–16886

Ritter JA, Barrick JDW, Sachse GW, Gregory GL, Woerner MA, Watson CE, Hill GF, Collins JE Jr (1992) Airborne flux measurements of trace species in an arctic boundary layer. J Geophys Res 97:16601–16625

Ritter JA, Barrick JDW, Watson CE, Sachse GW, Gregory GL, Anderson BE, Woerner MA, Collins JE Jr (1994) Airborne boundary layer flux measurements of trace species over Canadian boreal forest and northern wetland regions. J Geophys Res 99:1671–1685

Rodriguez-Iturbe I, Mejia JM (1974) The design of rainfall networks in time and space. Water Resour Res 10:713–728

Rotach MW, Gryning S-E, Tassone C (1996) A two-dimensional Lagrangian stochastic dispersion model for daytime conditions. Q J Roy Meteor Soc 122:367–389

Roujean J-L (1999) Two-story equations of transmission of solar energy (TSETSE) in open boreal conifer tree stands. J Geophys Res 104D:27869–27879

Saïd F, Attié JL, Bénech B, Druilhet A, Durand P, Marciniak MH, Monteny B (1997) Spatial variability in airborne surface flux measurements during HAPEX-Sahel. J Hydrol 188/189:878–911

Saugier B, Granier A, Pontailler JY, Dufrene E, Baldocchi DD (1997) Transpiration of a boreal pine forest measured by branch bag, sap flow and micrometeorological methods. Tree Physiol 17:511–519

Scala JR, Garstang M, Tao W-K, Pickering KE, Thompson AM, Simpson J, Kirchhoff VWJH, Browell EV, Sachse GW, Torres AL, Gregory GL, Rasmussen RA, Khalil MAK (1990) Cloud draft structure and trace gas transport. J Geophys Res 95:17015–17030

Schmid HP (1997) Experimental design for flux measurements: matching scales of observations and fluxes. Agr Forest Meteorol 87:179–200

Schmid HP, Lloyd CR (1999) Spatial representativeness at the location bias of flux footprints over inhomogeneous areas. Agr Forest Meteorol 93:195–209

Schmid HP, Oke TR (1990) A model to estimate the source area contributing to turbulent exchange in the surface layer over patchy terrain. Q J Roy Meteor Soc 116:965–988

Schotanus P, Nieuwstadt FTM, DeBruin HAR (1983) Temperature measurement with a sonic anemometer and its application to heat and moisture fluctuations. Bound-Lay Meteorol 26:81–93

Schuepp PH, Leclerc MY, Macpherson JI, Desjardins RL (1990) Footprint prediction of scalar fluxes from analytical solutions of the diffusion equation. Bound-Lay Meteorol 50:355–376

Schulze R (1961) Über die Verwendung von Polyätylen für Strahlungsmessungen. Arch Meteor Geophys Bioklim B 11:211–223

Sellers PJ, Hall FG (1992) FIFE in 1992: results, scientific gains, and future research directions. J Geophys Res 97(D17):19091–19019

Sellers PJ, Shuttleworth WJ, Dorman JL, Dalcher A, Roberts JM (1989) Calibrating the simple biosphere model for Amazonian tropical forest using field and remote sensing data: Part 1: Average calibration with field and remote sensing data. J Appl Meteorol 28:727–759

Sellers PJ, Hall F, Asrar G, Strebel DE, Murphy RE (1992) An overview of the First International Satellite Land Surface Climatology Project (ISLSCP) Field Experiment (FIFE). J Geophys Res 97(D17):18345–18371

Sellers PJ, Hall F, Margolis H, Kelly RD, Baldocchi D, den Hartog G, Cihlar J, Ryan MG, Goodison B, Crill P, Ranson KJ, Lettenmaier DP, Wickland DE (1995a) The Boreal Ecosystem-Atmosphere Study (BOREAS): an overview and early results from the 1994 field year. B Am Meteorol Soc 76:1549–1577

Sellers PJ, Heiser MD, Hall FG, Goetz SJ, Strebel DE, Verma SB, Desjardins RL, Schuepp PM, MacPherson JI (1995b) Effects of spatial variability in topography, vegetation cover and soil moisture on area-averaged surface fluxes: a case study using the FIFE 1989 data. J Geophys Res 100:25607–25629

Sellers PJ, Meeson W, Closs J, Collatz J, Corprew F, Dazlich D, Hall FG, Kerr Y, Koster R, Los S, Mitchell K, McManus J, Meyers D, Sun K-J, Try P (1996a) The ISLSCP initiative I global datasets: surface boundary conditions and atmospheric forcings for land-atmosphere studies. B Am Meteorol Soc 77:1987–2005

Sellers PJ, Los SO, Tucker CJ, Justice CO, Dazlich DA, Collatz GJ, Randall DA (1996b) A revised land-surface parameterization (SiB-2) for atmospheric GCMs. Part 2: The generation of global fields of terrestrial parameters from satellite data. J Climate 9:676–705

Sellers PJ, Hall FG, Kelly RD, Black A, Baldocchi DD, Berry J, Ryan M, Ranson KJ, Crill PM, Lettenmaier DP, Margolis H, Cihlar J, Newcomer J, Fitzjarrald D, Jarvis PG, Gower ST, Halliwell D, Williams D, Goodison B, Wickland DE, Guertin FE (1997) BOREAS in 1997: Experiment overview, scientific results and future directions. J Geophys Res 102:28731–28769

Shao Y, Henderson-Sellers A (1996) Validation of soil moisture simulation in land-surface parameterization schemes. Global Planet Change 13:11–46

Shuttleworth WJ (1988a) Evaporation from Amazonian rainforest. P Roy Soc Lond B Bio 233:321–346

Shuttleworth WJ (1988b) Macrohydrology – the new challenge for process hydrology. J Hydrol 100:31–56

Shuttleworth WJ (1991) Insight from large-scale observational studies of land/atmosphere interactions. Surv Geophys 12:3–30

Shuttleworth WJ, Dickinson RE (1989) Tropical deforestation: comment on the discrepancy between recent GCM and observational results. Q J Roy Meteor Soc 115:1177–1179

Shuttleworth WJ, Gash JHC, Lloyd CR, Roberts JM, Marques A de O, Fisch G, de Silva P, Ribeiro MNG, Molion LCB, de Abreu Sa LD, Nobre CA, Cabral OMR, Patel SR, de Moraes JC (1984) Eddy correlation measurements of energy partition for Amazonian forest. Q J Roy Meteor Soc 110:1143–1162

Shuttleworth WJ, Gash JHC, Lloyd CR, McNeil DD, Moore CJ, Wallace JS (1988) An integrated micrometeorological system for evaporation measurement. Agr Forest Meteorol 43:295–317

Smith EA, Hodges GB, Bacrania M, Cooper HJ, Owens MA, Chappell R, Kincannon W (1997) BOREAS net radiometer engineering study. Final Report (NASA Grant NAG5-2447), Florida State University, Tallahassee, USA, 51 p

Soegaard H, Boegh E (1995) Estimation of evapotranspiration from a millet crop in the Sahel combining sap flow, leaf area index and eddy correlation technique. J Hydrol 166:265–282

Soegaard H, Nordstroem C, Friborg T, Hansen BU, Christensen TR, Bay C (2000) Trace gas exchange in a high-arctic valley. 3. Integrating and scaling CO_2 fluxes from canopy to landscape using flux data, footprint modeling, and remote sensing. Global Biogeochem Cy 14:725–744

Staebler RM, Fitzjarrald DR, Moore KE, Czikowsky MJ, Acevedo OC (2001) Topographic effects on flux measurements at Harvard Forest. American Meteorological Society. 14th Symposium on boundary layer and turbulence. 7–11 August 2000. Aspen, CO, pp 390–393

Stenberg P (1996) Correcting LAI-2000 estimates for the clumping of needles in shoots of conifers. Agr Forest Meteorol 79:1–8

Stephens BB, Wofsy SC, Keeling RF, Tans PP, Potosnak MJ (2000) The CO_2 budget and rectification airborne, study: strategies for measuring rectifiers and regional fluxes. In: Inverse methods in global biogeochemical cycles. Geophysical Monog. 114, Amer. Geophys Union, 324 p

Sun J, Lenschow DH, Mahrt L, Crawford TL, Davis KJ, Oncley SP, MacPherson JI, Wang Q, Dobosy RJ, Desjardins RL (1997) Lake-induced atmospheric circulations during BOREAS. J Geophys Res 102D:29155–29166

Swinbank WC (1951) The measurement of vertical transfer of heat and water vapour by eddies in the lower atmosphere. J Appl Meteorol 8:135–145

Taylor GI (1914) In: Report on the work carried out by the S.S. Scotia (1913). H. M. Stationery Office, London, pp 55–62

Taylor CM (1995) Aggregation of wet and dry surfaces in interception schemes for general circulation models. J Climate 8:441–448

Taylor CM (2000) The influence of antecedent rainfall on Sahelian surface evaporation. Hydrol Process 14:1245–1259

Taylor CM, Blyth EM (2000) Rainfall controls on evaporation at the regional scale: an example from the Sahel. J Geophys Res 105:15469–15479

Taylor CM, Harding RJ, Thorpe AJ, Bessmoulin P (1997a) A mesoscale simulation of land-surface heterogeneity from HAPEX-Sahel. J Hydrol 188/189:1040–1066

Taylor CM, Saïd F, Lebel T (1997b) Interactions between the land surface and mesoscale rainfall variability during HAPEX-Sahel. Mon Weather Rev 125:2211–2227

Taylor CM, Lambin EF, Stephenne N, Harding RJ, Essery RLH (2002) The influence of land-use change on climate in the Sahel. J Climate 15(24):3615–3629

Tennekes H (1973) A model for the dynamics of the inversion above a convective boundary layer. J Atmos Sci 30:558–581

Thom AS (1975) Momentum, mass and heat exchange. In: Monteith JL (ed) Vegetation and the atmosphere. Academic Press, Chichester, pp 57–109

Tsvang LR, Aligusseynov AK, Perepelkin VG, Sulev MA, Meolder ME, Zeleny J (1987) Experiments on heat-balance closure in the atmospheric surface-layer and on the earth surface. Izvestiya AkademiI Nauk SSSR Fizika Atmosfery I Okeana 23:3–13 (in Russian)

Tsvang LR, Fedorov MM, Kader BA, Zubkovskii SL, Foken T, Richter SH, Zeleny YA (1991) Turbulent exchange over a surface with chessboard-type inhomogeneities. Bound-Lay Meteorol 55:141–160

Turner CL, Seastedt TR, Dyer MI, Kittel TGF, Schimel DS (1992) Effects of management and topography on the radiometric response of a tallgrass prairie. J Geophys Res 97D:18855–18866

Tuzet A, Castell J-F, Perrier A, Zurfluh O (1997) Flux heterogeneity and evapotranspiration partitioning in a sparse canopy: the fallow savanna. J Hydrol 188/189:482–493

Valentini R, de Angelis P, Matteucci G, Monaco R, Dore S, Scarascia Mugnozza GE (1996) Seasonal net carbon dioxide exchange of a beech forest with the atmosphere. Global Change Biol 2:199–207

Van den Hurk B, Malhi YS (1993) Net radiometer comparison experiments during the EFEDA-campaign. Memorandum. Department of Meteorology, Agricultural University of Wageningen, the Netherlands, 26 p

Van den Hurk BJJM, Viterbo P, Beljaars ACM, Betts AK (2000) Offline validation of the ERA40 surface scheme. ECMWF Technical memorandum 295, January 2000

Van Leeuwen WJD, Huete AR (1996) Effects of standing litter on the biophysical interpretation of plant canopies with spectral indices. Remote Sens Environ 55:123–138

Verma SB (1990) Micrometeorological methods for measuring surface fluxes. Remote Sensing Reviews 5:99–115

Verma SB, Baldocchi DD, Anderson DE, Matt DR, Clement RJ (1986) Eddy fluxes of CO_2, water vapour and sensible heat over a deciduous forest. Bound-Lay Meteorol 36:71–91

Verma SB, Ullman FG, Billesbach DP, Clement RJ, Kim J (1992) Eddy correlation measurements of methane flux in a northern peatland. Bound-Lay Meteorol 58:289–304

Viterbo P, Beljaars A (1995) An improved land-surface parameterization scheme in the ECMWF model and its validation. J Climate 8:2716–2748

Viterbo P, Betts AK (1999) The forecast impact of the albedo of the boreal forests in the presence of snow. J Geophys Res 104D:27803–27810

Walter-Shea EA, Blad BL, Hays CJ, Mesarch MA, Deering DW, Middleton EM (1992) Biophysical properties affecting vegetative canopy reflectance and absorbed photosynthetically active radiation at the FIFE site. J Geophys Res 97D:18925–18934

Wang JR (1995) An overview of the measurements of soil moisture and modeling of moisture flux in FIFE. J Geophys Res 100D12:18955–18959

Wang JR, Gogineni SP, JAmpe (1992) Active and passive microwave measurements of soil moisture in FIFE. FIFE Special Issue, J Geophys Res-Atmos 97D17:18979–18997

Wang Q, Lenschow DH, Pan L-L, Schillawski RD, Kok GL, Prevot AS (1999) Characteristics of the marine boundary layers using two Lagrangian measurement periods: 2. Turbulence structure: First Aerosol Characterization Experiment (ACE 1). J Geophys Res 104D:21767–21784

Webb EK (1970) Profile relationships: the log-linear range and extension to strong stability. Q J Roy Meteor Soc 96:67–90

Webb EK, Pearman GI, Leuning RL (1980) Correction of flux measurements for density effects due to heat and water vapour transfer. Q J Roy Meteor Soc 106:85–100

Weseley ML, Eastman JA, Stedman DH, Yalvac ED (1982) An eddy correlation measurement of NO_2 flux to vegetation and comparison to O_3 flux. Atmos Environ 16:815–820

Wichura B, Buchmann N, Foken T (2000) Fluxes of the stable isotope ^{13}C above a spruce forest measured by hyperbolic relaxed eddy accumulation method. American Meteorological Society. 14th Symposium on boundary layer and turbulence, 7–11 August 2000. Aspen, CO, pp 559–562

Wilson KB, Baldocchi DD (2000) Seasonal and interannual variability of energy fluxes over a broadleaved temperate deciduous forest in North America. Agr Forest Meteorol 100:1–18

Wilson KB, Hanson PJ, Mulholland PJ, Baldocchi DD, Wullschleger SD (2001) A comparison of methods for determining forest evapotranspiration and its components: sap flow, soil water budget, eddy covariance and catchment water balance. Agr Forest Meteorol 106:153–168

Wofsy SC, Harriss RC, Kaplan WA (1988) Carbon dioxide in the atmosphere above the Amazon basin. J Geophys Res 93:1377–1387

Wofsy SC, Goulden ML, Munger JW, Fan S-M, Bakwin PS, Daube BC, Bassow SL, Bazzaz FA (1993) Net exchange of CO_2 in a midlatitude forest. Science 260:1314–1317

Wojcik GS, Fitzjarrald DR (2001) Energy balances of curing concrete bridge decks. J Appl Meteorol 40:2003–2025

Wood EF, Lettenmaier DP, Zartarian VG (1992) A land-surface hydrology parameterization with sub-grid variability for general circulation models. J Geophys Res 97:2717–2728

Wood R, Stromberg IM, Jonas PR (1999) Aircraft observations of sea-breeze frontal structure. Q J Roy Meteor Soc 125B:1959–1995

Wright IR, Gash JHC, da Rocha HR, Roberts JM (1996) Modelling surface conductance for Amazonian pasture and forest. In: Gash JHC, Nobre CA, Roberts JM, Victoria RL (eds) Amazonian deforestation and climate. John Wiley & Sons, Chichester, pp 437–457

Wyngaard JC (1988) Flow-distortion effects on scalar flux measurements in the surface layer: Implications for sensor design. Bound-Lay Meteorol 42:19–26

Wyngaard JC, Brost RA (1984) Top-down and bottom-up diffusion of a scalar in the convective boundary layer. J Atmos Sci 41:102–112

Yarosh ES, Ropelewski CF, Berbery EH (1999) Biases of the observed atmospheric water budgets over the central United States. J Geophys Res-Atmos 104(D16):19349–19360

Part C

The Value of Land-surface Data Consolidation

Edited by Paul A. Dirmeyer

Chapter C.1

Motivation for Data Consolidation

Paul A. Dirmeyer · Holger Hoff

The land surface is an important component of the global climate system. Evidence of this, as well as portents of future change, is buried in the growing morass of data that has been collected during the last several decades, and that continues to be collected today. The puzzle of environmental change can be solved but many of the pieces remain hidden from obvious sight, scattered far and wide among the many existing scientific datasets.

The main obstacle to a complete understanding of our environment is the fractious nature of the data collected. There exist many short and incomplete time series of observations of our natural world. Spatial coverage is spotty – only over very small fractions of the globe do we have relatively complete samplings of weather and climate, biology, geochemistry or soils, and typically where one component is well observed, others are practically unsampled. Many of the existing *in situ* observational networks are slowly decaying through lack of maintenance. Many records are recorded on outdated digital media or exist only as paper records, and are badly in need of rescue before they become unusable or are destroyed (cf. Vörösmarty et al. 2001). Ironically, as historical and traditional sources of data are threatened by gradual loss, remote sensing from orbital satellites is creating a huge new stream of information that threatens to drown us in data.

Most scientists are not adequately trained in issues of computational science like database structure and management. Most research grants provide little or no support for such data infrastructure. The result is data chaos. A wealth of potentially useful data gathered in the field, generated by numerical methods, or in some combination, goes underutilised because of technical roadblocks. The solution to reconciling these issues and creating a unified understanding of the Earth system is data consolidation. The problems and barriers to consolidation include issues of data volume, scaling, description/documentation, standardisation, preservation, and compensation of data providers (National Research Council 1995).

Figure C.1 illustrates in a schematic way the path to data consolidation. In the *top panel*, various independent surface observing networks are represented by the various coloured dots scattered irregularly over the sur-

Fig. C.1. Levels of data consolidation (for explanations see text)

face of this hypothetical country. A gridded model product is depicted in yellow. The red swath indicates data collected from a satellite passing over the area. All of these data coexist over roughly the same area at approximately the same time (say within an hour of each other). The first step of consolidation is to collect the data from these various networks and sources. A higher level of consolidation would be to coregister this data, giving all of the datasets uniform spatial and temporal resolutions, as illustrated in the *centre panel*. However, there may still be inconsistencies within each field and between the various data sources. The final step would be to generate a fully consistent, unified and well documented dataset. Using the methods of data assimilation, one can produce datasets that are not only complete and lacking gaps or voids, but also physically consistent internally in space and time. In Part C, we examine the need and utility of land-surface data consolidation, examine degrees to which this has been achieved in the past using the framework depicted in Fig. C.1 as a guide, and explore the directions that it should be pursued in the future. Chapter C.1 outlines the problems that prompt the need for data consolidation. Chapter C.2 describes the ways in which terrestrial scientific data have been organised during the past and present. Chapter C.3 puts forth guidelines for achieving a more useful degree of consolidation, and addresses real-world issues such as maintenance and human factors. Data assimilation for land-surface variables is discussed in Chapt. C.4, and conclusions are given in Chapt. C.5.

C.1.1 The Volume of Data

The Earth system sciences have been prolific in terms of data generation. Today there are more timely observational and field data available than ever before. In fact, the Earth science community is currently swimming in data. Over the past few decades there have been numerous meteorological, ecological and hydrological field studies and campaigns at scales ranging from point locations up to continental scale. Table C.1 shows a list of some of the land surface related field campaigns during the last 20 years. The list is far from complete, but gives some idea of the efforts that have been put forth.

Such field studies are usually of finite duration. There are also ongoing activities to collect data on an operational basis around the world. There exists global meteorological coverage between coordinated observing networks like the Global Observing System (GOS) network administered by the World Meteorological Organisation (WMO) and the near-real-time atmospheric analyses produced by the European Centre for Medium-Range Weather Forecasts (ECMWF) and the US National Centers for Environmental Prediction (NCEP). There is also continuous ocean monitoring, including the Tropi-

cal Ocean-Global Atmosphere (TOGA) Tropical Atmosphere Ocean (TAO) array of moored buoys monitoring the El Niño-Southern Oscillation in the tropical Pacific (Hayes et al. 1991). Fluxes of water, energy and carbon between the land surface and atmosphere are monitored by the ever-expanding network of measurement towers comprising FLUXNET (IGBP 1996). There is also a worldwide Baseline Surface Radiation Network (BSRN) of more than 30 stations that continuously measure radiative fluxes at the Earth's surface (Ohmura et al. 1998). The Long Term Ecological Research (LTER) network is a collaborative effort investigating ecological processes operating at long time scales and over broad spatial scales (Greenland and Swift 1991). The LTER network promotes synthesis and comparative research across site ecosystems and among other related national and international research programmes.

Satellites have given us the ability to monitor the environment on a global scale, providing coverage of temperature (AVHRR), precipitation (MSU, TRMM), cloud cover (GMS, GOES, Meteosat), vegetation and land use (Landsat, SPOT, AVHRR). New remote sensing initiatives promise to deliver orders of magnitude more data (e.g. the Earth Observing System, EOS), the Advanced Earth Observing Satellite (ADEOS) and Envisat). The growth in the volume of observational and remote sensing data is not slowing, but increasing. Add to this the results of simulations by models of the ocean, land, atmosphere and coupled combinations thereof, both for operational and research purposes, and there is seemingly no limit to our capability to generate data.

The vast majority of the data collected goes underutilised because too few people know what exists, and many of those that do know cannot spend the time and effort, often formidable, to comprehend and decode yet another data format. Volume and difficulty of access are great impediments. Often when data are widely used, the amount of hardware, computer time, and manpower spent reinventing or tailoring code to read the same dataset in multiple laboratories is enormous.

Every dataset has a format. These formats usually have conventions that aim to keep the datasets compact, easily addressable, and readable. These goals are often ideals – too frequently the storage, ordering, and distribution procedures for the data are arcane. The issues of data structure and documentation are often given a low priority, as the creators of the data are biased by their own familiarity with the data. This is not a problem if the data are not shared. Only when the broader community is given access to these datasets, and they begin bombarding the investigators with questions, is the realisation made of the need for careful forethought.

Beyond structure and documentation, there are other practical matters which may degrade the quality of a data archive. Unavoidably, there are periodic instrument failures. These create gaps in the data stream. More sin-

Table C.1. List of field experiments

Name	Location	Period	References
Hydrological and Atmospheric Pilot Experiment – Modélisation du Bilan Hydrique (HAPEX-MOBILHY)	Southern France	1985–1987	André et al.(1989)
First ISLSCP Field Experiment (FIFE)	Central Kansas, USA	1987–1989	Sellers and Hall (1992); Hall and Sellers (1995)
Regio-Klima-Projekt (REKLIP)	Middle and southern upper Rhine Valley	1989–	Parlow (1996)
Anglo-Brazilian Climate Observation Study (ABRACOS)	Manaus, Ji-Paraná and Marabá, Brazil	1990–1995	Gash and Nobre (1997)
Hydrological and Atmospheric Pilot Experiment in the Sahel (HAPEX Sahel)	Western Niger	1991–1993	Goutorbe et al. (1994)
European International Project on Climatic and Hydrological Interactions between Vegetation, Atmosphere and Land Surface (ECHIVAL) Field Experiment in Desertification Threatened Areas (EFEDA)	South-eastern Spain	1991–1995	Bolle et al. (1993)
Hei Ho River Basin Field Experiment (HEIFE)	Gansu Province, China	1992–1993	Wang et al. (1993)
Boreal Ecosystem-Atmosphere Study (BOREAS)	Central Canada	1993–1996	Sellers et al. (1995); Hall (1999)
Mackenzie GEWEX Study (MAGS)	Mackenzie River basin, Canada	1994	Stewart et al. (1998)
Observation at Several Interacting Scales (OASIS)	Murray-Darling basin, Australia	1994–1995	*http://www.clw.csiro.au/research/ waterway/interactions/oasis/*
Northern Hemisphere Climate Processes Land Surface Experiment (NOPEX)	Central Sweden	1994–1996	Halldin et al. (1999)
Baltic Sea Experiment (BALTEX)	Baltic Sea basin	1994–2001	Raschke et al. (1998)
Monitoring the Usable Soil Reservoir Experimentally (MUREX)	South-western France	1995–1997	Calvet et al. (1999)
GEWEX Continental-Scale International Project (GCIP)	Mississippi River basin, USA	1995–2000	Coughlan and Avissar (1996); Lawford (1999)
GEWEX Asian Monsoon Experiment (GAME)	Siberia, Tibet, Thailand, Huaihe River basin, China	1996–	Yasunari (1993)
Large Scale Biosphere-Atmosphere Experiment in Amazonia (LBA)	Amazon region of South America	1996–	LBA Science Planning Group (1996)
Inner Mongolia Grassland Atmosphere Surface Study (IMGRASS)	Xilinhot, Inner Mongolia	1997–2000	*http://www.iap.ac.cn/english/iap/ Divisions/LAGEO.htm*
Southern Great Plains (SGP)	Oklahoma and Kansas, USA	1997, 1999, 2001	*http://hydrolab.arsusda.gov/sgp97/ http://daac.gsfc.nasa.gov/ CAMPAIGN_DOCS/SGP99/*
Semi-Arid Land-Surface-Atmosphere Program (SALSA)	Upper San Pedro River basin, Mexico and USA	1997–	Goodrich et al. (2000)
Couplage de l'Atmosphère Tropicale et du Cycle Hydrologique (CATCH)	Niger, Benin	2000–	*http://www.lthe.hmg.inpg.fr/catch/*

ister are errors in instrument calibration, deployment or data transcription. Often the bulk of the effort in a field campaign is spent performing "quality control" on the data, identifying, and if possible, correcting errors and mistakes. These nagging problems, combined with the fact that there is little incentive for public release of datasets, combine to disinterest the field scientist. As Lundin et al. (1999) point out, the field of science grades its participants on their production of papers in peer-reviewed journals, and most journals are not designed to publish datasets or their documentation. This leaves the providers of environmental datasets in a quandary.

Finally, failures and deterioration of recording media are problems that have not been solved in our rapid advancement of computer technology. Once all information was written on paper. Many of the older records still exist only in this form. Paper and ink deteriorate over time with exposure to heat, humidity, chemicals, insects, and so forth. Magnetic recording media, such as magnetic tapes, are vulnerable to electromagnetic contamination, deterioration, and physical failure of the medium: breakage of the tape. Newer technologies such as optically-based media (CDs, DVDs) and solid state storage devices eliminate the problems associated with moving parts and reduce the threats from exposure to harsh environments. But while these new media may last a century or more, there is no guarantee that the machinery to read them will survive so long. This is already a problem with data recorded on magnetic media only a decade or so ago.

C.1.2 The Breadth of Data

The processes in the Earth system, be they chemical, ecological, geophysical, hydrological or meteorological, occur across a span of space and time scales. In fact, the variations evident on long time scales and large spatial scales are inevitably the accumulation of very local, short-lived events; chemical reactions occurring on the molecular level in fractions of a second, growth and death of individual plants, advection of trace gasses by turbulent eddies in the lower atmosphere, or the motion of water molecules from cloud to raindrop, to soil and river, and back again via evaporation as water vapour.

These processes are measured and monitored by discrete sampling at one or more locations over some period of time. From these discrete samples, bulk properties are assumed, and a continuous picture of the space-time evolution of the Earth system is compiled. In some cases, the Earth system itself performs the aggregation through accumulation in measurable "reservoirs" such as biomass (carbon) or river system discharge (water). Information from before the era of direct measurement

may be gleaned from geostratigraphic, palynological, dendrochronological, sediment or ice-core data. But whatever the approach to the gathering of data, information can only be extrapolated so far across space and time scales from the original scale of the measurement. The further one strays in space or time, the less confidence one has that the data remain representative. Thus, the breadth of data is also a concern for compiling an accurate picture of the Earth system.

The necessary data layers for the study of biospheric feedback are obviously a function of the specific questions asked. The more interdisciplinary and non-local the study, the broader the data requirements. Figure C.2 presents a schematic that suggests the range of space and time scales that are measured in the field, and that are simulated by models, for the Large Scale Biosphere-Atmosphere Experiment in Amazonia (LBA). Note that the space and time scales covered by observations invariably include gaps, and that there is not always agreement between the scales of models and observations. Granted, one of the utilities of models is to fill gaps in the observational record to help create a continuous picture of nature. However, observational data are necessary to calibrate, initialise, and validate the models. So at some level there must be scale agreement between models and observations, or the model results must be questioned.

Notice also that there is not always agreement between the scales covered by different disciplines. This presents another problem, as the Earth system operates

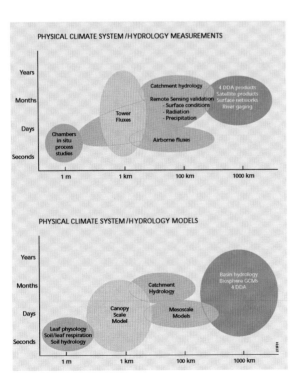

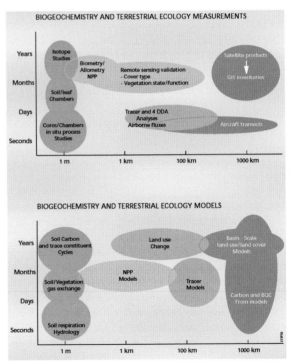

Fig. C.2. Space-time distribution of observations and model data for ecological and hydrometeorological research from the LBA concise experiment plan (LBA Science Planning Group 1996)

as an intradependent unit. For instance, soil carbon and nutrients are highly variable over small distances, evolve very slowly, and their uptake by plants is in fact highly dependent on the soil water content. But hydrological measurements typically do not exist on similar space-time scales. Thus, data gaps between disciplines can be as problematic as gaps within disciplines.

The previous example shows why a breadth of data is necessary. Usually scientists in one discipline need data collected or modelled by another discipline to complete their own studies. An obvious example is the hydrology community, who cannot model or understand the behaviour of catchment discharge and water-table variability without information on precipitation and near-surface atmospheric conditions provided by the meteorological community. This is an example of the necessity of a broad dataset to provide input or boundary conditions to a model. Broad datasets are also necessary to provide parameters for models (e.g. vegetation parameters for the land-surface component of a weather forecast model) and for validation and model development. Often the key to solving a scientific mystery in one discipline, or at one scale, lies in data from another discipline or scale. Stories of the solution of such scientific mysteries are a favourite of the media, providing the general public an entertaining look at the workings of science. But such situations are in fact real and they provide a highly visible confirmation of the need for a broad perspective on data collection and sharing.

C.1.3 The Trend toward Interdisciplinary Science

In our increasingly interdisciplinary scientific world, many find utility, and often absolute need, in using data from many sources. As the study of environmental sciences grows, focus on component studies is expanding to focus on coupled systems. It is increasingly necessary to bring together information and data across traditional discipline boundaries.

There exist numerous stumbling blocks: the use of jargon, poor or missing documentation, and a lack of recognition of the existence or applicability of the co-located data necessary to address certain problems. A complete dataset for a component study is typically an incomplete dataset for system studies.

Even within disciplines, there is often too little communication between the architects of data archives from separate but similar efforts. Separations by geography, politics, or factions within scientific communities, may leave large segments uninvolved in the work, and unaware of the results of others' work. For example, a communication gap persists between numerical modellers and those who collect data in the field. A modeller has to bring together many disparate datasets to provide parameters, initial conditions, boundary conditions, and

validation measures for the model. At the same time, a researcher involved in analysis or evaluation may have to contend with output from several different models, each in a unique format. These different specialties often act nearly independently.

Just as the volume of data has increased, so the breadth and depth of scientific research has increased, such that it becomes increasingly difficult for any individual to keep abreast of all of the progress in one's field. This problem is compounded as disciplines interact in systems studies.

There have been efforts to bridge the gap between different scientific interests and disciplines. The International Satellite Land Surface Climatology Project (ISLSCP) continues to lead a series of data initiatives to collect and co-register a wide range of information relevant to environmental science. Recent workshops on land-surface process modelling (IGPO 1998), modelling root water uptake (Feddes et al. 2001), and coupled land-surface/atmosphere interactions (INSU 2000) have brought together modellers and field scientists with interests in surface hydrology, weather forecasting, climate analysis and ecology. As different groups come together and combine their resources and strengths to address scientific issues in the terrestrial environmental system, a strong need is arising for data consolidation.

C.1.4 Who Needs Consolidated Data?

Anyone with more than a cursory need for datasets with relevance to Earth system sciences would benefit from a consolidated approach to data management. The potential benefits of structured, searchable, lasting archives include the facilitation of data transfer from experimenters to modellers which traditionally operated as disparate communities. Due to the evolving interdisciplinary nature of science and existing data structuring efforts, these groups are becoming a single integrated community. Consolidated data products are increasingly generated from an inseparable mix of experiments and modelling. Here we spell out some specific examples of propitious applications of data consolidation to specific user groups.

The developers and users of numerical models of the Earth system or its components probably handle a greater volume of data than anyone else, with the possible exception of those involved in operational data networks. Models may deal with the simulation of a single process (e.g. photosynthesis), a component of the Earth system (vegetation), an integrated set of components representing a fundamental element of the system (e.g. the land-surface), or a coupled set of elements (e.g. climate). Data requirements increase with scale and complexity. Data are needed to specify the physical and empirical parameters in the model that are not fundamental universal constants for calibration of the model. Data are needed to define the initial state of the system to be modelled, and possibly the time evolution of "boundary conditions" that are not directly simulated by the model but which have a direct impact on the simulated system. In addition, it is usually necessary to have observational information on the variables actually simulated by the model, as an independent means of validation.

Observational data are not as easy to produce as model-derived products. However, they are more important in that they represent some measure of the true conditions of the Earth system. Observational datasets are often produced from *in situ* measurements which may come from short-lived field campaigns, or from long-term observational networks. These data are monitored and processed to remove spurious or erroneous reports. In addition, remote sensing from space-based sensors contribute a large, continuous stream of data with global or near-global coverage. These data are typically in the form of radiances or brightness temperatures which must undergo processing and quality control to produce data in terms of more conventional quantities.

Typically, every observational dataset and every model has developed around it a format and method for data management. Virtually every plan is unique, and is developed with some degree of independence from other existing data management plans. Each has unique needs which must be addressed. But little thought is given to how the data can be shared beyond the immediate needs of those developing the plan. Inadequacies in the plan often are found well after it is in place. Indeed, not all problems can be foreseen. However, if a broad, general infrastructure for data exchange was defined across the Earth science community (as typically exists within engineering and computer science communities, e.g. those standards embodied by the Institute of Electrical and Electronics Engineers (IEEE)), it could aid these disparate data management planning activities in designing datasets that can be exchanged more easily across a wide spectrum of disciplines. In this regard, Earth scientists seriously trail their brethren in engineering and computer sciences.

C.1.5 What Is Consolidation?

Data consolidation means the collection, harmonisation and standardisation of data from different sources that are relevant to common applications. This involves bringing together data from different disciplines which have historically operated separately – to foster mutual understanding and future co-operation. Consolidation has several facets:

- Enhancing the net value of the data by bringing together data from different sources (e.g. *in situ* measurements, remote sensing and models);

- Linking and co-registering the data on matching spatial and temporal resolutions, projections, and gridding approaches;
- Dealing with gaps in the data, in coverage, in time, and across scales. This includes the development of methodologies for scaling, averaging, interpolation, aggregation and disaggregation;
- Promoting data and software standards, in the formatting of data and metadata and in documentation;
- Offering tools to accompany data that aid in the access and use of the consolidated datasets.

It should be acknowledged that there is likely no such thing as a single, universal standard that can accept and account for every contingency. That doesn't mean such an ideal should not be pursued. The goal of such standards is accurate documentation, ease of use and access. The key is co-registration (data on compatible spatial and temporal intervals), and spanning scales where a multi-scale spectrum of data exist.

An analogue to the concept of data consolidation can be found in the realm of scientific journals. A journal article is like a single dataset. Each is different, and usually very specific to a discipline. But, every journal paper should contain documentation complete enough so that any other scientist could reproduce the results. Also, every journal paper should be sufficiently annotated with references so an interested reader can work back through the bibliography and learn more about the subject. Many journal papers are esoteric and full of jargon specific to the field of research, but the best written papers crystallise their points so that they transcend their parochial corners of science. A specific journal brings together related work from different sources within a discipline in a standardised format, and presents it in volumes.

Journals and papers across all scientific disciplines have a common structure and organisation. Someone from atmospheric science can pick up a journal on ecology, and be capable of navigating it, easily finding the table of contents, recognise content from the titles, identify authors and their affiliations, locate articles by page number, browse abstracts or summaries, and read the articles in depth from introductory material through the results to conclusions. Any scientific reader should be able to recognise and understand the pertinence of footnotes, figures, tables and references, regardless of the topics and methods discussed.

Why should datasets not be similarly universal in their accessibility? Someone from any Earth science background should be able to easily access and comprehend the structure of any Earth science dataset. An interested researcher should be able to find information and references to gain a full understanding and appreciation thereof.

Quality enhancement is part of the data processing from raw data to the final product. It involves reformatting for consistency, uniform documentation and meta-data extraction for inter-operability and data screening, e.g. detection and removal of spurious elements. Different levels of quality assurance are defined, depending on the product specifications. Quality assurance and control is normally done by the original data collector; BAHC is focusing on facilitating links between different data providers regarding the uniformity and consistency in quality procedures.

The "aggregation problem" has been recognised as a troubling source of uncertainty since the early days of Earth system modelling. For many years, modellers working on numerical weather prediction, climate modelling and carbon-cycle modelling, have taken methods and paradigms developed and validated at very small spatial and short temporal scales, and applied them directly into regional or global-scale models. This procedure frequently involves a "scale jump" of several orders of magnitude; e.g. plot-level data are used to describe land-surface/atmosphere exchanges for grid cells of several hundred kilometres on a side. BAHC, in co-operation with ISLSCP, has had one of its main research foci on the issue of upscaling and aggregation in land-surface processes (e.g. Michaud and Shuttleworth 1997).

Scaling of data and information may be required either from small to large scales (both in space and time) or vice versa (downscaling and disaggregation). Spatial aggregation (upscaling) aims at a spatially averaged representation of landscape heterogeneities, e.g. of vegetation cover, soils, topography and physical properties. There are basically two different approaches to deal with sub-grid scale variability when modelling soil-vegetation-atmosphere fluxes. Either the spatially distributed parameter found within a particular grid box is replaced with an "effective" value, composed as a weighted average over the grid box, which allows the model to simulate a single aggregate flux, or the model simulates fluxes for a discrete set of parameter values found within a particular grid box, and then averages the fluxes weighted by their relative areas (so-called "tile" or "mosaic" approach). In both cases, a high resolution observed land-surface dataset is the starting point for scaling and aggregation procedures. The strategy in BAHC focuses on producing spatially averaged data fields at multiple scales, which allows for both "aggregated" and "distributed" modelling. Downscaling of information, e.g. to the scale needed for vulnerability and risk assessments, involves different approaches, i.e. stochastic and dynamic techniques. Within the stochastic approaches, empirically observed relationships are established between the large-scale circulations and the local atmospheric conditions. Local information is reproduced from the global data. Dynamic approaches use the output from the global circulation models or numerical weather prediction models as lateral boundary conditions for limited area models.

There is a cost for data consolidation. The main cost is in the extra human effort that goes into producing, checking and maintaining consolidated datasets. In some sense, cost is shifted from the clients, in the form of the time wasted trying to understand and navigate poorly documented and disorganised data, to the servers of the data. The benefit comes in the increased productivity of the users, who spend less time puzzling over the data, and more time using it. Thus, the financial benefit of data consolidation is proportional to the number of users of the data as well as to the volume of the data. The balance between cost and benefit should be weighed before any consolidation endeavour. There should be little additional equipment cost for data consolidation, as the volume of a consolidated dataset need not be significantly different than the conglomeration of original data.

In the remainder of Part C, we examine past efforts and the current status in accomplishing data consolidation in the Earth sciences. Then we delve more deeply into the necessary traits of consolidated datasets, including standards, tools and a maintenance plan to support consolidation. In addition, the very real human issues of motivating researchers to pursue data consolidation are discussed. We continue with a description of the great operational consolidator of disparate observational and model information – data assimilation – and finish with conclusions.

Chapter C.2

Existing Degrees of Consolidation

Paul A. Dirmeyer · Reinder A. Feddes · Forrest G. Hall · Sven Halldin · Holger Hoff · Paul Houser
Ronald W. A. Hutjes · Roy Jenne · John Leese · Timothy Kittel · Blanche W. Meeson · Richard J. Olson
Thomas J. Phillips · Andrew J. Pitman · Kiyotoshi Takahashi · Kristine Verdin

Examples from a range of degrees of data consolidation are presented here. They range from completely independent datasets (of which there are many), through various levels of consolidation. Examples of consolidated datasets include the products of operational data networks, both general and project specific data archives, and highly consolidated datasets which adhere strongly to standards of format, resolution and documentation.

C.2.1 Project-specific Data Collections

Data collection and distribution efforts focused on specific scientific projects or field campaigns were among the first to be consolidated to a useful degree. In small science projects, data management aspects are generally recognised as an integral part of the science because the investigator takes on these tasks directly within the science team. Quite often, such datasets are never intended for wide distribution, which can create problems if they are subsequently released to the community. Even if they are well documented, independent datasets almost always demand some degree of initial effort from each prospective user to read and use the data. Often such datasets are created such that they do not deliberately conform to an existing community data structure.

In large scientific projects involving many investigator teams (see Part B), the tasks of data management generally lie outside the purview of any one investigator, which creates the need for a dedicated group of individuals to take on these jobs under some management structure, usually institutionally separated from the investigators. This oversight is essential to data consolidation within a project. Data collections from field campaigns generally have a tight data structure, since more control can be exercised over the data collection, and fewer individuals are involved in data production.

Among the pioneers in data consolidation in the environmental sciences is the International Satellite Land Surface Climatology Project (ISLSCP). The First ISLSCP Field Experiment (FIFE), and the Boreal Ecosystem-Atmosphere Study (BOREAS; Chapt. A.7) are examples of coordinated field campaigns to gather *in situ* and remotely-sensed data at a range of spatial scales from point

measurements to 10^4–10^5 km^2 over the course of several years. This was the basis for construction of the FIFE Information System (FIS; Strebel et al. 1990) and later the BOREAS Information System (BORIS; BOREAS Science Team 1994). Infrastructure for FIS was provided by NASA's Pilot Land Data System (PLDS), a prototype distributed information management system designed to support field experiments. The PLDS provides services such as an on-line information system, science project support, a browse facility for analogue data, data publication on CD-ROMs, datasets documentation, distribution of and tracking of data, and user assistance.

Figure C.3 illustrates the nested approach to field observation which allowed these field experiments to

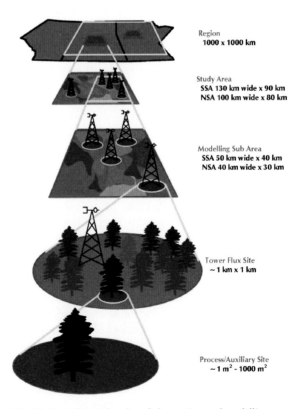

Region
1000 x 1000 km

Study Area
SSA 130 km wide x 90 km
NSA 100 km wide x 80 km

Modelling Sub Area
SSA 50 km wide x 40 km
NSA 40 km wide x 30 km

Tower Flux Site
~ 1 km x 1 km

Process/Auxiliary Site
~ 1 m^2 - 1000 m^2

Fig. C.3. Nested spatial scales of observation and modelling employed in the BOREAS field campaign. *SSA:* Southern Study Area, *NSA:* Northern Study Area

bridge spatial and temporal scales. In both the FIFE and BOREAS experiments, a dedicated staff was assembled and participated with project management and the participating scientists in project design, data collection, analysis and publication activities, helping to ensure consolidation and consistency.

The HAPEX-Sahel field campaign in Niger (see Chapt. A.5) built on the experience of FIFE to create an information system compatible with FIS (Goutorbe et al. 1997). This planned consistency greatly simplified the archival and use of HAPEX-Sahel data. Experiments such as FIFE, BOREAS, and HAPEX-Sahel made clear the advantages of data consolidation and standardisation of file systems across projects. As with other experiments, HAPEX-Sahel data are distributed electronically via the Internet, and on CD-ROMs.

The Northern Hemisphere Climate-Processes Land-Surface Experiment (NOPEX) is a BAHC-coordinated large-scale land-surface experiment. The objective of the NOPEX project is to study land-surface processes at a regional scale for a mixed land cover dominated by boreal forest. A basic design concept of the System for Information in NOPEX (SINOP) was to include not just scientific data and measurements in the database but all NOPEX information. Thus, SINOP is not merely a data archive but also a complete information management system for NOPEX (Lundin et al. 1999). A general principle in building a relational database is to keep the database tables focused and small, and concentrate on relations between tables. On the basis of this principle the layout of SINOP is now concentrated on two groups of tables; information and data; and the associations among and between members of those groups. This allows for complete consolidation of the experiment, and not just of the data.

The Global Energy and Water Cycle Experiment (GEWEX) is an international effort which has several Continental-Scale Experiments (CSEs). It includes the Mackenzie GEWEX Study (MAGS) as well as experiments past and present over the Baltic Sea region (Baltic Sea Experiment, BALTEX), the Mississippi River basin (GEWEX Continental-Scale International Project, GCIP; and the GEWEX Americas Prediction Project, GAPP), the Amazon Basin (LBA; Chapt. A.6), the Murray-Darling Basin Water Budget Project (MDB), and several sites in eastern Asia, ranging from the humid tropics to Siberia (GEWEX Asian Monsoon Experiment, GAME; Chapt. A.8). The CSEs include data and modelling efforts, and a Coordinated Enhanced Observing Period (CEOP) across all regions that is discussed later in this chapter. Each of the CSEs has a data management plan but they have been developed somewhat independently. For example, GCIP relies upon existing or planned observing programmes and data centres active over the Mississippi River basin. Much of the GCIP data effort

centres on the assembly of information about relevant datasets. Meanwhile, LBA and GAME involve the deployment of substantial additional enhanced observing campaigns and networks which allow those experiments more freedom to develop their own data standards and dissemination policies.

The GCIP Data Management Services System (DMSS) implementation strategy makes maximum use of existing data centres which are made an integral part of the GCIP-DMSS through three data source modules that specialise by data types (i.e. *in situ*, model output, and satellite remote sensing data). The atmospheric data needed for GCIP research is available through the existing systematic observing programmes operating over the continental US, including satellite observations. Such data are also readily available through existing data centres, particularly the National Climate Data Center operated by NOAA. This is also the case for streamflow data available through the National Water Information Service operated by the US Geological Survey. Thus GCIP-DMSS must perform consolidation on existing data from other sources outside the project in order to increase its utility within the project. However, the data needed by GCIP for land-surface process studies and coupling of a land-surface/atmospheric model are not available from a central data source and some variables needed are not collected and archived in a systematic manner. This is particularly the case for geographic data (topography, soils, geology, and river and basin location and characteristics) and for vegetation information (time-, season- and location-dependent datasets) combined from field surveys and remote sensing.

LBA has developed a data information system (LBA-DIS) that integrates data from different facets of the project. Aiming at integration across scales and disciplines, data from a variety of LBA modules, including field data, remote-sensing images and model results are integrated and made available for other groups. This integration of data and information supports co-locating field experiments and campaigns, coupled modelling, and assessment of global change effects in the Amazon region. The implementation of such a system for data integration does not need to be centralised. A distributed system, using available web technologies, can provide such integrated data products. The data information system's first release was a pre-LBA dataset, led by Brazil, which is a compilation of the major datasets of regional experiments that took place in Amazonia during the last 20 years. Since then, the LBA-DIS working group has developed Beija-flor, a web-based system for the submission and distribution of scientific data. The system includes a metadata editor that facilitates the entry, editing, and storage of metadata in standard formats (Fig. C.4).

Fig. C.4.
The Beija-flor data system of
LBA (Beija-flor is Portuguese
for hummingbird)

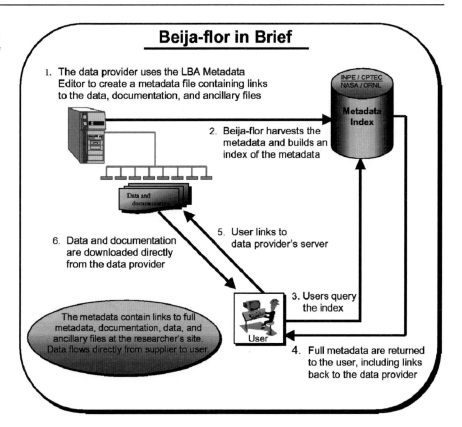

The GEWEX Asian Monsoon Experiment (GAME) has developed the GAME Archive and Information Network (GAIN) to provide the necessary data for the experiment's research activities, including process studies on the energy and water cycle, atmospheric and hydrological model development, and four-dimensional data assimilation. GAME data management is designed to satisfy the following conditions:

- the necessary data shall be collected in as consistent a manner as possible;
- the well-documented quality control shall be applied to the collected data;
- the data shall be archived under the appropriate condition to keep its integrity;
- the data shall be provided to the users in a smooth way and at nominal cost.

GAME data management applies the technique of assigning data levels to its products, to reflect the degree of processing and quality control that has been applied to the data. This is an approach commonly used in the management of large datasets that mix observations with modelling, particularly when a large fraction of the data stream comes from remote sensing that depends upon retrieval techniques. Attributing levels of processing to datasets is one strategy to consolidate data within a project. In GAIN there are three levels:

- Level 1: Original, direct instrumental reading (Data may not be referred to Earth coordinates and may require conversion to geophysical parameters).
- Level 2: Geophysical parameters obtained from instruments or derived from Level 1 data (Data are referred to Earth coordinate, usually as station data. Spatial distribution may not be homogeneous).
- Level 3: Homogeneous data fields derived from Level 2 data by an analysis or modelling techniques.

In addition to this type categorisation, GAIN also assigns one of several ranks of quality control, to help the user assess the calibre and usefulness of each dataset.

The United States Department of Energy Program for Atmospheric Radiation Measurement (ARM) implemented the Southern Great Plains (SGP) Cloud and Radiation Testbed (CART) site for extended measurement of the surface radiation budget and associated quantities (Stokes and Schwartz 1994). The site consists of *in situ* and remote-sensing instrument clusters arrayed across approximately 140 000 km^2 in north-central Oklahoma and south-central Kansas. Scientists are using the information obtained from the site to improve cloud and radiative models and parameterisations and, thereby, the performance of atmospheric general circulation models (GCMs) used for climate research. Deployment of the first instrumentation to the SGP site occurred in the spring of 1992. Additional instrumentation and data processing capabilities have been added incrementally in the succeeding years. Around this observational site have bloomed other monitoring and field projects, such as the Oklahoma Mesonet (Brock et al. 1995) and a biennial series of SGP field campaigns beginning in 1997.

Experimental projects also can be spatially distributed across the globe. FLUXNET is an integrated worldwide network of flux towers dedicated to long-term measurements of carbon dioxide, water vapour, and energy exchange from a variety of ecosystems. In 1996 the BAHC core project launched an international initiative aiming to coordinate a worldwide network of long-term carbon, water and energy flux measurements over terrestrial ecosystems. Regional collections of eddy covariance flux towers were formalised into the EUROFLUX and AmeriFlux networks in 1996. Although some towers had been in operation for many years, this was the start of a flux community effort to understand the controls on carbon fluxes. The FLUXNET project was established in 1997 to compile the long-term measurements from the regional networks into consistent, quality assured, documented datasets for a variety of worldwide ecosystems (Baldocchi et al. 1996; Running et al. 1999). The FLUXNET network is a "partnership of partnerships" formed by linking AmeriFlux (North and South Americas), EUROFLUX (Europe), MedeFlu (Mediterranean region), AsiaFlux (Asia), and OzNet (Australia) regional networks plus independent sites. New sites and regional networks and a more comprehensive scientific agenda have been added to FLUXNET so that by 2002 there were over 150 sites measuring continuous fluxes or are planning to become operational in the near future. The core variables for the FLUXNET database include net ecosystem exchange, sensible heat, and latent heat from eddy correlation, photosynthetic active radiation, net radiation, air temperature, precipitation, relative humidity, wind speed and direction above the canopy, barometric pressure, soil temperature, soil heat flux, and carbon dioxide concentration. Associated site information includes vegetation, edaphic, hydrological and meteorological characteristics.

Operational networks of routine observations are another example of distributed data gathering. Although not primarily for the purposes of scientific research, such networks are extremely useful for field campaigns and climate research. The World Meteorological Organisation is responsible for the design and implementation of the Global Observing Systems (GOS) of meteorological, oceanographic and terrestrial measurements (e.g. Kibby 1996; Summerhayes 1998). The network includes a system of telecommunications and data processing to ensure the timely distribution of weather and climate information to operational interests at public weather services around the world. The WMO establishes standards for data reporting and dissemination, which include formats for gridded data (GRIdded Binary; GRIB), point data (Binary Universal Form for the Representation of meteorological data; BUFR) and hourly station reports (Aviation Routine Weather Report; METAR). GOS, along with specific systems for the monitoring of climate, ocean, and the land surface (G3OS), comprise one of the broadest observational networks.

C.2.2 Subject-specific Archives

Subject-specific data collections may span a large range of space and time, but be bound by a common set of parameters and syntax because of the specialised nature of the branch of science that they represent. One example of this is the set of data that describe soil characteristics, as collected by soil scientists across the world. Global-scale spatial datasets of soil types and their allied soil hydrological properties have existed for many years. A number of regional and global soil databases relevant to global change research is currently available (Table C.2).

The largest is the National Soil Characterization Database (NSCD) of the United States Department of Agriculture (USDA 1994). It contains analytical data for more than 20 000 pedons of US soils (standard morphological descriptions are available for ~ 15 000 of these) and about 1 100 pedons from other countries.

A second useful resource is the World Inventory of Soil Emission potentials (WISE) database compiled by the International Soil Reference and Information Centre (Batjes and Bridges 1994; Batjes 1995). This database has a bias towards tropical regions but its primary purpose is estimating in a spatially explicit fashion the soil factors that control global change processes. The WISE database has information on the type and relative extent of the component soil units of each $0.5° \times 0.5°$ latitude by longitude grid cell of the world (derived from the revised 1 : 5 M scale FAO soil map of the world) with selected morphological, physical and chemical data for more than 4 350 soil profiles. A subset of 1 125 WISE profiles, along with some from the NSCD, is used in the Global Pedon Database (GPDB) of the International Geosphere-Biosphere Programme Data and Information System (IGBP-DIS) (Tempel et al. 1996). Existing soil maps and soil classification systems need to be translated into physical and hydrological properties for many IGBP activities, including global estimates of water-holding capacity needed by BAHC and data for predicting trace gas fluxes and global carbon cycling in GCTE. To date in the DIS soils activity, a pedon database has been produced, pedotransfer functions have been developed for soil thermal properties, a procedure for making the data spatially explicit has been implemented and, amongst others, global fields of C, N and thermal properties for surface and subsoil horizons are available.

Table C.2. Database names, soil profile numbers, contact and website information for six global soil databases available to the global change community. See text for references and additional information

Database	Institution	# Pedons	Contact	Website information
AMAZONIA	Embrapa (Brasil)/Woods Hole Research Center	1 153	D. Nepstad	*http://www.whrc.org/science/tropfor/LBA/WHRCsoilpr01.htm*
CANADA	Canadian Forest Service	1 462	M. J. Apps	*None available*
HYPRES	Alterra/MLURI	5 521	J. H. M. Wosten, A. Lilly	*http://www.mluri.sari.ac.uk/hypres.htm*
IGBP DIS	Data and Information System (IGBP)	Variable	G. Szejwach	*http://www.pik-potsdam.de/igbp-dis/igbp-site*
NSCD	USDA	>21 000	E. Benham	*http://www.statlab.iastate.edu/soils/ssl/natch_data.html*
WISE	ISRIC	4 350	N. H. Batjes	*http://www.daac.ornl.gov/SOILS/Isric.html*

Windows-compatible software allows users to select the location and spatial resolution for each variable of interest. Various pedotransfer functions to predict soil hydrological properties have also been tested using the IGBP-DIS dataset (Bisher et al. 1999)

Thirdly, the UNSODA database (Nemes et al. 1999) contains non-georeferenced soil hydrological data for approximately 800 soil profiles from around the world and, from which, a suite of pedotransfer functions have been derived (Schaap 1999).

Three other regional soil databases are also relevant to global change studies. A database of 1 153 soil profiles from the Amazon was developed as an input to a rooting depth model for producing a regional map of Amazonian plant available water (de Negreiros and Nepstad 1994). This database is now available electronically (Table C.2).

Another relatively new database from the Canadian Forest Service emphasizes soils data for Canadian forest and tundra sites (Siltanen et al. 1997). It has been used in analyses of carbon storage in boreal systems.

The third regional soil database is HYPRES (Hydraulic Properties of European Soils) established by 20 institutions from 12 European countries (Wösten et al. 1998; 1999). This database holds a wide range of both soil pedological and hydrological data and has a flexible relational structure that allows interrogation by a number of attributes or by a combination of attributes. Almost all records are geo-referenced and can be linked to soil profile descriptions. A common problem with many of these databases is that they are often not internally consistent, particularly where they have been developed over a number of years or from a variety of sources. This issue was addressed within HYPRES by standardising both the particle-size fractions according to the FAO clay, silt and sand ranges, and the hydraulic data by fitting the Mualem-van Genuchten parameters to the individual soil moisture retention and hydraulic conductivity curves stored in the database. A number of both class and continuous pedotransfer functions were also developed.

Closely related to soils, a root database (Jackson et al. 2000) of more than 1 000 profiles exists that covers various combinations of maximum rooting depth of fine and coarse roots, root length densities, root biomass (and surface area in a small subset of the data), as well as root nutrient concentrations by biome and plant life form (e.g. Jackson et al. 1996, 1997) There is however a lack of information on annual crops within the database. This root database currently has no spatial expression. Work is however under way to implement the data in a spatially explicit manner at 0.5° × 0.5° grid scales for use in climate models. This could be done by using the information within the root database to characterise the root profiles of the 12 major biomes identified within the global ISLSCP (International Satellite Land Surface Climatology Project) spatial dataset that will have a resolution of 0.5° square grid cells. This represents a first step in providing spatial root profile data at a global scale, but it is also recognised that further field observations of root profiles within particular, as yet unrepresented, vegetation types may be necessary to improve the overall predictions of the root profile.

C.2.3 Broad Data Archives

There are several examples of data collections which do little to standardise the data formats, yet are useful nonetheless by bringing together a range of datasets in one place, usually under a common documentation format. These archives act as a kind of general store for data. Often the datasets are accessible online and search tools are typically provided.

The International Council for Science (ICSU) established a system of World Data Centres (WDC) to serve the International Geophysical Year (IGY) of 1957–1958, and developed data management plans for each IGY scientific discipline. One of the first broad geophysical data archives, the WDC was so successful that it was made permanent and continues to operate to this day, collecting and disseminating geophysical and environmental data.

Another broad data archive is the National Center for Atmospheric Research (NCAR) Data Storage System (DSS). The DSS was established in 1965 to help provide data support for research at NCAR, to help university research, and also to aid the broader research community. This archive contains a broad range of datasets relevant to weather and climate research going back in some cases to the 19th century. More than 500 distinct datasets are catalogued in the DSS archive, ranging in size from less than one megabyte to over one terabyte. The total size of this archive has grown from 2.4 TBytes in 1990 to 11.9 TBytes in 1999. Many of the smaller datasets are accessible over the Internet. Larger datasets may be accessed directly by holders of computer accounts at NCAR (where the data are stored on the centre's mass storage system), or may be ordered on a variety of media at a nominal cost. The datasets are not in a standardised format, but the source and format of each dataset is recorded in easily accessible documentation.

The International Geosphere-Biosphere Programme (IGBP) has established a data information system (DIS) for its core projects. BAHC, in co-operation with IGBP-DIS, has defined a number of integrated data products that serve various IGBP programme elements. For these datasets, production methods are established and data access is facilitated. In order to provide access to distributed IGBP data and information, BAHC supports the IGBP-DIS information harvesting and IGBP metadata-base activities. This system is now operational and indexes a number of BAHC web sites.

The National Aeronautic and Space Administration (NASA) maintains a set of Distributed Active Archive Centers (DAACs) for remote sensing and other related Earth system science datasets. These datasets are largely the product of NASA or NASA-supported projects, and thus there exists a larger degree of standardisation than, for example, the NCAR-DSS. Each DAAC has one or more discipline responsibilities (e.g. land processes, biogeochemical dynamics, or snow and ice), and certain sets of products like the Pathfinder datasets have a well defined structure and documentation standards. Ultimately, NASA will provide a data information system for the Earth Observing System (EOS) suite of remote sensing products that will come from the next generation of research satellites. Much effort has been invested in designing a management structure for this potentially enormous volume of data.

Because of the timing of the launches of new satellites, the readiness of a critical number of researchers, and the maturity of a number of global models, the 2001–2003 time period has been identified as an optimal period for a Coordinated Enhanced Observing Period (CEOP) among the CSEs of GEWEX (BALTEX, MAGS, GCIP, GAME and LBA). The CEOP activities planned for the 2001–2003 period are considered an initial pilot attempt at coordinating across different regional experiments with a modest and focused set of objectives and is being managed primarily by the GEWEX Hydrometeorology Panel (GHP) as an International Project.

In CEOP, it is also critical that all aspects of the data collection effort be carried out simultaneously to realise its full potential in an effective manner:

- the same data assimilated products can be applied to all regions with the same model capabilities. Dynamic linkages between regions, validation of satellite and model products, as well as process studies, can then be carried out much more effectively and efficiently;
- the simultaneity also leads to a greater utilisation of the scientific effort, significant efficiencies may be made in the acquisition, processing and analysis of observational and model data; and transferability studies will be much more complete.

A multi-scale approach to the analysis, diagnostic and model development activities of the CEOP carries the scaling approaches pioneered in FIFE and BOREAS through the global scale. Three principal scales have been identified:

- Reference Sites: Well instrumented locations of small to intermediate scale areas (10^4 km^2 or less) distributed around the globe in different climatic regimes which can provide the data needed on a mesoscale or smaller for research in land area and hydrology processes and model validation.
- Regional Scale Areas: Large scale areas ranging from the size of a sub-area of a Continental Scale Experiment ($> 10^5$ km^2) to a whole continent and surrounding seas as a function of the research objective in the specific region.
- Global Scale: Co-operative research which is either in continental areas, distributed over the Earth, or includes all land and ocean areas of the Earth.

The primary emphasis in the CEOP data collection and management is on the Reference Site datasets contributed by the CEOP participants and distributed over the different continental land masses. The benefits to research on land area/hydrology/atmosphere coupling issues as well as coupled modelling evaluation will be greatly enhanced by a wide participation in compiling the Reference Datasets during the CEOP. It is recognised that not all Reference Sites can provide a complete list of the observation types in the composite dataset, yet the value of these datasets prevails. The Reference Sites Datasets are intended to be exchanged widely among the CEOP Research Community.

C.2.4   Co-registration

One step beyond the standardisation of format and documentation described above involves actual co-registration of data on identical spatio-temporal grids. Often at this stage compromises must be made to ensure uniformity across the final products from all data sources.

However, when well executed, the benefits of standardisation outweigh the price of compromise.

The first attempt to create a broad co-registered dataset of land-surface data on a global scale was the ISLSCP Initiative I dataset (Sellers et al. 1996). Data on topography, vegetation, soils, radiation, meteorology and hydrology were collected from many sources of observational data, and hybrid data-model analyses for a two-year pe-

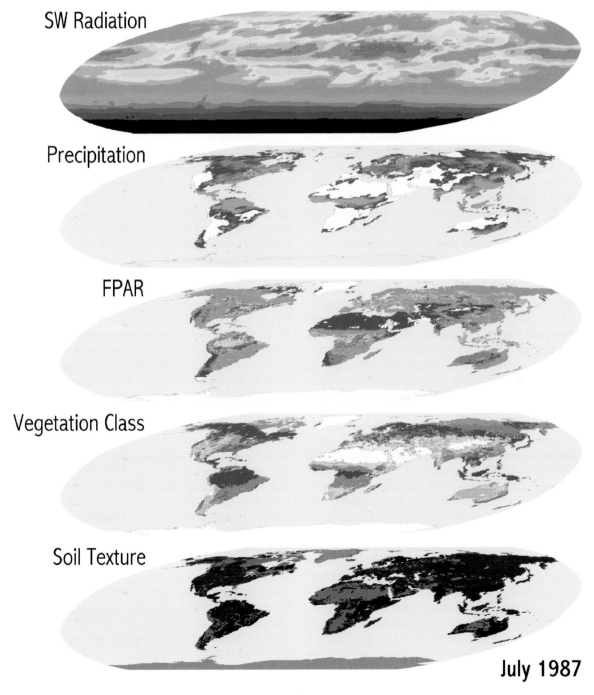

Fig. C.5. An example of data layers within the ISLSCP Initiative I dataset

riod were rendered on the same global 1° × 1° grid (Fig. C.5). A limited number of time scales were represented depending on the data source; single stationary fields, monthly, monthly-mean diurnal cycle, and 6-hourly data. All data are available in flat ASCII files, and all datasets are uniformly and comprehensively documented. Initiative I represented a triumph in data consolidation, and the datasets produced have been very popular and useful to a variety of users including modellers of land surface and atmosphere interactions, and a very diverse collection of other scientific disciplines and educators (e.g. agronomy, computer science, fluid dynamics, zoology). Since its release, over 9 000 CD-ROM sets have been ordered and over 120 000 files pulled from the on-line FTP library.

The creation of the ISLSCP Initiative I collection was a community effort. The collection was developed and reviewed using a peer review process that spanned the period from conception of the project to beyond completion. The collection editors and their team coordinated the overall effort, established methods to ensure that the agreed upon standards and conventions were followed, oversaw the peer review process, and documented and restructured a handful of the parameters in the collection. Participating members of the research community provided most of the parameters and documentation to the editors in final form. All data and documentation passed through the multi-phase peer review and all were revised and improved as a result of the review process.

All this additional work made a significant difference in the quality of the final product and in its utility. This is evident from several surveys that have been done since its release in 1995. These surveys show that this data collection has fostered research among the intended community and among a much broader audience that has acquired this collection. For many of these individuals the data collection made their research possible: without it, one quarter of survey respondents said they would not have performed the research. Many others responded that the data collection saved them a significant effort on a project that they were doing (91% of survey respondents). This is a key tenet of data consolidation; that a well-planned and executed effort can result in large net savings of time and resources in the end. This effort is being superseded by ISLSCP Initiative II, which is covering a core 10-year period (1986–1995) with data at spatial scales as high as 0.5°, and time scales as high as 3-hourly. BAHC is a key supporter of ISLSCP Initiative II.

The Global Soil Wetness Project (GSWP; Dirmeyer et al. 1999) is producing multi-year datasets of soil moisture, temperature and surface fluxes by integrating a variety of different land-surface models in a stand-alone mode using externally specified surface forcing and standardised soil and vegetation distributions at identical resolutions from the ISLSCP Initiatives I and II.

Because they are model products, there are no problems with gaps, and the project is designed to produce co-registered datasets from each model. GSWP has been designed around the ISLSCP datasets, and thus the model data produced in GSWP are likewise registered on the ISLSCP grids. This has allowed easy comparison of the land-surface models, land-surface model improvement, and ultimately prompting advances in weather and climate modelling as scientific focus turns from the role of oceans more toward the role of the land surface.

Whereas horizontal spatial co-registration across land-surface models was easily achieved, vertical and temporal registration was more problematic. Meteorological forcing data from ISLSCP Initiative I were supplied as 6-hour means or aggregates. This interval was too long for most land-surface models to operate effectively, so the data had to be interpolated down to intervals of an hour or less. Simple linear interpolation is inappropriate for some quantities such as downward shortwave radiation (which must conform to local times of sunrise and sunset) and convective precipitation (which contains high temporal variability that is essential in determining the partitioning between soil infiltration and surface runoff). In addition, different models had different characterisations of the vertical soil profile, using anywhere from one to 20 layers. This provided challenges in the specification of standards for diagnostic output. These trials illuminate some of the challenges to achieving full data consolidation.

The Vegetation/Ecosystem Modelling and Analysis Project (VEMAP) was an ongoing multi-institutional, international effort addressing the response of ecosystem structure and terrestrial biogeochemistry to changes in climate and other drivers across the conterminous United States. This project compared the sensitivity of a set of biogeochemistry and dynamic vegetation models to changing climate, elevated atmospheric CO_2 concentration, and other sources of altered forcing for both current (and historical) periods and under scenarios of future change. These analyses have been of simulations under both equilibrium conditions (Phase 1; VEMAP Members 1995; Pan et al. 1998; Yates et al. 2000) and transient forcing (Phase 2; Schimel et al. 2000). Like GSWP, keys to the project have been the development of a common, integrated model input database (Kittel et al. 1995, 1997) that assures that model comparisons are the result of model differences rather than differences in drivers and standardisation of model input and output files to facilitate dissemination and analysis.

The VEMAP dataset is an example of a dataset that is integrated in terms of both structure and content. With respect to structure, the dataset has a common ½° grid, common time steps, and common file format (netCDF, for Phase 2) for model inputs, outputs and validation data. Content of the input dataset was dictated by two requirements: that all input requirements of the ecological models be met in terms of time step (both daily and monthly time steps) and driving variables (climate, soil parameters and, for some models, vegetation type) and that climate inputs faithfully represent those aspects of climate that control ecological processes. In this last regard, specific objectives in developing the climate dataset were that it be spatially and temporally complete, reflect topographic and aspect controls over climate, and accurately represent climate variability. In addition, physical consistency among variables must be maintained in space and time (e.g. consistency in daily minimum and maximum temperature and solar radiation input) (Kittel et al. 1995). These requirements were needed to ensure that resulting simulation of ecological processes would be realistic enough to be compared with validation datasets, such as satellite-based observations (Schimel et al. 1997). They were accomplished through the use of station climate data where and when available and the application of statistical and empirical models to interpolate these data in both space and time in a physically realistic manner.

GTOPO30 provides the first global coverage of moderate-resolution (30 arc-second) elevation data. It was developed to meet the needs of the data user community for regional and continental scale topographic data. The initial release of GTOPO30 was completed in 1996 at the United States Geological Survey (USGS) Earth Resources Observation Systems (EROS) Data Center with broad international collaboration. GTOPO30 is the result of co-registration of data at one level, in that it was developed using multiple sources of elevation information. The majority of the elevation values for the globe were obtained by sub-sampling pre-existing, higher resolution digital elevation data. Yet elevations for more than 40% of the globe had to be derived through computer techniques which convert vector topographic data to hydrologically connected grids using the ANUDEM elevation gridding programme (Hutchinson 1989). Existing 1 : 1 000 000 scale digital mapping was used as the source data wherever possible. In areas of the globe with inadequate map coverage, additional data layers were contributed and digitised by co-operators. Following creation of the 30 arc-second digital elevation models (DEMs) from the two data sources, the resulting DEMs were then merged in a way that minimised visual and numerical discontinuities.

The HYDRO1k dataset was made possible by the completion of GTOPO30 in 1996. This project had the goal of developing global hydrological derivatives from the GTOPO30 DEM. The rationale for the development of HYDRO1k is that the GTOPO30 dataset, with its nominal cell size of 1 km, has been and will continue to be applied by many scientists and researchers to hydrologi-

cal and morphological studies. Inevitably, these studies require processing of the DEM and development of derivative products. The HYDRO1k project systematically developed hydrologically sound products for all the landmasses of the globe in a consistent fashion. The basis of the HYDRO1k data layers is a hydrologically correct version of the GTOPO30. The correction of the GTOPO30 DEM involves an iterative procedure utilising standard geoprocessing tools. Digital mapped hydrography, at 1 : 1 000 000 scale, along with existing drainage basin boundaries, were used to compare with DEM-derived streamlines and drainage divides. Areas of discrepancy between the mapped and derived sources were corrected in the DEM and the derivatives regenerated until a satisfactory match was obtained.

Following completion of the hydrologically correct DEM, derivative data layers were generated, including river flow directions, flow accumulations, slope, aspect, and a compound topographic (wetness) index. Additionally, topologically sound representations of the Earth's rivers and basins are being developed which can be used for comparison among rivers at corresponding scales. The structure of the HYDRO1k river and basin datasets has been developed by applying the Pfafstetter logic (Pfafstetter 1989) for subdividing hydrological units into sub-catchments (Verdin and Verdin 1999).

The Atmospheric Model Intercomparison Project (AMIP), an initiative of the WCRP for testing global atmospheric models under common specifications of radiative forcings and ocean boundary conditions, involves nearly all of the world's modelling centres (Gates et al. 1999). An international scientific panel directs AMIP, with logistical support provided by the Program for Climate Model Diagnosis and Intercomparison (PCMDI). From the perspective of land-surface specialists, the AMIP affords a unique opportunity to analyse climate simulations produced by more than 30 combinations of atmospheric models and land-surface schemes.

In the second phase of the intercomparison that is now under way (designated as AMIP II), the required output variables permit comprehensive analysis of the model climates. For instance, the evolution of land-surface climate is captured by 17-year time series of atmospheric forcings, continental hydrology and boundary fluxes, which are supplied at spatial resolutions of a few degrees latitude/longitude. Most of these fields are provided as monthly means, but some are sampled at six-hourly intervals (see Table C.3). The model output must adhere to a common data structure and be transmitted in one of two formats: 1) the WMO standard gridded-binary (GRIB) format or 2) a version of the netCDF format that conforms to a self-describing metadata standard known as COARDS. AMIP II imposes terabyte-scale data management tasks that include ingestion, organisation, storage, quality control, and distribution of model output.

Table C.3.
Standard model output of surface variables from AMIP II climate simulations of the period 1979–1995

Required monthly mean variables	Required six-hourly data
Ground and surface air temperatures	Total precipitation rate
Surface and mean sea-level pressure	Mean sea-level pressure
Zonal/meridional winds and stresses	**Optional six-hourly data**
Specific humidity	Surface air temperature
Evaporation and sublimation	Surface pressure
Up/down-ward shortwave and longwave fluxes	Zonal/meridional winds and stresses
Latent and sensible heat fluxes	Specific humidity
Convective and total precipitation rates	Up/down-ward shortwave and longwave fluxes
Snowfall: rate, depth, cover, melt	Latent and sensible heat fluxes
Soil moisture: total and at 10 cm depth	Snow depth
Surface and total runoff	Total soil moisture and runoff

C.2.5 Tools and Data

AMIP serves as an example of the pairing of standardisation of data formats and diagnostic tool development. AMIP specifies precise descriptions of climate model output variables and data output formats. These standards allow PCMDI to perform diagnosis and analysis of model simulations easily, and to report back to the modelling groups quickly. From AMIP I to AMIP II, the analysis response time for submitted datasets is being reduced from 6 months or more to 2 weeks, even though the reported data volume has increased by more than an order of magnitude for most model groups. PCMDI organises the model output according to a standard file structure, then stores the data in netCDF format following a metadata convention known as GDT, which is especially tailored for climate applications. Data quality control then follows, entailing the use of a suite of analysis codes and visualisation utilities that enable rapid identification of problems. PCMDI rectifies simple data anomalies (e.g. incorrect units) by semi-automated procedures, while enlisting the assistance of the modelling groups for non-routine problems. Quality-controlled data are then compressed and distributed to the diagnostic sub-projects via file transfer protocol servers or archival tapes.

These data management tasks have been rendered tractable by PCMDI's development of specialised software tools. For example, the formatting of data is facilitated by the Library of AMIP Data Transmission Standards (LATS), which is linkable with the modelling centres' C or Fortran data processing codes. The object-oriented Climate Data Management System (CDMS) organises the multidimensional gridded datasets characteristic of model intercomparisons. PCMDI's data management procedures are also subject to ongoing refinement. For instance, the GDT and NCAR netCDF metadata conventions are currently being merged to produce a new standard called netCDF Climate-Forecast (CF) that is designed to meet the needs of climate modelling and weather forecasting centres. A host of other tools exist to analyse and manipulate NetCDF data, which can be generally applied in other projects (e.g. VEMAP).

The Project for the Intercomparison of Land Surface Parameterization Schemes (PILPS) (Henderson-Sellers et al. 1995, 1996) coordinated the running of a suite of land-surface schemes (LSSs) at point and plot scale. PILPS has developed standard formats and complementary software tools to allow for consistency checks of the products of land-surface models driven by certain observed datasets. For example, interpolation codes were provided with the 3-hourly meteorological forcing data so that 20–30 minute data could be obtained if required by LSSs. Standards and tools aid in model development and code debugging.

The main difficulties experienced by PILPS were in the collection of results of simulations generated by the modellers. There were four problems: the format of the data, the ability of modellers to report precisely the required quantities, data management and accessibility to the data by modellers and by the central group of PILPS. The format of datasets sent to PILPS was specified precisely in each experiment, but this format changed over time. A thorough consideration of this issue was performed by Polcher and Shao (1996) who proposed a specific format that should have aided exchange of information between the modellers and PILPS. Since different people led each phase of PILPS, this format was not always used.

The ability of modellers to report precisely the required quantities was a major problem. Many output quantities are difficult to define exactly and, despite considerable effort, PILPS could not always be sufficiently exact on specifications. Apparently trivial issues such as sign convention and units of quantities caused major problems and a great deal of effort had to be spent in post-processing model results to ensure consistency. PILPS developed "consistency checks" which attempted to ensure that energy and water were conserved by models which enabled identification of some model errors and failures of modellers to report required quantities properly. The consistency checks are now in the form of a semi-automated tool that is part of the process of submitting model results to PILPS.

Data management was a major problem for PILPS. Twenty to thirty modelling groups submitting data on average three times each; coupled with internal post-processing iterations, this led to about five versions of data per model. Given that PILPS has about ten specific experiments, each including about 20 models reporting five sets of results each, the number of datasets to be managed became a significant concern and requests for copies of data could be difficult. New tools for data management, and the increased utility of the Internet have helped deal with this problem and aided accessibility to the data by modellers and by the central group of PILPS. The experiences of PILPS have served to demonstrate the potential utility of consolidated tools for data management and standardisation for both the participants and managers of a large modelling project.

There are two projects to standardise Earth system modelling: the European Programme for Integrated Earth System Modelling (PRISM) and the American Earth System Modelling Framework (ESMF). ESMF and PRISM both promote a modular interoperable approach to climate modelling, emphasising the need for standards and tools to ease implementation and exchange of component models and the data among models. Thus, the environmental modelling community is beginning down the road that will aid consolidation where model products are concerned.

The internet has the potential to serve as an ideal distribution network for scientific data. In order for this potential to be realised, the problems of large dataset size and dataset analysis must be addressed. The Distributed Ocean Data System (DODS) is a software framework used for data networking. DODS provides tools for making local data accessible to remote locations regardless of local storage format. It also provides metadata query, parsing and subsetting of data before delivery, which is accomplished by library extensions to most popular analysis and display packages as well as common programming languages like FORTRAN and C. Simply stated, one opens a file on a DODS server just as one opens a file on one's local disk, except that instead of a pathname, a URL is given as the file name. Subsetting is a particularly powerful function. Even if one only wishes to analyse a particular time or spatial subset of the data, data access methods such as File Transfer Protocol (FTP) require the entire dataset be transferred to a local machine. This can become problematic, as the transfer of remotely stored large datasets to local machines may be impractical or impossible. DODS allows specific spatial, temporal, and variable subsetting before transfer over the network. The GrADS (Grid Analysis and Display System)-DODS Server (GDS) extends this capability by allowing server-side analysis. A fully featured research tool which can access such data types as netCDF, binary and GRIB, GrADS offers an ideal way to visualise gridded datasets. By contrast, the GDS allows the product of mathematical operations performed on one or more remote datasets to be transmitted. This makes the transfer of full datasets unnecessary and shifts the computational burden off the local machine and onto the remote server; actions which greatly improve efficiency.

These examples illustrate the promise of data consolidation. Achieving it on a large scale is another matter. The next chapter describes some of the experiences and lessons of past and present attempts, and proposes requirements for full data consolidation.

Chapter C.3

Achieving Full Data Consolidation

Paul A. Dirmeyer · Sven Halldin · Holger Hoff · Ronald W. A. Hutjes · Roy Jenne · Pavel Kabat
John Leese · Richard J. Olson · Jan Polcher

The ultimate goal of data consolidation is to create a scientific and computational environment which will enable the addressing of scientific issues raised in the research community, and facilitate the development of models, analysis techniques, and observational approaches by promoting scientific exchange. In other words, to allow scientists to concentrate on the science with minimum distraction by making the necessary data easy to share, use and understand.

Because the specific needs of different users or different disciplines are not identical, any single form of consolidated data is likely to be less useful and useable to some than others. Different scientists may require the same data sampled at different space or time intervals, or different levels of correction. So can any single consolidated dataset meet the needs of all potential users? Probably not. But it should be more useful, in the net, than unconsolidated data. Users with special needs can, as they do now, approach the original data providers for raw data or alternative forms of it. If a significant fraction of the user base evolves the need for the data in an alternative form, they should have the sway necessary to help garner the resources to produce it. Inclusivity is also a goal.

C.3.1 Necessary Elements

Many of the elements of data consolidation have already been applied in the Earth sciences, but nowhere has there been a complete implementation. Full consolidation must include:

1. Complete sets of standardised data from different sources and disciplines brought together in one or more places.
2. Tools and documentation to facilitate use and understanding of the data.
3. An over-arching management and data structure.

In order to accomplish these points, it is essential to have collaboration between modellers, analysts, and the community organising observational campaigns and networks. With a structure to eliminate as much of the software re-invention as possible, it is practical to begin addressing community-wide issues without distractions that have hindered scientific productivity. With such a structure, model advancement and scientific inquiry should be able to accelerate.

Completeness may mean simply having all available data at hand. But, in fact, gaps exist because of a number of reasons including instrument failures, human error, political or budget constraints. Even a complete dataset of, for example, components of the water cycle may still be incomplete in some sense if the terms do not sum to zero – that is, if the budget does not balance. Calibration or precision errors may easily lead to such a situation, greatly reducing the utility of the data (see Chapt. B.9).

History has shown that large, centrally planned efforts at data standardisation and software development often flounder under their own weight and do not achieve full success. The biggest successes often begin as grassroots efforts started by one or a few clever, far-sighted individuals. However, for every computational "entrepreneur" that succeeds in changing the field, many do not. There is danger in an effort that is too big, or too small. The best approach may be through an initial small effort that illustrates a prototype, and proves the concept to the larger community, engendering its support.

Any attempt at full consolidation must take advantage of the lessons learned in previous efforts by various disciplines, expanding upon methodologies that have been found to work well, and avoiding the pitfalls exposed in previous projects. Although these elements will likely require greater initial effort to enact, they should pay dividends in the end. These elements are:

- Collection and provision of high-quality data at a range of scales (point, plot, regional and global) from existing sources;
- Provision of a consistent means for evaluation, comparison and validation of the data;
- A framework that facilitates participation by modellers, analysts and instrumentalists, and that benefits the development and research aims of all groups.

There is no sense in having a neat, complete dataset if it is not in a useful form for those who need it. Active solicitation of participation by modelling groups and representatives of field projects and remote sensing teams is tantamount to success. Any effort must bring data providers and users together under common causes. It is also important that any effort at data consolidation fold the results of participant modellers back into the data stream, as well as into the data development and maintenance efforts. The results of participating model experiments should be consolidated as well, allowing for complete end-to-end validation and application. A standard and uniform approach to data formatting and model input/output should be designed to make participation in the consolidation effort as simple as possible, with the goal of requiring minimal training and tailoring of code by participants.

Once in place, any consolidated data system should exhibit three essential qualities, namely retrievability, usability, and queriability:

- Retrievability is the absolute minimum requirement. It is prerequisite for getting hold of any data. Good retrievability reduces the time spent on searching for data and overcoming formatting problems, such that the usefulness of the data for a given problem can be assessed.
- Usability is the second level of requirement. An assessment of measurement locations and techniques, instruments and calibration methods used, and data-processing details, can only be done if a dataset adheres to minimum usability criteria, i.e. is accompanied with detailed and easily accessible metadata. It is a basic scientific requirement that information about retrievable data is such that measurements can be repeated and results interpreted.
- Queriability represents the highest ambition level for a data archive. It is provided most simply as accompanying documentation with stand-alone datasets. Real database approaches are needed when data from several datasets are combined, e.g. to obtain the spatial distribution of variables collected by several research groups, and to create conditional datasets. It is easy to lose information on the origin of individual components of a dataset when it is consolidated with many others. Yet, that information may be crucial in assessing its applicability to a specific investigation, or in shedding light upon a curious result. Metadata packaged with the datasets in self-describing archives minimises the chance that data and its documentation will become separated.

These qualities and elements of consolidated datasets must be present to maximise the usefulness of the data.

C.3.2 Tools

A critical component necessary for the success of data consolidation will be the establishment of a set of tools that facilitate participation by data analysts, modellers, and diagnosticians. An important aspect is the standardisation of datasets and software to facilitate the ingestion and processing of data. Any consolidation effort should not set out to invent yet another standard. An existing standard format or formats should be accepted, and well-designed tools for interfacing the data may make the differences in storage formats invisible to the user. The kinds of tools necessary for a full data consolidation effort include:

- A computing/data archive facility, possibly mirrored in a few locations, each housing a complete set of data, a library of routines for data input and output, documentation, etc.;
- Benchmarking tools and assistance for consistency assessment of data and codes, also provided as a tool for model developers;
- A catalogue of data and guidelines for model coupling, to simplify model testing and evaluation; and data access and application for analysis, validation and comparison.

By standardising the interfaces between data, models and analysis tools, preparation by any new participant is performed only once, and information can flow easily between researchers.

An example of this approach is the effort by the GEWEX Global Land-Atmosphere System Study (GLASS) called ALMA (Assistance for Land-surface Modelling Activities; *http://www.lmd.jussieu.fr/ALMA/*; Polcher 2001). To facilitate the exchange of data (observations, model analyses and remotely-sensed data) for driving and validating land-surface schemes as well as the resulting data produced by these schemes, ALMA has proposed a set of standards (Fig. C.6). The aim is to have a data exchange format that is stable but still general and flexible enough to evolve with the needs of land-surface schemes. This should ensure that the implementation of procedures for data exchange needs to be done only once, and that future comparisons of land-surface schemes can be conducted more efficiently.

Choices have been made for the description of the variables to be included in datasets, the names, units and sign conventions as well as the numerical format chosen for the files. A key feature to this approach is that the set defined is not closed but expandable, in acknowledgment of the fact that all potential contingencies in the evolution of models cannot be anticipated at the outset. This

Fig. C.6.
The ALMA model for data structure and consolidation with respect to land-surface model applications

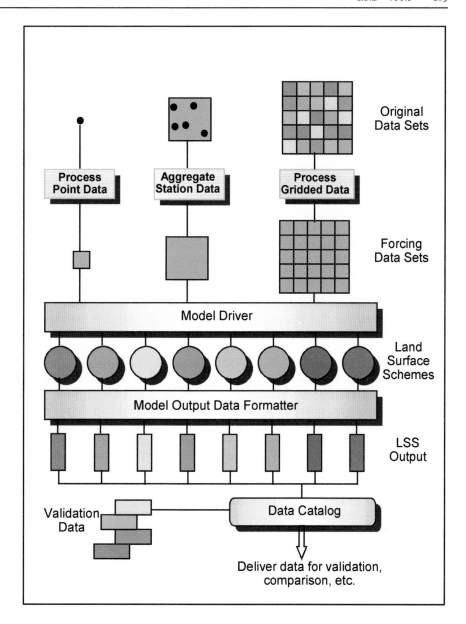

helps prevent restriction of the approach's usefulness to only the community in which it originated. It can evolve to serve wider and diverse scientific communities.

Another element of the ALMA approach that fosters community involvement and evolution is the concept of a "software bazaar" where users can download and exchange the source code for programmes and subroutines that apply and use the ALMA standards. This helps minimise re-invention of utilities for data handling, and exposes the code to more eyes, aiding debugging and improvement of code in an open-source environment. Again, this allows the scientist to spend more time on science.

The ALMA data exchange convention should not change the working habits of the groups participating in intercomparisons. It should only provide a stable data exchange protocol around which working methods can be developed. An important element of the ALMA approach is the maintenance of a forum for the exchange of expertise and software for the manipulation of data used to force and generated by land-surface schemes. This should be particularly useful for the groups who manage comparison projects or operate multiple models. The ALMA standards are being adopted rapidly across not only the GEWEX land-surface modelling studies such as PILPS and GSWP, but in other modelling activities such as the Snow models Intercomparison Project (SnowMIP) and the Land Data Assimilation System (LDAS) project (see Sect. C.4.5).

C.3.3 Data Maintenance

It is not enough to collect and redistribute data. Data must be updated, controlled for quality, and back-up copies must be made for insurance against hardware and media failures. Simply put, data maintenance is also required. New additions to datasets must be incorporated in a consistent manner with existing components, and cross references must be updated. As media degrade or become obsolete, data must be moved to more modern media. Examples of the challenges of data maintenance from a selected field campaign, global data network, and broad archive are presented in this section.

SINOP was successfully started in 1994 as the data and information system for NOPEX (Halldin and Lundin 1995). Since that time, it has been necessary to maintain an advanced technical level to keep up with the exploding development of computer and networking technology. It is a general objective for any information system to guarantee access to data, both in time and between scientists. Development of the NOPEX information system has focused on the qualities of retrievability, usability, and queriability mentioned previously. These attributes will degrade without continuing support and effort by the data management team, and a commitment from the agencies that support the work. This is true not only in NOPEX but also for any field campaign, particularly as its funding cycle draws to an end.

A solution frequently employed is to migrate experimental datasets to archival centres whose mission is specifically to maintain accessibility to these data long after the funding for the original experiment has run out. This is true of operationally collected data as well as data collected in field campaigns. One such archival site is the NCAR-DSS. The DSS (Data Support Section) has a small fraction of its staff dedicated to specific data archival projects but the bulk of the staff and resources go towards handling everything else in NCAR's vast store for atmospheric science and related disciplines. Most of their effort is toward updating the archive to ever increase its content, documentation and accessibility, and to help users. It is necessary for NCAR to update its existing archives continually and obtain new datasets to meet the ever-growing demand for data services. There is also the need to constantly migrate data to new hardware as old systems are retired and replaced by larger and faster ones. Both recent data and long-time series are important for research and are very popular.

A tremendous amount of DSS data is transferred out of archive into the community. Scientists within NCAR and the academic consortium it serves can access the archive directly via their accounts computers at NCAR. For those without direct access, data are sent by tape, CD-ROM and over the Internet to many users worldwide. In 1995, 5.6 terabytes of DSS archived data were distributed. In 1997 the use was 2.5 terabytes sent by tape, 0.8 terabytes on CD-ROM, and 8.4 terabytes accessed online, for a total of 11.8 terabytes. By 1999, the total use had increased to 16.6 terabytes – a factor of three in four years. There is no sign that the trend is abating. Clearly, such growing demands on large data archives demonstrate that they have an enormous need for maintenance.

The increasing amount of ecosystem carbon and energy exchange data presents new challenges to FLUXNET to provide qualified flux data to policy-makers dealing with global change issues, and modellers interested in regional scaling or validation of soil-vegetation transfer models and biogeochemical cycling models (Running et al. 1999). These users have identified the need to have estimates of Net Ecosystem Exchange (NEE) for monthly and annual time periods from a variety of ecosystems. Data from eddy covariance measurements are usually reported at half-hour intervals, with the objective to collect data 24 hours a day, 365 days a year. However, the average data coverage during a year is only about 65% due to system failures or data rejection. Therefore, gap-filling procedures have been established for providing complete datasets. Standardisation of the procedure allows defensible fillings and the creation of comparable datasets, which form the basis for inter-site comparisons. These efforts have been evolving over time, leading to more consistent and easily accessible datasets.

Research scientists perform initial data processing, quality assurance and documentation of the hourly or half-hourly flux data based on the unique focus of each site. Regional networks (e.g. AmeriFlux, EUROFLUX, etc.) acquire data from sites, review data and metadata, convert to common formats and distribute to the user community through an FTP or Web server. FLUXNET gets data and metadata from regional networks, compiles site characteristics, generates value-added products and provides access to the data. At each step, data and metadata are reviewed, processed, and standardised in co-operation with the site principal investigators.

The emphasis throughout FLUXNET is on high quality, reliable and credible information and value-added products. Measurements and terminology from disparate sites and networks are brought together into a common framework and harmonised, thereby increasing substantially the usage and value of the flux data and information for the global change research community. Given the constant stream of new data, maintenance of the data system at every level is of the utmost importance.

These brief examples illustrate the maintenance problems involved in collection and preparation of consolidated datasets, and the challenges in keeping them useful and available to the scientific community after the individual field efforts have drawn to a close. These data management and maintenance problems only multiply as the volume of data increases. Thus, the burgeoning stream of Earth system data coming from remote sensing will do nothing to ease the burden. Now more than ever, careful planning is needed to develop frameworks built on flexible, open-ended, self-describing formats that allow for growth, evolution, and survival of datasets.

C.3.4 Motivating Data Providers

The institution of science is structured to reward publication of peer-reviewed papers but does not place much currency on data production, which has typically been achieved with considerable expense and manpower. Because of this, factors other than the scientific and technological often play a major role in how data information systems are operated. They must, if the data providers are not to be dissuaded from contributing the necessary time and effort to make their datasets widely useful and useable.

The technical problems of running a database should not be neglected totally but it should be clear that the hardware and software of today is so easy and cheap that technology is a relatively small problem. The main technical problem is, furthermore, closely interlinked with organisation of the research and database groups. It is never possible to get a well-functioning database unless the underlying research group accepts a certain order in its way of functioning. Determining and establishing the management structure is not difficult given trained personnel and support. In science, however, getting the trained personnel and support for data management is difficult.

Because most funding agencies are specifically supporting science, data and information systems are considered to be part of an infrastructure that someone else is assumed to furnish. In practice, few institutes provide this infrastructure. Personnel with skills in database management and data communication are in such high demand in the private sector that they are not likely to remain or even seek employment at scientific institutes. This shifts the burden to the scientists. Researchers are not rewarded for taking time and intellectual effort to document and quality-control a dataset because datasets are generally not considered publishable and citable in peer reviewed journals. If the effort includes consolidation into a central database and loss of control over the dataset, then the motivation is that much less. Spending much time on such activities is often counterproductive to one's career. Any attempt to build truly consolidated, centralised databases must take this into account, provide adequate infrastructure for proper maintenance, and ideally find a means to reward the data provider for their efforts.

In essence, the dataset provider must be motivated to do the work. Researchers are motivated not only by curiosity and a desire to help society, but often by the practical expectations of funding and recognition. It is possible to force researchers to provide documented datasets if this is a condition for securing funding. They will, however, not do it willingly unless they get proper credit for their work. This means that they should be able to have their datasets (strictly speaking, their dataset documentation) published in internationally recognised journals such that they contribute to the researcher's citation index. This could actually create a way to promote standards and consolidation, by promoting such a publication. Publication of a documented dataset may be refused if it does not fulfill relevant standards. Of course, international standards for dataset publication do not now exist. A suggested suite of remedies to this situation is given by Vogel (1998).

Clearly, if tools are in place and standards are well defined, the burden on data providers is greatly relieved. Provided proper credit is attributed to the creators of the dataset whenever it is used (a provision for access of consolidated databases), where there once was a burden a true motivation arises.

When datasets are not the domain of individual scientists but rather national or international agencies, then serious negotiations are often needed to come to agreements with various authorities on the use and distribution of the data. Strangely, whereas the scientific establishment may drive researchers to undervalue their datasets, often government agencies overvalue theirs. Frequently, national governments will restrict or prevent the distribution of data on the basis of threats to their national security should the data fall into the wrong hands. Almost without exception, this form of restriction reflects an inflated sense of self importance and nationalism on the part of the national authorities. Such practices run counter to the position of the World Meteorological Organization, which clearly states in Resolution 40 (WMO 1996) that it "commits itself to broadening and enhancing the free and unrestricted international exchange of meteorological and related data and products."

Another impediment that some national and international agencies place on the ability of scientists to access their data is to charge for it. Charges for the cost of media or shipping are understandable. But many nations and organisations request exorbitant fees for data that is in fact collected with public funding. That is, the data are paid for twice – once by the taxpayers and again by the scientists (who are, typically, funded by agencies whose revenues come, again, from the taxpayers). Justification is often made by nations who are solicited from abroad for their data that they are merely trying to prevent a sort of "trade deficit" in exporting data for free. Yet such practices do more harm than good, by stymieing research and promoting poor international relations, and raise only minuscule capital for government coffers. The free exchange of scientifically valuable data is an issue of data ethics and should be instituted as an accepted international *modus operandi*.

Chapter C.4

Terrestrial Data Assimilation

Paul Houser · Michael F. Hutchinson · Pedro Viterbo · Hervé Douville · Steven W. Running

Accurate assessment of the spatial and temporal variation of terrestrial system storages (energy and mass) is essential for addressing a wide variety of highly socially relevant science, education, application, and management issues. Improved land-surface state estimates find direct application in agriculture, forest ecology, civil engineering, water resources management, crop system modelling, rainfall-runoff prediction, atmospheric process studies, and climate and ecosystem prediction. Data assimilation is a method by which observations and modelling are combined to create a continuous dataset in space and time, devoid of gaps. Pioneered for the atmosphere in operational Numerical Weather Prediction (NWP) centres (the National Centers for Environmental Prediction, Kalnay et al. 1996; the European Centre for Medium-Range Weather Forecasts, Gibson et al. 1994; the NASA Data Assimilation Office, Schubert et al. 1993), the method is now being applied at the land surface. Spatially and temporally variable rainfall and available energy, combined with land-surface heterogeneity, cause complex variations in all processes related to surface hydrology between the scales of conventional measurement networks. The characterisation of the spatial and temporal variability of water and energy cycles are critical to improve our understanding of land-surface/atmosphere interaction and the impact of land-surface processes on climate extremes. Because the accurate knowledge of these processes and their variability is important for weather and climate predictions, most NWP centres have incorporated land-surface schemes in their models. However, errors in the NWP forcing accumulate in the surface water and energy stores, leading to incorrect surface water and energy partitioning and related processes. This has motivated the NWP centres to impose *ad hoc* corrections to the land-surface states to prevent this drift.

Land data assimilation entails the use of uncoupled land-surface models forced with near-surface meteorological observations, and is therefore not affected by NWP forcing biases. In practice, gaps in the meteorological observing network mean that land models are forced with output from atmospheric analyses produced with their own data assimilation techniques for gap-filling, as well as satellite data and radar precipitation measurements. Existing high-resolution vegetation and soil coverage data can be used to specify land-model parameters. By virtue of the use of a gridded regional or global land-surface model in the assimilation process, co-registration of all output data is achieved automatically. The land model, run at high resolution, produces results that can be aggregated to various scales to assess water and energy balances and validated with various *in situ* observations. Ultimately, observations of land-surface storages (soil moisture, temperature, snow) and fluxes (evaporation, sensible heat flux, runoff) can be used to further validate and constrain the land data assimilation predictions. By continuously confronting theoretical and observational knowledge, data assimilation presents a rich opportunity to better understand physical processes and observation quality in a structured, iterative and open-ended learning process.

Data assimilation is also an important tool to help us make sense of voluminous and disparate data types that are becoming available from new space-based Earth observation platforms, as well as traditional *in situ* observations. Inconsistencies between observations and predictions are easily identified in a data assimilation system, and demand explanation, providing a basis for observational quality control and validation. Finally, the data assimilation system can extend or "advect" the available observation information in time and space (e.g. surface soil moisture observations vertically into the root zone) to provide continuous fields for use in subsequent research and application. Essentially, data assimilation is used to consolidate disparate observational and model information operationally into a unified, complete description of the terrestrial system that can be used community-wide by scientists to study important phenomena, evaluate models and data, and enhance prediction.

As with oceanic or atmospheric data assimilation, there exist both statistical and dynamic approaches to fill the gaps between observations in space and time. Statistical methods apply assumed or estimated properties of a given variable to derive continuous fields or improve resolution. These may be as simple as linear interpolation or regression, or may employ sophisticated techniques such as empirical consideration of the physical terrain or use of neural networks. Stochastic techniques also may be used to generate "weather" where only time-averaged data exist, thereby downscaling in time and producing more realistic forcing for individual modelling efforts.

Dynamic data assimilation employs physically-based models to fill gaps between observations, and produces a physically-consistent estimate of the space-time evolution (Fig. C.7). For land data assimilation, either stand-alone land-surface schemes or coupled land-atmosphere models may be used. However, both statistical and dynamic land data assimilation represent only an approximation to the true evolution of the physical state – the

more ground truth data that can be brought to bear on the problem, the better the result.

When compared to its atmospheric counterpart, terrestrial data assimilation is in its infancy (Mahfouf and Viterbo 1998). This is due to a lack of an operational exchange of observed data, the small-scale structure of many land-surface variables, and the relatively crude specification of the surface in the atmospheric host models used by the operational centres. Moreover, the operational data assimilation centres, mainly supporting weather forecasts, were slow to recognise the importance of the land surface. Finally, surface observables are re-

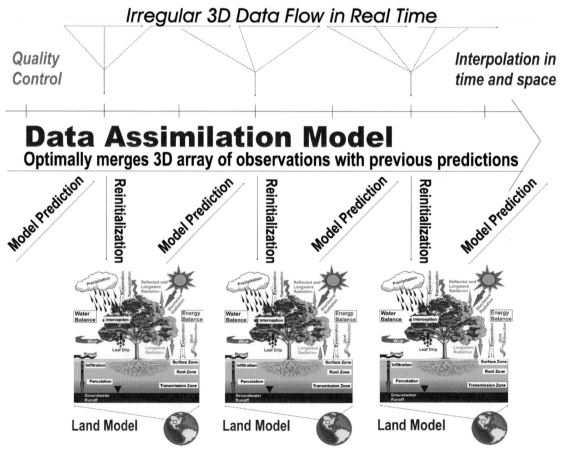

Fig. C.7. The land-surface data assimilation process

lated in a non-linear way to the model state variables and the structure of model and observations errors is poorly known.

C.4.1 Topographic Coherence of Weather – The Role of Statistical Assimilation

C.4.1.1 Stochastic Weather Models for Scenario Generation

A major requirement of many IGBP Programmes, including BAHC, GCTE, LUCC and GAIM, is the development of techniques for the simulation of spatially and temporally detailed atmospheric inputs to hydrological and ecological models. These are required at the spatial and temporal scales needed by real management systems. The spatial scales asked for are normally not coarser than a few kilometres. The temporal scale usually required is at least the daily time scale, incorporating mean behaviour as well as measures of variation and extremes. The spatial scale is much finer than the spatial resolution of general circulation and regional atmospheric models. A commonly accepted way of generating fine-scale scenarios is to perturb parameters of stochastic models of the observed current weather, for example, in accordance with broad-scale simulations of general circulation models (IGBP 1993). This approach has been used widely in climate impact research, both for the current weather and for projected future climates (Zorita et al. 1992; Kittel et al. 1995; Mearns et al. 1997; Semenov and Barrow 1997; Wilby et al. 1998). It has been called semi-empirical downscaling by Giorgi and Mearns (1991) and rests on three principal assumptions:

1. The validity of the broad-scale scenarios generated by GCMs;
2. The maintenance in changed climates of the observed links between the broad-scale atmospheric behaviour and stochastic weather model parameters;
3. The ability of stochastic weather models, with parameters essentially consisting of first- and second-order summary atmospheric statistics, to simulate accurately spatially and temporally detailed atmospheric inputs to ecological and hydrological models.

These assumptions are ordered by their degree of validity. Assumption 1 is most open to question, particularly with regard to significant interactions between the atmosphere, the land-surface and the ocean system and therefore a very significant source of uncertainty in scenario generation. This is underlined by significant differences in the broad scale precipitation scenarios provided by different GCMs (CSIRO 1996). Nevertheless, GCMs provide a starting point for one type of generation of a range of climate change scenarios. Given the

uncertainties associated with GCM predictions it is now recognised that broad scale scenarios should also be generated in the light of perceived vulnerabilities of land-surface systems. This is illustrated below in Fig. C.8 and further discussed in Part E.

Assumption 2 is also open to question but is supported by the observed broad scale temporal and spatial coherence of the current atmosphere system. This coherence appears to be founded on physical principles, and is therefore likely to be maintained in changed climates. The validity of the links identified under assumption 2 will be enhanced if the stochastic weather models are simply parameterised, so that they can be calibrated robustly from minimal data. This can facilitate the identification of the parameters most likely to respond to climate change, especially if the stochastic models incorporate physically-based structures that can be identified from observed data (Hutchinson 1995a).

The validity of assumption 3 is generally accepted. This has formed the basis for the development of stochastic weather models to provide inputs to ecological and hydrological models, beginning with the work of Jones et al. (1972), Richardson (1981) and Srikanthan and McMahon (1984), through to more recent work by Racsko et al. (1991), Shah et al. (1996) and Wilks (1999b). First- and second-order long-term weather statistics, from which stochastic simulation is a direct corollary, can well calibrate atmospheric variability, including probabilities of extreme events for the current weather. The main ongoing issue here is the practical one of identifying and calibrating space-time stochastic models that incorporate adequately observed spatial and temporal behaviour. This is made difficult by the relative sparseness of measured surface weather data and the need to calibrate observed longer term weather variations, over decades or more.

There has been steady progress in the development of methods for observing and interpreting spatially detailed atmospheric inputs by remote sensing and consequent potential for integrating remotely-sensed data with ground-based data (Georgakakos and Kavvas 1987; Stewart and Finch 1993; Fo and Crawford 1999). However, there are ongoing difficulties in calibrating remotely-sensed surface weather data accurately, particularly rainfall data (O'Connell and Todini 1996). Radar estimates of rainfall amounts can be in error by as much as a factor of two (Barros and Kuligowski 1998). As recognised in IGBP (1993), the main role for remotely-sensed data in the context of stochastic weather model development appears to be in providing insight into the nature of the spatial variability of surface weather. The ground-based meteorological data network remains the only source of daily surface weather data with extensive temporal coverage over the 20th century and near complete global land coverage. It is important that this network be maintained and, where appropriate, upgraded.

C.4.1.2 Scheme for Generating Stochastic Weather Scenarios

If stochastic models with global coverage are to be developed from the ground-based meteorological data network, there is a critical need for accurate spatial interpolation of appropriate weather statistics across the Earth's land surface. This has been achieved by incorporating topographic dependencies into the interpolation process. The task of developing stochastic space-time weather models from standard meteorological networks can then be conveniently divided into three steps (Hutchinson 1995a), as shown in Fig. C.8. This approach conveniently isolates different aspects of weather model development. In particular, it helps to identify the differing spatial dependencies and spatial scales of various model parameters and weather anomalies, with consequent differing requirements with regard to model complexity and observation network density.

The first step in this scheme is the development of stochastic models at single points where recorded weather data are available. It is at this stage that questions of appropriate temporal scale for ecological and hydrological models are addressed, as well as the incorporation of simple physically-based model structures, as discussed above. Both daily and monthly time scales have been used, although the daily time step is the most common. It is sufficient to model rainfall extremes, soil erosion and time-critical temperature-dependent events in plant growth and yield. The monthly time step is sufficient to model drought, natural vegetation, and broad scale hydrology. The variables most commonly required for ecological and hydrological models are precipitation, daily maximum and minimum temperature, solar radiation, atmospheric humidity and potential evaporation (IGBP 1993).

The second step in Fig. C.8 is the development of techniques to extend the parameters of point stochastic models spatially across the landscape. This is usually done by incorporating the effects of fine scale topography, as provided by the GTOPO30 and other elevation models. It is after parameters have been interpolated across the landscape that broad scale long-term weather change perturbations are applied. As both daily and monthly point stochastic weather models are normally calibrated on a month-by-month basis (Richardson 1981; Georgakakos and Kavvas 1987), the spatial interpolation of model parameters is related strongly to the interpolation of summary monthly weather statistics.

The most critical parameters in stochastic weather model calibration are the means of the different variables. Measures of variance and serial dependence are often less well defined by the data, and less critical to the performance of the fitted weather model. Variances are normally required to calibrate probabilities of extreme events, but Hutchinson (1995a), building on the work of Stidd (1973), has shown that daily rainfall distributions can be calibrated accurately with a truncated normal distribution using essentially just two first-order statistics, the mean rainfall amount and the mean number of dry days. Similarly, Chia and Hutchinson (1991) have shown that the variance of daily sunshine duration can be estimated reasonably from the daily mean and Geng (1986) has demonstrated empirical relationships between mean and variance parameters of the Richardson (1981) weather model. Mean weather parameters typically display the most complex spatial patterns, often strongly modulated by topography. Thus effective methods for the spatial interpolation of monthly mean weather statistics should be sufficient to enable the interpolation of all of the parameters of suitably parameterised stochastic point models.

Fig. C.8.
Scheme for fine scale stochastic space-time simulation of weather from ground-based data and broad scale atmospheric scenarios for climate impact and vulnerability assessments

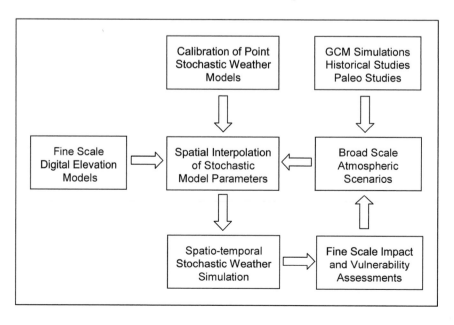

Developments in the first two steps in Fig. C.8 are guided by the requirements of the third step, the development of coordinated space-time models that adequately respect both spatial and temporal dependencies of daily or monthly weather anomalies. These anomalies are essentially normalised differences between actual daily (or monthly) weather values and the monthly mean values. Normalised anomalies show a high degree of spatial and temporal coherence and can be modelled effectively by a first-order multivariate autoregressive model. Richardson (1981) showed that such a model was appropriate for normalised daily temperature and solar radiation anomalies, and that the parameters defining the first-order temporal correlation structure could be assumed to be constant across all sites examined. Similarly, Schneider and Griffies (1999) found that a first-order autoregressive model was sufficient to model GCM output and could be used as a basis for predictability studies of area-averaged variables. Autoregressive structures can also be simply extended to account for interannual variability, such as in the twofold autoregressive Markov model described by Karner and Rannik (1996). More complex autoregressive integrated moving average models have also been used to model space-time precipitation (Shah et al. 1996).

C.4.1.3 Stochastic Models

Stochastic weather model development has been almost totally dominated by developments in stochastic models of precipitation, for which there is a vast literature (Georgakakos and Kavvas 1987). This reflects both the dominant control exerted by precipitation on ecological and hydrological processes and the difficulty in developing complete weather models. Only relatively simple daily and monthly precipitation models, however, can be used for long-term weather-scenario applications. More complex precipitation models have been developed for restricted types of precipitation and have complex structures that are acknowledged to be difficult to calibrate and validate over large areas (Valdes et al. 1985; Sivapalan and Wood 1987). However, there has been progress in simplifying the parameters of rainfall models based on cluster processes and in regionalising these parameters for environmental impact studies (Cowpertwait et al. 1996).

Two classes of daily stochastic precipitation models have achieved common use in scenario generation. The traditional approach uses a first-order two-state Markov chain to model rainfall occurrence and models rainfall amounts on wet days independently using a gamma distribution. This approach has been extended by fitting more complex models, but the basic separation between rainfall occurrence and rainfall amount has remained (Wilks 1999a). The occurrence structure has been ex-

tended by fitting different distributions to lengths of wet and dry spells and by extending the order of the Markov chain. Models for rainfall amount have been extended to include the mixed exponential distribution and by conditioning the parameters of the gamma distribution on the wet or dry status of preceding days. These extensions are obtained at the expense of fitting additional parameters. A simple occurrence model based on the normal distribution was suggested by Hutchinson (1995a). This model uses no additional parameters but was found to simulate dry spell lengths better than a first-order Markov chain model at locations across the whole USA.

The second approach consists of conditional stochastic precipitation models that incorporate physically-based controls by having model parameters conditioned on classified broad scale atmospheric circulation patterns. Such models have been developed by Bardossy and Plate (1992), Hay et al. (1991) and Wilson et al. (1992). These models have been found to match various observed mean and extreme rainfall statistics and to respect observed spatial dependencies. These models offer fairly direct links to large-scale circulation patterns. Their utility in generating future weather scenarios, with respect to Assumption 2 above, depends on the maintenance of the observed classes of atmospheric circulation patterns, and the constancy of their links with the rainfall model parameters. These models have provided valuable insight into the non-linear sensitivity of hydrological processes to postulated changes in global weather.

Complete stochastic weather models, such as the point model proposed by Racsko et al. (1991) and the multi-site model proposed by Wilks (1999b), have typically been constructed along the lines of the model proposed by Richardson (1981). These models have been used widely in scenario applications. In these models, weather variables, including daily maximum and minimum temperature and daily total solar radiation, are conditioned on the daily occurrence and non-occurrence of precipitation. However, this conditioning is weakly defined, since the *process* that affects temperature and solar radiation directly is the occurrence of significant cloud. This is not always associated with precipitation. More direct process-based relationships between the commonly required surface weather variables need to be explored. Such developments may overcome some of the reservations expressed by Katz (1996) about the use of such conditional stochastic weather models in generating long-term weather change and variability scenarios.

These models also need to be extended to better account for observed interannual variability. As widely recognised, and discussed by Katz and Parlange (1998), stochastic weather models fitted to daily statistics tend to underestimate interannual variability. This can be partially addressed by fitting more complex models, but these models tend to vary with geographic location and this can make the construction of multi-site models difficult Wilks (1999a). It is also likely that this phenomenon is due in part to non-stationarities in the weather system, such as those associated with ENSO (Phillips et al. 1998). More generic methods are called for to address this issue.

C.4.1.4 Topographic Dependent Interpolation of Weather Model Parameters

Statistical interpolation techniques appear to be best suited to the task of spatially extending the parameters of point simulation models. The techniques include kriging (Cressie 1991) and thin plate smoothing splines (Wahba and Wendelberger 1980). These methods have similar accuracy although splines tend to be more easily calibrated (Hutchinson and Gessler 1994). Thin plate smoothing splines have been used to interpolate monthly mean weather parameters across the Australian continent (Hutchinson 1991), England (Semenov and Brooks 1999) and Canada (Price et al. 2000), at spatial resolutions of a few kilometres. The PRISM method (Daly et al. 1994) fits local elevation-based regressions to weather data. It has been used to interpolate long-term weather statistics across the USA, also at a spatial resolution of a few kilometres. Thin plate smoothing splines have also been used to interpolate weather means at coarser resolution across Europe (Hulme et al. 1995) and all continents except Antarctica (Leemans and Cramer 1991).

The major factor in the accuracy and spatial resolution of these interpolated weather surfaces has been the incorporation of dependences on elevation as indicated in Fig. C.8. This is well known in the case of temperature, where the dependence on elevation is almost linear. Monthly mean precipitation is also modulated strongly by topography, but its influence varies spatially. Both thin plate smoothing splines and the PRISM method can incorporate this spatially varying dependence. Hutchinson (1995b) and Running and Thornton (1996) have shown that the relative impact of elevation on precipitation patterns is two orders of magnitude greater than the impact of horizontal position. Thus precipitation patterns can be influenced significantly by relatively modest topographic features (Barros and Kuligowski 1998). The spatial resolution of this dependence has been estimated as 4–10 km (Daly et al. 1994; Thornton et al. 1997; Hutchinson 1998).

The number of data points and approximate standard errors of monthly mean weather surfaces fitted across Australia are given in Table C.4. These errors are typical for elevation-dependent surfaces derived from standard meteorological networks. As for most regions of the world, the number of stations that record pre-

Table C.4. Number of data points and approximate standard errors of fitted monthly mean climate surfaces across Australia (Hutchinson 1991)

Climate variable	Number of data points	Standard error
Solar radiation	150	3%
Daily maximum temperature	900	0.2–0.4 °C
Daily minimum temperature	900	0.3–0.5 °C
Precipitation	10 000	5–15%
Pan evaporation	300	5%

cipitation exceeds by about an order of magnitude the number of stations that record other weather variables. This gives a reasonable indication of the relative spatial complexity associated with each variable. An example of a final interpolated surface for precipitation is shown in Fig. C.9. Further investigation is needed to clarify the scales of various topographic effects on precipitation (Hutchinson 1998) and to better define the topographic effects on temperature, such as topographically controlled inversions in winter and proximity to large water bodies (Hutchinson 1991).

C.4.1.5 Spatial Structure and Topographic Dependencies of Weather Anomalies

Since hydrological responses can be required simultaneously across larger regions, the spatial covariance structure of the weather anomalies generating these responses needs to be accommodated. Two applications of this analysis may be distinguished. The first is the spatial interpolation of values in real time from observation networks. This has particular relevance for calibration and validation of models of observed hydroecological response. Monthly rainfall anomalies can be interpolated reasonably well from standard meteorological networks (Lyons 1990), but additional remotely-sensed data are usually required to interpolate daily rainfall satisfactorily (Fo and Crawford 1999).

The second application is the statistical simulation of spatially distributed weather anomalies for scenario applications. Normalised weather anomalies tend to display broad spatial patterns that are controlled by broad synoptic patterns. These patterns tend to be independent of topography, unlike the corresponding non-normalised quantities. This can simplify the task of analysing spatial covariance structure, which can be addressed using standard multivariate statistical interpolation and analysis methods (Bell 1987; Daley 1991). However, while relatively simple models can be used, rainfall anomaly structure can be complicated by anisotropy (Obled and Creutin 1986), by systematic differences between inten-

Fig. C.9.
January mean precipitation
interpolated from 16 000 stations across Australia across a
2.5 km grid

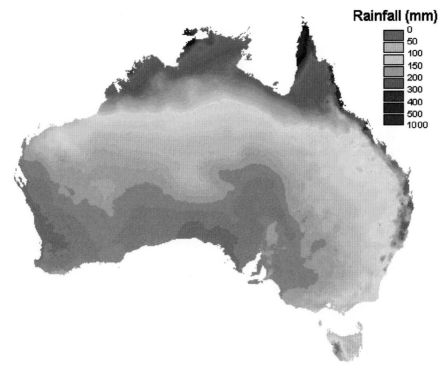

Rainfall (mm)

0
50
100
150
200
300
400
500
1000

sity-based and occurrence-based correlations (Hutchinson 1995a) and by non-stationarity of spatial correlations over time (Jones and Wendland 1984). Thus, accurate space-time simulation of daily rainfall remains an
active and challenging subject for hydrology (O'Connell
and Todini 1996).

The scheme described here for stochastically generating spatially and temporally detailed weather values
from ground-based meteorological data has been widely
adopted for environmental impact research. Its strength
is its reliance on well-established methods for spatial
and temporal analysis. These methods reflect a high
degree of spatial and temporal coherence in the surface
weather, particularly when dependences on topography
are incorporated. The methods are subject to ongoing
development and are critically dependent on the maintenance, and upgrade where appropriate, of the ground-
based meteorological network. Equally important issues
are the identification and calibration of simple, physically-based model structures, and whether these structures can be used with confidence in changed weather
(Hutchinson 1995a). There is a similar search for parsi-
monious, physically-based model structures in hydrology (O'Connell and Todini 1996). That such stochastic
model structures can be used in a changed weather can
only be truly ascertained after such a change occurs.
Nonetheless, these structures are well founded in existing methods for calibrating current and past weather,
and are an appropriate assumption for scenario generation.

C.4.2 Terrestrial Model Prediction

An essential component of any terrestrial assimilation
strategy is realistic simulation; the model physics[1] will
allow for the spatial and temporal extrapolation of observed information into unobserved regions. Recent
advances in understanding the soil water dynamics,
plant physiology, micrometeorology and hydrology that
control biosphere-atmosphere interactions have spurred
the development of Land Surface Schemes (LSSs), whose
aim is to represent simply yet realistically the transfer
of mass, energy and momentum between a vegetated
surface and the atmosphere (e.g. Dickinson et al. 1993;
Sellers et al. 1986). LSS predictions are regular in time
and space but these predictions are influenced by model
structure, errors in input variables and model parameters, and inadequate treatment of sub-grid scale spatial variability. Several studies have shown that models
with physically observable parameters and states are
preferred in data assimilation studies (Entekhabi et al.
1994; Houser et al. 1998). Therefore, LSSs must be developed that replace the current conceptual parameters and
states that are not observable with more physically-re-
alistic representations. An important development in this

---

[1]   Throughout this part, the term "physics" is used to represent all
physical processes, including chemical and biological as well as
thermodynamic, hydrodynamic, and other "diabatic" processes
that produce or consume non-mechanical energy.

regard is the coupling of radiative transfer models for the direct prediction of multi-frequency brightness temperatures by LSSs (Burke et al. 1997).

LSSs generally simulate the diurnal dynamics of soil moisture (both liquid and frozen), soil temperature, skin temperature, snowpack water equivalent, snowpack density, canopy water content, and the traditional energy flux and water flux terms of the surface energy and water balance. LSSs are evolving toward the inclusion of carbon and nitrogen physics, dynamic vegetation and ecosystem physics, groundwater interaction, runoff routing, and highly refined vertical and horizontal heterogeneity parameterisations. LSSs are increasingly addressing the problem of sub-grid heterogeneity by subdividing each GCM grid cell into a mosaic of tiles (after Avissar and Pielke 1989), each tile having its own vegetation/soil representation and hence water and energy balance. LSSs are also being explored that simulate catchment-based topographic processes, rather than the traditional grid-based approach (Koster et al. 2000; Ducharne et al. 2000). Another new frontier in land-surface modelling is the explicit coupling of carbon and nitrogen dynamics with water and energy dynamics on timescales ranging from minutes to centuries. This enables the prediction of ecosystem responses and feedbacks to weather and climate variability, as well as allowing for the assimilation of carbon, nitrogen, and vegetation observations.

There are strong justifications for studying LSSs, both uncoupled and coupled with atmospheric and ocean models. Coupling the LSS to an atmospheric model allows for the study of the interaction and feedbacks between the atmosphere and land surface. However, coupled modelling also imposes strong land-surface forcing biases predicted by the atmospheric model on the LSS. These biases in precipitation and radiation can overwhelm the behaviour of LSS physics (Dirmeyer 2001). In fact, several weather prediction centres must "correctively nudge" their LSS soil moisture toward climatological values to offset its drift. An uncoupled LSS can use observed land-surface forcing, use less computational resources, and still address many relevant scientific questions. Arguably, the most critical terrestrial coupling with the atmosphere is through precipitation. There are currently large efforts to derive precipitation (e.g. Global Precipitation Climatology Project) and to assimilate precipitation estimates into coupled models (Hou et al. 2000), which may improve the forcing of land models in coupled systems significantly.

PILPS has been responsible for a series of complementary experiments that focuses on identifying parameterisation strengths and inadequacies in about 30 land-surface process models. PILPS is a project designed to improve the parameterisation of the continental surface, especially hydrological, energy, momentum and carbon exchanges with the atmosphere. This is an important exercise because there are significant differences in the formulation of individual processes in the available land-surface schemes. These differences are comparable to other recognised differences among current global climate models such as cloud and convection parameterisations. PILPS emphasizes sensitivity studies with and intercomparisons of existing land-surface codes and the development of areally extensive datasets for their testing and validation (Henderson-Sellers et al. 1993).

Recognising the importance of soil moisture in the climate system, the International Satellite Land Surface Climatology Project (ISLSCP), which is a contributing project of GEWEX, began the Global Soil Wetness Project (GSWP) in 1994 (Dirmeyer et al. 1999). The initial efforts of the GSWP were to implement a land-surface modelling effort using a CD-ROM set of land-surface data developed by ISLSCP (Meeson et al. 1995). The CDs contain, in addition to other information, meteorological observations and parameter datasets sufficient to obtain soil moisture estimates for 1987–1988 for a $1° \times 1°$ grid. Ten groups, using various LSSs, including BATS, Mosaic, multiple versions of SiB, and others, produced soil moisture fields for these two years. Through the GSWP, Entin et al. (1999) attempted to validate these soil moisture fields using various soil moisture observations from the Northern Hemisphere mid-latitudes. They found that no model was able to recreate the actual soil moisture for all the areas studied. They also discovered that no model was able to recreate the seasonal cycle of soil moisture in Illinois and Russia, though all the models were deficient in recreating the changes of soil moisture in Mongolia and China, some of the few locations where routine soil moisture observations are made. The quality of the forcing data vary greatly from place to place, and may be a factor in the poor performance of the LSSs over certain regions (Oki et al. 1999). A second CD-ROM set is planned by ISLSCP, which will contain data for at least ten years (1986–1995). The GSWP will use these data to force LSSs which should then address another of the main issues raised when citing the difficulty of performing soil moisture validation, namely that of simulating interannual variability.

C.4.3 Terrestrial Observations

Another essential component of terrestrial data assimilation is the regular provision of land observations with known error characteristics. The data assimilation problem is best posed when the state observations being assimilated have similar physical complements in the LSS, and these states have significant memory or inertia so that an improvement is preserved and (hopefully positively) impacts subsequent predictions. Observations of significance to terrestrial data assimilation include temperature (air temperature, surface skin temperature, canopy temperature, and soil temperature), moisture (near-surface humidity, surface and profile soil moisture content, surface saturation, total water storage, plant water content, depression storage, lakes and rivers), snow (aerial extent, snow water equivalent, depth), carbon and nitrogen (plants and soil), and vegetation biomass (height, leaf area index, greenness). Land-surface fluxes, such as runoff, latent and sensible heat flux, carbon and nitrogen flux, and radiative fluxes, can be used in terrestrial data assimilation in the context of backing out a mass or energy state correction through conservation equations. Generally, it is more robust to perform a multi-variate analysis or assimilation, where an observation is used to constrain multiple relevant LSS states (this is further improved when observations of several different states are used). Data assimilation methods are designed to merge predictions and observations depending on the perceived errors of each. Establishing these errors can be the most complex and subjective part of data assimilation. Therefore, it is critical that observation error characteristics be well established through instrument calibration and validation. Large-scale terrestrial data assimilation development has lagged behind atmospheric data assimilation, primarily due to a lack of suitable observations available regularly in time and space. However, with the deployment of several new Earth system remote sensing platforms, this situation is quickly changing. The status of a few particularly critical terrestrial observations is described in more detail below.

Remote sensing of surface temperature is a relatively mature technology (see Chapt. B.8). The land surface emits thermal infrared radiation at an intensity directly related to its emissivity and temperature. The absorption of this radiation by atmospheric constituents is smallest in the 3–5 and 8–14 µm wavelength ranges, making them the best windows for sensing land-surface temperature. Some errors due to atmospheric absorption and improperly specified surface emissivity are possible, and the presence of clouds can contaminate or obscure the signal. Generally, surface-temperature remote sensing can be considered an operational technology, with many spaceborne sensors making regular observations (i.e. Landsat TM, AVHRR, MODIS, and ASTER) (Lillesand and Kiefer 1987). The evolution of land-surface temperature is linked to all other land-surface processes through physical relationships, so it is an ideal observation to assimilate.

Remote sensing of near-surface soil moisture content is a developing technology, although the theory and methods are well established (Eley 1992). Long-wave passive microwave remote sensing is ideal for soil moisture observation, but there are technical challenges in correcting for the effects of vegetation and roughness. Microwave soil moisture remote sensing has been limited previously to aircraft campaigns (e.g. Jackson 1997a). There are several current or future space-borne passive

and active (radar) microwave sensors that may be useful to derive soil moisture information in a data-assimilation context, including the Defense Meteorological Satellite Program (DMSP) SSM/I (Engman 1995; Jackson 1997b), the EOS-AMSR (Advanced Microwave Sounding Unit), the Tropical Rainfall Measurement Mission – Microwave Imager (TRMM-TMI), and the European Space Agency Soil Moisture and Ocean Salinity (ESA-SMOS) instruments. All of these sensors have adequate spatial resolution for land-surface applications but have a very limited quantitative measurement capacity, especially over dense vegetation and topographic relief. Because of the large error in remotely-sensed microwave observations of soil moisture, there is a real need to maximise its information by using data assimilation algorithms that can potentially account for this error.

An important and emerging technology with respect to terrestrial data assimilation is the potential to monitor variations in total water storage (ground-water, soil water, surface waters, water stored in vegetation, snow and ice) using satellite observations of the time variable gravity field. The Gravity Recovery and Climate Experiment (GRACE), an Earth System Science Pathfinder mission, will provide highly accurate estimates of changes in terrestrial water storage in large basins when it is fully operational after it has been launched successfully in 2002. Wahr et al. (1998) note that GRACE will provide estimates of variations in water storage to within 5 mm on a monthly basis (Rodell and Famiglietti 1999). Birkett (1995, 1998) demonstrated the potential of satellite radar altimeters to monitor height variations over inland waters, including climatically sensitive lakes and large rivers and wetlands. Such altimeters are currently operational on the ERS-2, ENVISAT and TOPEX/POSEIDON satellites, and are planned for the JASON-1 satellites.

Finally, snow aerial coverage and snow water equivalent can be monitored routinely by many operational platforms, including the Advanced Very High Resolution Radiometer (AVHRR), Geostationary Operational Environmental Satellites (GOES) and SSM/I. Recent algorithm developments even permit the determination of the fraction of snow cover within Landsat-TM pixels (Rosenthal and Dozier 1996). Cline et al. (1998), describe an approach to retrieve snow water equivalent from the joint use of remote sensing and energy balance modelling.

C.4.4 Data Assimilation Concepts and Methods

Charney et al. (1969) first suggested combining current and past data in an explicit dynamic model, using the model's prognostic equations to provide time continuity and dynamic coupling amongst the fields. This concept has evolved into a family of techniques known as *four-dimensional data assimilation* (4DDA). "Assimilation is the process of finding the model representation which is most consistent with the observations" (Lorenc 1995). In essence, data assimilation merges a range of diverse data fields with a model prediction to provide that model with the best estimate of the current state of the natural environment so that it can then make more accurate predictions (see Fig. C.7). The application of data assimilation in land-surface studies has been limited to a few one-dimensional, largely theoretical studies (i.e. Entekhabi et al. 1994; Milly 1986) primarily due to the lack of sufficient spatially-distributed hydrological observations (McLaughlin 1995). However, the feasibility of synthesising distributed fields of soil moisture by the novel application of 4DDA applied in a hydrological model was demonstrated by Houser et al. (1998). Most land data assimilation schemes include the following steps:

- An *error checking* procedure is used to correct or eliminate erroneous data (Bengtsson 1985). Observations can contain different types of error, including errors due to faulty instruments, improper processing, or unsatisfactory communication of the data.
- Observations can rarely be assumed to be physically and spatially identical to the modelled state. Therefore, an *observation operator* is often employed to facilitate bias correction, space and time interpolation, and range matching.
- The actual *analysis or merging* of observations with model predictions is performed using a data assimilation algorithm. Common data assimilation methods include direct insertion, Newtonian nudging, optimal or statistical interpolation, Kalman filtering, and variational approaches (often using an adjoint model).

"The process of replacing model values by 'observed' ones is called direct insertion" (Daley 1990) or updating. This method assumes "perfect" observations, or observations with no error. Thus, model predictions that are known to contain error are totally rejected and replaced with the perfect observation. Any spatial or temporal information advection is performed entirely through the model physics.

Newtonian nudging continuously adds a forcing function to the model's prognostic equations to "nudge" the model state gradually toward the observations. These small forcing terms, based on the difference between the simulated and observed state, gradually correct the model fields, which are assumed to remain in approximate balance at each time step (Stauffer and Seaman 1990).

Statistical interpolation is a minimum variance method that is closely related to kriging (Bhargava and Danard 1994). The technique can be traced back to Kolmogorov (1941) and Wiener (1949), who applied it to various areas of science and engineering. With the development of computer power, and through the inspiration of Gandin's publication, *Objective Analysis of Meteorological Fields*

(Gandin 1963), most major western meteorological services were using statistical interpolation operationally by the mid-1970s.

The Kalman filter has been extensively utilised in data assimilation research (Ghil et al. 1981; Cohn 1982). The Kalman filter assimilation scheme is a linearised statistical approach that provides a statistically optimal update of the system states, based on the relative magnitudes of the covariances of both the model system state estimate and the observations. The principal advantage of this approach is that the Kalman filter provides a framework within which the entire system is modified, with covariances representing the reliability of the observations and model prediction.

Variational methods were first introduced by Sasaki in 1958, and their use proved effective because they can incorporate many constraints easily (Ikawa 1984). "The variational algorithm requires the computation of the gradient of the distance function to be minimised with respect to the model state at the beginning of the assimilation period" (Courtier and Talagrand 1990). Thus, variational assimilation is principally very simple. One first defines a scalar function that describes the distance between the observations and the model prediction. Then, one simply seeks the model solution that minimises this function (Courtier and Talagrand 1990). The complexity comes from the generally large size of the minimisation problem.

One of the major components of any assimilation system is quality control of the input data stream. Quality control (QC) refers to the process by which observational data and their attributes are analysed to identify data items which are likely to contain gross errors and the attempts to correct or remove such errors. Observation errors are usually of two types: *natural error* (instrument or representativeness error), and *gross* or *rough errors* (improperly calibrated instruments, incorrect spatial/temporal registration, incorrect coding of observations, or telecommunication errors). These errors can be either random or spatially and/or temporally correlated with each other. Clearly, QC for any single observation must involve information other than the observational datum itself. Common QC algorithms can be categorised as follows:

- *Theory, realism,* or *sanity checks* see if the observation absolute value or time rate of change is physically realistic. This check filters such things as observations outside the expected range, unit conversion problems, etc.
- *Buddy checks* compare the observation with comparable nearby (space and time) observations of the same type and reject the questioned observation if it exceeds a predefined level of difference.
- *Background checks* examine if the observation is changing similarly to the model prediction.

According to a 1991 National Research Council report, "to produce research-quality data from a new satellite mission, the observed data should be subjected to a critical evaluation by an assimilation system in order to identify error characteristics of the instruments and the algorithms" (National Research Council 1991). The assimilation system can provide a systematic and powerful means of merging new, remotely-sensed observations with all earlier and current *in situ* and remotely sensed measurements. In a real-time context, data assimilation can provide quality assurance and validation of the observations, and can provide rapid identification and diagnosis of problems that might otherwise go unnoticed for longer periods. The data assimilation system can extend the available observations in time and space to provide continuous fields for use in subsequent research and application.

The continuous confrontation of theoretical and observational knowledge in a data assimilation system presents a rich opportunity to better understand physical processes and observation quality in a structured, iterative, and open-ended learning process. Data assimilation is also an important tool to help us make sense of voluminous and disparate data types that are becoming available from new space-based Earth observation platforms. Inconsistencies between observations and predictions are easily identified in a data assimilation system, providing a basis for observational quality control and validation. Modern data assimilation techniques use relevant observations and a state-of-the-art land-physics model to estimate the state of the land surface. For each observation, a background value is derived from the model forecast for comparison. Systematic differences between observations and model predictions can identify systematic error. Thus, the consistency of the model provides guidance to identify observation problems in a data assimilation context. This methodology clearly illustrates the importance of a good quality forecast and an analysis that is reasonably faithful to the observations. If the land model makes reasonably good predictions, then the analysis must only make small changes to an accurate background field (Hollingsworth et al. 1986). In many cases the analysis fields can provide guidance for identifying observational problems that can be compared with carefully chosen *in situ* observations to provide conclusive proof.

C.4.5 Current Projects

Subsurface moisture and energy stores exhibit persistence on various time scales that have important implications for extended climatic and hydrological predictions. Because these stores are time-integrated, errors in NWP forcing accumulate in them, which leads to incorrect surface water and energy partitioning. Land Data Assimilation Systems (LDAS) which are uncoupled LSSs that are forced primarily by observations and are therefore not affected by NWP forcing biases, are currently under development (Brutsaert 1998). The implementation of a LDAS also provides the opportunity to correct the model's trajectory using remotely-sensed observations of soil temperature, soil moisture and snow using data assimilation methods.

A multi-institutional LDAS research effort involving NASA, NOAA, Princeton University, the University of Washington, Rutgers University and the University of Maryland is currently under way. This LDAS operates in both retrospective and real-time modes at a $1/8°$ resolution over the continental United States using several different land-surface models. Project information and a real-time image generator are located at the LDAS web site: *http://ldas.gsfc.nasa.gov/*. Model parameters are taken from the high-resolution AVHRR-derived vegetation and soil survey classifications (Mitchell et al. 1999). Figure C.10 shows July average downwelling surface short-wave derived from GOES and the Eta model, total monthly precipitation derived from NEXRAD radar, gauges and the Eta model, experimental LDAS average skin temperature predictions, and experimental average near-surface soil moisture. A more complete description is given above.

NASA and NOAA are currently extending the North American LDAS project described above to all global land. This high-resolution, near real-time Global Land Data Assimilation Scheme (GLDAS) will use all relevant remotely-sensed and *in situ* observations within a land data assimilation framework. This development will increase greatly our skill in land surface, weather, and climate prediction, as well as provide high-quality, global land-surface *assimilated data fields* that are useful for subsequent research and applications.

Loosely linked to GLDAS are other projects, like the European project called ELDAS (Development of a European Land Data Assimilation System to predict Floods and Droughts), which is supported by the European Union 5th Framework Programme (*http://www.knmi.nl/samenw/eldas/*). ELDAS has been designed to develop a general data assimilation infrastructure for estimating soil moisture fields on the regional (continental) scale, and to assess the added value of these fields for the prediction of the land-surface hydrology in models used for numerical weather prediction and climate studies.

ELDAS uses a common infrastructure implemented at three participating institutes: ECMWF, DWD and CNRM/Météo France. The procedure followed will be able to generate soil moisture fields for a suite of land-surface schemes – respectively TESSEL (Tiled ECMWF Surface Scheme of Exchange processes at the Landsurface), TERRA (SVAT from the DWD) and ISBA (Interaction Sol Biosphère Atmosphère of CNRM) – but also others. The analysis method follows on the work by Rhodin

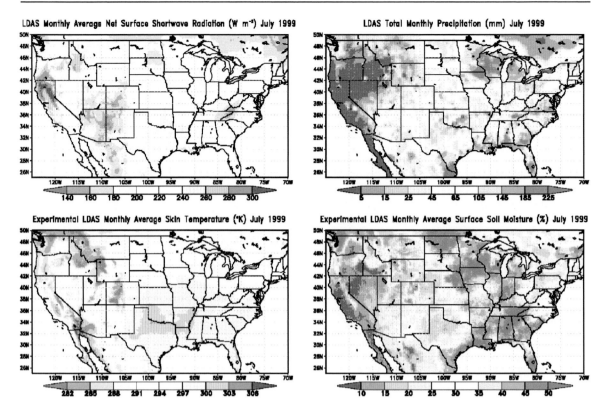

Fig. C.10. An example of atmospheric forcing and land-surface state fields from the North American LDAS project

et al. (1999) and Hess (2001). Simulations will be carried out with the full atmospheric model, but with model precipitation and radiation replaced by the observed data. Daily soil moisture field will be generated for a grid covering Europe in a sequential, cycled way, updating the atmospheric initial fields using analyses, and propagating the soil fields as first guess. The model grid will be different for different case- and validation studies, but use a common set of up/down scaling procedures.

Integral to the project are validation studies meant to assess the quality of the soil moisture fields using independent data from the GSWP 2000 dataset (Global Soil Wetness Project of GEWEX), from SSM/I (Special Sensor Microwave/Imager) or AMSR (Advanced Microwave Sounding Unit) validation (comparing these to computed top-of-the-atmosphere microwave radiation, generated from the surface state in the data assimilation modules), from GLDAS output (GLDAS forcings will be fed to the ELDAS data assimilation system) and from French river basin datasets. The usefulness of the dataproduct will be assessed in flood forecasting systems for UK rivers and the Rhine, and in European numerical weather predictions.

At the European Centre for Medium-Range Weather Forecasts, observations of screen-level temperature (SLT) and humidity are assimilated using an optimal interpolation technique (Douville et al. 2000). Over land, screen-

level winds are not included in the assimilation because it is felt that the observations reflect local circulations, poorly described in the assimilating model. Snow-depth observations are combined with a model snow density field and a short-term forecast background to produce an analysis of snow mass (water equivalent). An analysis of soil water is performed, based on the SLT analysis (Fig. C.11). The errors in short-term forecasts of SLT are combined linearly in an optimal way to produce soil water corrections (Douville et al. 2000). Note that, in sharp contrast to the atmosphere, no remote sensing information is used in the surface data assimilation, hampering the quality of the analysed products in data-void areas. All the variables described above are analysed in the 40-year ECMWF re-analysis.

Terrestrial data assimilation systems are also under development outside the immediate meteorological context. The CAMELS (Carbon Assimilation and Modelling of the European Land-Surface) project recently started the development of a prototype carbon cycle data assimilation system (CCDAS) in order to produce operational estimates of "Kyoto sinks". To produce a best estimate of carbon uptake CAMELS will use all of the constraints implied by the different data sources, as well as of the physiological and ecological constraints embodied in terrestrial ecosystem models (see Fig. C.12). This is essentially a data assimilation problem, requiring a system similar to those used to initialise weather fore-

Fig. C.11.
Hydrological budget simulated from 1 June to 9 October 1987 by the ECMWF single-column model forced with observed precipitation and radiative fluxes; *top*: nudging technique, *bottom*: OI technique. *P*, *E*, *R*, and *I* stand for the integrated precipitation, evaporation, runoff, and soil moisture increments respectively. The increments are represented by *vertical bars*. The observed soil moisture variation and evaporation are represented by *black disks and circles*

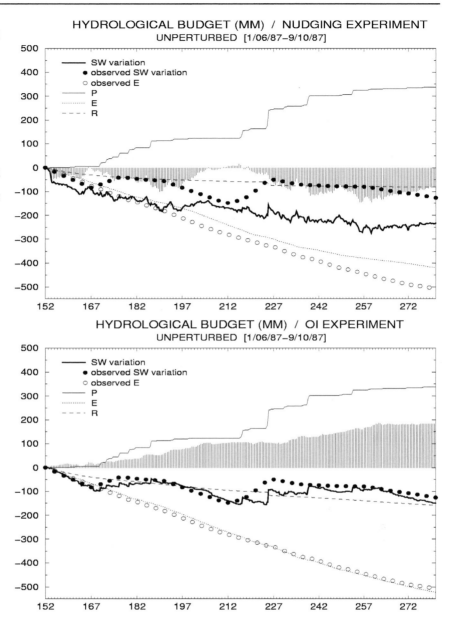

cast models. In this case observations are used to constrain the internal parameters of the terrestrial ecosystem models, while they are used to interpolate the observations.

The CCDAS scheme will use existing data sources (e.g. flux measurements, carbon inventory data, satellite products) and the latest terrestrial ecosystem models to produce operational estimates of the European land carbon sink. The terrestrial ecosystem models TRIFFID (terrestrial carbon cycle model from Hadley Centre, Cox et al. 2000), BETHY (Biosphere Energy Transfer Hydrology model from the Max Planck Institute for Meteorology, Knorr 2000; Knorr and Lakshmi 2001), ORCHIDEE (ORganizing Carbon and Hydrology In Dynamic EcosystEms from Laboratoire des Sciences du Climat et de l'Environnement, Viovy et al. 2001; Friedlingstein et al.

2001) will simulate the European land carbon sink at high resolution using operational analyses plus remote sensing products (e.g. seasonally-varying fAPAR). Atmospheric transport models will link terrestrial ecosystem model-predicted carbon fluxes to atmospheric CO_2 observations. Inverse models (e.g. Bousquet et al. 2000) will then be used to adjust both terrestrial ecosystem model parameters and prior estimates of carbon fluxes based on a 20–25 year simulation period. Eventually an online implementation of the CCDAS in a high-resolution atmospheric General Circulation Model (AGCM) will use the existing data assimilation structure in the AGCM plus the terrestrial ecosystem inverse model to nudge internal model variables, such as respiring soil carbon and leaf nitrogen, based on the atmospheric measurements.

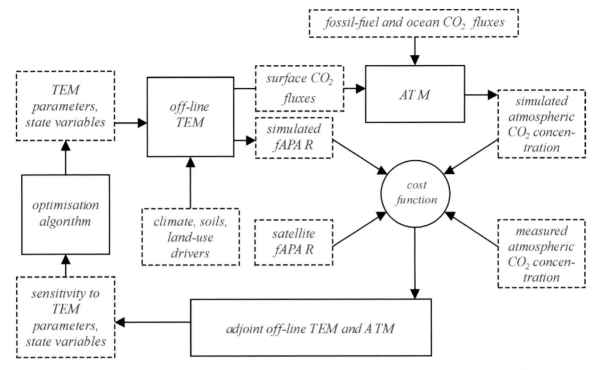

Fig. C.12. Off-line carbon data assimilation system to be developed in CAMELS (Carbon Assimilation and Modelling of the European Land Surface; *ATM*: Atmospheric Transport Model; *TEM*: Terrestrial Transport Model; *fAPAR*: fraction of Absorbed Photosynthetically Active Radiation)

C.4.6   Future Opportunities

With new land-surface state observations, and the recognised importance of the land surface on weather and climate prediction, terrestrial data assimilation will come of age this decade. However, there are many challenging issues to address to realise this goal, which are summarised as follows:

C.4.6.1   Terrestrial Observation

Profile soil moisture is arguably the most critical land state, and it remains largely unmeasured. Castelli et al. (1999) showed there is some soil water information in infrared radiative temperature measurements, but the best hope for operational soil moisture observations is with passive L-band microwave sensors. They provide information on soil water in the top 5–10 cm of soil, in areas free of contamination from the canopy water. An assimilation model can be used to solve the inversion problem and obtain a profile of water in the root zone (Calvet et al. 1998).

Subsurface soil temperatures have time scales of the order of days to years that can serve as very valuable diagnostics for anomalies in the surface forcing, in particular those related to cold season dynamics (Viterbo

et al. 1999). An effort is needed to disseminate in near-real time soil temperatures, observed regularly down to a depth of 1 m in many WMO stations.

A number of technological capabilities are maturing that make real-time analysis of land-surface hydroecological systems possible – an ecological equivalent to meteorological data assimilation. Key ecological data include land-cover type, vegetation phenological status, and leaf area index are measured globally by satellite systems such as the Earth Observing System and are available weekly. Energy balances and carbon flux measurements from FLUXNET eddy covariance towers are now becoming available continuously (see Chapt. B.4).

Independent, continuous and comprehensive validation is critical to the success of any terrestrial data assimilation system. The structure of the assimilation scheme ensures the accurate reproduction of assimilated observations, but it is recognised that data assimilation constraints can cause other model predictions to diverge from reality.

C.4.6.2   Terrestrial Simulation

Comprehensive, physically-based terrestrial simulation models must be developed that include standard and accepted processes such as water and energy balance, evapotranspiration, soil moisture depletion, stream runoff routing, groundwater interaction, surface water and wetland processes, cold season dynamics, urban processes, photosynthesis, carbon and nitrogen cycling, plant and ecosystem dynamics, vegetation primary production, and coupling of all these processes with the overlying atmosphere. These models must also parameterise realistically the processes arising from sub-grid heterogeneity in precipitation, radiation, vegetation, soils, topography and atmospheric turbulence. These ecohydrological process models must have physically realistic states and parameters to ease their use in terrestrial data assimilation systems. Finally, parameter calibration methods must be fully developed and tested to specify robust parameters for large-scale terrestrial simulation.

C.4.6.3 Terrestrial Data Assimilation

Multi-variate terrestrial data assimilation methods must be further refined for operational use. These methods should include use of both *in situ* and remote measurements of soil temperature and moisture, and snow cover and depth. Longer-term goals should be the assimilation of runoff, groundwater and vegetation characteristics. The emphasis should be on combined use of observations and coupled models.

Hollingsworth et al. (1986) showed that assumptions on the bias and horizontal correlation structure of the model and observations can have a significant impact on error estimations. In practice, data assimilation is often implemented with the assumption that observations and predictions are unbiased and uncorrelated in space. These assumptions work reasonably well for *in situ* observations, but satellite observations are usually biased by inaccurate algorithms, and their errors are usually correlated horizontally because the same sensor is making all the observations. These assumptions must be evaluated, and observation and model errors must be defined better.

Terrestrial models have very little physics that act in the horizontal dimension, therefore it may be possible to limit the terrestrial data assimilation to one dimension. The benefits and drawbacks of using a one-dimensional (vertical) or a multi-dimensional (vertical, horizontal, and time) assimilation method must be evaluated. Multi-dimensional assimilation methods are much more computationally intensive, but maybe able to extend observations better into data-sparse regions. The exploration of sub-grid scale terrestrial data assimilation (i.e. into a tiled or mosaic model) and the ability of the data assimilation algorithm to downscale observations to fine resolutions also remain unexplored issues.

Data assimilation systems can include a radiative transfer observation operator, to allow the assimilation of radiances directly, rather than derived quantities. However, non-linearity in the forward model may make this method non-workable for some land-surface variables. The benefits and drawbacks of direct radiance assimilation must be explored more thoroughly.

Snow cover is a readily available observation that may be of great use in terrestrial data assimilation systems. However, terrestrial models usually use a snow water equivalent state variable. Assimilation and modelling methods need to be able to use snow cover information in order to infer snow mass (i.e. Liston 1999).

The development of off-line land data assimilation systems must be further developed and included in coupled prediction systems. A surface data assimilation using information from a variety of ground-based and remote-sensing observations will better produce realistic soil temperature, soil water and snow mass global fields. Moreover, short- to long-term forecasts in such a system will provide the community with the most reliable global estimates of surface fluxes on a daily basis, together with a realistic diurnal cycle.

An interesting aspect of LDAS is the possibility of multiple land-surface predictions made by several different LSSs, with various initialisations. These various land-surface prediction ensembles and super-ensembles will be intercompared and explored. It is quite possible that through this land-surface ensemble prediction strategy, spatially-varying confidence limits on predictions could be established and new ways to evaluate and improve predictions could be developed.

Finally, there is a need to perform additional long-term coupled land-atmosphere re-analyses that include improvement in terrestrial simulation, observation, and data assimilation. The resulting datasets would improve greatly our understanding, prediction, and practical application of terrestrial systems.

Chapter C.5

Conclusions

Paul A. Dirmeyer

Data consolidation is the bringing together of related datasets from disparate sources, in differing spatial and temporal resolutions and in various formats, into an organised, standardised, co-registered and fully documented database for redistribution. The morass of data in the Earth sciences hampers interdisciplinary understanding and discovery. To improve navigability, some degree of organisation and indexing of data is necessary. There are various degrees of organisation, from simply collecting data in one physical or virtual place, to co-registering the data on identical grids and similar formats, to creating datasets that are entirely self-consistent via data assimilation with physically-based models. All of these efforts reflect some level of data consolidation.

The goals of data consolidation are to maximise the usefulness, continuity, and completeness of a range of data useful to a broad community. Within the Earth sciences, data consolidation brings together information from a range of disciplines, including hydrology, ecology, meteorology, soil sciences and geography. Consolidation reduces the burden of interpretation on scientists, freeing them from the need to unlock the idiosyncratic nature of a suite of individual datasets for various sources. Thus, data consolidation is a service to improve efficiency, and gives the data user more time to spend on science. The data providers benefit by having their data more easily and widely used, rewarding them with greater recognition and adding value to the overall science endeavour. At the same time, consolidation can only work if the providers of data are rewarded for their efforts, in the form of recognition and citation whenever their data are used.

Consolidation has two main facets. The first is the synthesis of disparate datasets across space and time to produce continuous, high-quality data. This process includes gap filling, aggregation and/or downscaling, quality control, and co-registration. When the quantities to be synthesised are physically related, methods of data assimilation within a physically-based model can be applied to ensure consistency across the state variables. Coherence in space and time due to factors such as topography can be exploited to fill gaps statistically. Data assimilation can be performed in near-real time with the latest models available, or in a retrospective mode with frozen versions of the models and historical observations so as to produce a dataset that is consistent across time as well as space. Well-integrated consistent data products, whether from pure observational data or model products, have an enhanced intrinsic value.

The other aspect of consolidation is the standardisation of formats, documentation, attribution, and tools to generate, access and distribute the consolidated datasets. Whereas synthesis makes the data useful, standardisation makes it useable. The net result is an approach to geophysical data that, when institutionalised, maximises quality and accessibility. However, innovation and progress must be recognised, and any approach to data consolidation must be flexible enough to be adaptable to new data streams. To be viable, a consolidation standard must be built to evolve as necessary.

A top-down dictum on data consolidation would almost certainly fall short of the mark, and be resisted by the research community. A grass-roots approach would likely spawn better ideas, but would suffer from a lack of broad perspective and backing from the large institutions in the scientific community. The best path to data consolidation is a joint effort including representation from both the largest and smallest data providers and users. It is essential that no one be left out of the process, or the data stream. This includes institutions that have historically restricted the free dissemination of data.

Key to the success of any attempt at data consolidation is a change in the mentality of scientific funding organisations. Large operational centres have the infrastructure to dedicate to data consolidation efforts. But non-operational data streams stemming from individual scientific projects with one or few researchers must have motivation and support to participate in consolidation. Funding agencies must recognise that in terms of support, infrastructure does not grow on trees. The interdisciplinary and inter-project exchange that today's programme managers advocate so strongly would be much easier to attain if they specifically supported data management in each grant. This could be done by the establishment and support of a data archive centre to serve projects that do not fall under the aegis of large programmes with existing data systems, or by specifically supporting data consolidation on a grant-by-grant basis. Regardless of the approach, it is also necessary that all programmes co-operate and support a common set of standards across both national and disciplinary boundaries.

Consolidation of data through standardisation, co-registration, and complete documentation that follows the data everywhere will ease a major burden for researchers. It will help to accelerate scientific progress as surely as standardisation of machine parts accelerated industrial efficiency in the early 20th century, or as networking standards have helped beget the Internet and World Wide Web today. So much of science continues to operate like a scattering of peasant farms. Achieving data consolidation within the terrestrial sciences would be like mechanisation a century ago – nothing short of revolutionary.

References Part C

André J-C, Goutorbe JP, Schmugge T, Perrier A (1989) HAPEX-MOBILHY: results from a large-scale field experiment. In: Rango E (ed) Remote sensing and large-scale global processes. Wallingford, UK, International Association of Hydrological Sciences, pp 13–20

Avissar R, Pielke RA (1989) A parameterization of heterogeneous land surfaces for atmospheric numerical models and its impact on regional meteorology. Mon Weather Rev 117:2113–2136

Baldocchi DD, Valentini R, Running SR, Oechel W, Dahlman R (1996) Strategies for measuring and modelling carbon dioxide and water fluxes over terrestrial ecosystems. Global Change Biol 2: 159–168

Bardossy A, Plate EJ (1992) Space-time model for daily rainfall using atmospheric circulation patterns. Water Resour Res 28:1247–1259

Barros AP, Kuligowski RJ (1998) Orographic effects during a severe wintertime rainstorm in the Appalachian Mountains. Mon Weather Rev 126:2648–2672

Batjes NH (ed)(1995) A homogenized soil data file for global environmental research: a subset of FAO, ISRIC and NRCS profiles (Version 1.0). Working Paper and Preprint 95/10, ISRIC, Wageningen

Batjes NH, Bridges EM (1994) Potential emissions of radiatively active gases from soil to atmosphere with special reference to methane: development of a global database (WISE). J Geophys Res 99:16479–16489

Bell TL (1987) A space-time stochastic model of rainfall for satellite remote-sensing studies. J Geophys Res 92:9631–9643

Bengtsson L (1985) Four-dimensional data assimilation. International Conference on the Results of the Global Weather Experiment and Their Implications for the World Weather Watch, Geneva, Switzerland, 27–31 May 1985, GARP Publications Series No. 26, pp 187–216

Bhargava M, Danard M (1994) Application of optimum interpolation to the analysis of precipitation in complex terrain. J Appl Meteorol 33:508–518

Birkett CM (1995) The contribution of the TOPEX/POSEIDON to the global monitoring of climatically sensitive lakes. J Geophys Res 100:25179–25204

Birkett CM (1998) Contribution of the TOPEX NASA radar altimeter to the global monitoring of large rivers and wetlands. Water Resour Res 34:1223–1239

Bisher I, Sorooshian S, Mayr T, Schaap M, Wosten H, Scholes R (1999) Comparison of pedotransfer functions to compute water holding capacity using the Van Genuchten model in inorganic soils, IGBP-DIS Working Paper No. 22

Bolle HJ, André JC, Arrue JL, Barth HK, Bessemoulin P, Brasa A, de Bruin HAR, Cruces J, Dugdale G, Engman ET, Evans DL, Fantechi R, Fiedler F, van de Griend A, Imeson AC, Jochum A, Kabat P, Kratzsch T, Lagouarde J-P, Langer I, Llamas R, Lopez-Baeza E, Melia Miralles J, Muniosguren LS, Nerry F, Noilhan J, Oliver HR, Roth R, Saatchi SS, Sanchez Diaz J, de Santa Olalla M, Shuttleworth WJ, Søgaard H, Stricker H, Thornes J, Vauclin M, Wickland D (1993) EFEDA: European Field Experiment in a Desertification-threatened Area. Ann Geophys 11:173–189

BOREAS Science Team (1994) BOREAS Experiment Plan. Available from *http://www-eosdis.ornl.gov/BOREAS/bhs/Acrobat.html*, 735 p

Bousquet P, Peylin P, Ciais P, Rayner P, Friedlingstein P, Lequere C, Tans P (2000) Interannual CO_2 sources and sinks as deduced by inversion of atmospheric CO_2 data. Nature 290:1342–1346

Brock FV, Crawford KC, Elliott RL, Cuperus GW, Stadler SJ, Johnson HL, Eilts MD (1995) The Oklahoma mesonet: a technical overview. J Atmos Ocean Tech 12:5–19

Brutsaert W (1998) Land-surface water vapor and sensible heat flux: spatial variability, homogeneity and measurement scales. Water Resour Res 34:2433–2442

Burke EJ, Gurney RJ, Simmonds LP, Jackson TJ (1997) Calibrating a soil water and energy budget model with remotely sensed data to obtain quantitative information about the soil. Water Resour Res 33:1689–1697

Calvet J-C, Noilhan J, Bessemoulin P (1998) Retrieving the root-zone soil moisture from surface soil moisture or temperature estimates: a feasibility study based on field measurements. J Appl Meteorol 37:371–386

Calvet J-C, Bessemoulin P, Noilhan J, Berne C, Braud I, Courault D, Fritz N, Gonzalez-Sosa E, Goutorbe J-P, Haverkamp R, Jaubert G, Kergoat L, Lachaud G, Laurent J-P, Mordelet P, Olioso A, Péris P, Roujean J-L, Thony J-L, Tosca C, Vauclin M, Vignes D (1999) MUREX: a land-surface field experiment to study the annual cycle of the energy and water budgets. Ann Geophys 17:838 –854

Castelli F, Entekhabi D, Caporali E (1999) Estimation of surface heat flux and an index of soil moisture using adjoint-state surface energy balance. Water Resour Res 35:3115–3125

Charney JG, Halem M, Jastrow R (1969) Use of incomplete historical data to infer the present state of the atmosphere. J Atmos Sci 26:1160–1163

Chia E-H, Hutchinson MF (1991) The beta distribution as a probability model for daily cloud duration. Agr Forest Meteorol 56: 195–208

Cline DW, Bales RC, Dozier J (1998) Estimating the spatial distribution of snow in mountain basins using remote sensing and energy balance modeling. Water Resour Res 34:1275–1285

Cohn S (1982) Methods of sequential estimation for determining initial data in numerical weather prediction. PhD thesis, Courant Institute of Mathematical Sciences, New York University, New York, NY, 183 p

Coughlan M, Avissar R (1996) The Global Energy and Water Cycle Experiment (GEWEX) Continental-Scale International Project (GCIP): An overview. J Geophys Res 101:7139–7148

Courtier P, Talagrand O (1990) Variational assimilation of meteorological observations with the direct and adjoint shallow-water equations. Tellus 42A:531–549

Cowpertwait PSP, O'Connell PE, Metcalfe AV, Mawdsley JA (1996) Stochastic point process modelling of rainfall; 1: Single site fitting and validation. J Hydrol 175:17–46

Cox PM, Betts RA, Jones CD, Spall SA, Totterdell IJ (2000) Acceleration of global warming due to carbon cycle feedbacks in a coupled climate model. Nature 408:184–187

Cressie NAC (1991) Statistics for spatial data. John Wiley & Sons, New York

CSIRO (1996) Climate change scenarios for the Australian region. Climate Impact Group, CSIRO Division of Atmospheric Research, Melbourne Australia, *http://www.dar.csiro.au/publications/scenarios.htm*

Daley R (1990) Basic theory of data assimilation, part II: Continuous data assimilation through direct insertion methods. Meso-scale Data Assimilation, 1990 Summer Colloquium, National Center for Atmospheric Research, Boulder, Colorado, 6 June–3 July

Daley R (1991) Atmospheric data analysis. Cambridge University Press, New York

Daly C, Neilson RP, Phillips DL (1994) A statistical-topographic model for mapping climatological precipitation over mountainous terrain. J Appl Meteorol 33:140–158

De Negreiros GH, Nepstad DC (1994) Mapping deeply rooting forests of Brazilian Amazonia with GIS. Proceedings of ISPRS Commission VII Symposium – Resource and Environmental Monitoring, Rio de Janeiro, Brazil, 7, pp 334–338

Dickinson RE, Henderson-Sellers A, Kennedy PJ (1993) Biosphere-Atmosphere Transfer Scheme (BATS) Version 1e as coupled to the NCAR Community Climate Model. NCAR Tech. Note, NCAR/TN-387+STR

Dirmeyer PA (2001) Climate drift in a coupled land-atmosphere model. J Hydrometeorol 2:89–100

Dirmeyer PA, Dolman AJ, Sato N (1999) The global soil wetness project: a pilot project for global land-surface modeling and validation. B Am Meteorol Soc 80:851–878

Douville H, Viterbo P, Mahfouf J-F, Beljaars ACM (2000) Evaluation of the optimum interpolation and nudging techniques for soil moisture analysis using FIFE data. Mon Weather Rev 128:1733–1756

Ducharne A, Koster RD, Suarez MJ, Stieglitz M, Kumar P (2000) A catchment-based approach to modeling land-surface processes in a general circulation model, 2: Parameter estimation and model demonstration. J Geophys Res 105:24823–24838

Eley J (1992) Summary of workshop, soil moisture modeling. Proceedings of the NHRC Workshop held March 9–10, 1992, NHRI Symposium No. 9 Proceedings

Engman ET (1995) Recent advances in remote sensing in hydrology. Rev Geophys 33(suppl. 2):967–975

Entekhabi D, Nakamura H, Njoku EG (1994) Solving the inverse problem for soil moisture and temperature profiles by sequential assimilation of multifrequency remotely sensed observations. IEEE T Geosci Remote 32:438–448

Entin JK, Robock A, Vinnikov KY, Zabelin V, Liu S, Namkhai A, Adyasuren TS (1999) Evaluation of global soil wetness project soil moisture simulations. J Meteorol Soc Jpn 77:183–198

Feddes RA, Hoff H, Bruen M, Dawson TE, de Rosnay P, Dirmeyer P, Jackson RB, Kabat P, Kleidon A, Lilly A, Pitman AJ (2001) Modelling root water uptake in hydrological and climate models. B Am Meteorol Soc 82:2797–2809

Fo AJP, Crawford KC (1999) Mesoscale precipitation fields. Part I: Statistical analysis and hydrologic response. J Appl Meteorol 38:82–101

Friedlingstein P, Bopp L, Ciais P, Dufresne J-L, Fairhead L, LeTreut H, Monfray P, Orr J (2001) Positive feedback between the carbon cycle and future climate change. Geophys Res Lett 28(8):1543–1546

Gandin LS (1963) Objective analysis of meteorological fields. Israel Program for Scientific Translations

Gash JHC, Nobre CA (1997) Climatic effects of Amazonian deforestation: some results from ABRACOS. B Am Meteorol Soc 78:823–830

Gates WL, Boyle JS, Covey C, Dease CG, Doutriaux CM, Drach RS, Fiorino M, Gleckler PJ, Hnilo JJ, Marlais SM, Phillips TJ, Potter GL, Santer BD, Sperber KR, Taylor KE, Williams DN (1999) An overview of the results of the Atmospheric Model Intercomparison Project (AMIP I). B Am Meteorol Soc 80:29–55

Geng S (1986) A simple method for generating daily rainfall data. Agr Forest Meteorol 36:363–376

Georgakakos KP, Kavvas ML (1987) Precipitation analysis, modeling and prediction in hydrology. Rev Geophys 25(2):163–178

Ghil M, Cohn S, Tavantzis J, Bube K, Isaacson E (1981) Application of estimation theory to numerical weather prediction. In: Bengtsson L, Ghil M, Källén E (eds) Dynamic meteorology: data assimilation methods. Springer-Verlag, New York, NY, pp 139–224

Gibson JK, Kallberg P, Nomura A, Uppala S (1994) The ECMWF re-analysis (ERA) project – plans and current status. 10th Intl. Conf. on Interactive Information Processing Systems (IIPS), American Meteorological Society

Giorgi F, Mearns LO (1991) Approaches to the simulation of regional climate: a review. Rev Geophys 29:191–216

Goodrich DC, Chehbouni A, Goff B, MacNish B, Maddock T, Moran S, Shuttleworth WJ, Williams DG, Watts C, Hipps LH, Cooper DI, Schieldge J, Kerr YH, Arias H, Kirkland M, Carlos R, Cayrol P, Kepner W, Jones B, Avissar R, Begue A, Bonnefond J-M, Boulet G, Branan B, Brunel JP, Chen LC, Clarke T, Davis MR, DeBruin H, Dedieu G, Elguero E, Eichinger WE, Everitt J, Garatuza-Payan J, Gempko VL, Gupta H, Harlow C, Hartogensis O, Helfert M, Holifield C, Hymer D, Kahle A, Keefer T, Krishnamoorthy S, Lhomme J-P, Lagouarde J-P, Lo Seen D, Luquet D, Marsett R, Monteny B, Ni W, Nouvellon Y, Pinker R, Peters C, Pool D, Qi J, Rambal S, Rodriguez J, Santiago F, Sano E, Schaeffer SM, Schulte M, Scott R, Shao X, Snyder KA, Sorooshian S, Unkrich CL, Whitaker M, Yucel I (2000) Preface paper to the Semi-Arid Land-Surface-Atmosphere (SALSA) program special issue. Agr Forest Meteorol 105(1–3):3–20

Goutorbe J-P, Lebel T, Tinga A, Bessemoulin P, Brouwer J, Dolman AJ, Engman ET, Gash JHC, Hoepffner M, Kabat P, Kerr YH, Monteny B, Prince S, Saïd F, Sellers P, Wallace JS (1994) HAPEX-Sahel: a large scale study of land-atmosphere interactions in the semi-arid tropics. Ann Geophys 12:53–64

Goutorbe J-P, Lebel T, Dolman AJ, Gash JHC, Kabat P, Kerr YH, Monteny B, Prince S, Strickler JNM, Tinga A, Wallace JS (1997) An overview of HAPEX-Sahel: a study in climate and desertification. J Hydrol 189:4–17

Greenland D, Swift LW Jr (1991) Climate variability and ecosystem response: opportunities for the LTER network. Bull Eco Soc Amer 72:118–126

Hall FG (1999) Introduction to special section: BOREAS in 1999: experiment and science overview. J Geophys Res 104:27627–27639

Hall FG, Sellers PJ (1995) First International Satellite Land Surface Climatology Project (ISLSCP) Field Experiment (FIFE). J Geophys Res 100:25383–25396

Halldin S, Lundin L-C (1995) SINOP: system for information in NOPEX, 2nd ed. NOPEX Technical Report No. 1. NOPEX Central Office, Uppsala University, 24 p

Halldin S, Gryning S-E, Gottschalk L, Jochum A, Lundin L-C, Van de Griend AA (1999) Energy, water and carbon exchange in a boreal forest landscape – NOPEX experiences. Agr Forest Meteorol 98/99:5–29

Hay LE, McCabe GJ Jr, Wolock DM, Ayers MA (1991) Simulation of precipitation by weather type analysis. Water Resour Res 27:493–501

Hayes SP, Mangum LJ, Picaut J, Sumi A, Takeuchi K (1991) TOGA-TAO: a moored array for real-time measurements in the tropical Pacific ocean. B Am Meteorol Soc 72:339–347

Henderson-Sellers A, Yang Z-L, Dickinson RE (1993) The Project for Intercomparison of Land-Surface Parameterization Schemes. B Am Meteorol Soc 74:1335–1349

Henderson-Sellers A, Pitman AJ, Love PK, Irannejad P, Chen T (1995) The Project for Intercomparison of Land-Surface Parameterizaton Schemes (PILPS) Phases 2 and 3. B Am Meteorol Soc 76:489–503

Henderson-Sellers A, McGuffie K, Pitman A (1996) The Project for the Intercomparison of Land-Surface Parametrization Schemes (PILPS): 1992–1995. Clim Dynam 12:849–859

Hess R (2001) Assimilation of screen-level observations by variational soil moisture analysis. Meteorol Atmos Phys 77:145–154

Hollingsworth A, Shaw DB, Lonnberg P, Illari L, Arpe K, Simmons AJ (1986) Monitoring and observation of analysis quality by a data assimilation system. Mon Weather Rev 114:861–879

Hou AY, Ledvina DV, da Silva AM, Zhang SQ, Joiner J, Atlas RM (2000) Assimilation of SSM/I-derived surface rainfall and total precipitable water for improving the GEOS analysis for climate studies. Mon Weather Rev 128:509–537

Houser PR, Shuttleworth WJ, Gupta HV, Famiglietti JS, Syed KH, Goodrich DC (1998) Integration of soil moisture remote sensing and hydrologic modeling using data assimilation. Water Resour Res 34:3405–3420

Hulme M, Conway D, Jones PD, Jiang T, Barrow EM, Turney C (1995) Construction of a 1961–1990 European climatology for climate change modelling and impact applications. Int J Climatol 15:1333–1363

Hutchinson MF (1989) A new procedure for gridding elevation and streamline data with automatic removal of spurious pits. J Hydrol 106:211–232

Hutchinson MF (1991) The application of thin plate smoothing splines to continent-wide data assimilation. In: Jasper JD (ed) Data assimilation systems. BMRC Research Report No. 27, Bureau of Meteorology, pp 104–113

Hutchinson MF (1995a) Stochastic space-time weather models from ground-based data. Agr Forest Meteorol 73:237–264

Hutchinson MF (1995b) Interpolation of rainfall means. Int J Geogr Inf Syst 9:385–403

Hutchinson MF (1998) Interpolation of rainfall data with thin plate smoothing splines: II. Analysis of topographic dependence. J Geog Inf Decis Anal 2(2):168–185

Hutchinson MF, Gessler PE (1994) Splines – more than just a smooth interpolator. Geoderma 62:45–67

IGBP (International Geosphere-Biosphere Programme) (1993) Biospheric aspects of the hydrological cycle. The Operational Plan. IGBP Global Change Report No. 27. International Council of Scientific Unions, Stockholm

IGBP (International Geosphere-Biosphere Programme) (1996) Current status and perspectives of FLUXNET. Global Change Newsletter 28:14–16

IGPO (International GEWEX Project Office) (1998) Land-surface parameterizations/soil vegetation atmosphere transfer schemes workshop: conclusions and working group reports. IGPO Publication Series No. 31, 77 p

Ikawa M (1984) An alternative method of solving weak constraint problems and a unified expression of weak and strong constraints in variational objective analysis. Pap Meteorol Geophys 35:71–79

INSU (Institut National des Sciences de l'Univers) (2000) GEWEX/INSU International Workshop on Modelling Land-surface Atmosphere Interactions and Climate Variability. 134 p

Jackson TJ (1997a) Southern Great Plains 1997 (SGP97) hydrology experiment plan. *http://hydrolab.arsusda.gov/sgp97/*

Jackson TJ (1997b) Soil moisture estimation using special satellite microwave/imager satellite data over a grassland region. Wat Resour Res 33:1475–1484

Jackson RB, Canadell J, Ehleringer JR, Mooney HA, Sala OE, Schulze ED (1996) A global analysis of root distributions for terrestrial biomes. Oecologia 108. 389–411

Jackson RB, Mooney HA, Schulze ED (1997) A global budget for fine root biomass, surface area, and nutrient contents. P Natl Acad Sci USA 94:7362–7366

Jackson RB, Schenk HJ, Jobbágy EG, Canadell J, Colello GD, Dickinson RE, Field CB, Friedlingstein P, Heimann M, Hibbard K, Kicklighter DW, Kleidon A, Neilson RP, Parton WJ, Sala OE, Sykes MT (2000) Below-ground consequences of vegetation change and their treatment in models. Ecol Appl 10:470–483

Jones DMA, Wendland WM (1984) Some statistics of instantaneous precipitation. J Clim Appl Meteorol 23:1273–1285

Jones JW, Colwick RF, Threadgill ED (1972) A simulated environmental model of temperature, evaporation, rainfall and soil moisture. T Am Agr Eng 15:366–372

Kalnay E, Kanamitsu M, Kistler R, Collins W, Deaven D, Gandin L, Iredell M, Saha S, White G, Woollen J, Zhu Y, Chelliah M, Ebisuzaki W, Higgins W, Janowiak J, Mo KC, Ropelewski C, Wang J, Leetmaa A, Reynolds R, Jenne R, Joseph D (1996) The NCEP/NCAR 40-year reanalysis project. B Am Meteorol Soc 77:437–471

Karner O, Rannik U (1996) Stochastic models to represent the temporal variability of zonal average cloudiness. J Climate 9:2718–2726

Katz RW (1996) Use of conditional stochastic models to generate climate change scenarios. Climatic Change 32:237–255

Katz RW, Parlange MB (1998) Overdispersion phenomenon in stochastic modeling of precipitation. J Climate 11:591–601

Kibby H (1996) The global climate observing system. World Meteorological Organization Bulletin 45:140–146

Kittel TGF, Rosenbloom NA, Painter TH, Schimel DS, VEMAP Modeling Participants (1995) The VEMAP integrated database for modeling United States ecosystem/vegetation sensitivity to climate change. J Biogeogr 22:857–862

Kittel TGF, Royle JA, Daly C, Rosenbloom NA, Gibson WP, Fisher HH, Schimel DS, Berliner LM, VEMAP2 Participants (1997) A gridded historical (1895–1993) bioclimate dataset for the conterminous United States. In: Proceedings of the 10th Conference on Applied Climatology, 20–24 October 1997, Reno, NV. Amer Meteorol Soc, Boston, pp 219–222

Knorr W (2000) Annual and interannual CO_2 exchanges of the terrestrial biosphere: process-based simulations and uncertainties. Global Ecol Biogeogr 9:225–252

Knorr W, Lakshmi V (2001) Assimilation of fAPAR and surface temperature into a land-surface and vegetation model. In: Lakshmi V, Albertson J, Schaake J (eds) Land-surface hydrology, meteorology, and climate observations and modeling. American Geophysical Union, Washington, D.C., pp 177–200

Kolmogorov A (1941) Interpolated and extrapolated stationary random sequences. Izvestia an SSSR seriya mathematicheskaya 5:85–95

Koster RD, Suarez MJ, Ducharne A, Stieglitz M, Kumar P (2000) A catchment-based approach to modeling land-surface processes in a general circulation model, 1: Model structure. J Geophys Res 105:24809–24822

Lawford RG (1999) A midterm report on the GEWEX Continental-scale International Project (GCIP). J Geophys Res 104: 19279–19292

LBA Science Planning Group (1996) The large-scale biosphere-atmosphere experiment in Amazonia (LBA): concise experiment plan. Available from the LBA project office, CPTEC/INPE, Rodovia Presidente Dutra, km 40, Caixa Postal 01, 12630-000 Cachoeira Paulista, SP, Brazil, 41 p

Leemans R, Cramer W (1991) The IIASA database for mean monthly values of temperature, precipitation and cloudiness of a global terrestrial grid. International Institute for Applied Systems Analysis (IIASA). RR-91-18. Laxenburg, Austria

Lillesand TM, Kiefer RW (1987) Remote sensing and image interpretation, 2nd ed. John Wiley & Sons, Chichester

Liston G (1999) Interrelationships among snow distribution, snowmelt, and snow cover depletion: implications for atmospheric, hydrologic, and ecologic modeling. J Appl Meteorol 38:1474–1487

Lorenc AC (1995) Atmospheric data assimilation. Meteorological Office Forecasting Research Division Scientific Paper No. 34

Lundin L-C, Halldin S, Nord T, Etzelmüller B (1999) System of information in NOPEX – retrieval, use, and query of climate data. Agr Forest Meteorol 98/99:31–51

Lyons SW (1990) Spatial and temporal variability of monthly precipitation in Texas. Mon Weather Rev 118:2634–2648

Mahfouf J-F, Viterbo P (1998) Land-surface assimilation. Proceedings of the Seminar on Data Assimilation, Sep 1996, ECMWF, Reading, pp 319–347

McLaughlin D (1995) Recent developments in hydrologic data assimilation. Rev Geophys, Suppl. 977–984

Mearns LO, Rosenzweig C, Goldberg R (1997) Mean and variance change in climate scenarios: methods, agricultural applications, and measures of uncertainty. Climatic Change 35:367–396

Meeson BW, Corprew FE, McManus JMP, Myers DM, Closs JW, Sun KJ, Sunday DJ, Sellers PJ (1995) ISLSCP initiative I – global datasets for land-atmosphere models, 1987–1988. Vol. 1–5, published on CD-ROM by NASA (USA_NASA_GDAAC_ISLSCP_001 – USA_NASA_GDAAC_ISLSCP_005)

Michaud JD, Shuttleworth WJ (1997) Executive summary of the Tucson Aggregation Workshop. J Hydrol 190:176–181

Milly PCD (1986) Integrated remote sensing modeling of soil moisture: sampling frequency, response time, and accuracy of estimates. Integrated Design of Hydrological Networks (Proceedings of the Budapest Symposium, July 1986). IAHS Publ. No. 158, pp 201–211

Mitchell K, Houser P, Wood E, Schaake J, Lettenmaier D, Tarpley D, Higgins W, Marshall C, Lohmann D, Lin Y, Ek M, Cosgrove B, Entin J, Duan Q, Pinker R, Robock A, Van den Dool H, Habets F (1999) The GCIP Land Data Assimilation System (LDAS) project – now underway. GEWEX News 9(4):3–6

National Research Council (1995) Finding the forest in the trees: the challenge of combining diverse environmental data. National Academy Press, Washington, DC, 129 p

National Research Council (US) Panel on Model-assimilated Data Sets for Atmospheric Oceanic Research (1991) Four-dimensional model assimilation of data: a strategy for the Earth system sciences. National Academy Press, Washington, DC

Nemes A, Schaap MG, Leij FJ (1999) The UNSODA unsaturated soil hydraulic database, version 2.0. US Salinity Laboratory, Riverside, California

Obled C, Creutin JD (1986) Some developments in the use of empirical orthogonal functions for mapping meteorological fields. J Clim Appl Meteorol 25:1189–1204

O'Connell PE, Todini E (1996) Modelling of rainfall, flow and mass transport in hydrological systems. J Hydrol 175:3–16

Ohmura A, Gilgen H, Hegner H, Müller G, Wild M, Dutton EG, Forgan B, Fröhlich C, Philipona R, Heimo A, König-Langlo G, McArthur B, Pinker R, Whitlock CH, Dehne K (1998) Baseline surface radiation network (BSRN/WCRP): new precision radiometry for climate research. B Am Meteorol Soc 79:2115–2136

Oki T, Nishimura T, Dirmeyer P (1999) Assessment of annual runoff from land-surface models using Total Runoff Integrating Pathways (TRIP). J Meteorol Soc Jpn 78:235–255

Pan Y, Melillo JM, McGuire AD, Kicklighter DW, Pitelka LF, Hibbard K, Pierce LL, Running SW, Ojima DS, Parton WJ, Schimel DS, VEMAP Members (1998) Modeled responses of terrestrial ecosystems to elevated atmospheric CO_2: a comparison of simulations by the biogeochemistry models of the Vegetation/Ecosystem Modeling and Analysis Project (VEMAP). Oecologia 114:389–404

Parlow E (1996) The Regional Climate Project REKLIP – an overview. Theor Appl Climatol 53(1–3):3–7

Pfafstetter O (1989) Classification of hydrographic basins: coding methodology. Departamento Nacional de Obras de Saneamento, August 18, 1989, Rio de Janeiro, available from Verdin JP, US Geological Survey, EROS Data Center, Sioux Falls, South Dakota 57198 USA, (unpublished manuscript)

Phillips JG, Cane MA, Rosenzweig C (1998) ENSO, seasonal rainfall patterns and simulated maize yield variability in Zimbabwe. Agr Forest Meteorol 90:39–50

Polcher J (2001) The Global Land-Atmosphere System Study (GLASS). BAHC News 9:5–6

Polcher J, Shao Y (1996) A standard format for reporting PILPS experiments. Global Planet Change 13:217–223

Price DT, McKenney DW, Nalder IA, Hutchinson MF, Kesteven JL (2000) A comparison of two statistical methods for spatial interpolation of Canadian monthly mean climate data. Agr Forest Meteorol 101:81–94

Racsko P, Szeidl L, Semenov M (1991) A serial approach to local stochastic weather models. Ecol Model 57:27–41

Raschke E, Karstens U, Nolte-Holube R, Brandt R, Isemer H-J, Lohmann D, Lobmeyr M, Rockel B, Stuhlmann R (1998) The Baltic Sea experiment BALTEX: a brief overview and some selected results of the Authors. Surv Geophys 19:1–22

Rhodin A, Kucharski F, Callies U, Eppel DP, Wergen W (1999) Variational analysis of effective soil moisture from screen-level atmospheric parameters; application to a short-range forecast. Q J Roy Meteor Soc 125:2427–2448

Richardson CW (1981) Stochastic simulation of daily precipitation, temperature and solar radiation. Water Resour Res 17:182–190

Rodell M, Famiglietti JS (1999) Detectability of variations in continental water storage from satellite observations of the time dependent gravity field. Water Resour Res 35:2705–2723

Rosenthal W, Dozier J (1996) Automated mapping of montane snow cover at subpixel resolution from the Landsat Thematic Mapper. Water Resour Res 32:115–130

Running SW, Thornton PE (1996) Generating daily surfaces of temperature and precipitation over complex topography. In: Goodchild MF, Steyaert LT, Parks BO, Johnston C, Maidment D, Crane M, Glendinning S (eds) GIS and environmental modeling: progress and research issues. GIS World, Fort Collins, USA, pp 93–98

Running SW, Baldocchi DD, Turner D, Gower ST, Bakwin P, Hibbard K (1999) A global terrestrial monitoring network integrating tower fluxes, flask sampling, ecosystem modeling and EOS satellite data. Remote Sens Environ 70:108–127

Sasaki Y (1958) An objective analysis based on the variational method. J Meteorol Soc Jpn 36:77–88

Schaap MG (1999) Rosetta Version 1.0. US Salinity Laboratory, Riverside, California

Schimel DS, VEMAP Participants, Braswell BH (1997) Spatial variability in ecosystem processes at the continental scale: models, data, and the role of disturbance. Ecol Monogr 67:251–271

Schimel D, Melillo J, Tian H, McGuire AD, Kicklighter D, Kittel T, Rosenbloom N, Running S, Thornton P, Ojima D, Parton W, Kelly R, Sykes M, Neilson R, Rizzo B (2000) Contribution of increasing CO_2 and climate to carbon storage by ecosystems of the United States. Science 287:2004–2006

Schneider T, Griffies SM (1999) A conceptual framework for predictability studies. J Climate 12:3133–3155

Schubert SD, Rood RB, Pfaendtner J (1993) An assimilated dataset for earth science applications. B Am Meteorol Soc 74:2331–2342

Sellers PJ, Hall FG (1992) FIFE in 1992 results, scientific gains, and future research directions. J Geophys Res 97:19091–19109

Sellers PJ, Mintz Y, Sud YC, Dalcher A (1986) A simple biosphere model (SiB) for use within general circulation models. J Atmos Sci 43:505–531

Sellers PJ, Hall F, Margolis H, Kelly B, Baldocchi D, den Hartog G, Cihlar J, Ryan MG, Goodison B, Crill P, Ranson KJ, Lettenmaier D, Wickland DE (1995) The Boreal Ecosystem-Atmosphere Study (BOREAS): an overview and early results from the 1994 field year. B Am Meteorol Soc 76:1549–1577

Sellers PJ, Meeson BW, Closs J, Collatz J, Corprew F, Dazlich D, Hall FG, Kerr Y, Koster R, Los S, Mitchell K, McManus J, Myers D, Sun K-J, Try P (1996) The ISLSCP initiative I global datasets: surface boundary conditions and atmospheric forcings for land-atmosphere studies. B Am Meteorol Soc 77:1987–2005

Semenov MA, Barrow EM (1997) Use of a stochastic weather generator in the development of climate change scenarios. Climatic Change 35:397–414

Semenov MA, Brooks RJ (1999) Spatial interpolation of the LARS-WG stochastic weather generator in Great Britain. Climate Res 11:137–148

Shah SMS, O'Connell PE, Hosking JRM (1996) Modelling the effects of spatial variability in rainfall on catchment response; 1: Formulation and calibration of a stochastic rainfall field model. J Hydrol 175:67–88

Siltanen RM, Apps MJ, Zoltai SC, Mair RM, Strong WL (1997) A soil profile and organic carbon data base for Canadian forest and tundra mineral soils. Nat. Resour. Can., Can. For. Serv., Noth. For. Cent., Edmonton, Alberta, Canada. 50 p

Sivapalan M, Wood EF (1987) A multidimensional model of nonstationary space-time rainfall at the catchment scale. Water Resour Res 23:1289–1299

Srikanthan R, McMahon TA: (1984) Synthesizing daily rainfall and evaporation data as input to water balance-crop growth models. J Aust I Agr Sci 51–54

Stauffer DR, Seaman NL (1990) Use of four-dimensional data assimilation in a limited-area mesoscale model. Part I: Experiments with synoptic-scale data. Mon Weather Rev 118:1250–1277

Stewart JB, Finch JW (1993) Application of remote sensing to forest hydrology. J Hydrol 150:701–716

Stewart RE, Leighton HG, Marsh P, Moore GWK, Ritchie H, Rouse WR, Soulis ED, Strong GS, Crawford RW, Kochtubajda B (1998) The Mackenzie GEWEX study: the water and energy cycles of a major North American river basin. B Am Meteorol Soc 79:2665–2683

Stidd CK (1973) Estimating the precipitation climate. Water Resour Res 22:2107–2110

Stokes GM, Schwartz SE (1994) The Atmospheric Radiation Measurement (ARM) program: programmatic background and design of the cloud and radiation test bed. B Am Meteorol Soc 75:1201–1221

Strebel DE, Newcomer JA, Ormsby JP (1990) The FIFE information system. IEEE T Geosci Remote 28:703–710

Summerhayes C (1998) The Global Ocean Observing System (GOOS). World Meteorological Organization Bulletin 47:32–39

Tempel P, Batjes NH, van Engelen VWP (1996) IGBP-DIS soil dataset for pedotransfer function development. Working paper and preprint 96/05, International Soil Reference and Information Centre (ISRIC), Wageningen

Thornton PE, Running SW, White MA (1997) Generating surfaces of daily meteorological variables over large regions of complex terrain. J Hydrol 190:214–251

USDA (United States Department of Agriculture) (1994) National soil characterization data. Soil Survey Laboratory, National Soil Survey Center, Soil Conservation Service, Lincoln, USA

Valdes JB, Rodríguez-Iturbe I, Gupta VK (1985) Approximations of temporal rainfall from a multidimensional model. Water Resour Res 21:1259–1270

VEMAP Members (1995) Vegetation/Ecosystem Modeling and Analysis Project (VEMAP): comparing biogeography and biogeochemistry models in a continental-scale study of terrestrial ecosystem responses to climate change and CO_2 doubling. Global Biogeochem Cy 9:407–437

Verdin KL, Verdin JP (1999) A topological system for delineation and codification of the Earth's river basins. J Hydrol 218: 1–12

Viovy N, Francois C, Bondeau A, Krinner G, Polcher J, Kergoat L, Dedieu G, de Noblet N, Ciais P, Friedlingstein P (2001) Assimilation of remote sensing measurements into the ORCHIDEE/STOMATE DVGM biosphere model. In: Recueil des actes du 8 Symposium International "Mesures Physiques et Signatures en Télédetection", Aussois, France, 8–11 Janvier 2001, pp 713–718

Viterbo P, Beljaars ACM, Mahfouf J-F, Teixeira J (1999) The representation of soil moisture freezing and its impact on the stable boundary layer. Q J Roy Meteor Soc 125:2401–2426

Vogel RL (1998) Why scientist have not been writing metadata. EOS Trans Am Geophys Union 79:373–380

Vörösmarty CJ, Askew A, Grabs W, Barry RG, Birkett C, Döll P, Goodison B, Hall A, Jenne R, Kitaev L, Landwehr J, Keeler M, Leavesey G, Schaake J, Strzepek K, Sundarvel SS, Takeuchi K, Webster F (2001) Global water data: a new endangered species. EOS Trans Am Geophys Union 82:54–58

Wahba G, Wendelberger J (1980) Some new mathematical methods for variational objective analysis using splines and cross validation. Mon Weather Rev 108:122–1143

Wahr J, Molenaar M, Bryan F (1998) Time-variability of the Earth's gravity field: hydrological and oceanic effects and their possible detection using GRACE. J Geophys Res 103:30205–30230

Wang J, Gao Y, Hu Y (1993) An overview of the HEIFE experiment in the People's Republic of China. Exchange processes at the land surface for a range of space and time scales. In: Bolle H-J, Feddes RA, Kalma JD (eds) Proceedings of a symposium held during the joint meeting of the International Association of Meteorology and Atmospheric Physics and IAHS at Yokohama, July 1993, IAHS Pub. 212:397–406

Wiener N (1949) Extrapolation, interpolation, and smoothing of a stationary time series. John Wiley & Sons, Chichester, 84 p

Wilby RL, Wigley TML, Conway D, Jones PD, Hewitson BC, Main J, Wilks DS (1998) Statistical downscaling of general circulation model output: a comparison of methods. Water Resour Res 34:2995–3008

Wilks DS (1999a) Interannual variability and extreme-value characteristics of several stochastic daily precipitation models. Agr Forest Meteorol 93:153–169

Wilks DS (1999b) Simultaneous stochastic simulation of daily precipitation, temperature and solar radiation at multiple sites in complex terrain. Agr Forest Meteorol 96:85–101

Wilson LL, Lettenmaier DP, Skyllingstad E (1992) A hierarchical stochastic model of large-scale atmospheric circulation patterns and multiple station daily precipitation. J Geophys Res 97:2791–2809

WMO (World Meteorological Organization) (1996) Exchanging meteorological data: guidelines on relationships in commercial meteorological activities. WMO Policy and Practice. WMO Report 837, Geneva, 24 p, available from the WMO, P.O. Box 2300, CH-1211 Geneva 2, Switzerland

Wösten JHM, Lilly A, Nemes A, Le Bas C (1998) Final report on the EU funded project using existing soil data to derive hydraulic parameters for simulation models in environmental studies and in land-use planning. (CHRX-CT94-0639), DLO Winand Staring Centre, Report 157, Wageningen, The Netherlands

Wösten JHM, Lilly A, Nemes A, Le Bas C (1999) Development and use of a database of hydraulic properties of European soils. Geoderma 90:169–185

Yasunari T (1993) GEWEX-related Asian monsoon experiment (GAME). Adv Space Res 14:161–165

Yates DN, Kittel TGF, Cannon RF (2000) Comparing the correlative Holdridge model to mechanistic biogeographical models for assessing vegetation distribution response to climatic change. Climatic Change 44:59–87

Zorita E, Kharin V, von Storch H (1992) The atmospheric circulation and sea surface temperature in the north Atlantic area in winter: their interaction and relevance for Iberian precipitation. J Climate 5:1097–1108

Part D

The Integrity of River and Drainage Basin Systems: Challenges from Environmental Change

Edited by Michel Meybeck and Charles J. Vörösmarty

Chapter D.1

Introduction

Michel Meybeck · Charles J. Vörösmarty

The foregoing parts demonstrate that the dynamics and biophysical character of land-atmosphere interactions are intimately connected to the dynamics and biophysical character of the land-based water cycle. The hydrological cycle has been shown to play a central role in our analysis of climate change, the impacts of land-use and land-cover change, and vegetation dynamics.

There are additional, significant issues that must be considered to more completely define the full dimension of global change with respect to the hydrological cycle. These collectively define a central role for humans in shaping the character of the terrestrial water cycle, not only at local scales, but over regional and even global domains as well.

Humans exert an influence on the water cycle not only through the highly-publicised greenhouse effect but also through the forces of land-cover change, land-management practices, urbanisation, and the construction and operation of water engineering facilities. These factors all dramatically alter hydrological dynamics and form the key focal points of this section. These issues will be addressed in the context of our primary scientific question:

- Over a decades-to-century time frame, what are the relative impacts on the terrestrial hydrological cycle of (*a*) climatic variation and greenhouse warming, (*b*) land-cover change and land management, and (*c*) direct alterations due to water resource management?

The global change research community has arguably focused its attention on climate change and much less on these other factors. Is this the correct focal point for our collective efforts? An objective answer to this central question will colour the progress of our science over the next several years, and we contend it is a necessary starting point as we look toward the future.

It is clear that movement toward a global picture of hydrospheric change – one in which humans figure prominently – will require us to identify appropriate hydrological and socio-economic principles and to combine these within a common framework. A central goal of this chapter is to summarise recent findings and to explore their use in moving us toward a global synthesis. Our emphasis will be on the biogeophysical aspects of this question, but considering socio-economic issues as they become relevant.

Building on a long and rich history of small-scale catchment-scale studies dating back more than 100 years, there are exciting new opportunities in the water sciences as we move toward a global view of environmental change and its impact on the water cycle. More traditional hydrological research has uncovered the mechanics of the water cycle describing such processes as evapotranspiration, soil physics, groundwater dynamics and runoff generation. Process-based knowledge has also accumulated on the mobilisation and transport of constituents – including pollution – which are entrained in runoff and river flow. This work provides us with the fundamental principles necessary to detect and interpret the ongoing forces of environmental change. It thus merits an important place in this synthesis chapter and we treat it explicitly in several sections.

Early on, scientific hydrology was turned toward a pragmatic goal of providing sufficient understanding to predict, or at least better manage, catastrophic flooding, drought, erosion and sedimentation, and pollutant source areas and eutrophication. In fact, much of what prompted hydrological analysis was driven by the needs of hydrological engineers. Humans are thus hardly passive when it comes to hydrological events and we have done much to transform the terrestrial hydrosphere into a highly managed biogeochemical cycle (Fig. D.94). This is certainly true at the local scale, and we contend that these changes are now certainly of regional importance, and ultimately pandemic in extent. With population growth and economic development will come increasing pressures to control water supplies in service to humanity. It is thus important to articulate the role of humans and to prepare for the wise management of what are in many parts of the world increasingly scarce water resources. Integrated Water Resources Management (IWRM) will constitute a key emphasis in our discussion.

The community is poised for major progress toward global synthesis. This results from the wide availability, relatively recently, of state-of-the-art datasets and analysis tools including GIS (Geographic Information System) and remote sensing. Analogous to the paired catchment study which has served as the mainstay of hydrological research at the small scale, the conceptual framework of the drainage basin as a functioning hydrological unit permits us to analyse how the spatial organisation of whole river systems conditions continental runoff. This perspective will be critical to our success in progressing upwards in scale from the small catchment to the meso-scale catchment to continents, and ultimately the globe.

We have several specific goals for this section:

- to articulate the role of humans in the terrestrial water cycle by assessing the relative importance of different sources of anthropogenic impact: climate change, land-cover and land-use change, and water engineering;
- to define a strategy for moving across time and space scales;

- to summarise recent developments in the field over the last 10–20 years and explore how these might be used to move us toward a more synthetic view of a rapidly changing water cycle, ultimately to the global scale; and,
- to identify appropriate water management principles that could be applied in the face of these ongoing environmental changes.

This part is structured according to a scaling framework which permits us to place recent findings into a common context. Detailed sections on local to small-catchment scale processes are followed by a regional analysis. We turn next toward an analysis of emerging trends at the global scale. At each stage we re-visit our central hypothesis, assess its validity, and identify key areas for future progress. Several case studies of specific river basins are completing the part. A concluding section identifies key steps forward.

Chapter D.2

Responses of Hydrological Processes to Environmental Change at Small Catchment Scales

Alfred Becker · Mike Bonell · Reinder A. Feddes · Valentina Krysanova · Jeffrey J. McDonnell
Roland E. Schulze · Christian Valentin

Alfred Becker

D.2.1 Introduction

Chapter D.2 deals with fundamental hydrological processes and their modelling at "small catchment scales". We specifically define such catchments as having areas from $\sim 10^{-1}$ km^2 to 10^3 km^2, known as the hydrological micro- to meso-scale. Since the exchange processes between the land surface and the atmosphere (energy, water etc.) at small scales are already treated in Chapt. A.2, the primary focus in this chapter is on so-called "wet hydrology", i.e. soil moisture dynamics, runoff generation and resulting lateral flows of water and associated transports of sediments, chemicals and nutrients. The processes at and below the land surface in soils and aquifers represent an important part of the terrestrial phase of the hydrological cycle and associated biogeochemical cycles.

The dynamics of individual hydrological processes and their spatial differentiation is highly complex, leading to significant uncertainties. For example, the quantification of the different contributions to catchment-level runoff of landscape units, such as vegetated in contrast to non-vegetated (bare or sealed), sparsely vegetated or mixed, built-up areas; or dried out in contrast to moist areas (wetlands, shallow groundwater areas), is always problematic and often not sufficient in accuracy. The main reason is the enormous spatial and temporal variability of infiltration capacities dependent not only on soil and vegetation type, but also on current soil moisture. Accordingly, various simplifications are applied in modelling. These are often acceptable in large-scale modelling. They may cause problems, however, in smaller scale simulation studies and in special investigations of, for example, the effects of changing land use (see Sect. D.2.6.1).

With this in mind, the primary aim of this Chapt. D.2 is:

- to present an overview of fundamental hydrological processes and their spatial and temporal variability;
- to summarise recent improvements in our understanding of these processes;
- to provide specific information on the different component processes of runoff generation and lateral flows along various pathways, especially below the land surface;
- to provide a review of the utility of comprehensive field studies in small catchments;
- to review the movement towards high-resolution distributed hydrological modelling using GIS-based parameterisations; and,
- to highlight the rapid development and degree of application of "integrated" ecohydrological models which serve to describe the complex links and interaction between energy and water and associated biogeochemical fluxes at micro- and meso-scales.

Most parts of the chapter are descriptive by intention. Equations and modelling details are generally not presented due to space limitations. However, relevant references are given to available textbooks, review papers and selected papers.

Alfred Becker

D.2.2 Terrestrial Hydrological Processes – Overview, Definitions, Classification

D.2.2.1 Fundamental Hydrological Processes

Fundamental hydrological processes occur roughly similar in character at all spatial scales. However, from a practical standpoint they are generally best studied at small scales, such as the plot, hillslope or small catchment (headwater) scale. An overview of these processes and their typical chronological sequence in the terrestrial phase of the hydrological cycle is given in Fig. D.1.

The left part of Fig. D.1 is focused essentially on "vertical processes" which basically define the water balance of a landscape unit. These specifically include the major processes of precipitation, evaporation, transpiration and runoff generation. An additional set of more specific component processes includes interception, snow cover dynamics, depression storage at the land surface, including initial wetting, infiltration, soil moisture dynamics in the unsaturated zone, including percolation and root water uptake, groundwater recharge and capillary rise, overland flow and subsurface stormflow generation. These are represented diagrammatically in Fig. D.2.

Most of these processes are treated extensively in available text books on hydrology (Maidment 1993; Dyck and Peschke 1995; Dingman 2001) and will not be discussed here. This chapter therefore concentrates on their relevance to global change. A fundamental element of this relevance is the remarkable temporal and spatial variability of these processes. Selected processes are discussed in detail in the sections below, including treatment of soil moisture dynamics (Sect. D.2.3), overland flow and erosion (Sect. D.2.4), subsurface stormflow (Sect. D.2.5), and ecohydrological processes (Sect. D.2.6).

D.2.2.2 Spatial Differentiation of Vertical Hydrological Processes

Fundamental vertical hydrological processes can best be studied and understood using elementary areal units (patches) characterised by similarities in a wide array of attributes. These may include similarity in terms of topographic characteristics (elevation, slope class), land use and land cover, soil type and texture, hydrogeology (especially depth of the groundwater table or impervious layers), proximity to river networks and catchment boundaries (water divides).

Elementary areal units belonging to the same category or sharing similar hydrological behaviours are variably referred to as hydrotopes or Hydrological Response Units (HRUs) (Becker et al. 2002; Flügel 1995; Becker and Braun 1999). Natural landscapes and river basins are composed of a variety of hydrotopes, which may markedly differ from each other in essential hydrological characteristics. Accordingly, landscapes show a well known "mosaic structure", or landscape patchiness, with variably sized and shaped polygons when mapped. This is illustrated, for example, by the mixed use landscape in Fig. D.3. The mosaic structure represents an appropriate disaggregation scheme for hydrological studies, at least with regard to the vertical processes, to which runoff generation belongs.

Concerning the spatial differentiation of water balance components it should be emphasized that, for example, wet surfaces such as water surfaces (AW), wetlands and various shallow groundwater areas (AN; cf.

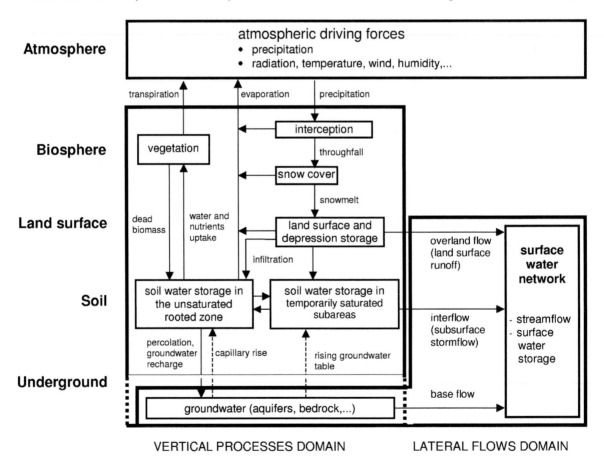

Fig. D.1. Schematic representation of the typical sequence of terrestrial hydrological processes with indication of vertical processes (fluxes in the *left block* marked by *up-* and *downward arrows*) and lateral flows (*lower right part, horizontal arrows*), after Becker et al. (2002)

Fig. D.3 and Fig. D.6), where evapotranspiration (ET) occurs at or near the potential rate, often exist adjacent to dry areas (e.g. dry vegetated areas, sealed areas, bare soils) where ET during dry periods is equal to or near zero. Such wet/dry contrasts add the further complication of advective processes. Analogously, during rainfall and snowmelt events, overland flow (RO) is normally generated only on limited areas, in particular from sealed or other impervious or less permeable areas (AIMP, e.g. uncovered rocks, clay and gleyic soils or parts of urban areas), as well as from saturated areas, whereas adjacent vegetated areas, especially those with deep groundwater (AG), do not generate any direct runoff (for comparison see Fig. D.6 in Sect. D.2.2.3).

Such, sometimes drastic, differences also exist between different environments, for example, high mountains versus lowlands, dry areas versus wetlands, and different climate zones (arid, semi-arid, mediterranean, temperate, humid etc.). This is illustrated by the examples in Table D.1 and Fig. D.4 (taken from L'vovich 1979 and Falkenmark and Chapman 1989, respectively). The significant differences in the partitioning of the different annual amounts of precipitation (*upper left downward arrows* in the three diagrams in Fig. D.4) into evapotranspiration (*upper right upward arrows*) with indication of both real (*thick arrows*) and potential (*thin arrows*) evapotranspiration and runoff (streamflow) are clearly visible in both the figure and the table. Additional information can be found in Dawdy (1991) and Schulze (1998).

An important feature of hydrotopes is their internal uniformity ("homogeneity" or, more strictly, "quasi-homogeneity") in essential process characteristics and related parameters which can be approximated by aver-

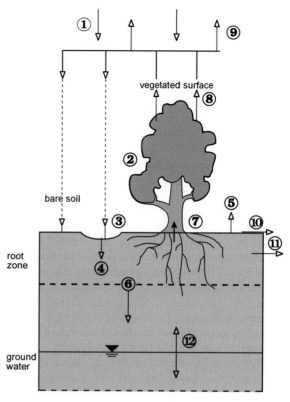

Fig. D.2. Water balance components (vertical processes) in a patch (or hydrotope). *Legend:* 1 Precipitation; 2 canopy interception storage; 3 depression storage at the land surface (including initial wetting); 4 infiltration; 5 evaporation from bare soil; 6 soil moisture recharge and percolation; 7 root water uptake; 8 transpiration; 9 total evapotranspiration; 10 infiltration excess/surface runoff/overland flow generation; 11 subsurface stormflow generation; 12 groundwater recharge and abstraction (by capillary rise or root water uptake); 13 snow cover dynamics (if snow cover exists; not represented in the figure)

Fig. D.3.
Example of the disaggregation of a landscape (or a subcatchment) into hydrotopes

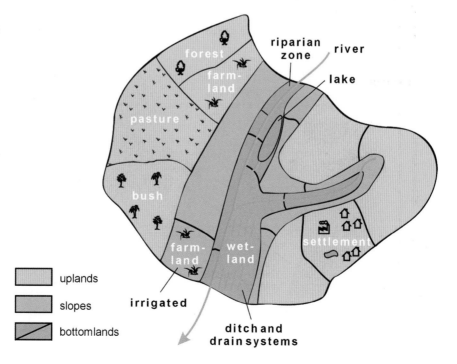

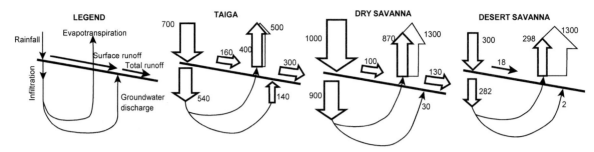

Fig. D.4. Land-surface water budgets in contrasting humid and arid climatic regimes (see text for explanation) (after Falkenmark and Chapman 1989)

age values, each of which is representative for the hydrotope under consideration. A necessary prerequisite for application of such a hydrotope-related parameter estimation is a practical delineation of individual hydrotopes. This has been facilitated in recent years by the increasing availability of GIS-based maps of land-surface attributes such as topography (DEMs, Digital Elevation Models), land use and land cover, soil type and texture, hydrogeology, especially depth of the ground-water table or impervious layers. By overlaying (or geo-referencing) maps of these or other attributes, as listed in the previous chapter, the land surface can be disaggregated into hydrotopes (e.g. Becker et al. 2002; Lahmer

et al. 1999) as is illustrated in the left part of Fig. D.5. How these characteristics and parameter values can be estimated from GIS-based information is described further briefly in Sect. D.2.6.3.

Among the attributes that specifically determine land-surface heterogeneity in terms of patchiness are land use and land cover. They, in addition to water use and management practices, are often subject to changes which include two broad categories:

1. direct changes in land cover due to urbanisation, industrialisation and mining, deforestation or afforestation;

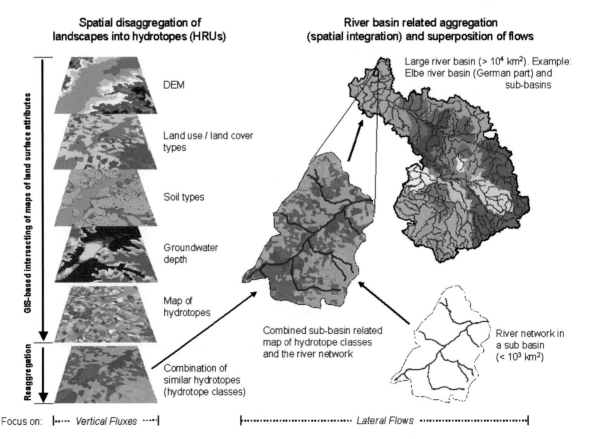

Fig. D.5. GIS-based disaggregation, re-aggregation and integration in river basin-related hydrological and ecological studies

2. agricultural management practices, including crop rotation, application of fertilisers and pesticides, tillage, soil conservation measures, ecological farming, animal husbandry and grazing, irrigation and drainage.

In the case of irrigation, which is often combined with surface water diversions and/or groundwater abstractions from elsewhere, water-saturated and wet areas may occur even in climatically dry periods. On the other hand, artificial drainage by ditch systems and/or pipes in the soil always generate drier and at least temporarily "dried out" areas.

These changes generally propagate from the patch, hydrotope, or hillslope scale, for example from the non-irrigated and irrigated farmlands in Fig. D.3. The distinctive hydrological behaviour of contrasting land areal units and management techniques thus makes it imperative to disaggregate (discretise) the land surface into hydrotopes or similar small areal units and to apply high resolution distributed models, as discussed in Sect. D.2.6.

D.2.2.3 Runoff Generation and Runoff Components

Runoff is generated at a hydrotope whenever "excess water" occurs as the difference between interval rainfall and/or snowmelt minus evapotranspiration and soil water recharge (1 and 13, 5, 8 and 9, as well as 6 in Fig. D.2; right lower block in Fig. D.1). Lateral flows are then generated through one or more of the following flow pathways (Buttle 1998; Becker et al. 1999a):

- land-surface runoff (overland flow RO = rainfall or snowmelt intensity minus infiltration capacity; 1, 13, 4 and 10 in Fig. D.2);
- interflow (subsurface stormflow RI; 11 in Fig. D.2); and,
- baseflow (through groundwater recharge (12 in Fig. D.2), which becomes groundwater storage and thus generates increases in baseflow RG).

These flows are first routed downslope through the runoff generating subcatchments along different surface and subsurface pathways to the nearest channel, then downstream along the river network and in the end constitute, in aggregate, the total basin outflow or "basin runoff" (discharge) in the final channel cross-section of the river basin under consideration (Fig. D.1, lower right block; Becker et al. 2002). This process includes a conceptual change from the primary disaggregation scheme of hydrotopes (or rasters of a regular grid) useful for the analysis of vertical processes (left in Fig. D.5), to one involving the integration of lateral flows from various contributing areas in the river network, the geography of which is defined by rivers, river sections, lakes, aquifers and wetlands. This is illustrated schemati-

cally on the right of Fig. D.5 in terms of re-aggregation, spatial integration and superposition.

It should be emphasized that among all hydrological processes runoff generation in catchments is most variable in space and with time, depending on the combinations of three main controlling factors: (1) climate, (2) soil and geology and (3) vegetation. The combination of these three factors determines the amounts, and thus the relative streamflow contributions, of surface and subsurface runoff which may differ considerably (Buttle 1998). A brief summary of these runoff components is given in Table D.1, with reference to Fig. D.6. More information about subsurface stormflow processes and the displacement of "old" pre-event water by "new" event water is provided in Sect. D.2.5.2.

In arid and semi-arid environments, infiltration excess overland flow (also called Hortonian overland flow from Horton's pioneering work in 1933) is the dominating runoff component (for comparison see RO_{Hor} in Fig. D.6). This is so because high intensity rain storms represent the most important runoff generating events, with rainfall intensities generally exceeding the soil infiltration capacity. A special type of this direct surface runoff generation is from impervious areas, such as bare rock (RO_{imp} in Fig. D.6) and sealed areas (built up, paved, etc.).

In more humid climates, RO_{Hor} is less prevalent and saturation excess overland flow dominates (RO_{sat} in Fig. D.6). It is generated over areas of surface saturation, which occur predominantly in near-stream zones and on shallow soils due to rising groundwater tables. This is illustrated in Fig. D.6 by the dashed line in the valley floor aquifer, which temporarily intersects the soil surface and thus produces dynamically growing saturated areas in the riparian zone (AN) during heavy or long-lasting rainfall and snowmelt. These areas generate not only saturation excess overland flow, but also an increase in subsurface stormflow (RN) into the channel. Dunne and Black (1970) showed how this type of direct rapid surface runoff (RO_{sat}) into the channel is produced on the time-scale of events, and McDonnell et al. (1999) argued that these saturated areas seem to scale directly with catchment area since topographic gradient decreases as basin scale increases. Consequently, saturation excess overland flow (RO_{sat}) is a main runoff producing mechanism across scales, but plays an increasing role over larger spatial domains. The variable source area concept of Hewlett and Hibbert (1967) can be considered as the best formulation of this kind of catchment-scale runoff generation process.

In steeper terrain, near-stream contributions are matched (or in mountainous terrain exceeded) by rapid subsurface stormflow in the form of transmissivity feedback, lateral preferential flow (e.g. through macropores and pipes and highly permeable layers: RI_1 in Fig. D.6), or pressure wave translatory flow, including piston flow (RI_2 in the figure) and groundwater ridging (RI_3) (Buttle

Fig. D.6.
Representation of a valley
cross-section indicating
(*i*) typical landscape sub-units
similar in their runoff generation and evaporation behaviour and (*ii*) preferred runoff
generation areas of the different flow components (*GW*:
groundwater; other abbrevia-
tions see Table D.1)

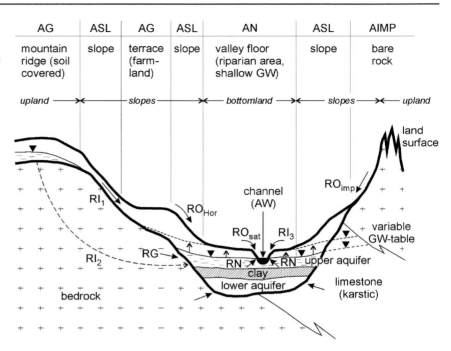

1998; Bonell 1998). For a more detailed treatment see Sect. D.2.5.1. In steep humid areas where subsurface stormflow is the dominant source of storm runoff into the channel, this may be viewed as a subsurface satu-rated area, variable source area process.

The third runoff component is baseflow, the long-term, slowly varying runoff-generating mechanism, which is more or less stable even during longer dry periods (RG in Fig. D.6, complemented by RN) (Sklash and Farvolden 1979). It is fed from aquifers in the catchment and some-

Table D.1. Storm runoff components and essential characteristics of landscape sub-units (see Fig. D.6) (Becker et al. 1999a)

Overland flow RO	
RO_{Hor}	Infiltration excess overland flow from soils when rainfall or snowmelt intensity exceeds infiltration capacity ("Horton" flow, high spatial variability). Preferred conditions: bare soil and cropland, esp. in arid and semi-arid regions and high intensity rainstorm events.
RO_{imp}	RO from impervious areas such as bare rocks, sealed areas (paved, built-up, etc.) in all climate zones (nearly constant areal extent). After an initial loss of very few millimetres, RO_{imp} amounts to 100% of rainfall or snowmelt in each event.
RO_{sat}	Saturation excess overland flow ("Dunne" flow) from dynamically varying saturated areas due to rising groundwater tables intersecting the land surface, with RO_{sat} – amounting also nearly equalling to rainfall or snowmelt. Preferred conditions: near-stream riparian areas, flat valleys with gentle concave slopes, and shallow groundwater areas, mainly in humid and semi-humid regions, even with low intensity long lasting rain or snowmelt.
Subsurface stormflow (interflow) occurring as short-term exfiltration of subsurface water to the land surface in depressions or at lower slopes, or directly into channels:	
RI_1	Subsurface stormflow through preferential flow pathways such as macropores, pipes, highly permeable layers, e.g. at the soil bedrock interface, often induced by transmissivity feedback.
RI_2	Piston flow (subsurface pressure wave transmission especially in mountainous terrain).
RI_3	Groundwater ridging (subsurface pressure wave transmission in lowland and riparian zone aquifers).
RN	Direct subsurface flow (quick return base flow) into the channel system from the riparian zone.
RG	Groundwater outflow or long-term baseflow
Typical landscape sub-units (hydrotopes)	
AG	Areas with the groundwater table deep below the surface so that plant roots cannot reach it.
AN	Areas with shallow groundwater table, e.g. wetlands, near-stream riparian areas.
AW	Open water surfaces.
ASL	Slope areas with increased potential for infiltration excess overland flow generation.
AIMP	Impervious or less permeable areas, e.g. uncovered rocks, clay and gleyic soils, sealed areas.

Fig. D.7.
Streamflow hydrograph measured at the outlet of a small
mountainous catchment in the
Harz middle mountain range
in central Germany (Becker
and McDonnell 1998)

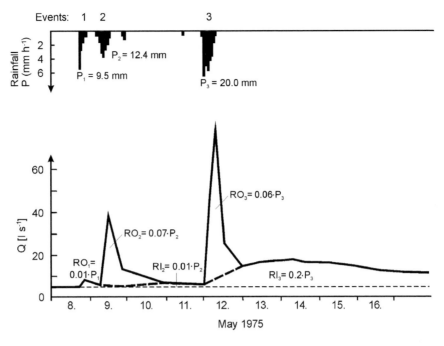

times beyond, i.e. from neighbouring catchments (depending on the location of the subsurface water divides).
As is illustrated in Fig. D.6, RG generally flows first into
the valley floor aquifers and then, by passing through
them, into the river system (in connection with or as RN).

During dry periods when the transpirational water
demand of the vegetation is increasing, water can be supplied from groundwater in the riparian zone (area AN)
or from associated wetlands, as illustrated in Fig. D.6. This
groundwater (RG) may then become re-distributed into
the unsaturated rooting zone of the soil and is then
transpired by the riparian zone vegetation. As has been
shown by Becker and Pfützner (1986) this redistribution
plays an important role in catchments with large lowlands having shallow groundwater. As a consequence large
reductions in baseflow and, hence, in streamflow can be
observed during dry periods within the growing season.

Further aspects of runoff generation and resultant
lateral flows at small catchment scales will be discussed
in Sect. D.2.5 on recent improvements in understanding
these processes. How lateral flows are or may be modified when moving downstream into larger complex river
systems with meanders, inundation plains, lakes and
reservoirs is discussed in Chapt. D.3.

D.2.2.4 Time Behaviour of Runoff Components

The numerous flow components as characterised before
are not all considered explicitly in runoff modelling.
Normally, at the catchment scale, only the three main
runoff components are described (Becker et al. 1999a; see
Fig. D.6): (*i*) surface flow (RO; overland flow), (*ii*) "interflow" (RI; considered as the superimposed subsurface

stormflow in different components as discussed before),
and (*iii*) baseflow (RG).

These three components differ characteristically in
their average *response time* T_r, i.e. in the time between a
runoff generating event, for example a rainstorm, and
the corresponding increase in streamflow (Dyck and
Peschke 1995). T_r is correctly defined as the first moment
of the component-related response function $h(t)$. It is
slightly longer than the time to peak flow. In the streamflow hydrograph in Fig. D.7, recorded at the outlet of the
1.6 km^2 mountainous Schaefertal catchment in the Harz
mountains in central Germany, T_r was estimated to be
about 2–3 hours for the surface flow component RO, and
about 2 days (50 h) for the interflow component RI (third
peak RI$_3$ in Fig. D.7; Becker 1989; Becker and McDonnell
1998). The RO responses to the three rainfall events (P_1,
P_2, P_3) are clearly visible in the hydrograph after each
rainfall. They amount to 1%, 7%, and 6%, respectively of
rainfall (event water). The interflow response is well-re-
flected only after the more intense third rainfall event P_3.
It amounts here to 2% of P_3. The recorded high peaks
marked by RO represent pulse response functions of the
surface flow system (RO) in the catchment. In the case of
RO$_3$ it is that of a 5 h pulse input. From it the Instantaneous Unit Hydrograph (IUH) can easily be derived.

From hydrograph analyses of this type and from various other modelling studies, T_r was found to be in the
ranges as listed in Table D.2 (central column). These response times and the related response functions are essential for describing streamflow and its temporal variation, especially during flood periods.

In addition, average *transit or travel times* T_t and corresponding residence times of the moving water molecules and associated solutes and substances are of in-

Table D.2.
Response times T_r and transit times T_t of the three main runoff components for the example shown in Fig. D.7

Runoff component	Response time T_r	Transit time T_t
Surface flow	<1 day down to <1 hour	Nearly equal T_r
Interflow	About 1 day up to a few weeks	Some weeks up to >1 year
Baseflow	About 100 days up to >1 year	Several years and more

terest. They are accordingly defined as the 1st moment of the "age spectrum" $g(t)$ of these molecules. An example of such an age spectrum is represented and discussed in Sect. D.2.5.2 (Fig. D.16). Field investigations, especially tracer studies, have shown that for the two subsurface flow components T_t is larger than T_r by at least one order of magnitude (e.g. Sklash and Farvolden 1979; Kendall and McDonnell 1998; McDonnell et al. 1999). This is expressed by the generalised ranges of T_t listed in the right column of Table D.2. Some discussion can be found in Sect. D.2.5.2. T_t needs to be known for specific water quality, hydroecological and environmental impact assessment studies.

Despite our capacity to assess many of the surface and subsurface flow components, deficiencies still exist concerning their generation, pathways and time-related behaviours. This requires further research, especially in small catchments, which should include experimental studies to improve the understanding of the mechanisms controlling runoff generation.

Mike Bonell

D.2.2.5 Unresolved Understanding of Processes of Subsurface Flow and Limitations in Assessing the Controlling Subsurface Characteristics and Parameters

The introduction into runoff processes and runoff components in the previous chapter represents a brief overview of our recent general understanding (Fig. D.6 and Table D.1). A principal obstacle to the characterisation of subsurface flows is that they are not visible and thus cannot be observed directly. Thus for lateral storm pathways our understanding is still rather limited, in particular in terms of

- the special pathways of these flows, which at the catchment scale often lead through a sequence of different flow systems (soil matrix, macropores, highly permeable soil horizons, "flow pipes", aquifers and other local or large groundwater systems);
- the associated different forms of water movement, i.e. flow of water particles, on the one hand, combined, at least in some systems (soil matrix, valley and bedrock aquifers …), with "pressure wave transmission" inducing the displacement of "old" water by "new" (event) water, and thus causing the observed remarkable differences between measured subsurface flow

response times and transit times in catchments, as listed in Table D.2;
- the resulting overall (integrated) transit times of water molecules through the catchment under study; and,
- the determination or derivation of catchment or subsystem related response and transit times and related model parameters from measurable subsurface characteristics.

Some of these aspects need to be discussed in the following for the development of a generally useable "bottom-up" approach to determine or derive catchment related response or transit times or related model parameters from measurable subsurface characteristics (e.g. soil and bedrock hydraulic and other properties). Such times and parameters have been and still need to be determined indirectly either from flow records (response times, mainly through model calibration and validation) or by rather expensive and laborious tracer measurements (transit times), the application of which is often limited to smaller catchments. Reference to these types of measurements and their analysis will be made in Chapt. D.2.5.

Both, response and transit times, are directly dependent on the permeability features of soil and rock hydraulic characteristics. It is generally agreed that field saturated (satiated) hydraulic conductivity, K^* (where K^* maybe as low as $0.5\,K_{sat}$, as defined by Bouwer 1966) is most essential here. However, K^* is also understood to be one of the most sensitive and problematic parameters in physically-based modelling (Freeze 1972; Beven et al. 1995; Bonell 1998a). The principal issue is the representation of macropores and pipes (using the specifications of Beven and Germann 1982) which are now recognised as means for rapid transfer of subsurface stormflow. There is a mismatch in terms of what available *in situ* field methods are measuring, which are biased towards matrix K^* mainly because of the small areas or volumes being sampled, and the K^* required for modelling at larger scales. By using environmental tracers (e.g. Bazemore et al. 1994; Bonell et al. 1998) and artificial ones (e.g. Lange et al. 1996) K^* values are determined for the transit of subsurface stormflow, which are up to three orders of magnitude higher in comparison with *in situ* measurements (Bonell 1998a).

Also evident from the use of tracers is that during and immediately after storms only a small proportion of the total available porosity (and corresponding available water storage) participates and drives the hillslope transmission of subsurface flow (Bronswijk et al. 1995; Lange et al. 1996). An additional problem is that knowledge of these subsurface networks of lateral transfer, most notably pipes, is virtually unknown (O'Loughlin et al. 1989; Vertessy and Elsenbeer 1999). A principal message is that one cannot import into models soil core K_{sat} estimates as a substitute for representing this parameter at the hillslope scale, let alone at very small catchment scales of about 1–2 ha (Molicova et al. 1997; Bonell 1998b). The lack of a macropore flow algorithm associated with subsurface stormflow and the invalidity of Richards' equation under transient saturation where macropore flow is the principal conveying mechanism, represents a major obstacle which produces apparent discontinuities in the hydraulic properties near saturation and thus need to be overcome (van Genuchten and Leij 1992).

The problems of characterising soil hydraulic properties apply also to the near-surface soil horizons linked with infiltration excess (Hortonian) and saturation excess (Dunne) overland flow. This has been confirmed in the various field measurement and modelling studies of Loague and co-workers on the R-5 catchment (0.1 km^2) near Chickasha, Oklahoma, in rolling prairie grassland (Loague and Freeze 1985; Loague and Kyriakidis 1997). Their conclusions warrant some attention as they are contrary to expectation: in spite of having values for permeability or field saturated hydraulic conductivity for every element of 1 m^2 for a grid size of 100 m^2, the results of stochastic-conditional simulations proved to be better than modelling with estimates of K^* from such a high resolution of field measurement. Furthermore, this very detailed spatial variability of the surface K^* could not take into account runoff-runon (reinfiltration effects). And finally, the same higher resolution of measurements caused a marginally lower overall mean K^*, thus causing an underestimation of overland flow and the total storm runoff in the hydrographs. Thus, despite the remarkably high resolution of field parameterisation, the performance of their model "is still fairly wretched", because it did not capture the physics of the storm runoff generation process sufficiently well (Loague and Kyriakidis 1997). An additional factor is that the scale of measurements taken could have been below the repetitive unit of Bear (1979) or the minimum length scale of observation to accommodate spatial variabilty of soil hydraulic properties.

Work elsewhere in the open eucalypt woodland of tropical semi-arid Queensland in Australia (Bonell and Williams 1986a,b; Williams and Bonell 1988) also illustrated the considerable temporal as well as spatial variability of K^* in response to the alteration of the surface through compaction and sealing by raindrop impact; and then breakage of the surface seal by dessication from the sun's heating coupled with biological activity (ants, termites).

Considering these and other similar results Bronstert (1999) realistically states that in most situations the observed sensitivity of K^* (to small changes in soil conditions) is "not a model artefact but a representation of the variability and stochasticity of the natural conditions at the hillslope and microcatchment" scale and that "… small variations in state variables cause large changes in runoff generation".

Accordingly, Bronstert proposed and applied a double porosity approach (macropores and matrix) in the parameterisation of hydraulic conductivities for use in his hillslope model HILLFLOW, rather than a single porous medium as adopted in other models (e.g. TOPOG). The problematic representation of K^* such as in TOPOG (O'Loughlin et al. 1989) during the prediction of catchment water yield causes modellers to inflate their K_{sat} or K^* estimates obtained from laboratory or field techniques. During the interstorm periods, however, these inflated estimates cause matrix percolation velocity to be much higher than in reality. Consequently, predicted soil moisture recessions grossly underestimate those measured in the field. In contrast, the high macroporosity and moderate matrix conductivity of HILLFLOW simulated soil moisture recession much better. Considering problems in the direct estimation of catchment related K_{sat} values described above, calibration techniques are still of special interest.

As indicated by Bonell (1998), the dynamic response characteristics of storm hydrograph modelling using hybrid metric-conceptual models (e.g. IHACRES; Jakeman et al. 1993; Jakeman and Hornberger 1993) such as hydrograph recession time constants could provide a catchment integrated K^* estimate if compared with independent hillslope estimates using experimental designs at the hillslope scale (e.g. cascade troughs, Bonell and Williams 1986a; e.g. hillslope pulse-wave experiment; Chappell et al. 1998), and standard pumping test methodologies associated with hydrology. This is a fertile area for future field experimental hydrology linked with model testing. The objective here is to establish closer links between hy-drograph parameters and independent field measurements.

Another critical issue is the understanding developed mainly from tracer studies that the storm hydrograph is generally dominated by pre-existing or "old" water (pre-event water). This has been observed especially in catchments in the humid temperate latitudes and it has raised questions on the source and on identifying the mechanisms for displacing old water from catchments (see Chapt. D.2.5). It is understood that deeper groundwater, including that from fractured bedrock, plays a role (Montgomery et al. 1997; Anderson et al. 1997; Wilson et al. 1991a,b; Durand et al. 1992) although most attention in hillslope hydrometric studies have been focused on the role of the shallow subsurface stormflow. It can be concluded that hillslope hydrology studies should always be coupled with complementary hydrogeological studies in order to assess and quantify the role of bedrock groundwater contributions to the storm hydrograph. A detailed review of the various mechanisms on the displacement of old (pre-event) water by new (event) water has been provided by Bonell (1998): an extract is included in Chapt. D.2.5.

Nevertheless, many questions remain as key research issues in hillslope hydrology, in particular the precise role of pressure waves leading to displacement of pre-existing soil water vis-à-vis the physical propagation of subsurface water by preferential flow (both of which can occur during transient saturation of porous media within storm events).

Reinder A. Feddes

D.2.3 The Unsaturated Zone and Its Interaction with the Atmosphere through the Biosphere

D.2.3.1 The Role of Soil Moisture in Coupled Land-surface/Atmosphere Modelling

The atmospheric modelling community is not entirely confident that the role of land surface is important in defining the character of weather and climate. However, work from the ABRACOS study (see Gash and Nobre 1997, as well as Chapt. A.6 and B.2) and the intercomparison of different Land Surface Parameterization Schemes (LSS) in the Project for Intercomparison of Land-surface Parameterization Schemes (PILPS; see review of Pitman et al. 1999 and Chapt. A.2.3) is beginning to provide evidence that the land surface directly affects the atmospheric boundary layer. Deforestation experiments show that the climate at regional scales is affected, with perturbations leading to geographically remote changes in temperature and precipitation via atmospheric teleconnections (see also Kleidon and Heimann 2000). Climate-system models also demonstrate that land-cover changes during the past 21 000 years have amplified climate variations regionally and even globally (e.g. Claussen and Gayler 1997; Kubatzki and Claussen 1998).

Soil moisture, liquid or frozen, is important in climate and climate change (IPCC 1995), as it represents an essential part of the memory of the land-atmosphere system. In atmospheric model simulations, anomalies in soil moisture affect the precipitation, temperature and motion fields of the atmosphere at regional and global scales. Correct partitioning of precipitation between evaporation and runoff (including surface runoff and root zone drainage) is an important reason for including sophisticated Land Surface Schemes (LSS) in GCMs and Numerical Weather Prediction (NWP) models (see Fig. D.1).

PILPS demonstrated that alternative SVAT/LSS schemes, driven by the same meteorological forcings of air temperature, humidity, wind speed, incoming solar radiation, long-wave radiation and rainfall can produce remarkably different surface energy and water balances (SVAT field experiments are discussed in Sect. A.7.2.9). Most studies in climate modelling have focused on the role of liquid soil moisture, but the role of seasonally frozen soil moisture and its influence on the surface energy balance may also be important, although it has received little attention so far.

There is no controversy about the importance of soil moisture in atmospheric modelling. It controls the partitioning of available energy into latent (evapotranspiration) and sensible heat and thus represents a key for climate modelling. This has been confirmed by retrospective studies using NWPs and GCMs. For example, the initialisation of the soil moisture field strongly influences subsequent precipitation and evaporation (see also Sect. A.4.5).

Given the spatial scales implicit in climate models, any notion of an observable soil moisture or an "average" soil moisture is problematic. A gap therefore exists between what climate modellers mean by soil moisture and what the observational community means by the same term. This is limiting our progress towards better models of large-scale soil moisture. Modellers need a soil moisture definition which can be used in a hierarchy of models and which is actually translatable into an observable quantity. In addition, land-surface modellers want to compare their simulations to observed soil moisture, which can only be done if the modelled and observed soil moisture are representative of similar spatial scales. Observed soil moisture representative of a point cannot be compared to soil moisture simulated by a land-surface scheme. One key product needed by the community is a clear conversion from climate-model soil moisture into measureable soil moisture. It is not clear how to do this currently, but it is a key theme for future research.

Only microwave satellite remote sensing with its potential global coverage has the potential to provide the necessary soil moisture information for use in hydrological models to assess water resources from local up to the continental scale. Active microwave remote sensing of soil moisture relies on the large contrasts between the dielectrical constants of water and dry soil, and provides areal information on soil moisture content with depth and its variability. Passive microwave radiometers measure the thermal emission from the surface. However, the different parts of the electro-magnetic spectrum, reflected or emitted by the Earth's surface, as measured by satellite sensors, are often difficult to interpret. Therefore, intermediate algorithms are required. These algorithms, unfortunately introduce additional sources of error and need further refinement if we are ever to realise the potential of remote sensing (Stewart et al. 1998).

Many hydrological models were not designed to assimilate the type of spatial data provided by remote sensing. For example, even spatially-distributed hydrological models make essentially calculations at a single point, having averaged multi-point data such as for rainfall. At present, the accuracy of variables/parameters derived from remote sensing data is often insufficient. Even if the accuracy of remote sensing products were to be good enough for use with large-scale models, there are still problems in visible band sensors with the temporal availability of the data, e.g. because of infrequent sampling or presence of clouds.

D.2.3.2 Soil Water Flow and Root Water Uptake at the Field Scale

Soil water uptake by plants through their root systems represents a key link between the land surface and atmosphere. This includes linkages between the soil (where water and nutrients reside) and the organs and tissues of the plants, where these resources are used by the canopy. Fluxes along the soil-plant-atmosphere continuum are regulated by above-ground plant structures and properties, like the leaf stomata, which can regulate plant transpiration when interacting with the atmosphere. Fluxes are also controlled by the below-ground plant properties depth, distribution and activity of roots, and the soil properties water potential, water content and hydraulic conductivity (Jackson et al. 2000).

Plant root systems show a remarkable ability to vary over soil depth and to adapt to changes in the availability of water and chemical properties (e.g. nutrients, salinity) in soils. Root response to soil properties, in turn, affects the uptake of soil water and nutrients and the storage of carbon below ground. Root distribution may change when ecosystems respond to greenhouse warming and carbon dioxide fertilisation. For example, at higher atmospheric CO_2 concentration, stomata can contract somewhat for a given influx of CO_2. Transpiration thus decreases, and, coupled with generally higher photosynthesis in higher CO_2 concentrations, water-use efficiency can increase dramatically (Field et al. 1995). Increased water-use efficiency will potentially feed back to changes in root characteristics, with the possibility of further, substantial changes in the local water (and energy) balances. Exploration of such feedbacks has only just begun.

We explore now how information about plant root characteristics such as rooting depth, distribution and function has been and could be used in land-surface modelling from the perspective of improving our depiction of the hydrological cycle and climate. *Within this context, one objective is to explore the level of detail necessary to properly parameterise models of water and energy flux on local, regional and global scales.* As a step toward this goal we examine the existing databases on plant rooting depth, distribution and dynamic water uptake behaviour and some of the key models which use this information.

Any reasonably accurate mathematical description of water uptake by vegetation with heterogeneous soil and root properties is complicated. One promising approach, however, is to regard the root system as a diffuse sink which penetrates each depth layer of soil uniformly, though not necessarily with a constant strength throughout the root zone as a whole. Water balance considerations of an infinitely small soil volume result in the continuity equation for soil water:

$$\frac{\partial \theta}{\partial t} = -\frac{\partial q}{\partial z} - S \tag{D.1}$$

where θ is the soil water content ($cm^3 cm^{-3}$), t is time (d), z is the vertical coordinate (cm) taken positively upward, q is the soil water flux density ($cm\ d^{-1}$) and S is the actual root water uptake rate ($cm^3 cm^{-3} d^{-1}$).

Darcy's equation is used to quantify the soil water fluxes, which for one-dimensional vertical flow can be written as:

$$q = -K(h)\frac{\partial(h+z)}{\partial z} \tag{D.2}$$

where K is the hydraulic conductivity ($cm\ d^{-1}$) and h is the soil water pressure head (cm).

Combination of Eq. D.1 and D.2 results in the well-known and widely applied Richards' equation:

$$\frac{\partial \theta}{\partial t} = C(h)\frac{\partial h}{\partial t} = \frac{\partial\left[K(h)\left(\frac{\partial h}{\partial z}+1\right)\right]}{\partial z} - S(z) \tag{D.3}$$

where C is the differential water capacity ($d\theta/dh$) (cm^{-1}), being the slope of the soil moisture characteristic $h(\theta)$. Van Genuchten (1980) has provided analytical expressions for the strongly non-linear hydraulic functions $h(\theta)$ and $K(h)$.

Under *optimal* soil moisture conditions the maximum possible root water extraction rate, $S_p(z)$, integrated over the rooting depth, is equal to the potential transpiration rate, T_p ($cm\ d^{-1}$), which is governed by atmospheric conditions. $S_p(z)$ may be determined by the root length density, $L_{root}(z)$ (cm root length cm^{-3} soil). This root length density at depth z is taken as fraction of the total root length density over the rooting depth D_{root} (cm) according to:

$$S_p(z) = \frac{L_{root}(z)}{\int_{-D_{root}}^{0} L_{root}(z)\partial z} T_p \tag{D.4}$$

Stresses due to overly dry or wet conditions and/or high salinity concentrations may reduce $S_p(z)$. The *water stress* may still be described by, for example, the function proposed by Feddes et al. (1978) and, for *salinity stress*, the response function as suggested by Maas and Hoffman (1977).

In order to simplify parameter calibration and use of existing experimental data, one may assume the water and salinity stress to be multiplicative. This means that the *actual* root water flux density, $S(z)$ (d^{-1}), can be calculated from the expression:

$$S(z) = \alpha_{rw} \alpha_{rs} S_p(z) \qquad (D.5)$$

where α_{rw} (–) and α_{rs} (–) are the reduction factors due to water and salinity stresses, respectively. Integration of $S(z)$ over the rooting depth then yields the actual transpiration rate T.

To obtain a solution of Eq. D.3 one has to supplement it with conditions for the initial situation and for the top and bottom boundaries of the flow system. At the top, apart from the governing meteorological conditions, the crop/natural vegetation plays a dominant role in the partitioning of the various fluxes. Hence for the assessment of land-surface properties like leaf area index, potential transpiration, potential soil evaporation and rainfall interception, one needs a coupling of the soil water balance model with a daily crop/vegetation growth model (see e.g. Spitters et al. 1989). Only in this way, can a proper prediction of crop/vegetation development and growth in dependency of the actual prevailing soil conditions be obtained. In current state-of-the-art simulation models of soil and transport processes, this feedback is often missing.

The bottom boundary of the flow system may be situated either in the unsaturated zone or in the saturated zone. At such a lower boundary, one can specify either a pressure head, a flux, or a relation between the two (e.g. Belmans et al. 1983). With the lower boundary condition the connection with the saturated zone can be established. The coupling between the two systems is possible by considering the phreatic surface as an internal moving boundary. One-dimensional root system models, however, may fail when lateral transports of water by subsurface or overland flow occur, as shown schematically in Fig. D.1. Becker (1995), Hatton et al. (1995), and Kim (1995) show that for catchments with complex sloping terrain and groundwater tables, a vertical domain SVAT has to be coupled with either a process or a statistically based scheme that incorporates lateral transfer (see Sect. D.2.5). A SVAT-model based on Eq. D.3, has been developed by Van Dam and Feddes (2000). For details see Van Dam (2000), as well as the website (*http://www.alterra.nl/models/swap/index.htm*).

A crucial question is now: "*Are land-surface and climate models indeed sensitive to the representation of roots?*" To answer this question, one can apply the SWAP model of Van Dam and Feddes (2000) to evaluate the effect of root distribution on the course of actual transpiration in time. As an *example of one application* we take a grass vegetation (covering the soil completely) with a rooting depth of 80 cm growing on a loamy sand (containing 10% clay) of 2 m depth. At the bottom boundary, a free drainage is assumed to prevail. An initial condition throughout the soil profile applies a soil water pressure head of $h = -200$ cm, implying a rather wet soil. At the soil surface, a potential transpiration rate T_p of 4 mm d^{-1} is applied for two different relative root density distributions, for example, Root1 where most of the roots are located in the top soil and Root2 where most of the roots are in the lower soil (Fig. D.8a).

The result of the simulation is shown in Fig. D.8b. Transpiration is more sensitive to the moisture content θ of the densely-rooted soil layer than to that in the remainder of the root zone. Hence Root1 produces an earlier onset of moisture stress than Root2, after 30 days showing an actual transpiration rate that is about half of that of Root2. Similar results were reported by Desborough (1997). *This is a clear demonstration that roots can influence the behaviour of a land-surface model considerably, the role of roots being particularly important when soil moisture limits evapotranspiration.*

A potentially important aspect that occurs in the real world, but that models overlook, is the marked influence that plant roots can have on the distribution and redistribution of soil water via the processes of "*hydraulic lift*". Through hydraulic lift, deep rooted herbs, grasses, shrubs and trees take up water from lower moist soil layers (e.g. close to a groundwater table) and exude that water during the night into the drier upper soil layers and thus provide benefit to themselves and neighbouring plants (e.g. Dawson 1996; Jackson et al. 2000b). For example, sugar maple trees can hydraulically lift 100 litres of water through their root systems into the upper soil layer each night. The result of hydraulic lift is usually a decline in groundwater table depth as well as stream discharge, as compared with vegetation systems where hydraulic lift is absent. Please note that "*capillary rise*", which means upward movement of water from the groundwater table through the soil, is distinctly different from hydraulic lift, where water is transported through the roots!

D.2.3.3 Effects of Frozen Soil Moisture

Another phenomenon often neglected in modelling is the freezing of soil water which seasonally covers up to 35% of the Earth's surface. In addition, the melting of permafrost in tundra regions has been identified as a potentially large source of atmospheric methane as a result of warming in these regions (IPCC 1990). Given

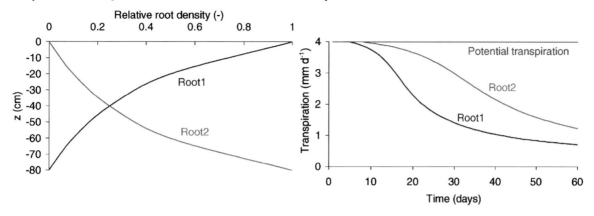

Fig. D.8. a Two different relative root density distribution functions adopted for grass; **b** simulation with the 1D-SWAP model of Van Dam and Feddes (2000) of the effect of these two different root distribution functions on the actual transpiration rate in time, taking $T_p = 4$ mm d^{-1} as upper boundary condition

the lack of validation of land-surface schemes in simulating frozen soil moisture regimes, and the resulting uncertainty underlying simulations of high latitude warming and possible methane release, it is surprising that little attention has been given to this area by the climate modelling community. Where efforts have been made to look at high latitude soil moisture, it is clear that models in their standard mode do not resolve all soil moisture variations in these regions properly. The main impact of frozen soil moisture is probably on runoff, but it is not clear whether the impact would actually affect a climate model simulation significantly. However, the need to simulate pan-Arctic land runoff into the oceans as part of the driving mechanism of deep water formation in the North Atlantic and the thermohaline circulation suggests it is an important variable to simulate accurately.

Models of water infiltration and movement in frozen soils require computationally-expensive, iterative techniques or high-resolution finite difference schemes. This complexity cannot be justified in climate models given the computational expense and general lack of detailed soil-based input information. Given these problems, it is understandable that far simpler models have evolved to represent freezing and thawing.

Qualitatively, it may be adequate to simulate the timing of spring melt and thus the beginning of freshwater runoff into the oceans to a temporal accuracy of a couple of weeks. However, to achieve this temporal accuracy in atmospheric models will likely require a simulation of infrared flux from the atmosphere to a level of accuracy not currently achievable in climate models. There is, then, a need to establish the role played by frozen soil moisture in affecting the timing and magnitude of spring runoff. Effectively, we believe the question "*does frozen soil moisture need to be included in climate models?*" requires a positive answer but the required level of accuracy needs yet to be determined. In any case more attention should be paid to the role of seasonally-frozen soil moisture.

D.2.3.4 Representation of Available Soil and Root Information in Land-surface Models

The sensitivity of the soil water balance to "*plant-available water capacity*" and, by implication, to plant root characteristics more generally, has been addressed by several investigators. Milly (1994) formulated a simple supply-demand-storage model where the sensitivity is a function of the capacity itself such that if capacity is so large as to equalise all temporal variability of water supply and demand, then further increases in capacity have no effect. If capacity is small enough to be fully utilised, then the long-term evaporation/runoff partitioning is very sensitive to capacity. Some of the results of Milly (1994) suggest that a doubling or halving of the available water capacity produces about a 25% decrease or increase of runoff (with compensating changes in evapotranspiration). It was also noted that the actual land system seems to operate close to the point of maximum available water capacity. This was taken as an indication that actual rooting depths reflect ecologically optimised responses to weather and climate variability. A similar near-maximisation of evaporation may occur on the global scale. The results of Milly and Dunne (1994) suggest a strong sensitivity of continental evaporation to water capacity, hence improved root modelling in GCMs could substantially improve soil-vegetation control instead of uncontrolled continental evaporation that extracts water from a deep soil moisture reservoir.

Recent studies (Dunne and Willmott 1996; Jackson et al. 1996) suggest that the *preponderance of surface roots might create effective rooting depths in water budget* models that are much smaller than those typically used in GCMs. Vörösmarty et al. (1998a) tested this notion on several hundred well-monitored catchments in the conterminous USA, and demonstrated that the use of a 0.5 m rooting depth uniformly (in contrast to using rooting depths that for some forest well exceeded 2 m) across all land-cover types and several evapotranspiration functions yielded improved water budget closure.

Kleidon and Heimann (2000) investigated the effects of *larger rooting depths* (i.e. increased soil water storage capacity and availability) on the surface energy balance and the atmosphere using a GCM. They derived a global dataset of deep-rooted vegetation assuming that vegetation adapts to its environment in an optimum way, i.e. maximising net primary productivity. The incorporation of deep-rooted vegetation into the GCM led to large-scale differences in the simulated surface climate and the atmospheric circulation, mostly in the seasonal humid tropics. Hence, deep-rooted vegetation seems to form an important part of the tropical climate system. This conclusion appears to be true for some temperate forest ecosystems as well (Dawson 1996).

Finally, Dirmeyer et al. (2000) found that most of the sensitivity of surface evaporative fluxes to soil moisture exists when soil wetness is low, and roots greatly constrain transpiration by conveying soil moisture stress to the plants. The parameterised roots in some LSSs (e.g. SSiB, Dirmeyer and Zeng 1997; BATS, Dickinson et al. 1993; Mosaic, Koster and Suarez 1992) appear to be quite efficient at extracting moisture from the soil and maintaining steady transpiration rates for moderately wet and wet soils, but restricting water fluxes when soil wetness is low.

Concerning *root density* De Rosnay and Polcher (1998) have explicitly taken into account sub-grid scale variability of vegetation and root profiles in computing root water uptake in their LSS SECHIBA, which receives climatic forcing from the GCM. Normalised (compare with Eq. D.4) root density

$$L'_{root} = \frac{L_{root}(z)}{\int_{-D_{root}}^{0} L_{root}(z)\partial z}$$

is here assumed to depend exponentially on soil depth as

$$L'_{root} = e^{cz} \tag{D.6}$$

where c is a fitting constant depending on the biome considered and z is taken positively upwards (see Fig. D.9).

The use of a relatively easily observable parameter such as c makes the development of a global dataset for root water uptake parameterisation feasible. Taking into account root profiles (and not only rooting depth) the authors conclude that it improves the representation of

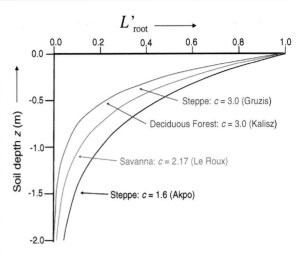

Fig. D.9. Observed normalised root density profiles $L'_{root}(z)$ depending on the biome-specific constant c (after De Rosnay and Polcher 1998)

transpiration and of continental evaporation by the soil-plant system. In a new version of the SECHIBA model (De Rosnay et al. 2000), Equations D.1–D.3 are combined with the macroscopic root modelling approach of Eq. D.4 and specification of subgrid scale variability of the surface, which is new for a GCM. This opens a wide range of possibilities for future applications in large-scale modelling, including simulating groundwater flow and soil-plant-atmosphere interactions at various scales. It also permits representation of root water uptake at different soil depths depending on the seasons.

Finally, Hallgren and Pitman (2000), who performed a sensitivity assessment of the BIOME3 model, found that, if root distribution in the model was varied within observational uncertainty, the model predicted significantly different distributions of plant functional types and that root distribution was one of the most important parameters within that model.

Because of the sensitivity of simulated transpiration in climate models, *global datasets of root and soil properties* are increasingly needed. The depth at which plants are able to grow roots has important implications for the hydrological balance, as well as for carbon and nutrient cycling. Canadell et al. (1996) show that across biomes the maximum rooting depth is 7.0 ± 1.2 m for trees, 5.1 ± 0.8 m for shrubs and 2.6 ± 0.1 m for herbaceous plants. Hence, deep root habit is far more prevalent than the traditionally held view would have it. Cumulative root distribution Y from the soil surface down to rooting depth d, here defined as the depth where Y reaches an arbitrary value (e.g. 99%) can for various biomes be fitted with a vegetation-dependent coefficient β to the following asymptotic equation (Gale and Grigal 1987; Jackson et al. 1996, 1997; Zeng 2001):

$$Y = 1 - \beta^d \tag{D.7}$$

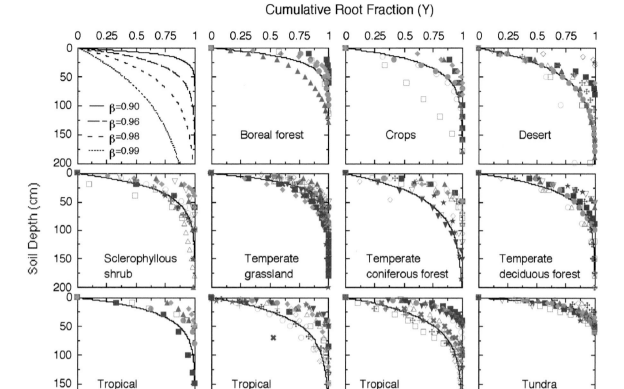

Fig. D.10. Cumulative root fraction Y as a function of soil depth d for 11 terrestrial biomes according to Eq. D.7. The curve in each biome panel is the least squares fit for all studies with data to at least 1 m depth in the soil (after Jackson et al. 1996)

Values for β, and properties like fine/total root biomass ratio, root length, maximum rooting depth, root/shoot ratio and nutrient content of different terrestrial biomes (Fig. D.10) can be found in the above cited literature references.

To date, a root database (Jackson et al. 2000a,b) of more than 1000 profiles exists that covers various combinations of maximum rooting depth of fine and coarse roots, root length densities, root biomass (and surface area in a small subset of the data), as well as root nutrient concentrations by biome and plant life form (e.g. Jackson et al. 1996, 1997) There is, however, a lack of information on annual crops within the database. This root database currently has no spatial expression. Work is however underway to implement the data in a spatially explicit manner at 0.5° × 0.5° grid scales for use in climate models. This could be done by using the information within the root database to characterise the root profiles of the 12 major biomes identified within the global ISLSCP (International Satellite Land Surface Climatology Project) spatial dataset which will have a resolution of 0.5° square grid cells. This represents a first step in providing spatial root profile data at a global scale, but it is also recognised that further field observations

of root profiles within particular, as yet unrepresented, vegetation types may be necessary to improve our overall understanding of the importance of contrasting root profiles.

Global-scale spatial datasets of soil types and their allied soil hydrological properties have existed for many years. A number of regional and global soil databases relevant to global change research are currently available (see Table C.2).

Within the modelling community however, there is an arguable *over-reliance on the empirical correlation between soil texture and soil hydrological properties* for derivation of model parameters. Although such relationships can be demonstrated and used (for example, Wösten et al. 1999) they generally lack a high degree of statistical confidence. The use of only soil texture inevitably leads to a high degree of variability in soil hydrological properties within each textural class. This variability can be reduced by considering other basic soil factors such as organic matter content, soil type and pedogenesis in determining soil porosity which, in turn, should lead to improvements in model parameterisation.

It is also important to note that as soil type (taxonomic unit) and soil characteristics (e.g. waterlogging or the presence of mechanically impeding layers) often modify the form of the plant root profile, there is a need to maintain referential integrity between the biome type and the soil taxonomic unit. There is a substantial risk of providing inappropriate or highly unlikely combinations of soil type and biome type as a consequence of overlaying these as two independent spatial datasets. Thus prior to providing global-scale rooting parameters and soil hydrological data, there must be a process of validation to ensure that the dataset provides realistic estimates.

Mesoscale parameterisation of soil hydraulic characteristics for application of Eq. D.3 in mesoscale hydrological modelling (e.g. Kabat et al. 1997) is gaining increasing importance. This is mainly due to the need to introduce at this scale plant-interactive soil water depletion and stomatal conductance parameterisations and to improve the calculation of deep percolation and runoff. Covering a grid of several hundreds of square kilometres, Richards' parameterisation in such SVAT schemes is assumed to be scale-invariant. The parameters describing the non-linear, area-averaged soil hydraulic functions in this scale-invariant equation are being treated as calibration parameters, which do not necessarily have a physical meaning. The saturated hydraulic conductivity is one of the soil parameters to which the models show very high sensitivity. It is shown that saturated hydraulic conductivity can be scaled in both vertical and horizontal directions for large flow domains.

Generally, a distinction is made between *effective* and *aggregated* soil parameters. Effective parameters are defined as area-averaged values or distributions over a domain with a single, distinct textural soil type. These can be obtained by "similar media" scaling or inverse modelling. The effective behaviour of the reference soil water retention and soil hydraulic conductivity curves can be obtained from similar media scaling. To obtain effective mesoscale soil hydraulic parameters there is a considerable potential in combining large-scale inverse modelling of unsaturated flow in combination with *remotely sensed* areal evapotranspiration and areal surface soil moisture (e.g. Feddes et al. 1993a,b). The *inverse* technique is found to provide effective soil parameters which perform well in predicting both the area-averaged evaporation and the area-averaged soil moisture fluxes, such as percolation.

Aggregated soil parameters represent grid-domains *with several textural soil types*. In soil science dimensional methods have been developed to scale up soil hydraulic characteristics. With some specific assumptions, these techniques can be extrapolated from classical field-scale problems in soil heterogeneity to larger domains, compatible with the grid-size of large-scale models. Aggregated soil parameters obtained from regression relationships between soil textural composition and hydraulic characteristics, predict evaporation fluxes well, but fail to predict soil water balance terms such as percolation and runoff. This is a serious drawback that could eventually hamper improvement in the representation of the hydrological cycle in mesoscale atmospheric models and in GCMs.

D.2.3.5 Conclusions

Putting soils and roots in a wider global change context poses questions as to predictions of changes in roots and root water uptake due to land-use and climate change. A research strategy for modelling root water uptake and the consolidation of root datasets might be possible through a combination of approaches (Feddes et al. 2001).

From this review, we suggest that a high priority is to *refocus efforts from vertical complexity in LSS towards horizontal complexity and the parameterisation of sub-grid scale heterogeneity.*

What is the best approach to parameterise and model root water uptake in hydrological, climate and weather prediction models? A systematic evaluation of the role of roots and root dynamics in water fluxes between the land surface and the atmosphere is required. A first priority is to firmly establish relationships between root biomass, rooting depth, root distribution and root functions with vegetation type, soil type, soil texture, topography and climate. Synergies are to be gained from a combination of this information.

- Two different modelling approaches should be pursued to improve root water uptake descriptions: increased detail/complexity of existing physically-based models. For this approach it was hypothesised that root water uptake can be modelled when more complete information on vertical distribution of roots and a physical description of root functioning is available;
- keeping root water uptake models as simple as possible (since it is not clear if modelling and prediction improves with more explicit description of roots, soils, and vegetation) and an implicit description of roots which assumes that water in the root zone is available to the plants.

Both approaches are needed to ensure that relevant processes are considered and understood, but that appropriate computational weight is paid to each, depending on its importance. Complexity really means completeness; both in terms of data and in terms of the relevant processes modelled. The goal here is accuracy and a proper scientific understanding of the physical processes. An "optimising systems" perspective could guide the modelling of the vegetation/soil system with respect to the use of available resources, and the degree of detail required.

With respect to the above modelling strategies, data needs from the communities of global climate modelling, numerical weather prediction and mesoscale modelling need to be specified clearly. Potentially relevant root data are: rooting depth (maximum and 90% value), root distribution over depth, root surface area or "active roots", proportion of fine/coarse roots, proportion of live/dead roots, root biomass, and possibly also nutrient content (N, P, K, Ca, Mg) of roots.

This "wish list" needs to be matched with an inventory of existing root data on regional and global scale, including levels of uncertainty in data. Regional priorities for root data are to be established from the climate perspective, like in monsoon regions such as India, or regions of strong climatic gradients or transition zones. Synergies in determining root functionality could be gained by linking existing datasets functionally, e.g. rooting depth and soil texture with vegetation/biome information. Root data collection should include existing field studies. One important initiative that provides data from a network of harmonised measurements is FLUXNET. Currently more than 80 stations in various biomes and climates provide additional site information, e.g. on water use and net primary production relationships for certain plant types – see *http://daacl.ESD.ORNL.Gov/FLUXNET*. More examples of soil data archives can be found in Sect. C.2.2.

Christian Valentin

D.2.4 Overland Flow, Erosion and Associated Sediment and Biogeochemical Transports

Environmental changes are expected to exacerbate current problems on the continental water cycle through changes in climate regimes and changes in land use and land cover. Also changes in atmospheric composition may have some impact on runoff production and soil erosion through changes in vegetation and soil organic matter. Because land degradation is not only a consequence but also a cause of environmental change through the alteration of albedo and evaporation, it is critical to predict the plausible impacts of changing climatic regimes and land use on overland flow, flood generation and erosion.

D.2.4.1 Impact of Climate Change

D.2.4.1.1 *Soil Organic Matter, Aggregate Stability and Crusting*

Changes in soil surface structure play a crucial role in the partition of rainfall into infiltration and runoff (e.g. Valentin 1996) and in the size and the quantities of the erodible particles (e.g. Le Bissonnais 1996). Because aggregate stability is greatly influenced by soil organic

matter and its dynamics, small changes in the supply and mineralisation of organic matter can have a major impact on soil structure and greatly influence soil and hillslope hydrology (e.g. Piccolo and Teshale 1998). Higher CO_2 concentrations stimulate photosynthesis through a "fertilisation" effect (Melillo et al. 1993) which should enhance plant growth, soil organic matter and soil structure. For instance, the soil concentration of the glycoprotein glomalin and thereby soil aggregation has been shown experimentally to increase with atmospheric carbon dioxide (Rillig et al. 1999). Most authors consider, however, that these positive consequences will be notably offset by higher temperature which would intensify the mineralisation of the humified organic carbon, thus enhancing the risk of soil crusting and runoff generation (e.g. Piccolo and Teshale 1998). This effect would be particularly pronounced where increased soil evapotranspiration due to global climate change would largely exceed rainfall. This would lead in semi-arid regions to increased soil organic matter degradation. In West Africa, the data collected along a climatic transect suggest that any decrease in precipitation would induce an increase in the runoff coefficient due to the decline in natural vegetation cover and the change in the dominant crust (Table D.3, Valentin et al. 1994) with a maximum of annual overland production under natural conditions occurring for a mean annual rainfall of about 600 mm (Valentin and d'Herbès 1999).

D.2.4.1.2 *Overland Flow, Flood Generation and Erosion*

Because runoff is threshold-dependent, slight changes in precipitation patterns (amount, frequency, intensity, seasonal distribution) can have large impacts on overland flow. The combined effect of more evaporation and less rainfall would lead to less runoff in some parts of the world as, for example, in Central America. Other regions would face increased runoff, like the monsoon regions of north-east India and south China (Lal 1994). In Europe, climate change scenarios result in a strong north-south gradient of change for runoff (Arnell 1999) although the relative impacts of natural climate variability may be greater than the impact of human-induced climate change (Hulme et al. 1999). The reliability of any estimation of the impact of climate change on local overland flow remains questionable because of the limitations in currently-applied hydrological models and the difficulties of downscaling GCM results to the catchment scale.

Despite these limitations, a number of researchers have attempted to simulate the change in overland flow and erosion in catchments using climatic scenarios in various regions of the world. Chiew et al. (1995) simulated the impact of climate change on soil moisture and runoff in 28 Australian catchments using a hydrological

Table D.3. Variations of main hydrological parameters as influenced by natural surface conditions along a climatic gradient in West Africa (Valentin et al. 1994)

Ecological zone	Guinean	Sudanian	Sahel	Desert
Mean annual rainfall (mm)	>1 200	600 – 1 200	200 – 600	<200
Vegetation cover (%)	>50	> 50	< 50	0
Crust type	No	Structural	Erosion	Gravel
Crdry (%)	4	25	50	83
Crwet (%)	5	30	59	87
Prdry (mm)	30	19	8	4
Prwet (mm)	15	8	4	3
FIR (mm h^{-1})	23	14	6	1

Crdry: runoff coefficient under dry initial conditions, *Crwet:* runoff coefficient under moist initial conditions, *Prdry:* pre-runoff rainfall under dry conditions, *Prwet:* pre-runoff rainfall under moist conditions, *FIR:* final infiltration rate at saturation.

daily rainfall runoff model and five global climate model scenarios. Estimated changes in rainfall by the years 2030 and 2070 were always amplified in runoff, especially in drier catchments. In the wet and temperate catchments, the percentage change in runoff was twice as much as the percentage change in rainfall, while in the arid areas, increases in rainfall enhanced runoff by more than five times the change in rainfall. Compared to precipitation, temperature increases alone had negligible impacts on soil moisture and runoff. Catchments in arid or semi-arid regions are especially sensitive to climate change because the annual runoff is naturally highly variable.

Where annual rainfall is known to have decreased significantly over the last decades (Sahel, Western Australia), no concurrent decrease has been observed in the frequency of extreme events (Albergel 1987; Yu and Neil 1993). Under these conditions, no decrease in the frequency of flooding is to be expected from a reduction of mean annual rainfall. In China, Guo and Ying (1997) used a monthly water balance model to simulate soil moisture and runoff. These authors calculated the change of maximum runoff QMAX under different climatic scenarios including temperature increases of 0.5 and 1.0 °C and three levels of precipitation increase (5, 10 and 15%). In both catchments, runoff was found to be more sensitive to precipitation variation than to temperature increase (Fig. D.11). At 5% significance level and a likely climatic scenario (temperature increase of 1 °C and rainfall increase by 10%), the simulated peak flood discharge at two studied catchments increased by 38 and 47%. Such increases might seriously affect flood protection works and water resources systems.

By contrast, in the Sacramento basin, California, hydrological modelling showed that with 20% increased rainfall and a 4 °C regional temperature rise, summer runoff remained well below normal (Gleick 1987). The crop-growth and farm-scale hydrological model erosion-productivity impact calculator (EPIC; Williams and Singh 1995) was used by Eheart and Tornil (1999) to simulate

the effects of global change on streamflows in midwestern states of USA. Their analysis showed that the coupled effects of decreases in runoff and increases in irrigation might lead to an increase in the annual mean occurrence of low flow from 3 days yr^{-1} to 13 days yr^{-1}, despite the streamflow accretion from groundwater-supplied irrigators. The EPIC model was also run for 100 sites in the US corn belt selected randomly from the National Resources Inventory site (Lee et al. 1996). The results showed that mean water erosion would vary approximately linearly with mean precipitation, with about a 40% change for a 20% change in mean precipitation.

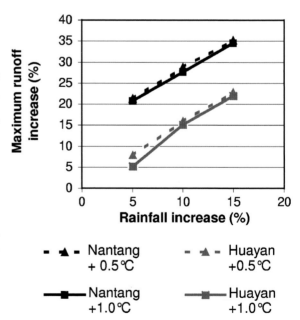

Fig. D.11. Change in maximum runoff (QMAX) under different climatic scenarios including two levels of temperature change (+0.5 °C, +1.0 °C), and three levels of precipitation change (+5%, +10%, +15%) in two Chinese basins. Nantang: area 1 080 km^2, mean annual rainfall 1 631 mm, Guangdong province; Huayuan: area 2 601 km^2, mean annual rainfall 1 114 mm, Hubei province (after the data of Guo and Ying 1997)

In Greece, Panagoulia and Dimou (1997) simulated the impact of changing climate patterns through a set of hypothetical and monthly GISS (Goddard Institute for Space Studies) scenarios of temperature increase coupled with precipitation changes. All climates yielded larger flood volumes and greater mean values for flood peaks associated with precipitation increases. The combination of higher and more frequent flood events could lead obviously to greater risks of flooding. The winter swelling of streamflow could increase erosion of the river bed and banks and hence modify the river profile.

In Belgium, Gellens and Roulin (1999) applied a step conceptual model and the outputs of seven global climate models to eight catchments. For all but two scenarios, the catchments presented an increase in flood frequency. For all the scenarios, catchments with prevailing surface flow experienced an increase in flood frequency during winter months. Similarly, Boorman and Sefton (1997) applied two climate scenarios and eight climate sensitivity tests to three UK catchments using two conceptual hydrological models. They showed that under climate change, floods would increase in magnitude while low flows would be reduced. In England, Roberts (1998) simulated that climate change would cause an increase of runoff from areas receiving high rainfall and a decrease from low rainfall areas. Under a rainfall scenario on the UK South Downs, with a 10% increase in winter rainfall, annual erosion increased by up to 150% (Favis-Mortlock and Boardman 1995). Analysis of synoptic flood generation suggested that any future increase in the frequency and intensity of cyclonic atmospheric circulations might result in a higher frequency of extreme floods and, consequently, increased sediment fluxes (Longfield and Macklin 1999).

In high latitudes and in cold mountainous regions, a warmer climate would result in more precipitation falling as rain rather than as snow. Springtime high water would thus be decreased due to reduced maximum snow water-equivalents (Vehviläinen and Lohvansuu 1991 in Finland; Tanakamaru and Kaodya 1993 in Japan). In these regions with snowmelt-dominated flow regimes, the high flow period is predicted to shift from spring to winter (Lettenmaier et al. 1992). Under climate scenarios and accompanying precipitation changes, the predicted annual runoff in a snow-dominated region in Japan was more sensitive to precipitation changes than to temperature increases (Tanakamaru and Kaodya 1993). In Scandinavian mountains, rainstorm-triggered floods and debris flows in summer and autumn would increase in frequency during climatic warming (Nyberg and Rapp 1998). In subartic Canada, a $2 \times CO_2$ climate warming scenario with an annual temperature increase of 4 °C and no precipitation change indicated lesser snow amounts and a shorter snow cover period. On average, the soil would remain at field capacity for a shorter time (Rouse 1998).

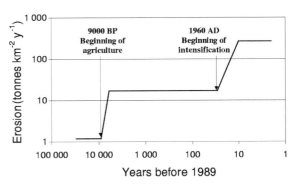

Fig. D.12. Change in erosion rate from 30 000 years B.P. to 1987 in the Kuk catchment (Papua New Guinea), area 6.2 km², mean annual rainfall 2 700 mm (after the data of Hughes et al. 1991)

D.2.4.2 Impact of Land-use Change

More than climate change, land-use changes are expected to reduce the land cover and deplete the soil organic carbon pool, thus enhancing the risk of soil crusting, overland flow and erosion. As a consequence of deforestation, biomass burning, cultivation and overgrazing, there has already been a significant decline in soil carbon content in both temperate and tropical lands associated with standing biomass (Flint and Richards 1991; Walker 1994).

Tropical Forest

Deforestation is broadly considered to be one of the major causes of increased flooding and accelerated erosion. The study of the palaeoenvironmental record extending back more than 30 000 years in Papua New Guinea (Hughes et al. 1991) showed a clear and sharp increase in erosion rates at 9 000 years B.P. due to forest clearance for agriculture (Fig. D.12). Another sharp increase has been observed recently as a result of agricultural intensification. The hydrological responses to deforestation include not only a change in total water yields but also dry weather flow (delayed flow) and storm runoff (quickflow; Bonell and Markham 1998). For instance, Lørup and Hansen (1997) showed that annual runoff from two cultivated catchments in south-western highlands of Tanzania was 30–36% higher than from a similar catchment under evergreen forest but the specific low flows from the cultivated catchments were about 220 and 330% of the corresponding low flows from the forest catchment. Recorded rainstorm hydrographs in the three catchments were similar. In Malaysia, after forest conversion to cocoa and oil palm, water yield from small catchments increased by 157 and 470% during the first years of plantation (Abdul Rahim 1988). The same paired catchment experiment showed a corresponding increase in specific peak dis-

Fig. D.13.
Soil erosion from runoff plots under different land-management practices and slope gradient conditions in five monsoon Asiatic countries (after the data of Sajjapongse and Syers 1995)

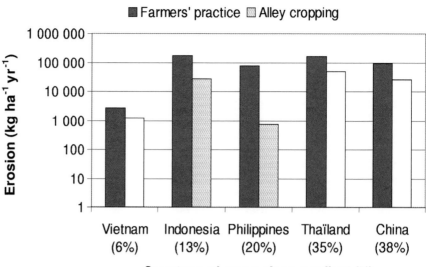

charge of 37% after clear-cutting (Abdul Rahim 1992). While soil erosion from forested catchments remains generally very limited (e.g. 0.55 t ha^{-1} yr^{-1} in French Guyana), it increases dramatically after mechanised clearing (2–8 t ha^{-1} yr^{-1}; Fritsch 1994). Some techniques tend to reduce erosion from cultivated land under monsoon climate. For example, experiments conducted in five Asiatic countries (Sajjapongse and Syers 1995; Fig. D.13) showed that agroforestry can reduce soil erosion significantly as compared to current farming practice.

Semi-arid and Arid

In the tropics, soil organic carbon content has been reported to decline to about 60% of that under natural vegetation after 3–5 years of cultivation for sandy soils and after 5–10 years for finer-textured soils (Feller and Beare 1997). As a result, the proportion of unstable aggregates and the runoff coefficient nearly doubles after 2–5 years of cultivation in the savanna zone of West Africa (Valentin et al. 2004). In the Sahel, severe crusting due to cultivation (Valentin 1996) and overgrazing (Hiernaux et al. 1999) can decrease soil infiltration capacity from 40 mm h^{-1} to 1 mm h^{-1} (Casenave and Valentin 1992). In West Africa, when the soil is left bare, the potential erosion increases gradually from less than 2 t ha^{-1} yr^{-1} for an annual rainfall of 150 mm in the fringe of the Sahara to nearly 81 t ha^{-1} yr^{-1} under 2 000 mm in southern Ivory Coast (Valentin et al. 1994). Population growth is still considered as the main driver of land-use change and increased water erosion (Planchon and Valentin 2004) in spite of some opposite findings under very special conditions (Tiffen et al. 1994). Using the present relations between soil erosion and population density, and the mean annual rate of population growth in the region,

2.63%, Planchon and Valentin (2004) calculated that degraded area should increase by 202 000 km^2 within the next 30 years in West Africa. This would lead to an increase of 13% of the area occupied by degraded soils. The area affected by water erosion would increase by 26%, mainly in the moist savanna zone (1 000–1 500 mm of annual rainfall). Although far less densely populated than the wettest region, many areas in the wet savanna may have a population density exceeding 70 inhabitants km^{-2}, which is a critical threshold in the region for severe water erosion hazard. In the wettest zone (> 1 500 mm of annual rainfall), areas not yet eroded are expected to be abruptly eroded as a result of land clearing using heavy machinery (Valentin 1996). In the semi-arid zone, an increased runoff coefficient due to more severe soil crusting should foster rill and gully development.

Mediterranean Region

Overland flow and soil erosion would result in changes of major magnitude in the Mediterranean region where they would be influenced also far more by land-use change than by the temperature increases due to climate change (Le Houérou 1990). For instance, in southern France, floods occur more rapidly and are twice as intense in a river basin with 90% forest-grassland and 10% vineyards than in a neighbouring basin with 30% forest-grassland and 70% vineyards (Galea et al. 1993). In deforested and intensively grazed portions of the Segura basin in Spain, soil losses as high as 200–300 t ha^{-1} yr^{-1} are reported (Lopez Bermudez 1990). But accelerated rural depopulation associated with abandonment of low productivity upland areas can also result in accelerated erosion, at least during the first year after abandonment. For instance, in south-eastern Spain, the infiltration ca-

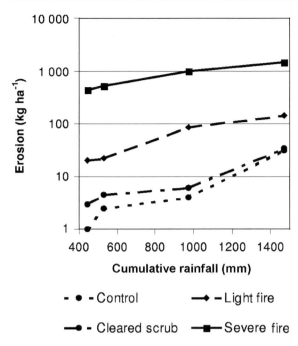

Fig. D.14. Soil erosion from runoff plots near Pisa (Italy) under four treatments: (*1*) control with natural Mediterranean vegetation, (*2*) vegetation regrowth after clearing prior to the experiment, (*3*) light fire after the removal of the litter, maximum temperature at the soil surface 180 °C, and above 100 °C during 7 minutes, (*4*) severe fire with the addition of woody litter –13 kg m^{-2}, maximum temperature at the soil surface 475 °C, and above 100 °C during one hour (after the data of Giovannini and Luccchesi 1992)

pacity was decreased from 55 to 19 mm h^{-1} and the erosion rate increased from 0 to 334 g m^{-2} h^{-1}, three years after abandonment. Ten years after abandonment, almost no runoff and erosion was observed (Cerda 1997). Also in Mediterranean Spain, a substantial regeneration of organic matter was observed 20 years after land abandonment and the estimated erosion rate was reduced from 2.2 t ha^{-1} yr^{-1} in the first years to 0.6 t ha^{-1} yr^{-1} after establishment of vegetative cover (Ruecker et al. 1998). But land that is no longer cropped is often grazed which tends to favour surface stoniness and concentrated overland flow (Ruiz Flano et al. 1992). Abandoned terraces become more vulnerable to piping, especially on illite and smectite-rich marls (Lopez Bermudez and Romero Diaz 1989).

Land abandonment tends also to favour wildfires which can have great impacts on overland flow and erosion. Severe fires that produce a temperature near 500 °C at the soil surface cause soil hydrophobicity and promote runoff production and soil erosion (e.g. Giovannini and Lucchesi 1992; Fig. D.14). After a summer forest fire in a catchment in southern France, the total runoff coefficient was 88% with a rainfall of 88 mm during the subsequent December, with 89% of the peak flow being generated by surface flow as shown by sodium and chloride tracers (Martin and Lavabre 1997). Severe ero-

sion was recorded also during the first winter after the Mt. Carmel fire, Israel (Ne'eman et al. 1997). In California, the response to winter storms in the first three years following a June fire which burned 1 214 ha was such that sediment not trapped in the debris basins accumulated to reach 108 000 m^3, reducing channel capacity and increasing the flood hazard (Keller et al. 1997). Immediately after disturbances by fire, agriculture or grazing, *Brachypodium retusum* grassland seems to be the best option for protecting soil, at least for Spanish conditions. For rangeland management, establishing natural *Quercus ilex* woodland produces the most resistant soil aggregates followed by *Quercus coccifera* and *Pistacea lentiscus* scrubland (Cerda 1998).

Temperate Zone

The decreasing use of organic manure and the growing use of inorganic fertilisers have promoted the depletion of soil organic matter and soil crusting. The thresholds at which runoff begins is now becoming as low in Western Europe as in the Sahel (Valentin and Bresson 1992). Already, crusting on silty loam, cultivated soils can lead to an infiltration capacity as low as 1 mm h^{-1} (Le Bissonnais et al. 1998). Besides the decline in aggregate stability, Boardman et al. (1994) listed among the causes of recent increases in the flooding of agricultural lands in north-western Europe (*i*) changes in land use, (*ii*) changes in the area of arable cropping, (*iii*) increase in the size of fields, (*iv*) compaction by farm vehicles; and (*v*) the expansion of urban areas into valley bottoms. These authors distinguished two main types of flooding in the loess European regions: winter flooding associated with wet soils and the cultivation of winter cereals, and summer flooding due to thunderstorm activity and runoff from sugarbeet, maize and potato crops.

Deforestation can also result in increased flooding as illustrated by simulations of a small lowland catchment in central Poland where a deforestation scenario of 10% of the catchment area increased the flood hydrograph volume by 13%, increased the peak flow rate by 17% and produced a 74% increase in sediment yield (Banasik 1989). Mining has also been an important cause of flooding. Studying a 119-year flood stage record at York, UK, Longfield and Macklin (1999) showed that the late 19th century was a period characterised by low flood frequency and magnitude, but high contaminant concentrations and downstream fine sediment delivery as a consequence of upland metal mining. A decline in flood frequency and magnitude as well as contaminant fluxes was observed between 1904 and 1943, resulting from the cessation of base metal mining. Between 1944 and 1968 significantly enhanced high flood frequency and magnitude as well as sediment fluxes resulted from changes in agricultural practices. Rates of flux declined between 1969 and 1977 owing to extremely low flood frequencies and magnitudes. Over the last two decades, the series of extreme magnitude floods have remobilised mining-contaminated alluvium, inducing high pollutant loads from metal mining. Two pairs of basins, one rural and one urbanised, were selected in the Midwestern USA for in-pair comparisons during the period 1940–1990 (Changnon and Demissie 1996). Anthropogenic changes affecting runoff in both rural basins accounted for two-thirds of the fluctuations in mean flows, while precipitation changes accounted for the remainder. The urbanised area doubled within one of the urban basins from 1940–1990, and these land-use changes explained more of the increase in mean flows and peakflows than in the urban basin with less change in land use. By 1990 precipitation accounted for 69% of the upward trend in mean flows since 1941 in the heavily developed urban basin, as compared to 37% in the less settled urban basin.

In temperate regions, the expansion of forests and fallow lands may reduce flooding hazards. After a 3-year calibration period, 207 ha of one catchment (total area of 310 ha) in New Zealand were converted from tussock grassland to pine plantation in 1982 (Fahey and Watson 1991). No change in water yield was observed until late 1988. In 1989, annual runoff from the planted catchment was 100 mm less than that from the adjacent control tussock catchment (218 ha). The peak flow rates of small storms (< 10 litres $s^{-1} ha^{-1}$) were most affected by afforestation and showed an average reduction of up to 50% for the 1988–90 period. Storm quickflow volumes showed a 29% reduction over the same period. Greater interception through increased evaporation rates from a wetted forest canopy is believed to be the main reason for reduced water yields after almost eight years of tree growth. Alternatively, reducing upstream soil erosion and river flows can in the future have some detrimental impacts on the sediment balance of the deltas and favour coastline erosion (McCully 1996; Brierley and Murn 1997).

High Latitudes

Major deforestation in Sweden from 2 500 B.P. led to an increase in sediment load associated with increased runoff and changes in channel hydrodynamics (Dearing et al. 1990). The period 500–300 B.P. showed a change to greater losses of topsoil and finer particles. More generally, changes in flooding hazards and erosion are expected to result from climate change or climate-change induced land-use change. In cold countries, altered hydrological regimes due to climate change could further exacerbate encroachment of agricultural land use into wetlands. As a result, their functions such as flood control, pollution filtration, nutrient recycling, sediment accretion, groundwater recharge and water supply, erosion control, and plant and wildlife preservation (Hartig et al. 1997) might be lost. As evidenced from ^{210}Pb-chronology in lake sediments in north-western Ontario, climatic changes have had a greater effect on erosion and sediment accumulation over the past 20 years than human disturbances such as clearcutting or fire (Blais et al. 1998).

High Altitudes

Land degradation in headwaters can have harmful effects in the regions located downstream (e.g. Krecek et al. 1996). And, compared to other ecoregional zones, mountains are more vulnerable to land-use change, particularly deforestation. In Himalaya, increasing population pressure has led to a greater use of marginal lands and to the conversion of forest to agriculture. As a result, streamflow increases in the rainy season (Rai and Sharma 1998) as well as soil erosion. However, the impact of Himalayan deforestation on the disastrous floods in the lower Ganges plain is still questionable because most of the catastrophic floods in the plains are the result of heavy monsoonal precipitation in the plains or in the sub-Himalayan zone (Messerli et al. 1995).

Conclusions

Major changes in overland flow and soil erosion are expected in the developing countries as a result of rapid land-use changes. High altitude sites may show particular sensitivities. Changes should also be observed in mid- and high latitudes regions which should gradually experience the impacts of global climate change and its attendant effects on soil organic matter. Overland flow and erosion in the temperate zone is also dependent on some forms of land-cover conversion. A typology of vulnerability to overland flow and soil erosion based on broad climatic zones is suggested by this review, and may provide a useful approach to future studies of these dynamics across human-occupied lands.

Jeffrey J. McDonnell

D.2.5   Subsurface Stormflow and Lateral Flow Processes

D.2.5.1   Rapid, Shallow Subsurface Stormflow Processes

Understanding of lateral flow processes and the generation of subsurface stormflow is important in the context of global change research since one-dimensional lumped models such as SVATs often fail when lateral transport of water by subsurface flow occurs (see Sect. D.2.3). Subsurface stormflow (RI_1 and RI_2 in Fig. D.6 earlier in this report) is a generic term, representing different below-ground runoff generation mechanisms which respond rapidly enough to a rainfall or snowmelt event to contribute to flood streamflow (Anderson and Burt 1990). This contribution depends on the type and duration of the event as well as topographic and other characteristics of the catchment (Hewlett and Hibbert 1967; Bonell 1998).

Our perception of rapid subsurface flowpaths has evolved greatly over the past two decades (McGlynn et al. 2002). Early studies focused on how rainfall and snowmelt (event water) moved rapidly into the channel during episodes (e.g. Mosley 1979). With the advent of conservative isotopic tracer studies, there is now consensus that pre-event water stored in the catchment before the episode is the dominant contributor to stormflow emerging in the stream – averaging 75% world-wide (Buttle 1998). Another consensus is that preferential flow is a ubiquitous phenomenon in natural soils, particularly in steep catchments (Clothier 2002). Today, research focuses largely on mechanisms to explain rapid movement and/or effusion of old water into stream channels. The main processes identified include:

- *Transmissivity feedback*, i.e. rapid subsurface flow where water tables rise vertically into more transmissive layers and result in rapid lateral flux.

In glaciated till-mantled terrain or in more temperate or sub-tropical areas where saprolite is found, the process known as transmissivity feedback (Rodhe 1987) may dominate the generation of rapid subsurface stormflow. In these instances, vertical recharge of the till or saprolite must first occur before water tables rise into the more transmissive mineral soil. Once water table rises into this zone, lateral flow begins – and the timing of well response into the mineral soil has been observed by many to coincide with rapid streamflow response (Kendall et al. 1999).

- *Lateral pipe flow*, i.e. transient saturated flow at a soil-bedrock interface where pipes enable water to move rapidly downslope.

Another commonly observed form of rapid subsurface stormflow production is by way of a vertical bypass flow lateral pipe flow response, as described by McDonnell (1990). Several recent studies have observed this process in Canada (Peters et al. 1995), Japan (Tani 1997), USA and New Zealand (McDonnell et al. 1999). In steep terrain with relatively thin soil cover, water moves to depth rapidly as vertical bypass flow (McIntosh et al. 1999). Since matrix storage declines rapidly with depth, the addition of only a small amount of new water (rainfall or snowmelt) is required to produce saturation at the soil-bedrock or soil-impeding layer interface. These aforementioned studies have each documented rapid lateral flow through pipes or openings at the soil-bedrock interface. Other studies have pointed to this in the form of rapid transmission of pressure waves, as discussed below. Whatever the mechanism, rapid lateral flow occurs at the permeability interface through the transient saturated zone. Once rainfall inputs cease, there is a rapid dissipation of positive pore pressures and the system reverts back to a slow drainage of matrix flow. Burns et al. (1998) showed that in addition to controlling rapid flowpaths, these bedrock flow channels also control the degree of solute flushing and consequently the base cation concentration of the lateral flow. They compared the composition of the mobile pipeflow with the slowflow in the matrix and found significant differences between the concentrated matrix water and relatively dilute bedrock flowpath, high mobility water.

Recent work by McDonnell et al. (1996), McDonnell (1997) and Freer et al. (1997) has shown that by mapping the impeding layer surface, one may be able to model the spatial pattern of transient water table development and thus the location of the mobile water flow path (Fig. D.15). This represents an advance in our ability to translate complex, non-Darcian flow processes into simple surfaces that can be incorporated into catchment model structures (Beven 2001).

- *Shallow interflow*, i.e. lateral flow through the organic or litter layer where either hydrophobicity or large conductivity contrast might enable water to move laterally on the timescale of hydrograph rise and fall.

A less widely cited example of rapid subsurface flow production is rapid lateral flow through the litter layer, sometimes called the "thatched roof effect" (Ward and Robinson 1990) or pseudo-overland flow (as reported by McDonnell et al. 1991). Recent studies by Brown et al. (1999) and Buttle and Turcotte (1999) using chemical end member mixing and isotopic tracing approaches have shown that this rapid lateral litter layer flow (perched on the mineral soil surface) may be a dominant mechanism in upland forested catchments during summer rainstorms. This is a combination of the high short-term rainfall intensities and water repellency that may develop at these sites during dry periods.

- *Pressure wave translatory flow*, i.e. where displacement occurs when kinematic wave speed arrives ahead of the advected water flux.

While pressure wave translatory flow was proposed back in the early 1960s by Hewlett and colleagues as part of the Variable Source Area concept, recent papers have rejuvenated interest in this area (Rassmussen et al. 2000). Torres et al. (1998) found that "a pressure head signal advanced through the soil profile on average 15 times greater than

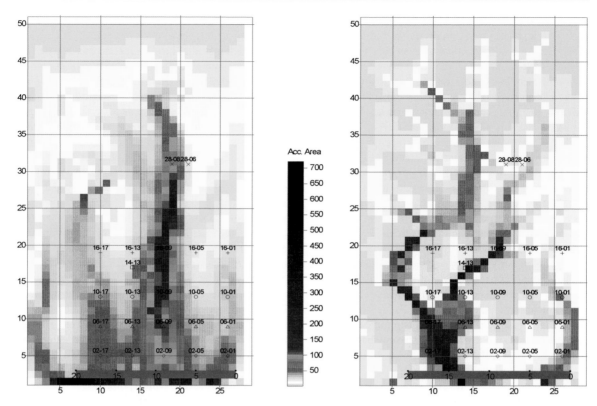

Fig. D.15. Two topographically-inferred flowpaths – one from surface topography (*left*) and one from bedrock topography (*right*), Panola Mountain Research Watershed (from Freer et al. 2002, ©American Geophysical Union). Since transient saturation occurs at the soil-bed-rock interface and saturation builds from the "bottom-up", the bedrock topography is the surface that controls the spatial distribution of mobile flow at the hillslope scale. *Acc. Area* = Accumulated Area (ground surface defined by surface contours contributing water to a point)

the estimated water and wetting front velocities." Hence initial pressure head response appears to be driven by the passage of a pressure wave rather than the advective arrival of "new water". This pressure water was thought to be the cause of the rapid effusion of old stored water from the deeper sandstone groundwaters on their hillslopes. Much work remains to be done to fully elucidate these processes in different environments. Notwithstanding, the research to date does remind all working in the area of subsurface storm-flow processes that rapid travel times of fluid pressure head or water content through the unsaturated zone could be interpreted, mistakenly, as preferential or macropore flow (or as very high hydraulic conductivities) (Smith 1983). The notion that soil water pressure velocities are often much faster than tracer velocity is an important point to keep in mind.

D.2.5.2 Separating Event Water and Subsurface Stormflow in the Storm Hydrograph

On the time scale of storm events, the mixture of waters contributing to flow in the channel often includes rainfall and subsurface flow from various positions and depths within the catchment. The simple two component hydrograph separation technique (Fig. D.16) is often used:

$$Q_t = Q_p + Q_e \tag{D.8}$$

$$C_t Q_t = C_p Q_p + C_e Q_e \tag{D.9}$$

$$X = \frac{(C_t - C_e)}{(C_p - C_e)} \tag{D.10}$$

where Q_t is streamflow, Q_p and Q_e are contributions from pre-event subsurface stormflow (soil water, groundwater) and event (rainfall, snowmelt) water, C_t, C_p and C_e are the conservative tracer composition in streamflow, pre-event and event waters, respectively; and X is the pre-event fraction of streamflow.

Buttle (1998) notes that the use of these mixing equations to solve for the event and pre-event components of streamflow rests on several assumptions:

- there is a significant difference between the isotopic composition of the event and pre-event components;
- the event water signature, as well as the pre-event water signature, are constant in time and space, or any variations that are found can be quantified;
- soil water contributions from the unsaturated zone are negligible, or if not, they must be similar to the groundwater signature; and,
- surface storage contributions to the channel are negligible.

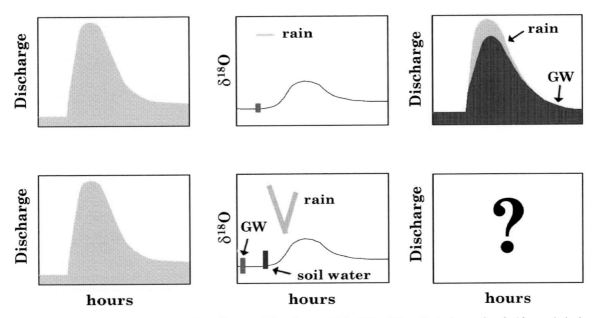

Fig. D.16. The hydrograph separation problem (from Kendall et al. 2001, © John Wiley & Sons Limited, reproduced with permission). Classical assumptions of spatial and temporal homogeneity of end members (rainfall and groundwater) allows for a clear and unambiguous separation of the hydrograph. However, temporal variations in rainfall and often different concentrations for soil water and groundwater pose problems for simple separation of the storm hydrograph

While several authors have discussed the implications of these assumptions (Sklash 1990 and several others), Genereaux (1998) quantified how some of these assumptions affect the uncertainty in the 2-component separation. Regardless of how one might deal with assumptions implicit within the technique, the approach itself only distinguishes between inputs of "new water" (rainfall) and "old" stored water (i.e. that water in the subsurface prior to the rainfall event). Since volume of the subsurface is not considered in this approach, neither water nor tracer mass balance is really considered (Harris et al. 1995). Ultimately, a distribution function representing the water age spectra for streamwaters is what we seek – to be able to associate percentages of water of various ages to volumes in the channel at any given time. This will likely necessitate development of new age-dating isotopic techniques and more clever statistical analyses of water chemistry records (like, for example Kirch-ner et al. 2000).

Age Components of Subsurface Stormflow

Subsurface flow represents a distribution of water ages and flowpath lengths that collectively contribute to flow in the channel. Between rainfall and snowmelt events, the mean residence time is often quantified using a transfer function based on the assumption of steady-state well mixed flow. This approach was outlined by Kreft and Zuber (1978) and used widely in the hydrological literature for assessing residence times of soil water (Stewart and McDonnell 1991) and groundwater (Maloszewski

et al. 1983). Implicit in these approaches is that the isotopic composition of waters in the catchment are ideal tracers since they themselves represent the water molecule (Kendall and McDonnell 1998). The tracer concentration of the output can be related to the tracer concentration of the input by the convolution inegral

$$c(t) = \int_{-\infty}^{t} c_{in}(\tau)\, g(t - \tau)\, dt \qquad (D.11)$$

where $c(t)$ is the output isotope concentration, t is the exit time from the system, τ is the entry time to the system, $t - \tau$ is the transit time, $c_{in}(\tau)$ is the input isotope concentration and $g(t - \tau)$ is the system response function that describes the transit time distribution of the conservative tracer particles through the system (Turner and Barnes 1998). Stewart and McDonnell (1991) found that a dispersion model led to the best system response function for reproducing reasonable residence time parameters based on physical measurements in steep wet catchments.

Perhaps the best contribution to this area of research was that of Uhlenbrook et al. (2002). They studied a 40 km² catchment in the Black Forest of Germany and documented the mean residence time of subsurface flow (RI_1 in Table D.4), shallow groundwater flow (RI_2 in Table D.4) and deep groundwater (RG in Table D.4) and computed both the mean residence time of these components and their contribution to low flow at the catchment outlet. Their data show that shallow groundwater had the largest contribution (69%) to the low flow total and a mean residence time of 2–3 years (see Table D.4).

Table D.4.
Results for mean residence time computation for the Brugga catchment, Germany (Uhlenbrook et al. 2002)

Subsurface flow type	Percentage of the low flow	Mean residence time
Subsurface stormflow[a]	11	Hours–days
Shallow groundwater[b]	69	2–3 years
Deep groundwater[c]	20	6–9 years

[a] Using the convolution integral method with ^{18}O.
[b] Using CFC tracing.
[c] Using the tritium approach and the system response function of Eq. D.11.

To date, the mean residence time approaches have been limited to a consideration of subsurface flow contributions to low flow and not stormflow. Turner et al. (1987) were the first to examine how these techniques might be used to quantify the mean age of subsurface stormflow contributions to the channel hydrograph. More recently, McDonnell et al. (1999) used this approach to quantify the age spectra of event water at the Maimai catchment in New Zealand.

Pathways of Rapid Deeper Subsurface Flow

For the most part, hydrometric studies in hillslope hydrology have focused attention on the role of shallow subsurface stormflow as the primary source of subsurface contributions to the storm hydrograph. The role of deeper groundwater (RG in Fig. D.6) has generally been regarded as passive and disregarded in topographic-wetness studies and topographically-based rainfall-runoff models. Work in Oregon by Montgomery et al. (1997) and Anderson et al. (1997) has demonstrated that saturated water flow through deep permeable bedrock can be an important contributor to the runoff volume and chemistry in steep topography. They found lateral subsurface stormflow movement into and out of the underlying sandstone bedrock, resulting in bedrock flow chemistry signatures in headwater stream channels in the Oregon Coast range. Similarly, Onda et al. (2001) found that bedrock contributes to channel stormflow in steep Japanese catchments. Here, large contributions of water from bedrock formations formed the largest fraction of water appearing in the channel on the time scale of a storm runoff event.

D.2.5.3 Modelling Lateral Flow at the Catchment Scale

The balance between practical simplifications and justifiable model complexity is unresolved in subsurface stormflow modelling. In most cases the available data motivates the use of simple, conceptual model approaches rather than the use of a fully-distributed, physically-based model with a large number of parameters. The steady-state assumption is the hallmark of conceptual runoff models incorporating subsurface stormflow routines. Here, an unambiguous, monotonic function between the groundwater storage and runoff is the basic underlying structure (Seibert et al. 2003). Consequently, the dynamics of the simulated runoff from subsurface stormflow always follows the simulated rise and fall in groundwater levels. TOPMODEL (Beven et al. 1995) is an example of such a conceptual model. While TOPMODEL simulates spatially-distributed groundwater levels using a topographic index, these groundwater levels always go up and down in parallel. The simulated runoff from the subsurface follows the same dynamic. Thus it is assumed implicitly that the groundwater storage and runoff can be described as a succession of steady-state flow conditions.

Recent work by Seibert et al. (2003) has tested the steady state assumption with an analysis of detailed groundwater level data along two opposing hillslopes along a stream reach in a Swedish till catchment. Groundwater levels in areas close to the stream followed the dynamics of the runoff (Fig. D.17a). The correlation between groundwater level and runoff decreased markedly for wells farther than c. 40 m from the stream (Fig. D.17c–d). The levels were often independent of streamflow, with upslope area groundwater capable of rising when riparian groundwater and runoff were falling, and vice versa. There was a high degree of correlation between groundwater levels at similar distances from the stream.

Despite the widespread acceptance of the steady-state assumption previously, the work of Seibert et al. (2003) shows that it is not valid for the hillslope tested, and a growing number of investigations from around the world are beginning to yield similar results. The implications for modelling subsurface stormflow in light of this is to move to more "box-like" model structures, like the new Dynamic TOPMODEL (Beven and Freer 2001) and the SOFTMODEL approach of Seibert and McDonnell (2002).

D.2.5.4 Subsurface Flow and Catchment-scale Nutrient Dynamics

Terrestrial-aquatic boundaries can be considered environmental transition zones, and are normally character-

Fig. D.17.
Relation between runoff and
depth to groundwater for four
different locations in Sweden.
a Well J3G1, 14 m from stream;
b well TG3, 26 m from stream;
c well WG4, 78 m from stream;
d well J6G1, 103 m from stream
(from Seibert et al. 2003,
© American Geophysical Union)

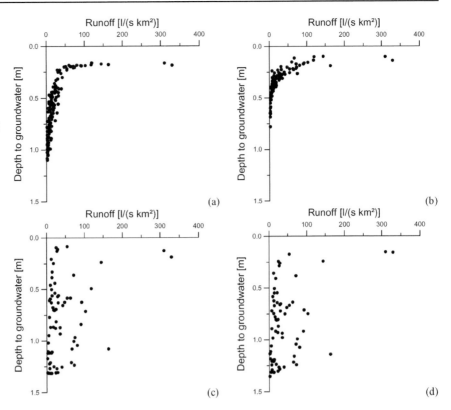

ised by abrupt changes in subsurface hydrological flowpaths and biogeochemical environments. Abrupt transitions in soil moisture conditions leads to corresponding
changes in subsurface biogeochemical conditions (e.g.
redox potential, pH). A gradient of the vertical and horizontal zones can be defined within the soil profile and
spatially within the catchment.

Cirmo and McDonnell (1997) reviewed the linkages
between the hydrological and biogeochemical environments in temperate forested catchments. In general, the
hillslope-riparian zone interface and wetlands may be
important ecotonal boundaries in the catchment. Recent
work has shown that the degree of topographic convergence in the landscape may have a large effect on the
flushing of mobile constituents like DOC (Dissolved Organic Carbon) and NO_3^- (Creed et al. 1996). Where water tables are close to the surface in hollows and in the
headwater sections of perennial channels, these constituents may be flushed into the stream during episodes.
This flushing response appears to be the dominant form
of export of dissolved forms of carbon and other nutrients into the channel. Over agricultural lands, some of
this flushing may occur in near-stream zones where saturation excess overland flow mobilises chemical constituents on the flood plain. Much of this nutrient load may
also be associated with fine sediment entrainment and
washoff.

McHale et al. (2002) argue that the initial step in identifying sources of NO_3^- and the flowpaths by which it

reaches the catchment outlet is to identify the geographic
end members (sources) of stream water and the subsurface water flowpaths. Creed et al. (1996) used the Regional Hydro-Ecological Simulation System model in an
effort to identify mechanisms responsible for the release
of NO_3^- from the Turkey Lakes catchment in central Ontario, Canada. They identified two release mechanisms
(1) NO_3^- flushing, where NO_3^- that has accumulated in
upper soil layers is flushed to the stream by a rising water table during storms, and (2) a draining mechanism,
where NO_3^- rich snowmelt water recharges deep groundwater via preferential flow pathways and is subsequently
released slowly over the year. These mechanisms suggest that the timing of the release of NO_3^-, although influenced by catchment biota, is controlled mainly by
catchment hydrology. More recently, Inamdar et al.
(2004) presented a conceptual model that couples the
flushing and draining of subsurface nutrient concentrations with the subsurface stormflow processes simulated
within TOPMODEL. Figure D.18 shows four stages
through a hypothetical hydrograph where the spatial
patterns of runoff are related to old/new water fractionation of the stormflow (with new water plots above the
dotted line in each hydrograph in Fig. D.18) and the
mixing and movement of water in the subsurface (shown
as transmissivity feedback as described earlier). These
"stages" of flushing for labile constituents like DOC and
NO_3^- provide a linkage to the process understanding of
how lateral flow occurs throughout the catchment. The

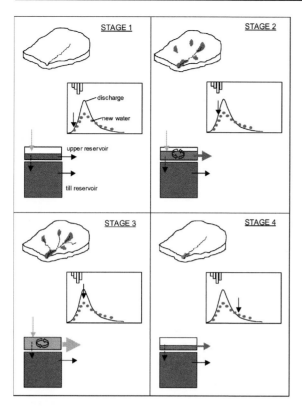

Fig. D.18. A conceptualisation of nutrient flushing in relation to subsurface stormflow (from Inamdar et al. 2004, © American Geophysical Union)

precise coupling between hydrological and biogeochemical processes remains an important focus of new research.

D.2.5.5 Conclusions

Research needs for the future have recently been discussed by Beven (2001), McDonnell and Tanaka (2001) and Uhlenbrook et al. (2003). How to parameterise preferential flow continues to be a primary concern. New ideas such as momentum dissipation (Clothier 2002) may be a way forward. Bedrock flow contributions to channels is being recognised more widely (Onda et al. 2001). Similarly, the critical importance of the hillslope-riparian (near stream zone) linkage is key to understanding nutrient transformations and the connectedness-disconnectedness of hillslope inputs to the channel. While most hillslope hydrologists have ignored channel processes, the hyporheic zone may be one of the most important final processing zones for subsurface stormflow before it enters the stream (Bencala 2000). This may be true as well at the large river scale (Fernald et al. 2001). Future work should explicitly address how vegetation water use and uptake influences soil moisture distribution and threshold responses of subsurface stormflow.

Valentina Krysanova · Alfred Becker · Roland E. Schulze

D.2.6 Integrated Ecohydrological Modelling Considering Nutrient Dynamics in River Catchments

D.2.6.1 General issues

We begin our discussion with a general consideration of appropriate spatial resolutions for modelling. It generally is accepted that coarser resolution schemes and simplified models are more appropriate for use in larger scale modelling, although the overall degree of complexity generally increases with an increasingly larger spatial domain. This is sometimes called the "Hydrological Paradox", the notion that many small-scale processes and spatial heterogeneities that differentiate the land-surface in nature "are smoothed out" when viewed over larger spatial domains. In addition, the integration, aggregation and superposition of various flow components in catchments and along the river courses contribute to this effect. This means that, in principle, modelling at small scales must be considered to be more demanding in terms of a required higher spatial resolution and more explicit modelling of individual processes.

In this context, a set of key issues emerges:

- The most appropriate approach toward assessing land-surface heterogeneity in terms of land-surface patchiness is distributed hydrological modelling with subdivision (alternatively termed "discretisation" or "disaggregation") of the land surface into elementary land-surface areal units characterised by an internally "uniform" or near-homogeneous behaviour ("patches", "hydrotopes", "hydrotypes" or "Hydrological Response Units", HRU). This approach introduced in Sect. D.2.2.2 (Fig. D.3, Fig. D.5) is discussed briefly in Sect. D.2.6.3.
- A special advantage of the disaggregation approach is the potential for directly deriving or estimating model parameters for individual distributed patches or hydrotopes from now generally available GIS data. This constitutes a major step forward in operational distributed hydrological modelling, since it allows modellers to estimate certain model parameters directly, at least for those component models describing the "vertical processes". The direct parameter estimation approach is outlined briefly in Sect. D.2.6.4. Serious limitations have been observed in using point data and information in the estimation of soil hydraulic properties controlling runoff generation and the parameters of related models (see Sect. D.2.3.4 and D.2.2.5).
- Model calibration is still necessary for some of the dynamic attributes of component models, as well as for lateral flow processes. The latter dynamics are

related to catchments, i.e. normally to larger land-surface areas (up to the mesoscale) delineated by water divides. Parameter calibration of catchment models may benefit from the use of catchment related "effective" parameters. Our initial remarks on this (Sect. D.2.2.3) are now supplemented by discussion in Sect. D.2.6.4.

- Development of multi-purpose "integrated" ecohydrological models has been at the scale of both small catchments and mesoscale river basins. The aim of integrated ecohydrological modelling is to unite the dynamics of energy (including evapotranspiration) and water fluxes (runoff generation and resulting lateral flows) and associated material fluxes (transports of sediments, chemicals, nutrients and pathogens). An introduction to this type of modelling is given in Sect. D.2.6.6 with reference to two multi-purpose models, viz. the SWIM and ACRU models, applications of which are illustrated in Chapt. D.3 and in the case studies described in Chapt. D.6 and D.7 as well as in vulnerability studies in Chapt. E.6.

D.2.6.2 Structure of Integrated Ecohydrological Models

Ecohydrology combines the study of hydrological, biogeochemical and ecological processes and their interrelations in soil, plants and water. It aims to impove understanding of those hydrological factors determining the development of natural and human-driven terrestrial ecosystems, and of ecological factors controlling water fluxes and resources. River catchments represent an appropriate scale for ecohydrological modelling owing to their hierarchical structure and natural boundaries. They can be considered as integrators of the effects of many forces, including topography, climate, soils, land use and direct use of water resources by humans and natural ecosystems.

In principle, an ecohydrological model should account for the numerous interrelated ecological and hydrological processes evident in catchments, including water fluxes in the soil, streams and waterbodies, vegetation growth, and nutrient cycling. Usually such models consider all related processes in an integrated manner and, as a rule, are based on the mathematical description of physical and biogeochemical processes.

A physically-based model describes the natural system using the basic mathematical representations of flows of mass, momentum and energy. Usually at the catchment scale, a physically-based model has to be fully distributed (see below). However, the fact that a model is physically based does not necessarily mean that it is based on fundamental physical laws alone. Conceptual approaches taking into account the general process behaviour at an appropriate level of representation may

also be included. It has been shown that the inclusion of physical laws in a model does not, by itself, guarantee an enhanced level of output. Even if physical laws included in a model are proven to represent a sound mathematical description, for example depicting a soil column in laboratory conditions, where soil has been well mixed, may not automatically be an appropriate guide to behaviours at the scale of grid elements typically used in distributed hydrological models (Beven 1996). These equations usually require parameters and variables assumed to be uniform over a spatial scale of hundreds of metres or even kilometres. Models that include mathematical descriptions of physical, biogeochemical and hydrochemical processes, and combine significant elements of both a physical and conceptual semi-empirical nature, can be termed process-based ecohydrological models. These have been shown to adequately represent natural processes at this scale.

Ecohydrological models for catchments contain as a basic element a hydrological submodel. Another necessary component is a vegetation submodel. Also, such a model usually includes the submodels for biogeochemical cycles (e.g. C, N, P), each at a certain level of complexity. The hydrological, vegetation and biogeochemical submodels are usually coupled (e.g. see Fig. D.19, Fig. D.20 and Fig. D.21) to include important interactions and feedbacks among the processes, such as water and nutrient drivers for plant growth, water transpiration by plants or nutrient transport with water. Usually, vertical and lateral fluxes of water and nutrients in catchments are modelled separately (cf. Sect. D.2.2).

An important question is the level of detail an ecohydrological model should exhibit in its process representation. This is not a trivial question. Model complexity is not an end in and of itself. If a complex phenomenon, or process, can be described mathematically in a simplified form and parameterised using available information, this is arguably preferable to the case, where the level of detail is high, parameterisation of the model is problematical, and control of the model behaviour is difficult. Some users and less experienced modellers believe that the more detail is included in the model, the better, and that more complex models guarantee better representation of reality. However, experience of using complex process-based models during the past decades has led to the conclusion that model complexity is generally defined as a compromise solution as follows: *include only submodels which are essential and necessary, parameters which are physically meaningful and can be estimated with confidence, and interrelations which can be understood and validated in simulation experiments.* Also, modelling results should not be interpreted as exact predictions, but in the first place behave with qualitative similarity to the real world, for example, as indicators of possible trends, or as a measure of differences in management scenarios.

Spatially distributed, or semi-distributed, models are usually required in the field of ecohydrological modelling in view of the inherent land-surface heterogeneity faced at the catchment scale. This is particularly the case for land-use change impact studies. The lumped, i.e. spatially averaged, models are not optimal for integrated ecohydrological modelling. The simplest way to overcome the lumped structure of a model is to subdivide a catchment into subcatchments. This enables one to take into account the differences in topography or land use in different parts of the catchment and to account for spatial variations in the model variables and parameters considered. This is usually done by first simulating all the processes in the subcatchments, and then aggregating the outputs to the level of the whole catchment.

The next step is to further subdivide the land surface into either regular grid cells, or irregular units, using the principle of similarity. In the latter case maps of, for example, subcatchments, land use, soil and groundwater table are overlain to create irregular Elementary Areal Units (EAUs), or Hydrological Response Units (HRUs), which can later be combined into classes inside subcatchments called hydrotopes (as introduced in Sect. D.2.2.1 and D.2.2.2, cf. Fig. D.3). Then the HRUs are not modelled separately, but every hydrotope class is modelled once in a time step. This method of spatial disaggregation takes into account the landscape heterogeneity and is computationally efficient.

In the case where all major vertical and lateral flows between regular grid cells or irregular units (e.g. HRU polygons) are considered in the model, and the model accounts for spatial variations in all variables and parameters, it is called a "fully-distributed" model. There are other ways to take into account spatial variability and reduce the level of complexity in comparison with the fully-distributed models. This can be done by considering the lateral processes for a subset of aggregated units, e.g. subcatchments. If the model subdivides the catchment into subcatchments only, or if it considers EAUs, but the lateral flows are first aggregated at the subcatchment level and then routed downstream, the model is called "semi-distributed".

Dynamic models represent time-dependent processes, whereas steady-state or static models describe stationary or equilibrium processes. Dynamic models can be continuous in time in contrast to event-based models, which represent some discontinuous or abrupt phenomena. Ecohydrological models usually belong to the class of continuous dynamic models.

The spatial and temporal resolution of the model should be appropriate for its use. The spatial resolution, scale of application and objective of the study are connected: a model developed for a small catchment for research purposes may have a fine spatial resolution (e.g. 50 m grid cell or less) in order to study flow components and their pathways. It may be a lumped model when only "precipitation-runoff" relations are of interest in a nearly homogeneous catchment. Also, a model for a mesoscale catchment developed for predictive purposes can have a coarser resolution (e.g. 200 m).

D.2.6.3 Assessment of Land-surface Heterogeneity in Modelling

As follows from the previous chapter, spatial variability of land-surface processes must be taken into account for effective modelling. Delineation of subareas with different characteristics related to runoff generation, evapotranspiration or nutrient uptake is thus critical. Further, distributed or at least semi-distributed models are required, especially if changes in land use and management and their effects on hydrology, water resources and ecology have to be simulated (cf. Sect. D.2.2.2). Such changes normally take place at relatively small landscape units and need to be considered in our models.

However, a basic problem in distributed modelling is the determination of model parameters for the various sub-areas. An efficient approach is to use available GIS-based maps, for example of land use, vegetation cover, soil types, topography (Digital Elevation Model) and hydrogeology, as a basis for the parameter estimation. These maps can serve first to delineate areal units within which land-surface characteristics are more or less uniform (cf. Fig. D.5 in Sect. D.2.2.2). They represent homogeneous or quasi-homogeneous HRUs, or classes of them in the form of hydrotopes.

An important step in the application of this procedure is to define the number and type of hydrotopes to be modelled separately in each application. The general rule is to keep the number of hydrotopes as small as possible, but as large as necessary, to address the given modelling task at hand.

The number of hydrotopes depends on the given landscape patchiness. It may well be that in a relatively uniform landscape as, for example, the Amazon forest or the North American prairies, a smaller number of hydrotope classes is sufficient, whereas in most parts of Europe or in mountainous landscapes a larger number of hydrotope classes, differing in essential characteristics, must be distinguished. A special sensitivity study has been performed in the Stör River basin located in the northern part of the Elbe River drainage area in Germany to attempt to answer this question (Becker and Braun 1999). It was shown that at least six types of land-surface units (hydrotopes) should be distinguished: open water surfaces; shallow groundwater areas (including wetlands, riparian zones) and areas with deep groundwater, both in the two sub-categories of forest and non-forested areas (cropland, grassland etc.) and impervious areas.

Whatever the specific case may be, the following land-surface features and related attributes should be considered in the general definition and application of the hydrotope approach (for the symbols see Fig. D.6):

- topography in terms of position (elevation, exposition etc.) and in relation to the drainage network (for example upland AG, slope ASL, and valley floor AN);
- land-use type including open water surfaces (AW), urbanised areas and settlements and sealed areas (AU), bare land and permeable vegetated areas, which may be further subdivided into forests, arable land, pasture, savannas, bushland, etc.;
- soil type with a number of hydrophysical and geochemical characteristics; and,
- hydrogeology including the depth to the groundwater table, saturated thickness, character of impervious or less permeable layers.

The three types of sub-areas mentioned under 1) are represented schematically in Fig. D.6. They are particularly important in hydrological modelling and may, therefore, be characterised more specifically with reference to the dominant processes, especially of runoff generation and resulting lateral flows:

- in upland areas (AG) groundwater is often deep below the surface and below the root zone, so that during long dry periods soil moisture is continuously decreasing and evapotranspiration is reduced in comparison with potential evapotranspiration;
- slopes (ASL), particularly in mountain environments, often have impervious or less permeable layers such as the soil-bedrock interface, clay layers, etc. where subsurface storm flow RH (often called interflow) may be generated during heavy rainfall and snowmelt events;
- valley floors (AN) are generally flat and have shallow groundwater tables, so that evapotranspiration is often near its potential rate and saturated areas may occur during heavy rainfall and snowmelt, a condition which directly generates overland flow RO.

D.2.6.4 GIS-based Estimation of Land-surface Characteristics and Related Model Parameters

Geographic Information Systems (GIS) play an increasingly important role in ecohydrological modelling. However, these systems are typically not used directly for ecohydrological modelling, because there is no explicit representation of time in the data structures, and there is no direct implementation of the differential equations describing the conservation of mass in the GIS. Nevertheless, GIS is useful both in the preparation of the input data for model parameterisation and for the visualisation of the model results as maps, summary statistics and other visualisation products. Some models use so-called GIS interfaces for both pre-processing and post-processing of model inputs and outputs. In addition, GIS can be used effectively during the verification stage of the modelling, e.g. for comparing the patterns of simulated outputs and observed data.

GIS may be vector- or raster-based. Vectors are defined as quantities having a starting coordinate and an associated directionality. Vectors are used in GIS to specify precisely the position of points, lines and polygons. Another way of representing spatial data is as a raster (or cellular) dataset. Each cell in a raster structure is assigned only one geographically-referenced value, and different attributes can be stored separately. Operations on multiple raster files involve the retrieval and processing of the data from corresponding cell positions in the different data layers. An example of vector-based GIS is ARC/INFO, whereas GRASS (Geographic Resources Analysis Support System; 1993) represents a raster-based GIS.

From a modelling perspective, vector and raster data structures both have advantages and disadvantages. High resolution raster models are well suited for analysing spatial variability, whereas network analysis is facilitated by vector-based data structures. By reducing cell size, the user can approximate polygons by raster cells quite successfully. The advantages of raster data structures are: easy overlays, combination with other layers, and spatial analysis. The advantages of vector data structures are: compact data structures (related to, for example, hydrotopes as polygons), good representation of networks (like stream channel systems) and accurate graphics. Both types of GIS are used widely as interfaces to hydrological and ecohydrological models.

If a raster data structure is used, a mesh of cells is laid over the catchment, and all processes in the soil are simulated for each cell. This is usually based on a digital elevation model, which provides a raster-based structure for calculations. However, if the mesh size is small, the number of grid cells can become very large, and simulation times prohibitive. The other option is to use a vector data structure, or irregular cells (polygons) and to use a principle of similarity in order to increase the polygon sizes and thus decrease simulation time.

At the pre-processing stage, GIS can be used for a number of purposes, such as:

- delineation of basin and sub-basin boundaries;
- delineation of hydrotopes (sets of polygons);
- assigning model parameters;
- delineation of the basin structure, e.g. as a combination of sub-basins and hydrotopes; and,
- delineation of the routing structure.

Essential characteristics such as topographic parameters (slope, aspect), water storage capacities (e.g. interception and depression storage capacity, soil porosity, field capacity, water holding capacity), permeabilities (e.g. infiltration, hydraulic conductivities), and vegetation characteristics (e.g. maximum leaf area index (LAI), root depth, albedo) can be assigned to the areal units using GIS.

Many of these characteristics can be used directly as model parameters in ecohydrological models, in particular those for the vertical processes domain related to each of the delineated EAUs or hydrotopes, for example, water holding capacity and porosity in the unsaturated soil zone. Other characteristics can serve to derive necessary model parameters by using available general relations, such as pedotransfer functions.

The following standard procedure for quantifying model parameters for distributed or semi-distributed hydrological models can be used:

- acquisition of GIS-based maps;
- delineation of the basin and sub-basin areas based on DEMs;
- conjunction of a relevant subset of these maps to generate Elementary Areal Units, characterised by an average elevation, slope gradient and aspect, definite type of land use, vegetation cover, soil type and depth to groundwater;
- combination of similar EAUs within basins or sub-basins into hydrotopes as areal units showing the same (or similar) hydrological behaviours, and definition of the basin structure;
- determination of hydrotope-related model parameters by selecting and transforming the mappable attributes by means of "related tables" or "look-up functions", such as pedotransfer functions; and,
- delineation of the river routing network.

Many of the approaches described above cannot be used in coarse-grid climate models. However, quite recently a major step has been made into a more hydrology-oriented direction by Koster et al. (2000) and Ducharne et al. (2000). They have re-derived a surface parameterisation for climate modelling that uses catchments as the basic units. This procedure may help hydrologists to play a more active role within climate modelling and to achieve more rapid improvements in the representation of hydrological processes in global change studies.

D.2.6.5 Calibration and Verification of Component Models (Modules) within Integrated Models

During the calibration procedure two different methods can be used for the model calibration, viz.

- manual calibration, as a trial-and-error method; or,
- automatic, computer-aided calibration.

The latter method is based on a numerical algorithm aimed at finding the best fit as the maximum or minimum of a given numerical objective function, e.g. minimum in the difference of water balance, maximum in the case of the Nash and Sutcliffe (1970) efficiency. The automatic calibration method is better suited for simplified models and has to be used with caution for more complex models. In cases where this method is used, special attention should be paid to the choice of a single optimisation criterion, choice of calibration parameters and specification of their ranges. Besides, the modeller has to be aware that the optimisation may lead to a local optimum instead of the global one, and that different sets of parameters may lead to equally good simulation results. The trial-and-error method is the most widely used and is recommended especially for more complex models. Remarkable progress in automatic parameter calibration has been achieved during the past years, even with multi-parameter models and areally-distributed validation data.

It is widely agreed that after model calibration, parameters must be validated using data different from those used for the calibration (e.g. by considering other time periods or other sub-basins). A prerequisite is that either the catchment conditions (e.g. climate, land use, groundwater abstraction) must be stationary during both the calibration and validation periods, or observed changes can be assessed by related parameter changes.

Verification of an ecohydrological model output implies that it has been matched against measured data, be it hydrological, biogeochemical, or biological in nature. This can be achieved by comparing simulated and measured water discharge or groundwater levels, nitrogen (N) and phosphorus (P) concentration or load, crop yields and spatial patterns of LAI. Successful verification of model output in several catchments under different environmental conditions increases its credibility.

D.2.6.6 Examples of Integrated Ecohydrological Models: SWIM and ACRU

Two examples of ecohydrological models are briefly described below: SWIM (Krysanova et al. 1998a) and ACRU (Schulze 1995). Case studies applying these models are described in later chapters (Chapt. D.6: SWIM, Chapt. D.7: ACRU).

D.2.6.6.1 The SWIM Modelling System

SWIM, i.e. Soil and Water Integrated Model (Krysanova et al. 1998a; Fig. D.19), represents a continuous-time spatially distributed model, integrating hydrological processes, vegetation growth (agricultural crops and natural vegetation), nutrient cycling (nitrogen, N and phos-

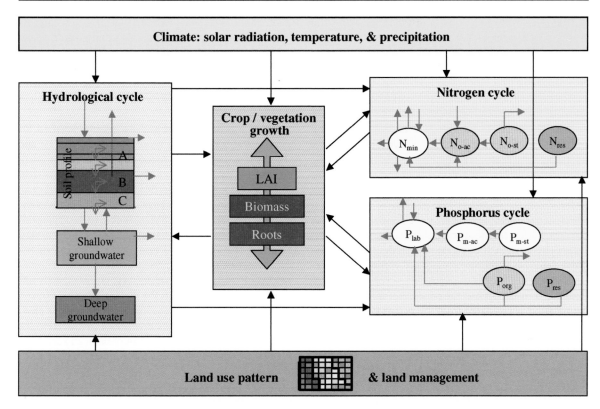

Fig. D.19. Flow chart of the SWIM model, integrating hydrology, crop/vegetation growth, and nitrogen and phosphorus cycles. The model is forced by climate and land use

phorus, P), and sediment transport at the river basin scale. In addition, the system includes an interface to the GRASS Geographic Information System (1993), which facilitates the extraction of spatially distributed parameters of elevation, land use, soil, vegetation and the routing structure for the basin under study (cf. Sect. D.2.6.4). The model can be applied to integrated modelling in mesoscale and large river basins with an area up to 100 000 km², or, after appropriate verification in representative sub-basins, for scenario analyses and impact studies in larger basins or regions.

A three-level scheme of spatial disaggregation into "basin – sub-basins – hydrotopes" or "region – climate zone – hydrotope", plus vertical subdivision of the root zone into a maximum of 10 layers is used in SWIM. As described earlier (see Sect. D.2.6.1) a hydrotope is a set of elementary units in a sub-basin or climate zone, which has the same or similar land uses and soils. In a typical model application for a river basin, water and nutrient balances are first calculated for every hydrotope, after which outputs from the hydrotopes are integrated (by area-weighting) to estimate the sub-basin outputs. The routing procedure is then invoked to estimate the sub-basin lateral flows of water, nutrients and sediments, taking into account transmission losses.

The simulated hydrological system consists of four control volumes: the soil surface, the root zone, the shal-

low aquifer and the deep aquifer. The soil column is subdivided into several layers. The water balance for the soil column includes precipitation, surface runoff, evapotranspiration, percolation and subsurface runoff. The water balance for the shallow aquifer includes groundwater recharge, capillary rise to the soil profile, lateral flow, and percolation to the deep aquifer.

The module representing crops and natural vegetation is an important interface between hydrology and nutrients. A simplified EPIC approach (Williams et al. 1984) is included in SWIM for simulating arable crops (like wheat, barley, rye, maize, potatoes, etc.) and aggregated vegetation types (like "pasture", "evergreen forest", "mixed forest"), using specific parameter values for each crop/vegetation type. A number of plant-related parameters are specified for 74 crop/vegetation types in the database attached to the model, i.e. biomass-energy ratio, harvest index, base and optimal temperature for plant growth, maximum LAI, fraction of growing season when LAI declines, maximum root depth, potential heat units required for maturity of crop, etc. Vegetation in the model affects the hydrological cycle by the cover-specific retention coefficient, which influences runoff, and indirectly the amount of evapotranspiration, which is simulated as a function of potential evapotranspiration and LAI.

Sediment yield is calculated for each sub-basin with the Modified Universal Soil Loss Equation (MUSLE,

Williams 1975; Williams and Berndt 1977) using the stormflow, the peak runoff rate, the soil erodibility factor, the crop management factor, the erosion control practice factor, and the slope length and steepness factor for every hydrotope inside the sub-basin. The stochastic parameter taken from a gamma-distribution is included to allow realistic representation of peak flow rates, given only daily rainfall and monthly rainfall intensity. The sediment routing model consists of two components operating simultaneously – deposition of sediments and degradation of bottom sediments in the streams. The estimation of deposition and degradation in the stream channel is based on the stream velocity in the channel, which is calculated as a function of the peak flow rate, the flow depth, and the average channel width. The sediment delivery ratio through the reach is defined as a non-linear function of the stream velocity. If the delivery ratio is less than 1, sediment deposition occurs and degradation is zero. Otherwise, the deposition is zero, and the degradation is estimated from the delivery ratio taking into account the channel bottom erodibility factor.

The nitrogen and phosphorus modules include the following pools: nitrate nitrogen (N_{min} in Fig. D.19), active and stable organic nitrogen (N_{o-ac} and N_{o-st}, respectively), organic nitrogen in the plant residue (N_{res}), labile phosphorus (P_{lab} in Fig. D.19), active and stable mineral phosphorus (P_{m-ac} and P_{m-st}, respectively), organic phosphorus (P_{org}), and phosphorus in the plant residue (P_{res}), while the following flows are represented: fertili-

sation, input with precipitation, mineralisation, denitrification, plant uptake, leaching to groundwater, losses with surface runoff, interflow and erosion. The interaction between vegetation and nutrient supply is modelled by the plant consumption of nutrients and using nitrogen and phosphorus stress functions, which affect plant growth.

SWIM has been tested and its outputs verified for hydrological responses in more than 20 mesoscale and large river basins (with 64 to 100 000 km^2 drainage area); for nitrogen dynamics in two basins; for crop growth in a region; and for erosion in two basins (Krysanova et al. 1998a,b; Krysanova et al. 1999a,b). All these model verifications have shown that SWIM is able to describe with reasonable accuracy the basic hydrological dynamics, including the spatial and temporal variability of the main water balance components; the cycling of nutrients in soil and their transport; vegetation growth (including agriculture crops); the dynamic features of soil erosion and sediment transport under different environmental conditions. This provides a justification for studying the effects of changes in climate and land use on the interrelated set of basin processes.

D.2.6.6.2 *The ACRU Model*

ACRU (derived originally from Agricultural Catchments Research Unit) is a daily time step, process-based and

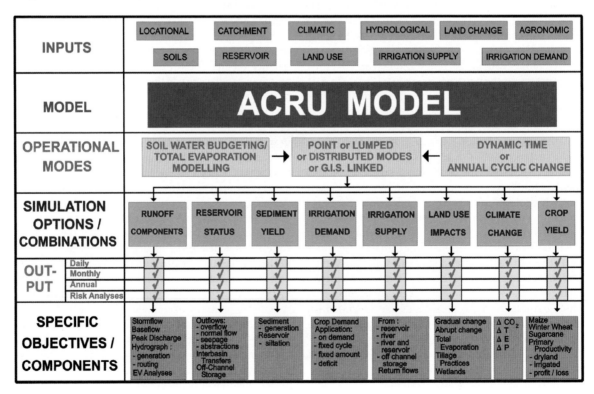

Fig. D.20. The ACRU modelling system and its major conceptual elements (from Schulze 1995)

multi-purpose distributed ecohydrological model (Fig. D.20 and Fig. D.21) with options to output, *inter alia*, daily values of runoff (i.e. event stormflow plus baseflow components), peak discharge, recharge to groundwater, sediment yield, phosphorus yield, reservoir status, irrigation water supply and demand as well as seasonal crop yields at a specific location or for a catchment. Conceptually the ACRU model is structured to represent processes over small catchments, where "small" designates a relatively homogeneous area (hydrologically speaking) over which any one of the catchment's representative daily rainfall, soils, land uses or antecedent soil moisture status can dominate the magnitude of stormflow. In practice, such small catchments are, ideally, from 1–50 km^2 in area.

For simulating large catchments, the individual small and relatively homogeneous subcatchments are hydrologically interlinked, cascading from upstream to downstream, with daily stormflow and baseflow from the land segment of each subcatchment computed separately and summed as the overall catchment runoff. This runoff then cascades into the next downstream catchment, is added to that catchment's runoff, and total flows are then attenuated by routing them downstream along channels, with their attendant hydraulic characteristics, and through reservoirs.

The land-based segment of ACRU revolves around multi-layer soil water budgeting in which partitioning and redistribution of saturated as well as unsaturated soil water takes place (Fig. D.21). Because of its strong process representation of interception, soil and surface manipulation, and dynamic above- and below-ground vegetation attributes affecting the evaporation and transpiration processes (including CO_2 transpiration feedback mechanisms), the model is structured to be hydrologically sensitive to changes in catchment land uses and management strategies as well as to climate change. Details, including all sequences assumed and equations used, are given in Schulze (1995).

The generation of catchment stormflow in ACRU, in addition to that from adjunct impervious areas, is based on the premise that after initial abstractions (through depression storage and the infiltration which occurs before runoff commences, both influenced strongly by tillage practices, land uses and seasonal rainfall intensity patterns), the stormflow produced is a function of

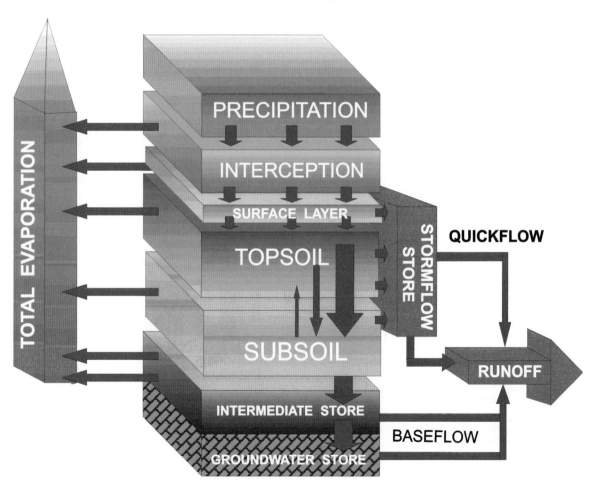

Fig. D.21. ACRU: Schematic of process representations (from Schulze 1995)

the magnitude of the rainfall and the dryness/wetness of the soil. This in turn is defined by the soil water deficit from a critical response depth of the soil. The soil water deficit antecedent to a rainfall event is simulated by ACRU's multi-layer soil water budgeting routines on a daily basis. The critical response depth results in different dominant runoff-producing mechanisms, predominantly surface v. subsurface v. baseflow responses, the proportions of which vary under different climatic, edaphic, topographic, vegetative and climatic regimes. Not all the stormflow generated by a rainfall event exits the catchment's outlet on the same day; stormflow is therefore split into quickflow (i.e. same day response) and delayed stormflow (Fig. D.21), with this "lag" (which may be conceptualised as a surrogate for simulating interflow) being dependent, *inter alia*, on the catchment's soil properties, its area, slope and drainage density.

The embedded sediment yield module in ACRU is based on the widely verified MUSLE, in which daily event-based stormflow and peak discharge (related respectively to the detachment and sediment yield processes) drive sediment production at the catchment level, while the source erosion components (dependent upon soil erodibility, slope length/gradient, surface cover/management and support/conservation practice) are derived from the Revised Universal Soil Loss Equation, RUSLE (Renard et al. 1991). A feature of ACRU's sediment yield module is that it can simulate the "first flush"

effect of extraordinarily high sediment yield from the first event of the rainy season – a phenomenon observed to be very important in semi-arid/sub-humid catchments.

Phosphorus in ACRU is modelled in two separate, but interacting, states representing the different sources of P and the mechanisms of P transport within a catchment (Kienzle et al. 1997). Relevant processes are depicted schematically in Fig. D.22. A relatively unique feature is the P derived from human sources from populations living in areas of inadequate sanitation with pit latrines which are liable to discharge to receiving waters. Loadings are differentiated between those from humans living within a 250 m buffer zone of streams, where elevated P impacts occur, and those from humans living beyond 250 m. Faeces-related *E. coli* concentrations in rivers, derived from livestock and humans and used as an indicator of the health status of streams, are modelled by processes similar to those of phosphorus.

ACRU's output from the catchment outlet (e.g. runoff, peak discharge, sediment yield) as well as that of internal state variables (e.g. soil moisture) and individual processes (e.g. interception, wetland responses, flow routing) have been verified in many catchments and field experiments covering many land uses and hydroclimates in South Africa, Swaziland, Zimbabwe, USA and Germany (e.g. Schulze 1995), including the Mgeni catchment (Kienzle et al. 1997), on which a case study is presented in Chapt. D.7.

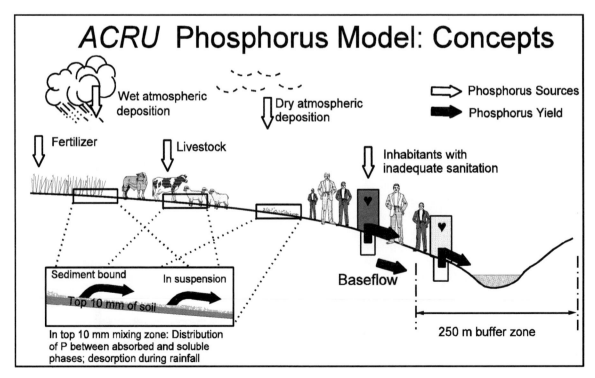

Fig. D.22. Major processes represented within the ACRU phosphorus module (after Kienzle et al. 1997)

Alfred Becker

D.2.7 Conclusions

Remarkable progress has been achieved during the past two decades in understanding hydrological processes and their physical, chemical and biological interactions. Comprehensive field studies in small catchments have played an important role here, as also to some extent the land-surface experiments discussed in Part B. The latter have combined, in different degrees, conventional measurements and observations with the application of advanced monitoring techniques such as:

- continuous monitoring of energy and gas fluxes at the land-surface/atmosphere interface;
- remote sensing to assess areal patterns of land-surface features and characteristics (land-surface heterogeneity) as well as their temporal variation;
- tracer techniques to investigate runoff generation processes and subsurface flows, in particular their different pathways and related transit times/residence times;
- detailed high-resolution modelling with distributed hydrological models.

Reference to some of these studies has been made in previous chapters in connection with the presentation and discussion of some of their results, the most important of which may be summarised as follows:

1. *Soil moisture patterns* and their dynamic variation with time clearly affect the atmospheric boundary layer and regional climate. Soil moisture represents the essential "water memory" of the land surface since with the biosphere soil water and groundwater can be supplied for evapotranspiration even from deeper soil horizons and from groundwater. Depending on root length and groundwater depth, water can also be transported to the atmospheric boundary layer through the plants during long dry periods. This is often not taken into account adequately in land-surface parametrisation schemes applied in atmospheric models. Most of these models try to represent soil moisture just by a single average number considered as representative of large areas. However, this does not reflect reality. Furthermore, the *seasonal freezing and thawing* of soil moisture during the cold season over about 35% of the Earth's surface has not so far received enough attention and is not yet taken into account in land-surface schemes. These issues are discussed in Sect. D.2.3, in particular Sect. D.2.3.1.
2. Additionally, soil moisture complements other land-surface features such as topography, land use, soil type, depth of impervious or saturated zones in controlling runoff generation and the associated *mobilisation of sediments and biogeochemical fluxes* (erosion and nutrients) during rainfall and snowmelt events. These processes and their dependence on the controlling land-surface characteristics differ between different environments and climatic zones (cf. Sect. D.2.4). Some of the controlling characteristics, for example soil organic matter, aggregate stability and crusting, are subject to change due to climate change and direct human impacts, and therefore should receive particular attention.
3. A significant step forward has been achieved in the understanding of *subsurface lateral flow processes*. Some flow pathways were identified only as early as in the 1960s, namely subsurface stormflows along preferential pathways in the ground, which include macropores, "pipe flows" and, especially in mountain environments, highly permeable soil layers at the soil bedrock interface. They are described in Sect. D.2.5, taking into account special phenomena such as the displacement of "pre-event water" by new incoming "event water" induced also by pressure wave propagation in the ground, including piston flow and groundwater ridging, and which play an important role here. In spite of this new knowledge, subsurface stormflow remains poorly conceptualised in many environments and crudely parameterised in most hydrological models. Those processes need to be classified and categorised for different climate and hydrological settings.
4. The high percentages of *"pre-event" water and relatively small contributions of "event" water in flood streamflow* are due to the significant differences between the response times of streamflow rises (flood flow) as a consequence of runoff generation in the catchment. The hydrograph composition and geographic source of stormflow in catchments vis-à-vis displacement of "pre-event water" by new incoming "event water" is not described in our current hydrological models. Most notably, the differences between particle flux and pressure wave propagation are not clearly separated in our current models. They need to be known for modelling subsurface biogeochemical fluxes through catchments and can only be explained by the aforementioned displacement processes and pressure propagation. More specific information is provided in Sect. D.2.2.3 and D.2.5.
5. Upscaling hydrological information is a crucial issue because some *hydrological processes and modelling approaches are strongly scale-dependent*. More attention should be paid to the definition of the spatial thresholds above which extrapolations are no longer valid.
6. Similarly, particular attention should be paid to the effect of *foreseeable extreme climatic events upon hydrological processes*. Since these events have rarely

been recorded in the past, they have not been used for calibration of predictive models. Sensitivity analysis should be thus conducted, considering that most relations are not linear. "Over-reaction" of hydrological processes to extreme events needs to be explored.

It should be emphasized that many of these improvements in understanding have been achieved by joint cooperative studies of BAHC and its main partner programmes, the International Hydrological Programme (IHP) of the UNESCO and both the World Climate Research Programme (WCRP) and the Operational Hydrological Programme (OHP) of the WMO. Several improvements have already been taken into account in modelling, at least at small catchment scales and in mesoscale river basins (see Chapt. D.2 and D.3). This concerns in particular the assessment of heterogeneity in terms of land-surface patchiness (see Sect. D.2.2) and related direct parameter estimation techniques, as discussed in Sect. D.2.6.4. Others, especially those on runoff generation mechanisms and subsurface stormflow, are usually not yet considered or only partly reflected in current modelling practices.

Chapter D.3

River Basin Responses to Global Change and Anthropogenic Impacts

Roland E. Schulze

D.3.1 Introducing the River Basin Scale and Its Response to Anthropogenic Change

Natural heterogeneities occur across a range of spatial scales from small plots to the globe, but dominate hydrological responses over a narrower spectrum, as already described in Chapt. D.1. If the river basin scale is defined spatially as spanning areas in the range of 10^2 to 10^5 km^2, then the conceptual depiction in Fig. D.96 shows that neither soil nor local topography is the dominant natural hydrological driver at this scale. On the other hand, physiography, vegetation, regional climate, the waterscape and, to some extent, macro-climate become key variables.

Similarly, anthropogenic influences also occur across a range of scales, but generally dominate hydrological responses most strongly over a relatively narrow range of spatial domains. Tillage and cropping practices, for example, might have major impacts on local hillslopes, but not typically a major one on larger river basins unless there is a nearly complete conversion of the catchment to agriculture. Water engineering, such as farm ponds, small impoundments, and irrigation can also generate considerable local effects. Collectively, these can

have substantial impacts even in very large river systems, as will be shown in Chapt. D.4. Further, water pollution generated from point and non-point sources in small catchments can persist up to the domain of a larger river basin.

If any one aspect characterises the river basin scale it is the juxtapositioning of human impacts on the hydrological regime, and vice versa, which is encapsulated by the so-called DPSIR model (Table D.5; i.e. Driving forces, Pressures, States, Impacts and Responses; McCartney et al. 2000). The elements are defined as:

- *Driving forces*, which are the broad features of change, such as anticipated changes through greenhouse forcing, demands of rising populations, expectations of food security (in developing economies) and water security, manifesting themselves as management and governance responses
- *Pressures* implying the specific, more proximal causes of hydrological change, including regional-scale climate change, variability such as caused through ENSO and changes in the frequency and intensity of extreme weather events, land-use change through both rural/urban migrations (particularly in developing economies) and extensive and intensive agri-

Table D.5. Changing hydrology at the river basin scale structured in terms of the DPSIR (Driving forces, Pressures, States, Impacts and Responses) model (adapted from McCartney et al. 2000, by Schulze 2001)

Driving forces	Pressures (i.e. causes of hydrological changes)	States (on hydrosystems: past, present, future)	Impacts (+ or – results of change)	Responses (international, national, local institutional)
Inter-seasonal climate variability	Regional climate change	Rivers: quantity	Degradation of ecosystems	Agenda 21
Greenhouse gas forcing	Local land use change	Rivers: seasonality	Loss of water rights	Dublin Statement
Rising population	Channel manipulation (dams, channel modifications)	Rivers: quality	Increased need for reliable water supply	ICM/IWRM as legal instrument
Rising security expectations	River basin water management	Groundwater	Amplification of extremes	New management strategies
State subsidies and directives	Rural-urban migration	Wetlands		New research directions
International market forces		Reservoirs		Ecosystem rehabilitation
		Lakes		Modelling

culture as well as hydraulic engineering including artificial impoundment, irrigation, channel modifications, and interbasin transfers;

- *States* implying instantaneous conditions of the hydrological cycle, resulting from proximal pressures and the history of these pressures. *State* embodies not only the quantity of water and its seasonal distribution, but also its quality in regard to suspended solids, water chemistry and the biological health of the river water. *State* also includes an accounting of the presence and condition of wetlands, lakes, artificial impoundments, and groundwater; and,

- *Impacts* which are the positive or negative environmental, social and/or economic consequences arising from changes in the state of the hydrological system or ecosystems resident within a river basin. *Impacts* include the degradation of terrestrial and aquatic ecosystems, loss of water rights, the amplification of extreme events or the loss of reliable water supply.

The DPSIR concept can be applied to all hydrological scales. However, there are a number of aspects to the river basin scale which set it apart from others and make it particularly amenable to analysis using DPSIR.

- *Biogeophysical attributes typical of small and large hydrological scales merge at river basin scale.*

  This is the scale at which all three of vertical, lateral and larger horizontal water movements attain more or less equal importance, as against the smaller scales where vertical (e.g. precipitation, infiltration) and, to a degree, lateral fluxes (e.g. interflow) often dominate. Furthermore, the river basin may be viewed as the meeting point of spatial upscaling (from point measurements to heterogeneous river basin fluxes) and climatic downscaling (from GCMs to regional climate models to distributed landscape process models).

- *Fundamental regional differences in hydrological responses attain importance at the river basin scale.*

  Differences in hydrological responses occur over space and time, and clear distinctions arise within a basin under contrasting climatic and/or physiographic regimes. River basins, by providing the potential for aggregated fluxes to be monitored at major tributary and mainstem mouths, allow researchers to distinguish responses both within and across river basins. They thus lend themselves to understanding contrasts between, for example, high *v.* low altitudes, flat *v.* mountainous terrain or humid *v.* arid basins. This has the complementary advantage of providing us with a framework for rejecting what Falkenmark et al. (1999) term "temperate zone imperialism", i.e. assuming that the way hydrological processes are represented in temperate zones may be transferred readily to other parts of the world in model algorithms.

- *At the river basin scale anthropogenic pressures reshape hydrological responses and produce aggregate impacts often far downstream of their origin.*

  In many parts of the world effectively entire river basins are experiencing massive manipulation of the land- and waterscapes in support of humans and "the four fs", i.e. food, fibre, fodder and fuel (Falkenmark et al. 1999; Schulze 2001). Development on the catchment includes urbanisation (both formal and of the informal shantytowns without proper services), intensification of agriculture (and associated pesticide/herbicide problems), extensification of agriculture (into climatically marginal areas, and including deforestation and overgrazing), or mining, all of which alter water quantity, its seasonal distribution and water quality. More direct water engineering includes dam construction (and associated changes to downstream flows of water and matter), inter-basin water transfers, major groundwater abstractions, land/wetland drainage, irrigation and in-channel modifications (e.g. dredging, channelisation). These multiple, engineered alterations to the natural hydrological system have become intermingled and/or diluted downstream through the patchwork of landscape changes, thus making it difficult to isolate individual causes of hydrological change or to lay individual blame on the impacts. Clearly a basin-scale perspective is essential to disentangle such complexities.

- *Policy relevancy and water management are applied at the river basin scale.*

  The river basin scale is the "action scale" (Falkenmark et al. 1999) for coping mechanisms with respect to water problems, including water scarcity, flooding, pollution, and public health issues. Because it is the action scale at which there is simultaneously a demand for public water supply, public sanitation, reduced flood risk, cheap hydropower or secure food supplies through irrigation, the river basin has also become a convenient management unit. It is the scale at which sound policy should be (but often is not) carried out in an environmentally sustainable manner by balancing the direct needs of people – often short term – with the indirect needs of a healthy environment – often longer term (Acreman 2000). Because there is a continuum of water users throughout the catchment, this is the scale where upstream actions can have impacts on (both negative and/or positive), or cause conflict with, downstream users. This is of particular importance in international drainage basins. With competition between sectors for a finite amount of water and the quest for an equitable allocation of this resource to both environment and development, multipurpose integrated water resources management (IWRM) becomes a major tool for managing river basins. In essence, the river basin scale is where "real people/communities with real land use and water decisions have to operate on real catchments" (Schulze 1999) and where the issues of stakeholders meet those of water policy-makers. Since this is the scale at which problems have their origin in planned human activities, but often with limited (or even erroneous) understanding of linkages between water, land and vegetation, it is also the scale at which many of these problems could be avoidable, provided driving forces were correctly identified and understood (Falkenmark et al. 1999) and scientists, managers and policy-makers appreciated more fully the interrelationships and feedbacks of the DPSIR concept (cf. Table D.5).

- *Attempts are made to "Right Previous Wrongs".*

It is within river basins *per se* that concepts such as channel restoration, wetland rehabilitation and ecological integrity are being enacted. The river basin is thus the scale at which biotic/abiotic linkages are researched and at which controlled floods are used as a management tool to mimic more closely pristine catchment flows in order to sustain aquatic habitats (Jewitt and Görgens 2000).

- *Hydrological models become operational tools for real-world decision-making.*

Hydrologists develop and/or apply models through the range of scales from point to small catchment to river basins to the global scale (Schulze 2000). At smaller experimental catchment levels, insights into micro- and meso-scale understanding of processes are usually the focus (cf. Chapt. D.2) while at continental to global levels, Earth system scale interlinkages and feedbacks are now being understood more fully than in the past. At neither of these extremes are major operational decisions made, however. The river basin, on the other hand, uses hydrological simulation models to aid in the engineering design and sizing of hydraulic structures, forecasts of high or low flows, water quality assessments, and estimation of sediment delivery and siltation of reservoirs. These often use the scenario approach to answer "what if" questions applied to water resources planning, design and operating problems, and design of alternative and mitigating options. The models thus become part of the water manager's armoury of decision support systems.

In light of the aforegoing discussion and the general introduction to Part D, the core of this chapter will initially address aspects of five focal areas, specifically:

- natural landscape processes and their role in shaping the character of the river basin;
- anthropogenic modifications to the river basin landscape;
- the combined character of river basins, represented today by a composite of natural and engineered landscapes;
- integrated water resources management; and,

- re-naturalisation of the contemporary hydrological landscape.

A full discussion of each of these foci would be immense in scope. Thus, for each focus, a few selected, key topics only will be covered.

Following the chapter on global scale are three river basin case studies. These cover a range of climates, levels of development and problems, and are

- the Amazon in South America, a large, humid and tropical basin with relatively little anthropogenic disturbance (Chapt. D.5);
- the Elbe, in Germany, typifying river basin problems in a highly developed country located in the temperate zone (Chapt. D.6); and,
- the Mgeni in South Africa, characterising land-use impacts in a sub-humid climate with a mixed developed/developing world economy (Chapt. D.7).

D.3.2 Natural Landscape Processes at the River Basin Scale

D.3.2.1 Introduction

Important natural landscape processes and factors influencing them which are relevant to the river basin scale are summarised in Table D.6. These processes and their landscape determinants occur largely as responses to atmospheric drivers of precipitation and other climatic characteristics, physiography and available energy. As a consequence of environmental pre-conditions, hydrological pathways at the river basin scale which determine the partitioning of precipitation into components of streamflow and evaporation are very different, dependent upon prevailing climatic and physiographic regimes. These aspects have already been discussed in Chapt. D.2. Three examples of natural landscape processes at the river basin scale follow. The first focuses on natural watercourses and their resident ecosystems, the second on evaporation and transmission losses from riverine systems and the third on wetlands.

D.3.2.2 Natural Watercourses and Aquatic Ecosystems

Natural Rivers: Attributes

Natural stream channels are central elements in most landscapes, being effectively interspersed into the landscape through dendritic drainage patterns (Nilsson and Jansson 1995). They are important natural corridors for the flows of energy, matter, plant and animal species, and are often key elements in the regulation and main-

Table D.6. Drivers of hydrology and water resources at different spatial scales (Schulze 2001)

Driver	Land- and waterscape Process, action structure	Crosscutting theme	Scale		
			Small catchment	River basin	Continental
Atmospheric drivers (including climate variability, forecastability and change)	**Natural landscape**	Integrated water resources management / Rehabilitation and renaturalisation of riverscapes and their catchments			
	Topography and its heterogeneity		×	×	
	Soils and their heterogeneity		×	×	
	Natural cover and its heterogeneity		×	×	
	Overland flow generation		×	×	
	Unsaturated zone processes		×	×	
	Subsurface stormflow		×	×	
	Groundwater recharge		×	×	
	Soil loss and sediment yield		×	×	
	Biogeochemical processes		×		
	Flow routing – channel			×	×
	Transmission losses			×	×
	Floodplain/wetland exchanges			×	×
	Sediment routing			×	×
	Estuarine processes			×	×
Anthropogenic drivers (including socio-economic development)	**Anthropogenic landscape**				
	Dryland cropping		×	×	×
	Plantation forestry		×	×	×
	Nutrient dynamics		×	×	×
	Urbanisation/industry		×	×	
	Mining		×	×	
Direct water management	**Water and river engineered landscape**				
	Groundwater abstraction		×	×	
	Irrigation		×	×	
	Channel modifications		×	×	
	Water storages and releases			×	×
	Water diversions (incl. transfers)			×	×

tenance of landscape biodiversity (Nilsson and Jansson 1995). The primary driving forces within natural rivers and their ecosystems are fundamentally annual discharge and the seasonal dynamics of river flows.

Rivers unmodified by human actions constitute complex and dynamic systems, displaying great spatial variability as well as morphological and hydrological uniqueness, with many self-regulating functions for flood peak attenuation, sediment storage and nutrient recycling (Sear et al. 2000). Channel forms and processes are dominated by interactions between the climatic and physiographic characteristics unique to each river basin, e.g. available energy of a river is a function of slope and discharge, while channel boundary conditions are determined by bank sediments, bank material and riparian vegetation. Natural history of the setting is also important. A good example is that slope and channel sedimentation are often the consequences of past glaciation, tec-

tonic movement, or high magnitude floods. Channel form may alter in response to changes in discharge and/or changes in sediment delivery (Sear et al. 2000).

The River Continuum and Pulse Concepts

Rivers and their ecosystems are the result of adaptations to the natural hydrological regime, which includes floods for exchanges of water, nutrients, sediments and organisms (Ward and Stanford 1995). The spatial and temporal variations in water depth and flow patterns in the channel, and the frequency and duration of inundation result in the inherent ecological diversity of river systems. Through this strong relationship between channel and floodplain, two important concepts have been identified by river ecologists, contributing to a holistic view of the river/floodplain complex (Acreman et al. 1999):

- *The River Continuum Concept* (Vannote et al. 1980) emphasizes the longitudinal connectivity from up-stream to downstream (i.e. source to mouth), associated with natural changes in intra- and inter-seasonal river flows, of water quality and aquatic species, distributed according to the gradients of sediment, nutrients, and organic matter which change progressively in the downstream direction; and,
- *The River Pulse Concept* (Junk et al. 1989) stresses the lateral connectivity between rivers and their floodplains. Floodplain inundation is regarded as the main driving force regulating the life of the river, with the channel supplying floodplains with nutrients and sediments and the floodplain supplying the river with improved water quality through settlement of sediments, processing of nutrients and detoxification of pollutants.

The longitudinal and lateral diversity embedded in these two concepts of natural watercourses stands in stark contrast with modified river systems where channel form is simplified, its diversity and its adjustment to dynamism is controlled, and riparian vegetation is managed or removed from floodplains, hydrologically disconnecting these important subsystems from the channel (Sear et al. 2000).

Riverine Ecosystems in Arid Zones: An Example of Adaptation

In few hydroclimatic zones are adaptations by riverine ecosystems to flow regimes as rapid as in arid regions. These regions are dissected by near-permanently dry, ephemeral river networks. A striking feature of rivers dissecting the Namibian desert, for example, are the "linear oases" of riparian forests adapted to the high variability in flow regimes (Jacobson et al. 1995). "Average annual" floods (if and when any flows occur in a given year) fill the river to its bankfall stage, maintaining the forests by providing essential nutrients and water. In the long periods of no flow the water table drops, older trees die, and open spaces are created for younger trees to fill. The massive episodic floods with return periods of tens of years (equivalent to the life expectancy of riverine trees) demolish whole forest reaches, create new channels within floodplains, recharge groundwater, deposit seeds and deposit logs/trees/branches which are sometimes visible for tens to hundreds of years (Jacobson et al. 1995). In the Namib, as in other arid riverine landscapes, forest composition along the length of the river varies in response to a close relation with water status. Where flow velocities are high, only trees rooting in rock fissures are found (e.g. *Ficus* spp.). Where river gradients are lower, clumps of flood resistant anas (*Faidherbia albida*) often line the channel reaches, with deep rooted

perennial woody species along less frequently flooded middle reaches and even more drought resistant camelthorn (*Acacia erioloba*) flanking the outer reaches where floodwaters are only experienced infrequently (Jacobson et al. 1995).

D.3.2.3 Evaporation and Transmission Losses from Riverine Systems

River Evaporation Losses

In humid, temperate or cold regions where either potential evaporation (E_p) is relatively low or precipitation exceeds E_p in most months, evaporation losses from riverine systems do not impact net streamflows significantly. However, when rivers, especially large ones having their source waters in higher rainfall areas, flow through arid regions where rainfall is low and E_p very high, evaporation losses from the water surface in the channel and surrounding riparian vegetation zones can become a significant factor in the water budget at river basin scale. In highly developed river systems, flows are highly regulated and releases or withdrawals from reservoirs may constitute a large fraction of basin water balance, for example, when used to satisfy irrigation (Vörösmarty and Sahagian 2000). Water management in such areas is thus challenging, in order to sustainably satisfy downstream ecological and societal water requirements without shortage or excess waste.

A good example of where evaporation losses from channels and the adjacent riparian zone needs to be factored into releases carefully is on the 2 300 km long Orange River in South Africa. Over the final 1 400 km of the river system downstream of the Vanderkloof Dam, mean annual precipitation averages only 145 mm (30–300 mm) while average annual potential evaporation is 2 850 mm (2 660–2 940 mm). With verifications against Bowen ratio energy balances and manual flow gaugings, McKenzie and Craig (1998) show that evaporation from the channel water surface is well approximated by class A-pan evaporation. Using findings of riparian water use by reeds and riparian trees (Birkhead et al. 1996), and aerial photograph estimates of the surface area of water and riparian vegetation, McKenzie and Craig (1998) calculated evaporation losses for the final 1 400 km of the Orange River at 575×10^6 m^3 yr^{-1} for flow releases of 50 m^3 s^{-1} and 990×10^6 m^3 yr^{-1} when flow releases are 400 m^3 s^{-1} (the larger number represents the impact of increased water surface area at higher flows). This equates to a 5.1% and 8.8% loss, respectively, of the natural mean annual runoff of the Orange River – a significant fraction for an already stressed river basin system. Assuming mean annual net river evaporation to be the sum of the differences between mean monthly A-pan evaporation and mean monthly rainfall for each of the 12 months, Fig. D.23

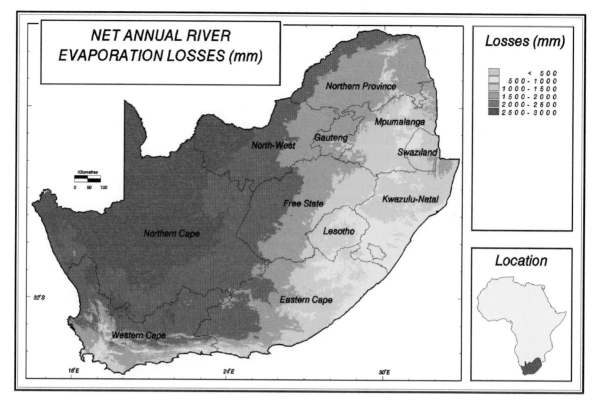

Fig. D.23. Estimated net annual evaporation from river surfaces over South Africa, Lesotho and Swaziland (after Schulze 2001)

shows that for South Africa net annual river evaporation can range from under 400 mm to over 2 500 mm per unit area of river surface. This large range establishes an important water resources challenge for the region.

Channel Transmission Losses

In addition to evaporation losses from riverine systems, transmission losses can take place by vertical recharge of water into the channel bed, lateral infiltration into the banks and, especially, infiltration occurring when high flows inundate the floodplain. Such losses can dominate upstream-downstream flow relationships in semi-arid and arid regions characterised by abundant ephemeral streams. These streamflow reductions also manifest themselves in regions with perennial but highly seasonal flows, especially at the onset of the runoff season when water levels in the channel are rising.

Two examples from southern Africa illustrate the increasing importance of riverine transmission losses with aridity. The first involved a phased, controlled release from the Sterkfontein Dam in South Africa into the Wilge River to Vaal Dam, 443 km downstream. Releases were initiated at 21 m^3 s^{-1} for two days followed by a stable 45 m^3 s^{-1} discharge for 19 days and then a 60 m^3 s^{-1} flow during the following nine days. Muller et al. (1985) show that the higher the flow, the relatively lower the losses.

Thus, for the 45 m^3 s^{-1} sustained flow, losses were 7% while for the 60 m^3 s^{-1} flow losses were reduced to 5.5% (they were expecting a 15–20% loss). Of these losses, 90% were from infiltration and the remaining 10% from evaporation.

The second example is from the arid 15 500 km^2 Kuiseb River basin in Namibia where annual E_p is around 3 000 mm. At its source in the east, mean annual precipitation is around 300 mm, but as the Kuiseb River winds 420 km westwards, mean annual precipitation decreases along a steep gradient to ~ 25 mm, while inter-annual variability increases. Episodic runoff is thus generated in the east and ephemeral flows move westwards into the generally rainless zone (Jacobson et al. 1995). The significance of rapid transmission losses downstream is clearly illustrated in Table D.7 and Fig. D.24. The table shows that at flow recording stations from east (360 km inland) to west (near the delta), the number of flood events per year declines from 5 to 1, average runoff volumes decrease (despite an increasing catchment area) and first recorded floods occur progressively later (December to February). Although total volume of water varies between rainy seasons, Fig. D.24 illustrates the persistency of the pattern of rapid downstream transmission losses year by year, with larger floods travelling farther downstream and, by inference, floods travelling further if antecedent conditions provide sufficient moisture (Jacobson et al. 1995).

Table D.7. Characteristics of floods and channel transmission losses from the Kuiseb River in Namibia (after Jacobson et al. 1995)

Recording station	Average month of first flood	Average no. of floods per year	Average runoff (m³)	Distance from delta (km)	Estimated[a] MAP (mm)
Us	December	5	6 000 000	360	300
Schlesien	January	3	16 000 000	250	200
Gobabeb	January	2	4 500 000	85	40
Swartbank	February	1	1 750 000	50	25
Rooibank	February	1	570 000	18	25

[a] Estimated from mapping of mean annual precipitation (MAP) of Namibia in Jacobson et al. (1995).

D.3.2.4 Wetlands

Wetlands embrace a diverse range of habitats, occupying the transitional environment between permanently wet (aquatic) and generally drier (terrestrial) areas of the landscape, while functioning differently to either aquatic or dry habitats. Their transitional status can apply both to their long-term state or their temporary status due to events or the progression of the seasons.

The Hydrological Significance of Wetlands

In a river basin context, wetlands provide many hydrological functions and public service functions that are "free of charge". These include supply of water, groundwater discharge/recharge, flow attenuation and flood protection, retention of sediments and of nutrients, recycling of nutrients including pollutants, and bank stabilisation. In addition, wetlands provide areas for recreation and are often a mode of water transport. Wetlands'

products include forest resources, wildlife, fish, forage resources, agricultural land (e.g. for rice production), peat for energy and water supply. They, furthermore, provide biological diversity.

- *Wetlands as suppliers of water.* Because of their topographically low position in the landscape, wetlands, with their higher soil water content in the topsoil than more upland sites, tend to be providers/generators of surface water as well as groundwater (McCartney and Acreman 2003);
- *Wetlands as flow attenuators.* Wetlands act as important water storage sites during wet periods and as a reserve during dry periods, when they tend to maintain recession flows, especially from headland wetlands (McCartney and Acreman 2003). This is illustrated well by Schulze (1979) by the differences in flow characteristics between the Ntabamhlope wetland-dominated catchment in South Africa (33.6 km²) and its adjacent non-wetland DeHoek catchment (14.6 km²). Ntabamhlope is characterised by slower recessions in terms of unit flows (i.e. m³ s⁻¹ km⁻²) in the post rainy

Fig. D.24.
Reductions in downstream flows through channel transmission losses in the Kuiseb River, Namibia (after Jacobson et al. 1995)

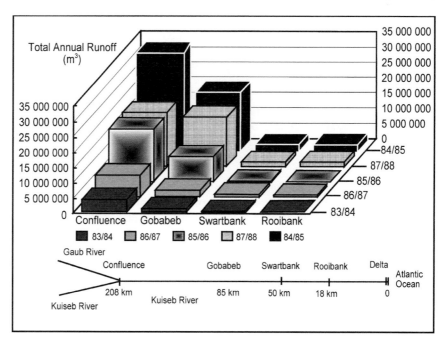

Fig. D.25.
Comparison of mean monthly streamflows and their inter-annual coefficients of variation in adjacent wetland (Nta-bamhlope, 33.6 km²) and non-wetland catchments (DeHoek, 14.6 km²) in South Africa (after Schulze 1979)

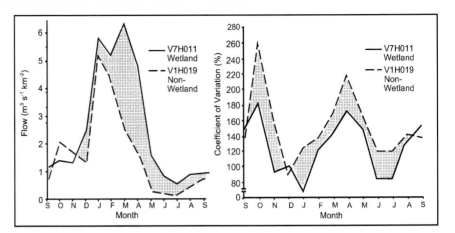

season (January to April), as well as the lower distinct peaks of flows (Fig. D.25, *left*). This is corroborated by much lower interannual coefficients of variation in this wetland-dominated catchment for 10 months of the year (Fig. D.25, *right*).

- *Wetlands as maintainers/improvers of water quality.* Because of reduced flow velocities sediments are deposited within wetlands. The reduced sediment load, as well as the storage and removal of heavy metals, pesticides and herbicides carried by the sediment, enhances downstream water quality (McCartney and Acreman 2003). Furthermore, a large variety of aerobic and anaerobic processes take place in wetlands (e.g. nitrification, denitrification, ammonification and volatilisation). Wetlands are also responsible for the removal of certain chemicals from the water column, in some cases converting soluble forms of heavy metals to insoluble forms, or removing nutrients from agricultural runoff, which could otherwise result in eutrophication of receiving water bodies. Russel and Maltby (1995) have shown in the Torridge catchment in the UK, for example, that wetlands reduced soluble P concentrations by 50–80% while also finding that 95% of the nitrates discharging into the same wetland were being stored and denitrified. The nutrient retention capacity of wetlands is especially effective during low flows, but relatively ineffective during high flows (McCartney and Acreman 2003). Because of their use as a buffer zone in areas vulnerable to pollution, wetlands are increasingly being protected, rehabilitated, and reconstituted.
- *Wetlands as rechargers/dischargers of groundwater.* Because impermeable soils or geological formations (e.g. sills) typically underlie wetlands and restrict water percolation, most wetlands do not recharge aquifers. If they do, it is usually very slowly, as in the Hadejia-Nguru wetlands of Nigeria (Hollis et al. 1993). Subterranean water stored in wetlands is, however, a very important retainer of elevated soil moisture during dry periods. McCartney (1998) showed that the dambo wetlands of Zimbabwe remain wet into the

dry season by maintaining a water table close to the ground. This acts as an important water resource for cultivators and pastoralists.

- *Wetlands as evaporators of water.* Water stored in many wetlands may be dominated by evaporation losses, with often only a small proportion of stored water supporting recession flows (McCartney 1998). Such high evaporative demands by wetlands can manifest themselves also by strong diurnal fluctuations in the wetland's water table hydrograph, as has been illustrated by Donkin et al. (1995) for the Nta-bamhlope wetland. Donkin et al. (1995) and Birkhead et al. (1996) have, in very different hydroclimatic zones within South Africa, shown that in summer months evapotranspiration from wetlands was often 1.5 to > 2 times in excess of Penman open water estimates, but only ~ 0.5 times as high in winter, when dormant reeds shade the evaporating surface.

Threats to Wetlands

Hydrology is the single most important determinant in the establishment and maintenance of wetlands and their inherent properties. Overt drainage and destruction of wetlands is pandemic. It is estimated that 1.9×10^6 km² of the world's natural wetlands have been lost (Meyer and Turner 1992). The extent of threats to wetlands are illustrated by the estimate that since 1930 in England 64% of the wet grassland in the Thames Valley has been lost (English Nature 1997) and that in the upper Mgeni river basin in South Africa 66% of the wetland area had been lost to man-induced causes (Jewitt and Kotze 2000).

In addition to outright destruction, threats to wetlands also arise through the removal of water and modification of the delicate interconnectedness between flow and sediment/chemical regimes. When hydrological conditions in wetlands change, even slightly, biotic and abiotic responses may be massive (Mitsch and Gosselink 1993), with potentially major detrimental consequences

Table D.8. Comparison of percentages of selected hydrologically important land uses in South Africa, Lesotho and Swaziland as a whole, and in some of South Africa's provinces (after Schulze 2001, based on Thompson's National Land Cover images from 1996 satellite imagery)

Land-use category	Country or province[a] (area in km^2)						
	RSA, Lesotho, Swaziland (1 267 681)	Mpumalanga Province (78 238)	Northern Cape (362 393)	Gauteng Province (18 610)	Free State (129 833)	KwaZulu-Natal (92 285)	Western Cape (129 578)
Plantation forest	1.47	9.19	0.01	1.29	0.16	6.75	0.83
Urban	1.29	1.95	0.16	19.30	0.80	1.57	0.83
Cultivated – rainfed	10.87	15.80	0.28	19.14	28.34	15.83	13.53
Irrigated	1.23	1.74	0.46	0.92	0.53	1.47	3.52
Severely degraded	4.79	2.11	0.88	0.13	0.68	7.97	2.76
Wetland	0.48	0.13	0.81	0.40	0.89	0.83	0.18
Grassland	21.30	40.70	12.00	42.30	50.80	39.20	0.94
Reservoir	0.38	0.51	0.18	0.47	0.63	1.11	0.50
Others[b]	58.19	27.87	85.22	16.05	17.17	25.27	76.91

[a] For location of provinces refer to Fig. D.23.
[b] "Others" will be mostly vegetation still in a relatively near-natural state.

to the integrity of wetland ecosystems (McCartney and Acreman 2003). This can take place by drainage and conversion to intensive arable land for agriculture, groundwater abstraction (Acreman and José 2000), permanent flooding due to small dam construction, alien plant infestation (Jewitt and Kotze 2000), irrigation or construction of dams upstream of the wetland. In the case of the latter, Schulze et al. (1987) report that regulated flows from proposed dams upstream of the Franklin wetland in South Africa would, through flow regulation, reduce influent discharges into the wetland for nine months of the year and decrease water storage within the wetlands significantly during low flow periods.

D.3.3 Anthropogenic Modifications of the River Basin Landscape

D.3.3.1 Land-Use Change and Its Impacts on Hydrological Responses

D.3.3.1.1 *Extent of Land-Use Changes*

Our recent history is one of dramatic increases in the incidence of anthropogenic disturbance to the natural landscape. Humans are utilising a growing proportion of the Earth's surface and appropriating increasing amounts of the Earth's productive capacity and natural resources (Turner et al. 1991). Both the extent of their modification and the predominance of different types of disturbances are determined by prevalent land-use patterns.

Worldwide only about 25% of the Earth's 130×10^6 km^2 surface area remains in near-natural conditions (Turner et al. 1993). Losses of natural wetlands have already been

mentioned. Approximately 4×10^6 km^2 of the total land mass has been urbanised (Douglas 1994), 14.3×10^6 km^2 is under cropland and 2×10^6 km^2 irrigated (Richards 1990). Natural closed forests have been reduced by 7.2×10^6 km^2 (Meyer and Turner 1992).

Historically, in the "old world", especially in temperate climates, land-use change has been going on for centuries, and in the UK 70% of the land is now under agriculture and grazing, 10% urban, 10% derelict land and mines while only 10% remains in natural woodland (DOE-UK 1993). In "newer world" countries the levels of urbanisation and agriculture may be proportionately lower, but other hydrologically sensitive land-use issues such as soil and land degradation (e.g. through overgrazing or deforestation of woodlands) become important, as in the case of South Africa (Table D.8).

D.3.3.1.2 *The Interplay between Society and Nature in Land-Use Change, and Its Hydrological Significance*

The drastic changes in land use have their roots in the interplay between society and the natural landscape. Societal requirements for food, water, energy or hazard prevention have to be satisfied by the natural resources of water, biomass, energy and minerals (Falkenmark et al. 1999). Through society's needs, the natural landscape is therefore manipulated for purposes of livelihood in both physical terms (e.g. land conversion) and chemical terms (e.g. waste and pollutant production). This, in turn, has environmental side effects on, for example, air, water, land and ecosystems. The natural system, in turn, generates with reactive responses (both passive – usually in the developing world, and active –

more characteristic of developed countries) which can feed back into the landscape again and into a changed capacity to provide free environmental services. Falkenmark et al. (1999) distinguish between two types of land-use activities that have a fundamental bearing on the hydrological cycle, i.e.

- land use which is dependent on water, i.e. where water poses limitations on societal production from land, an example being when there is too little water or too much water – both of which are limitations/disturbances caused by nature; and
- land use which impacts directly on water, i.e. when water balances and flows change through altered partitioning of precipitation into evaporation and runoff as a consequence of disturbances created by societies (e.g. urbanisation, overgrazing, tillage practices), or when flows of matter are altered by new land-management scenarios (e.g. sediments or nutrients). In either case, every land-management decision becomes a water resources decision (Falkenmark et al. 1999).

D.3.3.1.3 Types of Land-use Change Based on the Degree of Alteration from Natural Ecosystems

Hobbs and Hopkins (1990) have categorised land uses into four divisions based on the degree of alteration from a natural ecosystem state:

- Conserved ecosystems, i.e. where no deliberate modifications to the natural landscape has occurred, either by design (e.g. wilderness/conservation areas; unutilised government owned land; catchments reserved for water production) or by default (e.g. harsh environments). Hydrological responses are not altered from baseline responses;
- Utilised ecosystems, i.e. exploitation of indigenous ecosystems (e.g. by non-plantation forestry; pastoralism; recreation), with negligible hydrological impacts;
- Replaced ecosystems, i.e. with intensively managed removal of native ecosystems and replacement by simpler, less biodiverse systems geared towards agricultural, horticultural and plantation forestry production, with hydrological consequences in regard to water quantity and its seasonality, as well as to water quality, ranging from minor to major; and
- Completely removed ecosystems, i.e. destruction of native ecosystems, with each land use encompassing a suite of deliberate and/or inadvertent impacts of varying severity (e.g. urban and industrial development, mining, transport), with major local hydrological (e.g. attributes of peak flow) and water quality (e.g. chemical pollution) impacts.

D.3.3.1.4 Land Transformations and Ecohydrological Changes of State

In a manner akin to that of ecologists who view a land-use transformation as a change in "ecosystem state" (in terms of structure, composition and/or function) so, to hydrologists, changes in land use also represent hydrological response transformations (Schulze 2001). These may be represented as follows:

- Attribute changes (Schulze 1995, 2001) include changes in:
  - above ground characteristics, such as biomass and aerodynamic roughness, implying changes in canopy and litter interception losses, in consumptive water use by plants and in shading the soil, thereby partitioning evaporation of water from the soil surface and from plant tissue (transpiration) differently;
  - ground/surface character changes, such as the extent of compaction or imperviousness, surface crusting and sealing, surface roughness, surface cover by litter/mulch or conservation structures, all of which alter pathways of water entry into the soil and consequently runoff/erosion from the surface; and
  - below-surface attributes, such as bulk density and hence water transmissivity following various types of tillage practices.
- Ecological response changes are expressed by Hobbs (2000) in terms of decline and recovery over time as rapid and permanent changes of state (e.g. following urbanisation or permanent clearing for agriculture), gradual degradation (e.g. from overgrazing), sudden disturbance but with full recovery (e.g. following a fire), sudden disturbance but with only partial recovery (e.g. urbanisation followed by a re-greening of the urban landscape), recovery following cessation of stress (e.g. rehabilitation of rangeland following removal of cattle) and re-establishment following deliberate action (e.g. ecosystem restoration or wetland rehabilitation).
- Hydrological response changes, from the river basin perspective, are changes of ecosystem states and the associated attribute changes are accompanied by changes in flows of water and/or sediments and/or nutrients. The more purely hydrological flows may be manifested as increases in flows (e.g. following urbanisation), decreases in flows (e.g. following afforestation), changes in seasonality of flows (e.g. after dam construction), changes in peakedness/responses of flows (e.g. higher peaks and shorter lag times following a fire), changes in the partitioning of flows into baseflow or interflow or stormflow (e.g. following institution of agricultural conservation practices) or re-naturalisation of flows, after rehabilitation (Schulze 2001).

D.3.3.1.5 Hydrological Considerations Regarding Land-use Change

The potential impact of land use on hydrological responses is beyond dispute. In considering land-use impacts, however, a number of factors need to be borne in mind:

- Land-use change often leads to ecosystem degradation. This view is held strongly by ecologists (e.g. Hobbs 2000). However, the feedbacks to hydrology (and to society) may be either negative (e.g. through population pressure/drought in a subsistence agricultural setting), or positive (e.g. the effects of community gardens within a subsistence agricultural setting);
- Land-use change usually takes place slowly, and in a largely piecemeal fashion as individual farmers respond to market forces and/or legislation (Acreman and Adams 1998; Robinson et al. 2000) or as cities expand. This is especially important in a regional context;
- Land transformation is easily measurable at local scale, but not easily distinguishable regionally. At point, plot or experimental catchment scale process studies often provide unequivocal evidence of direct hydrological effects associated with a particular land use. However, at progressively larger river basin scales these impacts are difficult to detect as they become attenuated by aggregation with other stable land uses and/or as their effects are diluted downstream (e.g. Acreman and Adams 1998; Schulze et al. 1998; Robinson et al. 2000);
- Some land-use impacts display considerable lag. Impacts, particularly those reacting through the groundwater system, such as nitrate pollution, or the effects of plantation/afforestation on low flows, or those related to sediment transfers may only manifest themselves several years to decades later. Such potential impacts require careful consideration of the time dimension of the larger problem;
- The impact of land use often depends on its intensity. The intensity of urbanisation is a case in point, as is the application of artificial fertiliser on agricultural land; and,
- Land-use management often has more significant hydrological effects than land-cover change. Hydrologists tend to place great emphasis on land-cover change. It is often, however, the management practice (e.g. grazing-controlled or overgrazed, depth and type of tillage, rate/amount of fertiliser application) and mechanical practice (e.g. terraces, contours and other conservation structures) that partition precipitation into hydrological responses (e.g. overland flow, recharge, sedimentation, nitrate leaching) to a far greater degree than, for example, changing from crop A to crop B (e.g. Moerdyk and Schulze 1991; Schulze 2001).

The conclusion drawn from these six points is that it is difficult to quantify hydrological responses to the impacts of land-use change at the river basin scale when considering that extensive areas of a basin are owned or used by many individuals and, hence, are the result of a large number of individual management decisions. Even if a time trend in a hydrological variable can be identified, its link to land use is often complicated by intra- and interannual climatic variability (Acreman and Adams 1998). This alone can already mask/obliterate any statistical significance one may wish to ascribe to the impact of land-use change by itself (e.g. examples in Schulze et al. 1998).

What follows is a brief overview of selected examples of the hydrological impacts of plantation afforestation, urbanisation, fertiliser application and salinisation.

D.3.3.2 Plantation Afforestation Effects

Planned conversion of land cover to plantation forests is widepread and has important hydrological consequences. For example, in Wales the aim is to expand area under forests by 50% (Robinson et al. 2000), in the UK a proposed doubling in area was announced by HMSO (1995) and in South Africa already 15 000 km^2 are under fast-growing exotic forest plantations (Dye and Bosch 2000). Afforestation of, for example, a native grassland to monocultured plantation forest constitutes potentially the most drastic agricultural land-use change in hydrological terms (Robinson et al. 2000), particularly if it is areally extensive.

D.3.3.2.1 Characteristics of Plantation Forests and their Hydrological Impacts

Plantation forests constitute tall and aerodynamically rough trees with typically deep root systems, a dense and closed canopy cover which is often evergreen, and thick litter layers. In contrast, the vegetation they typically replace in many parts of the world consists of shorter and aerodynamically smoother land covers with shallower root systems, a less dense and more open canopy cover (often seasonally dormant) and thin litter layers. Hydrologically, these characteristics translate into reduced availability of incident rainfall due to elevated canopy and litter interception and enhanced evaporative losses from dense, evergreen and aerodynamically rough trees having deep root systems. As a result, runoff, erosion and water quality are altered.

Not only do forests intercept more precipitation than grasses because of their greater canopy biomass, but their higher aerodynamic roughness (r_a) allows the forest to transport stored heat from its surface to the air more readily under non-rainy conditions. This enhances

Fig. D.26.
Increases in annual evapotranspiration in South Africa for three tree species as an index of reduction in annual runoff in afforested catchments in different climatic regions and with varying extents of forest coverage (after Dye and Bosch 2000)

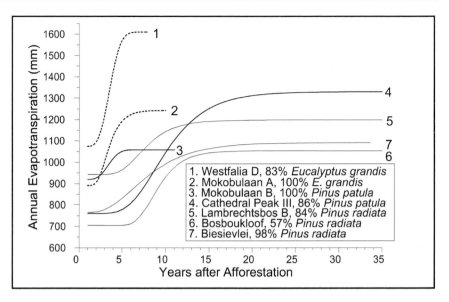

1. Westfalia D, 83% *Eucalyptus grandis*
2. Mokobulaan A, 100% *E. grandis*
3. Mokobulaan B, 100% *Pinus patula*
4. Cathedral Peak III, 86% *Pinus patula*
5. Lambrechtsbos B, 84% *Pinus radiata*
6. Bosboukloof, 57% *Pinus radiata*
7. Biesievlei, 98% *Pinus radiata*

dry canopy transpiration, which can also utilise soil moisture extracted from deeper layers in the soil profile, especially during dry periods. The higher r_a furthermore evaporates intercepted water from the wet canopy at rates up to 10 times that from shorter vegetation by assisting advective processes, which increase the rate at which heat can be supplied from the atmosphere to the forest surface (Calder 1999).

Progressive increases in evapotranspiration arise in conjunction with forest growth and maturity through a rotation, as illustrated from experimental evidence in South Africa (Fig. D.26). Evapotranspiration increases most notably in the earlier years of canopy closure. However, species differences are apparent (with eucalypts exhibiting faster canopy development than *Pinus* species), and climatic influences can speed up or slow down canopy development (e.g. for *P. patula* the site at Mokobulaan is notably warmer than that at Cathedral Peak), while the degree of afforestation influences evapotranspiration (cf. curve 6 *v.* 7 in Fig. D.26).

The increases in annual "observed" evapotranspiration illustrated above, determined as the residual between observed precipitation and runoff in gauged catchment experiments, represents the reduction in runoff as the trees grow. The values of 150–500 mm in Fig. D.26 concur with those of Bosch's and Hewlett's (1982) worldwide review and with the 200 mm difference observed in the Plynlimon research catchments in Wales between a grassed and forested catchment (Kirby et al. 1991).

Both literature and theory support the idea that total annual streamflow from regrowing forested catchments tend to become progressively reduced (although there is evidence that in mature forests this trend may be reversed as canopy biomass decreases again), but also that the reductions generally take place both in stormflow generation (because of increased interception losses and infiltrability) and in baseflow generation (because of the

deeper rooting), with dry season flows usually being impacted relatively more severely than wet season flows (e.g. Scott et al. 1998; Calder 1999). There are always some locally determined exceptions to these generalities, however (Calder 1999). The changing composition of plant communities and their respective water use efficiencies also regulates water budgets (Shiklomanov and Krestovsky 1988).

Not only do the extent of afforestation, ambient climate and site condition determine streamflow reduction, but also the topographic position/location of forests within a catchment. In South Africa there is evidence from clearfelling experiments in riparian *v.* non-riparian forests that because of increased soil water availability through lateral drainage, trees growing in the riparian zone have a considerably higher impact on runoff generation than those growing on more upslope positions (Table D.9).

D.3.3.2.2 *Modelling Impacts of Plantation Afforestation at River Basin Scale*

Many models have been developed to assess the impacts of afforestation at ungauged locations within river basins (Calder 1999). These models are generally of two types, empirical and process-based. Examples of both of these, which are applied to environmental decision-making in South Africa, are illustrated below.

The empirical Scott curves (Scott et al. 1998), based on relatively limited experimental evidence from selected gauged catchments, distinguish between total annual and low (dry season) flows for two major genera of forest plantations (eucalypts and pines) at optimum *v.* sub-optimum (based on climate) sites, assuming the sensitive riparian zones in a catchment not to be planted with trees. Figure D.27 shows marked differences in to-

Table D.9. First-year runoff responses between forest clearfelling in riparian *v.* non-riparian zones recorded in three experimental catchments in South Africa. Response is expressed in terms of volume increase per unit area of catchment treated, and as a percentage flow increase per 10% of the catchment cleared (after Dye and Bosch 2000)

Catchment	Mean annual precipitation (mm)	Experimental treatment	First year increase in runoff (% per 10% cleared)
Westfalia D	1 611	Clearfelling indigenous forest	9
		Clearfelling non-riparian indigenous forest	3.5
Witklip 2	996	Cut and poison riparian scrub	44
		Clearfelling non-riparian pines	37
Biesievlei	1 427	Clearfelling riparian pines	44
		Clearfelling non-riparian pines	14

tal and critical low flow reductions between faster growing eucalypts and slower growing pines at different ages and between optimum and sub-optimum sites. While easy to use, the Scott curves do not account adequately for the diversity of rainfall and temperature regimes, nor for variations in soils properties or management practices within afforested river basins (Dye and Bosch 2000).

Process-based models such as the ACRU (Schulze 1995 and updates) are now being used more frequently under water conflict situations in South Africa. This daily time step, physical-conceptual and multi-layer soil water budgeting model can account for a variety of explicit water cycle processes. These include enhanced wet canopy interception, deep rootedness, *in situ* climatic and soil conditions, forest rotation length, decline of forest leaf biomass and water use with age, stomatal control, management scenarios (e.g. thinning; site preparation) and riparian zone influences on subsoil water availability (Schulze et al. 1997; Meier et al. 1997). Model output has been extensively verified under plantation afforestation conditions (e.g. Schulze and George 1987; Jewitt and Schulze 1999), with model enhancement and verification studies continuing. Results with the ACRU model in potentially afforestable areas with mean annual precipitation ≥ 650 mm show that the wetter the area, the higher the runoff reduction will be in absolute terms,

while the drier the area, the higher will be the runoff reduction in relative (i.e. %) terms. ACRU has also been used to show how more severe local (i.e. individual subcatchment) impacts of land use on flows become attenuated and diluted in a river basin as flows accumulate and cascade downstream (Schulze et al. 1998). This is illustrated in Fig. D.28 for median annual flows in the Pongola-Bivane River basin of ~ 4 000 km upstream of Subcatchment 17. Individual subcatchments display highly variable degrees of plantation afforestation and local impacts, but the accumulated flows are highly smoothed downstream.

D.3.3.2.3 *Plantation Forests and Water Conflict*

Conflicts surrounding forests and water resources abound at river basin scale, particularly where afforestation to exotic species is practised in drier climates. The conflicts are often of an upstream:downstream nature (cf. Fig. D.28), where downstream users may be impacted by upstream afforestation (e.g. Schulze et al. 1998). The steeper reductions in dry season flows (cf. Fig. D.27), which are vital for maintenance of aquatic habitats and to local irrigators, as well as afforestation to exotic species within riparian zones and the invasion of alien tree species there, are

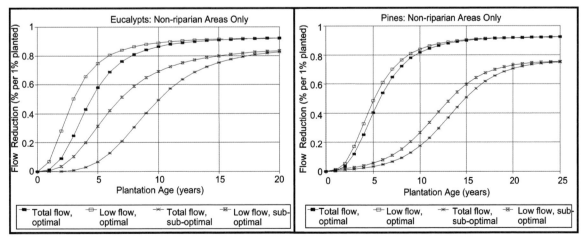

Fig. D.27. Flow reduction curves for eucalypts and pines in South Africa, assuming riparian zones unplanted (after Scott et al. 1998)

Fig. D.28.
Impacts of varying percentages of plantation afforestation per subcatchment (% afforestation in brackets) on median annual streamflows in the Pongola-Bivana river basin in South Africa, and its attenuation downstream by flow accumulations (after Schulze et al. 1998)

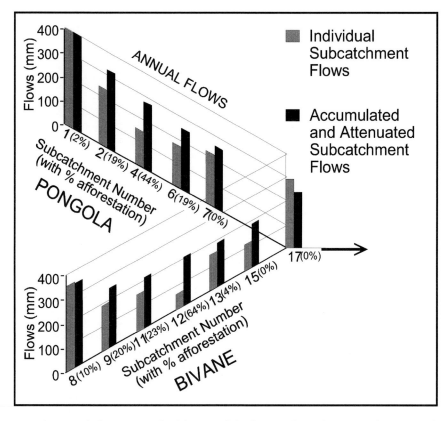

major concerns. In fact, plantation forest and alien invasive clearance of the riparian zone has been used as a compensatory mechanism for permitting more trees to be planted upslope. From a purely water resources perspective, countries such as South Africa, which are hydrologically sensitive to plantation afforestation, have a considerable body of legislation in place controlling which catchments may be afforested, to what extent, and where, within the catchment, afforestation may take place. There is even a levy now placed on afforestation because it is a "stream flow reduction activity" (NWA 1998).

D.3.3.3 Urban Influences on Hydrological Responses

While some 50% of the world's inhabitants live in urban areas which cover 4×10^6 km^2 of the Earth's surface (Douglas 1994), the intensity of urbanisation varies greatly among and within countries. For example, 88% of England's population is urbanised and urban expansion there is increasing by 15 000 ha yr^{-1} (Robinson et al. 2000), while in South Africa as a whole 48% of the population resides in urban areas (covering < 1.3% of the country), but this varies widely by province between a mere 9.1% of population urbanised in Limpopo Province against 96.0% in Gauteng Province (Schulze 1997).

Relative to its extremely modest overall area (~ 3% worldwide), the impacts of urban zones on water quantity and water quality are disproportionately high and

highly varied (Falkenmark et al. 1999; Robinson et al. 2000; Schulze 2001). The impacts depend on:

- interactions with the local environment;
- the level and history of development;
- the degree of "formality" *v.* "informality" of urbanisation (i.e. type of water infrastructure and services);
- the degree and type of associated industrialisation;
- technical strategies adopted for water supply to meet water demands; and,
- measures to minimise downstream water quality degradation.

This brief overview of urban influences on river basin level hydrology touches on issues of urban water supply systems, the urban drainage system and flood protection – all artificial hydrological systems.

Starting in the 1800s and extending into much of the 20th century, for many "old world" cities and towns water was originally supplied from local surface and/or groundwater sources. There was generally poor waste control, generated by surface or combined sewer systems. For most cities with their high demands for potable water (with a characteristic, rhythmic daily and often also a seasonal pattern) clean water was a chronic concern, until modern water treatment and sanitation systems were installed. Although this is seldom of present-day concern in the developed world, these problems persist in developing world cities and the progress

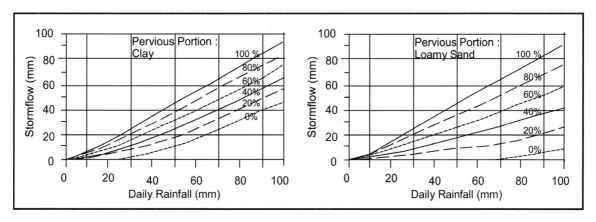

Fig. D.29. Conversion of daily rainfalls to stormflows for different percentages of imperviousness on different soil textures, assuming pervious areas to be well grassed and the impervious portion to be connected to stormwater drainage (from Schulze 2001)

of economic development can be traced to improvements in water supply and sanitation (Meybeck and Helmer 1989). Cities today rely increasingly on imported sources of water from remote sources. If remotely supplied water is from the upstream part of the basin in which the city resides, natural river flows, after passing through the urban area, are controlled and often reduced. If the sources are external to the basin local river flows emerging from the city can be increased through leakage of bulk water supply, return flows, groundwater flooding and/or leakage of stormwater drains.

A major hydrological response of urbanisation is the effect of increasing impervious areas (e.g. paved areas, roofs). This serves typically to reduce the time from the onset of stormflow to peak discharge and to increase the peak. Figure D.29 illustrates the conversion of daily rainfall into stormflows for different percentages of imperviousness for a system "connecting" stormwater drains to a stream and for contrasting soil textures that make up the pervious portion, which is assumed to be well grassed. However, not all impervious areas are adjunct, i.e. connected directly to a stormwater system, and some paved/roofed areas are disjunct, i.e. disconnected and discharging onto surrounding pervious areas where they contribute to enhanced runoff response. These factors all increase local flood peaks, as does channelisation which, by straightening and converting natural channels to concrete, can reduce the times to peak by a factor of 3–4. Local urban flooding can be particularly severe with rapid convergence of flood surges through the drainage system when restricted by culverts or with a high urban water table prevailing (Robinson et al. 2000). Some of these hydrograph changes are illustrated in Fig. D.30.

Urban flood protection relies on engineering techniques that reduce runoff rates and to either settle or break down pollutants. Innovations include underground storm tanks, oversize sewers with restricted outlets and "hydrobrakes", flow diversions or controls of runoff response at the source (e.g. porous pavements, gravel drains, diversions to pervious areas, contraction of artificial

Fig. D.30.
Urban hydrograph characteristics typified by flows upstream and downstream of urban outfalls in the Cut at Binfield, UK (after Robinson et al. 2000)

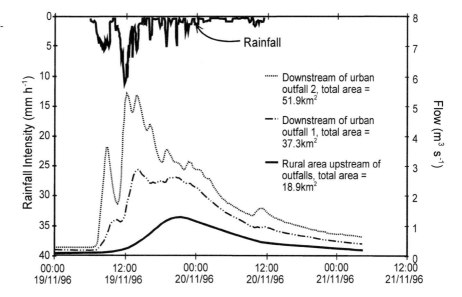

wetlands). Flood protection by channel diversions, embankments or provision of floodways all offer protection, but impact the downstream aquatic environment. A further protection measure is the construction of flood storage reservoirs upstream of flood prone urban areas, with these reservoirs purposely kept empty and with controlled flow releases during a flood event.

D.3.3.4 Water Quality Degradation Resulting from Agricultural Pollution by Nitrogen and Phosphorus

Water quality problems arise as a direct consequence of both intensified land management and water use. Agricultural systems today constitute a major source of sediments, nutrients, pesticides as well as coliform bacteria into river systems, which have become increasingly degraded in recent decades. This degradation can be measured as loss of natural aquatic systems, their component species, the amenities they provide, a reduction in supply of useable water and increases in the costs of treating the water for societal use (Carpenter et al. 1998). The USEPA (1989) estimated that 76% of pollution entering rivers and lakes in the USA was from non-point sources, of which agriculture contributed 64%. The remaining 24% was made up of almost equal proportions of point source pollution from urban, mining and industrial areas and from natural sources.

Although eutrophication, i.e. nutrient enrichment of water bodies by nitrogen (N) and phosphorus (P), is a natural phenomenon, accelerated cultural eutrophication through the supply of surplus nutrients – predominantly from non-point agriculture sources – is a feature of drainage basins which has developed only over the past 50 years. Eutrophication of inland waters may result in excessive production of plankton and rooted plants, which can lead to a marked turbidity in water, a reduction in biodiversity through loss of habitat, changes in dissolved oxygen with resultant fish kills, growth of blue/green algal blooms which are toxic and lethal to fish and render drinking water unpalatable, and increased growth of aquatic weeds. All these consequences make it expensive to purify water in order to make it fit for human consumption. Eutrophication accounts for ~ 60% of impaired river reaches in the USA (Carpenter et al. 1998). Sediments are the most prevalent direct water pollutant, especially since many important agricultural nutrients (in particular phosphorus) and pesticides are adsorbed by sediment particles and therefore travel downstream with sediments in storm runoff.

This section briefly reviews non-point agricultural pollution by nitrogen and phosphorus, in both cases addressing causes and processes as well as consequences at river basin scale. A similar overview of non-point pollution by pesticides, as well as of pollution abatement and mitigating factors, is provided by Schulze (2001). Non-point agricultural inputs of N and P can be continuous, intermittent, or linked to seasonal agricultural activities and extreme rainfall/flood producing events. Unlike point sources, which can be relatively easy to identify and control, non-point sources are difficult to measure and regulate (Carpenter et al. 1998).

D.3.3.4.1 *Processes Contributing to Nitrogen Pollution and Examples of Its Repercussions*

Nitrogen is naturally present as organic N in the living systems residing within drainage basins, even without any human activities. The nitric form that pollutes water results from the activity of soil micro-organisms which transform organic matter, while the nitrates derive mostly from agricultural activities such as application of mineral fertilisers and animal effluent (Gouy et al. 1999). In addition to its nutrient properties, nitrate can also be considered as a toxic substance for humans, with a WHO standard for drinking water at 50 mg NO_3^- per litre, while ammonium (NH_4^+) is a potential toxic compound for fish in the upper pH range when transformed into NH_3.

The causes of agricultural N pollution in fresh waters are manifold and complex (Williams et al. 2000). Intensification of agriculture means that existing fields are ploughed more frequently, and higher livestock densities are found on grasslands. Intensification also means that N fertilisers are being used with both higher amounts per application and with more frequent applications. Agricultural extensification also brings more new lands into production and hence eligible for N losses to aquatic ecosystems. To these sources must be added diffuse atmospheric deposition on soils, direct inputs of urban sewage effluent either from stormwater sewers or combined sewer overflows and animal effluent from storage installations (Bach et al. 1999; Gouy et al. 1999).

Emissions of N into water systems from agricultural fields (Fig. D.31) result from an imbalance between plant requirements and the instantaneous availability of N available from soils. This arises from a mismatch due to over-application of fertiliser, notably during non-cropped periods when N mineralisation is high. In the USA, for example, only ~ 18% of N input by fertilisers is used (Carpenter et al. 1998) with the remaining 82% lost as the greenhouse gas N_2O or NH_3 into the atmosphere (only to be redeposited again later) or accumulating in the soil to be leached to surface waters or to groundwater. In the near-surface soil horizons the fate of N is influenced by soil-crop interactions and turnover processes between organic and mineral N pools. Transport, by leaching, of dissolved N species into groundwater is influenced climatically by the control of microbial processes through temperature and water availability as well as redox conditions of the unsaturated zone.

Fig. D.31.
Inputs, outputs and processes of transport of N and P from agricultural land (after Carpenter et al. 1998; modified by Schulze 2001)

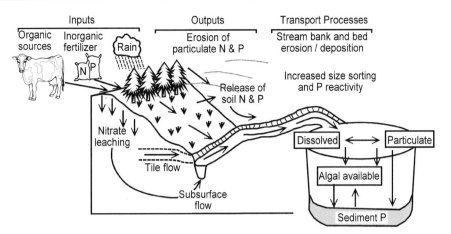

Important climatic variables controlling N chemistry include the timing of rainfall, number of soil wetting cycles by natural rainfall or irrigation application and, in colder climates, the length of time between crop uptake of N and soil freezing, with longer periods more conducive to accumulations of mineral N in the soil. Climatic factors are modulated by the permeability of the soil and its stratification. Soil texture is an important factor. The lighter the soil, the higher the rate of leaching. Carpenter et al. (1998) cite N exports from agricultural land to water at 10–40% for heavier loams and clays, but 25–80% for sandy soils. Measurements of the N budget in three sandy and three clayey catchments in Denmark show sandy catchments to leach 45% of total N application, with only 4.3% entering water courses, while clayey catchments leached only 33%, but with 12.3% entering water courses (NERI 1997). Because of its mobility, the pathway of nitrates into surface waters is largely through groundwater and tile drainage, when present. Because of the time lags involved, the phenomenon constitutes a long-term problem (Fig. D.31). Washoff of N also occurs, exacerbated in temperate Northern Hemisphere countries by the major N applications occurring during planting season when runoff is still high from the largely saturated catchments (Graham et al. 1999a).

The literature abounds with examples of nitrogen accumulation in inland waterways, derived solely from fertilisation of intensively cultivated crops. For example, in 3 300 catchments in France from which water is abstracted for human consumption, 12% were N-polluted above the EU potability limit of 50 mg l^{-1} (Gouy et al. 1999), while in the Burren River of south-east Ireland the increase in nitrates from 15 to 35 mg l^{-1} from 1979 to 1996 is attributed to higher proportions of ploughed land (Sherwood 1999). In Ireland, channel lengths of Class A (i.e. unpolluted) streams have decreased from 84% in 1971 to 57% in 1994, while streams of Class B (slightly polluted) increased from 6% of stream lengths to 27% and the moderately polluted Class Cs from 5% to 14%. Verstappen et al. (1999) report that in the Netherlands 3/4 of all lakes, canals and ditches fail to meet water quality standards.

In Austria, which obtains 98% of its potable water from groundwater, NO$_3^-$ in groundwater exceeded European Union thresholds of 50 mg l^{-1} in 15% of samples taken countrywide and 52% in Vienna province (BMLF 1996). Trends of NO$_3^-$ concentrations in five aquifers in north-eastern France, shown in Fig. D.32, clearly illustrate the increases over time.

The above examples reflect predominant influent flows of polluted groundwaters into streams. A comprehensive river basin scale GIS-oriented modelling system, MONERIS (**MO**delling of **N**utrient **E**missions into **RI**ver **S**ystems; Werner et al. 1991; Werner and Wodsak 1994) was applied by Behrendt et al. (1999) to analyse nutrient inputs in the four largest river basins within Germany, i.e. the Elbe (83 700 km^2), Weser (38 400 km^2), Rhine (102 500 km^2) and Danube (56 400 km^2), which together constitute 80% of Germany's total area. They show that agricultural sources constitute 64% (range: 56–77%) of the total N input into rivers (Fig. D.33). Significant, however, is that of the agricultural inputs, groundwater at 74% dominates over input via tile drain-

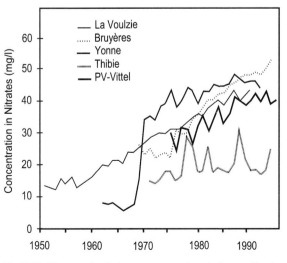

Fig. D.32. Time trends of nitrate concentrations in five aquifers in north-eastern France (after Gouy et al. 1999)

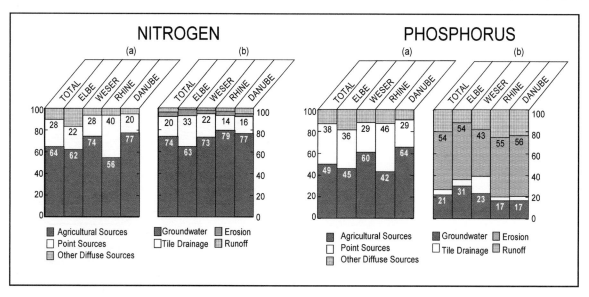

Fig. D.33. Nitrogen (*left*) and phosphorus (*right*) impacts from different sources for the in-country portions of the four main rivers in Germany in 1995. In each case (**a**) shows percentages from major sources and (**b**) percentages of different pathways into the river from agricultural sources (original values modelled by Behrendt et al. 1999; diagram from Schulze 2001)

age (20%), which in any event may be considered leaching en route to the groundwater table, while surface runoff plays a minor role as a source for N in river systems. Regional variations which reflect differences in agricultural practices are displayed, with groundwater sources making up 63% in the Elbe as against 79% in the Rhine (Fig. D.33).

D.3.3.4.2 Causes and Processes of Phosphorus Enrichment and Examples of Hydrological Repercussions

Unlike nitrogen, phosphorus (P), when added as a fertiliser or when it is released through organic matter decomposition, has a low solubility and is readily fixed (adsorbed) onto soil particles. Therefore, P is normally associated with a river's sediment load, with particles containing adsorbed P compounds mobilised off the land and into the stream as part of the soil erosion process (Fig. D.31). Being so closely related to erosion, nutrient contamination by P, apart from its availability through excessive fertilisation, is influenced largely by those factors conducive to high soil losses. These factors include the generation of overland flow, which acts as a transport mechanism from land to stream, soil cover by plants and their residues, topography and soil properties (particle size, structure, organic matter content). Because surface erosion depends on soil aggregate stability, P mobilisation in drier regions will thus often be related to soil crusting which, during aggregate breakdown (cracking) through soil shrinkage, enhances ero-

sion (Gustafson et al. 1999). In cold climates, erosion can be strongly related to aggregate breakdown through freeze-thaw processes. If predictions of future climate with possibly higher convective rainfall activity, longer dry periods between rainfall events, and more freeze thaw cycles, are fulfilled, losses of P through enhanced erodibility are likely to increase.

The mobilisation and processing of P is complex and along three major pathways (Graham et al. 1999a):

- *Movement of P-carrying sediments across the catchment surface.* This occurs during storm events and depends on rainfall intensity, vegetation cover (canopy, residue), surface topography (including gradient and roughness), soil hydrological and mechanical properties and conservation practices. These vary from place to place according to the level of management, and for some of the factors also from month to month.
- *Movement of material from the catchment surface to the channel.* This can be expressed by the "delivery ratio" (DR), i.e. the ratio between material mobilised locally across the catchment surface and that passing downstream through the system. Walling (1990) has shown that rates of sediment transport can be ~ one order of magnitude less than catchment erosion rates. DR depends, *inter alia*, on proximity to watercourses, slope gradient, riparian vegetation, soil moisture content and field drains. Phosphorus-carrying sediment can be either from the stream bank or from field surfaces, and up to 80% of sediment load can derive from but 10% of the catchment (Wall-

ing 1990). What is also important is that P fixes more readily to finer soil grains, because the suspended sediment size in a river is usually finer than that on the catchment.

- *Within channel movement.* Phosphorus-carrying sediment includes channel storage (which may "bury" the P) and floodplain deposition. Such deposition constitutes a major feedback of in-channel P to the catchment's floodplain during large floods, as 30–50% of the river suspended load can be deposited again (Graham et al. 1999a).

In contrast to N, only small amounts of P make their way into groundwater, and only when the soil has become "overloaded" and its fixation capacity has been fully saturated (Gouy et al. 1999). Dissolved P is largely available for biological uptake to algae, while particulate P is less immediately so. Particulate P can, therefore, be a source of biological production in the eutrophication process over a much longer time scale than dissolved P or from N through groundwater leaching. P is therefore often a limiting element, controlling rates and longer-term durations of eutrophication.

At the river basin scale, the main contributor of P to smaller rivers is often livestock waste followed by eroded soil from cultivated land (Kienzle et al. 1997). In developing nations the contribution of P from inhabitants living adjacent to streams in informal settlements without proper sanitation facilities can be dominant (e.g. Chapt. D.7, Mgeni Case Study). In larger rivers, dissolved P from sewage effluent and other diffuse sources is often an important contributor.

There is a considerable literature on P increases in river systems as well as on their sources and pathways. In an example from Ireland, Sherwood (1999) reports that of 19 Irish river basins studied, only two export < 0.4 kg total P ha^{-1} at the river mouth, while three export a loading > 1 kg to sea and 11 between 0.5 and 1.0 kg total P ha^{-1}. The problem is much larger than the figures above indicate, however, because upstream of the river mouth P has already been absorbed by rooted plants and algae, and P has already been deposited in sediments.

Figure D.33 shows that P input in Germany's four largest river basins in 1995 displays very different dominant processes when compared to those of N. Agriculture at 49% (42%–64%) remains the major source of P in the rivers, compared to point sources from sewage and industry at 38% and other diffuse sources such as atmospheric deposition and urban areas at 13%. The source of agricultural P input is overwhelmingly through erosion at 54% (range: 43%–56%), with input via groundwater and surface runoff contributing 21% (17%–31%) and 20% (10%–26%), respectively. Tile drainage constitutes only 5% (2%–16%) of the input.

D.3.3.4.3 Conclusions

Two trends of importance to river basin scale responses with regard to nutrient loading can be traced to the use of fertilisers. The first is the tremendous overall increase in N and P (and pesticide) usage since the 1950s. Carpenter et al. (1998) have calculated that of $\sim 600 \times 10^6$ Mg P applied as fertiliser from 1950–1995, 58% has accumulated in agricultural soils which, when considering a standing stock of P in the top 0.1 m of soil of $1\,300 \times 10^6$ Mg, implies a P content increase of 25% in that soil horizon – the most vulnerable to pollution. Many other examples may be cited of general increases, such as that in France where mineral N application between 1970 and 1997 has risen by a factor of 2.65. The second trend, stemming from new environmental legislation, has been a reduction in fertiliser applications of late in certain parts of the world. For example, in Denmark the N leaching from the root zone has decreased by 17% in the decade from 1985 following a 24% reduction in N application (Olesen 1999). Much remains to be done in the agricultural sector's contribution to reducing N and P loadings from river systems, however. Thus, while Germany's four major rivers show a 26% decrease in overall N input from all sources from 1985–95, that from agricultural sources was only 10% and, indeed, total P inputs increased by 7% (most notably in the Elbe at 11% and Danube at 12%) over the same 10-year period in which total P input from all sources has shown a 62% decline (Behrendt et al. 1999).

This brief review of some of the major causes and processes of agricultural pollution and their repercussions amply suggests the seriousness of soil, surface water and groundwater contamination by nitrates and phosphates. Multiple management strategies which include legislation, abatement measures and landscape management, either on-site or off-site, have been proposed (e.g. Carpenter et al. 1998; Gouy et al. 1999; Gustafson et al. 1999; Williams et al. 2000; summarised in Schulze 2001). However, a major perceived need is that of appropriate scaling-up of models describing the fate and behaviour of agricultural pollutants from soil profile and plot scale – at which processes are reasonably well understood – to the river basin scale. Relatively few models exist for river basin scale simulations that are not empirically based, or static and that are sufficiently process-oriented to simulate realistically in-catchment heterogeneities for a range of landscapes, management scenarios and climates.

D.3.3.5 Salinisation

River salinity has become a major water quality issue in many, especially semi-arid, regions of the world under heavy irrigation, including the Murray-Darling basin

(MDB) in Australia, the Indus River basin in Pakistan, the Colorado in the USA or the Breede in South Africa. The rises in salinity levels in these basins, with their high dependence on consumptive water use from river systems, is symptomatic of intensified current dryland and irrigation land-use practices. This anthropogenic water use has taken the place of natural systems, resulting in a "massive hydrological imbalance" which may take centuries to re-establish (MDBMC 1999). The magnitude of the problem is exemplified well by the Indus basin, in which 35–40% of the 16 Mha irrigated suffer from salt-affected soils. Of the 6 Mha affected, more than 2 million are severely affected (Sarwar 2000) to the extent that through the application of saline and sodic groundwater, soils have become almost impervious (Table D.10). Two Mha have already had to be abandoned as a result of excessive salinity (Wolters and Bhutta 1997). Because of widespread salinisation, the average annual salt inflow into the Indus River waters is now 33.0×10^6 t, while the natural outflow to sea is < 50% or 16.4×10^6 t, i.e. an addition of 16.6×10^6 t is added to the salt load of the river annually. Of this additional load, 87% are salts accumulated in soil profiles from irrigated lands and their underlying geological formations/aquifers (Nespak/ MMI 1993). By implication, 1 t of salts per ha irrigated land is being added annually to the Indus, the waters of which, when used for irrigation downstream, causes further salinisation of land.

This chapter will focus briefly on sources of salts, manifestations of salinity, salt mobilisation and controls on salinity, with examples taken mainly from the Murray-Darling and Indus basins.

D.3.3.5.1 Primary Sources of Salinisation

A major primary source of salinisation is in groundwater where salts occur naturally over geological time spans and, in the case of Australia, are manifest as salt lakes or salinised lands (MDBMC 1999). Naturally salinised landscapes support salt-tolerant native vegetation, though in most areas, the groundwater is stored deep enough not to affect plants (Fig. D.34). In the MDB, for example, the largest store of salt occurs in sedimentary fractured rocks up to 600 m below the present surface, but with important other source areas being thin lenses of saline groundwater, salts stored in the capillary fringe zone above the water table, and in alluvial sediments.

D.3.3.5.2 Secondary Sources of Salinisation

Deforestation. Over vast tracts of the 1 Mkm² MDB, clearing of about 50% of the deep-rooted dense natural evergreen forest, and in places up to 90% of the native vegetation, followed European settlement starting about 170 years ago. Natural vegetation was replaced with shallow-rooted and low rainfall-intercepting annual crops and pastures, causing decreased losses through transpiration (including canopy interception) and increased

Table D.10.
Classes of salt-affected soils in the Indus basin (after Qureshi and Barrett-Lennard 1998)

Class	Area (10^6 ha)	Characteristics
Slightly saline-sodic	0.7	In patches in cultivated fields
Porous saline-sodic	1.9	Throughout root zone; pervious
Severely saline-sodic	1.1	High groundwater tables; dense; nearly impervious
Sodic tubewell water	2.3	Application of sodic groundwater; almost impervious

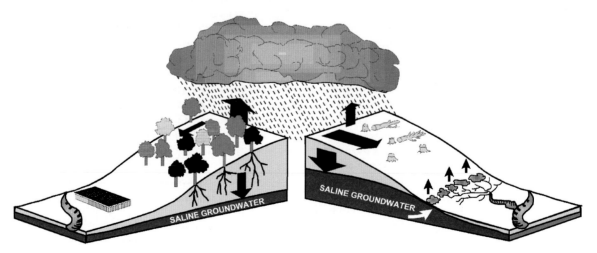

Fig. D.34. Deforestation impacts on salinisation processes in a catchment with naturally occurring saline groundwater in the Murray-Darling basin (after MDBMC 1999)

recharge to the groundwater table. With enhanced aquifer recharge (Fig. D.34), the water table rises, and saline natural groundwater intersects the land surface, with the resultant surface wash-off enhancing stream salinity. Waterlogging near the soil surface further affects soil structure and its chemistry.

Regulation of rivers by dams and diversion of water for irrigation. Salinisation in rivers is compounded by water use for irrigation. Flows are reduced, natural salts are added through return flows and local groundwater table levels rise through persistent water losses through deep percolation from unlined earthen canals and over-irrigation of fields. This brings salts to within a critical 1.5–2.0 m of the surface where they may be drawn upwards by capillary action, evaporation from the soil surface and transpiration by plants. All of this is compounded by then irrigating with poor quality water downstream (cf. Table D.10), adding further salts. A classic example of such groundwater table rise occurs in the heavily irrigated Punjab in Pakistan's Indus basin where, at the time of the extensive canal construction phase some 100 years ago, the groundwater table depth originally ranged from 20–30 m below the soil surface. However, persistent recharge through seepage over the years raised groundwater levels to within the critical 1.5–2.0 m from the surface in many places, and it has intersected major rivers, as illustrated in Fig. D.35. Again, problems in this semi-arid area are rendered more serious than elsewhere because of a natural saline groundwater.

Soils saturated with saline water can fluctuate seasonally with the rains. Thus, in the Indus, water table levels are deepest in the pre-monsoon dry period May–June and shallowest post-monsoon immediately after the rainy season in September. Sarwar (2000) cites that in the Indus just prior to the monsoon 13% of irrigated lands ($\sim 2 \times 10^6$ ha) have water tables within 1.5 m of the surface, while after the rains 25% of the irrigated area ($\sim 4 \times 10^6$ ha) is waterlogged. Because of the complex regional dynamics of salinisation, conflicting areas of land abandonment in the Indus due to secondary salinisation appear in the literature. One estimate (Sarwar 2000) cites 40 000 ha being abandoned annually for this reason.

D.3.3.5.3 Impacts of a Saline Water Table within 2 m of the Soil Surface

Of the many impacts of a shallow saline water table, MDBMC (1999) and Sarwar (2000) highlight the following:

- *Agricultural land.* Salt affected land, usually defined as land on which productivity is reduced by $\geq$ 50% of its natural potential, is characterised by crop failures in wetter years and on most waterlogged areas. On pastures/grazing lands in the MDB a vegetation succession has been observed with salinisation, by which N-fixing clovers disappear first, thereafter barley-type grasses, followed by volunteer (natural) salt-tolerant species to finally the bare soil. Typically, farmers underestimate the extent of salt-affected land.
- *Regional infrastructure.* Rising water tables under roads and buildings cause damage through sulphates which occur in saline groundwater. This destroys the strength of concrete structures, e.g. reducing the "life" of a road by up to a factor of four (MDBMC 1999).
- *Urban areas* may suffer from decay of house foundations, plumbing corrosion and septic tank damage. Additionally, high water tables can enhance runoff.
- *Environment.* Both aquatic and terrestrial ecosystems may be unable to cope with salt increases associated with secondary salinity, with impacts on ecological communities, particularly those in wetlands, being particularly severe.
- *Socio-economic* impacts of salinisation in lesser developed communities which are heavily dependent

Fig. D.35.
Enhanced salinity through rising of the groundwater tables after the introduction of canal irrigation in the Punjab region of the Indus basin, Pakistan (after Wolters and Bhutta 1997)

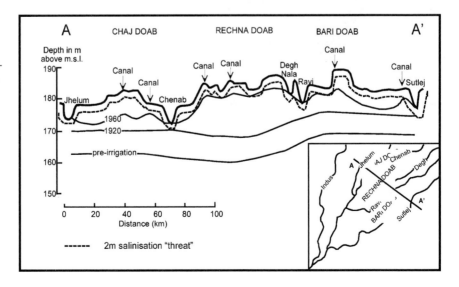

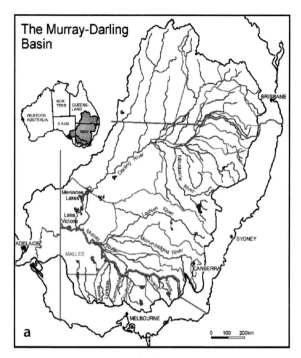

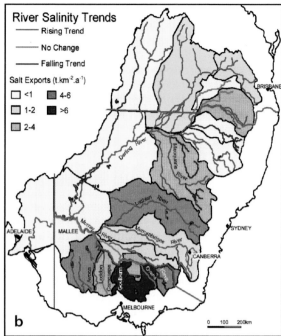

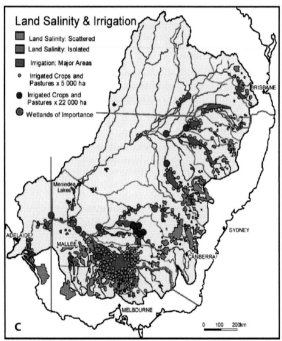

Fig. D.36. The Murray-Darling basin. **a** Location; **b** river salinity and its trends, as an example of changes in state (after MDBC 1997); **c** dryland salinity and irrigated areas in the mid-1990s as a manifestation of pressures (after LWRRDC 1998)

on irrigated agriculture (e.g. in the Punjab 75% of the population's activities are related to agriculture – Sarwar 2000), include health problems in both humans and animals and may cause major migrations of affected populations (Sarwar 2000).

D.3.3.5.4 *Manifestations of Salinity: A Case Study in the Murray-Darling Basin (MDB)*

Within river basins, salinity manifests itself both as a land-based and aquatic phenomenon. Examples are presented below, taken from the MDB (Fig. D.36). These also include estimates of the quantity of salt expected over the next 100 years.

- *Land salinisation.* Secondary salinisation resulting from land-use change (mainly deforestation and irrigation) has accelerated, with salinity some 200 years ago having been restricted to particular locations (e.g. the salt lakes of Victoria and the Mallee region; cf. Fig. D.36) and to those years with heavy rainfall and consequent groundwater level rise. Land salinisation affects vegetation and the rate of soil erosion through soil structure decline and stream banks becoming less stable. Because it manifests itself with a time lag land salinisation, once it becomes evident as a rising water table which breaks the surface, is often very late for remedial action. It is estimated that ~ 300 000 ha in the MDB are affected by land salinisation, with the major distribution distinctly around the eastern and southern perimeter of the basin in the upper Macquarie, Lachlan and Murrumbidgee catchments in the east and the Campaspe, Lodden and Avoca catchments, together with the Mallee region, in the south (cf. Fig. D.36).

- *Present river salinity and its trends.* Salt enters rivers by two processes: it is either washed off the land during stormflows or percolates through the irrigated soil profile and then enters into the river via baseflow when the groundwater level intersects the stream channel. Not all the salt that is mobilised is immediately exported out of the system by rivers. For example,
  - large scale re-storage of salt occurs in lower sections of a river basin through diversions (e.g. for irrigation) and entrapment;
  - sections with poor drainage retain much of the salt delivered to the surface by groundwater; and,
  - much of the salt load through land-use change or over-irrigation may manifest itself in the river only several decades later.

  At present, ~ 45% of salt mobilising in MDB catchments is exported to its rivers, similar to the Indus with ~ 50%. According to the MDBC (1997), mid-1990s river salt concentrations at yields > 6 t km^{-2} yr^{-1} are highest in the southern tributaries of the Murray River (i.e. the Ovens, Goulburn and Campaspe basins; Fig. D.36), with the equivalent of 4–6 t salt km^{-2} yr^{-1} in the Avoca, Lachlan and Klewa river basins, while the mainstem of the Darling River and its northern tributaries display some of the lowest salt concentrations in the MDB. There clearly is a strong spatial relationship between present levels of salinity in rivers and the combination of its two major causes in the MDB, i.e. dryland salinity and irrigation (cf. Fig. D.36). Currently 1.1 Mha in the lower basin are under irrigation, making up 75% of Australia's irrigated lands, and ~ 80% of the MDB's water is committed to irrigation. Significantly, many of these very rivers (and numerous others) currently show rising trends in salt loads. Equally significantly, some of the highest yielding southern tributaries of the Murray are already displaying falling river salinity levels (Fig. D.36).
- *Trends in salt mobilisation.* It is anticipated that over the next century salt mobilisation in the MDB will more than double (104%) from current values. This will, however, vary considerably by State, with the highest estimated increases projected for South Australia (135%) ~ three times as high as the projected increase in Queensland (38%). Furthermore, predicted rates of increase in mobilisation are not uniform over time. Overall, however, the salt export into the rivers is predicted to decrease from the present 45% to 38% in 2100 (MDBMC 1999).

D.3.3.5.5 *Controls on Salinity: Examples from the Past, the Present and the Future*

A paradigm shift is occurring in controlling river basin salinity, with a move from physical/structural measures favoured in the past to more integrative, non-structural measures proposed for the future. Thus, prior to the 1960s, measures taken in the Indus had a focus on controlling groundwater table depth by preventing canal leakage. The idea was to contain the capillary salinisation process, through closure of canals in the monsoon season and lowering full supply levels (Sarwar 2000). This gave only local/temporary relief, and so in the 1960s to 1970s the strategy evolved into one employing vertical tubewell drainage (over 14 000 were constructed, each with a capacity of 80 l s^{-1} and covering 2.6×10^6 ha) with the aim of combatting salinity by lowering the groundwater table and supplementing irrigation supplies from the pumped groundwater, either directly or mixed with canal water. However, tubewells depleted fresh groundwater, resulting in saline groundwater intrusion into aquifers, potentially aggravating the salinity problem. Post-1970 measures focused on expansive horizontal tile drainage, assuming that only small volumes of saline effluent would be produced. Eurocentric criteria which were applied in this semi-arid area (based on steady state assumptions regarding moisture and solute fluxes occurring in humid areas) resulted in significant (3-fold) over-design, followed by an increase in irrigation water demands (Vlotman et al. 1990). At present, the shift is towards local, on-farm management measures, both physical (e.g. canal lining which can cut seepage losses by up to 40%, and reclamation of salt-affected soils by physical and chemical measures) and non-structural (e.g. adoption of irrigation schedules appropriate to crop and climate, applying guidelines for safe use of different quality water and reclamation by biotic means).

The action plan for the Murray Darling basin is, in many ways, fundamentally different to that of the Indus. Instead of utilising solely established technologies, a co-ordinated approach was adopted in 1988 through the MDB Ministerial Council, striving for equitable, efficient and sustainable management of water, land and environmental resources as an integrated entity. Their "whole" catchment approach to salinity mitigation has included physical measures like groundwater interception of ~ 135 000 t salt each year along downstream rivers with help of upstream States' funding, as well as drainage diversion schemes diverting ~ 500 000 t salt each year, which has reduced salinity fluctuations and electrical conductivity by 76 µS. However, their focus has been on bottom-up and community-led participatory approaches. These included conflict resolution between the competing needs of downstream river protection (i.e. optimising water quality) and upstream land management (e.g. irrigation return flow generation) as well as economic evaluation of feasible alternative schemes, together with their environmental effects. Furthermore, obligations and rights with respect to salinity can be defined for State and individual stakeholders alike, setting salinity baselines (i.e. thresholds, at 1980s levels). Additional management mechanisms include salinity credits, saline groundwater interception before it reaches the stream by re-vegetation to deep rooted forests, improvements aimed at enhanced water use efficiency in irrigation (e.g. scheduling), and market instruments. Market instruments include the exploiter pays principle, subsidies and tax breaks for good management practice which include water trading within strict environmental guidelines.

D.3.4 Water and River Engineered Landscape

The three major drivers of hydrological responses which transcend the range of hydrological scales are

- atmospheric drivers (including natural climate variability and human induced changes in climate trends);
- anthropogenic drivers (mainly land-use changes impacting quantity and quality of river flows, which reflect largely also levels of socio-economic development); and,
- intervention in the hydrological regime by direct water management.

It is our contention that the latter generally dominates in importance, often by a wide margin. Through irrigation water supply and demand, river channel modifications, water storages in dams, and/or water diversions through inter-basin transfers, humans have succeeded in directly ameliorating the vagaries of spatially and temporally inconsistent water supplies by controlled abstractions, releases and flow attenuation to satisfy their demands for direct use and provision of safety. These direct, usually channel-based manipulations stand in stark contrast to the indirect ones of land-use and land-management changes, which essentially partition precipitation into new pathways in a less drastic way. Climate variability and change provide even less direct impacts as a source of anthropogenic change to the water cycle. This chapter will focus only on river channel modification and dams, together with their impacts.

D.3.4.1 River Channel Modification

In the developed world there is hardly a river that is entirely natural any longer and river channel modifications may be defined as those management activities that alter the form of the river channel, specifically affecting the form, cross-section and longitudinal profile (Sear et al. 2000). The difference between natural and modified water courses can be characterised in four ways, i.e. changes in channel form, changes in the rates of proc-

esses, the role of vegetation and connectivity with floodplains. Each of these is interrelated and contributes to modifying the hydrological response of a river basin by changing channel capacity and flow resistance.

There are six main types of channel modification: straightening, resectioning, reinforcing, embanking, culverting and weir/sluice construction (Sear et al. 2000). Major reasons for channel modification include extensive land drainage schemes which require improvement in the efficiency of the river network and protection from floods – either to confine high flows within the river network or to speed up the passage of flow through the channel.

Closely associated with these modifications is their maintenance in order to facilitate conveyance of water at design optima. Maintenance represents continual or periodic disturbance of the channel to remove sediment accumulation, reinforcement of the river banks, removal of debris and weeds, enhancement to instream habitats and fostering morphological diversity through river rehabilitation and restoration. Maintenance is, however, seldom based on the estimation of real need and therefore many channels remain either over- or under-maintained.

Using the UK as an example, the long history of river channel modification started by the Romans with drainage dyke and aqueduct construction and undertaken on a larger scale in the 17th–18th century for drainage of lowlands subject to inundation, continued with realignment of upland streams for milling and later demand for increased agricultural output. Initially, these channel modifications centered on design for a certain discharge rather than a particular return period. Table D.11 expresses the current percentages of river channel modification in the UK.

Geomorphological impacts, which depend on the location of the reach, its extent and the type of modification, can either be direct or indirect. Direct impacts such as lowering of the river bed and straightening can eliminate meandering reaches completely, eliminate natural pools and shorten river length. They often steepen slopes, reduce channel roughness and increase flow ve-

Table D.11. Extent of river channel modification (as a percentage) in the UK (after Environment Agency 1998)

Modification[a]	In UK upland	In UK lowland
Straightened	0.0	6.2
Resectioned	12.5	44.0
Reinforced	35.4	51.9
Embanked	5.1	14.9
Culverted	3.5	9.3
Weired/sluiced	8.4	15.0

[a] Percentages can add to >100% because more than one modification may have been undertaken along a single reach of a river.

locities to the extent that enhanced stream power can destroy engineered protection works, causing major bank failure with consequent increases in sediment supplies from the exposed banks. Indirect impacts imply both up- and downstream channel adjustments (Brooks 1987). Upstream of modified reaches the increase in water surface slope caused by drawdown can initiate incision and bank failure, while elevated flood levels resulting from embankments may induce backwatering and sediment accumulation. On the downstream side in high energy river systems (e.g. with gravel beds) channels may become enlarged with bank instability. In low energy river systems, on the other hand, morphological adjustments are characterised by accumulations of sediment, reducing channel capacity in the modified reach and raising bed elevations upstream. Such impacts may vary between 1–5 km downstream.

D.3.4.2 Dams and their Impacts

D.3.4.2.1 Purposes, Definitions and General Impacts of Dams

Dams (i.e. reservoirs, impoundments) are amongst the most significant human engineering achievements. Collectively, they also represent what are arguably the most significant interventions in the terrestrial hydrological cycle and have been an influential part of societal development – at least on the regional scale – for many centuries (Acreman et al. 2000). The main purposes of dams are threefold (Acreman et al. 1999, 2000):

- to *regulate flows*, with water impounded during high flows released for downstream use to meet human needs as well as industrial, environmental and/or agricultural demands during times of inadequate flows, especially where runoff displays high seasonality;
- to *create hydraulic head*, providing potential energy for hydropower production or for gravity-fed irrigation canals; and,
- for *control purposes*, of which the most common is for protection from floods (with 8% of single purpose and 39% multi-purpose dams built specifically for flood control, according to the International Commission on Large Dams, ICOLD 1994).

Globally, approximately 40 000 registered "large" dams exist with a combined full supply storage of 7.6×10^3 km^3, i.e. 7.6×10^{12} m^3 (Vörösmarty et al. 2003). Large dams, registered by the International Commission on Large Dams (ICOLD), have to fulfil at least one of the following criteria, i.e. wall > 15 m in height, wall > 10 m in height, but with a crest length of ≥ 500 m, storage capacity > 10^6 m^3 and/or spillway capacity > 2 000 m^3 s^{-1}.

From an holistic perspective the impacts of dams, as summarised by McCartney et al. (1999), include

- *ecosystem changes*, i.e. changes in natural capital through alterations in flow regimes and soil erosion as well as loss of habitat and ecosystem disturbance, with both direct impacts (e.g. species loss, changes in water courses and low flows) and indirect impacts (e.g. changes in species or soil fertility) occurring;
- *environmental changes*, by which facets of natural, physical and human capital are changed through reduction in water quality, loss of biodiversity, increases in waterborne diseases or changes in infrastructure, again with both direct and indirect impacts; and,
- *social changes*, which include changes in societal, human and financial capital through migration and demographic change, stakeholder development, as well as employment and savings opportunities, once more with direct impacts (e.g. increased cashflows or access to services) and indirect ones (e.g. macroeconomic change) possible.

Dams provide many positive benefits by increasing food yields from irrigated agriculture which feeds growing populations, promoting economic growth and new employment, assuring water supplies to rural and urban areas, providing cheap, regular and renewable hydropower, and controlling medium sized floods. While these benefits are clear to see, the value and effectiveness of dams to human society has come into question on environmental, political, and economic grounds (Gleick 1998). This short overview will address some impacts of dams on floods, in-channel processes and downstream river ecosystems.

D.3.4.2.2 Impacts of Dams on Flow Regimes, Channel Morphometry and Riparian Zone Processes

Dams replace natural flows with an artificial regime that typically dampens high flows and stabilises and elevates low flows. Impacts on individual flood flows by large dams depend on both the storage capacity of the dam relative to the flood volume and the way in which the dam is operated. Floods of low recurrence intervals (1–2 years) frequently have their total volume stored, whereas floods of medium return periods (< 10 years) allow floodwater to pass through the dam, but usually result in a greatly reduced peak and attenuation of the hydrograph. Floods of high return period (> 50 years), however, remain mostly unaffected by reservoir storage and only a small attenuation will be evident.

Potential impacts of dams on flow regimes are illustrated well by the Glen Canyon Dam on the Colorado River in the western USA. With a capacity of 30×10^9 m^3

and a depth of 150 m, the dam was completed in 1963 as a major flow regulator to downstream states Arizona, Nevada and California and for use in hydroelectric production. Streamflow gauging and suspended sediment monitoring on the Colorado commenced in the 1920s at Lees Ferry, 25 river km downstream of today's Glen Canyon Dam and at Grand Canyon, an additional 140 river km downstream from Lees Ferry. The reservoir has a highly stratified hypolimnion in the summer months (April to October) of cold, dense water with high specific conductance (~ 1100–1500 μS) as a result of high salt inflows.

Webb et al. (1999) report that natural snowfed peak flows from May–July have been converted to seasonally uniform flows. The two-year flood peak of $2160 \text{ m}^3 \text{s}^{-1}$ is now only $850 \text{ m}^3 \text{s}^{-1}$, i.e. ~ 40% of the pre-dam value, while the 10 year flood peak has been attenuated from $3950 \text{ m}^3 \text{s}^{-1}$ to only $1440 \text{ m}^3 \text{s}^{-1}$ (~ 36%). The highly variable, unregulated flow which ranged between a highest mean flow for the year of $761 \text{ m}^3 \text{s}^{-1}$ (in 1924) to a lowest mean annual flow of $171 \text{ m}^3 \text{s}^{-1}$ now fluctuates only mildly around the mean annual flow of $476 \text{ m}^3 \text{s}^{-1}$, while a steep pre-dam flow duration curve is now flat, with the 10th percentile flow reduced to only half its previous value (Webb et al. 1999).

Such drastic changes to the natural flow regime have manifested themselves through changes to in-channel morphology and riparian zone processes, often in amplified form. For example, Webb et al. (1999) notes the following alterations:

- *Aggradation of debris fans* since dam closure has been reduced to 25%, with the flow competence downstream of Glen Canyon Dam now greatly reduced and pre-dam reshaping of debris fans virtually ceased.
- *The sediment mass balance* has changed completely, with formerly "prodigious" sediment loads now being deposited into the dam and post-dam releases consisting of essentially clear water, with suspended sediment loads at ~ 5% of pre-dam values at Lees Ferry, increasing to 25% at Grand Canyon by which stage tributary contributions have become significant again. Immediately downstream of Glen Canyon Dam sediments coarser than 0.5 mm diameter now comprise < 1% of their pre-dam load.
- *Bed scour* began downstream of Glen Canyon Dam after closure, with only the fine sediment being scoured and the bed becoming armoured with coarser sediments and cobbles. Since the mid-1960s permanent scour of ~ 4.6 m below that of pre-dam levels has been observed at Lees Ferry. From 1963–1975 an estimated $8.5 \times 10^6 \text{ m}^3$ sediment was scoured between Glen Canyon Dam and Lees Ferry, 25 km downstream.
- *Sand bar erosion* takes place below dams because there is little replenishment of sand by floods. Downstream of Glen Canyon Dam 52% of sand bars on the Colorado showed reductions in size by 1990 (Webb et al. 1999).

- *Colonisation of riparian species*, both native but especially non-native species, has taken place on sand-mantled banks due to the absence of annual flooding. Along the Colorado growth rates of older trees in old high water zones have decreased, while seedling establishment has moved more towards the stream.
- *Temperature, clarity and aquatic productivity* have all changed markedly. Before construction of the Glen Canyon Dam the Colorado was mainly heterotrophic, with little primary production in sediment laden waters. However, cold water (pre- *v.* post-dam seasonal water temperature ranges changed from 0–29 °C to 7–15 °C) and low sediment concentrations have altered aquatic ecosystems drastically. An historically heterotrophic system is now autotrophic, resulting in high photosynthetic productivity downstream because of stabilised flows and a consistently large wetted perimeter. However, productivity decreases again with distance from the dam. Species composition has changed "dramatically" (Webb et al. 1999) with numerous species disappearing, others emerging. Native fishes have faced substantial competition and predation from non-native species.

While the Glen Canyon Dam example was from a perennial river, Jacobson et al. (1995) emphasize that along ephemeral rivers such as the Kuiseb in Namibia the downstream riparian zone impacts of dams are even more amplified, primarily because riparian vegetation along ephemeral streams is structured and maintained almost exclusively by infrequent flooding. Floods, furthermore, transport nutrient-rich organic matter to downstream ecosystems and infrequent floods play a major role in recruitment of new riparian forests. Along the Kuiseb, for example, a massive dieback of ana trees has been triggered by an absence of flooding.

D.3.4.2.3 Impacts of Dams on Downstream River Ecosystems

The downstream consequences of impoundments should be viewed as part of an integrated system of whole-river ecology. Such impacts should thus be considered within an hierarchical framework of 1st, 2nd and 3rd order inter-connected effects along the lines first proposed by Petts (1984), but subsequently modified and amplified upon, as in Fig. D.37.

- *1st order impacts* make up the immediate and abiotic effects associated with dam closure, influencing transfers of energy and material into, and within, the downstream river and connected ecosystems, with changes in water quantity (reductions in flow volume, variability and magnitudes), water quality (in regard to thermal regulation and changes to the water chemistry)

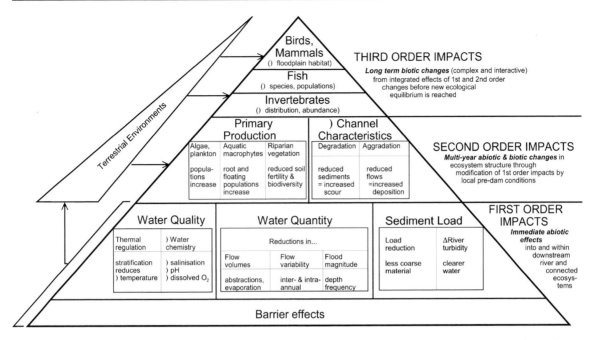

Fig. D.37. A framework for assessing impacts of dams on river ecosystems (after Petts 1984, but modified by Schulze 2001 from information in McCartney et al. 1999 and Webb et al. 1999)

and sediment load (load reduction and changes in river turbidity.

- *2nd order impacts* take place over several years and are both abiotic and biotic in nature, with changes in upstream and downstream ecosystem structure (changes in channel characteristics) and production (of algae, aquatic macrophytes and riparian vegetation). Second order impacts result from modifications of 1st order impacts by local conditions and are dependent upon river characteristics prior to dam closure.

- *3rd order impacts* constitute the complex and interactive long-term biotic changes in birds, mammals, fish and invertebrate populations and their distributions. They result from the integrated effects of all 1st and 2nd order changes before a new ecological equilibrium is reached.

D.3.4.2.4 *Conclusions*

To date, most studies on the environmental impacts of dams have been conducted on river systems in temperate climates (McCartney et al. 1999). A great deal more knowledge is required of possible downstream impacts of dams in tropical climates (where biological processes proceed faster and ecological changes become apparent more quickly) as well as in semi-arid to arid climates (e.g. Jacobson et al. 1995; Webb et al. 1999) where the converse is true. There is no doubt that priorities need to be established for improved management and policy formulation for more effective dam management. In this

regard McCartney et al. (1999) summarise effective dam management as a reconciliation between the *demand for water* (e.g. for domestic, agricultural and industry/ power determined by population, level of development and available resources) and the *sustainable supply of water* (at different scales and from different sources). Ideally, water demand will be determined subject to ecological constraints in regard to biodiversity/habitats, water quantity/quality, human health and climate change impacts, while sustainable supply is sustained through constraints/controls imposed by institutions (again at scales ranging from local to national).

D.3.5 The Road Ahead 1: Integrated Water Resources Management (IWRM)

D.3.5.1 Integrated Water Resources Management as a Response to an Inheritance of Damaged River Basins

Up until a few decades ago, and in some cases only a few years ago, the conventional approaches in most parts of the world to water and its utilisation were highly simplistic, largely because of a mindset of ideas and practices from managing small human populations and low-intensity human activities. For example, it could at one time be reasonably concluded that any waste produced would be effectively evacuated by rivers without any long-term/significant damage to water quality downstream. Furthermore, society's approach tended to be one of conquer, develop land and water and, when a problem was created, to migrate. Technologies such as dam building or inter-catchment transfers were applied to manage adequate supplies of water for society's and agriculture's needs and water quality problems were often solved by chemical treatment downstream of waste rather than upstream at its source (Falkenmark et al. 1999).

What we have inherited as a consequence, are damaged ecosystems (Newson et al. 2000), as illustrated in Fig. D.38, in which the spontaneous regulatory functions of rivers and their contributing catchments have been disturbed (e.g. through deforestation, increased erosion, stream fragmentation from dam construction) or removed (e.g. wetlands drainage). The manner of exploit-

ing water and the land from which it is generated has also changed through intensification of water use (e.g. irrigation, dryland cropping, urbanisation) and destruction of traditional extensive exploitation (e.g. marginalisation of more traditional land-use systems, exploitation of marginal lands).

Examples from the UK of what we have inherited include the 61% of the UK's agricultural land that is now drained, urban areas in England that have grown by 58% from 1945 to 1990, the 66% of channels in England and Wales that have been modified profoundly and the 83% of river channels that are maintained regularly (Newson et al. 2000).

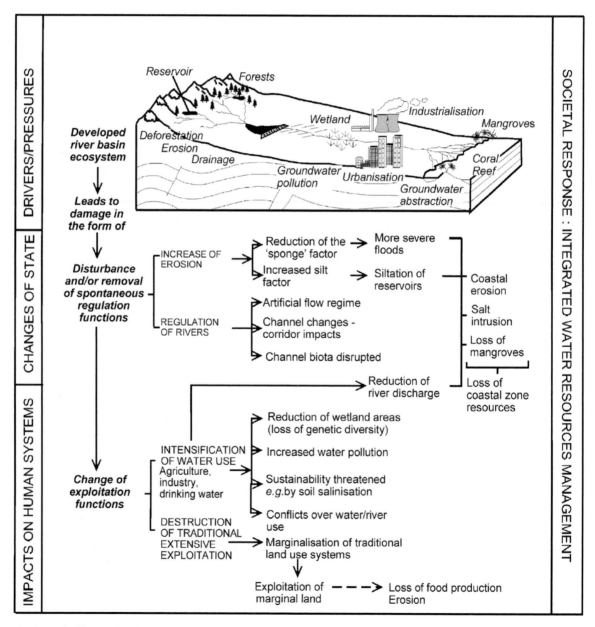

Fig. D.38. The "damaged" inherited catchment ecosystem within a framework of the DPSIR model, resulting in IWRM as the societal response (after Marchand and Toornstra 1986, and Newson 1997)

Remediation efforts associated with the "damaged" catchment ecosystems can be through reactive responses, for example, by making recommendations for precautionary actions (e.g. Newson 1997) or by promoting rehabilitation of damaged elements of the system (e.g. Brooks and Shields 1996). Alternatively, remediation can be afforded through proactive responses, in seeking to prevent the destructive causes and adopting a "least regrets" approach in water management by conserving the natural capital of the broader land/water environment through Integrated Water Resources Management, IWRM (Brooks and Shields 1996; Newson 1997; Newson et al. 2000).

D.3.5.2 What Is Integrated Water Resources Management (IWRM)?

Of the plethora of definitions of IWRM which abound in the literature, three have been selected:

- IWRM is the co-ordinated planning and management of land, water and other environmental resources for their equitable, efficient and sustainable use (from the UK; Calder 1999);
- IWRM may be viewed as a framework for planning, organising and controlling water systems to balance all relevant views and goals of stakeholders (from the USA; Grigg 1999); and,
- IWRM is seen as a philosophy, a process, and a management strategy to achieve sustainable use of resources by all stakeholders at catchment, regional, national and international levels, while maintaining the characteristics and integrity of water resources at the catchment scale within agreed limits (from South Africa; DWAF 1998).

Implicitly or explicitly these definitions place IWRM beyond simply the management of water quantity and quality, or being a catchment manager's "wish list". At the same time, it is not as overarching and broad a socio-economic or politico-institutional concept as Integrated Catchment Management (ICM), which UNESCO (1993) defines as:

> "the process of formulating and implementing a course of action involving natural and human resources in a catchment, taking into account social, economic, political and institutional factors operating within the catchment and the surrounding river basins to achieve specific social objectives"

and to which may be added

> "… to achieve a sustainable balance between utilisation and protection of all environmental resources in a catchment, and to grow a sustainable society through stakeholder, community and government partnerships in the management process" (DWAF 1998).

IWRM is thus a vital, albeit incomplete, subset of ICM, particularly if socio-political aspects do not receive the same emphasis as biophysical factors in management scenarios (Schulze 1999). Diagrammatically IWRM may be placed into a context of levels of management complexity and levels of integration needed (Fig. D.39). The distinction between IWRM and ICM thus becomes apparent in the figure.

Of the many approaches to IWRM that have been proposed, six are viewed as embodying its major goals and strategies (DWAF 1996, 1998; Calder 1999; Schulze 1999; Ashton 2000; Frost 2001; Schulze 2001):

- *A systems approach.* IWRM recognises individual components as well as linkages between them and addresses the needs of both human and natural systems (DWAF 1996). It recognises water and land manage-

Fig. D.39.
The relationship of IWRM and ICM to the level of management complexity and level of integration needed (after Ashton 2000). Their value to society is also indicated

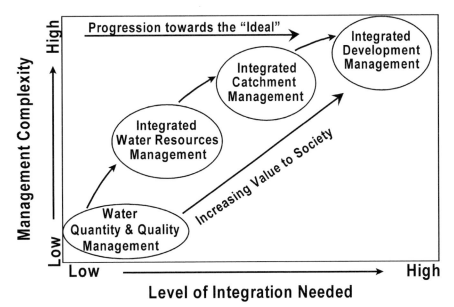

ment at the local catchment level to be mutually dependent, thus enabling an upward integration of strategic water management at scales beyond the local. IWRM seeks solutions by an incrementally evolving and iterative process rather than by attaining one optimal solution, using a blend of "soft system" tools focusing on the human dimension together with "hard system" methodologies such as models and their decision support systems (Calder 1999; Schulze 1999). IWRM concepts recognise that solutions should focus on underlying causes and not merely their symptoms.

- *An integrated approach.* Integration implies "holistic" and "with integrity" (Schulze 1999), beyond being only "comprehensive" (DWAF 1998), with integration of the biophysical system, the socio-political and the anthropogenic (water engineered) system, together with the aquatic system and its variable instream flow requirements. While integration is not achieved easily where social, political, administrative and natural boundaries coincide poorly, two types of integration are required (Jewitt and Görgens 2000; Frost 2001). The first is *horizontal integration*, which takes place within the same hierarchical level (at macro-scale or micro-scale), and where integration could be either between nations sharing a river, or between different water use sectors within the same river basin between upstream v. downstream users, or between activities of adjacent land uses/users within a catchment (Fig. D.40). The second type of

integration is *vertical*, where collaboration and/or coordination cross a range of political, legislative or management sectors, of types of modelling systems, or components of a natural system (such as river basins or aquatic ecology) which function at different vertical scales within the same sector (Fig. D.40).

- *A management approach.* In more generic terms management in IWRM implies maximising use of resources, minimising deleterious consequences over the long term, and even reversing the consequences of previously damaged systems (cf. Sect. D.3.6), while seeking the well-being, and enhancing the quality of life, of the inhabitants within an affected area and finding equitable solutions that are fair and just to all concerned (Schulze 1999). Management should recognise the intrinsic value and importance of water and not merely its availability to satisfy economic needs. In more pragmatic terms, management in IWRM requires that land and water be managed together, for every land-use decision becomes a water resources decision (Falkenmark et al. 1999), that water be managed at the lowest appropriate level using a bottom-up rather than a top-down approach, and that water allocation takes full account of all affected stakeholders, including the non-vocal poor as well as the environment. Furthermore, water has to be recognised as an economic good and principles of demand management need to be applied, with appropriate pricing policies to encourage efficient usage of water between

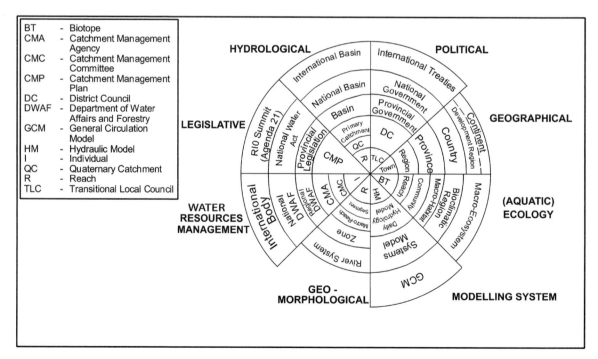

Fig. D.40. Examples of horizontal and vertical integration, taken from an IWRM approach in the Kruger National Park, South Africa (after Jewitt and Görgens 2000)

competing sectors such as domestic, agriculture, industry and the environment. Pricing should also embody notions of fairness and equity.

- *A stakeholder approach.* This recognises the importance of the involvement of individuals, landowners and government agencies, in a participatory process where all decisions around the management and sustainable use of land and water resources are made.
- *A partnership approach.* This emphasizes the commonality of objectives as well as the definition of collective rules, responsibilities and accountabilities of every individual who, and every water use and administrative agency which, participates in the process of decision making on use and management of land and water resources. This partnering must take place at all levels, down to that of the village and even the individual household (DWAF 1996; Calder 1999). It thus reflects a commitment to the principle of stewardship at all levels of management (Ashton 2000).
- *A balanced, sustainable approach.* Decisions must be designed to achieve compromise, between long-term and viable economic development for all the catchments' dependants (local to international), equitable access of water resources to them and protection of resource integrity (DWAF 1996; Calder 1999). Sustainability is not to be confused with zero growth (Ashton 2000). The relationship illustrated in Fig. D.41 shows that while sustainable resource utilisation requires a high level of resource management, it is also highly vulnerable to degradation.

D.3.5.3 The River Basin as the Fundamental Unit for IWRM

The river basin has been promoted and advocated as a basic, appropriate and ideal spatial unit by many prominent experts and international fora (Newson et al. 2000).

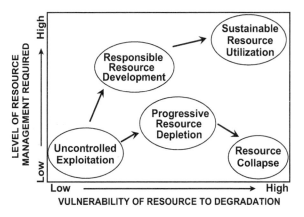

Fig. D.41. The relationship between sustainable resource utilisation in terms of the level of resource management required and the vulnerability of the resource to degradation (after Ashton 2000)

Within the river basin one can organise human activities for regional planning (Smith 1969), integrate patterns and processes of both natural and social systems (Young et al. 1994) and, particularly, integrate ecosystems-based land, environment and water resources management (Agenda 21 1992; Johnson 1993; Falkenmark 1997). Calder (1999) cites numerous indicators of "catchment consciousness" in science, society and policy making, for example, in:

- *Hydrology*, with identification of hydrological and environmental capacities and impacts of critical land use/development on water through catchment-scale research;
- *Hydraulic engineering*, focusing on imaginative uses of technology to assist in water management on a catchment basis (e.g. metering, leakage control, control of water pollution at source or recycling);
- *Economics*, through use of resource and environmental economics to assess new water schemes, water pricing and water as a tradeable commodity;
- *Society*, through a rise in public awareness of water issues within catchments, stakeholder involvement in decision making, and checks and balances on development and land-use change with respect to the water environment; and,
- *New policy frameworks* which use the river basin as a basic unit (e.g. Agenda 21) and entrench catchment based IWRM firmly in recent revisions to water laws (e.g. South Africa's National Water Act) with built-in legal structures for concepts such as Catchment Management Agencies (NWA 1998).

The factors listed above already make for compelling arguments in favour of using the river basin as the spatial management unit in IWRM. Other advantages (Schulze 1999) include the fact that

- it is a topographically clearly bounded unit through which to study inputs and outputs of the water-related biophysical system, particularly if the focus is on surface water and water quality;
- it is a natural "integrator" of interdependent water issues such as water supply, flood control and water quality (sediment, chemical and biological pollution production); and,
- it is the natural unit of analysis for river navigation; furthermore,
- the catchment may be hierarchically subdivided into modular, relatively homogeneous and hydrologically cascading sub-units in cases where issues of strong local interest prevail, or where the catchment is physiographically or socio-economically complex/diverse, or where a sub-unit adequately addresses particular stakeholder needs.

The river basin as a management unit is not without disadvantages, however. Even in terms of water, management may encompass more than one contiguous and topographically defined catchment, particularly where inter-basin transfers of water take place, or groundwater is a major component of usable water, as major aquifers may not coincide with surface water catchments (Schulze 1999). Furthermore, administrative units of countries, provinces or local authorities within which decision-making is generally effected, seldom follow natural catchment boundaries. Indeed, the river itself is often a political boundary acting as a management divide rather than an integrator. Also, a river basin may straddle a number of countries, with complex international water security issues at stake as well as complex cultural historical backgrounds, with countries possibly at different levels of development or covering major (but different) hydroclimatic regions (Schulze 1999; Newson et al. 2000). Additionally, neither the social world of different tribes, ethnic/linguistic groupings, nor the economic world of different levels of development, regional collaboration, trade, industry and capital flows follow natural catchment boundaries. It is also argued that major regional problems such as air pollution and the potential impacts of climate change are determined by factors beyond the natural catchment, as well as "natural" regions (e.g. of agricultural production) often being defined by climatic, physiographic or edaphic boundaries which do not coincide with the river basin (Schulze 1999).

Despite these disadvantages, the typology of river basins is ideally matched to issues relating to the integrated management of water and affiliated aquatic ecosystem resources.

D.3.5.4 At What Space and Time Scales Should IWRM Be Carried Out?

The appropriate temporal and spatial scales for IWRM are those at which policy-makers, catchment managers and stakeholders believe that they can achieve their objectives associated with the problem(s) at hand. This will depend on the life expectancy of a planning or management option, or the time it takes for such a plan to become operational. Within an overarching "scale of operation" of an IWRM plan, however, there should be embedded a hierarchy of intermediate and internal smaller space and shorter time scales, to define interim stages of implementation, which we can identify as goals or milestones (Schulze 1999).

With regard to spatial scales, IWRM must take stock of issues cast at a multitude of scales – global scale issues (e.g. water conventions, climate change), international scale problems (e.g. international rivers), national

issues (e.g. water management agendas), catchment scale issues, local government scale initiatives, community scale issues and household scale problems (especially in poorer countries where household water and food security can become major issues of existence). Spatial scale considerations in IWRM often reflect the level of development of a country. In poorer countries the space scale tends to be much smaller, determined by factors such as the distance range over which one can mobilise communities, or land availability around a village, or access to local water sources (Schulze 1999). When focusing at broader scales, it is often impossible to keep in view the "fine grained variation" embodied in all the various processes and there is a risk of overlooking local features, needs, circumstances, and/or aspirations, especially of the poor within the catchment (Frost 2001). On the other hand, when focusing on too fine a scale, there is a danger of losing sight of the wider context of IWRM and of losing sight of the overall governing processes of IWRM (Frost 2001).

Similarly, temporal scales in IWRM should not be viewed as static, but rather as a hierarchy of overlapping scales. Several classes of time scales are identified and need to be considered in juxtaposition within IWRM (Schulze 1999):

- *climate scales*, at intra-seasonal, inter-seasonal and decadal time frames;
- *river flow scales*, driven by the climate scales, which for surface waters range from high flow/drought "cycles", frequently related to the El Niño/Southern Oscillation phenomenon at multiple year scales, and associated inter-seasonal variability, superimposed upon which are the seasonality and concentration of streamflows within a year and intra-annual variability, the forecastability of river flows up to a season ahead and extremes such as floods. For groundwater, temporal recharge patterns and water table fluctuation are important time scale issues;
- *ecological time scales*, which are determined by magnitudes, frequencies and durations of low and high flows as biological triggers;
- *agricultural time scales*, where for crops the intra- and inter-seasonal timeframes are important whereas for forestry, inter-seasonal to decadal timeframes are of significance;
- *economic time scales*, ranging from longer term international to national, regional to local to shorter term individual rural subsistence household time scales;
- *political time scales*, which need to distinguish between essentially stable government structures *v.* potentially unstable government structures and inter-election time scales for national to local governance structures;

- *management and planning time scales*, often of the order of 10–20 years; and,
- *wealth/development level time scales*, wherein wealthy countries tend to have longer term planning horizons in contrast to poorer countries, which show relatively short-term response-based planning (Schulze 1999).

The scale at which IWRM is best initiated is the scale at which people are impacting on land and water resources. Thus, in Europe when the Rhine basin is impacted, the cause is large scale rather than localised factors, and large scale IWRM thus has to be effected, whereas in much of rural Africa, anthropogenic activities are often dependent upon, and in turn impact on, smaller 1st and 2nd order tributaries, and hence effective IWRM generally takes place there at smaller spatial scales and with shorter time horizons.

D.3.5.5 Differences in IWRM between Developing Countries and Developed Countries

Additional major distinctions between the IWRM characteristics in more developed countries (MDC) and lesser developed countries (LDC) are summarised in Table D.12.

Because of MDCs' high levels of expectation from IWRM, their proactive perspective and a generally non-life-threatening environment and infrastructure, IWRM there usually focuses more on quality of life and environment as well as on long-term issues. These include preservation of the environment, with a focus on aquatic ecosystems, the re-naturalisation and rehabilitation of the catchment and its receiving streams, water quality issues, demand management of water and potential impacts of climate change on water resources (Schulze 1999).

Table D.12. Characteristics influencing IWRM in more developed *v.* lesser developed countries (after Schulze 1999)

More developed countries	Lesser developed countries
Infrastructure	
High level of infrastructural development, with infrastructure generally improving	Infrastructure often fragile and frequently in a state of retrogression
Infrastructure enables coping with natural disasters (e.g. floods, drought)	More often at mercy of natural disasters; heavy damage and high death toll
High ethos of infrastructure maintenance	Low ethos of infrastructure maintenance
High quality data and information bases available, well co-ordinated	Data and information bases not always readily available
Capacity	
Scientific and administrative skills abundantly available	Limited scientific and administrative skills available
Expertise developed to local levels	Expertise highly centralised
Flexibility to adapt to technological advances	Often in survival mode; technological advances may pass by
Economy	
Mixed, service-driven economics buffered by diversity, highly complex interactions	High dependence on land, i.e. agricultural production; at mercy of the vagaries of climate
Economically independent and sustainable	High dependence on donor aid, NGOs
Multiple planning options available	Fewer options available in planning
Long-term planning perspective	Shorter term planning perspective
Countries wealthy, money available for planning and IWRM	Wealth of countries limited, less scope for planning and IWRM
Socio-political	
Population growth low or even negative	High population growth rates and demographic pressures on land, water, and other resources
Generally well informed public with good appreciation of planning	More poorly informed public, less appreciation of science/planning
High political empowerment of stakeholders	Stakeholders often not empowered, afraid to act or to exert pressure
Decision making decentralised	Decision making centralised
Environmental awareness and management	
High level of expectation of planning and IWRM	Lower level of expectation and attainment of goals
Desire for aesthetic conservation	Need for basic sustenance
Multiple planning options available	Fewer options available in planning
Long-term planning perspective	Shorter term planning perspective
Countries wealthy, money available for planning and IWRM	Wealth of countries limited, less scope for planning and IWRM

In contrast, and as a consequence of poorer infrastructure, higher vulnerability to natural events and often operating in survival mode, IWRM in LDCs frequently has to address more fundamental issues (Schulze 1999) such as creating basic water supplies (*v.* demanding water of highest quality), managing the water supply (*v.* demand management), poverty alleviation (*v.* quality of life enhancement), harnessing the environment (*v.* sustaining it), short-term needs (*v.* long-term perspectives), climate variability, both intra- and inter-seasonal (*v.* progressive climate change) and creating a sufficient infrastructure (*v.* maintaining and improving it).

D.3.5.6 Conditions for the Success of IWRM

By its very definition, the multi- and inter-disciplinary IWRM requires inputs from the biophysical and socio/politico/economic sciences, as well as from the engineering and computer sciences. Hydrologists have a special role to play (Schulze 1999). Because of hydrologists' understanding of the inter-connectivities of the terrestrial water cycle at catchment scale – natural, anthropogenically altered and water-engineered – they will be called upon to:

- quantify the states and fluxes of the water system over time and space (by observation, data handling and simulation modelling skills);
- assess the impacts of catchment and channel modification on resource and environment;
- consider both water quantity (identifying source areas within the catchment, seasonality, availability, probability) and water quality (physical, chemical and biological), including the impact of present and future land management;
- forecast availability and the utility of water resources by providing information on assurance of supply;
- assess the risks associated with specific conditions and in specific geographic areas (including flooding and drought potential, quantified in terms of magnitude, duration, frequency and location); and,
- disseminate and make useful technical information available in a format that can be used by managers, in order to reduce the subjectivity of their decisions (Schulze 1999).

While the hydrologists' role in IWRM is undoubtedly significant, real integration will, however, only be achieved when all disciplinary perspectives are brought together to produce truly holistic options and an understanding of interdependent consequences. Hydrologists should thus have an important responsibility in becoming more proactive in initiating an interplay with the non-biophysical sciences associated with IWRM.

If IWRM is a philosophy, a process and the act of putting principles into practice, then the last two are the more difficult of the three steps. Preconditions for successful IWRM include close involvement of all stakeholders, political support at all levels of governance, effective cooperation of land and water management, recognition that while IWRM is a long-term process, realistic short and medium-term goals need to be set and audited, and acceptance that each catchment is unique with respect to IWRM, thus enabling specific institutional arrangements to be adaptable for each catchment's situation (Farrington and Lobo 1997; Ashton 2000; Frost 2001; Schulze 2001).

Despite the enthusiasm for, logical appeal of, and belief in whole catchment water management – reinforcing common messages via Agenda 21 and new water legislation – initiatives have commonly not lived up to their expectations or reputations (Frost 2001). Factors inhibiting the success of IWRM, particularly in LDCs (but not exclusively so) include sectoralism within and between government departments and the generally fragmented nature of institutional structures; lack of clearly defined overall management strategies; absence of research to assess the resource availability and the value of water; difficulties in appreciating water as a source of conflict between sectors (e.g. rural *v.* urban) and within a sector (e.g. dryland *v.* irrigated agriculture; commercial *v.* subsistence agriculture) as well as between upstream and downstream users; deficiencies in information, in human capacity and in land and water management options; poor stakeholder involvement; and, a lack of critical evaluation of the performance of actions during and after the process of IWRM (Calder 1999; Falkenmark et al. 1999; Frost 2001). These issues are elaborated upon in detail by Schulze (2001).

D.3.5.7 Conclusions

IWRM aims at finding long-term, sustainable means to successfully cope with environmental management and simultaneously satisfying society's water needs, given particular environmental preconditions (climate, soil, topography). It accomplishes this through the balancing of the different functions of water between competing sector (e.g. environmental, agricultural, industrial) and stake-holder needs, entraining the involvement of numerous groups, from scientists to policy-makers to local landowners (Falkenmark et al. 1999). This brief overview of selected aspects of IWRM has shown that this philosophy and process is not easily put into practice because of the many conceptual, scale, disciplinary and practical problems encountered. IWRM does, however, provide a framework within which to research and evaluate a range of policy options, and it offers the opportunity to assist in assessing the risks and options available to environmental, social and economic policy-makers. The social dimension of the process is particularly important, and IWRM re-connects people to water issues within their catchment through consultative processes, stakeholder participation and partnership options (Newson et al. 2000). Although most people relate well to IWRM, it paradoxically has not been easy to translate it into operational terms.

D.3.6 The Road Ahead 2: Restoration of Riverine Ecosystems

The "human footprint" on riverine ecosystems has been significant. Impacts are evident in response to purposeful actions, ignorance, or maladapted management. One management response is to "undo" these actions through a holistic approach such as IWRM, which has to include evaluating past unsustainable actions and undertaking restorative action, often at considerable cost.

Fundamental questions relate to the condition to which the river basin should be restored – should it mimic pristine conditions? what were such pristine conditions? should it reflect conditions centuries ago, 100 years ago or at a certain point in more recent time? One cannot turn back the hands of time, but a clear management goal for development of flowing waters is that they mimic dynamic environmental conditions close to those of their natural state, which in turn can support high structural/morphological diversity by creating a range of habitats in which natural ecological processes can be supported. This is often achieved by softer, bio-engineered procedures (Gunkel 2000).

Many terms are used for modes of restoration to near-natural conditions. Gunkel (1994) identifies three such procedures, the first being re-naturalisation:

- *Re-naturalisation* aims at consciously reproducing near-natural streams with natural flows within the landscape, a riparian zone with high biodiversity and dynamic channel morphological diversity.

On the one hand, this can be achieved in a relatively short space of time by the "de-construction" of man-made hydraulic structures followed by artificial re-construction to a more natural riverine landscape. This requires interdisciplinary co-operation across conventional water engineering sciences, engineering biology and the ecological sciences. Re-naturalisation of this type includes restoration of meanders to their original locations (e.g. in the lower Geul Valley, Netherlands), regaining water retention (inundation) space on a river's floodplain by relocating economic activity and removing buildings/obstacles (e.g. in the river Sieg, a tributary of the Rhine, Germany) and restoration of former river branches previously isolated from the main river (e.g. along the French side of the Rhine bend between Kunheim and Marckolsheim, Alsace).

Costs associated with re-naturalisation efforts are typically high: Gunkel (2000) cites $150–200 per running metre. Despite that, results are not as yet always satisfactory, partially because projects may be poorly conceived, and partially because of inadequate understanding of suitable methods for re-naturalisation.

On the other hand, re-naturalisation of riverine ecosystems downstream of large impoundments can be achieved without de-construction through the concept of managed flows. Flow management achieves ecological instream flow requirements by simulating key aspects of the the natural flow regime by compensatory reservoir releases at certain times of the year. These can simulate widely contrasting conditions such as major flood pulses, minor freshets or sustained simulated baseflows. By operational definition (Acreman et al. 2000), managed flows are controlled water releases to inundate a specific area of downstream floodplain or delta to restore and maintain ecological processes and natural resources. This should be achieved through collaboration with, and with warning to, stakeholders, while at the same time retaining sufficient impounded water reserves for hydroelectric production generation, irrigation or inter-basin transfers. This trade-off maximises benefits for both flood dependent and reservoir dependent development. Examples of environmental, social and economically managed flow releases cited by Acreman et al. (2000) include maintenance of flood recession agriculture (e.g. the Tana in Kenya and Kafue in Zambia), restoration of biodiversity of a delta (e.g. of the Senegal and Indus Rivers, in both cases sustaining mangroves), flushing of sediments (e.g. Logone in Cameroon, Colorado in the USA), control of disease vectors (e.g. Mahaweli in Sri Lanka), restoration of wetlands and their resources (e.g. Hadeji in Nigeria) and maintenance of floodplain ecosystems (e.g. Pongola in South Africa).

While managed flows work well in perennial rivers, they may be largely ineffectual in arid-zone ephemeral streams such as those in western Namibia (Jacobson et al. 1995). This is because it is seldom feasible to match dam releases to the magnitude of natural floods in hydro-climates where artificial floods are typically but a fraction of the average natural flood peak. Furthermore, in such areas, floods usually only travel downstream for a few 10s of km, with little overbank flooding. In addition, natural floods carry heavy loads of silt, organic matter and nutrients, all deposited in, and enriching, the downstream riparian ecosystem. In contrast, artificial floods emerging from large impoundments (cf. the managed flood of the Grand Canyon) carry almost no organic material, nutrients or sediments compared with natural ones, but pick up sediments below the dam and deposit them further downstream (Jacobson et al. 1995).

The second mode of restoration to near-natural conditions is re-vitalisation:

- *Re-vitalisation* implies a change in the use of the entire upstream river catchment which can then, under proper use, convert the river system into a natural one in the long term.

Re-vitalisation is thus a process which evolves over many decades. It does not apply regulatory and re-constructive interventions in the channel, except at very specific points of disturbance. As such, re-vitalisation can only be effected in areas of extensive land use.

By Gunkel's (1994; 2000) definition, there is a third mode of restoration, i.e. rehabilitation:

- *Rehabilitation* affords the cessation of every artificial use and/or maintenance of the stream as well as the contributing catchment.

This "wilderness", "do-nothing" or "laissez-faire" philosophy is a long process, with the area being left entirely to nature and protected from human land use. As in the case of re-vitalisation, rehabilitation can only be applied in thinly populated areas. It cannot be applied in densely populated areas with continued utilisation of the stream system. There only re-naturalisation can achieve positive change.

D.3.7 Conclusions

The river basin scale is not simple to define by lower and upper bounds of areal extents, for it occurs within that continuum of natural heterogeneities which spans scales from point to global, superimposed upon which are anthropogenic influences which transcend scales from the individual agricultural field to major cross-basin engineered features.

It does, however, constitute that intermediate range of scales at which responses to those processes from the hillslope to small catchments (see Chapt. D.2) can manifest themselves as hydrological concerns when under the influence of anthropogenic drivers and pressures. On the other hand, the accumulated effects from individual river basins can modify responses at the subcontinental to global scales, as reviewed in Chapt. D.4.

The river basin is that scale at which different physiographic and climatic attributes can produce fundamentally different hydrological responses. In this chapter this is illustrated in the sections on salinisation and channel evaporation/transmission losses in semi-arid zones. In particular, however, the river basin is that scale at which anthropogenic actions can reshape responses of water quantity and water quality through changes to the landscape component of the basin (e.g. afforestation, urbanisation or degradation resulting from agricultural pollutants) or to alterations to the channel component of the basin (e.g. channel modifications, dams), both with potentially severe hydro-ecological consequences.

The river basin scale has, accordingly, become the spatial unit of action and water management to the damaged catchment ecosystem, in particular through the concept of Integrated Water Resources Management with its emphasis on systems, integration, management, stakeholder, partnership and sustainability approaches. Characteristics influencing IWRM are shown to be fundamentally different in more developed *v.* lesser developed countries. It has also become the scale at which attempts are made to "right previous wrongs" through re-naturalisation, re-vitalisation and re-habilitation programmes.

This chapter has provided a perspective on the above largely by describing how the elements of the DPSIR model (i.e. drivers, pressures, state changes, impacts and responses) have intersected with each other at the river basin scale. The chapter has set the scene for the presentation of three river basin scale case studies in Chapt. D.5 to D.7, each under the broad theme of integration in water management, but with regional emphasis focusing on differences in hydroclimatic regimes and levels as well as types of anthropogenic disturbances.

Acknowledgments

The contributions of Prof. Michel Meybeck of the University of Paris in France, who provided a comprehensive chapter review, and of Drs. Mike Acreman and Mathew McCartney of the Centre for Ecology and Hydrology at Wallingford in the UK for providing much literature and comment, are acknowledged gratefully, as are those of Jürgen Surendorff, Kent Thornycroft, Cynthia O'Mahoney and Mark Horan of the University of KwaZulu-Natal in Pietermaritzburg, South Africa who helped with preparation of diagrams. This work has been partially funded by the Water Research Commission of South Africa.

Chapter D.4

Responses of Continental Aquatic Systems at the Global Scale: New Paradigms, New Methods

Charles J. Vörösmarty · Michel Meybeck

D.4.1 Introduction

Water figures prominently in the science of the Earth system and in the international agenda on global change. As a key component of the Earth's climate and biogeochemistry the global hydrological cycle has received significant attention with respect to its role in land-atmosphere exchanges of water, energy and CO_2. This subject has, for this reason, constituted a major portion of this book. Water is also a key vehicle in the global mobilisation and transport of carbon, nutrients and suspended sediment, and it is these horizontal fluxes that orient major interconnections between the continental land mass and the world's oceans. We shall focus in this chapter on the terrestrial water cycle and its role in the horizontal transport of land-derived materials which has been voiced as an Earth system and global change issue

several times within the IGBP (Pernetta and Milliman 1995; Vörösmarty et al. 1997a). We refer here to "terrestrial aquatic systems" for rivers, lakes, reservoirs, wetlands and groundwaters while using "continental aquatic systems" to include the former as well as deltas, coastal lagoons, estuaries, fjords, etc.

Despite the enormous emphasis placed on climate change in the international research agenda on global change (e.g. Houghton et al. 2001; US National Assessment 2000; Lemellä and Helenius 1998) the dialogue with respect to continental aquatic systems (CAS) must necessarily extend to other human dimensions issues including land-cover change, population growth, urbanisation, economic development, and water resources management (Fig. D.42). Not only are issues within each conceptual domain important to understand, but their interactions are as well. As an example of the complexities which may arise, consider a major change in land cover from forest to agriculture that

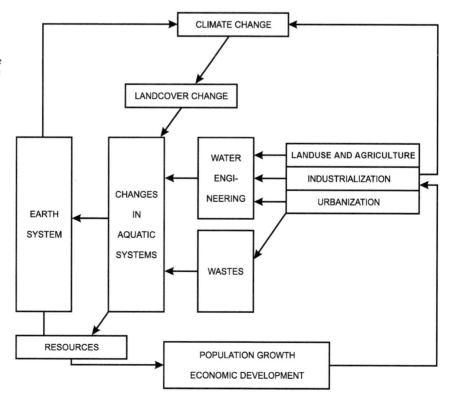

Fig. D.42.
Elements of global change with respect to global river systems for hydrology and nutrient biogeochemistry. The interactions of climate change and variability, land-cover change, population growth and industrialisation, and water engineering are shown

creates elevated erosion upstream which then translates into siltation of downstream reservoirs, forcing construction of additional reservoir systems. These impoundments promote consumptive use of water through new irrigation schemes that further fragment the landscape and change local weather patterns. Changes in the reliability of weather for crops then prompts additional reservoirs to be constructed, which begins to compromise water quality, interfere with the migration patterns of economically important fisheries, and distort the seasonality of nutrients delivered to the coastal zone and so forth.

Both hydraulic engineering and extensive land-cover change makes it increasingly difficult to detect the signature of climate change on global water resources (Vörösmarty 2002a). Nonetheless, we hypothesise that while their aggregate effect is difficult to quantify with precision, such contemporary changes are now pandemic in nature. We arrive at this conclusion from documentary evidence and modelling studies available in the recent literature which will be presented in this section of Part D.

Over continental and global domains, the aggregate impact on the terrestrial water cycle of direct human interventions such as water resource management, land-use change/intensification, and urbanisation exceeds that arising from climate change in the relatively recent past and will continue to do so over the next several decades (i.e. ±50 years from present).

It is important to note that this hypothesised effect is global in its domain. Thus, while bona fide global-scale studies will be summarised, we also will consider the aggregate effect indicated by results from more site-specific research. *Recent climate change* here refers to anthropogenically-induced changes due to the enhanced greenhouse effect; natural climate variability and climate change are distinct, but also considered. We also recognise that the absolute and relative impact of individual forcings will be site- and region-specific.

The goal of this part of the text is to assess the validity of our contention given above, and to review emerging subsidiary hypotheses, tools, and datasets. We also will review in detail the evidence for our overall assessment of the relative importance of different key agents of global change – climate change, land-cover/land-use change, and water engineering.

We will attempt to identify key unknowns and offer our assessment of likely changes into the future. We will use, as appropriate, global-scale observations and key geospatial biophysical datasets currently available. There is a growing body of knowledge now formalised into a set of aggregate relationships linking constituent fluxes and biophysical attributes of continental water systems. Together these methods constitute a new state-of-the-art in river basin analysis for water and constituent fluxes. These merit our due attention and, in fact, are essential to formulating a new paradigm regarding global-scale changes to river systems.

D.4.2 Terms of Reference

D.4.2.1 Relevant Time and Space Scales Associated with Global Change and Continental Aquatic Systems

The hydrosphere is influenced by global change over a variety of scales and is closely tied to the state of the climate system. Thus, its natural variability is associated with periodicities that span several temporal domains – from long-term oscillations associated with the Milankovic cycle of eccentricity in Earth orbits around the sun ($\sim 10^4$ yr) to much shorter quasi-periodic events such as El Niño/Southern Oscillation, North Atlantic Oscillation, Arctic Oscillation ($\sim 10^0$–10^1 yr). Still higher-frequency and episodic events characterise hydrology, especially over local domains. Additional concerns surround the impact of progressive climate change over the decades-to-century time domain, together with associated increases in climate extremes that are postulated to occur. These drive debates on the very nature and thresholds associated with glaciation and de-glaciation, as well as the vulnerability of human population to flooding and drought.

During the Holocene, climate and tectonics were the key determinants of the state of river basins. The situation has recently changed due to the introduction of significant human impacts on the Earth system. Throughout history, humans have pursued very direct and growing roles in shaping the character of the terrestrial water cycle, as they harness and use elements of continental aquatic systems in service to society. Recent changes have moved us rapidly toward a global-scale impact that is only now being articulated. For instance, it is now recognised that humans control and use a significant proportion (> 50%) of continental runoff to which they have access (Postel et al. 1996). And, the impact of the hydraulic engineering needed to afford such control has substantially distorted continental runoff (Vörösmarty and Sahagian 2000; Rosenberg et al. 2000). Large reservoir construction alone has doubled or tripled the residence time of river water, with the mouths of several large rivers showing delays on the order of months to years (Vörösmarty et al. 1997b). Such regulation has enormous impacts on suspended sediment and carbon fluxes, waste processing, and aquatic habitat (Dynesius and Nilsson 1994; Vörösmarty et al. 1997b,c; Stallard 1998). Land-cover change also has been shown to influence local patterns of runoff and feedbacks to the atmosphere as discussed earlier.

The emphasis in this section will be on macro-scale domains. However, because our knowledge of aquatic system change is better articulated at more local and regional domains, we will identify sensitive systems using the more traditional literature as required through-

Fig. D.43.
The Earth system showing key components of the terrestrial water cycle and its biogeochemistry. Although an aggregate view of the Earth system is important for understanding global-scale processes, river-related phenomenon are spatially-distributed in nature. This chapter will articulate the emergence of a new class of datasets and models that can be used to articulate global change on river systems in a spatially-explicit, yet fully global context (after Mackenzie et al. 2001)

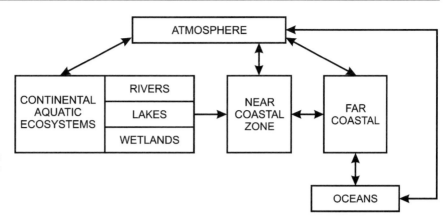

out the text. We also emphasize a time horizon of the near contemporary, but we will use a palaeo-perspective to support our understanding of present conditions and to shed light on long-term system behaviour.

D.4.2.2 Emerging Techniques for Analysing Continental Aquatic Systems and Global Change

With the recent advent of spatially-discrete, and often high-resolution Earth system datasets that depict the continental land mass, together with satellite remote sensing and GIS-based tools for analysis of river basins, the community finds itself poised to go well beyond descriptions of spatially-isolated vignettes of aquatic system change (Meybeck and Helmer 1989; Fraser et al. 1995) to now assemble a truly global picture of these progressive system changes. The macro-systemic view of continental aquatic systems within larger Earth system models, for example (Mackenzie et al. 2001; Ver et al. 1994; Rabouille et al. 2001) need not be limited to aggregated functional units (Fig. D.43). Using river networks of varying spatial resolutions from 30' (latitude × longitude) (Vörösmarty et al. 2000a,b) down to 1-km (USGS-EDC 1998) and many other resolutions in between (Fekete et al. 2001; Graham et al. 1999b), we can now depict the actively discharging as well as potential river networks of the world, with great geographic specificity.

A fundamental accounting unit is the drainage basin and its river networks. Our emphasis will thus be on surface waters, although we will also consider groundwater fluxes. We will treat water, sediment, organic matter and inorganic carbon, major nutrients (in both dissolved and particulate form)[1]. We will consider the mobilisation, storage, transformations, and routing (transport) of this matter from small upland catchments (dis-

cussed in prior chapters) through the continuum of continental aquatic systems with final delivery to the coastal oceans or inland seas.

D.4.3 Changes in River Connectivity and Basin Characteristics: Palaeo to Present

The organisation of rivers, and the character of their associated drainage basins, are anything but static and have changed dramatically over several time horizons. The spatial organisation of rivers represent at the global scale enormous, ordered tree structures. These topologies change, for example, in response to the availability of water, creating a contrast between potential and actively discharging channels. Such changes can take place over tens of thousands of years associated with climate shifts between glaciation and alti-thermal conditions. They also occur, but to a much smaller degree, from year-to-year, and even sub-annually in many parts of the world under arid and semi-arid conditions. These are areas on the brink of their capacity to sustain perennial flow and often are of enormous importance to local inhabitants in arid and semi-arid regions (Adewale 2001). Such changes also take place in response to the impact of tectonism such as in the rift lakes of Africa, or as a result of major glaciations which simultaneously redirect the river flows when ice sheets are encountered and increase the exposure of the continental shelf to land-surface hydrological processes. Rivers and drainage basins are thus highly dynamic entities. We are interested not only in the linkages of river systems within the continental land mass but also the linkages between the continental land mass and world's oceans which will also vary in response to such major forcing factors.

D.4.3.1 A Global Classification System for Flow Connectivity in River Systems

The structure of the global system of rivers can be viewed from the standpoint of potential flow pathways

---

[1] Toxic substances including heavy metals and pesticides are still not adequately measured at the global scale. The associated fluxes to the oceans are very difficult to assess or model and are therefore not covered here.

Fig. D.44.
Summary of the distribution of global exo- and endorheic area, average runoff and the associated human population density for the four main relief types (after the data of Meybeck et al. 2001). Mountains are true water towers in endorheic areas, runoff is less linked to relief over the exorheic domain

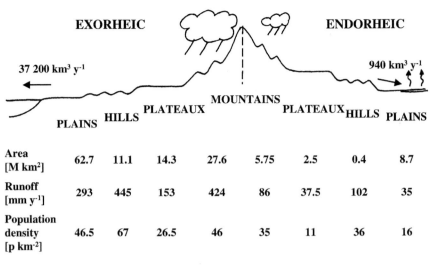

	PLAINS	HILLS	PLATEAUX	MOUNTAINS	PLATEAUX	HILLS	PLAINS
Area [M km²]	62.7	11.1	14.3	27.6	5.75	2.5	0.4	8.7
Runoff [mm y⁻¹]	293	445	153	424	86	37.5	102	35
Population density [p km⁻²]	46.5	67	26.5	46	35	11	36	16

(mediated by topography), actively discharging channels (mediated by available runoff) as well as their exit points that empty to either an ocean (exorheism) or inland receiving body (endorheism). If the whole surface of the continents was characterised by an excess of runoff, rivers would fill any depression and the entire continental surface would be exorheic. This is not the case and many river networks end in terminal depressions that are dry or as lakes that fail to overflow to the oceans, i.e. corresponding to endorheic basins. In a 30' (latitude × longitude) simulated river network (Simulated Topological Network, STN-30p; Vörösmarty et al. 2000a,b), the majority of potential flow pathways (draining 87% of the land mass) connects the interior of the continents to one of four oceans (i.e. Arctic, Atlantic, Pacific, Indian) or the Mediterranean/Black Sea. An additional set of river systems empty into major land-locked receiving waters (e.g. Caspian and Aral Seas, Great Salt Lake, Lake Chad) or large topographic depressions in extremely dry regions (e.g. Altiplano in the Andes, Takla Makan in western China). STN-30p thus defines the global network of rivers based fundamentally on topographic control. An idealised summary of the characteristics of exorheic and endorheic basins and their associated receiving bodies under a contemporary or near-contemporary condition is shown in Fig. D.44. The majority of the world's runoff and population are in exorheic basins.

Using contemporary Africa as an example, the Zambezi would be considered exorheic/flowing. The Tamanrasett/Saoura and the Araye drainage basin systems in the Sahara (Fig. D.45) would be considered exorheic/non-flowing. The Araye is known from palaeogeographic studies to have maintained an active discharge system over much of its area at 6 000 years before present (Williams and Faure 1980). Endorheic basins can also be flowing (Okawango, Chari) and non-flowing (Tafassassett, north of Lake Chad).

Thus, the STN-30p working definition of the endorheic domain, and the higher resolution HYDRO1K stream line data file (USGS-EDC 1998), is conditioned upon runoff for the recent past as well as the next several decades. Alternative definitions of global river connectivity to the ocean exist (Graham et al. 1999b; Alcamo et al. 2000; Oki et al. 1999).

Within basins, an operational threshold of 3 mm yr⁻¹ runoff has been proposed by Meybeck (1994) and Vörösmarty et al. (2000b) to define all types of active flow (intermittent episodic or occasional, seasonal, perennial), and a threshold of 30 mm yr⁻¹ to define the perennial flow only. The 3 mm yr⁻¹ (± 1 mm yr⁻¹) threshold is verified in the extremely arid Lake Eyre basin (Australia) where some eastern tributaries flow only once every 10 years, with a long-term average runoff of 1-2 mm yr⁻¹ (Kotwicki and Isdale 1991). It is now possible to map the patterns resulting from the conjunction of river networks, conditioned fundamentally on geology and topography, with climate-based controls which will limit the potential flow pathways to some subset of the global "potential". Collaborative efforts between the University of New Hampshire and the WMO Global Runoff Data Centre (Koblenz, Germany) enabled global-scale tests of this hypothesised threshold to be made. The work has generated global maps of runoff at 30' spatial resolution (latitude × longitude) conditioned on the spatial distribution of meteorological forcings, while at the same time "anchoring" these runoff values by measurements of stream discharge (Fekete et al. 1999, 2002). The result is a spatially-resolved mapping with mass balance conditioned upon the observational record.

To identify portions of river networks eligible for intermittent, seasonal or perennial flow (i.e. exceeding the 3 mm yr⁻¹ runoff threshold) requires an aggregate runoff to be computed upstream of any point within a simulated river network. This can be done by accumulating the spatially-distributed runoff and converting this to

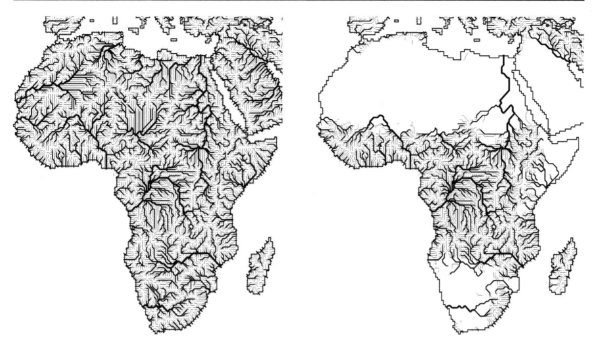

Fig. D.45. The concept of potential versus active river networks defined by climate. Results are shown for Africa contrasting a 30-minute potential river routing system (from Vörösmarty et al. 2000a, © with permission from Elsevier Science) to that defined by contemporary climate. A 3 mm yr^{-1} runoff threshold (after Meybeck 1994, and Vörösmarty et al. 2000b) for non-flowing river systems was employed to differentiate active river systems from potential systems

discharge within the river networks. Next, the corresponding cumulative drainage is recorded and the ratio of discharge to area (with suitable scaling parameters, Fekete et al. 2001) can be established. In this way, one can generate, for example, flow fields at the tops of mountain ranges and track the loss within alluvial fans. A simulated picture of allogenic rivers such as the Nile can also emerge, through which huge amounts of river water discharge through an otherwise desertic region.

In reality, at least three types of surface flow must be considered, operationally defined as: arheic or non-flowing (< 3 mm yr^{-1} runoff), episodic and seasonal runoff, (>3, <30 mm yr^{-1}), and permanent runoff (>30 mm yr^{-1}). Drainage systems like the Tamaransett Saoura represent today arheic river systems. Despite their dryness, they exhibit well-defined potential flow pathways to the ocean. Episodic and seasonal runoff systems can be defined both in terms of weather-related events or snow-melt and freeze-thaw dynamics in cold regions. The former is exemplified by the Wariner and Margaret creeks, two western lake Eyre tributaries which only flow a few times per hundred years (Kotwicki 1986). The Lena River which flows into the Arctic is an excellent example of the latter, with the maximum: minimum daily discharge ratio of more than 50:1 with the peak associated with spring runoff and the minimum with wintertime freezing.

These categories are naturally highly dynamic and particular river systems can show all three types of flow. For example, in the Nile catchment the White Nile and Blue Nile have permanent and highly seasonal flows, the Atbara is non-permanent and seasonal, the El Miklwaddi is occasional. During seasonally dry or drought conditions small headwater streams have little buffering capacity against such variability and often fail to discharge during dry spells. In contrast, downstream reaches benefit from a larger regional groundwater component which can sustain discharge. River networks in the Sahel have been shown to undergo such changes on a decade-to-decade time frame (Adewale 2001). Using high resolution maps, Landsat imagery, and geomorphometric analysis over a period of 30 years, these studies have shown a sustained and progressive loss of river networks from the 1960s into the 1990s. This work reveals the highly plastic nature of river systems. Shifts that are occurring today also occurred in the past and ultimately are a reflection of the major wet-dry transition zones of Africa (Fig. D.45). As interesting as these dynamics are to the curiosity of a geoscientist, it should be remembered that this variability has a human dimension as well, with river channels serving as an important water resource system to humans in this region.

By applying the four unique combinations of flow systems (i.e. of exorheic, endorheic, flowing, and non-flowing) we obtain a new division of the continental land mass (Table D.13). As mentioned above, the majority of the global land mass is exorheic (115.6 Mkm^2) and nearly 75% of this land shows flowing river systems (84.2 Mkm^2). Endorheic basins are quantitatively less important and

Table D.13. Division of the Earth's land mass by the new classification system presented in this study. The individual entries represent the four combinations of exorheic (external), endorheic (intenal), rheic (flowing), and arheic (non-flowing) land mass. The partitoning applies to all non-glaciated land area. All entries in Mkm²

Continent	Exorheic		Endorheic		Total
	Rheic	Arheic	Rheic	Arheic	
Africa	15.0 (50%)	11.9 (39%)	1.40 (5%)	1.74 (6%)	30.1
Asia	26.2 (59)	9.29 (21)	4.92 (11)	3.93 (9)	44.4
Australasia	2.28 (28)	3.64 (44)	0.008 (≪1)	2.27 (28)	8.19
Europe	8.09 (80)	0.072 (1)	1.89 (19)	0.043 (<1)	10.1
North America	17.3 (78)	4.48 (20)	0.37 (2)	0.18 (<1)	22.3
South America	15.3 (85)	1.99 (11)	0.28 (2)	0.34 (2)	17.9
Globe	84.2 (63)	31.4 (24)	8.86 (7)	8.51 (6)	133.0

have areas represented as flowing and non-flowing that are nearly equal. This pattern varies greatly over individual continents. Europe, with relatively wet conditions, shows 80% of its land mass flowing to the oceans, while nearly all of its remaining land mass discharges endorheically to the Caspian Sea. In contrast, Australia dominated by arid conditions shows only 28% of its area with active discharge to the oceans. More than 60% (3.64 Mkm²) of its total land area that would discharge to the ocean (5.92 Mkm²) is rendered inactive due to insufficient runoff. Further, most of Australia's endorheic areas are dominated by regions of non-flow. Other continents show relatively minor endorheic zones that are split roughly evenly between rheic and arheic regimes. Asia is the most typical of the global pattern with a roughly similar division of land mass into each of the four categories.

Finally, endorheic water bodies themselves serve as endpoints for a long continuum of hydrological processes operating across the land mass, and time series of volumetric storages make them useful harbingers of climate variability and progressive climate change signals. Present-day internal lakes such as Lake Chad indeed show multi-year trends due to climate variability (Birkett et al. 1999; Birkett 1995) (Fig. D.46). These inland waters are excellent targets for satellite remote sensing (Vörösmarty et al. 1999). Recent work using ocean radar altimetry re-focused on inland targets showed an enormous rise in the water stage – and therefore volume – of Lake Victoria that could be traced directly to a significant precipitation event with associated massive flooding (Birkett et al. 1999). Based on observations of past lake behaviour, the impacts of such a rise may last for a considerable length of time as the lake empties its excess water.

D.4.3.1.1 *Role of Groundwater in Land-Ocean Linkage*

Most of the renewable source of continental runoff is not generated by the deep (regional) groundwater flow mechanism. In areas of modest relief, the majority of groundwater flows are delivered to stream networks over relatively short lateral distances (Freeze and Witherspoon 1967; Dingman 2001). For example, in the Contoocook River catchment in New Hampshire, 93% of catchment runoff is discharged by rivers through the outlet of this basin which has an area of 2 000 km² (approximate size of our 30' grid cell). In the karstified Floridan aquifer in the SE United States, the vast majority of available water is discharged into streams and only about 3–5 cm yr^{-1} of the total 35 cm of annual runoff recharges the groundwater (Fetter 1988). The High Plains aquifer in the central US flows from W to E at a rate of about 0.3 m d^{-1} but ultimately discharges into streams and springs or is evaporated (Fetter 1988). These results are not atypical of the land mass as a whole – L'vovich and White (1990) and Shiklomanov (1996) indicate that at most only about 5% of continental runoff from groundwater discharges directly to the ocean suggesting that any replenishable deep groundwater either (*i*) constitutes a relatively minor flux to begin with or (*ii*) re-emerges into stream channels on the continental land mass.

Despite their relatively minor role supplying freshwater discharge to the ocean globally, direct groundwater inputs at local and regional scales may be quite important. Groundwater can constitute the dominant form of freshwater flow to the coastal zone (e.g. karstic regions in Florida, Yucatan, China, Mediterranean) (Church 1996). Recent use of ^{226}Ra as a tracer by Moore (1996) also pointed out the high proportion of direct groundwater inputs from the Georgia coastal plain. Given the dense settlement of the coastal zone by humans, the sensitivity of aquifer systems to global change should, nonetheless, be addressed. Furthermore, since groundwaters are generally more mineralised than rivers, direct groundwater flux of dissolved constituent loads is likely to be disproportionate in its importance relative to very modest fluxes for water *per se*. The influence of direct input of groundwaters to the coastal zone remains to be firmly established at the global scale.

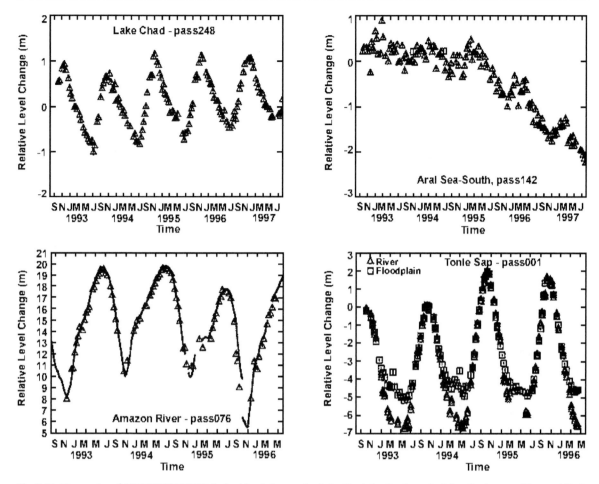

Fig. D.46. Time series of TOPEX/POSEIDON-derived level changes for Lake Chad, the Southern Aral Sea, the Amazon River and Tonle Sap. Note the comparison between the Amazon altimetry results and gauge data at Manacapuru (*solid line*). The satellite altimeter is both able to monitor rivers and their floodplains (e.g. Tonle Sap). Remarkable is the rising levels of Lake Chad, while the Aral Sea has decreased by 2.5 m between 1993 and 1997 (compiled from Birkett 1995, 1998, © American Geophysical Union)

D.4.3.2 Major Earth System Processes Controlling Land-to-ocean Coupling: Glaciation/De-glaciation, Climate Variability and Recent Tectonics

During the last 10^5 years, individual drainage basins and some regional river systems have been exposed to drastic changes of natural origin, which can be considered as very slow by human standards, but extremely rapid in terms of geological processes, for example on the order of a few 100s to 1000s of years. These changes are primarily linked through the major glaciation/de-glaciation cycles as described above occurring over the last 20 000 years. A complete picture must also include the influence of well-documented changes in precipitation patterns – with general increases – during the Holocene (last 10 000 years). Tectonic activity such as faulting, subsidence and volcanic eruptions may also locally or regionally modify river networks, but the influence is typically low, with high rates of uplift for example being of the order of 1 cm yr^{-1}.

D.4.3.2.1 *Glaciation/De-glaciation Controls of River Networks*

Over the long-term, large areas (16×10^6 km^2 in North America, 7.2×10^6 km^2 in Europe, 2.7×10^6 km^2 in Ural-Siberia; Flint 1971) are controlled by major onsets of glaciation and, later, de-glaciation in response to global climate change. This effect includes the impact of sea level change which is associated with differential exposure of the ocean shelves as land mass. The period since the Last Glacial Maximum to the present will constitute our time domain, with two sub-domains: the Holocene and the period of sedentary agriculture (corresponding to the IGBP PAGES-LUCIFS (Land-Use Change in Fluvial Systems) time frame: 2 000 years B.P.).

These changes are important from the standpoint of the horizontal transport of water, sediment, carbon, nutrient, sulphur over the non-glacierised land mass of the Earth. It is surprising that over the last 18 000 years the total area of non-glacierised land mass has changed little overall from the contemporary area of about 130 Mkm² (Gibbs and Kump 1994; Vörösmarty et al. 2000a). The reason for this is that an approximately equal area of land lost to ice sheets is acquired by the exposure of continental shelf associated with the lowering of sea level. Further, the non-frozen area subject to water flow, both endo- and exoheric, has changed little. Where changes have occurred is in the actively discharging portion of exo- and endorheic basins, in other words, areas manifesting the differences in climate between maximum glaciation and the contemporary. An approximate 15% increase is tabulated in land classified as actively discharging runoff (Table D.14). This is consistent with the notion that the Earth's climate was substantially different with cold/dry conditions prevailing (Frenzel 1992; see also Part A).

Examples of major changes to the network composition of drainage basins due to ice cap obstruction and to sea level changes can be found on all continents, but changes were most pronounced for North America, central Asia and South-east Asia.

North America

During the late Wisconsin glaciation 20 000 B.P., nearly all of Canada and Alaska were covered by a single ice cap of 16 Mkm² (Flint 1971). Major contemporary river systems such as the Yukon, Mackenzie, Churchill, Nelson, St. Lawrence, the northern part of the Mississipi as well as part of the Columbia were completely buried under a thousand metres of ice. During the ice melt period, giant catastrophic events due to ice dam breakups similar to present-day Icelandic Jökulhaupts were common at the edge of retreating ice sheets, particularly in Montana, Saskatchewan, Alberta, Ontario. Peak discharges have been estimated to well exceed 10^5 m³ s^{-1}, exceeding the discharge from the present-day Amazon, and having major consequences on corresponding fluxes to downstream channels and to the coastal zone (Benito et al. 1998).

After the complete retreat of glaciers by about 6 000 B.P. the land eroded by glacial scouring was left with millions of lakes from 1 ha to 100 000 km². A chain of North American Great Lakes, from Great Bear Lake to Lake Ontario at the edge of the hard rock Canadian shield, was created and now constitutes the most prominent hydrological artifact from the glaciation remaining on the continent. These features maintain an important and persistent role in the contemporary hydrology of North America. These lakes have very long residence times, on the order of 10s to 100s of years, thereby affecting the storage and timing of runoff as well as of sediment and nutrients in the Mackenzie, Nelson, Churchill and St. Lawrence Rivers, which are among the world's top 20 in size (Meybeck 1994).

Eurasia

In northern Europe and Asia, maximum glacial extent during the Valdai period was limited to Scandanavia, parts of northern Siberia (e.g. Taymir Peninsula), and to central Asian mountains. Northern European river basins were also ice-covered or blocked by the ice caps as for the Neva, N. Dvina, Kolyma, and Pechora which probably were diverted and forced to flow westwards to the Baltic and/or (backwards) to the Caspian Sea (Grosswald 1984; Maslenikova and Mangerud 2001). The Ob river might also have been blocked by the northern ice caps and overflowed to the Aral Sea through the Turgay Channel according to Grosswald (1984, 1998). The exact nature and timing of such connexions is still debated (Velichko 1984; see a more complete description in Gataullin et al. 2001 and Thiede et al. 2001). During melting of the Eurasian ice cap, river discharges were much higher than at present particularly for the Volga River. As a result the Caspian Sea was over-flowing to the Azov and Black Seas through the Manych depression north of the Caucasus. Cataclysmic emptying of ice-dammed lakes with discharge up to 1 000 000 m³ s^{-1} have been reported in the Altai mountains (Baker et al. 1993; Rudoy 1998). During the last Valdai glaciation, the maximum water discharge was much higher, up to eight times, than today according to palaeo-landscape analysis made in the East European Plain by Sidorchuk et al. (2001). In many other continents the lowering of sea level at the last glaciation re-

Table D.14.
Land area in ice sheets and additional area created by the lowering of sea level at Last Glacial Maximum at 18 000 B.P. Although we see a nearly equal loss of runoff-producing area from ice sheets and gain in exposed continental shelf, the spatial distribution and land-to-ocean connectivity has changed. Drainage patterns are from the work by Gerasimov (1964) quoted by Gibbs and Kump (1994). Internal area may differ here from more recent estimates (Table D.13)

	18 000 BP area (× 10⁶ km²)	Present area (× 10⁶ km²)
Total ice-free area	129.9	133.2
Exposed shelf area	15.9	–
Additional land covered by ice	18.4	–
Total ice free area	129.9	133.2
Ice-free area with external drainage	105.1	108.2
Ice-free area with internal drainage	24.8	24.9
Total ice-free area with runoff	104.7	121.3
Ice-free area with external drainage and runoff	90.4	104.1
Ice-free area with internal drainage and runoff	14.3	17.2

sulted in wide areas of continental platform being exposed and in extended or new drainage systems as in the Adriatic Sea, the Persian Gulf, the Argentina Coast, and particularly in South-east Asia where the Sunda and Java river systems, now mostly drowned, reached one Mkm^2 (Vanney 1991; Gibbs and Kump 1994).

D.4.3.2.2 *Progressive Changes in Past Climate*

Water balance changes are another important factor regulating the natural evolution of river systems, particularly in the centre of the continents where natural large-scale depressions occur. If the water balance is positive the depression will gradually fill to the point where it can overflow to the next set of river channels downstream (Coe 1998). When the excess is marginal, the corresponding lake or wetland can be occasionally cut-off. Over human time scales, such variations are rarely seen. Two examples are the Okavango Delta – Zambezi River connection in southern Africa and the Kerulen – Amur linkage in eastern Asia. These systems, both of the order of 10^6 km^2 in area, generate intermittent overflow that links and then severs any connections to downstream reaches. This has the effect of making the Okavango and Kerulen systems functionally endorheic or exorheic. Actually, all such depressions are potentially exorheic should sufficient runoff be made available, but these systems are at the margin of active discharge and actually do discharge water downstream but for very short periods of time (see Sect. D.4.3.1).

There is also an historical record of a connection linking the Aral and Caspian Seas through the Uzboi channel at least until A.D. 1000 (Létolle and Mainguet 1993), today completely dry except for the brackish Sary Kamich remnant lake. Other well-documented examples of Holocene connections/disconnections are known for the Omo-Turkana basin and for Lake Victoria and the Nile (although for the latter tectonic subsidence has also played a role). This is true as well during former Quaternary periods for the Tanganyika-Zaire River, Death Valley-Colorado River, and Lake Bonneville (now Great Salt Lake)-Columbia basin connection through the Snake river (Flint 1971). Major changes in river system connectivity have also been studied within the Bolivian Altiplano (Titicaca-Poopo-Uyuni), which has ultimately remained endorheic at least throughout the Holocene. There are also likely connections between the Mars Chiquita system (Argentina) and the Parana River.

The occurrence of lakes has been extremely variable during recent Earth history. From the standpoint of aggregate freshwater volumes, tectonic lakes are major and relatively long-lasting storage features at the Earth's surface (10^5 to 10^6 years). The Caspian Sea, African rift lakes, Lake Baikal, Titicaca and Aral Sea are examples and by far dominate the total volume of lake water stored on the surface of the Earth. However, from the standpoint of open water, lakes originating from glacial activity which are typically only 6 000 to 10 000 years old correspond to 50% of the present global lake area (2.43 Mkm^2, excluding Caspian Sea) (Meybeck 1994). This figure may have been much different during the more humid periods of the Holocene with the extension of present-day lakes such as Torrence, Gardner, Eyre in Australia, Chad in Africa, Lake Bonneville in the North American Great Basin, and the development of multiple lakes now dry in the Sahara, Southern Africa, central Asia, and the Altiplano.

The key period of 6 000 B.P. probably corresponds to a maximum extension of lakes at the global scale, a reflection of higher runoff and correspondingly higher lake volumes. This period is the focal point for ongoing analysis by the PAGES BIOME 6000 (Global Palaeo Vegetation) Project. One of its specific aims is to depict the status of the Sahara which was much more humid than under present-day conditions (Fig. D.45). Seasonal river runoff from high relief (Hoggar, Air, Tibesti) was very likely and the resulting allogenic river flow across the Libyan Desert has been investigated. Yet, we do not know exactly how this system was connected to the global ocean. How much reached the Libyan Gulf or the Niger basin? How far west could the Sahoura and Tamanrasset River traverse? Were there active tributaries of the lower Nile, north of the 5th Cataract? Fundamental studies have yet to be completed.

From an Earth systems perspective, there have been few pioneering studies (Gibbs and Kump 1994) and this short review has clearly shown that at least since the Last Glacial Maximum, land-ocean connections through rivers and its related fluxes (of water, carbon, nutrients, particulates) have been greatly variable as have been the global extension of inland waters. Yet the spatial distribution of these changes and their detailed timing and distribution remain to be established at the global scale.

D.4.3.2.3 *Conditioning of River Networks by Recent Climate and Impending Climate Change*

More recently and for areas between 10^5 and 10^6 km^2, we find important effects on river systems due to climate variability observed in the recent monitoring record (last 50–100 years). What evidence do we see for very recent changes in basin runoff and river discharge that we can explain through climate variation?

Long-term normalised runoff data (1900–1990 period) of major rivers with little relative land-cover change and limited direct water management as the Athabaska, N. Dvina, Lena, Niagara, Parana, Zaire and Amazon do not show definite trends over this period (Fig. D.47). Specific ENSO impacts have been searched for in such systems (Amarasekera et al. 1997; Cluis 1998). They are par-

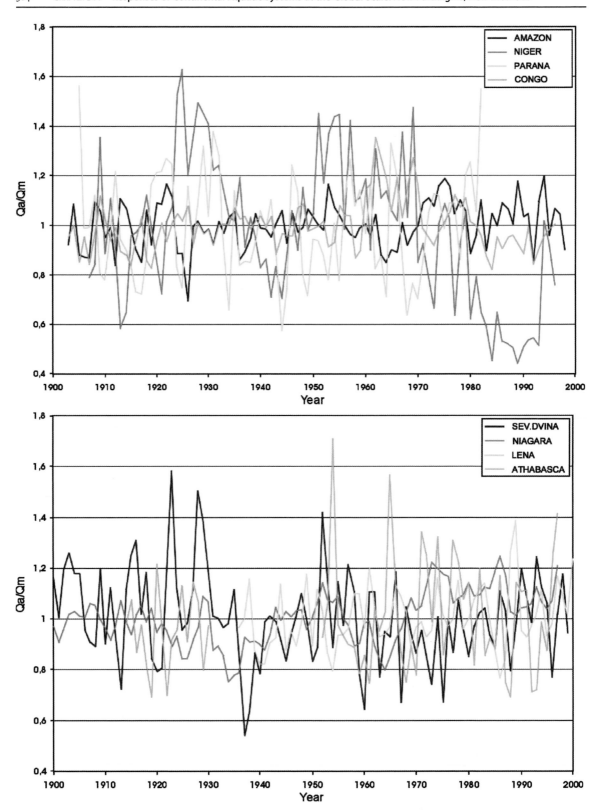

Fig. D.47. Annual hydrograph time series for large drainage systems are determined by an interplay between land-surface properties, climate regime, and ENSO/AO-type events. Normalised time series from rivers representing most latitudinal/climate zones are presented here. *Upper panel*: Amazon, Niger, Parana, Congo; *lower panel*: Northern Dvina, Niagara, Lena, Athabasca. (*Qa:* annual value, *Qm:* mean of the documented period). Data provided by: J.L. Guyot, IRD Toulouse (Amazon, Niger, Congo); A.I. Shiklomanov, University New Hampshire (Northern Dvina, Lena); S. Kempe (Parana); GRDC Koblenz (Niagara, Athabasca)

ticularly well marked for the Parana River in 1982 which doubled its normal annual flow in relation to an ENSO event according to Depetris et al. (1996). However, this year has been identified as an outlier in this series, even if there is some positive correlation between the annual flow at Corrientes and the ENSO index, according to Amarasekera et al. (1997). These authors also found weak negative correlations for the Amazon (confirming the results of Richey et al. 1989b), for the Nile (Blue and White Nile), and Congo and reported other reduced discharges for the Murray-Darling. The upper Niger river discharge presents a marked decrease between the late 1960s and 1990 (Fig. D.47), which is a direct reflection of the decade-to-decade progression into and then out of drought (Laraque et al. 2001). The Amazon and Zaire year-to-year runoff variability is among the world's lowest due to their position across the Equator, and their enormous size (6.4 and 4.0 Mkm^2 respectively) (Richey et al. 1989b; Laraque et al. 2001). The Niagara river variability (basin area 0.7 Mkm^2) is also extremely low due to the smoothing influence of the Great Lakes and does not present a long-term trend.

Following the pioneering work of Probst and Tardy (1989) on river runoff fluctuations at the global scale, IPCC impacts assessments on water systems (Arnell and Liu 2001), as well as the results from the 2nd Conference on Climate and Water (Lemmelä and Helenius 1998) and Shiklomanov (2000) demonstrate aptly the difficulty in piecing together a coherent picture of recent climate change. Over the last decade, a commonly applied practice among hydrologists has been to use meteorological or climate outputs from GCMs as direct inputs into stand-alone hydrology models of varying sophistication. This has resulted in a bewildering assortment of results based on no standard set of GCM models, spatial resolutions, or land-surface hydrology models. The resulting chaos has meant that a global synthesis has remained elusive using this approach.

Furthermore, the sought-after climate change signal – at least for runoff – is only partially supported by the historical record. While some studies show a possible intensification of the water cycle, for instance through more extreme precipitation in the US (Karl et al. 1996; Karl and Knight 1998) and other parts of the world (Easterling et al. 2000), the effect of these increases on the rest of the hydrological cycle is quite variable. Lins and Slack (1999) used approximately 400 stream gauging time series from the US with 50-years of continuous records (from unregulated systems) to conclude that the variability of streamflow has actually decreased while mean runoff increased. This was later confirmed by McCabe and Wollock (2002) who found statistically significant increases in annual moisture surplus (moisture that eventually becomes runoff) over the contiguous US as a whole, but especially in the east. And while Yue et al. (2003) found similar increases in minimum and mean

daily flows in northern Canada, they found the opposite to be true (significant decreases in minimum, mean and maximum daily flows) in southern Canada. A similar situation has been recorded for soils (Robock et al. 2000) using data from more than 600 sites from around the world with 6-to-55 year time series. The time series indicate a modest increase in summer wetness for the majority of sites examined. Contrary to the expectation (by GCM modelling) of drier conditions, especially in mid-continental areas, there is evidently a beneficiary effect due to climate change. And the effect of increased precipitation extremes on floods is still debated (Douglas et al. 2000; Groisman et al. 2001; Robson 2002; McCabe and Wolock 2002). Taken together, these results are paradoxical with respect to the findings on precipitation, and constitute an interesting and important new area of research as to why the observed precipitation intensification has not translated into a more coherent signal „downstream".

In contrast, the Arctic may be the harbinger of global climate and water cycle change. There is mounting evidence for an increase in storm tracks passing through the region of the polar front and increased precipitation (in particular for winter over Eastern Siberia) associated with the Arctic and North Atlantic Oscillation (Serreze et al. 2000). This has translated into a change in the absolute amount of winter-time runoff, co-located in Siberia with the advection of warm, moisture-laden air masses from the lower latitude (Lammers et al. 2001). Major Arctic river discharges, in addition, have been slowly rising (Semiletov et al. 2000; Peterson et al. 2002), leading to concerns about how changes in the Arctic land-based water cycle will influence North Atlantic deep water formation which is dependent on gradients of salinity and water temperature (SEARCH SSC 2001; Vörösmarty et al. 2001, 2002).

D.4.4 Human Conditioning of Continental Runoff

More recently and for areas between 10^5 and 10^6 km^2 we also find important direct effects due to water engineering activities associated specifically with impoundment, interbasin transfers, and irrigation water consumption. Such hydraulic manipulation to stabilise flow and to make use of otherwise inaccessible water resources constitutes a major human endeavour that is likely to continue well into the future. In this chapter we present some of the major types of distortion to hydrographs, together with some emerging evidence that such alterations – and their potential impacts – are indeed global in domain.

Alongside terrestrial vegetation – through evapotranspiration (ET) – humans are major users of terrestrial water. However, the relative role of humans in shaping continental-scale water balance has increased over time through (a) activities such as deforestation and wetland

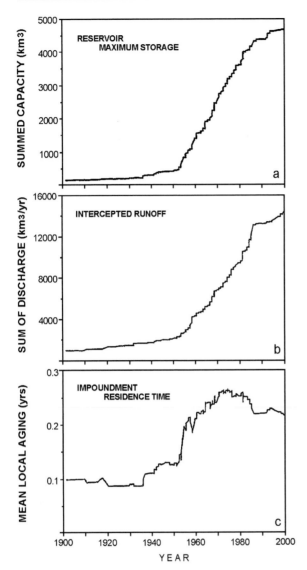

Fig. D.48. Time series of impounded volume (**a**), intercepted runoff (**b**), and local residence time change (**c**) associated with a subset of approximately 600 geo-referenced large reservoirs (> 0.5 km³ storage capacity). The basins in which these large reservoirs reside constitute 53% of continental runoff. Local residence time change persists downstream of these facilities and translates into changes to natural discharge regime, sediment flux, and river chemistry. These large reservoirs together constitute only about 60% of the presumed capacity of all registered dams and an unknown fraction of all impoundments when small farm ponds and other unregistered impoundments are considered (from Vörösmarty and Sahagian 2000, reprinted by permission of Bioscience)

drainage, which typically reduce ET and increase runoff, and (*b*) direct control and use of surface and groundwater supplies.

The spectrum of water engineering is broad and supports agricultural (irrigation), industrial, and domestic uses. These uses can either be considered sustainable from the standpoint of exploiting renewable supplies of water (such as river discharge for cooling power plants) or non-sustainable (such as in the mining of deep aqui-

fers for irrigation). When such non-sustainable uses are considered, there can be major associated losses of water stocks that collectively may even have an impact on the apparent rise of global sea level (Sahagian et al. 1994; Vörösmarty and Sahagian 2000).

Large reservoir systems, interbasin transfers, and consumptive water uses such as for irrigation are capable of significant and easily identifiable restructuring of natural river flow regimes. There can be substantial changes in long-term net runoff (i.e. precipitation minus evaporation) as well as in the timing and magnitude of downstream peak and low flows. Large impoundments also change the character of natural hydrographs. Discharge time series of pre- and post-regulated rivers unequivocally show this effect, even with relatively modest local residence time change. Figure D.50a shows a 35-yr time series of river discharge for the Nile River below the Aswan High Dam. The maximum-to-minimum discharge ratio decreased from a natural condition of 12:1 to 2:1 after construction of the dam, and the occurrence of high and low flows was shifted by several months. Similar levels of distortion are reflected in hydrographs from basins with significant irrigation water loss and inter-basin transfers (Fig. D.50b,c). Collectively, such water engineering works have strongly influenced and "fragmented" natural aquatic ecosystems throughout the Northern Hemisphere (Table D.15).

D.4.4.1 A Focus on Reservoirs

Large dam and reservoir construction is one of the better-documented water resources activities, although by no means is sufficient information available for exhaustive global-scale assessment. A large worldwide effect can be demonstrated from the more than 40 000 registered reservoirs (Rosenberg et al. 2000). Figure D.48 shows three measures of interaction of large impoundments with the global water cycle over the 20th century. Much of the last half of the century is characterised by an explosion of dam-building and associated impoundment of continental runoff, which has recently subsided due to a variety of technical, socio-economic, and environmental constraints (Gleick 1998). Both maximum storage and locally-intercepted discharge rose dramatically between 1940 and 1980 and mean runoff age within reservoirs more than doubled to over 3 months. Since 1980 the flattening of the maximum storage time series has contributed to the stabilisation and recent reduction in local ageing, but the global mean residence time is still substantial.

The geography of this dam-building and its potential impact on river systems has been recently assessed at the global scale by combining geographically-referenced water balance models (Federer et al. 1996; Vörösmarty et al. 1996a, 1998a), observed discharge/runoff (Vörös-

Table D.15. The control and fragmentation of Northern Hemisphere river systems showing the level of distribution by biome type (from Dynesius and Nilsson 1994)

Biome	Number of river systems	Impact class distribution of the river systems		
		Not affected (%)	Moderately affected (%)	Strongly affected (%)
Tundra and barren arctic	42	79	10	12
Subtropical and temperate rain forests	10	70	10	20
Temperate needle-leaf forests	85	53	13	34
Temperate broad-leaf forests and subpolar deciduous thickets				
Northern provinces	5	40	20	40
Southern provinces	37	3	38	59
Mixed mountain and highland systems	29	14	24	62
Temperate grasslands	11	0	27	73
Evergreen sclerophyllous forests	11	0	18	82
Cold-winter (continental) deserts	8	0	0	100
Lake systems	5	0	0	100
Warm deserts	1	0	0	100
All river systems		39	19	42

marty et al. 1996a; Fekete et al. 1999, 2002), a river networking topology (Vörösmarty et al. 2000a,b), and a global digital database of large reservoir statistics (Vörösmarty et al. 1997b,c). The conjunction of these datasets permits a mapping of the distribution of runoff and reservoir attributes and quantitative measures of the uneven, though nearly global distribution of human impact from this hydraulic engineering. Dam building we denote as "neo Castorisation", a term which refers to the human regulation of river flow through the creation of artificial impoundments, akin to the activities of another prolific dam builder, the North American Beaver (*Castor canadensis*) (Vörösmarty et al. 1997c).

A subset of 633 large reservoirs (> 0.5 km³ potential storage capacity) having sufficient information to be geographically-referenced and characterised using the approach described above can be used in the analysis (Table D.16). The likely total volume of impounded water from all registered reservoirs is about 7 000 km³, and the 633 analysed represent about 70% of this total. The value of 7 000 km³ is unremarkable except for the fact that it represents a 600% increase in the standing stock of water in river systems globally, an important metric of global change by any measure, especially considering the fact that most large reservoirs have been created during the last 50 years (Fig. D.48). This recent recalcula-

Table D.16. Key attributes of the geographically-referenced large reservoir systems (Vörösmarty et al. 1997b). The mean residence time change ($\Delta\tau_R$) is derived from mean annual conditions estimated locally for reservoirs in each continent or the globe

Continent[a]	n	Sum of maximum capacities (km³)	Mean of maximum capacities (km³)	Sum of intercepted discharge (km³ yr⁻¹)	Discharge-weighted mean $\Delta\tau_R$ (yr)
Africa	42	912	21.7	736	0.83
Asia					
Endorheic	19	102	5.4	58.3	1.17
North[b]	14	569	40.6	903	0.42
South	176	827	4.7	2 560	0.22
Australia/Oceania	16	47	3.0	44.2	0.71
Europe[c]	88	430	4.9	1 770	0.16
North America	180	1 195	6.6	3 500	0.23
South America	98	807	8.2	6 190	0.09
Total	633	4 889	7.7	15 762	0.21

[a] Defined by rivermouths within the STN-30 (see Vörösmarty et al. 2000a,b).
[b] Drainage into Arctic Ocean.
[c] Area west of the Ural Mountains and north of Caucasus Mountains.

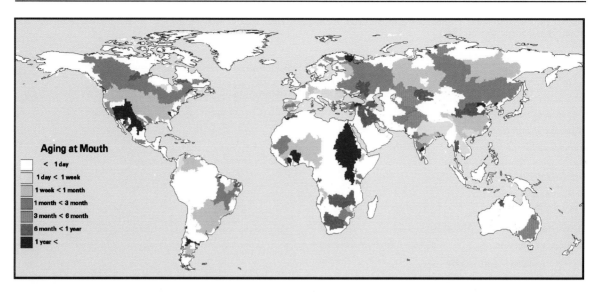

Fig. D.49. Ageing of continental runoff at the mouths of 236 regulated drainage basins due to impoundment from large reservoir systems (> 0.5 km³ storage capacity). Ageing is the difference between the average residence time of water in river channels (groundwater not included) without reservoirs and with reservoirs. Many of the world's river systems have an additional ageing from impoundment exceeding one month. To place these estimates into perspective, natural residency times are but from 16–26 days (see Covich 1993, Vörösmarty et al. 2000a,b; from Vörösmarty et al. 1997b)

tion (Vörösmarty et al. 2003) is based on $n = 45\,000$ registered dams and does not consider small impoundments totalling several hundreds of thousands, suggesting that the findings presented here are conservative.

An index of reservoir-induced impact on continental runoff is the discharge-weighted change in residence time of otherwise free-flowing rivers, which in turn serves as a measure of the anthropogenic "ageing" of continental freshwater in reservoir systems. This index can be computed with reasonable assurance using currently available datasets. The calculation will demonstrate a significant "imprint" from reservoir construction on global runoff.

The ageing of runoff within individual river segments, subcatchments and entire drainage basins is a function of water retention in reservoirs and river discharge. The calculation can be made using spatially-distributed, gridded datasets that simulate the continental land mass. Within each grid element, all reservoirs can be assumed to contribute to the net delay in downstream transport of river water. Mean local ageing ($\Delta\tau_R$) is defined as the aggregate reservoir volume divided by mean annual discharge. A value of 0.67 represents the fraction of potential volume utilised on average during the operation of a reservoir (USGS 1984). Water emerging downstream of an impoundment maintains its characteristic $\Delta\tau_R$ until modified by tributary inflows (which themselves may be influenced by upstream dams) or by encountering the next downstream reservoir. Through an appropriate discharge-weighting that avoids artificial ageing when water is lost to evaporation and/or consumptive withdrawals, the local change in residence time in a grid-based stream segment i ($\Delta\tau_{si}$) is calculated as:

$$\Delta t_{si} = \frac{\sum_{1}^{n} Q_j\, \Delta t\, \tau_{sj}}{\max\left(\sum_{1}^{n} Q_j, Q_i\right)} + \Delta t\, \tau_{Ri} \qquad (D.12)$$

where Q is discharge, j represents the adjacent upstream cell index (1 to n). $\Delta\tau_{Ri}$ is the local reservoir residence time for a reservoir with storage volume V_{Ri} and discharge Q_i, with

$$\Delta\tau_{Ri} = \frac{V_{Ri}}{Q_i}$$

The resulting distribution of Δt_s can be mapped onto digitised river networks (Fig. D.49). The index can be used to identify river reaches having a persistent imprint of impoundment-induced ageing. Contributing drainage areas and their apparent degree of river regulation can also be mapped. This approach obviates the need to compute explicitly natural residence times for river water, a capability that only recently has begun to be developed at continental and global scales (Hagemann and Dumenil 1998; Olivera et al. 2000; Fekete et al. 2001).

The imprint of reservoir storage is initiated at each reservoir and its effects persist downstream. Globally, using the set of 633 large reservoirs, we see that the mouths of several large rivers show a reservoir-induced ageing of continental runoff that exceeds three months and for river basins with large impoundments at the global scale the discharge-weighted mean is nearly 60 days (Fig. D.48). The local ageing can be of the order of years for the larger reservoirs and months for those that are intermediate in size. The average residence time for continental runoff in free-running river channels likely varies between only 16 days (Covich 1993) and 26 days (Vörösmarty et al. 2000a). Thus, mean age of river water has likely doubled or tripled to well over one month worldwide in regulated basins.

Impacts Associated with Runoff Ageing

From case studies as well as global synthesis we find that this ageing can lead to significant changes in net water balance, flow regime, reoxygenation of surface waters, temperature regime, and sediment transport (Vörösmarty et al. 1997b,c; 2003).

Dammed river systems affect drainage basin water balance and riverine hydrographs by changing the magnitude and timing of evaporation, net runoff, and downstream peak and low flows. The effects are highly reservoir-specific and dependent on size, geometry in relation to local topography, natural river flow, and climate. Some of these changes are related to the residence time changes brought about through impoundment.

Relative to natural rivers, impoundments can show greatly increased evaporative losses and thereby reduce net basin runoff. These losses are a function of increases in surface area related to reservoir filling and hence residence time change. Such losses are especially important in arid and semi-arid areas where freshwater resources are already scarce. Several examples demonstrate this effect. Reservoirs with residence times of the order of one to three years lose significant amounts of water. Lake Nasser on the Nile (maximum capacity 162 km^3; surface area 6 500 km^2) annually loses approximately 7% of its total capacity and 13% of its inflow to evaporation and seepage (Said 1993). Lake Kariba on the Zambezi River (185 km^3; surface area 5 200 km^2) loses a larger proportion of its inflow, about 20% (Vörösmarty and Moore 1991) as does the much smaller Tiga Reservoir (2.0 km^3; 178 km^2) in Nigeria which evaporates 26% of its upstream inputs (Oyebande 1995).

Impoundments with long residence times also produce significant distortion of natural hydrographs (Fig. D.50). Alternative statistics can also be applied linking the overall impact of regulation by multiple upstream reservoirs to exceedence probabilities of streamflow as outlined in Dingman (1981) to help explain hydrograph distortion. From a global change perspective, these are significant and instantaneous changes to the apparent behaviour of continental runoff, with large changes in the seasonal pattern of discharge apparent within only a few months or years. Since large, impounded river corridors accumulate flow over thousands if not millions of km^2, equivalent changes from other agents of global change such as the greenhouse

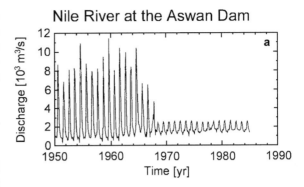

Nile River at the Aswan Dam

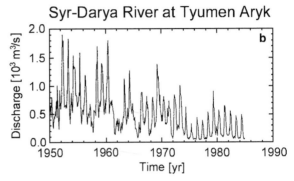

Syr-Darya River at Tyumen Aryk

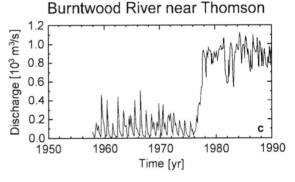

Burntwood River near Thomson

Fig. D.50. History of hydrograph distortion by water engineering works in three heavily-regulated rivers. **a** Nile River discharge is recorded just below the Aswan High Dam. The stabilisation of flow is apparent and it is not difficult to identify the time at which the dam was constructed and Lake Nasser reservoir filled. The post-impoundment Nile shows reduced overall discharge, substantially truncated peak flows, higher low flows, and a many-month shift in the timing of the natural hydrograph. **b** Sustained decreases in flow for the Syr-Darya River are associated with expanding water use for irrigation and the well-known, corresponding contraction of the Aral Sea. **c** Dramatic increases in discharge for the Burntwood River in Manitoba exemplify inter-basin transfer schemes, here between the Churchill and Nelson Rivers which empty into Hudson Bay. (Monthly data from Vörösmarty et al. 1996b, see *http://www.watsys.sr. unh.edu*)(from Vörösmarty and Sahagian 2000, reprinted by permission of Bioscience)

effect or progressive land-use change are to us implausible, especially over such brief time scales.

In addition to the obvious difficulties such hydrographic changes create for modelling studies of the terrestrial water cycle, they also bear direct environmental impacts downstream. Flow stabilisation below dams reduces floodplain size and flooded extent, leading to a wide array of effects including diminished evaporation and disruption to riparian ecosystems whose biogeochemistry, trophic structure, and diversity are all highly adapted to periodic inundation with its associated input of sediment and nutrients (Sparks 1995; UNEP/UNESCO 1986; Petts 1984; ICOLD 1994).

These varied impacts are of more than simple academic interest, as a multitude of environmental impacts, often very costly to society. Flow stabilisation and fragmentation, for example, interfere with fisheries, often of great economic value. Reduced delivery of nutrients to coastal systems compromises coastal food chains, including the potential for stimulating toxic and nuisance algal blooms. Reservoir silting results in a decline in flood regulation and hydroelectric capacity, downstream scouring of streambeds and instability of river deltas resulting in the failure or costly reinforcement of engineering structures.

D.4.4.2 Impacts of Land-cover Change on Water Budgets

In addition to the land-surface/atmosphere interactions documented in Part A of this book, there have also been recent studies of the impact of land-cover change on water balance which ultimately can be translated into changes in river basin exports of runoff. Land-cover changes influence the local dispensation of precipitation into either evapotranspiration, soil recharge and/or surface and subsurface runoff. Typically, reduction in biomass and deforestation will increase net runoff (see Chapt. D.3), though the characteristics of the timing of such runoff will vary given the condition of soil water and infiltration potential which serves as a trigger mechanism for groundwater recharge or surface water runoff. Fragmented landscapes generate changes in local weather (Avissar and Liu 1996; Pielke et al. 1997; see for example Fig. D.15). At drainage basin scales, the shifting mosaic of land cover and secondary regrowth of vegetation creates a complex hydrological signature as different age classes and successional species are present with unique water use efficiencies (Shiklomanov and Krestovsky 1988). Over still larger domains, land-cover change can

weaken the water cycle and distort global teleconnections (Chase et al. 1996; Costa and Foley 2000; Pitman and Zhao 2000; compare Chapt. A.8). Nonetheless, the impact of 110 000 km^2 of net annual deforestation (FAO 1999) on continental runoff has yet to be fully quantified and contributes greatly to our difficulties in interpreting other sources of impacts on drainage basin hydrology, such as the impacts of climate change. A series of now classic studies (Dickinson and Henderson-Sellers 1988; Lean and Warrilow 1989; Shukla et al. 1990) exploring the reduction in intensity of atmospheric water re-cycling in Amazonia associated with wholesale deforestation has set the stage for this question. Although the focus of these studies was on precipitation and evaporation, it can be demonstrated that reductions in runoff can also be traced to this weakening of water re-cycling. For example, in the Lean and Warrilow (1989) modelling studies of deforestation, mean annual P and E were reduced by 20% and 27%, respectively, from the natural condition, while runoff declined by 12%. We have come a long way since such studies and can now consider the impacts of more realistic land-cover changes (see Part A and Part B).

For the tropics in January, Chase et al. (2000) showed, relative to natural vegetation, a net heating of the land-surface associated with contemporary land covers, with a corresponding decrease in evaporation, but with little change in precipitation (Table D.17). This yielded an approximate 10% overall *increase* in net convergence (P − E) (i.e. water available for soil recharge or continental runoff), despite the overall deceleration of the water cycle due to land-cover change. They also noted large anomalies propagated to regions far from the origin of the land-surface change. Such results suggest that feedbacks between the land surface and the atmosphere are important in defining LUCC-induced water balance changes. Simple cause-and-effect in predicting the impacts of deforestation, while of practical significance for management at the local scale, yields an incomplete picture – at least over broader geographic regions and time domains.

D.4.5 Global Sediment Flux

The flux of sedimentary materials, either as suspended solids or as bedload, constitutes an important part of

Table D.17.
Change in heat and water balance for 10 composite Januaries over the Tropics (30° S to 30° N) due to land-cover change as predicted from a GCM. Results are averaged only over areas where land-cover differences exist between the cases (from Chase et al. 2000)

	Natural	Current	Difference
Sensible heat flux (W m^{-2})	36.34	42.90	+6.56
Latent heat flux (W m^{-2})	77.62	73.83	−3.79
Temperature (K)	292.39	292.83	+0.44
Precipitation (mm d^{-1})	3.78	3.75	−0.03
Evaporation (mm d^{-1})[a]	2.74	2.61	−0.13
Net convergence (P − E) (mm d^{-1})	1.04	1.14	+0.10

[a] Calculated here using a latent heat of vaporisation of 585 calories g^{-1}.

the planetary geological cycle and one with which fluvial systems are intimately connected. It is not simply the sediments which are to be considered, but also the associated chemical signatures of this material, be they sorbed nutrients or organic fractions containing carbon as well as nutrients.

The mobilisation of sediment through local erosion, its movement to fluvial systems, its transport, deposition, and re-mobilisation is a complex amalgam of processes. All of the domains through which sediment moves are highly sensitive to global change – both natural and anthropogenic. The response, however, is highly dependent on the scale considered and the full spectrum of continental aquatic systems through which water and sediment flows, from source areas to receiving waters.

D.4.5.1 The Continuum of Fluxes from Field Erosion to River Mouth Export

Factors controlling local erosion include land cover and management, the energy structure of precipitation, slopes, soils and their susceptibility to mechanical weathering. These factors are mechanistically treated by the Universal Soil Loss Equation (USLE) and the Modified MUSLE (Williams and Berndt 1977). For basin and ultimately macro-scale (continental or global-scale) application, field-scale sediment flux is hardly the same as flux at river mouth. Field-mobilised sediments take a somewhat tortuous and long journey that can include redeposition at the bottom of the field slopes, remobilisation and introduction into fringing riparian zones, movement into a river channel, redeposition and then remobilisation along floodplains, in lakes and artificial impoundments, and in deltas (Meade 1988).

This concept is articulated quantitatively in the so-called sediment delivery ratio (Walling 1983; Hadley et al. 1985) which charts the efficiency of a river system to transport sediment horizontally toward its endpoint. A topology of source-to-sink areas is implicit in this concept. It is not uncommon to find in many, even modestly-sized basins (c. 10^4 to 10^5 km^2), that the delivery through the mouth of the river is but a modest fraction of the field-level erosion, only a few percent. This ratio is a function of area as the numerical influence of source areas is "diluted" by the preponderance of sink areas that develop in the lower reaches of the river.

There are of course exceptions to this generalisation. An important consideration is the type of sediment transported. The slope of the sediment delivery ratio as a function of area is, therefore, much shallower for the Yellow River with its fine-grained loess sediments.

The hypsometric relationship between position in the drainage basin and elevation relative to maximum height is also important (Vörösmarty et al. 2000b). Several patterns of average basin elevation profiles along river courses can be described (Fig. D.51a). These can be expressed as a relationship between two normalised variables, the relative elevation (elevation at given location/maximum elevation) and fractional area drained (sub-basin area at given location/total basin area) (Strahler 1957). The Amazon River illustrates the common concave profile through which only a small proportion of total area shows a relatively high altitude and the vast majority of the basin is lowland. 90% of the Amazon Basin has an altitude lower than 20% of its maximum altitude; 50% of this basin's area has less than 5% of the maximum elevation. The Colorado basin profile is quite different and convex. Due to the Colorado Plateau about 50% of the basin is still at an altitude higher than 50% of maximum elevation. The difference between the convex and concave types is explained here by the occurrence of the deeply incised Colorado canyon (up to 1600 m). Such convex profiles are relatively infrequent among large basins. The Irrawaddy has an intermediate profile.

Such hypsometric relationships bear important consequences on suspended sediment transport. For example, concave profiles are typically linked with sediment mobilisation occurring first in the periphery of a basin followed by effective retention inside depositional areas which can include extensive floodplains (Walling 1983). These systems bear the imprint of a long-term history of erosion and of deposition in stable basins. In contrast, convex systems contain more potential headwater landscape that is vulnerable to erosion and relatively less lowland available for sediment retention. A convex basin is relatively efficient at mobilising sediment from headwater areas and transporting it through river networks to the basin mouth. Convex profiles in mid-basins characterise deep plateau incision corresponding to channel erosion as in the Colorado (Fig. D.51a).

D.4.5.2 Approaches toward Estimating Basin Fluxes

D.4.5.2.1 Typologies

Because so much of the Earth's land mass is poorly monitored with respect to hydrography and constituent fluxes (IAHS Ad Hoc Group on Global Water Data Sets 2001; Meybeck and Ragu 1997; Vörösmarty et al. 1997a), a sensible methodology is needed to apply information from well-monitored basins to the vast majority of river systems that are not routinely sampled. A typology approach for riverine flux (Meybeck 1993a; Vörösmarty and Meybeck 1999), takes a set of characteristics describing well-monitored basins and then maps these to similarly classified basins in poorly monitored parts of the world. Modern application of this approach uses static or dynamic GIS or remote sensing-based datasets to depict

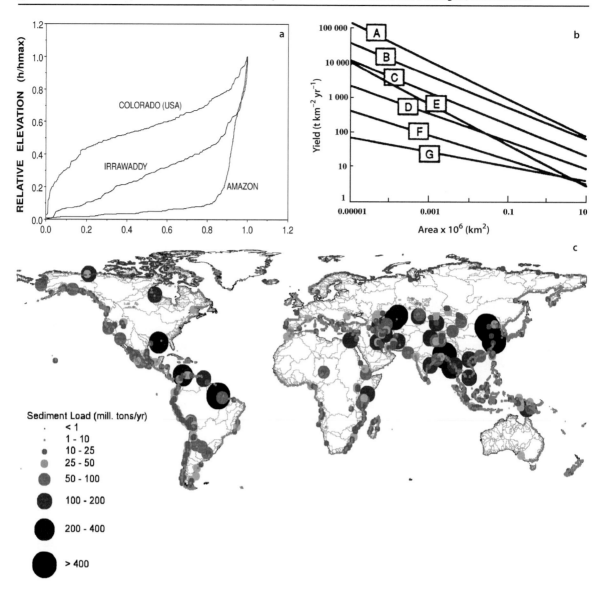

Fig. D.51. a Hypsometric curves for large rivers showing various stages of relief maturity: active mountain uplift (Irrawaddy), stability (Amazon), plateau incision (Colorado) (from Vörösmarty et al. 2000a, reprinted by permission of American Jounal of Science). b Global negative relationships between sediment yield and basin area for seven different biophysical environments used for the global sediment yield mapping (Milliman and Syvitski 1992, © 1992 by the University of Chicago). A = High Mountain, B = Mountain Asia-Oceania, C = Mountain N/S America, Africa, Alpine Europe, D = Mountain Non-Alpine Europe, High Arctic, E = Upland, F = Lowland, G = Coastal Plain (note the log-log scale). c Sediment load at river mouth (Mt yr^{-1}) calculated at a resolution of 30 minutes from regressions of Milliman and Syvitski (1992)

the geospatial character of individual drainage basins and to derive easily-mappable attributes. Milliman and Syvitski (1992) analysed suspended sediment flux in many different drainage basins distributed across the globe. Their analysis represents a simple but robust typology that applies maximum elevation drained and geographic (climate) domain by continent (Fig. D.51b) to classify drainage basins. Basin-scale responses are then cast as type-specific functions of drainage area. A mapping of this typology at the global scale based on digital topography (Meybeck et al. 2001) and a simulated network of global rivers (Vörösmarty et al. 2000a,b;

Fig. D.51c) completed for this review shows a global total of about 23 billion metric tons.

Though the capacity to map these patterns globally is satisfying, the approach is limited by the baseline data from which the typological relationships are derived. In the case of the global sediment typology, results are an admixture of modern rates erosion but purposely-chosen, pre-impoundment conditions from reviews of Milliman and Meade (1983) and Milliman and Syvitski (1992). Independent analysis (Vörösmarty et al. 2003) indicates that perhaps 30% of the natural global flux is now intercepted by modern reservoir systems which the typol-

Table D.18. Estimates of global sediment fluxes (F_{tss}) based on multifactor regression for sediment yields (from Ludwig and Probst 1998)

Study	Parameters used for sediment yield prediction	Global (Regional) F_{tss}, Gt yr^{-1}	Remarks
Langbein and Schumm 1958	APPT	(10.8)	Hand-fitted relationship. The given sediment flux corresponds only to about 42% of global APPT (250 to 1 250 mm range).
Douglas 1967	APPT	(11.5)	Hand-fitted relationship. The given sediment flux corresponds only to about 42% of global APPT (250 to 1 250 mm range).
Wilson 1973	APPT	(19.3)	Hand-fitted relationship. The given sediment flux corresponds only to about 42% of global APPT (250 to 1 250 mm range).
Ohmori 1983	APPT	56.6	As above, but the relationship was also extrapolated to the remainder 58% of global APPT (incl. >1 250 mm range) – see text.
Fournier 1960	Fournier index, relief	64.0	Sediment yields were correlated with a seasonality index for precipitation. Different relationships for different relief types.
Pinet and Souriau 1988	Elev, Orogeny type	16.2	2 linear relationships with elevation depending on the orogeny type. Important sedimentation (17.6 Gt yr^{-1}) in young orogenies.
Ahnert 1970	LR	9.3	Linear relationship with local relief (which is the difference between maximal and minimal elevation in a given sector).
Jansen and Painter 1974	QA, Elev, Slope, AT, Lith Mi, Veg1	26.7	Multiple correlation models depending on the climate type to which the drainage basins belong. Variable basin sizes.
Probst 1992, model I	Slope, Q, APPT, Veg1	22.9	Multiple correlation model on the basis of data from large river basins. Best model including 4 variables.
Probst 1992, model II	Slope, Lith Mi, APPT, Q, Veg1	21.7	Multiple correlation model on the basis of data from large river basins. Best model including 5 variables.

tss = total suspended solids; *APPT* = annual precipitation; *Four* = Fournier's seasonal precipitation index; *LR* = local relief; *QA* = annual runoff; *AT* = annual temperature; *Lith Mi* = mechanical erodibility indices; *Veg1* = index characterising the soil protection capacity of vegetation.

ogy would thus fail to accommodate. A more complete global typology would explicitly consider a pre-disturbance condition (i.e. with natural vegetation, no artificial impoundment) and constrast that against modern sediment delivery (i.e. with land-cover change and elevated erosion, dam construction). The global-scale impact of progressive river regulation, as suggested by Fig. D.48 and D.49, awaits analysis.

D.4.5.2.2 Spatially-explicit Multiple Regression

Typology-based multiple regressions have been employed in sediment transport studies (Mozzherin 1992; Sidorchuk 1991, 1994; Jansen and Painter 1974; Pinet and Souriau 1988, and others), where total suspended solids flux is expressed as a function of numerous variables including discharge, climate, relief, lithology, land use/cover, area, and the presence of lakes and impoundments (see also Table D.18). A broad set of forcings has been used, from precipitation alone to ensembles of lithology, vegetation, area, runoff, elevation, and/or slope. Major differences are apparent, representing in our view a non-standardisation of terminology, an amalgam of disturbance *v.* predisturbance conditions, and spatial and temporal incongruities, failing to take into consideration, among other factors, the sediment storage in basins.

As an example, Fig. D.52 shows the results from an approach used by Ludwig and Probst (1998) who obtained a global flux of 16 billion metric tons based on a multiple regression including logarithm of annual runoff, basin area, elevation, slope, annual temperature, a mechanical erosion index, and a vegetation protection index. This map is the first of its kind to simulate the river sediment yield at the global scale (0.5° resolution). Previous maps by Milliman and Meade (1983), Walling and Webb (1983) and Probst (1992) assigned river fluxes at the regional scale. Although a high degree of spatial heterogeneity is apparent within this map, the original data from which the multiple regression has been derived represents an admixture of different size classes of rivers. Sediment delivery is thus impacted by the delivery ratio concept but difficult to disentangle from the other driving variables in such regressions. In some sense this is a difficulty with all such models that are limited by a relatively small database (n = 50 to 100 in most cases), necessitating the use of all size classes of rivers. A call for a standardised resolution for sediment yields in such macro-scale models was made by Jansson (1982), but this plea has not been heeded. Last but not least, the analysis does not include the impact of modern reservoir sedimentation, which as shown above, can intercept 50% or more of riverine sediment flux in regulated basins (Vörösmarty et al. 2003; Meade and Parker 1985). Once again, despite the promise of global-scale, mappable typologies,

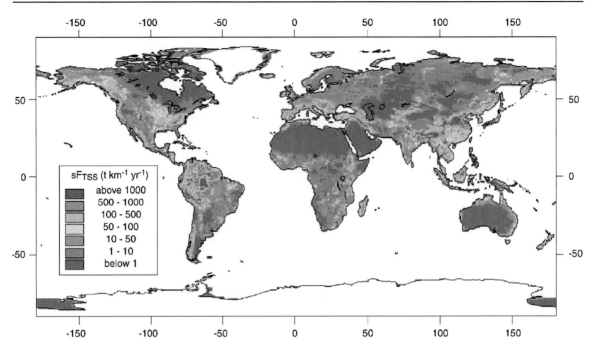

Fig. D.52. Global map of yields of total suspended solids (sF_{TSS}) in river systems (from Ludwig and Probst 1998, reprinted by permission of American Journal of Science)

we can be greatly restricted by the inherent assumptions, interpretation, and source data from which those characterisations are built.

D.4.5.2.3 Global Extrapolation from Archival Data

Major efforts aimed at compiling global databases on observed river basin hydrography (Vörösmarty et al. 1996b; Fekete et al. 1999, 2002), river chemistry and sediment flux (Meybeck and Ragu 1997; Milliman et al. 1995) have recently been completed. These databases can be used to construct estimates of continental sediment (and other constituent) fluxes over global, continental, or ocean basin domains. One such example is given in Fig. D.53 which attempts to derive a mean concentration

of total suspended solids (TSS) for the entire globe using the GEMS/GLORI (Global River Inputs to the Ocean) database prior to damming (Meybeck and Ragu 1997). The first 100 river basins are ranked with respect to size and a mean, discharge-weighted mean TSS is computed by including information up to and including the currently ranked river. The rivers examined represent an admixture of all types of different river systems with respect to climate, size (from 6.4 to 0.05 Mkm^2), source area, etc. Mean TSS starts at about 180 mg l^{-1} for the Amazon alone and then increases with progressive inclusion of other rivers into the calculation of the mean due to higher yields by some of the world's largest rivers. Large steps upward are noted for major sediment-laden rivers draining either large moutain systems (Chang Jiang #10 and Ganges #17) or highly erodible soils (Huang He #26). As

Fig. D.53.
Mean total suspended solids (*TSS*) concentration extrapolated from the GEMS-GLORI (Meybeck and Ragu 1996, 1997) constituent database. Ranked basins according to size and average cumulated means (Amazon is number 1, then Congo, Ob, Mississippi, Nile, etc.). The ultimate TSS compares favourably with the global average calculated from results shown in Fig. D.51c

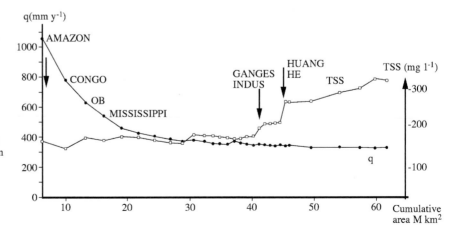

Fig. D.54.
The total quantity of discharge intercepted by large reservoir systems (> 0.5 km³ storage capacity) at contrasting levels of local residence time change. The local residence time change is a good predictor of the fraction of inflowing sediment deposited within a reservoir. Worldwide, 70% of the intercepted runoff flows through reservoir systems that can trap more than 50% of incident sediment flux (from Vörösmarty et al. 1997b)

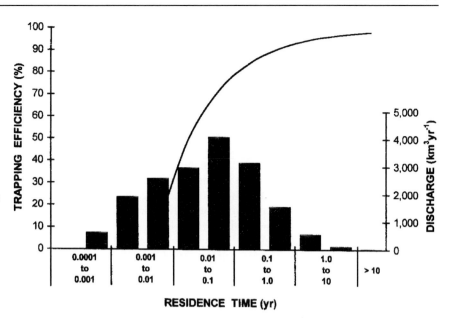

the river rank increases, smaller and smaller rivers are encountered which have relatively lower capacities to trap sediment and hence the mean rises. While conveying interesting global patterns such an analysis can give only general trends since it represents an incomplete accounting of the world's rivers. The database used in the figure has about 100 entries corresponding to more than 70% of global exorheic area. For comparison, a 30' spatial resolution (latitude × longitude) database of potential rivers will have about 6 000 basin entries while a 10' dataset would show something on the order of 20 000.

D.4.5.2.4 Explicit Consideration of Impoundment

From our foregoing discussion, it should be recognised that a more complete picture of global sediment flux should explicitly consider the interception and deposition of riverborne sediment within reservoirs. The methodology is similar to that used in computing runoff ageing. Large reservoirs (> 0.5 km³ maximum storage) can be geographically co-registered to STN-30 river segments bearing observed or estimated discharges, and a set of mathematical formula applied that estimate sediment trapping efficiencies for impounded and unimpounded sections of individually regulated drainage basins. We can apply an approximation to the relationship originally developed by Brune to predict individual reservoir trapping efficiencies (TE) as a function of τ_R, the reservoir residence time (Ward 1980): TE = 1 – $(0.05/\tau_R^{0.5})$. Within each of the 236 large reservoir-regulated drainage basins, we can identify all sub-basins which contain large reservoirs. For these we determine an aggregate impounded volume, which together with discharge yields a sub-basin residence time and an aggregate siltation capacity based on the Brune/Ward equation.

Whole basin sediment trapping can be adjusted by a discharge-weighting that accounts for the effect of unimpounded interfluvial areas. For these calculations, we take the generally accepted assumption that suspended sediment flux within a given river basin is proportional to discharge (Milliman and Meade 1983; Walling and Webb 1983; Milliman and Syvitski 1992) and make corrections to individual basins when contrary data is available. The resulting distribution of residence times and trapping efficiencies can be mapped onto simulated river systems and summarised at individual reservoir, continental, and global scales.

Encouraging results are obtained when comparing simulated results to independent compilations of pre- and post-impoundment sediment fluxes (Milliman and Syvitski 1992; Meybeck and Ragu 1997). The median and mean absolute disparities between observed and predicted in trapping efficiencies were 4% and 9%, respectively.

Continental totals for discharge intercepted by large reservoirs and expressed as a function of computed τ_R are given in Fig. D.54. Much of the runoff intercepted by large reservoirs globally has a residence time of ≥ 0.01 years (i.e. only three days), corresponding a 50% or greater sediment trapping efficiency. The most significant sediment trapping at large reservoirs is in endorheic Asia (mean = 91%), North Asia (90%), Australia/Oceania (90%), and Africa (86%). However, no region of the globe shows a discharge-weighted mean of less than 50%. The discharge-weighted, mean global trapping efficiency is 62%.

For the globe, approximately 70% of discharge flowing locally through large dams has a sediment trapping efficiency of 50% or more. A key question is how river systems downstream of such reservoirs counteract the loss of suspended load by clear water erosion of river banks. However, due to flow stabilisation, in particular loss of energy at high flows, it is anticipated that the impact of this trapping persists below the impoundment and these calculations provide a reasonable estimate of net loss in sediment flux. The true significance of such statistics becomes apparent when placed into a drainage basin context. Thus, when we remove the tabulation of sequential downstream interception, we find that approximately 9 000 $km^3 yr^{-1}$ or 24% of total continental runoff is intercepted by the most downstream of large reservoirs in each regulated basin, suggesting an important impact on global sediment retention. Furthermore, regulated basins represent more than 50% of total global runoff and their mean discharge-weighted residence time is one month, a τ_R associated with substantial sediment trapping. When the effect of dilution by unimpounded subcatchments is considered, our estimated retention of sediment within basins regulated by large reservoirs globally is 30%.

The current estimate of anthropogenic interference with the global suspended cycle is twice that given earlier by Meybeck (1988), 1.5 $Gt yr^{-1}$ or 8% of total flux, but this earlier estimate was based only on the trapping of a few major basins as reported by Milliman and Meade (1983). In a recent paper (Vörösmarty et al. 2003), this analysis of large reservoirs was extented to the > 44 000 smaller reservoirs. The discharge-weighted mean $\Delta\tau_R$ for individual small reservoirs (0.011 yr) was more than an order of magnitude smaller than for large reservoirs. Despite their modest individual contributions, small reservoirs collectively intercept an additional 23% of global sediment flux, or a little under half of the total anthropogenic flux in regulated basins which is approximately 50%. Taking into account unregulated basins, the interception of global sediment flux arising from all registered reservoirs ($n \sim 45\,000$) is conservatively placed at 25–30%, with an additional but unknown impact due to still smaller unregistered impoundments ($n \sim 800\,000$).

The framework presented here is a precursor to more detailed analysis of this problem wherein spatially and conceptually explicit models of sediment source and sink areas are applied. These results demonstrate that river impoundment should now be considered explicitly in global elemental flux studies, such as for water, sediment, carbon and nutrients. From a global change perspective, the long-term impact of such hydraulic engineering works (which are generally designed to last 50 to 200 years) on the world's coastal zone appears to be significant but has yet to be fully elucidated.

D.4.5.3 Additional Temporal Complexities

An important characteristic of drainage systems has not yet been mentioned. This refers to the timing of the movement of sediment through basins. Both seasonal and episodic discharges have a tremendous impact on the sus-

pended sediment flux that fails to be captured by the annual time step used in the models described above. For example, only a few days of discharge may be required to discharge an entire year's sediment flux or many years' worth (Meade and Parker 1985; Meybeck et al. 2003). An excellent example of this is the Eel River in coastal California which discharged the equivalent of eight years of its suspended load during three days of extreme river flow (Syvitski and Morehead 1999). Recent work (Morehead et al. 2003; Syvitski 2003), based on a model considering the equivalence between potential energy and gravity-driven sediment movement, offers promise toward capturing the timing of riverborne sediment flux.

There are also long-term timing issues as demonstrated by Trimble (1977) and Meade (1988) and more recently by Sidorchuk (1991, 1994). In the case of the Trimble study over the Piedmont region of the eastern US, he was able to trace sediment moving in multidecadal "waves" defined by several sequential historical events, from initial breaking of sod for agriculture starting in the 1700s, subsequent acceleration of erosion through the early part of the 1900s, reservoir construction in the early 1900s with associated siltation of these impoundments, and later re-mobilisation as sediments deposited within these impoundments were re-mobilised. The Mount St. Helens eruption also gave evidence of non-linear evolution of the Toutle river sediment transport after such a catastrophic event, when sediment yields reached 50 000 $t km^{-2} yr^{-1}$ during the first years after the blast compared to 300 $t km^{-2} yr^{-1}$ before this event (Major et al. 2000). The amount of highly erodible material generated by such volcanic eruption is enormous, billions of tons generated by debris avalanche and lahars[2]. Such material will slowly make its way to the coastal zone depending on its distance to the ocean. These events have been so far completely absent from global budgets although their local to regional importance in tectonically active regions (e.g. circumpacific fire belt) may be enormous.

Another factor in sediment transfer at the Earth's surface is inherited from the last glacial period. This enormous amount of stored sediments in formerly glacierised basins from 10 000 to 6 000 B.P. is still gradually moving downstream in river systems. Part of it is stored in the multiple lakes of glacial origin as mentioned previously. Detailed study of lake basin filling over 10 000 years allows reconstruction of sediment movement. In Jura lakes (Bichet et al. 1998) the first 2 000 years after deglaciation showed a rapid decrease of sediment transfer, followed by a relative low and stable transfer until the Bronze Age, when soil erosion increased after the first forest clearing. Similar trends are found in the French Massif Central from the Lake Chambon sediment filling (Macaire et al. 1997).

---

[2] Landslide or mudflow of pyroclastic material on the flank of a volcano.

D.4.5.4 Globally, What Is the Net Change in Riverborne Sediment Flux due to Humans?

At the river basin scale, it is difficult to assess the balance between processes that simultaneously accelerate sediment transport and those that restrict it. It is well known that land-cover change typically results in huge increases in sediment loads as in China (see summaries in Vörösmarty et al. 1998b), while at the same time sediment interception by reservoirs can also be significant. Given limits in the available information on reservoir systems (Vörösmarty and Sahagian 2000) and the availability of catchment models it will therefore be extremely challenging to piece together a global picture. New work by Walling and Fang (2003) offers a tantalising picture of what might emerge at the global scale. Of the 142 time series of major river basin suspended sediment flux, only four showed increases while 68 showed significant downward trends. About half of the basins with declining sediment flux also had declining water flow. The declines in water flow and in sediment flux reflect the increased level of control of land-surface hydrology, contrary to any increases in sediment flux due to increased erosion. This work offers important observational evidence for the presence of a large sink for sediment and possibly carbon in artificial impoundments as postulated by others (Smith et al. 2001; Stallard 1998).

The history of complex man-river sediment relationships can now be more completely understood from palaeo-reconstruction based on floodplain or delta cores. In China the Huang He (Yellow river) sediment flux to its delta was 1/10 of the present level (one billion tons per year) from 6 000 to 1 000 B.P. An abrupt increase occured at $c.$ 1 000 years ago as a consequence of human activities on the Loess Plateau – cultivation and deforestation (Saito et al. 2001). The degradation of vegetation cover may also have increased the flood frequency over the last thousand years (Vörösmarty et al. 1998b), thus enhancing erosion and transport.

Humans have also deeply modified the global landscape by moving earth in construction and mining activities which have also increased exponentially during human history (Hooke 2000). This author estimates that the total earth intentionally moved per capita is now 6 000 kg yr^{-1} (global average) corresponding to 35 billion t yr^{-1}, i.e. twice as much than the present river sediment load to oceans. Huge differences are observed between cultures: from 260 kg cap^{-1} yr^{-1} for Pascuan people in Easter Island to 31 000 for present US citizens. The unintentional accelerated erosion from agriculture could be twice as much as intentional earth moving (Hooke 2000). This enormous amount of sediment is mostly stored as colluvium and alluvium. It then moves into fluvial rivers with potential reservoir storage, at rates still unknown.

Humans now control more than one half of all water to which they have access (Postel et al. 1996). Given increased pressure on land and water resources through the coming decades (Rijsberman 2000) we anticipate the impact of both accelerated erosion and reservoir siltation to remain important. A major challenge thus faces us as we seek to quantify which of these two processes will predominate. Complicating these efforts will be a broad spectrum of effects, for which we have little information, even today. These include land-management techniques (some of which are now being adopted to reduce erosion), an accounting of reservoir capacities in the face of progressive siltation, engineering techniques to reduce that siltation, and climate change with its presumed increase in episodic flooding and high runoff.

D.4.6 Global River Transfer of Carbon and Its Alteration and Storage

Despite its limited importance in the yearly fluxes of carbon at the global scale – a few percent of, for instance, terrestrial gross primary production – it is necessary to explore river carbon transfers from land-to-ocean in order to assess (*i*) the long-term evolution of atmospheric CO_2 at the geological time scales (10^7 to 10^8 years), which is believed to be stabilised essentially by chemical weathering (see discussion in Broecker and Sarryal 1998); (*ii*) the present-day storage of organic carbon on land, and (*iii*) the present-day uptake of atmospheric CO_2 by soils during weathering reactions. On the other hand, riverine C is one of the least affected elements by direct human activities, particularly when compared to other bio-reactive elements such as N and P or to metals.

D.4.6.1 Sources, Sinks and Re-cycling

Riverine carbon originates from several sources. First is the action of carbonic acid (soil CO_2 in combination with water) on rock-forming minerals. For carbonated rocks half of the resulting HCO_3^- (also referred to as dissolved inorganic carbon [DIC]) originates from rock carbonate materials like calcite, half from soil atmospheric CO_2. For non-carbonated minerals such as aluminosilicates, 100% of the HCO_3^- originates from soil/atmospheric CO_2. Soil leaching of dissolved organic carbon (DOC) and erosion of topsoils containing particulate organic carbon (POC) is another pathway. Third, mechanical erosion of carbon-containing surficial sedimentary rocks yields particulate inorganic carbon (PIC) and, for shale, limestone and marls, POC. Wastewater from domestic and agro-industrial sources also produces POC (Servais et al. 1998). The relative importance and ages of the carbon sources are given in Table D.19 (Meybeck 1993a). Autochthonous (instream) river carbon such as precipitation of calcite (lake

Table D.19. Riverine carbon transfer and global change (TAC = Total Atmospheric Carbon) (from Meybeck and Vörösmarty 1999)

	Sources	Age (y)[a]	Flux[b] 10^{12} g C yr^{-1}	Sensitivity to global change[c]					
				A	B	C	D	E	F
PIC	Geologic	$10^4 - 10^8$	170	•					•
DIC	Geologic	$10^4 - 10^8$	140		•	•			•
	Atmospheric	$0 - 10^2$	245		•	•			•
DOC	Soils	$10^0 - 10^3$	200		•				•
	Pollution	$10^{-2} - 10^{-1}$	(15)?					•	
CO$_2$	Atmospheric	0	(20–80)		•	•	•		
POC	Soil	$10^0 - 10^3$	(100)	•					•
	Algal	10^{-2}	(<10)			•			•
	Pollution	$10^{-2} - 10^0$	(15)?					•	
	Geologic[d]	$10^4 - 10^8$	(80)	•					•

[a] Ages are counted since original CO$_2$ fixation. Recent ^{14}C analyses of river organic matter would give average ages closer to 10^3 years for both DOC and POC (Raymond and Bauer 2001).
[b] Present global flux to oceans mostly based on Meybeck 1982 and Meybeck 1993a.
[c] A = land erosion, B = chemical weathering, C = global warming and UV changes, D = eutrophication, E = organic pollution, F = basin management.
[d] The amount of geologic POC is still debated.

whitening) at pH exceeding 8.2 and production of algal biomass may also represent uptake of atmospheric carbon in continental aquatic systems.

One of the currently-debated river carbon issues is the recycling of rock-derived POC from shales hypothesised by Meybeck (1993a), based on recognition that minimum POC content in river suspended matter is about 0.5%, as in the Huang He, a value very close to the global shale average (Ronov 1976) or the loess POC content. Recent analysis of POC isotopic ratios and export rates on high erosion basins in Taiwan were conducted by Kao and Lui (1996, 1997) where it was learned that a recycling of fossil POC of up to 75% was apparent in the systems monitored. The proportion of recycled fossil POC could therefore be at its maximum in those sedimentary river basins characterised by high rates of rock erosion and limited soil erosion. However, high erosion basins on volcanic islands have zero recycled POC. In limestone-dominated basins, however, characterised by low mechanical erosion DiGiovanni et al. (2001) found a high proportion of fossil POC resulting from long-term de-calcification, although these basins do not provide much sediment at the global scale. Masiello and Druffel (2001) did not find evidence of such fossil POC in Santa Clara River particulates but most POC was aged soil organic matter. Raymond and Bauer (2001) performed ^{14}C age analysis of both POC and DOC in NE USA rivers and in the Amazon, and found average ages of around 670 years for the total organic carbon (TOC) discharged to the North Atlantic. The exact age distribution of organic carbon fluxes to the ocean remains to be established (Ludwig 2001). At the global scale, Meybeck's recent estimate (unpublished) for the recycled fossil POC is 30%, mostly from South-east Asia, but this is still debated (Ludwig 2001; Raymond and Bauer 2001).

A second issue relates to the potential influence of climate change on surficial rock weathering. An increase in temperature is likely to increase the weathering rates of aluminosilicates which are thought to be responsible for 67 to 70% of global DIC fluxes (Meybeck 1987; Amiotte-Suchet and Probst 1995). Another important control factor is the partial pressure of CO$_2$ in soils which is dependent on soil biogeochemistry and biomass production. Finally, the export rates of soil DIC, DOC, soil POC that constitute what can be called the riverine atmospheric carbon (Meybeck 1993a) are all directly linked to runoff (Degens et al. 1991; Meybeck 1993a; Bluth and Kump 1994; Amiotte-Suchet and Probst 1995; Ludwig et al. 1996).

The river export of total atmospheric carbon per unit of land area ranges from less than 0.5 g C m^{-2} yr^{-1} in the crystalline basin of the Sahel to more than 20 g C m^{-2} yr^{-1} in very humid tropical limestone basins (Meybeck 1993b). Lithology and runoff are therefore the two major driving forces of atmospheric carbon export, much ahead of temperature. Drainage pattern also plays an important role. In poorly-drained acidic waters, most of the atmospheric carbon occurs as dissolved organic carbon and not as DIC, as for the Rio Negro tributary of the Amazon and for many Arctic rivers where DOC exceeds 10 mg C l^{-1} (Artemyev 1996).

The spatial distribution of river carbon fluxes should also take into account sink terms. Particulate carbon can be stored in many sites along fluvial corridors, in slope deposits, floodplains, lakes and over the last few decades in artificial impoundments (Stallard 1998; Smith et al. 2001). Finally, much of the particulate riverborne

Table D.20. Global distribution of total atmospheric carbon (TAC) partitioned within river systems, by biome. The table presents results for individual chemical species from Meybeck (1993a)

	DOC	atm DIC	atm DC	soil POC	TAC	A (Mkm²)	Q (km³ yr⁻¹)	q (m yr⁻¹)
Total input to oceans								
C_{exp} (g m⁻² yr⁻¹)	1.99	2.44	4.43	0.99	5.42	99.9	37 400	0.375
M_{atm} C (Tg yr⁻¹)	199	244	443	99	542			
% (1)	37	45	82	18	100			
Total cold								
C_{exp} (g m⁻² yr⁻¹)	1.31	2.5	3.8	0.42	4.25	23.35	5 500	0.235
M_{atm} C (Tg yr⁻¹)	30.5	59.2	89.7	9.9	99.6			
% carbon species (1)	30.6	59.4	90.0	10.0	100			
% climatic zones (2)	15.3	24.2	20.3	10.0	18.4	23.3	14.7	
Total temperate								
C_{exp} (g m⁻² yr⁻¹)	1.5	4.5	6.0	1.5	7.5	22.0	10 250	0.465
M_{atm} C (Tg yr⁻¹)	32.2	100	132.2	33.7	165.9			
% carbon species (1)	19.4	60.3	79.7	20.3	100			
% climatic zones (2)	16.2	41	29.9	33.8	30.7	22.0	27.4	
Total arid + savanna								
C_{exp} (g m⁻² yr⁻¹)	0.15	0.6	0.75	0.23	0.98	28.7	2 430	0.085
M_{atm} C (Tg yr⁻¹)	4.45	17.1	21.5	6.75	28.3			
% carbon species (1)	15.7	60.4	76.1	23.9	100			
% climatic zones (2)	2.2	7.0	4.9	6.8	5.2	28.7	6.5	
Total humid tropics								
C_{exp} (g m⁻² yr⁻¹)	5.1	2.6	7.7	1.9	9.6	25.8	19 210	0.745
M_{atm} C (Tg yr⁻¹)	131.5	67.6	199.1	48.9	248			
% carbon species (1)	53	27.3	80.3	19.7	100			
% climatic zones (2)	66.2	27.7	45.0	49.4	45.8	25.8	51.4	

M_{atm}C = Total mass of riverine carbon carried to ocean; A = Exorheic drainage area of climatic zone; Q = Total river discharge; q = Average river runoff; DOC = Dissolved organic carbon; DIC = Dissolved inorganic carbon; DC = DOC + DIC; POC = Particulate organic carbon; (1) Percentage of carbon species in total carbon budget; (2) Percentage of climatic zone parcipitation in carbon budgets.

carbon that reaches the coastal zones of the world (90%) is likely to be sequestered within the coastal zone as for other particulates (Milliman 1991). In each of these sites the organic and inorganic carbon can be actively processed, exchanged, and recycled.

D.4.6.2 Estimates of Riverborne Carbon Flux to the Oceans

First estimates of carbon flux to the oceans were determined from an average water quality based on a few dozen world rivers and a global estimate for water runoff (Livingstone 1963; Meybeck 1979 for DIC; Meybeck 1982 for DIC, PIC, DOC, and POC). Since these first studies, our general knowledge on water quality has expanded to hundreds of rivers and tributaries (Meybeck and Ragu 1997) and to water balances defined spatially at relatively fine resolutions (Fekete et al. 1999, 2002; Oki

et al. 1995, 1999). From such databases two major approaches can be applied today: (*i*) a global typology of river basins, generally based on climatic or morphoclimatic characteristics (Meybeck 1993a, DiGiovanni et al. 2001) or (*ii*) an estimate of surface fluxes in individual grid cells based on multiple regression analysis using data from documented basins of various sizes in which runoff then temperature, lithology, soil organic carbon are applied as factors (Bluth and Kump 1994; Amiotte-Suchet and Probst 1995; Ludwig and Probst 1998; Ludwig et al. 1996).

An example of the first approach for total atmospheric carbon and its different origins is given in Table D.20. The table is organised to provide the relative sources of material for four global scale biomes (cold, temperate, arid and dry, humid tropics) which then can be compared to average land biomass productivity.

The second approach goes much further since the surface fluxes can be mapped provided that all driving

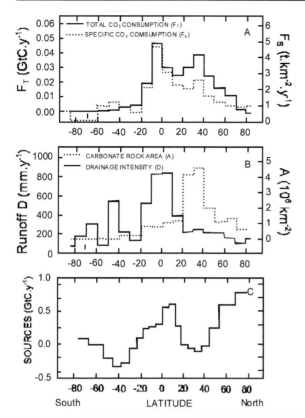

Fig. D.55a. Elements of carbon cycle transport from the continental land mass to the world's oceans: Latitudinal CO_2 weathering (from Amiotte-Suchet and Probst 1995)

factors are available as geographically-referenced files at appropriate resolutions, and can be reaggregated into various categories (such as by biomes, ocean basins, runoff classes).

An example of such a mapping is given in Fig. D.55b for DOC export over the exorheic part of the ocean (Ludwig et al. 1996; Amiotte-Suchet and Probst 1995). The DOC export (g C m^{-2} yr^{-1} or t km^{-2} yr^{-1}) = 0.0040 Q – 8.76 slope + 0.095 soil C, where Q is annual runoff in mm, slope is expressed in radians and soil C is in kg m^{-3}.

The maximum DOC export rate, more than 5 t C km^{-2} yr^{-1} (equivalent to g C m^{-2} yr^{-1}), is noteworthy first in the wet tropics (Amazonia, Zaire, South-east Asia) due to very high runoff and primary production, then in the Arctic Ocean basin despite a low runoff. In this case part of the contemporary DOC export may originate from peat-bog drainage of thousand-year old DOC (Artemyev 1996) (i.e. "sub-fossil" compared to the average age of DOC in most soils which is one order of magnitude lower). Under greenhouse warming the melting of sub-Arctic permafrost would result in a noticeable increase of sub-fossil DOC fluxes to the Arctic Ocean.

Ludwig et al. (1996) also provide the first mapping of POC export from soil erosion at the 0.5° × 0.5° resolution. POC yield is calculated for each cell combining a runoff estimate and a sediment yield estimate (see Fig. D.52) from which an average TSS concentration is derived. An associated percentage of POC in TSS is then calculated from a general POC% v. TSS regression al-

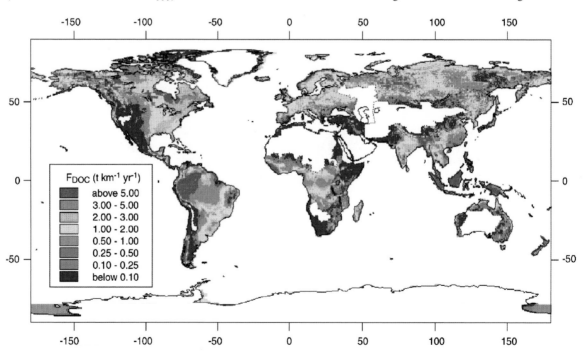

Fig. D.55b. Elements of carbon cycle transport from the continental land mass to the world's oceans: Map of dissolved organic carbon export rate in t km^{-2} yr^{-1} (from Ludwig et al. 1996, © American Geophysical Union)

ready set by Meybeck (1982) and Ittekot (1988) and re-validated on their database. Global hot spots of POC and DOC yields are not always located in the same regions: the circumarctic catchment has a very low POC yield while maximum values are observed for Northern China and Kenya coast. South-east Asia yields are very high for both POC and DOC. When re-aggregated for different biomes the average DOC and POC concentrations or yields may differ from those presented by Meybeck (1993a,b) (Table D.20), particularly for the wet tropics where Meybeck's average DOC is about 50% higher than that from Ludwig et al. (1996). This illustrates the limits of a typology approach based on one or two factors only. The wet tropics data used by Meybeck (1993a,b) are mostly based on the poorly-drained Negro and Congo, while the wet tropics considered by Ludwig et al. (1996) are greatly influenced by the whole Amazon, which is relatively better-drained.

Aitkenhead and McDowell (2000) based their global DOC budget on a linear regression between DOC export $(g\ m^{-2}\ yr^{-1})$ and the C:N ratio for 15 aggregated biomes, thus combining the regression and the typology approach. The C:N ratio accounts for 99.2% of the variance in annual DOC flux in their dataset. Their numbers are somewhat similar to those found by Schlesinger and Melack (1981) and are in sharp contrast to those of Meybeck (1981): their tundra DOC export rates, much affected by peat bogs and poor lateral drainage, are much higher than for the taiga while Meybeck postulated the reverse order, based on runoff intensity. Such differences in DOC

yields may in part result from different biome definitions used by different authors: the "polar" biome of Ludwig et al. (1996) is close to the "tundra" of Aitkenhead and McDowell (2000). The river DOC inputs for the oceans have been estimated as such over the last 20 years: 411 (Schlesinger and Melack 1981), 383 (Meybeck 1981), 205 (Ludwig et al. 1996), and $363 \times 10^{12}\ g\ yr^{-1}$ (Aitkenhead and McDowell 2000).

The specific fluxes of CO_2 consumed by rock weathering have been computed on 2 degree cells for which the dominant lithology has been determined, with carbonate rocks and all other types differentiated (Amiotte-Suchet and Probst 1995) from general regressions established between DIC export, rock type and runoff for small monolithologic and unpolluted catchments in France based on Meybeck (1986). Despite some limitation in this approach (one single region may not be representative of the global diversity of biogeochemical conditions, the runoff resolution is not the same as for lithology) these authors have provided the first mapping of net atmospheric carbon uptake through chemical weathering, that was then re-aggregated into latitudinal zones (Fig. D.55a) showing the combined effects of lithology and runoff and the dominance of carbonate rocks in the Northern Hemisphere.

The budget of natural and anthropogenic carbon sinks on land already addressed by Kempe (1984) has been very carefully revisited by Stallard (1998) by latitudinal bands (Fig. D.56). In addition, Stallard has considered a set of 864 scenarios for this carbon storage to bound the magnitude of the potential fluxes, a highly innovative ap-

Fig. D.56.
Carbon sequestration by terrestrial sinks within river basins in 864 scenarios (from Stallard 1998, © American Geophysical Union). **a** Human-induced carbon sink predicted by all scenarios; **b** averages of scenarios falling within $0.2 \times 10^{15}\ g\ C\ yr^{-1}$ intervals; **c** all scenarios within 10% of $1.0 \times 10^{15}\ g\ C\ yr^{-1}$; **d** all scenarios within 10% of $2.0 \times 10^{15}\ g\ C\ yr^{-1}$

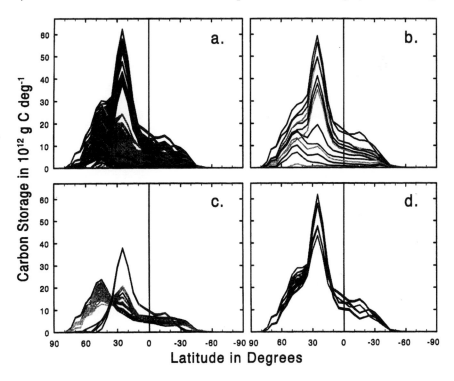

proach in global river studies. A critical finding was that human-induced burial (e.g. reservoir construction) may total up to 1.5 Pg C yr^{-1}, mostly in the 30–50° N latitudes where the land mass is maximum and independent studies have placed the so-called Northern Hemisphere "missing carbon" sink.

The stability of riverine carbon transfers to oceans is difficult to assess despite the relatively good data for DIC and even DOC in world rivers. When documented, long-term trends (> 30-year records) of DIC levels in major systems are relatively stable compared to other major ions like SO_4^{2-}, Na^+, Cl^-, as in former Soviet rivers (Tsirkunov 1998). DOC and POC measurements in waters by infrared techniques, more reliable than previous chemical analysis, are only available for most rivers over the last 15–20 years, and no trends can yet be observed. In heavily impacted and carbonate-rich river basins such as the Seine River, the DIC level is maintained close to the calcite saturation level most of the time and the increase of DOC is relatively limited due to processing of treated domestic wastes (Servais et al. 1998). On the other hand, N and P levels can be increased by more than 20-fold. However, direct release of domestic wastes to the coastal zone may represent a significant carbon flux increase at local to regional scales. Probably the least studied carbon transfer is river-dissolved CO_2 which is directly linked to the algal production and bacterial respiration, which can be expressed as a ratio. A good example has been documented for the Amazon (see Chapt. D.5). However, despite the pioneering study by Kempe on the Rhine (Kempe 1979, 1982), this question has not received much attention at the global scale.

D.4.7 Global Riverine Nutrient Flux to the Oceans

The assessment of pollution fluxes to inland and coastal waters is of broad international concern (Vörösmarty and Peterson 2000). Shifts in nutrient : nutrient delivery ratios contribute to coastal eutrophication, toxic blooms, hypoxia and anoxia in many parts of the world (Humborg et al. 1997; Justic et al. 1995a,b; Turner and Rabalais 1994; Shumway 1988). Land-based activities generating point and non-point sources in upstream catchments have been implicated (Hopkinson and Vallino 1995; Jordan et al. 1997; Hession et al. 2000; Russell et al. 1998; Goolsby and Battaglin 2001). Such changes can also encompass deceleration of geochemical fluxes as associated with the widespread retention of both solutes and particulates in reservoirs. Analysis of long time series of sediment flux on many rivers shows a majority with declining or stable fluxes to the coastal zone (Walling and Fang 2003), supporting the presence of a large sink for sediment and possibly carbon in artificial impoundments (Smith et al. 2001; Stallard 1998). Upstream reservoir retention robs coastal ecosystems of important sources of sediment and

nutrients, leading to the disintegration of deltas, shoreline erosion, and altered productivity upon which important fisheries are based (Milliman and Meie 1995; Vörösmarty et al. 1997b; Nixon 2003).

A growing body of evidence indicates that land-based human activities impart a biogeophysical signal for river fluxes at the global scale, and argue for appropriate monitoring and analysis tools to understand these progressive changes. It has been estimated that only a minority of the world's drainage basins (~ 20%) have nearly pristine water quality and that the riverine transports of inorganic N and P have increased several-fold over the last 150 to 200 years (Seitzinger and Kroeze 1998; Meybeck and Helmer 1989; Vörösmarty et al. 1986). Despite the apparent magnitude of these changes, we currently lack the observational support and synthetic perspective to address linkages between land-based activities and coastal ecosystems (Hobbie 2000).

Depicting land-to-ocean fluxes in Earth system models has thus generally not been well articulated (Vörösmarty et al. 1997a). However, important new work has recently emerged which considers land-based fluxes of water (Fekete et al. 1999, 2002; Oki et al. 1999), biogeochemical constituents (e.g. Seitzinger and Kroeze 1998; Ludwig and Probst 1996), and suspended sediments (Ludwig and Probst 1998) in a mutually consistent and geographically-specific manner. These recent studies have taken a pragmatic approach, combining global, biogeophysical datasets with relatively simple (e.g. multiple regression) models that predict a spatially-varying flux of material delivered from the land to the coastal zones of the world. Many provide a static, steady-state estimate of flux. There exists a major need for datasets and models to understand the primary factors regulating mobilisation, processing, and transport of constituents through the full continuum of inland aquatic systems from source area to ocean (Alexander et al. 2000; Vörösmarty and Peterson 2000). Quantifying the ongoing changes to contemporary river basins in the face of economic development, climate variability (e.g. ENSO, North Atlantic Oscillation, Arctic Oscillation), and the rapid discharge of water and materials during episodic floods remains an important, unresolved issue.

D.4.7.1 Inventory Methods

Nitrogen is an important biogeochemical constituent in both terrestrial and aquatic ecosystems. In a broad array of terrestrial ecosystems it serves as a limiting nutrient to ecosystem primary production and decomposition (Schlesinger 1997). Although not strictly limiting in most freshwater ecosystems, nitrogen nonetheless participates actively in the metabolism of river, lake and wetland ecosystems (Arheimer and Wittgren 1994; Jansson et al. 1994; Peterson et al. 2001; Wollheim et al. 2001).

Nitrogen in the coastal zone, however, plays an important role in both wetland and open water processes which ultimately support the maintenance of ecosystem health and fisheries of importance to humans (Bricker et al. 1999). Inorganic N is also the source of radiatively important trace gases such as nitrous oxide, whose emission is regulated by redox state which itself is sensitive to water availability.

We will take N as an example of a successful application of available techniques for global-scale analysis of biogeochemical fluxes. Recent studies show that a wide array of human activities have already accelerated the natural fluxes of nitrogen through catchments and over the full continuum of transport into the coastal zone. The focus of this change involves the main inorganic N forms, nitrate and ammonium, with less of an impact on organic forms (Meybeck and Ragu 1997). The evidence for an increasing anthropogenic loading of inorganic N is ubiquitous and it has been estimated (Galloway et al. 1995) that the combination of all sources of anthropogenic N fixation now equals or exceeds natural rates.

Much has been learned from descriptive studies which have taken observed patterns of river N flux and established (*i*) general tendencies for different types of rivers to export N (e.g. Lewis et al. 1999 for pristine catchment fluxes), (*ii*) time series following disturbance (e.g. Jaworski 1997), and (*iii*) seasonal fluxes of nitrogen through rivers (Alexander et al. 1996). Budgeting approaches have also been used to establish basin-wide N export estimates (Meybeck 1993b).

One such important effort was recently organised by the Scientific Committee on Problems of the Environment's International Nitrogen Project (SCOPE-N) as a complement to an earlier effort by SCOPE to quantify riverine fluxes (Degens et al. 1991). Although based on an inventory approach, results of the study elucidated several important points that are of direct relevance to the construction of models that predict the sources and fates of N at the macro-scale. Results were reported in Howarth et al. (1996) for both N and P and we summarise here the work on nitrogen.

The domain of the study was the entire contributing drainage area of the North Atlantic which was broken into 14 distinct regions. Individual sub-basins represented a broad cross-section of anthropogenic disturbance, from the Amazon to the highly eutrophied basins

Fig. D.57.
Relation between total accounted load and observed riverine flux for total N in the North Atlantic region. Note the sequestration and/or loss of total N loaded within drainage basins, on the order of 80% (from Howarth et al. 1996)

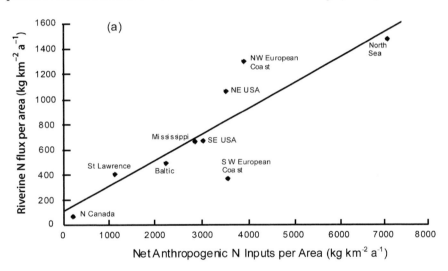

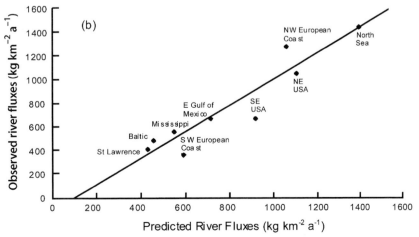

Table D.21. Some examples of the statistical approach used in estimating river material transport at the global scale (updated from Vörösmarty et al. 1997c)

Total Suspended Solids (TSS, see also Table D.18)	
Mozzherin (1992)	Runoff, area, relief, soils, typology
Jansen and Painter (1974)[a]	Climate, relief, land cover, lithology
Milliman and Syvitski (1992)[a]	Area, runoff, location[a]
Meybeck (1993a)	Runoff/temperature; biome typology
Ludwig and Probst (1996)[a]	Runoff, relief, landcover and others
NO$_3^-$	
Caraco and Cole (1999)	Land use, fertiliser, point sources deposition a. other factors; reused by Seitzinger and Kroeze (1998)
Meybeck (1982)[a]	Demophoric (population × energy consumption) index reused by Seitzinger and Kroeze (1998)
Seitzinger and Kroeze (1998)[a]	Loading, runoff, denitrification
TSS/TDS	
Meybeck (1976)	Runoff/river size
PO$_4^{3-}$	
Caraco (1994); Dillon and Kirchner (1975)	Geology and land use
Meybeck (1982)[a]	Demophoric index
DIC	
Amiotte-Suchet and Probst (1995)[a]	Lithology, runoff
POC/DOC	
Meybeck (1993a)	Runoff/temperature; biome typology
Ludwig et al. (1996)[a]	Runoff, slope, soil org. C and others, biome typology
Aitkenhead and McDowell (2000)[a]	Runoff, soil C/N ratio, biome typology
SiO$_2$	
Meybeck (1979)	Runoff, temperature
Bluth and Kump (1994)[a]	Runoff/lithology[a]

[a] Regression approach.

of Northern Europe. The work elucidated the importance of large river systems in the overall budget. The Amazon alone contributed 25% of the 13 Mt of total N exported to the ocean, while the Mississippi alone loaded an additional 14%. It also demonstrated the importance of considering smaller but highly impacted rivers. Collectively, the many basins of western Europe which have half of the area but one-fifth the discharge of the Amazon nonetheless generate two-thirds of the total N flux of the Amazon.

The work also elucidated the importance of different sources of N loadings across these basins. For all of the regions, the non-point sources of N dominate the tabulation of river flux. Sewage inputs for the whole North Atlantic basin, for example, account for only about 10% of total inputs. The bulk of the loading is from atmospheric deposition, industrial fertilisation and legume fixation. It was discovered that the temperate zone drainage basins have shown a 2 to 20-fold increase in river N fluxes, expressed on a per km^2 basis and compared to the pre-industrial condition. Northern Canada is considered to have remained relatively pristine as has the Amazon. Specific fluxes for the tropical Amazon indi-

cate much larger natural background levels than those from undisturbed temperate zone river basins, reflecting a relative abundance of N (and reduced role in nutrient limitation to growth) in tropical systems. The use of tropical systems, considered pristine (e.g. Meybeck 1982; Lewis et al. 1999), as a benchmark for the absence of anthropogenic impact, thus could seriously underestimate the degree of change occurring in the temperate zone and other parts of the world.

The Howarth et al. (1996) study was also able to correlate river N specific flux to net anthropogenic loads from fertiliser, atmospheric deposition, fixation, and net food and animal feeds imported/exported (kg N km^{-2} yr^{-1}). A critical finding was that only about 20–25% of the incident loading is actually transported through rivers to the ocean (Fig. D.57). There is, then, either a large storage (i.e. biomass and organic detritus storage in terrestrial ecosystems and soils) or loss (i.e. denitrification in wetlands and aquatic systems) within the basin. In contrast, Green et al. (2003) found a wide range globally, from 0–100%. The quantification and understanding of controls on these sinks remains a major research question.

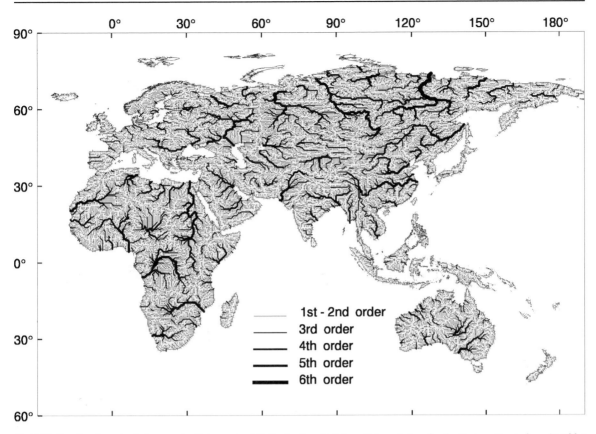

Fig. D.58. Simulated topological network of rivers at 30′ (latitude × longitude) spatial resolution for the Eastern Hemisphere (Strahler 1957). The order at 30′ resolution is shown from Vörösmarty et al. 2000a, © with permission from Elsevier Science

D.4.7.2 New Regression and Multiple Regression Models

Single or multi-factor statistical relationships represent a straightforward means of predicting constituent fluxes. Such relationships have been developed for river nutrient fluxes across a wide variety of scales, from small catchments, regions, and continents (Jordan et al. 1986; Smith et al. 1993; Frink 1991; Rekolainen 1990; Neill 1989) to the domain of the entire globe (e.g. Meybeck 1982; Cole et al. 1993). The approach has been used as well to predict changes in nutrient : nutrient ratios (Caraco 1994). Table D.21 gives examples of this approach for dissolved and particulate constituents across a spectrum of spatial scales.

Empirical relationships, up to the global scale, are developed from existing stream gauge information, topographical maps, digital elevation models, geological maps, remote sensing of land use, and inventories of dams and other water engineering works. These relationships are then distributed spatially using the drainage basin typology and classification systems discussed earlier. This could be achieved by first developing class-specific statistical functions; next, categorising the river systems of the globe according to the typology; mapping the biophysical drivers necessary to implement the functions spatially; and, finally, applying the statistical functions onto a set of base maps depicting individual drainage basins of the world (Fig. D.58).

The exercise yields region-to-region differences in terrestrial constituent loadings, which in turn establishes geographically-specific estimates of inputs to regional coastal zones and ocean basins. The regression-based approach for data-poor regions, in conjunction with actual data on major rivers, permits global budgets to be constructed for the contemporary setting, and when suitable forcings are applied, for past and future conditions. In this context, it is valuable to explore the issue of time lags between changing drainage basin attributes and riverine flux, to elucidate the impact of changing forcing factors on resulting constituent flux (see Jaworski et al. 1997).

The recent work of Seitzinger and Kroeze (1998) is instructive of applications using this class of model. The model is built around a statistically-derived non-linear equation:

$$DINexp_{riv} = 1.19 \times NO3exp_{riv} \times WS \qquad (D.13)$$

$$\begin{aligned}
NO3exp_{riv} &= EC_{riv} \times [Psources + EC_{ws} \times (Ppt_{ws} + Fert_{ws})] \\
&= 0.7 \times [1.85 \times Popd \times Urb \\
&\quad + 0.4 \times WaterRunoff^{0.8} \\
&\quad \times (Ppt_{ws} + Fert_{ws})]
\end{aligned} \qquad (D.14)$$

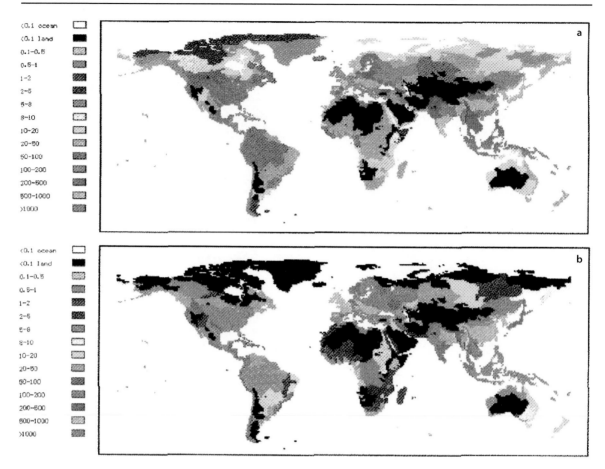

Fig. D.59. Examples of contemporary dissolved inorganic nitrogen flux from rivers predicted by a multiple regression model. **a** Total DIN export (kg N km^{-2} yr^{-1}); **b** export attributed to fertiliser use. Note the several areas of increased loading associated with high population and industrial fertiliser use (from Seitzinger and Kroeze 1998, © American Geophysical Union)

where DINexp$_{riv}$ is the export of dissolved inorganic N from a river (kg N km^{-2} yr^{-1}), 1.19 a scaling factor (converting the original relationship from NO$_3$ to DIN), EC$_{riv}$ an export coefficient, Psources point sources, EC$_{ws}$ a catchment export coefficient, Ppt$_{ws}$ atmospheric deposition of N to soils, and Fert$_{ws}$ the non-point inputs from fertilisers. The constant 0.7 represents the efficiency of transport through rivers *per se*, 1.85 is a human emission factor (1.85 kg cap^{-1} yr^{-1}), Popd the population density, and Urb the fraction of population that is urban, WaterRunoff is runoff in m yr^{-1}, and WS is the watershed area in km^2.

The structure of these equations is interesting from the standpoint of identifying and parameterising major drainage basin nutrient delivery processes. It distinguishes point from nonpoint loading terms, the latter of which are regulated by a variable efficiency of delivery to rivers (0.4 × WaterRunoff$^{0.8}$). Both point and nonpoint loads are together modified by a transport parameter which estimates the incomplete efficiency of river transfer of material.

The equation is not strictly mass-conserving on several counts. For instance, N fertilisers can be counted more than once as sewage load (point sources) and as diffuse sources. The equation does not include other important inputs such as fixation, loading from food-derived nitrogen in rural populations and point and nonpoint inputs from animal husbandry. Forage should not be counted twice as N fixation and as animal manure. The equation also does not account for transformations between organic inputs and inorganic outputs. However, it does represent a significant step forward, as it permits a relatively high resolution geography of N loads through rivers to be estimated. From their predictions, Seitzinger and Kroeze (1998) were able to map global patterns of DIN export, concluding that contemporary global river exports total 21 Tg DIN yr^{-1}, of which 75% is of anthropogenic origin (i.e. a 4-fold increase). Of this anthropogenic transport, more than 50% is attributable to fertiliser inputs. The authors mapped the geography of total riverine DIN export (Table D.22, Fig. D.59), together with exports from fertiliser, atmospheric deposition, and sewage. Globally they find that 8% of the total N inputs to the terrestrial environment can be accounted for as DIN (i.e. nitrate + ammonia) export by rivers, in rough agreement with Howarth et al. (1996) for North Atlantic catchments.

Table D.22. Ocean basin fluxes of dissolved inorganic N (from Seitzinger and Kroeze 1998)

Ocean	N₂O emissions		DIN export by rivers (10³ t N yr⁻¹)	Catchment area (10³ km²)	Runoff (km³)	Population (million)	Fertiliser use (10³ t N yr⁻¹)	NOy deposition (10³ t N yr⁻¹)
	Rivers (t N yr⁻¹)	Estuaries (t N yr⁻¹)						
Northern Hemisphere								
Arctic	3734 (<1)	1369 (1)	415 (2)	20730	4466	68	1618	2681
North Atlantic	179472 (17)	56524 (25)	4395 (21)	23743	10273	754	18726	6173
North Indian	296345 (28)	57529 (26)	4198 (20)	9980	5129	1260	11251	1681
North Pacific	455206 (43)	86333 (39)	6650 (32)	17662	10183	1529	23211	3113
European Seas	92836 (9)	11861 (5)	2539 (12)	13098	2772	598	14451	4269
Total	1027592 (98)	213616 (96)	18197 (87)	85213	32822	4209	69257	17918
Southern Hemisphere								
South Atlantic	8902 (1)	3264 (1)	989 (5)	19414	11299	294	1112	3210
South Indian	3565 (0)	1307 (1)	396 (2)	7215	2391	140	991	790
South Pacific	11228 (1)	4117 (2)	1248 (6)	10091	9282	266	2248	582
Total	23695 (2)	8688 (4)	2633 (13)	36720	22972	699	4351	4582

Model results are N₂O emissions and DIN export by rivers. Inputs to N model (does not include endorheic regions; see Tables D.13, D.14) are catchment area, runoff, population, fertiliser use, and NOy deposition. Numbers in parentheses are the percent of total.

The success in mapping contemporary patterns of river N flux will promote additional, more sophisticated statistical modelling studies. Although an ultimate goal is the construction of dynamical models over the full continuum of inland fluvial systems, it may still be feasible to gain additional predictive capacity through a subcatchment version of this type of model. In addition, the development of time-varying – intra- and inter-annual – statistical models is a logical next step.

D.4.7.3 "Hot Spots" at the Global Scale

A basic geography of nutrient inputs to the oceans must necessarily reflect the influence of non-uniform distri-

butions of nutrient content across world rivers. The distribution of measured river nitrogen is indeed skewed toward very high concentrations for both ammonium and nitrate. Such high levels are only observed in a few highly polluted basins of limited size, but their probable weight on the global budget is such that they could be termed "hot spots". From our present knowledge of nutrient quality in 250 rivers in the GEMS-GLORI register (Meybeck and Ragu 1997), these rivers are mostly located in western Europe (Ems, Weser, Thames, Seine, Scheldt) but they can also be found in some Chinese and Indian river basins. If the GEMS-GLORI results are considered to be unbiased and also representative of the 50% undocumented river discharge the following conclusions can be made (Fig. D.60):

Fig. D.60.
Global distribution of nitrate and ammonia fluxes in three classes of basins, for 250 major world rivers (data from Meybeck and Ragu 1996). The highest impacted basins (> 1 mg N-NO₃⁻ l⁻¹ and > 0.3 mg N-NH₄⁺ l⁻¹) correspond to 7 and 10% of the river runoff but to 35% and 55% of nitrate and ammonia fluxes, respectively. The pristine rivers (< 0.3 mg N-NO₃⁻ l⁻¹ and < 0.07 mg N-NH₄⁺ l⁻¹) account for 84% and 80% of the river runoff and to 48% and 22% of nitrate and ammonia fluxes, respectively

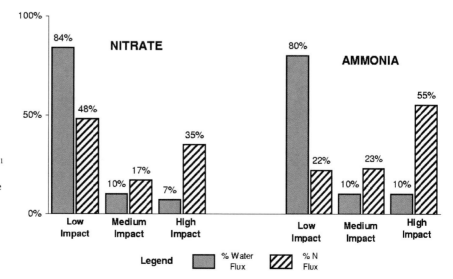

- The highest class of nitrate concentrations (> 1.0 mg N l^{-1}) corresponds to only 7% of global river discharge while the middle class (0.3 to 1.0 mg N l^{-1}) accounts for 10%, thus leaving 84% of river discharge with pristine or near-pristine nitrate levels (< 0.3 mg N l^{-1}). The relative weight of large rivers such as the nearly pristine Amazon and Orinoco (Lewis et al. 1999) or Zaire Rivers is apparent;
- The corresponding nitrate fluxes to oceans are quite different: 35%, 17%, and 48% for the highest, medium, and pristine concentration classes, respectively;
- For ammonium, the distribution is very similar to that of nitrate: the highest (N-NH$_4$ > 0.3 mg l^{-1}) medium, and lowest (near pristine) concentration class with < 0.07 mg N l^{-1} corresponds to 10%, 10%, and 80% of discharge, respectively;
- For ammonia flux to the ocean, the corresponding estimates for high, medium, and near pristine classes are 55%, 23% and 22%.

Most of the largest 100 rivers by area are now documented for nitrate and the few unknown concentrations can be estimated reasonably well, based on GEMS-GLORI (Meybeck and Ragu 1997) and several regional budgets. Rivers can be ranked by area, from the Amazon (6.4 Mkm^2) to smaller basins (< 20 000 km^2) and their discharge-weighted mean nitrate concentration and area-weighted specific flux (see Fig. D.61) computed at any point along the ranking axis as for the TSS fluxes (shown in Fig. D.53). The three largest rivers (Amazon, Zaire, Lena) do not appear to be appreciably affected by anthropogenic activities and are therefore characterised by low nitrate levels (mean of 0.125 mg N l^{-1}). The addition of the fourth largest river, the much more highly impacted Mississippi, increases the mean to 0.20 mg N l^{-1} which more or less remains stable until the European rivers are encountered at the 30th rank where the nitrate level gradually increases to 0.250 mg N l^{-1}. Ironically, the mean nitrate export per unit of basin area starts high in the Amazon at 150 and decreases to 70 kg N km^{-2} yr^{-1} upon reaching rank 30, and then slowly increases to 80 kg N km^{-2} yr^{-1} for the smallest rivers in this database, around 20 000 km^2 by rank 100. This apparent discrepancy reflects the high export rate of the Amazon and Zaire basins produced by the combination of high productivity and high runoff, compared to the other rivers that follow, mostly in the temperate and subarctic regions for the 14 first rivers ranked by area.

These examples shed light on the difficulties of computing global-scale means when relying only on major basins and suggests the need for a mechanism to include smaller, but potentially as important smaller river systems. Given a substantial population living within small coastal regions and considering those constituents that are heavily influenced by humans, a need for inventory or estimation of the role of smaller rivers is imperative.

D.4.7.4 Stoichiometric Changes of N : P : Si

The approaches outlined above have demonstrated that anthropogenic influences can substantially increase the efflux of inorganic N compared with the pre-industrial state. These increases are an important signal of global change, since N is limiting to both natural terrestrial ecosystems and aquatic systems in the coastal zone and is the source of radiatively important trace gas emission for nitrous oxide. Although our focus has been on river nitrogen, the biogeochemistry of this element is strongly coupled to the behaviour of other nutrients as well, most importantly phosphorus and silicon. We now turn our attention from some recent human-forced changes to the character of biogeochemical balances in rivers.

The productivity of natural assemblages of coastal zone phytoplankton is regulated by not only the level of inorganic nutrients but also demands for these nutrients

Fig. D.61.
Global average nitrate export rate Y_N (in kg N km^{-2} yr^{-1}) and nitrate concentration C_N (mg N-NO$_3$ l^{-1}) for the 250 largest rivers documented in the GEMS-GLORI database (Meybeck and Ragu 1996, 1997). Discharge-weighted averages are from 6.4 Mkm^2 (Amazon river, ranked #1, then Zaire, ranked #2, Mississippi, Lena etc.) to 65 Mkm^2, the total area documented

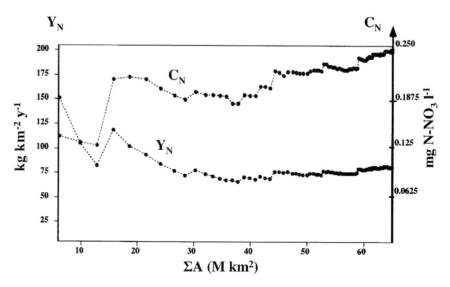

in particular ratios. Diatomaceous algae are a critical primary producer in coastal waters upon which a variety of higher organisms, including fish species of importance to humans, are dependent. These algae require inorganic silicon together with nitrogen and phosphorus in particular ratios in order to survive. A Si:N ratio of 1:1 (mole:mole) is required for proper algal nutrition. When this ratio exceeds 1:1 the N is limiting, as is the typical condition in natural ecosystems. However, when the ratio falls below 1:1, Si becomes limiting, invoking stress on diatom populations and favouring alternate primary producers. In the case of coastal waters this has been demonstrated to enhance the competitiveness of nuisance and toxic algae that place normal coastal food webs in jeopardy (Turner and Rabalais 1991; Justic et al. 1995a,b).

Analysis by Humborg et al. (1997) has suggested a shift in Si:N due to the trapping of Si by large reservoir systems. In addition, the anthropogenic loading of inorganic N in terms of industrial fertiliser application, atmospheric deposition, point source sewage, and rearing of livestock has the potential for further distorting this ratio. Table D.23 summarises Si:N ratios in several river systems of the world. These rivers can be categorised by their level of anthropogenic influence. It is important to note that the Si concentration will be determined fundamentally by the weathering of silicon-containing rock and soils with very few anthropogenic influences. From pristine through highly impacted systems Si ranges from 2 to 25 mg SiO_2 l^{-1}. In contrast, rivers show inorganic N concentrations (DIN) that span orders of magnitude when passing from pristine (DIN < 10 µM) to highly fertilised basins (DIN > 100 µM).

There is thus a strong contrast not only in N loading but in Si:N ratios across the spectrum of anthropogenic influence on river systems. In Table D.23, the highly eutrophied basins are in areas with high population density, with loadings from industrial fertiliser, point sources, and atmospheric deposition of N from regional air pollution. It is interesting to note that as China has recently embarked on a campaign of economic development, its major rivers appear now to be progressing to a state intermediate between pristine and highly eutrophied (Vörösmarty et al. 1998b). The three Gorges Dam is likely to modify the C:N:P ratio of the Chang Jiang inputs to the China Sea, in a way limiting the nitrate flux which has doubled over the last 10–20 years (Zhang et al. 1999). These rivers will thus be important to monitor as basins that are representative of the broader issue of cultural eutrophication of inland waterways.

D.4.8 Future Trends

D.4.8.1 Pressure on Inland Water Systems

Although an exact inventory of global water withdrawal has been difficult to assemble, the general features of anthropogenic water use are known. Reviews of the recent literature (Shiklomanov 1996; Gleick 2000) show a range in estimated global water withdrawals for the year 2000 between about 4 000 and 5 000 km^3 yr^{-1}. Despite reductions in the annual rate of increase in withdrawals from 1970 (Shiklomanov 1996, 2000; Gleick 1998), global water use has grown more-or-less exponentially together with human population and economic development over the industrial era. Aggregate irretrievable water losses (consumption), driven mainly by evaporation from irrigated land increased 13-fold during this

Table D.23. Ambient concentration and stoichiometric ratios for several key nutrients in pristine, rapidly developing and industrialised drainage basins (from Vörösmarty et al. 1998b)

River	Period	Si (µM)	N (µM)	P (µM)	Si:N	Si:P
Pristine/near pristine						
Amazon	Before 1972	187	3.2	0.4	58.4	468
	May–June 1976	111 – 121	7 – 11	0.3 – 0.75	12.9	221
Mackenzie	1981–1983	143	7.14	0.19	20.0	752
Yukon	1978–1985	275	8.35	0.35	32.9	786
Zaire	Nov. 1976	161	7.3	0.72	22.1	224
	May 1978	171	5.9	0.89	29.0	192
Industrialised						
Mississippi	1981–1987	108	114	7.7	1.0	14
Po	1981–1984	120	147	4.6	0.7	26
Rhine	1976–1978	130	310	14	0.4	9
Seine	1976–1982	120	372	20	0.3	6
Chinese rivers						
Yangtze	June 1980	100 – 105	65	0.5 – 0.9	1.6	146
Yellow	August 1986	128	64	1.1	2.0	116

same period. Global consumption for 1995 has been estimated at approximately 2 300 km³ yr⁻¹ or 60% of total water withdrawal (Shiklomanov 1996).

This use must be examined in relation to supply. Using recent estimates of long-term average runoff from the continents totaling approximately 40 000 km³ yr⁻¹ (Fekete et al. 1999; Shiklomanov 2000) and an estimated withdrawal of 4 000 to 5 000 km³ yr⁻¹, humans already exploit 10–15% of current renewable water supplies. Water withdrawal over the entire globe is therefore but a small fraction of continental runoff. However, upon closer examination, of the 31% of global runoff that is spatially and temporally accessible to society more than half (54%) is withdrawn or maintained for instream uses (Postel et al. 1996). Contemporary society is thus highly dependent on, and in many places limited by, the terrestrial water cycle defined by contemporary climate.

This dependency is likely to intensify in the future. From 1950 to 1998, water availability had already decreased from 16 000 to 6 700 m³ yr⁻¹ *per capita* (WRI 1998; Fekete et al. 1999). If we assume no appreciable change in global runoff over the next several decades, a projected increase in global population by 2025 to about 8.0 billion people (WRI 1998) means that *per capita* supplies will continue to decline to about 5 000 m³ yr⁻¹ (WRI 1998). Tabulating these statistics from the standpoint of accessible water, *per capita* availability would be reduced to about 1 500 m³ yr⁻¹. Given an estimate of mean global water use of 625 m³ yr⁻¹ *per capita* for 2025 (Shiklomanov 1996, 1997), withdrawals could therefore exceed 40% of the accessible global water resource even with presumed increases in use efficiency. The primary application of water is to irrigated cropland in the many regions of the world where rain-fed agriculture is limited or where specific crops such as paddy rice typically are inundated during growth. Irrigation today produces more than 40% of global food and agricultural commodity output (Shiklomanov 1996, 1997; United Nations 1997) on but 15–20% of all agricultural land worldwide. Providing adequate irrigation water to a growing population constitutes a major international security concern well into the future (United Nations 1997).

These statistics have obvious implications both for human society and natural ecosystems that are both highly dependent on renewable supplies of water. Figure D.62 shows the change in *per capita* water availability from the contemporary state to 25 years ahead. Many areas of the globe, already stressed by water shortages such as the Middle East, the Aral Sea basin, northern China, north-eastern Brazil and the arid south-western USA (to name but a few), will likely feel increased water stress well into the future. Associated with such stresses are a broad set of environmental and social challenges including:

- Increased levels of competition for water among humans and between humans and ecosystems which themselves require sustainable supplies;
- Non-sustainable uses of water or the need to develop costly new sources (e.g. from desalinisation);
- Limits to economic growth, including the curtailment of activities;
- Massive imports of food;
- Public health and pollution problems;
- Political instability and international conflicts in the roughly 250 international river basins.

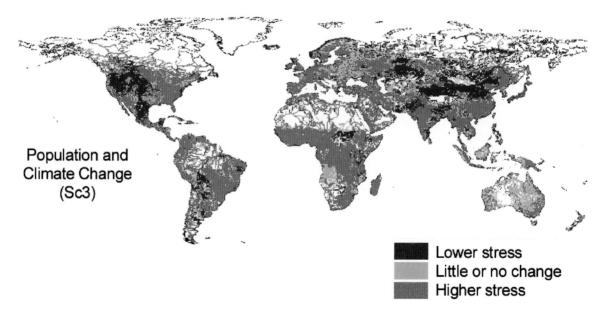

Population and Climate Change (Sc3)

■ Lower stress
▨ Little or no change
▨ Higher stress

Fig. D.62. Change in demand per unit discharge (a measure of water stress) from 1985 to 2025, showing areas of greater than 20% increase in stress (*red*), greater than 20% decrease in stress (*blue*), and relatively small change (*green*) (from Vörösmarty et al. 2000c, © American Association for the Advancement of Science, reprinted with permission)

D.4.8.2 The Future State of Riverine Carbon Loads

Projections of the future of river carbon are based on: (*i*) the simulation of weathering under a double CO_2 scenario; (*ii*) the simulation of TOC (Total Organic Carbon) export in regions most affected by runoff change and by temperature increase as in subarctic basins; (*iii*) the simulation of new sinks for DOC and POC particularly through the development of impoundments, from small hill slope reservoirs to large-scale reservoirs; (*iv*) the regionalisation of carbon inputs to oceans coupled to ocean models.

In the Duke Forest (North Carolina) the FACE fumigation experiment (Andrews and Schlesinger 2001) increased the atmospheric CO_2 by 200 ppmv. As a result the annual soil respiration increased by 27% which accelerated the rates of soil acidification and mineral weathering resulting in an increase in soil solution by 271% for cations and 162% for alkalinity and an increase of DIC flux to groundwaters by 33% over two years. As for many biogeochemical experiments the soil system may not yet have reached its steady state in this experiment. The DOC export under climate change with a doubled CO_2 regime has also been modelled by Clair et al. (1999) for Canada. They estimate an increase of DOC fluxes by 14% mostly owing to runoff increase particularly during the summer period in northern Canada. The three Gorges Dam on the Chiang Jiang is also attracting much attention, including the processing and retention of organic matter and nutrients (Chen 2002).

Aumont et al. (2001) have coupled for the first time the regionalised organic carbon budget made by Ludwig and Probst (1996), and Amiotte-Suchet and Probst (1995) (see Fig. D.55) with ocean carbon models. This is a promising direction which could eventually lead to a fully regionalised coupling between coastal basins and ocean biogeochemical provinces (see Ducklow's map in Schellnhuber 2001).

D.4.8.3 The Future State of Inorganic Nitrogen Loads in Rivers

The new global nitrogen models take into account the regionalisation, future scenarios, as well as developments in nutrient processing along stream orders. Seitzinger et al. (2002) and Kroeze et al. (2001) are exploring scenarios of river nitrogen inputs to the oceans such as: (*i*) Business as Usual (BAU) current nitrogen use trends continue between 1990 and 2050, i.e. the human population, fertiliser use, and atmospheric deposition of N oxides increase by 60, 145 and 70%, respectively; (*ii*) "low N diet" scenario only assumed for industrialised regions with a moderate shift in feeding habits of humans from animal to plant protein with a related de-

crease of the overall N inputs to soils in such industrialised regions to 40 kg N cap^{-1} yr^{-1} by 2050; (*iii*) low NO_y deposition in Europe, resulting from NO_x emission control. The first scenario results in more than a doubling of DIN export to oceans from most regions (Fig. D.63a), with the North Pacific and Indian Ocean receiving most of the related DIN, well ahead of the North Atlantic. The second scenario would significantly slow the DIN inputs to European seas. The third scenario would have only a moderate impact (–9% relative to BAU scenario) on riverine export of DIN because most N in European rivers originates from agriculture, not from fuel combustion.

The recent analysis of DIN and particulate nitrogen (PN) export to coastal seas made by Seitzinger et al. (2002) is spatially explicit and allows for the apportionment of N sources by world regions (Fig. D.63b). The BAU scenario predicts an increase of DIN export to the Ocean from 21 Tg N yr^{-1} to 47 Tg N yr^{-1} between 1990 and 2050. The largest increase is predicted from East Asia and from South Asia due to the population growth, agricultural and industrial development of the Chinese and

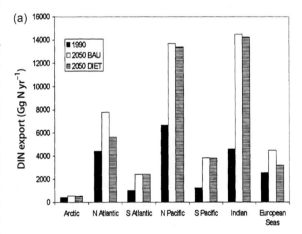

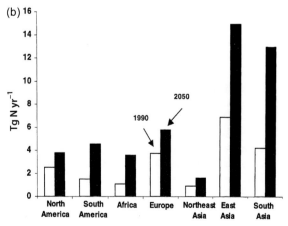

Fig. D.63. a DIN export rates by rivers to the world's oceans in 1990 and for two scenarios in 2050 (Kroeze et al. 2001); **b** DIN Export by rivers for world regions 1990 and 2050 Business as Usual Scenario (Seitzinger et al. 2002)

Indian subcontinents. This very detailed synthesis also takes into account a hypothetical removal of existing reservoirs which would increase the PN inputs to oceans from 23 to 30 Tg N yr^{-1}, thus pointing to their major influence as man-made nutrients sinks at the global scale.

In an Earth system perspective, the opening of the nitrogen cycle also results in a major increase in denitrification rates at the global scale. With a doubling DIN input scenario for 2050, Kroeze and Seitzinger (1998) also predicted that aqueous emissions of N_2O into the atmosphere will increase faster than DIN export rates: estuarine and river emission would increase by a factor of 3 and 4, respectively, while emission from continental shelves would only increase by 12.5%.

D.4.9 Future Research

In this review we discover that the issue of fluvial systems and its role in transporting water and constituents has recently moved toward a fully global-scale perspective. As a result, the transformation of river systems by climate variability and greeenhouse warming, land-cover change and water engineering is attaining legitimacy within the agenda on global change. We have provided what we believe to be mounting evidence of the predominance of human influences on the terrestrial water cycle.

Among these, the clearest imprint of human activities is the direct distortion of water and allied constituent fluxes, arising from land-cover change, agricultural management, urbanisation and hydraulic engineering. While this has been articulated amply at more local scales in earlier parts of this chapter (and will become apparent again among the case studies that follow), we are beginning to see quantitative evidence that these phenomena are now global in extent. The overall summary of Part D (Chapt. D.8) will take this argument further and place these changes into the context of a new geologic era – the Anthropocene (Meybeck 2002).

Our findings on the importance of direct human impacts on the terrestrial water cycle suggest the need to enlarge the scope of the global change dialogue, which historically has been motivated primarily by the issue of global climate change. For example, three quarters of the US Global Change Research budget is devoted to climate and carbon studies (Subcommitee on Global Change Research 2000). The climate issue also has motivated major assessment activities such as the IPCC and US National Assessments, which have themselves recognised some of the direct agents of change highlighted in this chapter. With mounting concerns about the state of freshwater resources it will not be difficult to justify consideration of the factors beyond climate which shape the global freshwater system (e.g. Global Water System Project, World Water Assessment Programme, Millennium Assessment).

This is not to say that climate change is unimportant. Its impact will manifest itself in combination with the more direct anthropogenic effects and do so in regionally unique ways (Fig. D.62). However, it is precisely at these scales that the current generation of GCMs do most poorly. And, a more complete picture of ongoing changes to the global water cycle must include a constellation of both natural variations and human impacts, both direct and climate-induced.

This realisation is being fueled by technical developments, which permit us to go beyond simple latitudinal, climate zone and continental averages, toward a fully geo-referenced and increasingly quantitative picture of the world's fluvial systems and their progressive change. We anticipate that the community will soon go beyond simple statistical approaches to the problem, and build time-varying flux models that can begin to address the tracking of source areas, sink areas and transport routes along the full continuum of inland water systems.

For instance, the well-known acceleration of erosion locally due to poor land management in many parts of the world apparently fails to be seen at river mouths (Walling and Fang 2003). Only a small fraction of anthropogenic N loads are transported to the coasts (Howarth et al. 1996). Extending this concept over still broader domains, we see that human activities in river basins can collectively impart a significant signature on continental runoff and also suspended sediment flux (Vörösmarty and Sahagian 2000). Fully global-scale assessment of water scarcity (Vörösmarty et al. 2000c) tabulating supply and demand along simulated river reaches ($n > 60\,000$) tripled the number of people previously assessed as experiencing severe water stress and competition for water using country-level statistics ($n \sim 200$).

Much of our knowledge about hydrology and the flux of materials out of drainage basins has been based on an extensive literature derived from small-scale catchment studies. Previous chapters of this book should demonstrate amply the progress of the science and management imperatives derived from such studies. There are characteristic behaviours of river systems which can only be articulated at the fine scale but, collectively, elicit unique drainage basin responses over intermediate and macro domains. Such dynamics could not otherwise be observed by focusing solely on small-scale catchment studies. A good example is the recent work using a stream-order approach in river basin biogeochemical models as in the Seine basin (Billen and Garnier 1999) where both diffuse N sources, point sources and major biogeochemical processes, as plankton uptake and, bacterial degradation in water and sediments, have been quantified and differentiated by stream order. In the Mississippi basin, Alexander et al. (2000) also showed progressive and unique contributions of different stream orders to the processing and sequestration of inorganic

Table D.24. Classes of catchment sensitivity to anthropogenic change and their representation within the three case studies of Chapt. D.5, D.6, and D.7

	Basin heterogeneity	Land-use change	Direct human impacts	Water demand	Climate variability (ENSO)	Basin area (10⁶ km²)	Runoff (mm yr⁻¹)	Population density P (km⁻²)
Amazon (Brazil)	X	X	O?	O	X	6.3	2460	2
Mgeni (South Africa)	XX	X	XX	XXX	XX	0.00408	162	1400
Elbe (Czech Republic, Germany	XX	X	XXX	X	X	0.148	170	169

nitrogen. Other types of nutrients modelling in medium sized basin, coupling TSS, nitrogen and phosphorus, are rapidly developing (see Chapt. D.3. A fully developed example is presented in the Elbe case study, Chapt. D.6).

The hydrological and biogeochemical processes should now be coupled in many basins with socio-economic interactions, and if possible an integrated model should be realised (see Chapt. D.6). We have chosen here to illustrate this approach with three case studies for the Amazon (Chapt. D.5), the Elbe (Chapt. D.6), and the Mgeni (Chapt. D.7) (Table D.24). The Amazon is typical of a large, still mostly-pristine, tropical basin in which river and flood plain interactions are dominant. The Mgeni catchment in South Africa is much smaller and characterised by dry climate with limited runoff and rapid development (population, agriculture). The Elbe catchment (Czech Republic, Germany) illustrates the multiple pressures on a medium-sized temperate catchment with a long history of human and river interactions. In these case studies we have addressed the global change issues raised at various scales in the previous chapters.

From our review we conclude that global water cycle analysis executed at the macro-scale *per se* is a valid approach in its own right, and that it is unlikely that much progress toward global synthesis will be made relying solely on more traditional case studies and place-based studies. At the same time, case studies are instructive of the dynamics that operate on individual basins that are part of the global system of rivers. For this reason we will present in the following section a series of detailed, well-documented case studies. Thus, we feel that it is imperative to continue focusing on sequentially larger domains in order to develop the requisite databases and modelling tools to perform fully global-scale assessments. In addressing the concept of "pristine-to-polluted" for rivers worldwide, what could only be articulated through vignettes a decade ago (Meybeck and Helmer 1989), can now be addressed over the global domain itself using rapidly emerging techniques and datasets. We maintain that in so approaching the problem the community will be fully able to identify dynamics and trends not otherwise apparent from more traditional, smaller scale catchment studies. This will constitute a major new avenue for future research.

Acknowledgment

This chapter would not have been possible without the countless, stimulating discussions we have had over the years with our colleagues in BAHC and affiliated Core Projects of the IGBP. We also wish to acknowledge our many collaborators and students on our funded research projects for important ideas and constructive critique of our work. In this context, we especially recognise members of the Water Systems Analysis Group (UNH) and the Laboratoire de Géologie Appliquée (Université Paris 6). We thank Ellen M. Douglas for valuable insights into recent climate-hydrology interactions. Grants from NASA, NSF, and NOAA helped to support this work.

Chapter D.5

Case Study 1:
Integrated Analysis of a Humid Tropical Region – The Amazon Basin

Jeffrey E. Richey · Reynaldo Luiz Victoria · Emilio Mayorga · Luis Antonio Martinelli · Robert H. Meade

D.5.1 Towards an Integrated Analysis of the Amazon Basin

The Amazon makes for an excellent test case for developing a model of how hydrological and biogeochemical cycles function at the land surface on regional to continental scales. With an area of 6 Mkm^2 containing the largest stand of tropical rainforest in the world, it contributes 15% of the global freshwater discharged to the oceans, and condensational energy release from convective precipitation is of sufficient magnitude to influence global climatic patterns. Its major tributaries represent different climate, soil, and topographic regimes. It is large scale and represents a series of hydrological and chemical regimes that are not atypical of world rivers (Stallard and Edmond 1983). As much of the Amazon is mostly undisturbed by anthropogenic activity, it provides one of the few opportunities left to develop models of how basins function, against which future change can be assessed. It is also relatively well-characterised, even over large scales. Given the heterogeneous nature of precipitation and precipitation instrumentation, river discharge is perhaps the most robust integrator of the long-term hydrological properties of a drainage basin.

In this chapter we use a heuristic model construct to develop a synthesis of the coupling between the hydrological and biogeochemical cycles of the Amazon River system. Our emphasis is on the large-mesoscale to continental scale features of the river system, which would be compatible in scale with the land-atmosphere interfaces represented by general circulation models. This synthesis is based in large part on the work of the CAMREX (Carbon in the AMazon River EXperiment) project (Richey and Victoria 1996; McClain et al. 2001).

D.5.2 Coupling Hydrology, Organic Matter and Nutrient Dynamics in Large River Basins

The central premise of a large river basin model is that the constituents of river water provide a continuous, integrated record of upstream processes whose balances vary systematically depending upon changing interactions between flowing water and the landscape, and the interplay of biological and physical processes (Richey and Victoria 1996). That is, the chemical signatures of riverine materials can be used to identify different drainage basin source regions, reaches or stages, and can be tied to landscape-related processes such as chemical weathering and nutrient retention by local vegetation. The compositions of the particulate and dissolved materials carried by the mainstem result from initially similar rainwaters that have been uniquely imprinted by contact with almost every plant, animal and mineral in the basin.

A strategy for capturing these dynamics is to build multiple time- and space-scale integrated models of changes in the water flow and biogeochemistry of river basins as a function of changes in land use and land cover and regional climatology over the basin.

There are three components to this strategy. The first is to build a spatial model of the physiography of a region. The next step is to model the flow of water across the landscape and down river channels. A tributary basin analysis is predicated on determining the interannual patterns in precipitation and runoff; runoff pathways and soil residence time can lead to differences in how chemical species are mobilised and particulates eroded. The hydrology of a regional-scale river system can be modelled as a geospatially-explicit water mass balance within the basin contributing to stream flow and downstream routing. As such, a model can be divided into two major components. A "vertical" component, which calculates the water balance at each individual grid cell, and a "horizontal" component, which routes the runoff generated by each grid cell to the ocean. The separation into vertical and horizontal components also allows for an easy interface to treat non-point source and in-channel chemical processes separately.

The final step is to couple the spatial and hydrological models to models of the origin and transport of selected chemicals, or the biogeochemical models. Models developed for terrestrial systems typically pay little attention to exports to rivers, and the common paradigm used for soil organic matter compartments is based on turnover rates. Mayorga et al. (2000) proposed an integrated modelling framework aimed at quantitatively describing the dynamics of organic matter cycling in mesoscale to large rivers by mechanistically tracking the evolution of organic matter from the land, through the river corridor, to the river (Fig. D.64). This model emphasizes the importance of physical and geochemical properties in controlling the preservation and fractionation of organic matter. The biogeochemical model is tightly coupled to spatially distributed models of water and sediment cycling and transport. The biological components of the border areas and the surface water act as filter systems to retain the organic carbon and nutrients for use in primary production and respiration. For instance, nitrogen leached as nitrate from terrestrial settings is often consumed in the process of denitrification when it enters the organic-carbon-rich environment of riparian zones. The organic matter from these producers is fuel for consumers in the water, sediments, and

soil. This matter may be consumed near its place of production or transported downslope or downstream before being consumed. Because oxygen transport is reduced in water, oxygen availability can become limited in many areas in these settings, leading to the development of anaerobic conditions. Usually there are low concentrations of electron acceptors such as nitrate and sulfate, so fermentation and methanogenesis are the main anaerobic means of organic-matter degradation. The distribution of aerobic and anaerobic conditions in these areas determines the relative production of CO_2 and CH_4 (and CH_4 oxidation). These degradation reactions also involve the mineralisation of organic N, with the potential for production of N trace gases. Contrasts between non-flooding and flooding conditions can create conditions for the sequential mineralisation of organic N, aerobic nitrification, and anaerobic denitri-

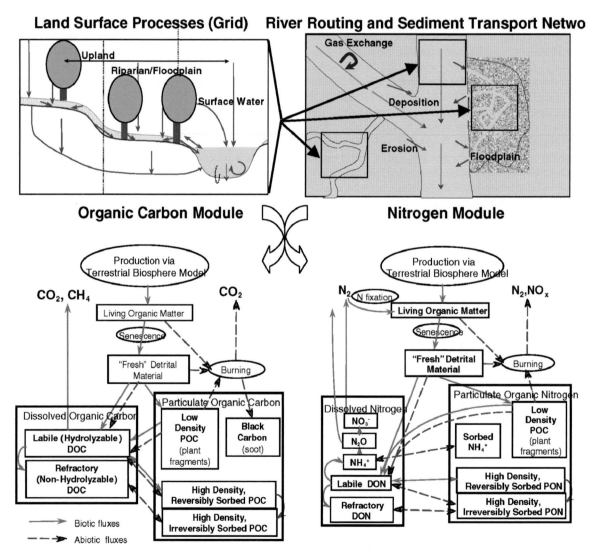

Fig. D.64. Mayorga et al. (2000) proposed an integrated modelling framework for describing the dynamics of organic matter (OM) cycling in mesoscale to large rivers, including *DOC* (dissolved organic carbon), *POC* (particulate organic carbon), *DON* (dissolved organic nitrogen), and *PON* (particulate organic nitrogen) (after Mayorga et al. 2000)

fication. These steps lead to the production of N_2 and N trace gases, removing N from cycling within an ecosystem.

Rivers and the water they carry thus provide a continuously flowing signal, recorded by isotopes, ions and molecules, of the cumulative effects of drainage basin processes such as weathering, oxidation/reduction, gas exchange, photosynthesis, biodegradation and partitioning. This recording is complementary to more classical methods of remote sensing-based on electromagnetic radiation, but is composited over a wider range of time and space scales and includes effects of subcanopy and subsurface processes. Hence an understanding of how these substances are routed from precipitation through their drainage systems to the oceans yields important information on the processes controlling regional-scale hydrological and biogeochemical cycles.

We will now use this framework to summarise the coupled hydrological and biogeochemical dynamics of the Amazon, in a manner that can lead to extrapolation to other systems.

D.5.3 The Amazon Basin: Vargem Grande to Óbidos

The Amazon drainage basin is characterised by areas of extremely high relief (Andes and sub-Andean trough) which transition into regions of low relief (shield areas and alluvial plain). The central Basin and the corridor to the Atlantic are composed of sediments from these formations as well as marine deposits. Over 90% of the sedi-

ment which eventually reaches the ocean is contributed by the Andes (Meade 1994). The geological stability of most of the Basin and the high amounts of rainfall have led to the development of a low-gradient topography and highly weathered soils, usually oxisols (mineral soils with an oxic horizon) and ultisols (mineral soils with an argillic horizon), in the region east of the Andes. Early on, the linkage between the geology, biology and hydrology of the Amazon River mainstem was recognised from visually observable features of the waters and was formalised into an Amazon river "typology" by Sioli (1950). This classification consists of visually distinct white-water, clear-water, and black-water rivers.

For the purposes of this paper, we divide the overall Basin into subbasins represented by the primary tributaries and the mainstem floodplain (Fig. D.65). The primary tributary basins range from about 50 000 km^2 to more than 1 Mkm^2, or from large mesoscale to continental scale in area (Table D.25). The Rio Madeira and the Amazon mainstem at Vargem Grande have predominantly Andean origin, with the greatest mean elevation, relief, and slope. The Rio Madeira originates in the Bolivian Andes and then passes across the Planalto Brasileiro and the central plain. The northern tributaries the Rios Içá and Japurá originate in the Andes and cross the sub-Andean trough and central plain to reach the main stem. Of the southern tributaries, the Juruá, and Purús drain the sub-Andean trough and the central plain. All of these rivers are considered white-water rivers, with high sediment concentrations. They drain relatively unweathered soils and geological formations and

Fig. D.65.
The Amazon drainage basin, with primary tributaries indicated. Vargem Grande (*VG*) and Óbidos (*OBI*) are the upstream and downstream ends of the river reach sampled, and Itapeuá (*ITA*) and Manacapurú (*MAN*) are mid-reach stations. The river network is derived from the Digital Chart of the World (DCW) river channel dataset. The attributes of the primary basins are summarised in Table D.25

Table D.25. Attributes of the primary tributary basins of the Amazon River, including the mainstem, tributaries draining from the north, and from the south (Fig. D.65). Vargem Grande (mainstem Solimões just above the confluence with the Rio Içá) represents primarily drainage from the Andes of Peru and Ecuador. Óbidos is the most downstream mainstem station that is routinely measured. To derive consistent river networks and drainage basin boundaries, a new algorithm was developed that creates flow-direction grids from a manually corrected vector cartographic dataset, the Digital Chart of the World; this topography-independend method was created in order to sidestep data problems present in the best available elevation dataset, GTOPO30. The new derived dataset was then used to calculate drainage basin areas, and combined with GTOPO30 to calculate mean basin elevation, relief, percent of basin > 500 m in elevation, and mean slope. Soil texture (mean soil organic C, clay content, and sand content) was calculated from Brazilian RADAM soil maps for the Brazilian Amazon and from the FAO $1° \times 1°$ gridded dataset for the non-Brazilian Amazon. These texture classes are used in hydrology models to estimate field capacity, wilting point, and available water capacity (field capacity minus wilting point) using the texture-based methods of Saxton et al. (1986). Rooting depth (which determines to what depth vegetation can access water) can be specified as a function of soil texture and vegetation type (*sensu* Nepstad et al. 1997)

Basin	Area (10^3 km^2)	Elevation (m)	Relief (m)	>500 m (%)	Slope (%)	Soil org. C (kg C m^{-2})	Clay (%)	Sand (%)	Forested (%)	Precipitation (mm yr^{-1})
Mainstem										
Vargem Grande	1 010	1 040	5 900	39	3.9	11.1	41	39	70	??
Óbidos	4 680	450	5 960	15	1.6	10.0	39	37	79	2 140
Northern										
Içá	120	210	3 470	7	0.4	10.6	45	30	89	2 640
Japurá	260	250	4 180	9	0.9	10.0	39	35	85	2 880
Negro	710	180	2 960	7	0.8	10.2	35	43	81	2 470
Southern										
Jutaí	50	90	150	0	0.1	9.7	35	32	94	2 890
Juruá	220	180	480	1	0.2	9.0	38	27	96	2 350
Purús	360	140	500	>1	0.2	9.3	37	29	93	2 210
Madeira	1 380	500	5 860	17	1.8	10.0	39	40	70	1 870

are characterised as "weathering-limited denudation" by Stallard and Edmond (1983), i.e. the transport processes are more rapid than the weathering processes that are creating materials for transport. Because the transported materials are still rich in more soluble components, the dissolved inorganic load of these waters is also high.

The Rio Negro, the classic black water river, lacks Andean headwaters and drains older, more weathered soils on lower relief surfaces of the Caatinga forest on the Planalto das Guianas, and its major tributary, the Rio Branco, drains a drier savanna region. The Rio Jutaí also has characteristics of black water. The loads that they carry reflect the capacity of transport processes to remove them, so they are characterised as "transport-limited denudation" (Stallard and Edmond 1983). These waters have very low suspended sediment concentrations and are usually tea-coloured in appearance due to a high load of organic acids. The clear water rivers are intermediate between the black and white, including the Trombetas, Tapajós, and Xingú.

The Amazon floodplain, or várzea, is an integral part of the river system, where floodwaters and local runoff regularly inundate the floodplain via an extensive network of drainage channels. Thousands of permanent lakes range in size from less than a hectare to more than 600 km^2, and are typically 6–8 m deep at high water. As the river falls, land is re-exposed and the lakes become isolated from the main channel, with depths decreasing

to 1–2 m. Numerous smaller tributaries drain exclusively lowland regions into the main channel or into the floodplain, while large "paranas" act as diversion canals between the main channel, floodplain, and tributaries, with the flow direction often depending on river stage. The upstream reach (Vargem Grande to Itapeuá) is characterised by rapid migration of mainstem and floodplain channels to produce an intricate scroll-bar topography with long, narrow lakes. The middle reach, between Itapeuá and Manacapurú, has a narrow floodplain with fewer lakes than upriver, and is controlled by structural features that constrain the river and allow almost no morphological change. The downstream reach, Manacapurú-Óbidos, has an incomplete levee system, which provides free access for overbank flows to a wide floodplain with a patchwork of wide, shallow lakes.

While large rivers define the fundamental character of the Amazon landscape, they owe their flow and chemical loads to a much denser network of small rivers and streams, at small to mesoscales. In the relatively low-gradient terrain of the Brazilian Amazon, streams follow slow and meandering courses through flat-bottomed valleys. The streams are often bounded by riparian forests whose elevation is similar to that of the streams and lower than that of the surrounding forests or grasslands by as much as several metres. In an area near Manaus, for example, stream density is approximately 2 km km^{-2} (Junk and Furch 1993); and the combined area of the streams and

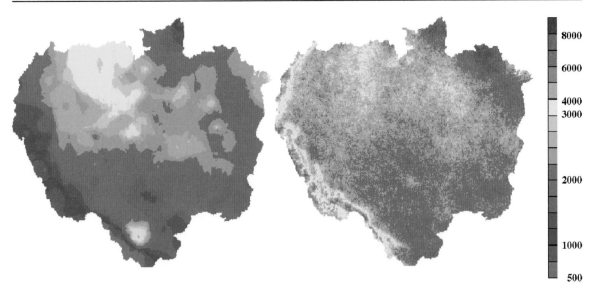

Fig. D.66. Annual rainfall fields (mm yr^{-1}), derived from the gauging network of the Agência Nacional de Aguas e Energia Elétrica (*on the left*) and from an AVHRR satellite index (*on the right*)

riparian zones in the Amazon Basin has been estimated at close to 1 Mkm^2 (Junk and Furch 1993). The hydrology of the small streams is strongly influenced by the surrounding land surface as well as by the upstream reaches. Most of the flow is fed by groundwater discharges, while stormflow occurs as saturation overland flow originating in near-saturation riparian soils. The chemistry of major ions of the streams falls within the general categories of black and clear waters, varying according to the geological characteristics and soil properties of their catchments.

D.5.4 Hydrology of the Amazon River System: A Mainstem Perspective

D 5.4.1 Patterns of Rainfall

As most recently summarised by Marengo and Nobre (2001) and Nobre (compare Chapt. A.6), the Amazon region is characterised by a strong annual cycle of rainfall, with varying patterns across the Basin (Fig. D.66). Annual variation in rainfall is linked to annual changes in the large-scale upper-air circulation. Although the sun largely controls the timing of the annual cycle of rainfall, rainfall in different parts of the Basin is triggered by different rain-producing mechanisms. Diurnal deep convection resulting from surface warming is most prevalent in central Amazônia, while deep convection in northern Amazônia is related to the Intertropical Convergence Zone and the moisture transports from the Atlantic. Instability lines originating near the mouth of the Amazon River can initiate rainfall in northern Amazônia, while convective activity at meso-scale and large-scale, associated with the penetration of frontal systems in south/south-east Brazil, can reach western southern Amazônia. Southern Amazônia has distinct dry and rainy seasons, with a maximum of rainfall in January-to-March. In the northern and central regions there is almost no dry season, but approaching the Equator there is a distinct maximum in spring and a "suppressed" maximum in the autumn. The north-west region (basins of the Rios Negro, Içá, and Japurá) has high rainfall, with more than 3 600 mm yr^{-1}, and peaks in April-to-June. The central part of Amazônia around 5° S averages about 2 400 mm yr^{-1}, while the region near the mouth receives more than 2 800 mm yr^{-1}. The extremely high and localised values of precipitation occur in narrow strips along the eastern side of the Andean slopes, reaching upwards of 7 000 mm yr^{-1}.

While these broad patterns of precipitation are accurate, a problem for modelling exists with the reconciling of predictions with sufficiently accurate measurements. Precipitation gauges in the north-west and especially along the eastern slopes of the Andes of Peru and Ecuador are so sparse as to render estimates of seasonality relative to observed river hydrographs problematic. Current gauge-based estimates of precipitation in the Solimões basin at Vargem Grande would require unrealistically-low evapotranspiration (cf. Salati et al. 1979) to match observed discharge at that station (Richey, unpublished data).

D.5.4.2 Mainstem and Tributary Hydrographs

The most striking features of Amazon River discharge are its magnitude and its highly damped hydrograph (Fig. D.67). The Amazon mainstem discharges between Vargem Grande and Óbidos are determined by inputs

Fig. D.67.
Amazon discharge at Óbidos
($m^3 s^{-1} \times 1000$). Óbidos, as the
last gauged station on the
Amazon mainstem, represents
the runoff of 4.6 Mkm² of
drainage basin, and is considered the defining point for
input to the ocean (with the
addition of the Rios Xingú
and Tapajós)

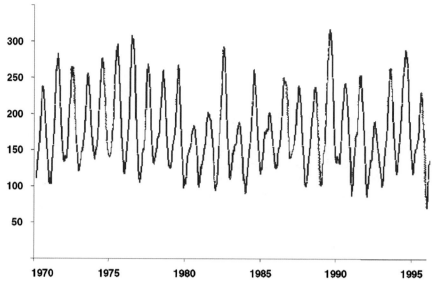

from upstream, from the primary tributaries and local rivers, and from exchange with the várzea. Though differences in stage of 7–10 m are common along the main stem, there is only a twofold to threefold difference between low and high discharges. The damped hydrograph of the main stem reflects in part the offset input from tributaries and along the channel. The peak flows from the northern and southern tributaries are typically three months out of phase as a result of the seasonal differences in precipitation (Fig. D.68).

As summarised from Richey et al. (1989a, 1990) and more recently by Carvalho and da Cunha (1998), the Amazon at Vargem Grande is primarily Andean water, with average minimum and maximum discharges of 20 000 $m^3 s^{-1}$ and 60 000 $m^3 s^{-1}$. The average annual discharge of the tributaries includes the Içá (9 000 $m^3 s^{-1}$), Japu-

rá (19 000 $m^3 s^{-1}$), Jutaí (3 000 $m^3 s^{-1}$), Juruá (8 000 $m^3 s^{-1}$), Purús (12 000 $m^3 s^{-1}$), Negro (28 000 $m^3 s^{-1}$), and Madeira (31 000 $m^3 s^{-1}$). Thus the main channel of the Amazon River at Óbidos receives input from major and minor tributaries draining an area of 4.2 × 10⁶ km², with low and high flows average 100 000 $m^3 s^{-1}$ and 220 000 $m^3 s^{-1}$, respectively. The total Amazon input to the Atlantic (Óbidos plus the Rios Tapajós and Xingú, represents a mean annual input of about 210 000 $m^3 s^{-1}$.

In-channel storage and especially floodplain storage and exchange influence the mainstem hydrograph and was shown to be essential to any modelling study of the river (Richey et al. 1989a; Vörösmarty et al. 1989). The increase in storage on the floodplain over the 8-month period of rising water is greatest during the mid-rising water phase in the upriver reach of the main stem, ap-

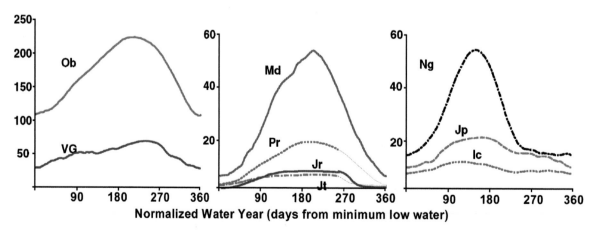

Fig. D.68. Composite discharge hydrographs ($m^3 s^{-1} \times 1000$) of the primary Amazon tributaries. Each hydrograph is for the water year of the respective tributary, which starts on the date indicated in parentheses (as calculated from mean discharge, 1972–1990). The Içá (*Ic*, Jan 31), Japurá (*Jp*, Jan 31), and Negro (*Ng*, Jan 31) are predominantly in the Northern Hemisphere, while the Jutaí (*Jt*, Oct 9), Juruá (*Jr*, Sep 18), Purús (*Pr*, Oct 19), and Madeira (*Md*, Sep 25) are in the Southern Hemisphere. Mainstem stations include Vargem Grande (*VG*, Sep 10; essentially representing Andean drainage) and Óbidos (*Ob*, Oct 19). The differences in timing represent the seasonality of the respective basins, established in large part by fluctuations in the Intertropical Convergence Zone

proaching 15 000 m^3 s^{-1}, with maximum rates in the midriver and downriver reaches of 3 000 m^3 s^{-1} and 9 000 m^3 s^{-1}, respectively. The decrease in storage during the 4-month falling water period is greatest during mid-falling water, ranging from –18 000 m^3 s^{-1} upriver to –8 000 m^3 s^{-1} at midriver. Water exchange between the várzea and main channel ranges from 3 000 m^3 s^{-1} during the dry season to 7 000 m^3 s^{-1} during the wet season in the up- and downriver sections; midriver flows are about half of these values. Overall, exchange is greatest during early-falling to mid-falling water in the upriver and downriver reaches, with a net flow from the floodplain to the main stem of about 20 000 m^3 s^{-1}. Net exchanges are generally lower in the midriver reach, where the area of the floodplain is relatively small. Therefore, water derived from local drainage constitutes a significant component of the water budget of the main stem. These flows correspond to about 30% of the flow at Itapeuá, and cumulatively to about 25% of the flow at Óbidos.

The predominant interannual variability in Amazon discharge occurs on the two to three year time scale, and oscillations in the hydrograph are coupled to the El Niño-Southern Oscillation cycle (ENSO), with a lag of about five months (Richey et al. 1989b). The oscillations of river discharge predate significant human influences in the Amazon Basin, and reflect both extra-basin and local factors. Climate records and general circulation model calculations suggest that interannual variations in the precipitation regime and hence discharge of the Amazon are linked to changes in the general circulation of the atmosphere over the tropical Pacific Ocean associated with ENSO. Qualitatively, the months of maximum pressure anomalies (Southern Oscillation negative phase) correspond to ENSO warm events and precede the negative flow anomalies in most cases. Major ENSO events, e.g. 1925–1926 and 1982–1983, are reflected in pronounced low discharges. The converse effect, high discharge associated with the Southern Oscillation positive phase, is also apparent. Only a portion of the variance in the discharge regime is linked to the ENSO phenomenon. Relations between runoff and precipitation are not straightforward, due in part to the carry-over storage ("basin-memory effects") typical of large catchments. Local climate influences, such as the boundary layer convergence mechanisms and the steady progression of individual fronts and air mass boundaries characteristic of the region, contribute to discharge variability.

In summary, the damped hydrograph of the mainstem reflects the large drainage basin area, the three-month phase lag in peak flows between the north- and south-draining tributaries related to seasonal differences in precipitation, and the large volume of water stored on or passing through the floodplain. Patterns of interannual variability indicate that considerable caution must be exercised in determining anthropogenic im-

pacts, particularly with the use of short-term records over large areas. Conversely, it would be difficult to uniquely identify a deforestation effect in the highly damped discharge regime of the Amazon River mainstem (with records dating back to 1903). The likelihood of linkages between the Amazon Basin and large-scale atmospheric circulation reinforces the importance of determining the factors controlling the hydrology of the Basin in the face of extensive land-use change. A rigorous analysis of the regional-scale hydrology of the Basin using field measurements and remote sensing integrated through realistic, physically-based modelling must be considered as the long-range goal for assessing and managing sustainable development in the Basin.

D.5.4.3 Models of Amazon Water Movement

The next major challenge is to relate the mainstem and tributary flow regimes to the antecedent conditions in the respective tributaries via spatially-explicit hydrological modelling. For example, a dynamic water balance model developed by T. Dunne (University of California Santa Barbara, pers. comn.; see also Vörösmarty et al. 1996a; Costa and Foley 1997) has been used to translate precipitation into runoff. As do many models of this type, it uses a tipping-bucket conceptualisation of moisture and moisture fluxes in the root zone. The tipping bucket assumes that all precipitation that enters the soil (assuming a fraction of precipitation runs off as overland flow) remains in the root zone until the moisture content reaches field capacity. At this point excess infiltrate is assumed to drain instantaneously to groundwater storage. Moisture content in the upper zone decreases only through transpiration, with transpiration ceasing when the moisture content drops to the wilting point. Evapotranspiration is calculated using a Priestley-Taylor formulation, where net radiation is the dominant forcing variable (under humid conditions). As such, the model is able to account for stomatal resistance to transpiration, as well as interception, canopy evaporation, and overland flow (which reduces possible additions to soil moisture). The translation of runoff from a cell to a downstream node that could be compared to existing gauging records is calculated using constant travel times and a convolution integral.

This model was applied to the Amazon at a spatial resolution of 0.05° × 0.05° (longitude by latitude). Input parameters were re-sampled to this resolution with a binary integer image format from data layers in both vector and raster data sources in a modelling environment managed by a GIS (Geographic Information System). For the relatively high-resolution models used here, surface meteorological stations by themselves are far too sparse and global datasets are far too coarse. Hence a combina-

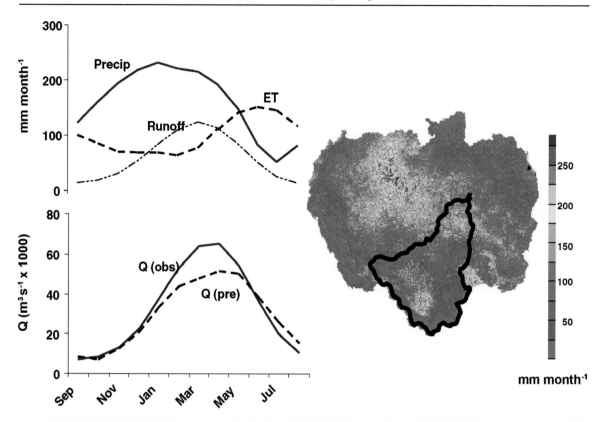

Fig. D.69. Hydrology model results for one time period for the whole basin, with reference to the specific discharge for the Rio Madeira (basin outline *indicated*)

tion of field point data and satellite-derived information. A time series of daily data from the Advanced Very High Resolution Radiometer (AVHRR) Global Area Coverage (GAC, about 5 km) from the NOAA-7, -9, and -11 satellites was used in conjunction with surface data to create fields of precipitation, solar radiation and temperature. The model performs reasonably well in representing the dynamics of these rivers. For example, the model hydrograph captures the phase and generally the amplitude of the Rio Madeira, which varies from a low of about $5\,000\;m^3\,s^{-1}$ to a high of about $70\,000\;m^3\,s^{-1}$ (Fig. D.69). The underestimate is thought to be due to difficulties in accurately estimating the amount of precipitation.

D.5.5   River Chemistry

A significant challenge in understanding the biogeochemical dynamics of the Amazon Basin in a systematic manner surrounds an optimal sampling strategy and means by which to sensibly integrate the results. The Basin is obviously very large, making it difficult to perform routine synoptic sampling. Because the hydrographs of the tributaries and mainstem stations may be out of phase with each other by up to three months, each

sample point occurs at a different point on the hydrograph of the respective river, even if the samples are collected on the same day. Furthermore, a volume of water travels downstream in channels from 1–$3\;m\;s^{-1}$, or 100–$300\;km\;d^{-1}$, so translocation of sampling points can be problematic. To solve these problems, the CAMREX (Carbon in the Amazon River Experiment) programme conducted a series of sampling cruises between Vargem Grande and Óbidos during the period 1982–1991, using the sampling protocol of Richey et al. (1986).

D.5.5.1   A Synoptic View of Chemical Profiles

We hypothesise that a good first-order descriptor of integrated riverine chemistry is the distribution of chemical concentrations with respect to discharge and, more specifically, to the rising and falling limbs of hydrographs. To account for differences in hydrograph timing on the respective tributaries, annual average discharge hydrographs from each station can be calculated, taking the mean discharge for each day from the period of record, starting from minimum low water for each tributary (Fig. D.70). These hydrographs represent the highly damped signal of multiple upstream events, aris-

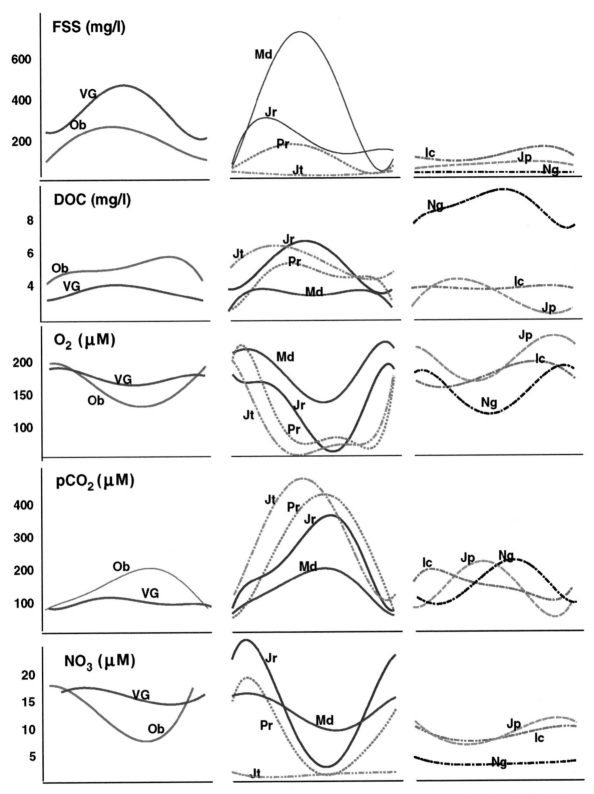

Fig. D.70. Composite chemical hydrographs for the primary tributaries and mainstem stations (as described in Fig. D.68). Chemical species include fine suspended sediments (*FSS*, < 64 m), dissolved organic carbon (*DOC*), dissolved carbon dioxide (*pCO₂*), dissolved oxygen (*O₂*), and nitrate (*NO₃*)

ing from the mixing of separate tributary inputs and along-channel processes. Concentrations of all chemical species were generally more variable over the hydrograph relative to the mainstem, and differed considerably from one tributary to the next.

Over 95% of the particulate material transported by the Amazon River is carried in suspension, with computed bedload transport rates only 1–2% of the total sediment load (Dunne, pers. comn.). Most of the suspended fraction is carried as fine sediments (< 64 µm). Peak fine suspended sediment concentrations occur about half way up the rising limb of the hydrograph. The mainstem at Vargem Grande has a peak of fine suspended sediment concentration of about twice that of Óbidos. The southern tributaries contribute by far the most sediment, led by the Rio Madeira (peak values approaching 700 mg l[-1]), followed by the Rios Purús and Juruá. In all these tributaries, the peak of fine suspended sediment occurs on the rising hydrograph. In contrast, the northern tributaries have lower concentrations, showing slight peaks on the falling limb. As the weight-percent of C and N to sediment is quite constant (Hedges et al. 1994, Richey et al. 1990), it may be inferred that the dynamics of the transport of particulate organic matter transport is governed primarily by the movement of the bulk mineral sediment.

Approximately 60–90% of the total dissolved inorganic carbon exists as bicarbonate at the average Amazon pH values of 6.5–7.2, with the balance being dissolved CO_2 gas; carbonate alkalinity is virtually zero (Richey et al. 1990). The mainstem is supersaturated with respect to CO_2 in the atmosphere by 10 to 20 times. Dissolved CO_2 increases almost precisely with the hydrograph, from 100 to 200 µM, with an overall mean of 130 µM. Because of the supersaturation, the total annual evasion of CO_2 from the river to the atmosphere is equivalent to about 50% of total dissolved inorganic carbon export, and is derived primarily from *in situ* respiration. Mainstem dissolved O_2 has the exact inverse distribution from pCO_2. It is 50% to 80% undersaturated relative to the atmosphere, and has its minimum at peak high water. Nitrate and dissolved organic N (not shown) together account for 50% of the total N in the mainstem, while NH_4 is present only at detection-limit concentrations. The overall mean NO_3 is 12 µM, with a seasonal distribution mirroring O_2. The annual average flux of total N out of the reach at Óbidos is 3 Tg yr[-1], of which 44% is fine particulate N, 27% is NO_3, 26% is dissolved organic nitrogen, and 3% is coarse particulate N.

Dissolved organic carbon at Óbidos is enriched by 1–3 mg l[-1] relative to Vargem Grande. The highest concentrations are found in the Negro, followed by the Jutaí, Juruá and Purús (Richey et al. 1990). The rise and fall of dissolved organic carbon generally tracks the hydrograph. The annual average export of dissolved organic carbon at Óbidos of 22 Tg yr[-1] is derived primarily from the Negro (31%) and Vargem Grande (20%). Várzea

drainage contributes about 17%. Dissolved CO_2 at Óbidos at high water is about twice that of Vargem Grande, while O_2 is lower. The southern tributaries have very elevated levels of pCO_2 and commensurately reduced levels of O_2 relative to the mainstem, especially the Jutaí, Juruá and Purús, all concurrent with the phase of the hydrograph. Values in the northern tributaries and the Madeira tend to be more consistent with the mainstem. NO_3 is reduced at Óbidos relative to Vargem Grande, and show dramatic depletions in the Juruá and Purús and to a lesser extent the Madeira over the course of the hydrograph. The Negro and the Jutaí both have little NO_3. Overall, inputs of NO_3 to the mainstream are more or less evenly divided between Vargem Grande (45%) and the sum of the tributaries (49%).

On the basis of these profiles, we can reconsider the basic groupings of the tributaries relative to the original typologies of water colour. There is a clear gradation in the relative "whiteness" of tributaries, from the high sediment Madeira to the mainstem Solimões at Vargem Grande, then to the Juruá, Purús, Içá and Madeira. The Jutaí often appears as "black" as the Rio Negro, has significant levels of dissolved organic carbon (though not as high as the Negro) and, like the Negro, has low NO_3. However it has higher levels of pCO_2 and lower O_2 than does the Negro. Perhaps the most intriguing feature in the tributaries is how dramatically and how in synchrony pCO_2 increases and O_2 and NO_3 decrease with the respective hydrographs. What has not been resolved at this time is whether this phenomenon is due to differences in inputs from land, in-channel processes, or local floodplain dynamics.

D.5.5.2 In-river Dynamics

The key to understanding biogeochemical transformations that occur in a large river system and can be summarised by the distributions such as those represented in Fig. D.70 is to understand the processes controlling the source, transport dynamics, and fate of organic matter and nutrients. Extensive work on the transport and mineralogy of sediment within the Amazon River system has shown that most river-borne particles are derived from the Andes but, due to erosion/deposition cycles, they clearly do not pass unaltered through the river system (Johnsson and Meade 1990; Meade et al. 1985; Meade 1994; Mertes et al. 1993, 1996; Martinelli et al. 1993). The hydrological and material transfer properties of a floodplain determine the degree to which biological and chemical processing imparts a signal on materials derived from either upland or main channel sources. At the most basic level, the hydrology of different water types – as they enter the floodplain from rainfall, local tributaries, and the main channel – is primarily a result of the timing of the supply with respect to other sources and the topography and soils of the floodplain.

Results of in-channel process modelling indicate even stronger ties between the mainstem and the floodplain than previously expected. Field measurements indicated that over 80% of the suspended sediment that is introduced to the floodplain from either the main channel or local tributaries is deposited on the floodplain surface (Dunne et al. 1998). On the other hand, substantial amounts of sediment enter the main channel via bank erosion. Estimates of the contribution of floodplain soils and sediment to the main channel flow indicate that approximately 1.5 billion tons of sediment are eroded annually (Dunne et al. 1998). Conservatively, these values indicate that the exchange of bank materials at least equals the discharge of sediment passing Óbidos annually (1.2 billion tons). It is thus unlikely that an individual particle travels through the study reach without intermediate storage. From ^{14}C and mineralogy data, it is possible to estimate that in the mainstem, a typical particle passes through the floodplain several times and that, given the pattern of channel migration, the floodplain in most reaches is recycled over a few thousand years. Measurements of the C, N, and ^{13}C of suspended sediments and floodplain sediments in different environments directly confirm the floodplain imprint (Martinelli et al. 1993; Quay et al. 1992).

Similarly, the distributions of the bioactive elements are strongly influenced by *in situ* and lateral processes, even in very large rivers. A sequence of processes controlling in-stream transformations, including in-stream metabolism, sediment transport, and gas exchange has been summarised (Richey et al. 1990; Richey and Victoria 1993; Devol et al. 1995; Devol and Hedges 2001; Quay et al. 1992, 1995; Bartlett et al. 1990; Benner et al. 1995; Amon and Benner 1996). These measurements indicate that although strongly heterotrophic, the mainstem Amazon is substrate-starved. Within-river respiration appears to be largely at the expense of a limited pool of moderately reactive biochemicals dispersed among highly degraded dissolved and particulate forms. The river is in quasi-steady state with respect to respiration balancing gas exchange.

To evaluate the relative importance of different factors on controlling these distributions, Richey and Victoria (1996) modelled the changes in O_2, pCO_2, NO_3, and dissolved organic carbon in a volume of water as it moves downstream, in terms of the respective source and sink terms. Observed dissolved organic carbon profiles can be predicted only with the floodplain as a source of dissolved organic carbon, and with about 60% of respiration being derived from dissolved organic carbon. Hence there is the clear implication that net dissolved organic carbon is not conservative. Downstream levels of NO_3 could be explained only through NO_3 released by *in situ* mineralisation. The conclusion is that exchanges between the floodplain and mainstem have a significant, if not dominant, impact on the metabolism of the mainstem.

D.5.5.3 Organic Geochemical Signatures

The previous discussion emphasized the distribution of the important and common "bulk" chemical parameters. Much more can be learned about the sources, chemical condition and fate of the material in transport by considering the more detailed chemical composition of these materials (Fig. D.71). The organic geochemical composition of the suspended sediments represents a sequence of processes originating with the fixation of carbon on land through degradation and mobilisation pathways.

Comparisons between the main fractions of organic matter (coarse, fine, and dissolved) and on the role of partitioning of organic materials between water and mineral surfaces within the river system (e.g. Hedges et al. 1986a,b, 1994) is illuminating. Comparison of the main fractions confirms that coarse particulate organic material appears to consist of relatively fresh tree leaves. Although both the fine particulate organic material and dissolved organic material fractions are highly degraded, fine particulate organic material is richer in nitrogen than dissolved organic material and contains 5–10 times more total amino acids. This trend extends to the molecular level, where fine particulate organic material is characterised by higher ratios of nitrogen-rich acidic amino acids versus carboxyl-rich acidic amino acids. This concentration of nitrogen in fine particulate organic material fits a scenario in which dissolved leaf degradation products are selectively partitioned onto negatively charged particles (as nitrogen imparts a local positive charge to organic molecules). In addition, measurements of ^{14}C age indicate that fine particulate organic material is the fraction with the longest maximum residence time and thus moves most slowly through the Basin. The ^{14}C age of bulk dissolved organic material is close to the modern atmospheric signature, which suggests that its diagenesis and transport times are surprisingly short. Comparison of tracers between the chemical fractions highlights the differences between the different size classes of organic matter, and the differences between basins. Although traditional humic extracts on Amazon samples have been done (Ertel et al. 1986), the information contained within the biochemical analyses provides a more coherent picture of the biogeochemical cycling of organic matter in the Amazon.

D.5.5.4 Dynamics of Floodplains

The várzea represents a complex and integral part of the biogeochemical cycles and foodwebs of the Amazon River system (Araújo-Lima et al. 1986). Portions of the white-water rivers remain in fixed channels while others meander, creating new surfaces for the colonisation of vascular plants. Biological activity is greatest on

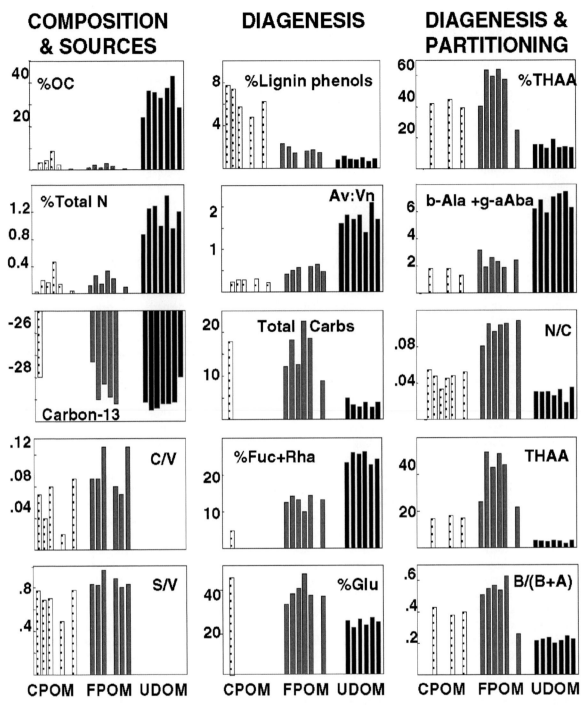

Fig. D.71. Significant trends in organic geochemical parameters (based on data from Hedges et al. 1994). *X*-axis parameters (each bar represents the average of one major tributary to the Amazon mainstem): *CPOM*: coarse particulate material (> 64 mm); *FPOM*: fine particulate material (0.2–64 mm); *UDOM*: ultrafiltered dissolved material (1000 Daltons to 0.2 mm). *Y*-axis parameters and the attributes they represent (*top* to *bottom*, *left* to *right*): %OC and %Total Nitrogen reflect bulk trends in carbon and nitrogen. ^{13}C-isotopes are source indicators that can be used in terrestrial systems to distinguish between organic matter produced at different altitudes, and sometimes between pollutants and natural compounds. Lignin Phenols (next 4 panels), released during CuO oxidation of terrestrial plants, allow discrimination between many trees and grasses. Cinnamyl/Vanillyl (*C/V*), Syringyl/Vanillyl (*S/V*), and Vanillic Acid : Vanillin (*Av : Vn*) plots indicate source changes and fungal degradation. Carbohydrates (next 3 panels) provide information on the diagenetic "age" of organic matter and discriminate between bacterial and plant sources. High percentages %Fucose + Rhamnose (*%Fuc + Rha*) are associated with highly degraded material. Of the amino acids (next 5 panels), total hydrolyzable amino acids (*THAA*) are general source and reactivity indicators, yield and compositional data are indicative of reworking, degradation and reactivity. High values of % b-Alanine + g-Aminobutyric acid (*%b-Ala + g-aAba*) and low values of *B/(B + A)* are indicative of diagenetic activity

the floodplains of white-water rivers, and the activity is tied to the cycle of flooding. Main communities on the floodplains are forests of flood-tolerant trees, non-forested lands that alternate between grasses during low-water and floating aquatic vascular plants during high-water, and open water lakes that expand as the flood-waters rise (Junk and Furch 1993).

Along the mainstem Amazon, flooding follows a three-stage pattern, with water entering the floodplain first through the deepest levee breaks, then through shallower levee breaks, and finally by overtopping the levees (Mertes et al. 1996). As these stages progress, water on the floodplain becomes increasingly similar to that of the main channel and less like the local drainage water. Gradients in water properties from river to local characteristics can often be observed across the floodplain. Trees on the floodplain grow mostly during low water periods and many drop their leaves during inundation. In contrast, phytoplankton in the lakes flourish after river water enters the floodplain and supplies nutrients for growth (Fisher and Parsley 1979). Phytoplankton communities in various floodplain lakes have been found to be limited by N only and by N and P (Zaret et al. 1981; Forsberg et al. 1988). The productivity of these systems can be quite high, as much as 25 t C ha^{-1} yr^{-1} in the herbaceous communities (Junk and Piedade 1993), ~ 16 t ha^{-1} yr^{-1} (Junk 1985) in the floodplain forest, and 3 t ha^{-1} yr^{-1} in the phytoplankton community of a lake (Schmidt 1973).

Much of the organic matter produced on the floodplain is also degraded there. Decomposition during the inundation phases includes a significant anaerobic component. For instance, floodplain lakes are often stratified with an anoxic lower layer, and low oxygen conditions also develop in the waters of the flooded forests and macrophyte beds. Methane fluxes have been measured during inundation in all the main vegetation communities on the Amazon floodplain. The magnitude of methane production within the floodplains has been characterised by numerous researchers, who estimate that it is the source of 5–10% of the total global CH$_4$ flux from wetlands (Devol et al. 1990; Bartlett et al. 1990). The majority of the flux is carried by ebullition, so that the areas of methane oxidation in the waters are bypassed.

D.5.6 Potential Impact of Anthropogenic Change on the River System

The diverse natural ecosystem of Amazônia supports a large resource-based economy. Deforestation for lumber and agriculture has made perhaps the greatest impact on Amazon ecosystems. Extensive road-building during 1960–1980 converted large areas of forest to crop land and cattle pasture. Between 1978 and 1988 the amount of deforested land in Amazônia increased from 78 000 to 230 000 km^2 (Skole and Tucker 1993). Mining and extraction of gold and aluminum have impacted the chemistry of certain Amazon River subcatchments, resulting in elevated concentrations of mercury in the river, sediments, fish and some humans (Martinelli et al. 1988, Roulet et al. 1998). Another developmental activity that has directly impacted the Amazon River system is dam building to produce inexpensive electricity to attract investment (Bunyard 1987). The reservoirs they created also produce methane gas that is eventually emitted to the troposphere.

As discussed by Richey et al. (1997), changes in land use would have indirect consequences through modification of uplands and through direct impact of the river corridors themselves. Changes in the uplands can affect the lower-lying river corridors by altering the fluxes of water, sediment, and biological materials transferred to them. Intact riparian zones and floodplains could buffer upland land-use changes. But the riparian zones and floodplains in the river corridors of the Amazon are also undergoing changes that will affect their nutrient dynamics and buffering ability. The floodplains of the Amazon and some of its major tributaries support important fisheries and are a source of easily accessed timber and fertile soils that can be used during the dry season for agriculture and cattle ranching. Deforestation and pasture creation has already occurred to a large extent in the floodplain of the eastern Amazon mainstem. The riparian areas of small streams are often subjected to the same deforestation and agricultural practices as the upland areas. The general response of these systems is decreased retention or filtering of water and materials. This may include decreased anaerobic processing and trace gas production. Without this filtering system, the rivers may carry larger loads of organic carbon and nutrients to downstream areas. One possibility is the enhancement of primary production downstream due to the increased nutrient loads. In general, rivers will respond with differing magnitudes and lags to natural or human perturbations depending upon the processes involved and the downstream transfer rates.

Even if the Basin undergoes massive deforestation, it will still be very different in character than many of the river basins of the developed world. Many of these other basins have drainage networks that have been channelised and cut off from the floodplains, agricultural lands that have been "reclaimed" from wetlands, and large loads of nutrients from fertilisers and anthropogenic chemicals. Over time, the Amazon may take on some of these characteristics.

D.5.7 Towards a Synthetic Model of Drainage Basins

Overall, we can hypothesise that the quantity and composition of dissolved and particulate bioactive materials in a parcel of water at any downstream node in the river system is predictable as the product of a common set of processes which occur differentially according to up-stream conditions of relative topography, soil organic content and texture, water residence time, and floodplain extent. That is, it should be possible to represent the chemical attributes of any geographic element in the Basin based on the mix of fairly simple properties, re-gardless of whether it is in the upper Rio Madeira or lower Rio Negro. The downstream chemical hydrograph at the mouth of that tributary is then a function of the chemical source and downstream routing. Within-channel proc-esses can significantly modify the river chemical profile.

Although the model of Fig. D.64 is not yet complete, it is a vision of how to:

- Evaluate our understanding of how biogeochemical and physical processes integrate in a river basin;
- Quantitatively constrain the importance of geochem-ical and physical processes and predict compositions and fluxes of bioactive elements at inaccessible loca-tions and for different land-use and climate change scenarios; and,
- Improve our ability to infer upstream landscape en-vironments and sequence of processes from observed biogeochemical signatures within a parcel of water.

Chapter D.6

Case Study 2:
Integrated Ecohydrological Analysis of a Temperate Developed Region: The Elbe River Basin in Central Europe

Valentina Krysanova · Alfred Becker · Frank Wechsung

D.6.1 General Outline of the Elbe River Basin

During the last decade integrated studies of the impacts of global change on the environment and society have been initiated in the Elbe River basin (Becker et al. 1999b; Krysanova et al. 1999a). The Elbe basin covers large parts of the Czech Republic and Eastern Germany. The Elbe River has a total length of 1 092 km, a drainage area of 148 268 km², and about 25 million human inhabitants. About two-thirds of the drainage basin is in Germany (our case study area – Fig. D.72) and one-third in the Czech Republic. Many tributaries of the Elbe River are controlled by dams and weirs whereas the Elbe mainstream in Germany is in a semi-natural state. The most downstream reaches of the Elbe (last 140 km) are affected by tides. Since large portions of the basin are located in the temperate zone under 1 000 m elevation, the river discharge is characterised by winter and spring high water.

Fig. D.72.
German portion of the Elbe drainage basin with three tributary subbasins the Mulde, the Saale and the Havel, two meso-scale subbasins the Stepenitz (1) and the Zschopau (2). The boundaries of the federal state of Brandenburg are indicated by the *thick line*

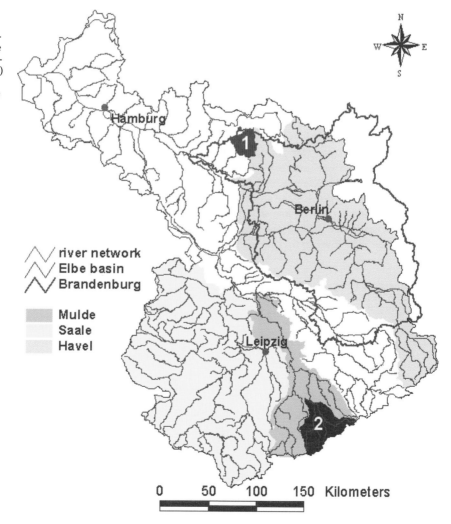

The buffering effect of high elevation snowmelt, which otherwise would affect flood discharge and low flows, is missing in the Elbe due to the absense of high mountains. Therefore, the ratio of monthly low flow and high flow in the Elbe is 1 : 21 (compared to 1 : 1.75 in the Rhine Valley). The long-term mean annual precipitation in the basin is 659 mm. Long-term mean discharge of the Elbe River is 712 m³ s⁻¹ at the last gauge not influenced by tides (Neu Darchau, 131 950 km²), the specific discharge is 5.4 l s⁻¹ km⁻², or a runoff of about 170 mm yr⁻¹. The basin discharges on average 22.5 × 10⁹ m³ annually, or 26% of the annual precipitation. A comparison of the observed and with SWIM simulated hydrographs of the Elbe in Neu Darchau in 1981–1990 is shown in Fig. D.73.

The German part of the Elbe drainage area is subdivided into three sub-regions based on relief and soils, namely, (a) the Pleistocene lowland, covering the Havel basin and the area to the west and north-west of it; (b) the loess sub-region in the lower parts of the Saale and the Mulde basins; and (c) the mountainous sub-region in the upper parts of the Saale and the Mulde basins (Fig. D.72). The agricultural lands that occupy about 56% of the total drainage basin area represent one of the most important sources of diffuse nutrient pollution. The Elbe and its tributaries are intensively used for fresh-water supply for domestic, industrial and agricultural purposes. Many large Czech and German nature reserves belong to the Elbe drainage area. The UNESCO-biosphere reserve "Middle Elbe" has been established to protect one of the largest continuous alluvial forests in central Europe.

A primary reason for selecting this river basin as a case study is its characteristic vulnerability to water stress during dry periods. Due to the position of the basin between the relatively "wet" maritime climate of western Europe and the more continental climate in eastern Europe with longer dry periods, the annual long-term average precipitation is relatively low, and in the lowland of the German part of the basin it is less than 600 mm yr⁻¹. Therefore, the Elbe River basin can be classified as the driest of the five largest river basins in Germany (Rhine, Danube, Elbe, Weser, Ems), with all the resulting problems and conflicts associated with such relative water scarcity.

Another important reason for using the Elbe as a case study is the excessive pollution of surface and groundwaters caused by high-intensity water use, excessive application of fertilisers and pesticides in agriculture, and discharge of domestic and industrial wastes. Excessive nutrient addition (of nitrogen and phosphorus) is one of the most widespread forms of water pollution. Despite the fact that emissions from point sources have notably decreased since the early 1990s due to reduction of industrial sources and introduction of new and better sewage treatment facilities, diffuse sources of agricultural pollution are still not sufficiently controlled. To evaluate and predict the effects of diffuse pollution control measures, process-based ecohydrological models are needed.

There is particular interest in the effects of presumed changes in climate and land use on hydrological processes, in terms of water balance, water quality, and on agriculture, especially crop yields. To investigate these effects the ecohydrological modelling system SWIM (cf. Sect. D.2.6.6.1) has been developed and applied in the German part of the Elbe River basin. Selected examples of application will be presented and briefly discussed below, namely:

- a comparative study of nitrogen dynamics in the lowland and mountainous parts of the basin (in the sub-basins of the Elbe Stepenitz and Zschopau);
- an attempt to regionalise the results with respect to nitrogen dynamics from sub-regional studies to the whole drainage area; and,
- land use and climate change impact studies performed in the federal state of Brandenburg, which largely overlaps with the north-eastern part of the Elbe basin.

All these studies provide an improved understanding of the complex interactions between climate, land use, hydrology and vegetation.

Fig. D.73.
Comparison of the observed and with SWIM simulated hydrographs of the Elbe River, gauge Neu Darchau in 1981–1990 (see Hattermann et al. 2003)

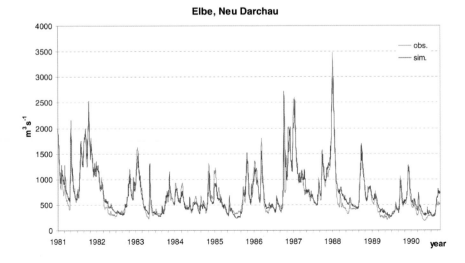

D.6.2 Integrated Analysis of Hydrological Processes and Nitrogen Dynamics

D.6.2.1 Comparison of Nitrogen Dynamics in the Lowland and Mountainous Sub-regions of the Elbe

Our objectives here are to validate the model for hydro-logical processes and nitrogen dynamics in lowland and mountainous sub-regions of the Elbe, and to analyse the major factors affecting nitrogen export from diffuse agricultural sources of pollution. Two mesoscale river basins with contrasting relief and soil types were selected: the Stepenitz in the Pleistocene lowland sub-region (gauge station Wolfshagen, 575 km^2) and the Zschopau (gauge station Lichtenwalde, 1504 km^2) in the mountainous sub-region (see Fig. D.72). Arable land is the dominant land use type in both basins (68 and 43% in the Stepenitz and the Zschopau, respectively), though the Zschopau also has a significant portion of forest (44%). Sandy and loamy-sandy soils prevail in the Stepenitz, while loess-type soils occupy 54% of the Zschopau basin. Long-term average precipitation is 630 and 940 mm yr^{-1}, runoff coefficients 0.28 and 0.42, and average specific runoff 6.3 and 13.8 l s^{-1} km^{-2} in the Stepenitz and the

Zschopau, respectively. The validation was performed under the conditions of uncertainty as regards crop rotations and crop management, as crop statistics are available at the district level only and are not geo-referenced.

As a first step, the hydrological module of SWIM (see Sect. D.2.6.6 and Fig. D.19) was validated for both mesoscale basins. The statistical evaluation of results was made by comparing the simulated and observed water discharge over a 5–6 year period, comparing means and coefficients of variations, and applying the common efficiency criteria after Nash and Sutcliffe (1970). The means and coefficients of variation of the observed and simulated time series were statistically comparable for both basins. The efficiency of runoff simulation was in a range from 0.67 to 0.88 with a daily time step (see e.g. a comparison of observed and simulated hydrographs for these basins in Fig. D.74).

An example of the modelled dynamics of NO$_3$-N in five soil layers and N flow components (mineralisation, plant uptake, losses with water and denitrification) is shown in Fig. D.75 for the dominant soil (Fahlerde: sandy loam covered by sandy layer) in the Stepenitz basin. As one can see, nitrate content in the soil usually decreases during the crop growth period and may drop to almost zero in late summer. Mineralisation is highest after harvesting and in September–October, and the lowest in

Fig. D.74.
Comparison of the observed and with SWIM simulated hydrographs of the Stepenitz River, gauge Wolfshagen (*upper part*) and Zschopau River, gauge Lichtenwalde (*lower part*) in 1983–1984

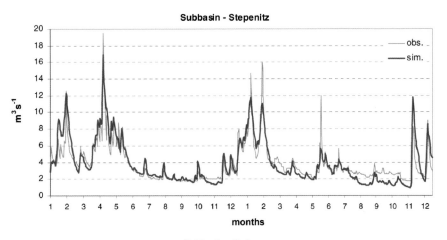

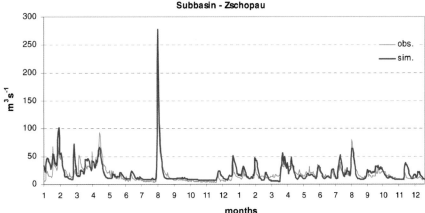

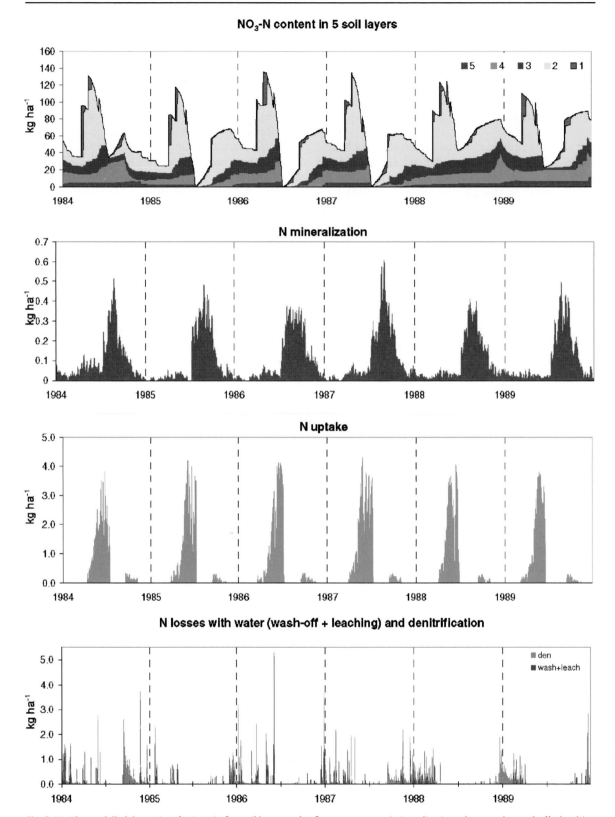

Fig. D.75. The modelled dynamics of NO_3-N in five soil layers and N flow components (mineralisation, plant uptake, washoff + leaching and denitrification) for the dominant soil (sandy loam with sandy covering) in the Stepenitz River basin in 1984–1989

winter. Denitrification and leaching occur mainly in winter time. The validation of nitrogen dynamics was performed using regional data on nitrogen balance components obtained from the literature, and the measured N concentrations at the basin outlet. Finally, the accumulated simulated NO_3-N load was compared with the accumulated measured NO_3-N load at the gauging stations Wolfshagen and Lichtenwalde. The annual accumulated loads were close to each other (usually 5 to 10% difference). In general, the validation was successful, despite uncertainties related to crop rotations and fertiliser application in the basins. More details on this study is given in Krysanova et al. (1999b).

Since the spatial distribution of crops in the basins was not known, the N dynamics in soils was analysed sequentially for typical crops and fertilisation schemes, considering a seven-year simulation period. The following crops were considered: summer barley, winter barley, winter wheat, summer barley and grass as a cover crop for winter. Different fertilisation schemes were applied. For example, we applied scheme FM, having three to four applications of N in spring under the condition that NO_3-N content is lower than a specified threshold, with a maximum amount equal to 150 kg ha^{-1} yr^{-1}, plus one application in autumn in the case of winter crops. Another test was on scheme FO, wherein the entire amount assigned to spring time is applied at once, in early spring, plus one application in autumn in the case of winter crops.

The analysis was performed on three representative soils in the Stepenitz basin: a sandy/sandy-loamy soil (53% of the area), a sandy-loamy/clay-loamy soil (12% of the area), and a sandy soil (9% of the area). In the Zschopau basin, two representative soils were analysed: sandy soil (40% of the area), and loess (37%). Table D.26 presents the annual amounts of the major nitrogen flow components in the chosen representative soils averaged over the seven-year simulation period and three crop combinations. The nitrogen wash-off with surface flow and interflow is assumed to contribute to surface water as quick and slower responses, as opposed to leaching to groundwater.

The analysis of differences in modelled nitrogen response in the different soils leads to the following conclusions:

- Nitrogen wash-off has different pathways in different soils. In sandy soils the leaching to groundwater prevails, while in the loamy and loess soils we mainly see lateral wash-off with interflow. The partitioning between both pathways is observed in soils of intermediate texture;
- Nitrogen losses with direct surface runoff are higher in the Zschopau basin due to its mountainous char-

acter and larger proportion of loamy soils. In contrast, they are practically missing in the Stepenitz;
- Simulated mineralisation is the highest in the loamy and loess soils mainly due to higher organic matter content and water availability in these soils;
- Application of the FO fertilisation scheme leads to higher N wash-off, and lower N utilisation (crop uptake divided by the sum of fertilisation and mineralisation) in comparison to the FM scheme.
- The overall utilisation of nitrogen is about 70–80% in the Stepenitz basin, and 55–65% in the Zschopau basin. As a consequence, fertilisation needs are higher in the Zschopau area because of more intensive runoff processes and correspondingly higher risk of losses.

In general, the simulated behaviour of nitrogen is in agreement with regional data reported in the literature and available measured data for the basins under study. The modelling results clearly reflect differences in the pathways of nitrogen loss from different soils, as well as the influence of climate and agricultural practices (fertilisation rates, application timing, rotation schemes) on nitrogen export. These differences serve as the basis for further scenario analysis and eventually for recommendations on improving agricultural practices aimed at reducing nutrient loads and improving water quality.

D.6.2.2 Regional Nitrogen Dynamics across the German Part of the Elbe Basin

After validation and comparison of results from the two sub-basins of the Elbe studies presented above, a first attempt was made to regionalise the SWIM application to the German part of the Elbe drainage area (see Sect. D.2.6). The lowland sub-region was subdivided into two climate zones according to the climate stations Hamburg and Potsdam with average temperatures of 8.7 and 8.8 °C, respectively. The loess and mountainous sub-regions (tributary basins Saale and Mulde in Fig. D.72) where the climate gradient is more significant, were subdivided into four climate zones based on meteorological stations Artern, Gera-Leumnitz, Hof-Hohensaas, and Bad Sachsa with average temperatures ranging from 6.5 to 8.6 °C.

The basic crop rotation and fertilisation schemes were constructed for simulation using information obtained from environmental agencies in the basin. The ten-year basic crop rotation scheme includes three years of winter wheat, two years of potatoes, one year of each of winter barley, winter rye, spring barley, and maize, and one year set-aside. The amount of mineral fertilisers applied for winter wheat is 120 kg N ha^{-1} yr^{-1}, for other grain crops 100 kg N ha^{-1} yr^{-1}, for potatoes 140 kg N ha^{-1} yr^{-1}, and for

Table D.26. Soil characteristics (BÜK-1000 in Hartwich et al. 1995) and nitrogen fluxes as average annual values for seven years simulation period in five representative soils in the Stepenitz and the Zschopau River basins

Basins	Soil number and mechanical structure		Soil characteristics (average over root zone)			Nitrogen fluxes (kg N ha^{-1} yr^{-1})							
			Clay (%)	Field capacity (mm m^{-1})	Saturated conductivity (mm h^{-1})	Fertili-zation	Minerali-zation	Crop uptake	Denitri-fication	Washoff with sur-face flow	Washoff with inter-flow	Leaching to ground water
Stepenitz	17	Sandy sediments	0.0	23.0	65.3	121.7	61.9	135.5	0.0	1.5	0.4	53.3
	26	Sandy upper layer over sandy loam	14.2	28.5	16.3	135.0	59.9	156.0	14.7	0.3	15.1	18.0
	19	Sandy loam over clay loam	25.6	34.0	6.5	140.0	75.8	154.4	2.5	0.7	67.0	0.1
Zschopau	57	Sand from magmatic and metamorphic rocks	4.5	24.0	27.1	144.3	25.7	113.7	0.0	7.6	1.5	62.1
	59	Loess mixed with weathering products	25.6	35.0	4.8	154.0	83.2	113.8	38.1	12.9	91.4	0.8

maize 180 kg N ha^{-1} yr^{-1}. The grain crops have mineral fertilisers applied three times, including once in the autumn for winter crops. For maize and potatoes, the total annual amount is applied once in the spring, together with seeds. In addition, organic fertilisers are added to soil under the following rules: for winter crops 30 kg ha^{-1} of organic N is applied 23–43 days before sowing, plus the same amount at the beginning of March, while for summer crops one application of 30 kg N ha^{-1} is performed at the end of October of the previous year, plus one more application of the same amount six weeks after seeding in the spring.

Simulation experiments were performed for all soils which occur on arable land in the six climate zones for the 30-year period 1961–1990, assuming minimal slope of 0.2%. The effect of higher slopes was investigated additionally through modelling experiments. This assumption increases only slightly the groundwater recharge and nitrogen leaching to groundwater for sandy soils, and more significantly for loess soils (which are otherwise rather small), and in a sense provides an upper estimate of nitrogen leaching.

In Fig. D.76 modelling results for the average annual nitrogen leaching to groundwater, denitrification, plant uptake and mineralisation, obtained under the assumption that the basic rotation and fertilisation scheme was applied for all soils and repeated three times during the simulation period, are mapped for the German part of the Elbe. The scale is 1 km × 1 km, using raster grids. Nitrogen leaching varies from very low in the loess sub-region and floodplain soils (due to high water capacity and low groundwater recharge) to 30–60 kg N ha^{-1} in sandy lowland soils, with a maximum of about 100 kg N ha^{-1}. When the real slope was taken into account, the leaching in the lowland was practically the same, and it decreased to almost zero in the loess soils.

The highest denitrification occurs in soils with the lowest leaching due to their higher field capacities and lower saturated conductivities which provide more optimal conditions for denitrification. In general, denitrification varies from zero in sandy soils to more than 100 kg ha^{-1} in the floodplain soils. Nitrogen uptake generally varies from 80 to 180 kg N ha^{-1} (minimum 26 kg N ha^{-1}), and it is clearly higher in the loess sub-region than in sandy and loamy-sandy soils in the Pleistocene lowland. Mineralisation of nitrogen is also higher in the loess sub-region compared to the lowland, with a maximum in the floodplain soils with a high organic N content (0.3% in the upper 15 cm layer). The modelling results presented in Fig. D.76 are comparable to the regional ranges indicated in the literature (DVWK 1985), and differences between the sub-regions appear plausible.

Simulation runs were performed for two other fertilisation schemes, a first increasing fertiliser application by 50%, and a second decreasing it by 50% for both min-

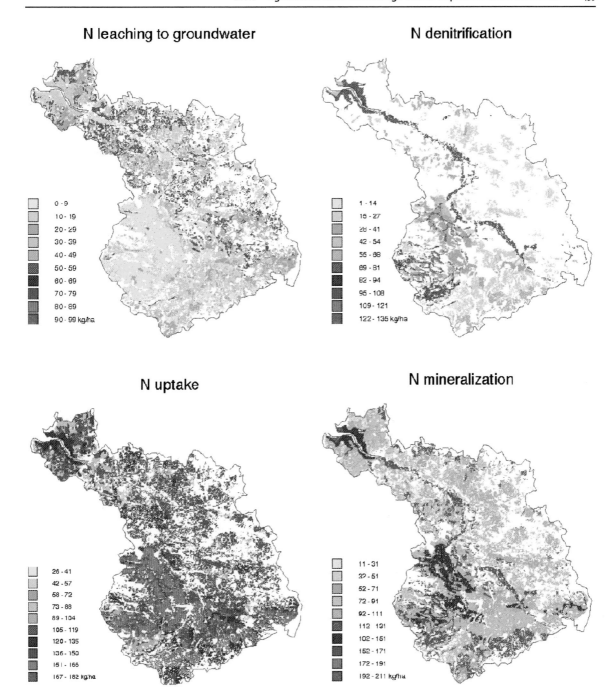

Fig. D.76. Modelling results for mean annual N flow components: leaching to groundwater, denitrification, plant uptake and mineralisation in agriculture soils of the Elbe basin (simulation period 1961–1990)

eral and organic N (Fig. D.77). Here the same tendencies are apparent as in Fig. D.76, namely, the highest N mineralisation and N uptake in loess soils, the highest N leaching in sandy soils, but the highest denitrification in loess soils, which have low saturated conductivities. Also, one can see how changes in fertilisation rates are translated into higher/lower uptake, mineralisation, leaching and gaseous N losses.

D.6.3 Agricultural Land-use Change and Its Impact on Water Resources

A land-use change impact study was performed for the federal state of Brandenburg, which overlaps with the lowland part of the Elbe basin (Fig. D.72). Besides general vulnerability to water stress due to natural climatic condi-

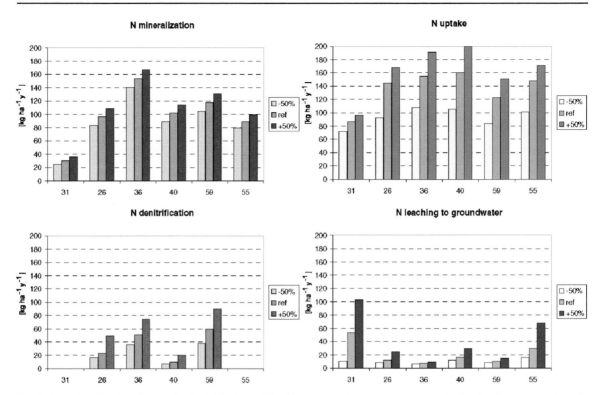

Fig. D.77. N mineralisation, plant uptake, denitrification and leaching to groundwater as average annual values for six representative soils (*31:* dry sand, *26:* sand over sandy loam, *36* and *40:* loess, *59:* loess mixed with weathering products, *55:* sand and gravel from rocks) in the Elbe for the period 1961–1990 obtained with SWIM for the reference rotation and fertilisation schemes (*ref*), and for the same rotation combined with 50% increased/decreased application of fertilisers

tions, water resources in Brandenburg are strongly affected by brown coal mining, which took place in the south-east of the federal state over the last decades. In an area of more than 2 000 km², groundwater was pumped out to depths of 100 m or more and released into rivers. Although the mining activities are now significantly curtailed, these water resources require replenishment. Surface waters are an obvious source for reconstituting the depleted groundwater pool. However, this results in decreasing river discharge, in particular in the Spree River, and may lead to general shortages in water availability elsewhere. As we will see, land-use change, through its impact on water budgets, is one possible option to counteract this development.

A tendency towards extensification in agricultural land use is observed in Brandenburg during the last decade. The increase in a temporary set-asides within specific crop rotations was the main means used to decrease the level of exising intensification on arable lands. Another effective way to extensify crop production and, at the same time, to protect environmentally sensitive areas, is to create buffer zones along river courses (river corridors) by converting the arable land there into grassland (i.e. permanent "set-asides"). The latter is also an efficient way to reduce nitrate pollution and sediment load in rivers. The primary objective of our study was to analyse the effects of these two alternative means of

agricultural extensification on water resources in Brandenburg using the SWIM model.

Six of the major crops grown in Brandenburg were considered: winter wheat, winter and summer barley, winter rye, silage maize, and potatoes. Temporary set-aside was interpreted in the model not as a black fallow, but as a cover crop (grass) with a maximum leaf area index (LAI) of 5, while all major crops had a maximum LAI between 6 and 6.5. Spatial and temporal distributions of crops across the study area are mainly determined by cropping preferences in response to soil qualities and sequences of cultivation. Existing relationships between soil quality and cropping patterns for northeastern Germany were used to define a reference scenario A. The scenario was based on three basic crop rotation schemes, which in turn were related to three soil classes: "poor" (sandy soils), "intermediate" (loamy-sandy soils) and "good" (loamy and loess soils). SWIM then distributed the chosen crops to locations ("fields") year by year by applying a stochastic crop generator. It was done in such a way that the crop distribution varied spatially according to the differences in soil quality and each year corresponded to the predefined mean crop distribution in the region.

Three types of land-use change scenarios were then formulated and applied:

Fig. D.78.
Assessment of impacts in percent compared to the reference scenario *A* of temporary set-aside (scenarios *B, C, D*), permanent set-aside as riparian buffer zones of 150 m and 500 m (scenarios *A150, A500*), and their combinations on water flow components in Brandenburg as computed by the SWIM simulation

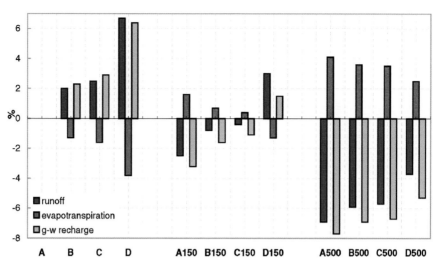

- modification of the basic rotation scheme by increasing the portion of temporary "set-aside" areas (scenarios B, C, and D). Scenario D has the largest portion of set-aside: 60, 40 and 30% of arable land for poor, intermediate and good soils, respectively;
- introducing river corridors of 150 and 500 m width onto the original land use map and converting cropland within them into meadows (permanent "set-aside", scenarios A150 and A500); and,
- combinations of temporary and permanent set-aside schemes as described above (scenarios B150, C150, D150, B500, C500, D500).

In scenarios B, C and D (Fig. D.78) the evapotranspiration decreased in the cropland area rather uniformly on all soils, and consequently the average soil moisture increased. This resulted in higher runoff and groundwater recharge. The regional impact of land-cover changes under scenarios B and C, however, was minor across the total area. Only the more significant changes in scenario D resulted in notable increases in the average regional runoff and groundwater recharge (6.7% and 6.4%, respectively).

The conversion of cropland into meadows within river corridors decreased the cropland portion of Brandenburg by 16.5% and 54%, respectively. The shift in the relative share of different soil qualities and deintensification as in B, C and D increased runoff *on cropland* between 5.3 (scenario A150) and 35.4% (scenario D500), depending on the assumed width of the river corridor and the share of temporary set-aside assigned to the scenario. However, the conversion of former cropland to meadows within river corridors decreased runoff and increased evapotranspiration on them quite significantly. For the entire area (scenario A500) it resulted in an increased evapotranspiration up to +4.1%, decreased runoff down to –6.9%, and decreased groundwater recharge down to –7.7%.

Thus, two opposite tendencies were identified through our simulation study. The temporary "set-aside" within

a crop rotation scheme would result in decreasing evapotranspiration and increasing runoff and groundwater recharge in the region, whereas the permanent set-aside within river corridors would reduce runoff and increase evapotranspiration. Land-use changes in terms of deintensification may compensate for the expected decrease of discharge in the river Spree over the coming period, if these changes assume increases in the portion of temporary set-aside areas, and do not involve conversion of arable land into permanent meadows (or forests). Runoff increases might be even higher with lower production intensity on the remaining area of arable crops, due to reduction in regional transpiration as a consequence of the lowered leaf area index. The land-use change impact study is described fully in Wechsung et al. (2000).

D.6.4 Climate Change Impacts on Hydrology and Crop Yields

An analysis of climate change impacts on hydrology and crop yields was also performed for Brandenburg (Fig. D.72) using the SWIM model. The reference scenario SB represents the observed climate over the period 1951–1990. To reflect uncertainties in the prediction of global climate change by current GCMs, two equilibrium scenarios (SE15, SE30) and two transient scenarios (ST15, ST30) assuming for temperature an endpoint increase of 1.5 and 3.0 °C, respectively, were used in our study. The scenarios were developed from ECHAMT21 results using a statistical downscaling method (Werner and Gerstengarbe 1997). Three periods were compared: 1980–1990 (control period A), 2020–2030 (period B), and 2040–2050 (period C).

As a first step, a hydrological validation was performed on three mesoscale river basins in the area. The crop module was then validated regionally for Brandenburg, using crop yield data for districts. Finally, a cli-

Table D.27. Climate change impact on hydrological processes and crop yield in Brandenburg, considering: Climate change only (*I*), in combination with adjustment of net photosynthesis to higher CO_2 (*II, III*), and with adjustment of transpiration to higher CO_2 (*III*) (the ranges exceeding ±5% are marked *bold*)

Simulation experiments		Hydrological flows/crop yield	Ranges of change to reference period 1980–1990 (%)			
			Period B: 2020–2030		Period C: 2040–2050	
			from	to	from	to
I	Climate change only	Evapotranspiration	–3.7	+6.3	–4.7	+1.3
		Groundwater recharge	–51.7	–24.0	–51.5	–30.7
		Runoff	–21.1	–2.3	–24.3	–7.1
		Winter barley yield	–7.8	+3.1	–13.8	–6.2
		Winter wheat yield	–20.6	–8.3	–22.3	–15.9
		Silage maize yield	+0.1	+8.2	–5.3	+1.5
II	Climate change combined with factor alpha	Winter barley yield	–1.8	+9.9	–5.5	+2.9
		Winter wheat yield	–15.2	–2.4	–14.5	–7.6
		Silage maize yield	+4.4	+12.9	+1.0	+8.3
III	Climate change combined with factors alpha and beta	Winter barley yield	+6.5	+16.1	+6.8	+14.0
		Winter wheat yield	–8.1	+4.5	–4.3	+3.6
		Silage maize yield	+15.9	+25.0	+18.8	+26.7

mate change impact study was performed, considering hydrological processes and crop yield. The cropping system was restricted to the major crops winter wheat, winter barley and silage maize.

The atmospheric CO_2 concentration for the reference period and two scenario periods were set to 346, 406 and 436 ppmv, respectively. Precipitation is lower than in the control period for all scenario periods. It is significantly lower for the scenario SE15 in periods B and C (–15.2%, –15.9%), and in period C for both transient scenarios (–13.4% and –12.1%).

The adjustment of net photosynthesis and evapotranspiration to altered atmospheric CO_2 concentration was studied considering two additional factors:

- adjustment of the potential growth rate per unit of intercepted photosynthetically active radiation (PAR) by a temperature dependent *correction factor alpha* based on experimental data for C3 and C4 crops; and
- assuming a CO_2 influence on transpiration at the regional scale (*factor beta*), which is coupled to the direct CO_2 effect of radiation use efficiency (*factor alpha*).

Different approaches toward adjusting of net photosynthesis and evapotranspiration to altered atmospheric CO_2 have been used in modelling studies (Goudrian and Ketner 1984; Rotmans and den Elzen 1993). Detailed studies on the interaction of higher CO_2 and water use efficiencies are described in (Easmus 1991; Kimball et al. 1999). In our study a semi-mechanistic approach to adjusting net photosynthesis (see Krysanova et al. 1999a) derived from a mechanistic model for leaf net assimila-

tion (Harley et al. 1992) was tested. The method takes into account the interaction between CO_2 and temperature. Additionally, a possible reduction of potential leaf transpiration due to higher CO_2 (factor beta), derived directly from the enhancement of photosynthesis, was considered.

Simulation runs, which included adjustment of net photosynthesis, have been carried out under two variants, without and with the factor beta. In this way, we accounted for current uncertainties regarding the significance of stomatal effects on higher CO_2 for regional evapotranspiration. Jarvis and McNaughton (1986) postulate that on the regional scale there is no control of stomatal resistance on evapotranspiration, because the humidity profiles are adjusted within the planetary boundary layer. This response would counter stomatal closure as a negative feedback. On the other hand, recent modelling studies (Kimball et al. 1995) suggest that stomata have far more control on regional and global evapotranspiration than postulated by Jarvis and McNaughton.

The simulation results (Table D.27) show that actual evapotranspiration is expected to increase or decrease slightly, depending on precipitation and temperature change. Despite the higher average temperature, it tends to decrease in period C for some of the four scenarios due to the lower precipitation. Runoff and groundwater recharge always decreased, whereas groundwater recharge responded more sensitively to the anticipated climate change (from –30.7 to –51.5% in period C). The crop yield of winter wheat and winter barley decreased in all four scenarios considering the "climate change only" case, whereas wheat was more sensitive: down to –22.3%. For

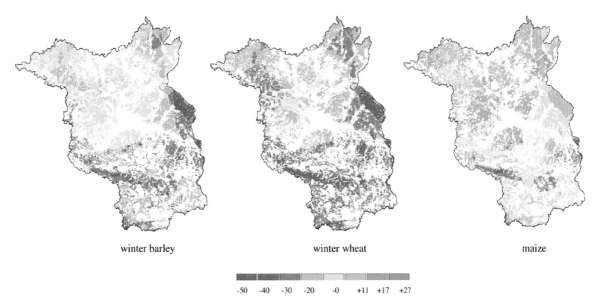

winter barley   winter wheat   maize

-50   -40   -30   -20   -0   +11   +17   +27

Fig. D.79. Changes in crop yield for equilibrium GCM scenario assuming 1.5 °C temperature increase (scenario SE15), period C in comparison with the reference scenario

maize, changes were not significant. In Fig. D.79 changes in crop yield for scenario SE15, period C are depicted. The impact of higher atmospheric CO_2 (factor alpha) compensated partly for climate-related crop yield losses. The assumption that additional stomatally-based reductions in transpiration are taking place at the regional scale (alpha and beta factors) lead to further increases in crop yield, which were larger for maize than for barley and wheat. In this case, positive changes, exceeding 5% for all four scenarios were predicted for winter barley and maize. Reduction in regional transpiration due to higher CO_2 (and hence greater water use efficiencies) may partly compensate for the decrease in runoff and groundwater recharge. This constitutes a feedback of vegetation on hydrological flows. Across the scenario runs, a high sensitivity of the main model outputs (three hydrological flows and crop yield) to precipitation was found. A full description of the climate change impact study for Brandenburg is given in Krysanova et al. (2000).

D.6.5 Conclusions

One of the major challenges for IGBP-BAHC has been the adequate description and modelling of the complex interactions between climate, hydrological and ecological processes across different scales. SWIM has been developed as a tool to serve this purpose at the mesoscale and regional scale. Model applications in a number of river basins in the range of about 100 to 10 000 km² drainage area have shown that an appropriately-scaled model can adequately describe basic hydrological processes (evapotranspiration, groundwater recharge, runoff generation), the cycling of nutrients (nitrogen and phosphorus) in soils and their transport in water. These phenomena can be treated consistently and used to articulate the influence of climate and other geophysical conditions, vegetation growth (especially agricultural crops), the dynamical features of soil erosion and sediment transport, as well as the effect of changes in climate and land use.

While results from SWIM model validation appear quite satisfactory, further development is foreseen for some processes, like nutrient retention in catchments and changes in atmospheric CO_2 and their influence on plant growth and evapotranspiration. This reflects the current level of maturity in our science. A significant challenge is thus to continue using ecohydrological models such as SWIM, despite their certain limits related to representation of complex ecohydrological processes in river basins, in a useful decision-making context in which scientists and catchment managers can jointly pursue creative new means of protecting the environment.

Chapter D.7

Case Study 3:
Modelling the Impacts of Land-use and Climate Change on Hydrological Responses in the Mixed Underdeveloped/Developed Mgeni Catchment, South Africa

Roland E. Schulze · Simon Lorentz · Stefan Kienzle · Lucille Perks

D.7.1 Setting the Scene

The "rainbow nation" tag is often used to depict the colourful natural, social and economic diversity of South Africa. It is also a fitting description of the physical, cultural and developmental diversity which prevails in the Mgeni catchment in the KwaZulu-Natal province and which has left a strong imprint of anthropogenic change on its varied hydrological systems. The Mgeni is an important catchment in South Africa: While its area of 4 079 km^2 upstream of Inanda Dam (15 km from its mouth into the Indian Ocean) constitutes a mere 0.33% of South Africa, it supplies water to nearly 15% of the country's 41 million inhabitants and to a region which, albeit small, produces around 20% of the country's gross domestic product.

The Mgeni is a catchment of contrasts:

- Physiographically the undulating landscape around New Hanover (Fig. D.80), with deep sandstone/dolerite derived soils, stands in stark hydrological contrast with the ruggedly incised topography with granite and gneiss derived, skeletal soils in the Valley of Thousand Hills between Nagle and Inanda Dams.

- Agricultural land uses are characteristically both intensive and extensive. High input and intensive first world cultivation of sugarcane and commercial timber plantations (Fig. D.81), together with irrigated crops/pastures and concentrated beef feedlots, cover substantial parts of the catchment (cf. Table D.28). There is also low input subsistence crop husbandry with less than one-third the yields and the overgrazed areas often overstocked with poor quality livestock. These two contrasting agricultural systems are often in close proximity to one another and affect hydrological responses in vastly different ways.

- The hydrology of westernised formal urban structures with planned residential, business and industrial sectors serviced by sophisticated water reticu-

Overview

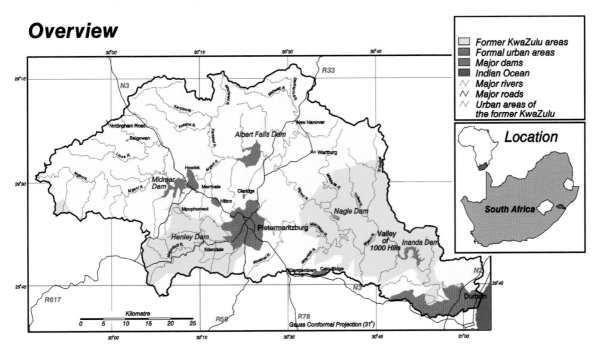

Fig. D.80. Mgeni catchment: overview

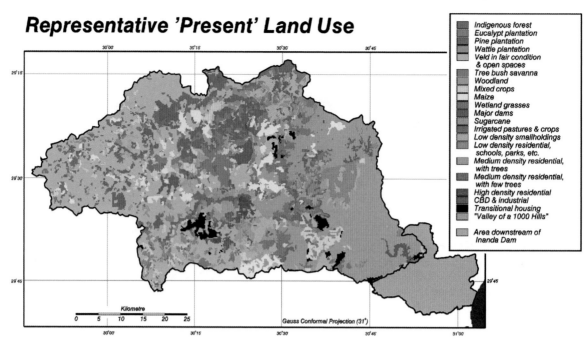

Fig. D.81. Representative "present" land use in the Mgeni catchment

Table D.28. Verification statistics for 1979–93 at six streamflow gauging stations (from Kienzle et al. 1997)

	Gauging station					
	Mpendle	**Lions**	**Karkloof**	**New Hanover**	**Nagle**	**Henley**
Management subcatchments attribute						
Area (km²)	296	362	337	430	439	178
Altitude Range (m)	1080 – 2013	1058 – 1997	1068 – 1725	633 – 1492	394 – 1097	928 – 1528
MAP (mm)	966	947	1032	889	851	915
MAR (% of MAP)	25.8	18.0	20.3	13.0	16.5	21.4
% difference $Q_S - Q_O$	1.5	0.9	1.1	0.5	6.2	2.8
Correlation coefficient	0.94	0.89	0.92	0.89	0.93	0.92
Slope	0.97	1.01	0.93	0.97	0.95	1.05
% difference $\sigma_S - \sigma_O$	−3.3	−13.5	−0.8	−8.9	−2.9	−14.3
Land-use proportions (% of total)						
Rural	0.3	3.4	4.7	17.0	0.3	2.5
Urban	0.0	0.6	46.6	0.9	0.0	0.0
Peri-urban	0.0	0.0	0.0	0.0	0.0	22.4
Commercial forest	11.3	13.0	32.7	40.0	11.3	7.3
Dryland agriculture	6.4	9.4	11.1	30.8	6.4	12.5
Irrigated agriculture	3.3	9.0	4.8	1.0	3.3	0.0
Natural vegetation	75.6	63.7	0.1	10.0	75.6	53.5
Farm dams	3.1	0.9	0.0	0.3	3.1	2.1

Q = streamflow; σ = standard deviation; subscripts: s = simulated, o = observed; MAP = mean annual precipitation; MAR = mean annual runoff.

lation systems contrasts markedly with that of the mushrooming informal shack dwelling settlements. The latter often have no potable water and domestic water supplies, which are still largely drawn directly from streams frequently infested with high concentrations of *Escherichia coli* pathogens emanating from poor sanitation facilities and high livestock densities.

This case study focuses on some present-day and projected future scenarios of anthropogenic actions, mainly through land use but also from potential climate change, on the spatial and temporal hydrological and water quality responses in the Mgeni catchment. The case study aims to highlight differences in developmental and management stages on the catchment's hydrological responses as well as the relative differences between land use and climate change impacts. A suitably structured hydrological simulation model operating at appropriately sensitive time steps and space scales is required. The daily time step, physical-conceptual and multipurpose ACRU agrohydrological model (Schulze 1995, and described briefly in Chapt. D.2), was selected because, in addition to outputting elements of streamflow, it can simulate reservoir yield, sediment and phosphorus yields, *E. coli* loadings, irrigation supply/demand and return flows and has been structured to represent processes of both land use and climate change impacts explicitly. The following sections

- firstly describe the attributes of the Mgeni catchment;
- outline the configuration of the Mgeni system for use with the ACRU model;
- present examples of model output verification;
- apply the model, first with a baseline land cover, against which the hydrology of present-day land use is compared, then to impacts of present-day land use on water quality in regard to its physical quality (i.e. sediment yield), chemical status (phosphorus loading) and its biological health (*E. coli* concentration); and,
- finally apply the model to selected scenario studies of land use and climate change on the Mgeni's water resources.

D.7.2 Attributes of the Mgeni Catchment and Human Pressures

The catchment is located between 29° 13' – 29° 46' S and 29° 46' – 30° 54' E (Fig. D.80) with an altitudinal range from 0–2 013 m. While the catchment mean annual precipitation is 902 mm (578 in 1992; 1 384 in 1987), the precipitation is spatially unevenly distributed and varies from 680 mm near the coast to 1 220 mm in the rugged western parts, with 80% of the inland rainfall occuring largely as convective storms in the summer months (October through March) while along the coast lower intensity general rains in summer make up 65–70% of the annual total. The distinct coastal *v.* inland climatic zonation also manifests itself in the temperature regime, with a 25 km frost free zone along the coast giving way to a May-to-September frost period inland, with up to 50 days of frost in places (Schulze 1997). Nevertheless, day-time midwinter (July) temperatures still rise to 16 °C (inland)

to 24 °C (coast) on average, while in midsummer (January) they average 25 °C (inland) to 28 °C (coast). Annual potential evaporation is 1 500 to 1 700 mm, ranging from 170 to 190 mm in January and to 90–100 mm in July, i.e. around 6 mm d^{-1} in summer and in excess of 3 mm d^{-1} in midwinter (Schulze 1997). Intra-seasonal and interannual variability of rainfall is high, with a coefficient of variation (CV) of annual precipitation around 25–30% while that of the annual runoff is 50–100% (Schulze 1997). As a consequence of the interplay between rainfall and evaporative demand, mean annual catchment runoff, at 162 mm, is only 18% of mean annual precipitation (Kienzle et al. 1997). The strong rainfall seasonality, low rainfall to runoff conversion and high ratio of annual evaporative demand to rainfall of nearly 2 results in large inland tracts of the Mgeni catchment being regarded as hydrologically semi-arid, although the coastal zone is sub-humid. Consequently, net supplementary irrigation requirements are high, at 680–900 mm yr^{-1} (Schulze 1997).

Umgeni Water, the Statutory Water Board responsible for the management and bulk supply of water, has over the past 15 years supplied a mean volume of around 200 Mm^3 annually to consumers living in and adjacent to the Mgeni catchment area. This volume represents 31% of the long-term annual water yield produced in the catchment area of the Mgeni upstream of, but including, Inanda Dam. In order to provide a secure water supply under the highly variable rainfall and runoff conditions, Umgeni Water manages five dams for water supply, with a combined capacity of 745.9 Mm^3. These are the Midmar, Albert Falls, Henley, Nagle and Inanda Dams (Fig. D.80). Their combined volume represents 113% of the mean annual runoff.

Some 6 million people are supplied by water captured in the Mgeni catchment. Without factoring in the present HIV/AIDS pandemic, the population in the Greater Durban and Pietermaritzburg metropolitan areas has been projected to expand to between 9 and 12 million by the year 2025 (Horne Glasson Partners 1989), with major population concentrations within the former homeland areas of KwaZulu in, and adjacent to, the Mgeni catchment (Fig. D.80). Apart from urban concentrations around Durban and Pietermaritzburg (Fig. D.80), the rural subsistence areas are amongst the most densely populated areas in the catchment with rural population densities from 200 to 1 000 inhabitants per km^2. It was recognised nearly 20 years ago that the rapidly accelerating water demand from the Mgeni catchment would exceed local raw water resources soon after the turn of the century (Breen et al. 1985). The increased occurrence of urban return flows, intensified agricultural practices and the present largely uncoordinated growth of sizeable informal urban and peri-urban settlements which are associated with population expansion and migration into the region are expected to lead to further deterioration of the water quality of local receiving streams, major rivers and reservoirs.

Monitoring Network

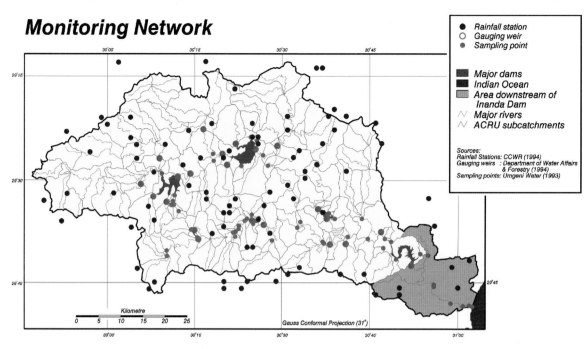

Fig. D.82. Mgeni catchment: monitoring network (after Kienzle et al. 1997)

During summer months, frequent convective thunderstorms of high intensity rainfall within the catchment result in the transport of suspended solid particles. These particles can carry appreciable concentrations of absorbed phosphorus and pathogens, which are indicated by the presence of *E. coli*. The phosphorus and pathogens which reach the receiving streams can result in serious chemical and biological water quality problems in rivers and reservoirs of the Mgeni system, affecting domestic, agricultural, industrial, ecological and recreational user groups. The anticipated decline in water quality will most probably be associated with increased purification costs and health risks in areas where untreated water is still widely used for domestic or recreational purposes. There is also the threat of potentially irreversible degradation of the riverine environment. Furthermore, many of the problems listed above could be substantially exacerbated by human-caused global climate change.

D.7.3 Configuration of the Mgeni System for Simulation Modelling

For purposes of simulation the Mgeni basin was delineated into interlinked, cascading subcatchments which, hydrologically, were considered to respond relatively homogeneously. These subcatchments were configured using the following requirements:

- they had to be relatively homogeneous in terms of climate, soils and land use;

- their outlets had to correspond to locations of streamflow gauging weirs, of instream flow requirement sites at which environmental water assessments are made, of water quality sampling points, as well as important water abstraction/return flow points and major impoundments (Fig. D.82);
- the 137 individual subcatchments had to be integrated into 12 management catchments (Fig. D.83) and six are described in more detail in Table D.28.

For the generation of simulated streamflows and their modifications by reservoirs, abstractions, return flows and irrigation practices, model input for each subcatchment included:

- locational information such as its area, links to up- and downstream subcatchments and mean elevations, the latter derived from a digital elevation model with a grid cell size of 250 m (250 m DEM);
- climate information, i.e. a representative quality controlled daily rainfall record for the 34-year period 1 January 1960 to 31 December 1993, monthly potential evaporation and temperature values derived from a 1' × 1' latitude/longitude grid (Schulze 1997), with these monthly values disaggregated within ACRU to daily values by Fourier Analysis;
- soils information, i.e. area-weighted texture values for critical soil water retention constants, plus thicknesses of top- and subsoils as well as values of saturated drainage rates;
- land-use information consisting of month-by-month above- and below-ground hydrological attributes of

Management Catchments

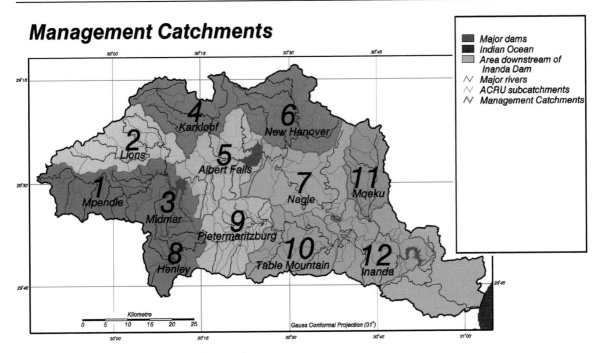

Fig. D.83. Management catchments within the Mgeni system

land cover/use (e.g. interception loss per rainday, leaf area index, water use coefficient, root mass distribution), area-weighted for both the baseline land cover classes (Fig. D.85) and the present land use classes (Fig. D.81);

- streamflow control variables such as baseflow recession constants, month-by-month coefficients of initial abstractions before runoff commences on a rainday (accounting for land use and seasonal rainfall intensity-related infiltrability rates), fractions of adjunct and disjunct impervious areas and effective soil depths from which stormflows are generated;
- dams (i.e. both the major water supply reservoirs as well as the 1 138 smaller farm dams located in the catchment), giving for each the full supply capacity, surface area at full supply capacity, dam area : volume relationships, monthly abstractions and return flows, legal and environmental flow releases, seepage rates and inter-catchment transfers; and,
- irrigated areas, inputting for each the area irrigated, monthly crop water demands, conveyance and other losses, source of irrigation water, mode of irrigation scheduling/cycle times and soil properties of the irrigated areas.

For daily sediment yield modelling additional information for each land use, soil and topographic unit included the following, which were each mapped on a digital terrain model with a grid cell size of 250 m and area-weighted per subcatchment:

- rainfall erosivity, computed month-by-month;
- soil erodibility indices derived at terrain unit (i.e. hill-slope element) level;
- topographic indices, with slope gradients computed from a 250 m DEM and flow accumulation calculated by a set of RUSLE-based equations (Renard et al. 1991);
- cover and management indices, computed month-by-month for each of the 21 land cover classes (Fig. D.85) from percentages of canopy and mulch cover, litter and root mass of the top 100 mm; and,
- support practice indices, as functions of land cover, slope and management practice for a set of rules assigned to each land cover.

Furthermore, for the phosphorus module the following were required per 250 m grid cell and, subsequently, per subcatchment:

- phosphorus fertiliser applications, according to locally determined frequencies and rates of application on the different crops grown in the Mgeni catchment, under further consideration of soil adsorption characteristics which depend on percentages of organic matter and clay as well as soil pH;
- wet deposition rates, locally determined at 0.145 g ha^{-1} mm^{-1} rainfall; and,
- dry deposition rates, locally estimated at 0.66 g ha^{-1} d^{-1} over rural areas, but doubled over urban areas.

For both the phosphorus and the *E. coli* modules two further sources of information were needed:

- livestock sources of phosphorus and *E. coli*, determined from local livestock densities of beef cattle, dairy cows and sheep with respective daily output of phosphorus of 11, 43 and 2.5 g; and,
- human sources of phosphorus and *E. coli* from pit latrines liable to discharge into natural water courses, determined from population densities in rural and informal urban settlements, with enhanced loadings for affected populations within 250 m of a receiving stream.

D.7.4   Verification of Simulated Hydrological Outputs

To generate confidence in the various scenarios of hydrological impacts within the Mgeni basin, especially at locations where no measurements are made, model outputs had to be carefully verified (history-matched) against observations for a wide range of physical and land use conditions. This verification required analysis of mean flows as well as of high and low flows. Verifications of simulated streamflows were undertaken for the 1979–93 period at the gauged outflows of six management catchments (Table D.28). These display wide altitude ranges and land-use combinations (Table D.28) with, for example, urban areas varying from 0 to 47%, peri-urban development from 0 to 22%, commercial af-

forestation from 7.3 to 40%, dryland agriculture from 6.4 to 31% and natural vegetation from 0.1 to 76%. Figure D.84 illustrates graphically the types of information used in verification analysis. Results of monthly totals of daily modelled *v.* observed streamflows show that simulations from all six management subcatchments produced correlation coefficients above 0.89 (Table D.28). In each case, simulated streamflow totals were within 6% of observed values, with five catchments simulating to within 3%. Differences of standard deviations were consistently low, with three management catchments' differences below 4% (Table D.28). Kienzle et al. (1997) showed that sediment yields could similarly be modelled realistically with ACRU, albeit with consistently slight overestimations, and also that *E. coli* relationships with sediment loads were mimicked well, but again with a tendency for some overestimation.

D.7.5   Modelling Impacts of Contrasting Land Uses on Streamflow Generation

Baseline streamflow is used in water resources analysis as an objective datum against which to compare hydrological responses to future land use or climate change. Baseline streamflow is defined here as that which would occur under climatic conditions identical to those at pre-

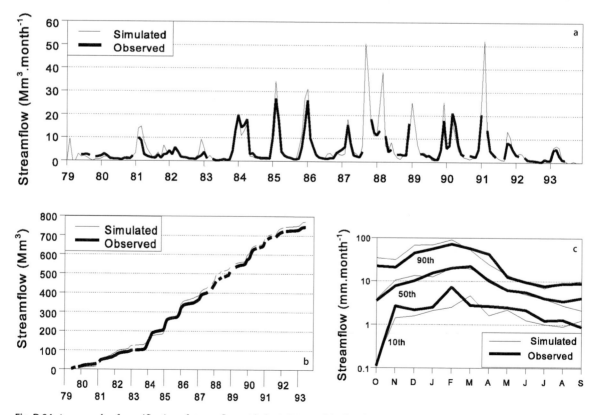

Fig. D.84. An example of a verification of streamflow with the ACRU model taken from the Lion's River management catchment (after Kienzle et al. 1997)

sent, but with the catchment assumed to be covered by natural vegetation (i.e. without anthropogenic interference and in equilibrium with contemporary climate). In this study the baseline land cover is represented by Acocks' (1988) Veld Types, of which six occur in the Mgeni catchment (Fig. D.85). Hydrological attributes such as month-by-month interception loss per rainday, water use and root distribution coefficients and those of initial abstraction for each of the Acocks' Veld Types were determined.

"Present" land use is represented in this study by field checked 1986 SPOT satellite imagery (hence verifications from 1979–93, i.e. seven years either side of 1986). Twenty-one hydrologically distinct land-use classes were identified, to each of which hydrological above-, ground- and below-ground attributes were assigned (Kienzle et al. 1997). Figure D.81 shows the patchwork of "present" land uses which, from the legend, indicates the potential for major land use driven impacts on subcatchment hydrological responses.

The impacts of present land use on runoff have been assessed by comparing model outputs produced under representative "present" land uses to those under baseline land covers for the 137 individual subcatchments of the Mgeni (Fig. D.86). Land-use impacts are seen to be highly significant within individual subcatchments, ranging from a > 100% increase (i.e. doubling) in mean annual runoff to a decrease by more than 60% when compared against runoff from the baseline land cover. The highest reductions of runoff are found in subcatchments which are under intensive agricultural use, in particular those with a high proportion of commercial forest or

sugarcane plantations (cf. Fig. D.81). On the other hand, subcatchments that are either highly urbanised or heavily overgrazed, show increases in water yield, as in the Pietermaritzburg area and in the Valley of Thousand Hills immediately upstream of Inanda Dam. The magnitude of the impact of present land use on hydrological responses becomes visually even more apparent in Fig. D.87, which shows that the near-linear relationship and statistically high correlation ($r^2 = 0.81$) between a runoff coefficient and mean annual precipitation (MAR/MAP) under baseline conditions is almost completely obliterated by the changes in land use to its present conditions ($r^2 = 0.09$).

D.7.6 Modelling Impacts of Land Uses on Water Quality Indicators

A series of spatially discrete, but interrelated, water quality processes presented by sediment, phosphorus and *E. coli* yields play themselves out in the Mgeni catchment primarily because of one important factor – anthropogenically induced changes in land cover. What is both scientifically and operationally challenging in the Mgeni system, however, is the mix of impacts of land use on water quality. The complexity arises due, in part, to the developed agricultural sector on the one hand, with its high phosphorus loadings from the intensive commercial farming and, on the other, the underdeveloped socio-economic sector with high sediment production through overgrazing or the high concentrations of human derived *E. coli* from informal peri-urban shack

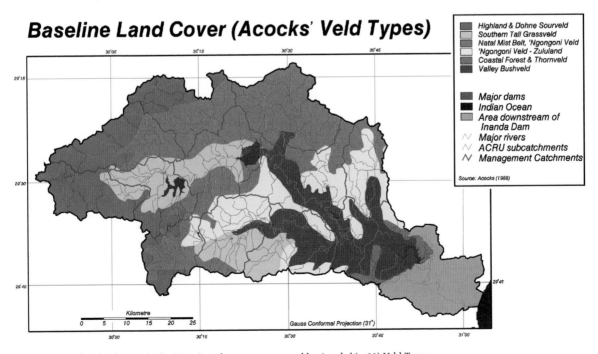

Fig. D.85. Baseline land cover in the Mgeni catchment, represented by Acocks' (1988) Veld Types

Impact of Present Land Use on MAR

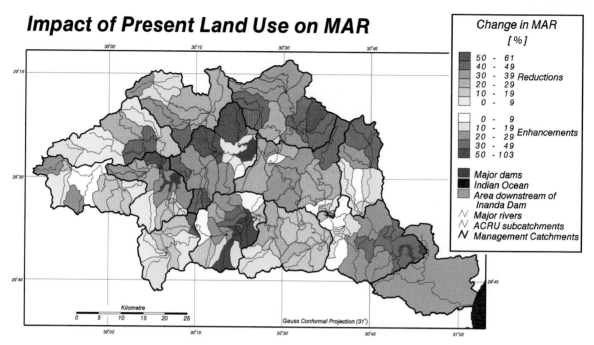

Fig. D.86. Impacts of "present" land use on mean annual runoff (*MAR*) in the Mgeni catchment (after Kienzle et al. 1997)

settlements. This impact study of land use on water quality indicators poses three key questions:

- In which subcatchments are sediment yields high *v.* low, and why?
- In which subcatchments are phosphorous yields high *v.* low, and why? and
- From where, within the Mgeni catchment, is non-point source *E. coli* derived, and why?

The simulations serve as an aid in defining catchment management practices aimed at maintaining acceptable sediment, phosphorus and pathogen loads to receiving waters.

Based on mean monthly rainfall erosivity derived from rainfall intensities, linked with grid values of soil erodibility *K*, the vegetal cover factor (ground + canopy) *C*, the topographic index *LS* and the management practice factor *P*, Kienzle et al. (1997) produced a map of poten-

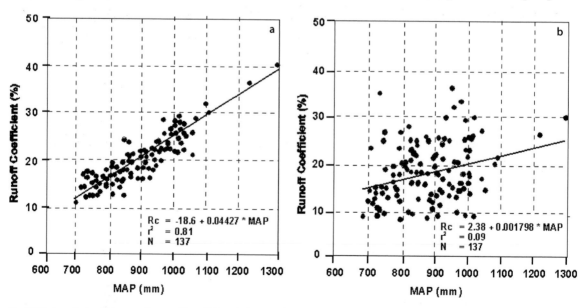

Fig. D.87. Associations between the runoff coefficient and mean annual precipitation (*MAP*) for the 137 subcatchments of the Mgeni for (**a**) baseline land cover and (**b**) present land-use conditions

Mean Annual Sediment Yield

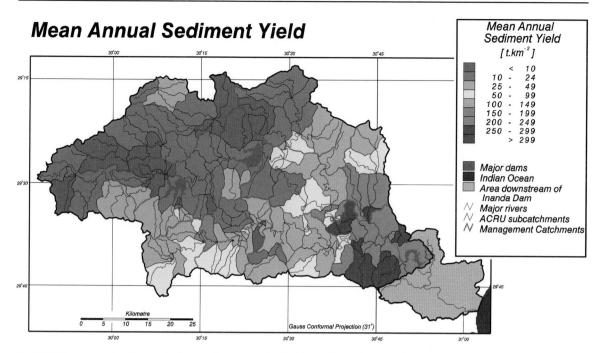

Fig. D.88. Mean annual sediment yields per subcatchment in the Mgeni (from Kienzle et al. 1997)

tial sediment source areas for the Mgeni catchment at a grid cell resolution of 250 m. Not all of the soil particles detached from the parent material in a catchment are transported to the stream network, however. Progressive deposition and re-entrainment of the sediments occur within a catchment. Hence, areas prone to erosion are not necessarily the areas which yield the highest sediment loads into receiving streams because of the inefficiencies inherent in the sediment delivery process. ACRU therefore estimates the amounts of sediment reaching the outlet of each subcatchment from daily stormflows and peak discharges rather than from mean rainfall erosivity indices. A unique approach was adopted with respect to the soil erodibility factors K, in that they were modelled to be dependent also on the prevailing soil moisture status. In the ACRU model, therefore, the soil erodibility potential is greater under dry soil moisture conditions, resulting in significantly enhanced sediment yield when a storm event occurs subsequent to a prolonged dry period – the so-called first flush effect, a phenomenon identified in the Mgeni from experimental catchment sediment yield data.

The averaged annual subcatchment sediment yields derived from simulated daily sediment loads for individual subcatchments ranged from 2 to 629 t km^{-2} (Fig. D.88), which compares well with those reported from reservoir sediment surveys in the Mgeni by Rooseboom et al. (1992), which ranged from 20 to 723 t km^{-2} yr^{-1}. Within the study area the degraded Valley of Thousand Hills delivers the highest sediment yields, as do areas around Edendale (cf. Fig. D.80), where large numbers of infor-

mal dwellings are located. Within the Mgeni catchment, sediment delivery ratios were found to vary from 9.6% of soil loss potential to 44.8%.

Time series were generated with ACRU to gain insight into temporal response patterns of sediment yields. Figure D.89 shows monthly totals of daily sediment yields for the 34-year simulation period for two subcatchments with, respectively, low and high sediment yields. What is apparent is that a single flood event with low recurrence probability can mobilise very large amounts of sediment. Simulations in a low sediment yielding subcatchment upstream of Midmar Dam resulted in 18.1% of the total 34-year sediment yield being transported during the single week in September 1987 when a flood with a recurrence interval of ~ 50 years occurred. In subcatchments with an already high sediment load, extreme events such as the September flood of 1987 play a relatively diminished role, because the background sediment load is higher. Thus, for a subcatchment near Inanda Dam, only 10.8% of the total 34-year sediment load was transported during the September 1987 floods.

Gridded inputs at 250 m resolution for wet and dry atmospheric deposition, numbers of beef cattle, dairy cows and sheep, amounts and application frequencies of fertilisers on sugarcane, maize and mixed crops as well as inputs from human faeces where inhabitants, especially those close to receiving streams, are without proper sanitation, were used to produce a mean annual phosphorus yield map (Fig. D.90), showing values ranging from 0.5 to 850 kg km^{-2} for the 137 subcatchments. The distribution of high and low annual phosphorus yield values is

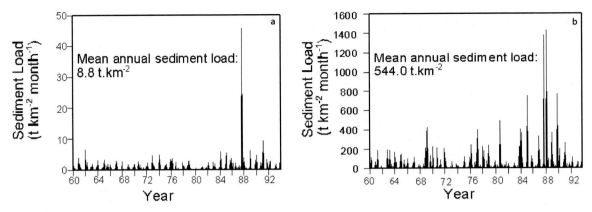

Fig. D.89. Time series of simulated monthly sediment loads for a subcatchment with a low and a high mean annual sediment load (after Kienzle et al. 1997)

quite dissimilar to that of annual sediment yields, despite adsorption of phosphorus to sediment particles. Thus, significant phosphorus loads emanate from the Albert Falls Dam area with its many cattle feedlots, whereas that subcatchment has a relatively low sediment yield.

While bearing in mind that ACRU tends to overestimate *E. coli*, areas of excessively high mean annual *E. coli* nevertheless occur (Fig. D.91; range 30–18 200 counts per 100 ml), associated with informal settlements having poor sanitary facilities, in particular along the Msunduzi river in its lower reaches and the central parts of the Valley of Thousand Hills. It is also evident that high *E. coli* concentrations are not always associated with areas of high sediment yield, despite the inherent link, particularly where impacts of large populations from informal settlements are the dominant cause.

D.7.7 Scenario Studies on Impacts of Land Use on Water Quantity and Quality

D.7.7.1 Effects of Individual Land Uses on Runoff at the Management Catchment Level

Impacts of present land use combinations at the management catchment scale were evaluated, as were impacts of two individual and hydrologically sensitive agricultural land uses, specifically irrigation and afforestation (Table D.29). All impacts were assessed against the mean annual runoff from baseline land cover (Scenario A) for the period 1960 through 1993. Scenario B shows that the present land-use combinations given in Table D.29 can already decrease an entire management

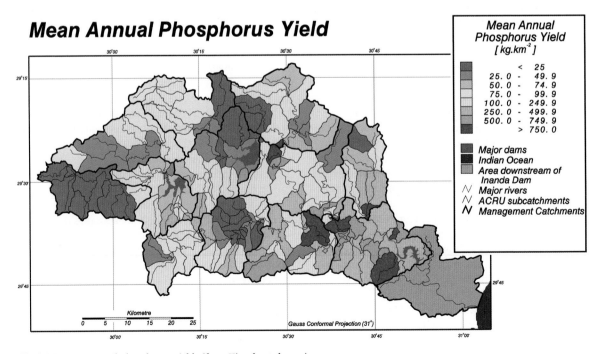

Fig. D.90. Mean annual phosphorus yields (from Kienzle et al. 1997)

Mean Annual E.coli Concentrations

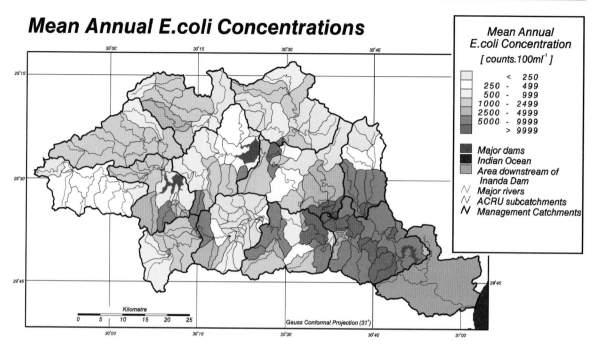

Fig. D.91. Mean annual *E. coli* concentrations from non-point sources per subcatchment in the Mgeni (from Kienzle et al. 1997)

catchment's annual runoff by up to 40%. In Scenario C the reduction in streamflow at the outlet of management catchments as a result of irrigation only (i.e. assuming all other areas to be under baseline land cover conditions) is shown to be severe. For example, with 9.0% of the area of the Lions management catchment under irrigated crops (cf. Table D.28), the resultant mean annual runoff reduction is nearly 23%, i.e. a streamflow reduction-to-area ratio of 2½ (22.8/9.0), while in the New Hanover management catchment this ratio is over 4 (4.4/1.0). Streamflow reduction-to-area ratios highlight the impact which irrigation abstractions, from both farm dams and direct pumping from streams, can have on water resources. Commercial afforestation to fast growing exotic species (e.g. eucalypts, pines, wattle – all harvested at 10–20 years), which is the agricultural land use most targeted as a streamflow reduction activity in the National Water Act of South Africa (NWA 1998), has run-

off reduction-to-area ratios ranging from only 0.65 in the Karkloof management catchment to 1.33 in the Lions management catchment.

Three features are highlighted by the above analysis:

- One cannot make simplistic statements about a given land use having a certain hydrological impact, for this is largely dependent on local climate and soils;
- Different land uses may have very different impact ratios when viewed from the basis of streamflow reduction to area of land use, as illustrated by the irrigation *v.* afforestation case above; and,
- Doubling the area under a given sensitive land use, e.g. commercial afforestation (Scenario E), does not double the runoff reduction from a catchment, but rather has a relatively diminishing impact. This arises because a streamflow reduction by a given land use depends on the original land cover it replaces, as well

Table D.29. Impacts of selected land-use scenarios on mean annual runoff in three management catchments (MCs) (*MAP:* mean annual precipitation)

Scenario		Mean annual runoff (mm)					
		Lions MC (MAP = 979 mm)		Karkloof MC (MAP = 1 081 mm)		New Hanover MC (MAP = 926 mm)	
A	Baseline land cover	233.4		345.6		218.5	
B	Present land use	204.5	(–12.4%)	277.6	(–19.7%)	130.1	(–40.5%)
C	Baseline + irrigation	180.2	(–22.8%)	319.7	(–7.5%)	208.9	(–4.4%)
D	Baseline + afforestation	192.9	(–17.4%)	272.0	(–21.3%)	156.7	(–28.3%)
E	Baseline + 2 × afforestation	178.4	(–23.6%)	241.6	(–30.1%)	134.4	(–38.5%)

as on the aggregation of individual runoff producing events – each of which responds differently over space in a non-linear hydrological system. The existing mix of other land uses on the impacted catchment is also important to consider.

D.7.7.2 Impacts of Subsistence Farming and Informal Settlements on Water Quality and Quantity

This scenario was modelled by altering relevant variables for the 16 subcatchments upstream of Pietermaritzburg to reflect a doubling in the area of subsistence agricultural smallholdings and a doubling of the informal settlement population, assuming no improvement in sanitary conditions. Hydrological responses were then simulated with the revised land uses and population densities and compared to simulated outputs under present conditions (Fig. D.92).

Water quality deterioration is immediately apparent due to the combination of informal settlements and subsistence agriculture. The relative increase in the different water quality indicators is significant, however. While sediment yield is simulated to increase by only 17%, the phosphorus yield increases by 76% and the *E. coli* concentrations by 40%. Phosphorus yield increase is by far the most significant due to its dependence on both of the projected changes (i.e. population influx as well as agricultural expansion). The *E. coli* result is affected by population increase, together with projected livestock increases associated with informal settlements. The sediment yield is affected predominantly by the increase in land under tillage and, to a lesser extent, the increase in runoff caused by the expansion of impervious and compacted areas associated with settlements. Water quantity is not as severely affected as the water quality. Baseflow decreases by some 20%, but the total streamflow increases by 6%. These changes are associated primarily with the increases in impervious and compacted areas.

D.7.7.3 Impacts of Potential Climate Change on Streamflows

Greenhouse warming is a plausible future scenario with potentially significant hydrological implications over southern Africa (Perks 2001; cf. Sect. E.6.4). Projected ratio changes in monthly rainfall magnitudes and absolute changes in monthly means of daily maximum and minimum temperatures were derived for the Mgeni catchment from the HadCM2+S GCM (i.e. including sulphate forcing, Murphy and Mitchell 1995) by interpolative downscaling of parameter values from the 2.50° latitude × 3.75° longitude GCM grid to a 0.25° grid by inverse distance weighting.

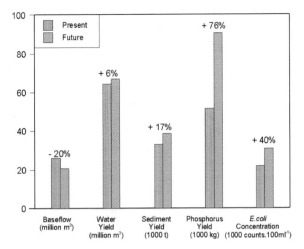

Fig. D.92. Comparison of the hydrology and water quality indices upstream of Pietermaritzburg between present and future land-use scenarios comprising a doubling of areas under subsistence agriculture and informal settlements (after Kienzle et al. 1997)

Over the Mgeni catchment, monthly T_{max} increases by 1.4 to 2.1 °C for a $2 \times CO_2$ climate change scenario with this GCM, T_{min} by 1.1 to 2.2 °C while summer month rainfalls decrease by 5 to 15% with virtually no rainfall changes in winter months. Daily values of present climate input were perturbed by their respective monthly changes while the $2 \times CO_2$ transpiration feedback option in ACRU was also invoked, assuming all vegetation to consist of C4 plants.

For this climate change scenario, impacts on streamflow from a baseline land cover showed individual subcatchments to decrease their mean annual runoff by 20–30% in the north-west of the Mgeni catchment, 10–20% in the centre and < 10% in the south-eastern subcatchments. The impact of this climate change on accumulated streamflow from upstream to downstream is illustrated in Fig. D.93, with the general west to east decrease in impact evident.

Five points of significance arise from this analysis:

- Given our present understanding of global climate change and techniques for downscaling climate driving variables to more local domains such as the Mgeni and its subcatchments, the trends of climate change induced streamflow changes (decreases in this case) are spatially relatively smooth compared with the highly localised and patchy impacts resulting from land-use change (cf. Fig. D.93 *v.* Fig. D.81);
- In a highly developed/perturbed catchment such as the Mgeni the changes in streamflows induced by present land uses are simulated to be far in excess of those from a typical $2 \times CO_2$ climate change scenario;
- The impacts of land use on streamflows can be both positive and negative (Mgeni: +100% to –60% change in mean annual runoff; Fig. D.86) while at the spatial scale of the Mgeni, climate change impacts on stream-

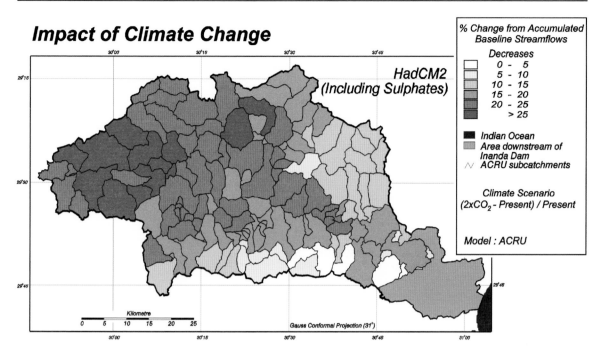

Fig. D.93. Accumulated changes to baseline streamflows in the Mgeni for a $2 \times CO_2$ climate change scenario (after Perks 2001)

flows are usually unidirectional only, i.e. either all positive or all negative (–5% to –30% mean annual runoff);

- For the above reasons, land-use impacts on streamflows can therefore be largely self-cancelling on accumulated flows downstream, while in the case of a unidirectional climate change, effects on accumulated flows downstream are amplified;

- In catchments where a general decrease in streamflows is predicted with climate change, as with the Mgeni, the non-linear runoff to rainfall relationship dictates that the biggest reductions generally occur in higher rainfall headwaters of a catchment (cf. Fig. D.93). The resulting large accumulated streamflow reductions cascade downstream to where human and industrial demand on water is usually highest (e.g. the Pietermaritzburg-Durban industrialised metropolitan areas). This aggregate change could, therefore, markedly influence future economic development in the area.

D.7.8 Conclusions

This case study set out to assess a select set of present day and future anthropogenic activities which influence the spatial and temporal character of the hydrology and water quality of the Mgeni catchment. The results have highlighted, *inter alia*, differences in the way the environmental system has reacted to the juxtapositioning of developed and underdeveloped sectors of the population and economy within one catchment.

To the modeller the elimination of uncertainties in simulation of land-use impacts remains a foremost challenge, as is the availability of suitable field observations to improve our understanding of processes and hence their representation in models. Furthermore, the modeller nowadays has to provide answers to pressing questions on land use and climate change impacts posed by legislators and operators alike, for example on streamflow reduction activities, pollution standards or environmental water requirements. The catchment manager, on the other hand, views land-use impacts in light of sustained water yields and purification costs of water contaminated by sediments, chemicals and pathogens, all of which link to socio-economic development within the catchment. To the manager this must be coupled with the prospect of rural to urban and peri-urban migration, uncertain population growth rates in the foreseeable future (because of the HIV/AIDS pandemic) and new uncertainties superimposed by the prospect of global climate change. Physiographically, socially and economically divergent catchments such as the Mgeni thus pose challenges to modellers and managers alike with respect to the impacts of land-use change, climate change, and their interactions well into the future.

Acknowledgments

This case study emanated from research funded by the Water Research Commission. Their financial support is gratefully acknowledged. The Computing Centre for Water Research (now closed) is also thanked for making available computing and GIS facilities.

Chapter D.8

Conclusions: Scaling Relative Responses of Terrestrial Aquatic Systems to Global Changes

Michel Meybeck · Charles J. Vörösmarty · Roland E. Schulze · Alfred Becker

D.8.1 Terrestrial Aquatic Systems and the Earth System under Pressure

From an even cursory reading of the material presented in Part D, it should be quite apparent that terrestrial aquatic systems encompass a broad set of biogeophysical landscape features and complex processes. Terrestrial aquatic systems include water, waterborne material, sediment and biota in vegetation, the soil unsaturated zone, groundwaters, wetlands, rivers, lakes and ar-

tificial water bodies such as reservoirs and canals. The fundamental drivers of water circulation and related material fluxes (for nutrients, carbon, particulate matter, pollutants) are multiple and combine physical, chemical and biological processes including open water evaporation, precipitation, infiltration, water runoff generation, water routing, erosion, leaching, weathering, silting, evapotranspiration, biological uptake and bacterial degradation. Together with their associated coastal zones, terrestrial aquatic systems constitute what we define as continental aquatic systems (CAS) (Fig. D.94).

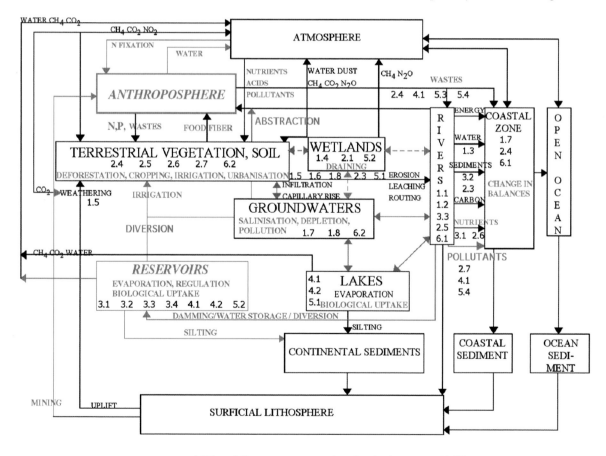

Fig. D.94. Continental aquatic systems (CAS) in the present-day Earth system (in *red*: major impacts of human activities; *grey*: water flux; *numbers* refer to Table D.30; modified from Meybeck 2003)

Some CAS fluxes have very short cycles of only days to weeks, such as those associated with major atmospheric cycles, while others have very long cycling times that span geological time scales. The time domain of water transferred from headwaters to the receiving bodies ranges from a few days when routed through river channels (Vörösmarty et al. 2000b), to years and even a century if large lakes and/or groundwater pools are present. At the global scale, river waters continuously carry enormous fluxes of material to the oceans (40 000 km^3 yr^{-1} of water with 20×10^{15} g of suspended matter, 4×10^{15} g of dissolved salt as Ca^{2+}, SO_4^{2-}, dissolved inorganic carbon and

0.4×10^{15} g of organic carbon). They also provide the coastal zone with essential nutrients as nitrogen, phosphorus and silica as well as with $25 \pm 10 \times 10^{15}$ g of particulates that regulate coastal morphology.

Human activities have greatly modified the Earth system through climate change, land-cover and land-use changes, water engineering and the release of wastes to aquatic systems (Turner et al. 1991). In the past 50 years, this anthropogenic influence has exceeded natural forcings in many parts of the world, or for some issues such as the nitrogen and phosphorus inputs to ocean, helping to define a new era, the *Anthropocene* (Crutzen and

Table D.30. Major global pressures on continental aquatic systems and the mapping of local-to-regional scale impacts to the global scale (also reported in Earth system dynamics, Fig. D.94, adapted from Meybeck 1998). *A*: human health, *B*: hydrological cycle balance, *C*: water quality, *D*: global carbon balance, *E*: fluvial morphology, *F*: aquatic biodiversity, *G*: coastal zone impacts. Only the major links between issues and impacts are listed here (*POPs* = Persistent Organic Pollutants)

Pressures	Local to regional changes of environmental states	Global impacts						
		A	B	C	D	E	F	G
1 Climate variability and climate change	1.1 Development of non-perennial rivers		•	•	•	•	•	•
	1.2 Segmentation of river networks					•	•	
	1.3 Changes in flow regimes		•			•	•	•
	1.3 Development of extreme flow events		•			•	•	
	1.4 Changes in wetland distribution/function	•	•		•		•	
	1.5 Changes in chemical weathering				•			•
	1.6 Changes in soil erosion				•	•		•
	1.7 Salt water intrusion in coastal groundwaters			•				
	1.8 Salinisation through evaporation		•	•			•	
2 Land-use change	2.1 Wetland filling or draining		•	•			•	
	2.2 Changes in water passways		•	•				
	2.3 Change in sediment transport					•	•	•
	2.4 Urbanisation	•	•	•			•	•
	2.5 Alteration of first order streams					•	•	
	2.6 Nitrate and phosphate increase	•		•	•			•
	2.7 Pesticide increase	•		•				•
3 River damming and channelisation	3.1 Nutrient and carbon retention				•			•
	3.2 Retention of particulates				•	•		•
	3.3 Loss of longitudinal and lateral connectivity						•	
	3.4 Creation of new wetlands	•		•	•			
4 Industrialisation and mining	4.1 Increases in heavy metals and POPs	•		•				•
	4.2 Acidification of surface waters			•			•	
	4.3 Salinisation	•		•				
	4.4 Sediment sources					•		
5 Urban wastes	5.1 Nitrate and phosphate increase	•		•				•
	5.2 Enhancement of water-borne diseases	•						
	5.3 Organic pollution	•		•			•	
	5.4 Heavy metals and POPs increase	•		•				•
6 Irrigation/water transfer	6.1 Partial to complete decrease of river fluxes					•	•	•
	6.2 Salinisation (evaporation and percolation)		•	•				

Stoermer 2000; Meybeck 2002). Few of these contemporary material fluxes are completely new. They represent major modifications of existing transfers of water and constituents. For example, mining can be considered to enhance the natural processes of erosion and weathering. Similarly, in the case of N_2 fixation from the atmosphere for nitrogen chemistry, the fixation by artificial fertiliser usage now exceeds the natural fixation. Most natural elemental cycles are accelerated, i.e. fluxes between Earth system reservoirs are increasing. Xenobiotic organic compounds (CFCs, PCBs, solvents or pesticides) that do not occur in nature are produced in increasing quantities and their fate in the environment represents a unique anthropogenic signature on the chemistry of the planet.

The Anthroposphere (Fig. D.94) is a new component of the Earth system, defined here by the land-use acceleration by humans in agriculture, urbanisation and industries by technical means, with their related energy and material transfers within and between these systems and releases of material to the environment. Some of the environmental impacts associated with the Anthroposphere can be regarded as permanent or irreversible, even on human time scales of a few generations. Examples abound, including dam/reservoir construction, river diversion, storage of pollutants in major continental aquatic system subcomponents, and inland and coastal water eutrophication. Some of the changes in water and dissolved matter fluxes may affect inland aquatic systems virtually immediately, with consequent impacts on the coastal zone delayed to varying degrees, up to several years for large basins. In contrast, coastal zone changes with respect to riverborne particulate fluxes can take decades or more, due to the much slower transfer of sediments through continental aquatic systems. The human impact on linkages between the continental land mass and the world's coastal zones thus depend on the specific elemental flux we are interested in, the degree of human disturbance, the level of socio-economic development and environmental governance, and the geophysical time constants involved (von Bodungen and Turner 2001).

The changes we note yield other important consequences, affecting also water availability and quality, fluvial morphology, aquatic biodiversity and human health (Table D.30). We acknowledge that these are of great importance to society, but have not been covered in this synthesis (see for example, Vellinga 1996; Rotmans and deVries 1997; Revenga et al. 2000; Vörösmarty and Sahagian 2000).

D.8.2 Spatial Organisation of Terrestrial Aquatic Systems and Their Responses to Anthropogenic Change

The spatial heterogeneity of terrestrial aquatic systems is always high, and any aquatic system can be regarded as "complex and very heterogeneous". Of course, the perceived degree of heterogeneity will depend on the scale of interest (Fig. D.95). Thus, what a small catchment process hydrologist might consider to be an overly simplistic representation of processes in a land-surface hydrology model embedded within a global circulation model, could be entirely appropriate for the task at hand.

At the plot scale (10^1–10^4 m^2), the distribution of root systems and soil water show a marked vertical heterogeneity (see Chapt. D.2). At such a scale, the lateral transfers of water and constituents are difficult to study except by using lysimeters or erosion plots. Here, human perturbation can arise from several sources such as direct land-cover change or diffuse atmospheric pollution (Fig. D.95) or the response of vegetation to atmospheric CO_2 increase.

At the small catchment scale (10^{-1} to 10^3 km^2) vertical heterogeneities still exist (Fig. D.95), in particular associated with groundwater (A), but important lateral transfers of water, particulate and dissolved material are also observed, both episodically or on a permanent basis. Vertical heterogeneities continue to be associated with the complex mosaic of land cover and its sensitivity to climate and CO_2 change (B), land use and management (C), use of agrochemicals (D), fine topography (E) and its related microclimate. Upstream-downstream flow structure is organised through a "waterscape" (F) that includes rills, gullies, brooks, small streams, and local wetlands, as well as artificial waterbodies, drainage works, ditches, rice paddies, ponds or farm reservoirs. Biogeochemical recycling of nutrients in wetlands (G) or within the riparian corridor (H) must be considered for understanding water and constituent fluxes at this scale.

At the larger river basin scale (10^2 to 10^5 km^2) heterogeneities arise both laterally and longitudinally. A first set of natural heterogeneities is apparent when comparing different subbasins in terms of river flow regimes, sediment supply and transfer, water chemistry linked to climate, relief, lithology and others (I, J, K). A second set of natural variations includes the longitudinal, downstream increase in stream order, variations in water velocities, and of the increased deposition of sedimentary materials. A final set of heterogeneities concerns the position and dynamics of the individual elements that constitute the natural and anthropogenic waterscape. These specifically include the floodplain (L), regional aquifers, swamps, individual lakes or lake provinces (M), reservoirs (N), surface and groundwater water use by humans (R), levees, navigation and irrigation canals (O), urbanisation, leaching of agricultural soils (P), point sources of pollution (Q), (R), damming and barrage operation (S). In the greatest basins shared by many countries the water policy and the economic development stage may add yet another type of heterogeneity. When rivers fi-

Fig. D.95.
Spatial organisation of terrestrial aquatic systems and key processes (see text for definitions of *individual letters*)

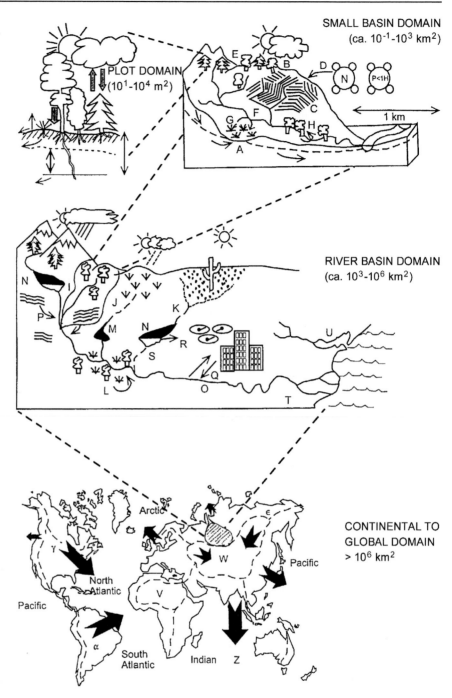

nally discharge water and materials to the ocean, their influence on the recipient coastal zone will depend on the geological, morphological, tidal, and hydrodynamical conditions at the local land/ocean interface (e.g. estuaries, deltas, lagoons) which may completely modify and/or weaken the river signal (T, U).

At the global scale (Fig. D.95) the overall heterogeneity of flux in terms of water, sediment, carbon, nutrients, pollutants can be assessed from direct field measurements. If the first 250 world exorheic rivers were monitored they would correspond to about half of world basins area discharging to oceans (Meybeck and Ragu 1997). These measurements can be used to set up the causal factors contributing to the global pattern of water and material fluxes, such as climate, lithology, relief, land use, population density, and socio-economic activities. These relationships permit us to then extrapolate riverine fluxes from better-known to nondocumented areas of the Earth. Global maps have now been established at various resolutions from 2' to 30' (latitude × longitude). From such

Fig. D.96.
Natural drivers (*top*) and anthropogenic pressures (*bottom*) occur across a range of spatial scales, but dominate hydrological responses over a narrower spectrum

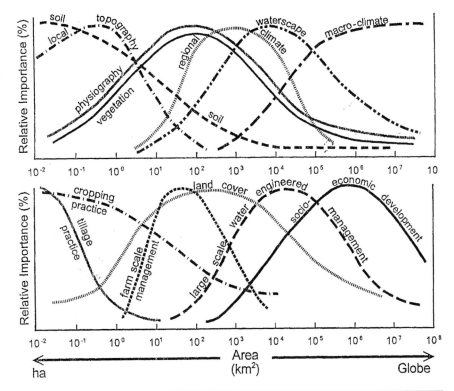

analysis (see Chapt. D.4) it is noted that the land mass is, again, highly variable, with specific riverine fluxes (per unit land area) commonly ranging over two to three orders of magnitude for the land area bearing permanent and occasional runoff (> 3 mm yr^{-1}). Moreover, these emerging models allow us to separate and map three types of land-ocean connections (Fig. D.95):

- arheism (V), which corresponds to the absence of surface runoff and riverine fluxes to receiving waters, as over the Sahara;
- internal river drainage or endorheism, as in the Volga basin and in central Asia (W); and,
- external river drainage to oceans or exorheism, as in South-east Asia (Z).

These models can also identify and apportion the sources of material originating from different areas within river basins, as for the Amazon (α), and the sinks of land-derived material, such as for carbon species in major lakes, floodplains, and reservoirs, as in the Saint Lawrence basin (γ). Oceans and regional seas freshwater and material budgets can also be facilitated once their drainage boundaries have been carefully setup. Past river connectivity to oceans due to climate change, as for the Kerulen/Amur system (ε) or new endorheism due to water use on basins (e.g. Colorado, Nile) illustrate the rapidly progressing nature of changes in river systems and their coupling to the coastal zone at the global scale.

D.8.3 Spatial Scale of Drivers Operating on Terrestrial Aquatic Systems

Spatial scales in hydrology (e.g. plot, field, small catchment, river basin or continental scales) are not fixed in their areal definitions, but rather merge into one another and therefore overlap. From the plot and field scales to the domain of whole large river basins, numerous types of dominant drivers can be identified that are both natural and anthropogenic (Fig. D.96).

D.8.3.1 Natural Drivers

At the plot scale (*c.* 10^{-5} to 10^{-2} km^2) natural drivers are related to soil (texture, structure, chemistry) and vegetation (biomass, leaf area index, root depth). At the crop field level (*c.* 10^{-2} km^2) and at the hillslope and/or elementary catchment level (*c.* 10^{-2} to 10 km^2) topography, microclimate, landslide occurrence, water routing within the soil toposequence figure prominently. For small to medium basins (*c.* 10^1 to 10^4 km^2) meso-scale landscape physiography (relief patterns, altitudinal gradients, wetland and small lake occurrence) as well as lithology must be considered. In large basins (*c.* 10^5 to 10^7 km^2) differences in regional climatic patterns (precipitation, temperature, evaporation) and fluvial system characteristics (channel types, floodplain extent, connections with major aquifers, estuarine types) are es-

Fig. D.97.
Nested and contrasting time scales in the responses of aquatic systems to natural and anthropogenic change from the individual hydrological response of representative basins (A) to the long-term Holocene evolution of catchments (D). Future evolution corresponds to different scenarios B2 and B3, C1 to C4, D1 to D3 (C: water quality; Q: river discharge; F: flux)

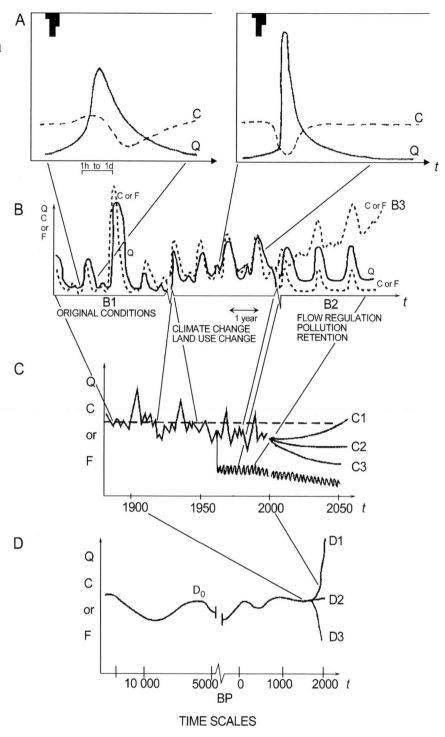

sential determinants of river flux regime and of net inputs to the coastal zone. At the global scale, the dominant features defining the overall pattern of runoff and material fluxes are macroclimatic attributes such as latitude, prevailing weather systems, continentality, endorheism, present-day tectonics, and past geological and climatic history.

D.8.3.2 Anthropogenic Drivers

Anthropogenic pressures can also be classified with respect to pertinent spatial scales. Over cultivated fields, water and material transfers are dependent on tillage and soil conservation (terracing, contouring), at the farm level on crop choice, planting dates, rotation patterns, plant densities, and fertiliser use. The presence of small dams and ponds, local irrigation, borehole abstraction, also should be taken into account. For whole basins (10^3 to 10^6 km^2), water management plays a major role on water and material fluxes through the operation of medium and large dams (storage from 10^8 to 10^9 m^3), large irrigation schemes (> 10 000 ha), overpumping of large aquifers, levees and lock operations, water diversions, urbanisation, industrialisation, and mining. Land-cover change resulting from deforestation, afforestation, plantation cropping, transition from pasture to cultivation also markedly affect water and its related fluxes at local to regional scales. Its accumulated effects on land-surface hydrology through feedbacks with the climate system, as discussed earlier in this volume, may also make such changes important to the water cycle at the global scale.

D.8.3.3 Integrated Water Management and Governance

Differences in demographics and stage of economic development, as well as the history of environmental protection, water policy and governance, add another dimension of complexity to identifying key drivers (Fig. D96). These transcend the realm of the natural system and are organised as a set of human dimension issues, particularly relevant at the country level up to the global scale. It must be noted that most of such anthropogenic factors are spatially distributed according to administrative and political boundaries, or sometimes to cultural and religious boundaries, which rarely correspond to those of the natural drivers. This discrepancy is one of the many difficulties we find at any scale for effectively applying integrated water management. In an increasing number of international river basins regional water management treaties have been set up (e.g. Colorado, Nile, Danube, Rhine, Baltic Sea basin). However, apart from drinking water standards set up by the World Health Organization, there is still no evidence for the international governance of the multiple water-related issues which have been identified here at the global scale. A harmonisation of conceptual and geographical boundaries to adequately manage both natural and human drivers is clearly needed.

D.8.4 Time Scales of Responses of Continental Aquatic Systems (CAS) to Imposed Changes

The time scales of continental aquatic systems' (CAS, i.e. all terrestrial aquatic systems plus the land-ocean interface) responses to changes is tentatively illustrated on Fig. D.97. The shortest time scale considered here is associated with stream and river hydrographs, which range from a few hours for the response of a small catchment to a single rainfall event, to a few weeks or months for seasonal high-water periods in the largest river basins. Two types of changes can be identified. In the first, the hydrograph response to a unit rainfall can be modified rapidly by a land-cover/land-use change or by a change in the regime of rainfall intensity, or duration may change the regime due to climate variability. In both cases, the pattern of water quality variation (C) with river discharge (Q) may also be quantitatively and qualitatively changed (Fig. D.97 A).

Over longer time horizons (for example, from 50 to 150 years), when long riverine records are available, we see that the year-to-year variability of hydrological fluxes can sometimes be modified by climate or land-cover changes, producing the more "spiky" hydrographs as in Fig. D.97 (A). However, the most striking changes at this time scale are those observed on regulated rivers (Fig. D.97, B2) or polluted rivers (Fig. D.97, B3).

When the hydrological and constituent fluxes are modelled and validated over the period of documentary records, they can be used to explore the future evolution of aquatic systems (Fig. D.97, C, C1 to C3), provided that reliable scenarios are available for both climate change and/or socio-economic pressures and water uses.

Other models, generally based on environmental archives, can be used to reconstruct the evolution of aquatic systems during the Holocene due to climate and land-use change (Fig. D.97, D0), then extrapolated to future conditions according to given socio-economic scenarios (Fig. D.97, D, D1 to D3). For some hydrological fluxes, such as for suspended matter, the human impact from the first wave of deforestation (Fig. D.97, D1) can largely exceed the variability induced by climate change only (Fig. D.97, D0). Such events are particularly well recorded in lake sediments and will be collected at the global scale within the LUCIFS programme of IGBP-PAGES (LUCIFS 2000). Recent evolution of aquatic systems shows various patterns including rapidly increasing fluxes (D1) such as for nitrates and some metals, decreasing fluxes to oceans (D3) in highly regulated river basins, or slight changes (D2) which are still difficult to differentiate from natural variability (D0). Table D.31 provides a summary of the various state changes of continental aquatic systems to natural drivers and human pressures with their related time scales.

D.8.5 Continental Aquatic Systems and Emergence of the Anthropocene

Major human pressures on continental aquatic systems have existed since the start of agriculture some 7 000 to 5 000 years ago. During this early period, impacts were

Table D.31. Major changes occurring on continental aquatic systems over different time scales (adapted from Acreman 2000 and Schulze 2001)

Time scale (years)	Drivers and pressures[a]	Changes of state of aquatic systems
10^6	Tectonic uplift	Stream incision
		Increased soil erosion
		River channel change
10^5	Glacial and inter-glacial periods (CV)	Changes in land/ocean connection
10^4	Changes in vegetation and partitioning of rainfall	Gradual evolution of river regime
		Changes in soil formation and erosion
10^3	Agricultural development (LU)	Transpiration, partitioning of water
	Urbanisation (LU)	Increased peak discharges
10^2	Industrialisation (DI)	Surface and groundwater contamination
10^1	Groundwater abstraction (DI)	Declining water table
	Fertiliser/pesticide applications (DI)	Surface and groundwater pollution
	Industrial decline, waste treatment (DI)	Cleaner rivers
	Land use/management change (LU)	Streamflow generation mechanisms
		Stormflows and baseflows changes
10^0	Interannual climate variability (CV, CC) (e.g. ENSO)	Droughts; floods; increased evaporation
	Change in crop types (LU)	Flow regime modification
	Large dam operation (DI)	Stormflow and baseflow regulation
10^{-1}	Seasonal climate variability (CV, CC)	Evapotranspiration
		Partitioning of water
	Tillage practices (LU)	Soil loss/sediment yield
10^{-2}	Extreme flood event (LU, CC)	Flood damage
	Small dam operation (DI)	Fewer spates; steadier flow
	Chemical spill and accident (DI)	Fish kills and damage to biota

[a] *CV:* Climate variability; *CC:* climate change; *DI:* direct human impacts (water use, waste releases); *LU:* land use change.

fairly localised as land use was effectively operating on much less than 1‰ of the Earth's surface. The development of major, centralised civilisations in Egypt, Mesopotamia, Indus Valley and China, some 3 000 years ago has often been accompanied by important land and water management and river engineering schemes. Notable examples exist in the Nile, Shatt El Arab, Indus and Huang He basins, then over the whole Mediterranean basin following expansion of the Roman Empire. Evidence of several mining impacts (e.g. Rio Tinto in Spain, Leblanc et al. 2000) and of general atmospheric pollution in the mid-northern latitudes around 2 500 B.P. in the ancient world has also been found in coastal marshes (Alfonso et al. 2001) and in peat bogs (Shotyk et al. 1998).

The development of such mining activities first in Eurasia, then in South America and the rest of the world, has resulted in a dotted distribution of hot spots of metal contamination sites. These sites now gradually leak contaminants to the downstream catchments. Careful studies of sedimentary archives of lakes, river deltas, estuaries (Valette-Silver 1992) and flood plains allow for the detailed reconstruction of these impacts which can last for a hundred years and more (Middelkoop 2002; Grosbois et al. 2001; Macklin et al. 1997; Hudson-Edwards 1999) and constitute a major threat for the future.

The progressive development in time and space of human pressures leading to the Anthropocene is presented in Fig. D.98 as a working hypothesis to be tested through future analysis (Meybeck 2002, 2003). We limited ourselves to gross categories of human pressure and postulated how an increasing fraction of the Earth's surface has been exposed to these. The progression to a global-scale impact can take two pathways. With the first, impacts are displayed locally, but due to the pandemic distribution of a particular class of change, the consequences are global in domain. A good example is the widespread conversion of land to agriculture and forestry. Global-scale impacts also arise from teleconnections operating over the planetary domain. For example, increased climate variability, hypothesised to be linked to greenhouse warming, has the potential to influence the entire planetary surface. Another example is the long-range atmospheric transport of pollutants as NO_x and SO_2, responsible for the acidification of surface waters, sometimes hundreds of kilometres away from emission sources. These statements should not imply that all impacts are now globally significant. In fact, most well-documented impacts on aquatic systems are essentially localised. And even for truly global phenomena, like greenhouse-induced climate change, some regional responses

Fig. D.98.
Working hypotheses on the occurrence of some major pressures on continental aquatic systems at the global scale and related environmental remediation responses (note the time acceleration; adapted from Meybeck 2003)

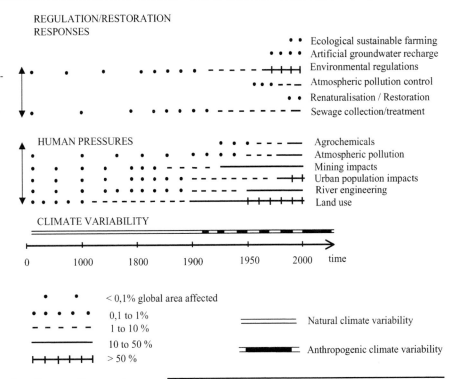

REGULATION/RESTORATION
RESPONSES

• • Ecological sustainable farming
• • • • Artificial groundwater recharge
Environmental regulations
• • • – – – Atmospheric pollution control
• • Renaturalisation / Restoration
Sewage collection/treatment

HUMAN PRESSURES

• • • – – — Agrochemicals
Atmospheric pollution
Mining impacts
Urban population impacts
River engineering
Land use

CLIMATE VARIABILITY

0   1000   1800   1900   1950   2000   time

•   •   < 0,1% global area affected
• • • •   0,1 to 1%
– – – – –   1 to 10 %
————   10 to 50 %
+–+–+–+–+   > 50 %

══════ Natural climate variability

▅▅▅▅▅ Anthropogenic climate variability

may be far more important than the overall mean change across the entire planet (e.g. Vörösmarty et al. 2000c).

Since the majority of human-induced sources of pressure on continental aquatic systems – such as population density, deforestation, urbanisation, use of agrochemicals, dam construction – have had an exponential rate of increase over the last two hundred years, the spatial distribution of these combined forces has now moved to the planetary scale. The continuing, fast rate of change thus necessitates an appropriate time scale adjustment on Fig. D.98. Only few indicators of pressures have been stabilised in recent decades (e.g. global mining production), but in most cases their impacts (e.g. of mines tailing) are still developing. Such delayed responses, termed environmental or ecological "time bombs", are still not addressed at the global scale.

Knowledge of the exact timing and extent of continental aquatic systems' pressures now requires in-depth analysis of human development at the global scale, including demographics and level of industrial development, over a long time horizon. Here, case studies can provide an important and practical mechanism for validating these hypotheses. For example, we have implicitly postulated that the impacts of mines on the aquatic environment or on human health, at the global scale, may have preceded those associated with urbanisation. Only through carefully assembled documentary evidence a sufficiently detailed environmental history can be developed. A multidisciplinary approach is critical. For example, the use of medical records to assess the history of lead poisoning versus cholera will yield important insights into the pressures at work in particular basins.

D.8.6 Continental Aquatic Systems Shared by Social Systems and the Biogeophysical Earth System: An Extension of the DPSIR Approach

The Driver-Pressure-State-Impact-Response or DPSIR concept (Sect. D.3.1) originally developed by OECD is now used in environmental resource studies linking the natural and social sciences (Salomons et al. 1999; von Bodungen and Turner 2001). While there are easily identified biogeophysical attributes of CAS (e.g. surface and groundwater reservoirs, sediment retention, nutrient cycling) these also can be viewed from a human perspective as a set of resources to provide drinking water, water for food and fibre, fish and game habitat, flood control, transportation, and instream processing of wastes. From this perspective CAS provide a set of free goods and services of great value to society.

The provision of CAS resources and services, their use and abuse leads to a complex interplay between physical and socio-economic phenomena as presented on Fig. D.94. The use of CAS for human benefit is driven by a broad array of socio-economic drivers, including demographic evolution, economic development, security needs, education level, conflict, social and economic crises. These operate in tandem with pressures on inland water resources such as damming, waste release, water diversion, abstraction and regulation, agrochemicals use. These pressures collectively modify the CAS state – a process generally regarded as environmental impact within the Earth sciences community. As described throughout this section, the resulting changes invoke a multitude of bio-

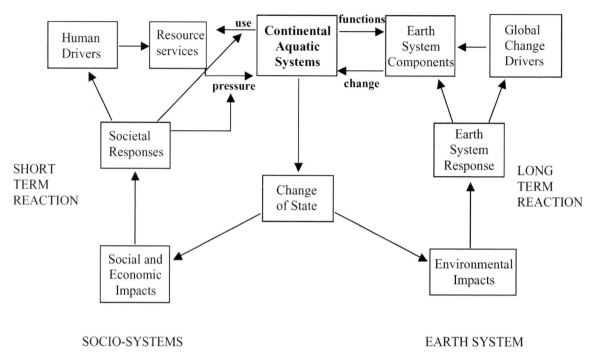

Fig. D.99. Continental aquatic systems shared by socio-economic and biogeophysical Earth systems during the Anthropocene (Meybeck 2003)

geophysical impacts. However, there are social and economic impacts as well, manifested as famine, insecurity, poverty, morbidity. These have been termed societal passive responses which, in most cases, lead to still further active responses aimed at decreasing these impacts through mitigation, rehabilitation, change of behaviour, conflict resolution, integrated management and resources sharing (Falkenmark 1997). When social and economic impacts are left unresolved or mismanaged, societal response to aquatic systems' degradation often include migration, nomadism and conflict (Fig. D.99). The socio-economic dimension of these issues extend to the historical, political, cultural and even religious realm.

We believe it is fair to state that contemporary anthropogenic impacts on continental aquatic systems have now reached a state at which regional and even global environmental systems are, or will soon be, modified in a quantitatively meaningful way. In Part D, we have documented this for water storage, sediment budgets, nutrient and other fluxes. These changes are just beginning to be taken into account in global Earth system analysis and models where they interact with other drivers such as climate change, land use/cover change. We are now at the point, both conceptually and technically, to simulate key interactions between aquatic systems and socio-economic systems (Vellinga 1996; Rotmans and de Vries 1997; Ehlers and Krafft 2001). These should now be entrained within the next generation of Earth system models (see Fig. D.94). This is an important step to take, as it will aid us in improving global-scale estimates of material budgets as well

as in elucidating transboundary interactions within the coupled Earth system, for example, land-use/cover change in one basin giving rise to modified rainfall patterns in an adjacent river system. Human pressures on continental aquatic systems therefore yield two types of consequences (Fig. D.99) that must be simulated: (*i*) short-term (10^{-1} to 10 yr) and localised impacts on aquatic resources, and (*ii*) long-term (10 to 10^2 yr) and sometimes broadscale impacts on Earth system components with corresponding Earth system response and changes.

A critical gap in our knowledge of fluvial systems concerns the internal dynamics of drainage basins and their response to anthropogenic change. A modelling effort emphasising process-level understanding could make a significant scientific contribution and provide the formal mechanism by which river basin simulations could be coupled to Earth system models. This work would rely heavily on case studies cast at the regional scale (*c.* 10^4 to 10^5 km^2). Several specific models can be envisioned, from relatively simple static material balance models to more complete biogeochemical process simulations based on various socio-economic scenarios, such as the change in diet in nitrogen flux simulation for 2050 (Kroeze et al. 2001; Seitzinger et al. 2002). The long-term goal of modelling material transformations along the entire continuum of fluvial systems from terrestrial mobilisation, through river corridor transport and transformation, with ultimate delivery to the coastal zone should now fully integrate the human dimension of Earth systems, characteristic of the new Anthropocene era.

References Part D

Abdul Rahim N (1988) Water yield changes after forest conversion to agricultural land use in Peninsular Malaysia. J Trop Forest Sci 1:67–82

Abdul Rahim N (1992) Impact of forest conversion on water yield in Peninsular Malaysia. Malays For 50(3):258–273

Acocks JPH (1988) Veld types of South Africa. Botanical Research Institute, Pretoria, South Africa. Mem Bot Surv S Afr 57, 146 p

Acreman MC (ed)(2000) The hydrology of the UK. Routledge, London, 303 p

Acreman MC, Adams B (1998) Low flows, groundwater and wetlands interactions – a scoping study, Part 1. Report to Environment Agency (W6-013), UKWIR (98/WR/09/1) and NERC (BGS WD/98/11), 97 p

Acreman MC, José P (2000) Wetlands. In: Acreman MC (ed) The hydrology of the UK. Routledge, London, pp 204–224

Acreman MC, Barbier EG, Birley MH, Campbell K, Farquharson FAK, Hodgson N, Lazenby J, McCartney MP, Morton J, Smith D, Sullivan CA (1999) Managed flood releases from dams – a review of current problems and future prospects. Institute of Hydrology, Wallingford, UK, Report to DFID-Project R7344, 57 p

Acreman MC, Farquharson FAK, McCartney MP, Sullivan CA, Campbell K, Hodgson N, Morton J, Smith D, Birley MH, Knott D, Lazenby J, Wingfield R, Barbier EG (2000) Managed flood releases from reservoirs: issues and guidance. Centre for Ecology and Hydrology, Wallingford, UK, Report to DFID and World Commission on Dams, 88 p

Adewale A (2001) River channel dynamics in response to climatic variations in the Sudano-Sahelian zone (SSZ). In: IGBP, IHDP, WCRP (eds) Challenges of a changing Earth. Global Change Open Science Conference, Poster P.3.0.021, Amsterdam

Ahnert F (1970) Functional relationships between denudation, relief and uplift in large mid latitude drainage basins. Am J Sci 268:243–263

Aitkenhead JA, McDowell WH (2000) Soil C:N ratio as a predictor of annual riverine DOC flux at local and global scales. Global Biogeochem Cy 14:127–138

Albergel J (1987) Genèse et prédétermination des crues au Burkina Faso. Du m^2 au km^2: étude des paramètres hydrologiques et de leur évolution. PhD, University of Paris

Alexander RB, Murdoch PS, Smith RA (1996) Streamflow-induced variations in nitrate flux in tributaries to the Atlantic coastal zone. Biogeochemistry 33:149–177

Alexander RB, Smith RA, Schwarz GE (2000) Effect of stream channel size on the delivery of nitrogen to the Gulf of Mexico. Nature 403:758–761

Alfonso S, Grousset F, Masse L, Tastet J-P (2001) A European lead isotope signal recorded from 6000 to 3000 years B.P. in coastal marshes (SW France). Atmos Environ 35(21):3595–3605

Amarasekera KN, Lee RF, William ER, Eltahir EAB (1997) ENSO and the natural variability in the flow of tropical rivers. J Hydrol 200:24–39

Amiotte-Suchet P, Probst JL (1995) A global model for present-day atmospheric/soil CO_2 consumption by chemical erosion of continental rocks (GEM-CO_2). Tellus 47B:273–280

Amon RMW, Benner R (1996) Bacterial utilization of different size classes of dissolved organic matter. Limnol Oceanogr 41:41–51

Anderson MG, Burt TP (1990) Subsurface runoff. In: Anderson MG, Burt TP (eds) Process studies in hillslope hydrology. John Wiley & Sons, New York, pp 365–400

Anderson SP, Dietrich WE, Montgomery DR, Torres R, Loague K (1997) Subsurface flow paths in a steep, unchanneled catchment. Water Resour Res 33(12):2637–2653

Andrews JA, Schlesinger WH (2001) Soil CO_2 dymamics, acidification and chemical weathering in a temperate forest with experimental CO_2 enrichment. Global Biogeochem Cy 15:149–162

Araújo-Lima CARM, Forsberg BR, Victoria R, Martinelli L (1986) Energy sources for detritivorous fishes in the Amazon. Science 234:1256–1258

Arheimer B, Wittgren HB (1994) Modelling the effects of wetlands on regional nitrogen transport. Ambio 23:378–386

Arnell NW (1999) The effect of climate change on hydrological regimes in Europe: a continental perspective. Global Environ Chang 9:5–23

Arnell NW, C Liu (eds)(2001) Hydrology and water resources. In: McCarthy JJ, Canziani OF, Leary NA, Dokken DJ, White KS (eds) Climate change 2001: impacts, adaptation, and vulnerability. Cambridge University Press, Cambridge

Artemyev VR (1996) Geochemistry of organic matter in river-sea systems. Kluwer Academic Publishers, Dordrecht, 190 p

Ashton P (2000) Integrated catchment management: balancing resource utilisation and conservation. CSIR, Pretoria, RSA, Aquatic Biomonitoring Course Notes, 11 p

Aumont O, Orr JC, Monfray P, Ludwig W, Amiotte-Suchet P, Probst JL (2001) Riverine-driven interhemispheric transport of carbon. Global Biogeochem Cy 15:393–405

Avissar R, Liu Y (1996) A three-dimensional numerical study of shallow convective clouds and precipitation induced by land-surface forcing. J Geophys Res 101:7499–7518

Bach M, Behrendt H, Huber A, Freude H-G (1999) Input paths and water load of nutrients and plant protection products in Germany. In: Van der Kraats JA (ed) Farming without harming: the impact of agricultural pollution on water systems. Drukkerij Belser, Lelystad, Netherlands, pp 71–82

Baker VR, Benito G, Rudoy A (1993) Paleohydrology of late Pleistocene superflooding. Altay mountains. Science 259: 348–350

Banasik K (1989) Estimation of the effect of land-use changes on storm-event sediment yield from a small watershed. In: Sediment transport modeling. American Society of Civil Engineers; New York; USA, pp 741–746

Bartlett KB, Crill PM, Banasi JA, Richey JE, Harriss RC (1990) Methane flux from the Amazon River floodplain: emissions during rising water. J Geophys Res 95:16733–16738

Bazemore DE, Eshleman KN, Hollenbeck KJ (1994) The role of soil water in stormflow generation in a forested headwater catchment: synthesis of natural tracer and hydrometric evidence. J Hydrol 162:47–75

Bear J (1979) Hydraulics of groundwater. McGraw-Hill, New York

Becker A (1989) Specific aspects of runoff formation. In: Proc. Int. Symp. on Headwater Control (Prague, November 1989), Vol. I. Prague University, Prague

Becker A (1995) Problems and progress in macroscale hydrological modelling In: Feddes (ed) Space and time scale variability and interdependencies in hydrological processes. Internat. Hydrology Series, Cambridge University Press, pp 1351–1443

Becker A, Braun P (1999) Disaggregation, aggregation and spatial scaling in hydrological modelling. J Hydrol 217:239–252

Becker A, McDonnell J (1998) Topographical and ecological controls of runoff generation and lateral flows in mountain catchments. In: Kovar K, Tappeiner U, Peters NE, Craig RG (eds) Hydrology, water resources and ecology in headwaters. Proc. HeadWater '98 Conf., Merano, April 1998, IAHS Publ. No. 248, pp 199–206

Becker A, Pfützner B (1986) Identification and modeling of river flow reductions caused by evapotranspiration losses from shallow groundwater areas. Proc. of the Budapest Symp., July 1986. IAHS Publ. No. 156

Becker A, Güntner A, Katzenmaier D (1999a) Required integrated approach to understand runoff generation and flow-path dynamics in catchments. In: Leibundgut C, McDonnell J, Schultz G (eds) Integrated methods in catchment hydrology. Proc. Int. Symp. at Birmingham, UK, July 1999, IAHS Publ. No. 258, pp 3–9

Becker A, Wenzel F, Krysanova V, Lahmer W (1999b) Regional analysis of global change impacts: concepts, tools and first results. Environ Model Assess 4(4):243–257

Becker A, Klöcking B, Lahmer W, Pfützner B (2002) The hydrological modelling system ARC/EGMO. In: Singh VP, Frevert DK (eds) Mathematical models of large watershed hydrology. Water Resources Publications, Colorado/USA

Behrendt H, Huber P, Ley M, Opitz D, Schmoll O, Scholz G, Uebe R (1999) Nährstoffbilanzierung der Flussgebiete Deutschlands. Institute for Freshwater Ecology and Inland Fisheries, Berlin, Germany, Report, 288 p

Belmans C, Wesseling JG, Feddes RA (1983) Simulation model of the water balance of a cropped soil: SWATRE. J Hydrol 63(3/4):271–286

Bencala KE (2000) Hyporheic zone hydrological processes. Hydrol Process 14:2797–2798

Benito G, Baker VR, Gregory KJ (eds) (1998) Paleohydrology and environmental change. John Wiley & Sons, Chichester, pp 215–264

Benner R, Osphal S, Chin-Leo G, Richey JE, Forsberg BR (1995) Bacterial carbon metabolism in the Amazon River system. Limnol Oceanogr 40:1252–1270

Beven KJ (1996) A discussion of distributed hydrological modeling. In: Abbott MB, Refsgaard JCH (eds) Distributed hydrological modelling. Kluwer Academic Publishers, Dordrecht Boston London, pp 255–278

Beven K (2001) How far can we go in distributed hydrological modelling? Hydrol Earth Syst Sc 5:1–12

Beven K, Freer J (2001) A dynamic TOPMODEL. Hydrol Process 15:1993–2011

Beven KJ, Germann PF (1982) Macropores and water flow in soils. Water Resour Res 18:1311–1325

Beven K, Lamb R, Quinn P, Romanowicz R, Freer J (1995) TOPMODEL. In: Singh VP (ed) Computer models of watershed hydrology. Water Resources Publications, Boulder, USA, pp 627–668

Bichet V, Campy M, Buoncristiani JF, Digiovanni C, Meybeck M, Richard H (1998) Variations in sediment yield from the upper Doubs River (Jura, France) since the late glacial period. Quaternary Res 51:267–279

Billen G, Garnier J (1999) Nitrogen transfers through the Seine drainage network: a budget based on the application of the "Riverstrahler" model. Hydrobiologia 410:139–150

Birkett CM (1995) The contribution of TOPEX/POSEIDON to the global monitoring of climatically sensitive lakes. J Geophys Res-Oceans 100(C12):25179–25204

Birkett CM (1998) The contribution of the TOPEX (NRA) radar altimeter to the global monitoring of large rivers and wetlands. Water Resour Res 34:1223–1239

Birkett CM, Murtugudde R, Allan T (1999) Indian Ocean climate event brings floods to East Africa's lakes and the Sudd Marsh. Geophys Res Lett 26(8):1031–1034

Birkhead AL, James CS, Olbrich BW (1996) Developing an integrated approach to predicting the water use of riparian vegetation. Water Research Commission, Pretoria, RSA. Report No. 474/1/97

Blais JM, France RL, Kimpe LE, Cornett RJ (1998) Climatic changes in northwestern Ontario have had a greater effect on erosion and sediment accumulation than logging and fire: evidence from ^{210}Pb chronology in lake sediments. Biogeochemistry 43:235–252

Bluth GJS, Kump LR (1994) Lithologic and climatologic controls of river chemistry. Geochim Cosmochim Ac 58:2341–2359

BMLF (Bundesministerium für Land und Forstwirtschaft) (1996) Gewässerschutzbericht

Boardman J, Ligneau L, de Roo A, Vandaele K (1994) Flooding of property by runoff from agricultural land in northwestern Europe. Geomorphology 10:183–196

Bonell M (1998a) Selected challenges in runoff generation research in forests from the hillslope to headwater drainage basin scale. J Am Water Resour As 34(4):765–785

Bonell M (1998b) Possible impacts of climate variability and change on tropical forest hydrology. Climatic Change 39(2–3):215–272

Bonell M, Williams J (1986a) The two parameters of the Philip infiltration equation: their properties and the spatial and temporal heterogeneity in a red earth of tropical semiarid Queensland. J Hydrol 87:9–31

Bonell M, Williams J (1986b) The generation and redistribution of overland flow in a massive oxic soil in a eucalypt woodland within the semiarid tropics in north Australia. Hydrol Process 1:31–46

Bonell M, Barnes CJ, Grant CR, Howard A, Burns J (1998) High rainfall response-dominated catchments: a comparative study of experiments in tropical north-east Queensland with temperate New Zealand. In: Kendall C, McDonnell JJ (eds) Isotope tracers in catchment hydrology. Elsevier, pp 347–390

Boorman DB, Sefton CEM (1997) Recognising the uncertainty in the quantification of the effects of climate change on hydrological response. Climatic Change 35:415–434

Bosch JH, Hewlett JD (1982) A review of catchment experiments to determine the effect of vegetation changes on water yield and evapotranspiration. J Hydrol 55:3–23

Bouwer H (1966) Rapid field measurements of air entry value and hydraulic conductivity of soils as significant parameters in flow system analysis. Water Resour Res 2:729–738

Breen CM, Akhurst EGJ, Walmsley RD (1985) Water quality management in the Umgeni catchment. Natal Town and Regional Planning Commission, Pietermaritzburg, South Africa. Supplementary Report 12

Bricker SB, Clement CG, Pirhalla DE, Orlando SP, Farrow DRG (1999) National estuarine eutrophication assessment: effects of nutrient enrichment in the nation's estuaries. National Oceanic and Atmospheric Administration

Brierley GJ, Murn CP (1997) European impacts on downstream sediment transfer and bank erosion in Cobargo catchment\New South Wales, Australia. Catena 31:119–136

Broecker WS, Sanyal A (1998) Does atmospheric CO_2 police the rate of chemical weathering? Global Biogeochem Cy 12:403–408

Bronstert A (1999) Capabilities and limitations of detailed hillslope hydrological modelling. Hydrol Process 13:21–48

Bronswijk JJB, Hamminga W, Oostindie K (1995) Field-scale solute transport in a heavy clay soil. Water Resour Res 31:517–526

Brooks A (1987) River channel adjustments downstream of channelisation works in England and Wales. Earth Surf Proc Land 12:337–351

Brooks A, Shields FF (1996) River channel restoration: guiding principles for sustainable projects. John Wiley & Sons, Chichester

Brown VA, McDonnell JJ, Burns DA, Kendall C (1999) The role of event water, rapid shallow flowpaths and catchment size in summer stormflow. J Hydrol 217L:171–190

Bunyard P (1987) Dam building in the tropics: some environmental and social consequences. In: Dickinson RE (ed) The geophysiology of Amazonia. John Wiley & Sons, Chichester, pp 63–68

Burns D, Hooper RP, McDonnell JJ, Freer J, Kendall C, Beven K (1998) Base cation concentrations in subsurface flow from a forested hillslope: the role of flushing frequency. Water Resour Res 34:3535–3544

Buttle J (1998) Fundamentals of watershed hydrology. In: Kendall C, McDonnell JJ (eds) Isotope tracers in catchment hydrology. Elsevier Science Publishers, p 816

Buttle J, Turcotte D (1999) Runoff processes on a forested slope on the Canadian shield. Nord Hydrol 30:1–20

Calder IR (1999) The blue revolution: land use and integrated water resources management. Earthscan, London, UK, 192 p

Canadell J, Jackson RB, Ehleringer JR, Mooney HA, Sala OE, Schulze ED (1996) Maximum rooting depth for vegetation types at the global scale. Oecologia 108:583–595

Caraco NF (1994) Influence of humans on P transfers to aquatic systems: a regional scale study using large rivers. In: Tiessen H (ed) Phosphorus cycles in terrestrial and aquatic ecosystems. SCOPE. John Wiley & Sons, Chichester

Caraco NF, Cole JJ (1999) Human impact on nitrate export: an analysis using world major rivers. Ambio 28:167–170

Carpenter SR, Caraco NF, Correll DL, Howarth RW, Sharpley AN, Smith VH (1998) Non-point pollution of surface waters with phosphorus and nitrogen. Ecol Appl 8:559–568

Carvalho NO, Baptista da Cunha S (1998) Estimatívá da carga sólida do Rio Amazonas e seus principais tributários para a foz e oceano: uma retrospectiva. A Agua em Revista 6(10):44–58

Casenave A, Valentin C (1992) A runoff capability classification system based on surface features criteria in semi-arid areas of West Africa. J Hydrol 130:231–249

Cerda A (1997) Soil erosion after land abandonment in a semiarid environment of southeastern Spain. Arid Soil Res Rehab 11:163–176

Cerda A (1998) Soil aggregate stability under different Mediterranean vegetation types. Catena 32:73–86

Changnon SA, Demissie M (1996) Detection of changes in streamflow and floods resulting from climate fluctuations and land use-drainage changes. Climatic Change 32:411–421

Chappell NA, Franks SW, Larenus J (1998) Multi-scale permeability estimation for tropical catchment. Hydrol Process 12:1507–1523

Chase TN, Pielke RA, Kittel TGF, Nemani RR, Running SW (1996) Sensitivity of a general circulation model to global changes in leaf area index. J Geophys Res 101(D3):7393–7408

Chase TN, Pielke RA, Kittel TGF, Nemani RR, Running SW (2000) Simulated impacts of historical land-cover changes on global climate in northern winter. Clim Dynam 16:93–105

Chen C-TA (2002) The impact of dams on fisheries: case of the Three Gorges Dam. In: Steffen W, Jäger J, Carson DC, Bradshaw C: Challenges of a changing Earth. Springer, Berlin, pp 97–99

Chiew FHS, Whetton PH, McMahon TA, Pittock AB (1995) Simulation of the impacts of climate change on runoff and soil moisture in Australian catchments. J Hydrol 167:121–147

Church TM (1996) An underground route for the watercycle. Nature 380:579–580

Cirmo C, McDonnell JJ (1997) Hydrological controls of nitrogen biogeochemistry and transport in wetland/near-stream zones of forested watersheds. J Hydrol 199:88–120

Clair TA, Ehrman JM, Higuchi K (1999) Changes in freshwater carbon exports from Canadian terrestrial basins to lakes and estuaries under a $2 \times CO_2$ atmospheric scenario. Global Biogeochem Cy 13:1091–1097

Claussen M, Gayler V (1997) The greening of Sahara during the mid-Holocene: results of an interactive atmosphere-biome model. Global Ecol Biogeogr 6:369–377

Clothier BE (2002) Rapid and far-reaching transport through structured soils. Hydrol Process 16(6):1321–1323

Cluis D (1998) Analysis of long runoff series of selected rivers of the Asia-Pacific region in relation with climate change and El Niño effects, GRDC-Report No. 21, WMO-Global Runoff Data Center, Koblenz, Germany, 58 p

Coe MT (1998) A linked global model of terrestrial hydrologic processes: simulation of rivers, lakes, and wetlands. J Geophys Res 103(D8):8885–8899

Cole JJ, Peierls BL, Caraco NF, Pace ML (1993) Human influence on river nitrogen. In: McDonnell M, Pickett S (eds) Humans as components of ecosystems: the ecology of subtle human effects and populated areas. Springer-Verlag, Berlin, pp 141–157

Costa MH, Foley J, J (1997) Water balance of the Amazon basin: dependence on vegetation cover and canopy conductance. J Geophys Res 102:23973–23989

Costa MH, Foley JA (2000) Combined effects of deforestation and doubled atmospheric CO_2 concentrations on the climate of Amazonia. J Climate 13:35–58

Covich AP (1993) Water and ecosystems. In: Gleick PH (ed) Water in crisis: a guide to the world's fresh water resources. Oxford University Press, Oxford, pp 40–55

Creed IF, Band LE, Foster NW, Morrison IK, Nicolson JA, Semkin RS, Jeffries DS (1996) Regulation of nitrate-N release from temperate forests: a test of the N flushing hypothesis. Water Resour Res 32:3337–3354

Crutzen PJ, Stoermer EF (2000) The anthropocene. IGBP Newsletter 41:17–18

Dawdy DR (1991) Problems of runoff modeling which are particular to the area or climate being modeled. In: Bowles DS, O'Connell PE (eds) Recent advances in the modeling of hydrologic systems. Kluwer Academic Publishers, Dordrecht, pp 541–547

Dawson TE (1996) Determining water use by trees and forests from isotopic, energy balance, and transpiration analyses: the role of tree size and hydraulic lift. Tree Physiol 16:263–272

Dearing JA, Alstrom K, Bergman A, Regnell J, Sandgreen P (1990) Recent and long-term records of soil erosion from southern Sweden. In: Boardman J, Foster IDL, Dearing JA (eds) Soil erosion on agricultural land. John Wiley & Sons, Chichester, pp 173–191

Degens ET, Kempe S, Richey JE (eds) (1991) Biogeochemistry of major world rivers. John Wiley & Sons, New York, 356 p

Depetris PJ, Kempe S, Latif M, Mook WG (1996) ENSO-controlled flooding of the Parana River (1904–1991) Naturwissenschaften 83:127–129

De Rosnay P, Polcher J (1998) Modelling root water uptake in a complex land surface scheme coupled to a GCM. Hydrol Earth Syst Sc 2:239–255

De Rosnay P, Bruen M, Polcher J (2000) Sensitivity of surface fluxes to the number of layers in the soil model used in GCMs. Geophys Res Lett 27(20):3329–3332 Desborough CE (1997) The impact of root-weighting on the response of transpiration to moisture stress in land surface schemes. Mon Weather Rev 125:1920–1930

Devol AH, Hedges JI (2001) The biogeochemistry of the Amazon River mainstem. In: McClain ME, Victoria RL, Richey JE (eds) The biogeochemistry of the Amazon Basin. Oxford University Press, New York, 365 p

Devol AH, Richey JE, Forsberg BR, Martinelli LA (1990) Seasonal dynamics in methane emissions from the Amazon River floodplain to the troposphere. J Geophys Res 95D:16417–16426

Devol AH, Forsberg BR, Richey JE, Pimentel TP (1995) Seasonal variation in chemical distributions in the Amazon (Solimões) River: a multiyear time series. Global Biogeochem Cy 9:307–328

Dickinson RE, Henderson-Sellers A (1988) Modeling tropical deforestation: a study of GCM land-surface parameterizations. Q J Roy Meteor Soc 114:439–462

Dickinson RE, Henderson-Sellers A, Kennedy PJ (1993) Biosphere-Atmosphere Transfer Scheme (BATS) version 1E as coupled to the NCAR community climate model. Tech. Note NCAR/TN-387+STR, 72 p

DiGiovanni C, Disnar JR, Macaire JJ (2001) Initial estimations of the annual organic productions of carbonated rocks. Eur. Union Geosci. Conf. Abst. J. Conf. Abstracts 4,1,187

Dillon PJ, Kirchner WB (1975) The effects of geology and land use on the export of phosphorus from watersheds. Water Res 9: 135–148

Dingman SL (1981) Planning level estimates of the value of reservoirs for water supply and flow augmentation in New Hampshire. Water Resour Bull 17:684–690

Dingman SL (2001) Physical hydrology, 2nd edn. Englewood Cliffs (NJ), Prentice-Hall

Dirmeyer PA, Zeng FJ (1997) A two-dimensional implementation of the Simple Biosphere (SiB) model. COLA Report 48, Center for Ocean-Land-Atmosphere Studies, Calverton, MD, USA, 30 p

Dirmeyer PA, Zeng FJ, Ducharne A, Morrill JC, Koster RD (2000) The sensitivity of surface fluxes to soil water content in three land surface schemes. J Hydrometeorol 1(2):121–134

DOE-UK (1993) Countryside survey, 1990 summary report. Department of the Environment, HMSO, London, UK

Donkin AD, Smithers JC, Lorentz SA, Lorentz RE (1995) Direct estimation of total evaporation from a southern African wetland. In: Campbell KL (ed) Versatility of wetlands in the agricultural landscape. American Society of Agricultural Engineers, St Joseph MI, USA, pp 501–513

Douglas I (1967) Man, vegetation and the sediment yield of rivers. Nature 215:925–928

Douglas I (1994) Human settlements. In: Meyer WB, Turner BL (eds) Changes in land use and land cover. Cambridge University Press, Cambridge, pp 149–170

Douglas EM, Vogel RM, Kroll CN (2000) Trends in floods and low flows in the US: impact of spatial correlation. J Hydrol 240: 90–105

Ducharne A, Koster RD, Suarez MJ, Stieglitz M, Kumar P (2000) A catchment-based approach to modeling land surface processes in a general circulation model. 2: Parameter estimation and model demonstration. J Geophys Res 105:24823–24838

Dunne T, Black RD (1970) Partial area contributions to storm runoff in a small New England watershed. Water Resour Res 6(5): 1296–1311.

Dunne KA, Willmott CJ (1996) Global distribution of plant-extractable water capacity of soil. Int J Climatol 16:841–859

Dunne T, Mertes LAK, Meade RH, Richey JE, Forsberg BR (1998) Exchanges of sediment between the floodplain and channel of the Amazon River in Brazil. Geol Soc Am Bull 110:450–467

Durand P, Robson A, Neal C (1992) Modelling the hydrology of submediterranean montane catchments. J Hydrol 139:1–14

DVWK (Deutscher Verband für Wasserwirtschaft und Kulturbau e.V.) (1985) Bodennutzung und Nitrataustrag – Literaturauswertung über die Situation bis 1984 in der Bundesrepublik Deutschland. Bonn

DWAF (Department of Water Affairs and Forestry) (1996) The philosophy and practice of integrated catchment management: implications for water resource management in South Africa. Water Research Commission, Pretoria, RSA, WRC Report, TT81/96, 140 p

DWAF (Department of Water Affairs and Forestry) (1998) Water law implementation process: a strategic plan for the department of water affairs and forestry to facilitate in the implementation of catchment management in South Africa. Water Research Commission, Pretoria, RSA, WRC Report, KV107/98, 66 p

Dyck S, Peschke G (1995) Grundlagen der Hydrologie. VEB Verlag für Bauwesen, Berlin

Dye PJ, Bosch JM (2000) Sustained water yield in afforested catchments – the South African experience. In: Von Gadow K, Putkala T, Tomé M (eds) Sustainable forest management. Kluwer Academic Publishers, Dordrecht, pp 99–120

Dynesius M, Nilsson C (1994) Fragmentation and flow regulation of river systems in the northern third of the world. Science 266: 753–762

Easmus D (1991) The interaction of rising CO$_2$ and temperatures with water use efficiency: commissioned review. Plant Cell Environ, special issue: Elevated CO$_2$ levels, 14(8):843–852

Easterling DR, Evans JL, Groisman PY, Karl TR, Kunkel KE, Ambenje P (2000) Observed variability and trends in extreme climate events: a brief overview. B Am Meteorol Soc 81(3): 417–425

Eheart JW, Tornil DW (1999) Low-flow frequency exacerbation by irrigation withdrawals in the agricultural midwest under various climate change scenarios. Water Resour Res 35(7):2237–2246

Ehlers E, Krafft T (eds) (2001) Understanding the Earth system. Compartments, processes interactions. Springer-Verlag, Heidelberg, 290 p

English Nature (1997) Wildlife and fresh water – An agenda for sustainable management. English Nature, Peterborough, UK

Environment Agency (1998) River habitat quality: the physical character of rivers and streams in the UK and the Isle of Man. Environment Agency, Bristol, UK

Ertel JR, Hedges JI, Devol AH, Richey JE, Ribeiro N (1986) Dissolved humic substances of the Amazon River system. Limnol Oceanogr 31:739–754

Fahey BD, Watson AJ (1991) Hydrological impacts of converting tussock grassland to pine plantation, Otago, New Zealand. J Hydrol New-Zealand 30:1–15

Falkenmark M (1997) Society's interaction with the water cycle: a conceptual framework for a more holistic approach. Hydrolog Sci J 42:451–466

Falkenmark M, Chapman T (1989) Comparative hydrology. UNESCO, Paris, France, 479 p

Falkenmark M, Andersson L, Carstensson R, Sundblad K (1999) Water: a reflection of land use. Swedish Natural Science Research Council, Stockholm, Sweden, 128 p

FAO (1999) State of the world's forests. UN Food and Agriculture Organization, Rome

Farrington J, Lobo C (1997) Scaling up participatory watershed development in India: lessons from the Indo-German Watershed Development Programme. Overseas Development Institute, London, UK. Natural Resource Perspectives 17

Favis-Mortlock D, Boardman J (1995) Nonlinear responses of soil erosion to climate change: a modelling study on the UK South Downs. Catena 25:365–387

Feddes RA, Kowalik PJ, Zaradny H (1978) Simulation of field water use and crop yield. Simulation Monographs, Pudoc. Wageningen, 189 p

Feddes RA, Menenti M, Kabat P, Bastiaanssen WGM (1993a) Is large-scale inverse modelling of unsaturated flow with areal average evaporation and surface soil moisture as estimated from remote sensing feasable? J Hydrol 143:125–152

Feddes RA, de Rooij GH, van Dam JC, Kabat P, Droogers P, Stricker JNM (1993b) Estimation of regional effective soil hydraulic parameters by inverse modelling. In: Russo D, Dagan G (eds) Water flow and solute transport in soils. Adv. Series in Agric. Sciences 20, Springer Verlag, Berlin, pp 211–231

Feddes RA, Hoff H, Bruen M, Dawson TE, de Rosnay P, Dirmeyer P, Jackson RB, Kabat P, Kleidon A, Lilly A, Pitman AJ (2001) Modelling root water uptake in hydrological and climate models. B Am Meteorol Soc 82:2797–2809

Federer CA, Vörösmarty CJ, Fekete B (1996) Intercomparison of methods for potential evapotranspiration in regional or global water balance models. Water Resour Res 32:2315–2321

Fekete BM, Vörösmarty CJ, Grabs W (1999) Global, composite runoff fields based on observed river discharge and simulated water balances. WMO-Global Runoff Data Center Report No. 22. Koblenz, Germany

Fekete BM, Vörösmarty CJ, Lammers RB (2001) Scaling gridded river networks for macroscale hydrology: development, analysis, and control of error. Water Resour Res 37(7):1955–1967

Fekete BM, Vörösmarty CJ, Grabs W (2002) High resolution fields of global runoff combining observed river discharge and simulated water balances. Global Biogeochem Cy 16(3): Art. No. 1042

Feller C, Beare MH (1997) Physical control of soil organic matter dynamics in the tropics. Geoderma 79:69–116

Fernald S, Wiggington P, Landers D (2001) Transient storage and hyporheic flow along the Willamette River, Oregon: field measurements and model estimates. Water Resour Res 37:1681–1694

Fetter CW (1988) Applied hydrogeology. Prentice Hall, Englewood Cliffs NJ, 691 p

Field CB, Jackson RB, Mooney HA (1995) Stomatal responses to increased CO$_2$: implications from the plant to the global scale. Plant Cell Environ 18:1214–1225

Fisher TRJ, Parsley PE (1979) Amazon lakes: water storage and nutrient stripping by algae. Limnol Oceanogr 24:547–553

Flint RF (1971) Glacial and quaternary geology. John Wiley & Sons, Chichester, 892 p

Flint EP, Richards JF (1991) Historical analysis of changes in land use and carbon stock of vegetation in south and South-east Asia. Can J Forest Res 2(1):91–110

Flügel WA (1995) Delineating hydrological response units by GIS analyses. In: Kalma JD, Sivapalan M (eds) Scale issues in hydrological modelling. John Wiley & Sons, Chichester, pp 181–194

Forsberg BR, Devol AH, Richey JE, Martinelli LA, dos Santos H (1988) Factors controlling nutrient concentrations in Amazon floodplain lakes. Limnol Oceanogr 33:41–56

Fournier F (1960) Climat et erosion. Presses Universitaires Paris, 201 p

Fraser AS, Meybeck M, Ongley ED (1995) Water quality of world river basins. UNEP Environment Library 14, UNEP, Nairobi, 40 p

Freer J, Beven KJ, Ambroise B (1996) Bayesian estimation of uncertainty in runoff prediction and the value of data: an application of the GLUE approach. Water Resour Res 32(7):2161–2173

Freer J, McDonnell JJ, Brammer D, Beven K, Hooper R, Burns D (1997) Topographic controls on subsurface stormflow at the hillslope scale for two hydrologically distinct catchments. Hydrol Process 11(9):1347–1352

Freer J, McDonnell JJ, Beven KJ, Peters NE, Burns DA, Hooper RP, Aulenbach B (2002) The role of bedrock topography on subsurface storm flow. Water Resour Res 38(12):1269, doi:1210.1029/2001WR000872

Freeze RA (1972) Role of subsurface flow in generating surface runoff; 2. Upstream source areas. Water Resour Res 8:1272–1283

Freeze RA, Witherspoon PA (1967) Theoretical analysis of regional groundwater flow; 2. Effect on water table configuration and subsurface permeability variation. Water Resour Res 3:623–634

Frenzel B (1992) Atlas of paleoclimates and paleoenvironments of the Northern Hemisphere: Late Pleistocene, Holocene. Geographical Res Inst, Hungarian Academy Sci., Budapest

Frink CR (1991) Estimating nutrient exports to estuaries. J Environ Qual 20:717–724

Fritsch JM (1994) The hydrological effects of clearing tropical rain forest and of the implementation of alternative land uses. IAHS Press, Wallingford, IAHS Publication No. 216, pp 53–66

Frost PGH (2001) Reflections on integrated land and water management. In: Gash JHC, Odada EO, Oyebande L, Schulze RE (eds) Freshwater resources in Africa. BAHC International Project Office, PIK, Potsdam, Germany, pp 49–56

Gale MR, Grigal DF (1987) Vertical root distributions of northern tree species in relation to successional status. Can J Forest Res 17:829–834

Galea G, Breil P, Ahmad A (1993) Influence du couvert végétal sur l'hydrologie des crues, modélisation à validations multiples. Hydrologie Continentale 8:17–33

Galloway JN, Schlesinger WH, Levy HI, Michaels A, Schnoor JL (1995) Nitrogen fixation: anthropogenic enhancement-environmental response. Global Biogeochem Cy 9:235–252

Gash JHC, Nobre CA (1997) Climatic effects of Amazonian deforestation: some results from ABRACOS. B Am Meteorol Soc 78(5):823–830

Gataullin V, Mangerud J, Svendsen JI (2001) The extent of the late Weichselian ice sheet in the south eastern Barents Sea. Global Planet Change 31:453–474

Gellens D, Roulin E (1999) Streamflow response of Belgian catchments to IPCC climate change scenarios. J Hydrol 210:242–258

Genereux D (1998) Quantifying uncertainty in tracer-based hydrograph separations. Water Resour Res 34(4):915–919

Gerasimov IP (ed) (1964) Fiziko-geograficheskiy atlas mira (physico-geographical atlas of the world). Soviet Acad of Science Moscow, 298 p

Gibbs MT, Kump LR (1994) Global chemical erosion during the last glacial maximum and the present: sensitivity to changes in lithology and hydrology. Paleoceanography 9:529–543

Giovannini G, Luccchesi S (1992) Effects of fire on soil physicochemical characteristics and erosion dynamics. In: Tabaud L, Prodon R (eds) Fire in Mediterranean ecosystems. Ecosystems Research Report No.°5, Commission of the European Communities, Brussels, Belgium, pp 403–412

Gleick PH (1987) Regional hydrologic consequences of increases in atmospheric CO_2 and other trace gases. Climatic Change 10:137–131

Gleick PH (1998) The world's water: the biennial report on freshwater resources (1998–1999). Island Press, Washington DC

Gleick PH (2000) Water futures: a review of global water resources projections. In: Rijsberman FR (ed) World water scenarios. EarthscanPages, London, pp 27–45

Goolsby DA, Battaglin W (2001) Long-term changes in concentrations and flux of nitrogen in the Mississippi River basin, USA. Hydrol Process 15:1209–1226

Goudrian J, Ketner P (1984) A simulation study of the global carbon cycle including man's impact on the biosphere. Climatic Change 6:167–192

Gouy V, Jannot P, Laplana R, Malé, J-M, Turpin N (1999) Country paper of France. In: Van der Kraats JA (ed) Farming without harming: the impact of agricultural pollution on water systems. Drukkerij Belser, Lelystad, Netherlands, pp 57–70

Graham D, House A, Hudson J, Leeks G, Williams R (1999a) Impact of agricultural pollution on water systems: country paper of the UK. In: Van der Kraats JA (ed) Farming without harming: the impact of agricultural pollution on water systems. Drukkerij Belser, Lelystad, Netherlands, pp 209–227

Graham ST, Famiglietti FS, Maidment DR (1999b) Five minute, 1/2° and 1° datasets of continental watersheds and river networks for use in regional and global hydrologic and climate system modeling studies. Water Resour Res 35:583–587

GRASS (1993) GRASS 4.1 Reference Manual. US Army Corps of Engineers. Construction Engineering Research Laboratories, Champaign, Illinois, 556 p

Green PA, Vörösmarty CJ, Meybeck M, Galloway JN, Peterson BJ, Boyer EW (2003) Pre-industrial and contemporary fluxes of nitrogen through rivers: a global assessmant based on typology. Biogeochemistry (in press)

Grigg NS (1999) Integrated water resources management: who should lead, who should pay? J Am Water Resour As 35:527–534

Grosbois CA, Horowitz AJ, Smith JJ, Febric KA (2001) The effect of mining and related activities on the sediment-trace element geochemistry of the Lake Coeur d'Alene, Idaho, USA. Part III: Downstream effects: the Spokane River basin. Hydrol Process 15:855–875

Groisman PY, Knight RW, Karl TR (2001) Heavy precipitation and high streamflow in the contiguous United States: trends in the twentieth century. Bull Am Meteorol Soc 82(2):219–246

Grosswald MG (1984) Glaciations of the continent shelves. Polar Geogr J 8:194–258, 287–351

Grosswald MG (1998) Late Weichselian ice sheets in Arctic and pacific Siberia. Quatern Int 45/46:3–18

Gunkel G (ed)(1994) Bioindikation. Gustav Fischer Verlag, Jena, 540 p

Gunkel G (2000) Das Ziel heisst: ein "naturnaher" Zustand. Technische Universität Berlin, Forschung Akuell 11:57–60

Guo S, Ying A (1997) Uncertainty analysis of impact of climate change on hydrology and water resources. In: Rosbjerg D, Boutayeb NE, Gustard A, Kundzewicz ZW, Rasmussen PF(eds) Sustainability of water resources under increasing uncertainty. Proceedings of an international symposium of the Fifth Scientific Assembly of the International Association of Hydrological Sciences (IAHS), Rabat, Morocco, 23 April to 3 May 1997, IAHS Press, Wallingford, UK, pp 331–338

Gustafsson A, Hoffmann M, Johnsson H, Krueger J, Kyllmar K, Ulén B (1999) N, P and pesticide pollution of water bodies by agricultural activities under Swedish conditions. In: Van der Kraats JA (ed) Farming without harming: the impact of agricultural pollution on water systems. Drukkerij Belser, Lelystad, Netherlands, pp 161–184

Hadley RF, Lal R, Onstad CA, Walling DE, Yair A (1985) Recent developments in erosion and sediment yield studies. Technical Documents in Hydrology, UNESCO/IHP, Paris

Hagemann S, Dumenil L (1998) A parametrization of the lateral waterflow for the global scale. Clim Dynam 14(1):17–31

Hallgren WS, Pitman AJ (2000) The sensitivity of a global biome Model (BIOME3) to uncertainty in parameter values. Global Change Biol 6:483–495

Harley PC, Thomas RB, Reynolds JF, Strain BR (1992) Modelling photosynthesis of cotton grown in elevated CO_2. Plant Cell Environ 15:271–281

Harris DM, McDonnell JJ, Rodhe A (1995) Hydrograph separation using continuous open-system isotopic mixing. Water Resour Res 31:157–171

Hartig EK, Grozev O, Rosenzweig C, Dixon RK (1997) Climate change, agriculture and wetlands in Eastern Europe: vulnerability, adaptation and policy. Climatic Change 36:107–121

Hartwich R, Behrens J, Eckelmann W, Haase G, Richter A, Roeschmann G, Schmidt R (1995) Bodenübersichtskarte der Bundesrepublik Deutschland 1 : 1 000 000. Hannover

Hattermann FF, Krysanova V, Wechsung F, Wattenbach M (2003) Macroscale validation of the eco-hydrological model SWIM for hydrological processes in the Elbe basin with uncertainty analysis. In: Fohrer N, Arnold J (eds) Regional assessment of climate and management impacts using the SWAT hydrological model. Hydrol Proc (in print)

Hatton TJ, Dawes WR, Vertessey RA (1995) The importance of landscape position in scaling SVAT models to catchment scale hydro-ecological prediction. In: Feddes (ed) Space and time scale variability and interdependencies in hydrological processes. Internat. Hydrology Series, Cambridge University Press, pp 43–53

Hedges JI, Clark WA, Quay PD, Richey JE, Devol AH, Santos U de M (1986a) Compositions and fluxes of particulate organic material in the Amazon River. Limnol Oceanogr 31:717–738

Hedges JI, Ertel JR, Quay PD, Grootes PM, Richey JE, Devol AH, Farwell GW, Schmidt FW, Salati E (1986b) Organic carbon-14 in the Amazon River system. Science 231:1129–1131

Hedges JI, Cowie GL, Richey JE, Quay PD (1994) Origins and processing of organic matter in the Amazon River as indicated by carbohydrates and amino acids. Limnol Oceanogr 39:743–761

Hession WC, McBride M, Bennett M (2000) Statewide non-point-source pollution assessment methodology. J Water Res Pl-Asce May/June 2000

Hewlett JD, Hibbert AR (1967) Factors affecting the response of small watersheds to precipitation in humid areas. In: Sopper WE, Lull HW (eds) International symposium on forest hydrology. pp 275–271

Hiernaux P, Bielders CL, Valentin C, Bationo A, Fernandez-Rivera S (1999) Effects of livestock grazing on physical and chemical properties of sandy soils in Sahel rangelands. J Arid Environ 41:231–245

HMSO (Her Majesty's Stationery Office) (1995) Rural England – a nation committed to a rural countryside. CM 3016, HMSO, London, UK

Hobbie J (ed) (2000) Estuarine science: a synthetic approach to research and practice. Island Press, Washington DC

Hobbs RJ (2000) Land-use changes and invasions. In: Mooney HA, Hobbs RJ (eds) Invasive species in a changing world. Island Press, Washington DC, pp 55–64

Hobbs RJ, Hopkins AJM (1990) From frontier to fragments: European impact on Australia's vegetation. Proceedings of the Ecological Society of Australia 16:93–114

Hollis GE, Adams WM, Kano MA (eds) (1993) The Hadejia-Nguru wetlands: environment, economy and sustainable development of a Sahelian floodplain wetland. IUCN, Gland, Switzerland, 244 p

Hooke RL (2000) On the history of humans as geomorphic agents. Geology 28:843–846

Hopkinson CS, Vallino J (1995) The nature of watershed perturbations and their influence on estuarine metabolism. Estuaries 18:598–621

Horne Glasson and Partners (1989) Water plan 2025. Umgeni Water, Pietermaritzburg, South Africa

Houghton JT, Ding Y, Griggs DJ, Noguer M, van der Linden PJ, Xiaosu D (eds) (2001) Climate change 2001: the scientific basis. Third Assessment Report, IPCC Working Group I, Cambridge University Press, Cambridge

Howarth RW, Billen G, Swaney D, Townsend A, Jaworski N, Lajtha K, Downing JA, Elmgien R, Caraco N, Jordan T, Berendse F, Freney J, Kudegarov V, Mardoch P, Zhao-Liang Z (1996) Regional nitrogen budgets and riverine N and P fluxes for the drainages to the North Atlantic Ocean: natural and human influences. Biogeochemistry 35:75–139

Hudson-Edwards KA, Macklin MG, Taylor MP (1999) 2000 years of sediment-borne heavy metal storage in the Yorkshire Ouse basin, NE England UK. Hydrol Process 13:1087–1102

Hughes PJ, Sullivan ME, Yok D (1991) Human-induced erosion in a highlands catchment in Papua New Guinea: the prehistoric and contemporary records. Z Geomorphol, Supl. 83:227–239

Hulme M, Barrow EM, Arnell NW, Harrison PA, Johns TC, Downing E (1999) Relative impacts of human-induced climate change and natural climate variability. Nature 397:688–691

Humborg C, Ittekkot V, Cociasu A, Bodungen BV von (1997) Effect of the Danube River dam on Black Sea biogeochemistry and ecosystem structure. Nature 386:385–387

IAHS Ad Hoc Group on Global Water Data Sets (2001) Global water data: a newly endangered species. Co-authored by Vörösmarty CJ (lead), Askew A, Barry R, Birkett C, Döll P, Grabs W, Hall A, Jenne R, Kitaev L, Landwehr J, Keeler M, Leavesley G, Schaake J, Strzepek K, Sundarvel SS, Takeuchi K, Webster F. An opinion editorial to AGU Eos Transactions 82(5):54–58

ICOLD (International Commission on Large Dams) (1994) Dams and the environment: water quality and climate. International Commission on Large Dams, Paris

Inamdar SP, Mitchell M, McDonnell JJ, McHale M, McHale P (2004) Use of new water ratios and surface saturated area estimates to test TOPMODEL applications to a forested headwater catchment. J Hydrol (in review)

IPCC (Intergovernmental Panel on Climate Change) (1990) Emission scenarios – IPCC special report. IPCC Secretariat, c/o WMO, Geneva, Switzerland

IPCC, Houghton JT, Meira Filho LG, Callender BA, Harris N, Kattenberg N, Maskell K (1995) Climate change 1995: the science of climate change. Contribution of working group I to the second assessment of the Intergovernmental Panel on Climate Change. Cambridge University Press, 572 p

Ittekot V (1988) Global trends in the nature of organic matter in river suspensions. Nature 332:436–438

Jackson RB, Canadell J, Ehleringer JR, Mooney HA, Sala OE, Schulze ED (1996) A global analysis of root distributions for terrestrial biomes. Oecologia 108:389–411

Jackson RB, Mooney HA, Schulze ED (1997) A global budget for fine root biomass, surface area, and nutrient contents. P Natl Acad Sci USA 94:7362–7366

Jackson RB, Schenk HJ, Jobbágy EG, Canadell J, Colello GD, Dickinson RE, Field CB, Friedlingstein P, Heimann M, Hibbard K, Kicklighter DW, Kleidon A, Neilson RP, Parton WJ, Sala OE, Sykes MT (2000a) Below-ground consequences of vegetation change and their treatment in models. Ecol Appl 10:470–483

Jackson RB, Sperry JS, Dawson TE (2000b) Root water uptake and transport: using physiological processes in global predictions. Trends Plant Sci 5(11):482–488

Jacobson PJ, Jacobson KM, Seely MK (1995) Ephemeral rivers and their catchments: sustaining people and development in Western Namibia. Desert Research Foundation of Namibia, Windhoek, Namibia, 160 p

Jakeman AJ, Hornberger GM (1993) How much complexity is warranted in a rainfall-runoff model? Water Resour Res 29(8): 2637–2649

Jakeman AJ, Chen TH, Post DA, Hornberger GM, Littlewood IG, Whitehead PG (1993) Assessing uncertainties in hydrological response to climate at a large scale. In: Proc. On Macroscale Modelling in Hydrosphere, 11–23 July 1993, Yokohama, IASH, 15 p

Jansen JMI, Painter RB (1974) Predicting sediment yield from climate and topography. J Hydrol 21:371–380

Jansson M (1982) Land erosion by water in different climates. UNGI Rapport No. 57, Uppsala University, Sweden

Jansson M, Andersson R, Berggren H, Leonardson L (1994) Wetlands and lakes as nitrogen traps. Ambio 23:320–325

Jarvis G, McNoughton KG (1986) Stomatal control of transpiration: scaling up from leaf to region. Adv Ecol Res 15:1–49

Jaworski NA, Howarth RW, Hetling LJ (1997) Atmospheric deposition of nitrogen oxides onto the landscape contributes to coastal eutrophication in the northeast United States. Environ Sci Technol 31:1995–2004

Jewitt GPW, Görgens AHM (2000) Scale and model interfaces in the context of integrated water resources management for rivers of the Krüger National Park. Water Research Commission, Pretoria, RSA. WRC Report, 627/1/00. 184 p

Jewitt GPW, Kotze DC (2000) Wetland conservation and rehabilitation as components of integrated catchment management in the Mgeni catchment, KwaZulu-Natal, South Africa. In: Bergkamp G, Pirot J-Y, Hostettler S (eds) Integrated wetlands and water resources management. Proceedings of a Workshop held at the 2nd International Conference on Wetlands and Development held in Dakar, Senegal, November 10–14 (1998) International Publication No. 56, Wageningen, The Netherlands

Jewitt GPW, Schulze RE (1999) Verification of the ACRU model for forest hydrology applications. Water SA 25:483–490

Johnson SP (1993) The earth summit: the United Nations Conference on Environment and Development (UNCED). Graham and Trotman, London, UK

Johnsson MJ, Meade RH (1990) Chemical weathering of fluvial sediments during alluvial storage: the Macuapanim Island point bar, Solimoes River, Brazil. J Sediment Petrol 60:827–842

Jordan TE, Correll DL, Peterjohn WT, Weller DE (1986) Nutrient flux in a landscape: the Rhode River watershed and receiving waters. In: Correll DL (ed) Watershed research perspectives. Smithsonian Inst. Press, Washington DC

Jordan TE, Correll DL, Weller DE (1997) Nonpoint sources of discharges of nutrients from Piedmont watersheds of Chesapeake Bay. J Am Water Resour As 33(3):631–645

Junk WJ (1985) Amazon floodplain – a sink or source of organic carbon? In: Degens ET, Kempe S, Herrera R (eds) Transport of carbon and minerals in major world rivers, Part 3. Mitt. Geol. Paläont. Inst. University Hamburg, SCOPE/UNEP Sonderbd. 58:267–283

Junk WJ, Furch K (1993) A general review of tropical South American floodplains. Wetlands Ecology and Management 2:231–238

Junk WJ, MTF Piedade (1993) Biomass and primary-production of herbaceous plant communities in the Amazon floodplain. Hydrobiologia 263:155–162

Junk WJ, Bayley PB, Sparks RE (1989) The flood pulse concept in river-floodplain systems. Can J Fish Aquat Sci 106:110–127

Justic D, Rabalais NN, Turner RE (1995a) Stoichiometric nutrient balance and origin of coastal eutrophication. Mar Pollut Bull 30:41–46

Justic D, Rabalais NN, Turner RE, Dortch Q (1995b) Changes in nutrient structure of river-dominated coastal waters – stoichiometric nutrient balance and its consequences. Estuar Coast Shelf S 40:339–356

Kabat P, Hutjes RWA, Feddes RA (1997) The scaling characteristics of soil parameters: from plot scale heterogeneity to subgrid parameterization J Hydrol 190(3–4):363–396

Kao SJ, Lui KK (1996) Particulate organic carbon export from a subtropical mountainous river (Lanyang Hri) in Taiwan. Limnol Oceanogr 41:1749–1757

Kao SJ, Lui KK (1997) Fluxes of dissolved and non-fossil particulate organic carbon from an Oceania small river (Lanyang Hsi) in Taiwan. Biogeochemistry 39:255–269

Kaplan DI, Bertsch PM, Adriano DC, Miller WP (1993) Soil-borne colloids as influenced by water flow and organic carbon. Environ Sci Technol 27:1193–1200

Karl TR, Knight RW (1998) Secular trends of precipitation amount, frequency, and intensity in the USA. B Am Meteorol Soc 79(2):231–241

Karl T, Knight RW, Easterling DR, Quayle RG (1996) Indices of climate change for the United States. B Am Meteorol Soc 77:279–292

Keller EA, Valentine DW, Gibbs DR (1997) Hydrological response of small watersheds following the Southern California Painted Cave fire of June 1990. Hydrol Process 11:401–414

Kempe S (1979) Carbon in the freshwater cycle. In: Bolin B, Degens ET, Kempe S, Ketner P (eds) Global carbon cycle. SCOPE rpt 13, John Wiley & Sons, New york, pp 317–342

Kempe S (1982) Long-term records of the CO_2 pressure fluctuations in freshwaters. In: Degens ET (ed) Transport of water and minerals in major world rivers. Mitt Geol-Paläntol Inst Univ Hamburg 52:91–332

Kempe S (1984) Sinks of the anthropogenically enhanced carbon cycle in surface fresh waters. J Geophys Res 89(D3):4657–4676

Kendall C, McDonnell JJ (eds)(1998) Isotope tracers in catchment hydrology. Elsevier Science Publishers, 839 p

Kendall K, Shanley J, McDonnell JJ (1999) A hydrometric and geochemical approach to testing the transmissivity feedback hypothesis during snowmelt. J Hydrol 219:188–205

Kendall C, McDonnell JJ, Weizu G (2001) A look inside black box hydrograph separation models: a study at the Hydrohill catchment. Hydrol Process 15(10):1877–1903

Kienzle SW, Lorentz SA, Schulze RE (1997) Hydrology and water quality of the Mgeni catchment. Water Research Commission, Pretoria, South Africa, WRC Report TT87/97, 88 p

Kim CP (1995) The water budget of heterogeneous areas: impact of soil and rainfall variability. PhD thesis, Wageningen Agricultural University, 182 p

Kimball BA, Pinter PJ Jr, Garcia RL, Lamorte RL, Wall GW, Hunsaker DJ, Wechsung G, Wechsung F, Kartschall T (1995) Productivity and water use of wheat under free-air CO_2 enrichment. Global Change Biol 1:429–442

Kimball BA, LaMorte RL, Pinter PJ Jr, Wall GW, Hunsaker DJ, Adamsen FJ, Leavitt SW, Thompson TL, Matthias AD, Brooks TJ (1999) Free-air CO_2 enrichment and soil nitrogen effects on energy balance and evapotranspiration of wheat. Water Resour Res 35(4):1179–1190

Kirby C, Newson MD, Gilman K (1991) Plynlimon research: the first two decades. Institute of Hydrology, Wallingford, UK, Institute of Hydrology Report 109

Kirchner JW, Feng X, Neal C (2000) Fractal stream chemistry and its implications for contaminant transport in catchments. Nature 403(6769):524–527

Kleidon A, Heimann M (2000) Assessing the role of deep rooted vegetation in the climate system with model simulations: mechanism, comparison to observations and implications for Amazonian deforestation. Clim Dynam 16:183–199

Koster RD, Suarez MJ (1992) Modeling the land surface boundary in climate models as a composite of independent vegetation stands. J Geophys Res 97:2697–2716

Koster RD, Suarez MJ, Ducharne A, Stieglitz M, Kumar P (2000) A catchment-based approach to modeling land surface processes in a general circulation model, 1: Model structure. J Geophys Res 105:24809–24822

Kotwicki V (1986) Floods of Lake Eyre. Engineering and Water Supply Dpt. Adelaide, South Australia

Kotwicki V, Isdale P (1991) Hydrology of Lake Eyre, Australia: El Niño link. Palaeogeogr Palaeocl 84:87–98

Krecek J Rajwar GS, Haigh MJ (eds) (1996) Hydrological problems and environmental management in highlands and headwaters. A. A. Balkema, Rotterdam, Netherlands, 196 p

Kreft A, Zuber A (1978) On the physical meaning of the dispersion equation and its solutions for different initial and boundary conditions. Chem Eng Sci 33:1471–1480

Kroeze C, Seitzinger SP (1998) Nitrogen inputs to rivers, estuaries and continental shelves and related nitrous oxides emissions in 1990 and 2050: a global model. Nutr Cycl Agroecosys 52:195–212

Kroeze C, Seitzinger SP, Domingues R (2001) Future trends in worldwide nitrogen transport and related nitrous oxide emissions: a scenario analysis. The Scientific World 1

Krysanova V, Becker A (1999) Integrated modelling of hydrological processes and nutrient dynamics at the river basins scale. Hydrobiologia 410:131–138

Krysanova V, Müller-Wohlfeil DI, Becker A (1998a) Development and test of a spatially distributed hydrological/water quality model for mesoscale watersheds. Ecol Model 106:261–289

Krysanova V, Becker A, Klöcking B (1998b) The linkage between hydrological processes and sediment transport at the river basin scale. In: Summer W, Klaghofer E, Zhang W (eds) Modelling soil erosion, sediment transport and closely related hydrological processes. IAHS Publ 249:13–20

Krysanova V, Wechsung F, Becker A, Poschenrieder W, Gräfe J (1999a) Mesoscale ecohydrological modelling to analyse regional effects of climate change. Environ Model Assess 4(4):259–271

Krysanova V, Gerten D, Klöcking B, Becker A (1999b) Factors affecting nitrogen export from diffuse sources: a modelling study in the Elbe basin. In: Heathwaite L (ed) Impact of land-use change on nutrient loads from diffuse sources. IAHS Publ 257:201–212

Krysanova V, Williams J, Bürger G, Österle H (2002) The linkage between hydrological processes and sediment transport at the river basin scale – a modelling study. In: Summer W, Walling DE (eds) Modelling erosion, sediment transport and sediment yield. International Hydrological Programme, Technical Documents in Hydrology, IHP-VI, pp 147–174

Kubatzki C, Claussen M (1998) Simulation of the global biogeophysical interactions during the last glacial maximum. Clim Dynam 14:461–471

Lahmer W, Becker A, Müller-Wohlfeil D-I, Pfützner B (1999) A GIS-based approach for regional hydrological modelling. In: Diekkrüger B, Kirkby MJ, Schröder U (eds) Regionalization in hydrology. IAHS Publ 254:33–43

Lal M (1994) Water resources of the South Asian region in a warmer atmosphere. Adv Atmos Sci 11:239–246

Lammers RB, Shiklomanov AI, Vörösmarty CJ, Fekete BM, Peterson BJ (2001) Assessment of contemporary Arctic river runoff based on observational discharge records. J Geophys Res 106(D4):3321–3334

Langbein WB, Schumm SA (1958) Yield of sediment in relation to mean annual precipitation. EOS Trans Am Geophys Union 39:1076–1084

Lange H, Lischeid G, Hoch R, Hauhs M (1996) Water flow paths and residence times in a small headwater catchment at Gardsjon, Sweden, during steady state storm flow conditions. Water Resour Res 32(6):1689–1698

Laraque A, Mahé G, Orange D, Marieu B (2001) Spatiotemporal variations in hydrological regimes within Central Africa during the XXth century. J Hydrol 245:104–117

Lean J, Warrilow DA (1989) Simulation of the regional climatic impact of Amazon deforestation. Nature 342:411–413

Le Bissonnais Y (1996) Aggregate stability and assessment of crustability and erodibility: 1. Theory and methodology. Europ J Soil Sci 47:425–437

Le Bissonnais Y, Benkahdra H, Chaplot V, Fox D, King D, Darousin J (1998) Crusting, runoff and sheet erosion on silty loamy soils at various scales and upscaling from m^2 to small catchments. Soil Till Res 46:69–80

Leblanc M, Morales JA, Borrego J, Felbaz-Poulichet F (2000) 4500 year old mining pollution in southwestern Spain: long-term implications for modern mining pollution. Econ Geol 95:655–662

Lee JJ, Phillips DL, Dodson RF (1996) Sensitivity of the US corn belt to climate change and elevated CO$_2$: II. Soil erosion and organic carbon. Agr Syst 52:503–521

Le Houérou HN (1990) Global change: vegetation, ecosystems, and land use in the southern Mediterranean basin by the mid twenty-first century. Israel J Bot 39:481–508

Lemmelä R, Helenius N (eds) (1998) Proc. of the Second International Conference on Climate and Water, Volume 1. Edita Ltd, Helsinki, Finland

Létolle R, Mainguet M (1993) Aral. Springer, Berlin, 385 p

Lettenmaier DP, Brettman KL, Vail LW, Yabusaki SB, Scott MJ (1992) Sensitivity of Pacific northwest water resources to global warming. Northwest Environ J 8(2):265–283

Lewis WM, Melack JM, McDowell WH, McClain M, Richey JE (1999) Nitrogen yields from undisturbed watershed in the Americas. Biogeochemistry 46:149–162

Lins HF, Slack JR (1999) Streamflow trends in the United States. Geophys Res Lett 26:227–230

Livingstone DA (1963) Chemical composition of rivers and lakes. Data of chemistry, US Geol Survey Prof Paper 440G, G1–G64

Loague KM, Freeze RA (1985) A comparison of rainfall-runoff modeling techniques on small upland catchments. Water Resour Res 21:229–248

Loague K, Kyriakidis PC (1997) Spatial and temporal variability in the R-5 infiltration dataset; Déjà vu and rainfall-runoff simulations. Water Resour Res 33(12):2883–2895

Longfield SA, Macklin MG (1999) The influence of recent environmental change on flooding and sediment fluxes in the Yorkshire Ouse basin. Hydrol Process 13(7):1051–1066

Lopez Bermudez F (1990) Soil erosion by water on the desertification of a semi-arid Mediterranean fluvial basin: the Segura basin, Spain. Agr Ecosyst Environ 33:129–145

Lopez Bermudez F, Romero Diaz MA (1989) Piping erosion and bad-land development in south-east Spain. Catena, Suppl 14:59–73

Lørup JK, Hansen E (1997) Effect of land use on the streamflow in the southwestern highlands of Tanzania. In: Rosbjerg D, Boutayeb NE, Gustard A, Kundzewicz ZW, Rasmussen PF (eds) Sustainability of water resources under increasing uncertainty. Proceedings of an international symposium of the Fifth Scientific Assembly of the International Association of Hydrological Sciences (IAHS), Rabat, Morocco, 23 April to 3 May 1997, IAHS Press, Wallingford, UK No. 240, pp 227–236

LUCIFS (2000) Land use and climate impacts on fluvial systems during the period of agriculture (LUCIFS). Pages Newsletter 8(3):10–19

Ludwig W (2001) The age of river carbon. Nature 409:466–467

Ludwig W, Probst JL (1996) Predicting the oceanic input of organic carbon by continental erosion. Global Biogeochem Cy 10:23–41

Ludwig W, Probst JL (1998) River sediment discharge to the oceans: present-day controls and global budgets. Am J Sci 298:265–295

Ludwig W, Probst JL, Kempe S (1996) Predicting the oceanic input of organic carbon by continental erosion. Global Biogeochem Cy 10:23–41

L'vovich MI (1979) World water resources and the future. AGU, Washington DC, (English translation edited by Nace RL)

L'vovich MI, White GF (1990) Use and transformation of terrestrial water systems. In: Turner BL, Clark WC, Kates RW, Richards JF, Mathews JT, Meyer WB (eds) The Earth as transformed by human action. Cambridge University Press, Cambridge, pp 235–252

LWRRDC (Land and Water Resources Research and Development Corporation) (1998) National dryland salinity program: management plan 1998–2003. Land and Water Resources Research and Development Corporation, Canberra, Australia

Maas EV, Hoffman GJ (1977) Crop salt tolerance – current assessment. J Irr Drain Div-Asce 103:115–134

Macaire JJ, Bossuet G, Choquier A, Cocirta C, DeLuca P, Dupis A, Gay I, Mathey E, Gueng P (1997) Sediment yield during late glacial and Holocene periods in the Lac Chambon watershed, Massif Central, France. Earth Surf Proc Land 22:473–489

Mackenzie FT, Lerman AB, Ver LMB (2001) Present, past and future of the global carbon cycle. A.A.P.G. Stud Geol 47:51–82

Macklin MG, Hudson-Edwards KA, Dawson EJ (1997) The significance of pollution from historic metal mining in the Pennine ore fields on river sediment contamination fluxes to the North Sea. Sci Total Environ 194/195:391–397

Maidment DR (ed) (1993) Handbook of hydrology. McGraw-Hill Inc.

Major JJ, Pierson TC, Dinehart RL, Costa JE (2000) Sediment yield following severe volcanic disturbance – A two-decade perspective from Mount Helens. Geology 28:819–822

Maloszewski P, Rauert W, Stichler W, Herrmann A (1983) Application of flow models in an alpine catchment area using tritium and deuterium data. J Hydrol 66:319–330

Marchand M, Toornstra FH (1986) Ecological guidelines for river basin development. Centrum voor Milieukunde, Rijksuniversiteit, Leiden, Netherlands, Report No. 28

Marengo JA, Nobre CA (2001) On the general characteristics and variability of climate in the Amazon Basin. In: McClain ME, Victoria R, Richey JE (eds) The biogeochemistry of the Amazon Basin. Oxford University Press, 384 p

Martin C, Lavabre J (1997) Estimation de la part du ruissellement sur les versants dans les crues du ruisseau du Rimbaud (Massif des Maures, Var, France) après l'incendie de forêt d'août 1990. Hydrolog Sci J 42:893–907

Martinelli LA, Ferreira JR, Forsberg BR, Victoria RL (1988) Mercury contamination in the Amazon: a gold rush consequence. Ambio 17:252–254

Martinelli LA, Victoria RL, Dematte JLI, Richey JE, Devol AH (1993) Chemical and mineralogical composition of Amazon River floodplain sediments, Brazil. Appl Geochem 8:391–402

Masiello CA, Druffel ER (2001) Carbon isotope geochemistry of the Santa Clara River. Global Biogeochem Cy 15:407–416

Maslenikova O, Mangerud J (2001) Where was the outlet of ice-dammed Lake Ksmi, Northern Russia. Global Planet Change 31:337–346

Mayorga E, Ballester V, Richey JE, Krusche A, Aufdenkampe A, Victoria R (2000) Towards a mechanistic, remote-sensing driven model of organic matter cycling linking the land surface and river system in the Amazon Basin. Abstract, LBA Science Conference, Belem

McCabe GJ, Wolock DM (2002) Trends and temperature sensitivity of moisture conditions in the conterminous United States. Climate Research 20(1):19–29

McCartney MP (1998) The hydrology of a headwater catchment containing a dambo. Unpublished PhD thesis, University of Reading, UK, 256 p

McCartney MP, Acreman MC (2003) Wetlands and water resources. In: Maltby E (ed) The wetlands handbook. Blackwells, Oxford, UK (in press)

McCartney MP, Sullivan CA, Acreman MC (1999) Ecosystem impacts of large dams. Centre for Ecology and Hydrology, Wallingford, UK, Report to IUCN. 78 p (Draft Report)

McCartney MP, Acreman MC, Bergkamp G (2000) Freshwater ecosystem management and environmental security. In: IUCN vision for water and nature: a world strategy for conservation and sustainable management of water resources in the 21st century. IUCN, Gland, Switzerland

McClain ME, Victoria TRL, Richey JE (ed) (2001) The biogeochemistry of the Amazon Basin. Oxford University Press, 364 p

McCully P (1996) Rivers no more: the environmental effects of dams. In: McCully P (ed) Silenced rivers: the ecology and politics of large dams. International Rivers Network, Zed Books, London, UK, pp 29–64

McDonnell JJ (1990) A rationale for old water discharge through macropores in a steep, humid catchment. Water Resour Res 26: 2821–2832

McDonnell JJ (1997) Comment on "the changing spatial variability of subsurface flow across a hillslide" by Ross Woods and Lindsay Rowe. J Hydrol New Zealand 36(1):97–100

McDonnell JJ, Tanaka T (2001) On the future of hydrology and biogeochemistry of forest catchments. Hydrol Process 15(10):2053–2056

McDonnell JJ, Bonell M, Stewart MK, Pearce AJ (1990) Deuterium variations in storm rainfall: implications for stream hydrograph separation. Water Resour Res 26:455–458

McDonnell JJ, Owens IF, Stewart MK (1991) A case study of shallow flow paths in a steep zero-order basin. Water Resour Bull 27(4): 679–685

McDonnell JJ, Freer J, Hooper R, Kendall C, Burns D, Beven K, Peters N (1996) New method developed for studying flow on hillslopes. EOS Trans Am Geophys Union 77:465–472

McDonnell JJ, Rowe L, Stewart M (1999) A combined tracer-hydrometric approach to assessing the effects of catchment scale on water flowpaths, source and age. International Association of Hydrological Sciences, Publication 258:265–274

McGlynn B, McDonnell JJ, Brammer D (2002) A review of the evolving perceptual model of hillslope flow in a steep forested humid catchment: a review of the Maimai catchment. J Hydrol 257:1–26

McHale MR, McDonnell JJ, Mitchell MJ, Cirmo CP (2002) A field-based study of soil water and groundwater nitrate release in an Adirondack forested watershed. Water Resour Res 38(4):1031, doi:10.1029/2000WR000102, 2002

McIntosh J, McDonnell JJ, Peters NE (1999) Tracer and hydrometric study of preferential flow in large undisturbed soil cores from the Georgia Piedmont, USA. Hydrol Process 13:139–155

McKenzie RS, Craig AR (1998) Evaporation losses from South African rivers. Water Research Commission, Pretoria, RSA, Report No. 638/1/99

MDBC (Murray-Darling Basin Commission) (1997) Salt trends: historic trend in salt concentration and saltload in stream flow in the Murray-Darling Drainage Division. Murray-Darling Basin Commission, Canberra, Australia, Dryland Technical Report 1

MDBMC (Murray-Darling Basin Ministerial Council) (1999) The salinity audit of the Murray-Darling Basin: a 100-year perspective (1999) Murray-Darling Basin Ministerial Council, Canberra ACT, Australia, 39 p

Meade RH (1988) Movement and storage of sediment in river systems. In: Lerman A, Meybeck M (eds) Physical and chemical weathering in geochemical cycles. Kluwer Academic Publishers, Dordrecht, pp 165–180

Meade RH (1994) Suspended sediments of the modern Amazon and Orinoco Rivers. Quatern Int 21:29–39

Meade RH, Parker RS (1985) Sediments in rivers of the United States. In: US Geological Survey (ed) National water summary, 1984. US Geol Surv Water-Supply Paper 2275, USGS, Reston, pp 49–60

Meade RH, Dunne T, Richey JE, dos Santos U, Salati E (1985) Storage and remobilization of sediment in the lower Amazon River of Brazil. Science 228:488–490

Meier KB, Brodie JR, Schulze RE, Smithers JC, Mngune D (1997) Modelling the impacts of riparian zone alien vegetation on catchment water resources using the ACRU model. Proceedings, 8th South African National Hydrology Symposium, Water Resources Commission, Pretoria, RSA, 13 p

Melillo JM, McGuire AD, Kikclighter DW, Moore B III, Vörösmarty CJ, Schloss AL (1993) Global climate change and terrestrial net primary production. Nature 363:234–240

Mertes LAK, Smith MO, Adams JB (1993) Estimating suspended sediment concentrations in surface waters of the Amazon River wetlands from Landsat images. Remote Sens Environ 43:281–301

Mertes LAK, Dunne T, Martinelli LA (1996) Channel-floodplain geomorphology of the Solimões-Amazon River, Brazil. Geol Soc Am Bull 108:1089–1107

Messerli B, Hofer T, Chapman GP (1995) Assessing the impact of anthropogenic land-use change in the Himalayas. In: Chapman G, Thompson M (eds) Water and the quest for sustainable development in the Ganges Valley. Mansell Publishing, London, Global Development and the Environment Series, pp 64–89

Meybeck M (1976) Total mineral transport by major world rivers. Hydrol Sci Bull 21:265–284

Meybeck M (1979) Concentration des eaux fluvials en éléments majeurs et apports en solution aux oceans. Rev Geol Dyn Geogr 21:215–246

Meybeck M (1981) River transport of organic carbon to the oceans. In: US Dep of Energy (ed) Flux of organic carbon by rivers to the oceans. NTIS Rep. CONF 8009140UC-11, Washington DC

Meybeck M (1982) Carbon, nitrogen, and phosphorous transport by world rivers. Am J Sci 282:401–450

Meybeck M (1986) Composition chimique naturelle des ruisseaux non pollués en France. Sci Geol Bull 39:3–77

Meybeck M (1987) Global chemical weathering estimated from river dissolved loads. Am J Sci 287:401–428

Meybeck M (1988) How to establish and use world budgets of river material. In: Lerman A, Meybeck M (eds) Physical and chemical weathering in geochemical cycles. Kluwer Academic Publishers, Dordrecht, pp 247–272

Meybeck M (1993a) Riverine transport of atmospheric carbon: sources, global typology and budget. Water Air Soil Poll 70:443–463

Meybeck M (1993b) C, N, P, and S in rivers: from sources to global inputs. In: Wollast R, Mackenzie FT, Chou L (eds) Interactions of C, N, P, and S biogeochemical cycles and global change. NATO ASI Series, Vol. 14, Springer-Verlag, Berlin, pp 163–193

Meybeck M (1994) Global lake distribution. In: Lerman A, Imboden D, Gat J (eds) Physics and chemistry of lakes. Springer-Verlag, Heidelberg, pp 1–35

Meybeck M (1998) The IGBP Water Group: a response to a growing global concern. Global Change Newsletters 36:8–13

Meybeck M (2002) Riverine quality at the Anthropocene: propositions for global space and time analysis, illustrated by the Seine River. Aquatic Sciences 64:376–393

Meybeck M (2003) Global analysis of river systems: from Earth system controls to Anthropocene syndroms. Philos T Roy Soc B (in press)

Meybeck M, Helmer R (1989) The quality of rivers: from pristine stage to global pollution. Palaeogeogr Palaeocl (Global Planet Change Section) 75:283–309

Meybeck M, Ragu A (1996) River discharges to the oceans. An assessment of suspended solids, major ions, and nutrients. Environment Information and Assessment Rpt. UNEP, Nairobi, 250 p

Meybeck M, Ragu A (1997) Presenting the GEMS-GLORI, a compendium for world river discharge to the oceans. Int Ass Hydrol Sci 243:3–14

Meybeck MM, Vörösmarty CJ (1999) Global transfer of carbon by rivers. IGBP Global Change Newsletter 37:18–19

Meybeck M, Green P, Vörösmarty CJ (2001) A new typology for mountains and other relief classes: an application to global continental water resources and population distribution. Mt Res Dev 21:34–45

Meybeck M, Laroche L, Dürr HH, Syvitski JPM (2003) Global variability of daily total suspended solids and their fluxes in rivers. Global and Planetary Change 39:65–93

Meyer WB, Turner BL (1992) Human population growth and global land-use/land-cover change. Annu Rev Ecol Syst 23:39–61

Middelkoop H (2002) Reconstructing flood plain sedimentation rates from heavy metal profiles by inverse modelling. Hydrol Process 16:47–64

Milliman JD (1991) Flux and fate of fluvial sediment and water in coastal seas. In: Mantoura RFC, Martin JM, Wollast R (eds) Ocean margin processes in global change. John Wiley & Sons, Chichester, pp 69–89

Milliman JM, Meade RH (1983) World-wide delivery of river sediment to the oceans. J Geol 91:1–21

Milliman JD, Meie R (1995) River flux to the sea: impact of human intervention on river systems and adjacent coastal areas. In: Eisma D (ed) Climate change: impact on coastal habitation. Crc Press Inc., pp 57–83

Milliman JM, Syvitski JPM (1992) Geomorphic/tectonic control of sediment discharge to the ocean: the importance of small mountainous rivers. J Geol 100:525–544

Milliman JD, Rutkowski C, Meybeck M (1995) River discharge to the sea. A global river index (GLORI). LOICZ repts and studies No. 2, Texel, The Netherlands, 125 p

Milly PCD (1994) Climate, soil water storage, and the average annual water balance. Water Resour Res 30, 2143–2156

Milly PCD, Dunne KA (1994) Sensitivity of the global water cycle to the water-holding capacity of land. J Climate 7:506–526

Mitsch WJ, Gosselink JG (1993) Wetlands, 2nd ed. Van Nostrand Reinhold, New York

Moerdyk MR, Schulze RE (1991) Impacts of forestry site preparation on surface runoff and soil loss. Agr Eng S Afr 23:319–325

Molicova H, Bonell M, Grimaldi M, Hubert P (1997) Using TOPMODEL towards identifying and modelling the hydrological patterns within a headwater humid tropical catchment. Hydrol Process 11(9):1169–1196

Montgomery DR, Dietrich WE, Torres R, Anderson SP, Heffner JT, Loague K (1997) Hydrologic response of a steep unchanneled valley to natural and applied rainfall. Water Resour Res 33(1): 91–109

Moore T (1989) Dynamics of dissolved organic carbon in forested and disturbed catchments, Westland, New Zealand, 1. Maimai. Water Resour Res 25(6):1321–1330

Moore WS (1996) Large ground water inputs to coastal waters revealed by 226Ra enrichments. Nature 380:612–614

Moore TR, Mark RK (1986) World slope map. AGU Eos Transactions 67:1353–1356

Morehead MD, Syvitski JP, Hutton EWH, Peckham SD (2003) Modeling the temporal variability in the flux of sediment from ungauged river basins. Global and Planetary Change 39(1–2):95–110

Mosley MP (1979) Streamflow generation in a forested watershed, New Zealand. Water Resour Res 15:795–806

Mozzherin VI (1992) The recent global suspended sediment yield and prognosis of its change. In: Chalov RS (ed) Problems of erosion, fluvial and mount. processes. Izhevsk, pp 63–85 (in Russian)

Muller AMM, Keuris H, Roux le F (1985) Die bepaling van rivierverliese in die Wilgerivier tussen Sterkfonteindam en Vaaldam. Department of Water Affairs, Pretoria, RSA. Directorate of Hydrology Report, 13 p

Murphy JH, Mitchell JFB (1995) Transient response of the Hadley Centre coupled ocean-atmosphere model to increased carbon dioxide. Part 2. Spatial and temporal structure of the response. J Climate 8:57–80

Nash JE, Sutcliffe IV (1970) River flow forecasting through conceptional models, 1. A discussion of principles. J Hydrol 10:282–290

Ne'eman G, Perevolotsky A, Schiller G (1997) The management implications of the Mt. Carmel research project. Int J Wildland Fire 7:343–350

Neill M (1989) Nitrate concentrations in river waters in the southeast of Ireland and their relationship with agricultural practices. Water Res 23:1339–1355

Nepstad DC, Klink C, Uhl C, Vieira IC, Lefebvre P, Pedlowski M, Matricardi E, Negreiros G, Brown IF, Amaral E, Homma A, Walker R (1997) Land-use in Amazonia and Cerrado of Brazil. Cienc Cult 49:73–86

NERI (Natural Environment Research Institute of Denmark) (1997) Technical report 210. Cited in: Olesen (1999) (in Danish)

Nespack/MMI (National Engineering Services Pakistan (Pvt) Ltd./ Mott MacDonald) (1993) Feasibility study on national drainage program 1. Executive summary. NESPAK and Mott MacDonald, Pakistan

Newson MD (1997) Land, water and development: sustainable management of river basin systems. 2nd ed. Routledge, London

Newson MD, Gardiner JL, Slater SJ (2000) Planning and managing for the future. In: Acreman MC (ed) The hydrology of the UK: a study of change. Routledge, London, pp 244–269

Nilsson C, Jansson R (1995) Floristic differences between riparian corridors of regulated and free-flowing boreal rivers. Regul River 11:55–66

Nixon S (2003) Replacing the Nile – Are anthropogenic nutrients providing the fertility once brought to the Mediterranean by a great river? Ambio 32(1):30–39

Nyberg R, Rapp A (1998) Extreme erosional events and natural hazards in Scandinavian mountains. Ambio 27(4):292–299

Ohmori H (1983) Erosion rates and their relation to vegetation from the viewpoint of worldwide distribution. Univ of Tokyo, Dept of Geography Bull 15:77–91

Oki T, Musiake K, Matsuyama H, Masuda K (1995) Global atmospheric water balance and runoff from large river basins. Hydrol Process 9:655–678

Oki T, Nishimura T, Dirmeyer P (1999) Validating land surface models by runoff in major river basins of the globe using Total Runoff Integrating Pathways (TRIP). J Meteorol Soc Jpn 77: 235–255

Olesen US (1999) Agriculture in Denmark, a source of nutrient pollution of the water environment. In: Van der Kraats JA (ed) Farming without harming: the impact of agricultural pollution on water systems. Drukkerij Belser, Lelystad, pp 33–44

Olivera F, Famiglietti J, Asante K (2000) Global-scale flow routing using a source-to-sink algorithm. Water Resour Res 36:2197–2207

O'Loughlin EM, Short DL, Dawes WR (1989) Modelling the hydrological response of catchment to land-use change. In: Hydrology and water resources symposium, comparisons in Austral hydrology. Inst. Engrs., Canberra, Australia, 28–30 Nov. 1989, Christchurch, New Zealand, pp 335–340

Onda Y, Komatsu Y, Tsujimura M, Fujihara J-I (2001) The role of subsurface runoff through bedrock on storm flow generation. Hydrol Process 15:1693–1706

Oyebande L (1995) Effects of reservoir operation on the hydrological regime and water availability in northern Nigeria. In: Petts G (ed) Man's influence on freshwater ecosystems and water use. IAHS Publ. No. 230, IAHS Press, Wallingford

Panagoulia D, Dimou G (1997) Sensitivity of flood events to global climate change. J Hydrol 191:208–222

Perks LA (2001) Refinement of modelling tools to assess potential agrohydrological impacts of climate change in Southern Africa. PhD thesis, School of BEEH, University of Natal, South Africa. 463 p

Pernetta JC, Milliman JD (eds) (1995) Land-ocean interactions in the coastal zone implementation plan. Global Change Report No. 33, IGBP, Stockholm, 215 p

Peters DL, Buttle JM, Taylor CH, LaZerte BD (1995) Runoff production in a forested, shallow soil, Canadian Shield basin. Water Resour Res 31(5):1291–1304

Peterson BJ, Wollheim WM, Mulholland PJ, Webster JR, Meyer JL, Tank JL, Marti E, Bowden WB, Valett HM, Hershey AE, McDowell WH, Dodds WK, Hamilton SK, Gregory SV, Morrall DD (2001) Control of nitrogen export from watersheds by headwater streams. Science 292:86–90

Peterson BJ, Holmes RH, McClelland JW, Vörösmarty CJ, Lammers RB, Shiklomanov AI, Rahmstorf S (2002) Increasing river discharge to Arctic Ocean. Science 298:2171–2173

Petts GE (1984) Impounded rivers: perspectives for ecological management. John Wiley & Sons, Chichester, 326 p

Piccolo A, Teshale AZ (1998) Soil processes and responses to climate changes. In: Climate change impact on agriculture and forestry. European Commission, EUR 18175, Luxembourg, pp 79–92

Pielke RA, Lee TJ, Copeland JH, Eastman JL, Ziegler CL, Finley CA (1997) Use of USGS-provided data to improve weather and climate simulations. Ecol Appl 7:3–21

Pinet P, Souriau M (1988) Continental erosion and large scale relief. Tectonics 7:563–582

Pitman A, Zhao M (2000) The relative impact of observed change in land cover and carbon dioxide as simulated by a climate model. Geophys Res Lett 27:1267–1270

Pitman AJ, Henderson-Sellers A, Yang Z-L, Abramopoulos F, Boone A, Desborough CE, Dickinson RE, Gedney N, Koster R, Kowalczyk E, Lettenmaier D, Liang X, Mahfouf J-F, Noilhan J, Polcher J, Qu W, Robock A, Rosenzweig C, Schlosser C, Shmakin AB, Smith J, Suarez M, Verseghy D, Wetzel P, Wood E, Xue Y (1999) Key results and implications from phase 1(c) of the Project for Intercomparison of Land-surface Parameterization Schemes. Clim Dynam 15:673–684

Planchon O, Valentin C (2004) Soil erosion in West Africa: present and future. In: Favis-Mortlock D, Boardman J (eds) Soil erosion and climatic change. Oxford University Press (in press)

Postel SL, Daily GC, Ehrlich PR (1996) Human appropriation of renewable fresh water. Science 271:785–788

Pretoria, RSA, NWA (1998) National Water Act of South Africa. Act No. 36 of 1998. Government Printer, Pretoria, RSA. 200 p

Probst JL (1992) Géochimie et Hydrochimie de l'érosion continentale. Mécanismes, bilan global actuel et fluctuations au cours des 500 derniers millions d'années. Sciences Géologique Memoirs 94. Strasbourg 161 p

Probst JL, Tardy Y (1989) Global runoff fluctuations during the last 80 years in relation to world temperature change. Am J Sci 289: 267–285

Quay PD, Wilbur DO, Richey JE, Hedges JI, Devol AH, Victoria RL (1992) Carbon cycling in the Amazon River: implications from the ^{13}C composition of particles and solutes. Limnol Oceanogr 37:857–871

Quay PD, Wilbur DO, Richey JE, Devol AH, Benner R, Forsberg BR (1995) The $^{18}O/^{16}O$ of dissolved oxygen in rivers and lakes in the Amazon Basin: a tracer of respiration and photosynthesis. Limnol Oceanogr 40:718–729

Qureshi RH, Barrett-Lennard EG (1998) Saline agriculture for irrigated land in Pakistan: a handbook. ACIAR Monograph 50, 142 p

Rabouille C, Mackenzie FT, Ver LM (2001) Influence of human perturbation on carbon, nitrogen and oxygen biogeochemical cycles in the global coastal ocean. Geochim Cosmochim Ac 65:3615–3641

Rai SC, Sharma E (1998) Land use/cover change and hydrology were studied in a watershed in the Sikkim Himalaya, India. Hydrol Process 12:2235–2248

Rassmussen T, Baldwin R, Dowd J, Williams A (2000) Tracer vs. pressure wave velocities through unsaturated saprolite. Soil Sci Soc Am J 64:75–85

Raymond PA, Bauer JE (2001) Riverine export of aged terrestrial organic matter to the North Atlantic Ocean. Nature 409:497–500

Rekolainen S (1990) Phosphorus and nitrogen load from forest and agricultural areas in Finland. Aqua Fenn 19:95–107

Renard KG, Foster GR, Weesies GA, McCool DK (1991) Predicting soil erosion by water. A guide to conservation planning with the Revised Universal Soil Loss Equation (RUSLE). USDA Agricultural Research Service, Tucson AZ, USA

Revenga C, Brunner J, Henninger N, Kassem K, Murray S (2000) Global freshwater ecosystem assessment. World Resources Institute, Washington DC

Richards JF (1990) Land transformation. In: Turner BL, Clark WC, Kates RW, Richards JF, Mathews JT, Meyer WB (eds) The Earth as transformed by human action. Cambridge University Press, Cambridge, pp 163–178

Richey JE, Victoria RL (1993) C, N and P export dynamics in the Amazon River. In: Wollast R, Mackenzie FT, Chou L (eds) Interactions of C, N, P, and S biogeochemical cycles and global change. Springer-Verlag, Berlin, pp 123–140

Richey JE, Victoria RL (1996) Continental-scale biogeochemical cycles of the Amazon River system. Verh Int Ver Theor Angew Limnol 26:219–226

Richey JE, Meade RH, Salati E, Devol AH, Nordin CF, dos Santos U Jr (1986) Water discharge and suspended sediment concentrations in the Amazon River: 1982–1984. Water Resour Res 22:756–764

Richey JE, Mertes LA, Victoria RL, Forsberg BR, Dunne T, Oliveira F, Tancredi A (1989a) Sources and routing of the Amazon River floodwave. Global Biogeochem Cy 3:191–204

Richey JE, Nobre C, Deser C (1989b) Amazon river discharge and climate variability: 1903 to 1985. Science 246:101–103

Richey JE, Hedges JI, Devol AH, Quay PD, Victoria R, Martinelli L, Forsberg BR (1990) Biogeochemistry of carbon in the Amazon River. Limnol Oceanogr 35:352–371

Richey JE, Wilhelm SR, McClain ME, Victoria RL, Melack JM, Araujo-Lima CARM (1997) Organic matter and nutrient dynamics in river corridors of the Amazon Basin and their response to anthropogenic change. Cienc Cult 49:98–110

Rijsberman FR (ed) (2000) World water scenarios. Earthscan, London

Rillig MC, Wright SF, Allen MF, Field CB (1999) Rise in carbon dioxide changes in soil structure. Nature 400:628

Roberts G (1998) The effects of possible future climate change on evaporation losses from four contrasting UK water catchment areas. Hydrol Process 12:727–739

Robinson M, Boardman J, Evans R, Heppell K, Packman JC, Leeks GJL (2000) Land-use change. In: Acreman MC (ed) The hydrology of the UK. Routledge, London, pp 30–54

Robock A, Vinnikov KY, Srinivasan G, Entin JK, Hollinger SE, Speranskaya NA, Liu SX, Namkhai A (2000) The global soil moisture data bank. B Am Meteorol Soc 81(6):1281–1299

Robson AJ (2002) Evidence for trends in UK flooding. Phil T Roy Soc A 1796:1327–1343

Rodhe A (1987) The origin of streamwater traced by oxygen-18. Doctoral Thesis, Dept. of Physical Geography, Uppsala University, 260 p

Ronov AB (1976) Global carbon geochemistry, volcanism, carbonate accumulation and life. Geochem Int 13:172–195

Rooseboom A, Verster E, Zietsman HL, Lotriet HH (1992) The development of the new sediment yield map of southern Africa. Water Research Commission, Pretoria, South Africa, WRC Report 297/2/92

Rosenberg DM, McCully P, Pringle CM (2000) Global-scale environmental effects of hydrological alterations: introduction. Bioscience 50:746–751

Rotmans J, den Elzen MGJ (1993) Modelling feedback mechanisms in the carbon cycle: balancing the carbon budget. Tellus 45B:1–20

Rotmans J, de Vries B (1997) Perspectives on global change. The TARGET approach. Cambridge University Press, Cambridge, 462 p

Roulet M, Lucotte M, Canuel R, Rheault I, Tran S, Gog YGD, Farella N, doVale RS, Passos CJS, daSilva ED, Mergler D, Amorim M (1998) Distribution and partition of total mercury in waters of the Tapajós River basin, Brazilian Amazon. Sci Total Environ 213:203–211

Rouse WR (1998) A water balance model for a subarctic sedge fen and its application to climatic change. Climatic Change 38: 207–234

Rudoy A (1998) Mountain ice-dammed lakes of Southern Siberia and their influence on the development and regime of the intracontinental runoff system of North Asia in the late Pleistocene. In: Benito G, Baker VR, Gregory KJ (eds) Paleohydrology and environmental change. John Wiley & Sons, Chichester, pp 215–234

Ruecker G, Schad P, Alcubilla MM, Ferrer C (1998) Natural regeneration of degraded soils and site changes on abandoned agricultural terraces in Mediterranean Spain. Land Degrad Dev 9: 179–188

Ruiz-Flano P, Garcia-Ruiz JM, Ortigosa L (1992) Geomorphological evolution of abandoned fields. A case study in the central Pyrenees. Catena 19:301–308

Russell MA, Maltby E (1995) The role of hydrologic regime on phosphorus dynamics in a seasonally waterlogged soil. In: Hughes JMR, Heathwaite AL (eds) Hydrology and hydrochemistry of British wetlands. John Wiley & Sons, Chichester

Russell MA, Walling DE, Webb RW, Bearne R (1998) The composition of nutrient fluxes from contrasting UK river basins. Hydrol Process 12:1461–1482

Sahagian D, Schwartz FW, Jacobs DK (1994) Direct anthropogenic contributions to sea level rise in the twentieth century. Nature 367:54–56

Said R (1993) The River Nile. Pergamon Press, Oxford

Saito Y, Yang Z, Hori K (2001) The Huang He (Yellow River) and Changjiang (Yangtze River) deltas: a review of their characteristics, evolution and sediment discharge during the Holocene. Geomorphology 41:219–231

Sajjapongse A, Syers JK (1995) Tangible outcomes and impacts from the ASIALAND management of sloping lands network. IBSRAM Proceedings No. 14, Bangkok, pp 3–14

Salati E, Dall'Olio A, Matsui E, Gat JR (1979) Recycling of water in the Amazon Basin: an isotopic study. Water Resour Res 15:1250–1258

Salomons W, Turner RK, de Lacerda LD, Ramachandran S (eds) (1999) Perspectives on integrated coastal zone management, Springer-Verlag, Heidelberg, 386 p

Sarwar A (2000) A transient model approach to improve on-farm irrigation and drainage in semi-arid zones. PhD Thesis, Wageningen University and Research Center, Wageningen, Netherlands. 147 p

Saxton KE, Rawls WJ, Romberguer JS, Papendick RI (1986) Estimating generalized soil-water characteristics from texture. Soil Sci Soc Am J 50(4):1031–1036

Schellnhuber H-J (2001) Earth system analysis and management. In: Ehlers E, Krafft T (eds) Understanding the Earth system. Compartments, processes and interactions. Springer-Verlag, Heidelberg, pp 17–55

Schlesinger WH (1997) Biogeochemistry: an analysis of global change. 2nd ed. Academic Press, San Diego

Schlesinger WH, Melack JM (1981) Transport of organic carbon in the world's rivers. Tellus 33:172–187

Schmidt GW (1973) Primary production of phytoplankton in the three types of Amazonian waters. III. Primary production of phytoplankton in a tropical floodplain lake of central Amazonia, Lago do Castanho, Amazonas, Brazil. Amazoniana 4:379–404

Schulze RE (1979) Hydrology and water resources of the Drakensberg. Natal Town and Regional Planning Commission, Pietermaritzburg, RSA, 179 p

Schulze RE (1995) Hydrology and agrohydrology – a text to accompany the ACRU 3.00 agrohydrological modelling system. Water Research Commission, Pretoria, RSA, WRC Report TT69/95, 552 p

Schulze RE (1997) South African atlas of agrohydrology and climatology. Water Research Commission, Pretoria, RSA, Report TT82/96, 276 p

Schulze RE (1998) Hydrological modelling: concepts and practice. IHE, Delft, 134 p

Schulze RE (1999) Integrated catchment management: summary of a CEH workshop on ICM. Centre for Ecology and Hydrology, Wallingford, UK, 11 p (available from author)

Schulze RE (2000) Transcending scales of space and time on impact studies of climate and climate change on agrohydrological responses. Agr Ecosyst Environ 82:185–212

Schulze RE (2001) Hydrological responses at river basin scale. School of Bioresources Engineering and Environmental Hydrology, University of Natal, Pietermaritzburg, South Africa, ACRUcons Report 38, 105 p

Schulze RE, George WJ (1987) A dynamic, process-based user-oriented model of forest effects on water yield. Hydrol Process 1: 293–307

Schulze RE, Chapman RA, Angus GR, Schmidt EJ (1987) Distributed model simulation of impacts of upstream reservoirs on wetlands. In: Walmsley RD, Botten ML (eds) Proceedings, Symposium on Ecology and Conservation of Wetlands in South Africa. Ecosystems Programme, FRD-CSIR, Pretoria, RSA, pp 115–123

Schulze RE, Summerton MJ, Meier KB, Pike A, Lynch SD (1997) The ACRU forest decision support system to assess hydrological impacts of afforestation practices in South Africa. Proceedings, 8th South African National Hydrology Symposium, Water Research Commission, Pretoria, RSA, 13 p

Schulze RE, Horan MJC, Shange SN, Ndlela RM, Perks LA (1998) Hydrological impacts of land use practices in the Pongola-Bivane catchment – Phase 3. University of Natal, Pietermaritzburg, RSA, School of Bioresources Engineering and Environmental Hydrology. ACRUcons Report 26, 58 p

Scott DF, Maitre Le DC, Fairbanks DHK (1998) Forestry and streamflow reductions in South Africa: a reference system for assessing extent and distribution. Water SA 24:187–199

Sear DA, Wilcock DN, Robinson M, Fisher K (2000) River channel modification in the UK. In: Acreman MC (ed) The hydrology of the UK. Routledge, London, pp 55–81

SEARCH SSC (2001) SEARCH: Study of Environmental Arctic Change, Science Plan. Seattle: Polar Science Center, University of Washington, 89 p, (available at: http://psc.apl.washington.edu/search/)

Seibert J, McDonnell JJ (2002) On the dialog between experimentalist and modeler in catchment hydrology: use of soft data for multicriteria model calibration. Water Resour Res 38(11): 1241, doi:1210.1029/2001WR000978

Seibert J, Bishop K, Rodhe A, McDonnell JJ (2003) Groundwater dynamics along a hillslope: a test of the steady state hypothesis. Water Resour Res 39(1):1014, doi:1010.1029/2002WR001404

Seitzinger SP, Kroeze C (1998) Global distribution of nitrous oxide production and N inputs in freshwater and coastal marine ecosystems. Global Biogeochem Cy 12:93–113

Seitzinger SP, Kroeze C, Bouwman AF, Caraco N, Dentener F, Styles RV (2002) Global patterns of dissolved inorganic and particulate nitrogen inputs to coastal ecosystems: recent conditions and future projections. Estuaries 25(4B):640–655

Semiletov IP, Savelieva NI, Weller GE, Pipko II, Pugach SP, Gukov AY, Vasilevskaya LN (2000) The dispersion of Siberian river flows into coastal waters: meteorological, hydrological, and hydrochemical aspects. In: Lewis EL, Jones EP, Lemke P, Prowse TD, wadhams P (eds) The freshwater budget of the Arctic Ocean. NATO Advanced Study Institute Series. Kluwer Academic Publishers, Dordrecht, pp 281–296

Serreze MC, Walsh JE, Chapin FS III, Osterkamp T, Dyurgerov M, Romanovsky V, Oechel WC, Morison J, Zhang T, Barry RG (2000) Observational evidence of recent change in the northern high latitude environment. Climatic Change 46:159–207

Servais P, Billen G, Garnier J, Idlafkih Z, Mouchel JM, Seidl M, Meybeck M (1998) Carbone organique: origines et biodegrabilité. In: Meybeck M, de Masily G, Fustec E (eds) La Seine en son bassin. Elsevier Paris, pp 483–529

Sherwood M (1999) Impact of agricultural pollution on water systems: Ireland. In: Van der Kraats JA (ed) Farming without harming: the impact of agricultural pollution on water systems. Drukkerij Belser, Lelystad, Netherlands, pp 137–149

Shiklomanov IA (1996) Assessment of water resources and water availability in the world: scientific and technical report. State Hydrological Institute, St. Petersburg, Russia, 127 p

Shiklomanov IA (1997) Comprehensive assessment of the freshwater resources and water availability in the world: assessment of water resources and water availability in the world. WMO, Geneva

Shiklomanov IA (2000) World water resources and water use: present assessment and outlook for 2025. In: Rijsberman FR (ed) World water scenarios. Earthscan, London, pp 160–203

Shiklomanov I, Krestovsky OI (1988) Influence of forests and forest reclamation practice on streamflow and water balance. In: Reynolds ERC, Thompson FB (eds) Forests, climate and hydrology: regional impacts. United Nations University Press, Tokyo

Shotyk W, Weiss D, Appleby P, Cheburkin A, Frei R, Glor M, Kramers J, Reese S, van de Knaap W (1998) History of atmospheric lead deposition since 12370 14C yr B.P. from a peat bog, Jura Mountains, Switzerland. Science 281:1635–1940

Shukla J, Nobre C, Sellers P (1990) Amazon deforestation and climate change. Science 247:1322–1325

Shumway SE (ed) (1988) Toxic algal blooms: hazards to shellfish industry. J Shellfish Res 7:587–705

Sidorchuk AY (1991) Sedimentation of the small rivers on the Russian Plain during the period of intensive agriculture. In: Larinov GA (ed) Erosional studies: theory, experiment, practice. Moscow University Publ, pp 140–42, (in Russian)

Sidorchuk A (1994) Modeling of sediment budgets through the fluvial system. In: Prelim. Proc. IGBP Inter-Core Project Workshop on Modeling the Delivery of Terrestrial Material to Freshwater and Coastal Ecosystems. UNH, Durham, NH

Sidorchuk A, Borisova O, Panin A (2001) Fluvial response to the Late Valdai/Holocene environmental change on the East European Plain. Global Planet Change 28:303–318

Sioli H (1950) Das Wasser im Amazonasgebiet. Forsch Fortschr 26: 274–280

Sklash MG (1990) Environmental isotope studies of storm and snowmelt runoff generation. In: Anderson MG, Burt TP (eds) Processes in hillslope hydrology. John Wiley & Sons, Chichester, pp 401–435

Sklash MG, Farvolden RN (1979) The role of groundwater in storm runoff. J Hydrol 43:45–65

Skole D, Tucker C (1993) Tropical deforestation and habitat fragmentation in the Amazon: satellite data from 1978 to 1988. Science 260:1905–1910

Smith CT (1969) The drainage basin as an historical basis for human activity. In: Chorley RJ (ed) Water, Earth and man. Methuen, London, pp 101–110

Smith RE (1983) Approximate soil water movement by kinematic characteristics. Soil Sci Soc Am J 47:3–8

Smith RA, Alexander RB, Tasker GD, Price CV, Robinson KW, White DW (1993) Statistical modelling of water quality in regional watersheds. Watershed '93, A National Conference on Watershed Management, pp 751–754

Smith SV, Renwick WH, Buddemeier RW, Crossland CJ (2001) Budgets of soil erosion and deposition for sediments and sedimentary organic carbon across the conterminous United States. Global Biogeochem Cy 15:697–708

Sparks RE (1995) Need for ecosystem management of large rivers and their floodplains. Bioscience 45:168–182

Spitters CJT, van Keulen H, van Kraalingen DWG (1989) A simple and universal crop growth simulator. In: Rabbinge R, Ward SA, van Laar HH (eds) Simulation and systems management in crop protection. Simulation Monographs, Pudoc, Wageningen, pp 147–181

Stallard RF (1998) Terrestrial sedimentation and the carbon cycle: coupling weathering and erosion to carbon burial. Global Biogeochem Cy 12:231–257

Stallard RF, Edmond JM (1983) Geochemistry of the Amazon: 2. The influence of geology and weathering environment on the dissolved load. J Geophys Res 88:9671–9688

Stewart MK, McDonnell JJ (1991) Modeling base flow soil residence times from deuterium concentrations. Water Resour Res 27: 2681–2693

Stewart JB, Engman ET, Feddes RA, Kerr Y (1998) Scaling up in hydrology using remote sensing: summary of a Workshop. Int J Remote Sens (19)1:181–194

Strahler AN (1957) Quantitative analysis of watershed geomorphology. Trans Am Geophys Union 38:913–920

Subcommittee on Global Change Research (SGCR) (2000) Our changing planet: the FY2001 US. Global Change Research Program. Supplement to the President's FY 2001 Budget. National Science and Technology Council, Washington DC, 74

Syvitski JPM (2003) Sediment discharge variability in Arctic rivers: implications for a warmer future. Polar Res 21(2):323–330

Syvitski JP, Morehead MD (1999) Estimating river-sediment discharge to the ocean: application to the Eel margin, northern California. Mar Geol 154:13–28

Tanakamaru H, Kaodya M (1993) Effects of climate change on the regional hydrological cycle in Japan. In: Exchange processes at the land surface for a range of space and time scales. Proceedings of the Yokohama Symposium, July 1993, IAHS Publ. No. 212, pp 535–542

Tani M (1997) Runoff generation processes estimated from hydrological observations on a steep forested hillslope with a thin soil layer. J Hydrol 200:84–109

Thiede J, Bauch HA, Hjort C, Mangerud J (eds) (2001) The late Quaternary stratigraphy and environments of Northern Eurasia and the adjacent Arctic seas. New contributions from QUEEN, Global Planet Change 31, No. 1–4, VII–X Sp. Iss.

Thompson MW (1996) A standard land-cover classificatin scheme for remote sensing applications in South Africa. S Afr J Sci 92: 34–42

Tiffen M, Mortimore M, Gichuku F (1994) More people, less erosion: environmental recovery in Kenya. African Centre for Technology Studies Press, Nairobi, 311 p

Torres R, Dietrich WE, Montgomery DR, Anderson SP, Loague K (1998) Unsaturated zone processes and the hydrologic response of a steep, unchanneled catchment. Water Resour Res 34(8):1865–1879

Trimble SW (1977) The fallacy of stream equilibrium in contemporary denudation studies. Am J Sci 277:876–887

Tsirkunov VV (1998) Salinisation. In: Kimstach V, Meybeck M, Baroudy E (eds) A water quality assessment of the former Soviet Union. E&FN Spon, London, pp 112–136

Turner JV, Barnes CJ (1998) Modeling of isotopes and hydrochemical responses in catchment hydrology. In: Kendall C, McDonnell JJ (eds) Isotope tracers in catchment hydrology. Elsevier, Amsterdam, pp 723–760

Turner RE, Rabalais N (1991) Changes in the Mississippi water quality during this century and implications for coastal food webs. Bioscience 41:140–147

Turner RE, Rabalais N (1994) Coastal eutrophication near the Mississippi River delta. Nature 368:619–621

Turner JV, Macpherson DK, Stokes RA (1987) The mechanisms of catchment flow processes using natural variations in deuterium and oxygen-18. J Hydrol 94:143–162

Turner BL, Clark WC, Kates RW, Richards JF, Mathews JT, Meyer WB (eds)(1991) The Earth as transformed by human action. Cambridge University Press, Cambridge

Turner BL, Moss RH, Skole DL (1993) Relating land use and global land-cover change: a proposal for an IGBP-HDP core project. IGBP, Stockholm, Sweden, IGBP Report 24, 65 p

Uhlenbrook S, Leibundgut CH (1999) Integration of tracer information into the development of a rainfall-runoff model. In: Leibundgut C, McDonnell J, Schultz G (eds) Integrated methods in catchment hydrology. Proc. Int. Symp. at Birmingham, UK, July 1999, IAHS Publ. No. 258, pp 93–100

Uhlenbrook S, Frey M, Leibundgut C, Maloszewski P (2002) Residence time based hydrograph separations in a meso-scale mountainous basin at event and seasonal time scales. Water Resour Res 38(6):1–14

Uhlenbrook S, McDonnell J, Leibundgut C (2003) Preface: runoff generation and implications for river basin modelling. Hydrol Proc 17(2):197–198

UNEP/UNESCO (United Nations Environment Program, United Nations Educational, Scientific Cultural Organization) (1986) The impact of large water projects on the environment. UNEP/EMINWA, UNESCO/IH, Paris

UNESCO (United Nations Educational, Scientific Cultural Organization) (1993) Integrated water resource management: meeting the sustainability challenge. IHP Humid Tropics Programme, Series 5. UNESCO Press, Paris

United Nations (1997) Comprehensive assessment of the freshwater resources of the world. UN, UNDP, UNEP, FAO, UNESCO, WMO, UNIDO, World Bank, SEI. WMO, Geneva, 33 p

United Nations (1999) Agenda 21: programme of action for sustainable development. Rio declaration on environment and development; statement of forest principles; the final text of agreements negotiated by governments at the United Nations Conference on Environment and Development (UNCED), 3 June 1992, Rio de Janeiro, Brazil. Department of Public Information, United Nations, Geneva (reprint)

US EPA (US Environmental Protection Agency) (1989) Report to congress: water quality of the Nation's Lakes. EPA 440/5-89-003. US Environmental Protection Agency, Washington DC

US National Assessment (2000) Climate change impacts on the United States: the potential consequences of climate variability and change: overview and foundation reports. Cambridge University Press, Cambridge

USGS (US Geological Survey) (1984) National water summary – hydrologic events and issues. US Geological Survey Water Supply Paper #2250. US Department of the Interior, Washington DC

USGS-EDC (US Geological Survey EROS Data Center) (1998) HYDRO 1K: elevation derivative database. US Geological Survey, Earth Resources Data Center, Sioux Falls SD, (see http://edcwww.cr.usgs.gov/landdaac/gtopo30/hydro/index.html)

Valentin C (1996) Soil erosion under global change. In: Walker BH, Steffen WL (eds) Global change and terrestrial ecosystems. Cambridge University Press, Cambridge, pp 317–338

Valentin C, Bresson L-M (1992) Morphology, genesis and classification of surface crusts in loamy and sandy soils. Geoderma 55:225–245

Valentin C, d'Herbès J-M (1999) Niger tiger bush as a natural water harvesting system. Catena 37:231–256

Valentin C, Collinet J, Albergel J (1994) Assessing erosion in West African savannas under global change: overview and research needs. 15th Intern. Congress of Soil Science, Acapulco, Mexico, Vol. 7a, pp 253–274

Valentin C, Rajot J-L, Mitja D (2004) Responses of soil crusting, runoff and erosion to fallowing in the savannas of West Africa. Agr Ecosyst Environ (in press)

Valette-Silver N (1992) Historical reconstructions of contamination using sediment cores: a review. NOAA Technical Memorandum NOS/ORGA 65. US Dept Commerce, 37 p

Van Dam JC (2000) Field-scale water flow and solute transport: SWAP model concepts, parameter estimation and case studies. Doctoral Thesis Wageningen University, ISBN 90-5808-256-3, 166 p

Van Dam JC, Feddes RA (2000) Numerical simulation of infiltration, evaporation and shallow groundwater levels with the Richards equation. J Hydrol 233:72–85

Van Genuchten MTh (1980) A closed form equation for predicting the hydraulic conductivity of unsaturated soils. Soil Sci Soc Am J 44:892–898

Van Genuchten MTh, Leij FJ (1992) On estimating the hydraulic properties of unsaturated soils. In: Van Genuchten MTh, Leij FJ, Lund LJ (eds) Indirect methods for estimating hydraulic properties of unsaturated soils. Proc. Int. Workshop, Riverside, California, October 11–13, 1989, University California, California, CA, pp 1–14

Vanney JR (1991) La Géographie de l'Océan. Oceanis Paris, 214 p

Vannote RL, Minshall GW, Cummins KW, Sedell JR, Cushing CE (1980) The river continuum concept. Can J Fish Aquat Sci 37:130–137

Vehviläinen, B, Lohvansuu J (1991) The effects of climate change on discharges and snow cover in Finland. Hydrolog Sci J 36:109–121

Velichko AA (ed) (1984) Late Quaternary environment of the Soviet Union. Longman Group Ltd., London, 327 p

Vellinga P (ed) (1996) The environment, a multidisciplinary concern. Inst for Environmental Studies, Vrije Universiteit, Amsterdam, 538 p

Ver LMB, Mackenzie FT, Lerman A (1994) Modeling preindustrial C-N-P-S biogeochemical cycling in the land coastal margin system. Chemosphere 29:855–887

Verstappen GGC, Steenvoorden JHAM, van Liere L (1999) The influence of agricultural nutrients and pesticides on Dutch surface waters. In: Van der Kraats JA (ed) Farming without harming: the impact of agricultural pollution on water systems. Drukkerij Belser, Lelystad, pp 199–208

Vertessy RA, Elsenbeer H (1999) Distributed modeling of storm flow generation in an Amazonian rain forest catchment: effects of model parameterization. Water Resour Res 35(7):2173–2187

Vlotman WF, Sufi AB, Sheikh IA (1990) Comparison of three subsurface pipe drainage projects in Pakistan. In: Proceedings of the Symposium on Land Drainage for Salinity Control in Arid and Semi-Arid Regions, Cairo, Egypt

Von Bodungen B, Turner K (eds) (2001) Science and integrated coastal management. Dahlem Conf Series, John Wiley & Sons, Chichester

Vörösmarty CJ (2002a) Global change, the water cycle, and our search for Mauna Loa. Hydrol Process 16:1335–1339

Vörösmarty CJ (2002b) Global water assessment and potential contributions from Earth System Science. Aquatic Sciences 64: 328–351

Vörösmarty CJ, Meybeck MM (1999) Riverine transport and its alteration by human activities. IGBP Global Change Newsletter 39:24–29

Vörösmarty CJ, Moore BI (1991) Modeling basin-scale hydrology in support of physical climate and global biogeochemical studies: an example using the Zambezi River. Studies in Geophysics 12: 271–311

Vörösmarty CJ, Peterson BJ (2000) Macro-scale models of water and nutrient flux to the coastal zone. In: Hobbie J (ed) Estuarine science: a synthetic approach to research and practice. Island Press, Washington DC, pp 43–80

Vörösmarty CJ, Sahagian D (2000) Anthropogenic disturbance of the terrestrial water cycle. Bioscience 50:753–765

Vörösmarty CJ, Gildea MP, Moore B, Peterson BJ, Berquist B, Melillo JM (1986) A global model of nutrient cycling: II. Aquatic processing, retention, and distribution of nutrients in large drainage basins. In: Correll D (ed) Watershed research perspectives. Smithsonian Institution Press, Washington DC

Vörösmarty CJ, Moore B, Gildea MP, Peterson B, Melillo J, Kicklighter D, Raich J, Rastetter E, Steudler P (1989) A continental-scale model of water balance and fluvial transport: application to South America. Global Biogeochem Cy 3:241–265

Vörösmarty CJ, Willmott CJ, Choudhury BJ, Schloss AL, Stearns TK, Robeson SM, Dorman TJ (1996a) Analyzing the discharge regime of a large tropical river through remote sensing, ground-based climatic data, and modeling. Water Resour Res 32:3137–3150

Vörösmarty CJ, Fekete B, Tucker BA (1996b) River discharge database, version 1.0 (RivDIS v1.0). Volumes 0 through 6. A contribution to IHP-V Theme 1. Technical Documents in Hydrology Series, UNESCO, Paris

Vörösmarty CJ, Wasson R, Richey JE (eds)(1997a) Modeling the transport and transformation of terrestrial materials to freshwater and coastal ecosystems. Workshop Report and Recommendations for IGBP Inter-Core Project Collaboration, IGBP Report 39, IGBP, Stockholm, Sweden, 84 p

Vörösmarty CJ, Sharma K, Fekete B, Copeland AH, Holden J, Marble J, Lough JA (1997b) The storage and aging of continental runoff in large reservoir systems of the world. Ambio 26: 210–219

Vörösmarty CJ, Meybeck M, Fekete B, Sharma K (1997c) The potential impact of neo-Castorization on sediment transport by the global network of rivers. In: Walling D, Probst J-L (eds) Human impact on erosion and sedimentation. IAHS Press, Wallingford, pp 261–272

Vörösmarty CJ, Federer CA, Schloss A (1998a) Potential evaporation functions compared on US watersheds: implications for global-scale water balance and terrestrial ecosystem modeling. J Hydrol 207:147–169

Vörösmarty CJ, Li C, Sun J, Dai Z (1998b) Emerging impacts of anthropogenic change on global river systems: the Chinese example. In: Galloway J, Melillo J (eds) Asian change in the context of global change: impacts of natural and anthropogenic changes in Asia on global biogeochemical cycles. Cambridge University Press, Cambridge, pp 210–244

Vörösmarty CJ, Birkett C, Dingman SL, Lettenmaier D, Kim Y, Rodriguez E (1999) HYDRA-SAT, HYDrological Radar Altimetry SATellite. NASA Post-2002 Land Surface Hydrology Mission Component for Surface Water Monitoring. NASA, Washington. (http://lshp.gsfc.nasa.gov/Post2002/ hydrasat/hydrasat2.html)

Vörösmarty CJ, Fekete BM, Meybeck M, Lammers R (2000a) Geomorphometric attributes of the global system of rivers at 30-minute spatial resolution (STN-30). J Hydrol 237:17–39

Vörösmarty CJ, Fekete BM, Meybeck M, Lammers R (2000b) A simulated topological network representing the global system of rivers at 30-minute spatial resolution (STN-30). Global Biogeochemical Cycles 14:599–621

Vörösmarty CJ, Green P, Salisbury J, Lammers R (2000c) Global water resources: Vulnerability from climate change and population growth. Science 289:284–288

Vörösmarty CJ, Hinzman L, Peterson BJ, Bromwich DL, Hamilton L, Morison J, Romanovsky V, Sturm M, Webb R (2001) The hydrologic cycle and its role in Arctic and global environmental change: a rationale and strategy for synthesis study. ARCUS, Fairbanks AK, 84 p

Vörösmarty CJ, Hinzman L, Peterson BJ, Bromwich DL, Hamilton L, Morison J, Romanovsky V, Sturm M, Webb R (2002) Arctic-CHAMP: a program to study Arctic hydrology and its role in global change. AGU Eos Transactions 83, 241, 144, 249

Vörösmarty CJ, Meybeck M, Fekete B, Sharma K, Green P, Syvitksi J (2003) Anthropogenic sediment retention: major global-scale impact from the population of registered impoundments. Global and Planetary Change 39:169–190

Walker BH (1994) Global change strategy options in the extensive agriculture regions of the world. Climatic Change 27(1): 39–47

Walling DE (1983) The sediment delivery problem. J Hydrol 65: 209–237

Walling DE (1990) Linking the field to the river: sediment delivery from agricultural land. In: Boardman J, Foster IDL, Dearing JA (eds) Soil erosion on agricultural land. John Wiley & Sons, Chichester

Walling DE, Fang D (2003) Recent trends in the suspended loads of the world's rivers. Global and Planetary Change 39:111–126

Walling DE, Webb BW (1983) Patterns of sediment yield. In: Gregory KJ (ed) Background to palaeohydrology. John Wiley & Sons, New York, pp 69–100

Ward PRB (1980) Sediment transport and a reservoir siltation formula for Zimbabwe, Rhodesia. Die Siviele Ingenier in Suid-Afrika, Januarie 1980, pp 9–15

Ward RC, Robinson M (1990) Principles of hydrology, 3rd ed. McGraw-Hill, London

Ward JV, Stanford JA (1995) Ecological connectivity in alluvial river ecosystems and its disruption by flow regulation. Regul River 11:105–119

Webb RH, Schmidt JC, Marzolf GR, Valdez RA (eds)(1999) The controlled flood in Grand Canyon. AGU, Washington DC, 367 p

Webb RH, Wegner DL, Andrews ED, Valdez RA, Patten DT (1999) Downstream effects of Glen Canyon Dam on the Colorado River in Grand Canyon: a review. In: Webb et al. (1999), pp 1–21

Wechsung F, Krysanova V, Flechsig M, Schaphoff S (2000) May land-use change reduce the water deficiency problem caused by reduced brown coal mining in the state of Brandenburg? Landscape Urban Plan 51/2–4:105–117

Werner PC, Gerstengarbe FW (1997) A proposal for the development of climate scenarios. Climate Change 8(3):171–182

Werner W, Wodsak H-P (1994) Stickstoff- und Phosphoreintrag in Fliessgewässer Deutschlands unter besonderer Berücksichtigung des Eintragsgeschehens im Lockergesteinbereich der ehemaligen DDR. Agrarspektrum 22, 241 p

Werner W, Olfs H-W, Auerswald K, Isermann K (1991) Stickstoff- und Phosphoreintrag im Oberflächengewässer über adiffuse Quellen. In: Hamm A (ed) Studie über Wirkungen und Qualitätsziele von Nährstoffen in Fliessgewässern. Academia Verlag, St Augustin, pp 665–764

Williams JR (1975) Sediment yield prediction with a universal equation using runoff energy factor. In: Present and prospective technology for predicting sediment yields and sources. USDA-ARS 40, pp 244–252

Williams JR, Berndt HD (1977) Sediment yield prediction based on watershed hydrology. Trans ASAE 20(6):1100–1104

Williams J, Bonell M (1988) The influence of scale of measurement on the spatial and temporal variability of the Philip infiltration parameters – an experimental study in an Australian savannah woodland. J Hydrol 104:33–51

Williams MAJ, Faure H (eds)(1980) The Sahara and the Nile – Quaternary environments and prehistoric occupation in northern Africa. A.A. Balkema, Rotterdam, 623 p

Williams JR, Singh VP (1995) The EPIC model. In: Computer models of watershed hydrology. Water Resources Publications, Colorado, pp 909–1000

Williams JR, Renard KG, Dyke PT (1984) EPIC – a new model for assessing erosion's effect on soil productivity. J Soil Water Conserv 38(5):381–383

Williams R, Burt T, Brighty G (2000) River water quality. In: Acreman MC (ed) The hydrology of the UK. Routledge, London, pp 134–149

Wilson L (1973) Variations in mean annual sediment yield as a function of mean annual precipitation. Am J Sci 273:335–349

Wilson GV, Jardine PM, Luxmoore RJ, Zelazny LW, Todd DE (1991a) Hydrogeochemical processes controlling subsurface transport from an upper Walker Branch Watershed during storm events. 1. Hydrologic transport processes. J Hydrol 123:297–316

Wilson GV, Jardine PM, Luxmoore RJ, Zelazny LW, Todd DE (1991b) Hydrogeochemical processes controlling subsurface transport from an upper Walker Branch Watershed during storm events. 2. Solute transport processes. J Hydrol 123:317–336

Wollheim WM, Peterson BJ, Deegan LA, Hobbie JE, Hooker B, Bowden WB, Edwardson KJ, Arscott DB, Hershey AE, Finlay J (2001) Influence of stream size on ammonium and suspended particulate nitrogen processing. Limnol Oceanogr 46:1–13

Wolters W, Bhutta MN (1997) Need for integrated irrigation and drainage management: example of Pakistan. Proceedings of ILRI Symposium Towards Integrated Irrigation and Drainage Management, Wageningen, The Netherlands

Wösten JHM, Lilly A, Nemes A, Le Bas C (1999) Development and use of a database of hydraulic properties of European soils. Geoderma 90:169–185

WRI (World Resources Institute) (1998) World resources: a guide to the global environment 1998–1999. World Resources Institute, Washington DC

Young GJ, Dooge JI, Rodda JC (1994) Global water resource issues. CUP, Cambridge

Yu B, Neil DT (1993) Long-term variations in regional rainfall in the south-west of Western Australia and the difference between average and high intensity rainfalls. International Journal of Climatology 13:77–88

Yue S, Pilon P, Phinney B (2003) Canadian streamflow trend detection: impacts of serial and cross-correlation. Hydrol Sci J 48(1): 51–63

Zaret TM, Devol AH, Santos AD (1981) Nutrient addition experiments in Lago Jacaretinga, central Amazon Basin, Brazil. Verh Internat Verein Limnol 21:721–724

Zeng X (2001) Global vegetation root distribution for land modeling. J Hydrometeorol 2:525–530

Zhang J, Zhang ZF, Liu SM, Wu Y, Xiong H, Chen HT (1999) Human impacts on the large world rivers: would the Chang Jiang (Yangtze River) be an illustration? Global Biogeochem Cy 13:1099–1105

Part E

How to Evaluate Vulnerability in Changing Environmental Conditions?

Edited by Roger A. Pielke, Sr. and Lelys Bravo de Guenni

Chapter E.1

Introduction

Roger A. Pielke, Sr.

The prevailing paradigm has been that climate variability and change can be projected decades or even a century or more into the future (e.g. IPCC 1996, 2001). However, for several reasons – e.g. imperfect representation of the full complexity of the Earth system, nonlinear spatial and temporal feedbacks, and imperfect foresight of human behaviour, it may not be possible to assess the range of potential future climate change accurately.

Therefore, a newer paradigm requires that our vulnerability to the entire spectrum of environmental threats be assessed, including the ranking of their seriousness. Once vulnerabilities have been so evaluated, whether or not they can be quantitatively predicted for the future can then be tested. This part of the BAHC synthesis discusses this perspective to assess risk associated with environmental variability and change. In particular, we should start the analysis by first assessing vulnerability, instead of projecting possible climate change scenarios, and then assessing vulnerabilities based on that subset of possible future climates.

This is a broader definition of vulnerability than that used by the impacts community. As discussed later, this approach involves first assessing all of the vulnerabilities of an environmental (or other) resource to environmental variability and change. Once these vulnerabilities are determined, estimates (with probabilities, if they can be quantified) of what scenarios would cause a vulnerability threshold to be exceeded are specified. Also, unlike the more narrowly defined concept of vulnerability, nonlinear feedbacks between the resource being affected and the forcings are included.

We will discuss first the relationship between prediction and vulnerability in Chapt. E.2 and E.3. The scenario approach will be reviewed later in Chapt. E.4 and its shortcomings illustrated. The reasons why a vulnerability approach as presented in Chapt. E.5 is more appropriate for environmental assessments is because it is regional and local in scale; it involves the evaluation of thresholds and the threat associated with extreme conditions; it can include a spectrum of threats, and can consider the effect of abrupt changes. In contrast, the existing paradigm for predicting the future starts from the global scale and then attempts to downscale to the regional and local scales; it focuses on averages, is long-term, and depends on skilful prediction of the future. It is therefore much less useful to policy-makers (Sarewitz et al. 2000). The in-depth discussion of why we need a vulnerability approach that then follows will be illustrated using water as an example.

Water is an essential component of life both in terms of its quality and quantity; however, this resource is threatened. For example, as reported in the Economist (May 29, 1999, p. 102), while 90% of the world's population have enough water at present, it is estimated that by 2050 more than 40% of the population will face some water shortage (see also Sect. E.6.1). The lack of access to safe water is even more serious. According to the same article in the Economist, developing countries often have very limited access to safe water supplies. Only about 30% of the residents in rural Brazil, for example, currently have access to safe water.

There is increasing recognition that threats to future water quality and quantity are influenced by a wide variety of environmental concerns. These concerns involve effects both from natural and human origin. Stahle et al. (2000), for example, document a 16th century megadrought in western North America that dwarfs any drought since then. Wilhite (2000) presents a series of articles that demonstrates the extensive effect of drought on human society. Kunkel et al. (1999) describe the importance of vulnerability to weather and climate extremes, and how this vulnerability changes in response to increases in human exposure, even when extreme weather statistics remain unchanged. Non-weather related influences include land-use change effects on runoff and stream flow, and deliberate engineering of water flow whereas human activities change the distribution and availability of water through engineering works. Examples of this kind of human activity include rerouting of rivers, creating artificial surfaces (reservoirs), changing surface water content through irrigation of drylands or drainage of wetlands, and through lifting groundwater to the surface.

In contrast with the current method for determining environmental risk by using scenarios based on general circulation model output, a vulnerability approach permits the inclusion of multiple environmental stresses associated with both short- and long-term threats, and involves local, regional, national and global scales. Chapter E.6 discusses this vulnerability perspective and provides specific case study examples.

Chapter E.2

Predictability and Uncertainty

Roger A. Pielke, Sr. · Gerhard Petschel-Held · Pavel Kabat · Brad Bass · Michael F. Hutchinson
Vijay Gupta · Roger A. Pielke, Jr. · Martin Claussen · Dennis Shoji Ojima

It is appropriate to consider water quantity and water quality as two facets of water resources. Both facets are intimately connected to the hydrological cycle, which itself is a component of the Earth's climate system. Since human activities and health are so connected to water resources, it is essential to determine how far into the future we can predict the condition of the water resources of a region with a sufficient level of confidence. When predictions are not possible, resilience must be built into a water system so that human (and natural) needs are not negatively affected. Even when skillful predictions are possible, they are seldom completely accurate. Thus uncertainty needs to be included when scientific analyses and predictions are used for water resource planning and management, particularly for issues such as adaptation or risk management. The prediction of weather and climate are essential aspects of planning for water resources in changing environmental conditions.

Lorenz (1979) proposed the concept of forced and free variations of weather and climate. He refers to forced variations as those caused by external conditions, such as changes in solar irradiance. Volcanic aerosols also cause forced variations. He refers to free variations as those which "are generally assumed to take place independently of any changes in external conditions". Day-to-day weather variation is presented as an example of free variations. He also suggests that "free climatic variations in which the underlying surface plays an essential role may therefore be physically possible".

However, if the ocean surface and/or land-surface changes over the same time period as the atmosphere changes, then the non-linear feedbacks (i.e. two-way fluxes) between the air, land and water, eliminate an interpretation of the ocean-atmosphere and land-atmosphere interfaces as boundaries. Rather than "boundaries", these interfaces become interactive media (Pielke 1998a, 2001). The two-way fluxes that occur between the atmosphere and ocean, and the atmosphere and the land-surface (as detailed in Part A of this book), must therefore necessarily be considered as part of the predictive system. On the time scale of what we typically call short-term weather prediction (days), important feedbacks include *biophysical* (e.g. vegetation controls on the Bowen

ratio), *snow cover*, *clouds* (e.g. in their effect on the surface energy budget), and *precipitation* (e.g. which changes the soil moisture) processes. This time scale is already considered to be an initial value problem (Sivillo et al. 1997) since operational numerical weather prediction models are routinely reinitialised twice daily. Seasonal and interannual weather prediction include the following feedbacks: *biogeochemical* (e.g. vegetation growth and senescence); *anthropogenic and natural aerosols* (e.g. through their effect on the long- and short-wave radiative fluxes and their effects on cloud microphysics and hence the hydrological cycle); *sea ice*; and *ocean sea surface temperature* (e.g. changes in upwelling such as those associated with an El Niño) effects. For even longer time periods (of years to decades and longer), the additional feedbacks include *biogeographical processes* (e.g. changes in vegetation species composition and distribution), *anthropogenic-caused land-use changes*, and *deep ocean circulation effects* on the ocean surface temperature and salinity. In the context of Lorenz's (1979) terminology, each of these feedbacks are free variations.

We begin to tackle this problem by using a hierarchy of models. We will consider two examples to illustrate this important point. First example is the 0-dimensional dynamical model (having no spatial dimensions, i.e. 0-th order in space) which fully and non-linearly couples radiation, biota, and the hydrological cycle, with other components of the Earth climate system. This step is necessary to obtain a fundamental theoretical understanding of the first-order effects on planetary climate. These first-order effects tend to be associated with both positive and negative feedbacks. It seems particularly important to include negative feedbacks, or the "homeostatic" mechanisms in the language of Watson and Lovelock (1983), in low-dimensional dynamical models. Negative feedbacks tend to be underrepresented in more complex infinite dimensional dynamical models, involving one or more spatial dimensions, and therefore are less well understood.

Insight from simple non-linear dynamical models should serve as foundations for the development of more complex models (Shackely et al. 1998; Ghil and Childress 1987). Negative feedbacks coming from coupling with biota have been explored in models such as Daisyworld (Meszaros and Palvolgyi 1990; Von Bloh et al. 1997, 1999; Nevison et al. 1999; Weber 2001), though full coupling with other components of an Earth-like climate system has yet to be explored. Still, even in simple models, new non-linear effects are only now being discovered (Nordstrom et al. 2004), a point which serves to emphasize the importance of understanding the role of positive and negative feedbacks on the climate system.

The second example is the so-called EMICs (Earth System Models of Intermediate Complexity; Claussen 2001). These models explicitly simulate the interactions among as many components of the natural Earth system as possible. They include most of the processes described in comprehensive models of atmospheric and oceanic circulation – usually referred to as "climate models" – albeit in a more reduced, i.e. a more parameterised form. Therefore, EMICs are considered to test scientific ideas in a geographically explicit model environment, not to make the most detailed and realistic prediction.

Regarding predictability we can distinguish between prediction, or forecast of the first and the second kind (Lorenz 1975). An example of a commonly known forecast of the first kind is short-term weather forecasts, i.e. the weather forecast for several days into the future is predicted given accurately monitored initial and boundary conditions. A prediction of the second kind occurs when boundary conditions determine the state of the system, and initial conditions are no longer important. Currently, longer term weather predictions of the first kind have been successful only in the case of forecasting seasonal weather such as a six month forecast of El Niño (Landsea and Knaff 2000) when the oceanographic monitoring system had already indicated an eastward moving Kelvin wave of tropical warm water.

The Intergovernmental Panel on Climate Change (IPCC) uses the term "projection" to indicate that a climate forecast, of the second kind is meant. However, the climate system of the future has not been shown to be independent, for example, of the initial (current) Earth's land cover. Moreover, we conclude that the term "projection" is misleading, because it suggests some more or less complete prediction of the future. However, most climate projections of the IPCC include only changes in the composition of the atmosphere, whereas other natural and anthropogenic forcings, such as solar variability, vegetation dynamics and land use (see Part A), are likely to affect future climate. We therefore propose to use the term "sensitivity experiments", when referring to the IPCC model results.

Predictability of climate can be limited owing to the non-linearity of the Earth system. Following Lorenz's (1968) terminology, non-linear systems, even without any external unsteady forcing, can be "transitive" or "intransitive", i.e. the statistics of the system can be sta-

tionary (ergodic) or can change with time, respectively. So far, all model simulations (e.g. Cubasch et al. 1994) have shown that the global climate system seems to respond almost linearly to greenhouse-gas forcing, if the next several decades, perhaps up to a century, are considered. However, since these are sensitivity model results, we do not know if this linearity will remain when the entire spectrum of natural- and human-forcings on these time scales are included. At the regional scale, the climate clearly exhibits intransitive behaviour as shown for the thermohaline circulation in the North Atlantic (e.g. Ganopolski and Rahmstorf 2001), Sahelian rainfall (Wang and Eltahir 2000), and Northern African deserts (e.g. Claussen et al. 1998). Thus at the global scale, intransitive behaviour cannot be excluded, because most models have not yet incorporated all feedbacks of the climate system. This argument further supports the use of the term "sensitivity experiment" instead of "projection", when referring to the IPCC results.

There are actually two types of prediction with respect to water resources. The first type of prediction involves an equilibrium impact of environmental change, δI. We can write this mathematically as

$$\delta I = f_1(\delta A, \delta B, \delta C, \ldots) \tag{E.1}$$

or in some cases as an implicit function

$$\delta I = f_2(\delta A, \delta B, \delta C, \ldots, \delta I) \tag{E.2}$$

where δA, δB, etc., represent a set of environmental perturbations from a reference state of basic variables A, B, etc. For example, δI could be the effect on river flow at a stream gauge (i.e. I is the river flow itself). δA could then be the radiative effect of increased CO_2 and other anthropogenically-emitted greenhouse gases with respect to the pre-industrial level. δB could be the biological effect of increased CO_2; δC could be human-caused land-use change; δD the direct radiative effect of anthropogenic aerosols; δE the indirect effect of these aerosols on cloud microphysics (cloud condensation nuclei, ice nuclei), etc. The choice of the perturbation depends on what is the specific impact of concern. δI can either represent a state variable or the statistics of a state variable such as a probability density function. δI can be time dependent (i.e. in nonequilibrium as a result of a change in f_1). When just one perturbation or a small number of perturbations on a subset of perturbations is imposed, the effect on δI is said to be a *sensitivity* experiment. When all significant (on δI) perturbations are included, the effect on δI is said to be a *realisation*. If a spectrum of perturbations are performed, which represent all possible situations, the experiment is referred to as an *ensemble*. Any one realisation selected from this ensemble is called a *scenario*. The uncertainty in terms of δI will be

Fig. E.1.
Schematic of different classes of prediction (for explanations see text; adopted from Pielke 2002)

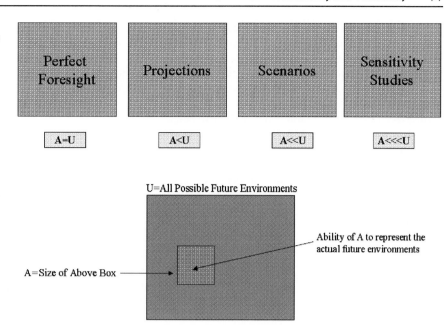

determined from the distribution (and probability of occurrence) of the scenarios. Figure E.1, adapted from Pielke (2002), illustrates this perspective of prediction.

Four classes of prediction are illustrated in the figure. In this schematic, the universe of all possible future environmental conditions are given by U, while the ability of a particular class of prediction to forecast the future is defined by A. If A covers a small area of U it is, of course, less likely to actually predict what the future will be. Indeed, with a sensitivity study A could lie outside of U. In the figure, a sensitivity study varies only a subset of environmental perturbations and/or does not include all important Earth system feedbacks. A scenario includes all important environmental perturbations and Earth system feedbacks, but is only a single realisation (or subset of realisations) from the spectrum of predictions possible from the non-linear, chaotic Earth system. Only when the envelope (the ensemble of all realisations) of the possible future conditions is obtained, is the prediction an actual projection (additional discussion of these classes of prediction are presented in Mac-Cracken 2002, and Pielke 2002).

As an example, the IPCC (1996) is actually a sensitivity study since not all anthropogenic effects on the Earth's climate system were considered. The IPPC report included the radiative effect of increased human-input greenhouse gases and aerosols, but did not include other important effects, such as land-use change, as discussed in Part A of this book. Moreover, since the IPCC then used downscaling to obtain regional estimates of climate change, the diversity of regional results among the GCM models produce impact estimates dI, dI/dt (as discussed in the following text) which makes the size of A even smaller for the regional scale, than for the global

scale. The vulnerability approach, in contrast, starts with the assessment of all values within U (as best as we can estimate the maximum realistic size of U), and, only then, seeks to determine which impacts are more likely than others.

The Earth system is considered as a dynamic system which includes the natural spheres (atmosphere, biosphere, hydrosphere, etc.) and the anthroposphere (economy, society, culture, psycho-social aspect, etc.) and the interactions between them (Schellnhuber 1998). The definitions of a sensitivity, realisation, ensemble, and scenario remain the same when the impact must be assessed by analysis of a dynamic system yet the assessment of $I(t)$, i.e. the impact as a function of time, becomes much more difficult in this case.

$$\frac{dI}{dt} = g(A(t), B(t), C(t), \dots, I(t)) \qquad (E.3)$$

The function given by Eq. E.3 represents a differential equation which is more difficult to solve than Eq. E.1. Note that equilibrium is determined by setting $dI/dt = 0$ and solving for I. Thus the dynamic description in Eq. E.3 is more general than the static equilibrium in Eq. E.1 or Eq. E.2. In particular, the description allows an assessment to be made of the stability or resilience properties of equilibrium points. The assessment becomes more difficult when the function g is non-linear, since features such as chaotic motion or complex bifurcation scenarios can arise. In these cases "surprises" can occur over time which are impossible to predict. For example, a stable equilibrium within an ecosystem can become unstable, which might lead to completely new structural properties in the system.

The accurate prediction of δI or dI/dt requires that f_1, f_2 and g be accurate representations of reality. If, however, there are large uncertainties in the specification of the perturbations and/or in the form of f_1, f_2 and g, the range of δI and dI/dt that results could be quite large. The choice of just one value of δI or dI/dt (or a limited subset of each) from a limited set of perturbations using f_1, f_2 and g will be incomplete, even if it is assumed that f_1, f_2 and g are accurate.

As an alternative, in the case of equilibrium considerations, the vulnerability of the water resource (or other environmental resource) can be determined by estimating what the maximum risks are. Equations E.1 and E.2 can be rewritten as

$$|\delta I|_{max} = |f_1(\delta A, \delta B, \delta C, \ldots)|_{max} \qquad (E.4)$$

$$|\delta I|_{max} = |f_2(\delta A, \delta B, \delta C, \ldots, \delta I)|_{max}$$

where $|\delta I|_{max}$ is the largest effect that results from the perturbations of the environmental conditions. In order to determine the maxima, however, it is necessary to know the ranges of possible values of the independent variables δA, δB, etc. Yet one might also determine the maximum possible values of δI when we assume only small changes in these variables. Mathematically this can be achieved by calculating the gradient of δI with respect to the input variables, i.e.

$$\vec{\nabla} A, B, \ldots, \delta I = \vec{\nabla} A, B, \ldots, f_1(\delta A, \delta B, \ldots) \qquad (E.5)$$

and analogously for f_2 (Lüdeke et al. 1999).

When considering a dynamic system, an analogous analysis is possible by considering the interval $G(A, B, C, \ldots, I)$ of possible values on the right hand side of Eq. E.3. We then obtain a so-called differential inclusion (Aubin and Cellina 1984), i.e.

$$\frac{dI}{dt} \in G(A, B, C, \ldots, I) \qquad (E.6)$$

which then allows a computation of the set of admissible "futures" of the possible trajectories $I(t)$ which are realisable by the assumed set of independent variables A, B, C, … This allows an evaluation of the maximum impact, I_{max}, at any arbitrary point in time.

Using these approaches, as long as f_1, f_2 and g are realistic representations of the climate system, policy-makers can concentrate their efforts at reducing the contributions of the perturbations that most contribute to δI or dI/dt. If these perturbations cannot be manipulated, this information also needs to be communicated to policy-makers.

The accurate determination of f_1, f_2 and g may not be possible. With respect to the future climate, general circulation models (GCMs) have been applied (e.g. IPCC 1996), however, they have been used as sensitivity experiments since, in general, only one or two perturbations have been initiated i.e. the radiative effect of an anthropogenic increase in CO_2 and other anthropogenic greenhouse gases or the radiative effect of an anthropogenic increase in aerosols. GCM and EMIC land-cover change simulations have also been per-

Fig. E.2.
Ecological vulnerability/ susceptibility links in environmental assessment as related to water resources (from Pielke and Guenni 1999)

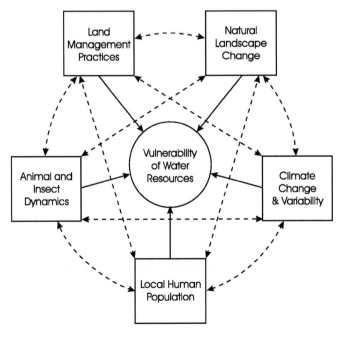

Predictability requires:
  - the adequate quantitative understanding of these interactions
  - that the feedbacks are not substantially nonlinear.

formed as sensitivity experiments (e.g. Brovkin et al. 1999; Claussen et al. 2001; Chase et al. 1996; Pitman et al. 1999; Bounoua et al. 2000).

An alternative approach is to determine what values of δI or dI/dt result in undesirable impacts. What are the thresholds beyond which we should be concerned? This approach involves starting with the impacts model (what is sometimes termed an "endpoint analysis" or "tolerable window approach") and, without using f_1, f_2 or g, determine the magnitude of δI or $\dot{I}$ that must occur before an undesirable effect occurs. The impact functions f_1, f_2 and g are then used to estimate whether such thresholds could be reached under any possible environmental variability or change. The estimates for what is realistic, with respect to climate, would include the GCM results but would also utilise palaeorecords, historical data, worst case combinations from the historical data, and "expert" estimates. Such an approach would provide a risk assessment for policymakers that is not constrained by uncertain predictions.

Figure E.2, from Pielke and Guenni (1999), illustrates a generic schematic as to how to assess δI and dI/dt for water resources.

Pielke and Uliasz (1998) and Pielke (1998a) discuss this type of an approach to estimate uncertainty in air quality assessments. Lynch et al. (2001) apply this technique to assess the sensitivity of a land-surface model to selected changes (plus and minus) of atmospheric variables such as air temperature and precipitation. Hubbard and Flores-Mendoza (1995) assess the effect on corn, soybean, wheat and sorghum production of positive and negative changes of precipitation and temperature in the United States. Tóth et al. (1997) and Petschel-Held et al. (1999a) have used this "inverse concept" in the form of the dynamically tolerable windows approach within an integrated assessment of climate change. Similarly, Alcamo and Kreilemans (1996) apply the general idea of the end-point analysis within their "safe landing analysis" of near term climate protection strategies. Figure E.3 illustrates an example where warmer and cooler condi-

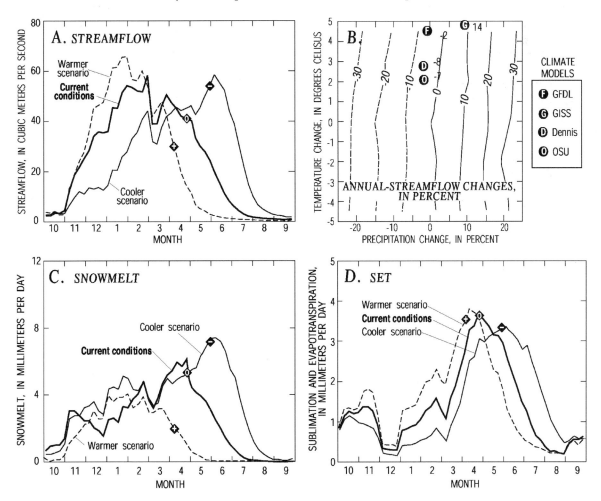

Fig. E.3. Simulated water budget responses to uniform change scenarios in the North Fork American river basin of California. **a, c,** and **d** illustrate mean changes under scenarios in which mean temperatures are changed, and **b** provides the percentage changes as function of changes in both mean temperature and mean precipitation (*contours* for uniform change scenarios – *dashed* where negative; *symbols* for GCM model sensitivity experiments; from Jeton et al. 1999)

tions are assessed in an impacts model to estimate the sensitivity of water resources in this region to this aspect of climate variability and change.

In recent years there has been a growing number of studies and modelling attempts on global environmental change which take uncertainty into account. Within these studies one might want to distinguish between (1) classical probabilistic approaches, (2) new, quantitative approaches based on theoretical frameworks such as cultural theory and (3) qualitative, yet formal approaches.

1. More traditional approaches within integrated assessments of climate change are taken, for example, in the PAGE 95 model (Plambeck and Hope 1996) or in the ICAM 2.0 and 2.5 models (Dowlatabadi and Morgan 1995) which use probability distribution functions to represent uncertainties in parameters and functional relationships. The implication of learning as the major process in reducing uncertainties is central to the studies by Kolstad and Kelly (1999). A recent overview on the issue of uncertainty in climate change assessments is given in Schellnhuber and Yohe (1998) or Dowlatabadi (1999).
2. An innovative approach is taken within the TARGETS modelling framework (Rotmans and deVries 1997; Hilderink et al. 1999) where uncertainties are related to three different world views, based on cultural

theory: individualistic, egalitarian and hierarchical. Different parameterisations of the model may then be determined. There is also a strong water component within this modelling framework from which to compute the basic supply and demand issues of water resources (Hoekstra 1996).
3. There is a recent attempt to apply qualitative modelling techniques to the analysis of environmental change. Originally suggested by the German Advisory Council on Global Change (WBGU 1994) and in cooperation with the Council, and further developed by Schellnhuber et al. (1997) and Petschel-Held et al. (1999b), the syndrome approach tries to identify major patterns of civilisation-nature interactions, which govern the dynamics of environmental change. Most interesting in the present context is the 1997 Annual Report of the Council (WBGU 1999) which focuses on the sustainable use of freshwater resources.

As discussed here, a vulnerability assessment provides a comprehensive framework within which to estimate environmental risk. This is in contrast to starting with a scenario approach which limits the spectrum of estimates to what can actually occur in the future. With the scenario approach, for example, environmental "surprises" (Canadell 2000) will be missed. These two divergent approaches are discussed further in the next chapter.

Chapter E.3

Contrast between Predictive and Vulnerability Approaches

Roger A. Pielke, Jr. · Thomas J. Stohlgren

While efforts to predict natural phenomena have become an important aspect of the Earth sciences, the value of such efforts, as judged especially by their capacity to improve decision making and achieve policy goals, has been questioned by a number of constructive critics. The relationship between prediction and policy making is not straightforward for many reasons. In practice, consolidative and exploratory models are often confused by both scientists and policy-makers alike, thus a central challenge facing the community is the appropriate use of such models and resulting predictions (Sarewitz and Pielke 1999).

Among the reasons for this criticism is that accurate prediction of phenomena may not be necessary to respond effectively to political or socio-economic problems created by such phenomena (for example, see Pielke et al. 1999). Indeed, phenomena or processes of direct concern to policy-makers may not be easily predictable or useful. Likewise, predictive research may reflect discipline-specific scientific perspectives that do not provide "answers" to policy problems since these may be complex mixtures of facts and values, and which are perceived differently by different policy-makers (for example, see Jamieson and Herrick 1995).

In addition, necessary political action may be deferred in anticipation of predictive information that is not forthcoming in a time frame compatible with such action. Similarly, policy action may be delayed when scientific uncertainties associated with predictions become politically charged as in the issue of global climate change, for example (Rayner and Malone 1997).

Predictive information may also be subject to manipulation and misuse, either because the limitations and uncertainties associated with predictive models are not readily apparent or because the models are applied in a climate of political controversy and high economic stakes. In addition, emphasis on predictive sciences moves both financial and intellectual resources away from other types of research that might better help to guide decision making as, for example, incremental or adaptive approaches to environmental management that require monitoring and assessment instead of prediction (see Lee and Black 1993).

These considerations suggest that the usefulness of scientific prediction for policy making and the resolution of societal problems depends on relationships among several variables, such as the timescales under consideration, the scientific complexity of the phenomena being predicted, the political and economic context of the problem, and the availability of alternative scientific and political approaches to the problem.

In light of the likelihood of complex interplay among these variables, decision makers and scientists would benefit from criteria that would allow them to judge the potential value of scientific prediction and predictive modelling for different types of political and social problems related to Earth processes and the environment.

Pielke et al. (1999) provide the following six guidelines for the effective use of prediction in decision making.

1. Predictions must be generated primarily with the needs of the user in mind. For stakeholders to participate usefully in this process, they must work closely and persistently with the scientists to communicate their needs and problems.
2. Uncertainties must be clearly articulated (and understood) by the scientists, so that users understand their implications. Failure to understand uncertainties has contributed to poor decisions that then undermine relations among scientists and decision makers. But merely understanding the uncertainties does not mean that the predictions will be useful. If policy-makers truly understood the uncertainties associated with predictions of, for example, global climate change, they might decide that strategies for action should not depend on predictions (Rayner and Malone 1997).
3. Experience is an important factor in how decision makers understand and use predictions.
4. Although experience is important and cannot be replaced, the prediction process can be facilitated in other ways, for example by fully considering alternative approaches to prediction, such as "no-regrets" public policies, adaptation, and better planning and engineering. Indeed, alternatives to prediction must be evaluated as a part of the prediction process.

5. To ensure an open prediction process, stakeholders must question predictions. For this questioning to be effective, predictions should be as transparent as possible to the user. In particular, assumptions, model limitations, and weaknesses in input data should be forthrightly discussed. Even so, lack of experience means that many types of predictions will never be well understood by decision makers.

6. Last, predictions themselves are events that cause impacts on society. The prediction process must include mechanisms for the various stakeholders to fully consider and plan what to do after a prediction is made.

Scenarios of the future such as projections of climate change, forest production, animal migration, or disease spread are – by definition – subsets of all possible outcomes (Cohen 1996). Most projections have much in common with the typical five-day weather forecast on the nightly news:

- they are based on data from the recent past;
- they incorporate relatively few response variables (e.g. temperature and precipitation);
- they are more accurate for some response variables than others (i.e. minimum temperature versus precipitation amounts);
- no level of certainty or probability is provided or implied;
- short-term projections are typically more accurate than long-term projections; and
- their use for planning is limited to a narrowly defined set of questions (e.g. should snow ploughs be positioned near highways in anticipation of a snowstorm; Sarewitz et al. 2000).

Seasonal forecasts, such as those predicting greater or less precipitation over the next few months, are based on larger-scale phenomena such as ocean temperatures and jet stream patterns, but have additional limitations and uncertainties about the spatial and temporal accuracy of the projections (Glantz 2001).

There is more uncertainty behind long-term climate change scenarios with 30- to 100-year timeframes compared to seasonal forecasts. Long-term climate scenarios have considerably more limitations and higher levels of uncertainty due to coarse-scale spatial resolution, natural spatial and temporal variation, insufficient understanding of multiple climate forcings, longer-term biological physical feedbacks in the models, and lack of stochastic extreme events. Globally averaged results would be expected to be more reliable than local and regional results. Reliance on scenarios is reliance on predictions with all their limitations and uncertainties (Sarewitz et al. 2000).

The value of predictions for environmental decision-making therefore emerges from the complex dynamics of the prediction process, and not simply from the technical efforts that generate the prediction product (which are themselves an integral part of the prediction process).

An alternative to the scenario-prediction approach is to evaluate vulnerability based on a more comprehensive assessment of multiple stresses, multiple response variables, and their interactions. For example, natural ecosystems are often affected by multiple stresses including land-use change, loss of biodiversity, altered disturbance regimes, invasive non-indigenous species, pollution, and rapid climate change (Stohlgren 1999b). Land managers are concerned about many natural and cultural resources, local economies, and stakeholder concerns for specific resources. The interactions of the multiple stresses and response variables are important and urgent considerations. A warming climate, for example, might alter the competitive advantage of exotic annual grasses only if nitrogen from air pollution were available, grazing were limited, and forest canopies remained fairly open. The effects of decreased precipitation on aquatic biota might be more of a problem where water was over-subscribed, low flows were adversely affected by water diversion, and exotic fish inhabited the streams. Human-dominated and natural ecosystems, and their interacting components and processes, can rarely be assessed by evaluating stresses independently. It is often possible, however, to identify vulnerabilities or sensitivities in cases where prediction is impractical or impossible.

Vulnerability assessments, of course, also have limitations. Detailed information on stresses, resources, and interactions is often scant, resulting in increased uncertainty. In general, the level of uncertainty is constrained by focusing on short-term planning horizons and high priority issues at local and regional scales. However, a vulnerability assessment is easily made consistent with the needs of decision makers. It is not a new concept, and is variously referred to in the literature as "adaptive management", "vulnerability assessment", "boundedly rational decision making", and "successive incremental approximations".

An important first step in considering the implications for research and policy is to recognise the role of models and predictions for purposes of both science and action. Bankes (1993) defines two types of models, consolidative and exploratory, differentiated by their uses.

- A *consolidative model* seeks to include all relevant facts into a single package and use the resulting system as a surrogate for the actual system.

The canonical example is that of the controlled laboratory experiment (Sarewitz and Pielke 1999). Other examples, include weather forecast and engineering design models. Such models are particularly relevant to decision making because the system being modelled can be treated as being closed, i.e. one "in which all the components of the system are established independently and are known to be correct" (Oreskes et al. 1994). The creation of such a model generally follows two phases: first, model construction and evaluation and second, operational usage of a final product. Such models can be used to investigate diagnostics (i.e. "what happened?"), process ("why did it happen?"), or prediction ("what will happen?").

- An *exploratory model* – or what Bankes (1993) calls a "prosthesis for the intellect" – is one in which all components of the system being modelled are not established independently or are not known to be correct.

In such a case, the model allows for computational experiments to investigate the consequences for modelled outcomes of various assumptions, hypotheses, and uncertainties associated with the creation of and inputs to the model. These experiments can contribute to at least three important functions (Bankes 1993). First, they can shed light on the existence of unexpected properties associated with the interaction of basic assumptions and processes (e.g. complexity or surprises). Second, in cases where explanatory knowledge is lacking, exploratory models can facilitate hypothesis generation. Third, the model can be used to identify limiting, worst-case, or special scenarios under various assumptions of and uncertainty associated with the model experiment. The limiting "worst cases" generated in such a model are explorations of the boundaries of the universe of model outputs, and may or may not have significance for understandings of the real world. Frequently consolidative and exploratory models are confused in this regard. Such experiments can be motivated by observational data (e.g. econometric and hydrological models), scientific hypotheses (e.g. general circulation models), or by a desire to understand the properties of the model or class of models independent of real-world data or hypotheses.

- A shift in the culture of research is also needed. Research must be truly interdisciplinary (rather than multidisciplinary).

Assessing multiple stresses requires a broad spectrum of expertise without losing sight of the complex interactions among the sciences. Scientists must also learn to work interactively with stakeholders so that the highest priority needs are met first. To date, climate change research has been heavily focused at the global-scale and longer timeframes, as evidenced by the reliance on GCMs, projected changes in mean temperature and precipitation, and 100-year or double-CO_2 model timeframes. There is often a huge "disconnect" with local and regional stakeholders who have demanded a greater emphasis on near-future responses associated with extreme events, information on a full range of climate characteristics, and interactions from multiple environmental stresses that they deal with on a daily basis.

This suggests a rebalancing of priorities towards adaptation and vulnerability/sensitivity assessments; away from consolidative modelling and toward more exploratory efforts. Such research has the potential to contribute to a range of important societal needs.

When the prediction process is fostered by effective, participatory institutions, and when a healthy decision environment emerges from these institutions, the products of predictive science may even become less important. Earthquake prediction was once a policy priority; now it is considered technically infeasible, at least in the near future. However, in California the close, institutionalised communication among scientists, engineers, state and local officials, and the private sector has led to considerable advances in earthquake preparedness and a much decreased dependence on prediction. On the other hand, in the absence of an integrated and open decision environment, the scientific merit of predictions can be rendered politically irrelevant, as has been seen with nuclear waste disposal and acid rain. In short, if no adequate decision environment exists for dealing with an event or situation, a scientifically successful prediction may be no more useful than an unsuccessful one.

- These recommendations fly in the face of much current practice where, typically, policy-makers recognise a problem, scientists then go away and do research to predict natural behaviour associated with the problem, and predictions are finally delivered to decision makers with the expectation that they will be both useful and well used. This sequence, which isolates prediction research but makes policy dependent on it, rarely functions well in practice.

E.3.1 Societal Needs

Recently, Lins and Slack (1999) published a paper showing that in the United States in the 20th century, there have been no significant trends up or down in the highest levels of streamflow. This follows a series of papers showing that over the same period "extreme" precipitation in the United States has increased (e.g. Karl and Knight 1998a; Karl et al. 1995).

The differences in the two sets of findings have led some to suggest the existence of an apparent paradox: How can it be that on a national scale extreme rainfall is increasing while peak streamflow is not? Resolving the paradox is important for policy debate because the impacts of an enhanced hydrological cycle are an area of speculation under the Intergovernmental Panel on Climate Change (IPCC 1996).

There does exist some question as to whether comparing the two sets of findings is appropriate. Karl and Knight (1998b) note that "As yet, there does not appear to be a good phys-ical explanation as to how peak flows could show no change (other than a sampling bias), given that there has been an across-the-board increase in extreme precipitation for 1- to 7-day extreme and heavy precipitation events, mean streamflows, and total and annual precipitation."

Karl's reference to a sampling bias arises because of the differences in the areal coverage of the Lins and Slack study and those led by Karl. Lins and Slack (1999) focus on streamflow in basins that are "climate sensitive" (Slack and Landwehr 1992). Karl suggests that these basins are not uniformly distributed over the United States, leading to questions of the validity of the Lins and Slack findings on a national scale (T. Karl 1999, pers. comm.). While further research is clearly needed to understand the connections between precipitation and streamflow, a study by Pielke and Downton (2000) on the relationship of precipitation and flood damages suggests that the relationship between precipitation and flood damages provides information that is useful in developing relevant hypotheses and placing the precipitation-streamflow debate into a broader policy context (cf. Changnon 1998).

Pielke and Downton (2000) offer an analysis that helps to address the apparent paradox. They relate trends in various measures of precipitation with trends in flood damage in the United States. The study finds that the increase in precipitation (however measured) is insufficient to explain increasing flood damages or variability in flood damages. The study strongly suggests that societal factors – growth in population and wealth – are partly responsible for the observed trend in flood damages. The analysis shows that a relatively small fraction of the increase in damages can be associated with the small increasing trends in precipitation. Indeed, after adjusting damages for the change in national wealth, there is no significant trend in damages. This would tend to support the assertion by Lins and Slack (1999) that increasing precipitation is not inconsistent with an absence of upward trends in extreme streamflow. In other words, there is no paradox. As they write, "We suspect that our streamflow findings are consistent with the precipitation findings of Karl and his collaborators (1995, 1998). The reported increases in precipitation are modest, although concentrated in the higher quantiles. Moreover, the trends described for the extreme precipitation category (> 50.4 mm d^{-1}) are not necessarily sufficient to generate an increase in flooding. It would be useful to know if there are trends in 24-hour precipitation in the > 100 mm and larger categories. The term "extreme", in the context of these thresholds, may have more meaning with respect to changes in flood hydrology."

Karl et al. (1995) document that the increase in precipitation occurs mostly in spring, summer, and autumn, but not in winter. H. Lins (1999, pers. comn.) notes that peak streamflow is closely connected to winter precipitation and that "precipitation increases in summer and autumn provide runoff to rivers and streams at the very time of year when they are most able to carry the water within their banks. Thus, we see increases in the lower half of the streamflow distribution."

Furthermore, McCabe and Wolock (1997) suggest that detection of trends in runoff, a determining factor in streamflow, are more difficult to observe than trends in precipitation: "the probability of detecting trends in measured runoff (i.e. streamflow) may be very low, even if there are real underlying trends in the data such as trends caused by climate change." McCabe and Wolock focus on detection of trends in mean runoff/streamflow, so there is some question as to its applicability to peak flows. If the findings do hold at the higher levels of run-off-streamflow, then this would provide another reason why the work of Lins and Slack is not inconsistent with that of Karl et al., as it would be physically possible that the two sets of analyses are complementary.

In any case, an analysis of the damage record shows that at a national level any trends in extreme hydrological floods are not large in comparison to the growth in societal vulnerability. Even so, there is a documented relationship between precipitation and flood damages, independent of growth in national population: as precipitation increases, so does flood damage.

From these results it is possible to argue that interpretations in the policy debate of the various recent studies of precipitation and streamflow have been misleading. On the one hand, increasing "extreme" precipitation has not been the most important factor in documented increases in flood damage. On the other hand, evidence of a lack of trends in peak flows does not mean that policy-makers need not worry about increasing precipitation or future floods. Advocates pushing either line of argument in the policy arena risk misusing what the scientific record actually shows. What has thus far been largely missed in the debate is that the solutions to the flood problems in the USA lie not only in a better understanding of the hydrological and atmospheric aspects of flooding, but also in a better understanding of the societal aspects of flood damage (see Pielke and Downton 2000, for further discussion).

E.3.2 Quantifying Uncertainty Using a Bayesian Approach

As new information on vulnerability and changes in the probability distribution of extreme values become better known, risk estimates should be updated. We discuss below a procedure for handling this problem by using a Bayesian approach to incorporate uncertainty.

Going beyond the consideration of a damaging event in the definition of hazard, let us consider, for example, the quantity of interest (from Chapt. E.2) to be δI, which is defined as the amount of environmental change in the context of water resources (e.g. amount of water recharge to an aquifer; toxic metal content or silting degree of a reservoir). We are interested in the thresholds of δI beyond which there are undesirable impacts, usually understood as undesirable effects on the human well-being. In probabilistic terms the values of interest are in the tail of the corresponding probability distribution. Since δI is the result of different perturbations δA, δB, $\delta \dot{I}$ and these perturbations have an inherent uncertainty associated with each, final uncertainty of δI will be the result of a combination of uncertainties as defined by functions f_1 and f_2 given above, as well as the inherent uncertainty associated with the validity of these functions. All these considerations imply that δI is a highly-dimensional quantity dependent on several parameters which might also be unknown. In probabilistic terms δI is a random variable in a highly dimensional space.

Despite these complexities, a vulnerability assessment of a water resource would be incomplete if a measure of uncertainty is not given to the event of δI being below or above a threshold value considered as a potentially hazardous situation. The reason for this is that policymakers are always interested in expected losses and these need to be quantified in a proper way.

Bayesian statistical methods are becoming increasingly important in evaluating uncertainty related with environmental change (Adams et al. 1984; Paté-Cornell 1996; Tol and de Vos 1998;Wikle et al. 1998). Given a statistical model for a variable or a set of variables depending on a given set of parameters, the Bayesian paradigm involves three main steps:

1. Consider a prior distribution for the model parameters based on prior (as for example subjective information by experts) knowledge and before using the data. When prior information is not available non-informative (diffuse) priors could be used.
2. Obtain an expression of the joint probability distributions of the observations conditioned on the model parameters (this is known as the likelihood function in classical statistics) which implies a proposed statistical model for the variable of interest.
3. Obtain the posterior probability distribution of the model parameters by combining prior information with the likelihood function using the Bayes rule.

This last step can be quite complex since it might involve the definition of a high dimensional joint probability distribution. However, modern computational techniques such as Markov chain Monte Carlo methods (Casella and George 1992) have made possible accurate representations of the joint posterior distributions of the parameters of complex statistical models.

The advantages of a Bayesian approach rely on the possibility of updating the joint posterior distribution of the model parameters when new information becomes available (something that is needed in transient risk estimation as proposed above). A hierarchical modelling approach can also be naturally implemented with-in the Bayesian framework by using the powerful tool of conditioning on the components or parameters of a previous step of the system. This is specially useful when uncertainties related to the spatial scale of scenarios need to be resolved. By adding a downscaling layer to the analysis, under some circumstances (e.g. numerical short-term weather prediction), it is sometimes possible to deal with the uncertainty of going from a grid box to a point value to better represent sub-grid scale variability.

In order to estimate realistic thresholds of δI associated with environmental perturbations, coherence of surface variability at the relatively fine space and time scales is of particular relevance. Since this estimation should be done in probabilistic terms, calibration of atmospheric variability by simply parameterised spatio-temporal stochastic models has proven to be a very useful tool for rainfall and temperature, as discussed in Sect. C.4.1. Extension of these methods within the Bayesian paradigm (Sansó and Guenni 1999) provide tools for including the uncertainties discussed above.

A good example of how to incorporate different sources of uncertainty by using a Bayesian framework is presented by Krzysztotowicz (1999). He proposes a Bayesian forecasting system (BFS) by which the total uncertainty about a hydrological predictant (as river stage, discharge or runoff volume) is decomposed into input uncertainty (e.g. time series of precipitation amounts needed as an input to a hydrological model) and hydrological uncertainty, which considers all sources of uncertainty beyond random inputs (e.g. model, parameters, estimation and measurement errors).

Van Noortijk et al. (1997) also use a Bayesian approach to quantify different sources of uncertainty in the process of taking optimal decisions to reduce flood damage along the Meuse river in the Netherlands (near the Dutch-Belgian border). Their approach attempts to quantify the expected economic losses due to flood damage at different discharge thresholds. This methodology fits very nicely with the vulnerability perspective proposed in this part of the book.

Chapter E.4

The Scenario Approach

Lelys Bravo de Guenni

The scenario approach uses a range of plausible future environmental conditions to assess their effects on water quality, river flows and other hydrological components. Examples of effects on water resource availability include the life span of a reservoir, a water supply system, and water management-related decisions. Most of these effects are usually assessed by using impact models which allow for the translation of changes in the components of the hydrological cycle (e.g. changes in river flow) into changes in the water resource system (e.g. hydropower generation). Changes in the environment (including climate and land-use changes) will affect the hydrological cycle and the vegetation patterns, which in turn will affect the environment including the climate system (Kite 1993). An integrated approach to assess the impact of environmental and land-use changes on water resources in different biomes requires a wide range of scenarios, hydrological models and impact models involving socio-economic aspects such as population growth and changed water management objectives.

Most of the scenario approaches up to the present that are used to assess impacts on water resources are numerical and stochastic modelling-based sensitivity experiments with a focus on atmospheric (weather) changes and variability. Moreover, these modelling experiments should more accurately be described as sensitivity studies, as discussed in Chapt. E.2, since the spectrum of human disturbances to the climate system is not included. Such experiments have been used for impact studies on water resources and agriculture and have been developed to investigate the impact of climate change where the radiative effect of anthropogenically-increased greenhouse gases is the dominant environmental perturbation of the climate system. Existing procedures to construct scenarios are more general and include arbitrary scenarios, historical analogues, palaeoclimatic records, as well as General Circulation Model (GCM) perturbation experiments, and Regional Climate Models (RCM) (Rosenzweig and Hillel 1998).

Arbitrary scenarios assume prescribed changes in climatic variables as for example temperature and/or precipitation. These changes might be applied to the mean (e.g. Rosenzweig et al. 1996), variance (e.g. Mearns et al.

1997) or other statistical characteristics and are normally used to identify sensitivities of the system to the prescribed changes. However, changes in one variable may be accompanied by changes in other variable(s) and some care must be taken to provide meteorologically plausible conditions.

Weather scenarios from historical records are used to construct possible realisations considering, for example, worst case situations (cool or warm periods, dry or wet periods). They can provide some insight on anomalous conditions and their usefulness is conditioned by the length of records available. Palaeoclimatic scenarios use records of pollen, ice cores, lake-levels and others. These records are useful to understand past atmospheric dynamics.

Weather change sensitivity from GCMs are produced under different forcing mechanisms (e.g. double-CO_2 scenarios). These experiments are appropriately referred to as sensitivity experiments, since only a subset of anthropogenic perturbations to the climate system have been evaluated using these models. Equilibrium or transient projections can be designed depending on whether the forcing conditions are instantaneous or are changing with time. In the last case GCMs are used to simulate the response of the climate system to a continuous increase in greenhouse gases, rather than to an instantaneous change. Although GCMs provide valuable insight into the dynamics of the atmosphere and the oceans, it is well known that they are not able to represent even current climate at local scales (Cubasch et al. 1996; Risbey and Stone 1996) and their spatial resolution is still very coarse for impact studies. Sub-grid scale topography, land cover and water bodies, for example, have major impacts on local maximum and minimum temperatures, precipitation, incident radiation and humidity. The variability in these factors needs to be well represented when used as inputs to water resource impact models.

Downscaling approaches have played a role as the main procedures to provide high spatial resolution data from the very coarse information from the GCMs (von Storch 1994). Different approaches are available: stochastic (Bárdossy and Plate 1992; von Storch et al. 1993; Conway and Jones 1998; Bellone et al. 2000); dynamic

(Giorgi et al. 1992; Jones et al. 1995); and regression-based methods which use regression techniques to predict local variables from the large-scale circulation. With the stochastic approaches, dependencies between the large-scale circulation and the local weather are modelled with stochastic weather generators by conditioning on large-scale covariates which introduce changes in the model parameters. Multiple year runs are an output of this approach. With the dynamic approach, output from GCMs or reanalysis data are used as lateral boundary conditions for RCMs in order to resolve more complex processes at the local scale (Giorgi et al. 1992).

If the GCMs had the ability to reproduce the climate system with the required spatial resolution, however, this dynamic downscaling step would not be necessary. Downscaling methods attempt to resolve a scale problem that enables comparison of GCM and RCM simulations with the point observations of environmental variables. However, neither the GCMs nor the regional climate models include all of the significant human effects on the climate system. The combined effects on the Earth's future environment of human land-use change, the biogeochemical effect on the atmosphere due to increased CO_2, and the microphysical effect of pollution aerosols, for example, have not yet been included in these models. Thus the existing GCM and RCM model runs with respect to the future should only be interpreted as sensitivity experiments, not forecasts, projections, or even scenarios (see Sect. E.2; Pielke 2001).

The application of statistical and dynamic downscaling as applied to numerical weather predictions, of course, is a very appropriate and valuable tool to improve the spatial and temporal skill of weather projections (please compare Sect. C.4.1). However, dynamic downscaling from the radiative effects of CO_2 and aerosol GCM sensitivity experiments cannot provide improved skill for several reasons.

First, with respect to dynamic downscaling, there is no feedback upscale to the GCM from the regional model, even if all of the significant large-scale (GCM scale) human-caused disturbances are included. The regional model runs themselves are incomplete in their representation of regional human changes (e.g. land-use change) and of biogeochemical effects. The GCM also has a spatial resolution that is inadequate to properly define the lateral boundary conditions of the regional model. As shown by Anthes and Warner (1978), the lateral boundary conditions are the dominant forcing of mesoscale atmospheric models as associated with propagating features in the polar westerlies. With numerical weather prediction, the observations used in the analysis to initialise a model retain a component of realism even when degraded to the coarser resolution of a global model. This realism persists for a period of time (up to a week or so) when used as lateral boundary con-

ditions for a regional numerical weather prediction model. This is not true with the GCMs where observed data do not exist to influence the predictions. A regional model cannot reinsert model skill when it is so dependent on lateral boundary conditions, no matter how good the regional model is.

If this conclusion is disagreed with, the first step to demonstrate that the regional climate model has predictive skill is to integrate an atmospheric GCM with observed sea-surface temperatures (SSTs) for several seasons into the future. The GCM output would then be downscaled using the regional climate model. There is expected to be some regional skill and this needs to be quantified. This level of skill, however, will necessarily represent the maximum skill theoretically possible with AOGCMs (Atmosphere-Ocean General Circulation Model) as applied to forecasts years and decades into the future since SSTs must also be predicted and not specified for these periods. Such experiments have not been systematically completed. Indeed, does the concept of predictive skill even make sense when we cannot verify the models until decades into the future?

The statistical downscaling, besides requiring that the GCMs are accurate predictions of the future, also requires that statistical equations used for downscaling remain invariant under changed regional atmospheric and land-surface conditions. There is no way to test this hypothesis. In fact, it is unlikely to be valid since the regional climate is not passive with respect to the larger-scale climate condition but is expected to change over time and feedback to the larger scales.

A critical issue in these methodologies, therefore, is how to handle the uncertainty associated with the climate system due to its complex non-linearity (Shukla 1998) and lack of predictability years and decades into the future. Because of the incomplete knowledge of the climate system and the "unknowable" knowledge inherent in the future of the human society, scenarios are meant to provide a range of possible climatic conditions (Hulme and Carter 1999). Extremes and "surprises" are an important part of this uncertainty since they can cause substantial damages to society. Stochastic approaches have a lot to offer on the grounds of uncertainty evaluation since global climate models are not actually formulated to perform an uncertainty analysis and in most cases this is carried out as an afterthought (Katz 1999).

Indeed, based on the different methodologies for scenario construction (including climate) that have been made, a quantitative assessment of what are the temporal and spatial limits of predictive skill is required. It is also concluded that none of the existing procedures to construct scenarios are able to represent the spectrum of plausible future environmental conditions on the local, regional (and even the global) spatial scales.

Chapter E.5

The Vulnerability Approach

Lelys Bravo de Guenni · Roland E. Schulze · Roger A. Pielke, Sr. · Michael F. Hutchinson

E.5.1 Risk, Hazard and Vulnerability: Concepts

Risk, in layman's terms, is the "chance of disaster" (Fairman et al. 1998). More formally

Risk is a quantitative measure of a defined hazard, which combines the probability or frequency of occurrence of the damaging event (i.e. the hazard) and the magnitude of the consequences (i.e. expected losses) of the occurrence.

Embedded within this definition is the term hazard, and implied in the phrase "consequences of the occurrence" is the concept of *vulnerability* to the hazard. These two terms are therefore described next, before re-visiting broader issues of risk.

A *hazard* is commonly described as the "potential to do harm". Defined more rigorously (Zhou 1995; Smith 1996; Fairman et al. 1998; Downing et al. 1999)

A hazard is a naturally occurring, or human induced, physical process or event or situation, that in particular circumstances has the potential to create damage or loss. It has a magnitude, an intensity, a duration, has a probability of occurrence and takes place within a specified location.

The above definition serves to highlight the concept that a physical process only becomes a hazard when it threatens to create some sort of loss (such as loss of life or damage to property) within the human environment (Smith 1996). This is therefore essentially an anthropocentric view of the concept of hazard and does not take into account the effect that an extreme natural event can have on an uninhabited area (Suter 1993). The assessment of losses and the determination of the detrimental effects on future overall sustainability in uninhabited areas are extremely difficult to undertake and they generally fall under the concept of ecological risk assessment (Suter 1993). Here, the magnitude of a hazard is thus determined by the extent to which the physi-

cal event can disrupt the human environment, i.e. a hazard is the combination of both the *active physical exposure* to a natural process and the *passive vulnerability* of the human system with which it is interacting (Plate 1996).

The physical exposure is essentially the damage-causing potential of the natural process and is a function of both its intensity and duration. The natural process becomes a hazard when it produces an event that exceeds

- *thresholds*, i.e. the critical limits (bounds) that the environment can normally tolerate before a negative impact is produced on a system or activity (Downing et al. 1999). In the case of rainfall, too much produces a flood hazard and too little a drought hazard. In Fig. E.4 the shaded area represents the tolerance limits of the variation about the average, within which a resource such as water can be used beneficially for social and economic activities within the human environment (Plate 1996). The magnitude by which an event exceeds a given threshold determines the damage-causing potential of such an event.
- *Intensity* refers to the severity, or damage-causing potential, of a natural process, e.g. rainfall at 20 mm h^{-1} is generally less damaging than 100 mm h^{-1} over the same time period. The hazard intensity is determined by the peak deviation beyond the threshold (vertical scale in Fig. E.4).
- *Duration*, the other variable determining the damage-causing potential of an event, implies exposure to an event, and the longer the exposure the greater the damage-causing potential (Zhou 1995; Plate 1996; Smith 1996). Hazard duration is determined by the length of time the threshold is exceeded (horizontal scale is Fig. E.4).

Response to a hazard, as discussed in more detail later, is either by

- *adaptation*, i.e. the long-term arrangement of human activity to take account of natural events (e.g. becom-

Fig. E.4.
The magnitude of environmental hazard expressed as a function of the variability of a physical element within the limits of tolerance (after Smith 1996)

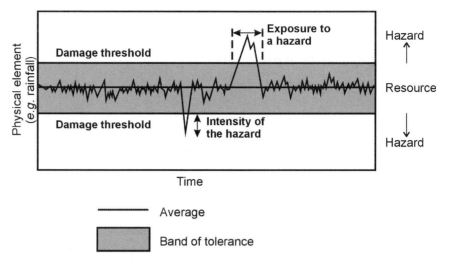

ing more dependent on groundwater than on more erratic surface water resources in more arid zones), or

- *mitigation*, i.e. the intentional response to cope with a hazard (e.g. only constructing buildings beyond a demarcated 1 : 50 year flood line).

Vulnerability implies the need for protection. From an anthropogenic viewpoint

Vulnerability is the characteristic of a person or group or component of a natural system in terms of its capacity to resist and/or recover from and/or anticipate and/or cope with the impacts of an adverse event (Blaikie et al. 1994; Downing et al. 1999).

Vogel (1998) describes vulnerability in terms of the resilience and susceptibility of a system, including its physical, social and human dimensions, while Plate (1996) adds the reliability of the system to its attributes.

- *Resilience* (Vogel 1998) is the capacity of a system (e.g. a dam) to absorb and recover from a hazardous event (e.g. a drought). Resilience therefore implies that there are thresholds of vulnerability.
- *Reliability*, on the other hand, is the probability that the system, or a component of a system, will perform its intended function for a specified period of time (e.g. what is the probability that the dam will be able to supply water to a city over the next 50 years?).

In terms of vulnerability, systems may be subjected to

- *assault events* (e.g. heavy rainfall; flood peak; pollution levels above a certain concentration), in which case the vulnerability threshold is determined by the system absorption and redirection capacities (e.g. a heavy rainfall saturating a soil and the soil then draining the excess water rapidly enough, or a dam filling to capacity and the spillway coping adequately with

the flood discharge). Systems may, according to Smith (1996), also be subjected to

- *deprivation events* (e.g. drought, soil erosion or leaching of fertilisers out of the soil), in which case the thresholds of vulnerability are determined by the retention and replacement capacities of the system (e.g. the buffer of deeper soil depth to storing moisture for a plant during a drought, or the rate of weathering to replace soil lost by erosion).

Vulnerability therefore invariably embraces an

- *external dimension* (Vogel 1998), i.e. the threat of an event, that may increasingly predispose people to risk (e.g. climate change and its impacts on water resources), as well as an
- *internal dimension*, i.e. the internal capacity to withstand or respond to an event, such as the defenselessness to cope with a hazard (e.g. poor people living on a floodplain) or the lack of means to cope with the aftermath of damaging loss.

People may thus face the same potential risk, but are not equally vulnerable because they may face different consequences to the same hazard.

Although intuitively simple, the vulnerability concept requires quantitative tools which are only now being developed. The issue of vulnerability is discussed in UNEP (2000), for example. One new aspect to the vulnerability perspective is that the affected resource can itself feedback to influence its environment. An attempt to quantify vulnerability in terms of the proportion of people affected under extreme rainfall conditions in South America has been presented by Guenni et al. (2001). They assumed that the proportion of people affected (or vulnerable) to extreme rainfall conditions is a random variable that can be calibrated by using available information on the conseqences of extreme events on the population in different South American countries.

Fig. E.5.
A schematic illustration in which risk changes due to variations in the physical system and the socio-economic system. In all the cases risk increases over time (with modifications after Smith 1996)

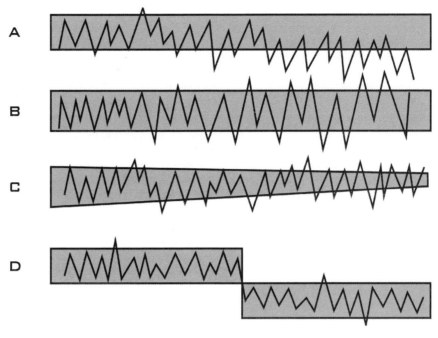

Risk Revisited

If risk is the probability of a specific hazard occurring and the loss caused by that hazard in regard to the level of vulnerability of the affected people or places, then several possibilities exist that give rise to increased risk. These are illustrated in Fig. E.5 (Smith 1996):

- Case A represents a scenario where the tolerance and the variability remain constant, but there is a gradual change over time in the mean value. In this particular case the frequency of extreme events at one end of the scale increases, as would be the case of a mean decrease in rainfall, or a decrease in runoff associated with upstream afforestation.
- Case B shows a scenario in which both the mean and the band of tolerance remain constant, but the variability increases. In this particular case the frequency of potentially damage producing events increases at both ends of the scale.
- In Case C the physical variable, e.g. runoff, does not change, but the band of tolerance narrows, i.e. the vulnerability of the human system increases (e.g. because of increased water demands on a river from people living directly in the floodplain). In this particular situation the frequency of damage-causing events increases at both ends of the scale (e.g. too little water during dry periods; vulnerability to flood damage during high flows).
- In Case D, there is an abrupt change in the mean, but the variability remains the same. Such a rapid transition provides little time to adopt to a change (such as might be possible with Case A).

In viewing risk as having a human component, in addition to its probabilistic one, three tiers of risk have been identified by Zhou (1995):

- a *lower band of risk*, which is acceptable to the affected people and where, for example, the benefits of doing nothing or little outweigh the disadvantages of carrying an unacceptable cost burden,
- a *middle band of risk*, where decisions have to be made which trade off the costs of reducing the risk *versus* the benefits of the risk reduction, and
- an *upper band of risk*, where doing nothing is completely unacceptable, irrespective of cost.

Each of these three tiers of risk is related to the balancing of benefits *v.* costs. This is usually done through risk management, which on the one hand has to be regulated by professional standards and legal measures while on the other hand it contains a large element of subjectivity.

The vulnerability approach described above provides a methodology by which the public in general and the policy-makers in particular will gain a better insight into which thresholds and limits in environmental changes might potentially cause stress or damage. They would have a better estimate of the uncertainty associated with the occurrence of these thresholds and they could better quantify the expected losses for specific hazards.

A cost-benefit analysis must necessarily be a next step for policy-makers to take the required actions in high vulnerability situations. In the following sections, specific examples of environmental variability and changes and multiple stresses to ecosystems with their expected impacts are presented.

Rik Leemans

E.5.2   Anthropogenic Land-use and Land-cover Changes

The Importance of Land-use and Land-cover Change

Issues related to water, water quality and availability, and water management are important on the international political agenda. But these topics do not operate in a vacuum. There are strong interactions between hydrological, atmospheric and biospheric processes on the one hand, and land use and water management on the other hand. Land and water are thus both systemic components of the Earth system, but simultaneously are strongly affected by environmental change. Land use and land-use change and its consequences for land cover (LUC-LCC) are therefore essential components in the development of comprehensive scenarios for the following reasons:

- Society strongly depends on the continued availability of renewable resources, such as food, fibre and fresh water. Humans already dominate the use of most biological productivity (Vitousek et al. 1997) and are a crucial factor in defining fresh-water availability and quality (Matthews et al. 1997; Hoekstra 1998; Arnell 1999). The continued supply of these resources can only be guaranteed by sustainable land uses but these are easily jeopardised by short-sighted or inappropriate human activities (Döös 1994). An integrated approach encompassing water, land and atmosphere with human behaviour is thus urgently needed;
- LUC-LCC alters the Earth's surface characteristics and thus climatic processes and patterns (e.g. Bonan 1997; Claussen and Gayler 1997, see also Chapt. A.5 or A.8). This also can alter water demand and supply or quality and availability;
- LUC-LCC are important determinants of carbon, water and energy fluxes (Braswell et al. 1997) and non-CO_2 greenhouse-gas emissions (e.g. Flint 1994; IPCC 1996). These fluxes and emissions directly influence atmospheric composition and radiative-forcing properties and, consequently, global and regional atmospheric and hydrological patterns;
- LUC-LCC are an important factor determining the response of species, ecosystems and landscapes to environmental change and variability (Peters and Lovejoy 1992; Leemans 1999). Land-cover modification and conversion through, for example, nitrogen addition by air pollution, drainage of wetlands and deforestation change the behaviour of ecosystems;
- Finally, LUC-LCC are, in their own right, an important component of global change (Turner et al. 1995).

All these LUC-LCC aspects, interactions and feedbacks are important and need to be considered and addressed in creating scenarios of Earth system dynamics.

LUC-LCC aspects, however, are difficult to define regionally and globally. LUC-LCC is strongly dependent on local environmental conditions and diverse societal, economic and cultural characteristics and evolves from diverse human activities (e.g. food and fibre production, mining, recreation and nature conservation) that are heterogeneous in their spatial (e.g. stand and landscape), temporal (e.g. diurnal, seasonal, annual and interannual) and societal dimensions (e.g. household, village, region and country). LUC-LCC dynamics are therefore fundamentally complex and amalgamate biogeochemical, hydrological, atmospheric and ecological processes with human behaviour and societal institutions. All these components have, by definition, their specific properties and act on a typical scale level (Fig. E.6). Regionally and globally, only the cumulative consequences of all these diverse local and regional LUC-LCC developments influence the Earth system in a systemic manner. But in (global) environmental-change or Earth-system studies it is these coarse aggregation levels that matter.

Existing Land-use and Land-cover Datasets and Scenarios

The starting point of LUC-LCC scenarios is always a description of current or historical LUC-LCC patterns. Land-cover databases are used to initialise the models required to develop the scenarios. However, there are large discrepancies among different land-cover data-bases (Townshend et al. 1991; Leemans et al. 1996). Some list only potential natural vegetation, others include one or more land-use classes, but none includes land use comprehen-

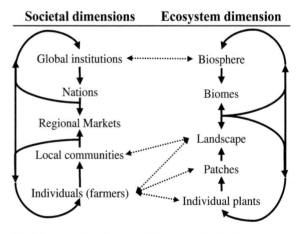

Fig. E.6. Interactions between different societal and ecosystems dimensions

sively. Further, most databases are compiled from diverse and heterogeneous sources and cannot be linked to a specific time period. The determination of change is therefore extremely difficult. Most of these datasets are of dubious quality. However, more recent statistical data sources on land use were improved and their internal consistency is enhanced (e.g. FAO 1999). Also the high-resolution spatially explicit global database, DISCOVER, has become available (Loveland and Belward 1997) and is frequently used in global-change research. This database is derived from satellite data from the early nineties and consists of land-cover classes useful for initialising land-cover scenarios. Further, several attempts have been made to develop historical land-use and land-cover databases (Klein Goldewijk and Battjes 1997; Ramankutty and Foley 1998). These attempts use historical proxy variables, such as historical maps, population-density estimates, and the location of cities and other infrastructure, to reconstruct likely historical land-cover patterns for the last centuries. Although it is difficult to determine the validity and reliability of these historical databases, all these improved databases are of utmost importance for initialising and validating regional and global models.

Many different scenarios for LUC-LCC exist. Many of them focus on local and regional issues and only a few are global in scope. However, most of the available LUC-LCC scenarios are not developed to determine global and regional environmental change, but more to evaluate the dynamics and environmental consequences of different agrosystems (e.g. Koruba et al. 1996), agricultural policies (e.g. Moxey et al. 1995), food security (e.g. Penning de Vries et al. 1997), and projections of agricultural production, trade and food availability (e.g. Alexandratos 1995; Rosegrant et al. 1995). Moreover, these studies do not well define the actual implications for land-cover patterns. At best they define an aggregated amount of arable land, pastures and other land uses.

LUC-LCC scenario studies use different approaches. Most of them are based on regression and process-based simulation models. Alexandratos (1995) has combined and expanded these approaches by interactively including expert judgement. Regional and national experts reviewed the model results. If such a panel determined inconsistencies with observed trends or likely trends, the scenarios were changed until a satisfactory solution emerged for all regions. The resulting scenario can therefore be interpreted as a consensus scenario. This approach generally leads to conservative estimates of possible trends, because the importance of new developments are underestimated.

Unfortunately, most of these LUC-LCC scenario studies do not consider changes in water supply. Some include weather change (e.g. Alcamo et al. 1998) and others assume that an increased demand for, for example,

irrigation water can easily be met when economic resources are available (e.g. Rosegrant 1997). Global-change assessments, however, require broader LUC-LCC scenarios than those do. Adequate scenarios should incorporate land-use activities and land-cover characteristics in order to make comprehensive estimates of the role of land and land use in defining the dynamics of the Earth system. Currently, the only approaches to deliver such capability are Integrated Assessment Models (IAMs: Weyant et al. 1996). Some of these IAMs are the Asian-Pacific Integrated Model (AIM: Matsuoka et al. 1994), Integrated Model to Assess the Global Environment (IMAGE 2: Alcamo 1994; Alcamo et al. 1998) and Integrated Climate Assessment Model (ICAM: Brown and Rosenberg 1999). These models already incorporate modules to simulate the consequences for LUC-LCC, which generate LUC-LCC scenarios on a resolution ranging from a coarse grid (IMAGE 2 and AIM) or for socio-economic regions (ICAM). The starting point for most of these models are crop-production models with different degrees of precision and focus. Most IAMs focus on arable agriculture and neglect pasturalism, forestry and other relevant land uses. None of them, however, satisfactorily incorporate regional water supply and demand issues. The feedback on land-use change on the overlying weather has also not been included in these studies. Only the IMAGE-2 group is currently developing such capability (Alcamo et al. 2000).

The Construction of Land-use Change and Land-cover Change Scenarios

Initially the consequences of land-use change were often only depicted as causing changes in the CO_2-emissions from tropical deforestation. The conversion of these forests is one of the important human sources of CO_2. Early carbon-cycle models used simply prescribed deforestation rates and emission factors to estimate future emissions. Land-use scenarios could only provide the relevant estimates. During the last decade a more comprehensive view emerged embracing the diversity of driving forces and regional heterogeneity. The current driving forces of most LUC-LCC scenarios are derived from population, income and productivity assumptions for agriculture and forestry. The first two are commonly assumed to be exogenous variables (i.e. scenario assumptions), while productivity is determined dynamically by the models used.

In most scenarios, population is generally split into urban and rural; each class is characterised by its specific needs and land uses. The demand for agricultural products (both the quantity and composition as specified by diet) is generally assumed to be a function of income and regional preferences. With increasing incomes there is a shift from grain-based diets towards more meat-based diets. Such a shift has large consequences for land use (Leemans 1999). Similar functional relations are assumed to determine the demand for non-food products, such as timber, fibre and biomass-based fuels. Productivity is determined by the locally or regionally prevailing environmental conditions. Soil and atmospheric conditions define potential yields of each individual crop. In the calculation of potential yields, changes in weather and CO_2 are frequently included. The actual yields are a fraction of the potential due to losses associated with improper management, limited water and nutrient availability, pests and diseases, and pollutants (Penning de Vries et al. 1997). During scenarios development, therefore, assumptions have to be made on the actual yield levels as a function of agricultural management, such as yield increases (e.g. fertilisation and irrigation) and protection measures (e.g. weeding and pest control). Irrigation is rarely considered explicitly or dynamically evaluated from realistic water supplies, nor is the effect of irrigation on the local weather, such as described by Stohlgren et al. (1998). Further, most yield calculations assume constant soil conditions. In reality, many land uses lead to land degradation, which affects yields and in turn limits and changes land use (Barrow 1991). Estimating future yield developments is thus difficult because of its dependence on management. Many contrasting perceptions therefore exist in the literature.

The demand for land-use products and the calculated productivities are then translated into the actual land use and corresponding land-cover patterns. In this step large methodological difficulties emerge. Agricultural land uses in certain regions intensify (i.e. increase yields), while in others they expand in area. The causal factors in different regions are different and the spatial and temporal consequences cannot be determined unambiguously. For example, deforestation is driven by timber extraction in Asia and by the conversion towards pasture in Latin America. In Europe, increasing productivity and a leveling demand drives the contraction of agricultural land and the subsequent expansion of forested land. Additionally, land-cover conversions are not a continuous process. Shifting cultivation is a common practice, but in many regions agricultural land has also been abandoned (e.g. Foster et al. 1998) or is abandoned regularly (Skole and Tucker 1993). Such dynamic and complex aspects of LUC-LCC make the development of comprehensive high-resolution LUC-LCC scenarios challenging, especially when simultaneously new land uses (e.g. biomass plantations) are emerging.

The result of most LUC-LCC scenarios is land-cover change. This is illustrated by an IMAGE-2 scenario (Fig. E.7). This scenario illustrates some of the complexities in the underlying land-use dynamics. Deforestation continues globally. Only in the latter half of this century is it expected that the forested area will increase again in all regions, except Africa and Asia. Pastures expand more rapidly than arable land. There are, however, large regional differences. One of the important assumptions in this scenario is that biomass will become a suitable energy source. The cultivation of such biomass crops requires additional land. Similar changes in future land use can develop if carbon sequestration forests are planted. Land use will thus respond to new needs for additional products and services.

Uncertainties in the Interpretation of Scenarios

Diverse applications of LUC-LCC scenarios have been developed. The different approaches are all very sensitive to the underlying assumptions of future changes in agricultural productivity. Estimating demand is relatively straightforward, because it depends on the number of people and their diet. Productivity in most scenarios is calculated through potential and actual yields (i.e. the interaction between environmental conditions and agricultural management; Rabbinge and van Oijen 1997). Large differences in productivity exist between scenarios and it is not always clear if potential or actual yields are used. This leads to large differences in the conclusion of different LUC-LCC scenario studies. For example, the FAO scenarios study (Alexandratos 1995) concluded that land availability was not a limiting factor in most countries, while IMAGE-2 scenarios (Alcamo et al. 1998) show that in Asia and Africa, land rapidly becomes limited over the same time period. In the IMAGE-2 scenario, relatively fast changes towards more meat-based diets lead to a rapid expansion of grazing systems on which productivity only slowly increases. This strongly influences the simulated extent of land use. The FAO study does not consider increases in grazing systems. This example shows that not clearly specifying the actual assumptions can lead to discrepancies and inconsistencies in scenario interpretation. Unfortunately most land-use scenario studies focus on agricultural production on arable land and few at the same time include grazing systems. Other important land uses, such as forestry, infrastructure (i.e. urban land uses) and nature conservation, and new land uses, such as biomass and carbon plantations, are generally not considered, nor is the influence of these changes in the climate system considered. In most scenario studies, integration of all land uses is not yet accomplished.

Reconciling demand and supply leads to different land-use and land-cover patterns (cf. Fig. E.7). This is often done by heuristic rules that link (inter)national socio-economic drivers with patterns of resource availability. Cellular automata, for example, explicitly resolve such rules by assigning probabilities for different land uses to each cell. These probabilities can change through time (e.g. White et al. 1997). The amount of land used is

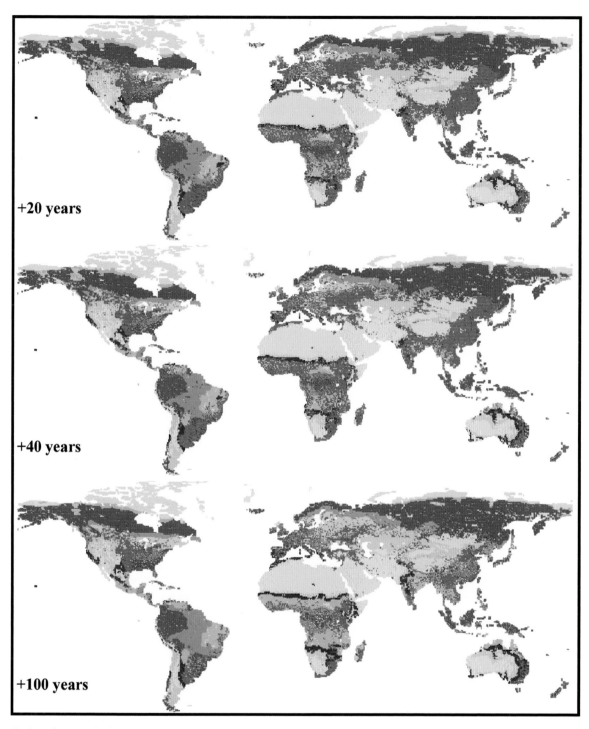

Fig. E.7. Changes in land cover for the IPCC-SRES B1 scenario, that depicts globalisation but with increased environmental consciousness (see scenario given in de Vries et al. 2000). The *red areas* indicate the changes in arable and grazing land. The other colours present the different biomes. The influence of these land-cover changes on the local, regional and global climate system, such as illustrated in Part A of the book, have not been included in the generation of this scenario

highly dependent on local crop productivity, land management (land quality, conservation or degradation) and accessible technology, water availability and quality, and competition with other land uses. Determining the actual patterns is therefore difficult. Only few regional and global studies actually simulate change in patterns (e.g. Alcamo et al. 1998; deKoning et al. 1998; Hall et al. 1995) but their results are always highly uncertain.

The purpose of the scenario development, its underlying assumptions and approaches used should be carefully and critically evaluated before the resulting land-cover patterns are applied in other studies. In addition the feedbacks between LUC-LCC and other components of the Earth system (e.g. the weather, the hydrological cycle, etc.) needs to be investigated, as described elsewhere in this volume. Despite the obvious limitations of current LUC-LCC scenarios, recent scenario developments show how useful organised thinking (i.e. detailed consistency checks between scenario assumptions and outcomes and explicit incorporation of interactions and feedbacks) about the future can be. However, a better perspective on how to use land-use and land-cover change as a driving force to environmental change is strongly required. This requires the further development of high-resolution integrated assessment models and comprehensive scenario studies, integrating other aspects of the Earth system, such as specifically land, land use and water issues.

Burkhard Frenzel

E.5.3 Procedures to Assess the Effect of Environmental Conditions on Water Resources: Natural Landscape Changes

Introduction

When dealing with the long-term palaeoecological processes that might influence present-day water budgets, it must be noted that climatic factors may have very long-lasting, but unexpected series of consequences. An example of this is the immigration history of forest plant-species into vast regions of higher latitude after the end of the last glaciation (Bernabo and Webb 1977; Gliemeroth 1995; Huntley and Birks 1983). These migrations, which were triggered by Late-glacial and early Holocene climatic changes, continued until other decisive factors such as stronger competitors or adverse geological or climatical situations finally stopped the immigration process.

This immigration and reorganisation process of the various forest communities influenced the water budgets of the regions into which they immigrated but the decisive climatic factor was at first the transition from the Late-glacial to the Holocene. Probably nearly all the other processes that took place during immigration were biotic consequences, which could not be predicted exactly if only the quality and intensity of the Late-glacial to Holocene climatic change was measured. It is scientifically very difficult to understand and to quantify these long-lasting, indirect climatic influences correctly, and demands reliable quantification of the climatic change triggers at a regional scale and with high and reliable

temporal resolution. The feedback of the surface environmental changes within the climate system also needs to be assessed. However, the normal physical dating procedures can give only approximate data. Therefore, warve-countings, dendrochronology or age determinations from annually-layered ice have to be used instead. This in turn means that much more comprehensive studies are needed. Moreover, it will often become important to look for marker horizons that can clearly identify synchronously-formed sediments within the ocean, in lakes and on the continents; in general, these will be wind-transported substances like volcanic ash, etc.

Vegetation Changes: Consequences for Water Budgets

It is evident that changes in the vegetation pattern and in the prevailing plant communities may be triggered by atmospheric change or by biotic and pedogenetic changes, or even by grazing and trampling animals (Gossow 1976; Peer 2000; Samjaa et al. 2000; Schaller 1998; Turner 1975). The approach is to study the vegetation and its ongoing or past changes on a broad regional scale, which helps to differentiate between more regional or local processes. This means, in general, that the various vegetation types within a landscape, e.g. a complete catchment, have to be studied in space and time in such a way that it opens up starting points for ecological investigations or calculations, i.e. one has to try to map or reconstruct the various important plant communities which existed or which had changed within the catchment area being studied. One has to determine whether these changes had been triggered by long-term changes in weather or by biotic processes only, e.g. by the immigration of stronger competitors or by pedogenesis, as far as this is independent from vegetation (Frenzel 1994).

Plant ecologists in general determine the water budgets of only certain plant species then, if necessary, extrapolate to whole plant communities. Hydrologists, on the other hand, generally investigate the water budgets of whole catchments, which might be covered by several, quite different, plant communities. Neither of these approaches yields reliable data for understanding what happens to the water budgets of a specific region, if certain plant communities are changing. To overcome these difficulties, at present it is only possible to rely on the evaporation data of identical plant communities at comparable sites, or at least to use data from related plant communities, even though the specific soil, water and atmospheric conditions may be different at individual sites. These difficulties increase when reconstructing former plant communities with no homologues (or at least analogues) to be found in modern vegetation; or if they do exist, do so under quite different environmental conditions. Such an example would be the early Holocene pine-birch forests, so characteristic of vast regions of the Northern Hemisphere, whose analogues are only found under special site conditions like dune fields, or under quite different environmental conditions to those of the region studied. It can be seen, therefore, that the degree of uncertainty in reconstructing the relevant water-budgets may increase in an often quite unknown way.

Changes in Permafrost and in Water Resources

The occurrence and distribution pattern of permafrost is strongly dependent on the average long-term weather and on its change over time. Permafrost also has a major influence on the water budgets of the regions where it occurs. There is a very comprehensive literature on these interrelationships (e.g. Washburn 1973), yet the interrelationships with the water budgets of those regions in which permafrost is disappearing are not investigated sufficiently. In view of the fact that the upper soil horizons in permafrost areas such as Siberia contain 40–80% of their volume as ice, a relative sudden melting of permafrost should have contributed huge masses of fresh water to lakes and rivers. This has never been estimated for the larger regions of central and eastern Siberia. On the other hand, if permafrost is formed in former river terraces, runoff from former lakes and rivulets should be impeded by the increase in volume, this in turn causing sediment accumulations and over-saturation in surface water which may also freeze. Each destruction of the former forest vegetation in a permafrost area will necessarily influence heat fluxes. Frost-heave will be the consequences (Fig. E.8), which will again hamper runoff. On the other hand, the litter of boreal coniferous forests, particularly pine and larch, produces a heat-flux impeding layer which prevents summer warmth penetrating deeper soil layers, thus favouring the expansion of permafrost.

Sea-level Changes and Their Influences on the Terrestrial Water Resources

Depending on their velocity and scale, and on the geological situation, sea-level changes would positively or negatively influence surface and groundwater runoff from larger regions (Liu et al. 1992; Bantelmann et al. 1984; Godwin 1952). Of course eustatic and isostatic sea-level changes have to be taken into account.

This is self-evident yet all the relevant factors are often very difficult to differentiate if the sediment input of larger rivers into their estuaries and the adjoining marine regions is complicating the situation. Sea-level changes caused, for example, by changes in the temperature of the ocean waters are another factor that can influence the water budgets of coastal regions. Moreover, off-shore coastal marine sediment transport can strongly influence the hydrology of adjoining terrestrial environments by changing both the submarine and shore topographies, and thus affecting groundwater circulation.

Other factors are the topogenous and ombrogenous peat-bogs that can form at the transition from ocean to land (Kossack et al. 1984). They are able to not only protect the coastline against waves and storms but also influence the terrestrial runoff, yet the vast palaeoecological literature on these peat-bogs demonstrates how difficult it is to differentiate between causes and consequences of their formation (Kossack et al. 1984).

Fig. E.8.
Permafrost region of central Yakoutia, Siberia. Forest has been cleared by man; frost-mounds are the immediate consequence

Migrating Dunes and Their Influence on Water Budgets

Two types of migrating dune fields have to be taken into consideration: those on the lowland coasts and those within the continents. Causes and consequences of the action of coastal dune fields are directly connected with the processes going on along the lowland coasts. Concerning the continental dune fields, one has to differentiate between those of arid to semi-arid climates against those of more temperate climates. The latter were mostly formed during glacial times and their migrations stopped by plant growth during the Holocene. In general, they will become active again if the protective vegetation cover is destroyed. This can occur due to landscape conversion by people. In colder climates, where plant growth is not rapid, fire and river action can cause terrestrial sand dunes to migrate again. In arid and semi-arid regions on the other hand, continental sand dunes will nearly always be able to migrate, provided the geological situation is favourable. Here, migrating sand dunes can block rivers and cause lakes to be formed. This is not restricted to modern arid to semi-arid regions only, but may also have happened during full-glacial or late-glacial times of various Pleistocene glaciations in what are now temperate climates.

During palaeoecological investigations these dune-made lakes or swamps may easily be taken as indicators of the weather becoming moister, though this has not always been the case. Comparable effects can be observed even today in temperate climates, as for example in the lowlands of the River Leba, northern Poland (Tobolski 1982; Fig. E.9). Here, migrating dunes are invading a peat-bog or an alder carr. Again, the palaeoecological impression would be that the weather had become moister but in fact the drowning of the peat bog and of the alder carr was only caused by migrating dunes impeding the run-off and pressing groundwater out of the sediments.

Such migrating dunes, both at the coast and within the continents, contain a lot of groundwater, thus favouring peat-bogs to grow between the dune ridges or even lakes to be formed (Pachur and Hoelzmann 1991). These water masses within the dune sands can also be influenced by permafrost. Moreover, it is inadequate to rely on only one study site or small region. The investigation of the history of the migrating dunes in the lowlands of the River Leba, as already mentioned, exemplifies this very well (Tobolski 1982). Only by a regional, comprehensive investigation, could it be convincingly shown that the middle to late Holocene start of the migration of these dunes was caused by man; this in turn influenced the morphology of the nearby sea coast, and the water budgets and vegetation types, without any atmospheric impact! A scientifically broad investigation of present-day ecological and hydrological conditions in homologue ecological situations is, therefore, of utmost importance for understanding what has happened and for avoiding incorrect palaeo- or ecological conclusions to be drawn.

Pedogenesis and Changing Water Resources

It is well known that pedogenesis is governed decisively by atmospheric conditions, by the hydrological situation and by the biosphere. With water resources, the most important processes are the formation of finer-grained, minerogenic particles and the narrowing of or even blocking of pores that might contain and conduct water. Soil formation is one long-duration process which is often influenced by much older climatic or geological

Fig. E.9.
Leba-lowland, northern Poland. A sanddune invades an ombrogenous peatbog, causing an inundation of the bog

processes. A broad literature informs about the wealth of difficulties in dating reliably the beginning and the various steps of pedogenesis in a given area (Morozova 1981). Where soils have been fossilised during the past by natural or anthropogenic processes (e.g. the erection of large burial mounds, etc.) the outlines of their formation-history can be traced to some extent. Yet it has always to be taken into consideration that even modern pedogenesis can penetrate deep into the soil, thus influencing former soil horizons. In this context it must be stressed that the spontaneous or anthropogenic changes to vegetation will influence soil formation intensively thus influencing water resources indirectly.

Final Remarks

The task is to assess the consequences of natural landscape changes on water resources. From the discussion here, it will be seen that these natural landscape changes are in general very complex and complicated processes, in which the triggering mechanisms may be simple (in general they are not) but the consequences and the interrelationships of processes are extremely non-linear and complicated. This being so, one should always restrain from simple, uni-directional explanations. Instead, the great variety of interacting processes has to be taken into consideration. The palaeoecological and ecological literatures show that, however, a uni-directional way of thinking is widely accepted. Very often we are a long way from a reliable understanding of what had happened palaeoecologically and thus what might have influenced – and still influences – the environment, including water resources. The only way to avoid these traps is to complete a scientifically and regionally very broad, very comprehensive, investigation of the palaeoecological and ecological processes acting in a given landscape. This is a very difficult task but is essential if we are to understand how human beings alter the natural environment.

Thomas J. Stohlgren

E.5.4 An Example of the Vulnerability Approach: Ecosystem Vulnerability

Rocky Mountain National Park and adjacent areas in the Front Range of the Rocky Mountains in Colorado, USA, like all bio-regions in the country, suffer from multiple stresses simultaneously including: rapidly changing cultural history, rapid human population growth, habitat loss and loss of species, altered disturbance regimes, invasive species, and air and water pollution (Stohlgren 1999a). Rapid climate change is an additional stress that may interact to a greater or lesser degree with these other stresses. There is a need to assess the risks associated with such changes, and the resilience of the ecosystem to resist negative impacts due to these changes.

Despite the recent attention that global climate change and potential long-term effects of increasing greenhouse gases in the atmosphere received, other environmental stresses are higher priority issues and concerns for land managers in the area (e.g. Rocky Mountain National Park Resources Management Plan 1999). The economy of the regions is transforming from extraction-based industries (forestry, mining) to tourism (natural areas and wildlife). The human population is increasing in the region at 2–3% yr^{-1} (a doubling rate in 20 years), and annual visitation to Rocky Mountain National Park now exceeds 3 million people (Stohlgren 1999a). Entire ponderosa pine (*Pinus ponderosa*) forests, decimated by turn-of-the-century logging and fire, have re-grown to create massive fuel loads, that due to a half-century of fire suppression, are ripe for catastrophic wildfire at the forest-suburb interface (Veblen et al. 2000). There is a very strong link of El Niño-Southern Oscillation on fire frequency and size throughout the Front Range (Veblen et al. 2000). Also demanding immediate management attention, are extra-large herds of elk, without native predators (grizzly bear, wolves). Many drainages have been altered by the near-complete extirpation of beaver, altering vegetation succession, nutrient cycling, soil erosion, and wildlife habitat. Invasive annual grasses and noxious weeds are invading these riparian zones, meadows, and low-elevation forests, which as a result, urgently require control actions. Air pollution adds more than 4 kg ha^{-1} yr^{-1} to terrestrial systems in this region (Baron et al. 2000), contributes to visibility problems, and potentially enhances the spread of invasive species. Even without changes in climate, these multiple stresses would remain top research and resource management priorities.

Assessing the vulnerability of Rocky Mountain National Park and adjacent lands to climate change requires a full understanding of ecosystem components and processes, existing stresses, and their interactions. Land managers are often bombarded by so many current and near-future problems and stakeholder demands, that long-term, subtle, chronic stresses like increasing greenhouse gases in the atmosphere are of less importance to them. Still, the long-term conservation of natural resources and healthy economies requires short- and long-term planning. A common sense approach might include: (1) direct and immediate attention to the most urgent and costly stresses; (2) additional research on the role of ecosystem components and processes within the climate system, and (3) some research attention on improving long-term scenarios. An alternative approach is to proportion research, management, and coping efforts at various scales (i.e. local, regional, national, global) to assure that various stakeholder needs are met at the appropriate scale. In any case, multiple stresses must be assessed simultaneously.

Introduction

The central Rocky Mountains, like all bio-regions in the USA, suffer from multiple stresses simultaneously including: rapidly changing cultural history, rapid human population growth, habitat loss and loss of species, altered disturbance regimes, invasive species, and air and water pollution (Stohlgren 1999b). Rapid large-scale atmospheric circulation change is an additional stress that may interact to a greater or lesser degree with these other stresses. Assessing the immediate and long-term needs of stakeholders, native biodiversity, the economy, and ecosystems requires that we understand all these stresses and set appropriate priorities for prevention, control, mitigation, and restoration. After a brief review of the individual stresses, potential interactions are discussed in this example, and strategies to assess priorities are presented.

It is often helpful to evaluate climate trends in a palaeoclimate context. The story that emerges in the central Rocky Mountains is one of tremendous temporal and spatial variability, and bi-directional changes in climate and vegetation (Nichols 1982; Alexander 1985; Whitlock 1993; Stohlgren 1999a,b; Schuster et al. 2000).

The varying treeline locations provide evidence of drastic changes in climate. During the Pinedale glaciation (22 400 to 12 200 years ago), the treeline in the southern Rocky Mountains in Colorado was 500 m (Nichols 1982) to 1 500 m (Fall 1988) below the present-day treeline, and annual temperatures were 7 to 13 °C cooler. The lower limit of permafrost was 1 000 m lower in Colorado (Fall 1988). During the late-glacial period (15 000 to 11 000 years ago), annual temperatures were 3 to 4.5 °C cooler than today, and the treeline was about 500 m to 700 m below the modern treeline (Fall 1988). During the Holocene transition (12 000 to 8 000 years ago), subalpine forests probably migrated upslope, and montane forests of lodgepole pine and Douglas-fir began to expand in the lower elevations. There was a climatic optimum from 9 000 to 7 000 years ago; mean annual temperatures increased 5 to 7.5 °C from the late-glacial into the Holocene. As the climate warmed between 9 000 and 4 500 years ago, the upper treeline advanced to as high as 300 m above its present position during the warmest period (Fall 1988). Cooling and treeline lowering occurred again between 4 500 and 3 100 years ago (Fall 1988) followed by another warming period from 3 000 to 2 000 years ago, and a cooling trend 2 000 years ago (Elias et al. 1986). The Little Ice Age from around 1550 to 1845 A.D. ended very abruptly (Schuster et al. 2000) and forced a descending of the upper treeline and perhaps lengthened intervals between fires. Tree rings evidence a warming trend in recent decades, but only in some areas (Weisberg and Baker 1995).

There are strong suspicions that the episodes of warming and cooling were abrupt rather than gradual changes (Fall 1988; Schuster et al. 2000). Tree-ring records nearby in the south-western US show numerous abrupt changes in precipitation between 800 and 2 000 years ago (Graumlich 1993; Hughes and Graumlich 1996), and the ice cores in Greenland (Groots et al. 1993; Alley et al. 1993) and Wyoming (Schuster et al. 2000) have documented changes in mean annual temperature by as much as 10 °C in a few years. Background rates of climate change in the topographically complex Rocky Mountains have been abrupt, bidirectional and unpredictable (Schuster et al. 2000). The present rates of change can be gauged against these background levels.

Rapidly Changing Cultural History

Cultural history has changed rapidly in the past 200 years having significant effects on the natural resources, landscapes, and ecosystems in the central Rocky Mountains (Buchholtz 1983). The discovery of gold in 1859 resulted in an exponential increase in the human population, which changed the regional economy and greatly altered nearly every ecosystem. Economic development began to centre on mining, forestry, agriculture, recreation, and the service industries that support these other economic activities (Lavender 1975). These activities required water storage and re-distribution projects proportionate with human population growth (Stohlgren 1999a).

The extraction-based culture in the Rocky Mountains gained momentum by mining copper, gold, lead, molybdenum, silver, tungsten and zinc, resulting in the continued contamination of many lands and waterways. Forestry was a major industry, though the rapid cutting of old-growth timber, relatively slow growth rates, and a shift to tourism has reduced the emphasis on forestry in many areas. Agriculture includes dryland and irrigated farming and livestock grazing. Livestock are frequently moved between high-elevation summer and low-elevation winter pastures. Tourism is now the major industry in the region – centered on Rocky Mountain National Park, several national forests, major ski resorts, and summer vacation use in nearly all mountain towns. It is against this backdrop of rapid cultural change that rapid human population growth has become a major concern in many areas (Riebsame 1997).

Rapid Human Population Growth

The human population grew rapidly between 1950 and 1990 with a 40-year increase of about 150% in Colorado. The population in Estes Park and visitor use of Rocky Mountain National Park has almost doubled since 1960

(Stohlgren 1999a). Current rates of human population change are 2 to 3% for many areas in the Rocky Mountains with urban sprawl and development in mountain communities. Population growth increases demands for water, power, and natural resources. Runoff and snow-melt from the peaks supply the Arkansas, South Platte, Colorado, and San Juan rivers with the water supply for most western states (Riebsame 1997). Human population growth and agriculture demands on water have many of these rivers "over-subscribed". Evaluating the effects of near-future weather and local land-use changes on the quality, quantity, and timing of water are a primary concern of stakeholders.

Habitat Loss and Loss in Biodiversity

Another top priority environmental issue in the Rocky Mountains, and one of the most urgent for land managers throughout the west, is habitat loss and loss of biodiversity. Most of the species on the endangered species list are there because of habitat loss (mostly associated with land exploitation) or invasive species (Flather et al. 1984). Beavers (*Castor canadensis*) that once played important roles in shaping vegetation patterns in riparian and meadow ecosystems in the Rocky Mountains are virtually absent in many areas (Chadde and Kay 1991; Knight 1994). Top predators such as grizzly bears and grays were hunted relentlessly by European settlers and have been extirpated from most of the Rockies. Amphibians have declined through habitat loss, predation by non-native sport fishes, timber harvest, increased ultraviolet radiation, and disease (Corn and Fogelman 1984; Bury et al. 1995). Nearly all native fisheries in the Rocky Mountains have been compromised by introduced fishes (Trotter 1987; Behnke 1992). Populations of bighorn sheep are at only about 2% to 8% of their populations at the time of European settlement (Singer 1995). Long-term changes in weather have not been implicated in the loss of any species in the US but abrupt changes in weather in an already fragmented landscape, and in combination with other stresses, may have a greater effect now than weather changes in the past. Shifting the focus of biodiversity preservation to the radiative effect of increased CO_2 in the atmosphere (see Sala et al. 2000) may be tangential given that the overriding causes of species endangerment are habitat loss (land-use change) and invasive species (Flather et al. 1984).

Altered Disturbance Regimes

Vegetation types that are adapted to frequent fire are a major concern in the foothills of the central Rocky Mountains. For example, ponderosa pine forests in the Front Range of Colorado show a tenfold increase in ponderosa pine biomass since 1890 in many stands (M. Arbaugh, US Forest Service, unpublished data; Veblen and Lorenz 1991). This has restored habitat for many wildlife species. However, more than 60 years of fire suppression has created hazardous fuels in a forest community that naturally burned at 20–60-year intervals in many areas (Veblen et al. 2000). Fire suppression has been effective: perhaps too effective for some species. For example, the number of three-toed woodpeckers increases greatly for three to five years in burned stands, as the birds feed on larval spruce beetles. These woodpeckers may have a profound affect on nearby forest stands by reducing the severity of spruce beetle epidemics. Herbaceous plant diversity skyrockets after a burn. Seventeen years after the 390-ha Ouzel burn in Rocky Mountain National Park, twice as many understorey species can be found in the burned area than can be found in adjacent unburned forest stands (Stohlgren et al. 1997). Species dependent on post-disturbance stands suffer habitat loss with each suppressed fire.

All rivers in the Rocky Mountain region have been altered by reservoirs or other water projects (transbasin canals, irrigation ditches, and small water impoundments). A major transbasin water import project in Colorado carries about 370 $Mm^3 yr^{-1}$ from the Colorado River (west of the Continental Divide) through a 7.8-km tunnel to the Big Thompson River (east of the Continental Divide). Ten reservoirs were built to support the project. In the catchments of the Arapaho-Roosevelt forests in Colorado are approximately 40 major reservoirs (C. Chambers, US Forest Service, pers. comn., February 1995). Reservoir-building directly tracks human population increases (Stohlgren 1999a). Domestic water use accounts for less than 6% of the total water use, agriculture accounts for about 90% of the water use (United States Geological Survey 1990).

Invasive Species

Invasive non-indigenous species are causing severe losses in agriculture, forestry and fishery resources in some regions and ecosystem processes throughout the US to the tune of $138 billion per year (Pimentel et al. 1999; Mack et al. 2000). Land managers, farmers, ranchers, and the public consider invasive species a top priority threat (Mack et al. 2000). European cheatgrass has invaded significant portions of the western pinyon-juniper woodlands, ponderosa pine, and Douglas-fir areas in the Rocky Mountains (Peters and Bunting 1994). Several rare plant species are being displaced by introduced plants in western rangelands (Rosentreter 1994). A dense cover of cheatgrass increased the fire frequency in many of these areas. With each fire, the dominance of non-native annual grasses is enhanced at the expense of native perennial grasses. Exotic plant species such as spotted knapweed, Kentucky bluegrass, common dandelion, and Russian thistle are rapidly invading many landscapes (Langner and Flather 1994; Stohlgren et al. 2000). Purple loosestrife, another European weed, is beginning to invade Rocky Mountain wetlands and streamsides. Purple loosestrife spreads quickly and crowds out native plants that animals use for food and shelter. Most invaders have no natural enemies in the United States and therefore spread unchecked (Thompson et al. 1987). The effects of these introduced plants and links to weather are poorly understood. The level of soil disturbance (road building, ploughing, small mammal burrowing, etc.) is most often implicated in the spread of invasive species.

Greenback cutthroat trout was near extinction by the early 1900s because of broad-scale stocking of non-native brown trout and rainbow trout, land and water exploitation, mining, and logging (Colorado Division of Wildlife 1986; Henry and Henry 1991). Three of the four other native subspecies of cutthroat trout are extinct (Greenback Cutthroat Trout Recovery Team 1983). Most aquatic ecosystems in the Rocky Mountains are now influenced by non-native brown trout from Europe and rainbow trout from the Pacific Coast. One of the subtle, yet devastating, trends in the Rocky Mountain fishery is loss of genetic diversity in native fishes from introductions of non-native fishes.

The Rocky Mountain goat and the moose, which are damaging to native plant species, were deliberately introduced into Colorado. Accidentally introduced mammals in Colorado and Wyoming include the house mouse and the Norway rat (Armstrong 1993). The potential effects of these introduced mammals on Rocky Mountain ecosystems are poorly understood.

Invasive exotic diseases are increasingly devastating to native flora and fauna. White pine blister rust is causing up to 50% mortality of white pines in areas of Montana. Lungworm-pneumonia complex is a bacterial disease that causes spontaneous mortality in the lambs of bighorn sheep in summer (Aguirre and Starkey 1994). Proximity to domestic sheep highly correlates with mortality in newly reestablished bighorn sheep populations (Singer 1995). Whirling disease, introduced from Europe, is a parasitic infection that attacks recently hatched trout. It is now affecting native and non-native trout populations in Colorado. At first, the disease was thought to affect only hatchery fishes; however, the native greenback cutthroat trout may also be susceptible.

While weather interacts with some of the disturbance mechanisms such as fire, dispersal and the human redistribution of species is the primary cause of invasion (Mack et al. 2000). The expense and difficulties of preventing, managing, and controlling invasive species may be only weakly linked to weather relative to other factors (land-use change, human-aided dispersal, pollution, trade, etc.).

Air and Water Pollution

Air and water pollution also continue to be major stresses throughout the region. For example, air pollution, primarily from the combustion of fossil fuels in the Denver-Boulder-Fort Collins metropolitan corridor, may diminish water quality, benefit exotic plant species, and affect nutrient cycling in Colorado (Baron et al. 1994). Chemical analyses of the high-elevation Colorado snowpack are revealing high concentrations (about 15 microequivalents/litre) of sulphate and nitrate in areas north-west of Denver (Turk et al. 1992). Remote areas of the world typically receive less than 0.5 kg ha^{-1} yr^{-1} of inorganic nitrogen, whereas the high-elevation sites in the Colorado Front Range now receive as many as 4.7 kg ha^{-1} yr^{-1} of inorganic nitrogen (Williams et al. 1996). In February 1995, the Colorado Air Quality Commission increased the Denver metropolitan area's particulate pollution limit from the current 41.2 t d^{-1} to 44 t d^{-1} in the next 20 years. Terrestrial biota of high-elevation areas may not have the ability to respond to this increased nitrogen loading (Nams et al. 1993). We can expect direct and indirect effects on ecosystem functions in forested catchments (Baron et al. 1994).

Water quality is a growing concern in the Rocky Mountains. The United States Geological Survey (1993) stated that all of Colorado's major drainages are affected to some degree by pollution. Historic mining operations still contribute toxic trace elements to more than 2 100 km of rivers and streams in Colorado. Of the 50 300 km of streams in Colorado, more than 900 km of streams do not meet water-quality criteria for fishing. Other Rocky Mountain states report similar problems (United States Geological Survey 1993). Although water developments affect riparian zones upstream and downstream from dams (Mills 1991), regional information on the biotic effects of water projects and pollution is extremely limited, fragmentary, or inaccessible. Air pollution and water pollution are immediate threats to human health, biodiversity, and local economies.

Potential Interactions

A vulnerability assessment requires an understanding of existing and potential interactions among stresses (Fig. E.10).

For example, air pollution and fertilisation may add high levels of nitrogen into rivers and streams (Baron et al. 1994; Williams et al. 1996). Nitrogen concentrations may be offset somewhat by increased precipitation and runoff, or they could be exacerbated by drought. The flow of nitrogen through the ecosystem might be further influenced by land-use change (e.g. deforestation, urbanisation, intensive grazing), invasive plant species (e.g. nitrogen fixers or nitrogen accumulators), or by the extirpation or re-introduction of beavers. Thus, assessing the vulnerability of water quality in a region to rapid change, multiple stresses and interactions must be considered. Setting appropriate priorities for prevention, control, mitigation, and restoration of water quality must consider all existing and potential stresses and their interactions. Local decisions on land use, control of invasive species, and fire management can have substantial effects on important resources.

As another example, increased CO_2, warmer temperatures, and increased precipitation may favour forest growth in several areas. However, increased forest growth may reduce understorey plant diversity, forage for wildlife, and riparian zone width, while increasing the forest fuel build-up and the threat of catastrophic wildfire in areas of increasing human population growth. Meanwhile, warmer night-time or winter temperatures may facilitate insect outbreaks by reducing the winter-kill of insect eggs and larvae, possibly counteracting the increased forest production with increased forest mortality (Robertus et al. 1991). Changes in forest composition, age structure, and biomass can greatly alter hydrology. Economic models of change must consider many possible interactions and outcomes from a suite of forces in complex systems.

Strategies to Set Priorities

Assessing the immediate and long-term needs of stakeholders, native biodiversity, the economy, and ecosystems requires that we understand all these stresses and set appropriate priorities for prevention, control, mitigation and restoration. It is common for land managers and stakeholders to address single stresses over the short-term (Stohlgren 1999a,b). Immediate issues related to rapid human population growth, habitat loss and loss of species, altered disturbance regimes, invasive species, and air and water pollution often take precedence over long-term responses to these same stresses. For example, rarely is long-term change recognised as the top research and management concern in city planning documents, county budget documents, or resource management plans in national parks and forests. Land managers are duly concerned and preoccupied over recent building permits on adjacent lands, loss of wetland habitat and recent amphibian decline, invasive noxious weeds and diseases, air-quality violations due to prescribed burning to bring forest-fuel levels under control, and current water-pollution problems from air pollution or pesticide use from nearby. Land man-

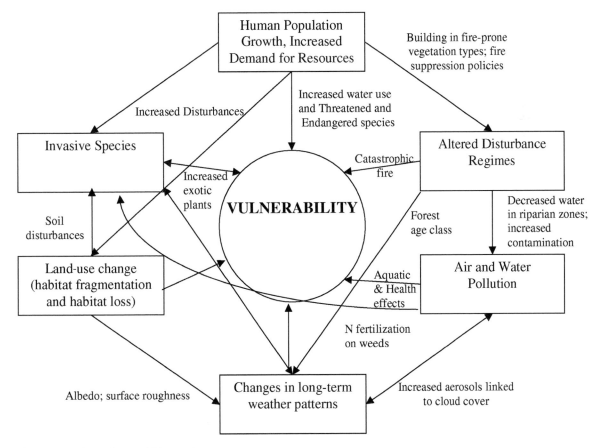

Fig. E.10. Existing and potential interactions among stresses

agers are often affected most by short-term and local issues.

Stakeholders with economic interests (e.g. local businesses, water managers and farmers) are often concerned with short-term economic returns and the immediate resource supplies that maintain the status quo. Thus, there may be significant pressure to manage for short-term needs that are easily addressed and matched to one- and two-year funding cycles, short-term public opinion and management plans, and average job longevity (c. three to five years). It follows that meagre research funds are frequently spent on short-term studies of problems.

There are several reasons why addressing only short-term stresses one at a time might be very shortsighted. First, as stated, there are many obvious interactions among the stresses that can have compounding and confounding effects. Second, there could be "threshold" effects, non-linear responses, or lag effects where the effects of single or multiple stresses go undetected until a given level is exceeded. Subtle changes are often the most difficult to detect and manage. Third, we do not fully understand, or even marginally understand, the long-term effects of multiple stresses in most ecosystems and regions. For example, rapid, but long-term, weather change or increases in variability or extreme events could make the management of the other stresses that much more difficult and expensive in the future.

Conclusion

The human population in the Rocky Mountain region will probably double in the next 20–40 years, creating increased demands for and pressures on natural resources, especially water. National parks, forests, and wilderness areas will probably become increasingly insular, and habitat fragmentation will increase in the na-ture reserves and the urban areas. Continued species decline (e.g. prairie dogs, amphibians, deer, old-growth and fire-dependent species), habitat loss (wetlands, riparian zones, and old-growth forests), increased air pollution and water developments, altered disturbance regimes, and introduced species and diseases will continue to affect many Rocky Mountain ecosystems. Vulnerability assessments must recognise modern humans as being both a primary stress factor and as stakeholders. Combined with standardised biotic resource inventories, predictive models, and long-term monitoring, vulnerability assessments are a logical approach to responsible land and water stewardship in the face of multiple stresses.

Setting priorities for the control and mitigation of multiple stresses requires detailed information on short- and long-term ecological and economic effects (Pielke 1998b). Some level of effort must be reserved for long-term, unpredictable responses. However, adequate funds must be allocated for the primary, short-term stresses that have an immediate effect on the region's resources and people. Current long-term GCM sensitivity experiments, with no known probability, may be of limited use to local land and water managers, but they may focus some attention on the need for a longer planning horizon and a more complete understanding of the Earth's climate system and potential human impacts to that system.

In the absence of adequate information and models, setting specific "coping" priorities is tenuous and should be an iterative process that should be improved as additional information becomes available. The palaeoclimate records warn us to be better prepared for rapid, abrupt changes – in any direction. Interdisciplinary research and monitoring of ecosystems, anticipating economic concerns, and clearly identifying society's immediate and long-term needs are prerequisites to vulnerability assessments and for setting and adjusting priorities.

Chapter E.6

Case Studies

Roger A. Pielke, Sr.

In order to illustrate that environmental vulnerability involves a spectrum of threats, several specific examples are provided in this chapter. These examples demonstrate why it is so important to start with the assessment of vulnerability when studying the interaction of humans with the rest of the environment.

Charles J. Vörösmarty · Jake Brunner
Carmen Revenga · Balazs Fekete · Pamela Green
Yumiko Kura · Kirsten Thompson

E.6.1 Population and Climate

Introduction

One aspect of the global change debate that has received relatively little attention from the Earth systems science community is the purposeful intervention of humans within the terrestrial hydrological cycle as we establish and use water resources. Recent studies have shown that this phenomenon is indeed global (Dynesius and Nilsson 1994; Vörösmarty et al. 1997; Postel et al. 1996) and the growth in water use by humans is likely to continue over the next several decades (Rijsberman 2000; United Nations 1997; Shiklomanov 1996). Water engineering will remain an important human endeavour over this time frame as both economic development and population growth will force future demands on the natural water system. The stabilisation of water supplies through control of terrestrial runoff and/or the translocation of water stocks from aquifers or through interbasin transfers will thus remain a fundamental feature of the land-based water cycle. These activities purposely seek to optimise water security and hence seek to reduce the degree of vulnerability to water shortages and/or excess.

In this contribution we document the emerging role of direct human usage of water. We explore the value of linking biophysical and socio-economic information within a common framework. We also highlight key areas of uncertainty and draw inferences about how our current knowledge base is related to the issue of water vulnerability. We summarise an emerging paradigm for successful water resources management and explain the role of the Earth systems science community in supporting this new strategy for water development.

Water Supply, Use and Emerging Patterns of Scarcity

Assessment of the global water supply has been an imprecise science. If we consider the renewable resource base to equal the long-term mean runoff we can find estimates which vary between 33 500 and 47 000 $km^3 yr^{-1}$ (Kourzoun et al. 1978; L'vovich and White 1990; Gleick 2000; Shiklomanov 1996; Fekete et al. 1999). One of the most recent such assessments, made using time series of observed discharges over approximately 70% of the discharging land surface of the globe and a model-based extrapolation, offer 39 600 $km^3 yr^{-1}$ (Fekete et al. 1999). A more classical accounting by Russian scientists cites 42 750 $km^3 yr^{-1}$ (Shiklomanov 2000). Although one could view these differences as being of simple academic interest only, they become amplified at regional or smaller scales and can give substantially different views of per capita water supplies and hence vulnerability to water scarcity. For example, in North Africa the former study gives a total of 362 $km^3 yr^{-1}$ while the latter shows 181 $km^3 yr^{-1}$. Assuming a population in 1995 of 150 million we get per capita supplies varying between 1 200 and 2 400 m^3 per capita. This estimate highlights the difficulty in attempting to assemble a coherent picture of water availability in the face of declining hydrographic monitoring (Vörösmarty et al. 2001). Further, in many parts of the world, groundwater is an important source of water supply (both renewable and non-renewable; Shiklomanov 1996). Our understanding of the hydrography of groundwater resources is even more limited, since well-logged, groundwater discharge/recharge, and aquifer property data are neither synthesised nor released to the global change community. Noteworthy exceptions exist for regional aquifers (e.g. High Plains in US; Weeks and Gutentag 1988; Nativ 1992), suggesting the feasibility of providing such information when suitable financial and technical resources are made available. Nonetheless, there is but one 1:10 M scale map (Dzhamalov and Zekster 1999) which is only now becoming available. High resolution global mapping of this important resource awaits further work and must be harmonised across the existing region-specific studies.

Larger uncertainties accompany aggregate levels of water use. Long term, there has been a more or less exponential rise in the withdrawal and consumption of water which supports human development (L'vovich 1990; Shiklomanov 2000). But significant uncertainties characterise the history of such studies. Recent reviews (Shiklomanov 1996; Gleick 2000) show that year 2000 water withdrawals as proposed in studies commencing in 1967 and ending in 1998 varied between 3 927 and 10 840 $km^3 yr^{-1}$ with the extreme high value made in 1974. Even assessments made as late as 1987 show a relatively large range of estimates from 3 927 to 5 186 $km^3 yr^{-1}$ for water withdrawals for year 2000. Country-level datasets (World Resources Institute 1996) for some countries are decades-old.

A prime example of such uncertainty is associated with the key component of water use globally, namely irrigation, which accounts for approximately 70% of contemporary withdrawals (Shiklomanov 2000). Irrigation area has been increasing steadily over the last several decades and it would be highly useful to have a systematic methodology to infer the land-area distribution, if not water use, by this water sector. The difficulty in assembling systematic data in part arises from the great diversity in how fresh water is obtained for farming, be it from groundwater resources in Arizona, from trapped or redirected local runoff in the rice paddies of South-East Asia, or from mainstream river water in the Nile floodplain or Columbia River valley. Figure E.11 shows a country-level accounting of irrigated land, contrasting reported statistics from FAO, the UN Food and Agriculture Organization, (World Resources Institute 1996; FAO 1998) with statistics derived from the 1-km land-cover

dataset of the US Geological Survey's Earth Resources Observation Systems Data Center (United States Geological Survey – EROS Data Center 1998) classified according to Olson (1994a,b). The disparity points to major uncertainties in our capacity to classify and inventory the area of irrigated lands. One important problem is that a standardised nomenclature is absent and "irrigated land" is interpreted differently across the globe (see also Frolking et al. 1999). In some cases, it is land that is potentially irrigable. In others, the term represents land that is now under irrigation or land that may have been retired from irrigation. The situation is further complicated when "irrigated land" represents an accounting of area multiplied by the number of crops grown each year (L'vovich 1990). A systematic nomenclature is clearly needed. A recent geo-referenced compilation (Döll and Siebert 2002) based on areas determined from maps generated primarily during the 1970s gives country-specific areas similar to FAO estimates for the mid-1990s, but the geographic updating has been optimised to match the national statistics.

Water scarcity has been defined in various ways, based on per capita supply, withdrawal-to-supply ratios, or coping ability (United Nations 1997; Falkenmark 1998). The inverse of per capita supply represents the water "crowding" and equals the number of people per flow unit. It has been postulated that a threshold of 1 000 people per $10^6 m^3$ yr of water supply is required for food self-sufficiency in the semi-arid tropics and subtropics (Falkenmark 1998). Although it can be argued that food self-sufficiency may not be an absolute necessity for the establishment of food security, it is standard policy in many countries as, for example, in India which has imported only a small fraction of its grain requirements for many years (FAO 1998). The level of economic development is important in coping with potential water stress. Thus, although southern Europe has a high level of relative water use (> 20% of supply), it is able to cope with such apparent scarcity. North Africa, on the other hand, is generally much less able to do so.

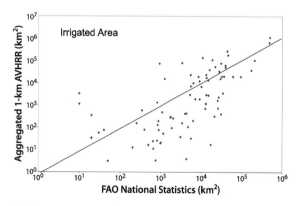

Fig. E.11. Comparison of two major compilations of irrigation area, showing country-level totals reported to the UN Food and Agricultural Organization (World Resources Institute 1996) and an aggregation to the national level of 1-km land cover (United States Geological Survey – EROS Data Center 1998;Olson 1994a,b) based on satellite remote sensing (NOAA-AVHRR)

Global Assessment of Water Vulnerability: A Biophysical Approach

To make estimates of global water scarcity, we used a newly-developed runoff database developed at the University of New Hampshire, in collaboration with the WMO Global Runoff Data Centre (GDRC) in Koblenz, Germany (Fekete et al. 1999, 2002). The dataset provides spatially-distributed renewable water supplies at 30-minute (50 km) grid increments. The runoff database combines available discharge monitoring data with a water balance model driven by atmospheric variables such as air temperature and precipitation (Vörösmarty et al. 1997). A total of 663 discharge gauging stations, which met strict selection criteria in terms of record length and consistency with neighbouring gauging stations, were selected from the GRDC archive for use in developing the database. These stations were co-registered to a 30-minute Simulated Topological Network (STN-30; Vörösmarty et al. 2000).

The Center for International Earth Science Information Network has produced a global population database using census data for 120 000 administrative units (CIESIN 2000). We compare conditions between 1995 and the year 2025, starting with a gridded dataset at 2.5-minute (4 km) grid increments. The 1995 cell counts were extrapolated exponentially to 2025 and then capped using the low growth projection for each country (United Nations Population Division 1998). The cell population counts were then aggregated to 30-minute grid increments to match the runoff database. Figure E.12 shows the population density in 1995. For this study, the low-growth projection is considered to be more realistic than either the medium or high projections (Seckler et al. 1999). But use of the low-growth projection, in conjunction with the maximum sustainable water supply, means that the water scarcity estimates should be considered as conservative. For the purposes of this study, we do not consider potential weather change impacts which add an additional layer of complexity and potential vulnerability. Future water scarcity may be more acute than this tabulation suggests.

The runoff and population databases were summed by river basin to calculate the renewable supply of water per person. We estimate scarcity using a set of standard rules (Hinrichsen et al. 1998). First, populations living in basins with water "crowding" of 590 to 1 000 people per $10^6 \, m^3$ yr of water supply are classified as water stressed, and tend to experience severe shortages in drought years. Second, people living in basins with 1 000–2 000 people per $10^6 \, m^3$ yr[1] are classified as water scarce, and can expect chronic shortages of freshwater that threaten food pro-

duction and hinder economic development. Finally, people living in basins with more than 2 000 people per $10^6 \, m^3$ yr face absolute water scarcity, beyond which development is highly constrained without access to alternative water sources.

River Basin Estimates

Figure E.13 and Fig. E.14 show river basins under varying degrees of water stress in 1995 and 2025. (Only basins larger than 25 000 km^2 are shown.) These maps show increased water scarcity in Southern and Eastern Africa, Pakistan and central Asia, but stable conditions in Latin America, South-east Asia, China and India. But this overview hides important basin-specific trends. Table E.1 shows total water supply, population, and water crowding in 1995 and 2025 for large basins that are currently water stressed or will be close to such stress by 2025. Together, they hold half the world's population. Average water crowding in these basins is projected to increase by 25%, from 546 people per $10^6 \, m^3$ yr in 1995 to 735 people per $10^6 \, m^3$ yr by 2025. Water crowding in some basins, notably those in China, will increase by less than 15%, but the Indus and the Nile will suffer declines in excess of 40%. In the case of the Indus, this will result from the growth in Pakistan's population, from 136 million in 1995 to 246 million in 2025. In the case of the Nile, the decline will mainly result from a doubling of Ethiopia's population, from 55 million to 108 million. Egypt, the second most populous country in the basin, will see its population grow more slowly, from 52 million to 63 million.

Global Estimates

The river basin estimates were summed to give the total number of people living in conditions of water scar-

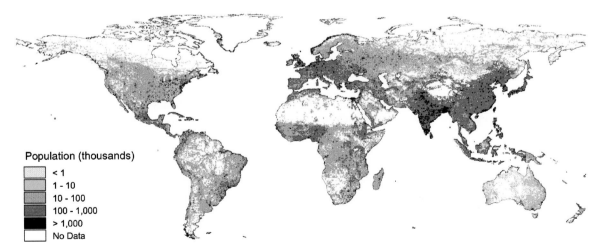

Fig. E.12. Population density from CIESIN (2000)

Population (thousands)

- < 1
- 1 - 10
- 10 - 100
- 100 - 1,000
- > 1,000
- No Data

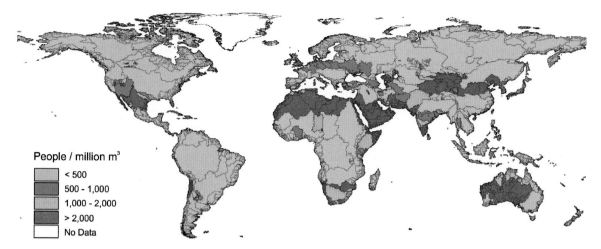

Fig. E.13. Water crowding computed by the conjunction of georeferenced data on population, water use, and water availability, organised by drainage basin for the year 1995

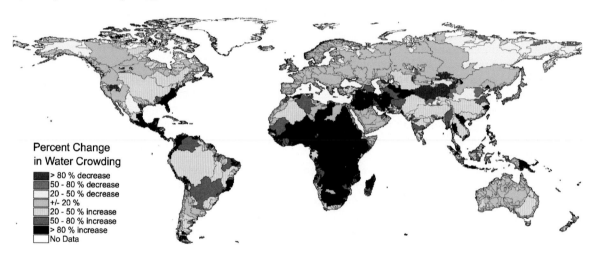

Fig. E.14. Change in water crowding computed by the conjunction of georeferenced data on population, water use, and water availability, organised by drainage basin and comparing conditions in 2025 to 1995

city in 1995 and 2025. Table E.2 shows that 2.3 billion people, or 41% of the world's population, lived in conditions of water scarcity (> 590 people per 10^6 m³ yr) in 1995, and 3.5 billion people, or 49% of the world's population, will live in these conditions by 2025. The number of people facing conditions of absolute water scarcity (> 2 000 people per 10^6 m³ yr) will increase from 1.1 billion to 1.8 billion.

These results require two qualifications. First, they are only based on population growth (which decline in a few parts of the world). They do not take into account the effects on water supply of pollution, weather change, or impoundment. They therefore probably underestimate the actual decline in water supply per person. Second, because the analysis assumes a constant water demand per person, the results inevitably misrepresent water scarcity in certain basins.

Effect of Tabulation Procedures and Their Contribution to Uncertainty

The total number of people living in basins that have an annual renewable water supply of less than 1 700 m³ per person in 1995 is 60% higher than some previous estimates (e.g. Shiklomanov 1996). Although more current population data and a relatively low global run-off are contributing factors, the main reason is that this study estimated water scarcity using river basins rather than countries. Basin-level analyses capture more of the inherent spatial variability in population and run-off (and hence water scarcity) and therefore global estimates that aggregate basin-level results will tend to be more accurate than those based on country averages.

Table E.1. Water scarcity by river basin, 1995 and 2025

River basin	Water supply (km³)	1995		2025	
		Population (millions)	Water crowding (people 10^6 m^{-3} yr^{-1})	Population (millions)	Water crowding (people 10^6 m^{-3} yr^{-1})
Ganges	1 228	483	393	669	544
Yangtze	921	407	441	464	504
Nile	280	127	453	218	781
Indus	152	183	1 210	307	2 020
Amu Darya	72	22	311	35	489
Krishna	52	66	1 270	86	1 680
Yellow	47	131	2 770	151	3 180
Syr Darya	26	22	854	28	1 080
Chao Phraya	18	15	808	17	927
Cauweri	12	28	2 380	32	2 660
Liao	12	35	3 050	39	3 330
Limpopo	11	15	1 400	17	1 640
Jordan	3	14	4 550	23	7 300

Table E.2.
Global water scarcity, 1995 and 2025

Water crowding (people 10^6 m^{-3} yr^{-1})	1995		2025	
	Population (millions)	Percent of total	Population (millions)	Percent of total
> 2 000	1 077	19	1 783	25
1 000 – 2 000	587	10	624	9
590 – 1 000	669	12	1 077	15
< 590	3 091	55	3 494	48
Unallocated	241	4	296	4
Total	5 665	100	7 274	100

Using the methodology presented above, we can also show that on a per capita basis, global renewable freshwater supply will fall from 6 900 m³ in 1995 to 4 500 m³ in 2050. This is a large drop, but it must be recognised that the reduction is in reality highly region-specific. Indeed, it has been estimated that the bulk of the human population is in spatial and temporal proximity to only 30% of the renewable supply of freshwater (Postel et al. 1996) and that we use 50% of this supply. This is especially important in the world's two most populous countries: China and India (Seckler 1999). Northern China is very dry, while southern China is very wet. Eastern India is very wet, while west and southern India are very dry. Because large numbers of people live in all these re-gions, aggregating them masks important within-country differences. Aggregate numbers thus understate the problem of potential water scarcity that is likely to be much more severe at more local scales.

To test this assertion, we tabulated relative water scarcity at four progressive levels of relative water use (i.e. total withdrawal/supply) using four different accounting units. We considered aggregations across 26 broad regions, all countries, drainage basins, and individual grid cells. Figure E.15 shows these results. First, our national level aggregation bears good resemblance to the work of Shiklomanov (1996) using his 26 composite regions. Our country-level aggregate also corresponds well at very low levels of water stress (< 0.1), although it differs substantially from the basin or grid accumulations. In turn, the basin aggregation differs substantially from the grid-based accounting at very low (< 0.1) levels of stress. In relative, but not absolute, terms it differs substantially from grids at moderate levels of stress. There is good correspondence between the basin-wide and grid-based estimates at high levels of stress. With an increasing number of accounting units (10^1–10^2 for regions-countries to 10^3 for basins), we see an increased separation of the classes. In particular, there is a large reduction in the importance of the intermediate classes, which are an apparent by-product of a large spatial accumulation of information.

Fig. E.15.
Assessments of water scarcity
as a function of accounting
unit. Here contrasting levels of
relative water demand (with-
drawal/supply) are used on the
horizontal axis. Levels beyond
0.4 indicate severe water stress.
All aggregations are made over
the global domain

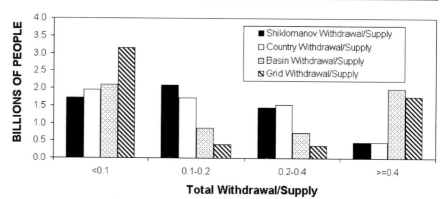

Although we can demonstrate sensitivity to the choice of accounting unit used, we cannot identify an optimal unit. This awaits further study by the water sciences community and a consideration of the broader issue of data availability and harmonisation across biophysical and socio-economic disciplines. The choice of account-ing unit is important from the standpoint of determin-ing the number of people at risk from water stress. In comparison to the regional or country-level tabulation which yields about 400 million under severe stress, the gridded or basin-level interpretation gives a substan-tially larger population, at about 2 billion. The conven-tional wisdom thus severely underestimates the degree to which the world's population is suffering from water scarcity.

Vulnerability and Essential Biophysical Data

The combination of biophysical and socio-economic datasets within a common geographic framework sug-gests that the world is a drier place than previously thought, and that population growth will result in a large and growing portion of the world's population living under conditions of water scarcity by 2025. The World Bank has long sought to help countries harness their water resources for economic development through its support to irrigation, water supply, sanitation, flood con-trol, and hydropower. Water has been one of the most important areas of World Bank investment: between 1985 and 1998, it lent more than $33 billion, or 14% of project lending (World Bank 1998a).

But these projects have encountered serious imple-mentation problems because they do not address the root causes of waste and mismanagement in the water sector. Several problems emerged. First, fragmented pub-lic investment programmes failed to optimise water al-location and emphasized supply development over de-mand management. Second, excessive reliance on over-extended government agencies resulted in poor finan-cial accountability and cost recovery. This led to a vi-cious circle of poor quality, unreliable service, low will-

ingness to pay, inadequate operating funds, and further deterioration in service. Third, damage to freshwater ecosystems inherent in some projects was treated, if at all, by limited corrective actions and rarely by substan-tive adjustments to project design.

In response to these problems, the World Bank is-sued a water sector strategy in 1993 (World Bank 1993). The principal goal of the strategy was to improve water sector performance by ensuring the provision of water services in an economically viable and environmentally sustainable manner. The strategy recognised that to im-prove the performance of the water sector, it is neces-sary to help countries reform their water management institutions, policies, and planning systems. The strat-egy is being implemented through a new generation of water resources management projects. Noteworthy ex-amples exist for Morocco and Mexico (World Bank 1996, 1998b).

To meet increased demand at reasonable cost, policy-makers are exploring better ways to allocate existing water supplies and encourage users to conserve water. Better water management requires a significant im-provement in the collection and analysis of hydrologi-cal data yet monitoring programmes in many countries have deteriorated sharply in recent years (Vörösmarty ans Sahagian 2000). Paradoxically, at a time of growing water scarcity, we know less and less about the challenges and opportunities we face.

The constraints on hydrological monitoring are both financial and institutional. For example, both China and India have dense monitoring networks and tremendous analytical capacity but the effective use of these assets is not possible for several reasons. First, very few sta-tions are linked by satellite or cell phone, resulting in lengthy delays as data are retrieved and transcribed by hand. Second, different agencies manage different gauges, often over different time periods, resulting in inconsistent data quality and fragmentary records. Third, hydrological data are not shared among agencies, let alone with the research community. Many data are considered state secrets. Fourth, limited data access is exacerbated by a move by many agencies toward full cost recovery. Finally, because of the focus on flooding, hydrological models are often only calibrated using data collected during the rainy season. More accurate predictions would be possible if the models were calibrated using data collected throughout the year.

Hydrological monitoring capacity in many countries is deteriorating because of inadequate cost-recovery and reduced government support. The number of stations reporting to the Global Runoff Data Centre peaked in the mid-1980s and then fell sharply (Fekete et al. 1999). The decline has been most marked in Africa, where a recent analysis shows that the number of gauging stations in most countries is well below WMO guidelines (Rodda 1998). Although several international initiatives (e.g. W-HYCOS, GRDC) have sought to compile and publish atmospheric, oceanic, cryospheric and hydrological data, they ultimately depend on the quality of the national networks. Even in the United States the hydrological monitoring network has degraded. The USGS has been monitoring river flow and water quality for over 100 years. The network coverage expanded throughout the 1960s and 1970s but contracted sharply during the 1990s (United States Geological Survey 1999). More than 100 river gauges with long-term records, the single most important class of stations for environmental monitoring and design engineering, are being lost each year (Lanfear and Hirsch 2000). As a result, the network has become vulnerable as collaborating agencies have cut costs to meet their own specific needs, for example by only monitoring high or low flows, or by limiting public access.

Many countries reject the principle of free data access. Some industrialised countries have introduced legislation that would restrict the open access that the research community has traditionally enjoyed. In response, the International Union of Geodesy and Geophysics, the International Association of Hydrological Sciences, and the World Meteorological Organization have all passed resolutions calling for the free and unrestricted access to hydrological data (compare Sect. C.3.4). In less developed countries, data access is primarily limited by high prices, which are justified on the basis of recovering operating costs. As a result, fewer data are available to analysts and policy-makers in developing countries to design and implement the new policies they need to cope with water scarcity.

The Promise of New Technologies

Ideally, a hydrological monitoring system will generate a real-time, continuous stream of discharge measurements from selected locations in the river basin. In practice, this goal is not attained: stage, not discharge, is measured (discharge is inferred from stage using a set of empirical relationships, which must be recalibrated on a regular basis), measurements are taken infrequently and irregularly, and the results are made available days or weeks after capture.

Advances in radar gauging, telemetry, the internet, and modelling can significantly improve monitoring performance. Satellite and cellular communications can transmit data in real-time from the gauging station to the internet. GIS-based hydrological models can integrate hydrological, meteorological, and elevation data to predict discharge and water quality across the river basin. For remote and inaccessible areas, Vörösmarty et al. (1999) propose the development of a hydrologically-oriented satellite to monitor river stage, surface velocity, and width. Using imaging radar and/or doppler lidar sensors, the satellite would measure the surface height of inland water bodies with an accuracy of 5–10 cm and river surface velocity with an accuracy of 20 cm per second, with a frequency of three to seven days. The data would be made freely available in near real-time to counteract the deterioration of terrestrial monitoring networks, data commercialisation, and long delays in proc-essing and distribution. But however advanced the technology, an adequate ground network is still required to calibrate these remotely sensed measurements.

Conclusions

The future management of water must emphasize demand management, efficiency improvements and conservation. This requires increasingly sophisticated information on the functioning of the hydrological system and how it is affected by water management decisions. To this end, the Earth systems science community can provide valuable insight into an important component of the analysis, namely, a quantitative description of the terrestrial water cycle. The analysis presented here suggests that when socio-economic and biophysical datasets are linked in a relatively high resolution setting, we see an even more urgent picture of potential vulnerability in world water resources than previously acknowledged. The challenge will be to harmonise the information needs of the two communities, that is of the natural sciences and the social sciences. The Earth systems modelling community provides high-resolution datasets on water availability which transcend political boundaries, a reflection of the integrative nature of the water cycle. While our knowledge of water supply continues to require improvement, our knowledge of population distribution and water use statistics is arguably much more primitive. In many countries, there has been no census for over a decade and water use statistics for several countries are many years old. Investment in hydrological monitoring and analysis should be balanced with increased support for socio-economic data collection. Socio-economists are well advised to adopt such a "boundary-free" perspective by standardising, both intra and internationally, the datasets necessary to quantify the drivers of water withdrawal and consumption. Finally, identification of water vulnerability will be based on vigilance and scientific hydrology. With international support, governments should seek to upgrade and extend their hydrological monitoring systems using the most cost-effective data collection, communication and analysis technologies available.

Acknowledgments

The authors would like to thank Uwe Deichmann (World Bank), who produced the population surfaces under contract to CIESIN and advised on the extrapolation of the data to 2025, and Greg Browder (World Bank), who provided the World Bank water sector project data and project papers. The authors would also like to acknowledge the contributions made by Richard Lammers (UNH) and Jose Simas, Danny Gunaratnam, and Nagaraja Rao

Fig. E.16.
The Lake Erhai basin

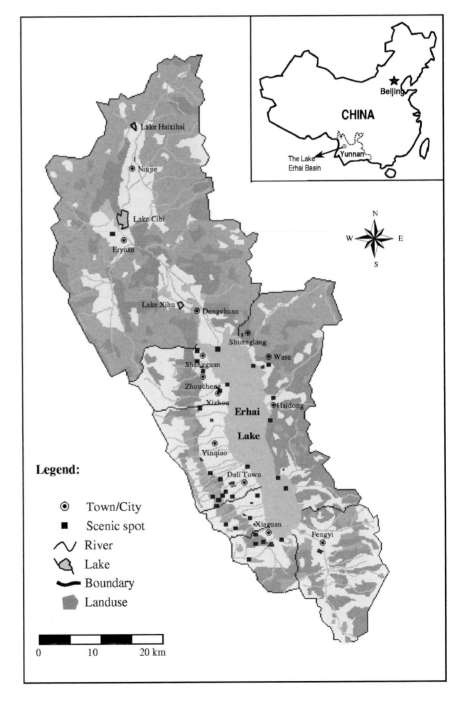

Harshadeep (World Bank). The US Agency for International Development and the Dutch Ministry of Foreign Affairs provided funding for this paper.

Brad Bass · Lei Liu · Gordon H. Huang

E.6.2   Water Resources in the Lake Erhai Basin, China

Introduction

One of the most critical concerns facing both the national and local governments in China is the environmental degradation associated with rapid economic development. Chinese governments have been in the process of designing region-specific environmental regulations, but the decision-making process requires a sound understanding of the significant drivers of regional environmental problems and the effects of policy changes on different sectors and decision variables. One sector that has drawn considerable interest is water resources; in particular, the impact of economic development on water resources and how these impacts are modified by other factors such as climatic variation and change.

The Lake Erhai basin provides an illustration of the integrated impacts of several factors, including weather, on water resources. The basin is host to agricultural and industrial production, a net-cage fishery, forestry and tourism, but its water resources are also required for municipal water and navigation. The rapid economic development and population growth of recent years has been accompanied by increased amounts of water pollution due to industrial wastewater emission, eutrophication in Lake Erhai due to nutrients from runoff and the fishery, and soil erosion due to deforestation. In order to address these concerns, a research project entitled "Integrated Environmental Planning for Sustainable Development in the Lake Erhai basin" was undertaken between 1996–1998 with support from the United Nations Environmental Programme.

The Lake Erhai basin is located at 25°25' – 26°10' N and 99°32' – 100°27' E and covers an area of 2565 km^2 within the jurisdiction of Yunan Bai National Autonomous Prefecture, China (Fig. E.16). Lake Erhai has an approximate area of 250 km^2, an average depth of 10.2 m and a storage volume of 28.2×10^8 m^3. There are 117 rivers and streams that drain into Lake Erhai, mostly from the north through several smaller lakes and which drain into the Mizu, Luoshijiang and the Yunganjiang Rivers and from the east. The Xier River is the only river flowing out of the lake and is the river most affected by industrial discharge. The basin also includes mountains, hills and alluvial plains. Land use includes forest (44.7%), grassland (20.3%), farmland (14.5%), human habitat (2.7%), abandoned land (1.7%), transportation (0.5%), and water bodies occupy 9.5% of the basin.

The vulnerability assessment diagram has been modified for this region and expanded to eight subsystems – population, agriculture, industry, tourism, water resources, pollution control, water quality and forestry – and their interrelations (Fig. E.17). For example, industrial development brings economic benefits but also consumes raw materials from forestry and agriculture and discharges wastewater into the Xier River. A system dynamics (SD) approach was used to model the interactions between the various components in the Lake Erhai basin. The relationships were quantified using Professional Dynamo Plus.

The governments that are involved want to consider how limits to economic development would impact on water quality and quantity, the sensitivity of these impacts to variations in weather and then determine an appropriate economic development policy. A vulnerability assessment was used to assess the impact of four different economic development policies on the available water resources in the Lake Erhai basin and the impact of variable precipitation. The results indicated that continued high levels of economic development carry the greatest risks to water quality, runoff from agriculture would become more of a threat under higher levels of precipitation, and a reduction in precipitation could lead to higher levels of eutrophication. However, the high levels of uncertainty in some key variables will increase the difficulty of justifying any particular planning alternative.

Verification and Sensitivity Analysis

The ErhaiSD model was verified using data from 1990–1994 for 14 variables. While the errors for most variables are quite low, the discrepancies between actual and simulated total industrial output are as high as 18% in one year and for industrial COD (Chemical Oxygen Demand) discharge the discrepancies were as high as 40% in 1994. These errors could be attributable to a number of factors. Important relationships may be missing from the model, or the relative weights of different relationships may be incorrect. A sensitivity analysis was used to examine the contributions of the model's parameters. A concept of sensitivity is defined as follows:

$$S_Q = \left\| \left[\frac{\Delta Q(t)}{Q(t)} \right] / \left[\frac{\Delta X(t)}{X(t)} \right] \right\| \tag{E.7}$$

where t is time, $Q(t)$ denotes the state at time t, $X(t)$ represents the parameter, S_Q is the sensitivity of Q to X, and $\Delta Q(t)$ and $\Delta X(t)$ represent the increments in Q and X at time t.

In assessing the sensitivity, 19 variables and 27 parameters were analysed. Each parameter was increased and decreased by 10% every four years from 1994 to 2010. The model was relatively invariable to these changes except for industrial COD discharge (ICODT) which was very

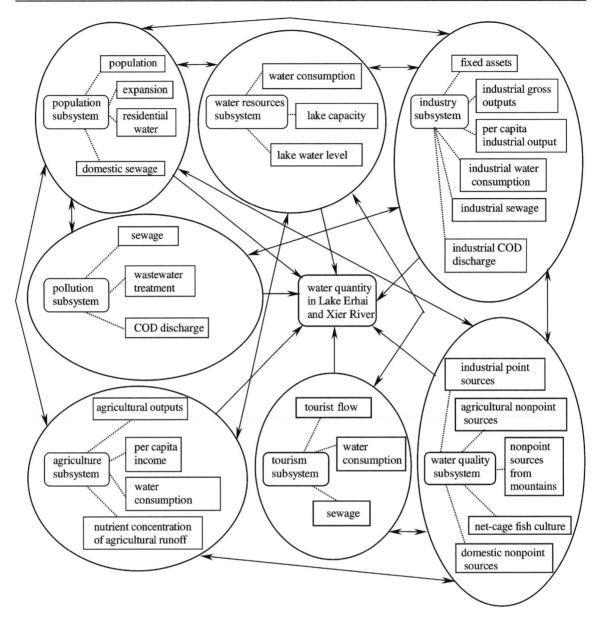

Fig. E.17. Interactive relationships among different subsystems

sensitive to several population parameters. To assess the seriousness of the problems with ICODT, planners, policy-makers and industrial stakeholders must agree on an acceptable uncertainty for this variable. If the model uncertainty is much higher than the acceptable uncertainty, then this decision variable should only be used in a restricted manner (ICODT increases or decreases relative to the previous time step) or not be used at all.

Vulnerability Assessment

The ErhaiSD model was run for a period of 15 years with 1996 identified as the year for generating alternative sce-

narios. The simulation exercise included a base run (BR) and three alternative scenarios (A1–A3), provided by local authorities based on a previous planning study (Wu et al. 1997), balancing economic and environmental objectives (1), emphasising industrial growth (2) and emphasising increasing water pollution control. The results are presented for gross industrial output (PIOT), industrial wastewater generation (ISWT), industrial COD emission and water pollution index (Fig. E.18a–d). The model suggests that water pollution will be reduced in all three alternatives. Alternative 2 will result in marginally higher amounts of economic growth than Alternative 1, yet with much larger amounts of industrial discharge. The COD concentrations could be reduced by

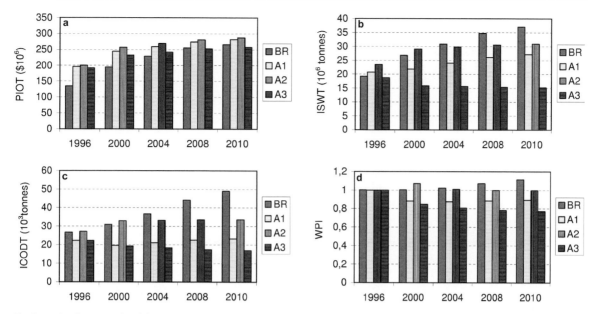

Fig. E.18. Simulation results of the ErhaiSD model with a base run (*BR*) and three alternative scenarios (*A1–A3*). **a** Gross industrial output values; **b** total industrial wastewater generation; **c** total industrial COD emissions; **d** water pollution index

50% through the development of regional wastewater treatment systems; the best improvement occurs in Alternative 3 due to the lower amounts of COD discharge.

Alternatives 1 and 3 produce substantially lower amounts of industrial discharge, but control of this variable is not sufficient as all of the alternatives contribute a similar amount of total dissolved N and P. Although Alternative 3 emphasizes agriculture, the amounts of total dissolved N and P in the back-flowing irrigation water (NAWUR and PAWUR) were only marginally higher than Alternatives 1 and 2.

Vulnerability to Weather

In addition to the direct discharge of industrial waste and nutrients, water quality is closely related to precipitation and runoff. To assess the vulnerability of the basin to seasonal weather variability – and perhaps climate change – three precipitation scenarios were considered, corresponding to dry, average and wet seasons in the region. Table E.3a–c present the impacts of the three precipitation scenarios on the Xier River, for the three planning alternatives, before and after the installation of wastewater treatment. The COD concentration is highest under the dry scenario, even with treatment facilities.

An additional simulation was conducted using the three precipitation scenarios under the assumption of increasing forest coverage. The results are presented for Alternative 1 only, as the impacts were similar across all three alternative futures (Table E.4a–c). Under the wet scenario, the N and P loadings are increased due to increased surface runoff, soil eroision and in-lake turbulence. In the dry scenario the N and P loadings are much lower. However, eutrophication could still be a problem due to intense sunlight and higher temperatures as occurred in August and September of 1996, for example, when the algae count and the COD reached their highest levels in ten years, most likely due to strong thermal stratification and weak dispersion.

Conclusion

The Lake Erhai basin's water supplies are vulnerable to a wide range of factors including the weather. The ErhaiSD model suggests that choosing an alternative management procedure, which balanced industrial growth with environmental concerns or one that favours agriculture, could reduce industrial discharge. This is particularly important under a scenario of reduced precipitation, as the COD levels are inversely proportional to water quantity. All of the planning alternatives were similar in terms of total dissolved N and P, but this could be reduced through increasing the forest coverage. The model suggested that N and P levels were proportional to water quantity, but it did not reflect the potential for eutrophication under a scenario of reduced precipitation.

These results are problematic for planners in that any precipitation scenario that deviates from average conditions will worsen water quality. The industrial discharge would have to be controlled through additional wastewater treatment and/or reduced industrial growth. If the future weather were to favour a drier precipitation regime, the region's development plan could be altered to

Table E.3. COD (Chemical Oxygen Demand) concentration in the Xier River before and after wastewater treatment processes

	Alternative 1			Alternative 2			Alternative 3		
	before	after	before/after	before	after	before/after	before	after	before/after
a Average season									
1998	73.02	63.14	1.2	91.76	78.85	1.16	73.23	61.72	1.19
2001	76.24	59.17	1.29	101.36	77.46	1.31	74.8	54.06	1.38
2004	81	56.9	1.42	104.42	70.2	1.49	76.04	47.21	1.61
2007	87.66	56.21	1.56	109.53	66.24	1.65	79.12	43.62	1.81
2010	91.89	55.1	1.67	112.08	60.94	1.84	79.77	38.56	2.07
b Dry season									
1998	97.16	83.63	1.16	122.85	105.19	1.17	97.53	81.77	1.19
2001	101.59	78.16	1.30	136.14	103.32	1.32	99.72	71.22	1.40
2004	108.04	74.91	1.44	140.30	93.24	1.51	101.31	61.67	1.64
2007	117.18	73.80	1.59	147.41	87.66	1.68	105.48	56.49	1.87
2010	123.09	72.25	1.70	151.03	80.33	1.88	106.40	49.44	2.15
c Wet season									
1998	61.68	53.49	1.15	77.19	66.50	1.16	61.83	52.3	1.18
2001	64.39	50.26	1.28	85.15	65.38	1.30	63.15	46.00	1.37
2004	68.37	48.44	1.41	87.73	59.43	1.48	64.24	40.41	1.59
2007	73.94	47.80	1.54	91.98	56.26	1.64	66.87	37.57	1.78
2010	77.43	47.08	1.64	94.07	51.89	1.81	67.41	33.43	2.02

Table E.4. Simulation results for Alternative 1 for normal, dry and wet season

	1998	2001	2004	2007	2010
a Normal season					
Forest coverage (FCR) (%)	44.7	50.5	56.2	62.0	67.7
N concentration in Lake Erhai (mg l^{-1})	0.31	0.30	0.28	0.27	0.25
P concentration in Lake Erhai (mg l^{-1})	0.0272	0.0265	0.0254	0.0251	0.0241
b Dry season					
Forest coverage (FCR) (%)	44.7	50.5	56.2	62.0	67.8
N concentration in Lake Erhai (mg l^{-1})	0.28	0.27	0.26	0.24	0.23
P concentration in Lake Erhai (mg l^{-1})	0.026	0.0255	0.0245	0.0244	0.0235
c Wet season					
Forest coverage (FCR) (%)	44.7	50.5	56.2	62.0	67.7
N concentration in Lake Erhai (mg l^{-1})	0.32	0.31	0.29	0.28	0.26
P concentration in Lake Erhai (mg l^{-1})	0.0279	0.0272	0.026	0.0256	0.0245

allow for more industrial growth, although it should be noted that industry requires significant amounts of water that might not be readily available. However, the amount of industrial discharge is the one variable that is highly uncertain in ErhaiSD. Without knowing the acceptable level of uncertainty, it is difficult to use the model as the sole basis for controlling industrial growth. Because of the high degree of uncertainty there are risks in accepting the model as a basis for controlling industrial discharge. This case study did not explicitly account for climatic uncertainties in the decision, an analysis which is presented in Bass et al. (1997).

Changming Liu

E.6.3   Yellow River: Recent Trends

Physical and Hydrological Characteristics

The Yellow River, or Huang He, the second largest river in China, is known for its high silt content. It originates from the Yueguzonglie basin, being about 4700 m in elevation on the north side of the Bayankala Mountain in the Qinghai-Tibet Plateau. It flows across nine provinces from west to east and, finally, it enters the Bo Sea (part of

Fig. E.19.
Upper and middle reach of
the Yellow River (Huang He)

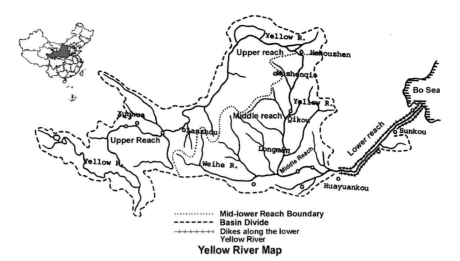

the Yellow Sea) at Kenli County, Shandong province. It lies between 96 to 119° E and 32 to 42° N, with a drainage area of 752 000 km² and a main course length of 5 464 km. Most of the Yellow River area is arid or semi-arid (Liu ans Liang 1989). The area of its drainage basin accounts for 8% of the national area while the runoff makes up only 2% of the national total.

The Yellow River is the most important water source of north-west and north China. It serves as the water supply for a population of 144 million, 16 Mha cropland, 50 large or middle sized cities, and as the energy base of Inner Mongolia, Shanxi, Shaanxi, and the Zhongyuan and Shengli oil fields. The upper and middle stream of the Yellow River covers an area of 730 000 km², which accounts for 97% of its total drainage area (Liu and Li 2000). Most of this area consists of mountains and high plateaus. The lower stream is quite flat with a length of 780 km flowing across the North China Plain.

Figure E.19 gives an overview of the Yellow River with its three sub-basins, e.g. upper, middle and lower reach.

The annual average runoff of the Yellow River is 58.0 billion m³ based on 56 years of data. The area above Lanzhou accounts for 55.6% of the total runoff of the river (Liu and Li 2000). The runoff changes greatly in different years and is distributed unevenly within a year. About 70% of the total runoff results from the rainy season (from June to September).

The prominent characteristic of the Yellow River is "short of water and rich in sediment load". The silt con-

tent of the Yellow River ranks the first among China's rivers (Liu and Liang 1989). The average annual sediment discharge (Table E.5) is 1.6 billion tons (taking the San-menxia station as representative), the average silt content is 35 kg m⁻³ the maximum sediment discharge is 3.92 billion tons per year (1933) and the maximum silt content reached 920 kg m⁻³). The sediment discharge during the flood season (July to October) provides 90% of the annual amount. The yearly fluctuation of sediment discharge is quite high. The sediment discharge in a high precipitation year is twice the average. The maximum and minimum sediment discharge ratio can reach even higher values. For example, the sediment discharge in 1933 was 3.92 billion tons, which was 8.2 times that of 1928 (0.48 billion t yr⁻¹). The hyperconcentrated sediment load in the Yellow River has resulted from environmental problems such as severe soil erosion in the Loess Plateau (Liu 1981; Liu and Liang 1989).

Change in Water Quality Indicators

The runoff of the Yellow River is decreasing which reduces the capacity of the river to cleanse itself by flushing. At the same time, pollution control of the Yellow River has lagged. Since there are no complete laws for water resource protection and a united management mechanism, the water use and sewage discharge increases have resulted in increased pollution load year

Table E.5. The yearly characteristic data of the Yellow River basin

River course	Catchment area (10⁴ km²)	Length of river course (km)	Natural runoff (10⁸ m³)	Sediment discharge (10⁸ t)	Precipitation (mm)	Average temperature (°C)
Upper reach	38.6	3 471.6	312.6	1.42	401.6	3.22
Middle reach	34.4	1 206.4	246.6	14.9	546.2	9.22
Down reach	2.2	785.6	21.0	0	675.3	14.46
Total	75.2	5 463.6	580.2	16	475.9	9.22

Table E.6. Statistical data of sediment yields after Chen et al. (1998) (empty spaces in the table are unmeasured values)

	River or reach	Catchment area (10^4 km^2)	Serious soil erosion area (10^4 km^2)	Annual average sediment discharge (10^8 t)	Silt content (kg m^{-3})	
					Flood	Average
High silt yield	Hekouzhen to Loumen	11.16	7.57	9.73	274	126
	Above Jiaokouhe of Beiluohe	1.72	0.66	1.0	492	250
	Above Tingkou of Jinghe	3.47	2.95	2.64	485	255
	Above Nanhechuan of Weihe	2.34	1.75	1.64	248	142
	Sub total	18.69	12.93	15.0	301	145
Low silt yield	Above Lanzhou of Yellow River	22.3		1.32		4.11
	Lanzhou to Hekou	16.3				
	Above Hejin of Fenhe	3.87		0.695		39.9
	Below Nanhechuan of Weihe	4.84				
	Above Wushe of Qinhe	1.29		0.118		7.95
	Above Heishiguan of Yi-Luohe	1.86		0.327		8.42
	Below Loumen	3.86		−0.73 (silting)		
	Sub total	54.3		1.96		5.24
Total		73.0		17.0		35.6

by year. The pollution of the Yellow River has become worse which has been attributed to different reasons with mining one of the main sources (Ongley 2000).

In Summer 1998, a water quality assessment was performed, based on monitoring data of 69 segments (26 of mainstream, 43 of tributaries) with a total length of 7 158 km in the Yellow River basin. The result illustrates the water problem of the Yellow River, i.e. reflects the serious degradation of the extensively used waters: grade[1] I–III made up only 26.1%, grade IV made up 38.6%, grade V made up 13.0% of the total length. The most seriously polluted water is unsuitable for any use, and does not even get a grade as it is inferior to grade V. 22.3% of the total length belonged to this category, that is below grade V. This means of the total length of the river with polluted water, i.e. belonging to grade IV, V and even worse than V, was 73.9% of the total length assessed in June to September 1998. Comparing the water quality of 1998 with that of 1985, the length of river with water quality worse than grade III has increased by 58.8%.

The long periods of low flow and drying-up in the lower reach of the Yellow River have led the wetland areas in the delta to decrease. The groundwater table has declined without enough fresh water recharge and the salinisation area has increased. The water environment of the wetland has changed and threatened the life of hundreds of wild plants and 180 kinds of birds. It is damaging the biological diversity of the delta.

Acknowlegement

This chapter was supported by the Chinese Key Project: G19990436-01.

Roland E. Schulze

E.6.4 Examples of Hazard Determination and Risk Mitigation from South Africa

The Approach, the Hydrological Model and the Spatial Databases Used

This section illustrates some of the concepts of risk, hazard and vulnerability within a context of hydrological risk management as described in Chapt. E.5, using examples from South Africa, defined here as the contiguous area of 1 267 681 km^2 made up of the nine provinces of the Republic of South Africa plus the landlocked Kingdoms of Lesotho and Swaziland. The country/regional scale is chosen to identify regions of similar hydrological hazard levels and thereby to distinguish between areas of higher and lower potential hydrological risk. Such a comparative view is important from the perspective that risk management is a national/regional responsibility and that this analysis may assist in identifying target areas for potential priority attention. Many of the hazard/risk indices presented by way of maps are in the form of ratios, rather than as absolute values, to highlight sensitive areas on a relative scale.

---

[1] In China, the water quality is classified in five grades, in descending order from the highest quality (I) to the poorest one (V).

A regional analysis requires a hydrological model to be operated with suitable hydrological spatial databases. The hydrological model selected was the ACRU agrohydrological modelling system (Schulze 1995, and see Sect. D.2.6.6.2), developed in South Africa for catchment scale simulation of runoff components, sediment yield, reservoir yield, irrigation supply/demand and crop yield analysis, and containing routines for the assessment of land use/management as well as climate impacts on hydrology. ACRU is a widely verified physical-conceptual and daily time step simulator operating on a multi-layer soil water budget described in detail in Schulze (1995). The ACRU model has been widely verified at catchment scale, both in South Africa under a range of hydroclimatic and physiographic conditions (e.g. Schulze 1995) and elsewhere (e.g. in the USA, in Germany and Zimbabwe).

For the production of South Africa scale maps shown in this chapter, quality controlled daily rainfall for the concurrent 44-year period 1950–1993 was input for each of the 1946 Quaternary catchments which have been delineated for South Africa, Lesotho and Swaziland. For each Quaternary catchment other atmospheric parameters (e.g. daily maximum and minimum temperatures; A-pan equivalent reference potential evaporation) were also input, as were hydrological soil parameters. For purposes of producing comparative hydrological hazard maps, land cover was assumed to be grassland in fair hydrological condition (i.e. 50–75% cover).

Examples of General Hydrological Hazard Indices

To set the scene, Fig. E.20 (*top*) shows mean annual precipitation (mm), which characterises the long-term quantity of available water to a region, to display a general westward decrease, with relatively low mean annual precipitation – a first indication of a largely semi-arid climate and potentially high risk natural environment. A simple aridity index expressed as the ratio of mean annual potential evaporation to mean annual precipitation (Fig. E.20, *middle*) emphasizes the hazard of hydrological semi-aridity, because it amplifies the effects of a low mean annual precipitation when that is evaluated in association with the region's high atmospheric demand. The aridity index is an already high 2–3 where mean annual precipitation is > 600 mm, increasing to > 10 and even > 20 in the west. A consequence largely of the high aridity index is that the conversion ratios of rainfall to runoff over most of South Africa are exceptionally low (Fig. E.20, *bottom*) with, overall, only 9% of rainfall manifesting itself as runoff. These low runoff ratios, simulated with the ACRU model (Schulze 1995), result in a high hydrological vulnerability over much of the region.

Example of a "Deprivation" Event, Hydrological Uncertainty and of Variable Sensitivity: The 1982/3 El Niño Event over South Africa

The 1982/1983 El Niño was one of the most severe experienced over South Africa. The manner in which such an event impacts on different hydrological responses is illustrated in Fig. E.21. Observed rainfall and simulated runoff and recharge into the groundwater zone through the soil profile are all expressed as ratios of their respective long-term (1950–1993) median values. For much of the region the El Niño season's rainfall was 60–75% of the median, however, with sizeable areas receiving within the range of expected rainfalls (i.e. 75–125%) while some others received only 20–60% of the norm (Fig. E.21, *top*). The corresponding runoff responses display much more complex patterns spatially and in the range of ratios. Much of the region yielded only 20–60% of the long-term runoff (Fig. E.21, *middle*), with considerable areas generating < 20% of the expected runoffs. This shows clearly once more the intensifying effects of the hydrological cycle on rainfall perturbations, as well as the dependence of hydrological responses not only on total rainfall amounts, but also on individual events, rainfall sequences and antecedent catchment wetness conditions, i.e. on the hydrological uncertainty created by meteorological and catchment conditions.

Some hydrological processes and responses display higher sensitivities, and thus higher potential vulnerabilities, than others. This is illustrated by the recharge to groundwater during this El Niño event, which was impacted even more severely than runoff (Fig. E.21, *bottom*). Generally only 0–20% of the expected recharge was simulated to take place, the reason being that a higher threshold has to be reached for recharge to commence than for stormflow to start occurring.

Example of an "Assault" Event as an Index of Potential Vulnerability and Stochasticity by Quantitatively Defined Endpoints in Regard to Depth, Duration, Frequency and Area Affected

Episodic flood generating events display considerable stochasticity (i.e. unknowable randomness). As an index of potential vulnerability, the flood hazard example presented below as an "assault" event illustrates the relative spatial differences over South Africa between the severity of the 1 : 50 year 1-day flood-producing rainfall, and consequent runoff, compared with what could be considered the annual expected 1-day values, i.e. the 1 : 2 year event. Ratios of 1 : 50 to 1 : 2 year rainfalls are generally between 2 and 4 (Fig. E.22, *top*), with lower ratios

Fig. E.20.
Indices of South Africa's largely semi-arid hydrological environment: (*top*) Mean annual precipitation (mm), (*middle*) aridity index expressed as the ratio of mean annual potential evaporation to precipitation and (*bottom*) the conversion ratio of mean annual runoff to rainfall (from Schulze 1997a)

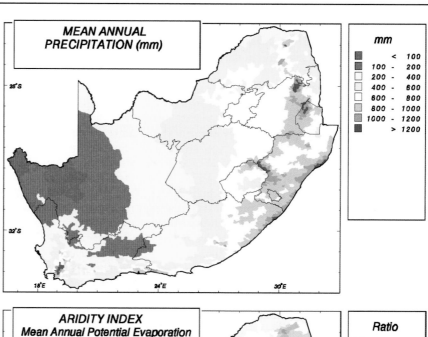

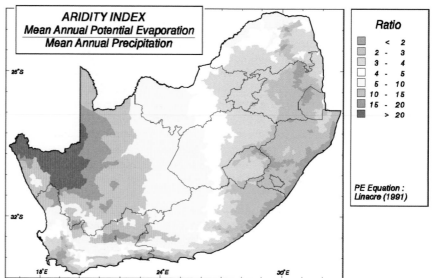

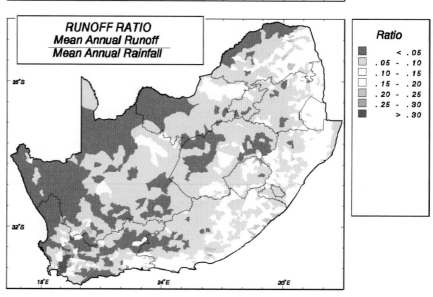

Fig. E.21.
Indices of hydrological amplifications of climate fluctuations: Ratios of the 1982/1983 hydrological year's rainfall (*top*), simulated runoff (*middle*) and recharge to groundwater (*bottom*) to long-term median values (after Schulze 1997a,b)

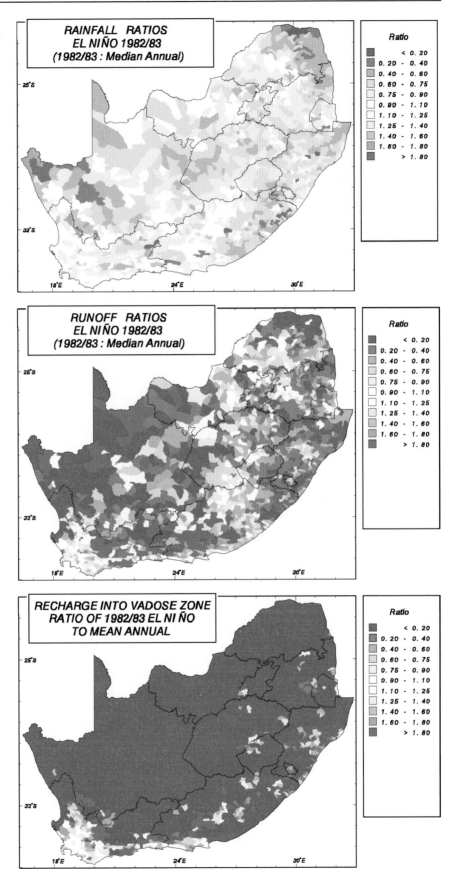

Fig. E.22.
Ratios of 50 year: 2-year 1-day rainfalls (*top*) and runoffs (*bottom*) as indices of potential vulnerability to flooding (from Schulze 2001)

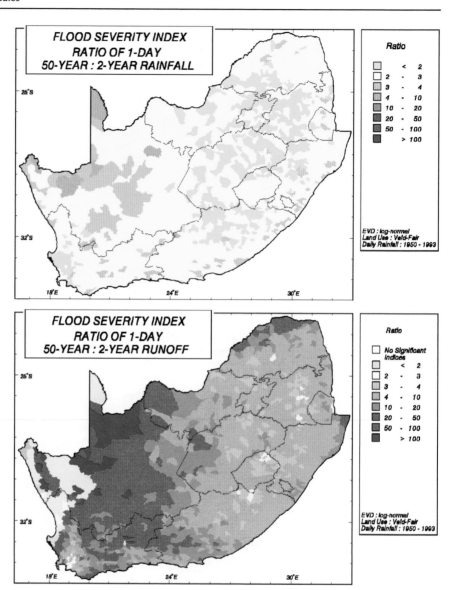

over central areas, but increasing to 10 in parts of the drier west. These rainfall ratios, however, manifest themselves as 1-day flood depths 4–10 times higher in the eastern areas of South Africa, and up to 50 times and higher over significant tracts of the drier west (Fig. E.22, *bottom*). While floods may be an infrequent occurrence in the west, this example illustrates that rare floods have the potential to do severe damage because of their unexpectedly high relative magnitudes.

Example of Uncertainty through Use of Short Datasets

Statistical hazard determination is frequently fraught with uncertainties as a consequence of using short datasets to determine high recurrence interval values of design rainfall or runoff. To illustrate this for hydrological design purposes, the 1:50 year 1-day rainfall and runoff estimated for a short 22-year period 1972–1993 was plotted as a ratio against 1:50 year 1-day rainfall and runoff estimated for double the period, i.e. the 44 years 1950–1993. In each case the log normal extreme value distribution was applied to the annual maximum series, in the case of runoff generated with the ACRU model. If the short records were representative of the expected population of the annual maximum series, the ratio would be around 1.

Figure E.23 shows this clearly not to be the case. Large tracts of South Africa display rainfall ratios between 0.75 and 0.95 and even < 0.75, while other areas show ratios in excess of 1.25 (and even 1.50). Estimating design rainfalls from short record lengths may thus result in severe underestimations or overestimations, with these errors amplified once design runoff is estimated from the rain-

Fig. E.23.
Ratios of "short" (1972–1993) to "longer" (1950–1993) design rainfall and runoff in South Africa for the 50-year return period 1-day event (from Schulze 2001)

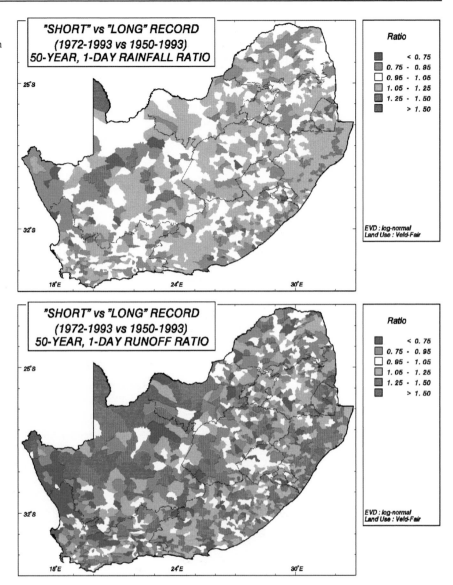

fall (Fig. E.23, *bottom*). The importance of record length can therefore not be overemphasized, particularly in light of the worldwide trend, certainly in developing countries and evident also in South Africa, of declining hydrometeorological recording networks.

Example of Secondary Hazard Modification Through Land-use Practices: The Case of Grazing Management

Land-use practices have already been shown to play a significant role in long-term average hydrological responses (cf. Mgeni Case Study, Chapt. D.7), but perhaps even more dramatically so at the extremities of frequency distributions (e.g. Schulze 1989, 2000). An example of secondary hazard modification by manipulating land-management practices is given below.

Much of South Africa's natural grassland (veld) has recently been shown to be heavily over-utilised (Hoffman et al. 1999). Hydrologically, the degradation through overgrazing of veld from good to poor condition, with its reduction in vegetal cover from > 75% to < 50%, implies enhanced stormflows through reduced interception, evaporation and transpiration potentials as well as infiltrability, shortened catchment lag times which increased peak discharges and greater exposure to soil erodibility through removal of mulch and shorter drop fall heights. These variables were changed for each of the 1946 Quaternary catchments covering South Africa in ACRU model simulations of stormflows, peak discharges and sediment yield to reflect veld in good *v*. poor management condition. Figure E.24 (*top*) shows that annual stormflows from veld in degraded condition are generally 1.5–2.5 times as high as those from

Fig. E.24.
Ratios of annual stormflows (*top*) and sediment yields (*bottom*) in South Africa from veld in poor *v.* good hydrological condition (from Schulze 2001)

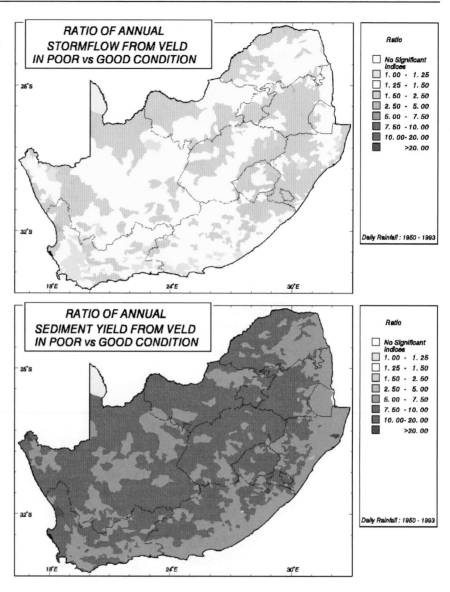

Fig. E.25.
Example of simple benefit analysis of seasonal runoff forecasts over South Africa (after Schulze et al. 1998)

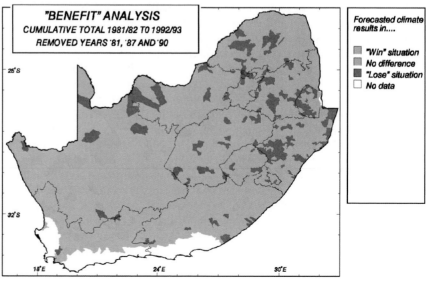

Fig. E.26.
Schematic illustration of the
reduction of uncertainty in a
reservoir operation through
application of forecasting
techniques

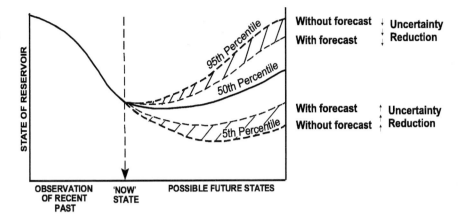

veld under good management. When converted to sediment yields, however, the factor difference becomes 2.5–7.5 times, and even > 7.5 times (Fig. E.24, *bottom*), clearly illustrating how a hazard, in this case stormflow and especially sediment yield, can be modified positively by good grazing management and/or rehabilitation of overgrazed lands.

Example of Vulnerability Modification through Seasonal Forecasting of Runoff

Vulnerability modification is a form of risk mitigation which includes, *inter alia*, assessing the benefits of forecasting streamflows for the rainy season ahead. Statistically derived categorical seasonal rainfall forecasts four months ahead are made for eight regions of South Africa by the South African Weather Service, for three categories, i.e. "above average", "near average" and "below average" seasonal rainfalls. If seasonal rainfall forecasts were a random process, such three-category forecasts would be correct 33% of the time. If seasonal categorical rainfall forecasts are "translated" into seasonal runoff forecasts, these could become very valuable reservoir operations and irrigation application planning tools for water resources managers. Seasonal categorical rainfall forecasts for the eight forecast regions in South Africa were downscaled to daily rainfall values using techniques described in Schulze et al. (1998b) for application with the ACRU modelling system to over 1 500 Quaternary catchments in South Africa. A simple benefit analysis of forecasting skill was undertaken, in which a "win" was recorded if, for the historical seasonal rainfall forecast, the simulated seasonal runoff was closer to the runoff simulated with actual historical rainfall than the median seasonal runoff, while a "loss" was recorded when median runoff was closer to the actual than the forecast runoff. "No difference" implies forecasted and median runoffs within 5% of one another. Figure E.25 illustrates that, when excluding three seasons out of 15 for which the rainfall forecast accuracy proved 100%

wrong, i.e. 1981/1982, 1987/1988 and 1990/1991, most of southern Africa scores more "wins" than "losses". The impacts of short term climate perturbations such as the El Niño phenomenon have already been illustrated. This forecast analysis indicates that even at the current level of seasonal forecast accuracy (around 62% if the 3 worst forecasts in 15 are omitted) these can potentially be "translated" into an operational tool for water resources managers which could prove statistically more accurate than the current practice of forecasting based on historical expected, i.e. median, runoffs with wide uncertainty bands, as shown in Fig. E.26.

Are Certain Areas in South Africa Hydrologically More Sensitive than Others to the Individual Forcing Variables of Climate Change?

Figure E.27 illustrates the relative sensitivities on mean annual runoff of ΔT (assumed to be a uniform increase of 2 °C over southern Africa) and ΔP (changed through –10% to +10% of the present). In each case the other two variables are held constant at present levels when running the daily ACRU model. The hydrological system is relatively insensitive to temperature changes that affect evaporation and hence runoff. The increase of 2 °C reduces mean annual runoff over most of summer rainfall areas in South Africa by only 5% (Fig. E.27, *top*). However, in the south-west winter rainfall region, the response to temperature becomes more dramatic, with a 2 °C increase by itself producing a simulated reduction in mean annual runoff in excess of 50%. The reasons for this are that under present climatic conditions evaporation losses there are relatively low from the moist soils in winter, but that with warming, faster drying soils between rainfall events significantly reduce runoff. The most significant sensitivity to climate change, however, remains that due to rainfall, with changes by one unit manifesting themselves as runoff changes by a factor of 2 to 5 (Fig. E.27, *bottom*), with the sensitivity more dominant in the extreme south-west.

Fig. E.27.
Sensitivities to selected atmospheric variables: Temperature (*top*) and rainfall (*bottom*) change impacts on mean annual runoff over South Africa (from Schulze and Perks 2000)

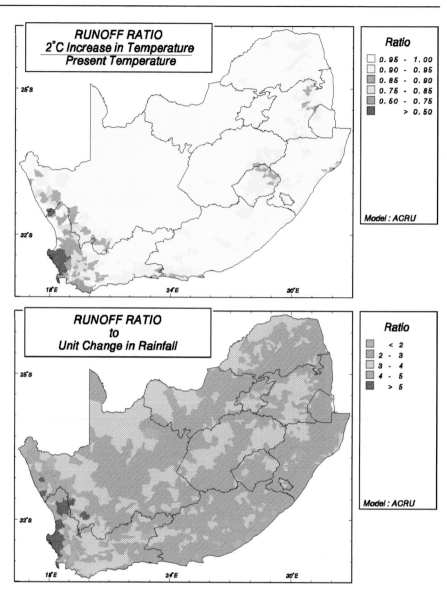

Conclusions

This chapter has provided examples of hydrological risk from South Africa. The examples bring to the fore two overarching issues in hydrological risk management. The first is the question of uncertainty in risk-related hydrological studies – uncertainties regarding meteorological and catchment conditions now and in the future, and uncertainties around input data and uncertainties emanating from the models used in hydrological risk management. The second revolves around the recurring theme of the hydrological system's potential for amplifying perturbations of the climate drivers, predominately of rainfall. It is the amplification and the uncertainty issues which will need to be stressed to practitioners and managers of hydrological risk and vulnerability time and again and to which researchers will need to focus more of their attention.

Acknowledgements

Results shown and discussed in this chapter derive from research funded by the Water Research Commission of South Africa and the US Country Studies for Climate Change Programme. They are thanked for their support, as are the Computing Centre for Water Research, Lucille Perks (climate change), Mark Horan (GIS) and Jason Hallowes (forecasting), all of the University of KwaZulu-Natal.

Chapter E.7

Conclusions

Roger A. Pielke, Sr. · Lelys Bravo de Guenni

We have provided a stark contrast between two approaches to assessing the potential effects of future environmental change. The first approach, the scenario approach, relies on computer model outputs, typically from General Circulation Models, to develop a set of scenarios for the future state of the Earth's environment. This scenario approach usually assumes that weather is the dominant, and sometimes only, forcing factor of concern, and focuses primarily on long-term change at global scales. The vulnerability approach, in contrast, focuses on multiple stresses at local and regional scales (primarily) and more immediate, shorter-term impacts. There are strengths and weaknesses in both approaches but the scenario approach, which is the most widely used, must start with vulnerability assessments. Only once we

know what are the specific threats to our environment do we need to estimate (using scenarios) whether environmental change is likely (or possibly) large enough to cause these threats to be realised, as well as to have a measure of the uncertainty attached to these scenarios. A vulnerability approach will better prepare the public for variable future weather, abrupt atmospheric shifts in any direction, and other urgent and costly environmental stresses. Table E.7 summarises the differences between the two approaches.

Specific examples of the vulnerability assessment procedure, with respect to water resources, were presented in this chapter. Vörösmarty and colleagues showed that the increase in human population in the next few decades is much more likely to threaten the availability of

Table E.7. General characteristics of the scenario and vulnerability approaches as typically used

Approach	Scenario	Vulnerability
Assumed dominant stress	Climate, recent greenhouse gas emissions t o th atmosphere, ocean temperatures, aerosols , etc	Multiple stresses: climate (historical climate variability), land use and water use, altered disturbance regimes, invasive species, contaminants/pollutants, habitat loss, etc.
Usual timeframe of concern	Long-term, doubled CO_2, 30 to 100 years in the future.	Short-term (0 to 30 years) and long-term research.
Usual scale of concern	Global, sometimes regional. Local scale needs downscaling techniques. However, there is little evidence to suggest that present models provide realistic, accurate, or precise climate scenarios at local or regional scales.	Local, regional, national and global scales.
Major parameters of concern	Spatially averaged changes in mean temperatures and precipitation in fairly large grid cells with some regional scenarios for drought.	Potential extreme values in multiple parameters (temperature, precipitation, frost-free days) and additional focus on extreme events (floods, fires, droughts, etc.); measures of uncertainty.
Major limitations for developing coping strategies	Focus on single stress limits preparedness for other stresses. Results often show gradual ramping of climate change-limiting preparedness for extreme events. Results represent only a limited subset of all likely future outcomes – usually unidirectional trends. Results are accepted by many scientists, the media, and the public as actual "predictions". Lost in the translation of results is that all models of the distant future have unstated (presently unknowable) levels of certainty or probability.	Approach requires detailed data on multiple stresses and their interactions at local, regional, national and global scales – and many areas lack adequate information. Emphasis on short-term issues may limit preparedness for abrupt "threshold" changes in climate some time in the short- or long-term. Requires preparedness for a far greater variation of possible futures, including abrupt changes in any direction – this is probably more realistic, yet difficult.

potable water than any of the climate change scenarios created by the general circulation models. Brad Bass and colleagues, and Changming Liu, documented the similar importance (and threat) associated with increasing population pressures in China. Roland Schulze shows hydrological risks exemplified for South Africa with special consideration of aridity and difficulties in management practices due to uncertainties in the prediction of seasonal weather such as ENSO events. Tom Stohlgren and colleagues document the range of threats to a pristine park area, in which invasive exotic plant species and air pollution are major threats. Other integrated assessments are also taking place (Becker et al. 1999; Stohlgren 1999b, and others). As shown in this chapter, the vulnerability assessment procedure is a more inclusive evaluation method than relying on scenarios at the start of an impact assessment. The scenario approach is only required once specific environmental threats are recognised. Once we identify environmental risk, the scenario approach can be used to determine if there is a possibility that the threat could be realised and, if possible, quantify the probability of the risk. Hutchinson shows how the strong topographic coherence of weather can be used to reduce some of the uncertainty in constructing fine scale scenarios and assessing vulnerability. Even without quantification and the ability to predict the future, however, the vulnerability perspective provides useful information for policy-makers (Sarewitz et al. 2000).

References Part E

Adams RM, Crocker TD, Katz RW (1984) Assessing the adequacy of natural science information: a bayesian approach. Rev Econ Stat LXVI:568–575

Aguirre AA, Starkey EE (1994) Wildlife disease in US national parks: historical and coevolutionary perspectives. Conserv Biol 8:654–661

Alcamo J (ed)(1994) IMAGE 2.0: integrated modeling of global climate change. Kluwer Academic Publishers, Dordrecht

Alcamo J, Kreilemans E (1996) Emission scenarios and global climate protection. Global Environ Chang 6:305–334

Alcamo J et al. (1998) Baseline scenarios of global environmental change. In: Alcamo J, Leemans R, Kreileman E (eds) Global change scenarios of the 21st century. Results from the IMAGE 2.1 model. Elseviers Science, London, pp 97–139

Alcamo J, Henrichs T, Rösch T (2000) World water in 2025: global modelling and scenario analysis for the world commision on water for the 21st century. University of Kassel, Kassel

Alexander RR (1985) Major habitat types, community types and plant communities in the Rocky Mountains. USDA Forest Service, Rocky Mountain Forest and Range Experiment Station, Fort Collins, Co

Alexandratos N (ed) (1995) Agriculture: towards 2010. John Wiley & Sons, Chichester

Alley RB, Meese DA, Shuman AJ, Gow AJ, Taylor KC, Grootes PM, White JWC, Ram M, Waddington ED, Mayewski PA, Zielinski GA (1993) Abrupt accumulation increase in the Younger-Dryas termination in the GISP2 ice core. Nature 362:527–529

Anthes RA, Warner TT (1978) Development of hydrodynamic models suitable for air pollution and other mesometeorological studies. Mon Weather Rev 106:1045–1078

Armstrong DM (1993) Lions, ferrets and bears. A guide to the mammals of Colorado. Colorado Division of Wildlife and University of Colorado Museum, Denver, Co

Arnell NW (1999) Climate change and global water resources. Global Environ Chang 9:31–49

Aubin J-P, Cellina A (1984) Differential inclusions. Springer, Berlin

Bankes SC (1993) Exploratory modeling for policy analysis. Oper Res 41:435–449

Bantelmann A, Hoffmann D, Menke B (1984) Veränderungen des Küstenverlaufs, Ursachen und Auswirkungen. In: Kossack G, Behre KE, Schmid P (eds) Archäologische und naturwissenschaftliche Untersuchungen an ländlichen und frühstädtischen Siedlungen im deutschen Küstengebiet vom 5. Jahrhundert v. Chr. bis zum 11. Jahrhundert n. Chr., Band 1: Ländliche Siedlungen. Deutsche Forschungsgemeinschaft, Weinheim, pp 51–68

Bárdossy A, Plate EJ (1992) Space-time model for daily rainfall using atmospheric circulation patterns. Water Resour Res 28:1247–1259

Baron JS, Ojima DS, Holland EA, Parton WJ (1994) Analysis of nitrogen saturation potential in Rocky Mountain tundra and forest: implications for aquatic systems. Biogeochemistry 27:61–82

Baron JS, Rueth HM, Wolfe AP, Nydick K, Allstott EJ, Minear JT, Moraska B (2000) Ecosystem responses to nitrogen deposition in the Colorado Front Range. Ecosystems 3:352–368

Barrow CJ (1991) Land degradation. Cambridge University Press, Cambridge

Bass B, Huang G, Russo J (1997) Incorporating climate change into risk assessment using grey mathematical programming. J Environ Manage 49:107–123

Becker A, Wenzel V, Krysanova V, Lahmer W (1999) Regional analysis of global change impacts: concepts, tools and first results. Environ Model Assess 4:243–257

Behnke RJ (1992) Native trout of Western North America. American Fisheries Society, Bethesda, Maryland

Bellone E, Hughes JP, Guttorp P (2000) A hidden Markov model for relating synoptic atmospheric patterns to precipitation amounts. Climate Res 15:1–12

Bernabo IC, Webb TI (1977) Changing patterns in the Holocene pollen record of northeastern North America: a mapped summary. Quaternary Res 8:64–96

Blaikie P, Cannon T, Davis I, Wisner B (1994) At risk: natural hazards, people's vulnerability and disasters. Routledge, London

Bloh W von, Block A, Schellnhuber H-J (1997) Self-stabilization of biosphere under global change: a tutorial geophysiological approach. Tellus 49B:249–262

Bloh W von, Block A, Parade M, Schellnhuber H-J (1999) Tutorial modeling of geosphere-biosphere interactions; the effect of percolation-type habitat fragmentation. Physica A 266:186–196

Bonan GB (1997) Effects of land use on the climate of the United States. Climatic Change 37:449–486

Bounoua L, GJ Collatz, Los SO, Sellers PJ, Dazlich DA, Tucker CJ, Randall DA (2000) Sensitivity of climate to changes in NDVI. J Climate 13:2277–2292

Braswell BH, Schimel DS, Linder E, Moore III B (1997) The response of global terrestrial ecosystems to interannual temperature variability. Science 278:870–872

Brovkin V, Ganopolski A, Claussen M, Kubatzki C, Petoukhov V (1999) Modelling climate response to historical land-cover change. Global Ecol Biogeogr 8:509–517

Brown RA, Rosenberg NJ (1999) Climate change impacts on the potential productivity of corn and winter wheat in their primary United States growing regions. Climatic Change 41:73–107

Buchholtz CW (1983) Rocky Mountain National Park: a history. Colorado Associated University Press, Boulder, Co

Bury RB, Corn PS, Dodd CK, McDiarmid RW Jr (1995) Amphibians. In: LaRoe ET (ed) Our living resources. National Biological Service, Washington DC

Canadell P (2000) Non-linear responses and surprises: a new Earth system science initiative. Global Change NewsLetter 43:1–2

Casella G, George EI (1992) Explaining the Gibbs sampler. Am Stat 46:167–174

Chadde SW, Kay CE (1991) Tall-willow communities on Yellowstone's northern range: a test of the natural regulation paradigm. In: Keiter RB, Boyce MS (eds) Greater Yellowstone ecosystem: redifining America's wilderness heritage. Yale University Press, New Haven, Conn., pp 231–262

Changnon SA (1998) Comments on "Secular trends of precipitation amount, frequency, and intensity in the United States." B Am Meteorol Soc 79:2550–2552

Chase TN, Pielke RA, Kittel TGF, Nemani R, Running SW (1996) The sensitivity of a general circulation model to global changes in leaf area index. J Geophys Res-Atmos 101:7393–7408

Chen J et al. (1998) Yellow River harnessing and water resources development. Yellow River Conservancy Press, 44 p

CIESIN (Center for International Earth Science Information Network) (2000) Gridded population of the world: provisional release of updated database of 1990 and 1995 estimates. Columbia University's Center for International Earth Science Information Network

Claussen M (2001) Earth system models. In: Ehlers EKT (ed) Understanding the Earth system: compartments, processes and interactions. Springer-Verlag, Heidelberg, pp 145–162

Claussen M, Gayler V (1997) The greening of Sahara during the mid-Holocene: results of an interactive atmosphere-biome model. Global Ecol Biogeogr 6:369–377

Claussen M, Brovkin V, Ganopolski A, Kubatzki C, Petoukhov V (1998) Modeling global terrestrial vegetation-climate interaction. Philos T Roy Soc B 353:53–63

Claussen M, Brovkin V, Petoukhov V, Ganopolski A (2001) Biogeophysical versus biogeochemical feedbacks of large-scale land-cover change. Geophys Res Lett 28:1011–1014

Cohen J (1996) How many people can earth support? WW Norton, New York

Colorado Division of Wildlife (1986) Wildlife in danger: the status of Colorado's threatened or endangered fish, amphibians, birds, and mammals, Fort Collins, Colo

Conway D, Jones PD (1998) The use of weather types and air flow indices for GCM downscaling. J Hydrol 213:348–361

Corn PS, Fogelman JC (1984) Extinction of montane populations of the northern leopard frog (Rana pipiens) in Colorado. J Herpetol 18:147–152

Cubasch U, Santer BD, Hellbach A, Hegerl G, Höck H, Maier-Reimer E, Mikolajewicz U, Stössel A, Voss R (1994) Monte Carlo climate change forecasts with a global coupled ocean-atmosphere model. Clim Dynam 10:1–19

Cubasch U, Storch H von, Waszkewitz J, Zorita E (1996) Estimates of climate changes in southern Europe using differnt downscaling techniques. Climate Res 7:129–149

de Vries HJM, Bollen J, Bouwman L, den Elzen MGJ, Janssen M, Kreilemann E (2000) Greenhouse-gas emissions in equity, environment- and service-oriented world: an IMAGE-based scenario for the next century. Technol Forecast Soc 63:137–174

deKoning GHJ, Veldkamp A, Fresco LO (1998) Land use in Ecuador: a statistical analysis at different aggregation levels. Agr Ecosyst Environ 70:231–247

Döll P, Siebert S (2002) Global modeling of irrigation water requirements. Water Resources Research 38(4):10.1029/2001 WR000355,2002

Döös BR (1994) Environmental degradation, global food production, and risk for large-scale migrations. Ambio 23:124–130

Dowlatabadi H (1999) Integrated assessment: implications of uncertainty. In: Encylopedia of Life Support Systems. Oxford University Press, Oxford

Dowlatabadi H, Morgan G (1995) Integrated assessment climate assement model 2.0, technical documentation. Department of Engineering and Public Policy, Carnegie Mellon University, Philadelphia

Downing TE, Olsthoorn AJ, Tol RSJ (eds) (1999) Climate change and risk. Routledge, London

Dynesius M, Nilsson C (1994) Fragmentation and flow regulation of river systems in the northern third of the world. Science 266:753–762

Dzhamalov RG, Zekster IS (eds) (1999) World map of hydrogeological conditions and groundwater flow and digital maps of hydrogeological conditions and groundwater flow. HydroScience Press

Elias SA, Short SK, Clark PU (1986) Paleoenvironment interpretations of the late Holocene, Rocky Mountain National Park, Colorado, USA. Review De Paleobiologie 5:127–142

Fairman R, Mead CD, Williams WR (1998) Environmental risk assessment: approaches, experiences and information sources. European Environment Agency, Copenhagen, 252 p

Falkenmark M (1998) Dilemma when entering the 21st century – rapid change but lack of a sense of urgency. Water Policy 1: 421–436

Fall PL (1988) Vegetation dynamics in the southern Rocky Mountains: Late Pleistocene and Holocene timber-line fluctuations. University of Arizona, 303 p

FAO (1998) FAO statistics on diskette. Food and Agriculture Organization of the United Nations, Rome

FAO (1999) FAOSTAT database collections, vol. 1999. Food and Agriculture Organization of the United Nations

Fekete BM, Vörösmarty CJ, Grabs W (1999) Global, composite runoff fields based on observed river discharge and simulated water balance. Global Runoff Data Centre, Koblenz

Fekete BM, Vörösmarty CJ, Grabs W (2002) High resolution fields of global runoff combining observed river discharge and simulated water balances. Global Biogeochem Cy 16(3):10.1029/ 1999GB001254,2002

Flather CH, Joyce LA, Bloomgarten CA (1984) Species endangerment patterns in the United States. USDA Forest Service, Rocky Mountain Forest and Range Experiment Station, Fort Collins, Co, 42 p

Flint EP (1994) Changes in land use in south and South-east Asia from 1880 to 1980: a data base prepared as part of a coordinated research program on carbon fluxes in the tropics. Chemosphere 29:1015–1062

Foster DR, Motzkin G, Slater B (1998) Land-use history as long-term broad-scale disturbance: regional forest dynamics in central New England. Ecosystems 1:96–119

Frenzel B (1994) Moore als Umweltindikatoren in klimatischen Grenzgebieten. Hohenheimer Umwelttagung 26:155–167

Frolking S, Xiao X, Zhuang Y, Salas W, Li C (1999) Agricultural land-use in China: a comparison of area estimates from ground-based census and satellite-borne remote sensing. Global Ecol Biogeogr 8:407–416

Ganopolski A, Rahmstorf S (2001) Rapid changes of glacial climate simulated in a coupled climate model. Nature 409:153–158

Ghil M, Childress S (1987) Topics in geophysical fluid dynamics: atmospheric dynamics, dynamo theory, and climate dynamics. Springer-Verlag, Heidelberg

Giorgi F, Marinucci MR, Visconti G (1992) A 2 × CO_2 climate change scenario over Europe generated using a limited area model nested in a general circulation model. 2. Climate change scenario. J Geophys Res 97:10011–10028

Glantz MH (2001) Currents of change: impacts of El Niño and La Niña on climate and society, 2nd edn. Cambridge University Press, Cambridge, New York

Gleick P (2000) Water futures: a review of global water resources projections. In: Rijsberman FR (ed) World water scenarios: analyses. Earthscan Publications, London, pp 27–37

Gliemeroth AK (1995) Paläoökologische Untersuchungen über die letzten 22 000 Jahre in Europa. S. Fischer, Stuttgart

Godwin H (1952) Recurrence-surfaces. Danmarks Geol Unders, II. raekke, nr. 80, pp 22–30

Gossow H (1976) Wildökologie. Begriffe, Methoden, Ergebnisse, Konsequenzen. BLV, München

Graumlich LJ (1993) A 1 000-year old record of temperature and precipitation in the Sierra Nevada. Quaternary Res 39:249–255

Greenback Cutthroat Trout Recovery Team (1983) Greenback cutthroat trout recovery plan revision. US Fish and Wildlife Service, Colorado Division of Wildlife, Fort Collins, Colo

Groots PM, Stuiver M, White JWC, Johnson SJ, Jouzel J (1993) Comparison of oxygen isotope records from the GISP2 and GRIP Greenland ice cores. Nature 366:552–554

Guenni L et al. (2001) A biogeophysical approach to the vulnerability concept: extreme rainfall and population distribution in South America during 1999. Global Change Open Science Conference, Amsterdam, The Netherlands, (poster paper)

Hall CAS, Tian H, Qi Y, Pontius G, Cornell J (1995) Modelling spatial and temporal patterns of tropical land-use change. J Biogeogr 22:753–757

Henry D, Henry M (1991) Greenbacks are back. Colorado Outdoors 40:23–25

Hilderink HBM et al. (1999) TARGETS – The CD-Rom

Hinrichsen D, Robey B, Upadhyay UD (1998) Solutions for a water-short world. Johns Hopkins University School of Public Health, Baltimore, Md

Hoekstra AY (1996) AQUA: a framework for integrated water policy analysis. Dutch National Institute of Public Health and the Environment (RIVM), Bilthoven, The Netherlands, 70 p

Hoekstra AY (1998) Perspectives on water. An integrated model-based exploration of the future. Technical University, Delft

Hoffman MT, Todd SW, Ntshona ZM, Turner SD (1999) Land degradation in South Africa. Pretoria, RSA, 245 p

Hubbard KG, Flores-Mendoza FJ (1995) Relating United States crop land use to natural resources and climate change. J Climate 8: 229–335

Hughes MK, Graumlich LJ (1996) Multimillenial dendroclimate records from the western United States. In: Bradley RS, Jones PD, Jouzel J (eds) Climatic variation and forcing mechanisms of the last 2000 years. Springer-Verlag, New York, pp 109–124

Hulme M, Carter TR (1999) Representing uncertainty in climate change scenarios and impact studies. ECLAT-2:12–54

Huntley B, Birks HJB (1983) An atlas of past and present pollen maps of Europe, 0–13 000 years ago. Cambridge University Press, Cambridge

IPCC (Intergovernmental Panel on Climate Change) (1996) Climate change 1995, the science of climate change. Contribution of Working Group I of the Second Assessment Report of the IPCC. Cambridge University Press, Cambridge

IPCC (Intergovernmental Panel on Climate Change) (2001) Summary for policymakers. A Report of Working Group I of the IPCC

Jamieson D, Herrick C (1995) The social construction of acid rain: some implications for science/policy assessment. Global Environ Chang 5:105–112

Jeton EE, Dettinger MD, Smith JL (1999) Potential effects of climate change on streamflow, eastern and western slopes of the Sierra Nevada, California and Nevada. USGS Water Resource Investigations, 44 p

Jones RG, Murphy JM, Noguer M (1995) Simulation of climate change over Europe using a nested regional climate model I. Assessment of control climate, including sensitivity to location of lateral boundaries. Q J Roy Meteor Soc 121:1413–1449

Karl TR, Knight RW (1998a) Secular trend of precipitation amount, frequency, and intensity in the United States. B Am Meteorol Soc 79:231–241

Karl TR, Knight RW (1998b) Reply. B Am Meteorol Soc 79:2552–2554

Karl TR, Knight RW, Plummer N (1995) Trends in high-frequency climate variability in the twentieth century. Nature 377:217–220

Katz RW (1999) Techniques for estimating uncertainty in climate change scenarios and impact studies. ECLAT-2:40–55

Kite GW (1993) Application of a land class hydrological model to climate change. Water Resour Res 29:2377–2384

Klein Goldewijk CGM, Battjes JJ (1997) A hundred year (1890–1990) database for integrated environmental assessments. National Institute of Public Health and the Environment, Bilthoven

Knight DH (1994) Mountains and plains: the ecology of Wyoming landscapes. Yale University, Thomson-Shore, Dexter, Mich

Kolstad CD, Kelly D (1999) Bayesian Learning, pollution, and growth. J Econ Dyn Control 23:491–518

Koruba V, Jabbar MA, Akinwumi JA (1996) Crop-livestock competition in the West African derived savannah: application of a multi-objective programming model. Agr Syst 52

Kossack G, Behre K-E, Schmid P (1984) Archäologische und naturwissenschaftliche Untersuchungen an ländlichen und frühstädtischen Siedlungen im deutschen Küstengebiet vom 5. Jahrhundert v. Chr. bis zum 11. Jahrhundert n. Chr. Deutsche Forschungsgemeinschaft, Weinheim

Kourzoun VI, Sokolov AA, Budyko MI, Voskresensky KP, Kalinin GP, Konoplyantsev AA, Korotkevich ES, L'vovich MI (1978) Atlas of the world water balance. UNESCO, Paris

Krzysztofowicz R (1999) Bayesian theory of probalistic forecasting via deterministic hydrologic model. Water Resour Res 35: 2739–2750

Kunkel KE, Pielke RA Jr, Changnon SA (1999) Temporal fluctuations in weather and climate extremes that cause economic and human health impacts: a review. B Am Meteorol Soc 80(6):1077–1098

Landsea CW, Knaff JA (2000) How much skill was there in forecasting the very strong 1997–1998 El Niño? B Am Meteorol Soc 81: 2107–2119

Lanfear KJ, Hirsch RM (2000) USGS study reveals a decline in long-record streamgauges. EOS Trans Am Geophys Union 80

Langner LL, Flather CH (1994) Biological diversity: status and trends in the United States. Rocky Mountain Forest and Range Experiment Station, Fort Collins, Colo

Lavender D (1975) The Rockies. Harper and Row, New York

Lee X, Black TA (1993) Atmospheric turbulence within and above a Douglas-fir stand. Part II: Eddy fluxes of sensible heat and water vapour. Bound-Lay Meteorol 64:369–389

Leemans R (1999) Modelling for species and habitats: new opportunities for problem solving. Sci Total Environ 240:51–73

Leemans R, Agrawala S, Edmonds JA, MacCracken MC, Moss RM, Ramakrishnan PS (1996) Mitigation: cross-sectoral and other issues. In: Watson RT, Zinyowera MC, Moss RH (eds) Climate change 1995. Impacts, adaptations and mitigation of climate change: scientific-technical analysis. Cambridge University Press, Cambridge, pp 799–797

Lins H, Slack JR (1999) Streamflow trends in the United States. Geophys Res Lett 26:227–230

Liu C (1981) The influence of forest cover upon annual runoff in the loess plateau of China. In: Ma LJC, Noble AG (eds) Water transfer: its environment: Chinese and American views. Methuen, New York, pp 131–142

Liu C, Li C (2000) Analysis on runoff series with special reference to drying up courses. Acta Geogr Sinica 55:257–265

Liu C, Liang J (1989) Some problems of flooding and its prevention in the lower Yellow River. In: Liu C, Liang J (eds) Taming the Yellow River: silt and floods. Kluwer Academic Publishers, Dordrecht, pp 223–242

Liu KB, Sun SC, Jiang XH (1992) Environmental change in the Yangtze River delta since 12000 years B.P. Quaternary Res 38:32–45

Lorenz EN (1968) Climatic determinism. In: Causes of climatic change, vol 30, 8. AMS, pp 1–3

Lorenz EN (1975) Climate predictability. In: Committee W-IJO (ed) The physical basis of climate and climate modelling, vol 16. GARÜ Publication Series

Lorenz EN (1979) Forced and free variations of weather and climate. J Atmos Sci 36:1367–1376

Loveland TR, Belward AS (1997) The IGBP-DIS global 1 km land-cover dataset, DISCover: first results. Int J Remote Sens 18:3291–3295

Lüdeke MKB, Moldenhauer O, Petschel-Held G (1999) Rural poverty driven soil degradation of climate change: the sensitivity of the disposition towards the Sahel syndrome with respect to climate. Environ Model Assess 4:315–326

L'vovich MI, White GF (1990) Use and transformation of terrestrial water systems. In: Turner BL, Clark WC, Kates RW, Richards JF, Mathews JT, Meyer WB (eds) The Earth as transformed by human action. Cambridge University Press, Cambridge, pp 235–252

Lynch AH, McIlwaine S, Beringer J, Bonan GB (2001) An investigation of the sensitivity of a land-surface model to climate change using a reduced form model. Clim Dynam 17:643–652

MacCracken MC (2002) Do the uncertainty ranges in the IPCC and US national assessments account adequately for possibly overlooked climatic influences? Climatic Change 52:13–23

Mack RN, Simberloff D, Lonsdale WM, Evans H, Clout M, Bazzaz FA (2000) Biotic invasions: causes, epidemiology, global consequences, and control. Ecol Appl 10:689–710

Matsuoka Y, Kainuma M, Morita T (1994) Scenario analysis of global warming using the Asian-Pacific Integrated Model (AIM). Energ Policy 23:1–15

Matthews E, Rotmans J, Ruffing K, Waller-Hunter J, Zhu J (1997) Critical trends: global change and sustainable development. UN Publication Department for Policy Coordination and Sustainable Development, New York

McCabe G, Wolock DM Jr (1997) Climate change and the detection of trends in runoff. Climate Res 8:129–134

Mearns LO, Rosenzweig C, Goldberg R (1997) Mean and variance change in climate scenarios: methods, agricultural applications, and measures of uncertainty. Climatic Change 35:367–396

Meszaros E, Palvolgyi T (1990) Daisyworld with an atmosphere. Idojaras 94:339–345

Mills JD (1991) Wyoming's Jackson Hole dam, horizontal channel stability, and floodplain vegetation dynamics. University of Wyoming, Laramie, 54 p

Morozova TD (1981) Razvitie pocvennogo pokrova Evropy v pozdnem plejstocene. Nauka, Moskva (in Russian: The development of European soils during the Late Pleistocene)

Moxey AP, White B, O'Callaghan JR (1995) CAP reform: an application of the NELUP economic model. J Environ Plan Manage 38: 117–123

Nams VO, Folkard NFG, Smith JNM (1993) Effects of nitrogen fertilization on several woody and non-woody boreal forest species. Can J Bot 71:93–97

Nativ R (1992) Recharge into southern high plains aquifer – possible mechanisms, unresolved questions. Environ Geol Water S 19(1):21–32

Nevison C, Gupta VK, Klinger L (1999) Self-sustained oscillations on Daisyworld. Tellus 51B:806–814

Nichols H (1982) Review of late Quaternary history of vegetation and climate in the mountains of Colorado. In: Halfpenny JC (ed) Ecological studies in the Colorado alpine: a festschrift for John W. Marr. Inst. Arctic and Alpine Res., occasional paper 37. University of Colorado, Boulder, Co, pp 27–33

Nordstrom K, Gupta V, Chase T (2004) Salvaging the Daisyworld parable under the dynamic area fraction framework. Scientist on Gaia-II. Cambridge, Ma, MIT Press (in press)

Olson JS (1994a) Global ecosystem framework-definitions. USGS EROS Data Center, Sioux Falls, SD

Olson JS (1994b) Global ecosystem framework-translation strategy. USGS EROS Data Center, Sioux Falls, SD

Ongley ED (2000) The Yellow River: managing the unmanageable. Water Int 25:227–231

Oreskes N, Shrader-Frechette K, Belitz K (1994) Verification, validation, and confirmation of numerical models in the earth sciences. Science 263:641–646

Pachur HJ, Hoelzmann P (1991) Paleoclimatic implications of Late Quaternary lacustrine sediments in Western Nubia, Sudan. Quaternary Res 36:257–276

Paté-Cornell E (1996) Uncertainties in global climate change estimates. Climatic Change 33:145–149

Peer T (2000) The highland steppes of the Hindukush Range as indicators of centuries old pasture farming. Marburger Geogr Schriften 135:312–325

Penning de Vries FWT, Rabbinge R, Groot JJR (1997) Potential and attainable food production and food security in different regions. Philos T Roy Soc B 352:917–928

Peters EF, Bunting SC (1994) Fire conditions pre- and post-occurrence of annual grasses on the Snake River plain. In: Monsen SB, Kitchen SG (eds) Proceedings – ecology and management of annual rangelands. USDA Forest Service Inter. Forest and Range Experiment Station, Ogden, Utah, pp 31–36

Peters RL, Lovejoy TE (eds) (1992) Global warming and biological diversity. Yale University Press, New Haven

Petschel-Held G, Bruckner T, Schellnhuber H-J, Tóth F, Hasselmann K (1999a) The tolerable windows approach. Theoretical and methodological foundations. Climatic Change 41:303–331

Petschel-Held G, Block A, Cassel-Gintz M, Kropp J, Lüdeke MKB, Moldenhauer O, Reusswig F, Schellnhuber H-J (1999b) Syndromes of global change: a qualitative modelling approach to assist global environmental management. Environ Model Assess 4:295–314

Pielke RA Sr (1998a) Climate predictions as an initial value problem. B Am Meteorol Soc 79:2743–2746

Pielke RA Jr (1998b) Rethinking the role of adaptation in climate policy. Global Environ Chang 8:159–170

Pielke RA Sr (2001) Earth system modeling – an integrated assessment tool for environmental studies. In: Matsuno T, Kida H (eds) Present and future of modeling global environmental change: toward integrated modeling. Terrapub, pp 311–337

Pielke RA Sr (2002) Overlooked issues in the US National Climate and IPCC assessments. Climatic Change 52:1–11

Pielke RA Jr, Downton MW (2000) Precipitation and damaging floods: trends in the United States, 1932–1997. J Climate 13:3625–3637

Pielke RA Sr, Guenni L (1999) Vulnerability assessment of water resources to changing environmental conditions. IGBP Global Change NewsLetter 39:22–24

Pielke RA Sr, Uliasz M (1998) Use of meteorological models as input to regional and mesoscale air quality models – limitations and strengths. Atmos Environ 32:1455–1466

Pielke RA Jr, Sarewitz D, Byerly R, Jamieson D (1999) Prediction in the earth sciences: Use and misuse in policy making. EOS Trans Am Geophys Union 80:309ff

Pimentel D, Lach L, Zuniga R, Morrison D (1999) Environmental and economic costs associated with non-indigenous species in the United States. College of Agriculture and Life Sciences, Cornell University, Ithaca, New York

Pitman A, Pielke RA Sr, Avissar R, Claussen M, Gash J, Dolman H (1999) The role of the land surface in weather and climate: does the land surface matter. IGPB Global Change NewsLetter 39:4–11

Plambeck EL, Hope C (1996) PAGE95: an updated valuation of the impacts of global warming. Energ Policy 24:783–793

Plate EJ (1996) Risk management for hydraulic systems under hydrological loads. In: 3rd IHP/IAHS Kovacs Colloquium on "Risk, Reliability, Uncertainty and Robustness of Water Resources Systems". UNESCO, Paris, 18 p

Postel SL, Daily GC, Ehrlich PR (1996) Human appropriation of renewable fresh water. Science 271:785–788

Rabbinge R, van Oijen M (1997) Scenario studies for future agriculture and crop protection. Eur J Plant Pathol 103:197–201

Ramankutty N, Foley JA (1998) Estimating historical changes in global land cover: Croplands from 1700 to 1992. Global Biogeochem Cy 13:997–1027

Rayner S, Malone EL (1997) Zen and the art of climate maintenance. Nature 390:332–334

Riebsame W (1997) Atlas of the New West. W.W. Norton, New York

Rijsberman FRe (2000) World water scenarios: analyses. Earthscan Publications, London

Risbey J, Stone P (1996) A case study of the adequacy of GCM simulations for input to regional climate change. J Climate 9:1441–1448

Robertus AJ, Burns BR, Veblen TT (1991) Stand dynamics of Pinus flexilis-dominated subalpine forests in the Colorado Front Range. J Veg Sci 2:445–458

Rocky Mountain National Park Resources Management Plan (1999) US National Park Service, Estes Park, Colorado, USA

Rodda JC (1998) Hydrological networks need improving! In: Water: a looming crisis. UNESCO, International Hydrological Programme, Paris

Rosegrant MW (1997) Water resources in the twenty-first century: challenges and implications for actions. International Food Policy Research Institute, Washington DC

Rosegrant MW, Agcaoili-Saombilla M, Perez ND (1995) Global food projections to 2020: implications for investment. International Food Policy Research Institute (IFPRI), Washington DC

Rosentreter R (1994) Displacement of rare plants by exotic grasses. In: Monsen SB, Kitchen SG (eds) Proceedings – ecology and management of annual rangelands. USDA Forest Service Inter. Forest and Range Experiment Station, Ogden, Utah, pp 170–175

Rosenzweig C, Hillel D (1998) Climate change and the global harvest. Oxford University Press, Oxford

Rosenzweig C, Phillip I, Goldberg R, Carroll I, Hodges T (1996) Potential impacts of climate change on citrus and potato production in the US. Agr Syst 52:455–479

Rotmans J, de Vries B (1997) Perspectives on global change: The TARGETS approach. Cambridge University Press, Cambridge

Sala OE, Chapin FS III, Armesto JJ, Berlow E, Bloomfield J, Dirzo R, Huber-Sanwald E, Huenneke LF, Jackson RB, Kinzig A, Leemans R, Lodge DM, Mooney HA, Oesterheld M, Poff NL, Sykes MT, Walker BH, Walker M, Wall DH (2000) Global biodiversity scenarios for the year 2100. Science 287:1770–1774

Samjaa R, Zöphel U, Peterson J (2000) The impact of the vole Microtus brandti on Mongolian steppe ecosystems. Marburger Geogr Schriften 135:346–360

Sansó B, Guenni L (1999) A stochastic model for tropical rainfall at a single location. J Hydrol 214:64–73

Sarewitz D, Pielke RA Jr (1999) Prediction in science and policy. Technol Soc 21:121–133

Sarewitz D, Pielke RA Jr, Byerly R (2000) Prediction, science, decision making and the future of nature. Island Press, Washington

Schaller GB (1998) Wildlife of the Tibetan steppe. University of Chicago Press, Chicago

Schellnhuber H-J (1998) Earth system management: the scope of the challenge. In: Schellnhuber HJ, Wenzel V (eds) Earth system analysis. Springer-Verlag, Berlin, pp 1–209

Schellnhuber H-J, Yohe G (1998) Comprehending the economic and social dimensions of climate change by integrated assessment. WCRP-Conference: Achievements, Benefits and Challenges, Geneva, 179 p

Schellnhuber H-J, Block A, Cassel-Gintz M, Kropp J, Lammel G, Lass W, Lienenkamp R, Loose C, Lüdeke M, Moldenhauer O, Petschel-Held G, Plöchl M, Reusswig F (1997) Syndromes of global change. GAIA 60:19–34

Schulze RE (1989) Non-stationary catchment responses and other problems in determining flood series: a case for a simulation modelling approach. In: 4th South African National Hydrological Symposium. Water Research Commission, Pretoria, RSA, pp 135–157

Schulze RE (1995) Hydrology and agrohydrology: a text to accompany the ACRU 3.00 Agrohydrological Modelling System. Water Research Commission, Pretoria, South Africa, 552 p

Schulze RE (1997a) South African atlas of agrohydrology and climatology. Water Research Commission, Pretoria, South Africa, 273 p

Schulze RE (1997b) Impacts of global climate change in hydrologically vulnerable region: challenges to South African hydrologists. Prog Phys Geog 21:113–136

Schulze RE (2000) Modelling hydrological responses to land use and climate change: a southern African perspective. Ambio 29:12–22

Schulze RE (2001) Risk, uncertainty and risk management in hydrology: a conceptual framework and a South African perspective. School of Bioresources Engineering and Environmental Hydrology, University of Natal, Pietermaritzburg, RSA., 52 p

Schulze RE, Perks LA (2000) Assessment of the impact of climate change on hydrology and water resources in South Africa. School of Bioresources Engineering and Environmental Hydrology, University of Natal, Pietermaritzburg, RSA, 118 p

Schulze RE, Hallowes JS, Lynch SD, Perks LA, Horan MJC (1998b) Forecasting seasonal runoff in South Africa: A preliminary investigation. Witwatersrand University, Johannesburg, RSA., p 19

Schuster PF, White DE, Naftz DL, DeWayne L (2000) Chronological refinement of an ice core record at upper Fremont Glacier in south central North America. J Geophys Res 105:4657–4666

Seckler D (1999) Revisiting the IWMI paradigm: increasing the efficiency and productivity of water use. Water Brief No. 2, IWWI

Seckler D, Barker R, Anarasunghe U (1999) Water scarcity in the twenty-first century. Int J Water Res Dev 15:29–42

Shackely S, Young P, Parkinson S, Wynne B (1998) Uncertainty, complexity and concepts of good science in climate change modelling: are GCMs the best tools? Climatic Change 38:159–205

Shiklomanov IA (1996) Assessment of water resources and water availability in the world. Stockholm Environment Institute, St. Petersburg, State Hydrological Institute

Shiklomanov I (2000) Water resouces and water use: present assessment and outlook for 2025. In: Rijsberman FR (ed) World water scenarios: analyses. Earthscan Publications, London, pp 160–203

Shukla J (1998) Predictability in the midst of chaos: a scientific basis for climate forecasting. Science 282:728–731

Singer F (1995) Bighorn sheep in the Rocky Mountain National Parks. In: La Roe Tea (ed) Our living resources: a report to the nation on the distribution, abundance, and health of US plants, animals, and ecosystems. USDI National Biological Service, Washington DC, pp 332–336

Sivillo JK, Ahlquist JE, Toth Z (1997) An ensemble forecasting primer. Weather Forecast 12:809–818

Skole D, Tucker C (1993) Tropical deforestation and habitat fragmentation in the Amazon: satellite data from 1978 to 1988. Science 260:1905–1910

Slack JR, Landwehr JM (1992) Hydro-climatic data network: a US Geological Survey streamflow dataset for the United States for the study of climate variations, 1974–1988. United States Geological Survey, Reston, VA 20192, 193 p

Smith K (1996) Environmental hazards. Routledge, London

Stahle DW, Cook ER, Cleaveland MK, Therrell MD, Meko DM, Grissino-Mayer HD, Watson E, Luckman BH (2000) Tree-ring data document 16th century megadrought over North America. EOS Trans Am Geophys Union 81:121–125

Stohlgren TJ (1999a) The Rocky Mountains. In: Mac MJ, Opler PA, Puckett Haecker CE, Doran PD (eds) Status and trends of the nation's biological resources, vol. 2. Biological Resources Division, US Geological Survey, Reston, VA, pp 473–504

Stohlgren TJ (1999b) Global change impacts in nature reserves in the United States. In: Aguirre-Bravo C, Franco CR (eds) North American symposium toward a unified framework for inventory and monitoring forest ecosystem resources, RMRS-P-12. USDA Forest Service Rocky Mountain Forest and Range Experiment Station, Fort Collins, CO, pp 5–9

Stohlgren TJ, Coughenour MB, Chong GW, Binkley D, Kalkhan MA, Schell LD, Buckley DJ, Berry JK (1997) Landscape-scale gap analysis: a complementary geographic approach for land managers. Landscape Ecol 12:155–170

Stohlgren TJ, Chase TN, Pielke RA Jr, Kittel TGF, Baron J (1998) Evidence that local land-use practise influences regional climate and vegetation patterns in adjacent natural areas. Global Change Biol 4:495–504

Stohlgren TJ, Owen AJ, Lee M (2000) Monitoring shifts in plant diversity in response to climate change: a method for landscapes. Biodivers Conserv 9:65–86

Storch H von (1994) Inconsistencies at the interface of climate impact studies and global climate research. Max Planck Institute for Meteorology Report 122:12

Storch H von, Zorita E, Cubasch U (1993) Downscaling of global change estimates to regional scales: an application to Iberian rainfall in wintertime. J Climate 6:1161–1171

Suter GW (1993) Ecological risk assessment. Lewis Publishers, Michigan

Thompson DQ, Stuckey RL, Thompson EB (1987) Spread, impact, and control of purple loosestrife (Lythrum salicaria) in North American wetlands. US Fish and Wildlife Service Research Report 2, Washington D.C., 55 p

Tobolski K (1982) Anthropogenic changes in vegetation of the Gardno-Leba lowland, North Poland. Preliminary report. Acta Palaeobot XXII:131–139

Tol RSJ, de Vos AF (1998) A Bayesian statistical analysis of the enhanced greenhouse effect. Climatic Change 38:87–112

Tóth F, Bruckner T, Füssel HM, Leimbach M, Petschel-Held G, Schellnhuber H-J (1997) From impact to emission: the tolerable windows approach. In: IPCC (ed) climate change and integrated assessment models – bridging the gaps. CGER-Report-1029-97. Center for Global Environmental Research, Tsukuba, Japan

Townshend JR, Justic GC, Li W, Gurney C, McManus J (1991) Global land-cover classification by remote sensing: present capabilities and future possibilities. Remote Sens Environ 35:243–355

Trotter PC (1987) Cutthroat: native trout of the west. Colorado Associated University Press, Boulder

Turk JT, Campbell DH, Ingersoll GP, Clow DA (1992) Initial findings of synoptic snowpack sampling in the Colorado Rocky Mountains. US Geological Survey, Denver, CO

Turner C (1975) Der Einfluß großer Mammalier auf die interglaziale Vegetation. Quartärpaläontologie 1:13–19

Turner BL, Skole DL, Sanderson S, Fischer G, Fresco L, Leemans R (1995) Land-use and Land-cover change: science/research plan. International Geosphere-Biosphere Programme, Human Dimensions of Global Environmental Change Programme, Stockholm, 132 p

UNEP (2000) Assessing human vulnerability due to environmental change; concepts, issues, methods and case studies. United Nations Environmental Programme, Nairobi, Kenya

United Nations (1997) Comprehensive assessment of the freshwater resources of the World. World Meteorological Organization, Geneva

United Nations Population Division (1998) World population prospects: the 1998 Revision. United Nations, New York

United States Geological Survey (1990) National water summary 1987 – Hydrologic events and water supply and use. In: USGS National Water Summary Series WSP 2350, Denver, Colorado

United States Geological Survey (1993) National water summary 1990–1991 – Hydrologic events and stream water quality. In: USGS National Water Summary Series WSP 2400, Washington D.C.

United States Geological Survey (1999) Streamflow information for the next century: a plan for the national streamflow information program of the USGS. USGS, pp 99–456

United States Geological Survey – EROS Data Center (1998) Global land-cover characterization based on 1-km AVHRR data for 1992/93. US Geological Survey Earth Resources Data Center, Sioux Falls, SD

van Noortijk J, Kok M, Cooke RM (1997) Optimal Decisions that Reduce Flood Damage along the Meuse: an Uncertainty Analysis. In: French S, Smith JQ (eds) The practice of Bayesian Analysis. Arnold, London, pp 151–172

Veblen TT, Lorenz DC (1991) The Colorado Front Range: A century of ecological change. University of Utah Press, Salt Lake City

Veblen TT, Kitzberger T, Donnegan J (2000) Climatic and human influences on fire regimes in ponderosa pine forests in the Colorado Front Range. Ecol Appl 10:1178–1195

Vitousek PM, Mooney HA, Lubchenco J, Melillo JM (1997) Human domination of earth's ecosystems. Science 277:494–499

Vogel C (1998) Vulnerability and global environmental change. LUCC Newsletter 3:15–19

Vörösmarty CJ, Sahagian D (2000) Anthropogenic disturbance of the terrestrial water cycle. Bioscience 50:753–765

Vörösmarty CJ, Sharma KP, Fekete BM, Copeland AH, Holden J, Marble J, Lough JA (1997) Storage and aging of continental runoff in large reservoir systems of the world. Ambio 26: 210–219

Vörösmarty CJ, Birkett C, Dingman L, Lettenmaier DP, Kim Y, Rodriguez E, Emmitt GD (1999) HYDRA-SAT: hydrological applications satellite. A report from the NASA Post-2002 Land Surface Hydrology Planning Workshop, Irvine, California, April 12–14, 1999

Vörösmarty CJ (lead), Askew A, Grabs W, Barry RG, Birkett C, Döll P, Grabs W, Hall A, Jenne R, Kitaev L, Landwehr J, Keeler M, Leavesley J, Schaake J, Strzepek K, Sundarvel S, Takeuchi K, Webster F (IAHS Ad Hoc Group on Global Water Data Sets) (2001) Global water data: a newly endangered species. An opinion editorial to AGU Eos Transactions 82:54–58

Vörösmarty CJ, Fekete BM, Meybeck M, Lammers R (2000) A simulated topological network representing the global system of rivers at 30-minute spatial resolution (STN-30). Global Biogeochem Cy 14:599–621

Wang G, Eltahir EAB (2000) Biosphere-atmosphere interactions over West Africa. 2. Multiple Equilibria. Q J Roy Meteor Soc 126: 1261–1280

Washburn AL (1973) Geocryology. A survey of periglacial processes and environments. E. Arnold, London

Watson A, Lovelock J (1983) Biological homeostasis of the global environment: the parable of Daisyworld. Tellus 49B:249–262

WBGU (German Advisory Council on Global Change) (1994) World in transition: the threat to the soils. Economica, Bonn

WBGU (German Advisory Council on Global Change) (1999) World in transition: ways towards sustainable management of freshwater resources. Springer-Verlag, Berlin

Weber SL (2001) On homeostasis in Daisyworld. Climatic Change 48:465–485

Weeks J, Gutentag E (1988) Region 17: high plains. In: Back W, Rosenshein J, Seaber P (eds) Hydrogeology. Geological Society of America, Boulder, CO, 524 p

Weisberg PJ, Baker WL (1995) Spatial variation in tree seedling and Krummholz growth in the forest-tundra ecotone of Rocky Mountain National Park, Colorado, USA Arctic Alpine Res 27:116–129

Weyant J, Davidson O, Dowlabathi H, Edmonds J, Grubb M, Parson EA, Richels R, Rotmans J, Shukla R, Tol RSJ, Cline W (1996) Integrated assessment of climate change: an overview and comparison of approaches and results. In: Bruce JP, Lee H, Haites EF (eds) Climate change 1995. Economic and social dimensions of climate change. Cambridge University Press, Cambridge

White R, Engelen G, Uljee I (1997) The use of constrained cellular automata for high-resolution modelling of urban land-use dynamics. Environ Plann B 24:323–343

Whitlock C (1993) Post-glacial vegetation and climate of Grand Teton and southern Yellowstone National Parks. Ecol Monogr 63:173–198

Wikle CK, Berliner LM, Cressie N (1998) Hierarchical Bayesian space-time models. Environ Ecol Stat 5:117–154

Wilhite DA (ed) (2000) Drought: a global assessment. Routledge Press, London

Williams MW, Baron JS, Caine N, Sommerfeld RA, Sanford RL (1996) Nitrogen saturation in the Rocky Mountains. Environ Sci Technol 30:640–646

World Bank (1993) Water resources management: a World Bank policy paper (prep. by Easter KW, Feder G, Le Moigne G, Duda AM). The World Bank, Washington DC, 141 p

World Bank (1996) Mexico – Water resources management project: staff appraisal report. The World Bank, Water Resources Management Project No. 15435, May 1996, 182 p

World Bank (1998a) Evaluating the Bank's water resources management policy: an approach paper (prep. by Bruce-Konuah A). The World Bank, Water Resources Management Program, October 1998, 6 p

World Bank (1998b) Morocco – Water resources management project: staff appraisal report. The World Bank, Water Resources Management Project No. 15760, January 1998, 76 p

World Resources Institute (1996) World resources: 1996–1997. Oxford University Press, New York

Wu SM, Huang GH, Guo HC (1997) An interactive inexact-fuzzy approach for multiobjective planning of water resources systems. Water Sci Technol 36:235–242

Zhou HM (1995) Towards an operational risk assessment in flood alleviation – theory, operationalization and application. Technical University, Delft, Netherlands, 204 p

Index

Symbols

30-minute Simulated Topological Network (STN-30) 378, 395, 516
4DDA (see *four-dimensional data assimilation*)

A

ABLE-2 (see *Amazon Boundary Layer Experiment-2*)
ABRACOS (see *Anglo-Brazilian Amazonian Climate Observation Study*)
absorption 21, 34, 35, 80, 97, 107, 113, 132, 183, 199, 281, 499
  –, of solar radiation 34, 80, 132
accelerometer 185
Acocks veld types 447
ACRU (see *Agricultural Catchments Research Unit*)
ADEOS (see *Advanced Earth Observing Satellite*)
Adour
  –, basin 226
  –, River 226
Adriatic Sea 383
Advanced Earth Observing Satellite (ADEOS) 248
Advanced Land Observing Satellite (ALOS) 211
Advanced Microwave Sounding Unit (AMSR) 281, 284
Advanced Very High Resolution Radiometer (AVHRR) 66, 80, 94, 207, 209, 210, 212, 248, 281, 283, 419, 422, 515
advection 31, 36, 52, 53, 119, 159, 165, 184, 187, 189, 190, 193, 194, 197, 250, 282, 385
aerodynamic
  –, gradient method 174
  –, resistance 10
  –, roughness length 11
aerosol 2, 80, 83, 84, 107, 123, 132, 167, 484–487, 497
  –, flux 2
  –, volcanic 484
afforestation 24, 36, 37, 77, 304, 322, 348–352, 374, 446, 450, 451, 461, 500
Africa 2, 7, 35, 37–39, 42–45, 53, 60, 62, 64, 65, 67–71, 73, 76, 80, 88, 130, 131, 134, 187, 215, 317, 318, 320, 336, 341, 343–347, 349–352, 358, 367–369, 371, 373, 374, 377–379, 383, 392, 395, 413, 441, 451, 452, 503, 514–516, 520, 527–529, 531–535, 537
African
  –, monsoon 35, 42, 43, 60, 61, 70, 75, 131
  –, rift lakes 383
African Monsoon Multidisciplinary Analysis (AMMA) 65
AGCM (see *atmospheric general circulation model*)
age spectrum 308
ageing 388, 389
  –, local 386, 388
aggregate stability 317
aggregating function 31
aggregation 31, 71, 218, 221–228, 233, 250, 252, 253, 289, 304, 305, 317, 328, 349, 452, 501, 515, 518
  –, problem 253
  –, spatial 253
Agricultural Catchments Research Unit (ACRU) 329, 332, 334–336, 351, 443, 444, 446, 449, 450, 452, 528, 531, 532, 534
  –, modelling system 334, 528, 534
  –, phosphorus module 336

agricultural
  –, expansion 44, 62, 70, 114, 452
  –, intensification 62, 319
agrochemicals 457, 463
AIM (see *Asian-Pacific Integrated Model*)
AIMP (see *impervious or less permeable areas*)
Aïr 383
air pollution 370, 409, 491, 501, 511–513, 537
aircraft
  –, sounding 194
  –, study 191
albedo 9, 10, 22, 23, 29, 34–38, 40–42, 45, 51, 53, 54, 57, 64–66, 68, 69, 73, 75, 76, 82–84, 87–90, 93, 96, 97, 101, 106, 107, 113–115, 118, 121–123, 127, 130, 131, 133, 202, 207, 211, 317, 332
  –, Charney's theory 35
  –, forests 106
  –, lakes 106
  –, planetary 36
  –, snow 113
  –, wetlands 106
Alberta 382
algal bloom 354, 390
allocation 204, 340, 368, 519
  –, coefficient 204
ALMA (see *Assistance for Land-surface Modelling Activities*)
ALOS (see *Advanced Land Observing Satellite*)
Altai Mountains 382
Altiplano 378
  –, dry lakes 383
Amazon Boundary Layer Experiment-2 (ABLE-2) 82–85, 87, 190, 194
Amazon Regional Micrometeorology Experiment (ARME) 82, 83, 85, 87, 88, 224
Amazon
  –, basin 79, 80, 82, 83, 85–89, 120, 131, 194, 218, 219, 233, 256, 391, 408, 415, 417, 419, 421, 422
    –, biogeochemical dynamics 422
    –, energy balance 85
    –, integrated analysis 415
    –, water balance 85, 86
  –, floodplain (várzea) 418
  –, forest 8, 80, 83, 195, 330
Amazon River 2, 79, 83, 87, 131, 381, 383, 384, 391, 415, 417–422, 424, 425, 427
  –, discharge 420, 421
  –, ecosystem 427
  –, hydrology 419
  –, nitrogen flux
    –, total 404
  –, particulate material transported 424
  –, profile 391
  –, sediment
    –, mineralogy 424
    –, transport 424
  –, total nitrogen export to ocean 404
  –, water movement model 421
Amazonia 1, 29, 30, 39, 46, 79–87, 92, 131, 132, 134, 157, 158, 164, 208, 229, 233, 250, 256, 390, 400, 419, 427

–, climate 79, 83, 131
–, deforestation 29
–, future climates 81
AMC (Atmospheric Mesoscale Campaign: see Wet Atmospheric
    Mesoscale Campaign, WETAMC)
AmeriFlux (measuring network) 189
amino acid 425, 426
AMIP (see *Atmospheric Model Inter-comparison Project*)
AMMA (see *African Monsoon Multidisciplinary Analysis*)
ammonia 406–408
–, flux 407
ammonium 354, 403, 407, 408
AMSR (see *Advanced Microwave Sounding Unit*)
Amsterdam Declaration on Global Change 1
analysis error 163
Andes 80, 378, 417–419, 424
–, headwaters 418
–, of Ecuador 418, 419
–, of Peru 418, 419
anemometer 109, 161, 162, 177, 178
–, sonic 109, 161, 177, 178
Anglo-Brazilian Amazonian Climate Observation Study
    (ABRACOS) 17, 81, 83–85, 87, 88, 132, 224, 310
–, experiments 17
anomaly approach 43
Anthropocene 412, 456, 461, 462, 464
anthroposphere 457
AOGCMs (see *atmosphere-ocean general circulation model*)
approach
–, dynamic 273, 497
–, integrated 182, 368, 496, 501
–, management 368
–, picket fence 214
–, stochastic 253, 497
–, tolerable window 488
aquifer 43, 301, 305–308, 346, 355, 358, 361, 370, 457, 459, 461, 514
–, deep 333, 386
–, shallow 333
Aral Sea 378, 381–383, 389, 410
–, basin 410
Araye 378
–, drainage basin system 378
ARC/INFO 331
Arctic
–, Ocean 378
–, Oscillation 96, 376, 385, 402
area
–, endorheic 378, 380
–, exorheic 378, 380
–, fractional, drained 391
–, impervious 305, 330, 335, 353, 445
Argentina coast 383
arheic 379, 380
arheism 459
ARM (see *Atmospheric Radiation Measurement*)
ARME (see *Amazon Regional Micrometeorology Experiment*)
ARNO model 16
ARPEGE model 221
Asia 37, 45, 54, 88, 113, 115, 116, 118, 120, 121, 123–126, 130, 133, 256, 258,
    320, 380, 382, 383, 392, 395, 398, 400, 401, 411, 459, 503, 515, 516
AsiaFlux (measuring network) 18, 258
Asian monsoon 8, 115, 116, 118, 121, 123, 134
–, climate 115, 133
Asian-Pacific Integrated Model (AIM) 502
aspen 19, 54, 94, 95, 103, 105, 106, 112, 187, 218
assault event 499, 528
assessment
–, global water vulnerability 515
–, vulnerability 3, 276, 489, 491, 494, 511, 513, 522, 536, 537
Assistance for Land-surface Modelling Activities (ALMA) 268,
    269
ASTER sensor 281
Aswan High Dam 386, 389
Atbara 379

Athabasca River 383, 384
Atlantic Ocean 79, 81, 131, 378
Atlas Mountains 44
atmosphere, chaotic dynamics 50
atmosphere-ocean general circulation model (AOGCMs) 497
atmospheric general circulation model (AGCM) 12, 46, 47, 257, 285
Atmospheric Mesoscale Campaign (AMC)
Atmospheric Model Inter-comparison Project (AMIP) 87, 263,
    264
atmospheric radiation measurement (ARM) 26, 28, 257
Australia 26, 196, 258, 278, 279, 309, 318, 358, 361, 378, 380, 383, 395
AVHRR (see *Advanced Very High Resolution Radiometer*)
aviation routine weather report 258
Avoca
–, basin 361
–, catchment 360
–, River 361
Azov Sea 382

B

background reflectance 207
BAHC (see *Biospheric Aspects of the Hydrological Cycle*)
balance, hydrological 2
balsam
–, fir 95
–, poplar 95
BALTEX (see *Baltic Sea Experiment*)
Baltic Sea 218, 256, 382, 461
Baltic Sea Experiment (BALTEX) 218, 256, 260
baseflow 305–307, 335, 336, 348, 350, 361, 445
baseline streamflow 446, 453
Baseline Surface Radiation Network (BSRN) 161, 170, 171, 248
basin
–, boundary 302, 370
–, characteristics 377
–, damaged ecosystem 367
–, dormant 378
–, elevation profile 391
–, endorheic 378, 382
    –, non-flowing 378
–, eutrophied 403, 409
–, export 406
–, flux, estimation 391
–, highly fertilised 409
–, hydrological measurement 232
–, rainfall 80
–, scale 232, 300, 301, 307–309, 317, 329, 330, 338, 349, 367, 370,
    372, 450, 457, 528
–, scale 305, 333, 339–341, 343, 349, 351, 354, 355, 357, 374, 390,
    397, 457
BATS (see *Biosphere-Atmosphere Transfer Scheme*)
Bayankala Mountain 525
Bayes rule 494
Bayesian
–, approach 493, 494
–, forecasting system (BFS) 494
–, statistical method 494
beaver (*Castor canadensis*) 108, 112, 387, 510
Beer's law 13
Belém 83
benchmarking 268
benefit analysis 500, 533, 534
BERMS (see *Boreal Ecosystem Research and Monitoring Sites*)
BETHY model (see *Biosphere Energy Transfer Hydrology model*)
BFS (see *Bayesian forecasting system*)
big leaf approach 31, 129
bighorn sheep 510, 511
biodiversity 3, 8, 63, 88, 115, 342, 354, 363, 365, 373, 456, 457, 491,
    509–512
–, loss 63, 363, 491, 510
biomass 8, 17, 33, 34, 46, 66, 73, 79, 80, 82–84, 93, 107, 131, 132, 159,
    163, 199–202, 204, 205, 211, 212, 250, 259, 281, 315–317, 319, 333,
    347–351, 390, 398, 399, 404, 459, 503, 510, 512

–, standing 199, 201, 319
biome 2, 18, 89, 114, 157, 207, 208, 233, 259, 314, 315, 317, 399–401, 496, 504
–, model 35, 40, 41, 43
–, paradox 37, 40, 41
BIOME 6000 (Global Palaeo Vegetation) Project 383
biosphere 1, 7, 8, 18, 33, 37, 48, 66, 68, 69, 71, 73, 76, 82, 113, 122, 129–131, 134, 136, 182, 199, 202, 205, 279, 337, 430, 486, 507
–, terrestrial 1, 48, 134, 199
Biosphere Energy Transfer Hydrology model (BETHY) 285
Biosphere-Atmosphere Transfer Scheme (BATS) 13, 67, 280, 314
Biospheric Aspects of the Hydrological Cycle (BAHC) 1
birch 94, 95, 506
Black Sea 378, 382
black spruce 95, 103–107, 110, 111, 160, 203
black water 418
–, river 418
Blue Nile 379
Bo Sea 525
Bolivian Andes 417
Boreal Ecosystem Research and Monitoring Sites (BERMS) 218, 220
Boreal Ecosystem-Atmosphere Study (BOREAS) 17, 54, 101, 103–106, 108, 110, 113, 135, 157, 161, 173, 186, 187, 193, 194, 196, 203, 208, 209, 212, 213, 216–218, 220–222, 224, 225, 228, 230, 255, 256, 260
boreal
–, climate 93, 132
–, ecosystem 93–95, 109, 111–114, 132–134, 211, 212
–, forest 2, 8, 40, 47, 54, 55, 93–97, 99–102, 104, 105, 107–110, 113, 114, 132, 133, 157, 187, 203, 211, 213, 216–218, 256
–, region 8
–, tree line 2
–, vegetation 94
BOREAS (see *Boreal Ecosystem-Atmosphere Study*)
boundary layer
–, budget 189, 190, 195, 231
–, method 190, 196
–, gradient method 191
–, structure 189
Bowen ratio 21, 25, 29, 53, 105, 118, 166, 173, 174, 176, 195, 215, 343, 484
–, (or energy balance) method 176
Brandenburg 429, 430, 435–439
Brazil 2, 25, 47, 79, 80, 83, 85–87, 131, 256, 410, 419, 482
breadth of data 250
brown coal mining 436
brown trout 511
Brucker period 126
Brune/Ward equation 395
BSRN (see *Baseline Surface Radiation Network*)
bucket model 12, 46, 49, 221
Buddy check 282
Budyko equation 52
buoyancy flux 162, 189, 191
Burntwood River 389
Burren River 355
business-as-usual scenario 81

C

Caatinga Forest 418
calculation 163
calibration 66, 67, 221
–, manual 332
–, model 328
California 161, 190, 318, 321, 364, 396, 421, 488, 492
CAMELS (see *Carbon Assimilation and Modelling of the European Land-Surface*)
Cameroon 373
Campaspe
–, basin 361
–, catchment 360
CAMREX (see *Carbon in the Amazon River Experiment*)
Canada 2, 17, 41, 54, 93–95, 97, 99, 101, 110–114, 132, 157, 211, 213, 216, 278, 319, 323, 327, 382, 404, 411

canopy
–, conductance 13, 47, 202
–, resistance 12, 13, 217
–, temperature 207, 280
–, turbulence 11
capacity, hydroelectric 390
capillary rise 302, 303, 312, 333
CarboEurope (measuring network) 18
carbohydrates 426
Carbon Assimilation and Modelling of the European Land-Surface (CAMELS) 284, 286
carbon cycle data assimilation system (CCDAS) 284, 285
Carbon in the Amazon River Experiment (CAMREX) 415, 422
carbon
–, balance 109, 111, 157, 182, 199, 231, 232, 456
–, biospheric exchange 109
–, cycle 8, 14, 18, 33, 36, 39, 45, 48, 49, 82, 93, 96, 112, 113, 136, 199, 212, 230, 284, 285, 400
–, data assimilation system (CCDAS) 284, 285
–, transport 400
–, dioxide
–, atmospheric concentration 37, 47, 93, 96, 110, 112, 113, 132, 212, 262, 311, 438
–, concentration 33, 46, 93, 109, 111, 113, 132, 201, 311, 317
–, dissolved 423
–, flux 17, 48, 83, 84, 109, 111, 157, 173, 174, 181, 193–196, 202, 204, 231, 232
–, flux measurement 17, 84
–, sinks 16
–, sources 16
–, uptake 13, 33, 187, 204
–, exchange 17, 19, 129, 133, 165, 204, 209, 212, 222, 231, 280
–, flux 2, 12, 16–19, 84, 93, 110, 111, 132–134, 212, 258, 285, 286, 376, 398, 399, 402
–, geographic variations 110
–, interannual variation 111
–, riverborne 399
–, seasonal variation 111
–, global river transfer 397
–, riverine 397, 398
–, load 411
–, transfer 398, 402
–, sequestration 93, 109–111, 503
–, sink 48, 112, 130, 133, 285, 401
–, soil organic carbon 319, 320, 399
–, stock methods 109
–, uptake 93, 109, 113, 132–134, 174, 196, 284, 401
carbon-14
–, age 398, 425
carboxylation capacity 13
carrying capacity 59, 130
CART (see *Cloud And Radiation Testbed*)
CAS (see *continental aquatic system*)
Caspian Sea 378, 380, 382, 383
Castor canadensis (see *beaver*)
cataclysmic emptying 382
CATCH project (see *Couplage de l'Atmosphère Tropicale et du Cycle Hydrologique*)
Caucasus 97, 382
C-band 210
CBL (see *convective boundary layer*)
CCDAS (see *carbon cycle data assimilation system*)
CDMS (see *climate data management system*)
Center for International Earth Science Information Network 516
Center for Ocean-Land-Atmosphere Studies (COLA) 68, 70, 71, 86, 87
central Asia 130, 382, 383, 459, 516
–, dry lakes 383
CEOP (see *Coordinated Enhanced Observing Period*)
Cerrado 80
CFCs 457
chamber method 111, 179, 189
Chang Jiang River 394, 409
change

–, ecohydrological 348
–, ecological response 348
–, environmental, at small catchment scales 301
–, global 376
channel
    –, morphometry 363
    –, transmission loss 344
Chari basin 378
Charney's theory of albedo 35
cheatgrass 510
chemical oxygen demand (COD) 522–525
Chiang Jiang basin 462
China 2, 116, 118–123, 125–127, 134, 280, 317, 318, 378, 380, 397, 401,
    409, 410, 462, 516, 518, 519, 522, 525–527, 537
Churchill River 382, 389
    –, system 382
circulation, mesoscale 24, 26, 28, 29, 31, 81, 132, 223
clear water river 418
climate
    –, $1 \times CO_2$ 46
    –, $2 \times CO_2$ 46, 47, 114, 319, 452, 453
    –, Amazonian 79, 83, 131
    –, anomaly 40, 59, 67–69, 76, 131
    –, Arctic, feedback 40
    –, Asian monsoon 115, 133
    –, boreal 93, 132
    –, change 45, 60, 81, 93, 96, 114, 317, 376, 383, 437, 441, 452, 485,
        492, 534, 535
        –, $2 \times CO_2$ scenario 452, 453
        –, driver 7
        –, high-latitude 96
        –, impact on crop yield 437
        –, impact on hydrology 437
        –, impacts on streamflow 452
        –, late glacial 505
    –, data management system (CDMS) 264
    –, definition 7
    –, feedback 113
    –, forced variations 484
    –, forcing 67
    –, free variations 484
    –, global 33, 129
    –, near ground 9, 129
    –, past, progressive changes 383
    –, regional 2, 8, 21, 32, 48, 60, 67, 71, 75–77, 82, 107, 115, 121, 123,
        130–132, 337, 339, 340, 497
    –, scenario 319, 491
    –, system 2, 7–10, 33, 36, 37, 60, 71, 82, 115, 123, 129–131, 134–136,
        205, 212, 216, 247, 280, 314, 376, 461, 484–487, 489, 496, 497,
        503–505, 508, 513
        –, interactions 71
    –, variability 381
        –, global 3
        –, regional 3
Climate Variability and Predictability (CLIVAR) 65
Cloud And Radiation Testbed (CART) 26, 28, 257
cloud
    –, data 207
    –, microphysics 484
clumping index 200
COD (see chemical oxygen demand)
COLA (see Center for Ocean-Land-Atmosphere Studies)
Colorado
    –, basin 3
        –, profile 391
    –, canyon 391
    –, Plateau 391
    –, River 3, 358, 363, 373, 383, 459, 461, 510
Columbia River 515
    –, system 382
compound, xenobiotic organic 457
condensation nuclei 107, 485
conductance 12, 13, 22, 23, 31, 47, 82, 83, 104–106, 112, 132, 201, 202,
    316, 364

–, model 13
–, stomatal 13, 22, 23, 31, 47, 106, 132, 201, 202, 316
    –, model 13
    –, relationship to photosynthesis 13
conductivity
    –, electrical 361
    –, hydraulic 15, 68, 69, 227, 259, 308, 309, 311, 316
    –, saturated hydraulic 15, 308, 309, 316
Congo River 384, 385
coniferous forest 54, 94, 95, 103, 106, 109, 218, 506
consolidation 2, 245, 247, 252, 255, 267, 289, 290
    –, degree 255
constant, psychrometric 176
continental aquatic system (CAS) 375–377, 391, 398, 455–457,
    461–464
    –, responses 461
continental aquatic system (CAS) 375–377, 391, 398, 455–464
    –, responses 461
continental-scale experiments (CSEs) 256, 260
continuity equation 189, 190, 231, 311
control, stomatal 103
convection 88, 90
    –, tropical 39, 40, 90
convective boundary layer (CBL) 24, 26, 189–197
    –, budget 193, 194, 196, 197
    –, gradient 191, 196
    –, thickness 189, 191
convergence 28, 30, 38, 42, 50, 75, 80, 84, 89–92, 116, 118, 131, 133, 189,
    190, 195, 196, 214, 218, 219, 327, 353, 390, 421
    –, net 390
Coordinated Enhanced Observing Period (CEOP) 65, 233, 256,
    260
correction factor 164, 438
corridor, riparian 457
cospectrum 178
Couplage de l'Atmosphère Tropicale et du Cycle Hydrologique
    (CATCH) project 63–65, 76
coupled modelling 66
Cretaceous 41
crop
    –, rotation 305, 431, 433, 436, 437
    –, yield 62, 332, 335, 430, 437–439, 528
        –, climate change impacts 437
crusting 317
cryosphere 7, 33
CSEs (see continental-scale experiments)
cultural history, rapidly changing 509
cumulus
    –, cloud 24, 25, 84, 85, 224
    –, convection 25, 30, 31, 115, 119, 120, 134
    –, frequency 25, 26
cycle hydrological 1, 3, 14, 26, 32, 89, 90, 129, 223, 227, 228, 299, 301,
    311, 316, 333, 340, 348, 363, 375, 456, 484, 489, 492, 496, 505, 514,
    528
Czech Republic 413, 429

D

DAACs (see Distributed Active Archive Centers)
Daihai Lake 126
dam 359, 363, 364, 443
    –, definition 363
    –, impact 363–365
    –, purposes 363
Danube basin 355
Darcy
    –, equation 311
    –, law 15
Darling River 361
data storage system (DSS) 260, 270
data
    –, assimilation 52, 248, 273, 281, 283, 287, 289
        –, methods 281–283, 287
    –, availability 85

-, broad archives 259
-, consolidation 2, 247, 248, 251–253, 255, 256, 262, 265, 267, 268, 289, 290
-, co-registration 2
-, distribution 2
-, maintenance 2, 270
-, meteorological 258
-, palaeobotanic 42
-, scaling 253
-, standardisation 2, 267
-, storage system (DSS) 260, 270
dataset, synthesis of disparate 2, 289
Death Valley 383
debris fan 364
deciduous forest 18, 25, 111, 160, 196
decomposition 93, 108–110, 113, 114, 133, 205, 356, 402
deep
-, aquifer 333, 386
-, groundwater 303, 325, 327, 330, 380
-, ocean circulation effects 484
Defense Meteorological Satellite Program (DMSP) 281
deforestation 2, 3, 24, 25, 29, 31, 35–37, 40, 44, 45, 48, 51, 62, 69, 70, 79–85, 88–94, 117, 118, 120, 127, 131, 132, 134, 233, 304, 319, 321, 322, 340, 347, 360, 366, 385, 390, 397, 421, 427, 461, 463, 501–503, 511, 522
-, arc of 79
-, experiment 36, 45, 88, 89, 92
-, in Amazonia 29
-, scenario 89, 321
degradation 59, 62, 63, 65, 68–70, 73–77, 111, 121, 122, 130, 131, 170, 317, 322, 334, 340, 347–349, 352, 354, 369, 374, 397, 412, 416, 425, 426, 444, 455, 464, 503, 504, 522, 527, 532
delineation of hydrotopes 331
delivery ratio (DR) 334, 356, 391, 393, 402, 449
delta values 324
DEM (see digital elevation model)
denitrification 334, 346, 404, 412, 416, 431–436
deposition 48, 83, 112, 126, 193, 212, 334, 354, 357, 391, 395, 404, 406, 409, 411, 424, 445, 449, 457
-, atmospheric 354, 357, 404, 406, 409, 411, 449
-, dry 445
-, wet 445
depression, tropical 116
deprivation event 499, 528
desert 8, 34, 35, 37, 39, 42, 53, 59, 60, 68, 72, 73, 102, 107, 121–123, 130, 133, 343, 379
-, equilibrium 73
-, green 102, 133
desertification 7, 35, 38, 44, 61–63, 66, 70, 75, 76, 134, 157
Deutscher Wetterdienst (DWD (Germany's National Meteorological Service)) 283
Deutsches Zentrum für Luft- und Raumfahrt (DLR) 187
DGVM (see dynamic global vegetation model)
DIAL (see Differential Absorption Lidar)
DIC (see dissolved inorganic carbon)
Differential Absorption Lidar (DIAL) 196, 197
digital elevation model (DEM) 263, 304, 330, 331, 405, 418, 444, 445
dimension
-, external 499
-, internal 499
DIN (see dissolved inorganic nitrogen)
disaggregation 252, 253, 302–305, 328, 330, 333
discharge
-, Amazon River 420, 421
-, Arctic rivers 385
-, cumulative 216
-, Elbe River 430
-, episodic 396
-, natural regime 386
-, seasonal 396
displacement height 11, 21, 22, 196
dissolved
-, inorganic carbon (DIC) 109, 397–399, 401, 402, 411, 424, 456
-, inorganic nitrogen (DIN) 406, 407, 409, 411, 412

-, export 406, 411, 412
-, organic carbon (DOC) 109, 327, 397–402, 411, 416, 423, 424
-, organic nitrogen (DON) 416, 424
-, oxygen 354, 423
-, phosphorous 357
Distributed Active Archive Centers (DAACs) 260
Distributed Ocean Data System (DODS) 265
disturbance 3, 17, 73, 79, 93, 95, 96, 111–113, 132, 184, 210, 212, 341, 347, 348, 362, 363, 374, 393, 403, 457, 491, 508–513
-, altered regime 510
DLR (see Deutsches Zentrum für Luft- und Raumfahrt)
DMSP (see Defense Meteorological Satellite Program)
DOC (see dissolved organic carbon)
DODS (see Distributed Ocean Data System)
DOLY vegetation model 47
DON (see dissolved organic nitrogen)
downscaling 253, 496, 497
-, approach 496
-, dynamic 497
-, statistical 437, 497
downstream peak 386, 389
DPSIR (see driving forces, pressures, states, impacts and responses)
DR (see delivery ratio)
drainage 15, 197, 297, 379, 382, 398, 427
-, basin 2, 16, 108, 263, 300, 340, 354, 377, 378, 381, 382, 388–392, 395, 396, 402–406, 409, 412, 415, 417, 418, 420, 421, 429, 430, 464, 517, 518, 526
-, concept 2, 300
-, Saharan 378
-, synthetic model 427
-, water balance 389
-, system
-, large 384
driver
-, anthropogenic 460
-, biophysical 62, 63, 405
-, hydrological 339
-, natural 459
driving forces, pressures, states, impacts and responses (DPSIR) model 339–341, 366, 374, 463
drought 25, 52, 59–63, 66, 67, 70, 75, 76, 81, 82, 115, 130, 131, 210, 276, 299, 343, 349, 370, 372, 376, 379, 385, 482, 498, 499, 511, 516
dry deposition 445
dryline 24, 25
DSS (see data storage system)
Duke Forest 411
dune, migrating 507
-, impact on water budget 507
Dunne overland flow 309
duration 96, 99, 498, 528
Durban 443, 453
DWD (see Deutscher Wetterdienst (Germany's National Meteorological Service))
dynamic global vegetation model (DGVM) 48, 71
dynamics, past atmospheric 496

E

Earth Observing System (EOS) 248, 260, 281, 286
Earth observing system advanced microwave sounding unit (EOS-AMSR) 281, 284
Earth Resources Observation Systems (EROS) 263, 515
Earth system measurement 155
Earth system modelling framework (ESMF) 265
Earth system models of intermediate complexity (EMICs) 48, 484
East Asia monsoon 116, 121, 123
East European Plain 382
Eastern Africa 516
EAU (see elementary areal unit)
ECHAM model (climate model based on the ECMWF model) 72
ECHAMT21 model (climate model based on the ECMWF model) 437
ECHIVAL Field Experiment in Desertification-threatened Areas (EFEDA) 157, 186, 208, 212, 221–224, 227

ECHIVAL (European International Project on Climatic and Hydrological Interactions between Vegetation, Atmosphere and Land-surfaces) 157, 186, 208, 212, 221–224, 227
ECMWF (see *European Centre for Medium-Range Weather Forecasts*)
ecohydrological
    –, change 348
    –, model 301, 329–332, 335, 430, 439
ecohydrology 329
Ecological Stratification Working Group (ESWG) 95
economics 369
ecosystem
    –, alteration 348
    –, aquatic 340, 341, 402
    –, boreal 93–95, 109, 111–114, 132–134, 211, 212
    –, conserved 348
    –, removed 348
    –, replaced 348
    –, riverine
        –, in arid zones 343
        –, restoration 373
    –, state 348
    –, terrestrial 340, 402
    –, utilised 348
    –, vulnerability 508
eddy
    –, correlation method 109, 176
    –, covariance 17, 159, 161, 163–166, 173, 174, 176–181, 193, 194, 202, 204, 230, 258, 270, 286
        –, fluxes 193, 194
        –, measurement 159, 163, 204, 270
        –, method 17, 163, 165, 176, 178
        –, system 164, 165, 173, 174, 176–178, 180, 202, 230
    –, diffusivity 11, 176
    –, flux 17, 163, 181, 182, 185, 189, 195, 230–232
    –, relaxed accumulation 178
Edendale 449
Edisol system 160
Eel River 396
Eemian warm period 41
EFEDA (see *ECHIVAL Field Experiment in Desertification-threatened Areas*)
Egypt, population grow 516
El Miklwaddi 379
El Niño-Southern Oscillation (ENSO) 18, 28, 39, 48, 70, 80, 85–87, 230, 248, 278, 339, 370, 376, 383–385, 402, 421, 484, 485, 508, 528, 534, 537
    –, forecast 486
        –, -Amazon 80, 385, 421
        –, -Congo 385
        –, -Murray Darling 385
        –, -Nile 385
        –, -Parana River 385
        –, -Sahel 70
Elbe
    –, basin 429, 430, 433, 435
        –, area 355
        –, nitrogen dynamics 433
    –, River 2, 330, 341, 355–357, 413, 429–431, 433–436
        –, discharge 430
ELDAS (see *European Land Data Assimilation System*)
elementary areal unit (EAU) 330, 332
EMICs (see *Earth system models of intermediate complexity*)
emissivity 10, 34, 35, 281
endorheic 215, 216, 378–380, 382, 383, 395
    –, area 378, 380
    –, basin 378, 382
        –, flowing 378
        –, non-flowing 378
    –, water body 380
endorheism 378, 459, 460
endpoint analysis 488
energy
    –, available 10, 15, 19, 35, 46, 52, 64, 75, 88, 129, 159, 160, 174, 176,

273, 310, 341, 342
    –, balance 9, 63, 85, 87, 159, 161, 166, 176, 210
        –, closure 159, 162, 165, 231
        –, closure problem 159
        –, global annual mean 9
        –, measurement 159, 166
        –, residual 159, 166
    –, balance (or Bowen ratio) method 176
    –, dissipation 101
    –, flux, turbulent 10
    –, global cycle 9
    –, storage 7
    –, transport 101
engineering, hydraulic 369
Enhanced Seasonal Observing Period for 1995 (ESOP-95) 26
ensemble, definition 485
ENSO (see *El Niño-Southern Oscillation*)
entrainment
    –, coefficient 191
    –, velocity 190, 194
environmental change
    –, risk assessment 482
EnviSat satellite 248, 281
EOS (see *Earth Observing System*)
ephemeral stream 344, 364, 373
EPIC model (see *Erosion-Productivity Impact Calculator model*)
Eppley PIR (pyrgeometer) 169–171
equilibrium
    –, green 73
    –, projection 496
    –, state 37, 67, 72, 73, 114
equivalent temperature 176
ErhaiSD model 522–525
EROS (see *Earth Resources Observation Systems*)
erosion 3, 62, 65, 93, 276, 299, 302, 317–322, 334, 336, 337, 348, 349, 356, 357, 360, 363, 364, 366, 376, 391–393, 395–398, 400, 402, 412, 424, 425, 439, 449, 455, 457, 499, 508, 522, 526
    –, local 391
    –, rate 319, 321, 356
Erosion-Productivity Impact Calculator (EPIC) model 318, 333
error
    –, checking 282
    –, gross 282
    –, natural 282
    –, propagation 232
ERS-1 satellite 211, 212
ERS-2 satellite 211, 281
ESA-SMOS (see *European Space Agency Soil Moisture and Ocean Salinity*)
Escherichia coli 336, 442–448, 450–452
ESMF (see *Earth system modelling framework*)
ESOP-95 (see *Enhanced Seasonal Observing Period for 1995*)
ESWG (see *Ecological Stratification Working Group*)
Ethiopia, population 516
Eurasia 51, 93–97, 101, 132, 133, 382, 462
    –, ice cap 382
EUROFLUX (measuring network) 160, 258, 270
Europe 17, 40, 44, 45, 52, 53, 56, 57, 70, 71, 87, 95, 110, 133, 157, 211, 216, 258, 278, 284, 317, 321, 330, 371, 380–382, 392, 404, 407, 411, 429, 430, 503, 511, 515
European Centre for Medium-Range Weather Forecasts (ECMWF) 52–56, 72, 86, 108, 113, 116, 194, 222, 228, 248, 283–285, 437
    –, forecast 116
    –, model 53, 54, 72, 194, 228, 437
European Land Data Assimilation System (ELDAS) 283, 284
European Space Agency Soil Moisture and Ocean Salinity (ESA-SMOS) 281
eutrophication 299, 346, 354, 357, 402, 409, 457, 522, 524
evaporation
    –, actual 49
    –, efficiency 49, 118
    –, losses from riverine systems 343
    –, potential 49, 113, 276, 343, 443, 444, 528, 529

–, representation of 11
evaporative fraction 46, 88, 103–105, 107, 166
evapotranspiration 11, 13, 14, 22, 65, 81, 86, 107, 114–116, 118, 120,
    179, 202, 209, 213–220, 286, 299, 303, 305, 310, 312–314, 316, 317,
    329–331, 333, 337, 346, 350, 385, 390, 419, 437–439, 455
    –, cumulative 216
event
    –, assault 499, 528
    –, deprivation 499, 528
    –, water 305, 307, 309, 323, 324, 326, 337
excess water 218, 305, 380, 499
exchange, radiative 211
    –, at surface 211
exorheic 378–380, 382, 383, 395, 400
    –, area 378, 380
    –, river 458
exorheism 378, 459
expansion, agricultural 44, 62, 70, 114, 452
experiment
    –, 2 × CO$_2$ sensitivity 114
    –, design 157, 163
    –, integrated 157, 233
    –, numerical 35, 46, 47, 51, 68, 121, 123, 126
        –, design 121
    –, transient 48
exposure, physical 498

F

FACE fumigation experiment 411
fAPAR (see *fraction of absorbed photosynthetically active
    radiation*)
fast component 14
feedback 33, 40, 45, 46, 59, 66, 71, 73–76, 113, 114, 121
    –, analysis 36, 37
    –, Arctic sea-ice-albedo 41
    –, biogeochemical 2, 33, 134, 484
    –, biogeographical 2, 34, 134
    –, biogeophysical 2, 33–37, 39, 41, 43, 134
    –, biophysical 13, 484
    –, climate-vegetation 47
    –, climatic 113
    –, clouds 484
    –, decadal biogeochemical 45
    –, desert-albedo 34, 53
    –, emissivity-vegetation 35
    –, hydrological 35, 36
    –, non-linear 482
    –, ocean sea surface temperature effects 484
    –, precipitation 484
    –, radiative 113
    –, sea ice 484
    –, snow cover 97, 113, 133, 484
    –, spatial 482
    –, temporal 482
    –, transmissivity 323
fen 94, 103–108, 112, 217
fertilisation 48, 112, 212, 311, 317, 334, 355, 356, 404, 433–436, 503, 511
fertiliser 62, 305, 321, 349, 354, 356, 357, 404, 406, 409, 411, 427, 430,
    433, 434, 436, 445, 449, 457, 461, 499
fetch 163, 176, 180, 181, 195, 230
field
    –, erosion 391
    –, experiment 2, 59, 63, 76, 79, 87, 131, 135, 182, 207–210, 212, 213,
        215, 218–220, 222, 249, 255, 256, 310, 336
        –, ABLE-2B 83, 87, 190
        –, ABRACOS 17, 81, 83–85, 87, 88, 132, 224, 310
        –, ARME 82, 83, 85, 87, 88, 224
        –, BALTEX 218, 256, 260
        –, BOREAS 17, 54, 101, 103–106, 108, 113, 135, 157, 161, 173,
            186, 187, 193, 194, 196, 203, 208, 209, 212, 213, 216–218,
            220–222, 224, 225, 228, 230, 255, 256, 260
        –, CAMREX 415, 422
        –, CATCH 63–65, 76

        –, EFEDA 157, 186, 208, 212, 221–224, 227
        –, FIFE 17, 135, 157, 159, 161, 193, 194, 208, 210–213, 215, 216,
            221, 222, 224, 225, 229, 230, 255, 256, 260
        –, FLUAMAZON 87
        –, GAME 1, 117, 119, 133, 213, 218, 256, 257, 260
        –, HAPEX-MOBILHY 23, 157, 214
        –, HAPEX-Sahel 17, 23, 63–66, 71, 76, 131, 157, 158, 173, 187,
            193, 200, 202, 204, 208, 210, 212–216, 221–225, 228, 230,
            256
        –, KUREX 159
        –, LBA-WET AMC 87
        –, MAGS 213, 218, 256, 260
        –, NOPEX 17, 103, 107–109, 113, 157, 168, 170, 193, 194, 208,
            212–214, 216, 218, 220, 221, 230, 256, 270
        –, RBLE 84, 85
        –, SALSA 208
        –, SALT 63–65, 76
        –, SEBEX 63, 64, 66, 76, 131
        –, SGP97 187
        –, TARTEX-90 159
        –, TOGA/COARE 170, 171
        –, WETAMC 85, 86
        –, WINTEX 213, 220
FIFE (see *First ISLSCP Field Experiment*)
FIFE Information System (FIS) 255, 256
fir 94, 95, 187, 509, 510
fire 95, 107, 510
First ISLSCP Field Experiment (FIFE) 17, 135, 157, 159, 161, 193, 194,
    208, 210–213, 215, 216, 221, 222, 224, 225, 229, 230, 255, 256, 260
FIS (see *FIFE Information System*)
fish backbone pattern 79
fitting constant 314
flood 45, 50, 53, 319, 320, 322, 341–345, 357, 362–364, 373, 402, 449,
    493, 531
    –, damages 493
    –, generation 317
    –, streamflow 333, 337
floodplain 83, 342–344, 357, 362, 373, 381, 390, 391, 398, 417, 418, 420,
    421, 424, 425, 427, 428, 434, 457, 459, 500, 515
    –, dynamics 425
flow
    –, accumulation 263, 352
    –, attenuator 345
    –, connectivity
        –, classification 377
        –, in river systems 377
    –, delayed 319
    –, distortion 161, 184
    –, Dunne 309
    –, Hortonian 309
    –, intermittent 378
    –, lateral 305, 322, 323, 326
        –, process 322, 337
    –, path 323, 324, 327
    –, perennial 377, 378
    –, pipe 308
    –, reduction curve 351
    –, saturation excess overland 305, 327
    –, seasonal 378
    –, subsurface
        –, rapid deep 326
        –, rapid shallow 322
Flux Source Area Model (FSAM) 181
flux
    –, aerosol 2
    –, airplane 186
    –, carbon 2
    –, buoyancy 162
    –, estimation 391
    –, footprint 163, 181, 230
    –, ground heat 101, 159
    –, heat 2
    –, long-wave radiative 484
    –, loss 178, 181, 182

–, measurement 183, 184, 186–188, 231
–, mesoscale 28, 53
–, nocturnal loss 182
–, momentum 166
–, short-wave radiative 484
–, terrestrial 9
–, tower site 17
–, trace gas 2
–, turbulent 25, 34, 88, 160–165, 177, 190, 191, 223, 224
–, water vapour 2
FLUXNET (measuring network) 17–19, 129, 135, 178, 182, 248, 258, 270, 286, 317
forcing
–, climate 67
–, radiative by greenhouse gas 81, 82
forecast, seasonal 491
forest
–, albedo 106
–, boreal 2, 8, 40, 47, 54, 55, 93–97, 99–102, 104, 105, 107–110, 113, 114, 132, 133, 157, 187, 203, 211, 213, 216–218, 256
–, coniferous 54, 94, 95, 103, 106, 109, 218, 506
–, deciduous 18, 25, 111, 160, 196
–, regrowth 48, 112
–, tropical 8, 47, 79, 81–83, 88, 89, 93, 131, 132, 158, 160, 164, 165
four-dimensional data assimilation (4DDA) 85, 257, 281, 282
Fourier analysis 444
FPAR (see fraction of photosynthetically active radiation)
fraction of absorbed photosynthetically active radiation (fAPAR) 209, 285
fraction of photosynthetically active radiation (FPAR) 66, 207, 209, 210
fraction, evaporative 46, 88, 103–105, 107, 166
fractional
–, area drained 391
–, vegetation coverage 121–123
France 90, 186, 187, 213, 214, 221, 222, 283, 320, 321, 355, 357, 374, 401
freezing 102, 230, 313, 337
friction velocity 181
FSAM (see Flux Source Area Model)
function, aggregating 31

G

GAC (see global area coverage)
GAIM (see Global Analysis, Integration and Modelling)
GAIN (see GAME Archive and Information Network)
GAME (see GEWEX Asian Monsoon Experiment)
GAME Archive and Information Network (GAIN) 257
GAME-HUBEX (see Huai-he River Basin Experiment)
Ganges 394
gap fraction 199
GAPP (see GEWEX Americas Prediction Project)
gas exchange, stomatal 11
GCIP (see GEWEX Continental-Scale International Project)
GCIP-DMSS (see GCIP Data Management Services System)
GCM (see general circulation model)
GCOS (see Global Climate Observing System)
GDRC (see Global Runoff Data Centre)
general circulation model (GCM) 1, 22, 31, 32, 47, 48, 68, 71, 75, 82, 87–90, 108, 130, 157, 223, 224, 230, 233, 257, 275, 285, 310, 313, 314, 316, 340, 385, 412, 437, 487, 492, 496, 497, 536
–, grid 64, 71, 157, 167, 222, 280, 452
–, simulation 68, 71, 81
geographic information system (GIS) 300, 301, 304, 328, 330–333, 355, 377, 391, 421, 453, 520, 535
Geographic Resources Analysis Support System (GRASS) 331, 333
Geostationary Operational Environmental Satellite (GOES) 27, 211, 248, 281, 283
German Advisory Council on Global Change (WBGU) 489
Germany 2, 162, 307, 325, 326, 330, 336, 341, 355–357, 373, 378, 413, 429, 430, 436, 515
Geul Valley (Netherlands) 373
GEWEX (see Global Energy and Water Cycle Experiment)
GEWEX Americas Prediction Project (GAPP) 256

GEWEX Asian Monsoon Experiment (GAME) 1, 117, 119, 133, 213, 218, 256, 257, 260
GEWEX Continental Scale Experiments 218
GEWEX Continental-Scale International Project (GCIP) 26, 218, 222, 256, 260
GEWEX Global Land/Atmosphere System Study (GLASS) 16, 268
GEWEX Hydrometeorology Panel (GHP) 260
GEWEX-CEOP 1 (see GEWEX Coordinated Enhanced Observing Period)
GEWEX-Coordinated Enhanced Observing Period (GEWEX-CEOP 1) 65, 233, 256, 260
GHP (see GEWEX Hydrometeorology Panel)
GIS (see geographic information system)
GISS (see Goddard Institute for Space Studies)
glacial maximum 38, 93
glaciation 381
–, controls of river networks 381
glacier 382
GLASS (see Global Land/Atmosphere System Study)
GLDAS (see Global Land Data Assimilation Scheme)
Glen Canyon Dam 363, 364
Global Analysis, Integration and Modelling (GAIM) 3, 275
global area coverage (GAC) 422
Global Change Open Science Conference 1
global circulation model (GCM) 157
Global Climate Observing System (GCOS) 65
Global Energy and Water Cycle Experiment (GEWEX) 1, 14, 26, 65, 117, 213, 214, 218, 228, 229, 256, 257, 260, 268, 269, 280, 284
Global Land/Atmosphere System Study (GLASS) 16, 268
Global Land Data Assimilation Scheme (GLDAS) 283, 284
global observing system (GOS) 248, 258
Global Pedon Database (GPDB) 258
global positioning system (GPS) 184–186, 231
Global Precipitation Climatology Project (GPCP) 280
Global River Inputs to the Ocean (GLORI) 394, 407, 408
Global Runoff Data Centre (GDRC) 515
Global Soil Wetness Project (GSWP) 14, 16, 49, 262, 269, 280, 284
global
–, model 229
–, scale 2, 3, 7, 16, 33, 36, 40, 47, 108, 130, 157, 207, 208, 248, 259–261, 300, 310, 311, 313, 315–317, 341, 370, 374, 377, 380, 383, 385, 386, 388, 392, 393, 395, 397–399, 402, 404, 405, 412, 456, 458–461, 463, 482, 483, 485, 486, 536
–, warming 36, 37, 117, 157
GLORI (see Global River Inputs to the Ocean)
Goddard Institute for Space Studies (GISS) 30, 319
–, GCM 30
Goddard Space Flight Center (GSFC) 66
GOES (see Geostationary Operational Environmental Satellite)
GOS (see global observing system)
Goulburn basin 361
GPCP (see Global Precipitation Climatology Project)
GPDB (see Global Pedon Database)
GPP (see gross primary productivity)
GPS (see global positioning system)
GRACE (see Gravity Recovery and Climate Experiment)
GrADS (see Grid Analysis and Display System)
GRASS (see Geographic Resources Analysis Support System)
grassland 37, 73, 87, 106, 120–123, 157, 164, 215, 309, 320–322, 330, 346, 349, 436, 522, 528, 532
Gravity Recovery and Climate Experiment (GRACE) 281
gray 511
grazing management 532
Great Bear Lake 382
Great Lakes 25, 382, 385
Great Salt Lake 378, 383
green
–, desert 102, 133
–, equilibrium 73
Greenback cutthroat trout 511
greenhouse gas 45, 84, 121, 123, 134, 485–487, 496, 508
–, increasing 508
–, radiative forcing 81, 82
greening 2, 35, 42, 43, 130, 348

–, of the Sahara 2, 35
GRIB (see *gridded-binary format*)
GRIB format (see *gridded-binary format*)
Grid Analysis and Display System (GrADS) 265
gridded-binary (GRIB) format 258, 263, 265
grizzly bear 508
gross primary productivity (GPP) 19, 109, 110
ground heat flux 101, 159
groundwater 514
  –, deep 303, 325, 327, 330, 380
  –, depth 337
  –, flow 109, 314, 325, 380
  –, net flux 213
  –, recharge 216, 302, 305, 322, 333, 390, 434, 437–439
  –, ridging 337
  –, shallow 301, 302, 307, 325, 330, 331
  –, storage 215, 305, 326
GSFC (see *Goddard Space Flight Center*)
GSWP (see *Global Soil Wetness Project*)
GTOPO30 digital elevation map 263, 276, 417, 418
Guangdong province
  –, area 318
  –, rainfall 318
Guiera senegalensis 204
Gulf of Mexico 24, 29, 52

H

habitat loss 510
Hadeji (Nigeria) 373
Hadley Centre
  –, AGCM 47
  –, regional climate model 230
Hadley Centre Ocean Carbon Cycle model (HadOCC model) 48
HAPEX (see *Hydrological Atmospheric Pilot Experiment*)
HAPEX-MOBILHY (see *Hydrological Atmospheric Pilot Experiment-Modélisation du Bilan Hydrique*)
HAPEX-Sahel (see *Hydrological Atmospheric Pilot Experiment Sahel*)
harvest measurement 200, 201, 204
hatched trout 511
hazard 320, 321, 347, 494, 498–500, 527, 528, 531, 532, 534
  –, definition 498
  –, determination 527
  –, index, hydrological 528
  –, response 498
  –, secondary modification 532
heat
  –, budget 193
  –, flux 2, 10, 12, 13, 19, 21, 22, 24–26, 28, 31, 32, 34, 36, 52, 66, 71, 75, 87, 101, 103, 107–109, 163, 165, 166, 202, 230, 506
    –, turbulent 21, 22, 34
  –, latent 9, 10, 12, 14, 19, 21, 22, 25, 28, 29, 34, 36, 52, 66, 67, 74, 75, 82, 87, 91, 101, 103, 108, 115, 118, 119, 129, 159, 163, 165, 166, 187, 202, 210, 214, 215, 217, 218, 221, 258
  –, sensible 9–11, 13, 15, 19, 21, 24, 25, 27, 28, 34, 36, 49, 52, 56, 64, 66, 74, 75, 81, 82, 87, 92, 101–103, 107–109, 118, 119, 129, 132, 159, 162, 164–166, 178, 187, 196, 202, 208, 210, 224, 230, 258, 273, 281, 310
    –, flux 10, 13, 21, 24, 25, 27, 28, 49, 56, 64, 74, 92, 103, 107–109, 119, 159, 162, 164, 178, 187, 196, 202, 210, 224, 230, 273, 281
height, geopotential 123, 125
High Plains 514
  –, aquifer 380
high
  –, emission scenario 81
  –, latitudes 322
HILLFLOW (hillslope model) 309
hillslope 301, 305, 308–310, 317, 324, 326–328, 374, 445, 459
  –, scale 305, 309, 324
HIV/AIDS pandemic 443, 453
Hoggar 383
hole, hyperbolic 179
Holocene 34, 35, 37–43, 62, 72, 130, 376, 381, 383, 460, 461, 505, 507, 509
  –, climate change 505
  –, late 62, 508

–, middle 40, 72
–, optimum 42
homogeneity 180, 190, 303, 325
Hortonian overland flow 305, 309
hot towers 29
house mouse 511
HRU (see *hydrological response unit*)
Huai-he River Basin Experiment (GAME-HUBEX) 119
Huang He (Yellow River) 3, 394, 397, 398, 525, 526
  –, basin 3
Hubei province
  –, area 318
  –, rainfall 318
Hudson Bay 95, 389
humidity
  –, relative 53, 102–107, 132, 195, 197, 258
  –, specific 23, 124, 125, 189
hybrid method 191
Hydra 160, 173
Hydraulic Properties of European Soils (HYPRES) 259
hydraulic
  –, conductivity 15, 68, 69, 227, 259, 308, 309, 311, 316
  –, engineering 369
  –, lift 312
HYDRO1k river and basin datasets 263
hydroelectric capacity 390
hydrograph 217, 307, 309, 310, 321, 323–327, 337, 346, 353, 363, 384, 389, 419–422, 424, 428, 461
  –, distortion 389
  –, separation problem 325
  –, urban characteristics 353
Hydrological Atmospheric Pilot Experiment (HAPEX)-Sahel 17, 23, 63–66, 71, 76, 131, 157, 158, 173, 187, 193, 200, 202, 204, 208, 210, 212–216, 221–225, 228, 230, 256
Hydrological Atmospheric Pilot Experiment-Modélisation du Bilan Hydrique (HAPEX-MOBILHY) 23, 157, 214
hydrological response unit (HRU) 302, 328, 330
hydrological
  –, balance 2
  –, catchment measurement 232
  –, cycle 1, 3, 14, 26, 32, 89, 90, 129, 223, 227, 228, 299, 301, 311, 316, 333, 340, 348, 363, 375, 456, 484, 489, 492, 496, 505, 514, 528
    –, terrestrial 299, 363, 514
  –, driver 339
  –, feedback 35, 36
  –, hazard index 528
  –, model 16, 228, 275, 276, 310, 311, 317, 319, 329, 332, 337, 415, 492, 496, 520
  –, monitoring 519–521
  –, paradox 328
  –, process 51, 64, 65, 69, 75, 76, 87, 228, 277, 301, 302, 305, 329, 332, 337, 338, 340, 377, 380, 430, 431, 438, 439, 528
  –, repercussion 356
  –, response 46, 76, 131, 278, 319, 334, 339, 340, 349, 362, 374, 441, 443, 446, 447, 459, 528, 532
    –, change 348
    –, of urbanisation 353
    –, transformation 348
    –, unit (HRU) 302, 328, 330
  –, risk management 527, 535
hydrology 63, 64, 107, 160, 285, 346, 369, 374, 415, 419, 422, 437
  –, climate change impacts 437
hydrosphere 1, 7, 33, 205, 299, 376, 486
hydrotope 302–305, 328, 330–333
hydrotype 328
hyperbolic hole 179
hyporheic zone 328
HYPRES (see *Hydraulic Properties of European Soils*)

I

IAM (see *integrated assessment model*)
IBIS (see *integrated biosphere simulator*)
ICAM (see *Integrated Climate Assessment Model*)

ice
-, age theory 7
-, cap 382
-, cover 207
-, sheets 42, 377, 382
ICM (see *integrated catchment management*)
ICODT (see *industrial COD discharge*)
ICOLD (see *International Commission on Large Dams*)
ICRISAT (see *International Crops Research Institute for the Semi-Arid Tropics*)
ICSU (see *International Council for Science*)
IEA (see *International Energy Agency*)
IFC (see *intensive field campaign*)
IFSARE (see *Interferometric Synthetic Aperture Radar for Elevation*)
IGBP (see *International Geosphere-Biosphere Programme*)
IGBP-DIS (see *International Geosphere-Biosphere Programme Data and Information System*)
IGY (see *International Geophysical Year*)
IHDP (see *International Human Dimensions Programme on Global Environmental Change*)
IHP (see *International Hydrological Programme*)
Illinois 25, 26, 50, 280
IMAGE 2 (see *Integrated Model to Assess the Global Environment*)
impact
-, 1st order 364
-, 2nd order 365
-, 3rd order 365
-, anthropogenic 62, 464
-, mesoscale 24
-, microscale 22
impervious or less permeable areas (AIMP) 303
imperviousness 348, 353
impoundment 339, 340, 363, 364, 373, 376, 386–389, 391, 393, 395–398, 402, 411, 444, 510
Inanda Dam 441, 443, 447, 449
index, topographic 326, 448
India 51, 70, 116, 317, 515, 516, 518, 519
Indian
-, Ocean 378
-, summer monsoon 36
Indonesia 39
Indus
-, basin 358, 359, 462
-, River 358, 373
industrial COD discharge (ICODT) 522, 523
inertial navigation systems (INS) 184
infiltration 2, 15, 62, 129, 216, 262, 301, 305, 309, 313, 317, 320, 321, 332, 335, 340, 344, 390, 455
-, capacity 305, 320, 321
-, excess 303
-, excess (Hortonian) overland flow 309
infrared radiation 9
inland ice 7
Inner Mongolia 76, 127, 133, 526
INPE (see *Instituto Nacional de Pesquisas Espaciais*)
INS (see *inertial navigation systems*)
insertion, direct 282
insolation 10, 34–37, 39, 40, 44, 107, 112, 130, 134, 211
-, change 44, 134
instantaneous unit hydrograph (IUH) 307
Instituto Nacional de Pesquisas Espaciais (INPE) 79, 85, 86
integrated
-, assessment model (IAM) 502
-, biosphere simulator (IBIS) 73
-, catchment management (ICM) 367
-, ecohydrological model 329, 332
-, experiment 157, 233
-, terrestrial experiment (ITE) 157, 167, 186, 202, 204, 221
-, water resource management (IWRM) 3, 299, 340, 365–374
    -, approach 368
    -, definitions 367
    -, spatial scale 370

-, temporal scale 370
Integrated Climate Assessment Model (ICAM) 489, 502
-, -2.0 model 489
-, -2.5 model 489
Integrated Model to Assess the Global Environment (IMAGE 2) 502
intensification, agricultural 62, 319
intensive field campaign (IFC) 214, 215
interaction 10, 21, 63, 66, 71, 82, 229, 501, 511
-, atmosphere-vegetation 34
-, mesoscale 71
Interaction Sol Biosphère Atmosphère model (ISBA) 221, 222, 226, 283
Interferometric Synthetic Aperture Radar for Elevation (IFSARE) 211
interflow 305, 307, 323, 331, 334, 336, 340, 348, 433
-, shallow 323
Intergovernmental Panel on Climate Change (IPCC) 81, 132, 310, 312, 376, 385, 482, 485–487, 492, 501, 504
International Commission on Large Dams (ICOLD) 363, 390
International Council for Science (ICSU) 259
International Crops Research Institute for the Semi-Arid Tropics (ICRISAT) 63
International Energy Agency (IEA) 170
International Geophysical Year (IGY) 259
International Geosphere-Biosphere Programme (IGBP) 1, 3, 14, 65, 167, 229, 233, 248, 258–260, 275, 276, 375, 381, 439, 461
International Geosphere-Biosphere Programme Data and Information System (IGBP-DIS) 258–260
International Human Dimensions Programme on Global Environmental Change (IHDP) 229, 233
International Hydrological Programme (IHP) 338
International Satellite Cloud Climatology Project (ISCCP) 88
International Satellite Land Surface Climatology Project (ISLSCP) 17, 66, 94, 122, 157, 181, 208, 213, 215, 229, 251, 253, 255, 259, 261, 262, 280, 315
-, Initiative I 66, 122, 261, 262
-, Initiative II 262
interpolation 187, 222, 252, 262, 264, 273, 276, 278, 282, 284
-, statistical 278, 282
Intertropical Convergence Zone (ITCZ) 60, 68, 70, 71, 80, 419, 420
inundation 211, 342, 343, 362, 373, 390, 427, 507
inventory method 402
inverse method 109, 110
inversion 16, 22, 52, 53, 84, 189, 190, 286
IPCC (see *Intergovernmental Panel on Climate Change*)
Ireland 355, 357
irradiance, solar 484
Irrawaddy River 391, 392
-, profile 391
irrigation 3, 62, 69, 305, 318, 335, 339, 340, 343, 347, 355, 357–363, 366, 373, 376, 385, 386, 389, 410, 443–445, 450, 451, 457, 461, 482, 502, 503, 510, 515, 519, 524, 528, 534
-, canal 359
ISBA (see *Interaction Sol Biosphère Atmosphère of CNRM*)
ISBA model (see *Interaction Sol Biosphère Atmosphère model*)
ISCCP (see *International Satellite Cloud Climatology Project*)
ISLSCP (see *International Satellite Land Surface Climatology Project*)
isotope concentration 325
Itapeúa 417, 418, 421
ITCZ (see *Intertropical Convergence Zone*)
ITE (see *integrated terrestrial experiment*)
IUH (see *instantaneous unit hydrograph*)
Ivory Coast 320
IWRM (see *integrated water resource management*)

J

jackpine 95, 101–106, 203
Java River system 383
jet, higher latitude zonal 39
Jura lakes 396

K

Kalman filter 282
Kelvin wave 485
Kenli County 526
Kentucky bluegrass 511
Kerulen system 383
key biome 2
Klewa
    –, basin 361
    –, River 361
Kuiseb
    –, basin 344
    –, River 344, 345
Kursk Experiment (KUREX-88) 159

L

La Niña 18, 85–87
Lachlan
    –, basin 361
    –, catchment 360
    –, River 361
Lagrangian
    –, model 180
    –, stochastic simulation 181
    –, theory 11
lahar 396
LAI (see *leaf area index*)
Lake Baikal 383
Lake Bonneville (Great Salt Lake) 383
Lake Chad 378, 380, 381, 383
Lake Chambon 396
Lake Erhai 521, 522, 524
    –, basin 521, 522, 524
        –, water resources 522
Lake Eyre 379, 383
    –, basin 378
Lake Gardner 383
Lake Kariba 389
Lake Nasser 389
Lake Ontario 382
Lake Torrence 383
Lake Victoria 380, 383
lake
    –, albedo 106
    –, dry 383
    –, extension 383
    –, level record 496
    –, tectonic 383
land
    –, abandonment 321, 359
    –, degradation 63, 65, 68, 69, 75–77, 121, 122, 131, 317, 322, 347, 503
    –, management 330, 457
    –, pastoral 62
    –, salinisation 360
    –, surface heterogeneity 273, 304, 328, 330, 337
    –, transformation 348
land-atmosphere
    –, feedback 49, 50
    –, interactions 21
land cover 62, 63, 68, 71, 76
    –, change 3, 39, 40, 207, 299, 310, 349, 375, 376, 390, 393, 412, 437, 457, 487, 504
    –, datasets 501
    –, scenarios 501
Land Data Assimilation System (LDAS) 222, 269, 283, 284, 287
land surface
    –, air temperature anomaly 96
    –, as carbon store 9
    –, change 45
    –, experiment 207, 337
    –, heterogeneity 330

    –, impact on weather 52
    –, model 15, 16, 108, 129, 157, 215, 221, 225, 232, 262, 264, 273, 283
    –, process 3, 10, 49, 52, 73, 75, 76, 87, 90, 129, 131, 221, 273, 281, 330
    –, runoff 305
    –, scheme (LSS) 14–16, 45–47, 221, 222, 228, 264, 273, 274, 279, 280, 283, 287, 313, 314, 337
land use 62, 76, 330, 457
    –, change 347, 349, 437
        –, Sahel 62
        –, scenario 436
        –, types 348
    –, dataset 501
    –, scenario 501
land-use change and land-cover change (LUC-LCC) 501
Land Use Change in Fluvial Systems programme (LUCIFS) 461
Land-Data Assimilation System (LDAS) 222, 269, 283, 284, 287
Landsat 94, 207, 248, 281, 379
    –, Thematic Mapper 94, 281
landscape
    –, mesoscale heterogeneity 26, 31
    –, natural 24, 25, 341, 347, 348, 508
        –, changes 505
        –, processes 341
large river basin
    –, model 415
    –, scale 349, 457
Large Scale Biosphere-Atmosphere Experiment in Amazonia (LBA) 1, 80, 81, 84–87, 132, 158, 208, 213, 218, 219, 229, 233, 250, 256, 257, 260
large-eddy simulation (LES) 24, 191
Last Glacial Maximum (LGM) 38, 39, 381, 382, 383
    –, climate 38
late glacial climate change 505
latent heat 9, 10, 12, 14, 19, 21, 22, 25, 28, 29, 34, 36, 52, 66, 67, 74, 75, 82, 87, 91, 101, 103, 108, 115, 118, 119, 129, 159, 163, 165, 166, 187, 202, 210, 214, 215, 217, 218, 221, 258
    –, flux 12, 14, 19, 21, 22, 25, 28, 29, 36, 52, 66, 74, 75, 82, 87, 101, 103, 108, 119, 129, 159, 163, 165, 166, 187, 202, 210, 214, 215, 217
    –, vaporisation 159
lateral
    –, flow 301, 302, 305, 307, 322, 326, 329–331, 333
        –, process 322, 337
    –, pipe flow 323
    –, preferential flow 305
Latin America 503, 516
LATS (see *Library of AMIP Data Transmission Standards*)
LBA (see *Large Scale Biosphere-Atmosphere Experiment in Amazonia*)
LBA Data Information System (LBA-DIS) 256
L-band 210
LCL (see *lifting condensation level*)
LDAS (see *Land-Data Assimilation System*)
LDC (see *lesser developed countries*)
leaching 334, 349, 354–357, 397, 432–436, 455, 457, 499
leaf
    –, area 7, 19, 22, 23, 25, 31, 34, 35, 40, 47, 53, 66, 83, 103, 106, 107, 113, 121–123, 199–202, 205, 207, 224, 281, 286, 312, 332, 436, 437, 445, 459
        –, index (LAI) 19, 22, 23, 31, 47, 53, 66, 68, 69, 74, 83, 103, 106, 107, 121–123, 199, 200, 202, 207, 217, 281, 286, 312, 332, 333, 436, 437, 445, 459
        –, specific 199, 202
    –, conductance 12
    –, cross section 11
    –, mass 34, 199, 200
    –, nitrogen concentration 201
    –, water
        –, content 201
        –, potential 12, 13, 106, 201
Lena
    –, basin 218
    –, River 379, 383, 384
LES (see *large-eddy simulation*)

Lesotho 344, 347, 527, 528
lesser developed countries (LDC) 371, 374
level of integration 367
LGM (see *Last Glacial Maximum*)
Library of AMIP Data Transmission Standards (LATS) 264
Libyan
  –, Desert 383
  –, Gulf 383
lichens 95
Lichtenwalde 431, 433
lift, hydraulic 312
lifting condensation level (LCL) 104, 195
light
  –, penetration 200
  –, use efficiency 19
lignin phenols 426
limestone 397, 398
Lions River 446, 451
lithosphere 7
litter 201, 321, 323, 348, 349, 445, 506
Little Ice Age 45, 509
livestock 60, 62, 63, 336, 354, 357, 409, 441, 442, 446, 452, 509
Lodden catchment 360
Loess Plateau 397, 526
logging 94, 95, 212, 508, 511
Logone 373
Long Term Ecological Research (LTER) network 248
LSS (see *land surface scheme*)
LTER network (see *Long Term Ecological Research network*)
LUCIFS programme (see *Land Use Change in Fluvial Systems*)
LUC-LCC (land-use change and land-cover change)
Luoshijiang River 522

M

Mackenzie GEWEX Study (MAGS) 213, 218, 256, 260
Mackenzie
  –, basin 218
  –, River 382
  –, river system 382
Macquarie catchment 360
macropore 232, 305, 308, 309, 337
MAGS (see *Mackenzie GEWEX Study*)
Mahaweli 373
Manacapurú 417, 418
management
  –, approach 368
  –, catchment 445, 446, 451
  –, of water systems 3
  –, practice factor 448
Manaus 82–85, 87, 232, 418
mangrove 373
Manych depression 382
MAP (see *mean annual precipitation*)
MAR (see *mean annual runoff*)
Margaret Creek 379
Markov chain 277, 494
  –, Monte Carlo method 494
Mars Chiquita system 383
mass flow controller 177
Massif Central 396
material flux 329, 455, 457, 458, 460, 461
MDB (see *Murray-Darling Basin*)
MDC (see *more developed countries*)
mean annual precipitation (MAP) 343, 344, 351, 430, 443, 447, 448, 451, 528
mean annual runoff (MAR) 343, 443, 447, 448, 450–453, 529, 534, 535
measurement
  –, airborne 184, 186
  –, accuracy 185
  –, flux 183, 186–188
  –, wind 185
  –, methods 109
  –, radiation 167
measuring network
  –, AmeriFlux 189
  –, AsiaFlux 18, 258
  –, CarboEurope 18
  –, EUROFLUX 160, 258, 270
  –, FLUXNET 17–19, 129, 135, 178, 182, 248, 258, 270, 286, 317
  –, MedeFlu 258
  –, OzNet 258
mechanism, homeostatic 484
MedeFlu (measuring network) 258
Mediterranean
  –, climate 45
  –, region 8, 258, 320
  –, Sea 378
Meiyu frontal activity 118, 120
memory
  –, properties 52
  –, time scales 75
Meteorological Office Surface Energy Scheme (MOSES) 67, 221
Meteosat 248
methane 93, 94, 108, 111, 112, 179, 193–195, 312, 313, 427
  –, flux 112
method
  –, aerodynamic gradient 174
  –, Bayesian 494
  –, blending-height 31
  –, boundary layer
    –, budget 190, 196
    –, gradient 191
  –, Bowen ratio (or energy balance) 174, 176
  –, hybrid 191
  –, inventory 402
  –, inverse 109, 110
  –, measurement 109
  –, micrometeorological 173, 174, 180
Mexico 24, 29, 52, 519
Mgeni 2, 336, 341, 346, 357, 413, 441–449, 451–453, 532
  –, basin 346
  –, catchment 336, 413, 441–445, 447–449, 452, 453
    –, human pressure 443
    –, land cover 447
    –, mean annual runoff 448
    –, present land use 442
  –, River 2
  –, system for simulation modelling 444
microclimate 201, 457, 459
micrometeorological method 173, 174, 180
micrometeorology 83, 166, 173, 279
microwave sensing 211
Middle East 71, 410
millet 64
mineralisation 112, 317, 334, 354, 416, 425, 431–436
mining 321, 427
Mississippi
  –, basin 25, 50, 52, 218, 256, 412
  –, River 50, 52, 256
  –, river system 382
  –, total nitrogen export to ocean 404
mitigation, definition 499
Mizu River 522
model
  –, ACRU 329, 334, 335, 351, 443, 446, 449, 528, 531, 532, 534
  –, ARPEGE NWP 221
  –, atmosphere-ocean 7, 36, 39, 41, 43
  –, atmosphere-only 37, 41, 43
  –, atmospheric 221
  –, Ball-Berry 13
  –, biome 35, 40, 41, 43
  –, BIOME-3 314
  –, biophysical 13, 18, 19, 202, 204
  –, Biosphere Energy Transfer Hydrology (BETHY) 285
  –, biosphere-atmosphere
    –, zonally-symmetric coupled 37
  –, bucket 12, 46, 49, 221

–, calibration 328
–, consolidative 491, 492
–, control m. 54
–, coupled atmosphere-ocean 7, 36, 43
–, coupled atmosphere-ocean-vegetation 36, 37, 40, 43
–, coupled land surface-atmosphere 310
–, CY48 53
–, DOLY vegetation m. 47
–, ECHAM 72
–, ECHAMT21 437
–, ECMWF 53, 54, 72, 194, 228, 437
–, ecohydrological 301, 329–332, 335, 430, 439
–, EMICs 48, 484
–, exploratory 490, 492
–, force-restore 15
–, fully coupled 41, 43
–, global 229
    –, scale 2
–, Hadley Centre
    –, AGCM 47
    –, coupled ocean-atmosphere GCM 48
    –, regional climate 230
    –, ocean carbon cycle 48
–, hydrological 16, 221, 228, 275, 276, 310, 311, 317, 319, 329, 332, 337, 415, 492, 496, 520
–, ICAM 2.0 489
–, ICAM 2.5 489
–, integrated ecohydrological 328
–, ISBA-MODCOU 226
–, Lagrangian 180
–, large river basin 415
–, macroscale hydrological 225
–, PAGE 95 m. 489
–, particle trajectory 180
–, photosynthesis 13, 47
–, physics 279, 282
–, prediction 279
–, regional scale 2
–, regression 405
    –, multiple 405
–, Simple Biosphere (SiB) 13, 86, 166, 280
–, simulation 129, 202, 278, 286, 312, 341, 502
–, stochastic 277
–, SWIM 333, 436, 439
Modelling of Nutrient Emissions into River Systems (MONERIS) 355
modelling
–, coupled 66
–, system SWIM 332
Modified Universal Soil Loss Equation (MUSLE) 333, 336
moisture
–, atmospheric storage 214
–, availability term 11
–, content 15, 26, 49, 75, 112, 210, 232, 281, 310, 312, 356, 421
–, flux 12, 22, 29, 50, 75, 131, 190, 214, 219, 316, 421
    –, convergence 50, 75, 131, 214
–, holding capacity 14, 15
–, sink term 15
–, state 49, 210
MONERIS (see Modelling of Nutrient Emissions into River Systems)
monitoring
–, hydrological 519–521
–, network 444
monsoon
–, Asian 8, 115, 116, 118, 120, 121, 123, 134, 320
–, circulation 35, 39, 60, 115, 123, 125, 126
    –, tropical 39
–, current 117, 120
–, East Asian 116, 121, 123
–, Indian summer m. 36
–, winter m. 125, 126, 134
Montana 382, 511
more developed countries (MDC) 371

Morocco 519
mosaic
–, of tiles approach 31
–, structure 302
MOSES (see Meteorological Office Surface Energy Scheme)
Mount St. Helens eruption 396
mountain, active uplift 392
Mulde basin 430
Murray-Darling Basin (MDB) 3, 256, 358–361
–, Water Budget Project 256
Murrumbidgee catchment 360
MUSLE (see Modified Universal Soil Loss Equation)

N

Namibia 344, 345, 364, 373
Nantang basin 318
NAO (see North Atlantic Oscillation)
NASA (see National Aeronautic and Space Administration)
NASA Data Assimilation Office 273
National Aeronautic and Space Administration (NASA) 66, 83, 85, 86, 193, 194, 211, 255, 260, 273, 283
National Center for Atmospheric Research (NCAR) 86, 186, 222, 260, 264, 270
National Centers for Environmental Prediction (NCEP) 73, 86, 190, 222, 248, 273
National Oceanic and Atmospheric Administration (NOAA) 97, 99, 100, 207, 256, 283, 422, 515
National Soil Characterization Database (NSCD) 258
National Water Act of South Africa 451
NBP (see net biome productivity)
NCAR (see National Center for Atmospheric Research)
NCAR Data Storage System (NCAR-DSS) 260
NCAR-DSS (see NCAR Data Storage System)
NCEP (see National Centers for Environmental Prediction)
NDVI (see Normalised Difference Vegetation Index)
NEE (see net ecosystem exchange)
needs, societal 492
Nelson
–, River 382, 389
–, system 382
net biome productivity (NBP) 112, 113
net ecosystem exchange (NEE) 109–112, 270
net primary production (NPP) 13, 63, 109–111, 133, 210, 317
netCDF, format 263, 264
network, topological 405
Neu Darchau 430
neutron probe 214, 216
New Hanover 441, 451
Newtonian nudging 282
Niagara River 383–385
Niamey 61, 63, 64
Niger
–, basin 383
–, River 61, 68, 384, 385
    –, discharge 385
Nigeria 346, 373, 389
Nile 3, 459, 461
–, basin 3, 462
–, catchment 379
–, floodplain 515
–, River 385, 386, 389
–, valley 44
nitrate 327, 334, 349, 354, 355, 403, 406–409, 416, 423–425, 431–433, 436, 511
–, concentration 355, 408
–, export rate 408
–, flux 407
nitrogen 13, 48, 110, 112, 170, 201, 205, 212, 280, 281, 285, 287, 332–334, 354–356, 402, 403, 406–409, 411–413, 416, 424–426, 430, 431, 433, 434, 439, 456, 457, 464, 491, 501, 511, 512
–, deposition 48, 112, 212
–, dynamics 431, 433
–, fixation 457

–, flow components 432, 433
–, in rivers 511
–, inorganic 402, 403
    –, load 411
    –, sequestration 413
–, leaching 357, 434, 435
–, loss 354, 435
–, mineral n. 355
–, mineralisation 354, 435, 436
–, organic 334, 354, 416, 424, 434, 435
–, pollution 354
–, ratio 401, 409
–, riverine flux 403
–, stoichiometric change 408
–, total
    –, Amazon River 404
    –, annual flux 424
    –, Mississippi 404
–, transport 355
–, uptake 435
NOAA (see *National Oceanic and Atmospheric Administration*)
nocturnal 181
NOPEX (see *Northern Hemisphere Climate-Processes
    Land-surface Experiment*)
Normalised Difference Vegetation Index (NDVI) 26, 72, 209, 210,
    223
North America 17, 29, 51, 54, 95–97, 99, 100, 108, 110, 113, 120, 133,
    190, 283, 284, 330, 381–383, 387, 482
    –, megadrought 482
North Atlantic
    –, basin 404
    –, deep water 385
    –, Oscillation (NAO) 45, 97, 376, 385, 402
North China Plain 526
Northern Dvina River 383, 384
Northern Hemisphere Climate-Processes Land-surface
    Experiment (NOPEX) 17, 103, 107–109, 113, 157, 168, 170, 193, 194,
    208, 212–214, 216, 218, 220, 221, 230, 256, 270
Norway rat 511
NPP (see *net primary production*)
NSCD (see *National Soil Characterization Database*)
Ntabamhlope 345, 346
numerical weather prediction (NWP) 106, 107, 218, 219, 221, 222,
    273, 283, 310
    –, centre 273
    –, model 221, 222
nutrient 1, 3, 8, 60, 65, 79, 93, 114, 131, 132, 205, 251, 259, 301, 311, 314,
    315, 317, 322, 327–330, 332–334, 337, 342, 343, 345, 346, 348, 354–357,
    364, 373, 375–377, 381–383, 390, 391, 396, 402, 404–409, 411–413,
    415, 416, 424, 427, 430, 433, 439, 455–458, 463, 464, 503, 508, 511,
    522, 524
    –, dynamics 326, 328, 415
    –, flux, riverine 402
    –, inputs to the oceans 407
    –, load 327, 357, 427, 433
NWP (see *numerical weather prediction*)

O

Ob River 382
Óbidos 417–422, 424, 425
observation 282
    –, operator 282, 287
    –, terrestrial 280, 286
    –, very small scale 195
observations of screen-level temperature and humidity (SLTH)
    284
observing period 216
OHP (see *Operational Hydrological Programme*)
Okavango 378
    –, system 383
Oklahoma 24–26, 28, 186, 257, 258, 309
OM (see *organic matter*)

Ontario 113, 114, 322, 327, 382
open-path analyser 177
Operational Hydrological Programme (OHP) 338
Orange River 343
orbital forcing 40–44, 130
ORCHIDEE (see *Organizing Carbon and Hydrology In Dynamic
    Ecosystems*)
organic matter (OM) 110, 113, 133, 315, 317, 321, 322, 337, 343, 354, 356,
    364, 373, 377, 398, 415, 416, 424–427, 433, 445
organisation, spatial 300, 377, 458
Organizing Carbon and Hydrology In Dynamic Ecosystems
    (ORCHIDEE) 285
Ovens basin 361
overgrazing 63, 319, 320, 340, 347, 348, 447, 532
overland flow 302, 303, 305, 307, 309, 312, 317, 319, 321–323, 327, 331,
    349, 356, 419, 421
oxygen, dissolved 354, 423
OzFlux (measuring network) 18
OzNet (measuring network) 258
ozone, deposition velocity 193

P

Pacific Ocean 378
Pacific-North America (PNA) 97, 99
paddy field 118–120, 133
PAGE 95 model 489
PAGES (Past Global Changes) IGBP Core Project 383
Pakistan 358, 359, 516
    –, population 516
palaeoclimate 40, 134
    –, reconstruction 33
    –, scenario 496
palsa 108, 112
Panola Mountain Research Watershed 324
Papua New Guinea 319
PAR (see *photosynthetically active radiation*)
paradox, hydrological 328
parameter, organic geochemical 426
Parana River 383–385
    –, runoff 385
particle trajectory model 180
particulate
    –, inorganic carbon (PIC) 397, 399
    –, nitrogen (PN) 411, 412
    –, organic carbon (POC) 397–402, 411, 416
    –, organic nitrogen (PON) 416
partnership approach 369
passive vulnerability 498
pastoral land 62
pastoralism 62, 348
patch 2, 22–24, 28, 31, 65, 84, 108, 223, 228, 302, 303, 305, 328
    –, scale 23, 24
PBL (see *planetary boundary layer*)
PCBs 457
PCMDI (see *Program for Climate Model Diagnosis and
    Inter-comparison*)
Peace River 113, 114
    –, region 113, 114
peat 93, 107, 108, 112–114, 345, 400, 401, 462, 506, 507
peat-bog 507
peatland 107, 108, 110–112, 132
PED (see *potential energy divergence*)
pedogenesis 315, 505, 507, 508
pedosphere 7, 33
pedotransfer function 258, 259, 332
Peltier element 169
percolation 302, 303, 309, 316, 333, 346, 359
perennial
    –, flow 377, 378
    –, rivers 373
permafrost 7, 93, 108, 111, 112, 114, 133, 312, 400, 506, 507, 509
Persian Gulf 383

perturbation 24, 33, 34, 37, 134, 136, 457, 485, 496
–, environmental 485, 486, 494
pesticide 305, 346, 354, 377, 430, 457
pH value 424
phosphorus 332–336, 354, 356, 408, 409, 413, 430, 439, 443–450, 452, 456
–, dissolved 357
–, enrichment 356
–, fertiliser 445
–, inorganic 402
–, input 357
–, labile 334
–, mean annual yields 450
–, mineral p. 334
–, mobilisation 356
–, non-point input 354
–, organic 334
–, particulate 357
–, pollution 354
–, processing 356
–, transport 355
–, yield map 449
photometer 169
photosynthesis 11, 13, 46–48, 109–112, 130, 132, 133, 160, 167, 181, 196, 199–202, 204, 209, 210, 231, 232, 251, 287, 311, 317, 417, 438
–, model 13, 47
–, net rate 47, 181, 204, 438
–, relationship to stomatal conductance 13
photosynthetically active radiation (PAR) 12, 13, 66, 87, 107, 110, 199, 202, 203, 208, 209, 211, 438
PIC (see *particulate inorganic carbon*)
picket fence approach 214
Piedmont region 396
Pietermaritzburg 374, 443, 447, 452, 453
Pilot Land Data System (PLDS) 255
PILPS (see *Project for Intercomparison of Landsurface Parameterization Schemes*)
pine 23, 83, 94, 95, 111, 114, 187, 223, 322, 350, 351, 451, 506, 508–511
Pinedale glaciation 509
pipe flow 323, 337
Pisa 321
piston flow 305, 337
Planalto Brasileiro 417
Planalto das Guianas 418
planetary boundary layer (PBL) 22–25, 28, 31, 32, 65, 129, 180, 189–191, 223, 224, 438
plant
–, physiology 7, 47, 48, 83, 130, 181, 279
–, uptake 334, 431, 432, 434–436
plantation
–, afforestation 349, 350
–, forest 349
plateau incision 391, 392
PLDS (see *Pilot Land Data System*)
plot scale 222
PN (see *particulate nitrogen*)
PNA (see *Pacific-North America*)
POC (see *particulate organic carbon*)
point source 406
policy framework 369
pollen record 496
pollution 1, 3, 182, 299, 322, 339, 340, 346, 348, 349, 354, 357, 369, 370, 402, 409, 410, 430, 431, 436, 453, 457, 462, 491, 497, 499, 501, 508, 509, 511–513, 517, 522–524, 526, 527, 537
–, air 370, 409, 491, 501, 511–513, 537
–, flux, assessment 402
–, water 511
PON (see *particulate organic nitrogen*)
ponderosa pine (*Pinus ponderosa*) 508, 510
Pongola 373
poplar 94, 95
population 320, 509, 510, 514, 516
–, density 516

–, human 378
–, growth 62, 299, 320, 375, 411, 453, 496, 508–510, 512, 514, 517, 519, 522
–, human 509
pore volume fraction 15
potential energy divergence (PED) 90, 91
prairie 113, 114
precipitation
–, anomaly 49, 52
–, cumulative 216
–, damage 493
–, pattern 419
–, recycling 53
–, transient development 44
–, variability 67, 72, 73
–, variance 49, 50, 130
predictability 484
–, of climate 485
–, of precipitation 49, 130
prediction 33, 49–52, 71, 85, 157, 194, 218, 233, 253, 273, 280, 282, 283, 286, 287, 309, 312, 316, 317, 437, 482, 484–487, 490–492, 494, 497, 537
–, classes 486
–, first kind 485
–, of climate 484
–, of weather 484
–, relationship to vulnerability 482
–, second kind 485
predictive
–, approach 490
–, information 490
pressure
–, anthropogenic 65, 340, 459
–, head 311, 312, 324
–, transducer 177
–, wave 305, 308, 310, 323, 324, 337
–, translatory flow 305, 323
–, transmission 308
PRISM (see *Programme for Integrated Earth System Modelling*)
process
–, biogeographical 484
–, hydrological 51, 64, 65, 69, 75, 76, 87, 228, 277, 301, 302, 305, 329, 332, 337, 338, 340, 377, 380, 430, 431, 438, 439, 528
–, terrestrial hydrological
–, definition 301
–, overview 301
–, vertical hydrological 302
production, net 201, 204
–, primary 13, 63, 109–111, 133, 210, 317
productivity, aquatic 364
profile
–, chemical 422
–, convex 391
Program for Atmospheric Radiation Measurement (ARM) 26, 28, 257
Program for Climate Model Diagnosis and Inter-comparison (PCMDI) 263, 264
Programme for Integrated Earth System Modelling (PRISM) 265, 278
–, method 278
Project for Intercomparison of Landsurface Parameterization Schemes (PILPS) 14–16, 45, 222, 225, 227, 264, 265, 269, 280, 310
projection
–, definition 485, 491
–, equilibrium 496
–, transient 496
propagation of errors 232
psychrometric constant 176
Punjab region 359
purple loosestrife 511
pyranometer 168, 169
pyrgeometer 161, 167–170

pyrheliometer 169
pyrradiometer 169

Q

QC (see *quality control*)
quality control (QC) 282
queriability 268
quickflow 319, 322, 336

R

RADARSAT (microwave space mission) 211
radiation
–, atmosphere-surface balance 207
–, atmospheric 34–36, 170
–, infrared 9
–, long wave 10, 64, 74, 87, 89, 208
–, measurement 167
–, net 9, 101, 161, 230
    –, sensor 161
–, short-wave 9, 10, 101, 169, 170, 262
–, solar 9, 13, 34, 35, 47, 73, 80, 84, 87, 105, 107, 132, 163, 169, 196,
    200, 217, 263, 276, 277, 310, 422
    –, absorption 34, 80, 132
–, thermal 35, 36
radiative
–, exchange 211
    –, at surface 211
–, feedback 113
–, forcing, greenhouse gas 81, 82
radiometer 161, 163, 167–171, 211, 230, 310
–, bolometric 169
–, design 169
–, net r. 161, 163, 168–171, 230
radiometry 167, 170
rainbow trout 511
rainfall
–, anomaly 60, 70, 80, 81, 278
–, distribution 61, 276
–, erosivity 445, 448, 449
–, variability, intra-seasonal 443
–, variation 60, 419
rain-use efficiency (RUE) 63
RAMS (see *Regional Atmospheric Modelling System*)
rate, photosynthetic 110, 201
RBLE (see *Rondônian Boundary Layer Experiment*)
RCM (see *regional climate models*)
realisation, definition 485
recovery ratio 159, 160, 164
region
–, arid 43, 71, 225, 317, 318, 343, 344, 377, 507
–, temperate 322
Regional Atmospheric Modelling System (RAMS) 26, 27, 117, 194
–, simulation 27
regional climate models (RCM) 496, 497
Regional Integrated Environmental Model System (RIEMS) 123
regression
–, model 405
–, multiple 393
    –, model 405
rehabilitation 341, 348, 362, 367, 371, 374, 464, 534
relationship, hypsometric 391
reliability, definition 499
remote sensing 207, 208, 211
–, input 208
repercussion, hydrological 356
reservoir 3, 15, 16, 93, 225, 313, 335, 363, 364, 373, 376, 386–389, 392,
    393, 395–397, 402, 409, 443, 449, 457, 494, 496, 528, 534
–, construction 376, 386, 388, 396, 402, 457
–, residence time 388, 395
–, trapping 395
residence time
–, local 386, 395

–, mean 325, 326, 386, 387
resilience 75, 484, 486, 499, 508
–, definition 499
resistance
–, aerodynamic 10
–, laminar 11
resource
–, management 77, 299, 369, 376, 508
–, renewable 501
respiration
–, autotrophic 109
–, heterotrophic 109, 130, 133, 179
response
–, function 307, 312, 325
–, hydrological 46, 76, 131, 278, 319, 334, 339, 340, 349, 362, 374,
    441, 443, 446, 447, 459, 528, 532
    –, change 348
    –, of urbanisation 353
    –, transformation 348
–, time 75, 178, 264, 307, 308, 337
–, scale 460
restoration 341, 348, 362, 373, 374, 509, 512
–, of riverine ecosystem 373
retrievability 268
Revised Universal Soil Loss Equation (RUSLE) 336, 445
Rhine
–, basin 371
    –, area 355
–, River 284, 355, 356, 371, 373, 402, 430, 461
–, valley 430
rice paddy field 118
Richard equation 15, 311
RIEMS (see *Regional Integrated Environmental Model System*)
Rio Branco 418
Rio Içá 417, 418
–, basin 419
–, discharge 420
Rio Japurá 417
–, basin 419
–, discharge 420
Rio Juruá 424
–, discharge 420
Rio Jutaí 418
–, discharge 420
Rio Madeira 417, 420, 422, 424, 428
–, sediment 424
Rio Negro 398, 418, 424, 428
–, discharge 420
–, basin 419
Rio Purús 424
–, discharge 420
Rio Tapajós 418, 420
Rio Tinto 462
Rio Xingú 418, 420
riparian
–, corridor 457
–, zone 305, 307, 327, 330, 343, 350–352, 364, 373, 391, 416, 419,
    427, 437, 508, 511–513
–, process 363
risk 498, 500
–, assessment 253
–, definition 498, 500
–, management 527, 535
–, mitigation 527
–, upper band of 500
river
–, ageing 388
–, bank reinforcement 362
–, carbon 397, 398, 411
    –, flux 398
    –, transfer 397
–, channel modification 362
–, chemistry 386, 394
–, connectivity 377

–, continuum concept 342, 343
–, corridor 79, 219, 389, 427, 436, 437
–, discharge 16, 68, 86, 215, 382, 383, 385, 386, 388, 407, 408, 415, 421, 429, 436, 460, 461
    –, Arctic 385
    –, measurement 86
    –, oscillation 421
–, engineered landscape 362
–, evaporation 343, 344
    –, loss 343
–, exorheic 458
–, flow 227, 228, 263, 299, 322, 342, 343, 353, 362, 370, 377, 383, 386, 387, 389, 396, 457, 485, 496, 520, 522
–, large, hypsometric curves 392
–, mouth 357, 391, 392, 412
    –, export 391
–, natural 341
–, navigation 369
–, network 302, 305, 343, 362, 377–379, 381, 387, 388, 391, 417, 418
    –, active 379
    –, conditioning 383
    –, deglaciation 381
    –, glaciation 381
    –, in the Sahel 379
    –, potential 379
–, nitrogen 407, 408, 411
    –, flux 404
–, nutrient flux 405
–, perennial 373
–, pulse concept 342, 343
–, regulation 359, 388, 393
–, rehabilitation 362
–, restoration 362
–, routing 16, 68, 207, 332, 379
    –, scheme 16, 68
–, salinity 357
–, scenarios 76, 81, 131, 262, 275–277, 317–319, 329, 348, 351, 357, 367, 401, 411, 428, 436–439, 443, 446, 451, 452, 460, 461, 464, 482, 483, 486, 488, 491, 492, 494, 496, 497, 501–503, 505, 508, 523, 524, 536, 537
–, section 305
–, segment 388, 395
–, surface evaporation 344
–, suspended solids 340, 390, 393, 394
–, system 300, 307, 339, 342, 343, 354, 356–358, 363, 365, 375–379, 381–383, 386–389, 391, 394–396, 399, 404, 405, 408, 409, 412, 459
    –, active 379
    –, dammed 389
    –, flow connectivity 377
    –, integrity 297
    –, potential 379
–, water 418
–, water demand 307, 352, 361, 365, 443, 445, 500, 501, 517, 519
river basin
    –, anthropogenic modification 347
    –, as fundamental unit 369
    –, biogeochemistry 415
    –, boundary 218
    –, drivers 339
    –, estimates 516
    –, hydrology 415
    –, hydrotopes 302
    –, impacts 340
    –, large 333, 334
    –, mesoscale 329, 333, 334, 338, 431, 437
    –, nutrient dynamics 415
    –, organic matter 415
    –, pressures 339
    –, responses 339, 340
    –, salinity 361
    –, scale 339, 341, 350
        –, DPSIR model 339, 374
    –, state changes 340

–, water flow 415
riverine system
    –, evaporation losses 343
    –, transmission losses 343
Rocky Mountain goat 511
Rocky Mountain National Park 508–510
    –, vulnerability 508
Rocky Mountains 24, 508–511
    –, Front Range 508
    –, water quality 511
Roman classical period 44, 45
Rondônian Boundary Layer Experiment (RBLE) 84, 85
root
    –, density 311–314
        –, distribution 312, 313
    –, distribution
        –, cumulative 314
        –, function 313
    –, fraction 315
    –, production 201, 204
    –, water
        –, extraction 311
        –, flux density 312
        –, uptake 251, 302, 303, 311, 314, 316
    –, zone 21, 22, 51, 103, 106, 223, 227, 273, 286, 310–312, 316, 331, 333, 357, 421
rooting depth 51, 83, 201, 259, 311–317
Rossby wave 39, 40, 80
roughness 10, 11, 21–23, 31, 34, 51, 66, 68, 74, 84, 88–90, 108, 115, 118, 121–123, 133, 176, 180, 181, 207, 210, 211, 224, 281, 348, 349, 356, 362
    –, length, aerodynamic 11
rubber boom 79
RUE (see *rain-use efficiency*)
runoff 14–16, 61, 68, 108, 129, 305, 307, 378, 385, 389, 437, 438, 450, 510, 515, 520, 534
    –, ageing 389
    –, coefficient 317, 320, 321, 431, 447, 448
    –, components 305
    –, continental 300, 376, 380, 386, 388–390, 396, 410, 412
    –, episodic 379
    –, generation 299, 301, 302, 305–309, 317, 323, 328–331, 337, 338, 350, 439, 455
    –, model 307, 318, 326
    –, net r. 386, 389, 390
    –, permanent 379
    –, response 46, 351, 528
    –, seasonal 379
    –, time behaviour of components 307
RUSLE (see *Revised Universal Soil Loss Equation*)
Russian thistle 511

S

Saale basin 430
Sacramento basin 318
SACZ (see *South Atlantic Convergence Zone*)
Sahara 2, 7, 35, 37–39, 42–44, 49, 59, 60, 72, 73, 130, 320, 378, 383, 459
    –, drainage basin system 378
    –, dry lakes 383
    –, greening 2, 35
Sahel
    –, climate 59, 68, 130
        –, anomaly 68
    –, region 2, 8, 17, 35, 39, 42, 45, 59–77, 130, 131, 134, 157, 159, 160, 173, 187, 193, 200, 202, 204, 208, 210, 212, 213, 215, 216, 221–225, 228, 230, 256, 318, 320, 321, 379, 398
Sahelian Energy Balance Experiment (SEBEX) 63, 64, 66, 76, 131
Sahoura River 383
Saint Lawrence basin 459
salinisation 357–360
    –, primary source 358
    –, secondary source 358
salinity 311, 312, 357–362, 385, 484
    –, control 361

–, stress 312
SALSA (see *Semi-Arid Land-Surface-Atmosphere*)
SALT (see *Savannas in the Long Term*)
salt, mobilisation 361
sand
    –, bar erosion 364
    –, desert 34, 35, 42
    –, dune 507
SAR (see *Synthetic Aperature Radar*)
Sary Kamich Lake 383
Saskatchewan 17, 111, 112, 187, 382
satellite
    –, data 63, 66, 76, 97, 100, 131, 182, 212, 218, 222, 273, 502
    –, Envisat 248, 281
    –, ERS-2 211, 281
    –, Jason-1 281
    –, remote sensing 207
    –, TOPEX/Poseidon 281, 381
saturated
    –, hydraulic conductivity 15, 308, 309, 316
    –, zone 65, 312, 323, 337
saturation, excess (Dunne) overland flow 305, 309, 327
savanna 35, 42, 62, 64, 65, 68, 73, 79, 131, 159, 202, 204, 222, 223, 225,
    230, 320, 418
Savannas in the Long Term (SALT) 63–65, 76
SBL (see *stable boundary layer*)
scale
    –, hillslope 305, 309, 324
    –, jump 253
    –, model
        –, global 2
        –, regional 2
    –, small catchment 301, 307, 309, 338, 457
    –, space 2, 18, 135, 300, 370, 417, 443
    –, time 7, 9, 15, 17–19, 33, 48, 50, 51, 60, 71–73, 75, 76, 87, 90, 93,
        109, 121, 126, 129, 132, 134–136, 158, 233, 248, 250, 251, 262, 275,
        276, 283, 286, 324, 326, 357, 370, 371, 383, 389, 397, 421, 456, 457,
        460–463, 484, 485, 494
scaling 221
Scandanavia 382
Scanning Multichannel Microwave Radiometer (SMMR) 211
scenario 76, 81, 131, 262, 275–277, 317–319, 329, 348, 351, 357, 367, 401,
    411, 428, 436–439, 443, 446, 451, 452, 460, 461, 464, 482, 483, 486,
    488, 491, 492, 494, 496, 497, 501–503, 505, 508, 523, 524, 536, 537
    –, 2 × CO_2 climate change 452, 453
    –, approach 341, 482, 489, 496, 536, 537
    –, business-as-usual 81
    –, construction 497
    –, definition 485, 491
    –, generation 275
    –, high emission 81
    –, interpretation uncertainties 503
    –, land use 501
    –, land-use change 502
    –, low emission 81
    –, palaeoclimatic 496
science, interdisciplinary 251
Scientific Committee on Problems of the Environment's
    International Nitrogen Project (SCOPE-N) 403
SD approach (see *system dynamics approach*)
sea
    –, ice 7, 41, 43, 96, 484
        –, albedo feedback 34, 130
        –, feedback 484
    –, surface temperature (SST) 35, 37, 40, 42–45, 49, 50, 59, 68,
        70, 73, 76, 80, 81, 131, 497
        –, anomaly 68, 70, 80
SEBEX (see *Sahelian Energy Balance Experiment*)
SECHIBA model 314
sediment
    –, annual discharge 526
    –, delivery ratio 334, 391, 449
    –, flux 319, 321, 386, 391–397, 402, 412
        –, global 390

–, riverborne 397
–, mass balance 364
–, mean annual yields 449
–, phosphorus-carrying 356, 357
–, simulated load 450
–, suspended flux 392, 395–397, 412
–, transfer 349, 396
–, transport 317, 333, 334, 356, 389–391, 393, 396, 397, 425, 439,
    506
–, trapping 395, 396
–, yield 321, 333, 335, 336, 392, 393, 396, 400, 443, 445, 446,
    448–450, 452, 527, 528, 532–534
Segura basin 320
Seine
    –, basin 412
    –, River 402
Semi-Arid Land-Surface-Atmosphere (SALSA) 208
semi-arid
    –, region 59, 71, 130, 225, 317, 318, 377, 507
    –, river basin 3
Senegal River 373
sensitivity 89, 90, 322, 491, 496, 528, 534
    –, analysis 522
    –, assessment 492
    –, experiment 24, 33, 48, 113, 114, 485, 487, 488, 496, 497, 513
    –, of convection 88, 90
    –, of precipitation 52, 91
    –, study 8, 36, 40, 66, 68, 76, 130, 131, 135, 280, 330, 486, 496
sensor, semi-conductor 169
seriousness, ranking 482
SGP (see *Southern Great Plains*)
SGP97 (see *Southern Great Plains Experiment of 1997*)
Shaanxi 526
Shaft El Arab basin 462
Shandong province 526
Shanxi 526
shelter coefficient 165
Shuttle Imaging Radar (SIR) 211
SiB model (see *Simple Biosphere model*)
Siberia 115, 256, 381, 382, 385, 506
Sieg River 373
signature, organic geochemical 425
silicon 169, 408, 409
Simple Biosphere (SiB) model 13, 86, 166, 280
Simplified Simple Biosphere (SSiB) model 66–68, 314
Simulated Topological Network-30 minute (STN-30) 378, 516
simulation
    –, Lagrangian stochastic 181
    –, terrestrial 286
SINOP (see *System for Information in NOPEX*)
SIR (see *Shuttle Imaging Radar*)
SLTH (see *observations of screen-level temperature and humidity*)
small catchment scale 301, 307, 309, 338, 457
SMMR (see *Scanning Multichannel Microwave Radiometer*)
Snake River 383
Snow Models Inter-comparison Project (SnowMIP) 269
snow
    –, albedo 113
        –, feedback 41, 96, 97
    –, cover 7, 34, 41, 51, 54, 93, 95–101, 113, 114, 132, 133, 207, 212, 213,
        281, 287, 301, 303, 319, 484
        –, feedback 97, 113, 133
    –, depth 41, 51, 54, 96, 99–101
    –, duration 96, 99
    –, extent 96, 97
    –, -free albedo 207
snowmelt 101, 106, 108, 114, 217, 303, 305, 319, 323–325, 327, 331, 337,
    379, 430, 510
SnowMIP (see *Snow Models Inter-comparison Project*)
snowpack 96, 97, 100, 214, 280, 511
    –, water balance 214
sod, initial breaking 396
SOFTMODEL 326
Soil and Water Integrated Model (SWIM) 329, 332–334, 430, 431,

433, 436, 437, 439
-, hydrograph 430, 431
soil
-, aggregated 316
-, carbon 48, 49, 113, 251, 285, 319
-, chemistry 459
-, erodibility 334, 336, 445, 448, 449, 532
-, erosion 62, 276, 317, 320–322, 334, 356, 360, 363, 396, 398, 400, 439, 499, 508, 522, 526
-, frozen 16, 55, 222, 310, 313
-, heat
-, flux 10, 21, 34, 36, 64, 160, 162–164, 174, 176, 222, 258
-, flux measurement 162
-, storage 162
-, storage measurement 162
-, hydrology 134, 222
-, matrix 308
-, moisture 14, 53, 210, 232, 281, 310, 312
-, change 484
-, characteristics 311
-, content 15
-, dynamics 59, 301, 302
-, frozen 312
-, state 210
-, organic matter 317
-, parameter, effective 316
-, profile 10, 258, 259, 262, 312, 323, 327, 333, 350, 357, 358, 361, 528
-, properties 68, 69, 74, 114, 207, 311, 314, 336, 356, 419, 445
-, structure 317, 359, 360, 459
-, texture 16, 226, 315–317, 353, 418, 459
-, type 102
-, water
-, balance 83, 312, 313, 316
-, content 227, 251, 311, 345
-, flux 311
-, freezing 56
-, pressure head 311, 312
-, vertical flux 15
-, wetness 15, 49, 50, 52, 211, 314
-, anomaly 50
soil-vegetation-atmosphere transfer (SVAT) 108, 222–224, 283, 310, 312, 316, 322
-, scheme 108
solar
-, flux 9
-, radiation 9, 13, 34, 35, 47, 73, 80, 84, 87, 105, 107, 132, 163, 169, 196, 200, 217, 263, 276, 277, 310, 422
-, variability 485
Solimões basin 419
solution, multiple equilibrium 37, 43, 73
solvent 457
sonic
-, anemometer 109, 161, 177, 178
-, temperature 161, 162, 178
sounding
-, by aircraft 194
-, sequences 193
South Africa 2, 336, 341, 343–347, 349–352, 358, 367–369, 373, 374, 413, 441, 451, 527–529, 531–535, 537
South America 45, 79–81, 88, 120, 208, 258, 341, 462, 499
South Atlantic Convergence Zone (SACZ) 80
South East Asia 382, 383, 398, 401, 459
Southern Africa, dry lakes 383
Southern Great Plains (SGP) 186, 187, 257, 258
Southern Great Plains Experiment of 1997 (SGP97) 187
Special Sensor Microwave/Imager (SSM/I) 211, 281, 284
species, invasive 510
specific
-, humidity 23, 124, 125, 189
-, leaf area 199, 202
SPOT, satellite imagery 447
spotted knapweed 511
Spree River 436

spruce 94, 95, 102–108, 110, 111, 114, 160, 179, 203, 510
Sri Lanka 373
SSiB model (see *Simplified Simple Biosphere model*)
SSM/I (see *Special Sensor Microwave/Imager*)
SST (see *sea-surface temperature*)
St. Lawrence River 382
stability 31, 38, 39, 75, 118, 135, 164, 165, 176, 177, 179–181, 218, 317, 321, 337, 356, 392, 402, 417, 486
stable boundary layer (SBL) 189, 193, 195
-, budget 195
-, thickness 189
stakeholder approach 369
standing biomass 199, 201, 319
Stepenitz
-, basin 431, 433
-, River 431, 432
Sterkfontein Dam 344
STN-30 (see *Simulated Topological Network-30 minute*)
stochastic
-, approach 253, 497
-, model 277
-, weather
-, model 275, 277
-, scenario 276
stomatal
-, conductance 13, 22, 23, 31, 47, 106, 132, 201, 202, 316
-, model 13
-, control 103
-, gas exchange 11
Stör River basin 330
storage
-, change 214, 217–220
-, component 14
-, volume 388, 522
storm hydrograph 324
stormflow 302, 303, 305–309, 322–324, 326–328, 335–338, 348, 350, 353, 361, 419, 445, 449, 528, 532–534
-, delayed 336
-, subsurface 302, 305–309, 322–328, 337, 338
stream
-, ephemeral 344, 364, 373
-, segment 388
streamflow 86, 87, 213, 215, 220, 227, 232, 256, 303, 305, 307, 318, 319, 322–324, 326, 337, 341, 343, 344, 346, 350, 352, 370, 385, 389, 442–446, 451–453, 492, 493, 534
-, baseline 453
-, generation 446
-, hydrograph 307
sublimation 101, 108, 214
subsidence 22, 75, 90, 91, 189–191, 193, 194, 381, 383
-, rate 190
subsistence
-, farming 452
subsurface
-, flow 308, 309, 323–326, 337
-, stormflow 302, 305–309, 322–324, 326–328, 337, 338
summer monsoon 35, 36, 39, 42, 51, 61, 115, 123, 125, 126, 134
Sunda River system 383
surface
-, aerodynamic roughness 207
-, albedo 34, 40, 41, 45, 51, 64–66, 68, 69, 75, 97, 106, 107, 113, 122, 131, 202
-, buoyancy flux 189
-, conductance to water vapour 105
-, energy
-, balance 9
-, budget 87, 132, 208, 210, 484
-, partition 102, 230
-, flow 216, 307, 319, 321, 379, 433
-, flux 22, 23, 28, 31, 36, 88–90, 92, 102, 107, 108, 121, 157, 158, 173, 176, 179, 182, 186, 187, 189–191, 193, 195, 196, 220, 221, 223, 224, 227, 228, 230, 231, 233, 262, 281, 287, 399
-, measurement 158, 173, 175, 182, 222
-, heat budget 21

–, heterogeneity 10, 25, 32, 102, 273, 304, 328, 330, 337
–, hydrology 107
–, layer 21, 24, 66, 174, 181, 189, 191, 195
–, modelling 221
–, moisture budget 22
–, radiation budget 10, 257
–, roughness 88, 122
–, runoff 2, 15, 68, 213, 216, 227, 228, 262, 303, 305, 310, 333, 334, 356, 357, 433, 459, 524
–, temperature 210
–, turbulent flux 173
–, water balance 14
susceptibility 487
SVAT (see *soil-vegetation-atmosphere transfer*)
Swaziland 336, 344, 347, 527, 528
Sweden 94, 108, 111, 194, 322, 327
SWIM (see *Soil and Water Integrated Model*)
synergism 33, 36
synergy 49, 204
Synthetic Aperature Radar (SAR) 211
System for Information in NOPEX (SINOP) 256, 270
system
–, approach 367
–, continental aquatic 376
–, dynamics (SD) approach 522
–, terrestrial aquatic 375, 455, 457, 458, 461
–, spatial heterogeneity 457
–, spatial organisation 457
–, spatial scale of drivers 459

T

Tabatinga 83
TAC (see *total atmospheric carbon*)
Tafassassett 378
taiga 34–36, 40, 41, 45, 93, 114, 130, 401
–, -tundra feedback 34, 36, 40, 41, 130
Taiwan 398
Takla Makan 378
Tamanrasset River 383
Tamanrrasett/Saoura drainage basin system 378
tamarack 94, 95, 107, 108
Tamaransett Saoura 379
TAO (see *Tropical Atmosphere Ocean*)
TARGETS modelling framework 489
Tartu experiment (TARTEX-90) 159
Taymir Peninsula 382
TE (see *trapping efficiencies*)
tectonic lake 383
tectonics 381
teleconnection 28, 29, 31, 33, 39, 40, 70, 129
–, regional 28
temperature
–, measurement 161
–, potential 22, 23, 35, 108, 189
–, sonic 161, 162, 178
–, sum 34
TERRA (SVAT from the DWD) 283
TESSEL (see *Tiled ECMWF Surface Scheme of Exchange processes at the Land surface*)
Thailand 2, 116–118, 120, 127, 133
thatched roof effect 323
thawing 102, 230, 313, 337
theory, Lagrangian 11
threshold 62, 321, 337, 355, 362, 376, 482, 488, 493, 494, 498–500
thunderstorm 24, 25, 321
Tibesti 383
Tibetan Plateau 115, 133, 218
Tiga Reservoir 389
tile approach 224, 228, 233
Tiled ECMWF Surface Scheme of Exchange processes at the Landsurface (TESSEL) 283
time
–, -domain reflectometry 214

–, transit t. 307, 308, 325, 337
–, travel t. 307
Titicaca Sea 383
TOC (see *total organic carbon*)
TOGA (see *Tropical Ocean-Global Atmosphere*)
TOGA/COARE (see *Tropical Ocean Global Atmosphere/Coupled Ocean-Atmosphere Experiment*)
tolerable window approach 488
TOPEX/Poseidon satellite 281, 381
TOPMODEL 16, 326, 327
TOPOG 309
topography 16, 25, 60, 107, 207, 210, 253, 256, 261, 276, 278, 279, 287, 289, 304, 316, 324, 326, 329–331, 337, 339, 356, 372, 377, 378, 389, 392, 417, 418, 424, 428, 441, 457, 459, 496
total
–, atmospheric carbon (TAC) 398, 399
–, organic carbon (TOC) 398, 411
–, runoff integrating pathways 68
–, suspended solids (TSS) 393, 394, 398, 400, 408, 413
–, flux 408
Toutle River 396
trace gas 2, 167, 178, 179, 182, 192, 194, 196, 205, 213, 250, 258, 403, 408, 416, 417, 427
–, flux 2
tracer 94, 212, 308, 309, 321, 323–325, 337, 380, 425
–, concentration 325
transfer, interbasin 340, 385, 386, 514
transient projection 496
transit time 307, 308, 325, 337
transmission
–, loss 333, 341, 344, 345, 374
–, from riverine systems 343
transmissivity feedback 323
transpiration 2, 13, 21, 25, 34–36, 46–48, 50, 51, 55, 75, 89, 101–103, 105, 106, 110, 113, 116, 131–133, 167, 199, 201, 202, 211, 222, 224, 225, 232, 301, 311–314, 329, 335, 348, 350, 358, 359, 421, 437–439, 452, 532
–, potential rate 201, 311, 312
–, rate, actual 312, 313
transport, biogeochemical 317
trapping efficiencies (TE) 395
travel time 307
tree line 2, 40, 95, 509
–, boreal 2
–, varying locations 509
trembling aspen 95
TRIFFID DGVM 48
TRMM (see *Tropical Rainfall Measurement Mission*)
TRMM-TMI (see *Tropical Rainfall Measurement Mission – Microwave Imager*)
Trombetas River 418
Tropical Atmosphere Ocean (TAO) 248
Tropical Ocean Global Atmosphere/Coupled Ocean-Atmosphere Experiment (TOGA/COARE) 170, 171
Tropical Ocean-Global Atmosphere (TOGA) 170, 171, 248
Tropical Rainfall Measurement Mission (TRMM) 85, 86, 107, 248, 281
Tropical Rainfall Measurement Mission – Microwave Imager (TRMM-TMI) 281
tropical
–, convection 39, 40, 90
–, convective heating centre 39
–, depression 116
–, forest 8, 47, 79, 81–83, 88, 89, 93, 131, 132, 158, 160, 164, 165
–, monsoon circulation 39
TSS (see *total suspended solids*)
tundra 34–36, 40, 41, 45, 54, 93–95, 114, 130, 183, 193, 259, 312, 401
turbulence
–, organised 31
Turgay Channel 382
typhoon 116

U

Umgeni water 443

uncertainty 484, 490, 493, 517, 523–526, 528, 531
–, quantifying 493
Universal Soil Loss Equation (USLE) 333, 336, 391
upper system boundary (USB) 180
upscaling 47, 221, 223, 224, 227, 253, 340
–, study 223
upwash contamination 184
upwelling, feedback 484
National Center for Atmospheric Research (NCAR) 86, 186, 222, 260, 264, 270
National Centers for Environmental Prediction (NCEP) 73, 86, 190, 222, 248
usability 268
USB (see *upper system boundary*)
USLE (see *Universal Soil Loss Equation*)
Uzboi channel 383

V

Vaal Dam 344
Valdai glaciation 382
validation 1, 14, 18, 76, 82, 85, 123, 131, 159, 184, 202, 204, 208, 209, 211, 222–224, 228, 232, 251, 252, 260, 262, 263, 267, 268, 270, 273, 278, 280, 281, 283, 284, 286, 308, 313, 316, 331–333, 431, 433, 437, 439
Valley of Thousand Hills 441, 449
value, critical 15, 54
vapour pressure 12, 104, 106, 110, 176
–, deficit (VPD) 12, 104, 106, 110
Vargem Grande 417–420, 422, 424
variability
–, interannual 39, 48–52, 73, 76, 85, 87, 110, 116, 173, 174, 212, 216, 277, 280, 344, 421, 443
–, seasonal 48
–, solar 485
Variable Infiltration Capacity (VIC) model 16, 219
várzea 418, 420, 421, 425
vegetation 459
–, -albedo-precipitation feedback 36
–, as dynamic state variable 7
–, boreal 94
–, carbon 49
–, change 2, 10, 40, 44, 46, 70, 130, 134, 505
–, cover 24, 25, 37, 39, 42, 45, 59, 60, 62, 65, 66, 79, 121–125, 131, 134, 163, 167, 224, 225, 253, 317, 330, 332, 356, 397, 507
–, data 122
–, degradation 75
–, density 34, 67, 71, 113
–, dynamics 7, 8, 28, 33, 59, 65, 67, 73, 76, 199, 204, 299, 485
–, simulation 7
–, fraction 44, 123
–, fractional coverage 121–123
–, function 201
–, global pattern 7, 43, 71
–, impact on climate change 7
–, measurement 202, 204, 232
–, mid-Holocene 38, 42
–, module, dynamic 43
–, natural 25, 51, 62, 64, 69, 74, 76, 121–123, 125, 126, 134, 222, 230, 276, 312, 317, 320, 332, 333, 390, 393, 446, 447, 501
–, physiology 199, 201
–, potential v. 122
–, productivity 200
–, root distribution 51
–, -snow-albedo feedback 41
–, structure 34, 39, 40, 43, 47, 199, 200, 202, 204, 210
–, types 37, 510
Vegetation/Ecosystem Modelling and Analysis Project (VEMAP) 262, 264
velocity scale 189
VEMAP (see *Vegetation/Ecosystem Modelling and Analysis Project*)
Venezuelan Llanos 87
verification 446

VIC (see *Variable Infiltration Capacity*)
volatile organic carbon (VOC) 109
Volga
–, basin 459
–, River 382
VPD (see *vapour pressure deficit*)
vulnerability 3, 118, 133, 136, 253, 276, 322, 329, 369, 372, 376, 430, 435, 482, 483, 486, 487, 489, 491–494, 498–500, 508, 511–514, 516, 520–522, 524, 527, 528, 531, 535–537
–, approach 136, 482, 483, 486, 490, 500, 536
–, assessment 3, 276, 489, 491, 492, 494, 511, 513, 522, 536, 537
–, concept 499
–, definition 482
–, modification 534
–, passive 498
–, potential v. 528
–, relationship to prediction 482

W

Wangara Experiment 24
Wariner Creek 379
washoff 327, 432
waste
–, domestic 430
–, industrial 430
water
–, atmospheric
–, balance 116, 120, 218, 219
–, cycle over monsoon Asia 115
–, availability 83, 118, 201, 350, 351, 354, 403, 410, 433, 436, 457, 501, 504, 514, 517, 520
–, balance 14, 86, 213, 214
–, concept 213, 214
–, equation 91, 214
–, soil 83, 312, 313, 316
–, body, endorheic 380
–, budget 86, 190, 215, 220, 256, 304, 350, 390, 436, 505–507
–, capacity
–, available 313, 418
–, differential 311
–, plant-available 313
–, conflict 351
–, crowding 516, 517
–, cycle 1, 2, 8, 14, 48, 65, 79, 81, 91, 115, 120, 125, 131, 132, 228, 233, 257, 267, 299, 300, 317, 362, 372, 375–377, 385, 386, 389, 390, 410, 412, 413, 461, 514, 520
–, global 14
–, terrestrial 299, 300, 372, 375–377, 389, 410, 412, 520
–, datasets 1
–, demand 365
–, density 15
–, event w. 305, 307, 309, 323, 324, 326, 337
–, excess 218, 305, 380, 499
–, exchange processes 2
–, integrated management 461
–, per capita availability 410
–, pollution 511
–, precipitable 81, 91
–, quality 308, 340, 341, 343, 346, 348, 349, 352, 355, 357, 361, 363–366, 369, 371, 372, 374, 376, 399, 402, 430, 433, 443, 444, 447, 448, 450, 452, 453, 456, 460, 461, 482, 484, 496, 501, 511, 512, 520, 522, 524, 527
–, degradation 354
–, indicator 447, 526
–, quantity 340, 348, 352, 364, 365, 367, 372, 374, 450, 484, 524
–, resources 3, 77, 115, 118, 228, 273, 299, 310, 318, 329, 330, 340–342, 344, 348, 351, 352, 367–369, 371, 372, 375, 376, 385, 386, 397, 436, 443, 446, 451, 463, 484, 485, 487–489, 494, 496, 499, 506–508, 514, 519, 520, 522, 534, 536
–, scarcity 515, 518
–, sector strategy 519
–, stress 12, 312, 410, 412, 430, 435, 515, 516, 518, 519
–, supply 313, 322, 335, 340, 345, 352, 353, 362, 369, 372, 430, 443,

445, 496, 502, 510, 514–517, 519, 520, 526
–, assessment 514
–, renewable 517
–, surface 302, 331
–, sustainable supply 365
–, system, management 3
–, table, saline 359
–, terrestrial 3, 281, 299, 300, 372, 375–377, 385, 389, 410, 412, 520
–, use 3, 46, 304, 317, 328, 340, 343, 348, 350, 351, 354, 358, 359, 362, 366, 368, 369, 386, 389, 390, 409, 410, 430, 438, 439, 445, 447, 457, 459, 461, 510, 514, 515, 517, 518, 520, 526
–, vapour 2, 10, 17, 18, 25, 47, 74, 81, 86, 105, 116, 119, 120, 129, 132, 133, 157, 161, 162, 177, 179, 190, 192, 194, 196, 197, 231, 250, 258
–, budget 193
–, flux 2, 17, 86, 116, 190, 196, 231
–, measurement 162
–, vulnerability, global assessment 515
–, withdrawal 409, 410, 515, 521
watercourse, natural 341
waterscape 339, 457
WBGU (see German Advisory Council on Global Change)
WCRP (see World Climate Research Programme)
WDC (see World Data Centres)
weather
–, anomaly 278
–, forced variations 484
–, free variations 484
–, model
–, parameters 278
–, stochastic 275, 277
–, prediction, numerical 52, 85, 157, 253, 283, 284, 317, 484, 497
–, scenario 496
–, stochastic 276
–, seasonal 48, 52, 485, 524, 537
Webb correction 162
Weser basin, area 355
West Africa 35, 37, 53, 60, 64, 65, 69, 215, 317, 318, 320
Wet Atmospheric Mesoscale Campaign (WETAMC) 85, 86
wet
–, deposition 445
–, hydrology 301
–, savanna 320
WETAMC (see Wet Atmospheric Mesoscale Campaign)
wetland 42, 94, 103, 106–108, 110, 121, 133, 195, 213, 217, 281, 301–303, 305, 307, 322, 327, 330, 336, 340, 341, 345–347, 354, 359, 366, 373, 375, 404, 427, 455, 457, 482, 501, 511, 513
–, albedo 106
–, hydrological significance 345
–, threats 346
wetting, initial 301, 303
whirling disease 511
White Gull Creek basin 217
White Nile 379, 385
white spruce 95
white-water river 417, 418, 425, 427
Wilge River 344
willow 95

wind
–, horizontal speed 190, 191
–, measurement, airborne 185
–, velocity 26, 82, 161, 163, 164, 174, 184, 185
–, measurement 161
Winter Experiment (WINTEX) 213, 220
winter monsoon 125, 126, 134
WINTEX (see Winter Experiment)
Wisconsin glaciation 382
WISE (see World Inventory of Soil Emission potentials)
WMO (see World Meteorological Organisation)
Wolfshagen 431, 433
wolve 508
World Bank 519, 521, 522
World Climate Research Programme (WCRP) 14, 167, 170, 213, 229, 233, 263, 338
World Data Centres (WDC) 259
World Inventory of Soil Emission potentials (WISE) 258
World Meteorological Organisation (WMO) 7, 161, 169, 248, 258, 263, 271, 286, 338, 378, 515, 520
World Radiation Centre 161, 169–171
World Radiation Reference (WRR) 169, 170
World Radiometric Reference 169
WRR (see World Radiation Reference)
Wyoming 509, 511

X

Xier River 522, 524, 525

Y

Yangtze River valley 119
Yellow River (Huang He) 121, 391, 525–527
–, basin 121, 527
–, runoff 526
Yellow Sea 526
Yueguzonglie basin 525
Yukon River system 382
Yunganjiang River 522

Z

Zaire 383, 385, 400, 408
–, basin 408
–, River 383
–, runoff 385
Zambezi River 383, 389
zero plane displacement height 11
zone
–, hyporheic 328
–, saturated 65, 312, 323, 337
–, temperate 321
–, unsaturated 302, 310, 312, 324, 354, 455
Zschopau
–, basin 431
–, River 431